Brain Mapping:
The Disorders

Brain Mapping
The Disorders

Edited by

John C. Mazziotta

Ahmanson–Lovelace Brain Mapping Center
UCLA School of Medicine
Los Angeles, California

Arthur W. Toga

Laboratory of Neuro Imaging
Division of Brain Mapping
Department of Neurology
UCLA School of Medicine
Los Angeles, California

Richard S. J. Frackowiak

Wellcome Department of Cognitive Neurology
Institute of Neurology
University College London
London, United Kingdom

ACADEMIC PRESS

A Harcourt Science and Technology Company

San Diego ▴ **San Francisco** ▴ **New York** ▴ **Boston** ▴ **London** ▴ **Sydney** ▴ **Tokyo**

Academic Press
A Harcourt Science and Technology Company
525 B Street, Suite 1900, San Diego, California 92101-4495, U.S.A.
http://www.academicpress.com

Academic Press
Harcourt Place, 32 Jamestown Road, London NW1 7BY, UK
http://www.hbuk.co.uk/ap/

Library of Congress Catalog Card Number: 99-67664

International Standard Book Number: 0-12-481460-3

PRINTED IN THE UNITED STATES OF AMERICA
00 01 02 03 04 05 PTP 9 8 7 6 5 4 3 2 1

Contents

Contributors   xi
Foreword   xv
Preface   xvii
Acknowledgments   xix

I Background and Technical Issues

1 The Study of Human Disease with Brain Mapping Methods

John C. Mazziotta and Richard S. J. Frackowiak

I. Diagnostic Methods   6
II. Surgical Strategies   19
III. Atlases   20
IV. Plasticity   24
V. References   26

2 Experimental Design and Statistical Issues

Karl J. Friston

I. Introduction   33
II. Functional Specialization and Integration   34
III. Spatial Realignment and Anatomical Normalization   36
IV. Statistical Parametric Mapping and Functional Specialization   37
V. Experimental Design   42
VI. Designing fMRI Experiments: A Signal-Processing Perspective   45
VII. Epoch- and Event-Related Studies   48

VIII. Inferences about Subjects and Populations   51
IX. Effective Connectivity and Functional Integration   53
X. Specialization, Integration, and the Lesion-Deficit Model   54
References   56

3 Preoperative Brain Mapping

John C. Mazziotta

I. Lesion Localization and Characterization   60
II. Targeting   67
III. Cortical Mapping   70
IV. Surgical Planning   72
References   73

4 Intraoperative Brain Mapping

Arthur W. Toga, George A. Ojemann, Jeffrey G. Ojemann, and Andrew F. Cannestra

I. Introduction   77
II. Electrical Stimulation Mapping (ESM)   78
III. Relation between ESM and Functional Imaging   83
IV. Electrophysiological Correlates of Cognition Derived Intraoperatively   83
V. Optical Imaging Intraoperatively   84
VI. Intraoperative Structural Maps   91
VII. Summary   100
References   100

5 Intraoperative Visualization

Ron Kikinis, Neerav R. Mehta, Arya Nabavi,
Emmanouel Chatzidakis, Simon Warfield, David Gering,
Neil Weisenfeld, Richard S. Pergolizzi, Jr., Richard B.
Schwartz, Nobuhiko Hata, William Wells III, Eric Grimson,
Peter McL. Black, and Ferenc A. Jolesz

I. Background and Introduction   107
II. Surgical Planning   108
III. Intraoperative MRI   116
IV. Concluding Remarks   126
    References   127

6 Disease-Specific Brain Atlases

Paul M. Thompson, Michael S. Mega,
and Arthur W. Toga

I. Challenges in Population-Based Brain
    Mapping   131
II. Types of Brain Atlases   134
III. Analyzing Brain Data   137
IV. Individualized Brain Atlases   138
V. Model-Driven Deformable Atlases   143
VI. Probabilistic Atlases   146
VII. Atlas-Based Pathology Detection   150
VIII. Cortical Modeling   155
IX. Cortical Averaging   157
X. Brain Averaging   164
XI. Dynamic (4D) Brain Atlases   168
XII. Conclusion   168
    References   170

II Neurological Disorders

7 Functional Imaging Studies of Aphasia

Cathy J. Price

I. Neurological and Neuropsychological Models
    of Language   181
II. Neuroimaging Studies of Aphasic
    Patients   184
III Examples of How Neuroimaging Experiments
    on Aphasic Patients Have Contributed to
    Normal and Abnormal Models of
    Language   186
IV. Conclusions   196
    References   198

8 The Functional Neuroimaging of Memory Disorders

P. C. Fletcher

I. Functional and Structural Studies: Major
    Conceptual Differences   202
II. Exploring Memory Impairment with PET and
    fMRI: Conceptual Difficulties   204
III. Overcoming the Difficulties: Making Functional
    Neuroimaging Useful to the Study of Memory
    Impairment   206
IV. Summary   214
    References   214

9 Brain Mapping in Dementia

Michael S. Mega, Paul M. Thompson, Arthur W. Toga, and
Jeffrey L. Cummings

I. Structural Imaging   218
II. Functional Imaging   226
III. Summary   232
    References   234

10 Movement Disorders: Parkinson's Disease

D. Eidelberg, C. Edwards, M. Mentis, V. Dhawan, and
J. R. Moeller

I. Introduction   241
II. Dopamine System Imaging in
    Parkinsonism   242
III. Functional Brain Imaging in the Resting
    State   245
IV. Brain Activation Studies: Motor Execution and
    Learning   249
V. Functional Brain Imaging in the Assessment of
    Therapeutic Interventions   250
VI. Conclusion   257
    References   257

11 Movement Disorders: Other Hypokinetic Disorders

David J. Brooks

I. Multiple-System Atrophy   263
II. Progressive Supranuclear Palsy   268
III. Corticobasal Degeneration   275
IV. Dopa-Responsive Dystonia   278
V. Akinetic-Rigid Huntington's Disease   279
VI. Conclusions   281
    References   281

12 Functional Imaging in Hyperkinetic Disorders

Guy Sawle

I. Tremor   285
II. Dystonia   287
III. Tics   290
IV. Chorea   292
V. Tardive Dyskinesia   294
VI. Restless Legs Syndrome   295
     References   295

13 Functional Imaging in Vascular Disorders

J. C. Baron and G. Marchal

I. Brief Overview of Methods Employed   299
II. Normal Physiology and Basic Pathophysiology of Brain Perfusion and Metabolism   300
III. Long-Standing Arterial Obstruction: Mapping Hemodynamic Failure   303
IV. Acute Ischemic Stroke: Mapping the Core, the Penumbra, and the Reperfused Tissue   303
V. Remote Metabolic Effects of Stroke   308
VI. Receptor Studies in Vascular Disorders   311
     References   312

14 The Epilepsies

John S. Duncan

I. Introduction   317
II. Positron Emission Tomography Studies of Cerebral Blood Flow and Glucose Metabolism   318
III. Positron Emission Tomography Studies of Specific Ligands   323
IV. Single Photon Emission Computerized Tomography   330
V. Functional MRI   334
VI. Diffusion-Weighted MRI   336
VII. Magnetic Resonance Spectroscopy   336
VIII. Electrophysiological Imaging   342
     References   343

15 MRI in Multiple Sclerosis

Guojun Zhao, David K. B. Li, and Donald Paty

I. Imaging in the Diagnosis of Multiple Sclerosis   357
II. MRI in Natural History Studies of Multiple Sclerosis   364
III. Clinical Correlations with MRI Findings   372
IV. MRI–Pathological Correlation   374
V. Application of MRI in Monitoring of Clinical Trials   375
VI. Evaluation of in Vivo Pathology with Newer MR Techniques   379
VII. Summary   381
     References   381

16 Structural and Functional Imaging of Cerebral Neoplasia

Jeffry R. Alger and Timothy F. Cloughesy

I. Introduction   388
II. The Clinical Challenges   388
III. Recent Neuroimaging Progress Related to Intracranial Neoplasms   403
IV. Summary and a Look into the Crystal Ball   411
     References   412

17 Neurodegenerative Disorders of the Cerebellum

Sid Gilman, Mary Heumann, and Larry Junck

I. Introduction   417
II. Friedreich's Ataxia   418
III. The Hereditary Cerebellar Degenerations   421
IV. The Sporadic Cerebellar Degenerations   432
     References   447

III Pediatric Disorders

18 Dyslexia and Related Learning Disorders: Recent Advances from Brain Imaging Studies

Michel Habib and Jean-François Démonet

I. Brain Mechanisms in Dyslexia: An Overview, with Special Emphasis on Morphological Brain Imaging   460
II. Event-Related Potentials and Developmental Language Disorders   465
III. Functional Brain Imaging in Dyslexia   466
IV. Conclusion   478
     References   479

IV Psychiatric Disorders

19 Depression
Helen S. Mayberg

I. Introduction   485
II. Depression in Neurological Disease   486
III. Idiopathic Depression   491
IV. Parallel Studies of Normal Sadness   497
V. Working Model of Depression   498
  References   500

20 The Neurobiology of Anxiety and Anxiety-Related Disorders: A Functional Neuroimaging Perspective
P. Chua and R. J. Dolan

I. Introduction   509
II. Conceptual Issues in Neurobiological Accounts of Anxiety   510
III. Fear and Emotional Processing in the Brain   511
IV. Neuroimaging Studies of Fear Processing   512
V. Classical Conditioning as a Model of Fear Learning   512
VI. Processing Learned Fear Responses with and without Awareness   514
VII. Amygdala Interactions Related to Processing Aware and Nonaware Fear-Relevant Stimuli   515
VIII. Induction of Anxiety in Volunteer Subjects   516
IX. Psychopathological Studies of Anxiety Disorders   517
X. A Neuroanatomical Model of Anxiety   518
XI. Outstanding Issues in Understanding Mechanisms of Anxiety   519
XII. A Model of Emotional Processing   519
XIII. Conclusions   520
  References   520

21 Functional Neuroimaging Studies of Schizophrenia
Sarah-J. Blakemore and Chris D. Frith

I. Introduction   523
II. Neuropathology   525
III. Structural Studies   525
IV. Neuroreceptor Imaging of Antipsychotics Using PET   528

V. Cognitive Activation Studies   529
VI. Imaging Symptoms   537
VII. Obstacles to Functional Neuroimaging and Schizophrenia Studies   539
VIII. Conclusion   541
  References   541

22 Addictive States
Frank W. Telang and Nora D. Volkow

I. Introduction   545
II. Pharmacological Properties of Drugs of Abuse in the Human Brain   546
III. Imaging and Addictive Processes: Evaluation of the Addicted Subject   549
IV. Studies with Abused Drugs   550
V. Conclusions   562
  References   562

V Therapeutics and Recovery of Function

23 Plasticity
Mark Hallett

I. Introduction   569
II. Methods   570
III. Peripheral Lesions   571
IV. Spinal Cord Lesions   575
V. Brain Lesions   575
VI. Activity/Learning   577
VII. Blind, Cross-Modal Plasticity   581
VIII. Severe Sensory Neuropathy   582
IX. Maladaptive Plasticity   582
  References   584

24 Recovery of Neurological Function
François Chollet and Cornelius Weiller

I. Evidence for Spontaneous Cerebral Reorganization during Recovery in Stroke Patients   588
II. Learning Processes and Recovery of Neurological Function   590
III. Pharmacological Approach to Recovery of Neurological Function: Neuromodulation of Brain Networks   592
  References   594

Contents

25 Therapeutics: Pharmacologic

William H. Theodore

I. Introduction   599
II. Imaging Studies of the Effects of Antiepileptic Drugs   600
III. Drugs in Basal Ganglia and Movement Disorders   606
IV. Cognitive Dysfunction and Dementia   606
V. Imaging Studies of "Antipsychotic" Drugs   607
VI. Conclusion: The Possibility of Individualizing Therapy   609
    References   609

26 Therapeutics: Surgical

Robert S. Turner, Thomas Henry, and Scott T. Grafton

I. Introduction   613
II. Surgical Therapies for Parkinson's Disease   615
III. Surgical Therapies for Tremor   632
IV. Surgical Therapies for Epilepsy   632
V. Summary   646
    References   646

Index   655

Contributors

Numbers in parentheses indicate the pages on which the authors' contributions begin.

Jeffry R. Alger (387)
Department of Radiological Sciences, Brain Research Institute, Jonsson Comprehensive Cancer Center, University of California, Los Angeles, California 90095

J. C. Baron (299)
INSERM U320, CYCERON, University of Caen, Caen 14074, France

Peter McL. Black (107)
Department of Neurosurgery, Brigham and Women's Hospital, Boston, Massachusetts 02115

Sarah-J. Blakemore (523)
Wellcome Department of Cognitive Neurology, University College London, London WC1N 3BG, United Kingdom

David J. Brooks (263)
MRC Cyclotron Unit, Hammersmith Hospital, London TW7 4QN, United Kingdom

Andrew F. Cannestra (77)
Laboratory of Neuro Imaging, Division of Brain Mapping, Department of Neurology, UCLA School of Medicine, Los Angeles, California 90095

Emmanouel Chatzidakis (107)
Department of Neurosurgery, Brigham and Women's Hospital, Boston, Massachusetts 02115

François Chollet (587)
INSERM U455 and Department of Neurology, Hôpital Purpan, Toulouse 31059, France

P. Chua (509)
Department of Neuropsychiatry and Department of Psychiatry, The Royal Melbourne Hospital, Parkville, Victoria 3050, Australia

Timothy F. Cloughesy (387)
Department of Neurology, Henry E. Singleton Brain Cancer Research Program, Jonsson Comprehensive Cancer Center, University of California, Los Angeles, California 90095

Jeffrey L. Cummings (217)
Department of Psychiatry and Biobehavioral Sciences, UCLA School of Medicine, Los Angeles, California 90095

Jean-François Démonet (459)
INSERM U455, Department of Neurology, CHU Purpan, 31059 Toulouse, France

V. Dhawan (241)
Functional Brain Imaging Laboratory, Department of Neurology, North Shore University Hospital, Manhasset, New York 11030

R. J. Dolan (509)
Wellcome Department of Cognitive Neurology, London WC1N 3BG, United Kingdom

John S. Duncan (317)
Institute of Neurology, London WC1N 3BG, United Kingdom

C. Edwards (241)
Functional Brain Imaging Laboratory, Department of Neurology, North Shore University Hospital, Manhasset, New York 11030

D. Eidelberg (241)
Functional Brain Imaging Laboratory, Department of Neurology, North Shore University Hospital, Manhasset, New York 11030; and Department of Neurology, The New York Hospital–Cornell Medical Center, New York, New York 10021

P. C. Fletcher (201)
Research Department of Psychiatry, Cambridge University, Addenbrooke's Hospital, Cambridge CB2 2QQ, United Kingdom

Richard S. J. Frackowiak (3)
Wellcome Department of Cognitive Neurology, Institute of Neurology, University College London, London WC1N 3BG, United Kingdom

Karl J. Friston   (33)
Wellcome Department of Cognitive Neurology, Institute of Neurology, London WC1N 3BG, United Kingdom

Chris D. Frith   (523)
Wellcome Department of Cognitive Neurology, University College London, London WC1N 3BG, United Kingdom

David Gering   (107)
Artificial Intelligence Laboratory, Massachusetts Institute of Technology, Cambridge, Massachusetts 02139

Sid Gilman   (417)
Department of Neurology, University of Michigan Medical Center, Ann Arbor, Michigan 48109

Scott T. Grafton   (613)
Departments of Neurology and Radiology, Emory University School of Medicine, Atlanta, Georgia 30322

Eric Grimson   (107)
Artificial Intelligence Laboratory, Massachusetts Institute of Technology, Cambridge, Massachusetts 02139

Michel Habib   (459)
Cognitive Neurology Laboratory, Department of Neurology, CHU Timone, 13385 Marseille, France

Mark Hallett   (569)
Human Motor Control Section, National Institute of Neurological Disorders and Stroke, National Institutes of Health, Bethesda, Maryland 20892

Nobuhiko Hata   (107)
Department of Radiology, Brigham and Women's Hospital, Boston, Massachusetts 02115

Thomas Henry   (613)
Departments of Neurology and Radiology, Emory University School of Medicine, Atlanta, Georgia 30322

Mary Heumann   (417)
Department of Neurology, University of Michigan Medical Center, Ann Arbor, Michigan 48109

Ferenc A. Jolesz   (107)
Department of Radiology, Brigham and Women's Hospital, Boston, Massachusetts 02115

Larry Junck   (417)
Department of Neurology, University of Michigan Medical Center, Ann Arbor, Michigan 48109

Ron Kikinis   (107)
Department of Radiology, Brigham and Women's Hospital, Boston, Massachusetts 02115

David K. B. Li   (357)
Department of Diagnostic Radiology, Vancouver Hospital and Health Sciences Center, University of British Columbia, Vancouver, British Columbia, Canada V6T 2B5

G. Marchal   (299)
INSERM U320, CYCERON, University of Caen, Caen 14074, France

Helen S. Mayberg   (485)
Rotman Research Institute, Departments of Psychiatry and Medicine (Neurology), University of Toronto, Toronto, Ontario, Canada M6A 2E1

John C. Mazziotta   (3, 59)
Ahmanson–Lovelace Brain Mapping Center, UCLA School of Medicine, Los Angeles, California 90095

Michael S. Mega   (131, 217)
Laboratory of Neuro Imaging and Alzheimer's Disease Center, Division of Brain Mapping, Department of Neurology, UCLA School of Medicine, Los Angeles, California 90095

Neerav R. Mehta   (107)
Department of Radiology, Brigham and Women's Hospital, Boston, Massachusetts 02115

M. Mentis   (241)
Functional Brain Imaging Laboratory, Department of Neurology, North Shore University Hospital, Manhasset, New York 11030

J. R. Moeller   (241)
Department of Biological Psychiatry, New York State Psychiatric Institute, New York, New York 10032

Arya Nabavi   (107)
Department of Neurosurgery, Brigham and Women's Hospital, Boston, Massachusetts 02115

George A. Ojemann   (77)
Department of Neurological Surgery, University of Washington, Seattle, Washington 98195

Jeffrey G. Ojemann   (77)
Department of Neurological Surgery and NeuroImaging Laboratory, Washington University School of Medicine, St. Louis, Missouri 63110

Donald Paty   (357)
Division of Neurology, Vancouver Hospital and Health Sciences Center, University of British Columbia, Vancouver, British Columbia, Canada V6T 2B5

Richard S. Pergolizzi, Jr.   (107)
Department of Radiology, Brigham and Women's Hospital, Boston, Massachusetts 02115

Cathy J. Price   (181)
Wellcome Department of Cognitive Neurology, Institute of Neurology, London WC1N 3BG, United Kingdom

Guy Sawle   (285)
Department of Neurology, Queens Medical Centre, Nottingham NG7 2UH, United Kingdom

Richard B. Schwartz   (107)
Department of Radiology, Brigham and Women's Hospital, Boston, Massachusetts 02115

Frank W. Telang (545)
Medical Department, Brookhaven National
Laboratory, Upton, New York 11973

William H. Theodore (599)
National Institutes of Health, Bethesda, Maryland
20892

Paul M. Thompson (131, 217)
Laboratory of Neuro Imaging, Division of Brain
Mapping, Department of Neurology, UCLA School of
Medicine, Los Angeles, California 90095

Arthur W. Toga (77, 131, 217)
Laboratory of Neuro Imaging, Division of Brain
Mapping, Department of Neurology, UCLA School of
Medicine, Los Angeles, California 90095

Robert S. Turner (613)
Departments of Neurology and Radiology, Emory
University School of Medicine, Atlanta, Georgia
30322

Nora D. Volkow (545)
Medical Department, Brookhaven National

Laboratory, Upton, New York 11973; and Department
of Psychiatry, State University of New York at Stony
Brook, Stony Brook, New York 11794

Simon Warfield (107)
Department of Radiology, Brigham and Women's
Hospital, Boston, Massachusetts 02115

Cornelius Weiller (587)
Department of Neurology, Universitätskrankenhaus
Hamburg-Eppendorf, D-20246 Hamburg, Germany

Neil Weisenfeld (107)
Artificial Intelligence Laboratory, Massachusetts
Institute of Technology, Cambridge, Massachusetts 02139

William Wells III (107)
Artificial Intelligence Laboratory, Massachusetts
Institute of Technology, Cambridge, Massachusetts
02139

Guojun Zhao (357)
Division of Neurology, Vancouver Hospital and Health
Sciences Center, University of British Columbia,
Vancouver, British Columbia, Canada V6T 2B5

Foreword

Each of these three remarkable volumes thoughtfully and thoroughly discusses different aspects of the functioning primate's brain. Nevertheless, as the reader must expect, they most extensively address the human brain in health and disease. The first volume, titled *Brain Mapping: The Methods,* is a perfect gem, constructed largely as a product of Arthur Toga's exact and imaginative mind as well as his remarkable capacity to enlist the talents of many of the best neuroscientists in the world to contribute relevant sections to the volume. The remarkably encyclopedic and well-presented contents of the *Methods* volume includes beautifully constructed, well-explained, and visualized diagrams, graphs, tables, cartoons, specific abstractions, and mini-analyses, all of which are dedicated to solving what seem to be infinitely complicated problems.

As might be expected, Drs. Toga and Mazziotta first and effectively justify the obvious concept that maps need not be confined to planar structures alone. They describe maps as large or small, as selecting one color against another, as looking at surfaces instead of what's inside, as being macroscopic or microscopic, as being electrical or structural, and as mappings of whatever other functions can be found to be potentially useful when related to each other in brain function. The sometimes witty outcome of all this is, "Give me a neurological circuit containing three or more points and I'll figure a way to convert it into a map."

Drs. Toga and Mazziotta and many other contributors have ingeniously taken over the formidable task of translating the brain into a functionally useful group of dynamic maps. Hundreds of thousands of years ago, formed by evolved selection and despite its almost infinite opportunities for permanent destruction, the human brain, millennium by millennium, started slowly to stitch itself together and gradually develop its present environmentally adapted and forward-seeing capacities. Concurrently, the brain has automatically developed and generated the orderly ways and logical thoughts that progressively create growing minds, which themselves will bring intellectual surprises as the planet opens the twenty-first century. These functional brain patterns currently can often be mapped by investigators as the recognized origins of many different forms of behavior. They also serve, to a limited degree, as generators of thoughts and expressions of mind. Understanding and describing these brain expressions and analyzing just how they generate ideas and actions provide humankind with the needed clues to understand incoming information and apply it to regulate both mind and generated behavior.

The second volume of the trilogy is *The Systems.* Lucid descriptions and great insight span the opening few chapters, which discuss the early history of gross neuroanatomy as well as summarizing the many modern days, months, and years that it took to construct an accurate functional brain map. The remaining parts of the second volume explain in literate, thorough terms the details and images that paint the quantitative as well as qualitative dimensions of the human brain's functional neuroanatomy. Additional discussions describe images of cognitive systems, ontology, aging, learning, and adult plasticity. What a treat!

The third volume brings us to *Brain Mapping: The Disorders,* formulated and edited by Drs. Mazziotta, Toga, and Frackowiak. It provides the meat of the trilogy for clinical neurologists. The third volume begins with chapters dedicated to background and technical issues. Drs. Mazziotta and Frackowiak initiate this discussion, focusing on the overall use of brain mapping to understand more about human disease and its potential treatments. Karl Friston describes the principles of experimental design and statistical issues, after which Drs. Mazziotta and Toga carry brain mapping into the neurosurgical suite. Notable neurosurgeons and neuroscientists then describe different parts of the technology and application of brain mapping as used in intraoperative procedures and intraoperative visualization. Dr. Paul Thompson provides the last chapter in this section in the form of disease-specific brain atlases.

The next 11 chapters devote their contents to semichronic and chronic neurologic disorders. The specific, collective topics include aphasia, memory disorders, and dementia; movement disorders; vascular disorders; epilepsy; multiple sclerosis; brain cancer; degenerative cerebellar disorders; and psychiatric disorders. The final section discusses therapeutics and recovery of function; it includes plasticity, natural recovery, and neuropharmalogic and surgical therapeutics.

It has been widely and wisely observed that during the past 20 years or so, two major processes have revolutionized the fundamental futures of medicine and neuroscience at both a fundamental and a clinical level. The first of these truisms traces to Watson and Crick's 1953 discovery of the double helix. The ultimate outcome of that finding has generated the increasing emphasis and success of today's discoveries of genetically influenced cell–molecular medicine. The second great appearance was Hounsfield's 1971 (Nobel-awarded) discovery of being able to image the gross anatomy of the living brain safely and painlessly, using computed axial tomography (CAT scanning). Shortly thereafter, other investigators employed isotope techniques to extend and modify dynamic positron emission tomography (PET) to identify the brain's patterns of metabolism during normal or abnormal functional activity. Earlier, Block and Purcell in 1973 (also Nobel winners) had developed nuclear magnetic resonance (NMR) as a

technique for identifying the behavior of protons in magnetic fields. Subsequently, Lautebur (1973) modified the technique in order to produce biologically adaptable magnetic resonance imaging (MRI) (see Brain Mapping: The Systems, Chapter 2, by Marcus Raichle).

The point of the preceding paragraph is that molecular genetics and brain mapping have produced two totally different but revolutionary approaches that have already improved patient care and survival in many instances. Nevertheless, brain function must be indispensably dependent upon cell–molecular mechanisms. Brain imaging, on the other hand, currently focuses on identifying the presence of normal or abnormal structural appearances in patients suffering somatic or psychological symptoms. Some current imaging pioneers seek abnormal brain patterns as markers for either bipolar disorder or schizophrenia but consistent abnormalities appear very few. What has prospered has been the strong surge of cognitive neuroscientists who can develop strict paradigms for challenging voluntary subjects and concurrently mapping their brains. The results, for the first time in history, reveal at least fragments of how the thinking brain may generate functional activities in response to selective stimuli. Also, we have begun to register brain patterns in response to certain kinds of problem solving. These most recent steps in mapping the brain, of course, have brought cognitive neuroscientists closer and closer to the study of not just how the brain works but how it expresses a great many aspects of the normal or abnormal human mind. I hope Drs. Mazziotta, Toga, and Frackowiak can be persuaded to assemble one more beautiful *Brain Mapping* volume addressing the brain mechanisms that generate normal and abnormal minds.

Fred Plum

Preface

The publication of this text comes at an important time: the beginning of a new era in neuroscience. It is quite predictable that during the first 50 years of this new era, clinical and cognitive neuroscience will capture the scientific imagination and the public interest. Just as the explosion in the discipline of physics dominated the first 50 years of the 20th century and molecular biology the past 50 years, so perhaps will the disciplines and directions of clinical and cognitive neuroscience point the way to the next 50 years. The methods and applications described in the *Brain Mapping* series will be an important part of that process.

Since the birth of clinical neuroscience, its discoveries have been largely observational, based on the description of patients' signs and symptoms. Similarly, microscopic data derived from the examination of tissue that was either surgically excised or obtained postmortem were qualitatively described. The advent of imaging, both structural and functional, changed this outlook in a vast and irreversible manner. No longer do we simply follow patients at the bedside or in the clinic but, rather, rapid decisions and progressively earlier treatments are administered based on imaging data, initially from x-ray computed tomography and later from magnetic resonance imaging, to augment clinical observations. After many decades of basic science explorations using functional imaging derived from positron emission tomography, single photon emission computed tomography, measurements with stable or radioactive xenon, and electrophysiological recordings, these methods have matured to the point where they are of value in the diagnosis, management, and therapeutic monitoring of patients with neurological, neurosurgical, and psychiatric disorders.

Clinical neuroscience has made the leap from post-mortem diagnosis to *in vivo* intervention. The advent of powerful new molecular, cellular, and pharmacological agents to alleviate or minimize neuropathological processes will make such studies all the more important. The appropriate selection of patients and their categorization by functional, chemical, and neuropharmacological criteria will fall into the domain of brain mapping ever more in the years to come. In addition, the disciplines of molecular neuroscience and those of cognitive neuroscience, neuropsychology, and clinical neuroscience will merge through brain mapping and neuroimaging techniques. The result will be to the benefit of patients and their families. This is critical, in that patients, families, and society at large bear a tremendous economic burden and personal grief from the high prevalence of disorders that affect the nervous system. Never before have so many methods been available to tackle these vexing problems.

The publication of this text brings together many of the most notable authors in the field today. While no text can be all-encompassing or convey every theory or data set ever collected, this text does provide a broad survey of disorders that affect functional systems in the human brain, the functional tools to explore them, and ways to monitor the natural reorganization of the brain after injury or as a result of surgical, pharmacological, or behavioral intervention. Organized to reflect those categories, topics are clustered around technical issues, neurological disorders, pediatric syndromes, and psychiatric disease in addition to therapeutics and recovery of function. The techniques discussed not only apply in the clinic, but extend all the way to the operating room, where the course of surgical procedures is increasingly guided by functional brain mapping techniques.

This third book in the *Brain Mapping* series comes full circle. The first text described the methods of the brain mapping technologies. The second addressed the normal brain, the functional organization of its systems, and the tools to study its dynamic changes. This latest book addresses the disorders that affect the brain and the monitoring of the therapies that attempt to minimize or ameliorate these neuropathological processes. The development of brain mapping methods, their ap-

plications in normal subjects, and their clinical use in patients will have a profound impact on our understanding of normal brain function in health and its aberration in patients with brain disorders. This book marks a point in time in that process.

John C. Mazziotta
Arthur W. Toga
Richard S. J. Frackowiak

Acknowledgments

This third book in the *Brain Mapping* series has afforded me the opportunity to work closely with a different group of scientists and friends. Here I want to make some personal comments of appreciation. Thanks to my colleagues who authored the chapters of this text, doing such an admirable job in the process. Many other collaborators also shared their data, insights, and wisdom with both the editors and the chapter authors. That input helped make the material complete and timely. In this same vein, the patients who participated in the research studies, along with their families, have provided the insights that have advanced this field. To all of them, your contributions are greatly appreciated.

I am grateful for the support of the University of California at Los Angeles School of Medicine, the Neuropsychiatric Institutes, and the Departments of Neurology, Radiological Sciences, and Pharmacology. Funding and support for such projects come from many sources. Thanks go to the Brain Mapping Medical Research Organization, the Pierson–Lovelace Foundation, the Ahmanson Foundation, the Tamkin Foundation, the Jennifer Jones-Simon Foundation, the Robson Family, and the North Star Fund, as well as donations from many other individuals and groups. This work and the scientific efforts described in this text were partially supported by a grant from the Human Brain Project through funding from the National Institute of Mental Health, the National Institute for Drug Abuse, and the National Cancer Institute (P01-NH/DA52176-7). My staff, Palma Piccioni, Laurie Carr, Leona Mattoni, and Martha Sanchez, have contributed enormously to the practical aspects of this effort as well as to both the quality and the timeliness of this production.

No text is ever produced with quality and value without those same attributes being found in the publishers. Thus, thanks go to Academic Press, in particular, Jasna Markovac, Ph.D., who sat through many a luncheon and dinner with us, keeping us on the time line, and to Marge Lorang for her diligent and tactful administrative assistance.

Finally, to my family, friends, students, teachers, and collaborators go my greatest admiration and thanks for being the source of our motivation to take on projects that I believe will be important and meaningful.

John C. Mazziotta

This third book in the *Brain Mapping* series describes the transition of brain mapping approaches from benchside to bedside. To participate in the development of a field that moves so rapidly and draws upon such a diverse set of sciences is truly a privilege. In designing a book project, one of the requirements is to select contributors that can communicate not only the science but also their excitement and enthusiasm. I think we have achieved that. Once again, thanks to the contributors who withstood our constant nagging, cajoling, and editorial heavy handedness. As with previous projects, many people participated, some of them even willingly. Thanks to Lidia Uce, Neli Hamlin, Andrew Lee, and members of the Laboratory of Neuro Imaging for all your help, direct and indirect. Thanks to the students and colleagues who provided images and other materials. I hope they are pleased with their inclusion and presentation. I would be remiss if I did not thank the supporters of the science that is included here. The National Science Foundation (DBI 9601356), the National Center for Research Resources (RR 13642), the National Institute for Neurological Diseases and Stroke (NS 38753), the National Library of Medicine (LM05639), the National Institute of Mental Health (MH52083), and the Human Brain Project funded jointly by the National Institute of Mental Health and the National Institute of Drug Abuse (MH/DA52176) all provided support.

Given that I still have lots of homework, particularly book chapters, book editing, and such, I want to acknowledge those with whom I share my home. My wife, Debbie, and my children, Nicholas, Elizabeth, and Rebecca, all provide the necessary support and acceptance of my life as an academic.

Arthur W. Toga

Working on this project has been richly rewarding for many reasons. First, I had the opportunity to work closely with my good friends John and Arthur on the other side of the Atlantic. Second, I had the pleasure of working on an important contribution to the literature. Third, I was provided with an excuse to work closely with many of my colleagues with whom I otherwise have too infrequent contact. The collective contributions of the authors are greatly valued by me and certainly will be appreciated by the readers.

I am grateful to my institution, the Institute of Neurology, University College London. Funding and support for such projects come from many sources. The Wellcome Trust provided the facilities and environment for work in functional imaging, as did the Leopold Muller Trust. My staff contributed enormously to many aspects of this effort from beginning to end. Thanks go to Terry Morris, Pat Forsdick, Marcia Bennett, Josephine Macauley, Chris Freemantle, and Peter Aston.

As with my other projects in cooperation with Academic Press, I want to personally thank them for everything. This text was handled expertly by Academic Press, in particular, Jasna Markovac, Ph.D. She and her staff kept this book on a tight and well-managed schedule.

Most important, I thank my family, friends, and students for their support and faith throughout this and other endeavors.

Richard S. J. Frackowiak

I
Background and Technical Issues

1

The Study of Human Disease with Brain Mapping Methods

John C. Mazziotta[*,1] and Richard S. J. Frackowiak[†]

*Ahmanson–Lovelace Brain Mapping Center, UCLA School of Medicine, Los Angeles, California 90095
†Wellcome Department of Cognitive Neurology, Institute of Neurology, University College London, London WC1N 3BG,
United Kingdom

I. Diagnostic Methods

II. Surgical Strategies

III. Atlases

IV. Plasticity

References

Disorders of the human nervous system are among the most debilitating and devastating of all human illnesses. Such disorders not only affect the physical abilities of patients but often severely compromise the quality of life, the ability to function in society, family relations, and the ability to maintain gainful employment. As such, neurological, neurosurgical, and psychiatric disorders affect not only the patients themselves but also their families and society at large because of the tremendous economic burden that results from their prevalence.

Undoubtedly, this is one of the reasons brain mapping methods have advanced so rapidly and hold such promise for improved diagnostics, monitoring of therapeutics, and providing insights into the basic mechanisms of brain disease. Never before have so many methods been available to tackle these vexing problems. This chapter provides an overview of the methods and their applications in human diseases that are discussed in more detail in subsequent chapters.

Throughout this chapter, and the text as a whole, it is important for the reader to keep in mind a number of important physical and physiological factors when trying to understand the approaches of different brain mapping applications to the study of human cerebral disorders (Table I).

First, it is important to keep in mind that specific methods address only one or a few aspects of underlying cerebral physiology and pathophysiology. For example, electromagnetophysiological techniques, such as electroencephalography (EEG), event-related potentials (ERP), and magnetoencephalography (MEG), provide information about large constellations of neurons and the net electromagnetophysiological vector

---

[1] To whom correspondence should be addressed.

Table I   Advantages and Limitations of Current Brain Mapping Techniques of Use in the Study of Patients with Neurological, Neurosurgical, and Psychiatric Disorders

Method	Advantages	Limitations
X-ray computed tomography (CT)	Excellent bone imaging	Ionizing radiation
	~100% detection of hemorrhages	Poor contrast resolution
	Short study time	
	Can scan patients with ancillary equipment	
	Can scan patients with metal/electronic devices	
Magnetic resonance imaging (MRI)	High spatial resolution	Long study duration
	No ionizing radiation	Patients may be claustrophobic
	High resolution	Electronic devices contraindicated
	High gray–white contrast	Acute hemorrhages problematic
	No bone-generated artifact in posterior fossa	Relative measurements only
	Can also perform chemical, functional, and angiographic imaging	
Positron emission tomography (PET)	Can perform hemodynamic, chemical, and functional imaging	Ionizing radiation
	Quantifiable results	High initial costs
	Absolute physiologic variables can be determined	Long development time for new tracers
	Uniform spatial resolution	Limited access
		Low temporal resolution
Single photon emission computed tomography (SPECT)	Can perform hemodynamic, chemical, and functional imaging	Ionizing radiation
	Widely available	Relative measurements only
		Nonuniform spatial resolution
		Low temporal resolution
Xenon-enhanced computed tomography (XeCT)	Uses existing equipment	Ionizing radiation
		High xenon concentrations have pharmacologic effects
Helical computed tomography (CT Angiography, CTA)	Provides high-resolution vascular images	Ionizing radiation
		Vascular and bony anatomy only
Electroencephalography (EEG)	No ionizing radiation	Low spatial resolution
	High temporal resolution	Weighted toward surface measurements
	Widely available	
	Can identify epileptic foci	
Magnetoencephalography (MEG)	No ionizing radiation	Low spatial resolution
	High temporal resolution	
	Can identify epileptic foci	
Transcranial magnetic stimulation (TMS)	No ionizing radiation	Low spatial resolution
	Potential for therapy	Has produced seizures in certain patient groups
	Can be linked to other imaging methods (PET, MRI)	
Optical intrinsic signal (OIS) imaging	No ionizing radiation	Complex signal source
	High temporal resolution	Invasive only (intraoperative)
	High spatial resolution	

that their firing produces (Gevins, 1996). The electrical techniques are weighted toward surface structures whereas the magnetic ones convey information about deeper brain structures with less distortion. This indicates another important aspect in assessing methods for the evaluation of disease states. That is, certain techniques are better suited to examination of particular sites in the brain than others. A detailed description of these methodological limitations can be found in the first volume in this series, *Brain Mapping: The Methods* (Toga and Mazziotta, 1996) and other texts (Mazziotta and Gilman, 1992; Toga and Mazziotta, 2000).

Second, a number of techniques are devoted specifically to describing brain structure. These include X-ray computed tomography (CT), conventional magnetic resonance imaging (MRI), and blood vessel imaging using either conventional angiography, magnetic resonance angiography (MRA), or helical CT. A number of techniques assess hemodynamic responses as a measure of function. These techniques include xenon-enhanced CT, functional MRI (fMRI), perfusion MRI (pMRI), and cerebral blood flow or blood volume measurements using positron emission tomography (PET) or single photon emission computed tomography (SPECT). All the techniques that evaluate cerebral function based on hemodynamic measurements are, by their very nature, at a physiological "distance" from the actual neuronal event. These methods assume that neuronal firing and blood flow increments or decrements are tightly coupled (Roy and Sherrington, 1890; Kety and Schmidt, 1945). In most cases, this holds true in the normal brain, a discussion of which is provided in detail in the second volume in this series, *Brain Mapping: The Systems* (Toga and Mazziotta, 2000), but may not always be true in pathologic states.

Third, the determination of chemical processes in the brain falls mainly in the domain of PET and magnetic resonance spectroscopy (MRS). The former can measure cerebral glucose metabolism (Sokoloff *et al.*, 1977; Phelps *et al.*, 1979; Reivich *et al.*, 1979; Huang *et al.*, 1991), protein synthesis (Hawkins *et al.*, 1989; Keen *et al.*, 1989), amino acid uptake, pH, and other variables and does so in a quantitative manner reported in physiological units when appropriate rate constants and other factors are incorporated into mathematical models for their estimation (Raichle, 1983; Phelps and Mazziotta, 1985; Mazziotta and Phelps, 1986). MRS provides relative measurements of chemical compounds relying primarily on hydrogen spectra but, at higher magnetic field strengths, can also estimate relative quantities of sodium, fluorine, carbon, and phosphorus containing molecules as well (Prichard and Shulman, 1986; Chance *et al.*, 1988; Chang *et al.*, 1991; Prichard *et al.*, 1991; Gruetter *et al.*, 1992; Prichard and Rosen, 1994).

Fourth, the evaluation of receptor systems in the brain, both transmitter molecules and receptor complexes, has been an active area of research in both health and disease using PET and SPECT (Frost, 1982). Most information has been derived for the dopaminergic system but data also exist for the cholinergic, serotonergic, opioid, and benzodiazepine (Mintun *et al.*, 1984; Farde *et al.*, 1986; Perlmutter *et al.*, 1986; Wong *et al.*, 1986; Arnett *et al.*, 1987; Fowler *et al.*, 1987; Savic *et al.*, 1988; Huang *et al.*, 1991; Madar *et al.*, 1997) systems.

Lastly, there are interactive approaches. The most time honored, of course, is the direct observation of signs and symptoms in patients with cerebral disorders. Such information can be obtained in the traditional clinical setting or in the highly unusual circumstance of awake surgical procedures where recording from or stimulation of cerebral tissue can be correlated with behavioral states in a conscious patient (Penfield and Rasmussen, 1950). These latter methods, while in use for more than 50 years, still provide unique information about structure–function relationships in the human brain. Recently, such measurements have been augmented through the use of optical intrinsic signal (OIS) imaging, in which changes in optical reflectance from the cortex are measured during surgery (Grinvald *et al.*, 1986; Frostig *et al.*, 1990; Haglund *et al.*, 1992; Toga *et al.*, 1995). Measurements can be made either in awake patients during behavior or with anesthetized patients receiving sensory stimulation or direct peripheral nerve stimulation. They provide measures of functional specificity for different brain regions in the operative field. The technique of transcranial magnetic stimulation allows the investigator to stimulate the cortex of the brain magnetically, resulting in an induced electrical discharge in the cortex and an observed or reported behavior from the subject (Merton and Morton, 1980; Mills, 1991; Wasserman *et al.*, 1992; Pascual-Leone *et al.*, 1994; Wasserman *et al.*, 1994). This technique has been used experimentally to map normal brain systems and in patients for experimental, diagnostic, and therapeutic purposes.

Each technique is capable of making measurements with a characteristic resolution in both the spatial and temporal domains. Tomographic imaging techniques produce the highest spatial resolution currently available whereas the electromagnetophysiological methods provide the highest temporal resolution. Although knowledge of resolution is important, it must be matched to the question of interest in a particular patient population—a decision that also requires knowledge of the sampling characteristics of the technique. This latter term refers to the volume of brain tissue that can be assessed with a particular measurement. Thus, whereas fMRI may survey functional responses throughout the entire brain, electrophysiological measurements from a depth electrode, despite producing data with exquisite spatial and temporal resolution, sample only a very small volume of brain tissue. As such, investigations aimed at trying to identify a functional disorder in a disease of unknown etiology, e.g., autism, would be better done with a global technique that surveys the entire cerebral landscape rather than with measurement at multiple sites with an electrophysiological method such as depth electrodes. The latter are better used to understand the local electrophysiology of a site that has a high probability of being abnormal, and possibly also causative of a given disorder.

With these principles in mind, we now consider a more specific examination of the different strategies that can be employed with modern brain mapping methods to assess patients, either as individuals or as groups, with neurological, neurosurgical, or psychiatric disorders (Table II).

I. Diagnostic Methods

A. Individual Processes

1. Structural Anatomy

a. Computed Tomography X-ray computed tomography (CT) was the first noninvasive imaging technique that allowed for direct visualization of the brain parenchyma (Newton and Potts, 1981; Williams and Haughton, 1985). It revolutionized the evaluation of patients with neurological and neurosurgical disorders because it could image bone and provided the first opportunity to see the brain directly (Brooks and Di Chiro, 1976). It has a small dynamic contrast range so that differentiation of gray and white matter is difficult (Fig. 1A). However, it is very sensitive for identifying cerebral hemorrhage and also lesions associated with an alteration in the blood–brain barrier, by virtue of leakage of iodinated contrast material into them (Williams and Haughton, 1985). These abilities make X-ray CT ideal for direct and immediate assessment of patients with cerebral hemorrhage, multiple sclerosis, brain tumors, and traumatic injuries. Contrast sensitivity and the time needed for scanning have improved since X-ray CT was introduced, but with the advent of MRI technology, many diagnostic studies that were formerly in the province of X-ray CT are now done with MRI. Nevertheless, X-ray CT continues to have an important role in resolving certain diagnostic questions and in particular patient circumstances (Mazziotta *et al.,* 1995).

X-ray CT has remained the imaging modality of choice for patients requiring urgent evaluation of suspected intracranial hemorrhage (Fig. 1C) and in patients with acute head trauma. In both these circumstances, the speed of the study and ease of patient access as well as availability of equipment are well matched to the ability of CT to evaluate such patients. X-ray CT is also the procedure of choice when evaluating abnormalities of bony structures of the head, particularly the skull base (Fig. 1B). Lastly, patients that cannot tolerate MR imaging because of claustrophobia or implanted ferromagnetic (Klucznik *et al.,* 1993; Kanal *et al.,* 1996; Shellock, 1997) or electronic devices or that need to be attached to ancillary equipment such as is frequently encountered in critical care situations (MacNamara and Evans, 1995) are also best scanned by X-ray CT if structural information is needed for clinical evaluation (Salcman and Pevsner, 1992; Moran *et al.,* 1994).

b. MRI Magnetic resonance imaging (Lauterbur, 1973; Mansfield, 1977) is the structural imaging modality of choice in all other situations. Its superior spatial resolution and contrast range, particularly useful in differentiating gray and white matter, are but two features of MRI that make it superior to CT for structural imaging (Atlas, 1991) (Fig. 2). The ability to image brain structures from any angle and avoidance of CT artifacts that result from soft tissue-dense bone boundaries in the field of view provide further arguments for the use of MRI in patients with cerebral disorders. This advantage is most notable in the posterior fossa, where artifacts from the dense petrous bones often obscure or obliterate relevant clinical information about the brainstem and cerebellum (see Chapter 3, Fig. 1). Gadolinium and other paramagnetic contrast agents can be used with MRI to provide the ability to detect blood–brain barrier defects in a manner analogous to that used in X-ray CT with iodinated contrast agents (Atlas, 1991).

2. Vascular Anatomy

a. MR Angiography The observation that protons leaving the field of view reduce the local signal in MR imaging studies has led to an entire field of MR angiography and associated flow-based MR imaging techniques (Atlas, 1991; Wilcock *et al.,* 1996). The so-called flow void occurs in the vascular system when blood that encounters a radiofrequency pulse leaves the field of view of the scanner and the resultant energy is therefore emitted outside the field of view. The result is a loss of signal within the lumina of blood vessels. When particular pulse sequences are utilized to optimize this effect, an image of vascular anatomy results. Like most angiographic procedures, the image depicts the contents of the blood vessel (within the lumen) as opposed to the blood vessel wall and associated structures. However, unlike conventional angiography, in which arterial, cap-

Table II Brain Mapping Methods of Use in the Study of Human Disease along with the Types of Measurements They Provide and Some of the Clinical Situations Where They May Be of Use

Method	Measurements provided	Disorders
X-ray computed tomography (CT)	Brain structure	Acute/chronic hemorrhages
	Blood–brain barrier integrity	Acute trauma
		General screening of anatomy
		Focal or generalized atrophy
		Hydrocephalus
Magnetic resonance imaging (MRI)	Brain structure	Acute ischemia
	Brain and cervical vasculature	Neoplasms
	Relative cerebral perfusion	Demyelinating disease
	Chemical concentrations	Epileptic foci
	Fiber tracts	Degenerative disorders
	Blood–brain barrier integrity	Infections
		Preoperative mapping
Positron emission tomography (PET)	Perfusion	Ischemic states
	Metabolism	Degenerative disorders
	Substrate extraction	Epilepsy
	Protein synthesis	Movement disorders
	Neurotransmitter integrity	Affective disorders
	Receptor binding	Neoplasms
	Blood–brain barrier integrity	Addictive states
		Preoperative mapping
Single photon emission computed tomography (SPECT)	Perfusion	Ischemic states
	Neurotransmitter integrity	Degenerative disorders
	Receptor binding	Epilepsy
	Blood–brain barrier integrity	Movement disorders
Xenon-enhanced computed tomography (XeCT)	Perfusion	Ischemic states
Helical computed tomography (CT Angiography, CTA)	Vascular anatomy	Vascular occlusive disease
	Bony anatomy	Aneurysms
		Arteriovenous malformations
Electroencephalography (EEG)	Electrophysiology	Epilepsy
		Encephalopathies
		Degenerative disorders
		Preoperative mapping
Magnetoencephalography (MEG)	Electrophysiology	Epilepsy
Transcranial magnetic stimulation (TMS)	Focal brain activation	Preoperative mapping
Optical intrinsic signal imaging (OIS)	Integrated measure of blood volume, metabolism, and cell swelling	Intraoperative mapping

illary, and venous phases are distributed in time and images of each phase can be produced independently, MR angiography provides a composite image of all medium- to large-diameter vessels, including arteries and veins.

Such studies have provided important opportunities for the evaluation of intracranial and cervical, medium and large vessels for abnormalities, including arteriovenous malformations, aneurysms, and occlusive disease (see Chapter 3, Fig. 2). Smaller caliber vessels still require conventional angiographic or helical CT (see below) evaluation for the assessment of disorders such as vasculitis.

b. Helical CT In this technique, also called spiral CT or CT angiography, conventional CT technology is modified to produce very rapid sequential images of the

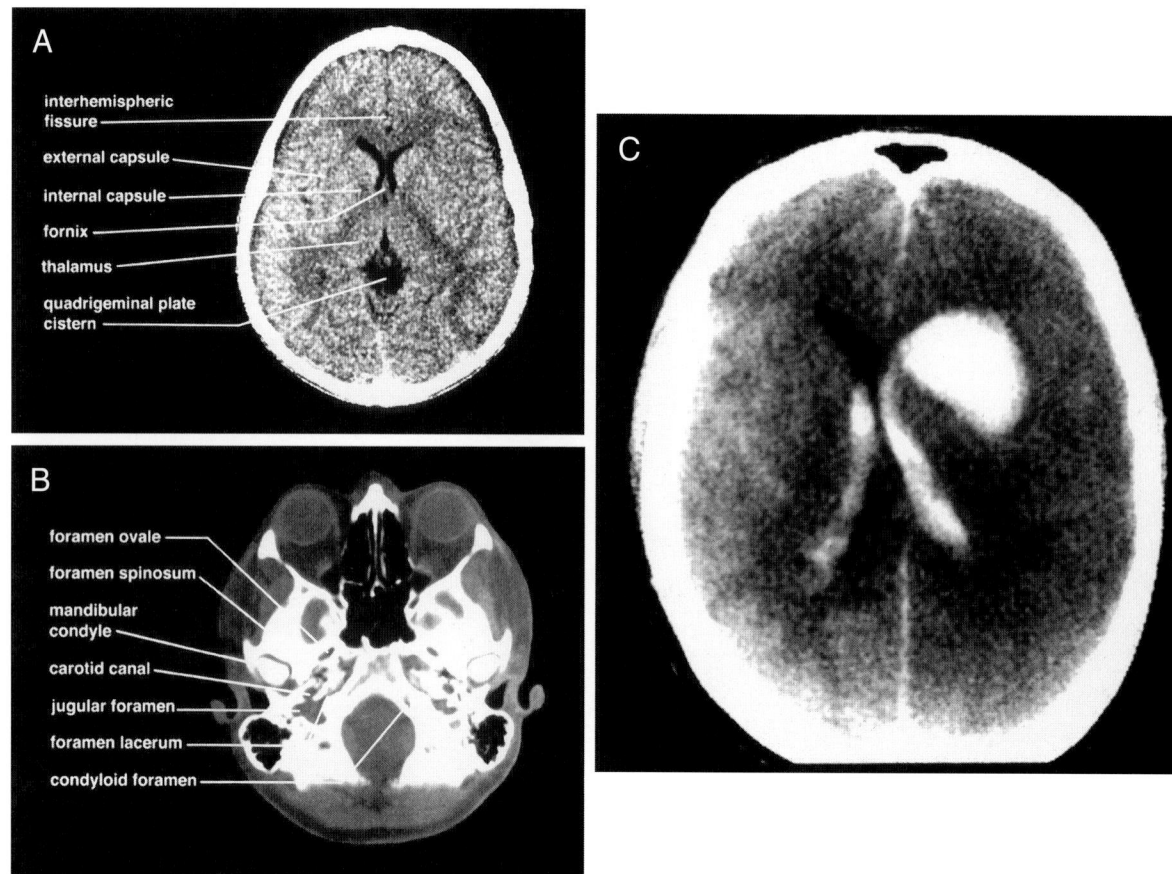

Figure 1 X-ray CT. (A) Images of the human brain from an X-ray CT device, demonstrating good anatomical detail, particularly of the skull and ventricular system as well as the subarachnoid CSF spaces. Note that there is less gray–white contrast than in MRI images (Fig. 2A). (B) X-ray CT provides very detailed images of bony structures that surround the central nervous system. This is particularly useful in evaluating pathologic states at the base of the skull, where conventional radiography is often difficult because of patient positioning and the overlap of bony structures in a two-dimensional radiograph. Further, in situations where trauma is a factor, often patients cannot be manipulated easily because of the possibility of fractures at the base of the skull or in the cervical spine. (C) Intracerebral hemorrhage demonstrated by X-ray CT. Sensitivity for detection of intracranial bleeds is effectively 100% with X-ray CT and it remains the imaging modality of choice in acute patients when identification of cerebral hemorrhage is urgent and important. This is typically the case in patients with an acute cerebral deficit when cerebral hemorrhage must be identified if thrombolytic or anticoagulant therapy is being contemplated.

head by having the relationship between the patient and the X-ray tube/detector system traverse a helical course through the tissue (Kaatee *et al.,* 1997; Achenbach *et al.,* 1998; Garg *et al.,* 1998). This process is rapid and so arranged that it is coincident with the delivery of a bolus of iodinated contrast material into the cranial and cervical vessels from a peripheral vein. The resultant images are high-resolution depictions of the intracranial anatomy that can be reconstructed in three dimensions (Fig. 3A). At present, limited information is available about the technique in terms of its clinical applications but it is likely that there are circumstances when it may be superior to conventional angiography. One of those is in the assessment of the local vascular anatomy of patients with aneurysms and arteriovenous malformations. In this case, conventional angiography, while high in spatial resolution, collapses the three-

dimensional structure of such lesions into a two-dimensional projection (Osborn, 1980; Morris, 1997). As such, the important relationship between aneurysm neck and a parent or daughter vessel may be difficult to ascertain or may require multiple intra-arterial contrast injections and radiation exposure for the patient. Helical CT allows for a true three-dimensional reconstruction of local anatomy that can be manipulated to assess such relationships in greater detail, while subjecting the patient to only a single radiation exposure and dose of iodinated contrast material (Fig. 3B). A similar situation exists when defining feeder vessels to arteriovenous malformations.

3. Blood Flow and Perfusion

a. PET The assessment of cerebral perfusion (i.e., cerebral blood flow per volume of tissue) can be determined quantitatively with PET using ^{15}O-labeled water,

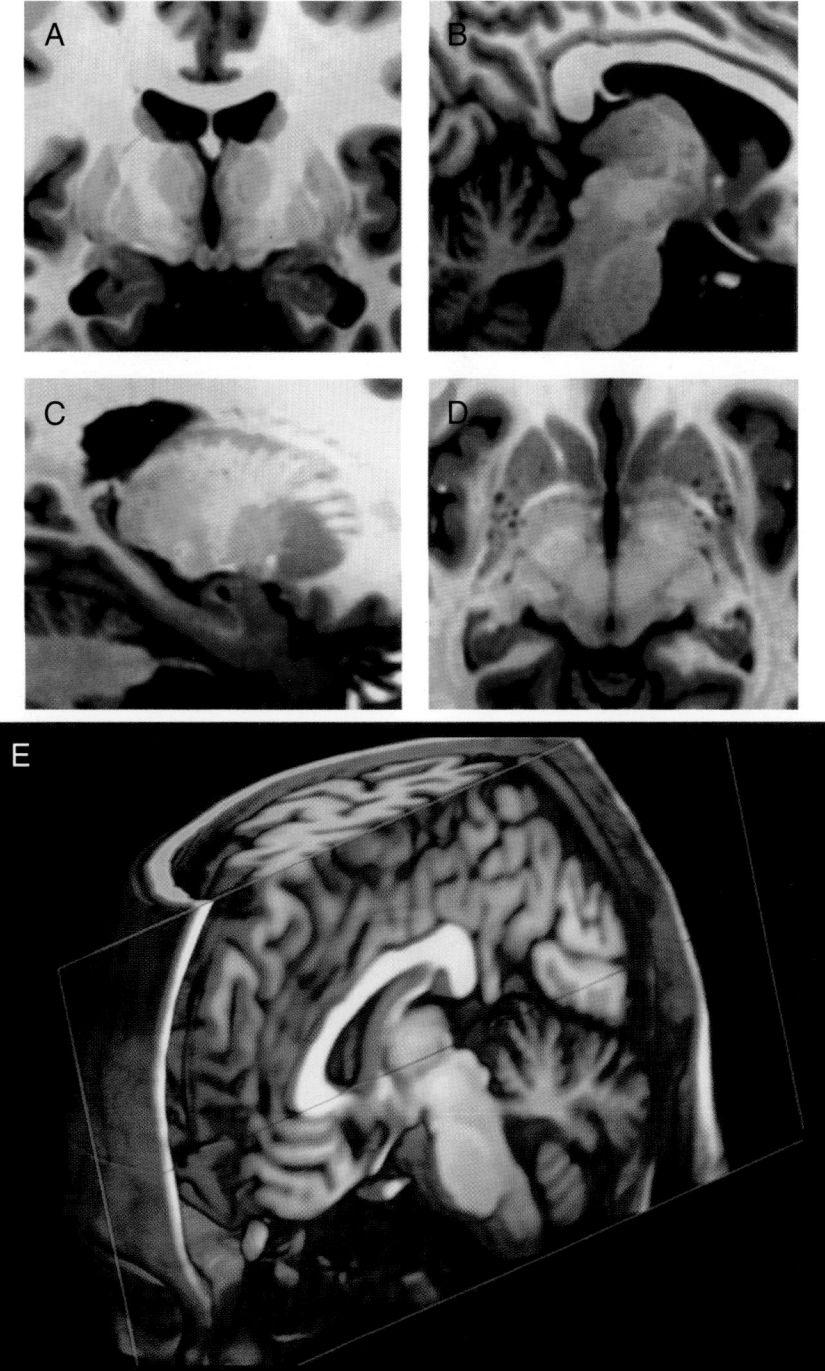

Figure 2 Magnetic resonance imaging. (A-D) Typical two-dimensional MRI images of the brain. Notice that the detailed anatomy of the brain parenchyma has better gray–white contrast than X-ray CT images (Fig. 1A). Notice also that there are none of the typical CT artifacts caused by the juxtaposition of dense bone and brain parenchyma. (A) Coronal view through the thalamus demonstrating the subnuclei of the thalamus, the mamillary bodies, and the internal, external, and extreme capsules as well as the two segments of the globus pallidus. Also note the detailed anatomy of the hippocampi. (B) Sagittal view demonstrating the colliculi of the midbrain, the midline of the thalamus, and the detailed anatomy of the midsagittal region of the cerebellum. (C) Sagittal view through the hippocampus and striatum. Note the fine bridges of gray matter between the caudate and the putamen. (D) Transverse section through the basal ganglia and upper midbrain. Note the periaqueductal gray matter, the detailed anatomy of the hippocampi, and the bilateral flow voids produced by the presence of the lenticulostriate arteries in the posterior portion of the putamen. (E) Three-dimensional reconstruction with cutaway of an MRI data set demonstrating the kind of anatomical detail that can be provided with three-dimensional MR images. Courtesy of Colin Holmes and colleagues, UCLA School of Medicine, Los Angeles, California. (A–D) Holmes *et al.* (1998). Enhancement of magnetic resonance images using registration for signal averaging. *J. Computer Assis. Tom.* **22**(1), 139–152.

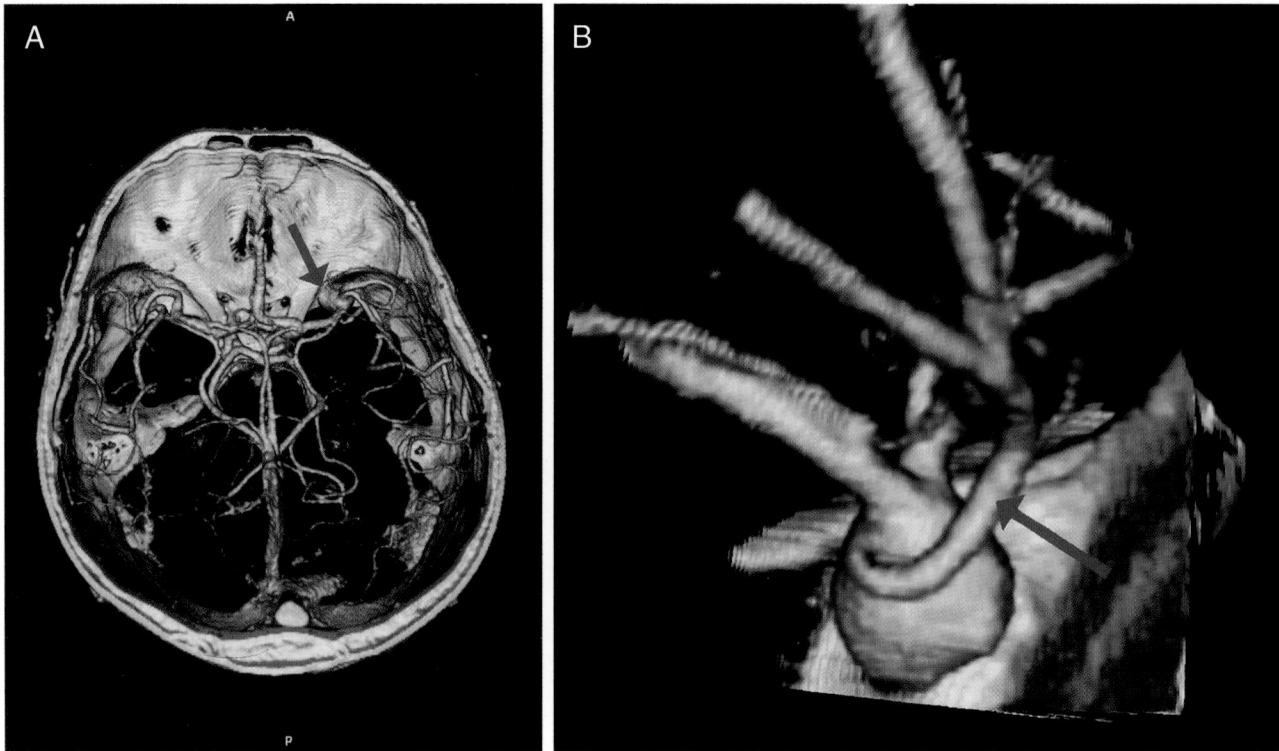

Figure 3 Helical CT angiography. (A) The cerebral vasculature is superimposed on this cutaway view of the skull seen from above. Note the fine detail provided for the vascular structures, including the Circle of Willis, the anterior, middle, and posterior cerebral arteries, and many of their branches. The red arrow indicates an aneurysm arising from the middle cerebral artery at the anterior edge of the middle cerebral fossa. (B) Close-up view of the aneurysm demonstrated in A. Note that in this three-dimensional reconstruction, it is possible to see all of the blood vessels that contribute to or arise from the aneurysmal sack (seen on the left). Unlike conventional angiography, where the overlap of the aneurysm and the parent or daughter vessels can be obscured because of the two-dimensional projection required in this technique, full three-dimensional images are possible with helical CT angiography. By digital reconstruction and rotation of the data set from helical CT, these complex and important relationships can be evaluated prior to surgery, whereas with conventional angiograms, multiple views would be required and still may not provide a sufficiently detailed view of these relationships. In addition, multiple views obtained with conventional angiography expose the patient to additional radiation and contrast risks. In this case, the angular artery (red arrow) arises from the aneurysm sack. Endovascular coil placement might lead to obstruction of this important vessel, thereby making such a patient a candidate for surgical rather than endovascular treatment of this lesion. Courtesy of Pablo Villablanca, UCLA School of Medicine, Los Angeles, California.

[11]C-labeled butanol, and potentially other agents (Jones *et al.*, 1976; Raichle *et al.*, 1983; Mazziotta *et al.*, 1985). These agents are freely diffusible and knowledge of the time–activity relationship of the tracer compound in arterial blood and the tissue concentration over time in the brain permits a calculation of cerebral perfusion with approximately $5\times5\times5$ mm spatial resolution. These methods have been used extensively to evaluate increments in perfusion associated with underlying neuronal activity such as that associated with the performance of behavioral tasks (Fox *et al.*, 1984; Mazziotta *et al.*, 1985) and, as such, can be used in patients to evaluate critical cortical and subcortical areas as part of the process of mapping brain regions adjacent to abnormal-

ities that are under consideration for surgical resection (see below).

Perfusion measurements are also of value in assessing patients with ischemic cerebrovascular disease (Baron, 1992) (see Chapter 13, Fig. 3), not only to determine baseline conditions but also in the assessment of "cerebral perfusion reserve" by challenging such patients with cerebral vasodilation drugs such as acetazolamide.

b. SPECT Perfusion measurements with SPECT (Kuhl and Edwards, 1964; Patton *et al.*, 1969; American Academy of Neurology (AAN), 1996) can be obtained in a semiquantitative way using xenon-133 (Obrist *et al.*, 1967, 1975) or in a relative fashion using a host of technetium-99m labeled agents (e.g., HMPAO) (Hung *et al.,* 1988; Lear, 1988; Nakano *et al.,* 1989; Rattner *et al.,* 1991). Studies have been used to evaluate patients with cerebrovascular disease (Chollet *et al.,* 1989; Oshima *et al.,* 1990; Devous *et al.,* 1993) as noted above and, more recently, in the assessment of patients with epilepsy (Devous *et al.,* 1990; Gzesh *et al.,* 1990; Rowe *et al.,* 1991a,b; Newton *et al.,* 1992; Berkowitz *et al.,* 1993; Ho *et al.,* 1995). In the latter circumstance, ictal, postictal, and interictal studies are made using these agents and subtracted to identify foci of epileptic discharges that increase cerebral perfusion during seizures and reduce perfusion postictally and interictally (O'Brien *et al.,* 1998) (Fig. 4).

c. Xenon-Enhanced CT Nonradioactive xenon-133 alters the X-ray attenuation characteristics of the tissues that absorb it (Drayer *et al.,* 1979). Xenon-133 can be readily inhaled. Using strategies similar to those discussed above, one can calculate changes in X-ray attenuation associated with the concentration of xenon in tissue to calculate an index of cerebral perfusion. This approach has been used to evaluate patients with cerebrovascular disease and head trauma. One confounding factor with this approach is that relatively high concentrations of xenon are needed in brain tissue to produce accurate perfusion estimates. At these concentrations, patients experience pharmacological effects, including sedation and, ultimately, anesthesia. These direct neuronal effects of xenon can, in turn, affect cerebral perfusion and contaminate physiological measurements.

Nevertheless, such studies have been used to assess patients in specific diagnostic categories when alternate techniques are not available.

d. Magnetic Resonance Imaging Relative and semiquantitative measurements of cerebral perfusion can be obtained with MRI using a variety of techniques (Moseley *et al.,* 1990; Warach *et al.,* 1992). The basic difference among these techniques is whether exogenous agents are infused into the patient or whether changes in endogenous signals from the vascular system are monitored. With the former, bolus contrast agents such as gadolinium are delivered to a subject intravenously (Villringer *et al.,* 1988; Belliveau *et al.,* 1991; Turner *et al.,* 1991; Kennan *et al.,* 1994). Their local concentration in the cerebral vasculature estimates local cerebral perfusion, which, in turn, is proportional to neuronal firing rates. The change in MR signal, induced by a local change in concentration of a contrast agent, reflects a relative change in local perfusion that can be estimated. A refinement of this approach requires information about the concentration of the contrast agent in the blood over a period of time. This can be measured directly from an arterial source, although such a method is rarely employed, or from a large-diameter blood vessel in the scanner's field of view, e.g., arteries in the Circle of Willis.

Alternatively, one can record endogenous local signal changes associated with alterations in cerebral perfusion induced by changing neuronal activity (Kwong *et al.,* 1992; Ogawa *et al.,* 1990, 1993; Turner *et al.,* 1993; Frahm *et al.,* 1994). When local neuronal firing increases,

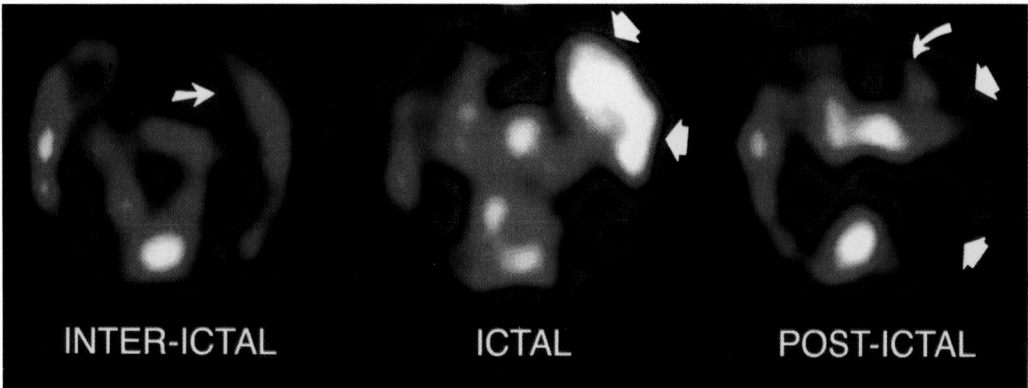

Figure 4 SPECT studies of a seizure focus in a patient with epilepsy, obtained using technetium-99m-labeled HMPAO. By comparing the ictal study (hyperperfusion), obtained by injecting the tracer on a telemetry ward, with the postictal and interictal scans (hypoperfusion), it is possible, with a high degree of accuracy, to identify seizure foci responsible for focal epilepsies. Although such SPECT studies of relative perfusion may be confusing or misleading when evaluated in only one of these states, the composite information obtained from ictal, postictal, and interictal studies, particularly when the studies are aligned, registered, and subtracted, is useful in the presurgical evaluation of patients with focal or complex partial epilepsy. Courtesy of Sam Berkovic, Austin Hospital, Melbourne, Victoria, Australia. From Berkovic, S. F., Newton, M. R., and Rowe, C. C. (1991). Localization of epileptic foci using SPECT. *In* "Epilepsy Surgery" (H. Luden, ed.), pp. 251–256. Raven.

there is an increment in cerebral blood flow to an area. There is, however, a proportionally smaller change in local tissue oxygen metabolism (Fox and Raichle, 1986). As a result, more oxygen is delivered per unit volume of tissue (because of decreased fractional extraction) but there is little change in the amount of oxygen used. Thus, the oxygen content (i.e., the concentration of oxyhemoglobin) in venous blood increases. When compared to images in a state of "baseline" neuronal firing, this local change in venous oxyhemoglobin results in an alteration of the MR signal since deoxygenated blood is more paramagnetic than oxygenated blood (the difference is about 0.2 ppm) (Pauling and Coryell, 1936). The change in signal provides an estimate of relative local cerebral perfusion changes associated with the change in underlying neuronal activity. This observation has resulted in the widespread use of this so-called BOLD (Blood Oxygen Level Dependent) technique (Ogawa *et al.*, 1990, 1993) in the evaluation of normal subjects performing behavioral tasks and in patients who are candidates for surgical resection of brain lesions such as tumors, vascular malformations, or epileptic foci. Such methods may ultimately supplant more invasive approaches to determining language dominance (Hunter *et al.*, 1999) and memory function such as those employing the intra-arterial injection of barbiturate compounds (e.g., WADA testing).

Diffusion imaging is also possible using MR scanning. Water, and hence protons, is freely diffusible. In the brain, however, the microscopic anatomy of the tissue compartmentalizes (with cells, along axons, etc.) this process (Warach *et al.*, 1992; Sorensen *et al.*, 1996; Baird and Warach, 1998). It is possible to obtain MRI images that are diffusion weighted (DW) and in which the signal is dependent on the ease with which protons diffuse in their local environment. Initially tested in cats (Moseley *et al.*, 1990), DW images were shown to demonstrate the boundaries of acute infarcts within minutes. Infarcted or ischemic areas are visible as regions of hyperintensity corresponding to local decreases in the apparent diffusion coefficient (ADC) of water. The use of DWI in screening patients with early ischemia who are candidates for thrombolytic therapy is especially relevant (Warach *et al.*, 1995a; Sorensen *et al.*, 1996; Lutsep *et al.*, 1997; Baird and Warach, 1998) (Fig. 5). When combined with MRI perfusion imag-

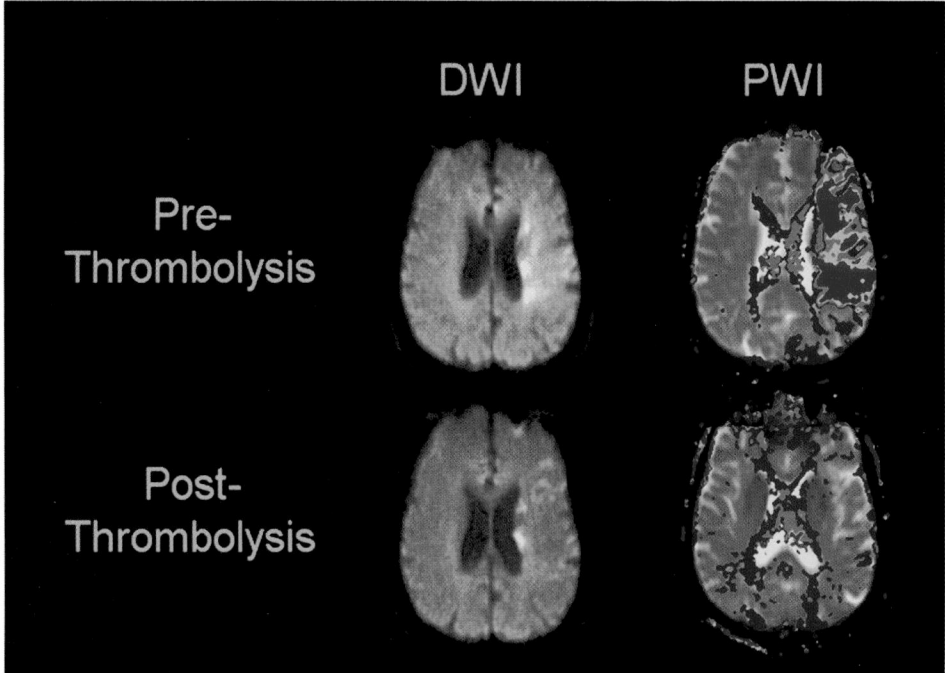

Figure 5 Diffusion-weighted (DWI) and perfusion-weighted (PWI) images from MRI are useful in selecting patients for cerebrovascular intervention therapies and monitoring their outcome. Here, a patient with a large region of hypoperfusion in the distribution of the middle cerebral artery (red and yellow areas of the PWI images) already has some evidence of tissue ischemia as demonstrated by the white areas that border the ventricle in the prethrombolysis DW image (top left). Following thrombolysis (bottom row), note that perfusion has been reestablished to the area of the middle cerebral artery and the patient is left with only a small ischemic injury on the posterior border of the lateral ventricle. Courtesy of Chelsea Kidwell, Jeffrey Saver, and Jeffrey Alger, UCLA School of Medicine, Los Angeles, California.

ing (Baird and Warach, 1998), an assessment of the proportion of jeopardized and infarcted tissue can be made (see Chapter 3, Fig. 4).

4. Metabolism

a. PET Glucose and oxygen metabolism can be assessed using PET and ^{18}F-labeled fluorodeoxyglucose (FDG), ^{11}C-labeled glucose, and ^{15}O-labeled oxygen (Jones *et al.,* 1976; Sokoloff *et al.,* 1977; Phelps *et al.,* 1979; Reivich *et al.,* 1979; Huang *et al.,* 1980; Raichle *et al.,* 1983; Mazziotta *et al.,* 1985). A common approach has been to use FDG to evaluate glucose metabolism. Thought to be primarily an assessment of synaptic activity (i.e., deoxyglucose uptake is maximal in the neuropil rather than in regions dominated by cell bodies), scanning of patients using FDG has provided useful observations in a wide range of disorders. Hypometabolic regions are found interictally

at epileptic foci (Chugani *et al.,* 1988, 1990; Engel *et al.,* 1990) and in specific cortical and subcortical regions in a wide range of degenerative (Kuhl *et al.,* 1982; Hawkins *et al.,* 1987; Mazziotta *et al.,* 1987; Mazziotta, 1989; Gilman *et al.,* 1990) and dementing (Mazziotta *et al.,* 1992) processes and also appear to reflect the malignancy grade of cerebral neoplasms (see Chapter 3, Fig. 6), particularly gliomas (Di Chiro, 1985) (Figs. 6–8). Oxygen metabolism and extraction measurements are of greatest utility in assessing patients with cerebrovascular disease, particularly that unique subset of patients for which extracranial-intracranial bypass or other surgical reperfusion approaches are contemplated (Baron, 1992).

b. MR Spectroscopy Concentrations of a wide variety of compounds can be estimated using MR spectroscopy (Prichard and Shulman, 1986; Hopkins and

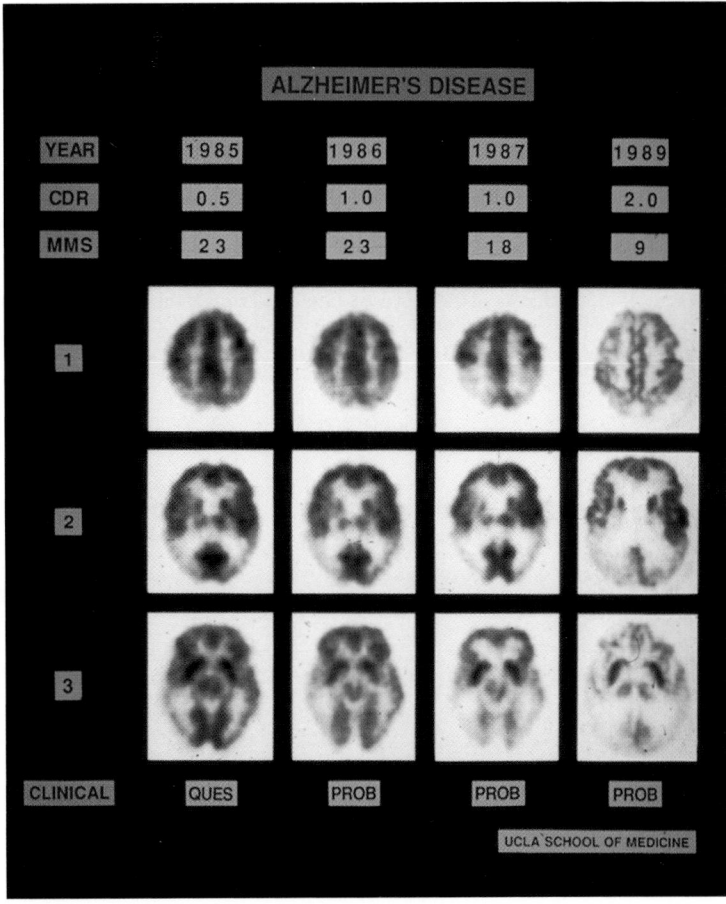

Figure 6 PET studies of cerebral glucose metabolism in a patient with Alzheimer's disease demonstrating the characteristic pattern of hypometabolism that occurs in the parietal and superior temporal cortices. As the disease progresses (columns from left to right), hypometabolism worsens in both spatial extent and magnitude, ultimately involving all neocortical structures.

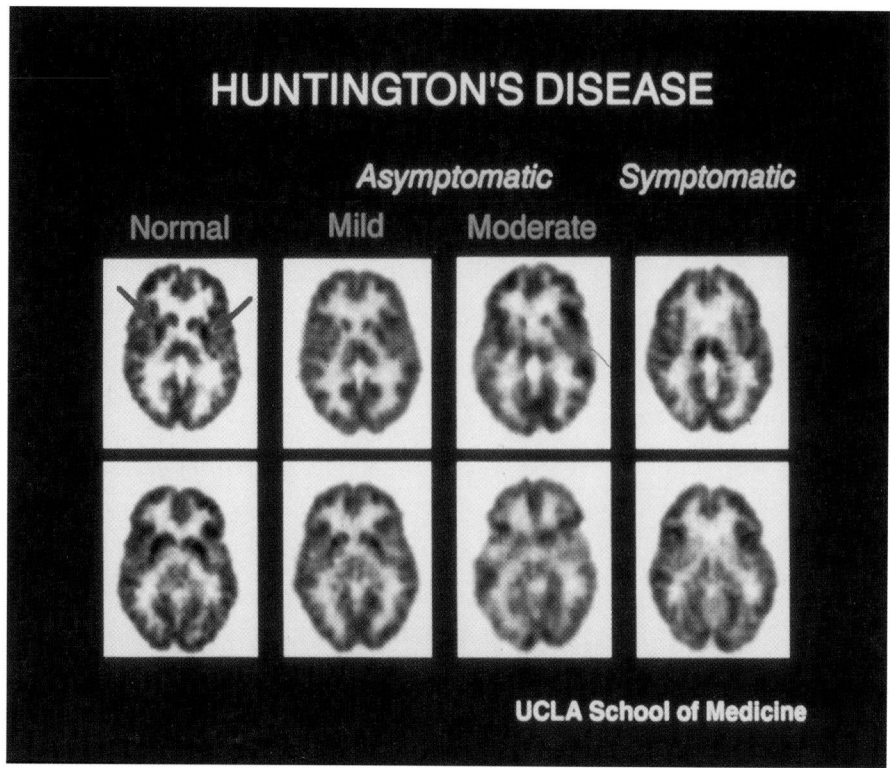

Figure 7 Glucose metabolism measured with PET in Huntington's disease. Two anatomical levels (rows) of a normal subject demonstrating glucose metabolism in the caudate and putamen as compared with three patients, two with mild or moderate asymptomatic Huntington's disease and one with frank symptoms. Note that in the symptomatic patient there is profound hypometabolism of the caudate and most of the putamen, whereas in the asymptomatic, gene-positive subjects, there is progressive loss of metabolism in the caudate. Such changes can be identified 5–7 years prior to the onset of symptoms in patients who carry an expanded triplet repeat sequence of the Huntington's disease gene. With advancing disease, hypometabolism extends throughout the striatum and can also involve the frontal cortices. Courtesy of Scott Grafton and John Mazziotta, UCLA School of Medicine, Los Angeles, California.

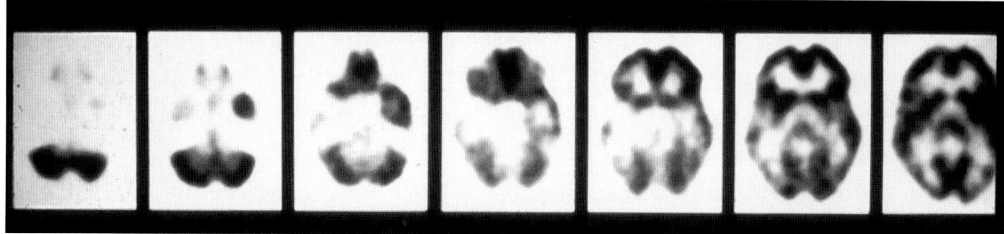

Figure 8 Interictal hypometabolism of the temporal lobe in a patient with complex partial epilepsy referable to that structure. Note the profound asymmetry of metabolism, particularly in the medial but also in the lateral portion of the temporal lobe, in six out of the seven axial images. Such studies are typically performed in the evaluation of patients who are candidates for surgical treatment of their epilepsy. In many patients, brain mapping techniques have obviated the requirement for depth electrodes and subdural grids in the presurgical evaluation of such individuals. Courtesy of Jerome Engel, Jr., *et al.*, UCLA School of Medicine, Los Angeles; J. Engel, P. H. Crandall, and R. Rausch (1983). The partial epilepsies. *In* "Clincal Neurosciences" (R. N. Rosenberg, ed.), pp. 1349–1380. Churchill Livingstone, New York.

Barr, 1987; Chance *et al.,* 1988; Barranco *et al.,* 1989; Chang *et al.,* 1991; Prichard *et al.,* 1991; Gruetter *et al.,* 1992; Sappey-Marinier *et al.,* 1992; McLaughlin *et al.,* 1993; Pritchard and Rosen, 1994; Wang *et al.,* 1996). Proton spectra provide information about lactate as a measure of carbohydrate metabolism. Other chemical species as well as spectra obtained from other isotopes can provide additional clues about various chemical pathways. In addition, "tracer" techniques can be employed, e.g., using $^{17}O_2$, ^{13}C-labeled glucose (Fox *et al.,* 1988; Pekar *et al.,* 1993; van Zijl *et al.,* 1993), analogous to those used with PET or SPECT (see Chapter 3, Fig. 8).

5. Ligands and Neuroreceptor Imaging

a. PET   Ligands have been developed for PET to image both pre- and postsynaptic neuroreceptors (Frost, 1982; Perlmutter *et al.,* 1986). Since one or both sides of a synapse can be affected by neuropsychiatric disease, the ability to image the integrity of both components of synaptic structure and function is useful. There are a large number of neurotransmitter systems in the human

brain but PET ligands have been developed for only some of these. Fewer still have been completely validated and are in clinical use.

The most extensively studied neurochemical system is the dopaminergic network. Presynaptic imaging of dopamine synthesis and reuptake have been evaluated with fluorinated (^{18}F) ligands. The uptake, metabolism, and flux of ^{18}F-labeled L-DOPA in presynaptic dopaminergic terminals have been used to evaluate movement disorders, particularly Parkinson's disease (Fig. 9), and a number of neuropsychiatric syndromes (Brooks *et al.,* 1992; Mazziotta, 1992; Schulz *et al.,* 1994). The same is true for the evaluation of presynaptic dopaminergic reuptake sites (e.g., with the WIN compounds). On the postsynaptic side, numerous ligands have been developed with a range of affinities for the different postsynaptic dopaminergic receptors (Farde *et al.,* 1986). These tracers include raclopride, spiperone, ethylspiperone, and others. They have provided data for the movement disorders, schizophrenia, pituitary tumors, and other disorders of the brain discussed in later chapters of this book. In addition, the la-

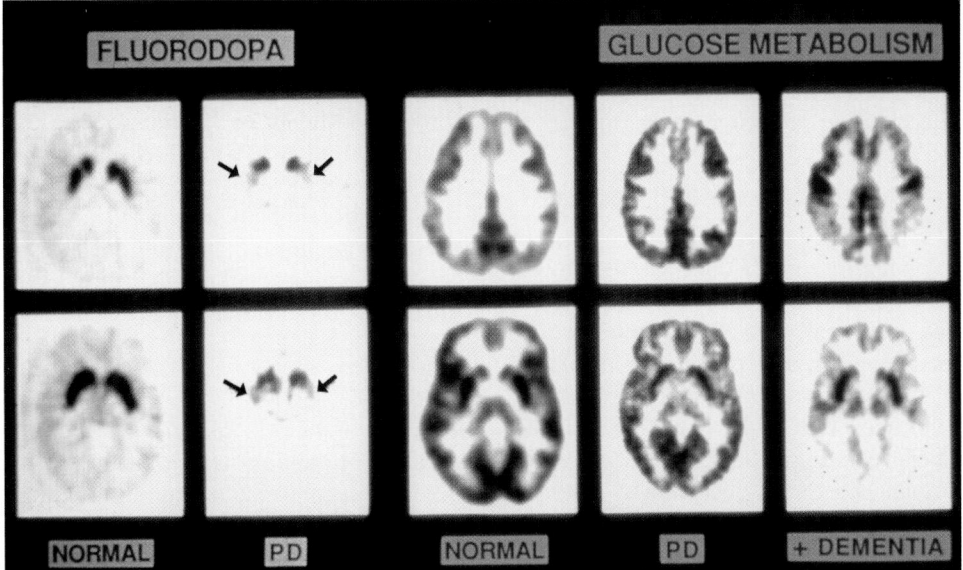

Figure 9 Evaluation of patients with Parkinson's disease using PET. Glucose metabolism in Parkinson's disease demonstrates a normal pattern in such patients (blue PD label) when compared to a normal control (middle column). When the dopamine-specific ligand [^{18}F]fluoro-L-DOPA is employed, however, a striking reduction in the uptake of this tracer is identified in the posterior putamen of patients with Parkinson's disease when compared to normals (columns 1 and 2). This is because the majority of the presynaptic dopaminergic terminals in Parkinson's disease are lost from the putamen at the onset of Parkinson's disease symptoms. Nevertheless, this population of synapses represents only a minority of all the synapses in this structure. As such, glucose metabolism remains normal but the uptake of fluoro-L-DOPA is dramatically reduced since this tracer images only that subpopulation of cells that have dopamine as their neurotransmitter. In patients with Parkinson's disease plus dementia, some patients also show hypometabolism of neocortex ("+DEMENTIA") similar to patients with Alzheimer's disease (see Fig. 6).

beling of drugs of abuse, such as ^{11}C-labeled cocaine, that bind to receptors in the dopaminergic system has been helpful as a means of exploring the neurobiology of chemical addiction.

Cholinergic tracers are in use for the study of neurodegenerative disorders such as Alzheimer's disease and markers of the serotonin system and have been employed in the evaluation of psychiatric disorders. The central benzodiazepine system can be scanned with flumazenil labeled with carbon-11 (Savic et al., 1988; Frey et al., 1991; Koepp et al., 1997). Important findings have been made with this tracer in patients with epilepsy (Fig. 10A). Finally, opioid receptors have been studied with a host of ligands that bind with varying affinities to the many subclasses of opiate receptor. These compounds have been used in the exploration of epilepsy, psychiatric disease, and movement disorders (Mayberg et al., 1991; Madar et al., 1997).

b. SPECT A number of iodinated compounds have been developed and serve as analogs of the PET tracers described above (Kung et al., 1990; Holman and Devous, 1992). Used in a similar fashion, relative estimates of binding and receptor uptake can be obtained with SPECT (van Huffelen et al., 1990) (Fig. 10B). Although the number of compounds of this type is few compared to the inventory of PET ligands, interest in their development and use will result in an ever-increasing set of SPECT tracers.

6. Electrophysiology

a. Electroencephalography (EEG) Electrophysiological techniques provide the best temporal resolution for studying neuronal activity in patients with neurosurgical, neurological, and psychiatric disorders. Although more limited in spatial resolution than the tomographic techniques, EEG data provide the investigator and clinician an opportunity to learn about the timing of events and their synchronicity (Berger, 1929; Abraham and Ajmone-Marsan, 1958; Chiappa, 1983; Gevins and Bressler, 1988; Daly and Pedley, 1990). Particularly important in the investigation of patients with seizures, EEG has been the mainstay of diagnostic evaluation of such patients (Engel, 1993). The limited spatial resolution of scalp EEG can be overcome by the use of more invasive techniques when patients are surgical candidates. In that setting, subdural grid electrodes, depth electrodes, and cortical surface electrodes can be used to obtain local extracellular field potential recordings and electrocorticograms (Engel, 1993).

System-specific information can be obtained with EEG when recording is accompanied by specific sensory, motor, or cognitive stimulation, a technique known as event-related potential (ERP) recording (Walter, 1963; Walter et al., 1964; Freeman, 1978; Chiappa, 1983; Coles et al., 1990). The resultant ERP maps are used to identify the general location and relative timing of a cortical representation of such functions in the brain (Gevins et al., 1979, 1987; Fender, 1987; Mesulam, 1990; Smith et al., 1990; Curran et al., 1993; Sanes and Donoghue, 1993). This approach has been used extensively to evaluate interruptions of the visual, auditory, or somatosensory systems (e.g., in multiple sclerosis) noninvasively and to provide more detailed cortical maps intraoperatively, or with subcortical or depth electrodes, in the evaluation of patients with seizure foci. Analysis of power spectra from scalp EEG and ERP recordings has also provided information of clinical use in neurodegenerative disorders and psychiatric syndromes.

b. Magnetoencephalography (MEG) The measurement of minute magnetic fields in the brain with MEG is analogous to the measurement of electrical fields with EEG. Requiring far more complex equipment, the MEG method may have greater spatial resolution and greater accuracy in identifying electrophysiological dipoles, both at the surface and in the depths of the brain. Clinically, the method has been used to identify seizure foci and in a research setting has been employed for the experimental investigation of a wide range of neuropsychiatric disorders. Data with this technique are limited in number due to the fact that only a small number of MEG installations are currently operational and evaluating patients clinically.

c. Transcranial Magnetic Stimulation (TMS) The creation of an intense focal magnetic field in the human cerebral cortex results in the induction of an electrical current that discharges cells lying tangentially within that volume of tissue (Merton and Morton, 1980; Barker et al., 1985; Mills, 1991; Jahromi et al., 1992; Wasserman et al., 1992, 1994; Pascual-Leone et al., 1992, 1994) (see Chapter 3, Fig. 12). The discharge can result in a pseudophysiologic response. That is, if the motor cortex is stimulated, the appropriate contralateral muscle groups will contract. If the visual cortex is stimulated, a subject or patient will see a flash or phosphene in the contralateral visual field. Transient high-frequency stimulation of the cortex by magnetic stimulation will temporarily and reversibly deactivate it

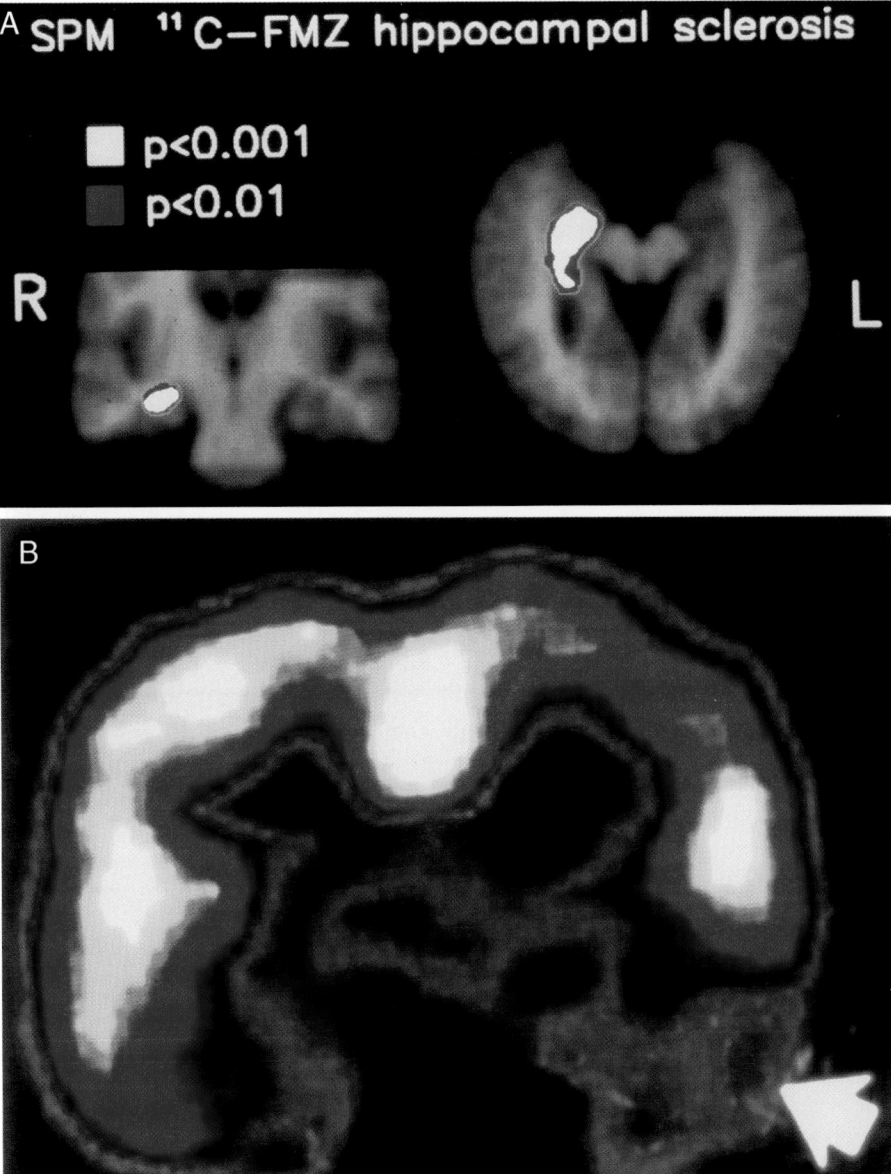

Figure 10 Benzodiazepine receptor imaging in patients with focal epilepsy. (A) PET evaluation with [^{11}C]flumazenil. In patients with focal temporal lobe epilepsy from hippocampal sclerosis producing complex partial seizures, flumazenil uptake is reduced in the medial temporal lobe. These changes are typically smaller in spatial extent than hypometabolism detected with FDG-PET. Comparison of the two studies may ultimately lead to more selective surgical resections of medial temporal lobe structures in such patients. Courtesy of John Duncan, National Hospital for Neurologic Disease, Queen Square, London. From *Brain* (1996), **119**, 1677–1687, with permission of Oxford University Press. (B) Similar imaging can be performed using SPECT and the iodinated compound iomazenil. Although slightly lower in spatial resolution, such studies provide information comparable to that discussed for flumazenil PET imaging in patients with focal epilepsy. Courtesy of Professor A. C. van Huffelen, University of Utrecht (van Huffelen *et al.*, 1990). From Berkovic *et al.* (1993). *In* "Surgical Treatment of the Epilepsies" (J. Engel, Jr., Ed.), 2nd Ed., p. 238. Raven, New York.

(Grafman *et al.*, 1994; Jennum *et al.*, 1994; Epstein *et al.*, 1996; Wasserman, 1998). This approach has been used to create reversible lesions and has also been used as a therapeutic maneuver in the treatment of chronic depression, in a fashion analogous to electroconvulsive shock therapy delivered focally to the frontal cortex (George *et al.*, 1996; Pascual-Leone *et al.*, 1996). When linked to a tomographic functional imaging technique, TMS can be used to map functional pathways directly in patients and normal subjects (Paus *et al.*, 1997). There are guidelines for the safe use of this technique, which must be used with caution in epileptic patients.

7. Optical Intrinsic Signal (OIS) Imaging

The newest brain mapping technique to be used in a clinical setting is optical intrinsic signal imaging. This method provides information about cortical blood flow, blood volume, and metabolism in an integrated fashion (Grinvald *et al.*, 1986; Frostig *et al.*, 1990; Haglund *et al.*, 1992; Toga *et al.*, 1995). The approach is straightforward. White light is shone onto the exposed cortex and the amount of light of different wavelengths reflected from the cortex is measured. The reflectance and wavelength composition of light changes as a function of the neural activity of the illuminated tissue as a function of blood volume, blood flow, cell swelling, and the oxidative state of the tissue (among other variables). An invasive technique, OIS is used in the operating room to provide functional maps in which neuronal activity is varied by stimulation of peripheral nerves or, in the awake patient, by the performance of behavioral (e.g., language) tasks (Haglund *et al.*, 1992; Toga *et al.*, 1995) (see Chapter 4, Figs. 9–12). The method has the best spatial and temporal resolution of all the functional imaging techniques, approaching 50 μm in the spatial domain and 50 ms in the temporal domain.

B. Multiple Modality Imaging

There are two types of multimodal integrative imaging studies that are important in the clinical evaluation of patients. The first is the within-subject integration of information from multiple brain mapping techniques, or the serial integration of multiple imaging studies in time using the same technique in the same individual (Fox *et al.*, 1985; Pelizzari *et al.*, 1989; Zhang *et al.*, 1990; Woods *et al.*, 1992, 1993; Collins *et al.*, 1994b; Strother *et al.*, 1994; Thurfjell *et al.*, 1994; Wieringa *et al.*, 1994; Turkington *et al.*, 1995). The second approach is the averaging or integration of information from multiple subjects, a much more difficult problem due to the

great anatomical and functional variability (Geschwind and Levitsky, 1968; Witelson, 1977; Galaburda *et al.*, 1978; Le May and Kido, 1978; Chui and Damasio, 1980; Weinberger *et al.*, 1982; Collins *et al.*, 1994a; Mazziotta *et al.*, 1995) that exists between individuals. The within-subject integration problem has yielded to a variety of excellent and elegant mathematical approaches for the alignment and registration of data. The between-subject problem is a more difficult one that is being resolved through the use of warping and morphing techniques that were described in the first volume of this series, *Brain Mapping: The Methods* (Toga and Mazziotta, 1996) (Talairach and Tournoux, 1988; Evans *et al.*, 1991; Friston *et al.*, 1991; Gee *et al.*, 1993; Thurfjell *et al.*, 1994).

1. Within-Subject Registration

A composite image of a patient derived from imaging using multiple imaging modalities or serial studies over time is a critical indicator of the clinical picture from an imaging perspective-for example, in a patient with a brain tumor in whom the natural history of the enlargement of the lesion can be evaluated quantitatively and objectively. Similarly, the integrated image of functional activation in cortical regions surrounding a lesion that is to be surgically resected can predict the relative risk of functional damage due to resection of normal cortex in the process of tumor ablation (Fig. 11).

Images of interictal spikes can be obtained by combining EEG and fMRI (Zhang *et al.*, 1990; Ives *et al.*, 1993; Wieringa *et al.*, 1994; Detre *et al.*, 1995; Warach *et al.*, 1995b). Such images capitalize on the excellent temporal resolution of EEG and the complementary high spatial resolution of MRI. The relationship between hypometabolism in a seizure focus, determined with PET measurements of glucose metabolism, can be compared with benzodiazepine receptor binding, thereby increasing the specificity and sensitivity of the combined result (Savic *et al.*, 1988). One can also combine electrophysiologic data sets with tomographic data from PET, MRI, and MRS to provide a composite preoperative assessment of patients with epilepsy (Theodore, 1992; Spencer, 1994; Knowlton *et al.*, 1997). PET measurements of cerebral blood flow, oxygen extraction, and oxygen metabolism in the same subject and comparison with diffusion-weighted MRI and MR angiography (or helical CT) provide a very complete picture of the supply–demand relationships of the brain parenchyma in patients with ischemic cerebrovascular disease. Such combined studies will undoubtedly become a clinical norm rather than an exception in the future.

Comparisons of scans between individual patients will become increasingly important. Such comparisons require that the scans be spatially normalized to account for individual differences in brain structure. The ability to normalize a scan to a standard brain space means that individual patient scans can be compared with a representative scan from a population of normal subjects that takes into account a realistic estimate of the anatomical variability in that population (Mazziotta *et al.*, 1995). In the structural domain, this ability may increase the sensitivity with which subtle heterotopias or other migrational abnormalities are identified in patients with focal epilepsy. Similarly, selected patterns of atrophy in neurodegenerative diseases should be detected in a more sensitive and specific fashion. The ability to compare representative scans across patient groups could also have importance for clinical trials where a patient group on an experimental therapy could be compared with a control group in an objective and quantifiable manner (see Chapter 6).

II. Surgical Strategies

A. Preoperative

The preoperative investigation of patients with cerebral lesions falls into two general categories. The first is targeting areas for the purpose of stimulation or ablation. These circumstances occur in patients with movement disorders in which parts of the basal ganglia or other subcortical regions are selected for lesioning (e.g., pallidotomy) as a means of improving symptoms. Lesioning using stereotactic focal radiation or direct surgical ablation, by heating or freezing, are of interest for the treatment of cerebral neoplasms and vascular malformations. In a similar fashion, stimulating electrodes are now being employed in the treatment of Parkinson's disease and certain types of tremors. The exact location for the placement of these electrodes requires knowledge of the structural anatomy and local electrophysiology or function to obtain maximal therapeutic benefit.

1. Targeting

Brain mapping techniques have already been employed for targeting. At present, these approaches are limited to obtaining better definition of a patient's structural anatomy and developing better atlases (see below) to identify selected portions of the brain, given the individual variability among patients. With high-resolution structural imaging, specific locations for potential lesions can be identified anatomically and a frameless stereotactic approach can be used to direct an ablation probe or stimulation electrode. Once located, electrophysiological recordings can be used to verify the local functional environment of a given anatomical site.

Functional activation of deep brain sites (e.g., medial globus pallidus) is an important area of current investigation. Ideally, such sites should be located both structurally and functionally to identify a surgical target more accurately preoperatively. Such a facility would mean less retargeting and repositioning of electrodes and probes, thus reducing operating room time and morbidity, resulting in a higher success rate. At present, there are no validated clinical examples of such an approach.

2. Differentiation of Normal from Abnormal Brain

Another important aspect of presurgical investigation of patients with brain mapping techniques is the identification of normal cortex or deep brain structures so that they may be avoided during surgical resection or ablation of cerebral lesions (Grafton *et al.*, 1994). The goal is to remove an abnormality in its entirety without removing normal brain tissue. Preoperative evaluation with PET, SPECT, fMRI, or TMS (Jennum *et al.*, 1994) may be employed to identify the functional anatomy of an individual's brain to determine the safety of surgery and a strategy for reaching an abnormal brain region to remove pathologic tissue.

Such functional imaging techniques may ultimately replace procedures such as the Wada test or reversible pharmacologic interruptions of brain function (Desmond *et al.*, 1995; Hunter *et al.*, 1999). The selective administration of barbiturates to brain regions that produce transient deficits has long been used to determine the relative safety associated with removal of a portion of the brain. Nevertheless, such tests are difficult to perform, particularly in younger children, and the exact distribution of the pharmacologic agent can be difficult to verify. If the same information can be obtained through presurgical use of functional imaging, it is hoped that both the accuracy and the ease of obtaining such data will be enhanced (Fig. 11).

The combined use of all of the noninvasive scanning methods, both current and experimental, resulting in integrated and composite images linked to interactive graphics stations in the operating room or the interventional neuroradiological suite, will surely become a part of such interventional procedures in the future. To be successful, these approaches need to be less costly, more accurate, and associated with lower eventual morbidity than conventional surgery without such ancillary noninvasive procedures.

B. Intraoperative Mapping

The most frequently used intraoperative brain mapping techniques are electrophysiological. These would include electrocorticography for the identification of epileptic foci. Where cortical resections are indicated, evoked potential recordings with cortical electrodes are combined with peripheral nerve stimulation in anesthetized patients. Such an approach is frequently used to identify sensory and motor cortices. When more complex information is needed, particularly about language areas of the cortex, anesthesia is reversed and the patient is awakened during surgery. Direct electrical stimulation of the cortex is then used to reversibly disrupt local neuronal function while the patient performs behavioral tasks (Penfield and Rasmussen, 1950). In this setting, a patient must be psychologically able to accept the disturbing aspects of awake neurosurgery. Ideally, the examiner and the patient should have good rapport so that behavioral testing can proceed in an efficient and cooperative manner.

Optical intrinsic signal imaging, currently used in a research setting, may soon augment intraoperative electrophysiological measurements by providing directly visualization of cortical maps and functional responses (Haglund *et al.*, 1992; Toga *et al.*, 1995) seen ultimately through the operating microscope in real time. The high spatial and temporal resolution of this approach can be used to validate preoperative images of functional anatomy from individual patients (see Chapter 4, Figs. 7–12). In addition, the functional brain maps can be updated as they become distorted from the preoperative state by changes in brain shape resulting from osmotic dehydration, ventricular drainage, or local edema during the course of the operation.

As more interventional neuroradiology and neurosurgery are performed within imaging devices, the reacquisition of structural and functional information as an invasive procedure unfolds will become possible and realistic. The most likely source of these data will be from MRI devices where either patients are moved in and out of the device to update imaging data or the interventional procedure is performed within the magnetic field directly. In either case, direct updates of structural and, potentially, functional information can be provided to the clinical team performing the procedure.

III. Atlases

The advent of modern mathematical and computational approaches to averaging imaging data across subjects has led to the generation of population-based probabilistic atlases (Mazziotta *et al.*, 1995). Such atlases are already in existence for the normal brain at different age ranges and for other regions of the body (Ackerman, 1991). Disease-based atlases may be useful for the differential diagnosis of human cerebral disorders.

The basic approach to generating such atlases is to obtain images from a large number of subjects (i.e., typically hundreds or even thousands) in a mathematical framework that produces a database that is probabilistic. Such an atlas allows the user to obtain relative information that takes into account the variance in structure and function in the human population (Fig. 12). Once established, such an atlas can interact with new data sets derived from individual subjects and patients or groups of subjects or patients. Thus, a clinician or investigator who performs an MRI scan of a single patient with focal epilepsy could call upon a digital probabilistic atlas of normal subjects and compare the patient with the average normal atlas. The atlas will use the normal variance information estimated from the population of normals from which it is generated to determine whether a patient's scan falls within or outside normal morphometric limits. If the atlas is constructed from sufficient subjects, a subpopulation could be selected that more closely resembles a patient's demographic profile. In such a case, one might ask for only those normal subjects from the atlas who are right-handed, of a particular racial origin, and of females, age 25–30. An increasing number of variables can be included in such a prior specification depending on the size of the data set constituting the atlas and the range of demographic information collected about the contributing subjects. As a result, it would become possible to detect subtle abnormalities of diagnos-

Figure 11 Within-subject registration techniques. (A) Three-dimensional reconstruction of a patient's brain with a cortical tumor in the region of the sensorimotor cortex (arrow). The structural data set, reconstructed from MRI, can then be combined with functional information about cerebral blood flow changes associated with motor tasks, indicated in color and derived from ^{15}O-labeled water PET studies when the patient moved the left leg (B), shoulder (C), or fingers (D). (E) Intraoperative view of the hand area of the motor cortex where the localization of sensorimotor hand function was identified intraoperatively with electrophysiologic techniques and labeled "1" and "2." (F) Operative view of site following resection of tumor. Because of the close proximity of the tumor to the activation site for finger movement, identified with preoperative PET imaging, it was predicted that this patient would have loss of fine motor control of the hand following complete resection of the lesion. This was in fact the case and is indicative of the accuracy and predictive power of preoperative mapping in patients with cortical lesions close to vital cortical structures. Courtesy of Roger Woods, UCLA School of Medicine, and Scott Grafton, Emory University School of Medicine.

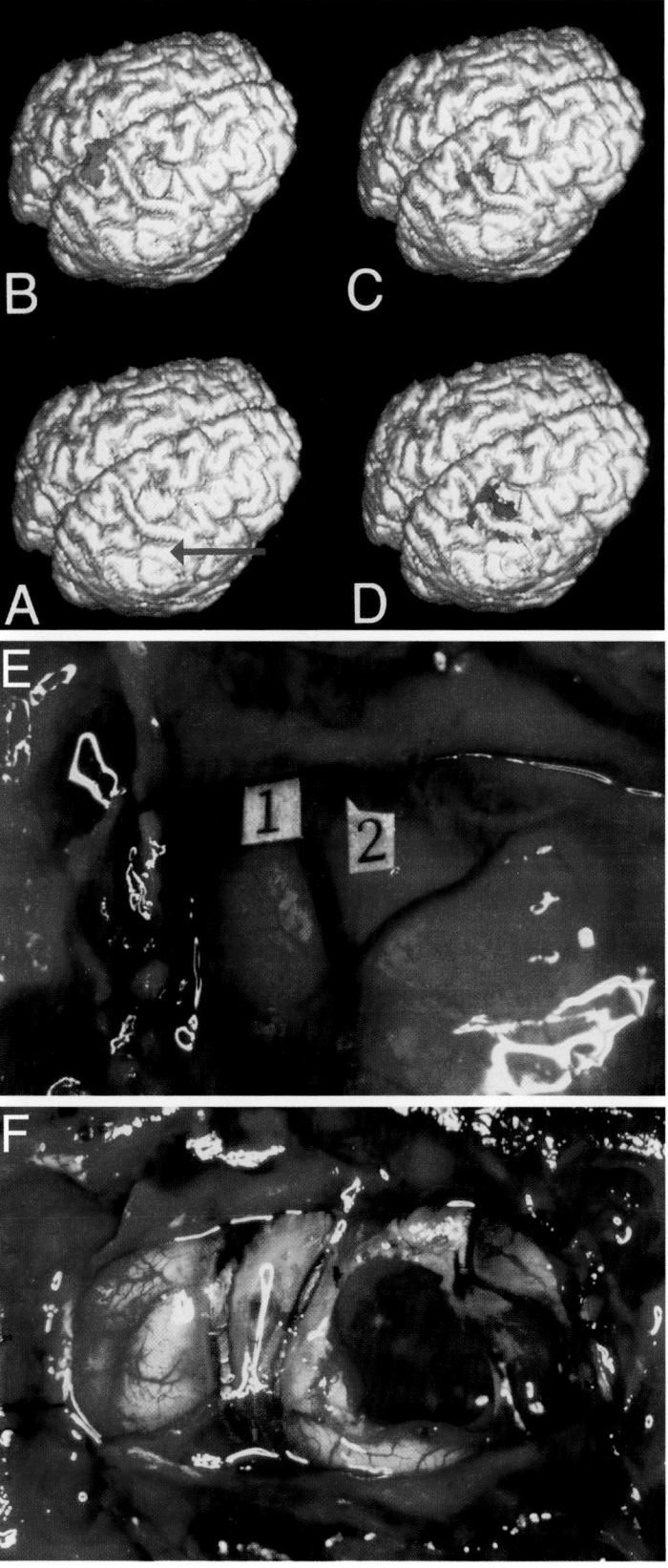

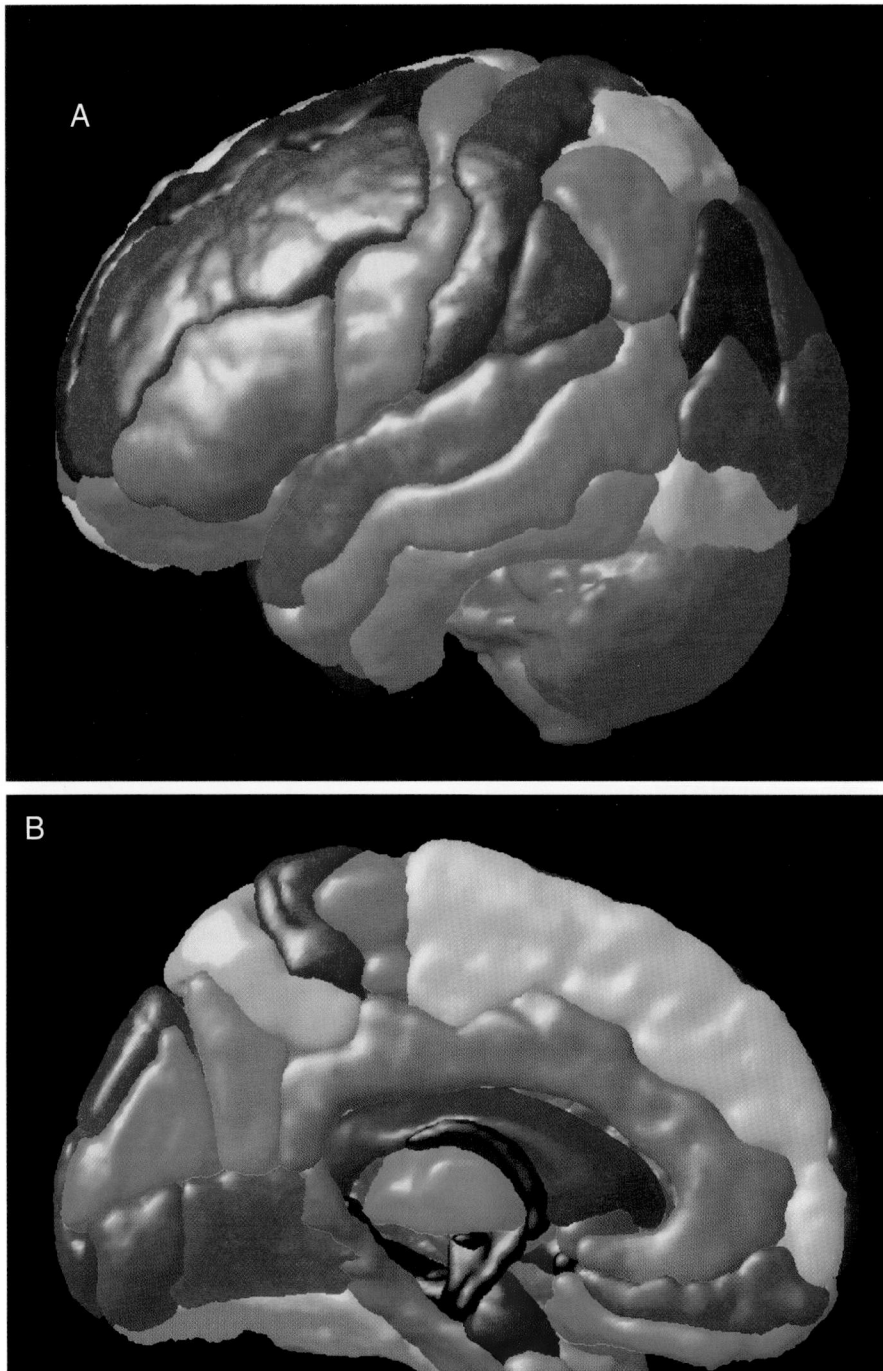

Figure 12 Probabilistic atlases. Population-based probabilistic atlas of the normal human brain derived from 67 subjects, ages 20–40 years, seen from the lateral (A) or midsagittal (B) views. The structures have been segmented to show cortical regions at a 50% confidence limit in the population. Courtesy of Alan Evans and colleagues, Montreal Neurologic Institute (Evans *et al.*, 1991).

tic importance that would not be identified by the less sensitive conventional approach of qualitatively examining two-dimensional image sets by eye. In addition, such an atlas-based approach will give an objective and quantifiable magnitude to any detected abnormality. The scan data from any patient can be added to an atlas database, increasing its value with regard to particular patient groups (Mazziotta *et al.*, 1995).

Disease-based atlases, thus generated, are currently being assessed as discussed in Chapter 6. It is possible to imagine morphometric or functional atlases for Alzheimer's (Fig. 13) and Parkinson's disease, schizophrenia, and other disorders. Such atlases would also provide a population- and disease-based opportunity to examine the natural history of morphometric or functional abnormalities as a function of disease progression, age of onset, or other variables. Such atlases could also be used to identify changes in natural course as a function of therapeutic intervention. Consider, for ex-

ample, a clinical trial with a new drug for Alzheimer's disease. The Alzheimer's disease population atlas would provide estimates of morphometric changes in focal atrophy as well as, for example, alterations in cerebral glucose metabolism as a function of disease progression. A population of patients at a certain stage of the disease could be divided into two groups, one given an experimental therapy and the other to be given placebo. Serial imaging of both groups with the appropriate techniques would then provide longitudinal imaging data. Comparisons of morphometric and metabolic changes as a function of time between the two groups would be undertaken to detect objective and quantitative differences between the two groups. Any differences would represent a measure of the effect of the therapeutic intervention on progressive atrophy or glucose metabolism due to the natural history of the disease. It is probable, although currently unproven, that such an approach will be more sensitive in detecting differences

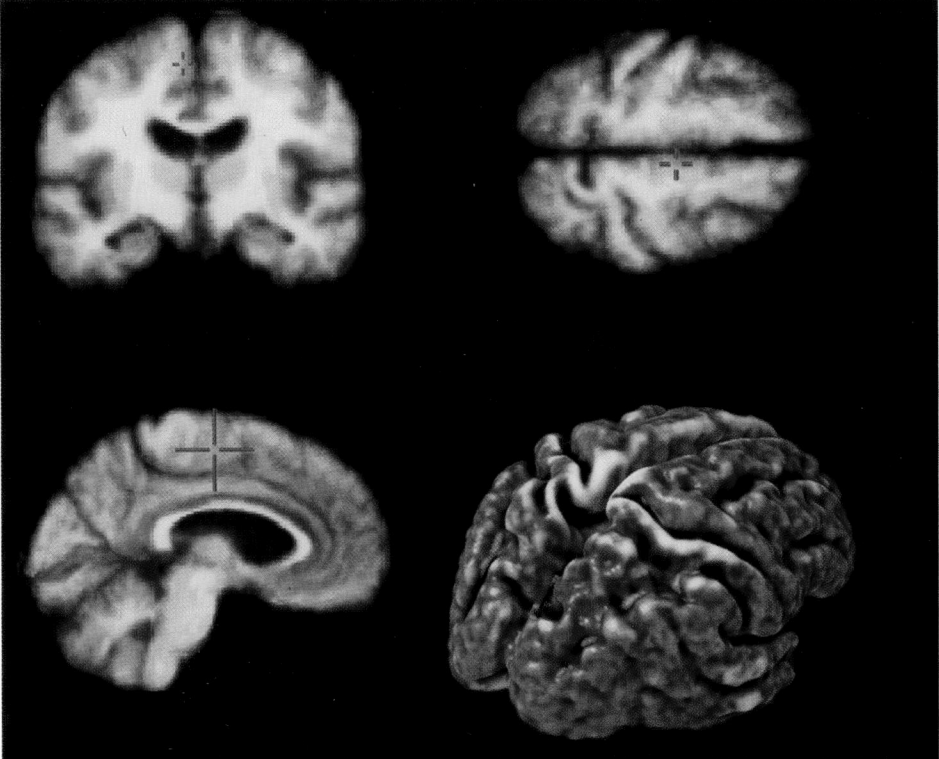

Figure 13 Probabilistic population atlas derived from nine individuals with Alzheimer's disease. This atlas is presented as a set of two-dimensional orthogonal views plus a three-dimensional rendering (bottom right) and is produced using a continuum-mechanical approach (see Chapter 6). Note the influence of atrophy on the composite image demonstrating widening of the major fissures of the brain as well as sulci in the neocortex. Such disease-based population atlases will be useful not only in tracking the natural history of cerebral disorders but also in providing objective and quantifiable information about structural and functional changes associated with experimental therapy for these disorders. Courtesy of Paul Thompson and colleagues, UCLA School of Medicine, Los Angeles, California.

between control and experimental groups, thereby requiring either fewer subjects or shorter time frames for therapeutic assessment, thus resulting in lower costs of clinical trials. A more complete discussion of these issues is provided in Chapter 6.

IV. Plasticity

One of the most exciting and dramatic observations to come from human brain mapping with a wide range of structural and functional techniques has been the dynamic plasticity of function in both normal brains and the brains of patients with neurological and neuropsychiatric disorders. Brain maps must therefore be viewed as dynamic, changing with development, disease progression, and normal learning and in the recovery of function after acute injury. The dynamic plasticity of functional brain maps provides an exciting opportunity to study these processes. It also means that the use of brain maps must take into account such variability in the design of brain mapping studies for patients with cerebral disorders.

For example, just as structural and functional studies must be normalized for spatial variability in the population, disease-based maps must be normalized in time to account for dynamic changes that occur with progression. Thus, a comparison of patients with Alzheimer's disease or other neurodegenerative disease should be stratified by time of onset, or other variables that take into account the pattern of changing functional maps. The same is true after an acute brain injury, such as trauma or cerebral infarction. The complex interaction and highly variable changes in blood flow, blood volume, water diffusion, oxygen, and glucose extraction and metabolism will all be more appropriately interpreted if they are stratified by time from onset of cerebral injury. So too will plastic changes associated with recovery and reorganization after irreversible damage. Compensatory properties of the human nervous system have been clearly demonstrated in studies of patients following stroke who recover motor

function (Fig. 14). The study of drug-induced, behaviorally associated and surgically promoted plasticity will, we predict, be an important part of brain mapping in the study of patients with neurological, neurosurgical, and psychiatric disorders.

The value of imaging data depends on an appreciation of the changing landscape of functional patterns. This is particularly true for techniques that make relative measurements. For example, relative cerebral blood flow measurements obtained with fMRI, SPECT, or PET may be misleading in patients with large cerebral infarctions. The evaluation of motor reorganization after cerebral infarction in a patient or group of patients must take into account a variable cerebral blood flow baseline in the setting of hemodynamically unstable tissue. Cerebral blood flow may be very low acutely, rise dramatically soon thereafter, and then reach some stable new level days, weeks, or months later. At present, it is uncertain how increments or decrements in blood flow associated with neuronal firing changes that are task-induced will behave under these different conditions of baseline blood flow. Is it valid to compare motor task activation in a cortical region when the "resting state" blood flow is altered from the normal value by 50–200%? Are there ceiling or floor effects in these responses? These issues need to be addressed before a proper interpretation of scans from such patients will be possible.

We predict that the ability to image plastic reorganization, both in normal and in pathologic states, will provide new insights, previously unavailable, about the constant reorganization of the human brain. Such information will be valuable in the design of behavioral, surgical, and pharmacological interventions in patients that facilitate and maximize the efficiency of the natural recovery processes. The imaging techniques should also provide a means to evaluate specific rehabilitation interventions, to determine their appropriateness, effectiveness, and timing, and to select patients for them. These abilities are currently lacking because the necessary information about the variables discussed has not been previously available.

Figure 14 Compensatory reorganization of the brain after acute injury induced by cerebral infarction. (A) Normal response of a group of control subjects performing a motor task with the right upper extremity. Relative increases in cerebral blood flow derived from PET measurements, demonstrating increased perfusion in the contralateral hand area of the motor cortex and the ipsilateral cerebellum. (B) The same motor task and methods were applied to a group of patients who had small subcortical cerebral infarctions associated with upper extremity paresis, all of whom recovered in the days to week following the acute ischemic injury. Note that when these subjects perform the same task, not only are there responses in the expected areas previously identified in the normal controls, that is, the contralateral hand area of the motor cortex and ipsilateral cerebellum, but there are also relative increases in cerebral blood flow in the ipsilateral hand area of the motor cortex and the contralateral cerebellum. Such studies provide useful insights into how the brain reorganizes following acute injury and can also be used to study compensation in more chronic states such as might be encountered in neurodegenerative disorders. These types of insights may be useful for designing more efficient, effective, and timely neurorehabilitation protocols employing behavioral, pharmacologic, or, potentially, surgical interventions for the restoration of function or its maintenance following acute or chronic injury to the brain. Courtesy of Francois Chollet and colleagues, Toulouse, France. From Chollet *et al.* (1991) *Ann. Neurol.* **29,** 63–71.

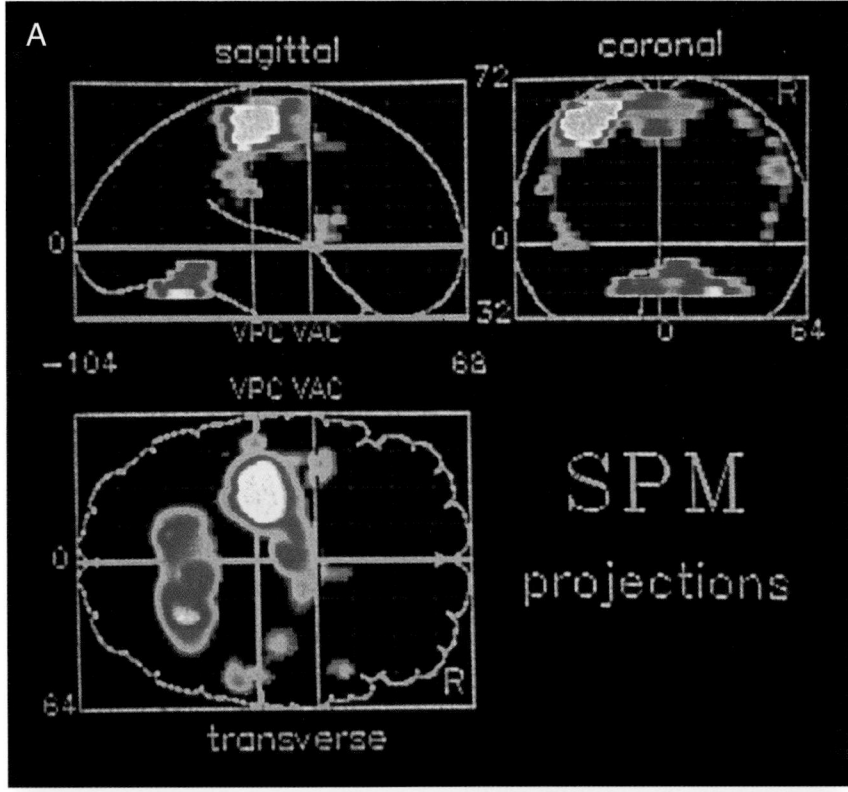

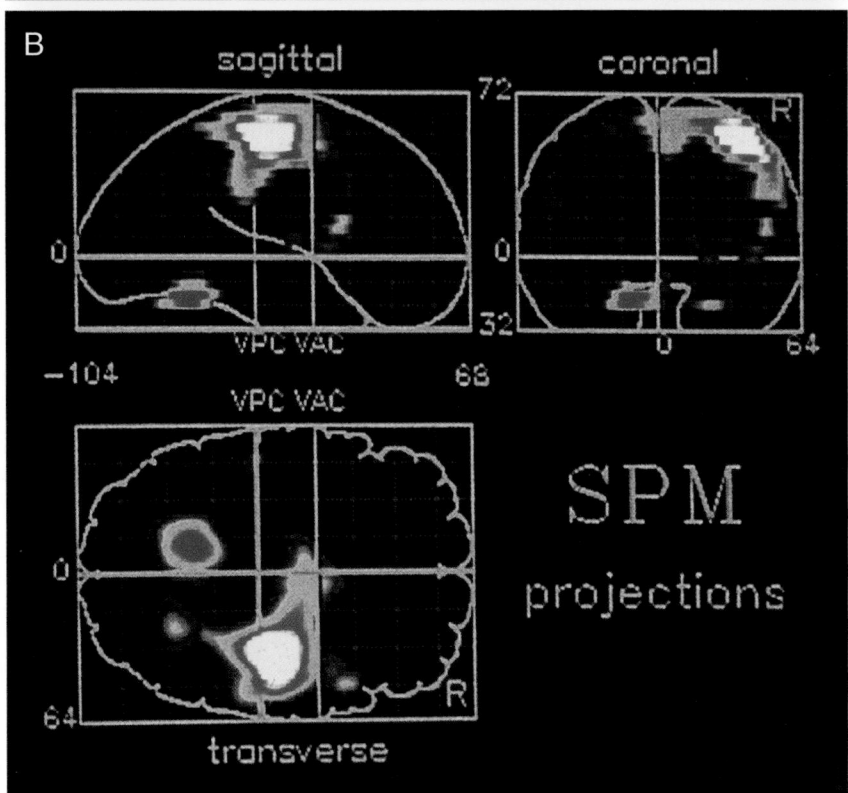

Acknowledgments

The authors thank contributing investigators for allowing their work to be reproduced in this chapter. This work was partially supported by a grant from the Human Brain Project (P01-MH52176-7) and generous support from the Brain Mapping Medical Research Organization, The Ahmanson Foundation, the Pierson–Lovelace Foundation, the Jennifer Jones–Simon Foundation, the Tamkin Foundation, the Northstar Fund, and the Wellcome Trust. The authors thank Laurie Carr for the preparation of the manuscript and Andrew Lee for assistance with the illustrations.

References

Abraham, K., and Ajmone-Marsan, C. (1958). Patterns of cortical discharges and their relation to routine scalp electroencephalography. *Electroencephalogr. Clin. Neurophysiol.* **10**, 447–452.

Achenbach, S., Moshage, W., Ropers, D., Nossen, J., and Daniel, W. G. (1998). Value of electron-beam computed tomography for the noninvasive detection of high-grade coronary-artery stenosis and occlusions. *N. Engl. J. Med.* **339**, 1964–1971.

Ackerman, M. J. (1991). The visible human project. *J. Biocommun.* **18** (2), 14.

American Academy of Neurology (AAN). (1996). Assessment of Brain SPECT: Report of the Therapeutics and Technology Assessment Subcommittee of the American Academy of Neurology. *Neurology* **46**, 278–285.

Arnett, C. D., Fowler, J. S., MacGregor, R. R., *et al.* (1987). Turnover of brain monoamine oxidase measured *in vivo* by positron emission tomography using L-[¹¹C]deprenyl. *J. Neurochem.* **49**, 522–527.

Atlas, S. W. (1991). "Magnetic Resonance Imaging of the Brain and Spine." Raven Press, New York.

Baird, A. E., and Warach, S. (1998). Magnetic resonance imaging of acute stroke. *J. Cereb. Blood Flow Metab.* **18**, 583–609.

Barker, A. T., Jalinous, R., and Freeston, I. L. (1985). Non-invasive magnetic stimulation of human motor cortex. *Lancet* **i**, 1106–1107.

Baron, J. C. (1992). In "Stroke: Pathophysiology, Diagnosis and Management" (H. J. Barnett, B. M. Stein, J. P. Mohr, and F. M. Yatsu, Eds.), 2nd ed., pp. 111–123. Churchill Livingstone, New York.

Barranco, D., Sutton, L., Florin, S., Greenberg, J., Sinnwell, T., Ligeti, L., *et al.* (1989). Use of ¹⁹F NMR spectroscopy for measurement of cerebral blood flow: Comparative study using microspheres. *J. Cereb. Blood Flow Metab.* **9**, 886–891.

Belliveau, J. W., Kennedy, D. N., McKinstry, R. C., Buchbindiner, B. R., Weisskoff, R. M., Cohen, M. S., *et al.* (1991). Functional mapping of the human visual-cortex by magnetic-resonance-imaging. *Science* **254**, 716–719.

Berger, H. (1929). Uber das elektroenzephalogramm des menschen. *Arch. Psychiatr. Nervenk.* **87**, 527–570.

Berkovic, S. F., Newton, M. R., Chiron, C., and Dulac, O. (1993). Single photon emission tomography. *In* "Surgical Treatment of the Epilepsies" (J. Engel, Jr., Ed.). Raven Press, New York.

Brooks, D. J., Ibanez, V., Sawle, G. V., Playford, E., Quinn, N., Mathias, C. J., *et al.* (1992). Striatal D₂ receptor status in patients with Parkinson's disease, striatonigral degeneration, and progressive supranuclear palsy, measured with ¹¹C-raclopride and positron emission tomography. *Ann. Neurol.* **31**, 184–192.

Brooks, R. A., and Di Chiro, G. (1976). Principles of computer assisted tomography (CAT) in radiographic and radioisotopic imaging. *Phys. Med. Biol.* **21**, 689–732.

Chance, B., Leigh, J. S., McLaughlin, A. C., *et al.* (1988). Phosphorus-31 spectroscopy and imaging. *In* "Magnetic Resonance Imaging" (C. L. Partain, J. A. Patton, *et al.*, Eds.). Saunders, Philadelphia.

Chang, L. H., Cohen, Y., and Weinstein, P. R. (1991). Interleaved ¹H and ³¹P spectroscopic imaging for studying regional brain imaging. *Magn. Reson. Imaging* **9**, 223–227.

Chiappa, K. H. (1983). "Evoked Potentials in Clinical Medicine." Raven Press, New York.

Chollet, F., Celsis, P., Clanet, M., Guiraud-Chaumeil, B., Rascol, A., and Marc-Vergnes, J. P. (1989). SPECT study of cerebral blood flow reactivity after acetazolamide in patients with transient ischemic attacks. *Stroke* **20** (4), 458–464.

Chugani, H. T., Shewmon, D. A., Peacock, W. J., Shields, W. D., Mazziotta, J. C., and Phelps, M. E. (1988). Surgical treatment of intractable neonatal onset seizures: The role of positron emission tomography. *Neurology* **38**, 1178–1188.

Chugani, H. T., Shields, W. D., Shewmon, D. A., Olson, D. M., Phelps, M. E., and Peacock, W. J. (1990). Infantile spasms. I: PET identifies focal cortical dysgenesis in cryptogenic cases for surgical treatment. *Ann. Neurol.* **27**, 406–413.

Chui, C., and Damasio, A. (1980). Human cerebral asymmetries evaluated by computed tomography. *J. Neurochem. Neurosurg. Psychiatry* **43**, 873–878.

Coles, M. G. H., Gratton, G., and Fabiani, M. (1990). Event-related brain potentials. *In* "Principles of Psychophysiology: Physical, Social, and Inferential Elements" (J. T. Cacioppo and L. G. Tassinary, Eds.), pp. 413–455. Cambridge Univ. Press, New York.

Collins, D. L., Peters, T. M., and Evans, A. C. (1994a). An automated 3D nonlinear image deformation for determination of gross morphometric variability in human brain. *Proc. Visual Biomed. Comput. VBC '94, Rochester, SPIE* **2359**, 180–190.

Collins, D. L., Neelin, P., Peters, T. M., and Evans, A. C. (1994b). Automatic 3D intersubject registration of MR volumetric data in standardized Talairach space. *J. Comput. Assist. Tomogr.* **18** (2), 192–205.

Curran, T., Tucker, D. M., Kutas, M., and Posner, M. I. (1993). Topography of the N400: Brain electrical activity reflecting semantic expectancy. *Electroencephalogr. Clin. Neurophysiol.* **88**, 188–209.

Daly, D. D., and Pedley, T. A. (1990). "Current Practice of Clinical Electroencephalography," 2nd ed. Raven Press, New York.

Desmond, J. E., Sum, J. M., Wagner, A. D., Demb, J. B., Shear, P. K., Glover, G. H., *et al.* (1995). Functional MRI measurement of language lateralization in Wada-tested patients. *Brain* **118**, 1411–1419.

Detre, J. A., Sirven, J. I., Alsop, D. C., O'Connor, J. J., and French, J. A. (1995). Localization of subclinical ictal activity by functional magnetic resonance imaging: Correlation with invasive monitoring. *Ann. Neurol.* **38**, 618–624.

Devous, M. D., Leroy, R. F., and Horman, R. W. (1990). Single photon emission computed tomography in epilepsy. *Semin. Nucl. Med.* **20**, 325–341.

Devous, M. D., Payne, J. K., Lowe, J. L., and Leroy, R. F. (1993). Comparison of technetium-99m-ECD to xenon-133 SPECT in normal controls and in patients with mild to moderate regional cerebral blood flow abnormalities. *J. Nucl. Med.* **34**, 754–761.

Di Chiro, G. (1985). Diagnostic and prognostic value of positron emission tomography using ¹⁸F-fluorodeoxyglucose in brain tumors. *In* "Positron Emission Tomography" (M. Reivich and A. Alavi, Eds.), pp. 291–309. A. R. Liss, New York.

Drayer, B., Gur, D., Wolfson, S. K., and Dujovuy, M. (1979). Regional blood flow of the posterior fossa: Xenon enhanced CT scanning. *Acta Neurol. Scan.* **60** (Suppl. 1), 218–219.

Engel, J. (1993). "Surgical Treatment of the Epilepsies," 2nd ed. Raven Press, New York.

Engel, J., Jr., Henry, T. R., Risinger, M. W., Mazziotta, J. C., Sutherling, W. W., Levesque, M. F., *et al.* (1990). Presurgical evaluation for partial epilepsy: Relative contributions of chronic depth electrode recordings versus FDG-PET and scalp sphenoidal ictal EEG. *Neurology* **40**, 1670–1677.

Epstein, C. M., Lah, J. K., Meador, K., Weissman, J. D., Gaitan, L. E., and Dihenia, B. (1996). Optimum stimulus parameters for lateralized suppression of speech with magnetic brain stimulation. *Neurology* **47,** 1590–1593.

Evans, A. C., Dai, W., Collins, D. L., Neelin, P., and Marrett, S. (1991). Warping of a computerized 3-D atlas to match brain image volumes for quantitative neuroanatomical and functional analysis. *SPIE Med. Imaging: Image Proc.* **1445,** 236–246.

Farde, L., Hakan, H., Ehrin, E., *et al.* (1986). Quantitative analysis of D_2 dopamine receptor binding in the living human brain by PET. *Science* **231,** 258–261.

Fender, D. H. (1987). Source localization of brain electrical activity. *In* "Handbook of Electroencephalography and Clinical Neurophysiology" (A. S. Gevins and A. Remond, Eds.), Vol. 1, pp. 355–403. Elsevier, Amsterdam.

Fowler, J. S., MacGregor, R. R., Wolf, A. P., *et al.* (1987). Mapping human brain monoamine oxidase A and B with ^{11}C-labeled suicide inactivators and PET. *Science* **235,** 481–485.

Fox, P. T., and Raichle, M. E. (1986). Focal physiological uncoupling of cerebral blood flow and oxidative metabolism during somatosensory stimulation in human subjects. *Proc. Natl. Acad. Sci. U.S.A.* **83** (4), 1140–1144.

Fox, P. T., Mintun, M. A., Raichle, M., and Herscovitch, P. (1984). A non-invasive approach to quantitative functional brain mapping with $H_2^{15}O$ and positron emission tomography. *J. Cereb. Blood Flow Metab.* **4,** 329–333.

Fox, P. T., Perlmutter, J. S., and Raichle, M. E. (1985). A stereotactic method of anatomical localization for positron emission tomography. *J. Comput. Assist. Tomogr.* **9** (1), 141–153.

Fox, P. T., Raichle, M. E., Mintun, M. A., and Dence, C. (1988). Nonoxidative glucose consumption during focal physiologic neural activity. *Science* **241,** 462–464.

Frahm, J., Merboldt, K.-D., Hanicke, W., Kleinschmidt, A., and Boecker, H. (1994). Brain or vein-Oxygenation or flow? On signal physiology in functional MRI of human brain. *NMR Biomed.* **7,** 45–53.

Freeman, W. J. (1978). Spatial properties of an EEG event in the olfactory bulb and cortex. *Electroencephalogr. Clin. Neurophysiol.* **44,** 586–605.

Frey, K. A., Holthoff, V. A., Koeppe, R. A., Jewett, D. M., Kilbourn, M. R., and Kuhl, D. E. (1991). Parametric *in vivo* imaging of benzodiazepine receptor distribution in human brain. *Ann. Neurol.* **30,** 663–672.

Friston, K. J., Frith, C. D., Liddle, P. F., and Frackowiak, R. S. (1991). Plastic transformation of PET images. *J. Comput. Assist. Tomogr.* **15** (4), 634–639.

Frost, J. J. (1982). Pharmacokinetic aspects of the *in vivo,* noninvasive study of neuroreceptors in man. *In* "Receptor Binding Radiotracers" (W. Eckelman, Ed.), pp. 25–39. CRC Press, Boca Raton, FL.

Frostig, R. D., Lieke, E. E., Ts'o, D. Y., and Grinvald, A. (1990). Cortical functional architecture and local coupling between neuronal activity and the microcirculation revealed by *in vivo* high-resolution optical imaging of intrinsic signals. *Proc. Natl. Acad. Sci. U.S.A.* **87,** 6082–6086.

Galaburda, A., Le May, M., Kemper, T., and Geschwind, N. (1978). Left-right asymmetries in the brain. *Science* **199,** 852–856.

Garg, K., Welsh, C. H., Feyerabend, A. J., Subber, S. W., Russ, P. D., Johnston, R. J., *et al.* (1998). Pulmonary embolism: Diagnosis with spiral CT and ventilation-perfusion scanning. Correlation with pulmonary angiographic results or clinical outcome. *Radiology* **208,** 201–208.

Gee, J. C., Reivich, M., and Bajcsy, R. (1993). Elastically deforming 3D atlas to match anatomical brain images. *J. Comput. Assist. Tomogr.* **17** (2), 225–236.

George, M. S., Wassermann, E. M., Williams, W. A., Steppel, J., Pascual-Leone, A., Basser, P., *et al.* (1996). Changes in mood and hormone levels after rapid-rate transcranial magnetic stimulation (rTMS) of the prefrontal cortex. *J. Neuropsychiatry Clin. Neurosci.* **8,** 172–180.

Geschwind, N., and Levitsky, W. (1968). Human brain: Left-right asymmetries in temporal speech regions. *Science* **161,** 181–187.

Gevins, A. (1996). Electrophysiological imaging of brain function. *In* "Brain Mapping: The Methods" (A. W. Toga and J. C. Mazziotta, Eds.). Academic Press, San Diego.

Gevins, A. S., and Bressler, S. L. (1988). Functional topography of the human brain. *In* "Functional Brain Imaging" (G. Pfurtscheller, Ed.), pp. 99–116. Huber, Bern.

Gevins, A. S., Zeitlin, G. M., Doyle, J. C., Yingling, C. C., Schaffer, R. E., Callaway, E., *et al.* (1979). Electroencephalogram correlates of higher cortical functions. *Science* **203,** 665–668.

Gevins, A. S., Morgan, N. H., Bressler, S. L., Cutillo, B. A., White, R. M., Illes, J., *et al.* (1987). Human neuroelectric patterns predict performance accuracy. Science **235,** 580–585.

Gilman, S., Junck, L., Markel, D. S., Koeppel, R. A., and Kluin, K. J. (1990). Cerebral glucose hypermetabolism in Friedreich's ataxia detected with positron emission tomography. *Ann. Neurol.* **28,** 750–757.

Grafman, J., Pascual-Leone, A., Alway, D., Nichelli, P., Gomez-Tortosa, E., and Hallett, M. (1994). Induction of a recall deficit by rapid-rate transcranial magnetic stimulation. *NeuroReport* **5,** 1157–1160.

Grafton, S. T., Martin, N. A., Mazziotta, J. C., Woods, R. P., Vinuela, F., and Phelps, M. E. (1994). Localization of motor areas adjacent to arteriovenous malformations. A positron emission tomographic study. *J. Neuroimaging* **4** (2), 97–103.

Grinvald, A., Lieke, E., Frostig, R. D., Gilbert, C. D., and Wiesel, T. N. (1986). Functional architecture of cortex revealed by optical imaging of intrinsic signals. *Nature* **324,** 361–364.

Gruetter, R., Novotny, E. J., Boulware, S. D., Rothman, D. L., Mason, G. F., Shulman, G. I., *et al.* (1992). Direct measurement of brain glucose concentration in humans by ^{13}C NMR spectroscopy. *Proc. Natl. Acad. Sci. U.S.A.* **89,** 1109–1112.

Gzesh, D., Goldstein, S., and Sperling, M. R. (1990). Complex partial epilepsy: The role of neuroimaging in localizing a seizure focus for surgical intervention. *J. Nucl. Med.* **31,** 1839–1843.

Haglund, M. M., Ojemann, G. A., and Hochman, D. W. (1992). Optical imaging of epileptiform and functional activity in human cerebral cortex. *Nature* **358**(6388), 668–671.

Hawkins, R. A., Mazziotta, J. C., and Phelps, M. E. (1987). Wilson's disease studied with FDG and positron emission tomography. *Neurology* **37,** 1707–1711.

Hawkins, R. A., Huang, S. C., Barrio, J. R., Keen, R. E., Feng, D., Mazziotta, J. C., *et al.* (1989). Estimation of local cerebral protein synthesis rates with L-[1-^{11}C]leucine and PET: Methods, model and results in animals and humans. *J. Cereb. Blood Flow Metab.* **9,** 446–460.

Ho, S. S., Berkovic, S. F., Berlangieri, S. U., Newton, M. R., Egan, G. F., Tochon-Danguy, H. J., *et al.* (1995). Comparison of ictal SPECT and interictal PET in the presurgical evaluation of temporal lobe epilepsy. *Ann. Neurol.* **37,** 738–745.

Holman, B. L., and Devous, M. D. (1992). Functional brain SPECT: The emergence of a powerful clinical method. *J. Nucl. Med.* **33,** 1888–1904.

Hopkins, A. L., and Barr, R. G. (1987). Oxygen-17 compounds as potential NMR T2 contrast agents: Enrichment effects of $H_2^{17}O$ on protein solutions and living tissues. *Magn. Reson. Med.* **4,** 399–403.

Huang, S. C., Phelps, M. E., Hoffman, E. J., *et al.* (1980). Noninvasive determination of local cerebral metabolic rate of glucose in man. *Am. J. Physiol.* **238,** E69–E82.

Huang, S. C., Yu, D. C., Barrio, J. R., *et al.* (1991). Kinetics and modeling of L-6-[^{18}F]fluoro-dopa in human positron emission tomographic studies. *J. Cereb. Blood Flow Metab.* **11**, 898–913.

Hung, J. C., Corlija, M., Volkert, W. A., and Holmes, R. A. (1988). Kinetic analysis of technetium-99m *d,l*-HM-PAO decomposition in aqueous media. *J. Nucl. Med.* **29**, 1568–1576.

Hunter, K. E., Blaxton, T. A., Bookheimer, S. Y., Figlozzi, C., Gaillard, W. D., Grandin, C., *et al.* (1999). ^{15}O-Water positron emission tomography in language localization: A study comparing positron emission tomography visual and computerized region of interest analysis with the Wada test. *Ann. Neurol.* **45**, 662–665.

Ives, J., Warach, S., Schmitt, F., Edelman, R. R., and Schomer, D. L. (1993). Monitoring the patient's EEG during echo planar MRI. *Electroencephalogr. Clin. Neurophysiol.* **87**, 417–420.

Jahromi, B. S., Robitaille, R., and Charlton, M. P. (1992). Transmitter release increases intracellular calcium in perisynaptic Schwann cells *in situ. Neuron* **8**(6), 1069–1077.

Jennum, P., Friberg, L., Fuglsang-Frederiksen, A., and Dam, M. (1994). Speech localization using repetitive transcranial magnetic stimulation. *Neurology* **44**, 269–273.

Jones, T., Chesler, D. A., and Ter-Pogossian, M. M. (1976). The continuous inhalation of oxygen-15 for assessing regional oxygen extraction in the brain of man. *Br. J. Radiol.* **49**, 339–343.

Kaatee, R., Beek, F. J. A., de Lange, E. E., van Leeuwen, M. S., Smits, H. F., van der Ven, P. J., *et al.* (1997). Renal artery senosis: Detection and quantification with spiral CT angiography versus optimized digital subtraction angiography. *Radiology* **205**, 121–127.

Kanal, E., Shellock, F. G., and Lewin, J. S. (1996). Aneurysm clip testing for ferromagnetic properties: Clip variability issues. *Radiology* **200**, 576–578.

Keen, R. E., Barrio, J. R., Huang, S. C., *et al.* (1989). *In vivo* cerebral protein synthesis rates with leucyltransfer RNA used as a precursor pool: Determination of biochemical parameters to structure tracer kinetic models for positron emission tomography. *J. Cereb. Blood Flow Metab.* **9**, 429–445.

Kennan, R. P., Zhon, J. H., and Gore, J. C. (1994). Intravascular susceptibility contrast mechanisms in tissues. *Magn. Reson. Med.* **31**, 9–21.

Kety, S. S., and Schmidt, C. F. (1945). The determination of cerebral blood flow in man by the use of nitrous oxide in low concentrations. *Am. J. Physiol.* **143**, 53–66.

Klucznik, R. P., Carrier, D. A., Pyka, R., and Haid, R. W. (1993). Placement of a ferromagnetic intracerebral aneurysm clip in a magnetic field with a fatal outcome. *Radiology* **187**, 855–856.

Knowlton, R. C., Laxer, K. D., Ende, G., Hawkins, R. A., Wong, S. T. C., Matson, G. B., *et al.* (1997). Presurgical multimodality neuroimaging in electroencephalographic lateralized temporal lobe epilepsy. *Ann. Neurol.* **42**, 829–837.

Koepp, M. J., Labbe, C., Richardson, M. P., Brooks, D. J., Van Paesschen, W., Cunningham, V. J., *et al.* (1997). Regional hippocampal [^{11}C]flumazenil PET in temporal lobe epilepsy with unilateral and bilateral hippocampal sclerosis. *Brain* **120**, 1865–1876.

Kuhl, D. E., and Edwards, R. Q. (1964). Cylindrical and section radioisotope scanning of the liver and brain. *Radiology* **83**, 926–936.

Kuhl, D. E., Phelps, M. E., Markham, C. H., *et al.* (1982). Cerebral metabolism and atrophy in Huntington's disease determined by FDG and computed tomography scan. *Ann. Neurol.* **12**, 425–434.

Kung, H. F., Ohmomo, Y., and Kung, M. P. (1990). Current and future radiopharmaceuticals for brain imaging with single photon emission computed tomography. *Semin. Nucl. Med.* **20**, 290–302.

Kwong, K. K., Belliveau, J. W., Chesler, D. A., Goldberg, I. E., Weisskoff, R. M., Poncelet, B. P., *et al.* (1992). Dynamic magnetic resonance imaging of human brain activity during primary sensory stimulation. *Proc. Natl. Acad. Sci. U.S.A.* **89**, 5675–5679.

Lauterbur, P. C. (1973). Image formation by induced local interactions: Examples employing nuclear magnetic resonance. *Nature* **242**, 190–191.

Le May, M., and Kido, D. (1978). Asymmetries of the cerebral hemispheres on computed tomograms. *J. Comput. Assist. Tomogr.* **2**, 471–476.

Lear, J. L. (1988). Quantitative local cerebral blood flow measurements with technetium-99m HM-PAO: Evaluation using multiple radionuclide digital quantitative autoradiography. *J. Nucl. Med.* **29**, 1387–1392.

Lutsep, H. L., Albers, G. W., DeCrespigny, A., Kamat, G. N., Marks, M. P., and Moseley, M. E. (1997). Clinical utility of diffusion-weighted magnetic resonance imaging in the assessment of ischemic *stroke. Ann. Neurol.* **41**, 574–580.

MacNamara, A. F., and Evans, P. A. (1995). The use of CT scanning by accident and emergency departments in the UK: Past, present and future. *Injury* **26**, 667–669.

Madar, I., Lesser, R. P., Krauss, G., Zubieta, J. K., Lever, J. R., Kinter, C. M., *et al.* (1997). Imaging of delta- and mu-opioid receptors in temporal lobe epilepsy by positron emission tomography. *Ann. Neurol.* **41**, 358–367.

Mansfield, P. (1977). Multiplanar image formation using NMR spin echoes. *J. Phys. C* **10**, L55–L58.

Mayberg, H., Sadzot, B., Meltzer, C. C., Fisher, R. S., Lesser, R. P., Dannals, R. F., *et al.* (1991). Quantification of mu and non-mu opiate receptors in temporal lobe epilepsy using positron emission tomography. *Ann. Neurol.* **30**, 3–11.

Mazziotta, J. C. (1989). Huntington's disease: Studies with structural imaging techniques and positron emission tomography. *Semin. Neurol.* **9**, 360–369.

Mazziotta, J. C. (1992). Movement disorders. *In* "Clinical Brain Imaging: Principles and Applications" (J. C. Mazziotta and S. Gilman, Eds.), pp. 244–293. Davis, Philadelphia.

Mazziotta, J. C., and Gilman, S. (1992). "Clinical Brain Imaging: Principles and Applications" Davis, Philadelphia.

Mazziotta, J. C., and Phelps, M. E. (1986). Positron emission tomography studies of the brain. *In* "Positron Emission Tomography and Autoradiography" (M. Phelps, J. Mazziotta, and H. Schelbert, Eds.), pp. 493–579. Raven Press, New York.

Mazziotta, J. C., Huang, S.-C., Phelps, M. E., Carson, R. E., MacDonald, N. S., and Mahoney, K. (1985). A noninvasive positron computed tomography technique using oxygen-15-labeled water for the evaluation of neurobehavioral task batteries. *J. Cereb. Blood Flow Metab.* **5**, 70–78.

Mazziotta, J. C., Phelps, M. E., Huang, S.-C., Baxter, L. R., Riege, W. H., Hoffman, J. M., *et al.* (1987). Reduced cerebral glucose metabolism in asymptomatic subjects at risk for Huntington's disease. *N. Engl. J. Med.* **316**, 357–362.

Mazziotta, J. C., Frackowiak, R. S. J., and Phelps, M. E. (1992). The use of positron emission tomography in the clinical assessment of dementia. *Semin. Nucl. Med.* **22**, 233–246.

Mazziotta, J. C., Toga, A. W., Evans, A., Fox, P., and Lancaster, J. (1995). A probabilistic atlas of the human brain: Theory and rationale for its development. *NeuroImage* **2**, 89–101.

McLaughlin, A. C., Pekar, J., Sinnwell, T., et al. (1993). ^{17}O and ^{19}F magnetic resonance imaging of cerebral blood flow and oxygen consumption, Syllabus, SMRM FMRI Workshop, Arlington, VA, p. 69.

Merton, P. A., and Morton, H. B. (1980). Stimulation of the cerebral cortex in the intact human subject. *Nature* **285**, 227.

Mesulam, M. (1990). Large-scale neurocognitive networks and distributed processing for attention, language, and memory. *Ann. Neurol.* **28** (5), 597–613.

Mills, K. R. (1991). Magnetic brain stimulation: A tool to explore the action of the motor cortex on single human spinal motoneurones. *Trends Neurosci.* **14,** 401–405

Mintun, M. A., Raichle, M. E., Kilbourn, M. R., Wooten, G. F., and Welch, M. J. (1984). A quantitative model for the *in vivo* assessment of drug binding sites with positron emission tomography. *Ann. Neurol.* **15,** 212–227.

Moran, S. G., McCarthy, M. C., Uddin, D. E., and Poelstra, R. J. (1994). Predictors of positive CT scans in the trauma patient with minor head injury. *Am. Surg.* **60,** 533–535.

Morris, P. (1997). "Practical Neuroangiography." Williams and Wilkins, Baltimore.

Moseley, M., Kucharczyk, J., Mintorovitch, J., Cohen, Y., Kurhanewicz, J., Derugin, N., Asgari, H., *et al.* (1990). Diffusion-weighted MR imaging of acute stroke: Correlation of T2 weighted and magnetic susceptibility-enhanced MR imaging in cats. *Am. J. Neuroradiol.* **11,** 423–429.

Nakano, S., Kinoshita, K., Jinnouchi, S., and Hoshi, H. (1989). Comparative study of regional cerebral blood flow images by SPECT using xenon-133, iodine-123 IMP, and technetium-99m HM-PAO. *J. Nucl. Med.* **30,** 157–164.

Newton, M. R., Berkovic, S. F., Austin, M. C., Rowe, C. C., McKay, W. J., and Bladin, P. F. (1992). Postictal switch in blood flow distribution and temporal lobe seizures. *J. Neurol. Neurosurg. Psychiatry* **55,** 891–894.

Newton, T. H., and Potts, D. G., Eds. (1981). "Radiology of the Skull and Brain: Technical Aspects of Computed Tomography," Vol. 5. Mosby, St. Louis.

O'Brien, T. J., So, E. L., Mullan, B. P., Hauser, M. F., Brinkmann, B. H., Bohnen, N. I., *et al.* (1998). Subtraction ictal SPECT co-registered to MRI improves clinical usefulness of SPECT in localizing the surgical seizure focus. *Neurology* **50,** 445–454.

Obrist, W. D., Thompson, H. K., King, C. H., and Wang, H. S. (1967). Determination of regional cerebral blood flow by inhalation of 133-xenon. *Circ. Res.* **20** (1), 124–135.

Obrist, W. D., Thompson, H. K., Jr., Wang, H. S., and Wilkinson, W. E. (1975). Regional cerebral blood flow estimated by 133-xenon inhalation. *Stroke* **6** (3), 245–256.

Ogawa, S., Lee, T. M., Nayak, A. S., and Glynn, P. (1990). Oxygenation-sensitive contrast in magnetic-resonance image of rodent brain at high magnetic fields. *Magn. Reson. Med.* **14,** 68–78.

Ogawa, S., Menon, R. S., Tank, D. W., Kim, S. G., Merkle, H., Ellermann, J. M., *et al.* (1993). Functional brain mapping by blood oxygenation level-dependent contrast magnetic-resonance-imaging-A comparison of signal characteristics with a biophysical model. *Biophys. J.* **64,** 803–812.

Osborn, A. G. (1980). "Introduction to Cerebral Angiography." Harper and Row, Hagerstown.

Oshima, M., Tadokoro, R., Makino, N., and Sakuma, S. (1990). Role of fast data acquisition method with ^{99m}Tc HMPAO brain SPECT in patients with acute stroke. *Clin. Nucl. Med.* **15,** 172–174.

Pascual-Leone, A., Valls-Sole, J., Wassermann, E. M., Brasil-Neto, J., Cohen, L. G., and Hallett, M. (1992). Effects of focal transcranial magnetic stimulation on simple reaction time to acoustic, visual and somatosensory stimuli. *Brain* **115,** 1045–1059.

Pascual-Leone, A., Cohen, L. G., Brasil-Neto, J. P., and Hallet, M. (1994). Non-invasive differentiation of motor cortical representation of hand muscle by mapping of optimal current directions. *Electroencephalogr. Clin. Neurophysiol.* **93,** 42–48.

Pascual-Leone, A., Rubio, B., Pallardo, F., and Catala, M.-D. (1996). Beneficial effect of rapid-rate transcranial magnetic stimulation of the left dorsolateral prefrontal cortex in drug-resistant depression. *Lancet* **348,** 233–238.

Patton, J. A., Brill, A. B., Erickson, J. J., *et al.* (1969). A new approach to the mapping of three-dimensional radionuclide distributions. *J. Nucl. Med.* **10,** 363.

Pauling, L., and Coryell, C. D. (1936). The magnetic properties and structure of hemoglobin, oxyhemoglobin and carbonmonoxyhemoglobin. *Proc. Natl. Acad. Sci. U.S.A.* **22,** 210–216.

Paus, T., Jech, R., Thompson, C. J., Comeau, R., and Evans, A. C. (1997). Transcranial magnetic stimulation during positron emission tomography: A new method for studying connectivity of the human cerebral cortex. *J. Neurosci.* **17,** 3178–3184.

Pekar, J., Sinnwell, T. M., Ligeti, L., et al. (1993). Double-label tracer experiments using multinuclear MRI: Mapping cerebral oxygen consumption and blood flow using ^{17}O and ^{19}F MRI. *Proc. 12th SMRM Annu. Mtg.,* New York, p. 1388.

Pelizzari, C. A., Chen, G. T., Spelbring, D. R., Weichselbaum, R. R., and Chen, C. T. (1989). Accurate three-dimensional registration of CT, PET, and/or MR images of the brain. *J. Comput. Assist. Tomogr.* **13** (1), 20–26.

Penfield, W., and Rasmussen, T. (1950). "The Cerebral Cortex of Man: A Clinical Study of Localisation of Function." Macmillan, New York.

Perlmutter, J. S., Larson, K. B., Raichle, M. E., Markham, J., Mintun, M. A., Kilbourn, M. R., *et al.* (1986). Strategies for *in vivo* measurement of receptor binding using positron emission tomography. *J. Cereb. Blood Flow Metab.* **6,** 154–169.

Phelps, M. E., and Mazziotta, J. C. (1985). Positron emission tomography: Human brain function and biochemistry. *Science* **228,** 799–809.

Phelps, M. E., Huang, S. C., Hoffman, E. J., Selin, C., Sokoloff, L., and Kuhl, D. E. (1979). Tomographic measurement of local cerebral metabolic rate in humans with (F-18)2-fluoro-2-deoxy-D-glucose: Validation of method. *Ann. Neurol.* **6,** 371–388.

Prichard, J. W., and Shulman, R. G. (1986). NMR spectroscopy of brain metabolism *in vivo. Annu. Rev. Neurosci.* **9,** 61–85.

Pritchard, J. W., and Rosen, B. R. (1994). Functional study of the brain by NMR. *J. Cereb. Blood Flow Metab.* **14** (3), 365–372.

Prichard, J. W., Rothman, D., Novotny, E., Petroff, O., Kuwabara, T., and Avison, M. (1991). Lactate rise detected by ^{1}H NMR in human visual cortex during physiological stimulation. *Proc. Natl. Acad. Sci. U.S.A.* **88,** 5829–5831.

Raichle, M. E. (1983). Positron emission tomography. *Annu. Rev. Neurosci.* **6,** 249–268.

Raichle, M. E., Martin, W. R. W., Herscovitch, P., Mintun, M. A., and Markham, J. (1983). Brain blood flow measured with intravenous H$_2$^{15}O. II. Implementation and validation. *J. Nucl. Med.* **24,** 790–798.

Rattner, Z., Smith, E. O., Woods, S., Dey, H., and Hoffer, P. B. (1991). Toward absolute quantitation of cerebral blood flow using technetium-99m-HMPAO and a single scan. *J. Nucl. Med.* **32** (8), 1506–1507.

Reivich, M., Kuhl, D. E., Wolf, A., *et al.* (1979). The (^{18}F)fluorodeoxyglucose method for the measurement of local cerebral glucose utilization in man. *Circ. Res.* **44,** 127–137.

Rowe, C. C., Berkovic, S. F., Austin, M. C., McKay, W. J., and Bladin, P. F. (1991a). Patterns of postictal cerebral blood flow in temporal lobe epilepsy: Qualitative and quantitative analysis. *Neurology* **41,** 1096–1103.

Rowe, C. C., Berkovic, S. F., Austin, M. C., Saling, M., Kalnins, R. M., McKay, W. J., *et al.* (1991b). Visual and quantitative analysis of interictal SPECT with technetium-99m-HMPAO in temporal lobe epilepsy. *J. Nucl. Med.* **32**(9), 1688–1694.

Roy, C. S., and Sherrington, C. S. (1890). On the regulation of the blood supply of the brain. *J. Physiol. (London)* **11,** 85–108.

Salcman, M., and Pevsner, P. H. (1992). Value of MRI in head injury: Comparison with CT. *Neurochirurgie* **38,** 329–332.

Sanes, J. N., and Donoghue, J. P. (1993). Oscillations in local field potentials of the primate motor cortex during voluntary movement. *Proc. Natl. Acad. Sci. U.S.A.* **90,** 4470–4474.

Sappey-Marinier, D., Calabrese, G., Fein, G., Hugg, J. W., Biggins, C., and Weiner, M. W. (1992). Effect of photic stimulation on human visual cortex lactate and phosphates using ^{1}H and ^{31}P magnetic resonance spectroscopy. *J. Cereb. Blood Flow Metab.* **12,** 584–592.

Savic, I., Roland, P., Sedrall, G., Persson, A., Pauli, S., and Widen, L. (1988). *In-vivo* demonstration of reduced benzodiazepine receptor binding in human epileptic foci. *Lancet* **ii,** 864–866.

Schulz, J. B., Klockgether, T., Petersen, D., Jauch, M., Müller-Schauenburg, W., Spieker, S., *et al.* (1994). Multiple system atrophy: Natural history, MRI morphology, and dopamine receptor imaging with ^{123}IBZM-SPECT. *J. Neurol. Neurosurg. Psychiatry* **57,** 1047–1056.

Shellock, F. (1997). "Pocket Guide to MR Procedures and Metallic Objects: Update 1997." Lippincott-Raven, Philadelphia.

Smith, M. E., Halgren, E., Sokolik, M. E., Baudena, P., Liegeois-Chauvel, C., Musolino, A., *et al.* (1990). The intracranial topography of the P3 event-related potential elicited during auditory oddball. *Electroencephalogr. Clin. Neurophysiol.* **76,** 235–248.

Sokoloff, L., Reivich, M., Kennedy, C., *et al.* (1977). The (^{14}C)deoxyglucose method for the measurement of local cerebral glucose utilization: Theory, procedure and normal values in the conscious and anesthetized albino rat. *J. Neurochem.* **28,** 897–916.

Sorensen, A. G., Buonanno, F. S., Gonzalez, R. G., Schwamm, L. H., Lev, M. H., Huang-Hellinger, F. R., *et al.* (1996). Hyperacute stroke: Evaluation with combined multisection diffusion-weighted and hemodynamically weighted echo-planar MR imaging. *Radiology* **199,** 391–401.

Spencer, S. S. (1994). The relative contributions of MRI, SPECT and PET imaging in epilepsy. *Epilepsia* **35,** S72–S89.

Strother, S. C., Anderson, J. R., Xu, X. L., Liow, J. S., Bonar, D. C., and Rottenberg, D. A. (1994). Quantitative comparisons of image registration techniques based on high-resolution MRI of the brain. *J. Comput. Assist. Tomogr.* **18** (6), 954–962.

Talairach, J., and Tournoux, P. (1988). Principe et technique des etudes anatomiques. *In* "Co-Planar Stereotaxic Atlas of the Human Brain—Three-Dimensional Proportional System: An Approach to Cerebral Imaging" (M. Rayport, Ed.), Vol. 3. Thieme, New York.

Theodore, W. H. (1992). MRI, PET, SPECT: Interrelations, technical limits, and unanswered questions. *In* "Surgical Treatment of Epilepsy" (W. H. Theodore, Ed.), pp. 127–134. Elsevier, Amsterdam/New York.

Thurfjell, L., Bohm, C., Greitz, T., Eriksson, L., and Ingvar, M. (1994). Accuracy and precision in image standardization in intra- and intersubject comparisons. *In* "Functional Neuroimaging: Technical Foundations" (R. W. Thatcher, M. Hallett, T. Zeffiro, E. R. John, and M. Huerta, Eds.). Academic Press, San Diego.

Toga, A. W., Cannestra, A. F., and Black, K. L. (1995). The temporal/spatial evolution of optical signals in the human cortex. *Cereb. Cortex* **5** (6), 561–565.

Toga, A., and Mazziotta, J. (1996). "Brain Mapping: The Methods." Academic Press, San Diego.

Toga, A., and Mazziotta, J. (2000). "Brain Mapping: The Systems." Academic Press, San Diego.

Turkington, T. G., Hoffman, J. M., Jaszczak, R. J., MacFall, J. R., Harris, C. C., Kilts, C. D., *et al.* (1995). Accuracy of surface fit registration for PET and MR brain images using full and incomplete brain surfaces. *J. Comput. Assist. Tomogr.* **19** (1), 117–124.

Turner, R., Le Bihan, D., Moonen, C. T. W., Despres, D., and Frank, J. (1991). Echo-planar time course MRI of cat brain deoxygenation changes. *Magn. Reson. Med.* **22,** 210–216.

Turner, R., Jezzard, P., Wen, H., Kwong, K. K., LeBihan, D., Zeffiro, T., and Balaban, R. S. (1993). Functional mapping of the human visual cortex at 4 and 1.5 Tesla using deoxygenation contrast EPI. *Magn. Reson. Med.* **29,** 277–279.

van Huffelen, A. C., van Isselt, J. W., van Veelen, C. W. M., van Rijk, P. P., van Bentum, A. M. E., Dive, D., *et al.* (1990). Identification of the side of the epileptic focus with ^{123}I-iomazenil SPECT. *Acta Neurochirurg. Suppl.* **50,** 95–99.

van Zijl, P. C. M., Chesnick, A. S., DesPres, D., Moonen, C. T., Ruiz-Cabello, J., and van Gelderen, P. (1993). In vivo proton spectroscopy and spectroscopic imaging of [1-^{13}C]-glucose and its metabolic products. *Magn. Reson. Med.* **30** (5), 544–551.

Villringer, A., Rosen, B. R., Belliveau, J. W., Ackerman, J. L., Lauffer, R. B., Buxton, R. B., *et al.* (1988). Dynamic imaging with lanthanide chelates in normal brain-Contrast due to magnetic-susceptibility effects. *Magn. Reson. Med.* **6,** 164–174.

Walter, D. O. (1963). Spectral analysis for electroencephalograms: Mathematical determination of neurophysiological relationships from records of limited duration. *Exp. Neurol.* **8,** 155–181.

Walter, W. G., Cooper, R., Aldridge, V., McCallum, W., and Winter, A. (1964). Contingent negative variation: An electrical sign of sensorimotor association and expectancy in the human brain. *Nature* **203,** 380–384.

Wang, Z., Zimmerman, R. A., and Sauter, R. (1996). Proton MR spectroscopy of the brain: Clinically useful information obtained in assessing CNS diseases in children. *AJR, Am. J. Roentgenol.* **167,** 191–199.

Warach, S., Li, W., Ronthal, M., and Edelman, R. R. (1992). Acute cerebral ischemia: Evaluation with dynamic contrast-enhanced MR imaging and MR angiography. *Radiology* **182,** 41–47.

Warach, S., Gaa, J., Siewert, B., Wielopolski, P., and Edelman, R. R. (1995a). Acute human stroke studied by whole brain echo planar diffusion-weighted magnetic resonance imaging. *Ann. Neurol.* **37,** 231–241.

Warach, S., Ives, J., Darby, D., Thangaraj, V., Edelman, R., and Schomer, D. (1995b). EEG triggered echo planar functional MRI: Localizing activation related to EEG discharges. *Hum. Brain Mapp. Suppl.* **1,** 420.

Wasserman, E. M. (1998). Risk and safety of repetitive transcranial magnetic stimulation: Report and recommendations from the International Workshop on the Safety of Repetitive Transcranial Magnetic Stimulation, June 5–7, 1996. *Electroencephalogr. Clin. Neurophysiol.* **108,** 1–16.

Wasserman, E. M., McShane, L. M., Hallett, M., and Cohen, L. G. (1992). Noninvasive mapping of muscle representations in human motor cortex. Electroencephalogr. *Clin. Neurophysiol.* **85,** 1–8.

Wasserman, E. M., Pascual-Leone, A., and Hallett, M. (1994). Cortical motor representation of the ipsilateral hand and arm. *Exp. Brain Res.* **100,** 121–132.

Weinberger, D., Luchins, P., Morihisa, J., and Wyatt, R. (1982). Asymmetrical volumes of the right and left frontal and occipital regions of the human brain. *Neurology* **11,** 97–102.

Wieringa, H. J., Peters, M. J., and Lopes da Silva, F. H. (1994). Integration of MEG, EEG, and MRI. *In* "Functional Neuroimaging: Technical Foundations" (R. W. Thatcher, M. Hallett, T. Zeffiro, E. R. John, and M. Huerta, Eds.). Academic Press, San Diego.

Wilcock, D., Jaspan, T., Holland, I., Cherryman, G., and Worthington, B. (1996). Comparison of magnetic resonance angiography with conventional angiography in the detection of intracranial aneurysms in patients presenting with subarachnoid haemorrhage. *Clin. Radiol.* **51,** 330–334.

Williams, A. L., and Haughton, V. M., Eds. (1985). "Cranial Computed Tomography: A Comprehensive Text." Mosby, St. Louis.

Witelson, S. (1977). Anatomical asymmetry in the temporal lobes: Its documentation, phylogenesis and relationship to functional asymmetry. *Ann. N.Y. Acad. Sci.* **299,** 328–354.

Wong, D. F., Gjedde, A., and Wagner, H. N. J. (1986). Quantification of neuroreceptors in the living human brain. I. Irreversible binding of ligands. *J. Cereb. Blood Flow Metab.* **6,** 137–146.

Woods, R. P., Cherry, S. R., and Mazziotta, J. C. (1992). Rapid automated algorithm for aligning and reslicing PET images. *J. Comput. Assist. Tomogr.* **16** (4), 620–633.

Woods, R. P., Mazziotta, J. C., and Cherry, S. R. (1993). PET registration with automated algorithm. *J. Comput. Assist. Tomogr.* **17** (4), 536–546.

Zhang, J., Levesque, M. F., Wilson, C. L., Harper, R. M., Engel, J., Jr., Lufkin, R., *et al.* (1990). Multimodality imaging of brain structures for stereotactic surgery. *Radiology* **175** (2), 435–441.

2

Experimental Design and Statistical Issues

Karl J. Friston

*Wellcome Department of Cognitive Neurology, Institute of Neurology,
London WC1N 3BG, United Kingdom*

I. Introduction

II. Functional Specialization and Integration

III. Spatial Realignment and
Anatomical Normalization

IV. Statistical Parametric Mapping and
Functional Specialization

V. Experimental Design

VI. Designing fMRI Experiments: A Signal-
Processing Perspective

VII. Epoch- and Event-Related Studies

VIII. Inferences about Subjects and Populations

IX. Effective Connectivity and
Functional Integration

X. Specialization, Integration, and the Lesion-
Deficit Model

References

I. Introduction

This chapter reviews experimental design and analysis issues that arise when using functional neuroimaging (PET and fMRI) to make inferences about functional anatomy in health and disease. These issues comprise those that are generic to all functional neuroimaging and some considerations that are specific to modeling the effects of pathophysiology. This chapter will consider, in turn, the basic principles of design and analysis and highlight their importance for, or role in, studies of brain disorders.

An important issue in design and analysis is the relationship between the neurobiological hypotheses and conceptual models of pathophysiology and how this relationship is realized in terms of the statistical models used to analyze neuroimaging time series. This chapter begins by reviewing the distinction between functional specialization and integration and how these principles serve as the motivation for most analyses of functional neuroimaging data. This distinction is critical from the point of view of clinical studies and relates closely to the distinction between organic brain syndromes that arise from regionally specific insults and those that can be framed in terms of disconnection syndromes. Indeed the distinction between specialization and connectionism arose historically from clinical studies (see below). This chapter will address the design and analysis of clinical studies from both these distinct perspectives but will conclude by noting that both have to be considered before definitive inferences about regionally specific abnormalities can be made.

Statistical parametric mapping is generally used to identify functionally specialized brain regions and is the most prevalent approach to characterizing functional anatomy and disease-related changes. The alternative perspective, namely that provided by functional integration, requires a different set of [multivariate] approaches that examine the relationship between changes in activity in one area and another. Statistical parametric mapping is a voxel-based approach, employing standard inferential statistics to make some comment about regionally specific responses to experimental factors. To assign an observed response to a particular brain structure or cortical area, the data must conform to a known anatomical space. Before considering statistical modeling, this chapter deals briefly with how a time series of images are realigned and mapped into some standard anatomical space (e.g., stereotactic space). The general ideas behind statistical parametric mapping are then described and illustrated with attention to the different sorts of inferences that can be made with different experimental designs. fMRI is special in the sense that the data lend themselves to a signal-processing perspective that can be exploited to ensure that both the design and analysis are as efficient as possible. Linear time-invariant models provide the bridge between inferential models employed by statistical mapping and simple signal-processing approaches. Temporal autocorrelations in noise processes represent another important issue, specific to fMRI, and approaches to maximizing efficiency in the context of serially correlated error terms will be discussed. Nonlinear models (Volterra series) will be considered because they provide constraints on when the assumptions behind linear models are likely to hold. fMRI can capture data very fast (in relation to other imaging techniques), engendering the opportunity to measure event-related responses. The distinction between event- and epoch-related designs will be discussed from the point of view of efficiency and the constraints provided by nonlinear characterizations. Before considering multivariate analysis and effective connectivity, we will close the discussion of inferences, about regionally specific effects, by looking at the distinction between fixed and random-effect analyses and how this relates to inferences about the subjects studied or the population from which these subjects came. The final section will deal with some conceptual issues about regional responses in disease by looking at abnormal regional responses in the context of functional [dis]integration.

II. Functional Specialization and Integration

The brain appears to adhere to two fundamental principles of functional organization: *functional integration* and *functional specialization,* where the integration within and among specialized areas is mediated by effective connectivity. The distinction relates to that between *localizationism* and *[dis]connectionism* that dominated thinking about cortical function in the nineteenth century. Since the early anatomic theories of Gall, the identification of a particular brain region with a specific function has become a central theme in neuroscience. However, functional localization per se was not easy to demonstrate: For example, a meeting that took place on August 4, 1881, addressed the difficulties of attributing function to a cortical area, given the dependence of cerebral activity on underlying connections (Phillips *et al.,* 1984). This meeting was entitled "Localisation of Function in the Cortex Cerebri." Goltz (1881), although accepting the results of electrical stimulation in dog and monkey cortex, considered that the excitation method was inconclusive, in that movements elicited might have originated in related pathways or current could have spread to distant centers. In short, the excitation method could not be used to infer functional localization because localizationism discounted interactions or functional integration among different brain areas. It was proposed that lesion studies could supplement excitation experiments; ironically, it was observations on patients with brain lesions some years later (see Absher and Benson, 1993) that led to the concept of *disconnection syndromes* and the refutation of localizationism as a complete or sufficient explanation of cortical organization. Functional localization implies that a function can be localized in a cortical area, whereas specialization suggests that a cortical area is specialized for some aspects of perceptual or motor processing and that this specialization is anatomically segregated within the cortex. The cortical infrastructure supporting a single function may then involve many specialized areas whose union is mediated by the functional integration among them. In this view, functional specialization is only meaningful in the context of functional integration and vice versa.

A. Functional Specialization and Segregation

The functional role played by any component (e.g., cortical area, subarea, or neuronal population) of the brain is largely defined by its connections. Certain patterns of cortical projections are so common that they could amount to rules of cortical connectivity. "These rules revolve around one, apparently, overriding strategy that the cerebral cortex uses—that of functional segregation" (Zeki, 1990). Functional segregation demands that cells with common functional properties be grouped together. This architectural constraint, in turn, necessitates both convergence and divergence of cortical connections. Extrinsic connections between cortical regions are not continuous but occur in patches or clusters. This patchiness has, in some instances, a clear relationship to functional segregation. For example, V2 has a distinctive cytochrome oxidase architecture, consisting of thick stripes, thin stripes, and interstripes. When recordings are made in V2, directionally selective (but not wavelength or color selective) cells are found exclusively in the thick stripes. Retrograde (i.e., backward) labeling of cells in V5 is limited to these thick stripes. All the available physiological evidence suggests that V5 is a functionally homogeneous area that is specialized for visual motion. Evidence of this nature supports the notion that patchy connectivity is the anatomical infrastructure that mediates functional segregation and specialization. If it is the case that neurons in a given

cortical area share a common responsiveness (by virtue of their extrinsic connectivity) to some sensorimotor or cognitive attribute, then this functional segregation is also an anatomical one. Challenging a subject with the appropriate sensorimotor attribute or cognitive process should lead to activity changes in, and only in, the area of interest. This is the model upon which the search for regionally specific effects is based and the basis of the analytic techniques adopted.

The analysis of functional neuroimaging data involves many steps that can be broadly divided into (i) spatial preprocessing, (ii) estimating the parameters of a statistical model, and (iii) making inferences about modeled effects using the parameter estimates and their associated statistics (see Fig. 1). We will deal first with spatial transformations: The relationship between regionally specific effects and homologous responses in other subjects (e.g., control subjects) has to be established to make inferences about abnormal responses. This necessitates some form of anatomical designation for regional responses that is facilitated by analyzing the data in relation to some standard anatomical space.

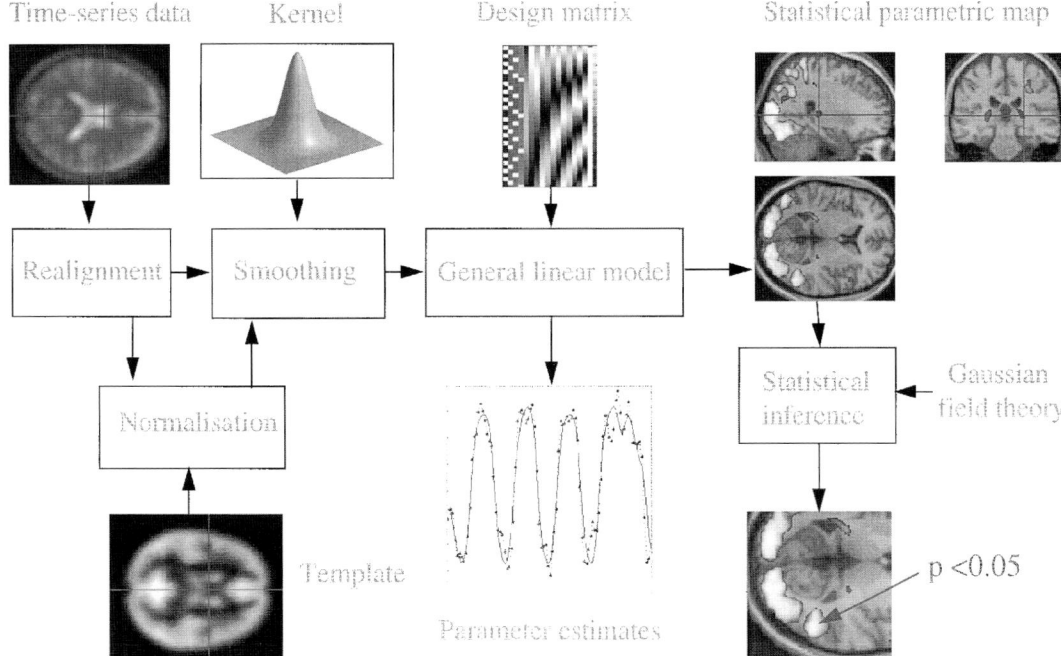

Figure 1 This schematic depicts the transformations that start with the fMRI time series and end with a statistical parametric map (SPM). SPMs can be thought of as "X-rays" of the significance of an effect. Voxel–based analyses require the data to be in the same anatomical space: This is effected by realigning the data (and removing movement-related signal components that persist after realignment). After realignment, the images are subject to nonlinear warping so that they match a template that already conforms to a standard anatomical space. After smoothing, the general linear model is employed to (i) estimate the parameters of the model and (ii) derive the appropriate univariate test statistic at each and every voxel (see Fig. 2). The test statistics that ensue (usually T or F statistics) constitute the SPM. The final stage is to make statistical inferences on the basis of the SPM (see Fig. 3) and characterize the responses observed using the fitted responses or parameter estimates.

III. Spatial Realignment and Anatomical Normalization

The analysis of PET and fMRI studies starts with a series of spatial transformations. These transformations aim to reduce artifactual variance components in the voxel time series. The first step is to realign the data to "undo" the effects of subject movement during the scanning session. After realignment the data are then transformed using linear or nonlinear mappings into a standard anatomical space. The data are usually spatially smoothed before entering the analysis proper.

A. Realignment

Changes in signal intensity from any one voxel can arise from head motion and this represents a serious confound, particularly in fMRI studies. Despite restraints on head movement, cooperative subjects still show displacements of up to a millimeter or so. Realignment involves (i) estimating the six parameters of an affine "rigid-body" transformation that minimizes the [sum of squared] differences between each successive scan and the first and (ii) applying the transformation by resampling the data using trilinear or sinc interpolation. Estimation of the affine transformation is usually effected with a first-order approximation of the Taylor expansion of the effect of movement on signal intensity using the spatial derivatives of the images. This allows for a simple iterative least-squares solution (that corresponds to a Gauss-Newton search) (Friston *et al.*, 1995a). However, in fMRI, even after perfect realignment, movement-related signals can still persist. This calls for a final step in which the data are adjusted for residual movement-related effects.

B. Adjusting for Movement-Related Effects

In extreme cases, as much as 90% of the variance in an fMRI time series can be accounted for by the effects of movement *after* realignment (Friston *et al.*, 1996a). Causes of these movement-related components include (i) subject movement between slice acquisition, (ii) interpolation artifacts, (iii) nonlinear distortion due to magnetic field inhomogeneity, and (iv) spin-excitation history effects. The latter can be pronounced if the TR (repetition time) approaches T_1, making the current signal a function of movement history. These multiple effects render the movement-related signal (y) a nonlinear function of displacement ($\mathbf{x}$) in the nth and previous scans $y = f(\mathbf{x}_n, \mathbf{x}_{n-1}, \ldots)$. By assuming a sensible form for this function, its parameters can be estimated using the observed time series and the estimated movement parameters $\mathbf{x}$ from the realignment procedure. The estimated movement-related signal is then simply subtracted from the original data. The form for $f(\mathbf{x})$ proposed in Friston *et al.* (1996a) was an autoregression-like model that was motivated by spin-excitation history effects and allowed displacement in previous scans to explain the current movement-related signal. Generally, for TRs of several seconds, interpolation artifacts supersede and an expansion of $f(\mathbf{x}_n)$ in terms of periodic basis functions of the current displacement appears to be sufficient.

This section has considered spatial realignment. In multislice acquisition, different slices are acquired at slightly different times. This engenders the possibility of temporal realignment to ensure that the data from any given volume were sampled at the same time. This is usually performed using sinc interpolation over time and only when the temporal dynamics of evoked responses are important and the TR is sufficiently small to permit interpolation.

C. Spatial Normalization

After the data are realigned, a mean image of the time series or some other coregistered (e.g., a T_1-weighted) image is used to estimate the warping parameters that map it onto some template that already conforms to a standard anatomical space (e.g., Talairach and Tournoux, 1988). This estimation can use a variety of models for the mapping, including (i) a 12-parameter affine transformation where the parameters constitute a spatial transformation matrix, (ii) low-frequency basis spatial functions (usually a discrete cosine set or polynomials), where the parameters are the coefficients of the basis functions employed, and (iii) a vector field specifying the mapping for each control point (e.g., voxel). In the final case, the parameters are vast in number and constitute a tensor that is bigger than the image itself. Estimation of the parameters of all these models can be accommodated in a simple Bayesian framework in which one is trying to find the deformation β that has the maximum posterior probability $p(\beta|\mathbf{Y})$ given the data $\mathbf{Y}$, where $p(\beta|\mathbf{Y}) \propto p(\mathbf{Y}|\beta) \cdot p(\beta)$. Put simply, one wants to find the deformation that is most likely given the data. This deformation can be found by maximizing the probability of getting the observed data, assuming the deformation is true, times the probability of that deformation. In practice, the deformation is updated iteratively to minimize the likelihood and prior potentials $H(\mathbf{Y}|\beta)$ and $H(\beta)$, where $p(\mathbf{Y}|\beta) = \exp\{-H(\mathbf{Y}|\beta)\}$ and $p(\beta) = \exp\{-H(\beta)\}$. The likelihood potential is generally taken to be the sum of squared differences between the template and deformed image and reflects the probability of actually getting that image if the transformation was correct. The prior potential can be used in the context of weighted least squares to incorporate prior information about the likelihood of a given transformation. Prior potentials can be determined empirically or motivated by constraints on the mappings and play a more essential role as the number of parameters specifying the mapping increases (Ashburner *et al.*, 1997).

In practice, most people use an affine or spatial basis function model and use iterative least squares to minimize the likelihood potential. As with realignment, this involves linearizing with Taylor expansions. A nice extension of this approach is that the difference between the template and index image can be minimized by using a [linear] combination of templates (e.g., depicting gray, white, CSF, and skull tissue partitions). This models intensity differences that are unrelated to registration differences and allows different modalities to be coregistered (Friston *et al.*, 1995a).

A special consideration here is the spatial normalization of brains that have gross anatomical pathology. This pathology can be of two sorts: (i) quantitative changes in the amount of a particular tissue compartment (e.g., cortical atrophy) or (ii) qualitative changes in anatomy involving the insertion or deletion of normal tissue compartments (e.g., ischemic tissue in stroke or cortical dysplasia). The former case is, generally, not problematic in the sense that changes in the amount of cortical tissue will not affect its optimum spatial location in reference to some template (and, even if it does, a disease-specific template is easily constructed). The second sort of pathology can introduce substantial "errors" in the normalization unless special precautions are taken. These usually involve imposing constraints on the warping to ensure the pathology does not bias the deformation of undamaged tissue. This involves "hard" constraints implicit in using a small number of basis functions or "soft" constraints by increasing the role of prior information in Bayesian schemes. An alternative strategy is to use another modality that is less sensitive to the pathology as the basis of the spatial normalization procedure.

D. Coregistration of Functional and Anatomical Data

It is useful to be able to coregister functional and anatomical images. However, with echo-planar imaging, geometric distortions of T_2^* images, relative to anatomical T_1-weighted data, are a particularly serious problem because of the very low frequency per point in the phase-encoding direction. Typically, for echo-planar fMRI, magnetic field inhomogeneity sufficient to cause dephasing of 2π through the slice corresponds to a voxel distortion in the plane. "Unwarping" schemes have been proposed to correct for the distortion effects (Jezzard and Balaban, 1995). Note that this is not an issue if one spatially normalizes the functional data.

E. Spatial Smoothing

The motivations for smoothing the data are fourfold: (i) By the matched filter theorem, the optimum smoothing kernel corresponds to the size of the effect that one anticipates. The spatial scale of hemodynamic responses is, according to high-resolution optical imaging experiments, about 2-5 mm. Despite the potentially high resolution afforded by fMRI, an equivalent smoothing is suggested for most applications. (ii) By the central limit theorem, smoothing the data will render them more parametric in their distribution and ensure the validity of inferences based on parametric tests. (iii) When inferences are made about regional effects using Gaussian field theory (see below), one of the assumptions is that the data are a reasonable lattice representation of an underlying and smooth Gaussian field. This necessitates smoothness to be substantially greater than voxel size. If the voxels are large, then they can be reduced by subsampling the data and then smoothing (with the original point spread function) with little loss of intrinsic resolution. (iv) In the context of intersubject averaging, it is often necessary to smooth more (e.g., 6 mm) to project the data onto a spatial scale where homologies in functional anatomy are expressed among subjects.

IV. Statistical Parametric Mapping and Functional Specialization

Functional mapping studies are usually analyzed with some form of statistical parametric mapping. Statistical parametric mapping refers to the construction of spatially extended statistical processes to test hypotheses about regionally specific effects (Friston *et al.*, 1991). Statistical parametric maps (SPMs) are image processes with voxel values that are, under the null hypothesis, distributed according to a known probability density function (usually Student's T or F distributions). The success of statistical parametric mapping is largely due to the simplicity of the idea. Namely, one analyzes each and every voxel using any standard (univariate) statistical test. The resulting statistical parameters are assembled into an image-the SPM. SPMs are interpreted as spatially extended statistical processes by referring to the probabilistic behavior of stationary Gaussian fields (Adler, 1981; Worsley *et al.*, 1992, 1996; Friston *et al.*, 1994a). Stationary fields model both the univariate probabilistic characteristics of a SPM and any stationary spatial covariance structure (stationary means not a function of position). "Unlikely" excursions of the SPM are interpreted as regionally specific effects, attributable to the sensorimotor or cognitive process that has been manipulated experimentally.

A. Different Approaches?

Statistical analysis corresponds to modeling the data in order to partition observed neurophysiological responses into components of interest, confounds, and error. A brief review of the literature may give the impression that there are numerous ways to analyze PET and fMRI time series with a diversity of statistical and conceptual approaches. This is not the case. With only a few exceptions, every analysis presented is a variant of the general linear model. This includes (i) simple T tests on scans assigned to one condition or another, (ii) correlation coefficients between observed responses and boxcar stimulus functions in fMRI, (iii) inferences made using multiple linear regression, (iv) evoked responses estimated using linear time-invariant models, and (v) selective averaging to estimate event-related responses in fMRI. Mathematically they are all identical. The use of the correlation coefficient deserves special mention because of its popularity in fMRI (Bandettini *et al.*, 1993). The significance of a correlation is identical to the significance of the equivalent T value testing for a regression of the data on some stimulus waveform. The correlation coefficient approach is useful but the inference is effectively based on a limiting case of multiple linear regression that obtains when there is only one regressor. In fMRI many regressors usually enter into a statistical model. For this reason, the T statistic provides a more flexible and general way of assessing the significance of regional effects and is preferred over the correlation coefficient.

Some researchers construct statistical maps based on nonparametric tests that eschew distributional assumptions about the data. These approaches may, in some instances, be useful but are generally less powerful (i.e., sensitive) than parametric approaches (see Aguirre *et al.*, 1998). Their original motivation in fMRI was based on the [specious] assumption that the residuals are not normally distributed.

B. The General Linear Model

The general linear model is an equation $\mathbf{Y} = \mathbf{X}\beta + \epsilon$ that expresses the observed response variable $\mathbf{Y}$ in terms of a linear combination of explanatory variables $\mathbf{X}$ plus a well-behaved error term (see Fig. 2 and Friston *et al.*, 1995b). The general linear model is variously known as "analysis of covariance" or "multiple regression analysis" and subsumes simpler variants, like the T test for a difference in means, to more elaborate linear convolution models such as finite impulse response (FIR) models. The matrix $\mathbf{X}$ that contains the explanatory variables (e.g., designed effects or confounds) is called the design matrix. Each column of the design matrix corresponds to some effect one has built into the

experiment or that may confound the results. These are variously referred to as explanatory variables, covariates, regressors, or, in fMRI, stimulus functions. The example in Fig. 1 relates to an fMRI study of visual stimulation under four conditions. The effects on the response variable are modeled in terms of functions of the presence of these conditions (i.e., boxcars smoothed with a hemodynamic response function) and constitute the first four columns of the design matrix. There then follows a series of terms that are designed to remove or model low-frequency variations in signal due to artifacts such as aliased biorhythms. The final column is whole-brain activity. The relative contribution of each of these columns is assessed using standard least squares and inferences about these contributions are made using T or F statistics, depending upon whether one is looking at a particular linear combination or all of them together. The operational equations are shown in Fig. 2. In this scheme the general linear model has been extended (Worsley and Friston, 1995) to incorporate intrinsic autocorrelations or serial correlation among the error terms and to allow for some specified filtering or smoothing of the data.

These equations can be used to implement a vast range of statistical analyses. The issue is therefore not so much the mathematics but the formulation of a design matrix $\mathbf{X}$ appropriate to the study design and inferences that are sought. The design matrix can contain both covariates and indicator variables. Each column of $\mathbf{X}$ has an associated unknown parameter. Some of these parameters will be of interest (e.g., the effect of a particular sensorimotor or cognitive condition or the regression coefficient of hemodynamic responses on reaction time). The remaining parameters will be of no interest and pertain to confounding effects (e.g., the effect of being a particular subject or the regression slope of voxel activity on global activity). Inferences about the parameter estimates are made using their estimated variance. This allows one to test the null hypothesis that all the estimates are zero using the F statistic to give an SPM{F} or that some particular linear combination (e.g., a subtraction) of the estimates is zero using a SPM{T}. The T statistic obtains by dividing a contrast or compound (specified by contrast weights) of the ensuing parameter estimates by the standard error of that compound. The latter is estimated using the variance of the residuals about the least-squares fit. An example of contrast weights would be $[-1\ 1\ 0\ 0\ .\ .\ .]$ to compare the differential responses evoked by two conditions, as modeled by the first two condition-specific regressors in the design matrix. If several parameter estimates are potentially interesting (e.g., using polynomial expansions (Büchel *et al.*, 1996) or basis functions of some parameter of interest), then the SPM{F} is usually employed. An important example, from the perspective of

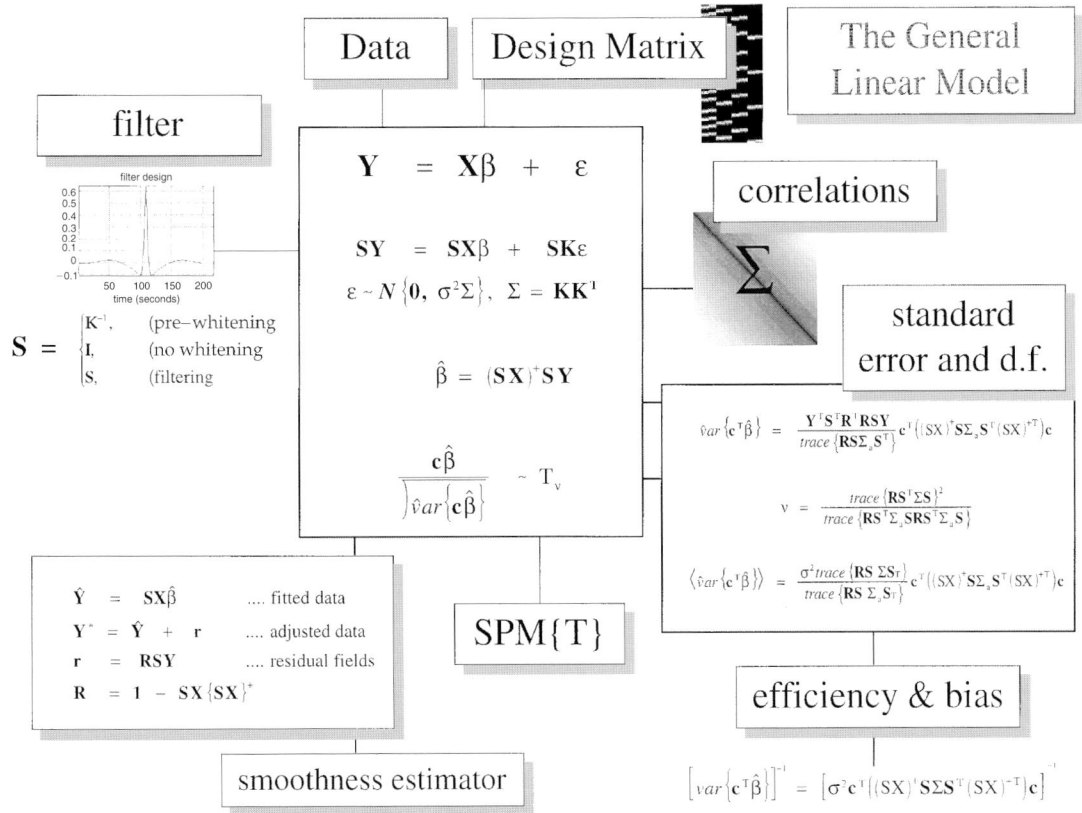

Figure 2 The general linear model. The general linear model is an equation expressing the response variable **Y** in terms of a linear combination of explanatory variables in a design matrix **X** and an error term with assumed or known autocorrelation Σ. In fMRI the data can be filtered with a convolution matrix **S**, leading to an augmented general linear model that includes [intrinsic] serial correlations and applied [extrinsic] filtering. Different choices of **S** correspond to different [de]convolution schema as indicated on the upper left. The parameter estimates obtain in a least-squares sense using the pseudoinverse (denoted by +) of the filtered design matrix. Generally, an effect of interest is specified by a vector of contrast weights **c** that give a weighted sum or compound of parameter estimates $c\hat{\beta}$, referred to as a contrast. The T statistic is simply this contrast divided by its estimated standard error (i.e., square root of the estimated variance). The ensuing T statistic is distributed with v degrees of freedom. The equations for the estimated variance of the contrast and the degrees of freedom associated with the error variance are provided in the right-hand panel. Efficiency is simply the inverse of the variance of the contrast estimate. Discrepancies between the actual and assumed intrinsic correlations (Σ and Σ_a, respectively) can result in systematic bias in the actual and estimated variance of the contrast. These expressions are useful when assessing the relative efficiency and robustness of a model. The parameter estimates can be either examined directly or used to compute the fitted responses (see lower left panel). Adjusted data refers to data from which estimated confounds have been removed. The residuals **r** obtain from applying the residual-forming matrix **R** to the data. These residual fields are used to estimate the smoothness of the component fields of the SPM used in Gaussian field theory (see Fig. 3).

fMRI, of the general linear model is the linear time-invariant model.

1. Linear Time-Invariant (LTI) Systems and Finite Impulse Response (FIR) Models

In Friston *et al.* (1994b) the form of the hemodynamic impulse response function (HRF) was estimated using a least-squares deconvolution and a FIR time-invariant (LTI) model, where evoked neuronal responses are convolved with the HRF to give the measured hemodynamic response (see also Boynton *et al.*, 1996). The estimated HRF resembled a Poisson or Gamma function peaking at about 5 s. This simple linear

framework is the cornerstone for making statistical inferences about activations in fMRI with the general linear model. The notion of temporal basis functions to model evoked responses in fMRI was subsequently introduced (Friston *et al.*, 1995c) and applied to event-related responses in Josephs *et al.* (1997) (see also Lange and Zeger, 1997).

2. Temporal Basis Functions

Temporal basis functions are important because they provide for a graceful transition between conventional multilinear regression models with one stimulus function per condition and FIR models with a parameter for each time point following the onset of a condition or trial type. Temporal basis functions offer useful constraints on the form of the estimated response that retain (i) the flexibility of FIR models and (ii) the efficiency of single-regressor models. In practice, the implementation of these constrained FIR models involves setting up stimulus functions that model expected neuronal changes [e.g., boxcars of epoch-related responses or spikes (delta functions) at the onset of specific events or trials]. These regressors are then convolved with a set of basis functions that model the HRF, in some linear combination, and are assembled into the design matrix. The basis functions can be as simple as a single canonical HRF, through to a series of delayed delta functions. The latter case corresponds to a FIR model proper and the coefficients are the impulse response function for the event or epoch in question. Selective averaging in event-related fMRI (Dale and Buckner, 1997) is mathematically equivalent to this limiting case.

The advantage of using temporal basis functions (as opposed to an assumed form for the HRF) is that one can model voxel-specific forms for hemodynamic responses and formal differences (e.g., onset latencies) among responses to different sorts of events. The advantages of using basis functions over FIR models are that (i) the parameters are estimated more efficiently and (ii) stimuli can be presented at any point in the interstimulus interval. The latter is very important because time-locking stimulus presentation and data acquisition gives a biased sampling over peristimulus time and can lead to differential sensitivities, in multislice acquisition, over the brain.

C. Statistical Inference and the Theory of Gaussian Fields

Inferences using SPMs can be of two sorts depending on whether one knows where to look in advance: With an anatomically constrained hypothesis about effects in a particular brain region, the uncorrected p value associated with the height or extent of that region in the SPM can be used to test the hypothesis. With an anatomically open hypothesis (i.e., a null hypothesis that there is no effect anywhere in the brain), a correction for multiple dependent comparisons is necessary. The theory of Gaussian fields provides a way of computing this corrected p value that takes into account the fact that neighboring voxels are not independent by virtue of smoothness in the original data. Provided the data are sufficiently smooth, the correction based on Gaussian field theory is less severe (i.e., is more sensitive) than a Bonferroni correction for the number of voxels.

The only assumptions underlying the use of Gaussian

field theory are that (i) the residual fields (but not necessarily the data) are a reasonable lattice approximation to an underlying Gaussian field, (ii) the fields have a twice-differentiable autocorrelation function (a common misconception is that the autocorrelation function has to be Gaussian-it does not), and (iii) their multinormal correlation structure is wide-sense stationary. The only way in which these assumptions can be violated is if (i) the data are not smoothed (with or without subsampling of the data to preserve resolution) or (ii) the statistical model is misspecified so that spatially coherent signals (e.g., global effects) end up in the residuals.

1. Anatomically Closed Hypotheses

When making inferences about regional effects (e.g., activations) in statistical maps, one often has some idea about where the activation should be. In this instance, a correction for the entire search volume is inappropriate. However, a problem remains in the sense that one would like to consider activations that are "near" the predicted location, even if they are not exactly coincident. There are two approaches to this: (i) prespecify a small search volume and make the appropriate correction (Worsley et al., 1996) or (ii) use the uncorrected p value based on spatial extent of the nearest cluster (Friston, 1997). This probability is based on getting the observed number of voxels, or more, in a given cluster (conditional on that cluster existing) and can be calculated using distributional approximations from the theory of Gaussian fields.

2. Anatomically Open Hypotheses and Levels of Inference

To correct for the volume of brain analyzed, the SPM is subject to thresholding, using some height and spatial extent thresholds, and p values are derived that pertain to (i) the number of activated regions (i.e., number of clusters above the height and volume threshold)-set-level inferences, (ii) the number of activated voxels (i.e., volume) comprising a particular region-cluster-level inferences, and (iii) the p value for each voxel within that cluster-voxel-level inferences. These p values are corrected for the multiple dependent comparisons and are based on the probability of obtaining c, or more, clusters with k, or more, voxels above a threshold t in an SPM of known or estimated smoothness. This probability has a reasonably simple form and is derived using distributional approximations from the theory of Gaussian fields (see Fig. 3 for details).

Set-level inference refers to the inference that the number of clusters comprising an observed activation profile is highly unlikely to have occurred by chance and pertains to the activation profile as characterized by its constituent regions. Cluster-level inferences are a

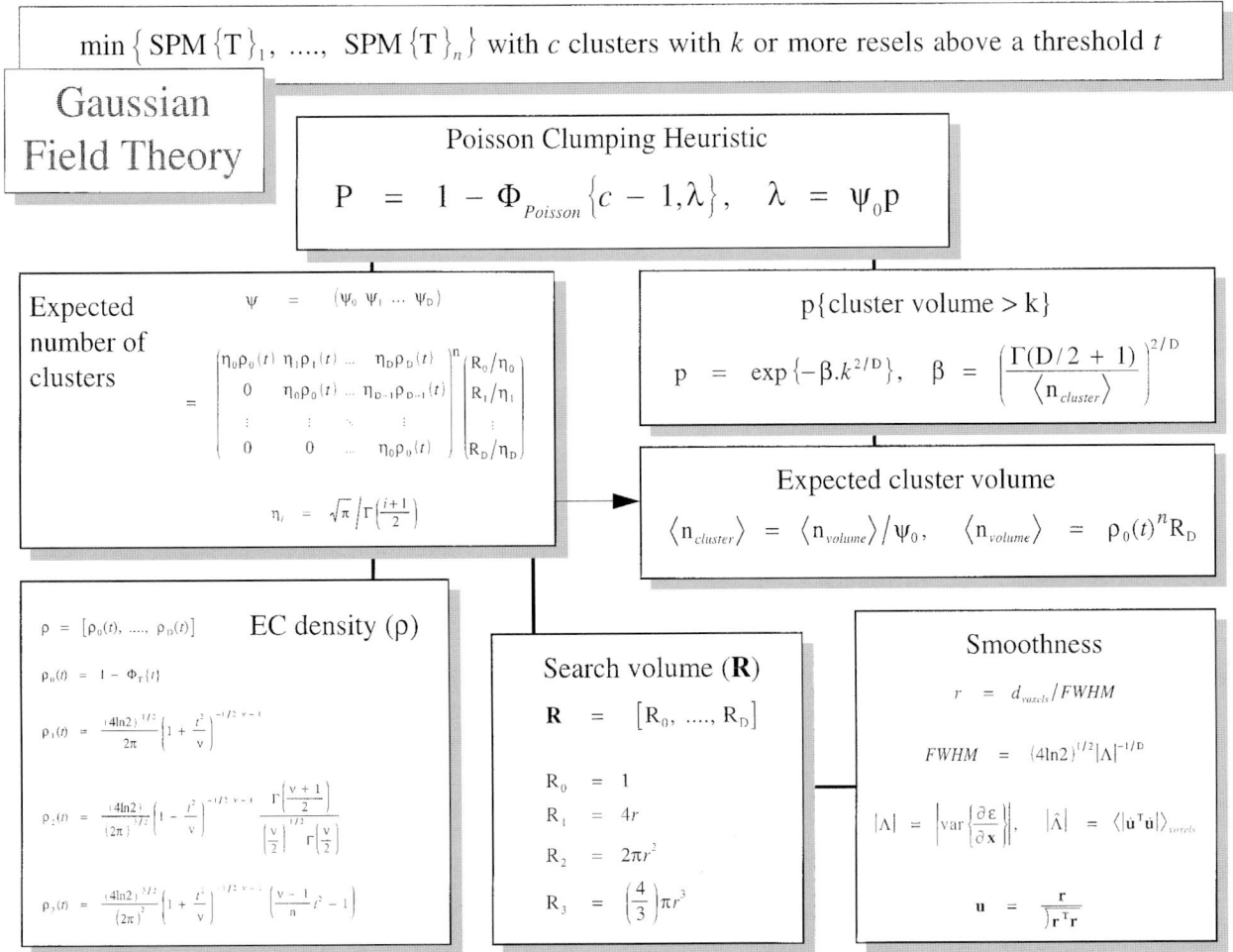

Figure 3 The theory of Gaussian fields: If one knew where to look beforehand, then inference can be based on the value of the statistic at the specified location in the SPM without correction. If, however, one did not have an anatomical constraint *a priori*, then a correction for multiple-dependent comparisons has to be made. These corrections are usually made using distributional approximations from the theory of Gaussian fields. This schematic deals with the most general case of n SPM{T}s whose voxels all survive a common threshold t (i.e., a conjunction). The central probability, upon which all voxel-, cluster-, or set-level inferences are made, is the probability P of getting c or more clusters with k or more resels (resolution elements) above this threshold. By assuming that clusters behave like a multidimensional Poisson point process (i.e., the Poisson clumping heuristic), P is simply determined given that the distribution of c is Poisson with an expectation λ. λ is the product of the expected number of clusters of any size and the probability that any cluster will be bigger than k resels. The latter probability is shown using a form for a single Z-variate field constrained by the expected number of resels per cluster (... denotes expectation or average). The expected number of resels per cluster is simply the expected number of resels in total divided by the expected number of clusters. The expected number of clusters is estimated with the Euler characteristic (EC) (effectively the number of blobs minus the number of holes). This estimate is, in turn, a function of the EC density for the statistic in question (with v degrees of freedom) and the resel counts. The EC density is the expected EC per unit of D-dimensional volume of the SPM, where the D-dimensional volume of the search space is given by the corresponding element in the vector of resel counts. Resel counts can be thought of as a volume metric that has been normalized by the smoothness of the SPM's component fields expressed in terms of the full width at half-maximum (FWHM). This is estimated from the determinant of the variance-covariance matrix of the first spatial derivatives of $\mathbf{u}$—the normalized residual fields $\mathbf{r}$. In this example, equations for a sphere of radius r are given. Φ denotes the cumulative distribution function for the subscripted statistic in question.

special case of set-level inferences that obtain when the number of clusters $c = 1$. Similarly, voxel-level inferences are special cases of cluster-level inferences that result when the cluster can be small (i.e., $k = 0$). Using a theoretical power analysis (Friston *et al.*, 1996b) of distributed activations, one observes that set-level inferences are generally more powerful than cluster-level inferences and that cluster-level inferences are generally more powerful than voxel-level inferences. The price paid for this increased sensitivity is reduced localizing power: Voxel-level tests permit individual voxels to be identified as significant, whereas cluster- and set-level inferences only allow clusters or sets of clusters to be so identified.

V. Experimental Design

In this section we will consider the different sorts of design that can be employed in neuroimaging studies. Generally speaking, modern studies of brain disorders rely upon factorial designs where the effect of pathology (one factor) is examined in terms of neuronal responsiveness (where responses are evoked by a second experimental factor).

A. Categorical Designs, Cognitive Subtraction, and Conjunctions

The tenet of cognitive subtraction is that the diference between two tasks can be formulated as a separable cognitive or sensorimotor component and that the regionally specific differences in hemodynamic responses identify the corresponding functionally specialized area. Early applications of subtraction range from the functional anatomy of word processing (Petersen *et al.*, 1989) to functional specialization in extrastriate cortex (Lueck *et al.*, 1989). The latter studies involved presenting visual stimuli with and without some sensory attribute (color, motion, etc.). The areas highlighted by subtraction were identified with homologous areas in monkeys that showed selective electrophysiological responses to equivalent visual stimuli.

Cognitive conjunctions (Price and Friston, 1997) can be thought of as an extension of the subtraction technique, in the sense that they combine a series of subtractions. In subtraction we test a hypothesis pertaining to the activation in one task relative to another. In conjunction analyses several hypotheses are tested, asking whether all the activations, in a series of task pairs, are jointly significant. Consider the problem of identifying regionally specific activations due to a particular cognitive component (e.g., object recognition). If one can identify a series of task pairs whose differences have only that component in common, then the region that activates in all the corresponding subtractions can be associated with the component in question. Conjunction analyses allow one to demonstrate this context-insensitive nature of regional responses. One important application of conjunction analyses is in multisubject fMRI studies where generic effects are identified as those that are conjointly significant in all the subjects studied (see below).

In relation to clinical neuroscience studies, subtraction designs are usually employed in a multifactorial context where regionally specific responses are assessed in the presence and absence of a specific pathology. The ensuing differences represent the condition × pathology interaction (see below).

B. Parametric Designs

The premise behind parametric designs is that regional physiology will vary systematically with the degree of cognitive or sensorimotor processing or deficits thereof. Examples of this approach include the PET experiments of Grafton *et al.* (1992), who demonstrated significant correlations between hemodynamic responses and the performance of a visually guided motor tracking task. On the sensory side, Price *et al.* (1992) demonstrated a remarkable linear relationship between perfusion in peri-auditory regions and frequency of aural word presentation. This correlation was not observed in Wernicke's area, where perfusion appeared to correlate not with the discriminative attributes of the stimulus, but with the presence or absence of semantic content. These relationships, or "neurometric functions," may be linear or nonlinear. Using polynomial regression, in the context of the general linear model, one can identify nonlinear relationships between stimulus parameters (e.g., stimulus duration or presentation rate) and evoked responses. To do this, one usually uses a SPM{F} (Büchel *et al.*, 1996).

The example provided in Fig. 4 illustrates subtraction, conjunction, and parametric approaches to design and analysis. These data were obtained from a fMRI study of visual motion processing using radially moving dots. The stimuli were presented over a range of speeds using isoluminant and isochromatic stimuli. To identify areas involved in visual motion, a stationary-dots condition was subtracted from the moving-dots conditions (see contrast weights on the upper right). To ensure significant motion-sensitive responses, using both color and luminance cues, a conjunction of the equivalent subtractions was assessed under both viewing contexts. Areas V5 and V3a are seen in the ensuing SPM{T}. The responses in left V5 shown in the lower panel speak to a very compelling inverted "U" relationship between speed and evoked response that peaks at around 8 degrees per second. It is this sort of relationship that parametric designs try to characterize. Interestingly, the form of these speed-dependent responses was similar using both stimulus types although luminance cues are seen to elicit a greater response. From the point of view of a factorial design, there is a main effect of cue (isoluminant vs isochromatic), a main [nonlinear] effect of speed, but no speed × cue interaction.

Clinical neuroscience studies can employ this approach by looking for the neuronal correlates of clinical (e.g., symptom) ratings over subjects. In many cases multiple clinical scores are available for each subject and the statistical design can usually be seen as a multilinear regression. In situations where the clinical scores are correlated, principal component analysis or factor analysis is sometimes applied to generate a new, and

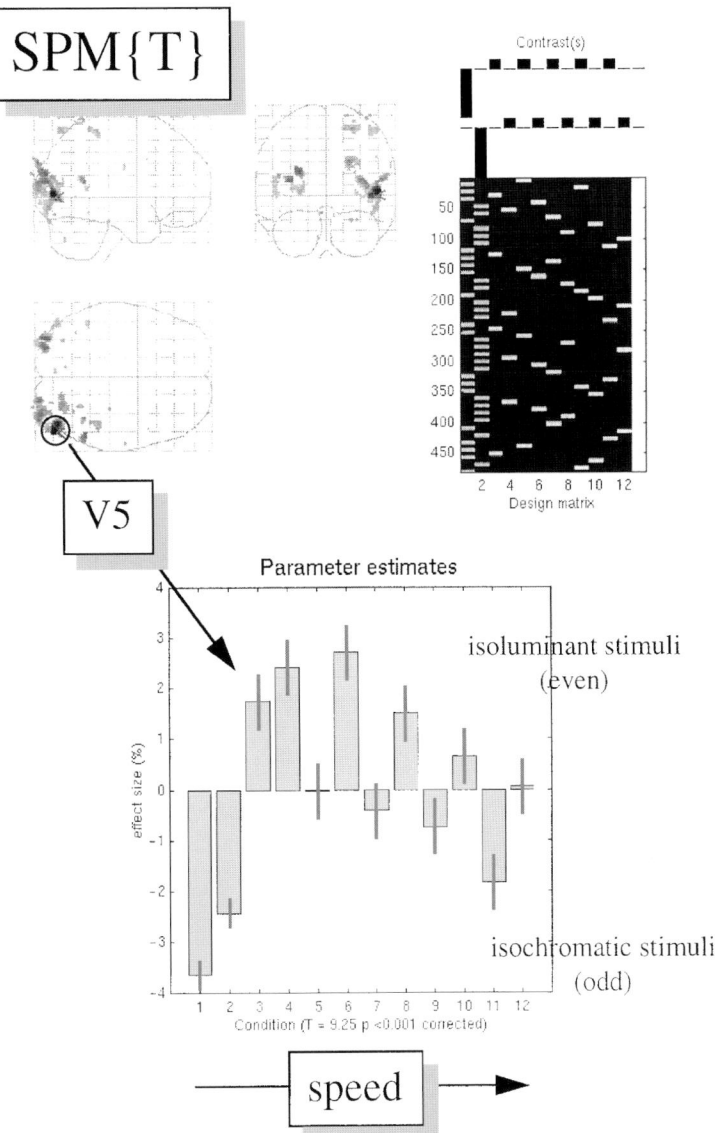

Figure 4 (Top right) Design matrix: This is an image representation of the design matrix. Contrasts: These are the vectors of contrast weights defining the linear compounds of parameters tested. The contrasts are displayed over the column of the design matrix that corresponds to the effects in question. The design matrix here includes condition-specific effects (box-cars convolved with a hemodynamic response function). Odd columns correspond to stimuli shown under isochromatic conditions and even columns model responses to isoluminant stimuli. The first two columns are for stationary stimuli and the remaining columns are for conditions of increasing speed. The final column is a constant term. (Top left) SPM*{T}*: This is a maximum-intensity projection of the SPM*{T}* conforming to the standard anatomical space of Talairach and Tournoux (1988). The *T* values here are the minimum *T* values from both contrasts, thresholded at p = 0.001 uncorrected. The most significant conjunction is seen in left V5. (Lower panel) Plot of the condition–specific parameter estimates for this voxel. The *T* value was 9.25 (p < 0.001 corrected; see Fig. 3).

smaller, set of explanatory variables that are orthogonal to each other. This has proved particularly useful in psychiatric studies, where syndromes can be expressed over a number of different dimensions (e.g., the degree of psychomotor poverty, disorganization, and reality distortion in schizophrenia; see Liddle *et al.,* 1992). In this way, regionally specific correlates of various symptoms can be revealed that may speak to their distinct pathogenesis in a way that transcends the syndrome itself. For example, psychomotor poverty is associated with left dorsolateral prefrontal dysfunction irrespective of whether the patient is suffering from schizophrenia or depression (see Chapter 20).

C. Factorial Designs

At its simplest, an interaction represents a change in a change. Interactions are associated with factorial designs where two or more factors are combined in the same experiment. The effect of one factor on the effect of the other is assessed by the interaction term. Factorial designs have a wide range of applications. An early application in neuroimaging examined physiological adaptation and plasticity (Friston *et al.*, 1992a) during motor performance by assessing time × condition interactions. Psychopharmacological activation studies are examples of factorial designs (Friston *et al.*, 1992b). In these studies cognitively evoked responses are assessed before and after a drug is given. The interaction term reflects the modulatory drug effect on the task-dependent activa-

tion. Factorial designs have an important role in the context of cognitive subtraction and additive factors logic by virtue of being able to test for interactions, or (context-sensitive activations) (i.e., to demonstrate the fallacy of pure insertion, see Friston *et al.*, 1996c). These interaction effects can sometimes be interpreted as (i) the integration of the two or more [cognitive] processes or (ii) the modulation of one [perceptual] process by another being manipulated (see Fig. 5).

From the point of view of clinical studies interactions are central. The effect of a disease process on sensorimotor or cognitive activation is simply an interaction and involves replicating a subtraction experiment in subjects with and without the pathophysiology being studied. Factorial designs can also embody parametric components. If one of the factors has a number of parametric levels, the interaction can be expressed as a difference in regression slope of regional activity on the parameter, under both levels of the other [categorical] factor. There is a special instance of these designs that is particularly relevant to clinical studies, where the categorical factor is the presence of a specific disease

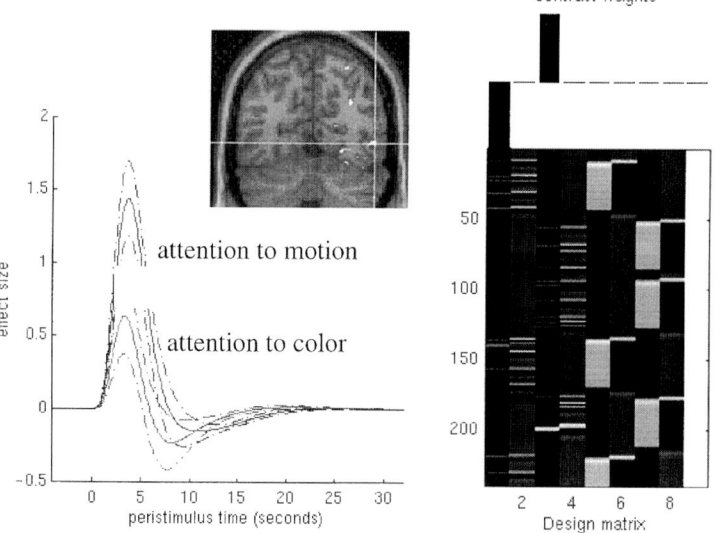

Figure 5 Analysis showing how to assess an interaction using an event-related design. Subjects viewed stationary monochromatic stimuli that occasionally changed color and moved at the same time. These compound events were presented under two levels of attentional set (attention to color and attention to motion). The event–related responses are modeled in an attention-specific fashion by the first four regressors (delta functions convolved with a hemodynamic response function and its derivative) in the design matrix on the right. The simple main effects of attention are modeled as similarly convolved box-cars. The interaction between attentional set and visually evoked responses is simply the difference in evoked responses under both levels of attention and is tested for with the appropriate contrast (upper right). Only the first 256 rows of the design matrix are shown. The most significant modulation of evoked responses under attention to motion was seen in left V5 (insert). The fitted responses and their standard errors are shown on the left as functions of peristimulus time.

process: In many pathology × task interactions it is difficult to say whether the abnormal activation is due to the fact that the patient cannot do the task or whether impaired performance is due to the abnormal physiological responses observed. This causal indeterminacy has, for example, bedeviled hypofrontality in schizophrenia research (see Chapter 8). Parametric variation in task difficulty can, however, be used to show differences in regression slopes (i.e., an interaction) that cannot be explained by the main effect of difficulty, providing a nice resolution of the performance confound. For example, premotor activity in Parkinson patients may be twice as sensitive to increases in the frequency of motor movements even though they cannot attain the rates achieved by normal subjects (or the corresponding premotor physiological activation).

VI. Designing fMRI Experiments: A Signal-Processing Perspective

In this section we consider fMRI time series from a signal-processing perspective, with particular reference to optimum experimental design and efficiency.

A. Temporal (Serial) Autocorrelations and Efficiency

fMRI time series can be viewed as a linear admixture of signal and noise. Signal corresponds to neuronally mediated hemodynamic changes that can be modeled as a (non)linear convolution of some underlying neuronal process responding to changes in experimental factors. Noise has many contributions that render it rather complicated in relation to other neurophysiological measurements. These include neuronal and nonneuronal sources. Neuronal noise refers to neurogenic signal not modeled by the explanatory variables and has the same frequency structure as the signal itself. Nonneuronal components have both white [e.g., RF (Johnson) noise] and colored components [e.g., pulsatile motion of the brain caused by cardiac cycles and local modulation of the static magnetic field (B_0) by respiratory movement]. These effects are typically low frequency or wide band (e.g., aliased cardiac-locked pulsatile motion). The superposition of all these components creates temporal correlations (Fig. 6) among the error terms in the statistical model (denoted by Σ in Fig. 2) that can have a severe effect on sensitivity to experimental effects. Sensitivity depends upon (i) the relative amounts of signal and noise and (ii) the efficiency of the experimental design. Efficiency is simply a measure of how reliable the parameter estimates are and can be defined as the inverse of the variance of a contrast of the parameter esti-

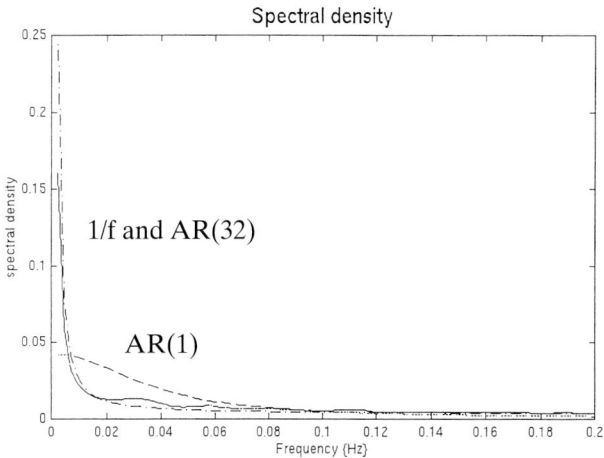

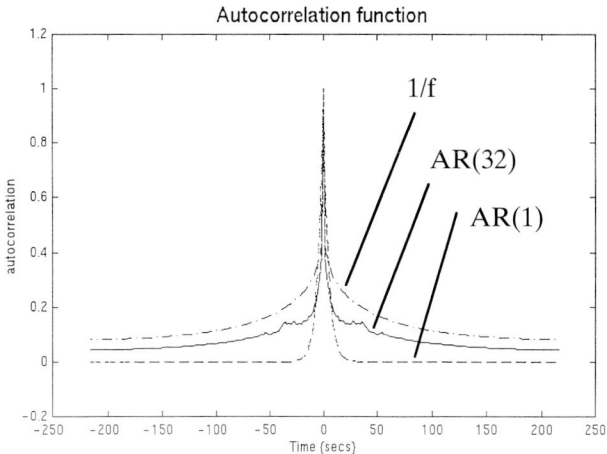

Figure 6 Spectral densities and corresponding autocorrelation functions for the residual terms of a fMRI time series. Three cases are shown: (i) an AR(32) model estimated using the Yule-Walker method (this is taken to be the closest to an empirical estimate), (ii) an AR(1) model estimate using the same method, and (iii) a model of the form ($q/f + 1$), where f is frequency in hertz and $q = 0.0178$.

mates (see Fig. 2). There are two important considerations that arise from this perspective on fMRI time series: The first pertains to optimal experimental design and the second to optimum (de)convolution of the time series to obtain the most efficient parameter estimates.

B. The Hemodynamic Response Function and Optimum Design

As noted above, an LTI model of neuronally mediated signals in fMRI suggests that, whatever the frequency structure of experimental variables, only those that survive convolution with the hemodynamic response function (HRF) can be estimated with any efficiency. By convolution theorem the experimental variance should therefore be designed to match the transfer function of the HRF. The corresponding frequency profile of this transfer function is shown in Fig. 7, (solid line in the upper left panel). It is clear that frequencies around 0.03 Hz are optimal, corresponding to periodic designs with 32-s periods (e.g., 16-s epochs). Generally, the first objective of experimental design is to comply with the natural constraints imposed by the HRF and to ensure that experimental variance occupies these intermediate frequencies.

C. The Noise Spectrum, Efficiency, and Bias

This is quite a complicated but important area. Conventional signal-processing approaches dictate that whitening the data engenders the most efficient parameter estimation. This corresponds to filtering with a convolution matrix $\mathbf{S}$ (see Fig. 2) that is the inverse of the intrinsic convolution matrix $\mathbf{K}$ ($\mathbf{K}\mathbf{K}^{T} = \Sigma$). There

are, however, two fundamental problems with this: (i) one generally does not know the intrinsic correlations Σ and (ii) even if they were known for any given time series, adopting the same assumptions for all voxels will lead to bias and loss of efficiency because each voxel has a different temporal autocorrelation structure (brain stem voxels will be subject to pulsatile effects, ventricular voxels will be subject to CSF flow artifacts, white matter voxels will not be subject to neurogenic noise, etc.). The first problem is highlighted in Fig. 6. Here the residuals from a long (1000 scan, TR = 1.7 s) time series were used to estimate the spectral density and associated autocorrelation functions (where one is proportional to the Fourier transform of the other) using the Yule-Walker method with an autoregression order of 32. Estimates using a commonly assumed AR(1) model (Bullmore *et al.*, 1996) and a modified $1/f$ model (Zarahn *et al.*, 1997) are also shown. The AR(1) is inad-

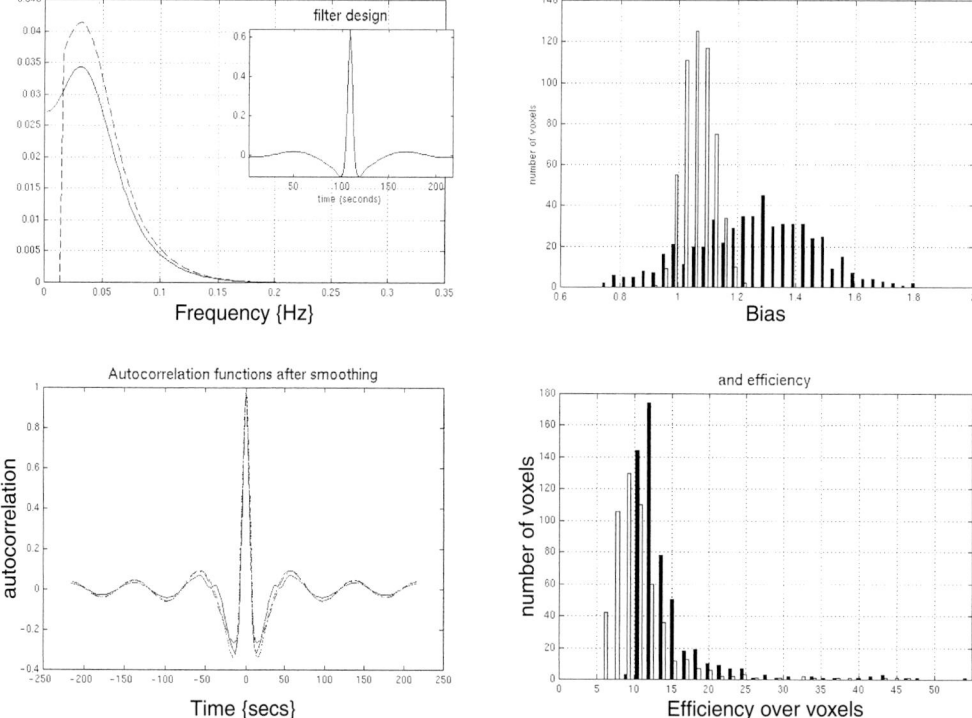

Figure 7 The effect of smoothing on bias when estimating the variance of contrasts. Upper left panel: The spectral density of the hemodynamic response function (a mixture of two gamma functions) (solid line) and after high–pass filtering at 1/64 Hz (broken line). The corresponding filter in time is shown in the insert. Lower left panel: The autocorrelation functions predicted on the basis of the three models shown in Fig. 6 after filtering with the kernel in the insert. Compare these autocorrelation functions with those in Fig. 6. The differences are now ameliorated. Right–hand panels: The distribution of bias (upper panel) and efficiency (lower panel) over voxels using an AR(8) model to estimate serial correlations at each voxel and assuming a stationary autocorrelation structure over voxels. Distributions are shown with (white bars) and without (filled bars) filtering. Note how filters reduce bias (at the expense of reduced efficiency). Bias is expressed as the ratio of estimated to actual contrast estimate variance and should, ideally, be unity. Efficiency is the inverse of the actual contrast estimate variance.

equate and fails to model long-range correlations (i.e., low frequencies). The $1/f$ model shown here is an extremely good approximation (even more so since the parameters of this model were not estimated using the data but taken from the literature). However, the $1/f$ model fails to model the short-term correlation structure as well as it could. The discrepancy between the assumed and actual form of temporal correlations means that, when the data are whitened in accord with the assumed models, the standard error of the parameter estimates (i.e., their variance) is overestimated [AR(1)] or underestimated ($1/f$). This leads directly to bias in the ensuing statistics (the T statistic is simply the quotient of the contrast and its estimated standard error; see Fig. 2). The effect is not trivial and can easily reach 50% for common experimental epoch- and event-related designs. Expected bias can be computed using the intrinsic and assumed autocorrelations and the expression for the expected estimate of parameter estimate variance in Fig. 2).

D. Temporal Filtering

One solution to this (bias problem) is described in Worsley and Friston (1995) and involves regularizing the temporal autocorrelation structure by filtering. This effectively imposes a structure on the temporal autocorrelations that renders the difference between the assumed and actual correlations less severe. Although less efficient, the ensuing inferences are less biased and more robust. The loss of efficiency can be minimized by appropriate experimental design and choosing a suitable filter **S**. The example in Fig. 7 combines the HRF and high-pass filtering at 1/64 Hz, which, when applied to the time series, markedly reduces the differences among the various models assumed for the autocorrelations (compare the lower panel of Fig. 6 with the lower panel of Fig. 7).

Appropriate temporal filtering not only reduces bias engendered by misspecification of the intrinsic correlations but addresses the second problem of nonstationariness of serial correlations over voxels. The upper right-hand panel of Fig. 7 shows bias over 512 voxels using a voxel-specific estimate for the intrinsic correlations and the average over all voxels for the assumed structure. Without filtering, the biases (solid bars) range from 80 to 150%. With filtering (open bars), they are reduced substantially. The effect on efficiency is shown in the lower right panel. It is interesting to note that temporal autocorrelations render the efficiency very variable over voxels (by nearly an order of magnitude). This is important because it means that fMRI is not homogeneous in its sensitivity to evoked responses from voxel to voxel.

E. Spatially Coherent Confounds and Global Normalization

Implicit in the use of high-pass filtering is the removal of low-frequency components that can be regarded as confounds. Other important confounds are signal components that are artifactual or have no regional specificity. These are referred to as global confounds and have a number of causes that can be divided into physiological (e.g., global perfusion changes mediated by changes in pCO_2) and nonphysiological (e.g., transmitter power calibration, B_1 coil profile, and receiver gain). The latter will generally scale the signal before the MRI sampling process. Other nonphysiological effects may have a nonscaling effect (Nyquist ghosting, movement-related effects, etc.). Instrumentation effects that scale the data motivate a global normalization by proportional scaling using the whole-brain mean. In fMRI spatially coherent confounds may, however, still persist and a more sensitive statistical model obtains when these can be identified and modeled.

In clinical studies the pathology itself may have globally coherent correlates. In this instance, it is important to ensure that global normalization does not compromise the efficiency with which regionally specific correlates can be estimated. If the pathophysiology is reflected in the global activity, the unconstrained use of this global variate as a confound will render regionally specific correlates inestimable. This is only the case for the main effect of pathology and does not affect task $\times$ pathology interactions. If one is interested in the main effect of pathology per se (e.g., SPECT scans of Alzheimer patients vs depressed patients), the global variate is usually computed using only parts of the brain that are minimally affected by that pathophysiology (e.g., the cerebellum).

F. Nonlinear System Identification Approaches

So far, we have only considered LTI models and first-order HRFs. Another signal-processing perspective is provided by nonlinear system identification (Vazquez and Noll, 1998). This section considers nonlinear models as a prelude to the next section on event-related fMRI, where nonlinear interactions among evoked responses provide constraints for experimental design and analysis.

We have described an approach to characterizing evoked hemodynamic responses in fMRI based on nonlinear system identification, in particular, the use of Volterra series (Friston *et al.*, 1998a). The approach enables one to estimate *Volterra kernels* that describe the relationship between stimulus presentation and the hemodynamic responses that ensue. Volterra series are essentially high-order extensions of linear convolution or FIR models. These kernels therefore represent a nonlinear characterization of the HRF that can model the responses to stimuli in different contexts and interactions among stimuli. In the context of fMRI, the kernel coefficients can be estimated by (i) using a second-order approximation to the Volterra series to reformulate the problem in terms of a general linear model and (ii) expanding the kernels in terms of temporal basis functions. This allows the use of the standard techniques described above to estimate the kernels (Fig. 8) and to make inferences about their significance on a voxel-specific basis using SPMs.

One important manifestation of the nonlinear effects, captured by the second-order kernels, is a modulation of stimulus-specific responses by preceding stimuli that are proximate in time. This means that responses at high stimulus presentation rates saturate and, in some instances, show an inverted U behavior (see Fig. 9, upper panel). This behavior appears to be specific to BOLD effects (as distinct from evoked changes in cerebral blood flow) and may represent a hemodynamic refractoriness. This effect has important implications for event-related fMRI.

VII. Epoch- and Event-Related Studies

A crucial distinction in experimental design is that between epoch- and event-related designs. In SPECT and PET, only epoch-related responses can be assessed because of the relatively long half-life of the radiotracers employed. In fMRI, however, there is an opportunity to measure event-related responses that may be important in some clinical contexts. An important issue in event-related fMRI is the choice of interstimulus interval, or, more precisely, stimulus onset asynchrony (SOA). The SOA, or the distribution of SOAs, is a critical factor in experimental design and is chosen, subject to some constraints, to maximize the efficiency of response estimation. The constraints on the SOA clearly depend upon the nature of the experiment but are generally satisfied when the SOA is small and derives from a random distribution. Rapid presentation rates allow for the maintenance of a particular cognitive or attentional set, decrease the latitude that the subject has for engaging alternative strategies, or incidental processing, and allows the integration of event-related paradigms using fMRI and electrophysiology. Random SOAs ensure that preparatory or anticipatory factors do not confound event-related responses and ensure a uniform context in which events are presented. These constraints speak to the well-documented advantages of event-

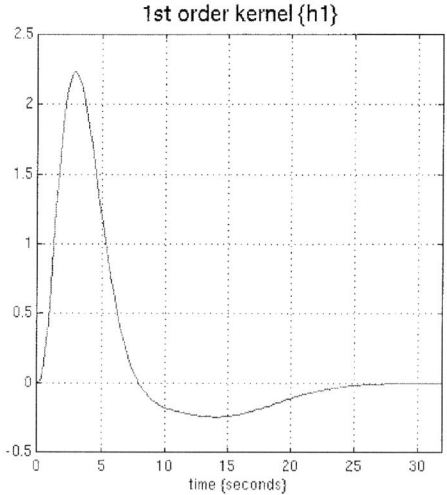

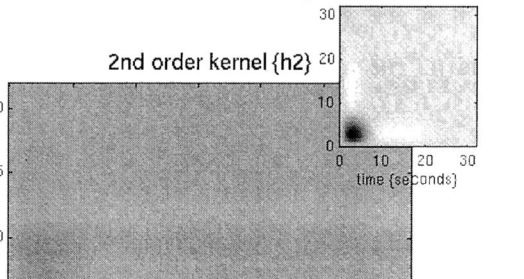

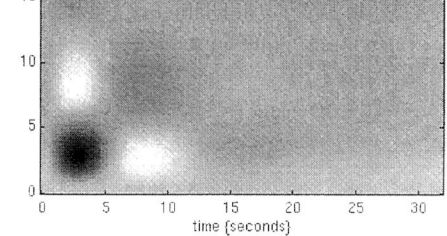

Figure 8 Volterra kernels **h1** and **h2** based on parameter estimates from a voxel in the left superior temporal gyrus at −56, −28, 12 mm. These kernel estimates were based on a single-subject study of aural word presentation at different rates (from 0 to 90 words per minute) using a second-order approximation to a Volterra series expansion modeling the observed hemodynamic response to stimulus input (a delta function for each word). These kernels can be thought of as a characterization of the second-order hemodynamic response function. The first-order kernel (upper panel) represents the (first order) component usually presented in linear analyses. The second-order kernel (lower panel) is presented in image format. The color scale is arbitrary; white is positive and black is negative. The insert on the right represents $[-\mathbf{h1} \cdot \mathbf{h1}^T]$, the second-order kernel that would be predicted by a simple model that involved convolution with **h1** followed by some nonlinear scalar function.

related fMRI over conventional blocked designs (Buckner *et al.,* 1996; Clark *et al.,* 1998).

In general, the explanatory variables are created by convolving a set of delta functions, indicating the presence of a particular event, with a small set of basis functions that model the hemodynamic response to those

Nonlinear interactions among stimuli

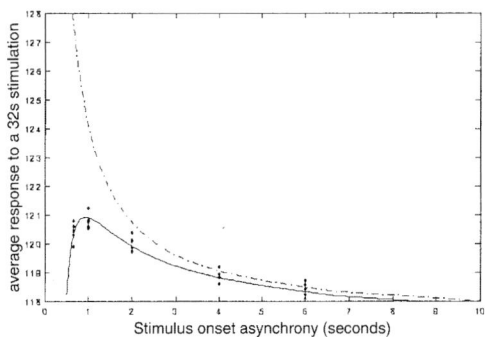

SOA and stationary stochastic designs

Stationary and nonstationary stochastic designs

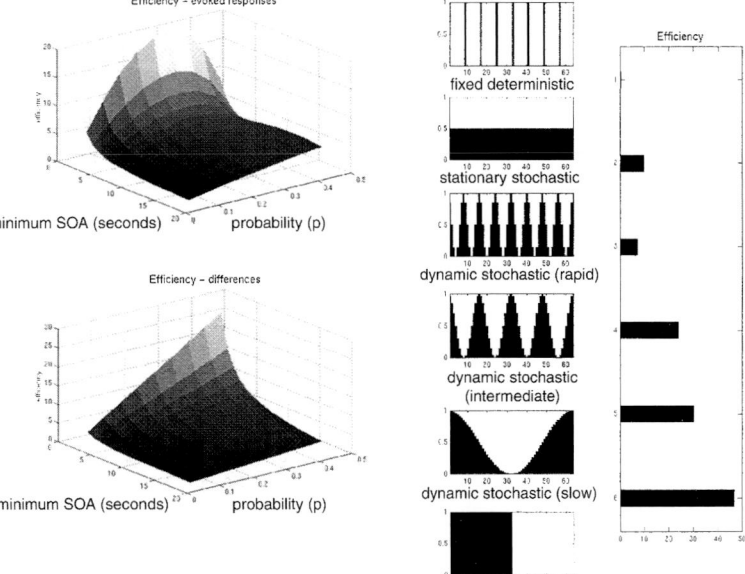

Figure 9 Upper panel: Plot of integrated responses over a 32-s stimulus train as a function of SOA. Solid line: Estimates based on a nonlinear convolution model (see Fig. 8). Dots: Empirical averages based on the presentation of actual stimulus trains. Broken line: The responses expected in the absence of second-order effects. Lower right panel: A comparison of some common (and some untried) designs. The left-hand column is a graphical representation of the occurrence probabilities **p** expressed as a function of time (seconds). The efficiency of each design is shown along the ordinate on the right assuming an SOA$_{min}$ of 1 s and a time series of 64 s for a single trial type design. The expected number of events (i.e., the mean value of **p**) was 0.5 in all cases (apart from the first), corresponding to an expected SOA of 2 s, or 32 events. Lower left panels: Efficiency in a stationary stochastic design with two event types, each presented with probability **p** every SOA$_{min}$. The upper graph is for a contrast testing for the response evoked by a particular trial type and the lower graph is for a contrast testing for differential responses.

events (Josephs *et al.,* 1997). A special case of this general approach obtains when the basis functions are delta functions placed at a discrete set of peristimulus times. This special case corresponds to selective averaging and requires stimulus presentation and data acquisition to be synchronized (which, in general, is not a good idea). We find the most useful basis set to be a canonical HRF and its derivatives with respect to its parameters (e.g., latency and dispersion). The nice thing about this approach is that it partitions differences among evoked responses into differences in magnitude, latency, or dispersion that can be tested for using specific contrasts and the SPM{T} [see Friston *et al.* (1998b) for details].

A. Stochastic and Deterministic fMRI Designs

fMRI designs can vary over a large number of parameters. To compare the efficiency of different designs, it is useful to have some common framework that accommodates them all. The efficiency can then be examined in relation to the parameters of the design. A general taxonomy of designs might be the following: Any design can be (stochastic) or (deterministic.) In stochastic designs (Heid *et al.*, 1997) one needs to specify the probabilities of an event occurring at all times those events could occur. In deterministic designs the occurrence probability is unity and the design is completely specified by the times of stimulus presentation or trials. The distinction between stochastic and deterministic designs pertains to how a particular realization or stimulus sequence is created. The efficiency afforded by a particular event sequence is a function of the event sequence itself, and not of the process generating the sequence (i.e., deterministic or stochastic). With a stochastic design, the design matrix **X,** and associated efficiency, are random variables. However, for finite length sequences the expected efficiency over an infinite number of realizations of **X** is easily computed:

In the framework considered here, the probability **p** of any event occurring is specified at each time it could occur (i.e., every SOA$_{min}$). Here, **p** is a vector with an element for every SOA$_{min}$. This formulation engenders the distinction between (stationary) stochastic designs, where the occurrence probabilities are constant, and nonstationary stochastic designs, where they change over time. For deterministic designs the elements of **p** are 0 or 1, the presence of a 1 denoting the occurrence of an event. An example of **p** might be the boxcars used in conventional block designs. Stochastic designs of the sort proposed by Dale and Buckner (1997) correspond to a vector of identical values and are therefore stationary in nature. Stochastic designs with temporal modulation of occurrence probability correspond to probability vectors with time-dependent probabilities varying between 0 and 1. With these probabilities the expected design matrices and expected efficiencies can be computed. The nice thing about this formulation is that by setting the mean of the probabilities **p** to a constant, one can compare different deterministic and stochastic designs given the same number of events (or, equivalently, same mean SOA). Some common examples are given in Fig. 9 (lower right panel) for an SOA$_{min}$ of 1 s and 32 expected events or trials over a 64-s period (except the

fixed-SOA deterministic example with 4 events). It can be seen that the least efficient is a fixed-interval deterministic design (despite the fact that the SOA is roughly optimal for this class), whereas the most efficient is a block design. A slow modulation of occurrence probabilities gives high efficiency while retaining the advantages of stochastic designs and may represent a very useful compromise between the high efficiency of block designs and the psychological benefits and latitude afforded by stochastic designs.

B. What Is the Minimum SOA?

This question can be addressed using Volterra series (see above). The results of a typical analysis are given in Fig. 9 (upper panel). This represents the average response, integrated over a 32-s train of stimuli as a function of SOA within that train. These responses were based on kernel estimates using data from a voxel in the left posterior temporal region of a subject obtained during the presentation of single words at a variety of rates. The solid line represents the estimated response and shows a clear maximum at just less than 1 s. The dots represent estimates based on empirical data from the same experiment. The broken line shows the expected response in the absence of nonlinear effects (i.e., that predicted by setting the second-order kernels to zero). It is clear that nonlinearities become important at around 2 s, leading to an actual diminution of the integrated response at subsecond SOAs. The implication of these data are that (i) SOAs should not really fall much below 1 s and (ii) at short SOAs the assumptions of linearity are violated. It should be noted that these data pertain to single-word processing in auditory association cortex. More linear behaviors may be expressed in primary sensory cortex, where the feasibility of using minimum SOAs as low as 500 ms has been demonstrated (Burock *et al.*, 1998), even when sampling at a lower rate (e.g., TR = 1 s).

C. Stationary Stochastic Designs and Efficiency

In this section, we consider why, in some instances, a very short average SOA is best whereas in others a longer SOA is more appropriate. Here we deal explicitly with multiple trial types and define the *trial onset asynchrony* (TOA) as the interval between onsets of a particular trial type. The best TOA depends upon the nature of the characterization of evoked responses that is required. For any given event type, the associated **p** determines the average or expected TOA for that event type. This is simply SOA$_{min}$/$\bar{\mathbf{p}}$, where $\bar{\mathbf{p}}$ is the mean probability for each trial type. By varying **p** and SOA$_{min}$, one can find the most efficient design depending upon whether one is looking for evoked responses themselves or differences among evoked responses. These two situations are depicted in the lower left panel of Fig. 9. It is immediately obvious that for both sorts of effects very small SOAs are optimal. However, the optimal occurrence probabilities are not the same. More infrequent events (corresponding to a smaller **p**) are required to estimate efficiently the responses themselves. This is equivalent to treating the baseline or control condition as any other condition (i.e., by including null events, with equal probability, as further event types). Therefore, by making the probability of null events and all other events equal to $1/(N + 1)$, where N is the number of event types, we get a mean TOA of $(N + 1)SOA_{min}$. Such designs result in optimal and equivalent efficiency for all comparisons (within stationary stochastic designs).

VIII. Inferences about Subjects and Populations

For a given group of subjects, there is a fundamental distinction between saying that the average response is significant in relation to the variability *of* the subjects' responses and that there is a significant response in relation to the variability *about* those subject-specific responses. This distinction relates directly to the difference between fixed and random-effect analyses. The following example tries to make this clear: Consider what would happen if we scanned six subjects during the performance of a single task, relative to a baseline. We then constructed a statistical model, where task-specific effects were modeled separately for each subject. Unknown to us, only one of the subjects activated a particular brain region. When we examine the contrast of parameter estimates, assessing the mean activation over all the subjects, we see that it is greater than zero by virtue of this subject's activation. Furthermore, because that model fits the data extremely well (modeling no activation in five subjects and a substantial activation in the sixth), the error variance, on a scan-to-scan basis, is small and the T statistic is very significant. Can we then say that the group shows an activation? On the one hand, we can say, quite properly, that the mean group response embodies an activation but clearly this does not constitute an inference that the group's response is significant (i.e., that this sample of subjects shows a consistent activation). The problem here is that we are using the *scan-to-scan* error variance and this is not necessarily appropriate for an inference about group responses. To make the inference that the group showed a significant activation, one would have to assess the

variability in activation effects from *subject to subject* (using the contrast of parameter estimates for each subject). This variability now constitutes the proper error variance. In this instance, the variance of these six measurements would be large relative to their mean and the corresponding T statistic would not be significant.

The distinction between the two approaches above relates to how one computes the appropriate error variance. The first represents a fixed-effect analysis and the second a random-effect analysis. In the former, the error variance is estimated on a scan-to-scan basis, assuming that each scan represents an independent observation (ignoring serial correlations). Here the degrees of freedom are essentially the number of scans (minus the rank of the design matrix). Conversely, in random-effect analyses, the appropriate error variance is based on the activation from subject to subject where the effect per se constitutes an independent observation and the degrees of freedom fall dramatically to the number of subjects. The term *random effect* indicates that we have accommodated the randomness of differential responses by comparing the mean activation to the variability in activations from subject to subject. Both analyses are perfectly valid, but only in relation to the inferences that are being made.

A. Random-Effect Analyses

The implementation of random-effect analyses in SPM is fairly straightforward and involves taking the contrasts of parameters estimated from a first-level (fixed effect) analysis and entering them into a second-level (random effect) analysis. This ensures that there is only one observation (i.e., contrast) per subject in the second-level analysis and that the error variance is computed using the subject-to-subject variability. The nature of the inference made is determined by the contrasts entered into the second level (see Fig. 10). The second-level design matrix simply tests the null hypothesis that the contrasts are zero (and is usually a column of ones implementing a single sample T test).

B. Conjunction Analyses and Population Inferences

There is, however, an alternative approach based on fixed-effect analyses that allows one to make qualitative inferences about population effects. This approach uses conjunctions as described in Section V.A. The motivation for conjunction analyses in the context of multi-subject studies is that they allow for (i) an inference at the level of the fixed-effect analysis based on the null hypotheses of no activation in any of the subjects studied and (ii) an inference at a second level about the

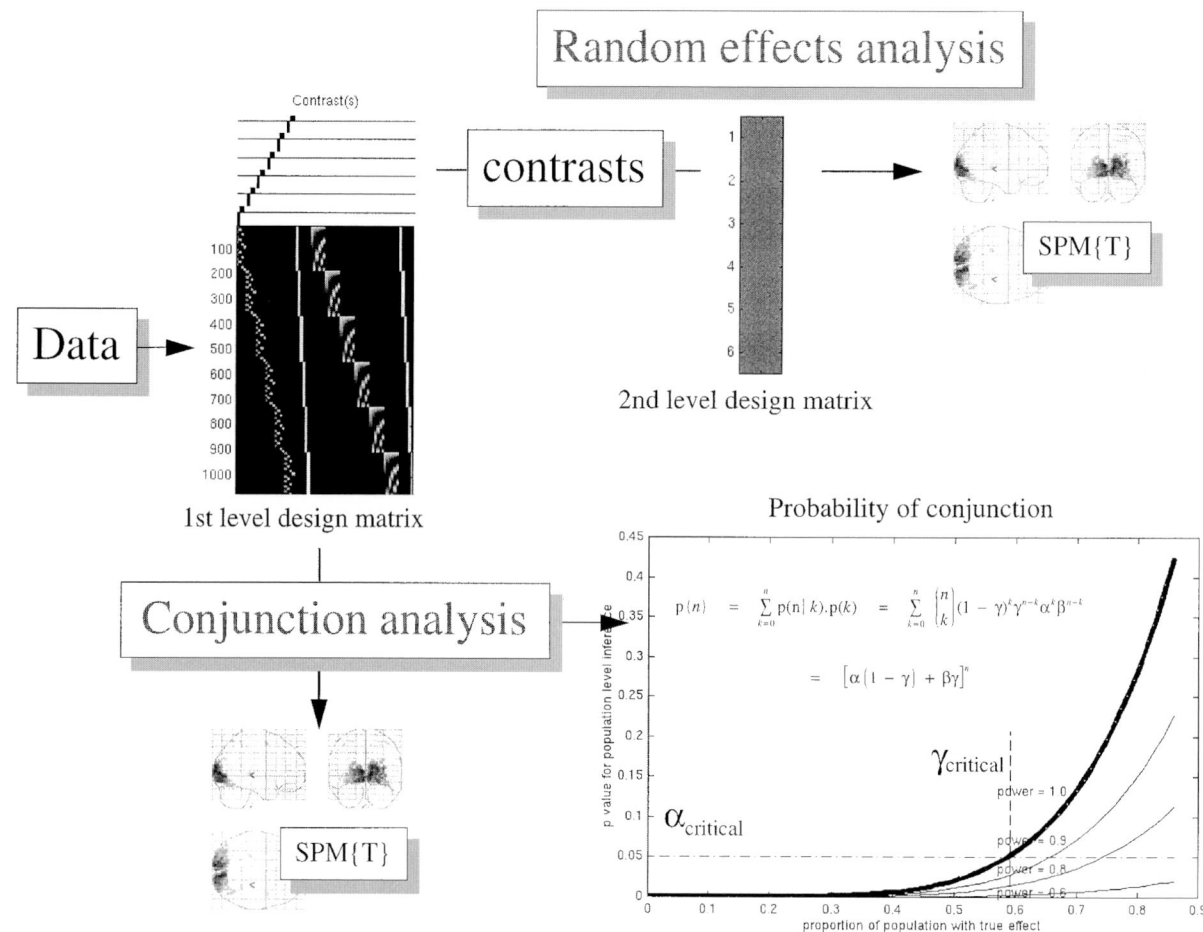

Figure 10 Schematic illustrating the implementation of random effect and conjunction analyses for population inference. The lower right graph shows the probability of obtaining a conjunction, conditional on a certain proportion γ of the population expressing the effect in question, for a test with specificity of $\alpha = 0.05$ (corresponding to a Z-variate threshold of 1.64) at several sensitivities ($\beta = 1, 0.9, 0.8,$ and 0.6). The critical specificity for population inferences $\alpha_{critical}$ and the associated proportion of the population $\gamma_{critical}$ are denoted by the broken lines. This probability $p(n)$ is given by the function of $\alpha, \beta,$ and γ.

population in terms of the proportion of the population that is likely to show the regionally specific effect. If, for any given contrast, one can establish a conjunction of effects over n subjects using a test with a specificity of α and sensitivity β, the probability under the null hypotheses of this occurring by chance can be expressed as a function of γ, the proportion of the population from which the subjects were sampled with the effect in question (see the equation in Fig. 10, lower right panel). Now this probability has an upper bound $\alpha_{critical}$, corresponding to a critical proportion $\gamma_{critical}$, that is realized when (the generally unknown) sensitivity is one. In other words, under the null hypothesis that the proportion of the population evidencing this effect is less than or equal to $\gamma_{critical}$, the probability of getting a conjunction over n subjects is equal to, or less than, $\alpha_{critical}$. In short, a conjunction allows one to say, with a specificity of $\alpha_{critical}$, that more than $\gamma_{critical}$ of the population show the effect in question. Formally, we can view this

analysis as a conservative $100(1 - \alpha_{critical})\%$ confidence region for the unknown parameter γ. Specifically, the confidence region is $\gamma > \gamma_{critical}$ if the conjunction occurs, and all values of γ if it does not. This approach retains the sensitivity of fixed-effect analyses yet still serves qualitative inferences about the population. These inferences can be construed as statements about how typical the effect is without saying that it is necessarily present in every subject.

In practice, a conjunction analysis of a multisubject study comprises the following steps: (i) A design matrix is constructed where the explanatory variables pertaining to each experimental condition are replicated for each session. This subject-separable design matrix implicitly models subject × condition interactions (i.e., different condition-specific responses among sessions). (ii) Contrasts are then specified that test for the effect of interest in each session to obtain a set of SPM{T}. (iii) These SPM{T} are thresholded at u (corresponding to some uncorrected specificity α) and combined to give an SPM of the conjunction. The ensuing SPM has two equivalent interpretations: First, it represents the intersection of the excursion sets, defined by the threshold u, of the subject-specific SPM$\{T\}$. Second, it is an SPM of the minimum value of the T values, thresholded at u. (iv) The corrected (for search volume) and uncorrected p values associated with each voxel are now computed as described in Fig. 2. These p values provide for inferences about the particular subjects studied. However, because we have demonstrated regionally specific conjunctions, one can now proceed to make an inference (at a second level) about the population from which these subjects came using the confidence region approach described above.

C. Random-Effect Analyses in Clinical Studies

Generally speaking, inferences about brain disorders will always depend upon random-effect analyses because the disease factor is expressed at the level of a subject, whether in terms of cross-sectional or longitudinal studies. Clearly an inference that only a certain proportion of patients show some effect will never be sufficient, if this effect is meant to be diagnostic. Conjunctions are sometimes employed to show that every patient in a cohort differs significantly (in terms of baseline activity or activation) from a large comparison group. The application of the confidence region approach above is, however, complicated by the fact that the contrasts constituting the conjunction are not independent (because they rely upon the same control subjects). As the size of the comparison group increases, the degree of collinearity between the subject-specific contrast does, however, reduce. Most PET studies have used fixed-effect analyses to date. These studies are entirely valid but it should be remembered that the inferences pertain to, and only to, the particular patients studied.

IX. Effective Connectivity and Functional Integration

There is a necessary relationship between approaches to characterizing functional integration and multivariate analyses because the latter are necessary to model interactions among brain regions. Multivariate approaches can be divided into those that are inferential in nature and those that are data led or exploratory. Most are based on the singular value or eigen decomposition of the between-voxel covariances in a neuroimaging time series.

A. Eigenimage Analysis and Related Approaches

In Friston *et al.* (1993) we introduced voxel-based principal component analysis (PCA) of neuroimaging time series to characterize distributed brain systems implicated in sensorimotor, perceptual, or cognitive processes. These distributed systems are identified with principal components or eigenimages that correspond to spatial modes of coherent brain activity. This approach represents one of the simplest multivariate characterizations of functional neuroimaging time series and falls into the class of exploratory analyses. Principal component or eigenimage analysis generally uses singular value decomposition (SVD) to identify a set of orthogonal spatial modes that capture the greatest amount of variance, expressed over time. As such, the ensuing modes embody the most prominent aspects of the variance-covariance structure of a given time series. Noting that the covariance among brain regions is equivalent to functional connectivity renders eigenimage analysis particularly interesting because it was among the first ways of addressing functional integration (i.e., connectivity) in the human brain. Subsequently eigenimage analysis has been elaborated in a number of ways. Notable among these are canonical variate analysis and multidimensional scaling (Friston *et al.*, 1996d,e). Canonical variate analysis was introduced in the context of ManCova (multiple analysis of covariance) and uses the generalized eigenvector solution to maximize the variance that can be explained by some explanatory variables relative to error. Canonical variate analysis can be thought of as an extension of eigenimage analysis that refers explicitly to some explanatory variables and allows for statistical inference.

In relation to fMRI data, eigenimage analysis (Sychra *et al.*, 1994) is generally used as an exploratory device to characterize coherent brain activity. These variance components may, or may not, be related to experimental design, and endogenous coherent dynamics have been observed in the motor system (Biswal *et al.*, 1995). Despite its exploratory power, eigenimage analysis is fundamentally limited for two reasons. First, it offers only a linear decomposition of any set of neurophysiological measurements and second, the particular set of eigenimages or spatial modes obtained are uniquely determined by constraints that are biologically implausible. These aspects of PCA represent inherent limitations on the interpretability and usefulness of eigenimage analysis of biological time series. These observations have motivated the exploration of nonlinear PCA and neural network approaches to imaging time series (e.g., Mørch *et al.*, 1995).

Two other important approaches deserve mention

here. The first is independent component analysis (ICA). ICA uses entropy maximization to find, using iterative schemes, spatial modes or their dynamics that are approximately independent. This is a stronger requirement than orthogonality in PCA and involves removing high-order correlations among the modes (or dynamics). It was initially introduced as spatial ICA (McKeown *et al.*, 1998), in which the independence constraint was applied to the modes (with no constraints on their temporal expression). More recent approaches use, by analogy with magneto- and electrophysiological time-series analysis, temporal ICA, where the dynamics are enforced to be independent. This requires an initial dimension reduction (usually using conventional eigenimage analysis). Finally, there has been an interest in cluster analysis (Baumgartner *et al.*, 1997). Conceptually, this can be related to eigenimage analysis through multidimensional scaling and principal coordinates analysis. In cluster analysis voxels in a multidimensional scaling space are assigned belonging probabilities to a small number of clusters, thereby characterizing the temporal dynamics (in terms of the cluster centroids) and spatial modes (defined by the belonging probability for each cluster). These approaches eschew many of the unnatural constraints imposed by eigenimage analysis and can be a very useful exploratory device.

B. Functional and Effective Connectivity

Imaging neuroscience has firmly established functional specialization as a principle of brain organization in man. The functional integration of specialized areas has proven more difficult to assess. Functional integration is usually inferred on the basis of correlations among measurements of neuronal activity. Functional connectivity has been defined as *correlations between remote neurophysiological events.* However, correlations can arise in a variety of ways: For example, in multiunit electrode recordings, they can result from stimulus-locked transients evoked by a common input or reflect stimulus-induced oscillations mediated by synaptic connections (Gerstein and Perkel, 1969). Integration within a distributed system is usually better understood in terms of effective connectivity: Effective connectivity refers explicitly to *the influence that one neural system exerts over another,* either at a synaptic (i.e., synaptic efficacy) or population level. It has been proposed that "the [electrophysiological] notion of effective connectivity should be understood as the experiment- and time-dependent, simplest possible circuit diagram that would replicate the observed timing relationships between the recorded neurons" (Aertsen and Preißl, 1991). This speaks to two important points: (i) Effective connectivity is dynamic, i.e., activity- and time-dependent, and (ii) it depends upon a model of the

interactions. The models employed in functional neuroimaging can be classified as (i) those based on regression models (Friston *et al.*, 1995d) or (ii) structural equation modeling (McIntosh and Gonzalez-Lima, 1994) (i.e., path analysis).

C. Characterizing Nonlinear Coupling among Brain Areas

Linear models of effective connectivity assume that the multiple inputs to a region are linearly separable. This assumption precludes activity-dependent connections that are expressed in one sensorimotor or cognitive context and not in another. The resolution of this problem lies in adopting nonlinear models that include interactions among inputs. These interactions can be construed as a context- or activity-dependent modulation of the influence that one region exerts over another, where that context is instantiated by activity in further brain regions exerting modulatory effects. These nonlinearities can be introduced into structural equation modeling using so-called moderator variables that represent the interaction between two regions in causing activity in a third (Büchel and Friston, 1997). From the point of view of regression models, modulatory effects can be modeled with nonlinear input-output models and, in particular, the Volterra formulation described above. In this instance, the inputs are not stimuli but activities from other regions. Because the kernels are high order, they embody interactions over time and among inputs. The influence of one region j on another i can therefore be divided into two components: (i) the direct influence of j on i, irrespective of the activities elsewhere, and (ii) an activity-dependent component that represents an interaction with inputs from the remaining regions. The example provided in Fig. 11 addresses the modulation of visual cortical responses by attentional mechanisms (e.g., Treue and Maunsell, 1996) and the mediating role of activity- or context-sensitive changes in effective connectivity. Figure 11 shows a characterization of this modulatory effect in terms of the increase in V5 responses to a simulated V2 input when posterior parietal activity is zero (broken line) and when it is high (solid lines). This sort of result suggests that parietal activity may be a sufficient explanation for attentional modulation of visually evoked extrastriate responses.

X. Specialization, Integration, and the Lesion-Deficit Model

In this section, the specialization of brain regions, underpinning neuropsychological constructs like category-specificity, is considered at a conceptual level to

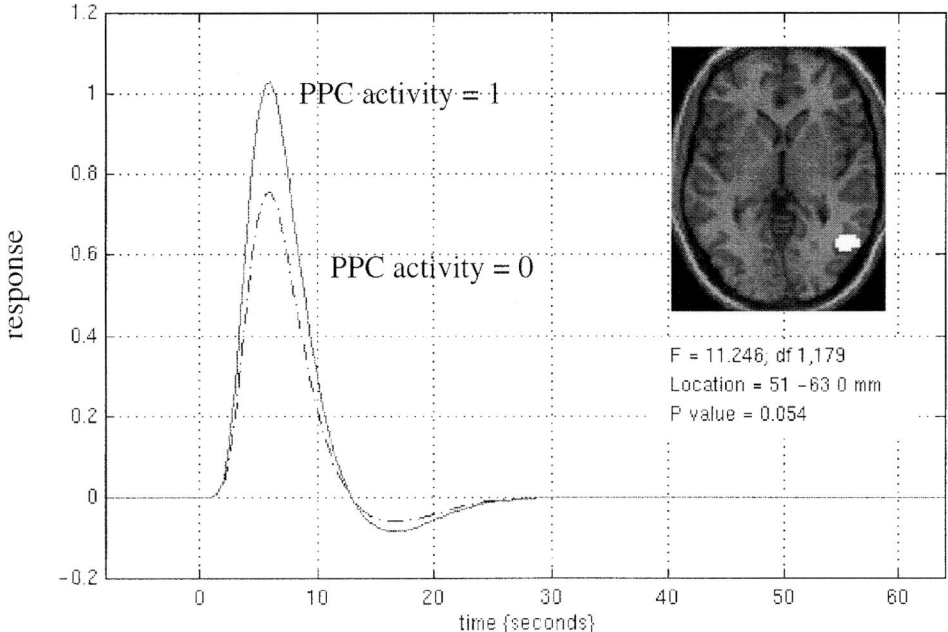

Figure 11 Characterization of effects of V2 inputs on V5 and the modulation of these effects by posterior parietal cortex (PPC) using simulated inputs at different levels of PPC activity. The broken lines represent estimates of V5 responses when PPC activity = 0 according to a second-order Volterra model with inputs based on the activity in V2, PPC, and the pulvinar (all normalized to zero mean and unit standard deviation). The simulated input, from V2, corresponded to a square wave of 500-ms duration convolved with a hemodynamic response function. The solid curves represent the same response when PPC activity is one. It is evident that V2 has an activating effect on V5 and that PPC increases the responsiveness of V5 to these inputs. The insert shows all the voxels in V5 that evidenced a modulatory effect ($p < 0.05$ uncorrected). These voxels were identified by thresholding SPMs of the F statistic testing for the contribution of second–order kernels involving V2 and PPC while treating all other components as confounds. Results for one subject are shown. Subjects were studied with fMRI under identical stimulus conditions (visual motion subtended by radially moving dots) while manipulating the attentional component of the task (detection of velocity changes).

show how different experimental designs emerge in the study of brain-damaged patients. To do this, we introduce a distinction between *latent representations* (the specific architecture and connection strengths that enable the detection of a specific stimulus) and *dynamic representations* (the expression of neuronal transients that occur when those stimuli are processed). Generally, lesion studies hope to make inferences about latent representations whereas neuroimaging deals with dynamic representations. The inferences from both fields of inquiry would be consistent if there were a simple relationship between them. However, by virtue of the nonlinear and modulatory (see above and Sandell and Schiller (1982), Girard and Bullier (1989), Treue and Maunsell (1996), and Hirsch and Gilbert (1991)) effect of extrinsic connections in the brain that can dynamically change the response properties of neurons, latent representations are clearly a function of activity in distant cortical areas. This poses a severe problem for the lesion-deficit model because a lesion may be anatomically distant

from specialized neuronal structures supporting the latent representations suggested by the psychological deficit.

The existence of long-range modulatory effects leads directly to the notion of two sorts of functional specialization in the brain: (i) specialization that depends only on "driving" feed-forward connections and that is context-insensitive and (ii) specialization that is conferred by modulatory interactions with other areas at the same level, or higher, in a cortical hierarchy. The latter specialization is context-sensitive and explicitly dependent upon functional integration and the effective connectivity that mediates it. The neuronal correlates of things like category specificity belong to the second, integrative class of specialization and, as such, rest upon latent representations that are labile, dynamic, and dependent upon influences from remote brain areas. These observations are closely related to the concept of (dynamic diaschisis) (see Chapter 7) and speak to the necessary role of neuroimaging in the study of brain-damaged patients.

A. Implications for Experimental Design

In terms of experimental designs that would distinguish between functional specialization that did and did not depend upon integration, it is helpful to consider how one might demonstrate functional specialization in the absence of integration. One experiment that comes to mind is the functional neuroimaging study of anesthetized subjects exposed to photic or aural stimuli. Although the effect of anesthesia is not specific in relation to removing functional integration, this thought experiment highlights the nature of the distinction. It may be the case that word stimuli are capable of eliciting neuronal responses in tertiary and higher sensory cortices, suggesting that there may be classical (receptive fields) for words. Alternatively the difference between word and nonword stimuli, in this context, would not be apparent because higher order (phonological or lexicosemantic) processing is required to realize any differential responses (i.e., category specificity for words) in putative word-form areas. There are more practical experiments that suggest themselves, using transcranial magnetic stimulation to disrupt higher order areas, allowing one to disambiguate between the two sorts of functionally specialized responses.

The central idea above is that category-specific responses may be constructed by interactions between the category-specific infrastructure and other neuronal systems. In other words, the specificity does not lie in the intrinsic response profiles of the neurons themselves but on the context in which these responses are elicited. This context is provided by afferents from other brain areas and the specificity may well lie in these connections. If this hypothesis is correct, then there are two approaches that are clear candidates for characterizing the neuronal correlates of phenomena like category-specific deficits. Both rely on neuroimaging and involve (i) an analysis of the effective connectivity among cortical regions and (ii) an explicit analysis of the context sensitivity of the responses elicited. In terms of analyses of effective connectivity (the influence that one neuronal system exerts over another), we are already in a position, using fMRI, to make inferences about modulatory or context-dependent changes in effective connection strengths (see Fig. 11). One can envisage experiments addressing the modulation of connections from V2 to the fusiform region by putative naming areas in the basal inferotemporal region (Brodmann's area 47). These experiments would point to the role of top-down modulation in configuring word-specific responses in early components of the ventral visual pathway. The second experimental approach depends upon measuring context-sensitive responses associated with dynamic representations. This effectively reduces to looking for

interactions and calls for multifactorial designs as discussed in Section V.C. This general approach has been refined to incorporate the influence of remote areas on regionally specific responses using "psychophysiological interactions" (Friston *et al.*, 1997). An example of this sort of experiment would test for interactions between activity in the basal inferotemporal region and the presence of (word and nonword) letter strings in the visual field when predicting activity in areas showing category specificity for words. This multifactorial approach is motivated, quite simply, by the presence of possible modulatory or nonlinear effects mediated by functional integration and would be impossible to implement using lesion studies. This line of argument suggests that it may no longer be sufficient to demonstrate, say, a face-specific area by simply presenting face and house stimuli to subjects. A multifactorial approach that emphasized the context-sensitive nature of this specificity would involve the presentation of houses and faces, both under two task conditions that emphasized house and face processing, respectively.

Acknowledgment

This work was funded by the Wellcome Trust.

References

Absher, J. R., and Benson, D. F. (1993). Disconnection syndromes: An overview of Geschwind's contributions. *Neurology* **43**, 862–867.

Adler, R. J. (1981). *In* "The Geometry of Random Fields." Wiley, New York.

Aertsen, A., and Preißl, H. (1991). Dynamics of activity and connectivity in physiological neuronal networks. *In* "Non Linear Dynamics and Neuronal Networks" (H. G. Schuster, Ed.), pp. 281–302. VCH, Weinheim/New York.

Aguirre, G. K., Zarahn, E., and D'Esposito, M. (1998). A critique of the use of the Kolmogorov-Smirnov (KS) statistic for the analysis of BOLD fMRI data. *Magn. Reson. Med.* **39**, 500–505.

Ashburner, J., Neelin, P., Collins, D. L., Evans, A., and Friston, K. (1997). Incorporating prior knowledge into image registration. *NeuroImage* **6**, 344–352.

Bandettini, P. A., Jesmanowicz, A., Wong, E. C., and Hyde, J. S. (1993). Processing strategies for time course data sets in functional MRI of the human brain. *Magn. Reson. Med.* **30**, 161–173.

Baumgartner, R., Scarth, G., Teichtmeister, C., Somorjai, R., and Moser, E. (1997). Fuzzy clustering of gradient-echo functional MRI in the human visual cortex. Part 1: Reproducibility. *J Magn. Reson. Imaging* **7**, 1094–1101.

Biswal, B., Yetkin, F. Z., Haughton, V. M., and Hyde, J. S. (1995). Functional connectivity in the motor cortex of resting human brain using echo–planar MRI. *Magn. Reson. Med.* **34**, 537–541.

Boynton, G. M., Engel, S. A., Glover, G. H., and Heeger, D. J. (1996). Linear systems analysis of functional magnetic resonance imaging in human V1. *J. Neurosci.* **16**, 4207–4221.

Büchel, C., and Friston, K. J. (1997). Modulation of connectivity in visual pathways by attention: Cortical interactions evaluated with structural equation modelling and fMRI. *Cereb. Cortex* **7**, 768–778.

Büchel, C., Wise, R. J. S., Mummery, C. J., Poline, J.-B., and Friston, K. J. (1996). Nonlinear regression in parametric activation studies. *NeuroImage* **4,** 60–66.

Buckner, R., Bandettini, P., O'Craven, K., Savoy, R., Petersen, S., Raichle, M., and Rosen, B. (1996). Detection of cortical activation during averaged single trials of a cognitive task using functional magnetic resonance imaging. *Proc. Natl. Acad. Sci. U.S.A.* **93,** 14878–14883.

Bullmore, E. T., Brammer, M. J., Williams, S. C. R., Rabe-Hesketh, S., Janot, N., David, A., Mellers, J., Howard, R., and Sham, P. (1996). Statistical methods of estimation and inference for functional MR images. *Magn. Reson. Med.* **35,** 261–277.

Burock, M. A., Buckner, R. L., Woldorff, M. G., Rosen, B. R., and Dale, A. M. (1998). Randomized event-related experimental designs allow for extremely rapid presentation rates using functional MRI. *NeuroReport* **9,** 3735–3739.

Clark, V. P., Maisog, J. M., and Haxby, J. V. (1998). fMRI study of face perception and memory using random stimulus sequences. *J. Neurophysiol.* **76,** 3257–3265.

Dale, A., and Buckner, R. (1997). Selective averaging of rapidly presented individual trials using fMRI. *Hum. Brain Mapp.* **5,** 329–340.

Friston, K. J. (1997). Testing for anatomical specified regional effects. *Hum. Brain Mapp.* **5,** 133–136.

Friston, K. J., Frith, C. D., Liddle, P. F., and Frackowiak, R. S. J. (1991). Comparing functional (PET) images: The assessment of significant change. *J. Cereb. Blood Flow Metab.* **11,** 690–699.

Friston, K. J., Frith, C., Passingham, R. E., Liddle, P. F., and Frackowiak, R. S. J. (1992a). Motor practice and neurophysiological adaptation in the cerebellum: A positron tomography study. *Proc. R. Soc. London, Ser. B* **248,** 223–228.

Friston, K. J., Grasby, P., Bench, C., Frith, C. D., Cowen, P. J., Little, P., Frackowiak, R. S. J., and Dolan, R. (1992b). Measuring the neuromodulatory effects of drugs in man with positron tomography. *Neurosci. Lett.* **141,** 106–110.

Friston, K. J., Frith, C., Liddle, P., and Frackowiak, R. S. J. (1993). Functional connectivity: The principal component analysis of large data sets. *J. Cereb. Blood Flow Metab.* **13,** 5–14.

Friston, K. J., Worsley, K. J., Frackowiak, R. S. J., Mazziotta, J. C., and Evans, A. C. (1994a). Assessing the significance of focal activations using their spatial extent. *Hum. Brain Mapp.* **1,** 214–220.

Friston, K. J., Jezzard, P. J., and Turner, R. (1994b). Analysis of functional MRI time series. *Hum. Brain Mapp.* **1,** 153–171.

Friston, K. J., Ashburner, J., Frith, C. D., Poline, J.-B., Heather, J. D., and Frackowiak, R. S. J. (1995a). Spatial registration and normalisation of images. *Hum. Brain Mapp.* **2,** 165–189.

Friston, K. J., Holmes, A. P., Worsley, K. J., Poline, J.-B., Frith, C. D., and Frackowiak, R. S. J. (1995b). Statistical parametric maps in functional imaging: A general linear approach. *Hum. Brain Mapp.* **2,** 189–210.

Friston, K. J., Frith, C. D., Turner, R., and Frackowiak, R. S. J. (1995c). Characterizing evoked hemodynamics with fMRI. *NeuroImage* **2,** 157–165.

Friston, K. J., Ungerleider, L. G., Jezzard, P., and Turner, R. (1995d). Characterizing modulatory interactions between V1 and V2 in human cortex with fMRI. *Hum. Brain Mapp.* **2,** 211–224.

Friston, K. J., Williams, S., Howard, R., Frackowiak, R. S. J., and Turner, R. (1996a). Movement related effects in fMRI time series. *Magn. Reson. Med.* **35,** 346–355.

Friston, K. J., Holmes, A., Poline, J.-B., Price, C. J., and Frith, C. D. (1996b). Detecting activations in PET and fMRI: Levels of inference and power. *NeuroImage* **4,** 223–235.

Friston, K. J., Price, C. J., Fletcher, P., Moore, C., Frackowiak, R. S. J., and Dolan, R. J. (1996c). The trouble with cognitive subtraction. *NeuroImage* **4,** 97–104.

Friston, K. J., Poline, J.-B., Holmes, A. P., Frith, C. D., and Frackowiak, R. S. J. (1996d). A multivariate analysis of PET activation studies. *Hum. Brain Mapp.* **4,** 140–151.

Friston, K. J., Frith, C. D., Fletcher, P., Liddle, P. F., and Frackowiak, R. S. J. (1996e). Functional topography: Multidimensional scaling and functional connectivity in the brain. *Cereb. Cortex* **6,** 156–164.

Friston, K. J., Büchel, C., Fink, G. R., Morris, J., Rolls, E., and Dolan, R. J. (1997). Psychophysiological and modulatory interactions in neuroimaging. *NeuroImage* **6,** 218–229.

Friston, K. J., Josephs, O., Rees, G., and Turner, R. (1998a). Non-linear event-related responses in fMRI. *Magn. Reson. Med.* **39,** 41–52.

Friston, K. J., Fletcher, P., Josephs, O., Holmes, A., Rugg, M. D., and Turner, R. (1998b). Event-related fMRI: Characterizing differential responses. *NeuroImage* **7,** 30–40.

Gerstein, G. L., and Perkel, D. H. (1969). Simultaneously recorded trains of action potentials: Analysis and functional interpretation. *Science* **164,** 828–830.

Girard, P., and Bullier, J. (1989). Visual activity in area V2 during reversible inactivation of area 17 in the macaque monkey. *J. Neurophysiol.* **62,** 1287–1301.

Goltz, F. (1881). *In* "Transactions of the 7th International Medical Congress" (W. MacCormac, Ed.), Vol. I, pp. 218–228. Kolkmann, London.

Grafton, S., Mazziotta, J., Presty, S., Friston, K. J., Frackowiak, R. S. J., and Phelps, M. (1992). Functional anatomy of human procedural learning determined with regional cerebral blood flow and PET. *J. Neurosci.* **12,** 2542–2548.

Heid, O., Gönner, F., and Schroth, G. (1997). Stochastic functional MRI. *NeuroImage* **5,** S476.

Hirsch, J. A., and Gilbert, C. D. (1991). Synaptic physiology of horizontal connections in the cat's visual cortex. *J. Neurosci.* **11,** 1800–1809.

Jezzard, P., and Balaban, R. S. (1995). Correction for geometric distortion in echo-planar images from B_0 field variations. *Magn. Reson. Med.* **34,** 65–73.

Josephs, O., Turner, R., and Friston, K. J. (1997). Event-related fMRI. *Hum. Brain Mapp.* **5,** 243–248.

Lange, N., and Zeger, S. L. (1997). Non-linear Fourier time series analysis for human brain mapping by functional magnetic resonance imaging (with discussion) *J. R. Stat. Soc., Ser. C* **46,** 1–29.

Liddle, P. F., Friston, K. J., Frith, C. D., and Frackowiak, R. S. J. (1992). Cerebral blood-flow and mental processes in schizophrenia. *J. R. Soc. Med.* **85,** 224–227.

Lueck, C. J., Zeki, S., Friston, K. J., Deiber, M. P., Cope, N. O., Cunningham, V. J., Lammertsma, A. A., Kennard, C., and Frackowiak, R. S. J. (1989). The colour centre in the cerebral cortex of man. *Nature* **340,** 386–389.

McIntosh, A. R., and Gonzalez-Lima, F. (1994). Structural equation modelling and its application to network analysis in functional brain imaging. *Hum. Brain Mapp.* **2,** 2–22.

McKeown, M., Jung, T.-P., Makeig, S., Brown, G., Kinderman, S., Lee, T.-W., and Sejnowski, T. (1998). Spatially independent activity patterns in functional MRI data during the Stroop color naming task. *Proc. Natl. Acad. Sci. U.S.A.* **95,** 803–810.

Mørch, N., Kjems, U., Hansen, L. K., Svarer, C., Law, I., Lautrup, B., and Strother, S. C. (1995). Visualization of neural networks using saliency maps. *In* "IEEE International Conference on Neural Networks, Perth, Australia," pp. 2085–2090. IEEE.

Petersen, S. E., Fox, P. T., Posner, M. I., Mintun, M., and Raichle, M. E. (1989). Positron emission tomographic studies of the processing of single words. *J. Cog. Neurosci.* **1,** 153–170.

Phillips, C. G., Zeki, S., and Barlow, H. B. (1984). Localization of function in the cerebral cortex: Past, present and future. *Brain* **107,** 327–361.

Price, C. J., and Friston, K. J. (1997). Cognitive conjunction: A new approach to brain activation experiments. *NeuroImage* **5,** 261–270.

Price, C. J., Wise, R. J. S., Ramsay, S., Friston, K. J., Howard, D., Patterson, K., and Frackowiak, R. S. J. (1992). Regional response differences within the human auditory cortex when listening to words. *Neurosci. Lett.* **146,** 179–182.

Sandell, J. H., and Schiller, P. H. (1982). Effect of cooling area 18 on striate cortex cells in the squirrel monkey. *J. Neurophysiol.* **48,** 38–48.

Sychra, J. J., Bandettini, P. A., Bhattacharya, N., and Lin, Q. (1994). Synthetic images by subspace transforms. I. Principal component images and related filters. *Med. Phys.* **21,** 193–201.

Talairach, P., and Tournoux, J. (1988). A stereotactic coplanar atlas of the human brain. Thieme, Stuttgart.

Treue, S., and Maunsell, H. R. (1996). Attentional modulation of visual motion processing in cortical areas MT and MST. *Nature* **382,** 539–541.

Vazquez, A. L., and Noll, C. D. (1998). Nonlinear aspects of the BOLD response in functional MRI. *NeuroImage* **7,** 108–118.

Worsley, K. J., and Friston, K. J. (1995). Analysis of fMRI time series revisited–again. *NeuroImage* **2,** 173–181.

Worsley, K. J., Evans, A. C., Marrett, S., and Neelin, P. (1992). A three–dimensional statistical analysis for rCBF activation studies in human brain. *J. Cereb. Blood Flow Metab.* **12,** 900–918.

Worsley, K. J., Marrett, S., Neelin, P., Vandal, A. C., Friston, K. J., and Evans, A. C. (1996). A unified statistical approach to determining significant signals in images of cerebral activation. *Hum. Brain Mapp.* **4,** 58–73.

Zarahn, E., Aguirre, G. K., and D'Esposito, M. (1997). Empirical analyses of BOLD fMRI statistics: I. Spatially unsmoothed data collected under null-hypothesis conditions. *NeuroImage* **5,** 179–197.

Zeki, S. (1990). The motion pathways of the visual cortex. *In* "Vision: Coding and Efficiency" (C. Blakemore, Ed.), pp. 321–345. Cambridge Univ. Press, Cambridge, UK.

3

Preoperative Brain Mapping

John C. Mazziotta

Ahmanson–Lovelace Brain Mapping Center, UCLA School of Medicine, Los Angeles, California 90095

I. Lesion Localization and Characterization

II. Targeting

III. Cortical Mapping

IV. Surgical Planning

References

An important use of current brain mapping methods is the presurgical evaluation of patients in terms of both their structural and their functional anatomy. The goal of such evaluations is to reduce the risk of injury to normal brain, to improve the accuracy and efficiency of the interventional procedure once underway, and to optimize the description of the pathologic lesion affecting the patient. Always used as a supplement to clinical data and standard interventional strategies, these approaches have and will take on increasing importance in the planning of neurosurgical procedures as well as in the arena of interventional neuroradiology. For the purpose of simplicity, both neurosurgical and interventional endovascular neuroradiological procedures will be referred to by the collective term "operative procedure" in the remainder of this chapter.

There are three basic types of preoperative brain mapping evaluations. First, brain mapping methods can be used to characterize and localize lesions in the central nervous system. Almost every brain mapping technique, a brief description of which was provided in

Chapter 1 (see Chapter 1, Tables I and II), with detailed methods provided in the first volume in this series, *Brain Mapping: The Methods* (Toga and Mazziotta, 1996), can be used for the structural and functional characterization of lesions. Typically, such abnormalities are suspected from a patient's history and confirmed by neurological examination. A single brain mapping method, typically a structural imaging technique, often provides the first confirmatory evidence that such a lesion exists. In the case of functional lesions, such as epileptic foci, electrophysiological techniques such as the electroencephalogram (EEG) are the most common initial confirmatory laboratory test. Such data typically provide information about where the abnormality is and what its effects may be on adjacent normal brain regions. The composite use of multiple brain mapping techniques can add information about the relationship of the lesion to adjacent or distant brain structures that are affected by its presence. These methods may also help to provide additional insights into the specific nature of the lesion by adding functional, chemical, and neurotransmitter information. By integrating data from multiple brain mapping methods, a composite description of the lesion and adjacent brain areas emerges that optimizes planning for the operative procedure, minimizes its risk, and expedites its execution.

The second approach to the use of brain mapping methods in the preoperative evaluation of patients with cerebral lesions is in lesion targeting. An ever-increasing number of disorders of the brain are being treated with stereotactic strategies that target selective brain lesions or normal brain centers for ablation or stimulation. In this situation, preoperative brain mapping has increased and will continue to increase the accuracy of identifying these sites, thereby increasing the success of the resultant procedures. The actual interventions may include external stereotactic radiation or the direct placement of probes or stimulators at specific sites in the brain. The treatment of movement disorders, where therapeutic lesions of the medial pallidum, subthalamic nuclei, or nuclei of the thalamus are employed to minimize or alleviate symptoms in advanced cases of Parkinson's disease, is a typical example of this approach. Electrical stimulation of the brain can also be used in these circumstances and requires accurate structural and functional placement of the stimulator electrode tip. It is not unreasonable to anticipate that in the near future, external stereotactic ablation methods will be developed that will obviate the need for direct insertion of probes into the brain parenchyma. In these situations, it will be essential to have very precise and accurate preoperative brain mapping data since electrophysiologic recordings from the targeted site, which are currently used, would not be available.

The final strategy for using preoperative brain mapping is in the evaluation of the normal brain, either adjacent to or remote from the lesion in question. Here the goal is to preserve normal tissue while, at the same time, removing the pathologic lesion in its entirety. A further value of this type of preoperative mapping is to provide guidance with regard to the strategy for reaching the pathologic lesion when such abnormalities are deep to the surface of the brain and require dissection or the passage of instruments through normal brain tissue. The rapidly growing body of knowledge about functional activation studies in normal subjects provides the necessary prelude for developing strategies for assessing patients with cerebral lesions prior to operative procedures. Through the development of large batteries of behavioral tests that produce reliable functional responses, specific approaches to individual patients can be selected from such inventories and delivered to provide individualized data about the functional organization of a given patient's brain. Such evaluations are typically performed using tomographic functional imaging techniques such as positron emission tomography (PET) or functional magnetic resonance imaging (fMRI) (Desmond et al., 1995). A major difficulty with this approach is that while detailed composite maps can be developed preoperatively, alterations in brain geometry during the surgery will make these preoperative maps progressively less accurate as the procedure progresses. Such distortions may occur as a result of therapeutic interventions, such as osmotic dehydration of the

brain or ventricular drainage, or they may occur secondary to local tissue edema that results from the procedure itself. As such, preoperative functional maps must be updated with structural and functional information provided during the procedure. Approaches to this intraoperative tracking include the use of optical intrinsic signal (OIS) (Grinvald et al., 1986; Frostig et al., 1990; Haglund et al., 1992; Toga et al., 1995) imaging and open or interventional MRI devices where the procedure is performed within an MRI device. Details about these strategies can be found in Chapter 4: "Intraoperative Brain Mapping."

I. Lesion Localization and Characterization

A. Structural Data

The preoperative localization of lesions has realized rapid improvements in the past 20 years. Prior to that time, the presence of nonvascular, soft tissue, intracranial lesions could only be inferred from the distortions that they caused in blood vessels or cerebrospinal fluid (CSF) spaces as demonstrated by conventional angiography or pneumoencephalography, respectively. With the advent of X-ray computed tomography (X-ray CT), brain parenchyma could be seen for the first time along with pathologic lesions (Brooks and Di Chiro, 1976; Newton and Potts, 1981; Williams and Haughton, 1985; Mazziotta and Gilman, 1992). The use of intravenous iodinated contrast material provided additional information about the integrity of the blood–brain barrier, typically altered in lesions of the brain. The introduction of magnetic resonance imaging (MRI) further enhanced the ability to identify intracranial abnormalities, particularly in regions where X-ray CT was poor (Lauterbur, 1973; Mansfield, 1977; Atlas, 1991). Thus, lesions that were smaller than a few millimeters or had subtle contrast differences between the X-ray attenuation characteristics of normal brain and the pathologic site were often identified with MRI whereas they went undetected with X-ray CT. Furthermore, while X-ray CT suffered from artifacts produced when the transmitted X-ray beam passed through dense bone (e.g., petrous bones), often obscuring structures at the frontal poles or in the brainstem and ventral cerebellum, MRI does not suffer from these difficulties (Fig. 1). The use of agents that incorporate gadolinium provides the same blood–brain barrier integrity survey with MRI that can be obtained using iodinated contrast materials and X-ray CT (Atlas, 1991).

MR imaging has its own set of constraints and difficulties, including the potential for artifacts in regions

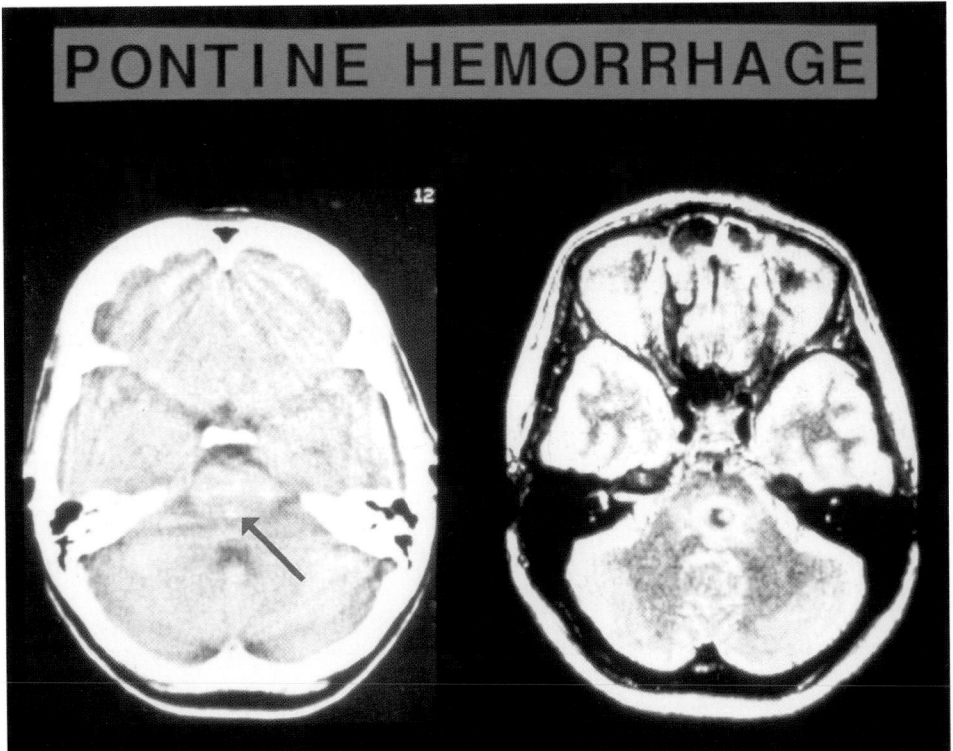

Figure 1 X-ray CT versus MRI in the evaluation of a small pontine hemorrhage. The patient is a 31-year-old female in her second trimester of pregnancy who experienced sudden bilateral sensory loss and a severe headache. An urgent X-ray CT study demonstrated the suggestion of a small punctate hemorrhage in the midportion of the pons. The artifacts through this region produced by the dense petrous bones on either side of the brainstem almost entirely obscure the abnormality, however. MRI demonstrated not only the hemorrhage but also associated surrounding edema and is devoid of artifacts since this methodology is not sensitive to variation in bone density of adjacent structures..

of the brain that are adjacent to sinuses (e.g., frontal poles and medial temporal lobes). Claustrophobia can be a problem for some patients and either makes the examination of a given patient impossible or contaminates the data because of movement artifact associated with patient agitation. Most important in the clinical setting is the fact that implanted electronic or ferromagnetic devices are a contraindication for MR imaging (Klucznik *et al.*, 1993; Kanal *et al.*, 1996; Shellock, 1997) as is the requirement for patients to be scanned who have ancillary life-support equipment or who are so critically ill that they cannot be positioned deep within an MR device for extended periods of time without direct observation by members of the health care team (Salcman and Pevsner, 1992; Moran *et al.*, 1994; MacNamara and Evans, 1995).

Despite these limitations, MRI is typically the structural imaging study of choice to identify the structural characteristics of cerebral lesions in the preoperative setting. One of the major exceptions to this general rule is in patients with acute cerebrovascular events where the exclusion of the diagnosis of intracranial hemorrhage is an urgent issue. In this setting, the speed of X-ray CT, its very high sensitivity for identifying blood, and the accessibility of the patient during the procedure make this modality the procedure of choice.

B. Vascular Anatomy

Conventional angiography has long been the mainstay of identifying abnormalities of blood vessels within the head. The advent of MR angiography has allowed a noninvasive way of visualizing medium- to large-vessel arteries and veins in both the brain and the neck (Atlas, 1991; Wilcock *et al.*, 1996). With this approach, a rapid survey of patients with suspected abnormalities of such vessels can be performed at the time of structural brain imaging. This provides the opportunity to identify aneurysms, arteriovenous malformations, or moderate-to high-grade stenotic lesions of the large to medium arteries (Fig. 2). Since MR angiography images all medium- and large-size vessels, the simultaneous depiction of both arteries and veins often requires multiple postacquisition reconstructions to avoid overlap in a

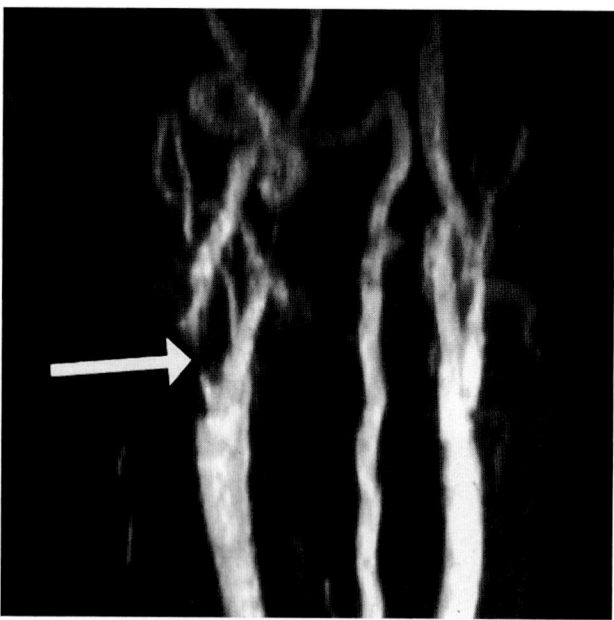

Figure 2 Magnetic resonance angiogram of the cervical vascula-ture demonstrating an occlusive arteriosclerotic lesion in the internal carotid artery. Noninvasive approaches such as MRA can be used to evaluate patients with known or suspected cerebrovascular disease in the same imaging session as that used to evaluate the brain parenchyma. Note that the venous structures are also imaged and, in certain orientations, can overlap the arterial structures of interest. Nevertheless, with digital displays, it is possible to manipulate the data set to provide additional views after the data acquisition has been completed and the patient has departed the imaging suite. Courtesy of Chelsea Kidwell and Jeffrey Saver, UCLA School of Medicine, Los Angeles, California.

two-dimensional projection. This is often the case for imaging of the carotid bifurcation where the jugular ve-nous system is in close proximity.

The development of helical CT is the latest advance in the ability to perform imaging of the vascular system of the brain and neck in a fashion that is less invasive than conventional angiography (Kaatee *et al.*, 1997; Achenbach *et al.*, 1998; Garg *et al.*, 1998). With this tech-nique, a bolus injection of iodinated contrast material is administered intravenously while a high-speed X-ray CT scanner employs a rapid movement of the patient relative to the X-ray tube such that the path of the X-ray beam is in a helical distribution through the body. With this approach, a single intravenous bolus ad-ministration of iodinated contrast provides a three-dimensional data set for the vascular anatomy of the brain. This strategy has been used to provide high-reso-lution detailed images of the three-dimensional struc-ture of aneurysms, arterovenous malformations, or other vascular abnormalities. Such data are of particular value in preoperative planning (see Chapter 1, Fig. 3). Conventional angiography results in a two-dimensional

projection of the lumens of blood vessels in the brain. Often the neck of an aneurysm and its relationship to its parent or daughter arteries can be hard to discern un-less a number of views are obtained at different projec-tion angles. This exposes the patient to multiple doses of X-ray radiation and an increased burden of iodinated contrast material. The use of helical CT using a single-bolus administration of contrast, allowing for the recon-struction of the entire complex anatomy of the aneurysm and its related vessels. The subsequent three-dimensional reconstruction of this complex can be rotated and dis-played from any angle, allowing the neurovascular surgi-cal team to closely examine these relationships and plan a strategy preoperatively. Such improved data collection occurs without the requirement for multiple radiation or contrast exposures by the patient.

C. Blood Flow and Perfusion

Preoperative lesion characterization that can benefit from perfusion imaging includes lesions associated with cerebrovascular disease and measurements in patients with epilepsy. In patients with ischemic vascular disease of the brain, PET, SPECT, and MRI have all been helpful in characterizing the underlying tissue conditions with re-gard to delivery of oxygen and glucose (see Chapter 13) (Kuhl and Edwards, 1964; Obrist *et al.*, 1967, 1975; Patton *et al.*, 1969; Jones *et al.*, 1976; Raichle *et al.*, 1983; Mazz-iotta *et al.*, 1985; Hung *et al.*, 1988; Lear, 1988; Chollet *et al.*, 1989; Nakano *et al.*, 1989; Oshima *et al.*, 1990; Rattner *et al.*, 1991; Devous *et al.*, 1993; American Academy of Neurology (AAN), 1996). Typically performed in con-junction with blood vessel imaging (described earlier) and, in the case of PET, with metabolic measurements, a complete picture of the supply and demand relationship is possible on the local level throughout the brain (Figs. 4 and 5; also see Chapter 1, Fig. 5 and Chapter 13, Fig. 3).

With PET, such measurements typically include the use of ^{15}O-labeled compounds to determine cerebral perfusion, oxygen extraction, and metabolism. These three variables along with cerebral blood volume have been shown to have a characteristic relationship as cerebral perfusion pressure falls (Jones *et al.*, 1976; Baron, 1992). That is, as perfusion pressure drops, blood volume increases as a compensatory means of main-taining adequate tissue perfusion. This is accomplished by venous dilation associated with a diminished periph-eral resistance in the cerebrovascular beds. If perfusion falls further, oxygen extraction increases from its typical 40% to values approaching 100%. Tissue in this state is in a highly precarious condition and in urgent need of measures to increase local perfusion pressure or reduce metabolic demands. If perfusion pressure falls further, no other compensatory hemodynamic mechanisms exist to maintain blood supply. At this point, metabolic demands exceed supply, that is, the definition of ischemia where cerebral oxygen metabolism falls and symptoms begin. If the interruption in perfusion is of a large enough magnitude and/or duration, cerebral infarction will occur and is heralded by a drop in oxygen extraction. At this point, cerebral perfusion can remain low, normalize, or actually increase. Using the combination of these PET studies, it is possible to evaluate, at a local level, the perfusion-metabolic demands of tissue (see Chapter 13, Fig. 3) and determine the appropriate interventional therapy for a given patient and major arterial feeding vessel (Fig. 3) (Baron, 1992). These issues are discussed further in Chapter 13: "Functional Imaging in Vascular Disorders." Analogous measurements of cerebral perfusion can be obtained with SPECT through the use of a wide variety of perfusion radiopharmaceuticals (Chollet *et al.*, 1989; Oshima *et*

al., 1990; Devous *et al.*, 1993). Associated measurements of oxygen extractions and metabolism, however, are not currently possible with SPECT.

Rapid advances in magnetic resonance imaging now provide a battery of variables that can be estimated with this technique. Images of apparent diffusion coefficients (ADC) and diffusion-weighted images (DWI) (Moseley *et al.*, 1990; Warach *et al.*, 1992, 1995a; Sorensen *et al.*, 1996; Lutsep *et al.*, 1997; Baird and Warach, 1998) accompanied by perfusion imaging, obtained by bolus administration of gadolinium-labeled intravascular compounds (Ogawa *et al.*, 1990, 1993; Kwong *et al.*, 1992; Turner *et al.*, 1993; Frahm *et al.*, 1994), can provide semiquantitative estimates about the perfusion conditions of local tissue beds (Fig. 4). While MR does not presently allow for the measurement of oxygen extraction or metabolism, MR spectroscopy has been used to evaluate lactate concentrations in normal

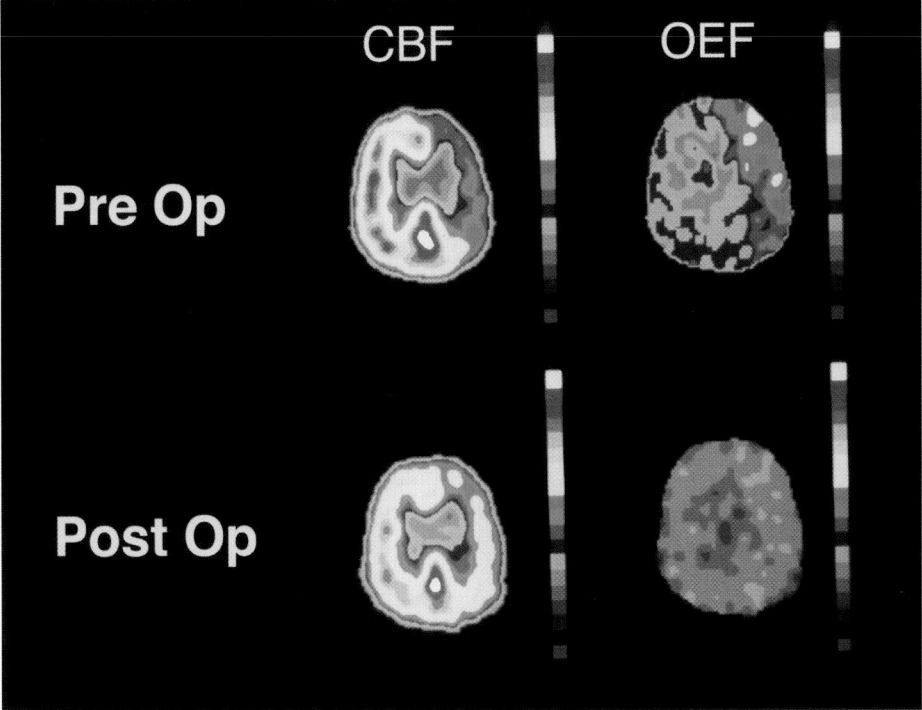

Figure 3 Pre- and postoperative PET studies in a patient with ischemic cerebrovascular disease who underwent extracranial-intracranial (EC–IC) bypass. This battery of studies was obtained with ^{15}O-labeled compounds to measure cerebral blood flow (CBF) and oxygen extraction (OEF) (also see Chapter 13, Fig. 3). In the preoperative study (top row), note that there is diminished blood flow (CBF) in the middle cerebral artery distribution on the right side of the image. In this same distribution, oxygen extraction is relatively increased, indicating that the tissue is underperfused but viable and compensated by enhanced oxygen extraction. Such patients are candidates for reestablishment of perfusion either by thrombolytic approaches or by surgical revascularization procedures. This patient underwent EC–IC bypass and postoperatively (bottom row) has near- normal cerebral blood flow (with the exception of the most anterior portion of the middle cerebral artery distribution) and normal oxygen extraction. Studies such as these can be used to select candidates for surgery, monitor their outcome, and help to determine the cost-effectiveness of such an approach. Courtesy of William Powers, Washington University School of Medicine, St. Louis, Missouri.

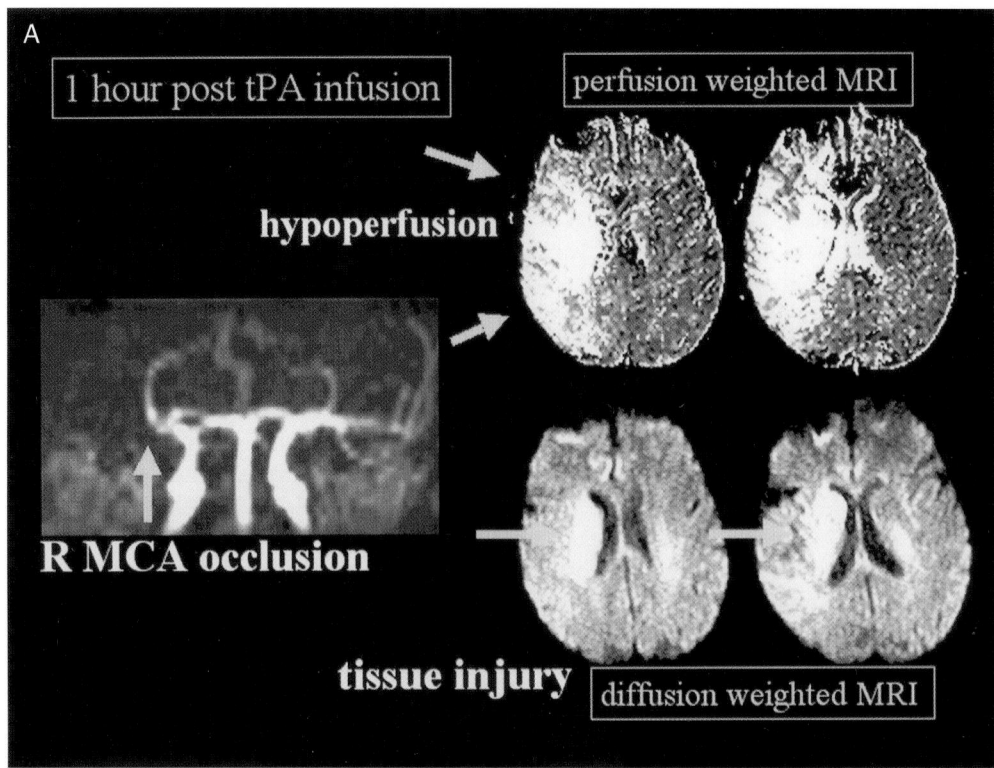

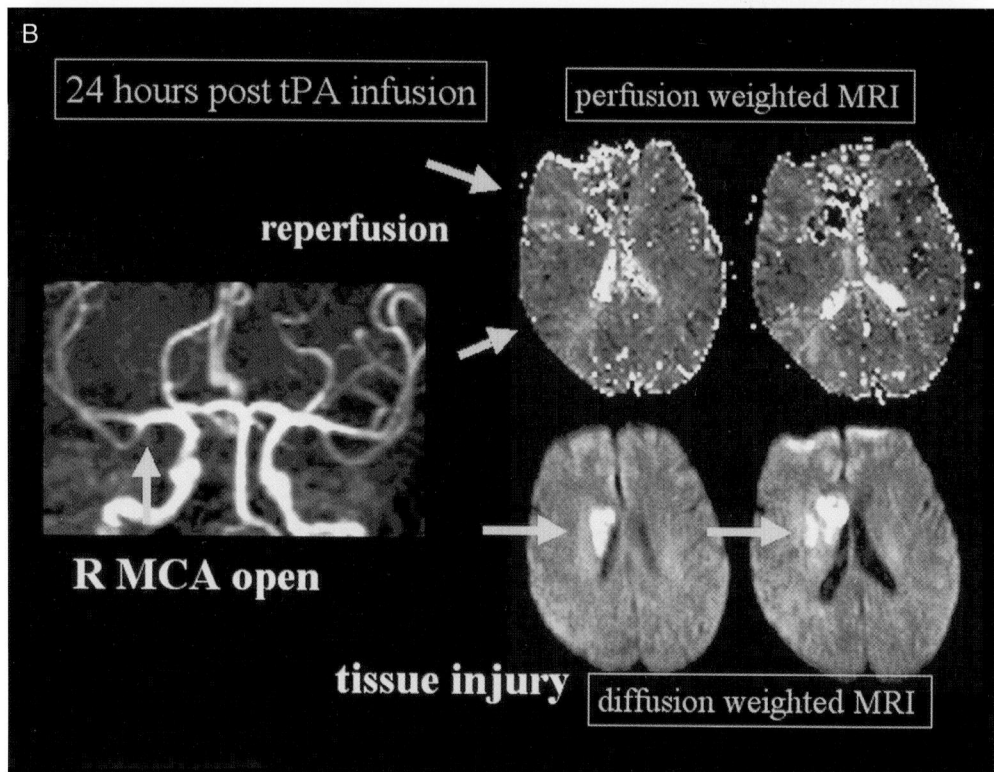

Figure 4 MRI studies of a patient with ischemic cerebrovascular disease prior to and following intra-arterial thrombolytic therapy. (A) Acute occlusion of the right middle cerebral artery demonstrated on MR angiography. Note that in the perfusion-weighted MRI, there is a large area, encompassing the entire right middle cerebral artery distribution, of hypoperfusion (seen as white) and extending all the way from the cortical surface to the lateral ventricle. In addition, the diffusion-weighted MRI demonstrates tissue injury and cell swelling in the central portion of this middle cerebral artery distribution closest to the ventricle (arrows, white region). (B) Twenty-four hours following thrombolytic therapy (tPA infusion), MR angiography demonstrates patency of the right middle cerebral artery and resolution of the perfusion deficit on the perfusion-weighted MRI study. There is

and ischemic tissue (Prichard and Shulman, 1986; Hopkins and Barr, 1987; Chance *et al.*, 1988; Fox *et al.*, 1988; Barranco *et al.*, 1989; Chang *et al.*, 1991; Prichard *et al.*, 1991; Gruetter *et al.*, 1992; Sappey-Marinier *et al.*, 1992; Pekar *et al.*, 1993; van Zijl *et al.*, 1993; Prichard and Rosen, 1994). The relationship between perfusion and diffusion measurements and how they are used in these circumstances is more completely described in Chapter 13: "Functional Imaging in Vascular Disorders."

Hemorrhagic lesions of the brain can be identified, their anatomy and volume determined, and their effect on adjacent or remote tissues characterized from a structural point of view with either X-ray CT or conventional MR imaging. When either of these techniques is combined with MR angiography, conventional angiography, or helical CT, the specific anatomy of the blood vessels in or near such hemorrhages can be evaluated. Thus, the composite battery of imaging studies can provide very specific information about the site and cause of intracranial hemorrhages and also suggest preoperative strategies for their correction, when possible.

Perfusion in patients with epilepsy, particularly those with focal epilepsy emanating from the medial temporal lobes and producing complex partial seizures, has been studied extensively with SPECT methods (Devous *et al.*, 1990; Gzesh *et al.*, 1990; Rowe *et al.*, 1991a,b; Newton *et al.*, 1992; Theodore, 1992; Berkovic *et al.*, 1993; Spencer, 1994; Ho *et al.*, 1995; O'Brien *et al.*, 1998) and less so with PET and fMRI. Invariably obtained with companion EEG data, interictal hypoperfusion is typical of the patients' focal temporal lobe epileptic foci. During seizures, ictal SPECT studies are performed by injecting subjects who are inpatients on epilepsy telemetry wards. In this state, the seizure focus typically shows hyperperfusion. Immediately following a seizure, postictal SPECT or PET studies have demonstrated profound hypoperfusion. Alignment and registration of such data sets and their subtraction provide enhanced anatomical accuracy and sensitivity for evaluating these relationships (Engel *et al.*, 1990; O'Brien *et al.*, 1998). Typically used as a prelude to surgical resection of medial temporal lobe structures, such studies have become an important adjunct to the standard clinical and electrophysiological evaluation of epilepsy patients who are surgical candidates (Berkovic *et al.*, 1993; Spencer, 1994).

More recently, EEG recordings obtained in MR devices capable of producing fMRI data have been employed to identify the site of epileptic spikes. By scanning subjects continuously as EEG data are recorded, fMRI images can be reconstructed at times when such spikes occur, adjusted for the appropriate hemodynamic delay (Ives *et al.*, 1993; Detre *et al.*, 1995; Warach *et al.*, 1995b). The same approach can be used for focal seizures, provided that the seizures do not produce head movements that would create artifacts in the resultant data. Once the MRI data are collected and adjusted for hemodynamic delay, images of relative hyperperfusion are produced associated with the site of interictal spikes or focal seizures (Fig. 5). The role of such studies in the preoperative evaluation of epilepsy patients who are surgical candidates remains to be determined but one of the most difficult problems with such patients is the rapid propagation of such seizures from the primary focus to adjacent or remote sites. It is conceivable that such fMRI–EEG data collections may help with this problem. A more complete discussion of the use of integrated brain mapping techniques in the evaluation of patients with epilepsy is provided in Chapter 14: "The Epilepsies" (Knowlton *et al.*, 1997).

D. Metabolism

PET measurements of glucose metabolism are obtained with the compound [^{18}F]fluorodeoxyglucose (FDG) (Sokoloff *et al.*, 1977; Phelps *et al.*, 1979; Reivich *et al.*, 1979; Raichle, 1983; Phelps and Mazziotta, 1985). Such studies have been extremely helpful in characterizing the early sites of abnormality in patients with neurodegenerative disorders (Kuhl *et al.*, 1982; Hawkins *et al.*, 1987; Mazziotta *et al.*, 1987, 1992; Gilman *et al.*, 1990). Typical patterns of hypometabolism are associated with Alzheimer's and Huntington's disease (Mazziotta and Phelps, 1986; Mazziotta *et al.*, 1987; Mazziotta, 1992), frontotemporal dementia, Creutzfelt–Jakob disease, and other dementing illnesses as described in Chapter 9: "Brain Mapping in Dementia" (see Chapter 1, Figs. 6, 7, and 9). Movement disorders have also been extensively studied with this approach. Hypometabolism occurs in the striatum in Huntington's disease (see Chapter 1, Fig. 7), in the frontal cortex and striatum in progressive supranuclear palsy, and in the cerebellum in Friedrich's ataxia, to name a few. Hypermetabolism is associated with cortical and subcortical regions in tardive dyskinesia and Meige syndrome. A more complete discussion of glucose metabolism in movement disorders can be found in Chapters 10–12.

---

a residual area of tissue injury, the white area on the diffusion-weighted MRI, but this residual tissue damage represents but a small portion of the large volume of tissue at risk when compared to the 1-h postthrombolytic therapy studies shown in A. Such an approach can be used to select patients for thrombolytic therapy, monitor their outcome, and determine the clinical effectiveness of such an approach. Courtesy of Steven Warach and colleagues, Beth Israel Deaconess Medical Center, Boston, Massachusetts. Steven Warach, all rights reserved.

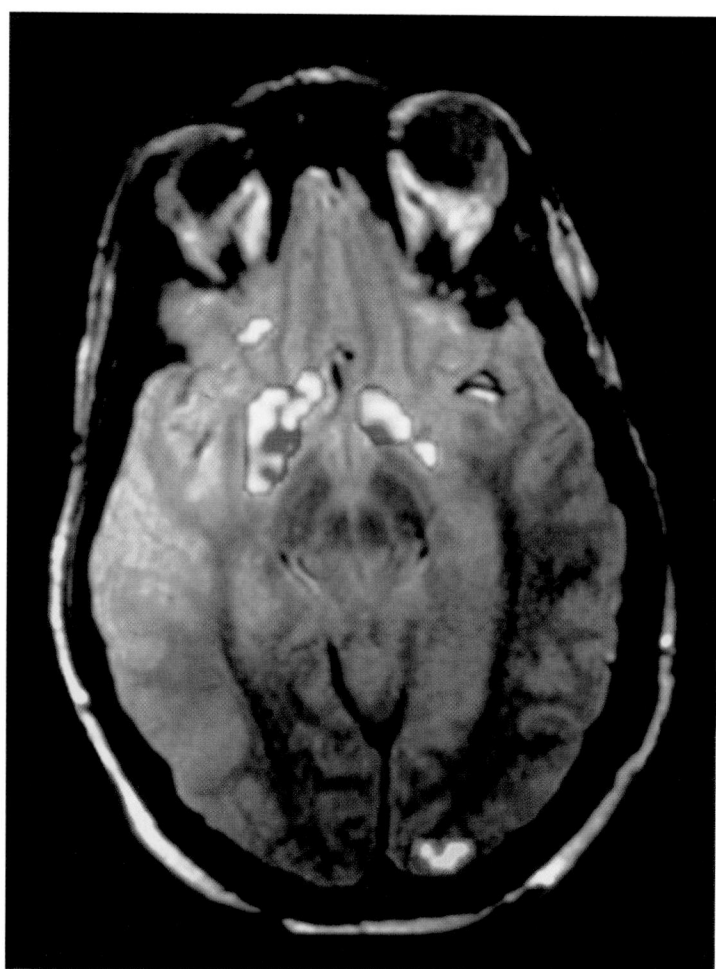

Figure 5 Electroencephalographically (EEG) triggered fMRI study of a patient with partial epilepsy. Thirty-five-year-old woman with partial epilepsy and EEG-documented epileptic discharges of the temporal lobes with left-sided onset. EPI-BOLD acquisition, of 2–3 s each, following the onset of the epileptic EEG discharges versus baseline (no discharges) state. Note the increased signal (color areas) in the anterior medial temporal lobes bilaterally associated with the EEG events. Courtesy of Steven Warach, John Ives, and Don Schomer, Beth Israel Deaconess Medical Center, Boston, Massachusetts. (From S. Warach and J. Ives (1996). EEG-triggered echo-planar functional MRI in epilepsy. *Neurology* **47**(1), 89–93.

Focal areas of hypermetabolism have been reported in high-grade cerebral neoplasms such as glioblastoma multiforme as well as in intracranial infections and have been useful in differentiating tumor recurrence from radiation necrosis (Fig. 6). The preoperative evaluation of patients with cerebral neoplasms using FDG-PET, MR spectroscopy (Fig. 8) (Prichard and Shulman, 1986; Chang *et al.*, 1991; Pekar *et al.*, 1993; Pritchard and Rosen, 1994), and perfusion measurements from MRI, SPECT, or PET has been used as an adjunct in surgical planning for such patients and in characterizing the degree of malignancy of a given lesion. As yet, it is not possible to determine a specific metabolic signature that accurately predicts neoplastic histologic type but it is quite plausible that the combination of multiple brain map-

ping techniques in such a fashion may ultimately produce such signatures in a noninvasive fashion. A more complete discussion of brain mapping techniques of use in cerebral neoplasms is provided in Chapter 16: "Structural and Functional Imaging of Cerebral Neoplasia."

E. Ligand, Neurotransmitters, and *in Vivo* Neuropharmacology

Brain mapping methods of use in the pharmacological characterization of brain lesions primarily fall in the domain of PET and SPECT. Using specifically designed tracers, it is possible to evaluate both the pre- and postsynaptic aspects of neurochemical systems, including the dopaminergic, cholinergic, serotonergic, benzodi-

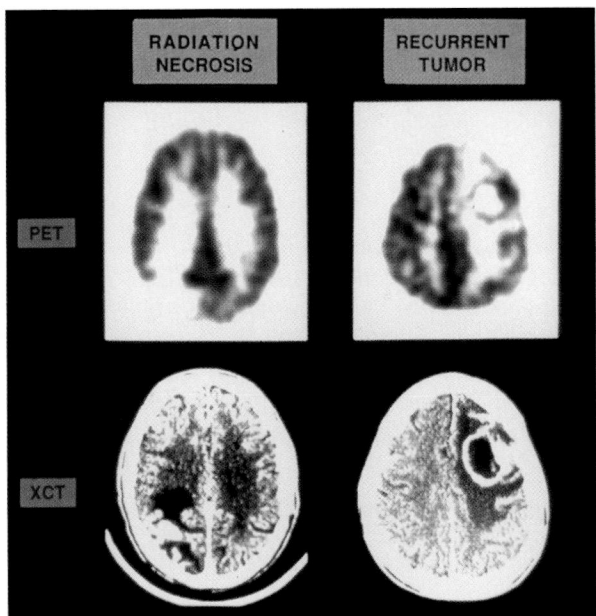

Figure 6 Glucose metabolism obtained with PET-FDG in two different patients, one with recurrence of a cerebral neoplasm and one with radiation necrosis following radiation therapy for a previously diagnosed cerebral neoplasm. Note that tumor recurrence typically is higher in malignancy grade than the original tumor at the time of first resection. As such, glucose metabolism tends to be increased proportional to the degree of malignancy and, as in this case, exceeds that of normal brain parenchyma (ring of hypermetabolism around the necrotic core). In radiation necrosis, glucose metabolism is absent due to the necrotic nature of the lesion. Nevertheless, both recurrence and necrosis tend to have a similar appearance on structural imaging studies with associated edema, contrast enhancement, and compression of adjacent structures.

azepine, and opioid systems. This approach has been used for lesion characterization in a wide range of disorders, including cerebral neoplasms, movement disorders, epilepsy, and neuropsychiatric disorders. The reader is referred to the appropriate chapters for details on these specific applications.

The application of composite brain mapping techniques using such an approach is illustrative and has been performed in patients with medial temporal lobe epilepsy producing complex partial seizures. As described above, such patients can be evaluated by using perfusion imaging with SPECT or PET, glucose metabolic imaging with PET, spike localization with fMRI, and traditional electrophysiologic techniques. Their evaluation with ligand imaging may provide additional insights and has been described using PET and SPECT methods (Fig. 7). For example, such patients have been studied using opiate ligands ([^{11}C]diprenorphine, carfentanil, and cyclofoxy) (Mayberg *et al.*, 1991), benzodiazepine agents ([^{11}C]flumazenil) (Savic *et al.*, 1988; Frey *et al.*, 1991; Koepp *et al.*, 1997), and others. Analogous studies have been performed with SPECT using an io-

dinated form of the benzodiazepine receptor ligand iomazenil (van Huffelen *et al.*, 1990; Berkovic *et al.*, 1993). Of particular interest has been the benzodiazepine studies, which seem to show more selective abnormalities in ligand binding with a smaller spatial extent than those identified with glucose metabolism PET studies or perfusion studies using either PET or SPECT (see Chapter 1, Fig. 10). In addition, these compounds identify cortical lesions in patients with neocortical forms of epilepsy that have been difficult to appreciate with either perfusion or metabolic imaging. All of these approaches provide additional anatomical and chemical characterization of these potentially operable lesions.

Ligand imaging has also been used in the evaluation of patients with cerebral neoplasms. The peripheral benzodiazepine receptors have been used as markers of neoplastic tissue, as has [^{18}F]fluorodeoxyuridine. Raclopride and spiperone have been used to evaluate pituitary tumors. All of these neuropharmacological approaches provide additional chemical specificity for the identification of these lesions as well as the determination of the spatial extent of these specific properties. When used in combination with structural imaging techniques and more generalized functional measures (Fig. 8), such as metabolism and perfusion, they can be useful adjuncts in the preoperative planning, characterization, and localization of cerebral lesions.

II. Targeting

The use of brain mapping methods to target sites in the brain of patients with cerebral disorders has two potential goals. The first is to identify sites in the brain for ablation therapy because they themselves are the cause of a pathologic condition, such as neoplasms or vascular malformations, and the second is to interrupt specific pathways that are contributing to a patient's symptom complex, such as in Parkinson's disease. In either case, optimal anatomical detail and, ideally, functional characterization should improve the operative results.

In the case of ablations, lesioning can be performed with stereotactic radiotherapy or with probes inserted into the substance of the brain that produce lesions by either heating or cooling. Since there is a tremendous degree of variability in the structural and functional organization of brains between individuals, a precise map of these locations is required within an individual patient to ensure that the ablative process affects only those tissues intended to help the patient and no additional brain regions. For stereotactic radiotherapy, all of the lesion localization and characterization tools that have been discussed above can be employed to provide the three-dimensional coordinates for ablating specific pathologic lesions in the

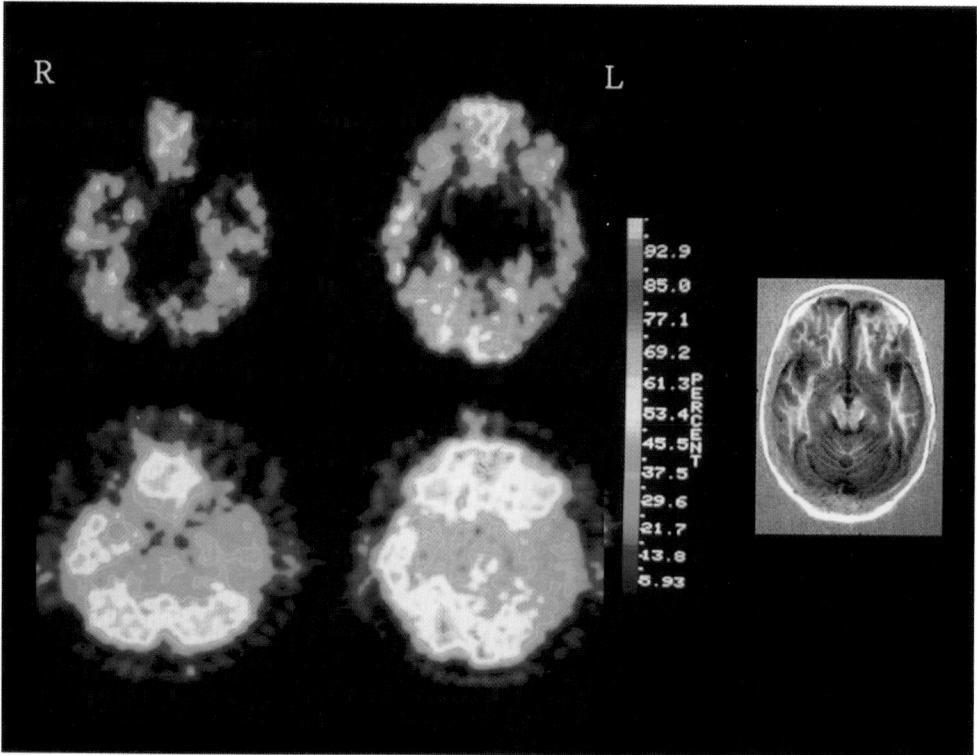

Figure 7 Benzodiazepine receptor imaging with PET and ¹⁸F-labeled flumazenil (top row, color). These studies are shown in comparison with conventional MRI (right-most image) for structural imaging and glucose metabolism (bottom row, color) obtained with PET. Notice that in this patient, with complex partial seizures of the medial temporal lobe, there is a degree of atrophy of the hippocampus as seen with conventional MR imaging. Widespread hypometabolism in the temporal lobe is seen interictally with FDG-PET. A much more selective abnormality is seen in the flumazenil image, which may reflect the selective loss of synaptic receptors associated with the primary site of the epileptic focus. Courtesy of Ivanka Savic, Karolinska Institute, Stockholm, Sweden. Reprinted with permission from I. Savic, P. Roland, G. Sedvall, A. Persson, S. Pauli, and L. Widen (1988). *In vivo* demonstration of reduced benzodiazepine receptor binding in human epileptic foci. *Lancet* **11,** 864–866. © by The Lancet Ltd.

brain (Fig. 9). This kind of treatment can be delivered with either a rigidly mounted frame that is bolted to the skull or frameless approaches that have been developed more recently.

For those procedures that require inserting probes or electrodes into the substance of the brain, past approaches have utilized relatively few intracranial landmarks and systems that stretch or compress individual brain anatomy to match an atlas. This approach was originally proposed and popularized by Talairach and Tournoux and is still in widespread use (Talairach and Tournoux, 1988). Once a target is identified by these approaches, with the anterior and posterior commissures as landmarks, a probe is inserted to the target coordinates. Typically, recordings are then made from that location using electrodes to record cell firing. Patients are usually awake during such procedures so that nonlethal cooling or heating of tissue can be performed in a reversible manner to determine the degree of beneficial effect or potential unanticipated deficits that might occur should a permanent lesion be placed at that location.

Modern brain mapping techniques are helpful in the targeting process in four ways. First, improved lesion localization and characterization can help better identify sites for ablative lesioning. Second, increased anatomical resolution with structural imaging techniques provides for better anatomical definition of landmarks in or near the target. Third, functional imaging approaches may allow for the physiologic activation of targeting sites, particularly in those situations where normal circuits are intended to be interrupted for a therapeutic purpose, as in Parkinson's disease (see Chapter 26). Fourth, the use of population-based probabilistic atlases (Mazziotta *et al.*, 1995) that account for variance in structure and function between subjects (Geschwind and Levitsky, 1968; Witelson, 1977; Galaburda *et al.*, 1978; Le May and Kido, 1978; Chui and Damasio, 1980; Weinberger *et al.*, 1982; Dupont *et al.*, 1994) can improve the probability of distorting a given subject's brain to match the atlas (see Chapter 6).

There is little doubt that the advances in MR imaging have produced anatomical data sets about individ-

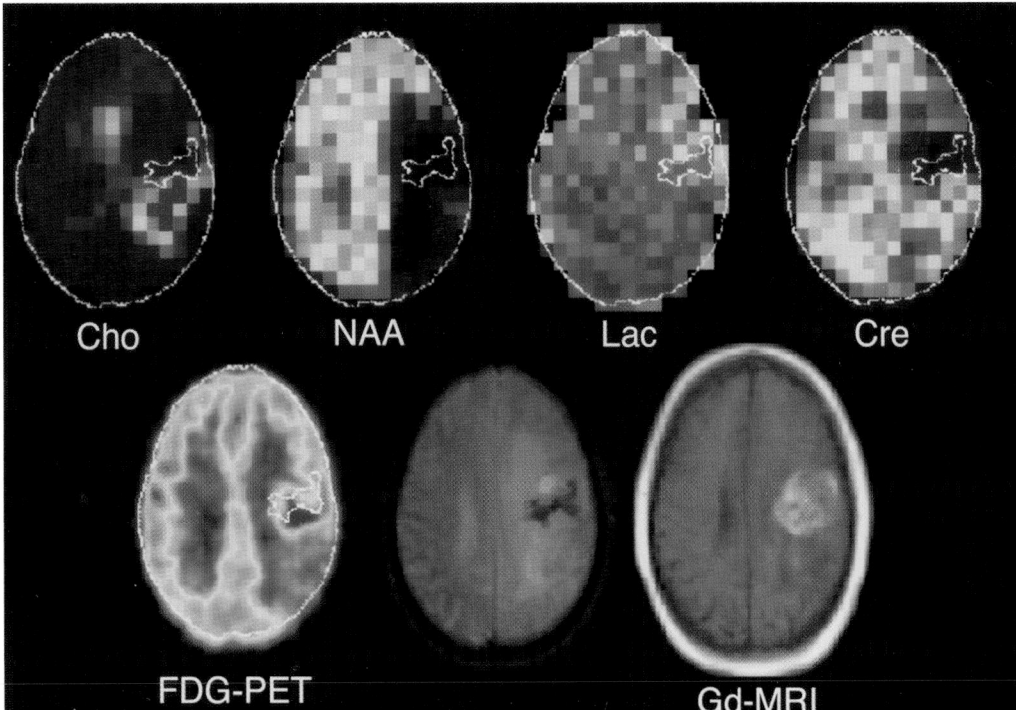

Figure 8 Magnetic resonance spectroscopy compared with FDG-PET and conventional MRI studies in a patient with a recurrent glioblastoma multiforme of the left superior temporoparietal region. The patient had been previously treated with partial surgical resection and radiation therapy. The images demonstrate magnetic resonance spectroscopy (MRS) (top row) of choline-containing compounds (Cho), N-acetyl-containing compounds, with N-acetylaspartate (NAA) as the prominent contributor, lactate (Lac), and creatine/phosphocreatine (Cre). These studies are compared with FDG-PET and conventional MRI obtained with and without gadolinium (Gd-MRI). Note that the high signal from choline in the proton-MRS corresponds to the highest metabolic values in the PET image. There is generally low NAA concentration in and around the lesion whereas the patterns for lactate and creatine/phosphocreatine are more complex. Such studies may help to better characterize the chemical nature of lesions such as cerebral neoplasms, particularly when used in comparison with PET and conventional structural imaging. Courtesy of Mario Quarantelli and Jeffrey Alger, Neuroimaging Branch, National Institutes of Neurologic Disorders and Stroke, NIH, Bethesda, Maryland, and the Department of Radiological Sciences, UCLA School of Medicine, Los Angeles, California, respectively.

ual patient brains that are beyond any previous level of detail in terms of both spatial resolution and contrast. In addition, high-field MR units or the approach of averaging multiple studies from a conventional field strength scanner and producing a low-noise composite image greatly improves the spatial resolution and allows for the direct visualization of the medial portion of the globus pallidus, substructures of the thalamus, details of the brainstem, the nuclei, and cortex of the cerebellum, and many other anatomical details previously unseen and of importance in targeting strategies.

Less progress has been made in the functional activation of sites that are potential sources of either stimulation or ablation. This is particularly important in situations where normal circuits are to be interrupted for a therapeutic purpose. There are only a few publications demonstrating activation, i.e., increases in relative cerebral perfusion, with fMRI in basal ganglia structures. Significant advances in both the paradigms and the imaging strategies will be required before reliable targeting of functional sites can be employed as preoperative methods. At present, the improvements in structural imaging help to better define the general target area and then a localized search with an electrode recording electrophysiological data is used for ultimate lesion or electrode placement. It is also possible that selective-ligand studies can identify specific neurotransmitter systems of interest for either ablative or stimulation therapeutic interventions (Frost, 1982; Mintun *et al.*, 1984; Farde *et al.*, 1986; Perlmutter *et al.*, 1986; Wong *et al.*, 1986; Arnett *et al.*, 1987; Fowler *et al.*, 1987; Huang *et al.*, 1991; Brooks *et al.*, 1992; Madar *et al.*, 1997).

Finally, the use of population-based probabilistic atlases can enhance the accuracy and planning of targets in the brain, particularly for deep structures. The ability to use a digital three-dimensional database of anatomical variance in comparison with the patient in question allows for an estimate of the local variability that might be encountered in the structure or structures of the brain slated for ablation or stimulation (Mazziotta *et al.*,

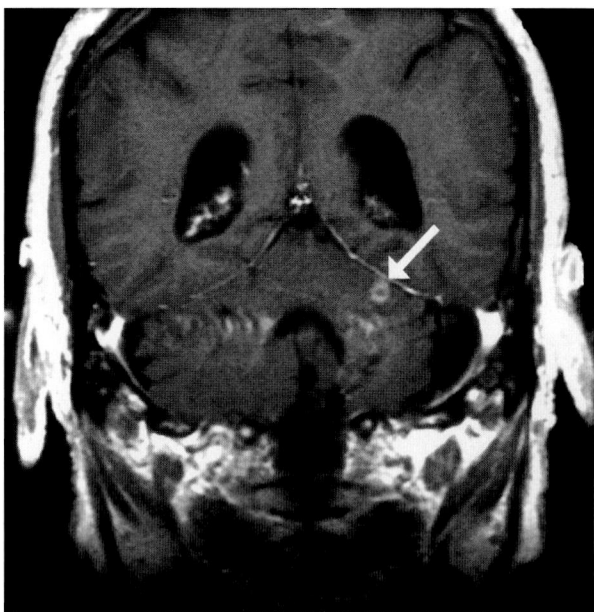

Figure 9 Magnetic resonance imaging and gadolinium enhancement of a small metastatic lesion on the superior surface of the left cerebellum in a patient with a primary lung cancer and metastases to the brain and liver. Such detailed images can then be entered into stereotactic coordinate systems for noninvasive and selective stereotactic radiotherapy.

1995). Through the use of appropriate three-dimensional warping algorithms, these atlases and reference systems have the capacity to distort an individual patient's brain to match that of a population in a far more sophisticated way than the linear stretching and compression originally popularized by Talairach and Tournoux (Talairach and Tournoux, 1988). These modern digital atlases also take into account the asymmetries of the cerebral hemispheres and make no assumptions about the presumed location of brain structures but rather, through the availability of hundreds, or potentially thousands, of entered subjects, calculate for a specific population the probability of a structure being at a certain location. Functional data will soon be added to such atlases, further extending their capabilities. A more complete discussion of atlas approaches in the evaluation of patients with cerebral disorders is provided in Chapter 6.

III. Cortical Mapping

Patients with lesions of the cerebral hemispheres that require surgical resection or neuroradiological intervention run the risk of losing normal functions controlled by adjacent tissues, particularly in the cerebral cortex. Such procedures are particularly worrisome in patients with lesions in or near motor or language cortices. Such patients, in the past, have been advised either not to have surgery or to anticipate a significant deficit. In other situations, patients have had their anesthesia reversed once the cortex of the brain is exposed and behavioral testing is conducted while sites of potential cortical resection are stimulated electrically to temporarily disrupt their functional activity (Penfield and Rasmussen, 1950). Such measures have been warranted because of the devastating consequences of injury or removal of cortex critical to movement or language function. In patients who have been candidates for resection of a temporal lobe, including the anterior and medial portions and anterior hippocampus, testing of memory function has also been part of their presurgical evaluation. In this situation, barbiturates are injected intraarterially, temporarily and reversibly inactivating the mesial temporal lobe or the majority of the entire hemi-

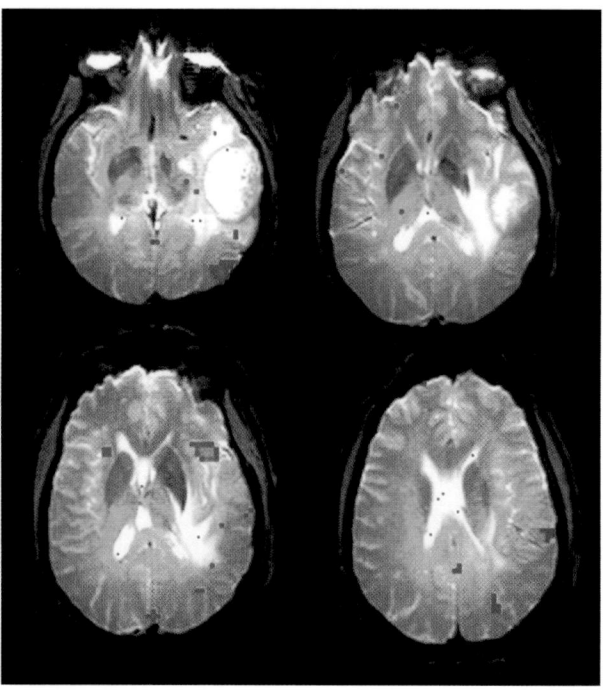

Figure 10 fMRI activation in a patient with a cerebral neoplasm of the left superior temporal region involving language cortex. Structural images demonstrate the neoplasm and its associated edema. The red regions demonstrate fMRI responses during a language task. Note that these responses are displaced by the presence of the tumor and edema but still occur. Such preoperative evaluations can be used to identify language or other system-specific functions of the brain so that they may be avoided during the planning and execution of resection procedures such as the one required in this patient with a cerebral neoplasm. Such studies are typically done for patients with lesions in or near a language, visual, or motor cortical regions. Courtesy of Susan Bookheimer, Brain Mapping Division, Neuropsychiatric Institute, UCLA School of Medicine, Los Angeles, California.

sphere, depending on which vessel is injected. During the pharmacological window of the drug, patients are behaviorally tested to determine whether they can learn new information. The assumption is that if the temporal lobe that is slated for removal is temporarily "anesthetized" by barbiturate administration, the remaining temporal lobe should be capable of memory function and the acquisition of new material for learning purposes. Thus, the Wada procedure, as it is called, substantiates that these functions will be retained after the surgical resection. Analogous approaches can be used for language areas in determining language dominance, particularly in children who are to have large cortical resections for the control of epilepsy. These studies are, in fact, difficult to perform and require a good rapport with the patient, who needs to be cooperative and motivated to provide data in an altered state and under the threatening conditions of an angiography suite. The barbiturate administrations are often accompanied by a transient hemiparesis, which, despite pretest warnings, is

anxiety-producing and adds to the stress and uncertainty of the situation.

For all these reasons, noninvasive preoperative mapping has been avidly pursued as a means of trying to avoid some of these more traumatic and potentially less accurate approaches to defining normal cortical areas and their relationship to lesions under consideration for surgical resection. PET, SPECT, fMRI, and transcranial magnetic stimulation have all been considered for these purposes (Merton and Morton, 1980; Chugani *et al.*, 1988, 1990; Engel, 1993; Engel *et al.*, 1990; Zhang *et al.*, 1990; Belliveau *et al.*, 1991; Mills, 1991; Pascual-Leone *et al.*, 1992, 1994, 1996; Wasserman *et al.*, 1992, 1994; Grafman *et al.*, 1994; Grafton *et al.*, 1994; Jennum *et al.*, 1994; Epstein *et al.*, 1996; Knowlton *et al.*, 1997; Paus *et al.*, 1997; Wasserman, 1998; Hunter *et al.*, 1999). Most data have been collected with PET and fMRI (Figs. 10 and 11). In these circumstances, batteries of behavioral paradigms are delivered to patients preoperatively to develop maps of relatively increased

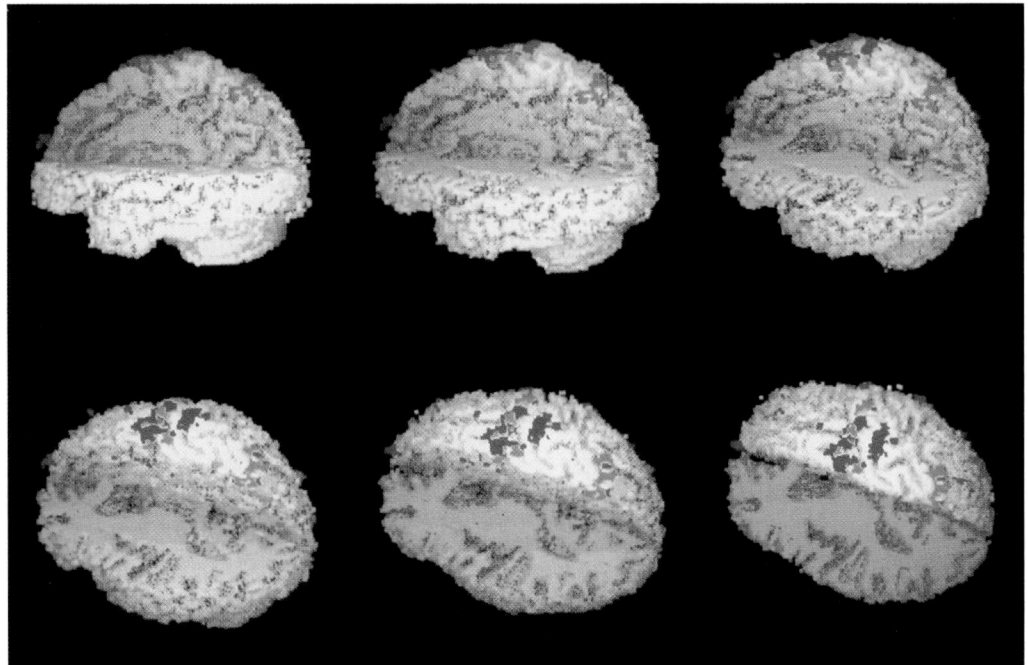

Figure 11 Presurgical mapping of a patient with a right parietal arterovenous malformation. In these images, the left hemisphere has been digitally removed to demonstrate the important findings in the right hemisphere. The structural images were obtained from conventional MRI and reconstructed in three dimensions. The lesion itself is demonstrated in orange. The blue color demonstrates the areas of increased relative cerebral blood flow when the subject followed a moving target on a video monitor with his eyes. Notice that areas in the premotor cortex, presumably the frontal eye fields, as well as the parietal cortex were activated with this task. The magenta color indicates the additional areas that were activated when the subject followed the moving target with both his eyes and his left index finger. A composite study such as this shows the actual lesion, its location in the three-dimensional organization of the cortex, and adjacent cortical areas that are activated by specific behavioral tasks, all of which can be displayed in the operating room to determine the boundaries of the lesion and the boundaries of normal cortical areas immediately adjacent. Courtesy of Scott Grafton, Emory University School of Medicine, Atlanta, Georgia.

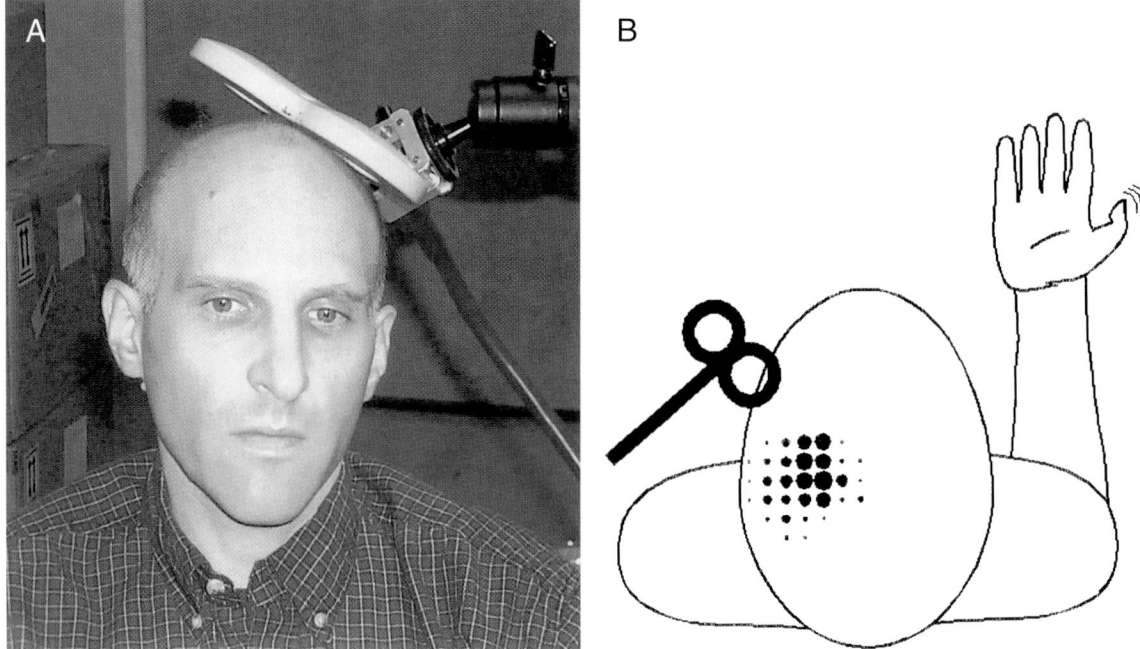

Figure 12 Transcranial magnetic stimulation of the human motor system. (A) A transcranial stimulator is positioned over the motor cortex. The stimulator itself consists of a coil of wire in a figure-eight configuration. When current flows through the wire, it induces an electrical current in the cortex immediately beneath the skull. This induced current discharges the neurons in this region and, depending on its target, will produce a wide variety of behavioral responses. In this case, the stimulator is positioned over the motor cortex and would result in a response consisting of motor movement of the contralateral upper extremity. (Subject is Dr. Wassermann.) (B) Maps of the responses can be generated demonstrating the distribution of these functions in the cortex of the human brain. Courtesy of Mark Hallett and Eric Wasserman, National Institute of Neurologic Disease and Stroke, NIH, Bethesda, Maryland.

perfusion associated with the performance of critical tasks thought to be near areas of surgical resection or surgical access to deeper lesions. Thus, patients with lesions in or near their sensory-motor cortices will be asked to perform tasks that activate the appropriate topographic regions of the motor cortex, and the relative distance between the lesion and the activation sites provides a measure of risk involved in the resection of the abnormality. A similar approach can be used for language areas (Desmond *et al.,* 1995; Hunter *et al.,* 1999). As such, these approaches will make it possible to determine language dominance and hemispheric memory function in medial temporal lobe structures. While still in an early stage of development, such strategies require that an ample amount of information is obtained in normal subjects to determine the consistency and reliability of such relative perfusion measurements for identifying memory, language, sensory, and motor functions within an individual. By combining these noninvasive brain mapping techniques with conventional Wada testing, intraoperative stimulation, and other forms of evaluation, it will be possible to validate the accuracy of the noninvasive approach and determine its applicability for direct use in presurgical patient populations.

IV. Surgical Planning

It is possible to combine all these data into composite images that can then be visualized in the operating room during the course of the operative procedure (Fox *et al.,* 1985; Pelizzari *et al.,* 1989; Evans *et al.,* 1991; Friston *et al.,* 1991; Woods *et al.,* 1992, 1993; Gee *et al.,* 1993; Collins *et al.,* 1994a,b; Mazziotta *et al.,* 1995; Turkington *et al.,* 1995). That is, data about lesion localization and characterization can be combined with targeting information as well as normal cortical responses of sites adjacent to and remote from the lesion to provide a composite picture defining the lesion itself and the surrounding brain structures in terms of both anatomy and physiology. Such maps can be linked to interactive displays that update as the surgeon or interventional neuroradiologist moves a linked probe relative to the patient. The displays can be multiplanar two-dimensional data sets or true 3D representations. In addition, the viewing apparatus can be integrated into goggles worn by the surgical team in which the operator can select an entirely synthetic image made up of preoperative mapping studies, an entirely pure view of the operative field, or a mixture of the two. Such systems have to update orientation so that the relative position of the surgeon's eyes and the patient's head is reflected in the rotation, translation, and scaling of the digital information obtained preoperatively. A more complete discussion of these approaches is described in Chapter 5. Irrespective of the type of preoperative mapping used or the methods employed for its ultimate visualization at the time of the procedure, the composite use of these brain mapping techniques to evaluate patients prior to invasive interventional procedures should increase their accuracy, safety, efficiency, and, ultimately, the success rate, resulting in a better outcome for the patients.

Acknowledgments

The author thanks contributing investigators for allowing their work to be reproduced in this chapter. This work was partially supported by a grant from the Human Brain Project (P01-MH52176-7) and generous support from the Brain Mapping Medical Research Organization, The Ahmanson Foundation, the Pierson–Lovelace Foundation, the Jennifer Jones Simon Foundation, the Tamkin Foundation, and the Northstar Fund. The author thanks Laurie Carr for the preparation of the text and Andrew Lee for assistance with the illustrations.

References

Achenbach, S., Moshage, W., Ropers, D., Nossen, J., and Daniel, W. G. (1998). Value of electron-beam computed tomography for the noninvasive detection of high-grade coronary-artery stenosis and occlusions. *N. Engl. J. Med.* **339,** 1964–1971.

American Academy of Neurology (AAN). (1996). Assessment of brain SPECT: Report of the Therapeutics and Technology Assessment Subcommittee of the American Academy of Neurology. *Neurology* **46,** 278–285.

Arnett, C. D., Fowler, J. S., MacGregor, R. R., *et al.* (1987). Turnover of brain monoamine oxidase measured *in vivo* by positron emission tomography using L- [^{11}C]deprenyl. *J. Neurochem.* **49,** 522–527.

Atlas, S. W. (1991). "Magnetic Resonance Imaging of the Brain and Spine." Raven Press, New York.

Baird, A. E., and Warach, S. (1998). Magnetic resonance imaging of acute stroke. *J. Cereb. Blood Flow Metab.* **18,** 583–609.

Baron, J. C. (1992). *In* "Stroke: Pathophysiology, Diagnosis and Management" (H. J. Barnett, B. M. Stein, J. P. Mohr, and F. M. Yatsu, Eds.), 2nd ed., pp. 111–123. Churchill Livingstone, New York.

Barranco, D., Sutton, L., Florin, S., Greenberg, J., Sinnwell, T., Ligeti, L., *et al.* (1989). Use of ^{19}F NMR spectroscopy for measurement of cerebral blood flow: Comparative study using microspheres. *J. Cereb. Blood Flow Metab.* **9,** 886–891.

Belliveau, J. W., Kennedy, D. N., McKinstry, R. C., Buchbindiner, B. R., Weisskoff, R. M., Cohen, M. S., *et al.* (1991). Functional mapping of the human visual-cortex by magnetic-resonance-imaging. *Science* **254,** 716–719.

Berkovic, S. F., Newton, M. R., Chiron, C., and Dulac, O. (1993). Single photon emission tomography. *In* "Surgical Treatment of the Epilepsies" (J. Engel, Jr., Ed). Raven Press, New York.

Brooks, D. J., Ibanez, V., Sawle, G. V., Playford, E., Quinn, N., Mathias, C. J., *et al.* (1992). Striatal D2 receptor status in patients with Parkinson's disease, striatonigral degeneration, and progressive supranuclear palsy, measured with ^{11}C-raclopride and positron emission tomography. *Ann. Neurol.* **31,** 184–192.

Brooks, R. A., and Di Chiro, G. (1976). Principles of computer assisted tomography (CAT) in radiographic and radioisotopic imaging. *Phys. Med. Biol.* **21,** 689–732.

Chance, B., Leigh, J. S., McLaughlin, A. C., *et al.* (1988). Phosphorus-31 spectroscopy and imaging. *In* "Magnetic Resonance Imaging" (C. L. Partain, J. A. Patton, *et al.,* Eds.). Saunders, Philadelphia.

Chang, L. H., Cohen, Y., and Weinstein, P. R. (1991). Interleaved ^{1}H and ^{31}P spectroscopic imaging for studying regional brain imaging. *Magn. Reson. Imaging* **9,** 223–227.

Chollet, F., Celsis, P., Clanet, M., Guiraud-Chaumeil, B., Rascol, A., and Marc-Vergnes, J. P. (1989). SPECT study of cerebral blood flow reactivity after acetazolamide in patients with transient ischemic attacks. *Stroke* **20** (4), 458–464.

Chugani, H. T., Shewmon, D. A., Peacock, W. J., Shields, W. D., Mazziotta, J. C., and Phelps, M. E. (1988). Surgical treatment of intractable neonatal onset seizures: The role of positron emission tomography. *Neurology* **38,** 1178–1188.

Chugani, H. T., Shields, W. D., Shewmon, D. A., Olson, D. M., Phelps, M. E., and Peacock, W. J. (1990). Infantile spasms. I: PET identifies focal cortical dysgenesis in cryptogenic cases for surgical treatment. *Ann. Neurol.* **27,** 406–413.

Chui, C., and Damasio, A. (1980). Human cerebral asymmetries evaluated by computed tomography. *J. Neurochem. Neurosurg. Psychiatry* **43,** 873–878.

Collins, D. L., Peters, T. M., and Evans, A. C. (1994a). An automated 3D nonlinear image deformation for determination of gross morphometric variability in human brain. *Proc. Visual Biomed. Comput. VBC '94, Rochester, SPIE* **2359,** 180–190.

Collins, D. L., Neelin, P., Peters, T. M., and Evans, A. C. (1994b). Automatic 3D intersubject registration of MR volumetric data in standardized Talairach space. *J. Comput. Assist. Tomogr.* **18** (2), 192–205.

Desmond, J. E., Sum, J. M., Wagner, A. D., Demb, J. B., Shear, P. K., Glover, G. H., *et al.* (1995). Functional MRI measurement of language lateralization in Wada-tested patients. *Brain* **118,** 1411–1419.

Detre, J. A., Sirven, J. I., Alsop, D. C., O'Connor, J. J., and French, J. A. (1995). Localization of subclinical ictal activity by functional magnetic resonance imaging: Correlation with invasive monitoring. *Ann. Neurol.* **38,** 618–624.

Devous, M. D., Leroy, R. F., and Horman, R. W. (1990). Single photon emission computed tomography in epilepsy. *Semin. Nucl. Med.* **20,** 325–341.

Devous, M. D., Payne, J. K., Lowe, J. L., and Leroy, R. F. (1993). Comparison of technetium-99m-ECD to xenon-133 SPECT in normal controls and in patients with mild to moderate regional cerebral blood flow abnormalities. *J. Nucl. Med.* **34,** 754–761.

Dupont, P., Orban, G. A., De Bruyn, B., Verbruggen, A., and Mortelmans, L. (1994). Many areas in the human brain respond to visual motion. *J. Neurophysiol.* **72,** 1420–1424.

Engel, J. (1993). "Surgical Treatment of the Epilepsies," 2nd ed. Raven Press, New York.

Engel, J., Jr., Henry, T. R., Risinger, M. W., Mazziotta, J. C., Sutherling, W. W., Levesque, M. F., *et al.* (1990). Presurgical evaluation for partial epilepsy: Relative contributions of chronic depth electrode recordings versus FDG-PET and scalp sphenoidal ictal EEG. *Neurology* **40,** 1670–1677.

Epstein, C. M., Lah, J. K., Meador, K., Weissman, J. D., Gaitan, L. E., and Dihenia, B. (1996). Optimum stimulus parameters for lateralized suppression of speech with magnetic brain stimulation. *Neurology* **47,** 1590–1593.

Evans, A. C., Dai, W., Collins, D. L., Neelin, P., and Marrett, S. (1991). Warping of a computerized 3-D atlas to match brain image volumes for quantitative neuroanatomical and functional analysis. *SPIE Med. Imaging: Image Proc.* **1445,** 236–246.

Farde, L., Hakan, H., Ehrin, E., *et al.* (1986). Quantitative analysis of D_2 dopamine receptor binding in the living human brain by PET. *Science* **231,** 258–261.

Fowler, J. S., MacGregor, R. R., Wolf, A. P., *et al.* (1987). Mapping human brain monoamine oxidase A and B with ^{11}C-labeled suicide inactivators and PET. *Science* **235,** 481–485.

Fox, P. T., Perlmutter, J. S., and Raichle, M. E. (1985). A stereotactic method of anatomical localization for positron emission tomography. *J. Comput. Assist. Tomogr.* **9** (1), 141–153.

Fox, P. T., Raichle, M. E., Mintun, M. A., and Dence, C. (1988). Nonoxidative glucose consumption during focal physiologic neural activity. *Science* **241,** 462–464.

Frahm, J., Merboldt, K.-D., Hanicke, W., Kleinschmidt, A., and Boecker, H. (1994). Brain or vein- Oxygenation or flow? On signal physiology in functional MRI of human brain. *NMR Biomed.* **7,** 45–53.

Frey, K. A., Holthoff, V. A., Koeppe, R. A., Jewett, D. M., Kilbourn, M. R., and Kuhl, D. E. (1991). Parametric *in vivo* imaging of benzodiazepine receptor distribution in human brain. *Ann. Neurol.* **30,** 663–672.

Friston, K. J., Frith, C. D., Liddle, P. F., and Frackowiak, R. S. (1991). Plastic transformation of PET images. *J. Comput. Assist. Tomogr.* **15** (4), 634–639.

Frost, J. J. (1982). Pharmacokinetic aspects of the *in vivo,* noninvasive study of neuroreceptors in man. *In* "Receptor Binding Radiotracers" (W. Eckelman, Ed.), pp. 25–39. CRC Press, Boca Raton, FL.

Frostig, R. D., Lieke, E. E., Ts'o, D. Y., and Grinvald, A. (1990). Cortical functional architecture and local coupling between neuronal activity and the microcirculation revealed by *in vivo* high-resolution optical imaging of intrinsic signals. *Proc. Natl. Acad. Sci. U.S.A.* **87,** 6082–6086.

Galaburda, A., Le May, M., Kemper, T., and Geschwind, N. (1978). Left–right asymmetries in the brain. *Science* **199,** 852–856.

Garg, K., Welsh, C. H., Feyerabend, A. J., Subber, S. W., Russ, P. D., Johnston, R. J., *et al.* (1998). Pulmonary embolism: Diagnosis with spiral CT and ventilation-perfusion scanning—Correlation with pulmonary angiographic results or clinical outcome. *Radiology* **208,** 201–208.

Gee, J. C., Reivich, M., and Bajcsy, R. (1993). Elastically deforming 3D atlas to match anatomical brain images. *J. Comput. Assist. Tomogr.* **17** (2), 225–236.

Geschwind, N., and Levitsky, W. (1968). Human brain: Left-right asymmetries in temporal speech regions. *Science* **161,** 181–187.

Gilman, S., Junck, L., Markel, D. S., Koeppel, R. A., and Kluin, K. J. (1990). Cerebral glucose hypermetabolism in Friedreich's ataxia detected with positron emission tomography. *Ann. Neurol.* **28,** 750–757.

Grafman, J., Pascual- Leone, A., Alway, D., Nichelli, P., Gomez-Tortosa, E., and Hallett, M. (1994). Induction of a recall deficit by rapid-rate transcranial magnetic stimulation. *NeuroReport* **5,** 1157–1160.

Grafton, S. T., Martin, N. A., Mazziotta, J. C., Woods, R. P., Vinuela, F., and Phelps, M. E. (1994). Localization of motor areas adjacent to arteriovenous malformations. A positron emission tomographic study. *J Neuroimaging* **4** (2), 97–103.

Grinvald, A., Lieke, E., Frostig, R. D., Gilbert, C. D., and Wiesel, T. N. (1986). Functional architecture of cortex revealed by optical imaging of intrinsic signals. *Nature* **324,** 361–364.

Gruetter, R., Novotny, E. J., Boulware, S. D., Rothman, D. L., Mason, G. F., Shulman, G. I., *et al.* (1992). Direct measurement of brain glucose concentration in humans by ^{13}C NMR spectroscopy. *Proc. Natl. Acad. Sci. U.S.A.* **89,** 1109–1112.

Gzesh, D., Goldstein, S., and Sperling, M. R. (1990). Complex partial epilepsy: The role of neuroimaging in localizing a seizure focus for surgical intervention. *J. Nucl. Med.* **31,** 1839–1843.

Haglund, M. M., Ojemann, G. A., and Hochman, D. W. (1992). Optical

imaging of epileptiform and functional activity in human cerebral cortex. *Nature* **358** (6388), 668–671.

Hawkins, R. A., Mazziotta, J. C., and Phelps, M. E. (1987). Wilson's disease studied with FDG and positron emission tomography. *Neurology* **37,** 1707–1711.

Ho, S. S., Berkovic, S. F., Berlangieri, S. U., Newton, M. R., Egan, G. F., Tochon-Danguy, H. J., *et al.* (1995). Comparison of ictal SPECT and interictal PET in the presurgical evaluation of temporal lobe epilepsy. *Ann. Neurol.* **37,** 738–745.

Hopkins, A. L., and Barr, R. G. (1987). Oxygen-17 compounds as potential NMR T2 contrast agents: Enrichment effects of $H_2{}^{17}O$ on protein solutions and living tissues. *Magn. Reson. Med.* **4,** 399–403.

Huang, S. C., Yu, D. C., Barrio, J. R., *et al.* (1991). Kinetics and modeling of L-6-[18F]fluoro-dopa in human positron emission tomographic studies. *J. Cereb. Blood Flow Metab.* **11,** 898–913.

Hung, J. C., Corlija, M., Volkert, W. A., and Holmes, R. A. (1988). Kinetic analysis of technetium-99m *d,l*-HM-PAO decomposition in aqueous media. *J. Nucl. Med.* **29,** 1568–1576.

Hunter, K. E., Blaxton, T. A., Bookheimer, S. Y., Figlozzi, C., Gaillard, W. D., Grandin, C., *et al.* (1999). ^{15}O-Water positron emission tomography in language localization: A study comparing positron emission tomography visual and computerized region of interest analysis with the Wada test. *Ann. Neurol.* **45,** 662–665.

Ives, J., Warach, S., Schmitt, F., Edelman, R. R., and Schomer, D. L. (1993). Monitoring the patient's EEG during echo planar MRI. *Electroencephalogr. Clin. Neurophysiol.* **87,** 417–420.

Jennum, P., Friberg, L., Fuglsang-Frederiksen, A., and Dam, M. (1994). Speech localization using repetitive transcranial magnetic stimulation. *Neurology* **44,** 269–273.

Jones, T., Chesler, D. A., and Ter-Pogossian, M. M. (1976). The continuous inhalation of oxygen-15 for assessing regional oxygen extraction in the brain of man. *Br. J. Radiol.* **49,** 339–343.

Kaatee, R., Beek, F. J. A., de Lange, E. E., van Leeuwen, M. S., Smits, H. F., van der Ven, P. J., *et al.* (1997). Renal artery senosis: Detection and quantification with spiral CT angiography versus optimized digital subtraction angiography. *Radiology* **205,** 121–127.

Kanal, E., Shellock, F. G., and Lewin, J. S. (1996). Aneurysm clip testing for ferromagnetic properties: Clip variability issues. *Radiology* **200,** 576–578.

Klucznik, R. P., Carrier, D. A., Pyka, R., and Haid, R. W. (1993). Placement of a ferromagnetic intracerebral aneurysm clip in a magnetic field with a fatal outcome. *Radiology* **187,** 855–856.

Knowlton, R. C., Laxer, K. D., Ende, G., Hawkins, R. A., Wong, S. T. C., Matson, G. B., *et al.* (1997). Presurgical multimodality neuroimaging in electroencephalographic lateralized temporal lobe epilepsy. *Ann. Neurol.* **42,** 829–837.

Koepp, M. J., Labbe, C., Richardson, M. P., Brooks, D. J., Van Paesschen, W., Cunningham, V. J., *et al.* (1997). Regional hippocampal ^{11}C]flumazenil PET in temporal lobe epilepsy with unilateral and bilateral hippocampal sclerosis. *Brain* **120,** 1865–1876.

Kuhl, D. E., and Edwards, R. Q. (1964). Cylindrical and section radioisotope scanning of the liver and brain. *Radiology* **83,** 926–936.

Kuhl, D. E., Phelps, M. E., Markham, C. H., *et al.* (1982). Cerebral metabolism and atrophy in Huntington's disease determined by FDG and computed tomography scan. *Ann. Neurol.* **12,** 425–434.

Kwong, K. K., Belliveau, J. W., Chesler, D. A., Goldberg, I. E., Weisskoff, R. M., Poncelet, B. P., *et al.* (1992). Dynamic magnetic resonance imaging of human brain activity during primary sensory stimulation. *Proc. Natl. Acad. Sci. U.S.A.* **89,** 5675–5679.

Lauterbur, P. C. (1973). Image formation by induced local interactions: Examples employing nuclear magnetic resonance. *Nature* **242,** 190–191.

Le May, M., and Kido, D. (1978). Asymmetries of the cerebral hemispheres on computed tomograms. *J. Comput. Assist. Tomogr.* **2,** 471–476.

Lear, J. L. (1988). Quantitative local cerebral blood flow measurements with technetium-99m HM-PAO: Evaluation using multiple radionuclide digital quantitative autoradiography. *J. Nucl. Med.* **29,** 1387–1392.

Lutsep, H. L., Albers, G. W., DeCrespigny, A., Kamat, G. N., Marks, M. P., and Moseley, M. E. (1997). Clinical utility of diffusion-weighted magnetic resonance imaging in the assessment of ischemic stroke. *Ann. Neurol.* **41,** 574–580.

MacNamara, A. F., and Evans, P. A. (1995). The use of CT scanning by accident and emergency departments in the UK: Past, present and future. *Injury* **26,** 667–669.

Madar, I., Lesser, R. P., Krauss, G., Zubieta, J. K., Lever, J. R., Kinter, C. M., *et al.* (1997). Imaging of delta- and mu-opioid receptors in temporal lobe epilepsy by positron emission tomography. *Ann. Neurol.* **41,** 358–367.

Mansfield, P. (1977). Multiplanar image formation using NMR spin echoes. *J. Phys. C* **10,** L55–L58.

Mayberg, H., Sadzot, B., Meltzer, C. C., Fisher, R. S., Lesser, R. P., Dannals, R. F., *et al.* (1991). Quantification of mu and non-mu opiate receptors in temporal lobe epilepsy using positron emission tomography. *Ann. Neurol.* **30,** 3–11.

Mazziotta, J. C. (1989). Huntington's disease: Studies with structural imaging techniques and positron emission tomography. *Semin. Neurol.* **9,** 360–369.

Mazziotta, J. C. (1992). Movement disorders. *In* "Clinical Brain Imaging: Principles and Applications" (J. C. Mazziotta and S. Gilman, Eds.), pp. 244–293. Davis, Philadelphia.

Mazziotta, J. C., and Gilman, S. (1992). "Clinical Brain Imaging: Principles and Applications." Davis, Philadelphia.

Mazziotta, J. C., and Phelps, M. E. (1986). Positron emission tomography studies of the brain. *In* "Positron Emission Tomography and Autoradiography" (M. Phelps, J. Mazziotta, and H. Schelbert, Eds.), pp. 493–579. Raven Press, New York.

Mazziotta, J. C., Huang, S.-C., Phelps, M. E., Carson, R. E., MacDonald, N. S., and Mahoney, K. (1985). A noninvasive positron computed tomography technique using oxygen-15-labeled water for the evaluation of neurobehavioral task batteries. *J. Cereb. Blood Flow Metab.* **5,** 70–78.

Mazziotta, J. C., Phelps, M. E., Huang, S.-C., Baxter, L. R., Riege, W. H., Hoffman, J. M., *et al.* (1987). Reduced cerebral glucose metabolism in asymptomatic subjects at risk for Huntington's disease. *N. Engl. J. Med.* **316,** 357–362.

Mazziotta, J. C., Frackowiak, R. S. J., and Phelps, M. E. (1992). The use of positron emission tomography in the clinical assessment of dementia. *Semin. Nucl. Med.* **22,** 233–246.

Mazziotta, J. C., Toga, A. W., Evans, A., Fox, P., and Lancaster, J. (1995). A probabilistic atlas of the human brain: Theory and rationale for its development. *NeuroImage* **2,** 89–101.

Merton, P. A., and Morton, H. B. (1980). Stimulation of the cerebral cortex in the intact human subject. *Nature* **285,** 227.

Mills, K. R. (1991). Magnetic brain stimulation: A tool to explore the action of the motor cortex on single human spinal motoneurones. *Trends Neurosci.* **14,** 401–405.

Mintun, M. A., Raichle, M. E., Kilbourn, M. R., Wooten, G. F., and Welch, M. J. (1984). A quantitative model for the *in vivo* assessment of drug binding sites with positron emission tomography. *Ann. Neurol.* **15,** 212–227.

Moran, S. G., McCarthy, M. C., Uddin, D. E., and Poelstra, R. J. (1994). Predictors of positive CT scans in the trauma patient with minor head injury. *Am. Surg.* **60,** 533–535.

Moseley, M., Kucharczyk, J., Mintorovitch, J., Cohen, Y., Kurhanewicz, J., Derugin, N., Asgari, H., *et al.* (1990). Diffusion-weighted MR imaging of acute stroke: Correlation of T2 weighted and magnetic susceptibility-enhanced MR imaging in cats. *Am. J. Neuroradiol.* **11,** 423–429.

Nakano, S., Kinoshita, K., Jinnouchi, S., and Hoshi, H. (1989). Comparative study of regional cerebral blood flow images by SPECT using xenon-133, iodine-123 IMP, and technetium-99m HM-PAO. *J. Nucl. Med.* **30,** 157–164.

Newton, M. R., Berkovic, S. F., Austin, M. C., Rowe, C. C., McKay, W. J., and Bladin, P. F. (1992). Postictal switch in blood flow distribution and temporal lobe seizures. *J. Neurol. Neurosurg. Psychiatry* **55,** 891–894.

Newton, T. H., and Potts, D. G., Eds. (1981). "Radiology of the Skull and Brain: Technical Aspects of Computed Tomography," Vol. 5. Mosby, St. Louis.

O'Brien, T. J., So, E. L., Mullan, B. P., Hauser, M. F., Brinkmann, B. H., Bohnen, N. I., *et al.* (1998). Subtraction ictal SPECT co-registered to MRI improves clinical usefulness of SPECT in localizing the surgical seizure focus. *Neurology* **50,** 445–454.

Obrist, W. D., Thompson, H. K., King, C. H., and Wang, H. S. (1967). Determination of regional cerebral blood flow by inhalation of 133-xenon. *Circ. Res.* **20** (1), 124–135.

Obrist, W. D., Thompson, H. K., Jr., Wang, H. S., and Wilkinson, W. E. (1975). Regional cerebral blood flow estimated by 133-xenon inhalation. *Stroke* **6** (3), 245–256.

Ogawa, S., Lee, T. M., Nayak, A. S., and Glynn, P. (1990). Oxygenation-sensitive contrast in magnetic-resonance image of rodent brain at high magnetic fields. *Magn. Reson. Med.* **14,** 68–78.

Ogawa, S., Menon, R. S., Tank, D. W., Kim, S. G., Merkle, H., Ellermann, J. M., *et al.* (1993). Functional brain mapping by blood oxygenation level-dependent contrast magnetic-resonance-imaging — A comparison of signal characteristics with a biophysical model. *Biophys. J.* **64,** 803–812.

Oshima, M., Tadokoro, R., Makino, N., and Sakuma, S. (1990). Role of fast data acquisition method with ^{99m}Tc-HMPAO brain SPECT in patients with acute stroke. *Clin. Nucl. Med.* **15,** 172–174.

Pascual-Leone, A., Valls-Sole, J., Wassermann, E. M., Brasil-Neto, J., Cohen, L. G., and Hallett, M. (1992). Effects of focal transcranial magnetic stimulation on simple reaction time to acoustic, visual and somatosensory stimuli. *Brain* **115,** 1045–1059.

Pascual-Leone, A., Cohen, L. G., Brasil-Neto, J. P., and Hallet, M. (1994). Non-invasive differentiation of motor cortical representation of hand muscle by mapping of optimal current directions. *Electroencephalogr. Clin. Neurophysiol.* **93,** 42–48.

Pascual-Leone, A., Rubio, B., Pallardo, F., and Catala, M.-D. (1996). Beneficial effect of rapid-rate transcranial magnetic stimulation of the left dorsolateral prefrontal cortex in drug-resistant depression. *Lancet* **348,** 233–238.

Patton, J. A., Brill, A. B., Erickson, J. J., *et al.* (1969). A new approach to the mapping of three-dimensional radionuclide distributions. *J. Nucl. Med.* **10,** 363.

Paus, T., Jech, R., Thompson, C. J., Comeau, R., and Evans, A. C. (1997). Transcranial magnetic stimulation during positron emission tomography: A new method for studying connectivity of the human cerebral cortex. *J. Neurosci.* **17,** 3178–3184.

Pekar, J., Sinnwell, T. M., Ligeti, L., *et al.* (1993). Double-label tracer experiments using multinuclear MRI: Mapping cerebral oxygen consumption and blood flow using ^{17}O and ^{19}F MRI. *Proc. 12th SMRM Annu. Mtg.,* New York, p. 1388.

Pelizzari, C. A., Chen, G. T., Spelbring, D. R., Weichselbaum, R. R., and Chen, C. T. (1989). Accurate three-dimensional registration of CT, PET, and/or MR images of the brain. *J. Comput. Assist. Tomogr.* **13** (1), 20–26.

Penfield, W., and Rasmussen, T. (1950). "The Cerebral Cortex of Man: A Clinical Study of Localisation of Function." Macmillan, New York.

Perlmutter, J. S., Larson, K. B., Raichle, M. E., Markham, J., Mintun, M. A., Kilbourn, M. R., *et al.* (1986). Strategies for *in vivo* measurement of receptor binding using positron emission tomography. *J. Cereb. Blood Flow Metab.* **6,** 154–169.

Phelps, M. E., and Mazziotta, J. C. (1985). Positron emission tomography: Human brain function and biochemistry. *Science* **228,** 799–809.

Phelps, M. E., Huang, S. C., Hoffman, E. J., Selin, C., Sokoloff, L., and Kuhl, D. E. (1979). Tomographic measurement of local cerebral metabolic rate in humans with (F-18)2-fluoro-2-deoxy-D-glucose: Validation of method. *Ann. Neurol.* **6,** 371–388.

Prichard, J. W., and Shulman, R. G. (1986). NMR spectroscopy of brain metabolism *in vivo. Annu. Rev. Neurosci.* **9,** 61–85.

Prichard, J. W., and Rosen, B. R. (1994). Functional study of the brain by NMR. *J. Cereb. Blood Flow Metab.* **14** (3), 365–372.

Prichard, J. W., Rothman, D., Novotny, E., Petroff, O., Kuwabara, T., and Avison, M. (1991). Lactate rise detected by ^{1}H NMR in human visual cortex during physiological stimulation. *Proc. Natl. Acad. Sci. U.S.A.* **88,** 5829–5831.

Raichle, M. E. (1983). Positron emission tomography. *Annu. Rev. Neurosci.* **6,** 249–268.

Raichle, M. E., Martin, W. R. W., Herscovitch, P., Mintun, M. A., and Markham, J. (1983). Brain blood flow measured with intravenous H$_2$^{15}O. II. Implementation and validation. *J. Nucl. Med.* **24,** 790–798.

Rattner, Z., Smith, E. O., Woods, S., Dey, H., and Hoffer, P. B. (1991). Toward absolute quantitation of cerebral blood flow using technetium-99m-HMPAO and a single scan. *J. Nucl. Med.* **32** (8), 1506–1507.

Reivich, M., Kuhl, D. E., Wolf, A., *et al.* (1979). The (^{18}F)fluorodeoxyglucose method for the measurement of local cerebral glucose utilization in man. *Circ. Res.* **44,** 127–137.

Rowe, C. C., Berkovic, S. F., Austin, M. C., McKay, W. J., and Bladin, P. F. (1991a). Patterns of postictal cerebral blood flow in temporal lobe epilepsy: Qualitative and quantitative analysis. *Neurology* **41,** 1096–1103.

Rowe, C. C., Berkovic, S. F., Austin, M. C., Saling, M., Kalnins, R. M., McKay, W. J., *et al.* (1991b). Visual and quantitative analysis of interictal SPECT with technetium-99m-HMPAO in temporal lobe epilepsy. *J. Nucl. Med.* **32** (9), 1688–1694.

Salcman, M., and Pevsner, P. H. (1992). Value of MRI in head injury: Comparison with CT. *Neurochirurgie* **38,** 329–332.

Sappey-Marinier, D., Calabrese, G., Fein, G., Hugg, J. W., Biggins, C., and Weiner, M. W. (1992). Effect of photic stimulation on human visual cortex lactate and phosphates using ^{1}H and ^{31}P magnetic resonance spectroscopy. *J. Cereb. Blood Flow Metab.* **12,** 584–592.

Savic, I., Roland, P., Sedrall, G., Persson, A., Pauli, S., and Widen, L. (1988). *In-vivo* demonstration of reduced benzodiazepine receptor binding in human epileptic foci. *Lancet* **ii,** 864–866.

Shellock, F. (1997). "Pocket Guide to MR Procedures and Metallic Objects: Update 1997." Lippincott-Raven, Philadelphia.

Sokoloff, L., Reivich, M., Kennedy, C., *et al.* (1977). The (^{14}C)deoxyglucose method for the measurement of local cerebral glucose utilization: Theory, procedure and normal values in the conscious and anesthetized albino rat. *J. Neurochem.* **28,** 897–916.

Sorensen, A. G., Buonanno, F. S., Gonzalez, R. G., Schwamm, L. H., Lev, M. H., Huang-Hellinger, F. R., *et al.* (1996). Hyperacute stroke: Evaluation with combined multisection diffusion-weighted and hemodynamically weighted echo-planar MR imaging. *Radiology* **199,** 391–401.

Spencer, S. S. (1994). The relative contributions of MRI, SPECT and PET imaging in epilepsy. *Epilepsia* **35,** S72–S89.

Talairach, J., and Tournoux, P. (1988). Principe et technique des etudes anatomiques. *In* "Co-Planar Stereotaxic Atlas of the Human Brain—Three-Dimensional Proportional System: An Approach to Cerebral Imaging" (M. Rayport, Ed.), Vol. 3. Thieme, New York.

Theodore, W. H. (1992). MRI, PET, SPECT: Interrelations, technical limits, and unanswered questions. *In* "Surgical Treatment of Epilepsy" (W. H. Theodore, Ed.), pp. 127–134. Elsevier, Amsterdam/New York.

Toga, A., and Mazziotta, J. (1996). "Brain Mapping: The Methods." Academic Press, San Diego.

Toga, A. W., Cannestra, A. F., and Black, K. L. (1995). The temporal/spatial evolution of optical signals in the human cortex. *Cereb. Cortex* **5** (6), 561–565.

Turkington, T. G., Hoffman, J. M., Jaszczak, R. J., MacFall, J. R., Harris, C. C., Kilts, C. D., *et al.* (1995). Accuracy of surface fit registration for PET and MR brain images using full and incomplete brain surfaces. *J. Comput. Assist. Tomogr.* **19** (1), 117–124.

Turner, R., Jezzard, P., Wen, H., Kwong, K. K., LeBihan, D., Zeffiro, T., and Balaban, R. S. (1993). Functional mapping of the human visual-cortex at 4 and 1.5 Tesla using deoxygenation contrast EPI. *Magn. Reson. Med.* **29,** 277–279.

van Huffelen, A. C., van Isselt, J. W., van Veelen, C. W. M., van Rijk, P. P., van Bentum, A. M. E., Dive, D., *et al.* (1990). Identification of the side of the epileptic focus with ^{123}I-iomazenil SPECT. *Acta Neurochirurg. Suppl.* **50,** 95–99.

van Zijl, P. C. M., Chesnick, A. S., DesPres, D., Moonen, C. T., Ruiz-Cabello, J., and van Gelderen, P. (1993). *in vivo* proton spectroscopy and spectroscopic imaging of [1-^{13}C]-glucose and its metabolic products. *Magn. Reson. Med.* **30** (5), 544–551.

Warach, S., Li, W., Ronthal, M., and Edelman, R. R. (1992). Acute cerebral ischemia: Evaluation with dynamic contrast-enhanced MR imaging and MR angiography. *Radiology* **182,** 41–47.

Warach, S., Gaa, J., Siewert, B., Wielopolski, P., and Edelman, R. R. (1995a). Acute human stroke studied by whole brain echo planar diffusion-weighted magnetic resonance imaging. *Ann. Neurol.* **37,** 231–241.

Warach, S., Ives, J., Darby, D., Thangaraj, V., Edelman, R., and Schomer, D. (1995b). EEG triggered echo planar functional MRI: Localizing activation related to EEG discharges. *Hum. Brain Mapp. Suppl. 1,* 420.

Wasserman, E. M. (1998). Risk and safety of repetitive transcranial magnetic stimulation: Report and recommendations from the International Workshop on the Safety of Repetitive Transcranial Magnetic Stimulation, June 5–7, 1996. *Electroencephalogr. Clin. Neurophysiol.* **108,** 1–16.

Wasserman, E. M., McShane, L. M., Hallett, M., and Cohen, L. G. (1992). Noninvasive mapping of muscle representations in human motor cortex. *Electroencephalogr. Clin. Neurophysiol.* **85,** 1–8.

Wasserman, E. M., Pascua-Leone, A., and Hallett, M. (1994). Cortical motor representation of the ipsilateral hand and arm. *Exp. Brain Res.* **100,** 121–132.

Weinberger, D., Luchins, P., Morihisa, J., and Wyatt, R. (1982). Asymmetrical volumes of the right and left frontal and occipital regions of the human brain. *Neurology* **11,** 97–102.

Wilcock, D., Jaspan, T., Holland, I., Cherryman, G., and Worthington, B. (1996). Comparison of magnetic resonance angiography with conventional angiography in the detection of intracranial aneurysms in patients presenting with subarachnoid haemorrhage. *Clin. Radiol.* **51,** 330–334.

Williams, A. L., and Haughton, V. M., Eds. (1985). "Cranial Computed Tomography: A Comprehensive Text." Mosby, St. Louis.

Witelson, S. (1977). Anatomical asymmetry in the temporal lobes: Its documentation, phylogenesis and relationship to functional asymmetry. *Ann. N.Y. Acad. Sci.* **299,** 328–354.

Wong, D. F., Gjedde, A., and Wagner, H. N. J. (1986). Quantification of neuroreceptors in the living human brain. I. Irreversible binding of ligands. *J. Cereb. Blood Flow Metab.* **6,** 137–146.

Woods, R. P., Cherry, S. R., and Mazziotta, J. C. (1992). Rapid automated algorithm for aligning and reslicing PET images. *J. Comput. Assist. Tomogr.* **16** (4), 620–633.

Woods, R. P., Mazziotta, J. C., and Cherry, S. R. (1993). PET registration with automated algorithm. *J. Comput. Assist. Tomogr.* **17** (4), 536–546.

Zhang, J., Levesque, M. F., Wilson, C. L., Harper, R. M., Engel, J., Jr., Lufkin, R., *et al.* (1990). Multimodality imaging of brain structures for stereotactic surgery. *Radiology* **175** (2), 435–441.

4

Intraoperative Brain Mapping

Arthur W. Toga,[*,1] George A. Ojemann,[†] Jeffrey G. Ojemann,[‡] and Andrew F. Cannestra[*]

Laboratory of Neuro Imaging, Division of Brain Mapping, Department of Neurology, UCLA School of Medicine, Los Angeles, California 90095
†Department of Neurological Surgery, University of Washington, Seattle, Washington 98195
‡Department of Neurological Surgery and NeuroImaging Laboratory, Washington University School of Medicine, St. Louis, Missouri 63110

I. Introduction
II. Electrical Stimulation Mapping (ESM)
III. Relation between ESM and Functional Imaging
IV. Electrophysiological Correlates of Cognition Derived Intraoperatively
V. Optical Imaging Intraoperatively
VI. Intraoperative Structural Maps
VII. Summary
References

I. Introduction

A variety of specialized techniques have been developed to identify functionally important brain areas during neurosurgical operations. Some of these methods provide perspectives on functional localization similar to those derived with techniques used outside the operating room: intraoperative electrical stimulation mapping identifies those brain areas whose function is necessary for a particular behavior at a point in time by interrupting local cerebral activity and interfering with the behavior. Behavioral changes after brain lesions and intraarterial amobarbital perfusion provide similar data in the extraoperative setting. Intraoperative optical imaging of "intrinsic signals" identifies brain areas with altered vascular and metabolic activity during a behavior, physiological alterations that are thought to reflect changes in neuronal activity. Imaging of blood flow and oxygen extraction changes by PET or fMRI provide somewhat similar data in the extraoperative setting. The intraoperative setting also provides opportunities for the direct recording of physiologic correlates of behavior in the electrocorticogram (ECoG) or of neuronal activity directly with microelectrodes. In general, intraoperative techniques provide greater local anatomic resolution than the analogous extraoperative techniques. However, such measurements are made over a more restricted area of brain, in special patient populations that come to neurosurgical procedures, and with the time constraints, nature and level of anesthesia (local or general), and other requirements of the operating room environment.

Many of these techniques were developed to reduce the morbidity of neurosurgical operations by identifying those areas where resections were likely to

---

[1] To whom correspondence should be addressed.

lead to functional deficits, areas that can then be avoided by the surgeon. The value of a technique in providing this information is thus an important clinical consideration, and is not necessarily the same for all methods. There is also considerable interest in relating extraoperative imaging to the intraoperative setting. This includes both methods for projecting preoperative structural imaging onto the surgical field and methods for relating preoperative functional imaging. However, it does not necessarily follow that sites with changes in blood flow or oxygen extraction on preoperative PET or fMRI will identify sites that the surgeon must spare to avoid a functional deficit. That relation must be established empirically for each technique. This chapter reviews the specialized techniques and their application for functional localization used in the operating room, methods for relating preoperative imaging to the surgical field, and the relation between them.

II. Electrical Stimulation Mapping (ESM)

The use of electrical stimulation to localize motor function in cortex dates back to the latter part of the nineteenth century (Fritsch and Hitzig, 1870). The technique was soon applied to human cortex (Bartholow, 1874) and by the early twentieth century, ESM was in regular use to identify sensory-motor cortex (Cushing, 1909; Foerster, 1931; Penfield and Boldrey, 1937). The technique depends on the observation that application of a relatively small electrical current to the surface of primary cortices (or to their subcortical pathways) evokes a response: localized movement from motor cortex, tingling dysesthesias from sensory cortex, phosphenes from calcerine visual cortex, and, rarely, a roaring noise from primary auditory cortex. Application of similar currents to the remaining association cortex usually does not evoke any response in the absence of stimulation-evoked epileptiform activity. The interpretive and experiential responses reported by Penfield after stimulation of temporal association cortex in patients with epilepsy (Penfield and Perot, 1963) are now thought to be associated with epileptiform activity (Gloor et al., 1982) and most likely represent partial seizures. However, Penfield noted that when applied to association cortex in a patient who was engaged in an ongoing language task, larger currents would disrupt that task, even though the same current had no observable evoked effects in the nonbehaving patient (Penfield and Roberts, 1959).

Thus, ESM utilizes two effects, the evoked responses at relatively small currents for identification of primary cortices and their subcortical pathways and, in the remainder of cortex, disruption of function with larger currents. Although the precise physiologic events responsible for these behavioral effects of stimulation are not known, it is likely that the smaller currents predominantly excite neurons and en passage fibers, whereas the larger currents block local function by depolarization. Most stimulation is with trains of pulses, commonly at frequencies of 30–60 Hz (with higher frequencies sometimes used for stimulation of thalamus or pallidum). Pulses are usually biphasic, to minimize tissue injury. Pulse duration is commonly 0.3–1 ms. Spatial resolution of stimulation depends on whether monopolar or bipolar electrodes are used, electrode size and spacing, and current level. With bipolar stimulation at currents below the level for afterdischarge, stimulation-evoked changes as assessed by optical imaging are confined to the area between the electrodes (Haglund et al., 1992, 1993a). Behaviorally, too, intraoperative stimulation effects are often quite localized, with electrode movements of a few millimeters altering responses. The safety of ESM is well established. Current levels that do not induce histologic changes in cortex of experimental animals have been determined (Babb et al., 1977). Extensive, repeated stimulation of human cortex in the current ranges used for mapping has been shown not to produce histologic changes (Gordon et al., 1990). Function returns to prestimulation levels almost immediately after removal of the current.

A. ESM Identification of Rolandic Cortex

Motor movements can be evoked from precentral rolandic cortex under local or general anesthesia in most adults. Under local anesthesia these effects usually require currents in the 2- to 4-mA range (between peaks of the biphasic pulses). They commonly have the classical homuncular arrangement, with face lateral inferior and leg mediosuperior (Penfield and Rasmussen, 1950). Sensory responses are evoked from the postcentral gyrus, often at a slightly lower threshold and usually in a similar homuncular pattern, though shifted slightly inferiorly compared to motor responses. In awake patients, both motor and sensory responses may be confined to individual digits and the large representation of the tongue and mouth easily identified, although the detailed representation of somatosensory cortex reported in primates (Kaas et al., 1979) is not usually evident in humans.

Motor responses can also be identified under light nonparalytic general anesthesia, but less reliably, at much larger currents (4–20 mA) and with a much less precise pattern. The tongue representation is also difficult to identify under general anesthesia. The subcortical motor pathways can also be reliably identified by stimulation under local or general anesthesia, using the threshold current for evoking movement from cortex. Although a reliable technique for identifying rolandic cortex in adults, adolescents, and older children, in younger children, under about age 5, motor responses to electrical stimulation are difficult to evoke, requiring long pulses at high currents (Jayakar *et al.*, 1992). Usually electrical stimulation mapping of rolandic cortex involves applying a small current (2 mA in the awake patient under local anesthesia, 4 mA under general anesthesia) and increasing the current at 2-mA increments until a visualized motor response, or in the awake patient, a sensory response, is evoked. Rarely, motor cortex stimulation will not elicit movement, but block volutional movements (Luders *et al.*, 1987). EMG motor responses after single pulses have also been used to identify motor cortex (Sutherling *et al.*, 1988). Sites of evoked motor or sensory responses are classically identified by sterile number tickets placed on the cortical surface, with their location recorded photographically.

Rolandic cortex has also been identified intraoperatively by somatosensory-evoked potentials (SSEPs) recorded in the ECoG. SSEP to median nerve stimulation is most often used to identify the central sulcus (Fig.

1). Motor cortex is identified by a maximal-amplitude negative peak. Phase reversal of the early negative (N_{2O}) peak occurs across the central sulcus, where the negative end of the dipole is on the somatosensory cortex (Nuwer *et al.*, 1992). However, this phase reversal may occur one sulcus away from the central sulcus as identified by stimulation mapping (Dinner *et al.*, 1986). Both short- and long-latency SSEPs are generated from somatosensory areas 1 and 3b, lying over the crest and within the central sulcus, respectively (Allison *et al.*, 1989). Motor cortex has not been implicated in these responses. However, increases in precentral gyrus root mean square voltage have been observed in the 40- to 108-ms latency range. These long latency changes may reflect activity in area 4 (Goldring *et al.*, 1970; Allison *et al.*, 1989). SSEPs are also dependent on anesthetic levels.

B. ESM Identification of Language

Intraoperative mapping of sites associated with disruption of language during electrical stimulation requires an awake patient under local anesthesia. The use of propofol intravenous anesthesia during the cran-

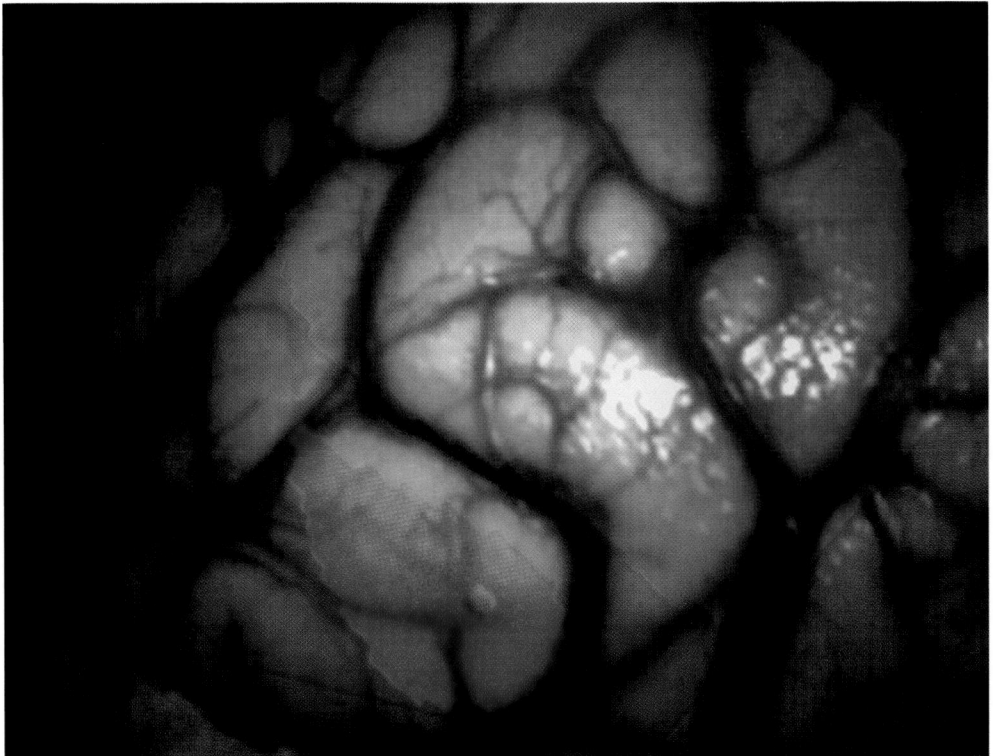

Figure 1 Somatosensory-evoked potentials are used to identify functionally active regions. Integrating the electrical potential results in a map that can be overlaid on preoperative cortical surface maps of the patient. These maps have significant utility when used to identify regions such as sensory versus motor strip and can help identify regions of the homunculus during surgery.

iotomy, with a block of dural sensation by intradural injection of local anesthetic adjacent to middle meningeal branches, has greatly facilitated this procedure. The propofol intravenous anesthesia rapidly reverses and the patient is then awake with all-pain sensitive structures blocked by the local anesthesia, for the surface of human cortex has no pain (or other sensory) receptors. With stimulation mapping of language, the patient engages in a language measure that involves multiple equivalent items. During every second or third item of this task, a site on the exposed cortical surface is stimulated, applying the current when the item is presented and maintaining it until a response is given or the next item presented. Items are commonly presented at 3- to 4-s intervals. Subsequent stimulation at other cortical sites continues until all sites on the exposed cortical surface have been stimulated (usually at least three times). The current used for stimulation is set immediately before mapping at the largest current that does not evoke after discharge in the ECoG recorded from the exposed area of cortex. Other stimulation parameters are similar to those used in motor mapping, biphasic pulses of 0.3- to 1-ms duration in trains of 30–60 Hz. Sites stimulated are identified by numbered tickets and the locations recorded photographically (Fig. 2). A site is related to

the measured function when performance is repeatedly disrupted during stimulation of that site. Performance in the absence of stimulation is used to ensure that any stimulation effects are unlikely to be random events. To ensure this with the relatively few samples of stimulation effect at each site, the language measure should be one where the patient makes few errors in the absence of stimulation. This becomes an issue in using stimulation mapping in the presence of a preexisting aphasia.

Penfield's initial studies with intraoperative stimulation mapping of language utilized object naming as the language measure (Penfield and Roberts, 1959). Object naming is a useful task for identifying language cortex, as it is the only language behavior that is disturbed in all aphasic syndromes. Penfield interpreted his stimulation mapping findings during naming as identifying the classical Broca and Wernicke areas in the dominant hemisphere, with performance also disrupted from stimulation of dominant supplementary motor area and face rolandic area of either hemisphere. Subsequent studies have shown that, in an individual patient, the sites where language performance is disturbed by dominant hemispheral stimulation are often localized to several very focal areas of 1–2 cm^2 in extent, usually with at least one posterior inferior frontal and one or two tem-

Figure 2 The exposed cortical surface is surveyed for functional representation. To keep track of which sites produce responses, numbered sterile paper labels are used to create a kind of functional map directly on the brain of the patient.

poral sites (Fig. 3) (Ojemann, 1994). Often these sites have sharp boundaries.

These focal sites are in somewhat different perisylvian locations in different patients, however, and are not always in the classical Broca or Wernicke locations, as illustrated in Fig. 4 (Ojemann *et al.,* 1989a). Remapping the same patient after periods of as long as 20 years has shown the sites to be in the same location (G. Grant, G. Ojemann, and E. Lettich, unpublished data). Focal sites related to language as measured by object naming have been identified in children as young as 4 (using extraoperative mapping techniques) and in adults as old as 80. Particular patterns of localization have been related to gender and preoperative verbal ability of these patients (Ojemann *et al.,* 1989a). The relationship between extent of a temporal resection close to temporal sites where stimulation disturbed object naming and postoperative language disturbances has been investigated in two studies, both showing that if a resection avoids these sites by a margin of 10–20 mm along a continuous gyrus, a postoperative language deficit was unlikely, whereas with closer resections, deficits occurred (Ojemann, 1983a; Haglund *et al.,* 1994a). These findings form the basis for considering intraoperative stimulation mapping of language in planning cortical resections: the

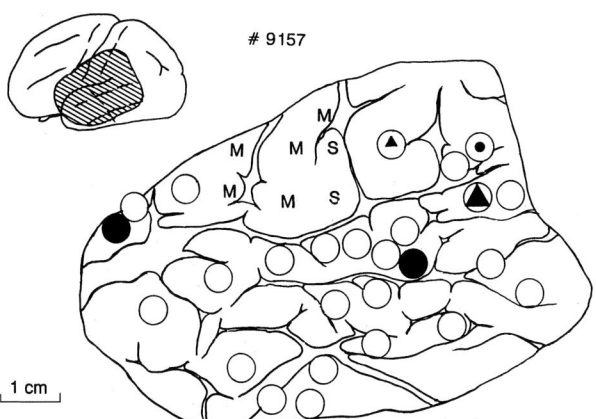

Figure 3 Location of sites in left inferior frontal temporal interior parietal cortex where affects of ESM on naming were assessed in English and Spanish in a 44-year-old bilingual ambidextrous male, left-dominant by intracarotid amobarbital assessment. Each circle is a site of stimulation, using 1-ms biphasic pulses at 60 Hz, 3 mA delivered through bipolar electrodes separated by 5 mm. No errors evoked at open circles. Repeated errors in English evoked at filled circles, with error on one of three items in English at site with small circle. Spanish naming was intact with stimulation of these sites. Repeated errors in Spanish evoked at site with large triangle, with error on one of three items in Spanish at site with small triangle. No errors in English evoked at these sites. Error rate in the absence of stimulation: 1.2% for English, 4.8% for Spanish. M and S identify motor and sensory responses to ESM. Note that repeated naming errors with ESM are localized to very focal areas that are separate for the two languages. From Ojemann (1994).

focal nature of areas where stimulation disrupts language in an individual patient, the variation in the location of these sites across the patient population, so that no anatomic landmark reliably indicates where these areas are located, and the evidence that these sites indicate what must be spared in a resection to avoid a deficit. Surface cortical stimulation predicts the effects of resections that include buried cortex and underlying white matter. This suggests that crucial sites for language are seldom only located in buried cortex away from surface sites. ESM of cortex buried in sulci has rarely shown disruption of language, and then only as an extension into the sulcus of surface sites (Ojemann *et al.,* 1989a; Ojemann, 1991b).

When several different language measures are assessed during ESM of an individual patient, the cortical sites where the different functions are altered are often somewhat separate. This separation has been shown for naming the same items in two different languages (illustrated in Fig. 3; also Ojemann and Whitaker, 1978; Ojemann, 1983a; Black and Ronner, 1987), for naming in sign compared to oral languages (Mateer *et al.,* 1982; Haglund *et al.,* 1993b), for naming compared to sentence reading (Ojemann, 1989), for object naming compared to verb generations (J. Ojemann *et al.,* 1993), and for recent memory for names (Ojemann, 1978; Ojemann and Dodrill, 1985; Perrine *et al.,* 1994). Resection of temporal cortical sites related to recent memory for names identified by ESM may contribute to memory deficits after dominant temporal lobectomy, with some evidence that preservation of these sites after their identification by ESM may reduce that risk (Ojemann and Dodrill, 1985, 1987).

ESM has also been used to investigate the relation of sequences of motor speech gestures to the detection of speech sounds, using one of the advantages of stimulation, that the "lesion" can be turned on and off and thus occurs only during part of the language task, in this case only when the speech sound was heard. That study found considerable overlap in the sites in dominant perisylvian cortex where sequences of motor speech gestures and detection of speech sounds were altered by stimulation (Ojemann and Mateer, 1979; Ojemann, 1983b).

Thus, stimulation mapping findings suggest overlap of areas important to speech production and perception, rather than the separation suggested by the classical Broca–Wernicke model. These stimulation mapping findings were the basis for a model of the organization of dominant-hemisphere lateral cortex, with the common area for language production and perception in posterior, frontal-superior, temporal-anterior, and inferior parietal perisylvian cortex, surrounded by areas related to recent verbal memory, with specialized sites

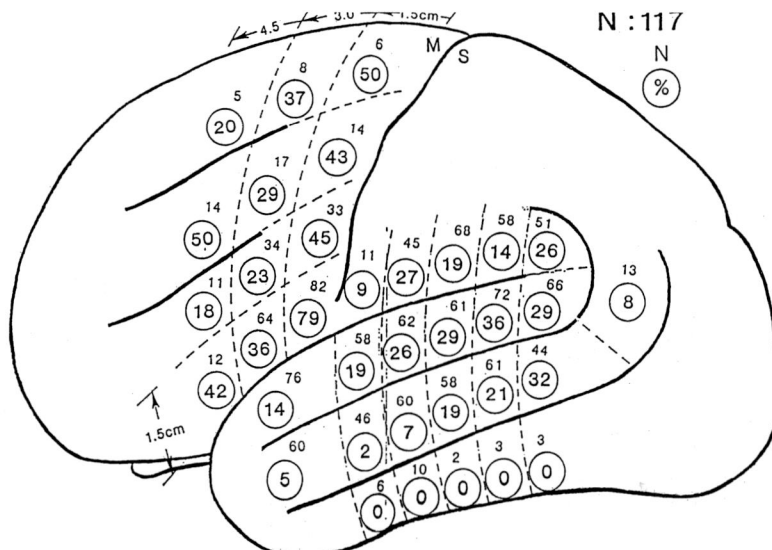

Figure 4 Variability in localization of sites where ESM significantly altered English naming in 117 patients, all left-dominant for language. Individual maps aligned to rolandic cortex and end of the sylvian fissure. Cortex divided into zones by dashed lines. Upper number in each zone is the number of patients with at least one site of ESM in that zone. Lower number is percentage of those patients with sites of significant evoked naming errors. M and S identify rolandic cortex. From Ojemann *et al.* (1989a).

specific to naming and syntax located between those functional areas (Ojemann, 1983a,b, 1991a). Stimulation mapping has also been used to investigate localization of visuospatial functions in the nondominant hemisphere, including memory for complex figures and identification of facial emotional expressions (Fried *et al.*, 1982).

Although perisylvian sites related to language by stimulation mapping identify areas that must be spared to avoid a postoperative language deficit, electrical stimulation can also evoke alterations in language from several dominant-hemisphere sites that can be resected with little risk of a persisting deficit. These include the supplementary motor area of posterior superior frontal lobe (Penfield and Roberts, 1959) and the "basal temporal" language area in the fusiform gyrus (Luders *et al.*, 1986).

Cortical ESM during language, memory, and motor functions has also been performed through chronic subdural grid electrodes. The relative merits of intraoperative ESM and extraoperative ESM through these chronic electrode arrays are discussed elsewhere (G. A. Ojemann *et al.*, 1993). Use of these chronic electrodes provides a technique to use ESM to assess language in young children and adults who cannot cooperate with an awake craniotomy. In general, the findings with ESM through these chronic electrodes have been similar to those observed with intraoperative ESM, with interfer-

ence with a specific behavior from localized sites, different sites for different behaviors, and variation between subjects in the location of these sites.

ESM during neurosurgical procedures for the treatment of dyskinesias has provided an opportunity to investigate localization of language and memory in the human thalamus (reviewed in Johnson and Ojemann, 2000). Stimulation of anterior lateral thalamus of the dominant hemisphere during the encoding phase of memory increases the likelihood of later correctly retrieving that material (Ojemann *et al.*, 1971). Stimulation of the same sites at the same currents at the time of retrieval decreases accuracy of retrieval. These effects are evident for verbal material with stimulation of the lateral thalamus in the dominant hemisphere and for visuospatial material in the nondominant hemisphere (Ojemann, 1979). They were attributed to activation of thalamocortical-specific attentional systems. Stimulation of the lateral thalamus of the dominant hemisphere also alters naming but in rather complex ways. Stimulation of the anterior lateral thalamus evokes repeated use of the same wrong name, often the last item correctly named at a subthreshold current. This was interpreted as reflecting the same effect in language as was seen with thalamic stimulation during the memory measure. More posteriomedially in the lateral thalamus, stimulation evoked perseveration on correct object names and slowed naming, reflecting the thalamic motor role (Ojemann and Ward, 1971; Mateer, 1978). Posterior to this area, in anterior lateral pulvinar, evoked naming changes were similar to those seen with cortical stimulation (Ojemann et al., 1968). Although studied in less detail, chronic stimulation of the globus pallidum has been associated with decreased verbal fluency and visuoconstruction abilities (Troster et al., 1997). A recent fMRI study of overt and silent word retrieval found left thalamic (as well as cortical) activation during both tasks and putamenal activation that was more pronounced for the overt task, especially in the nondominant hemisphere (Rosen et al., 2000).

III. Relation between ESM and Functional Imaging

The correlation between ESM localization of cortical function and functional imaging has been of longstanding interest. Initial correlations have focused on sensorimotor function and localization of the central (rolandic) sulcus. Using PET, Fox and colleagues (1987) advocated localizing sensorimotor activation in neurosurgical patients preoperatively using strong vibrotactile stimuli. These activations accurately localized the central sulcus, although precise determination of motor and sensory gyri was not possible. fMRI activations have also been correlated to ESM in localizing sensory cortex (Fried et al., 1995; Jack et al., 1994) and visual cortices (Fried et al., 1995). fMRI studies, using lighter sensory stimuli, have also correlated well with ESM and SSEP (Puce et al., 1995). These methods have been used successfully in patients with cerebral tumors (Mueller et al., 1996), although other evidence suggests that intrinsic tumors may alter regional activation (J. Ojemann et al., 1998). The precision of fMRI remains only partially predictive of ESM localization, with disagreement up to 20 mm reported in a large case series (Yetkin et al., 1997). Surgical resections that remain greater than 2.0 cm from fMRI activations were not associated with motor deficits postoperatively (Mueller et al., 1996). The use of fMRI for sensorimotor mapping is becoming increasingly common in the clinical setting.

This same correlation for language sites has proven to be more difficult. Localization of number counting in frontal sites has been reported to be accurate but with discrepencies of up to 2.0 cm (Yetkin et al., 1997). For more complex language tasks, several factors must be considered. In normal subjects, different tasks will activate different regions within the posterior temporal lobe (Fiez et al., 1996). However, object naming in normal subjects activated only inferior temporal (basal language) regions and inferior frontal sites (Bookheimer et al., 1995), with no activation in lateral temporal cortex,

where stimulation errors during object naming are routinely found (Ojemann et al., 1989a). Even during attempts simply to lateralize language function, object naming was considerably less reliable than verb generation in producing a lateralized response (Benson et al., 1999). More consistent frontal and lateral temporal activations have been reported in normals using a verb generation task where subjects produce verbs from visual or auditory noun stimuli (Raichle et al., 1994).

A promising strategy in determining language sites preoperatively has been to apply a battery of tests and combine the imaging results. Using auditory and object naming language tasks with PET in seven patients, Bookheimer and colleagues (1997) found blood flow changes in regions of ESM sites to be greater than in regions found to be uninvolved in the language task by ESM. In a different study (FitzGerald et al., 1997), the predictive value of functional imaging was directly measured using fMRI in a variety of language tasks including listening to words and text, reading, and visual and auditory verb generation. ESM localized language using object naming. Imaging was found to be 81% sensitive and 53% specific for identifying language areas in cortical regions adjacent to activations. Within 2 cm, the sensitivity was 92% but with no (0%) specificity in identifying cortical regions. A portion of this discrepancy could be explained by the somewhat different ESM maps found when using object naming or verb generation as the task (J. Ojemann et al., 1993). It remains unclear whether imaging methods can identify essential language sites with enough specificity to plan surgical resections. A better understanding of what behavioral tasks reliably activate frontal and especially temporal sites can contribute to correlations between fMRI or PET activations and ESM. Ultimately, the ability of imaging to predict and prevent postoperative deficits in language will determine the relevance of blood flow changes, even in the setting of discordant ESM data.

IV. Electrophysiological Correlates of Cognition Derived Intraoperatively

The intraoperative setting provides several opportunities to directly record physiologic events occurring in the human brain during various cognitive processes, including language, memory, and learning. The ECoG has been recorded from cortical sites related to naming by ESM during the same language measure but with overt response delayed to a later cue (Fried et al., 1981; Ojemann et al., 1989b). ECoG changes that were localized to those sites during that task included slow potentials at frontal language sites and local desynchronization at temporal language sites. These changes occurred simultaneously, beginning within 200 ms of presentation of the item to be named and lasting about 1 s, the time needed for a response. Thus, these studies identifed events occurring in parallel between frontal and temporal language sites. There was no evidence for serial processing. The nature of the temporal changes suggested that they might be generated by ascending thalamocortical circuits.

The intraoperative setting also provides an opportunity for direct recording of neuronal activity with extracellular microelectrodes. These studies provide a very different perspective on the neurobiology of cognition. Changes in neuronal activity that have statistical relationships to different aspects of behavior have been identified (reviewed in G. A. Ojemann *et al.*, 1998). Because of the invasive nature of microelectrode recording, many of the cortical studies with this technique have been confined to those regions that will be subsequently resected as part of the neurosurgical procedure; thus essential motor and language areas are not sampled. However, statistically significant changes were identified in 20–50% of the neurons sampled in the anterior temporal cortex. Those changes were equally likely in recordings from either dominant or nondominant hemisphere during a number of different language and verbal memory behaviors (Creutzfeldt *et al.*, 1989a,b; Schwartz *et al.*, 1996; Ojemann and Schoenfield-McNeill, 1999). Thus, direct recording of neuronal activity suggests widely distributed networks for these behaviors, in contrast to the strongly lateralized and focal nature of sites apparently crucial for these behaviors based on ESM or lesion effects. An individual neuron often changed activity in the same way for only one of the several behaviors sampled in different studies, including only naming or word reading but not both (Schwartz *et al.*, 1996; Ojemann and Schoenfield-McNeill, 1999), and only one of multiple languages (Ojemann, 1991b). Recordings obtained from dominant temporal cortex during naming were more likely to show relative inhibition, whereas those from nondominant temporal cortex relative excitation (Schwartz *et al.*, 1996). A few recordings have suggested specific patterns of activity for specific words (Creutzfeldt *et al.*, 1989a).

Changes in neuronal activity during recent memory for words or names have been frequently identified in temporal cortical neuronal recordings, also from either hemisphere (Ojemann *et al.*, 1988; Haglund *et al.*, 1994b; Weber and Ojemann, 1995; Ojemann and Schoenfield-McNeill, 1998, 1999). However, individual neurons usually changed activity in the opposite direction when the instruction was given to remember an item, compared to only identifying that same item, and individual neurons usually changed activity during only memory for names or words, but not both (Ojemann and

Schoenfield-McNeill, 1999). These changes were often sustained shifts in the frequency of activity, lasting seconds or minutes, changes suggestive of attentional mechanisms. Although these changes were proportionately large, the absolute frequency of activity of temporal cortical neurons was low, 1–3 Hz under control conditions and rarely beyond 20 Hz when activated. Some of the neurons changing activity with language tasks in temporal cortex may be important for verbal learning and actively inhibited in language identification tasks not requiring learning (Ojemann and Schoenfield-McNeill, 1998). Extracellular microelectrodes frequently record activity from several nearby neurons. Those nearby neurons frequently change activity with different behavior measures, a finding that suggests that human temporal association cortex is not organized in columns related to single functions, but rather has neurons that are part of the networks for different functions in close proximity (J. Ojemann *et al.*, 1992; Ojemann and Schoenfield-McNeill, 1999). Changes in neuronal activity have been identified during visuospatial measures, including identification and memory of faces and facial expressions, lines, and shapes (J. Ojemann *et al.*, 1992; Holmes *et al.*, 1996) and during music perceptions (Creutzfeldt *et al.*, 1989).

The findings from these microelectrode recordings have a number of implications for extraoperative functional imaging studies. fMRI and PET have rarely shown changes during language or memory measures in the anterior temporal regions where these recordings were obtained, although 20–50% of neurons sampled there in either hemisphere had significant changes in activity with these tasks. Most of those changes, though statistically significant and proportionally large, represent relatively low firing rates, so that little or no change may be induced in the metabolic processes that are imaged. Thus functional imaging may not indicate the full extent of neuronal activity. Most functional imaging techniques (including fMRI) do not have the spatial resolution to separate the nearby neurons that the microelectrode recordings relate to different functions. The microelectrode recordings also indicate that reduction in activity, relative inhibition, is an important component of the neurobiology of cognitive processes, especially language in the dominant hemisphere. The nature of any imagable metabolic correlate of this inhibition is not clear.

V. Optical Imaging Intraoperatively

A recently developed technique for intraoperative mapping is optical imaging. This approach capitalizes on the relationship between vascular and metabolic changes and neuronal activity. However, the spatial congruence between neuronal and vascular responses is not well defined (Lou *et al.*, 1987; Lindauer *et al.*, 1993), primarily due to differing etiologies of mapping techniques and resolution differences. For example, the blood oxygenation level dependent (BOLD) fMRI signal is thought to originate from venous capillary oximetry changes and therefore represents a venous blood dependent indicator of brain activation (Cohen *et al.*, 1994). In contrast, ^{15}O PET studies often measure changes in blood flow to produce functional maps (Phelps and Mazziotta, 1985). Optical imaging, on the other hand, produces complex activation maps (Frostig *et al.*, 1990; Narayan *et al.*, 1995) with several underlying etiologies (metabolic and vascular). Spatial maps, therefore, may differ across techniques. Indeed, Rao *et al.* (1993) indicated fMRI often centers over a sulcus or includes it, whereas optical signals often center over gyri (Toga *et al.*, 1995; Cannestra *et al.*, 1998a). Narayan *et al.* (1995) reported optical and intravascular fluorescent dye maps overspilled regions of electrophysiologic activity (using multi-unit recordings) by about 20%. However, few multimodality functional imaging studies exist.

As noted above, detection and measurement of functional activation often depend on perfusion-related signals. Studies of normal cognitive brain activity often take advantage of these hemodynamic changes to detect the site of the functional activity associated with specific behaviors. These methods assume a tight coupling between hemodynamic changes and neuronal activity. Tomographic techniques such as fMRI and PET have had tremendous impact on our understanding of normal brain function but the technology is difficult for the operating room setting. However, there are other approaches that are more compatible with an intraoperative environment. These are techniques that measure perfusion-related phenomena such as blood volume or flow changes, alterations in hemoglobin oxygenation ratios, or other manifestations of neurovascular coupling. Optical techniques, in particular, offer many distinct advantages intraoperatively.

Optical intrinsic imaging (OIS) currently offers the spatial (μm) and temporal (ms) resolution necessary to observe the rapid and transient changes that accompany functional activity. OIS measures optically active processes indirectly coupled to neuronal activity, including changes in blood volume, oxygenated hemoglobin concentrations, cellular swelling, and cytochrome activity. By altering the frequency of light measured, one or more of these contributing physiological processes can be emphasized.

Optical intrinsic signals are activity-related changes in the reflectance of cortex (and other tissues) to visible light. They were first demonstrated *in vitro* by Hill and Keynes (1949) and have been subsequently used *in vivo* (Orbach and Cohen, 1983; Grinvald *et al.*, 1986) for functional imaging in rodent (Blood *et al.*, 1995), cat (Ts'o *et al.*, 1990), primate (Frostig *et al.*, 1990), and human (Haglund *et al.*, 1992; Toga *et al.*, 1995; Cannestra *et al.*, 1996, 1998a,b, 1999a,b). The etiology of the intrinsic signals (Cohen, 1973) includes reflectance changes from several optically active processes indirectly coupled to neuronal firing (Cohen, 1973), including changes in blood volume (Frostig *et al.*, 1990; Narayan *et al.*, 1995), cellular swelling (Holthoff and Witte, 1996), hemoglobin concentrations (Malonek and Grinvald, 1996), and cytochrome activity (LeManna *et al.*, 1987), depending on the wavelength measured (Malonek and Grinvald, 1996).

The human intraoperative optical imaging system (Fig. 5) has been described in previous publications (Toga *et al.*, 1995; Cannestra *et al.*, 1996, 1998a,b). A slow-scan 16-bit CCD camera (Princeton Instruments, Trenton, NJ) is mounted via a custom adapter onto the video monitor port of a Zeiss operating microscope. Images are acquired through a transmission filter at 610 (600–620) nm. A circular polarizer and heat filters are placed under the main objective of the operating microscope to reduce glare artifacts from the cortical surface. White light illumination is provided by the Zeiss operating microscope light source through a fiber optic illuminator. Since the microscope view is determined by the surgeon's choice of magnifications and microscope placement, the experimental field of view (FOV) varies, depending on lens distance from the cortex and specific orientation. FOV often varies from 2.75 to 7.50 cm, resulting in spatial resolutions between 100 and 425 μm.

iOIS acquisitions consist of CCD exposure of the cortex during a control state (no stimulation) followed by a subsequent stimulated state. Control trials and experimental trials are interleaved. During sensorimotor mapping trials, baseline CCD images are acquired, and then physiologically synchronized PC software triggers the stimulations (2-s duration). Imaging protocols contain multiple (10–20) stimulation trials and include an equal number of nonstimulated interleaved controls. During language mapping, patients are asked to perform verbal and nonverbal tasks presented in a block paradigm (20 s of activation followed by a 10-s rest interval). All iOIS images are averaged across trials (4–20) to increase signal-to-noise ratios and maximize sensitivity.

Haglund *et al.* (1992) first demonstrated the use of OIS intraoperatively in their examination of epileptic afterdischarges in response to bipolar electrode stimulation of motor, premotor, and somatosensory cortex. They also utilized OIS to locate functional activity dur-

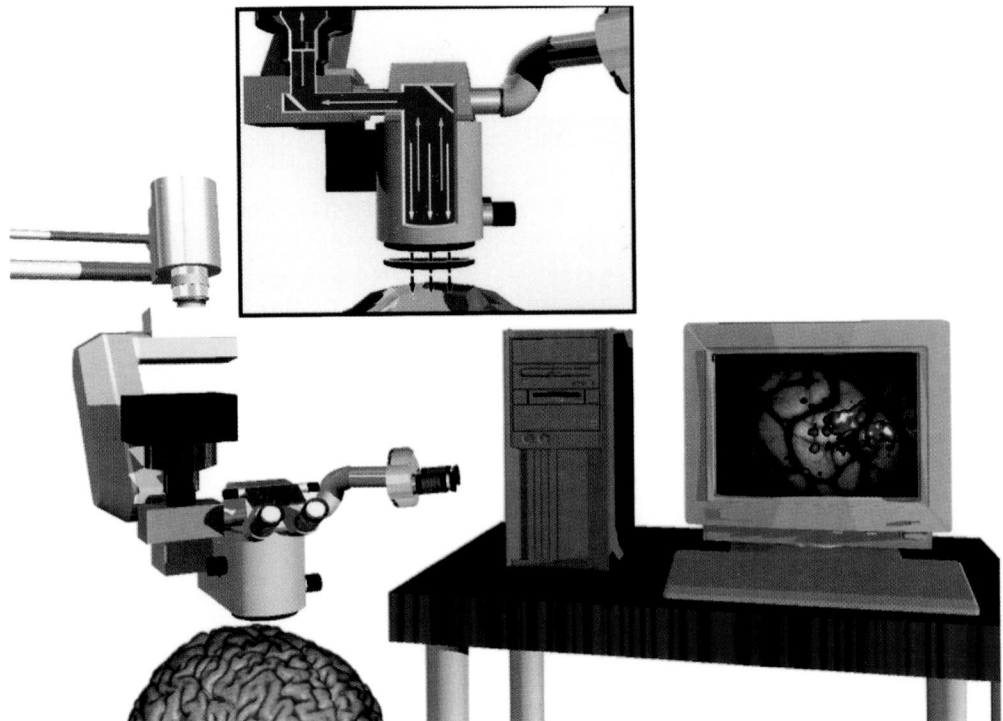

Figure 5 The intraoperative optical imaging system. A slow-scan CCD camera is mounted to the Zeiss operating microscope via the video monitor optics and custom-made adapter. White light generated by the microscope's fiber optic source is transmitted through the Zeiss optics and custom-made filter assembly (heat and polarizing filters) to illuminate the cortex. Reflected light passes back through the video optics, filtered at 610 nm and transmitted to the CCD camera. Acquisition, stimulation, and analysis is controlled by a PC present within the operating room. The inset describes the reflected light path through heat and polarizing filters, microscope optics, the transmission filter, and CCD assembly.

ing language tasks and compared the localization derived with OIS to that obtained with ESM during the same language task, object naming. OIS changes were recorded from a wider area than ESM interference, especially in the temporal lobe. This was a landmark paper as it demonstrated the use of optical methods, which up to this point had been employed only investigatively in monkeys, cats, and rodents. To stabilize the brain during the measurements and thus improve the signal to noise, they employed a glass plate. Toga *et al.* (1995) used the same approach but corrected for brain movement using image synchronization and postacquisition image manipulation rather than a glass plate.

Recent studies have examined the relationship between electrophysiological and hemodynamic responses in humans preoperatively and intraoperatively (Toga *et al.,* 1995; Cannestra *et al.,* 1998a). The purpose of the study was to define the temporal and spatial characteristics of activity-related cortical signals in humans using a multimodality imaging approach. Sensorimotor cortex was studied during 110-Hz index finger vibration in eight human subjects undergoing neurosurgical procedures for removal of intracranial pathology (tumors

or AVMs). Intraoperative optical imaging revealed rapid and transient signal changes by measuring changes in vascular and metabolic activity. Intraoperative cortical surface evoked potentials (somatosensory; SSEPs) were used to monitor electrical activity. Preoperative BOLD fMRI (at 3 T) was used with the same experimental parameters as the optical paradigm to interrelate findings to other noninvasive mapping studies. All three modalities were coregistered in 3D.

Different spatial patterns and temporal response profiles were observed. Although overall fMRI, optical, and SSEP activities were observed with very similar spatial distributions, fMRI and optical were separated by statistically significant distances. All modalities showed responses near the level of the superior genu of the central sulcus. SSEPs and optical activity were observed over the surface of pre- and postcentral gyri with very similar spatial distributions. The optical map colocalized with the SSEP map corresponding to a 13-mV potential change (within the 0.5-cm interpolation distance required for SSEPs). However, fMRI localized superior and deep to the iOIS and SSEP signals. Figure 6 details the colocalization of each modality. This colocal-

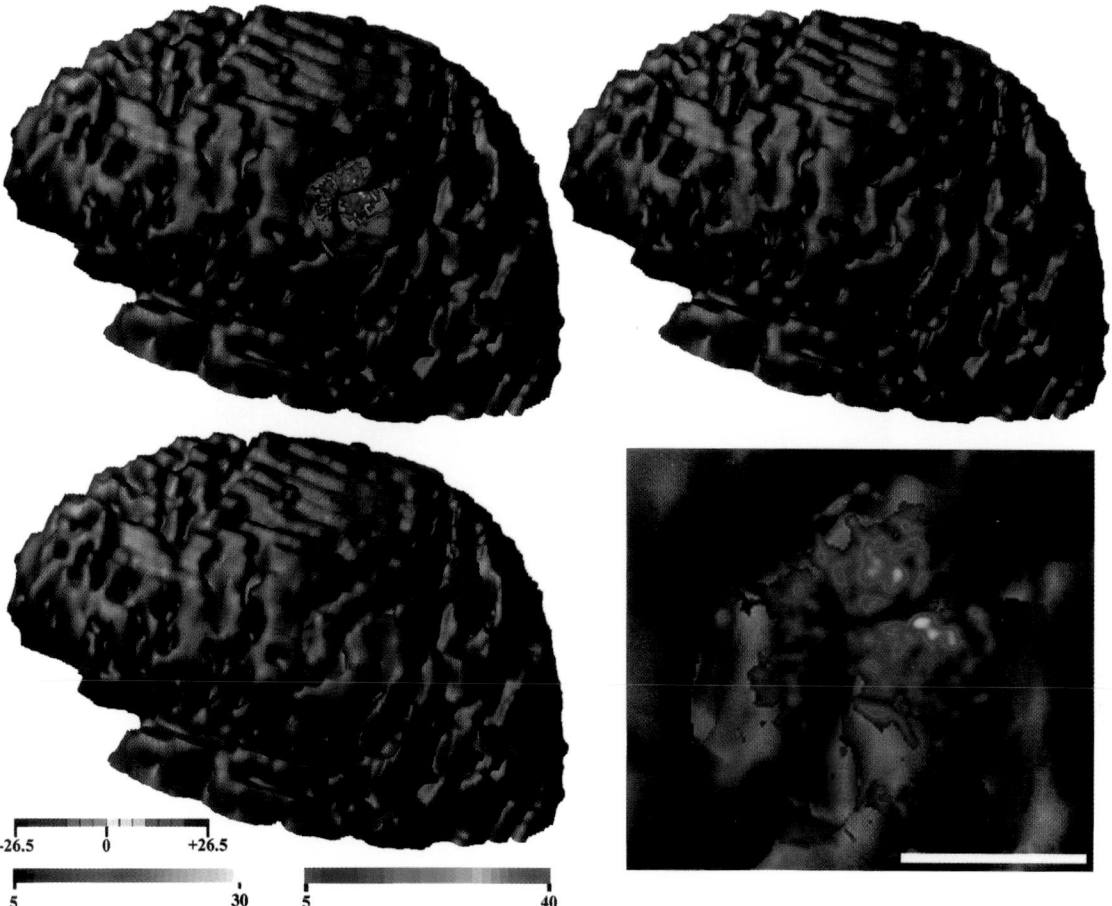

Figure 6 SSEP/iOIS/fMRI intrasubject studies reveal differences in spatial localization. SSEPs and iOIS were performed as in previous chapters. fMRI was performed preoperatively in a 3-T EPI scanner (TR = 2, flip = 79, N_{ex} = 1). Stimulations were provided by a 110-Hz pneumatically driven finger vibrator, synchronized and phase lagged to the scanner. iOIS utilized the same stimulator (SSEPs are from median nerve stimulation). Images are colocalized on a surface model obtained from fast spin-echo MRI. All signals localized at the level of the superior genu of the central sulcus. iOIS responses (upper right) were recorded over the crest of both sensory and motor cortex. SSEPs (lower left) also localized across this region. fMRI (upper right), however, localized within the central sulcus and did not include the gray matter included in the SSEP and iOIS responses. The iOIS and fMRI responses are colocalized in the zoomed window (lower right) and show colocalization at the superior genu. The fMRI signal clearly exists within the sulcus, whereas the iOIS signal localizes on the gyrus. iOIS hot (yellow) color is a 0.30% reflectance change (lower left scale, 5×10^{-4} to 30×10^{-4}) from baseline. The fMRI is a pseudocolored intensity map; hot is greater intensity (5–40 contrast units). The SSEP map indicates mV change (upper left scale, −26.5 to +26.5 mV). Scale bar is 1 cm.

izization pattern is consistent with close SSEP-optical spatial coupling observed in previous experiments (Toga *et al.*, 1995; Cannestra *et al.*, 1998a).

Temporal patterns (impulse response) also were different between the fMRI and optical techniques. The response curves were of similar shape and duration; however, the fMRI (average of all active pixels) response was delayed 2–3 s relative to the optical. However, this relationship became more similar when only pixels in the same location from each modality were examined.

Thus, in this multiple-modality mapping protocol, preoperative BOLD fMRI, intraoperative SSEPs, and optical methods measured different temporal–spatial characteristics of sensorimotor response in human brain. The observed spatial patterns and temporal profiles are not necessarily in disagreement, but rather may represent differing underlying etiologies between optical and BOLD fMRI. These findings clearly indicate different spatial-temporal patterns of functional activity are dependent upon technique and underlying physiology.

The important point here is that, as with any brain mapping, the maps collected intraoperatively will describe a specific aspect of the brain and multiple modalities may or may not colocalize in time and space.

A. Intraoperative Optical Imaging of Language

Building on previous human studies (Toga *et al.,* 1995; Cannestra *et al.,* 1996, 1998a,b), we investigated the topographical and temporal activation of language cortices within the surgical setting utilizing intraoperative OIS (iOIS) and ESM. We provide spatial compari-

son between disruption (ESM) and high-resolution activation (iOIS) maps, as well as paradigm-specific changes observed with iOIS (Figs. 7 and 8). This study characterizes spatial *and* temporal responses both between and within Broca's and Wernicke's areas. These observations demonstrate response profile differences dependent upon cortical region and language task (Fig. 9).

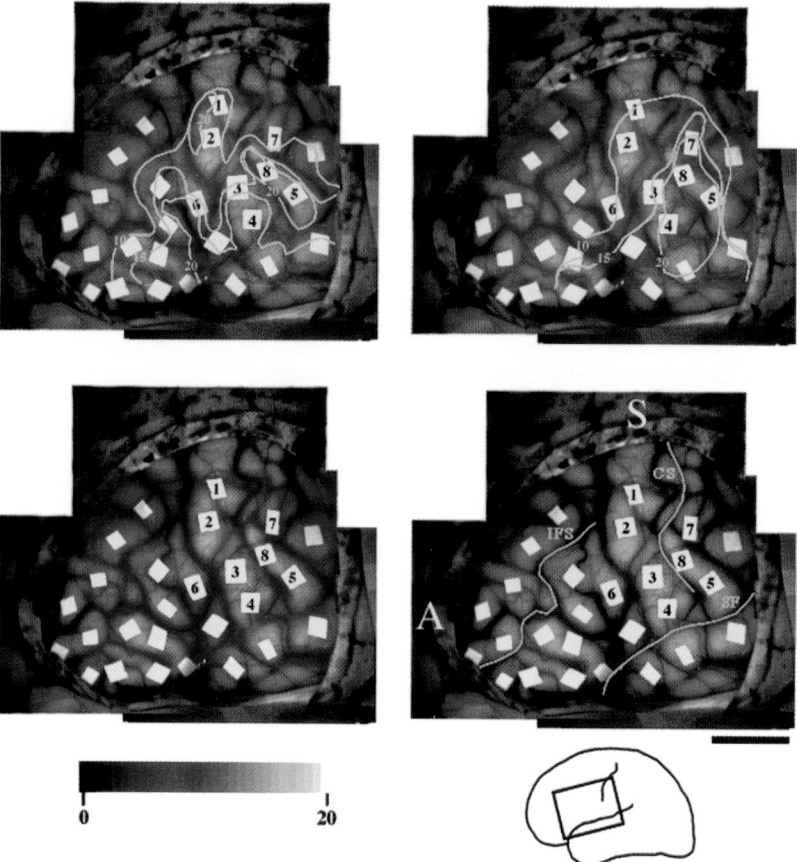

Figure 7 iOIS response over Broca's area demonstrates topographical specificity dependent upon task. iOIS was performed over frontal gyri during awake neurosurgical procedure. Images were acquired in a block paradigm scheme (10-s rest, 20-s activation) and averaged across blocks to obtain reflectance changes. Object naming using the Boston naming test (upper left) demonstrated a pattern of activity different from the nonvocal word discrimination task (upper right). Object naming revealed more activation over motor cortex (sites 1 and 2) and more activity anterior over the pars opercularis (region 6) and pars triangularis. The word discrimination task revealed less activity at 1, 2, and 6, but increased activity at region 7 and posterior cortex. Baseline activity is observed in the lower left panel. Markers indicate sampling regions for ESM. ESM at 1, 2, and 3 resulted in orofacial movements, whereas at 5, 7, and 8 the patient reported orofacial sensations (orofacial movement also occurred). ESM during the object naming protocol revealed speech arrest at regions 3, 4, and 6. In this patient (patient 1), the inferior frontal gyrus was noted to be shifted posterior and inferior due to the presence of a cavernous angioma deep to middle frontal gyrus. ESM was not performed in the temporal lobe. Magnitude contours (in yellow) indicate increasing intensity. Magnitude contours were obtained after applying a 5-pixel Gaussian blur. Images are averages of eight scans in one patient. A, anterior; S, superior. The left color bar is reflectance change $\times 10^{-4}$ for the object naming, word discrimination, and baseline iOIS images. Scale bar is 1 cm.

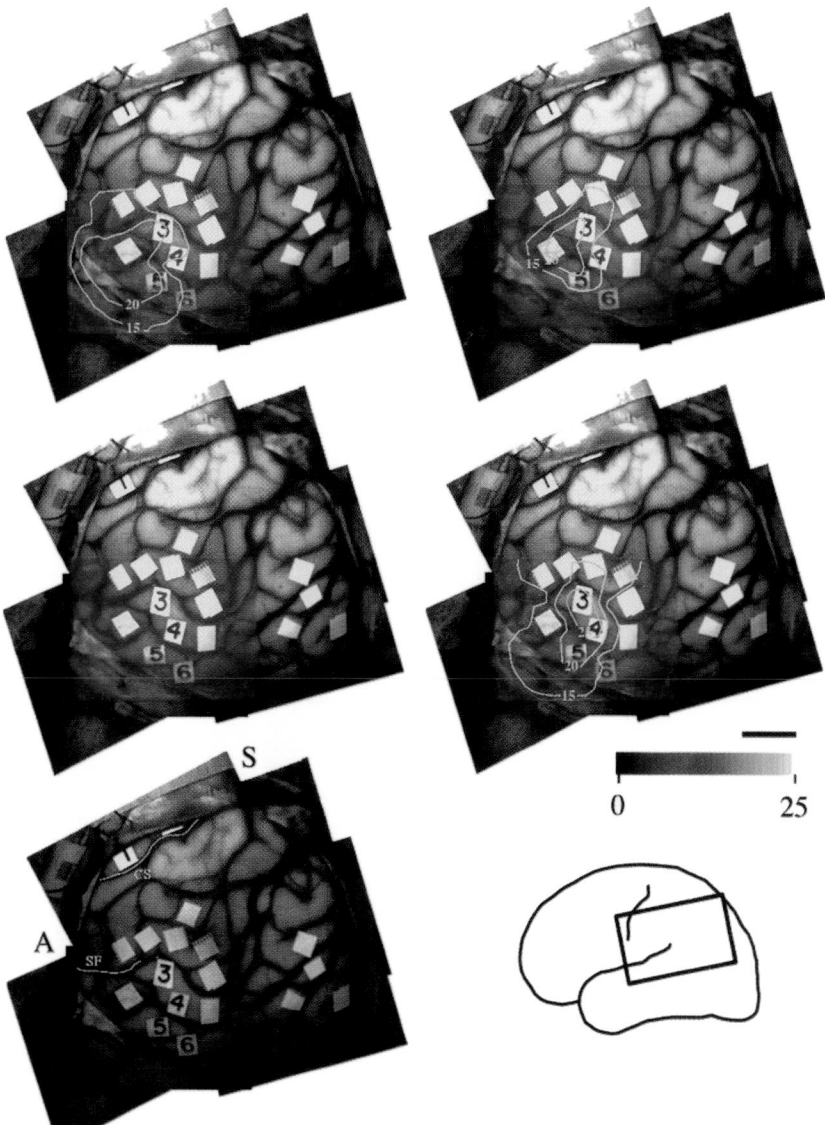

Figure 8 iOIS response over Wernicke's area demonstrates topographical specificity dependent upon task. iOIS was performed over the temporal-parietal junction during awake neurosurgical procedure. Images were acquired in a block paradigm scheme (10-s rest, 20-s activation) and averaged across blocks to obtain reflectance changes. Overall, regions 3 and 5 were activated by all paradigms. Comparison of object naming (upper left) and word discrimination (upper right) shows object naming selectively activated anterior and inferior cortices whereas word discrimination activated more superior and anterior anatomy (superior to region 3). Auditory-responsive naming tasks (lower right) activated all of the regions corresponding to both the object naming and the word discrimination paradigms and was more intense centrally. Additionally, the auditory-responsive naming discrimination paradigm also selectively activated more posterior cortex (near region 4). The magnitude contours demarcate these differences. Baseline activity is observed in the lower left panel. ESM revealed anomia for cortical regions 3, 4, 5, and 6. ESM at site 1 (at upper left cranial margin) revealed hand and arm movement, thereby localizing motor cortex. In this patient (patient 8), angular gyrus and STG were noted to be shifted anterior and inferior due to the presence of an AVM superior and deep to angular gyrus. Magnitude contours (in yellow) indicate increasing intensity. Magnitude contours were obtained after applying a 5-pixel Gaussian blur. Images are averages of eight scans in one patient. A, anterior; S, superior. The color bar is reflectance change $\times 10^{-4}$ for all paradigms. Scale bar is 1 cm.

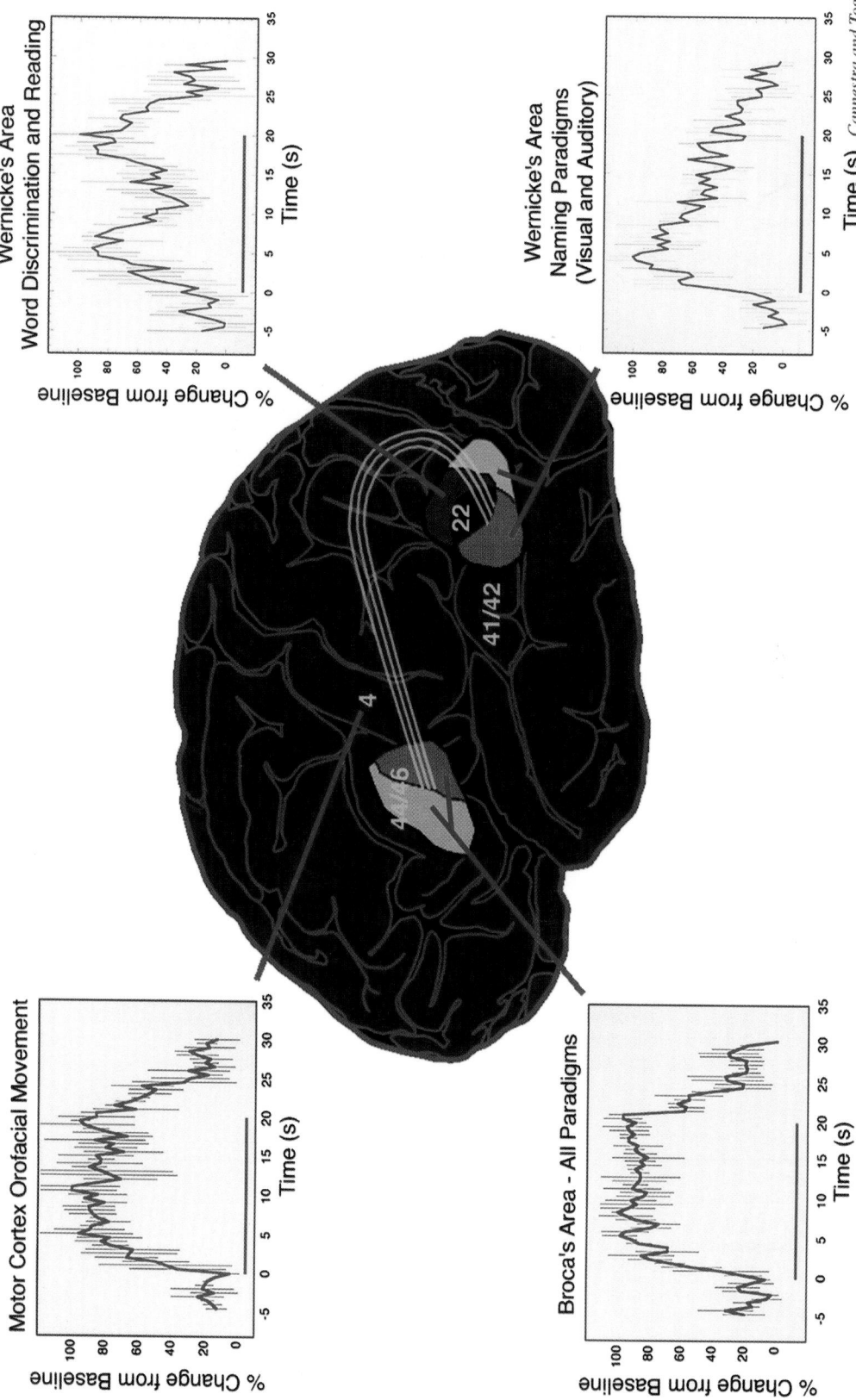

Figure 9 iOIS was performed during awake neurosurgical procedure to examine the temporal response over Broca's and Wernicke's areas. Images were acquired in a block paradigm scheme (10-s rest, 20-s activation) and averaged across blocks (4–8 trials per subject). Principal component analysis was then performed and the first eigenvector displayed as a time course measure. Response onset from both areas was similar, activating to peak within 3 s. Broca's area temporal response displayed a boxcar-type response profile (lower left graph) and was not dependent upon paradigm. Nonvocal orofacial movements (upper left) produced a similar boxcar profile. The response profiles from Wernicke's area displayed the initial peak for both naming (visual and auditory; lower right) and the word discrimination and reading paradigms (upper right). The word discrimination and reading paradigms also included an additional response peak midway through the response profile. Graphs were normalized and averaged across three subjects for each region. Error bars are standard deviations.

Cannestra and Toga

B. iOIS of Sensorimotor Cortex

iOIS results demonstrate a spatial correspondence between electrophysiological and optical signal response in the human sensorimotor cortex (Fig. 10). Where the optical signal was diffuse over sensory cortex and more focused over motor areas, the electrophysiological (EP) maps demonstrated similar spatial patterns. On the other hand, there was a larger contiguous optical signal focus over motor compared to sensory area that was less obvious in the EP map. EPs represent neuronal activity and optical signals represent vascular and metabolic factors (Frostig et al., 1990; Narayan et al., 1995). Nevertheless, the centers of highest response for EPs and optical signals do colocalize on both the sensory and motor strips.

During iOIS imaging, thumb, index, and middle fingers were stimulated individually to obtain separate cortical activation maps. Figure 11 describes iOIS responses relative to the classic homunculus representation (Penfield and Rasmussen, 1950). When maximal iOIS responses are examined over sensory cortex, middle, index, and thumb representations are oriented superior to inferior. These peak iOIS responses exist within the −13-mV (and larger) potential of the transcutaneous electric median nerve representation. The distances from the median nerve SSEP phase reversal (within the central sulcus) and maximal sensory iOIS responses are 10.7, 4.0, and 5.8 mm for thumb, index, and middle finger, respectively. These distances are similar to results obtained from single-digit electrocorticography (distances near 9, 6, and 5 mm for thumb, index, and middle finger, respectively (Sutherling et al., 1992).

In Fig. 11, the median nerve transcutaneous electrical SSEP response should encompass an area responsible for the thumb, index, middle, and portions of the ring finger (Wood et al., 1988). The observed result, finger (thumb, index, and middle) iOIS responses contained within the median nerve SSEP map, is consistent with such topography. Electrocorticography has revealed lateral inferior to medial superior order of thumb, index, and middle fingers, with overlap or reversal for adjacent digits (Baumgartner et al., 1993). The iOIS finger maps are consistent with these findings as well. Additionally, the superior/inferior orientation of middle, index finger, and thumb, and the small/diffuse representation of the middle finger, are consistent with traditional homunculus representations (Penfield and Rasmussen, 1950). In Fig. 11, the total area of response (union of all digits, 22.6 cm) is similar to areas reported by PET (22 cm; Rumeau et al., 1994) and the size of the homunculus representation drawn by Penfield and Rasmussen (1950). When individual digit electrocorticography data

were adapted from previous studies (Sutherling et al., 1992) and compared to perfusion-dependent responses, comparable localizations were observed. Individual finger SSEP localization (as observed by Sutherling et al., 1992) and iOIS finger responses were within 1.7, 2.0, and 0.8 mm for thumb, index, and middle fingers, respectively. Peak iOIS responses may therefore correlate to brain maps obtained by direct cortical stimulations (Penfield and Rasmussen, 1950) and single-dipole localizations in electrocorticography (Sutherling et al., 1992).

Figure 12 describes the relationship between iOIS, EPs, and electrocortical stimulation mapping (ESM). The thumb and face (cheek) were stimulated individually to obtain separate iOIS maps. Thumb localized within the region demarcated by ESM and large potential changes. The cheek response localized more inferior, consistent with the findings from ESM. Peak intensity iOIS responses accurately localized with stimulation sites. Thumb and facial responses activated large regions of motor cortex. Inferior postcentral activation also is observed during facial stimulation; however, it is probably distorted due to the location of arterial-venous malformation (AVM).

In Fig. 12 iOIS localizations were consistent with cortical maps obtained by ESM mapping and evoked potential measures. As in Fig. 11, the superior-inferior orientation of hand and face (from iOIS and ESM mapping) also is consistent with traditional homunculus representations (Penfield and Rasmussen, 1950). These results provide additional evidence that peak iOIS responses correlate with electrically defined cortical representations.

These studies begin to describe the morphology of perfusion-related responses and cortical activity. Optical reflectance imaging, combined with EP mapping of the sensorimotor cortex, illustrated the relationship between neuronal firing, vascular activity, and metabolic activity. Additionally, these experiments suggest the development and implementation of a noncontact, repeatable, intraoperative tool for functional mapping is of practical significance and may be used in conjunction with SSEPs to avoid eloquent brain during resection.

VI. Intraoperative Structural Maps

Structural maps to aid in the localization of specific anatomies historically have received considerable attention, as the ability to precisely target is essential for any neurosurgical procedure. The motivation for the development of accurate localization and visualization systems originates from the practical paradox of neurosurgical procedures (Black and Ronner, 1987; Nazzarro and Neuwelt, 1990; Black, 1991). The surgeon's goal is to

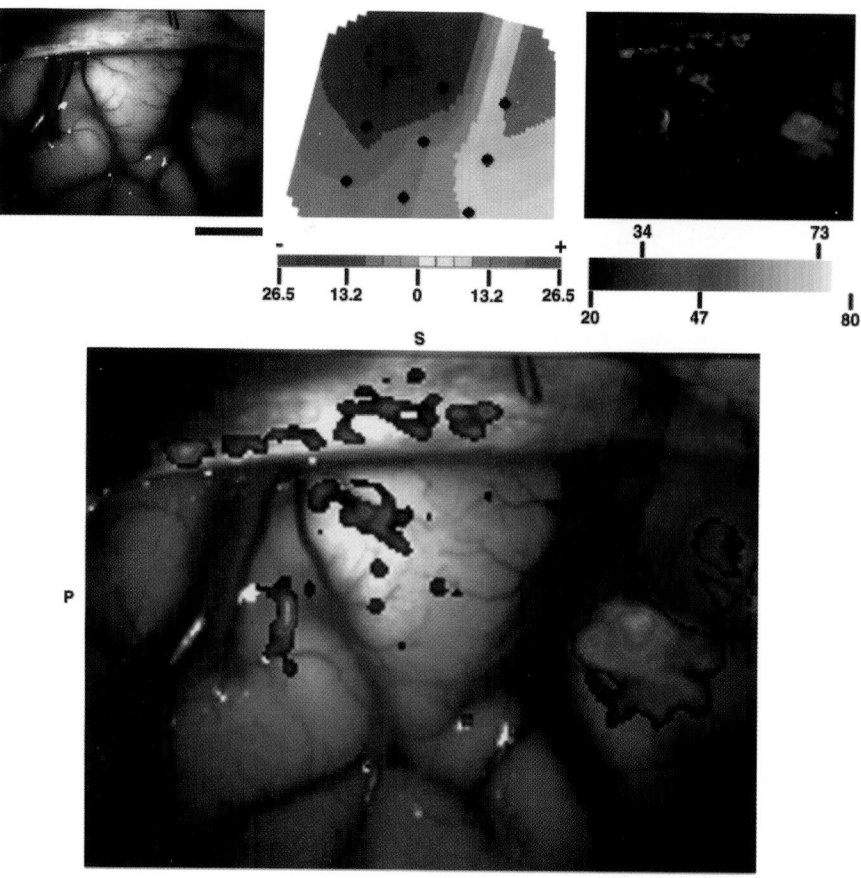

Figure 10 Colocalization of optical intrinsic signals and evoked potential maps over right sensorimotor cortex. Regions of highest measured electrical change correspond to those areas where optical signals were largest in size and intensity. Both sensory and motor strips were activated. The stimulus intensity was sufficient to cause involuntary hand and finger movements. Data collected from each method were obtained under identical stimulus conditions for each subject. Raw images of brain surface, EP maps, and optical signal images were superimposed by registering the fiducials observable from EP mapping with images obtained with the CCD camera. No distortion correction was employed to account for the brain curvature contributing to the evoked potential map that was not part of the optical signal image. Upper left panel shows raw image of cortical surface. Upper middle panel shows EP map of this region of cortex in response at 24.37 ms to median nerve stimulation. Black dots indicate grid electrode positions (9 of 20). Upper right panel shows optical signal image at 3 s post-stimulus onset. Bottom panel shows superpositioning of all three upper panels. Note relative absence of activation in sulcal margin. S, superior; P, posterior. EP color bar indicates mV; reflectance color bar indicates reflectance decrease $\times 10^{-4}$. Scale bar is 1 cm.

remove all intracranial pathology from the patient but at the same time avoid the removal of normal brain tissue that supports vital cerebral function. While much intracranial pathology is easily distinguishable upon visual inspection by the surgeon (abscesses, cysts, vascular malformations, and many neoplastic growths), some intracranial pathology necessitates additional intervention for delineation from normal tissue. For example, while tumor that is grossly necrotic, hemorrhagic, and visually distinct from cortical tissue may be removed in

a straightforward manner, tumor margins and tumor tissue buried in the depths of a tumor resection bed pose an extremely difficult practical problem. Tumor margins are often repeatedly biopsied to determine if the margin walls are free of viable tumor tissue. This painstaking process is typically random and often undersamples the tumor resection bed. It also is quite time-consuming since each biopsy requires frozen section preparation and analysis by a neuropathologist. The utilization of intraoperative structural brain mapping procedures can

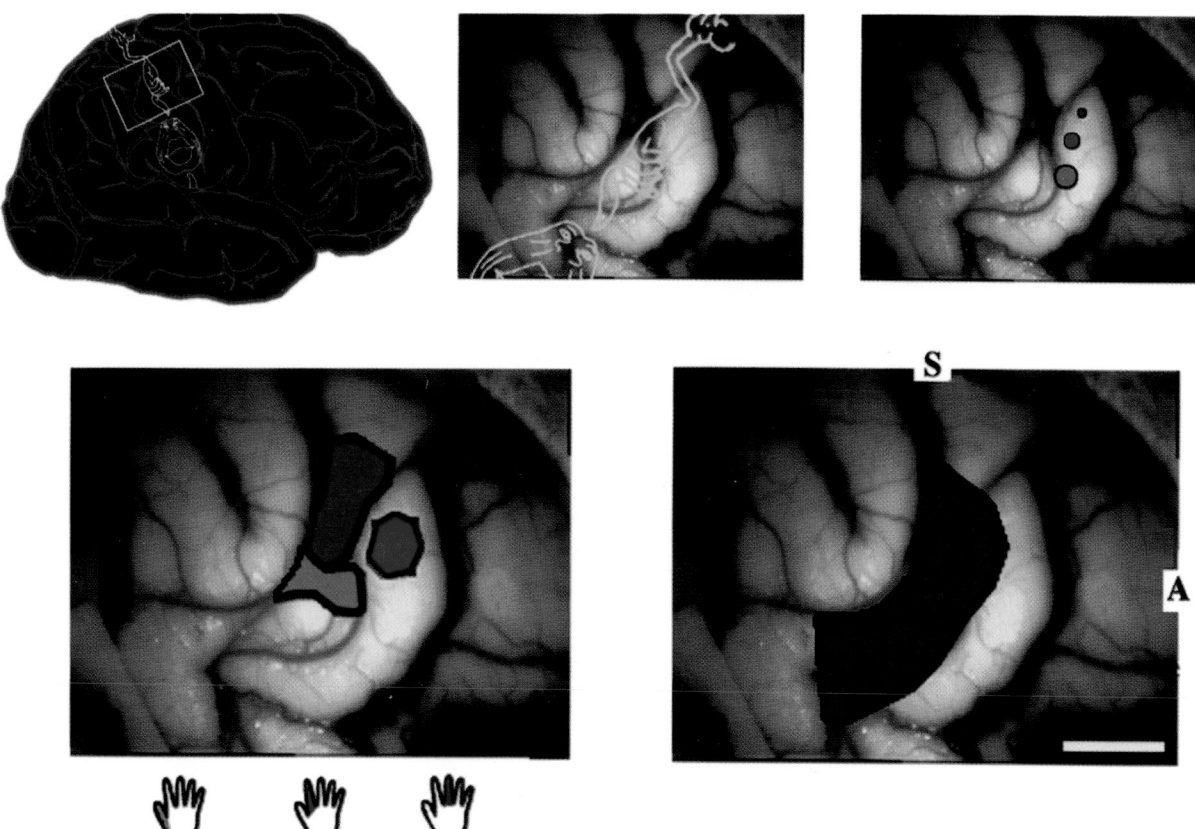

Figure 11 The orientation of individual-digit iOIS responses is consistent with the classical homunculus schematic (upper left, brain schematic, adapted from Penfield and Rasmussen, 1950). The middle upper panel applies this schematic over subject neuroanatomy. The upper right panel illustrates data from previous studies utilizing single-digit electrocorticography. Distances between median nerve SSEP and individual-digit dipoles were near 9, 6, and 5 mm for thumb (green), index finger (red), and middle finger (purple), respectively (adapted from Sutherling *et al.,* 1992). When peak iOIS responses are displayed over somatosensory cortex, a superior to inferior orientation is revealed for the middle finger (purple), index finger (red), and thumb (green), respectively (lower left panel). The distances from the median nerve SSEP phase reversal (within the central sulcus) and maximal sensory iOIS responses are 10.7, 4.0, and 5.8 mm for thumb, index finger, and middle finger, respectively. The peak iOIS digit responses corresponded to the region of −13-mV potentials obtained during median nerve transcutaneous electrical stimulation (lower right panel; blue). iOIS magnitude contours were obtained by tracing signals after a 5-pixel Gaussian blur (as observed in Fig. 2); only the peak contour is displayed over sensory cortex. The median nerve SSEP region corresponding to −13-mV and larger potentials (as in Fig. 2) is displayed only over sensory cortex. Localization of digit dipoles (adapted from Sutherling et al., 1992) utilizing published dimensions relative to the phase reversal of the median nerve SSEP (superior/inferior; Fig. 2) and distance from the central sulcus. Diameter of digit dipoles indicates ±2 standard errors of the mean localization. Scale bar is 1 cm. Spatial resolution is 335 μm/pixel. A, anterior; S, superior.

greatly influence the outcome of these surgical procedures.

Numerous approaches to intraoperative structural brain mapping have been devised, including stereotaxic frames where a coordinate system is indexed and referenced to either an atlas or preoperative neuroimaging studies. There are also frameless systems that index brain space using a probe or pointing device. While it can be argued that the historical origins of structural localization can be traced to phrenology (Franz Joseph Gall [1758–1828]), until the first Spiegel-Wycis stereotaxic apparatus was developed, the ability to precisely locate and treat neurological disease surgically was without reference to coordinates. Such maps consist of a comprehensive description of anatomy and a reference system to relate anatomy and a specific location.

To create such a map, image data must be acquired, a coordinate-based referencing system must be established, and a way to visualize the map must be available. Not only are these techniques essential for the intraoperative protocols but appropriate coordinate systems and methods for placing data there provide the opportunity to reference pre- and postoperative brain structure and function observations.

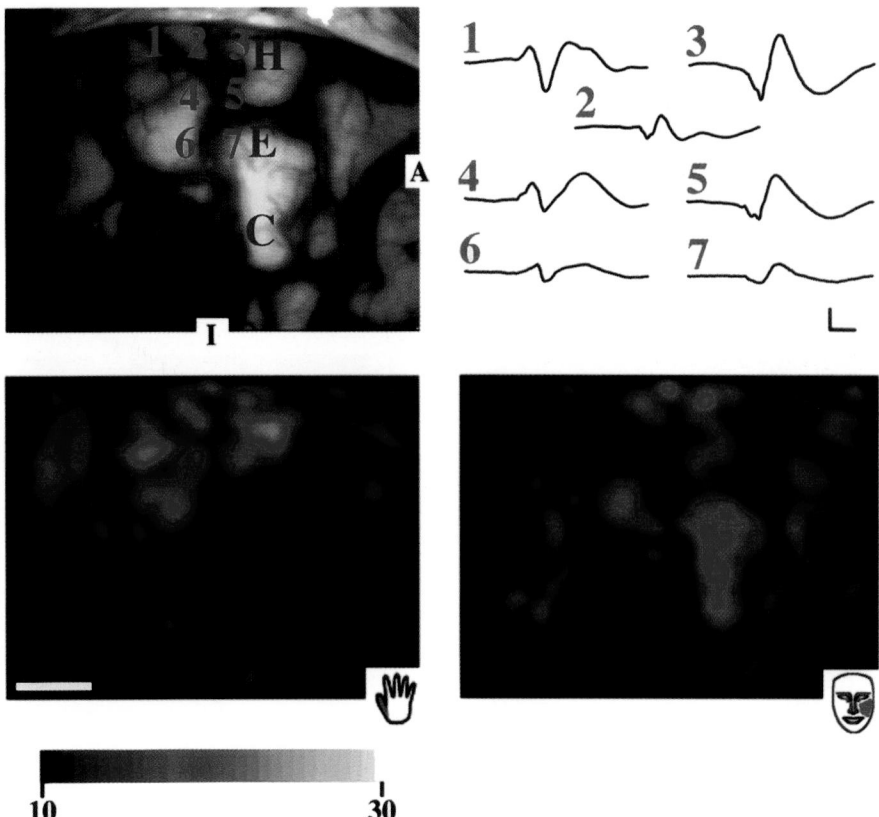

Figure 12 Thumb and facial (cheek) stimulations produced separate cortical activation maps (lower left and lower right, respectively). These regions correlated with intraoperative mapping of cortical representations by ESM (upper left, blue letters). Thumb also localized with large evoked potential changes observed from transcutaneous median nerve stimulation. Thumb localized more superior than cheek response. Thumb and facial responses activated large regions of motor cortex. Inferior postcentral activation is observed during facial stimulation but is probably distorted due to the location of the arterial-venous malformation (AVM). Images are from a single patient. Scale bar is 1 cm. Spatial resolution is 500 μm/pixel. A, anterior; S, superior.

A. Preoperative Maps

Preoperative brain mapping obviously plays a major role in neurosurgical treatment planning. Furthermore, preoperative data can be used as an anatomic framework for the mapping of other structure/function observations made intraoperatively (see below). Spatial resolution of structural data can be sufficient to enable the reconstruction and rendering of surface models that can present views similar to that exposed with the craniotomy (Fig. 13). Figure 13 illustrates a reconstructed model showing a craniotomy but the brain model is from preoperative MR. These preoperative maps (see Chapter 3) also can contain sufficient anatomic detail to facilitate the placement of the brain volume into a predefined coordinate system for comparing other modalities from the same subject or from multiple subjects (see below). MR pulse sequences that enable acceptable gray matter/white matter/cerebral spinal fluid dis-

crimination are the most useful. Multiple scans are the norm.

Intraoperatively, depending on the focus of the surgery or investigation, either 3D tomographic or cortical surface descriptions of anatomy are acquired. The advent of new (sometimes lower field strength) scanners capable of generating quality images of anatomy is revolutionizing intraoperative structural mapping (see below). Advances in MRI have made possible the use of closed- and open-configuration magnets within the confines of an operating room. While there are still significant practical limitations such as the physical proximity between the surgeon, the patient, and the scanner as well as the need for specialized gantries and MR-compatible instruments, this approach enables repeated surveys of structural changes to the brain resulting from the procedure.

Descriptions of the cortical surface anatomy, vascular system, or other features can be collected using dig-

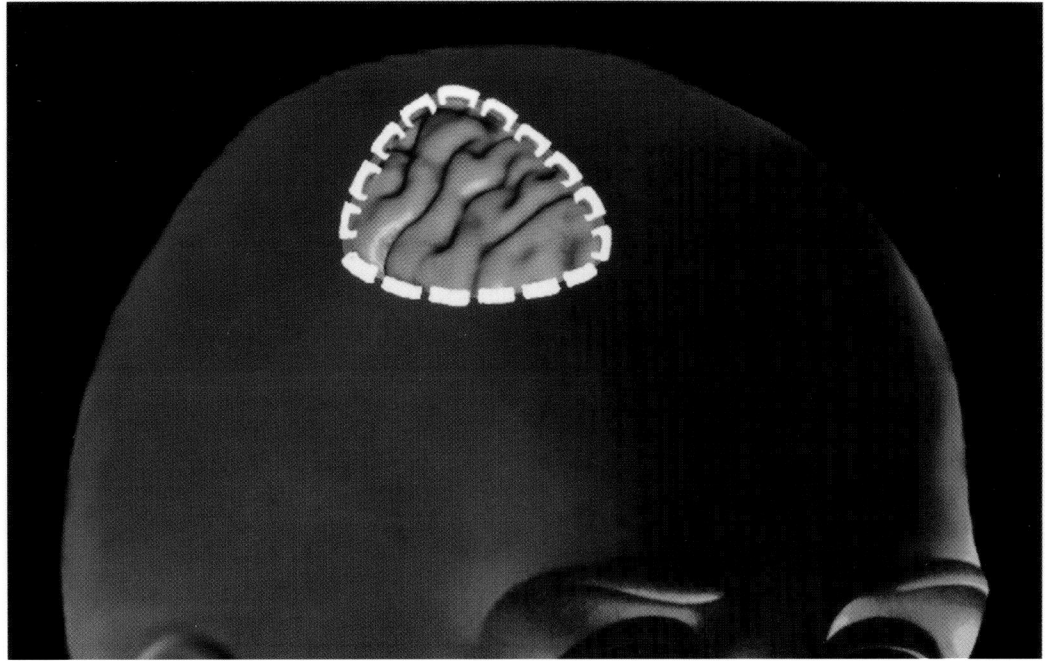

Figure 13 Using preoperative MR, this reconstruction illustrates a surface model of the cortex through a simulated craniotomy. It provides a view that is expected intraoperatively. The cortical model from the patient is also useful for mapping of functional data collected intraoperatively or even preoperatively.

ital camera systems and these data colocalized with the tomographic data collected preoperatively. This approach is most informative when the preoperative scans include functional measures from fMRI or PET spatially normalized with the structural scans. In this way, functional maps collected before and during the surgery can be compared. However, doing this with any degree of accuracy requires a spatial transformation of one or more of the data sets to make them equivalent in space. Often, due to the deformations of the brain as a result of the surgical procedures, the necessary spatial transformations are mathematically complex (Toga, 1998).

B. Spatial Normalization

Due to individual variations in anatomy among normal subjects, proportional scaling systems are typically employed to reference a given brain to an atlas brain (Talairach and Tournoux, 1988). More sophisticated elastic or fluid transformations, involving local matching, are becoming more commonplace (see below). These approaches locally deform a digital atlas to reflect the anatomy of new subjects. Commonly used human atlases include those of Talairach and Tournoux (1988) and the thalamic and brainstem anatomical maps of Schaltenbrand and Wahren (1977). Before describing the requisite transformations, we define appropriate indexes in the form of coordinate systems.

C. Stereotaxy and Coordinate Reference Systems

Indexing locations provides the relationship between location and structure and enables multiple-modality and multiple-subject comparisons. Although this topic is beyond the scope of this chapter and volume, there is a significant literature devoted to the definition, requisite transformation, and utilization of coordinate systems. The interested reader should refer to Toga and Mazziotta (1996) and Toga (1998). The intraoperative environment once again poses significant challenges for establishing a coordinate system, primarily because the shape and position of the brain can change so drastically due to the fact that the integrity of the cranium is compromised during the surgery.

Many approaches have been used to achieve a spatial relationship between data collected before the surgery and during it as well as to published atlases. These include the use of stereotaxic frames (there are many frame variants), using extrinsic fiducials placed on the scalp, or referencing external anatomic features such as the nasium, glabella, medial and lateral canthi, tragus, or other regions where there is a stable relationship between an external anatomic feature and the brain. The first known stereotactic device used for a neurosurgical procedure was the Spiegel–Wycis Stereoencephalotome, which was very similar to the

Horsley–Clarke device (1908) used in primates (notably, both Horsley and Clarke and Mussen designed human devices that never saw use). These systems provided an intracranial Cartesian coordinate system and defined specific $X, -Y, -Z$ translations to index the brain. Indeed the sole goal of stereotaxy is to provide a coordinate system to map out the internal anatomy of the head. The Spiegel–Wycis device utilized ear bars and infraorbital tabs to anchor a frame to the patient's head. Spiegel and Wycis (1952, 1950) used their device in conjunction with pneumoencephalograms to perform the first human stereotaxic neurosurgical procedure. Soon after their development, a multitude of neurosurgical stereotaxic devices were created. The first arc-based stereotaxic device was by Lars Leksell (1949). This system employed three coordinates to indicate the center of a semicircular arc, similar to rotational coordinates. Additional systems utilized a series of interlocking arcs (BRW; Brown *et al.,* 1980) or burr hole mounting (Austin and Lee, 1958).

The scope of stereotaxic neurosurgery is a field unto itself and far too extensive to be fully covered in this chapter. However, the use of stereotaxic devices was ubiquitous throughout the 1950s and 1960s. Indeed within 25 years after the introduction of human stereotaxic neurosurgery, over 37,000 patients had been treated (Speigel, 1982). The late 1970s saw the development of computerized tomography and the marriage of stereotaxic neurosurgery with tomographic imaging. The subsequent development of computer workstations and MRI led to creation of intraoperative visualization systems and frameless stereotaxy.

D. Registration

Many approaches have been used to achieve a spatial relationship between data collected before the surgery and during it as well as to published atlases. In existing atlases, spatial normalization systems are typically employed to reference a given brain with an atlas brain (Talairach and Szikla, 1967; Talairach and Tournoux, 1988). This allows individual data to be superimposed on the data in the atlas—in other words, to be transformed to match the space occupied by the atlas. In the Talairach stereotaxic system, piecewise affine transformations are applied to 12 rectangular regions of the brain, defined by vectors from the anterior and posterior commissures to the brain's extrema. These transformations reposition the anterior commissure of the subject's scan at the origin of the 3D coordinate space, vertically align the interhemispheric plane, and horizontally orient the line connecting the two commissures. Each point in the incoming brain image, after it is "warped" into the atlas space, is labeled by an (x, y, z)

address referable to the atlas brain. Although originally developed to help interpret brainstem and ventricular studies acquired using pneumoencephalography (Talairach and Szikla, 1967), the Talairach stereotaxic system rapidly became an international standard for reporting functional activation sites in PET studies, allowing researchers to compare and contrast results from different laboratories (Fox *et al.,* 1985, 1988; Friston *et al.,* 1989, 1991).

The use of spatial normalization schemes based upon deep white matter features (the AC and PC), such as outlined above, has yet to completely accommodate the most variable brain structure, the cortex. The cortex is also the site of interest for most functional activation studies and many neurosurgical procedures. Considerable normal variation in sulcal geometry has been found in primary motor, somatosensory, and auditory cortices (Missir *et al.,* 1989; Rademacher *et al.,* 1993), primary and association visual cortices (Stensaas *et al.,* 1974), frontal and prefrontal areas (Rajkowska and Goldman-Rakic, 1995), and lateral perisylvian cortex (Geschwind and Levitsky, 1968; Steinmetz *et al.,* 1989, 1990; Ono *et al.,* 1990). More recent 3D analyses of anatomic variability in postmortem, *in vivo* normal, and diseased populations have found a highly heterogeneous pattern of anatomic variation (Thompson *et al.,* 1996a,b, 1998a,b,c; Fig. 14). Figure 14 shows the mapping of cortical variability of the surface from a population of normal males. While the detail shown here is lower than that observed intraoperatively, it does illustrate the fact that the variability is greater in some cortical regions compared to others.

In view of the complex structural variability between individuals, a fixed brain atlas may fail to serve as a faithful representation of every brain (Roland and Zilles, 1994; Mazziotta *et al.,* 1995). Since no two brains are the same, this presents a challenge for attempts to rely on an atlas clinically. Even in the absence of any pathology, brain structures vary between individuals not only in shape and size but also in their orientations relative to each other. Such normal variations have also complicated the goals of comparing functional and anatomic data from many subjects (Rademacher *et al.,* 1993; Roland and Zilles, 1994).

The intraoperative brain is even more problematic. Due to the relatively extensive deformations of the geometry of the brain resulting from a compromised cranium, the presence of powerful pharmacologic agents, and the influence of the pathology, more sophisticated transformations that correct for these local, yet profound, influences are required. Local warping transformations (including local dilations, contractions, and shearing) can be used to adapt the shape of intraoperative views with the preoperative anatomy or of a brain

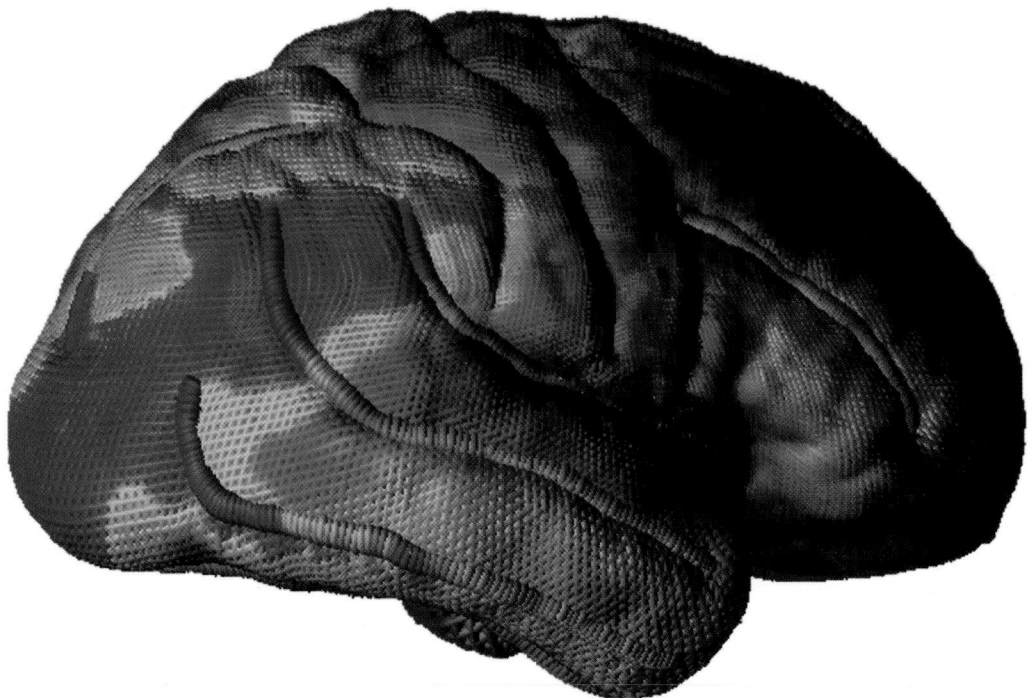

Figure 14 We have measured the variability of cortex by using measurements of surface features such as gyri and sulci. These enable the calculation of statistics that describe the variability of populations of subjects. See additional chapters in this volume. Cortical variability is greater in some regions compared to others but does make obvious the need for deformation strategies before single-subject published atlases can be relied upon for the identification and delineation of structures in the brain of any given patient or subject. In this figure, the regions of greatest variability can be found red. A temperature color scale codes the root-mean-square magnitude of variability at each point on the surface.

atlas. Pioneered by Bajcsy and colleagues at the University of Pennsylvania (Broit, 1981; Bajcsy and Kovacic, 1989; Gee *et al.*, 1993, 1995), this approach has been used in a variety of applications (Seitz *et al.*, 1990; Thurfjell *et al.*, 1993; Ingvar *et al.*, 1994), where warping transformations are applied to adapt one view to another or to establish coregistered functional scans.

Several deformable atlases have been designed according to the laws of continuum mechanics, which describe the deformational behavior of real materials. Recently, Christensen *et al.* (1993, 1995a,b, 1996) proposed a deformable MRI-based atlas driven by a viscous-fluid-based warping transform. The fluid model was motivated by capturing nonlinear topological behavior and large image deformations. The optimal deformation field maximizes a global intensity similarity function (defined on the deformed template and the target) while satisfying continuum-mechanical constraints that guarantee the topological integrity of the deformed template (Christensen *et al.*, 1996).

Because of the potential for extreme differences in cortical anatomy during surgery, basing the spatial transformation on anatomic features holds the most promise. Thus, to guide the mapping of an atlas onto an individual scan, higher level structural information can be invoked to guarantee the biological validity of the resulting transform (Collins *et al.*, 1996; Davatzikos, 1996; Thompson and Toga, 1996). In one approach (Thompson and Toga, 1996), anatomic surfaces, curves, and points are extracted (with a combination of automatic and manual methods) and forced to match (Fig. 15). The sulcal features in Fig. 15 not only provide guidance intraoperatively but are used to control the mathematics of the matching. The procedure calculates the volumetric warp of one brain image into the shape of another by calculating the deformation field required to elastically transform functionally important surfaces in one brain into precise structural correspondence with their counterparts in a target brain. The scheme involves the determination of several model surfaces, a warp between these surfaces, and the construction of a volumetric warp from the surface warp. Extremely complex surface deformation maps on the internal cortex are constructed by building a generic surface structure to model it. Connected systems of parametric meshes model primary sulci with deep trajectories. In advance,

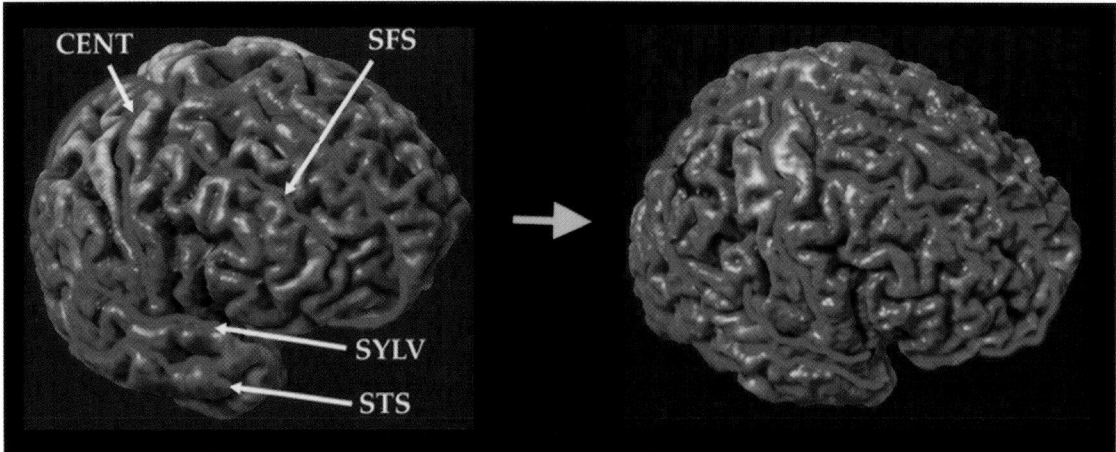

Figure 15 Using mathematical methods to map cortices from different individuals or from different time points of the same individual enables the calculation of the differences between them. Brain warping can be based on numerous different methods, but utilization of the cortical anatomy is the most useful in intraoperative applications because of the potential for extreme deformations during the procedure relative to image data collected preoperatively. Here specific sulcal features are identified and used in driving the warp to make the two equivalent in space.

a high-resolution model of the external cortex is automatically extracted from both scans with an active surface-algorithm (MacDonald *et al.*, 1993). These models are subsequently reparameterized to allow matching of specific cortical regions.

E. A Changing Morphology

One of the uniquely interesting aspects of intraoperative structural mapping is the potential to develop models that describe the way in which the geometry of the brain responds during surgery. Currently, structural brain imaging typically focuses on the analysis of 3D models of brain structure, derived from volumetric images acquired at a single time point. However, serial scanning, when combined with a powerful set of warping and analysis algorithms, can enable intraoperative responses to be tracked in their full spatial and temporal complexity.

Maps of anatomical change can be generated by warping scans acquired from the same subject over time (Thirion *et al.*, 1998; Thompson *et al.*, 1998a,c). Serial scanning of human subjects (Fox *et al.*, 1996; Freeborough *et al.*, 1996; Thompson *et al.*, 1998a,c) or experimental animals (Jacobs and Fraser, 1994) in a dynamic state of disease or development offers the potential to create 4-dimensional models of brain structure. These models incorporate dynamic descriptors of how the brain changes during maturation, disease, or intervention. For a range of patient populations, 4D models of the brain can be based on imaging and modeling its 3D structure at a sequence of time points. In a changing

morphology, warping algorithms enable one to model structural changes that occur over prolonged periods, such as developmental, aging, or disease processes, as well as structural changes that occur more rapidly, as in intraoperative procedures. A 4-dimensional approach can provide critical information on local patterns and rates of tissue growth, atrophy, shearing, and dilation that occur in the dynamically changing architecture of the brain (Toga *et al.*, 1996; Thompson *et al.*, 1998a,c).

F. Visualization and Frameless Stereotaxic Systems

Increasingly, the use of sophisticated visualization approaches is finding its way into the operating room. Atlases that describe a canonical brain or models of the individual patient can be rendered realistically. The Digital Anatomist project in Seattle (Sundsten *et al.*, 1991) and the Harvard Surgical Planning Laboratory (Kikinis *et al.*, 1996) have each synthesized digital anatomical models of extraordinary utility. These annotated models can be rotated and visualized interactively to illustrate complex spatial relationships among anatomic structures. Modeling strategies currently used to represent brain data have been motivated by the need to extract and analyze the complex shape of anatomical structures for high-resolution visualization and quantitative comparisons. Ray-tracing and surface-rendering techniques can then be applied to parameterized or triangulated structure models (Payne and Toga, 1990; Toga, 1994) to visualize complex anatomic systems. Because the digital models reside in the same stereotaxic space as the atlas data, surface and volume models are amenable to digital transformation, as well as geometric and statistical measurement (Thompson *et al.*, 1996a). An underlying 3D coordinate system is, therefore, central to all atlas systems, since it supports the linkage of structure models and associated functional data.

Furthermore, integrating preoperative maps describing brain structure and/or function of the patient with the exposed cortex is enormously advantageous. This has been utilized in a variety of image guidance systems. Chapter 5 describes such intraoperative visualization tools in greater detail.

Intraoperative guidance and visualization systems consist of a visualization/computer display with an external device to monitor position in three-dimensional space. The external positioning device often utilizes a wand or pointer or may be in the form of a probe (i.e., an ultrasound transducer). After acquisition of preoperative neuroimaging studies, scans are transformed into surface models and tissue is delineated for intraoperative display and manipulation. Once the patient is positioned for surgery (via cranial stabilization), the visualization system is then registered to the patient's surgical position to reflect the patient's anatomical position as observed by the surgeon. Once registered, the pointer may then be used to demarcate positions on the patient that are then displayed in the 3D images via the visualization system.

With the advent of computerized tomography in the late 1970s, the ability to localize and image neurosurgical pathology in 3D led to the development of a new neurosurgical field, image-guided stereotaxic surgery (Gildenberg, 1987). The early systems primarily used stereotaxic head holders that were worn during the preoperative scanning procedure. For example, one of the first systems utilized a modified Leksell head holder that attached to the CT scanner (Leksell and Jernberg, 1980). The patient was required to wear the apparatus from image acquisition through surgery. In the operating room, the head holder was attached to the head frame such that the image coordinates could be transposed to the operating room. Data from the subject were registered by simple rotation and translation. Later, the development of a fiducial system that contained all the information for three-dimensional reconstruction on each two-dimensional slice further facilitated the use of tomographic imaging with stereotaxic surgery (Brown *et al.*, 1980). Kelly introduced a practical method of registering computer-generated 3D volumetric images in the surgical field by using stereotaxic techniques (Kelly *et al.*, 1980, 1983, 1984). The subsequent introduction of practical computer workstations allowed the manipulation of preoperative images within

the stereotaxic coordinate system, and the first so-called frameless stereotaxic systems were created. One of the principles of this frameless stereotaxy is that stereotaxic positioning may be obtained without anchoring a head holder or frame on the patient prior to scanning. The application of more complex registration algorithms allowed for a best fit of a patient position to a series of surface-based fiducials. Registration was accomplished by the tagging of defined anatomical landmarks, including the medial and lateral canthi, tragus, occiput, and the external cranial margin. One of the first frameless systems (Watanabe *et al.*, 1987) utilized a multiarticulated mechanical arm attached to the operating room table or head holder. The angles of each of the arm's joints are accurately read by encoders, so that the computer can precisely calculate the location of the pointer in space. Two such commercially available systems are the ISG wand (ISG, Toronto, Canada) and the OAS wand (Radionics, Burlington, MA). The workstations are then able to reconstruct the preoperative scans into the real-life surgical position of the patient. A minimum of three and often more fiducial markers are used that can be identified accurately on both the patient and the preoperative image data set. As the pointer touches each of these positions on the patient, the point in space is registered in the workstation. When all of the points (or surface contours) are entered, the computer performs the registration and the system is ready for use. As the surgeon then points, the position of the pointer on the appropriate slice of the surface model is displayed. Advances in frameless stereotaxy have been rapid and newer systems use ultrasound (Roberts *et al.*, 1986, 1989), light-emitting diodes (Barnett *et al.*, 1993), magnetic fields (Kato *et al.*, 1991), or video tracking systems (Heilbrun *et al.*, 1992) that allow the use of a handheld pointer or instruments not attached to an articulated arm. One such commercially available system is the VectorVision2 system (BrainLab, Heimstetten, Germany) that utilizes stereoscopic optical tracking systems and triangulation principles. These systems use a handheld unattached wand that may be tracked in three dimensions by the use of infrared (IR) indicators at the ends. The use of a two-camera IR system provides the stereoscopic views necessary for three-dimensional placement. Another innovative approach provides digital overlay of the Schaltenbrand, Talairach, and Brodmann atlases onto individual patient MRI scans to assist in the identification of structures and facilitate surgical planning (Hardy, 1994; St.-Jean *et al.*, 1998).

Frameless stereotaxic systems have the enormous advantage of providing real-time feedback to the surgeon. As the probe is moved around the surgical space, the tip of the probe will correspond to the stereotaxic position if placed within the surgical opening. The surgeon can, therefore, see where the probe position lies at any time as well as what is near or beyond the probe's position. By providing an additional probe's eye view, these systems also provide information about guidance vectors to plan the surgical path. As such, these systems help to minimize the invasiveness of the planned surgery as well as increase the completeness of resection. Used preincision, they provide an excellent intraoperative tool for incision and craniotomy planning. Some systems provide input to a pointing device, integrating all imagery on a single screen; others include heads-up display. Inclusion of pointers that indicate the placement of a wand help blend all the available information into a unified presentation.

Functional data may also be melded to the anatomic navigation system to provide a functional guidance system. Preoperative functional mapping techniques such as positron emission tomography (PET) or functional MRI (fMRI) may be included with these systems to facilitate intraoperative functional identification. Recent studies have also provided a means for registering intraoperative functional maps to the visualization system. Modayur *et al.* (1998) has produced a system that can combine MRI, PET, ESM, and ECog data into a single neuronavigational tool. The application of functional data can minimize postoperative iatrogenic deficit.

Currently, the greatest limitation for these systems is the inability of preoperative scans to accurately reflect gross geometric changes during craniotomy and intraoperative manipulation. Cortical anatomy may have changed significantly due to patient position, craniotomy morphometry, swelling, drugs, or fluid drainage. There is still a need to combine the sophisticated deformable atlas approaches described above with real-time visualization and intraoperative procedures. Several bases to the deformable registration have been proposed. Efforts by G. Ojemann *et al.* have focused on the use of intracranial reregistration to the cortical surface to compensate for gross morphologic changes with craniotomy. This system reregisters the cortical surface based upon the position of the cortical blood vessels. Preoperatively, a MR angiogram is performed and integrated into the guidance system. After reflection of the dura, the brain surface is then reregistered to the branch points of the arteries with the MRA scan, subsequently minimizing cortical/navigational positioning errors. Vince *et al.* (1998) recently integrated intraoperative ultrasound with the BrainLab system. The ultrasound probe is tracked similarly to the pointer and tomographic images are reconstructed within the plane of the ultrasound. Future development may allow the realignment of the three-dimensional data set using ultra-

sound and positional data. Recently, intraoperative imaging technology has progressed sufficiently to allow the use of intraoperative MRI for practical operating room use. Ultimately, the use of a neuronavigational tool with an intraoperative MRI would provide the necessary means for constantly updating for gross morphologic and iatrogenic change.

VII. Summary

Intraoperative brain mapping has provided a rich collection of information about the structure and function of the human brain. Furthermore, it has helped fully characterize the complex, dynamic, transient, and distributed nature of brain function through direct observation using a diverse arsenal of mapping strategies. The unique opportunities for studying the brain unimpeded by other tissues make intraoperative mapping a critical contribution to both furthering basic scientific investigation and providing invaluable data to help guide and inform the surgeon. The advent of increasingly complex, sensitive, and integrative technology now enables the accurate combination of multiple-modality and multiple-subject data. While the most dramatic and oftentimes most poignant successes of intraoperative mapping have involved language, visual, and sensorimotor cortices, association and other brain regions stand to benefit from these well-proven and emerging technologies.

Acknowledgments

A.W.T. and A.F.C. thank the staff of the Laboratory of Neuro Imaging, without whom this work would not have been possible. A.W.T. and A.F.C. were supported, in part, by the National Institute of Mental Health (MH 52083), the National Center for Research Resources (P41-RR13642), the National Institute of Neurological Disorders and Stroke and the National Institute of Mental Health (NINDS/NIMH NS38753), the National Library of Medicine (LM/MH05639), the National Science Foundation (BIR 93-22434), and a Human Brain Project grant to the International Consortium for Brain Mapping, funded jointly by NIMH and NIDA (P20 MH/DA52176). G.A.O. and J. G.O. were supported, in part, by NIH Grants NS 36527 and DC/LM 02310 and the McDonnell-Pew Foundation.

References

Allison, T., McCarthy, G., Wood, C. C., *et al.* (1989). *J. Neurophysiol.* **62**, 711–722.

Austin, B., and Lee, A. (1958). A plastic ball and socket type stereotaxic detector. *J. Neurosurg.* **15**, 264–268.

Babb, T. L., Soper, H. V., Brown, W. J., Ottimo, C. A., and Crandall, P. H. (1977). Electrophysiological studies of long-term stimulation of the cerebellum in monkeys. *J. Neurosurg.* **47**, 535–565.

Bajcsy, R., and Kovacic, S. (1989). Multiresolution elastic matching. *Comput. Vision, Graph. Image Process.* **46**, 1–21.

Barnett, G. H., Kormos, D. W., Steiner, C. P., and Weisenberger, J. (1993). Use of a frameless, armless stereotactic wand for brain tumor localization with two-dimensional and three-dimensional neuroimaging. *Neurosurgery* **33,** 674–678.

Bartholow, R. (1874). Experimental investigations into functions of the human brain. *Am. J. Med. Sci.* **67,** 305–313.

Baumgartner, C., Doppelbauer, A., Sutherling, W. W., Lindinger, G., Levesque, M. F., Aull, S., Zeitlhofer, J., and Deecke, L. (1993). Somatotopy of human hand somatosensory cortex as studied in scalp EEG. *Electroencephalogr. Clin. Neurophysiol.* **88,** 271–279.

Benson, R. R., FitzGerald, D. B., LeSueur, L. L., Kennedy, D. N., Kwong, K. K., Buchbinder, B. R., Davis, T. L., Weisskoff, R. M., Talavage, T. M., Logan, W. J., Cosgrove, G. R., Belliveau, J. W., and Rosen, B. R. (1999). Language dominance determined by whole brain functional MRI in patients with brain lesions. *Neurology* **52,** 798–809.

Black, P. M. (1991). Brain tumors. *N. Engl. J. Med.* **324,** 1471–1476, 1555–1564.

Black, P. M., and Ronner, S. F. (1987). Cortical mapping for defining the limits of tumor resection. *Neurosurgery* **20,** 914–919.

Blood, A. J., Narayan, S. M., and Toga, A. W. (1995). Stimulus parameters influence characteristics of optical intrinsic responses in somatosensory cortex. *J. Cereb. Blood Flow Metab.* **15** (6), 1109–1121.

Bookheimer, S. Y., Zeffiro, T. A., Blaxton, T., Gaillard, W., and Theodore, W. H. (1995). Regional cerebral blood flow during object naming and word reading. *Hum. Brain Mapp.* **3,** 93–106.

Bookheimer, S. Y., Zeffiro, T. A., Blaxton, T., Malow, B. A., Gaillard, W. D., Sato, S., Kufta, C., Fedio, P., and Theodore, W. H. (1997). A direct comparison of PET activation and electrocortical stimulation mapping for language localization. *Neurology* **48,** 1056-1065.

Broit, C. (1981). "Optimal Registration of Deformed Images," Ph.D. Dissertation, University of Pennsylvania.

Brown, R. A., Roberts, T. S., and Osborn, A. G. (1980). Stereotaxic frame and computer software for CT-directed neurosurgical localization. *Invest. Radiol.* **15,** 308–312.

Cannestra, A. F., Blood, A. J., Black, K. L., and Toga, A. W. (1966). The evolution of optical signals in human and rodent cortex. *NeuroImage* **3,** 202–208.

Cannestra, A. F., Black, K. L., Martin, N. A., Cloughesy, T., Burton, J. S., Rubinstein, E., Woods, R. P., and Toga, A. W. (1998a). Topographical and temporal specificity of human intraoperative optical intrinsic signals. *NeuroReport* **9,** 2557–2563.

Cannestra, A. F., Pouratian, N., Shomer, M. H., and Toga, A. W. (1998b). Refractory periods observed by intrinsic signal and fluorescent dye imaging. *J. Neurophysiol.* **80** (3), 1522–1532.

Cannestra, A. F., Bookheimer, S. Y., O'Farrell, A., Sicotte, N., Martin, N. A., Becker, D., Rubino, G., and Toga, A. W. (1999a). Temporal and topographical characterization of language cortices utilizing intraoperative optical intrinsic signals (in press).

Cannestra, A. F., Bookheimer, S. Y., Martin, N. A., and Toga, A. W. (1999b). Temporal and spatial differences observed by functional MRI and human intraoperative optical imaging (in press).

Christensen, G. E., Rabbitt, R. D., and Miller, M. I. (1993). A deformable neuroanatomy textbook based on viscous fluid mechanics. *27th Annu. Conf. Inf. Sci. Syst.* 211–216.

Christensen, G. E., Miller, M. I., Marsh, J. L., and Vannier, M. W. (1995a). Automatic analysis of medical images using a deformable textbook. *Proc. Comput. Assist. Radiol.* 152–157.

Christensen, G. E., Rabbitt, R. D., Miller, M. I., Joshi, S. C., Grenander, U., Coogan, T. A., and Van Essen, D. C. (1995b). Topological properties of smooth anatomic maps. *In* "Information Processing in Medical Imaging" (Y. Bizais, C. Barillot, and R. Di Paola, Eds.), pp. 101–112.

Christensen, G. E., Rabbitt, R. D., and Miller, M. I. (1996). Deformable templates using large deformation kinematics. *IEEE Trans. Image Process.* **5** (10), 1435–1447.

Cohen, L. B. (1973). Changes in neuron structure during action potential propagation and synaptic transmission. *Physiol. Rev.* **53** (2), 373–418.

Cohen, M. S., and Bookheimer, S. Y. (1994). Localization of brain function using magnetic resonance imaging. *Trends Neurosci.* **17** (7), 268–276.

Collins, D. L., Le Goualher, G., Venugopal, R., Caramanos, A., Evans, A. C., and Barillot, C. (1996). Cortical constraints for non-linear cortical registration. *In* "Visualization in Biomedical Computing" (K. H. Höhne and R. Kikinis, Eds.). Springer-Verlag, Hamburg, Germany.

Creutzfeldt, O., and Ojemann, G. A. (1989). Neuronal activity in the human lateral temporal lobe. III: Activity changes during music. *Exp. Brain Res.* **77,** 490–498.

Creutzfeldt, O., Ojemann, G. A., and Lettich, E. (1989a). Neuronal activity in the human lateral temporal lobe. I: Responses to speech. *Exp. Brain Res.* **77,** 451–475.

Creutzfeldt, O., Ojemann, G. A., and Lettich, E. (1989b). Neuronal activity in the human lateral temporal lobe. II: Responses to the subject's own voice. *Exp. Brain Res.* **77,** 476–489.

Cushing, H. (1909). A note upon the faradic stimulation of the postcentral gyrus in conscious patients. *Brain* **32,** 44–54.

Davatzikos, C. (1996). Spatial normalization of 3D brain images using deformable models. *J. Comput. Assist. Tomogr.* **20** (4), 656–665.

Dinner, D. S., Luder, H., Lesser, R. P., and Morris, H. H. (1986). Invasive methods of somatosensory evoked potential monitoring. *J. Clin. Neurophysiol.* **3,** 113–130.

Fiez, J. A., Raichle, M. E., Balota, D. A., Tallel, P., and Petersen, S. E. (1996). PET activation of posterior temporal regions during auditory word presentation and verb generation. *Cereb. Cortex* **6,** 1–10.

FitzGerald, D. B., Cosgrove, G. R., Ronner, S., Jiang, H., Buchbinder, B. R., Belliveau, J. W., Rosen, B. R., and Benson, R. R. (1997). Location of language in the cortex: A comparison between functional MR imaging and electrocortical stimulation. *Am. J. Neuroradiol.* **18,** 1529–1539.

Foerster, O. (1931). The cerebral cortex of man. *Lancet* **109,** 309–312.

Fox, N. C., Freeborough, P. A., and Rossor, M. N. (1996). Visualization and quantification of rates of cerebral atrophy in Alzheimer's disease. *Lancet* **348** (9020), 94–97.

Fox, P. T., Perlmutter, J. S., and Raichle, M. (1985). A stereotactic method of localization for positron emission tomography. *J. Comput. Assist. Tomogr.* **9** (1), 141–153.

Fox, P. T., Burten, H., and Raichle, M. E. (1987). Mapping human somatosensory cortex with positron emission tomography. *J. Neurosurg.* **67,** 34–43.

Fox, P. T., Mintun, M. A., Reiman, E. M., and Raichle, M. E. (1988). Enhanced detection of focal brain responses using inter-subject averaging and change distribution analysis of subtracted PET images. *J. Cereb. Blood Flow Metab.* **8,** 642–653.

Freeborough, P. A., Woods, R. P., and Fox, N. C. (1996). Accurate registration of serial 3D MR brain images and its application to visualizing change in neurodegenerative disorders. *J. Comput. Assist. Tomogr.* **20** (6), 1012–1022.

Fried, I., Ojemann, G. A., and Fetz, E. E. (1981). Language-related potentials specific to human language cortex. *Science* **212,** 353–356.

Fried, I., Mateer, C., Ojemann, G., Wohns, R., and Fedio, P. (1982). Organization of visuospatial functions in human cortex: Evidence from electrical stimulation. *Brain* **105,** 349–371.

Fried, I., Nenov, V. I., Ojemann, S. G., and Woods, R. W. (1995). Functional MR and PET imaging of Rolandic and visual cortices for neurosurgical planning. *J. Neurosurg.* **83,** 854–861.

Friston, K. J., Passingham, R. E., Nutt, J. G., Heather, J. D., Sawle, G. V., and Frackowiak, R. S. J. (1989). Localization in PET images: Direct fitting of the intercommissural (AC-PC) line. *J. Cereb. Blood Flow Metab.* **9,** 690–695.

Friston, K. J., Frith, C. D., Liddle, P. F., and Frackowiak, R. S. J. (1991). Plastic transformation of PET images. *J. Comput. Assist. Tomogr.* **9** (1), 141–153.

Fritsch, G., and Hitzig, E. (1870). Uber die elektrische erregbarkeit des Grosshirns. *Arch. Anatom. Wissenschaft. Med.* **37,** 300–332.

Frostig, R. D., Lieke, E. E., Ts'o, D. Y., and Grinvald, A. (1990). Cortical functional architecture and local coupling between neuronal activity and the microcirculation revealed by *in vivo* high-resolution optical imaging of intrinsic signals. *Proc. Natl. Acad. Sci. U.S.A.* **87,** 6082–6086.

Gee, J. C., Reivich, M., and Bajcsy, R. (1993). Elastically deforming an atlas to match anatomical brain images. *J. Comput. Assist. Tomogr.* **17** (2), 225–236.

Gee, J. C., LeBriquer, L., Barillot, C., Haynor, D. R., and Bajcsy, R. (1995). Bayesian approach to the brain image matching problem. *Inst. Res. Cog. Sci. Tech. Rep.* 95–108.

Geschwind, N., and Levitsky, W. (1968). Human brain: Left-right asymmetries in temporal speech region. *Science* **161,** 186–187.

Gildenberg, P. L. (1987). Whatever happened to stereotactic surgery? *Neurosurgery* **20,** 983–987.

Gloor, P. A., Olivier, A., Quesney, L. F., Andermann, F., and Horowitz, S. (1982). The role of the limbic system in experimental phenomena of temporal lobe epilepsy. *Ann. Neurol.* **12,** 129–144.

Goldring, S., Aras, E., and Weber, P. C. (1970). *Electroencephalogr. Clin. Neurophysiol.* **29,** 537–550.

Gordon, B., Lesser, R. P., Rance, N. E., Hart, J., Webber, R., Uematsu, S., and Fisher, R. S. (1990). Parameters for direct cortical stimulation in the human: Histopathologic confirmation. *Electroencephalogr. Clin. Neurophysiol.* **75,** 371–377.

Grinvald, A., Lieke, E., Frostig, R. D., Gilbert, C. D., and Wiesel, T. N. (1986). Functional architecture of cortex revealed by optical imaging of intrinsic signals. *Nature* **324,** 361–364.

Haglund, M. M., Ojemann, G. A., and Hochman, D. W. (1992). Optical imaging of epileptiform and functional activity in human cerebral cortex. *Nature* **358,** 668–671.

Haglund, M. M., Ojemann, G. A., and Blasdel, G. G. (1993a). Optical imaging of bipolar cortical stimulation. *J. Neurosurg.* **78,** 785–793.

Haglund, M. M., Ojemann, G. A., Lettich, E., Bellugi, U., and Corina, D. (1993b). Dissociation of cortical and single unit activity in spoken and signed languages. *Brain Lang.* **44,** 19–27.

Haglund, M. M., Berger, M. S., Shamseldin, M. S., Lettich, E., and Ojemann, G. A. (1994a). Cortical localization of temporal lobe language sites in patients with gliomas. *Neurosurgery* **34,** 239–253.

Haglund, M. M., Ojemann, G. A., Schwartz, T. W., and Lettich, E. (1994b). Neuronal activity in human lateral temporal cortex during serial retrieval from short-term memory. *J. Neurosci.* **14,** 1507–1515.

Hardy, T. L. (1994). Computerized atlas for functional stereotaxis, robotics and radiosurgery. *SPIE* **2359,** 447–456.

Heilbrun, M. P., McDonald, P., Wilker, C., *et al.* (1992). Stereotactic localization and guidance using a machine vision technique. *Stereotact. Funct. Neurosurg.* **58,** 94–98.

Heilman, K., Wilder, B., and Malzone, W. (1972). Anomic aphasia following anterior temporal lobectomy. *Trans. Am. Neurol. Assoc.* **98,** 291–293.

Hill, D. K., and Keynes, R. D. (1949). Opacity changes in stimulated nerve. *J. Physiol. (London)* **108,** 278–281.

Holmes, M. D., Ojemann, G. A., and Lettich, E. (1996). Neuronal activity in human right lateral temporal cortex related to visuospatial memory and perception. *Brain Res.* **711,** 44–49.

Holthoff, K., and Witte, O. W. (1996). Intrinsic optical signals in rat neocortical slices measured with near-infrared dark-field microscopy reveal changes in extracellular space. *J. Neurosci.* **16** (8), 2740–2749.

Horsley, V., and Clarke, R. W. (1908). The structure and functions of the cerebellum examined by a new method. *Brain* **31,** 45–124.

Ingvar, D. H., and Lassen, N. A. (1994). Regional blood flow in the cerebral cortex determined by krypton 85. *Acta Physiol. Scand.* **54,** 325.

Jack, C. R., Jr., Thompson, R. M., Butts, R. K., Sharbrough, F. W., Kelly, P. J., Hanson, D. P., Riederer, S. J., Ehman, R. L., Hanagiandreou, N. J., and Cascino, G. D. (1994). Sensory motor cortex: Correlation of presurgical mapping with functional MR imaging and invasive cortical mapping. *Radiology* **190,** 85–92.

Jacobs, R. E., and Fraser, S. E. (1994). Magnetic resonance microscopy of embryonic cell lineages and movements. *Science* **263** (5147), 681–684.

Jayakar, P., Alvarez, L. A., Duchowny, M. S., and Resnick, T. J. (1992). A safe and effective paradigm to functionally map the cortex in childhood. *J. Clin. Neurophysiol.* **9,** 288–293.

Johnson, M. D., and Ojemann, G. A. (2000). Role of the human thalamus in language and memory: Evidence from electrophysiological studies. *Brain Cog.* (in press).

Kaas, J. H., Nelson, R. J., Sur, M., Lin, C. S., and Merzenich, M. M. (1979). Multiple representations of the body within the primary somatosensory cortex of primates. *Science* **204** (4392), 521–523.

Kato, A., Yoshimine, T., Hayakawa, T., *et al.* (1991). A frameless, armless navigational system for computer assisted neurosurgery. *J. Neurosurg.* **74,** 845–849.

Kelly, P. J., and Alker, G. J., Jr. (1980). A method for stereotactic laser microsurgery in the treatment of deep seated CNS neoplasms. *Appl. Neurophysiol.* **43,** 210–215.

Kelly, P. J., Kall, B., Goerss, S., and Alker, G. J. (1983). Precision resection of intra-axial CNS lesions by CT-based stereotactic craniotomy and computer monitored CO2 laser. *Acta Neurochir. (Wien)* **68,** 1–9.

Kelly, P. J., Kall, B. A., and Goerss, S. (1984). Transposition of volumetric information derived from computed tomography scanning into stereotactic space. *Surg. Neurol.* **21,** 465–471.

Kikinis, R., Shenton, M. E., Iosifescu, D. V., McCarley, R. W., Saiviroonporn, P., Hokama, H. H., Robatino, A., Metcalf, D., Wible, C. G., Portas, C. M., Donnino, R., and Jolesz, F. (1996). A digital brain atlas for surgical planning, model-driven segmentation, and teaching. *IEEE Trans. Vis. Comput. Graph.* **2** (3), 232–241.

Leksell, L. (1949). A stereotaxic device for intracerebral surgery. *Acta Chr. Scand.* **99,** 229–233.

Leksell, L., and Jernberg, B. (1980). Stereotaxis and tomography: A technical note. *Acta Neurochir.* **52,** 1–7.

LeManna, J. C., Sick, T. J., Pirarsky, S. M., and Rosenthal, M. (1987). Detection of an oxidizable fraction of cytochrome oxidase in intact rat brain. *Am. Physiol. Soc.* **253,** C477–C483.

Lindauer, U., Villringer, A., and Dirnagl, U. (1993). Characterization of CBF response to somatosensory stimulation: Model and influence of anesthetics. *Am. Physiol. Soc.* **264,** H1223–H1228.

Lou, H. C., Edvinsson, L., and MacKenzie, E. T. (1987). The concept of coupling blood flow to brain function: Revision required? *Ann. Neurol.* **22,** 289–297.

Luders, H., Lesser, R. P., Hahn, J., *et al.* (1986). Basal temporal language area demonstrated by electrical stimulation. *Neurology* **36,** 505–510.

Luders, H., Lesser, R. P., Marris, H. H., Dinner, D. S., and Hahn, J. (1987). Negative motor responses elicited by stimulation of the human cortex. *Adv. Epileptol.* **16,** 229–231.

MacDonald, D., Avis, D., and Evans, A. C. (1993). Automatic parameterization of human cortical surfaces. *Annu. Symp. Info. Proc. Med. Imaging* (IPMI).

Malonek, D., and Grinvald, A. (1996). Interactions between electrical activity and cortical microcirculation revealed by imaging spectroscopy: Implications for functional brain mapping. *Science* **272,** 551–554.

Mateer, C. (1978). Asymmetric effects of thalamic stimulation on rate of speech. *Neuropsychologica* **16,** 497–499.

Mateer, C. A., Polen, S. B., Ojemann, G. A., *et al.* (1982). Cortical localization of finger spelling and oral language: A case study. *Brain Lang.* **17,** 46–57.

Mazziotta, J. C., Toga, A. W., Evans, A. C., Fox, P., and Lancaster, J. (1995). A probabilistic atlas of the human brain: Theory and rationale for its development. *NeuroImage* **2,** 89–101.

Missir, O., Dutheil-Desclercs, C., Meder, J. F., Musolino, A., and Fredy, D. (1989). Central sulcus patterns at MRI. *J. Neuroradiol.* **16,** 133–144.

Modayur, B., Prothero, J., Ojemann, G., Maravilla, K., and Brinkley, J. (1997). Visualization-based mapping of language function in the brain. *NeuroImage* **6,** 245–258.

Mueller, W. M., Yetkin, F. Z., Hammeke, T. A., Morris, G. L., Swanson, S. J., Reichert, K., Cox, R., and Haughton, V. M. (1996). Functional magnetic resonance imaging mapping of the motor cortex in patients with cerebral tumors. *Neurosurgery* **39,** 515–521.

Narayan, S. M., Esfahani, P., Blood, A. J., Sikkens, L., and Toga, A. W. (1995). Functional increases in cerebral blood volume over somatosensory cortex. *J. Cereb. Blood Flow Metab.* **15,** 754–765.

Nazzarro, J. M., and Neuwelt, E. A. (1990). The role of surgery in the management of supratentorial intermediate and high-grade astrocytomas in adults. *J. Neurosurg.* **73,** 331–344.

Nuwer, M. R., Banoczi, W. R., Cloughesy, T. F., Hoch, D. B., Peacock, W., Levesque, M. F., Black, K. L., Martin, N. A., and Becker, D. P. (1992). Topographic mapping of somatosensory evoked potential helps identify motor cortex more quickly in the operating room. *Brain Topogr.* **5** (1), 53–58.

Ojemann, G. A. (1978). Organization of short-term verbal memory in language areas of human cortex: Evidence from electrical stimulation. *Brain Lang.* **5,** 331–348.

Ojemann, G. A. (1979). Altering memory with human ventrolateral thalamic stimulation. *In* "Modern Concepts in Psychiatric Surgery" (E. R. Hitchcock, H. T. Ballantine, Jr., and B. A. Meyerson, Eds.), pp. 103–109. Elsevier, Amsterdam/New York.

Ojemann, G. A. (1983a). Electrical stimulation and the neurobiology of language. *Behav. Brain Sci.* **6,** 221–230.

Ojemann, G. A. (1983b). Brain organization for language from the perspective of electrical stimulation mapping. *Behav. Brain Sci.* **6,** 189–206.

Ojemann, G. A. (1989). Some brain mechanisms for reading. *In* "Brain and Reading" (C. Von Euler, I. Lundberg, and G. Lennerstrand, Eds.), pp. 47–59. Macmillan, New York.

Ojemann, G. A. (1991a). Cortical organization of language. *J. Neurosci.* **11,** 2281–2287.

Ojemann, G. A. (1991b). Cortical organization of language and verbal memory based on intraoperative investigations. *Prog. Sensory Physiol.* **12,** 193–230.

Ojemann, G. A. (1994). Cortical stimulation and recording in language. *In* "Localization and Neuroimaging in Neuropsychology" (A. Kertesz, Ed.), pp. 35–55. Academic Press, San Diego.

Ojemann, G. A., and Dodrill, C. B. (1985). Verbal memory deficits after left temporal lobectomy for epilepsy. *J. Neurosurg.* **62,** 101–107.

Ojemann, G. A., and Dodrill, C. B. (1987). Intraoperative techniques for reducing language and memory deficits with left temporal lobectomy. *In* "Advances in Epileptology: The 16th Epilepsy International Symposium" (P. Wolf *et al.,* Eds.), pp. 327–330. Raven Press, New York.

Ojemann, G. A., and Mateer, C. (1979). Human language cortex: Localization of memory, syntax, and sequential motor-phoneme identification systems. *Science* **205** (4413), 1401–1403.

Ojemann, G. A., and Schoenfield-McNeill, J. (1998). Neurons in human temporal cortex active with verbal associative learning. *Brain Lang.* **64,** 317–327.

Ojemann, G. A., and Schoenfield-McNeill, J. (1999). Activity of neurons in human temporal cortex during identification and memory for names and words. *J. Neurosci.* **19,** 5674–5682.

Ojemann, G. A., and Ward, A. A. (1971). Speech representation in ventrolateral thalamus. *Brain* **94,** 669–680.

Ojemann, G. A., and Whitaker, H. A. (1978). The bilingual brain. *Arch. Neurol.* **35,** 409–412.

Ojemann, G. A., Fedio, P., and Van Buren, J. M. (1968). Anomia from pulvinar and subcortical parietal stimulation. *Brain* **91,** 99–116.

Ojemann, G. A., Blick, K. I., and Ward, A. A. (1971). Improvement and disturbance of short-term verbal memory with human ventrolateral thalamic stimulation. *Brain* **94,** 225–240.

Ojemann, G. A., Creutzfeldt, O., Lettich, E., and Haglund, M. M. (1988). Neuronal activity in human lateral temporal cortex related to short-term verbal memory, naming and reading. *Brain* **111,** 1383–1403.

Ojemann, G. A., Ojemann, J. G., Lettich, E., and Berger, M. S. (1989a). Cortical language localization in left, dominant hemisphere. *J. Neurosurg.* **71,** 316–326.

Ojemann, G. A., Fried, I., and Lettich, E. (1989b). Electrocorticographic (EcoG) correlates of language. I. Desynchronization in temporal language cortex during object naming. *Electroencephalogr. Clin. Neurophysiol.* **73,** 453–463.

Ojemann, G. A., Sutherling, W., Lesser, R., Dinner, D., Jayakar, P., and Saint-Hilaire, J. (1993). Cortical stimulation. *In* "Surgical Treatment of the Epilepsies" (J. Engle, Jr., Ed.), 2nd ed., pp. 399–414. Raven Press, New York.

Ojemann, G. A., Ojemann, J., and Fried, I. (1998). Lessons from the human brain: Neuronal activity related to cognition. *Neuroscientist* **4,** 285–300.

Ojemann, J. G., Ojemann, G. A., and Lettich, E. (1992). Neuronal activity related to faces and matching in human right nondominant temporal cortex. *Brain* **115,** 1–13.

Ojemann, J., Ojemann, G., and Lettich, E. (1993). Cortical stimulation during a language task with known blood flow changes. *Soc. Neurosci. Abstr.* **19,** 1808.

Ojemann, J. G., Neil, J. M., MacLeod, A. M., Silbergeld, D. L., Dacey, R. G., Jr., Petersen, S. E., and Raichle, M. E. (1998). Increased functional vascular response in the region of a glioma. *J. Cereb. Blood Flow Metab.* **18,** 148–153.

Ono, M., Kubik, S., and Abernathey, C. D. (1990). "Atlas of the Cerebral Sulci." Thieme, Stuttgart.

Orbach, H. S., and Cohen, L. B. (1983). Optical monitoring of activity from many areas of the *in vitro* and *in vivo* salamander olfactory bulb: A new method for studying functional organization in the vertebrate central nervous system. *J. Neurosci.* **3** (11), 2251–2262.

Payne, B. A., and Toga, A. W. (1990). Surface mapping brain function on 3D models. *Comput. Graph. Appl.* **10** (5), 33–41.

Penfield, W., and Boldrey, E. (1937). Somatic motor and sensory representation in the cerebral cortex of man as studied by electrical stimulation. *Brain* **60,** 389–443.

Penfield, W., and Perot, P. (1963). The brain's record of auditory and visual experience-A final summary and discussion. *Brain* **86,** 595–696.

Penfield, W., and Rasmussen, T. (1950). "The Cerebral Cortex of Man: A Clinical Study of Localization of Function." Macmillan, New York.

Penfield, W., and Roberts, L. (1959). "Speech and Brain Mechanisms." Princeton Univ. Press, Princeton, NJ.

Perrine, K., Devinsky, O., Uysal, S., Luciano, D., and Dogali, M. (1994). Left temporal neocortex mediation of verbal memory. *Neurology* **44,** 1845–1850.

Phelps, M. E., and Mazziotta, J. C. (1985). Positron emission tomography: Human brain function and biochemistry. *Science* **228,** 799–809.

Puce, A., Constable, R. T., Luby, M. L., McCarthy, G., Nobre, A. C., Spencer, D. D., Gore, J. C., and Allison, T. (1995). Functional magnetic resonance imaging of sensory and motor cortex: Comparison with electrophysiological localization. *J. Neurosurg.* **83,** 262–270.

Rademacher, J., Caviness, V. S., Jr., Steinmetz, H., and Galaburda, A. M. (1993). Topographical variation of the human primary cortices: Implications for neuroimaging, brain mapping and neurobiology. *Cereb. Cortex* **3** (4), 313–329.

Raichle, M. E., Fiez, J. A., Videen, T. O., MacLeod, A. K., Pardo, J. V., Fox, P. T., and Petersen, S. E. (1994). Practice-related changes in human brain functional anatomy during nonmotor learning. *Cereb. Cortex* **4**, 8–26.

Rajkowska, G., and Goldman-Rakic, P. (1995). Cytoarchitectonic definition of pre-frontal areas in the normal human cortex: II. Variability in locations of areas 9 and 46 and relationship to the Talairach coordinate system. *Cereb. Cortex* **5** (4), 323–337.

Rao, S. M., Binder, J. R., Bandettini, P. A., Hammeke, T. A., Yetkin, F. Z., Jesmanowicz, A., Lisk, L. M., Morris, G. L., Mueller, M. W., and Estkowski, L. D. (1993). Functional magnetic resonance imaging of complex human movements. *Neurology* **43**, 2311–2318.

Roberts, D. W., Strohbehn, J. W., Hatch, J. F., *et al.* (1986). A frameless stereotactic integration of computerized tomographic imaging and the operating microscope. *J. Neurosurg.* **65**, 545–549.

Roberts, D. W., Strohbehn, J. W., Friets, E. M., *et al.* (1989). The stereotactic operating microscope: Accuracy, refinement and clinical experience. *Acta Neurochir. Suppl. (Wien)* **46**, 112–114.

Roland, P. E., and Zilles, K. (1994). Brain atlases—A New Research Tool. *Trends Neurosci.* **17** (11), 458–467.

Rosen, H. J., Ojemann, J. G., Ollinger, J. M., and Petersen, S. E. (2000). Activation during word retrieval done silently and aloud using fMRI. *Brain Cog.* (in press).

Rumeau, C., Tzourio, N., Murayama, N., Peretti-Viton, P., Levrier, O., Joliot, M., Mazoyer, B., and Salamon, G. (1994). Location of hand function in the sensorimotor cortex: MR and functional correlation. *Am. J. Neuroradiol.* **15**, 567–572.

Schaltenbrand, G., and Wahren, W. (1977). "Atlas for Stereotaxy of the Human Brain," 2nd ed. Thieme, Stuttgart.

Schwartz, T. H., Ojemann, G. A., Haglund, M. M., and Lettich, E. (1996). Cerebral lateralization of neuronal activity during naming, reading, and line-matching. *Cog. Brain Res.* **4**, 263–273.

Seitz, R. J., Bohm, C., Greitz, T., Roland, P. E., Eriksson, L., Blomqvist, G., Rosenqvist, G., and Nordell, B. (1990). Accuracy and precision of the Computerized Brain Atlas Programme for Localization and Quantification in Positron Emission Tomography. *J. Cereb. Blood Flow Metab.* **10**, 443–457.

Speigel, E. A. (1982). History of human stereotaxy (stereoencephalotomy). *In* "Stereotaxy of the Human Brain: Anatomical, Physiological and Clinical Applications" (G. Schaltenbrand and A. E. Walker, Eds.), pp. 3–10. Thieme, Stuttgart.

Spiegel, E. A., and Wycis, H. T. (1950). Pallido-thalotomy in chorea. *Arch. Neurol. Psychiatry* **64**, 495–496.

Spiegel, E. A., and Wycis, H. T. (1952). "Stereoencephalotomy," Part 1. Grune & Stratton, New York.

Steinmetz, H., Furst, G., and Freund, H.-J. (1989). Cerebral cortical localization: Application and validation of the proportional grid system in MR imaging. *J. Comput. Assist. Tomogr.* **13** (1), 10–19.

Steinmetz, H., Furst, G., and Freund, H.-J. (1990). Variation of perisylvian and calcarine anatomic landmarks within stereotaxic proportional coordinates. *Am. J. Neuroradiol.* **11** (6), 1123–1130.

Stensaas, S. S., Eddington, D. K., and Dobelle, W. H. (1974). The topography and variability of the primary visual cortex in man. *J. Neurosurg.* **40**, 747–755.

St.-Jean, P., Sadikot, A. F., Collins, L., Clonda, D., Kasrai, R., Evans, A. C., and Peters, T. M. (1998). Automated atlas integration and interactive three-dimensional visualization tools for planning and guidance in functional neurosurgery. *IEEE Trans. Med. Imaging* **17** (5), 672–680.

Sundsten, J. W., Kastella, J. G., and Conley, D. M. (1991). Videodisc animation of 3D computer reconstructions of the human brain. *J. Biomed. Commun.* **18**, 45–49.

Sutherling, W., Crandall, P., Darcey, T., Becker, D., Levesque, M., and

Barth, D. (1988). The magnetic and electric fields agree with intracranial localization of somatosensory cortex. *Neurology* **38**, 1705–1714.

Sutherling, W. W., Leveaque, M. F., and Baumgartner, C. (1992). Cortical sensory representation of the human hand: Size of finger regions and nonoverlapping digit somatotopy. *Neurology* **42**, 1020–1028.

Talairach, J., and Szikla, G. (1967). "Atlas d'Anatomie Stereotaxique du Telencephale: Etudes Anatomo-Radiologiques." Masson & Cie, Paris.

Talairach, J., and Tournoux, P. (1988). "Co-planar Stereotaxic Atlas of the Human Brain." Thieme, New York.

Thirion, J.-P., Prima, S., and Subsol, S. (1998). Statistical analysis of dissymmetry in volumetric medical images. *In* "Medical Image Analysis" (in press).

Thompson, P. M., and Toga, A. W. (1996). A surface-based technique for warping 3–dimensional images of the brain. *IEEE Trans. Med. Imaging* **15** (4), 1–16.

Thompson, P. M., Schwartz, C., and Toga, A. W. (1996a). High-resolution random mesh algorithms for creating a probabilistic 3D surface atlas of the human brain. *NeuroImage* **3**, 19–34.

Thompson, P. M., Schwartz, C., Lin, R. T., Khan, A. A., and Toga, A. W. (1996b). 3D statistical analysis of sulcal variability in the human brain. *J. Neurosci.* **16** (13), 4261–4274.

Thompson, P. M., Giedd, J. N., Blanton, R. E., Lindshield, C., Woods, R. P., MacDonald, D., Evans, A. C., and Toga, A. W. (1998a). Growth Patterns in the Developing Human Brain Detected Using Continuum-Mechanical Tensor Maps and Serial MRI, 5th International Conference on Human Brain Mapping, Montreal, Canada.

Thompson, P. M., Moussai, J., Khan, A. A., Zohoori, S., Goldkorn, A., Mega, M. S., Small, G. W., Cummings, J. L., and Toga, A. W. (1998b). Cortical variability and asymmetry in normal aging and Alzheimer's disease. *Cereb. Cortex* (in press).

Thompson, P. M., Narr, K. L., Blanton, R. E., and Toga, A. W. (1998c). Mapping structural alterations of the corpus callosum during brain development and degeneration. *In* "The Corpus Callosum" (M. Iacoboni and E. Zaidel, Eds.). Kluwer Academic, Dordrecht/Norwell, MA (in press).

Thurfjell, L., Bohm, C., Greitz, T., and Eriksson, L. (1993). Transformations and algorithms in a computerized brain atlas. *IEEE Trans. Nucl. Sci., Part 1* **40** (4), 1167–1191.

Toga, A. W. (1994). Visualization and warping of multimodality brain imagery. *In* "Functional Neuroimaging: Technical Foundations" (R. W. Thatcher, M. Hallett, T. Zeffiro, E. R. John, and M. Huerta, Eds.), pp. 171–180.

Toga, A. W. (1998). "Brain Warping." Academic Press, San Diego.

Toga, A. W., and Mazziotta, J. C. (1996). "Brain Mapping: The Methods." Academic Press, San Diego.

Toga, A. W., Cannestra, A. F., and Black, K. L. (1995). The temporal/spatial evolution of optical signals in human cortex. *Cereb. Cortex* **5** (6), 561–565.

Toga, A. W., Thompson, P. M., and Payne, B. A. (1996). Modeling morphometric changes of the brain during development. *In* "Developmental Neuroimaging: Mapping the Development of Brain and Behavior" (R. W. Thatcher, G. R. Lyon, J. Rumsey, and N. Krasnegor, Eds.). Academic Press, San Diego.

Troster, A. I., Fields, J. A., Wilkinson, S. B., Pahwa, R., Miyawaki, E., Lyons, K. E., and Koller, W. C. (1997). Unilateral pallidal stimulation for Parkinson's disease: Neurobehavioral functioning before and 3 months after electrode implantation. *Neurology* **49**, 1078–1083.

Ts'o, D. Y., Frostig, R. D., Lieke, E. E., and Grinvald, A. (1990). Functional organization of primate visual cortex revealed by high resolution optical imaging. *Science* **249**, 417–420.

Vince, G. H., Krone, A., Woydt, M., and Roosen, K. (1998). Real-time ultrasound fusion in optical tracking MR/CT image guided surgery. *Eur. Soc. Stereotact. Func. Neurosurg.*, Freiberg, Germany.

Watanabe, E., Watanabe, T., Manaka, E., *et al.* (1987). Three-dimensional digitizer (Neuronavigator): New equipment for CT-guided stereotaxic surgery. *Surg. Neurol.* **27,** 543–547.

Weber, P. B., and Ojemann, G. A. (1995). Neuronal recordings in human lateral temporal lobe during verbal paired associate learning. *NeuroReport* **6,** 685–689.

Wood, C. C., Spencer, D. D., Allison, T., McCarthy, G., Williamson, P. D., and Goff, W. R. (1988). Localization of human sensorimotor cortex during surgery by cortical surface recording of somatosensory evoked potentials. *J. Neurosurg.* **68,** 99–111.

Yetkin, F. Z., Mueller, W. M., Morris, G. L., McAuliffe, T. L., Ulmer, J. L., Cox, R. W., Daniels, D. L., and Haughton, V. M. (1997). Functional MR activation correlated with intraoperative cortical mapping. *Am. J. Neuroradiol.* **18,** 1311–1315.

5

Intraoperative Visualization

Ron Kikinis,*,[1] Neerav R. Mehta,* Arya Nabavi,[†] Emmanouel Chatzidakis,[†] Simon Warfield,*
David Gering,[‡] Neil Weisenfeld,[‡] Richard S. Pergolizzi, Jr.,* Richard B. Schwartz,*
Nobuhiko Hata,* William Wells III,[‡] Eric Grimson,[‡] Peter McL. Black,[†] and Ferenc A. Jolesz*

*Department of Radiology, Brigham and Women's Hospital, Boston, Massachusetts 02115
[†]Department of Neurosurgery, Brigham and Women's Hospital, Boston, Massachusetts 02115
[‡]Artificial Intelligence Laboratory, Massachusetts Institute of Technology, Cambridge, Massachusetts 02139

I. Background and Introduction
II. Surgical Planning
III. Intraoperative MRI
IV. Concluding Remarks
    References

I. Background and Introduction

Conventional surgery has always relied upon the surgeon's "eye"—the surgeon's ability to determine anatomic relationships and discriminate pathology based on direct visualization. To distinguish tumor from normal parenchyma, abnormal from normal, is what gives the surgeon the unique ability to wield the scalpel and judge what is and what is not to be resected. Today, many surgeries are still performed as they were 50 years ago, despite the increased level of technology in imaging and minimally invasive procedures. The surgeon's vision is obstructed by the surfaces through which visible light cannot penetrate and restricted spatially by the size of incisions. Large incisions are made to facilitate direct visualization of the tissue in question but should be avoided for their invasiveness. The recent trend, minimally invasive surgery, however, has resulted in further limitation in surgical visualization. Through the use of small keyhole incisions and endoscopes, the visualization of the entire operational volume is impossible and the surgeon's eye should be complemented with image-based information.

On the other side of the fence lies radiology, which has always been tied to imaging technology. Both the diagnostic and therapeutic fields of radiology have been employing techniques that demonstrate both the normal anatomy and the pathology without the limitation of direct visualization. In the 1970s interventional treatment virtually exploded onto the medical arena, replacing some open surgical procedures with less invasive, technology-driven therapy. Percutaneous procedures (biopsies, drainages, thermal ablations, etc.) performed under X-ray fluoroscopic, ultrasound (US), and/or computed tomographic (CT) guidance have been introduced. Catheter-based techniques of angioplasty, stent deployment, and aneurysm coiling have all become part of everyday medical practice. These procedures are truly image guided and

---

[1] To whom correspondence should be addressed.

they are accomplished based on information only available from imaging modalities. Simultaneously, computerized image processing and visualization tools have been introduced for various surgical and radiation oncology applications. These algorithms process preoperatively acquired images to create anatomical models for localization, targeting, and visualization in three dimension (3D). These models were applied for radiation therapy procedures to calculate radiation doses and optimize trajectories for beams. The models were also employed for surgical planning and used for defining various access strategies and to simulate a planned surgery. In neurosurgery, the original frame-based stereotactic method has been replaced by frameless navigational systems to achieve intraoperative guidance utilizing the preoperatively created 3D models that are registered to the patient.

Now, these two fields, surgery and radiology, travel on the same course toward the ultimate goal of minimally invasive, image-guided procedures. Computers, image processing and visualization tools, advanced imaging technology, and modern therapeutic devices will all fuel the drive toward further integration between these two fields. Recently, at the Brigham and Women's Hospital at Harvard Medical School (Boston, Massachusetts), a full collaborative effort between radiologists, surgeons, and other professionals has been consolidated into an interdisciplinary Image Guided Therapy Program that aims for the refinement of image-guided techniques and the development of novel treatment approaches.

Minimally invasive therapy is a rather new concept that has recently appeared in modern-day medical practice. Image-guided surgery is a good example, especially when computers and "hi-tech" therapy devices are integrated in image-guided procedures. Unfortunately, the full utilization of these technological resources has only minimally influenced conventional, open surgery. However, the influence has taken hold, and minimally invasive surgical procedures are becoming more and more popular. Such procedures result not only in smaller incisions but also faster recovery times as well as lower morbidity and mortality.

A. Limitations of Conventional Surgery

Surgical visualization is inherently incomplete. The constraints of the surgical opening do not allow the therapist to fully appreciate the three-dimensional anatomy hidden behind the exposed surface. With minimally invasive surgeries being carried out through smaller incisions and endoscopic keyholes, this visualization is hampered even further. The need for image guidance is becoming more apparent with the ever decreasing surgical field.

Another limitation in surgical visualization is incomplete localization. Direct visualization of a tumor may not always ensure accurate identification of its borders. Radiologic information can complement this, allowing for more complete resection of pathology and minimizing damage to normal anatomy. Deep-seated lesions can also be reached with minimum damage to surrounding parenchyma with the help of image guidance (Jolesz, 1997).

B. Image-Guided Therapy

Image guidance provides a solution to the aforementioned limitations in conventional surgery. The volumetric information and the improved tissue characterization not only can help reduce the invasiveness of the procedure but can also help in the localization of the lesion. Targeting of lesions and delineation of normal anatomy surrounding the lesions can be accomplished by a variety of imaging modalities, including ultrasound, fluoroscopy, and magnetic resonance imaging (MRI). The first two of these imaging modalities are already in use in today's operating rooms. MRI, on the other hand, is just beginning to be used in this environment for intraoperative image guidance (Jolesz and Silverman, 1995; Black *et al.*, 1997; Nakajima *et al.*, 1997).

II. Surgical Planning

Surgical planning should include the generation of a 3D volumetric model, target definition, optimization of access routes, and the circumvention of functionally important anatomic structures. In diagnostic radiology, however, we display images in two-dimensional, cross-sectional formats. Such images offer superior views for the radiologist, who needs to identify subtle abnormalities. However, from the surgical perspective, these two-dimensional images are not as helpful. Mental reconstruction of three-dimensional anatomy from two-dimensional images has a marginal accuracy at best. The combination of these two-dimensional cross sections into a three-dimensional model via computer algorithms greatly enhances the therapeutic potential of these images. Views similar to the surgical point of view can be rendered on the computer screen, giving a more realistic depiction of anatomy. Using these models, the surgeon can select alternative approaches to the target. The simulation of access routes or trajectories, the optimization, and the subsequent decision making are the essential characteristics of surgical planning. Biopsies are based on a single trajectory whereas a surgical procedure consists of multiple trajectories. Therefore, open surgeries require a more preoperative planning, a more complex simulation method, and more painstaking decision making. In our hospital, surgical planning is performed in a highly computerized interdisciplinary research center (Surgical Planning Laboratory) where radiologists, surgeons, and computer scientists combine forces to develop and implement various surgical planning methods. The preoperative planning procedure involves an image-processing pipeline that filters the image data, classifies or segments the various tissue classes, and renders the image data in 3D. This pipeline eventually results in a 3D model of the anatomy that encompasses the entire operational volume. In this 3D model the organs and their normal or pathologic tissue components are distinctly separated or segmented from each other. The segmentation exploits the differences in signal intensity (contrast) between the tissue types as they are represented by various imaging modalities. Currently, digital anatomic atlases are applied to aid the segmentation process (template-driven segmentation). The segmented organs can be separately compiled and rearranged (removed, displaced, grouped, separated, cut out, made translucent or highlighted, etc.). These manipulations are necessary to expedite surgical simulation and planning. The models represent the "virtual patient" since they are created from actual patients' data. Because the real and the virtual patient comprise analogous information, they can be registered to each other. The registration process converts the image coordinates to the patient's frame of reference. Stereotactic frames around the patient's head can make this transformation between the imaging data and the patient's anatomy possible. Originally, stereotactic neurosurgery used plain X-ray-based projections. Now, CT or MR imaging is utilized for frame-based or computerized frameless stereotactic systems. The frameless devices use fiducial markers or anatomic landmarks to ascertain correspondence between the image-based coordinates and the patient's relevant anatomy. The position of the fiducial markers and the movement or actual location of the handheld surgical instruments can be tracked using optical, electromagnetic, or US sensors. These tracking methods are used for relating the positions of markers or instruments to the 3D image database and recall images for interactive display. These navigational systems, or interactive image guidance, are now available for neurosurgery, head and neck surgery, spine surgery, and orthopedic procedures. They can guide the surgeons to the predefined target, unless it has changed during the procedure. Surgical planning should indicate the likely changes in the anatomy during surgery, which are only perceptible through simulations actualized from intraoperative data.

To understand the surgical planning process we review its components separately:

A. 3D Preoperative Modeling

Three-dimensional modeling provides multiple advantages to both surgeons and radiologists. The most prevalent advantage is that of better visualization, particularly for those who are not accustomed to mental reconstruction of cross-sectional image slices. Reconstruction allows the therapist to have a firm grasp on the location of the lesion as well as potential risk (e.g., cutting vessels) that may be faced during resection. Proper surgical approaches and repeated rehearsals can be arranged prior to the patient even entering the operating room. This allows complex and little-known surgeries to be carried out not only with more safety but with the added benefit of training residents and fellows. Sometimes, the information acquired from 3D models can improve diagnosis. In one case, a patient who was initially thought to have a left MCA aneurysm (based on CT and MR imaging studies) was actually shown to have a coiling vessel upon 3D reconstruction, requiring no invasive therapy. Aside from the preoperative and diagnostic capabilities of models, the greatest power of this tool lies in intraoperative visualization. Surgical visualization is inherently incomplete and the surgeon must use his experience and judgment to help determine the location of vital structures and limits of pathology. The major purpose of large incisions and craniotomies is to facilitate surgical visualization. With the advent of minimally invasive procedures, with surgery being performed through small incisions and endoscopic "keyholes," limitations in direct visualization are at a maximum. In addition to this, surgical identification of indistinct tumor margins can prove to be difficult, if not impossible. What can be identified as pathology on an imaging study may not be as easily discernible in the operating room. By incorporating the imaging study into a useful surgical model, the pathology can be accurately estimated and resected, minimizing the removal of normal brain tissue. The preoperative planning, intraoperative localization, and targeting of lesions can all help realize the full potential of minimally invasive surgery (Hu et al., 1990; Kall et al., 1994; Kikinis et al., 1996; Nakajima et al., 1999).

Unfortunately, one cannot merely jump from an imaging study to a full three-dimensional model without entering through some specific steps. These include, but are not limited to, data acquisition, volume registration, and segmentation of relevant structures. Application of the model in the surgical setting requires the additional steps of model to patient registration and navigation. The first step, data acquisition, is simply the transfer of MR data, and any other studies to be incorporated into the model, to computer workstations. Next, multimodal volume registration is used to combine the data from the various imaging studies into one data set. For example, CT images can be combined with MR images, allowing for a model with excellent soft tissue (derived from the MR study) and bone (from the CT) visualization. However, all this information is still combined in one data set; hence segmentation must be performed to extract, from the singular data set, relevant structures into multiple sets on the computer. This is perhaps the most important step in the process. It literally transforms the two-dimensional images into a three-dimensional compilation of separate anatomic structures. Once this is accomplished, segmented structures can be combined into a working model (Kikinis *et al.*, 1996).

B. Data Acquisition

Currently, the imaging protocol used in neurosurgical cases at the Brigham and Women's Hospital includes a set of sagittal T1-weighted spin-echo, axial T1-weighted spin-echo, axial T2-weighted spin-echo, sagittal postcontrast SPGR ("spoiled GRASS"), and MRA data sets. All images are obtained using a 1.5-T superconducting MR imaging system (Signa; GE Medical Systems, Milwaukee, WI). The primary data from which three-dimensional models are created are the SPGR sequence, which consists of 124 1.5-mm-thick slices with a 256×256 matrix of pixels in a 240×240 field of view. The other image sets can be combined with the SPGR if visualization is still inadequate via the multimodal registration process. The magnetic resonance angiogram (MRA) is performed using a three-dimensional phase-contrast technique. The velocity encoding of the angiogram is variable, depending on which vessels need to be seen (typically 20 cm/s for veins, 60 cm/s for arteries—especially important for vascular cases). The data are transferred to a SUN computer workstation (SPARC-station 20, SUN Microsystems, Inc., Mountain View, CA) in the Surgical Planning Laboratory via a local network consisting of both conventional coaxial cables and fiber-optic cables (Dumoulin *et al.*, 1989; Permicone *et al.*, 1992; Edelman, 1993).

C. Filtering

One of the major drawbacks to using MR imaging is the relatively low signal-to-noise ratio (SNR) achieved with high-contrast images. Actually, one can obtain an image with a good SNR at the cost of having a low contrast level. Although this may occasionally be acceptable in the day-to-day diagnostic world of radiology, three-dimensional modeling currently requires a high contrast level. The segmentation algorithms in use are sensitive to noise and do not perform optimally in its presence. There are a number of ways in which improvements in

SNR can be accomplished, one of which is by simply increasing scanning time. A slice can be acquired repeatedly and averaged over the time domain. Of course, this sort of solution to the SNR problem has one major cost: precious scan time-hence the advent of postprocessing methods. The particular method currently in use at the Brigham and Women's Hospital Surgical Planning Laboratory is one based on anisotropic diffusion.

Conventional filtering techniques, including linear spatial filters ("low-pass filters") and nonlinear filters, have the disadvantages of edge blurring, image degradation, and generation of artifacts. These are unacceptable for medical images. The ideal filter should do three things: (1) minimize information loss by preserving image boundaries and detailed structures, (2) efficiently remove noise in regions of homogenous physical properties, and (3) enhance morphological definition by sharpening discontinuities. These criteria can be satisfied with anisotropic diffusion filtering. The details of this method are beyond the scope of this chapter, but suffice it to say that noise is efficiently reduced in areas of homogeneity while boundaries, object contours, and small structures (e.g., vessels) are not only preserved but even enhanced. As for MRA data depicting blood vessels, a 3D line filter is used to enhance visibility. This filter not only allows the discrimination of linear from nonlinear structures but also recovers original line structures from corrupted ones. Once the filter is applied, multimodal registration and segmentation can be initiated with the postprocessed images (Gerig *et al.*, 1992; Sato *et al.*, 1997).

D. Multimodal Volume Registration

The integration of various modalities into a single three-dimensional model is essential to optimize information. CT scans allow for excellent visualization of bone and acute blood. MR can currently delineate soft tissue better than any other modality. More importantly, the diversity of MR imaging schemes can produce an equally diverse set of images. T1-weighted images can produce an excellent view of normal anatomy, whereas T2-weighted images are better at picking out pathology. Hence, an edematous area of inflammation that "lights up" on a T2 image can be combined with a T1 data set. This would allow for a model with not only accurate anatomical information but also accurate pathological information.

One can also see how multimodal registration can be of benefit in cases where functional images are needed. Surgeries to resect epileptic seizure foci are currently performed with the benefit of functional mapping on the three-dimensional model. SPECT and fMRI images can also be registered and integrated into the model to facilitate the surgery.

Of even greater importance than the functional imaging is the vascular imaging. Whether by conventional angiography or magnetic resonance angiography, modeling with vascular structures in place is critical for the determination of optimal surgical approaches. Potentially dangerous hypervascular areas can be carefully traversed with the help of this information.

The maximization of mutual information algorithm is used to incorporate these various imaging modalities into one model. A base series of images is defined (usually sagittal SPGR images) and the other images are fused to this base set. To facilitate the description of this method, let us initially assume that two identical MR studies were to be registered together. If these two identical MRI studies of a patient were performed at different times and then combined, the simplest way to measure proper alignment would be to summate the differences in signal intensity. The lower the value, the more exact the alignment. However, the utility of registration is in the combination of different studies with different signals at given anatomical landmarks. Thus a combination of studies can be performed by aligning one signal format into another; aligning a CT signal to a MR equivalent, and vice versa. The algorithm optimizes the alignment of a data set to the base data set by positing an alignment transformation, evaluating the quality of the alignment by measuring the mutual information of the data (the entropy of the base data set and the aligned data set less the joint entropy of the fused data), and then refining the posited alignment transform until the mutual information is at a maximum. After proper registration, segmentation of the structures of interest can be performed (Vannier *et al.,* 1984; Levin *et al.,* 1989; Zhang *et al.,* 1990; Holman *et al.,* 1991; Wells *et al.,* 1996b).

E. Transcranial Magnetic Stimulation

An aside here is necessary to briefly explain transcranial magnetic stimulation (TMS). TMS is a noninvasive method of assessing cortical function via direct stimulation of nerve cells, something that was only previously achieved by direct electrode stimulation during open neurosurgical cases. Quite simply, a coil generates a magnetic field that penetrates the cranium and affects a given volume of cortex. Reponses are measured, both motor and visual, and recorded relative to a completed three-dimensional model. Hence, functional maps are generated in a noninvasive, yet accurate, way. Currently, motor and visual maps are generated by measuring muscle stimulation and vision suppression, respectively. When a region of visual cortex is stimulated, inhibitory neuronal populations activate and suppress the corresponding visual field. In the future, TMS not only will be used in a re-

search setting to help delineate functional neuroanatomy but it will also be used in the therapeutic setting. The maps can be used preoperatively and intraoperatively to more accurately predict functional outcomes after surgery and to help the surgeon spare important functional areas during tumor resection (Ettinger *et al.,* 1996, 1997).

F. Segmentation

Segmentation is the process of separating meaningful anatomic structures from original images, be it MRI, CT, or any other modality. The process is currently a semiautomated one, partially performed by the computer, followed by manual editing. This is perhaps one of the most important steps in three-dimensional modeling. The process of extracting pertinent anatomical and pathological information from images is one that is routinely performed by radiologists on a daily basis. Although computers are beginning to "see" relevant structures, there is still some way to go before the process becomes fully automated (although efforts to that end are currently underway).

The major tools used in the segmentation of structures are thresholding and connectivity algorithms. Thresholding is based on signal intensity differences between tissues. For example, white matter, on T1-weighted images, has a higher signal intensity than gray matter. By having the computer select a given range of signal intensities, excluding those signals outside the range, one can extract the relevant structure. To segment gray matter, one can threshold around a given range of signal intensities that include the signal of gray matter, excluding the higher white matter signal (which would fall outside of the threshold range). By selecting only a given range of signal intensities, the computer can discriminate between the various anatomic structures.

Of course, thresholding is not enough to properly segment structures. Another algorithm, the connectivity algorithm, is incorporated into the scheme to assist the thresholding. This takes advantage of the connections within an anatomic structure. For example, by thresholding gray matter, not only is brain gray matter included, but also little bits of skin and soft tissue that happen to have the same signal intensity as gray matter. To get rid of this "extraneous" signal, the connectivity algorithm selects only those points on the image, or voxels, that are connected to the structure of interest. The various structures are then displayed in different colors and the user can pick the structure of interest. Thus only brain gray matter is selected, with the extraneous soft tissue and skin excluded.

Unfortunately, thresholding and connectivity alone cannot always isolate the desired anatomical structure of interest. Three additional tools are then at the user's disposal: (1) volume of interest (VOI) definition, (2) morphological operations (dilation and erosion), and (3) cut plane. The VOI allows for a structure to be "boxed" in. For example, if only the cerebellum is to be isolated, a VOI box can be adjusted around the structure to include only the cerebellum and exclude other nearby structures (cerebral cortex, midbrain, pons, medulla, etc.). The morphological operations, dilion and erosion, are used to enlarge and shrink structures, respectively. Erosion is useful to break links between structures that are thinly connected by strands of tissue. Dilation reverses the effect of erosion, but without replacing the broken link. Finally, the cut plane is simply a triangular plane, which can be placed anywhere within the volume. It is used whenever VOI and erosion/dilation fail in separating objects (e.g., spinal bones and aorta).

The combination of these algorithms makes for a powerful set of tools used toward the formation of three-dimensional structures from the two-dimensional images. However, even with these algorithms, perfect automated segmentation has yet to be realized. A certain amount of manual editing must still be performed to achieve the degree of detail and accuracy necessary for surgical applications. This manual editing is performed on a slice-by-slice basis, fine-tuning what the computer generated. The process can take anywhere from 60 to 180 min. Finally, once every slice has the structure of interest (e.g., brain) highlighted, the slices can all be compiled into a three-dimensional image through the use of the marching cubes algorithm (Cline *et al.*, 1987, 1990, 1991; Lorensen and Cline, 1987; Warfield *et al.*, 1996; Wells *et al.*, 1996a).

G. Model to Patient Registration

Once the model has been successfully generated, it must be put to use in the operating room. Registration of the model to the patient allows for an enhanced level

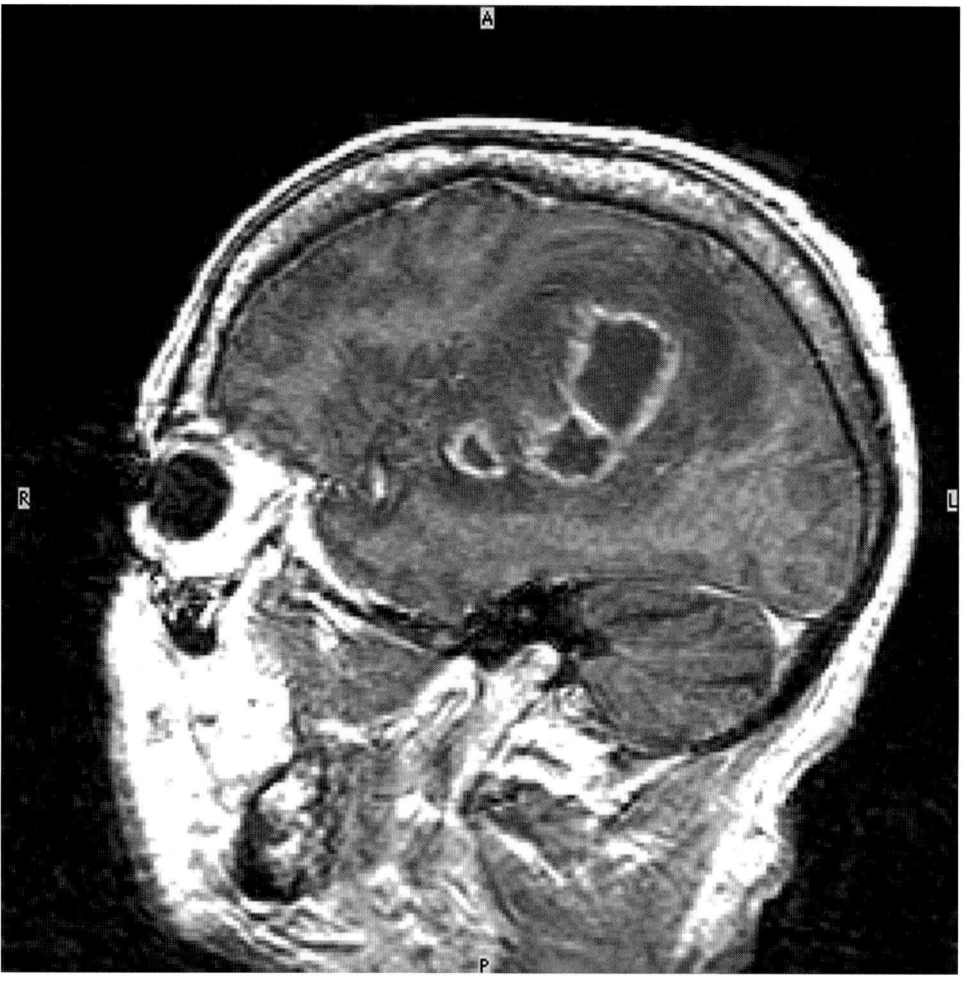

Figure 1 Sagittal T1–weighted postgadolinium image.

of interaction between the surgeon and the model. Currently, a variety of registration tools make it possible to map the model to the patient. These include (1) video registration, (2) laser scanning, and (3) LED probe surface scanning.

Video registration can be used to align either surface skin landmarks or cortical vessels. It essentially consists of superimposing a projection of the 3D model onto a video image of the actual surgical field. Anatomical fiducials such as the external auditory meatus, nasion, lateral canthus, and previous surgical scars are used for skin–skin alignment. Once an incision is made and the dura is opened, cortical vessels become visible and the registration can be refined by vessel–vessel alignment. As expected, the vessel–vessel registration provides a higher degree of accuracy than the skin–skin registration alone. The refinement has been shown, in a phantom study, to improve the registration accuracy from 8.9 to 1.3 mm. This method, however, does rely on human manipula-

tion, with the user magnifying and rotating the three-dimensional model to "fit" the video image. More automated registration techniques are now being utilized such as laser scanning and LED probe based surface scanning. These techniques automatically align the contours of the patient's skin with those on the model (Gleason *et al.*, 1994; Grimson *et al.*, 1996; Nakajima *et al.*, 1997).

H. Intraoperative Navigation

Intraoperative navigation is achieved with the use of a sterile probe equipped with an LED-based optical tracking system. There are three cameras positioned above the surgical field that can track two flashing LEDs on a probe. The computer display itself (see illustrative cases) has the three-dimensional model as well as the axial, coronal, and sagittal MR images. The tip of the probe is localized and displayed on the 3D model as well as on each of the three MR image planes being displayed. The

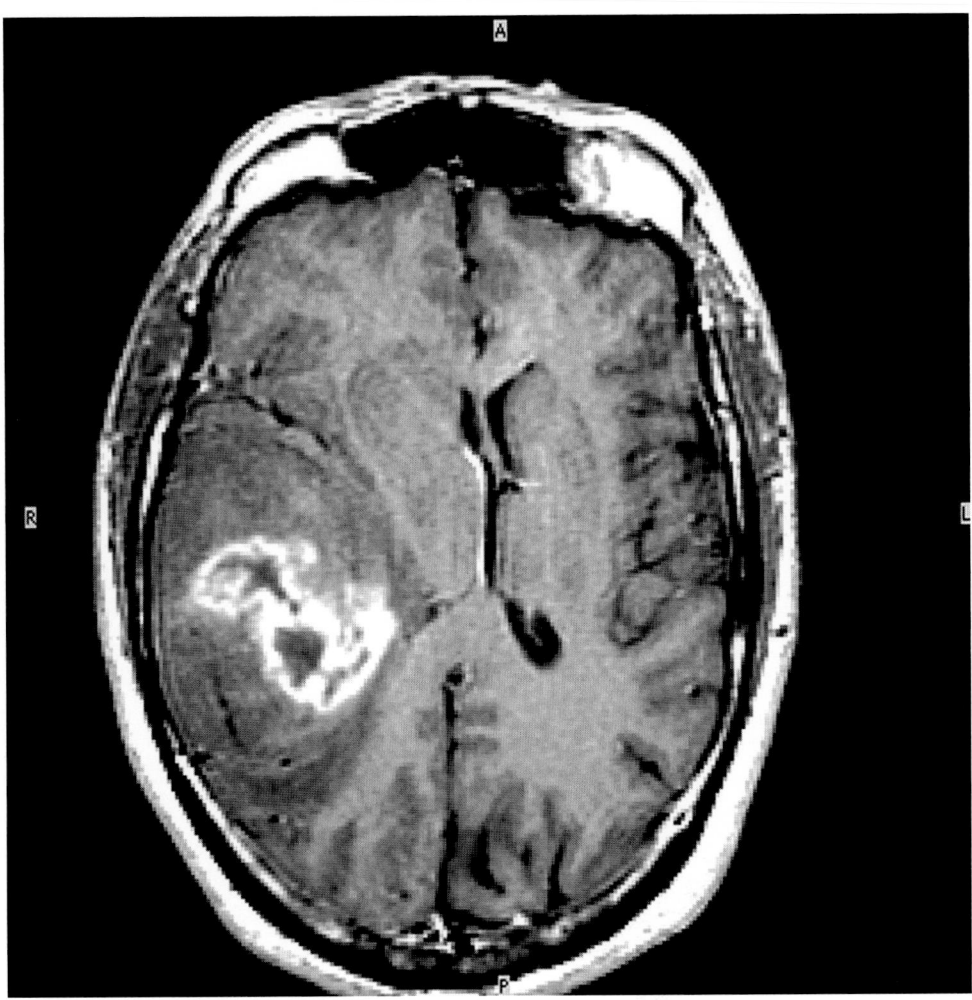

Figure 2 Axial T1–weighted postgadolinium image.

model can be translated, rotated, and magnified as desired while structures can be pointed out and outlined with the probe. The model can also be colored, with various structures being made translucent to help clarify relationships between tumor, vasculature, ventricles, and brain. Commonly, the outline of a tumor cavity is checked with the probe following tumor resection. The probe is also useful for intraoperative cortical mapping during epilepsy surgery, marking the electrodes placed on the patient's brain onto the 3D model itself (Kall *et al.*, 1986; Kato *et al.*, 1986; Kosugi *et al.*, 1988; Barnett *et al.*, 1993; Zamorano *et al.*, 1994; Golfinos *et al.*, 1995).

I. Illustrative Cases

1. Case 1: Low-Grade Glioma

The patient is a 45-year-old male with a history of unsteadiness, headache, and left-sided weakness with a left facial droop (began 1 month prior to presentation). MRI study showed a right-sided brain tumor. Figures 1 and 2 are sagittal and axial postgadolinium T1-weighted images. Figure 3 is an axial T2-weighted image. Note the presence of the tumor and surrounding edema. Three-dimensional reconstruction was performed with the tumor and edema separately segmented and reconstructed (Figures 4 and 5). Figure 5 shows the relationship of the tumor, vessels, and ventricles without the brain surface rendering. Intraoperatively, a right temporoparietal craniotomy was done under local anesthesia, followed by cortical sensory and motor mapping. Figure 6 shows the intraoperative display with a virtual probe pointing to the tumor on the model. Note the position of the virtual probe tip is indicated by crosshairs on all three MR cross sections (axial, coronal, and sagittal). The tumor was resected, histopathology showing glioblastoma multiforme, and the patient did not present with any deficits. In this case, the model proved very useful in localizing

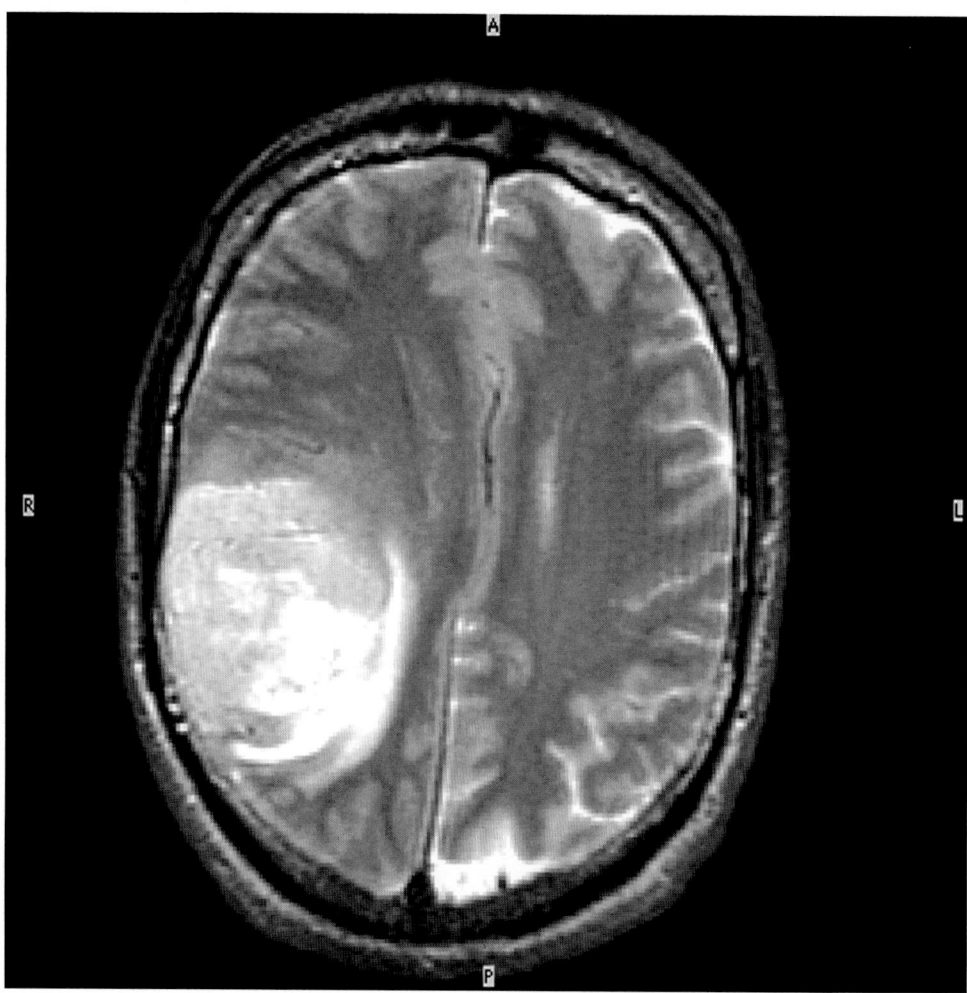

Figure 3 Axial T2–weighted image.

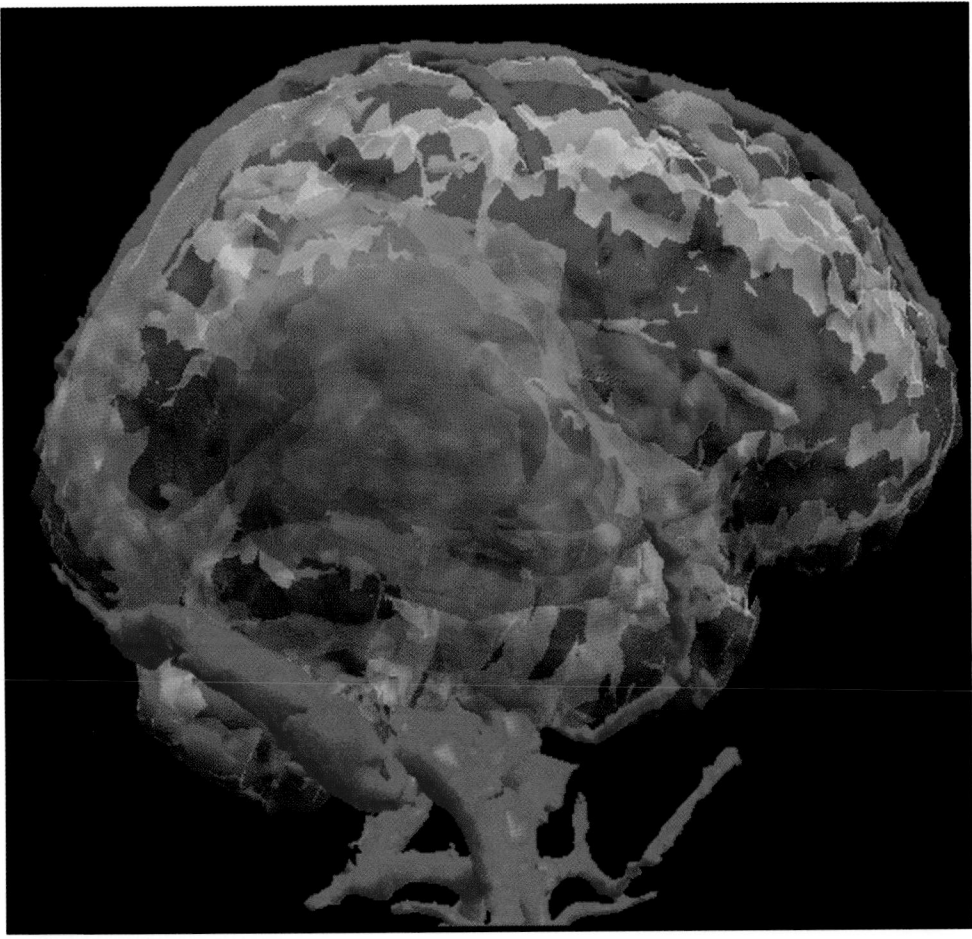

Figure 4 3D reconstruction showing lateral aspect of the transparent brain surface (white) with vessels (red), tumor (purple), edema (green), and ventricles (blue).

the lesion, especially in the midst of the surrounding edematous tissue.

2. Case 2: AV Malformation

The next case is that of a 20-year-old male who presented with sudden onset of headache while lifting weights. He had no neurological deficits or nuchal rigidity, but did have a mild photophobia. CT scan revealed a small hematoma in the posterior fossa. MRI study revealed an arteriovenous malformation in the cerebellar vermis with extension into the medial aspect of the right cerebellar hemisphere. The drainage was into the straight sinus. Figure 7 shows the sagittal postcontrast T1 image, and Figure 8 shows the axial T2 image. 3D reconstruction was carried out and the relationship of the cortical vessels to the malformation

is shown in Figure 9. A posterior fossa craniotomy was done under general anesthesia, feeding and draining vessels were identified with the surgical microscope, and the entire malformation was successfully removed.

3. Case 3: Intractable Epilepsy

The patient is a 2-year-old female with a history of tuberous sclerosis who presents with intractable seizures. This case demonstrates the utility of the three-dimensional model in mapping of seizure foci. A left fronto-parietal craniotomy was carried out under general anesthesia, after which an 8 × 8 grid of electrodes was placed over the seizure area. Each of the electrode points was then registered to the model with the surgical navigator (Figures 10 and 11). The grid of electrodes was then

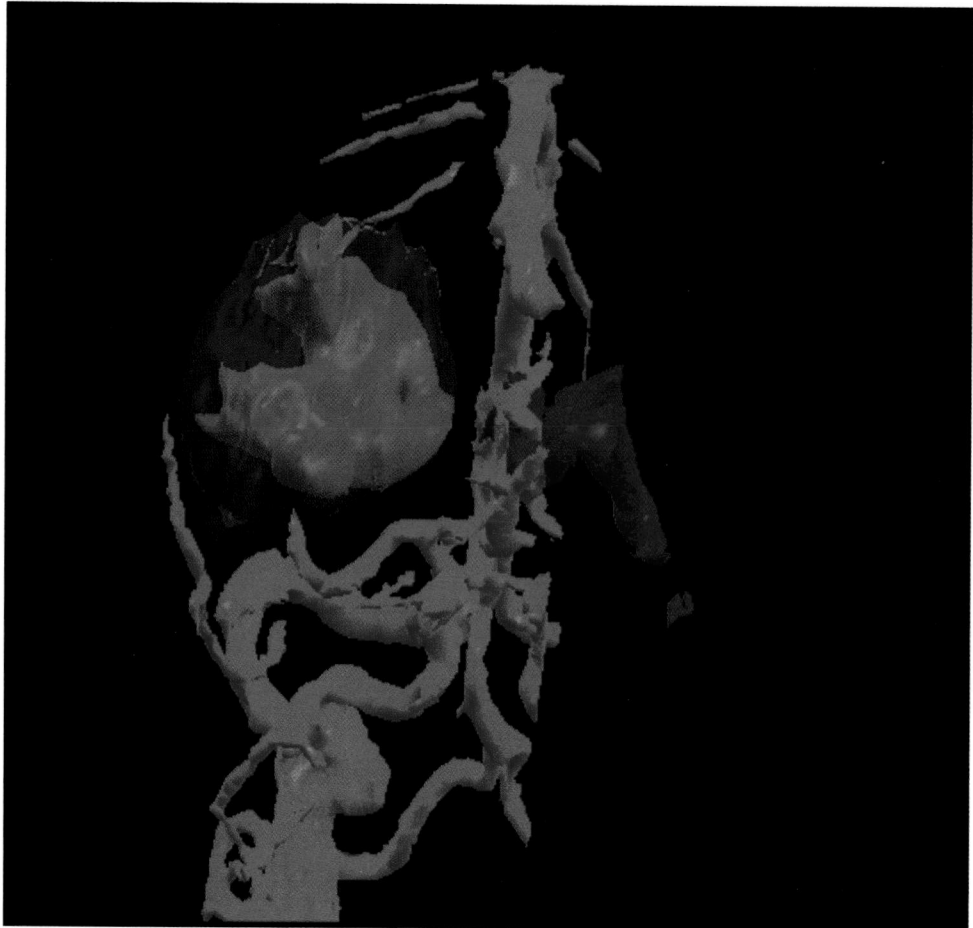

Figure 5 3D reconstruction showing anterior view of tumor (purple), edema (green), ventricles (blue), and vessels (red). Brain surface has been rendered invisible.

used to specifically identify exact points of seizure foci. Resection of these foci was carried out the following week.

III. Intraoperative MRI

One of the drawbacks to the use of preoperatively acquired image data is that it cannot depict the changes occurring during the operation. The surgical field is a dynamic environment, with anatomical changes being induced by the surgeon. A real-time, dynamic method of image acquisition is needed during the procedure to accurately gauge these anatomical changes and help assist the surgeon through the operation. Imaging methods of real-time visualization such as ultrasound and fluoroscopy are currently in use for such purposes. However, MRI has a clear advantage over these modal-

ities in its superior ability to define soft tissues, multiple planes of viewing, and diversity of scanning parameters (Jolesz and Zientara, 1996).

As discussed in the previous section, image guidance can be accomplished with the help of three-dimensional models with data obtained preoperatively. However, there are inherent conflicts with the use of images acquired preoperatively in the real-time surgical environment, which is wrought with anatomical changes. In neurosurgical cases, incomplete tumor resection, functional brain damage, and CSF leakage can result in inaccuracies within frame and frameless stereotactic systems. There is a need for an imaging modality that can not only characterize tissues but also obtain real-time images or frequent image updates. Open-configuration intraoperative MRI systems have the potential to serve in this role.

There are other modalities of real-time imaging that are currently in use. Ultrasound has commonly been

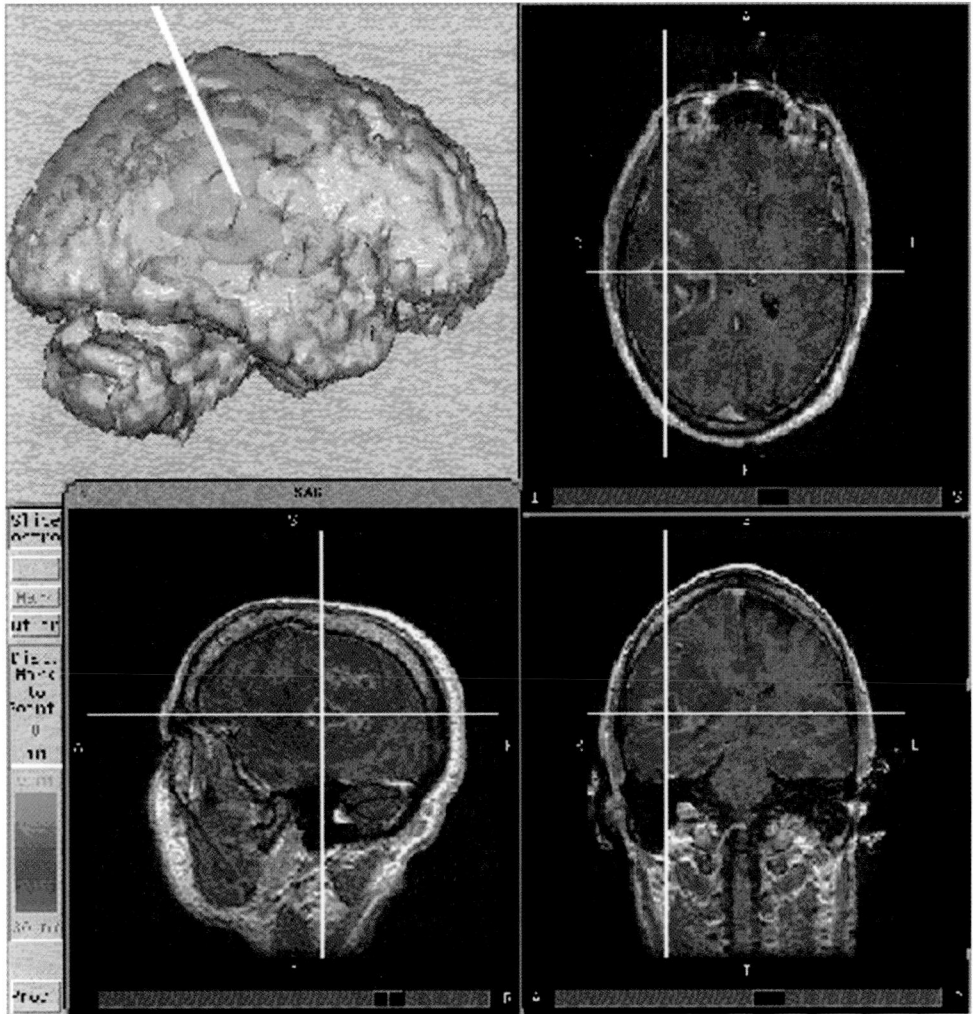

Figure 6 Intraoperative display with virtual probe pointing to tumor on 3D model. Axial, sagittal, and coronal sections display the tip of the virtual probe with crosshairs.

used in therapeutic interventions such as biopsies and fluid drainage procedures (e.g., pleural or ascitic fluid). Yet the contrast resolution of ultrasound, especially for tumor detection, is relatively poor. Although multiple planes of imaging can be acquired, one is restricted by probe placement. Another imaging modality currently in use for real-time imaging is that of fluoroscopic imaging. Aside the obvious disadvantage of flooding the patient with multiple doses of radiation, these "projection" images can only depict two-dimensional anatomy and multiple projections are required to fully appreciate three-dimensional relationships. Besides, X-ray fluoroscopy is, in most cases, insufficient for correct target localization unless the target is predominantly vascular.

Intraoperative MRI is superior to current methods of real-time imaging in its ability to characterize tissues. Accurate identification of pathology is essential in the resection of tumors and other lesions. This is of paramount importance in neurosurgery, where inaccurate resection may result in either incomplete tumor resection or destruction of normal brain tissue. MRI also has other benefits that can be exploited to guide therapy such as flow sensitivity, perfusion evaluation, and detection of temperature changes, the latter being essential in guiding thermoablative therapies such as laser interstitial therapy, focused ultrasound surgery, and cryotherapy (Jolesz and Blumenfeld, 1994; Gronemeyer et al., 1995; Jolesz and Silverman, 1995; Lufkin, 1995; Jolesz and Zientara, 1996).

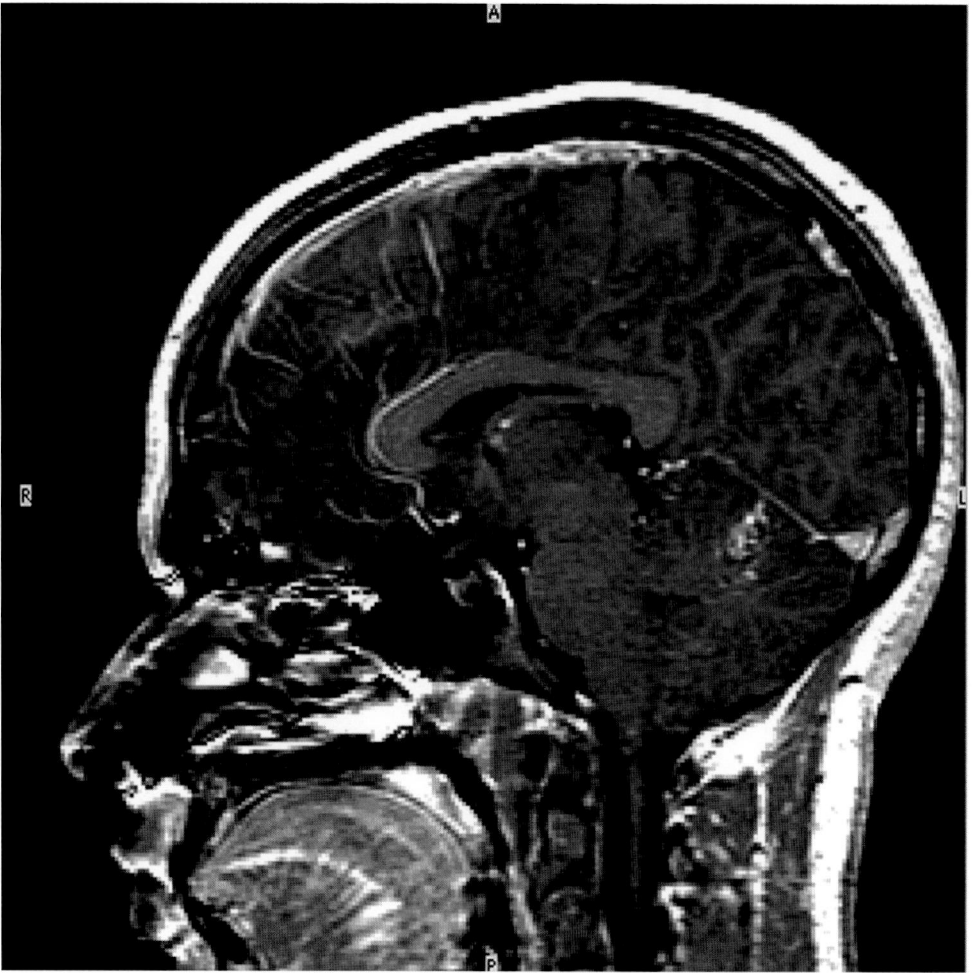

Figure 7 Sagittal T1–weighted postgadolinium image.

A. Equipment and Materials

Conventional MRI systems usually have a cylindrical configuration, precluding direct access to the patient. Open-configuration MR systems have been developed without much consideration to potential interventional or intraoperative use. These magnets have been resistive magnets with horizontal gaps that allowed access to the patient, but only enough for interventional percutaneous procedures. The first superconductive open magnet was developed for the specific purpose of interventional/intraoperative use. It was developed and tested via a collaborative effort between General Electric Medical Systems and the Brigham and Women's Hospital in Boston, Massachusetts. The configuration was that of two (double) "doughnuts" containing a 0.5-T superconducting magnet between which a vertical gap was created

for the physician to stand and carry out procedures. The patient can be positioned erect, sitting, or lying parallel or perpendicular to the axis of the magnet's bore. The gradient coils are concealed within the doughnuts, where they function without hampering the surgeon or interventionalist. Flexible radiofrequency (rf) transmit and receive coils, which are specifically designed for imaging a particular part of anatomy, can be adapted perfectly to the shape of the region of interest. These coils can also be sterilized and incorporated into the surgical drape. The coils significantly improve image quality and are of considerable strategic importance in interventional MRI. The unit itself is housed in what is a hybrid of an operating room and an interventional radiology suite. In addition, there are also computing facilities, display devices, and integrated therapy systems available.

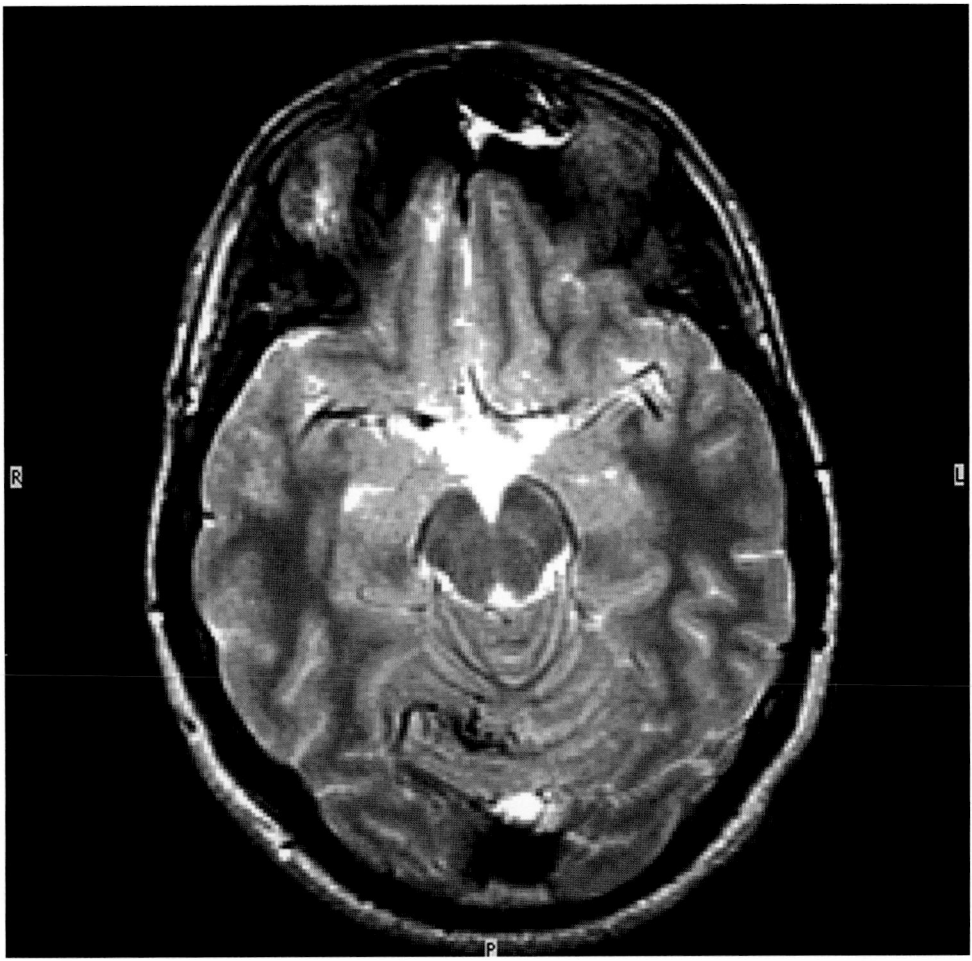

Figure 8 Axial T2–weighted image.

One of the major obstacles in the fruition of intraoperative MR was the development of MRI-compatible instrumentation. Conventional stainless steel surgical instruments clearly cannot be used in the MR environment secondary to the influence of the magnetic fields on them. However, nonferromagnetic instruments can also be deflected within the magnet, and all potential instruments must be rigorously tested. In addition, the creation of artifacts by the tools themselves when placed near a given target and electrical interference with image acquisition by electrocautery provided for quite a set of hurdles to be overcome. Thus although initially slow in development, MR tools have now come into being in the form of scalpels, drills, surgical drapes, and anesthesia equipment. A MR-compatible microscope was also successfully developed by Studor Medical Engineering (Rhen Paal, Switzerland) and implemented at Brigham and Women's Hospital in 1996.

Devices for intraoperative navigation were also integrated into the interventional MRI environment. As described in the previous section, an optical tracking system utilizing light-emitting diodes is used to track rigid probes and needles. Several modes can be selected and anatomy can be observed with respect to probe or needle position. A targeting mode can be accessed where the tip of a virtual needle is displayed, showing its expected trajectory through the patient's brain. However, the LED-based optical tracking method is not applicable to the tracking of flexible probes, endoscopes, and catheters within the body. For these, a tip-tracking method is used where miniature radiofrequency receive coils are attached to the tips of these instruments. The implementation of tip-tracking methods has been quite successful, with an accuracy of within 1 mm (Mueller *et al.*, 1986; Lufkin *et al.*, 1987; van Sonnenberg *et al.*, 1988; Martin *et al.*, 1992; Dumoulin *et al.*, 1993; Kandarpa *et al.*, 1993; Le-

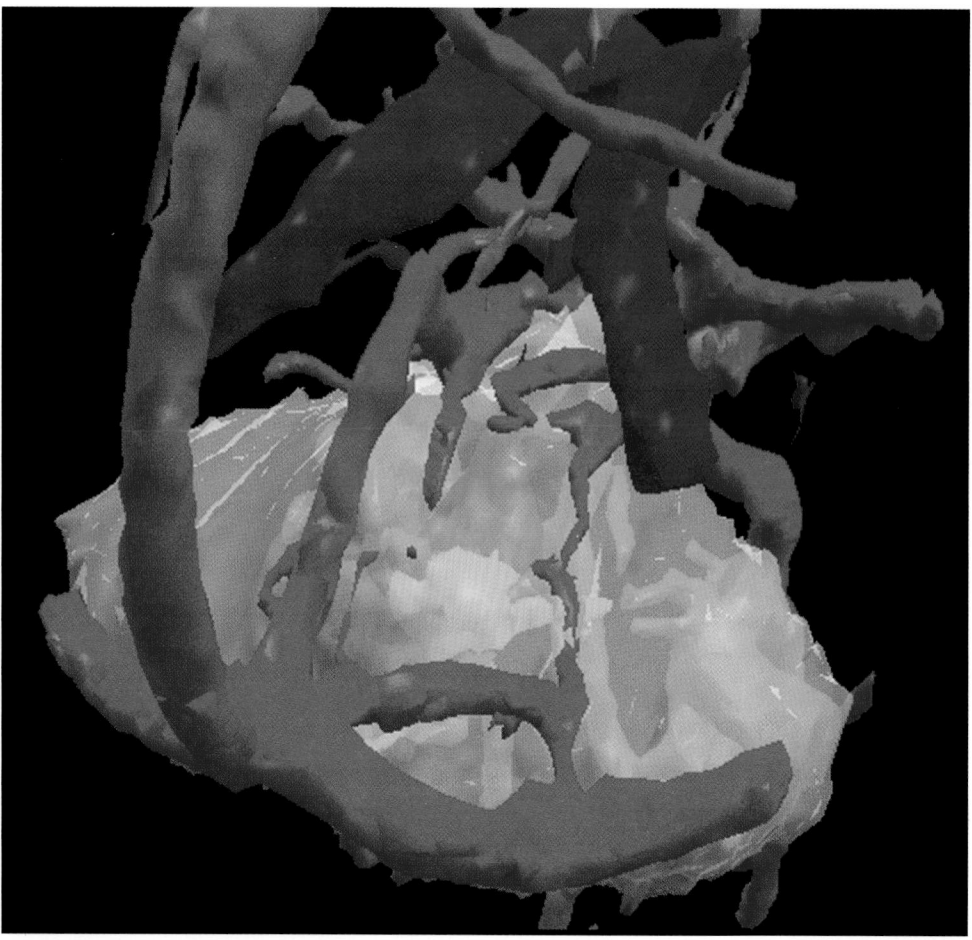

Figure 9 3D reconstruction showing transparent cerebellar surface (white) with vessels (red), ventricles (blue), and AV malformation (green).

ungh *et al.*, 1995; Schenck *et al.*, 1995; Silverman *et al.*, 1995).

B. Procedures Performed

Clinical implementation of intraoperative MRI involves a collaborative team effort between surgeons, radiologists, MR technologists, nurses, and engineers. The radiologist interprets images throughout the procedures and essentially guides the surgeon through the operation. Cases performed within the unit, thus far, include stereotactic biopsy, craniotomy, cyst drainage, spine surgery, and thermoablative procedures.

Many stereotactic biopsies have been performed at multiple locations. The first such procedure was performed in January 1996, with subsequent biopsies being performed on lesions located in brainstem, deep white matter, thalamus, basal ganglia, cerebellum, and cerebral cortex. Rare episodes of hemorrhage were immediately detected by the MRI and effectively treated, with no clinical sequelae.

The first craniotomy was performed in June 1996 after the development of the MRI-compatible drill and microscope. Thus far, the majority of procedures performed have been for low-grade glioma cases, but other lesions such as meningiomas, cavernous malformations, and arteriovenous malformations have been successfully resected within the unit. Transphenoidal pituitary resections were also implemented in this operating environment. So far, more than 300 open neurosurgical procedures have been performed.

Cysts, such as arachnoid cysts, have been decompressed and drained under MR guidance. Air or contrast can be administered directly within the cyst to help outline its cavity. Spine surgery had to await the develop-

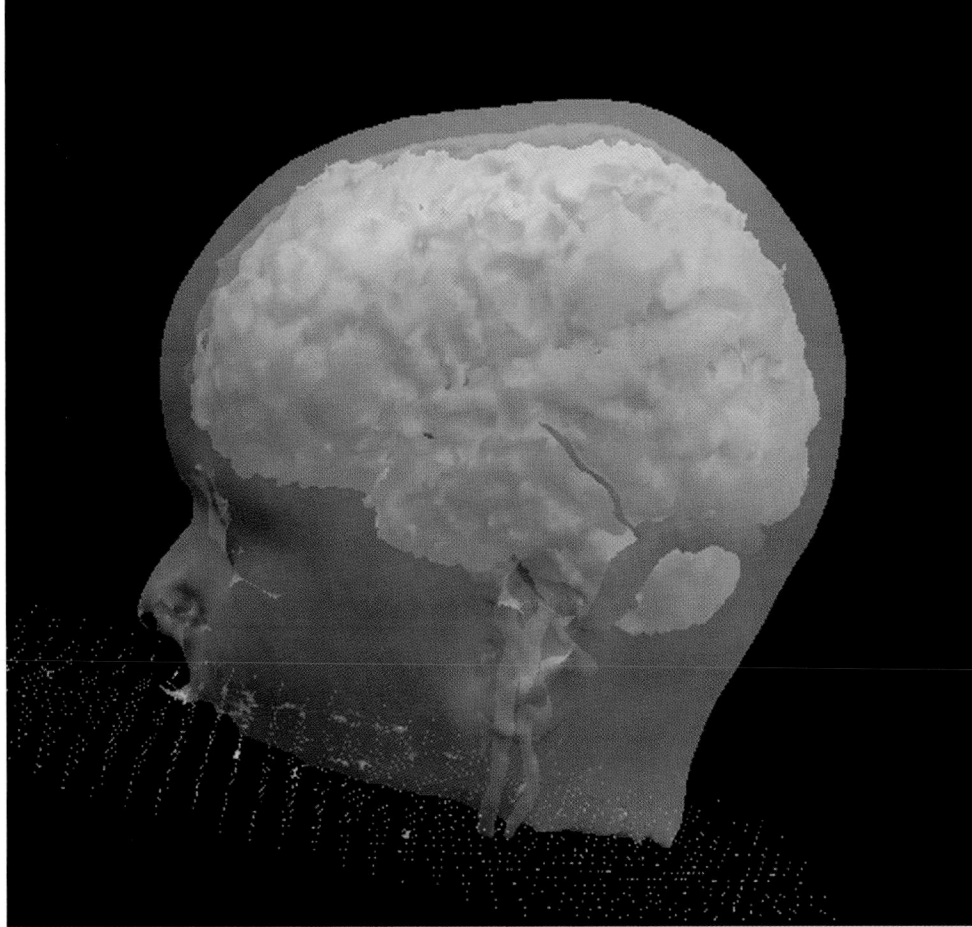

Figure 10 3D reconstruction showing brain surface (white) and vessels (red).

ment of spinal instruments compatible with the magnetic environment, but since its development, multiple procedures ranging from extra-axial cervical cyst evacuation to anterior cervical decompression for myelopathy and several discectomies have been carried out (Moriarty *et al.*, 1996; Black *et al.*, 1997). Currently, lumpectomies for breast cancer have been introduced into intraoperative MRI. Fat-suppressed contrast-enhanced images with high sensitivity for tumor detection are used. Since target definition (delineation of tumor margins) is essential for lumpectomy, MRI-guided tumor resection may have a substantial impact in breast cancer surgery.

We also introduced this technology for sinus endoscopy. MRI can complement endoscopic visualization by providing cross-sectional images related to the surface video displays by the tracked aspirator that is used as a "virtual pointer" (Nakajima *et al.*, 1998).

C. Thermal Ablations

There is an increased role of monitoring and control techniques with heat-sensitive advanced image-guidance methods used for thermal ablations. Although the biological mechanisms of freezing or heat-mediated tissue damage are well established, the 3D extent of the injury cannot be verified without volumetric imaging. In order to treat a neoplasm, a critical level of thermal destructive energy has to be delivered and at least 60°C should be reached within the treated tumor. The targeting should be accurate, but temperature-sensitive imaging methods should also be applied to limit the deposition of destructive energy within the mass and depict thermal diffusion into the adjoining normal tissue. Image-guided, monitored, and controlled thermal ablative treatment can be greatly effective and safe.

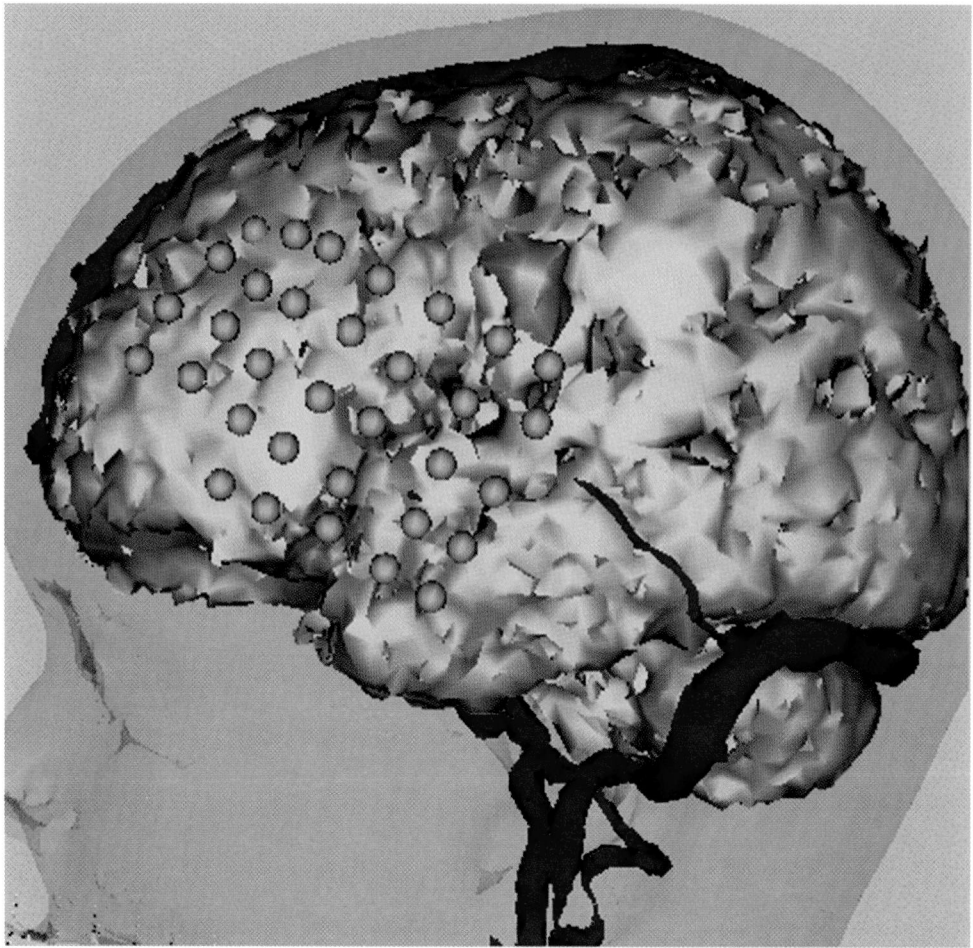

Figure 11 Intraoperative display with electrodes (yellow) mapped onto brain surface.

Most of the ablation methods use a relatively small-diameter probe, and therefore they are, in general, minimally invasive; these probes also limit the size and shape of tissue injury. The thermal diffusion from these probes is affected by the physical disposition of the tissue, and the shape of the treated volume may not exactly correspond to the shape of the treated tumor. The lesion size cannot be increased indefinitely because of the developing steady state between energy deposition and absorption. Tissue perfusion and blood flow can strongly influence the size of the ablation.

The advanced cryoablation systems use gases (i.e., argon). The probes are small (down to 1-mm diameter), the tubing is flexible, and the gas containers are light. These features make the percutaneous cryotherapy approach suitable for MRI guidance, with which the monitoring of the developing iceball is pos-sible. MRI provides clear definition of tumor margins and depicts a sharply outlined and easily interpretable signal void that represents the frozen tissue water (iceball). The signal void is due to the reduced mobility and the subsequently shortened T2 relaxation time. During the freezing, MR images confirm the slowly increasing size of the iceball and during the thawing process they confirm its diminishing size. Cryoablation results in tissue death within the treated volume. Because of the good definition of the iceball during the procedure, MRI is appropriate for monitoring and controlling freezing. The MRI-guided percutaneous cryoablations are now under clinical testing in our institution. The primary applications are colorectal metastases in the liver, deep musculoskeletal, retroperitoneal, or paraspinal masses, which are difficult to access with open surgery, breast cancer, and prostate cancer.

Interstitial laser treatment (ILT) uses optical fibers and infrared light for thermal coagulation. The optical fibers fit within relatively thin needles, and the laser treatment can follow a biopsy procedure. The ILT procedure is fully compatible with MRI, which can provide temperature-sensitive images with relevant temporal resolution. The real-time MRI monitoring is very effective in detecting the thermal diffusion and convection of heat within the inhomogeneous tissue, where flow and tissue perfusion can influence the relative amount of energy and the size of the treated tissue volumes. MRI-guided ILT has become an accepted minimally invasive treatment option for brain tumors, liver tumors, breast cancer, and head and neck malignancies (Cline *et al.*, 1993). It is a relatively easy method, which can be well adapted to the MRI environment.

Radiofrequency (rf) ablations have been tested for liver tumor treatment and for other benign neoplasms (uterine fibroids). Recently, the incompatibility of rf energy deposition and MR imaging has been resolved and MRI-guided rf ablations have become feasible.

The most tempting thermal ablation method is focused ultrasound surgery (FUS). This method is based on acoustic energy deposition and secondary thermal effects. FUS is noninvasive, and the well-focused high-energy US beam causes no tissue damage outside the target. Without visualizing the high-temperature focal spot, this technique was unsuccessful because of the lack of targeting and monitoring possibility. Instituting the temperature-sensitive MRI guidance into the targeting and energy-delivery process rehabilitated the old method. Under MRI guidance a relatively low-power energy deposition can be used for targeting. The small temperature elevation within the focal spot causes no permanent tissue damage but can be detected by MRI and, if necessary, relocated on the target. When targeting is concluded, the power can be raised to obtain a higher than 60° temperature, which causes irreversible tissue damage. This method is clinically tested in our hospital for the treatment of benign and malignant breast lesions. We outline the tumor and, within the contours, deposit sufficient number of overlapping focal spots to cover the entire tumor. There are other likely applications of FUS. Small power levels can transiently open the blood-brain barrier and chemo- or immunotherapy of brain tumors may be conceivable (Cline *et al.*, 1994, 1995).

When monitoring thermal therapies, there are a few fundamental requirements that must be fulfilled by the imaging modality. There should be sufficient, visually detectable image contrast dependent upon the temperature-induced changes within the tissue. In addition, the temporal resolution must be such that the images are acquired faster than the thermal changes. Currently, T1 and phase-sensitive images show the greatest sensitivity to temperature changes and, therefore, have been used extensively in monitoring therapy (Asher *et al.*, 1991; Bleier *et al.*, 1991; Bettag *et al.*, 1992; Fan *et al.*, 1992; Panych *et al.*, 1992; Cline *et al.*, 1993, 1994, 1995; Kahn *et al.*, 1994; Anzai *et al.*, 1995). Within the range of 37–50°C, T1 varies linearly with a relationship of 1% per °C. Beyond 50°C, changes in the tissue itself drastically affect the MR signal properties. Temperatures higher than 56–60° C result in irreversible tissue damage, protein denaturation, and coagulation necrosis (tissue phase changes).

D. Real-Time Imaging (Dynamic MRI)

The need for real-time imaging in interventional and functional MRI applications has fueled the development of dynamic and adaptive MRI. This is exemplified by the thermal therapies previously mentioned. Dynamic MRI is a distinct entity from standard MRI in that a large quantity of images must be continually and rapidly acquired, in order not to interrupt or slow down the procedure being performed. Dynamic MRI is separated into two major categories: nonadaptive and adaptive.

The nonadaptive dynamic MRI methods consist of either fast pulse sequences or reduced k-space sampling. Fast pulse sequences exploit the ability to quickly manipulate magnetic field gradients (as in echo planar imaging) and rf pulses to acquire as much information in as short a time as possible. Current fast pulse methods include RARE, fast GRE techniques, BURST, GRASE, echo planar imaging, ultrafast, and spiral scan methods. Acquisition rates for complete MRI images using these fast nonadaptive dynamic methods can be less than 1 s (echo planar, BURST). Drawbacks to these sequences include a lowered tissue contrast, reduced spatial resolution, or low SNR. Reduced k-space sampling methods include MR fluoroscopy, Fourier-enhanced keyhole imaging, and sampling using prior knowledge. MR fluoroscopy combines fast MR pulse sequences, an efficient imaging algorithm, and hardware for image reconstruction to provide a real-time visualization. Essentially, the method improves temporal resolution at the cost of acquiring fewer data (hence, less spatial resolution). Keyhole Fourier imaging begins by acquiring a high-resolution baseline image, with high-spatial frequency, and subsequently updating it with data at a lower spatial frequency. This technique assumes that there is no

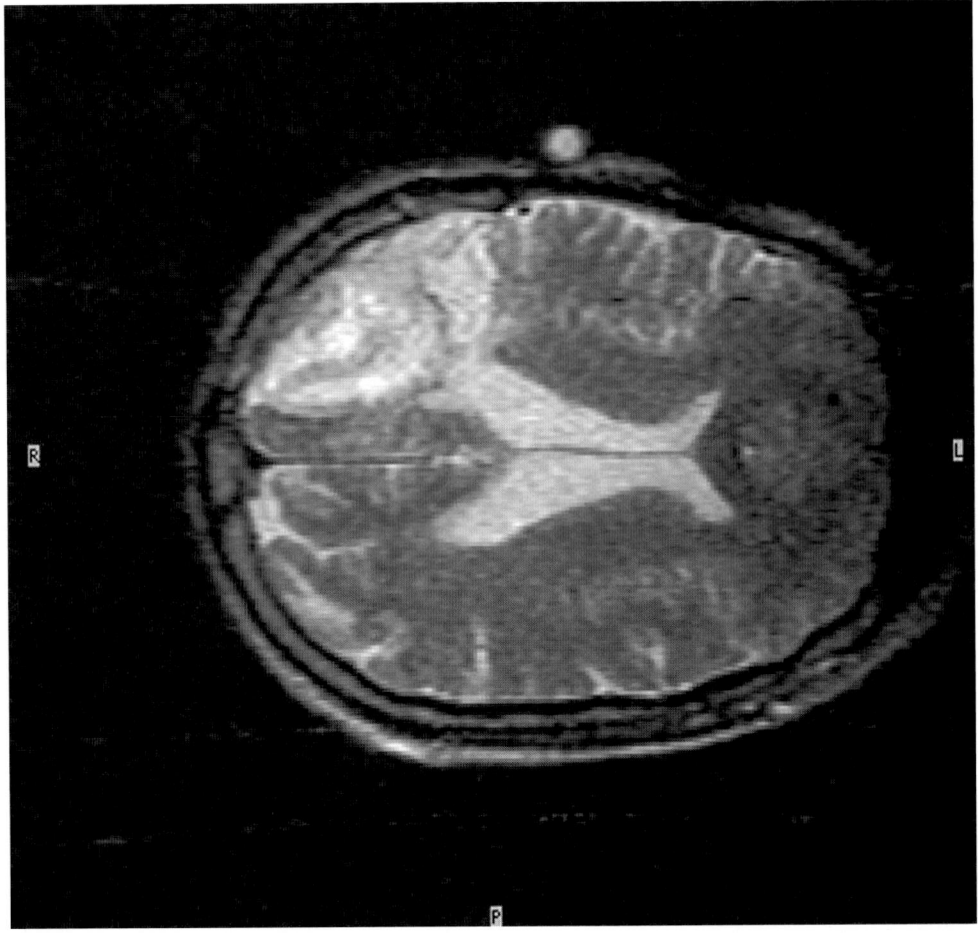

Figure 12 Axial T2–weighted image (before operation).

change in the underlying morphology responsible for the higher spatial frequency variation in the image (Haase *et al.*, 1986; Hennig *et al.*, 1986; Hennig and Friedburg, 1988; Haase, 1990; Bullock *et al.*, 1991; Cohen and Weisskoff, 1991; Oshio and Feinberg, 1991; Gowland *et al.*, 1992; Chenevert and Pipe, 1993; Moonen *et al.*, 1994; Spielman *et al.*, 1995).

Adaptive MRI methods are those that modify the data acquisition strategy based on the information in the most current image. There are currently three adaptive MRI methods that have been under development at Brigham and Women's Hospital: (1) wavelet-encoded MRI, (2) singular value decomposition (SVD) encoded MRI, and (3) adaptive Fourier transform encoded MRI. Wavelet-encoded MRI replaces the phase-encoding step of image acquisition with one based on wavelet encoding. Rf pulses are used to excite a spatial profile, shaped like wavelet functions, along an axis (for 2D images, perpendicular to the frequency encode axis). The profiles acquired are contractions, dilations, and translations of the same basic wavelet function that is spatially localized. The major advantages of the wavelet encode is its reduced sensitivity to motion and its multiresolution structure. This allows for adaptive imaging, where a lower resolution part of the FOV can be analyzed to determine the likelihood of changes in higher resolution data. This region can then be updated accordingly. Adaptive wavelet-encoded images also have less artifact than Fourier-encoded images but suffer from decreased SNR. Another method of non-Fourier-based encoding utilizes singular value decomposition. This is a highly efficient modality of encoding that essentially creates a compact set of encoding profiles in both the horizontal and vertical directions of a 2D image. Parts of the encode (termed eigenimages) can be recombined to form the full image, with resolution being proportional to the number of eigenimages used. This method allows for faster encoding, compared to Fourier-based schemes, and also paves the way for adaptive techniques to be applied (via the multiresolution features and even keyhole techniques). The spatial resolution of SVD-encoded MRI is equivalent to standard Fourier-encoded MRI, but there is a slightly decreased SNR. Adaptive Fourier-encoded MRI techniques have also been investigated. k-space data can be selectively acquired and analyzed for a change compared to data acquired previously. This method has thus far proved to be about five times faster than the nonadaptive Fourier method. Dynamic imaging will significantly evolve over the next decade, and it will be the basis for future minimally invasive therapies (Riederer *et al.*, 1988; Wright *et al.*, 1989; Holsinger *et al.*, 1990; Brummer *et al.*, 1992; van Vaals *et al.*, 1992; Cao and Levin, 1993; Panych

et al., 1993; Panych and Jolesz, 1994; Zientara *et al.*, 1994).

E. Illustrative Cases

1. Case 1: Glioma Resection

The first case is that of a 45-year-old woman with a right parietal glioma removed within the intraoperative MR unit. Figure 12 shows a T2 image acquired within the open MR unit before the operation began. Figure 13 is a postresection T1-weighted image showing resection of the tumor with a thin rim of postoperative enhancement. The major trade-off to note here is that of image quality for the sake of image acquisition time. Intra-

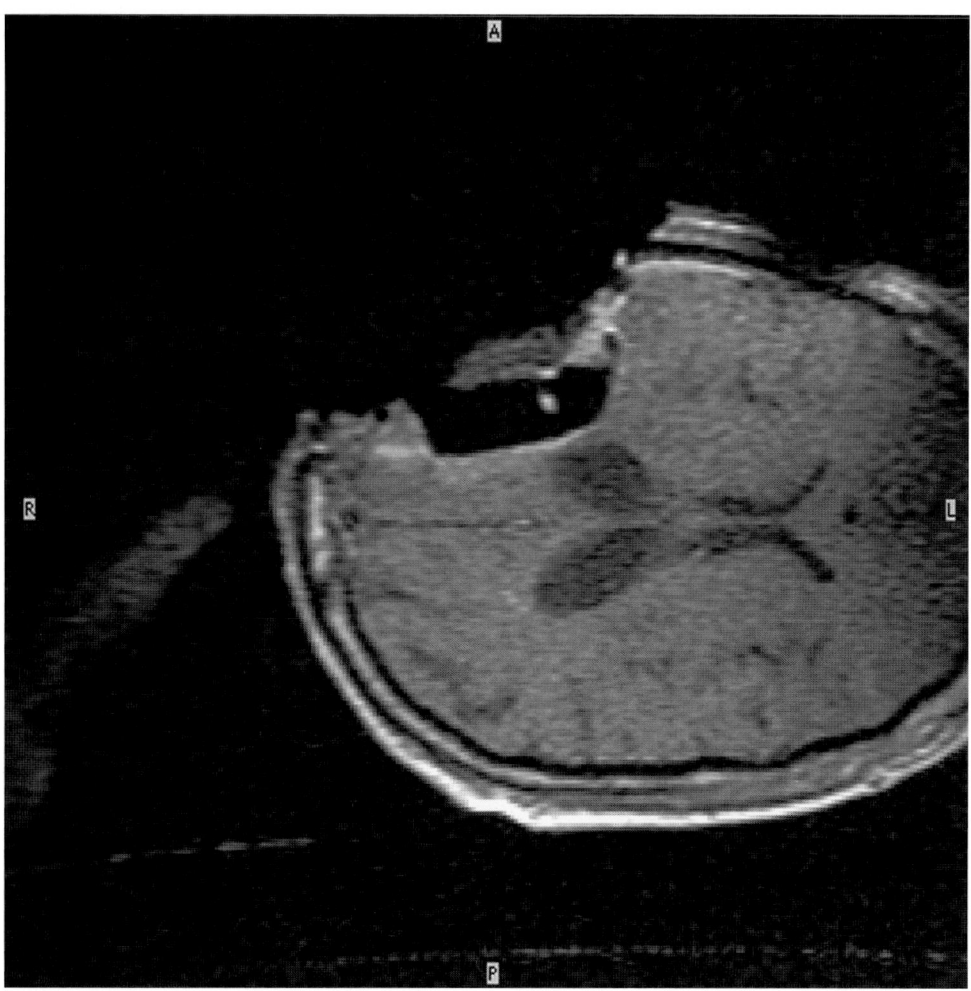

Figure 13 Axial T1–weighted image (postresection). Note peripheral rim of enhancement.

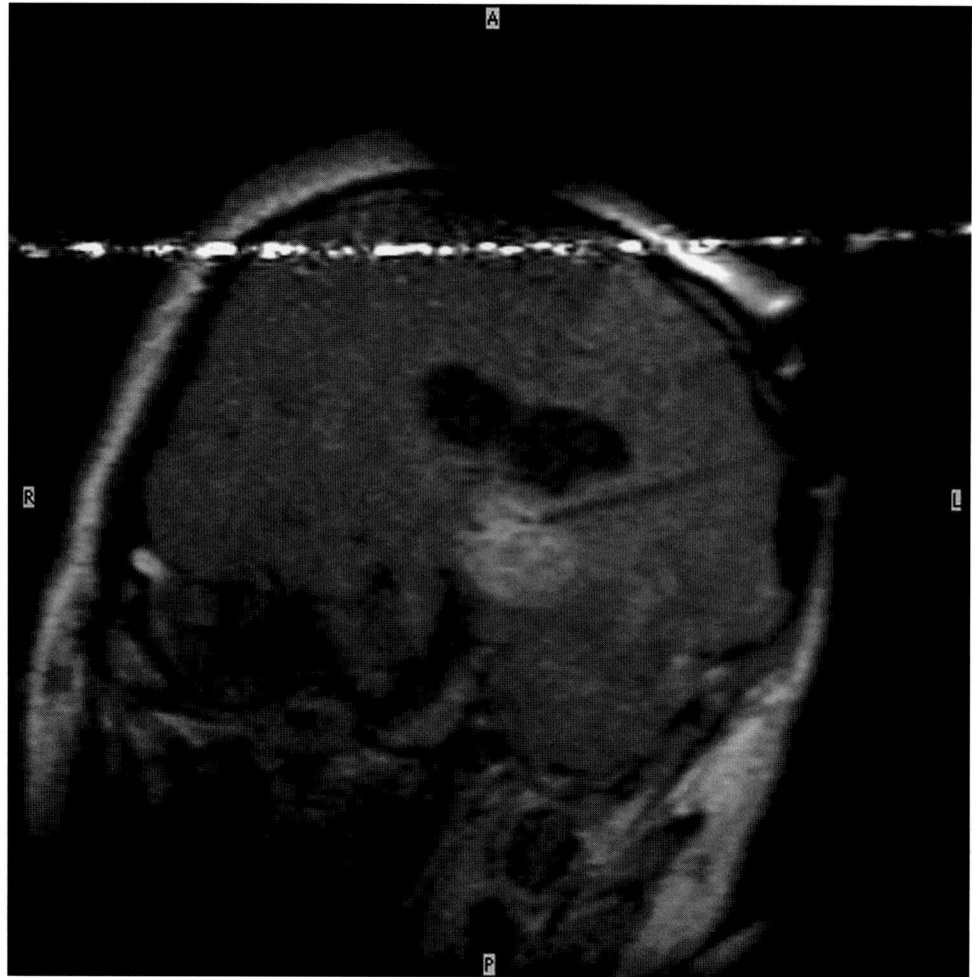

Figure 14 Coronal T1–weighted postgadolinium image showing optical fiber and lesion.

operatively, the images are preferably acquired as rapidly as possible to prevent delays in the surgery. The trick is to achieve a balance between optimal image quality and expediency of image acquisition, an issue that has been at the forefront of both open and conventional MR research.

2. Case 2: Laser Hyperthermia

The second intraoperative case is that of laser interstitial therapy. The patient is a male in his mid 50s who presented with a thalamic lesion. This case was ideal for the open MR system because of the deep and difficult location of the tumor. Figure 14 is a postcontrast image showing the optical fiber and lesion. Temperature distribution during therapy is seen by chemical shift image (Figure 15).

IV. Concluding Remarks

The "operating room of the future" is a recently coined term used to describe the cooperative venture between imaging, image processing, and surgery toward the singular goal of providing minimally invasive surgical procedures with improved localization, targeting, monitoring, and improved patient management. It is an environment that will utilize imaging, preoperative and intraoperative, to help guide the localization, targeting, and navigation within the confines of the patient's anatomy. Neurosurgery, in particular, is well suited toward this end, but minimally invasive, image-guided approaches will help to enhance therapy and reduce morbidity for other surgical fields as well. Although the future of healthcare is riddled with financial uncertainty, an enduring commitment toward improving patient care will nevertheless help drive image-guided therapy to the

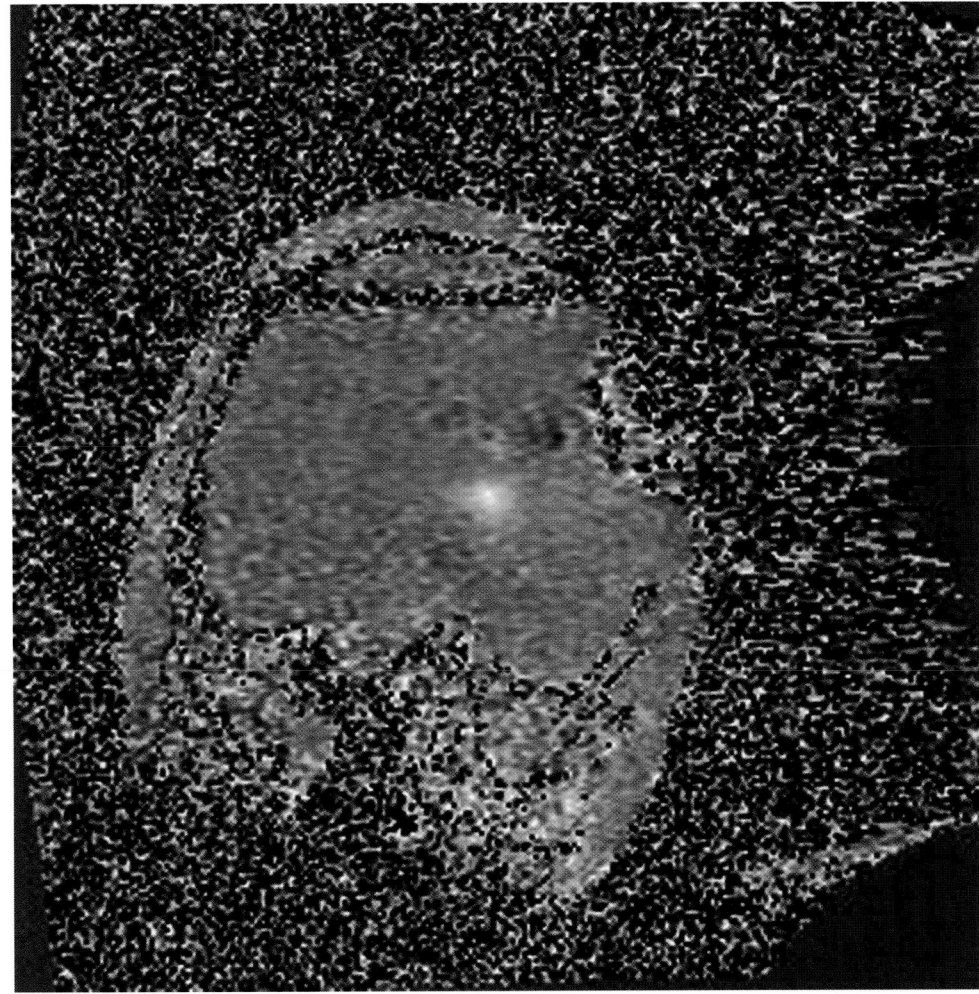

Figure 15 Chemical shift image demonstrating temperature distribution during therapy.

forefront of medicine (Jolesz and Shtern, 1992; Jolesz *et al.*, 1994).

Acknowledgments

The authors acknowledge support by the NIH and NSF, specifically PO1 CA67165-03, 1R01RR11747-01, and 1P41RR13218-01.

References

Anzai, Y., Lufkin, R., DeSalles, A., *et al.* (1995). Preliminary experience with a technique for MR-guided thermal ablation of brain tumors. *AJNR, Am. J. Neuroradiol.* **16,** 39–48.

Asher, P., Justich, E., and Schrottner, O. (1991). Interstitial thermotherapy of central brain tumors with Nd:YAG laser under real time monitoring by MRI. *J. Clin. Laser Med. Surg.* **9,** 79–83.

Barnett, G., Kormos, D., Steiner, C., and Weisenberger, J. (1993). Intraoperative localization using an armless, frameless stereotactic wand. *J. Neurosurg.* **78,** 510–514.

Bettag, M., Ulrich, F., Shober, R., Sabel, M., Kahn, T., and Bock, W. (1992). Laser-induced interstitial thermotherapy in malignant gliomas. *Adv. Neurosurg.* **22,** 253–257.

Black, P., Moriarty, T., Alexander, E., III, Stieg, P., Woodard, E., Gleason, P., Martin, C., Kikinis, R., Schwartz, R., and Jolesz, F. (1997). Development and implementation of intraoperative MRI and its neurosurgical applications. *Neurosurgery* **41,** 831–842.

Bleier, A., Jolesz, F., Cohen, M., *et al.* (1991). Real time magnetic resonance imaging of laser heat deposition in tissue. *Magn. Reson. Med.* **21,** 132–137.

Brummer, M., Dixon, W., Gerety, B., and Tuithof, T. (1992). Composite *k*-space windows (keyhole techniques) to improve temporal resolution in a dynamic series of images following contrast administration. *SMRM Conf. Abstr.* 4236.

Bullock, P., Mansfield, P., Gowland, P., Worthington, B., and Firth, J. (1991). Dynamic imaging of contrast enhancement in brain tumors. *Magn. Reson. Med.* **19,** 293–298.

Cao, Y., and Levin, D. (1993). Feature recognizing MRI. *Magn. Reson. Med.* **30,** 305–317.

Chenevert, T., and Pipe, J. (1993). Dynamic 3D imaging at high temporal resolution via reduced k-space sampling. *SMRM Conf. Abstr.* 1262.

Cline, H., Dumoulin, C., Lorensen, W., Hart, H., and Ludke, S. (1987). 3D reconstruction of the brain from magnetic resonance images using a connectivity algorithm. *Magn. Reson. Imaging* **5,** 345–352.

Cline, H., Lorensen, W., Kikinis, R., and Jolesz, F. (1990). 3D segmentation of MR images of the head using probability and connectivity. *J. Comput. Assist. Tomogr.* **14,** 1037–1045.

Cline, H., Lorensen, W., Souza, S., Jolesz, F., Kikinis, R., Gerig, G., and Kennedy, T. (1991). 3D surface rendered MR images of the brain and its vasculature. *J. Comput. Assist. Tomogr.* **15,** 344–351.

Cline, H., Schenck, J., Watkins, R., Hynynen, K., and Jolesz, F. (1993). Magnetic resonance guided thermal surgery. *Magn. Reson. Med.* **30,** 98–106.

Cline, H., Hynynen, K., Hardy, C., Watkins, R., Schenck, J., and Jolesz, F. (1994). MR temperature mapping of focused ultrasound surgery. *Magn. Reson. Med.* **31,** 628–636.

Cline, H., Hynynen, K., Watkins, R., Adams, W., Schenck, J., Ettinger, R., Freund, W., Vetro, J., and Jolesz, F. (1995). A focused ultrasound system for MRI guided ablation. *Radiology* **194,** 731–737.

Cohen, M., and Weisskoff, R. (1991). Ultrafast imaging. *Magn. Reson. Imaging* **9,** 1–37.

Dumoulin, C., Souza, S., Walker, M., and Wagle, W. (1989). Three-dimensional phase contrast angiography. *Magn. Reson. Med.* **9,** 139–149.

Dumoulin, C., Souza, S., and Darrow, R. (1993). Real-time position monitoring of invasive devices using magnetic resonance. *Magn. Reson. Med.* **29,** 411–415.

Edelman, R. (1993). MR angiography: Present and future. *AJR, Am. J. Roentgenol.* **161,** 1–11.

Ettinger, G., Grimson, W., Leventon, M., Kikinis, R., Gugina, V., Cote, W., Karapelou, M., Aglio, L., Shenton, M., Potts, G., and Alexander, E., III. (1996). Non-invasive functional brain mapping using registered transcranial magnetic stimulation. IEEE Workshop on Mathematical Methods in Biomedical Image Analysis.

Ettinger, G., Leventon, M., Grimson, W., Kikinis, R., Gugino, V., Cote, W., Sprung, L., Aglio, L., Shenton, M., Potts, G., and Alexander, E., III. (1997). Experimentation with a transcranial magnetic stimulation system for functional brain mapping. CVRMed/MRCAS '97, Grenoble, France.

Fan, M., Ascher, P., Shrottner, O., Ebner, F., Germann, R., and Kleinert, R. (1992). Interstitial 1.06 Nd:YAG laser thermotherapy for brain tumors under real-time monitoring of MRI: Experimental study and phase I clinical trial. *J. Clin. Laser Med. Surg.* **10,** 355–361.

Gerig, G., Kikinis, R., and Kuebler, O. (1992). Nonlinear anisotropic filtering of MRI data. *IEEE Trans. Med. Imaging* **11,** 221–232.

Gleason, P., Kikinis, R., Altobelli, D., Wells, W., Alexander, E., III, and Black, P. (1994). Video registration virtual reality for nonlinkage stereotactic surgery. *Stereotact. Funct. Neurosurg.* **63,** 139–143.

Golfinos, J., Fitzpatrick, B., Lawrence, R., and Spetzler, R. (1995). Clinical use of a frameless stereotactic arm: Results of 325 cases. *J. Neurosurg.* **83,** 197–205.

Gowland, P., Mansfield, P., Bullock, P., Stehling, M., Worthington, B., and Firth, J. (1992). Dynamic studies of gadolinium uptake in brain tumors using inversion recovery echo planar imaging. *Magn. Reson. Med.* **26,** 241–258.

Grimson, W., Ettinger, G., White, S., *et al.* (1996). An automatic registration method for frameless stereotaxy, image guided surgery, and enhanced reality visualization. *IEEE Trans. Med. Imaging* **15,** 129–140.

Gronemeyer, D., Seibel, R., Melzer, A., *et al.* (1995). Future of advanced guidance techniques by interventional CT and MRI. *Minimally Invas. Ther.* **4,** 251–259.

Haase, A. (1990). Snapshot flash MRI: Application to T1, T2, and chemical shift imaging. *Magn. Reson. Med.* **13,** 77–79.

Haase, A., Frahm, J., and Matthaei, D. (1986). FLASH imaging: Rapid NMR imaging using low flip angle pulses. *J. Magn. Reson. Imaging* **67,** 258–266.

Hennig, J., and Friedburg, H. (1988). Clinical applications and methodological developments of the RARE technique. *Magn. Reson. Imaging* **6,** 391–395.

Hennig, J., Nauerth, A., and Friedburg, H. (1986). RARE imaging: A fast imaging method for clinical MR. *Magn. Reson. Med.* **3,** 823–833.

Holman, B., Zimmerman, R., Johnson, K., Carvalho, P., Schwartz, R., Loeffler, J., Alexander, E., III, Pellizari, C., and Chen, G. (1991). Computer assisted superimposition of magnetic resonance and high resolution technetium-99m HMPAO and thallium-201 SPECT images of the brain. *J. Nucl. Med.* **32,** 1478–1484.

Holsinger, A., Wright, R., Riederer, S., Farzaneh, F., Grimm, R., and Maier, J. (1990). Real-time interactive magnetic resonance imaging. *Magn. Reson. Med.* **14,** 547–553.

Hu, X., Tan, K., Levin, D., *et al.* (1990). Three-dimensional magnetic resonance images of the brain: Application to neurosurgical planning. *J. Neurosurg.* **72,** 433–440.

Jolesz, F. (1997). Image-guided procedures and the operating room of the future. *Radiology* **204,** 601–612.

Jolesz, F., and Blumenfeld, S. (1994). Interventional use of magnetic resonance imaging. *Magn. Reson. Q.* **10,** 85–96.

Jolesz, F., and Shtern, F. (1992). The operating room of the future. *Invest. Radiol.* **27,** 326–328.

Jolesz, F., and Silverman, S. (1995). Interventional magnetic resonance therapy. *Semin. Intervent. Radiol.* **12,** 20–27.

Jolesz, F., and Zientara, G. (1996). MRI-guided interventions. *In* "Clinical MRI," 2nd ed., Chap. 12, pp. 380–390. W.B. Saunders.

Jolesz, F., Kikinis, R., and Shtern, F. (1994). The vision of image-guided computerized surgery: The high tech operating room. *In* "Computer-Integrated Surgery," pp. 717–721. MIT Press, Cambridge, MA.

Kahn, T., Bettag, M., Ulrich, F., *et al.* (1994). MR-imaging guided laser induced interstitial thermotherapy in cerebral neoplasm. *J. Comput. Assist. Tomogr.* **18,** 519–532.

Kall, B., Kelly, P., and Goerss, S. (1986). The computer as a stereotactic surgical instrument. *Neurol. Res.* **8,** 201–208.

Kall, B., Goerss, S., Kelly, P., and Stiving, S. (1994). Three-dimensional display in the evaluation and performance of neurosurgery without a stereotactic frame: More than a pretty picture? *Stereotact. Funct. Neurosurg.* **63,** 69–75.

Kandarpa, K., Jakab, P., Patz, S., Schoen, F., and Jolesz, F. (1993). Prototype miniature endoluminal MR imaging catheter. *J. Vasc. Intevent. Radiol.* **4,** 419–427.

Kato, A., Yoshimine, T., Hyakawa, T., Yomita, Y., Ikeda, T., Mitomo, M., Harada, K., and Mogami, H. (1986). Armless navigational system for computer assisted neurosurgery. *J. Neurosurg.* **65,** 545–549.

Kikinis, R., Gleason, P., Moriarty, T., Moore, M., Alexander, E., III, Steig, P., Matsumae, M., Lorensen, W., Cline, H., Black, P., and Jolesz, F. (1996). Computer-assisted interactive three-dimensional planning for neurosurgical procedures. *Neurosurgery* **38,** 640–649.

Kosugi, Y., Watanabe, E., and Goto, J. (1988). An articulated neurosurgical navigational system using MRI and CT images. *IEEE Trans. Biomed. Eng.* **35,** 147–152.

Leungh, D., Debatin, J., Wildemuth, S., *et al.* (1995). Real-time biplanar needle tracking for interventional MR imaging procedures. *Radiology* **197,** 485–492.

Levin, D., Hu, X., Tan, K., *et al.* (1989). The brain: Integrated three-dimensional display of MR and PET images. *Radiology* **172,** 783–789.

Lorensen, W., and Cline, H. (1987). Marching cubes: A high resolution 3D surface construction system. *Comput. Graphics* **21,** 163–169.

Lufkin, R. (1995). Interventional MR imaging. *Radiology* **197,** 16–18.

Lufkin, R., Teresi, L., and Hanafee, W. (1987). New needle for MR-guided aspiration cytology of the head and neck. *AJR, Am. J. Roentgenol.* **149,** 380–382.

Martin, A., Plewes, D., and Henkelman, R. (1992). MR imaging of blood vessels with intravascular coil. *J. Magn. Reson. Imaging.* **2,** 421–429.

Moonen, C., Duyn, J., and van Gelderen, P. (1994). Fast volume scanning with frequency shifted BURST MRI. *Magn. Reson. Med.* **32,** 429–431.

Moriarty, T., Kikinis, R., Jolesz, F., Black, P., and Alexander, E., III. (1996). Magnetic resonance imaging therapy: Intraoperative MR imaging. *Neurosurg. Clin. N. Am.* **7B,** 323–331.

Mueller, P., Stark, D., Simeone, J., *et al.* (1986). MR-guided aspiration biopsy: Needle design and clinical trials. *Radiology* **161,** 605–609.

Nakajima, S., Kikinis, R., Black, P., Atsumi, H., Leventon, M., Hata, N., Metcalf, D., Moriarty, T., Alexander, E., III, and Jolesz, F. (1997). Image-guided neurosurgery at Brigham and Women's Hospital. *Comput. Assist. Neurosurg.* 144–162.

Nakajima, S., Atsumi, H., Moriarty, T., Kikinis, R., Jolesz, F., and Black, P. (1997). The use of cortical surface vessel registration for image guided neurosurgery. *Neurosurgery* **40(6),** 1201–1210.

Nakajima, S., Atsumi, H., Bhalerao, A., Jolesz, F., Kikinis, R., Yoshimine, T., Moriarty, T., and Steig, P. (1997). Computer-assisted surgical planning for cerebrovascular neurosurgery. *Neurosurgery* **41(2),** 403–409.

Nakajima, S., Kikinis, R., Jolesz, F., Atusmi, H., *et al.* (1998). 3D MRI reconstruction for surgical planning and guidance. *Adv. Neurosurg. Nav.* (in press).

Oshio, K., and Feinberg, D. (1991). GRASE imaging: A novel fast MRI technique. *Magn. Reson. Med.* **20,** 344–349.

Panych, L., and Jolesz, F. (1994). Dynamically adaptive MRI by wavelet transform encoding. *Magn. Reson. Med.* **32,** 738–748.

Panych, L., Hrovat, M., and Bleier, A. (1992). Effects related to temperature changes during magnetic resonance imaging. *Magn. Reson. Imaging* **2,** 69–74.

Panych, L., Jakab, P., and Jolesz, F. (1993). An implementation of wavelet encoded MRI. *Magn. Reson. Imaging* **3,** 649–655.

Permicone, J., Siebert, J., Laird, T., Rosenbaum, T., and Potchen, E. (1992). Determination of blood flow direction using velocity-phase image display with 3D phase contrast MR angiography. *AJNR, Am. J. Neuroradiol.* **13,** 1435–1438.

Riederer, S., Tasciyan, T., Farzaneh, F., Lee, J., Wright, R., and Herfkens, R. (1988). MR fluoroscopy: Technical feasibility. *Magn. Reson. Med.* **8,** 1–15.

Sato, Y., Nakajima, S., Atsumi, H., Koller, T., Gerig, G., Yoshida, S., and Kikinis, R. (1997). 3D multi-scale line filter for segmentation and visualization of curvilinear structures in medical images. *In* "Proceedings of CVRMed II and MR-CAS III, Grenoble, France, 1997," pp. 213–222.

Schenck, J., Jolesz, F., Roemer, P., *et al.* (1995). Superconducting open configuration MRI system for image-guided therapy. *Radiology* **195,** 805–814.

Silverman, S., Collick, B., Figueira, M., Khorasani, R., Adams, D., Newman, R., Topulos, G., and Jolesz, F. (1995). Interactive MR-guided biopsy in an open configuration MR imaging system. *Radiology* **197,** 173–181.

Spielman, D., Pauly, J., and Metyer, C. (1995). Nonuniform k-space sampling techniques: Interleaved spirals. *Magn. Reson. Med.* **33,** 326–336.

Vannier, M., Marsh, J., and Warren, J. (1984). Three-dimensional CT reconstruction for craniofacial surgical planning and evaluation. *Radiology* **50,** 179–184.

van Sonnenberg, E., Hajek, P., Gylys-Morin, V., *et al.* (1988). A wire sheath system for MR-guided biopsy and drainage: Laboratory studies and experience in 10 patients. *AJR, Am. J. Roentgenol.* **151,** 815–817.

van Vaals, J., Tuithof, H., and Dixon, W. (1992). Increased time resolution in dynamic imaging. *J. Magn. Reson. Imaging* **2,** 44.

Warfield, S., Dengler, J., Zaers, J., *et al.* (1996). Automatic identification of gray matter structures from MRI to improve the segmentation of white matter lesions. *J. Image Guided Surg.* **1,** 326–338.

Wells, W., Grimson, W., Kikinis, R., and Jolesz, F. (1996a). Adaptive segmentation of MRI data. *IEEE Trans. Med. Imaging* **15,** 429–442.

Wells, W., Viola, S., Nakajima, H., *et al.* (1996a). Multi-modal volume registration by maximization of mutual information. *Med. Image Anal.* **1,** 35–51.

Wright, R., Riederer, S., Farzaneh, F., Rossman, P., and Liu, Y. (1989). Real-time MR fluoroscopic data acquisition and image reconstruction. *Magn. Reson. Med.* **12,** 407–415.

Zamorano, L., Jiang, Z., and Kadi, A. (1994). Computer assisted neurosurgery system: Wayne State University hardware and software configuration. *Comput. Med. Imaging Graphics* **18,** 435–441.

Zhang, J., Levesque, M., Wilson, C., Harper, R., Engel, J., Lufkin, R., and Behnke, E. (1990). Multimodality imaging of brain structures for stereotactic surgery. *Radiology* **175,** 435–441.

Zientara, G., Panych, L., and Jolesz, F. (1994). Dynamically adaptive MRI with encoding by singular value decomposition. *Magn. Reson. Med.* **32,** 268–274.

6

Disease-Specific Brain Atlases

Paul M. Thompson,[*,1] Michael S. Mega,[*,†] and Arthur W. Toga[*]

[*]Laboratory of Neuro Imaging and
[†]Alzheimer's Disease Center, Division of Brain Mapping, Department of Neurology, UCLA School of Medicine,
Los Angeles, California 90095

I. Challenges in Population-Based
Brain Mapping

II. Types of Brain Atlases

III. Analyzing Brain Data

IV. Individualized Brain Atlases

V. Model-Driven Deformable Atlases

VI. Probabilistic Brain Atlases

VII. Atlas-Based Pathology Detection

VIII. Cortical Modeling

IX. Cortical Averaging

X. Brain Averaging

XI. Dynamic (4D) Brain Atlases

XII. Conclusion
References

I. Challenges in Population-Based Brain Mapping

Advanced brain imaging technologies now provide a means to investigate disease and therapeutic response in their full spatial and temporal complexity. Imaging studies of clinical populations continue to uncover new patterns of altered structure and function, and novel algorithms are being applied to relate these patterns to cognitive and genetic parameters. Postmortem brain maps are also beginning to clarify the molecular substrates of disease.

As imaging studies expand into ever-larger patient populations, population-based brain atlases offer a powerful framework to synthesize the results of disparate imaging studies. These atlases use novel analytical tools to fuse data across subjects, modalities, and time. They detect group-specific features not apparent in individual patient scans. Once built, these atlases can be stratified into subpopulations to reflect a particular clinical group. The disease-specific features they resolve can then be linked with demographic factors such as age, gender, and handedness as well as specific clinical or genetic parameters (Mazziotta *et al.,* 1995; Toga and Mazziotta, 1996; Toga and Thompson, 1999a).

New brain atlases are also being built to incorporate dynamic data. Despite the significant challenges in expanding the atlas concept to the time dimension, dynamic brain atlases are beginning to include probabilistic information on growth rates that may assist research into pediatric disorders (Thompson and Toga, 1999). Imaging algorithms are also significantly improving the flexibility of digital brain templates. *Deformable brain atlases* are adaptable brain templates

---

[1]To whom correspondence should be addressed.

that can be individualized to reflect the anatomy of new subjects, and *probabilistic atlases* are research tools that retain information on cross-subject variations in brain structure and function. These atlases are powerful new tools with broad clinical and research applications (Fox *et al.*, 1994; Roland and Zilles, 1994; Kikinis *et al.*, 1996; Minoshima *et al.*, 1996; Toga and Thompson, 1998b).

A. Disease-Specific Atlases

This chapater introduces the topic of a *disease-specific* brain atlas (Fig. 1). This type of atlas is designed to reflect the unique anatomy and physiology of a particular clinical subpopulation (Mega *et al.*, 1997, 1998c, 1999; Thompson *et al.*, 1997, 1998a, 1999b; Narr *et al.*, 1999). Based on well-characterized patient groups, these atlases contain thousands of structure models as well as composite maps, average templates, and visualizations of structural variability, asymmetry, and group-specific differences. They act as a quantitative framework that correlates the structural, metabolic, molecular, and histologic hallmarks of the disease (Mega *et al.*, 1997). Additional algorithms are described that use information stored in the atlas to recognize anomalies and label structures in new patients. Because they retain information on group anatomical variability, disease-specific atlases are a type of probabilistic atlas specialized to represent a particular clinical group. The resulting atlases can identify patterns of altered structure or function and can guide algorithms for knowledge-based image analysis, automated image labeling (Collins *et al.*, 1994b; Pitiot *et al.*, 1999), tissue classification (Zijdenbos and Dawant, 1994), and functional image analysis (Dinov *et al.*, 1999).

We present data from several ongoing projects whose goal is to create disease-specific atlases of the brain in Alzheimer's disease, schizophrenia, and several neurodevelopmental disorders. Since current brain templates poorly represent the anatomy of these clinical populations, the resulting atlases offer a framework to investigate each disease. Pathological change can be tracked over time, and disease-specific features re-

solved. Rather than simply fusing information from multiple subjects and sources, new mathematical strategies are introduced to resolve group-specific features not apparent in individual scans.

B. Anatomical Templates

Central to the construction of a disease-specific atlas is the creation of averages, templates, and models to describe how the brain and its component parts are organized and how they are altered in disease. Statistical models are created to reveal how major anatomic systems are affected, how the pathology progresses, and how these changes relate to demographic or genetic factors. To create templates that reflect the morphology of a diseased group, specialized strategies are required for population-based averaging of anatomy (Thompson *et al.*, 1996a,b, 1998; Grenander and Miller, 1998). In one approach (Thompson *et al.*, 1999b), sets of high-dimensional elastic mappings, based on the principles of continuum mechanics, reconfigure the anatomy of a large number of subjects in an anatomic image database. These three-dimensional deformation fields are used to create a crisp anatomical image template with highly resolved structures in their mean spatial location. The mappings also generate a richly detailed local encoding of anatomic variability, with up to a billion parameters (Thompson and Toga, 1997; Grenander and Miller, 1998). The resulting variability parameters are stored as a tensor field (Section IX) and leveraged by pattern recognition strategies that automatically identify anatomical structures in new patients' scans and identify disease-specific characteristics (Pitiot *et al.*, 1999; Thompson *et al.*, 1999a).

C. Cortical Patterns

Cortical patterns are altered in a variety of diseases. Sulcal pattern anomalies have been identified in schizophrenia and epilepsy (Cook *et al.*, 1994; Kikinis *et al.*, 1994; Fuh *et al.*, 1997), and diffuse cortical atrophy is typical of Alzheimer's disease, Pick's disease, and other dementias (Schmidt, 1992). Gyral anomalies,

Figure 1 Elements of a disease-specific atlas. This schematic shows the types of maps and models contained in a disease-specific brain atlas. In the main text, we explain how these maps are created and review their applications. Examples are shown from an atlas that represents an Alzheimer's disease population. To construct the atlas, databases of structural imaging data are used to develop detailed models of cortical structure and anatomic subsystems. These models are statistically combined to create group average models that can be compared with a normal database. Patterns of variability, asymmetry, and disease-specific differences are also computed from the anatomic data. Specialized techniques create a well-resolved average image template for the patient population (continuum-mechanical template, center right). This template provides a coordinate framework to link *in vivo* metabolic and functional data with fine-scale anatomy and biochemistry. In recent studies (Mega *et al.*, 1997), histologic maps of postmortem neurofibrillary tangle (NFT) staining density were correlated with *in vivo* metabolism. 3D FDG-PET data, obtained 8 h before death, were compared with whole-brain cryosections acquired immediately postmortem and stained for NFTs by the Gallyas method. With the algorithm of Thompson and Toga (1996; warped image), distorted tissue sections were elastically warped back to their configuration in the cryosection blockface (top row). A further 3D registration projected the data into premortem MR and coregistered PET data (top right).

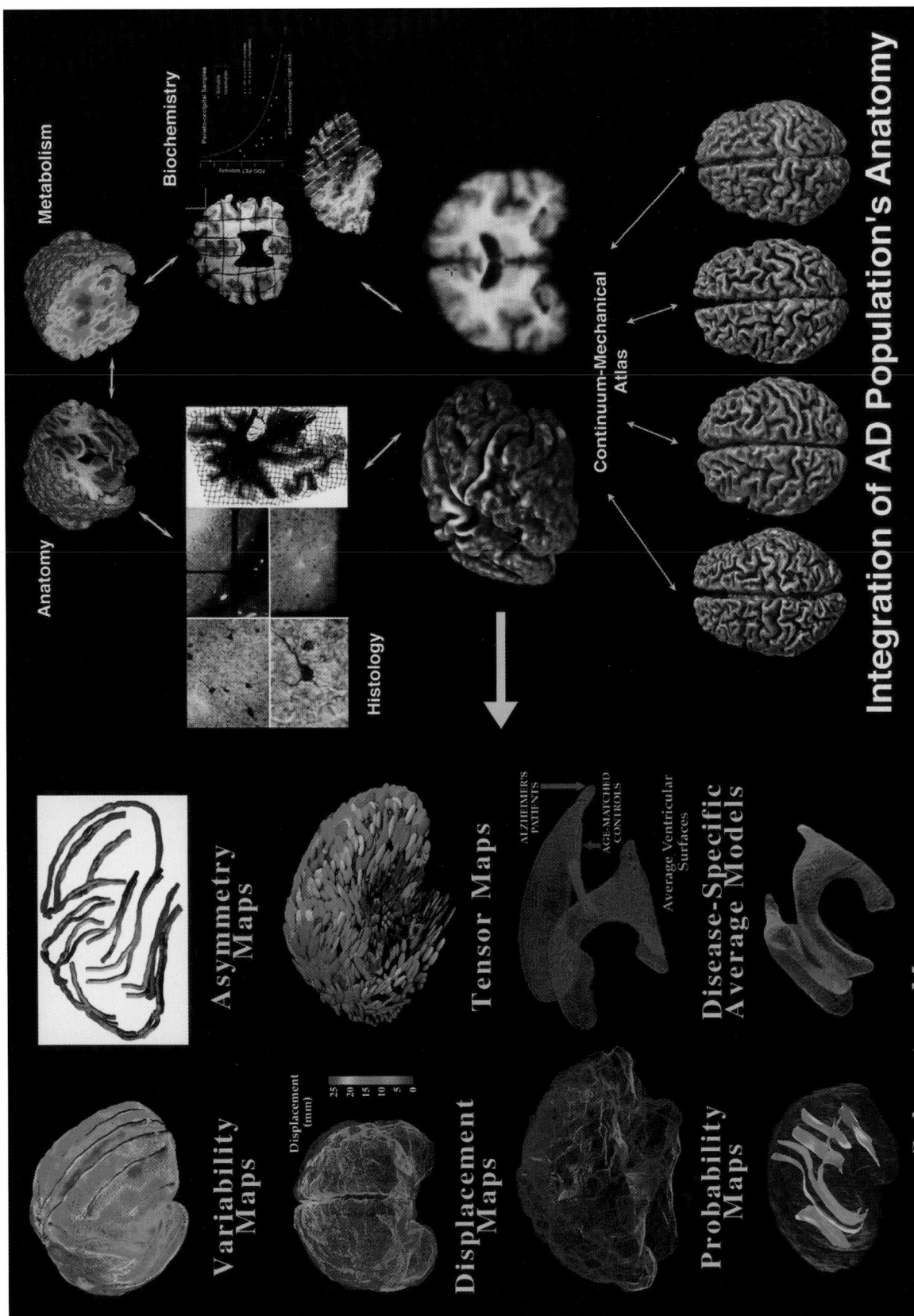

Metabolism

Biochemistry

Anatomy

Histology

Continuum-Mechanical Atlas

Integration of AD Population's Anatomy

Variability Maps

Displacement (mm)
25
20
15
10
5
0

Displacement Maps

Probability Maps

Asymmetry Maps

Tensor Maps

ALZHEIMER'S PATIENTS

AGE-MATCHED CONTROLS

Average Ventricular Surfaces

Disease-Specific Average Models

Subsystem Maps

such as cortical dysplasias, have also been linked with neurodevelopmental delay (Sobire *et al.,* 1995). Nonetheless, ratings of structural change in the cortex are still largely based on qualitative assessment (Berg *et al.,* 1993).

To clarify how diseases affect the cortex, specialized approaches are described for averaging cortical anatomy across subjects. Gyral pattern matching (Thompson and Toga, 1996; Thompson *et al.,* 1999b) is used to create average cortical models, to measure cross-subject differences, and to encode the magnitude and principal directions of anatomical variation at the cortex. In the resulting cortical templates, subtle features emerge. Regional asymmetries appear that are not apparent in individual anatomies. Population-based maps of cortical variation reveal a mosaic of variability patterns that are characteristic of each cortical region.

D. Pathology Detection

Normal anatomic complexity makes it difficult to design automated strategies that detect abnormal brain structure. Considerable research has focused on uncovering specific patterns of anatomic alterations in Alzheimer's disease and other dementias (Friedland and Luxenberg, 1988), schizophrenia (Kikinis *et al.,* 1994; Csernansky *et al.,* 1998), epilepsy (Cook *et al.,* 1994), attention deficit hyperactivity disorder (ADHD; Giedd *et al.,* 1994), autism (Filipek, 1996; Courchesne, 1997), and cortical dysplasias (Sobire *et al.,* 1995). Some of the results of these morphometric studies are summarized in Table I. At the same time, brain structure is so variable that group-specific patterns of anatomy and function are often obscured.

E. Demographic Factors

Reports of structural differences in the brain linked to gender and handedness are still controversial, and these factors may also interact with disease-specific abnormalities (Narr *et al.,* 1999; Thompson *et al.,* 1999a). Other factors that interfere with analysis include educational level, premorbid adjustment, treatment history and response, and the duration and course of illness (Carpenter *et al.,* 1993). Interactions of these variables make it harder to detect disease-specific patterns and relate them to clinical and genetic parameters.

The importance of these linkages has propelled computational anatomy to the forefront of brain imaging investigations. To distinguish abnormalities from normal variants, a realistically complex mathematical framework is required to encode information on anatomic variability in homogeneous populations (Grenander and Miller, 1998; Thompson and Toga, 1999a,b). As we

shall see, elastic registration, or *warping* algorithms, offer substantial advantages in creating brain atlases that encode patterns of anatomic variation and detect pathology.

F. Dynamic Brain Atlases

While brain atlases have traditionally relied on static representations of anatomy, many of the diseases that affect the human brain are progressive (e.g., dementia and neoplastic tumors). Other disorders may be relapsing-remitting (e.g., multiple sclerosis) or may display a normal phenotype with an aberrant time course (e.g., delayed neurodevelopment). The progression of a disease may also be modulated by therapy, ranging from drug treatment to surgery. In response to these challenges, *dynamic* brain atlases retain spatiotemporal information on patterns of neuroanatomic change. They offer a means to analyze the dynamics of disease. Later in this chapter, we describe how atlases can be expanded to incorporate quantitative (4D) maps of growth or degenerative patterns in the brain. These maps characterize local growth or atrophic rates in development or disease. Atlases that incorporate confidence limits on growth rates, in particular, offer a new type of normative framework to analyze aberrant brain development and childhood-onset disorders (Thompson and Toga, 1999a).

II. Types of Brain Atlases

A. Coordinate Systems

Rapid progress has been made in recent years by research groups developing standardized three-dimensional brain atlases (Talairach and Tournoux, 1988; Greitz *et al.,* 1991; Höhne *et al.,* 1992; Thurfjell *et al.,* 1993; Kikinis *et al.,* 1996). While few of these atlases aim to represent anatomy and function in disease, several commercially available atlases of pathology combine histologic data with illustrative metabolic or structural images. The *Harvard Brain Atlas* (Johnson, 1996) is a rich source of annotated CT, MRI, SPECT, and PET (single photon/positron emission computed tomography) images from a number of clinical populations. Cerebrovascular, neoplastic, and degenerative diseases are represented (including stroke, vascular dementia, and Alzheimer's disease), as are inflammatory, autoimmune, and infectious diseases (multiple sclerosis and AIDS). In a similar effort, the *Atlas of Brain Perfusion SPECT* has been produced by Brigham and Women's Hospital (Holman *et al.,* 1994). This atlas presents 21 SPECT images with coregistered scans (SPECT merged with CT or MRI), and all scans are annotated with rel-

Table I Morphometric Abnormalities in Disease

Although not an exhaustive list, this table summarizes the anatomic alerations found in major neurological diseases as well as confounding differences due to demographic factors. Most recent studies have used large archives of volumetric (3D) MRI scans to detect structural alterations in disease. Disease-specific atlases offer a computational framework to combine digital structure models from large clinical subpopulations, localizing group differences. Once models are in a conveniently parameterized format, they can be statistically combined to reveal group-specific features not observable in individual data. Patterns of average asymmetry and variation also emerge. Complementary to the more traditional volumetric descriptors, computational models and images can be combined to build probabilistic atlases that localize regional abnormalities.

Alzheimer's Disease   Sulcal and ventricular enlargement; gross cerebral atrophy, starting in temporal and parietal areas (Kido et al., 1989; Erkinjuntti *et al.*, 1993; Killiany *et al.*, 1993), entorhinal cortex (Arnold *et al.*, 1991; Braak and Braak, 1991), the basal nucleus of Meynert (Whitehouse *et al.*,1981), amygdala (Cuénod *et al.*, 1993), and hippocampus (West *et al.*, 1994); caudate, lenticular, and thalamic atrophy also reported (Jernigan *et al.*, 1991).

Schizophrenia   Enlarged lateral and third ventricles (median 40% increase; Lawrie and Abukmeil, 1998); reduced brain volume (averaging 3%) with disproportionate temporal lobe reductions and gray matter deficits (including hippocampus; Lawrie and Abukmeil, 1998; Csernansky *et al.*, 1998; Joshi *et al.*, 1998); shape difference (Casanova *et al.*, 1990; DeQuardo *et al.*, 1996; Bookstein, 1997) or increased bowing of corpus callosum, stronger in male patients (Narr *et al.*, 1999); thalamic and midline anomalies (Andreasen *et al.*, 1994, 1995); striatal enlargement secondary to antipsychotic medication (Chakos *et al.*, 1994; Harrison, 1999); loss or reversal of cortical pattern asymmetry (Falkai *et al.*, 1992; Shenton *et al.*, 1992; Bilder *et al.*, 1994)

Epilepsy   Abnormal hippocampal morphology, including unilateral or bilateral volume loss, in patients with temporal lobe epilepsy (TLE; Jack, 1994); left temporal lobe may atrophy faster in left unilateral TLE than right temporal lobe in right unilateral TLE

Autism   Regional cerebral enlargement in posterior, temporal, and occipital but not frontal cortices; enlarged total cerebellar volume (Piven *et al.*, 1997); controversy over localized hypoplasia at the cerebellar vermis, lobules VI and VII; possible hypoplastic and hyperplastic subgroups (Courchesne *et al.*, 1994; Filipek, 1995)

ADHD   4.7% smaller total cerebral volume ($p < 0.02$) and smaller cerebellum, with significant loss of normal right > left caudate asymmetry and reversal of lateral ventricular asymmetry (Castellanos *et al.*, 1996); reduced area of the callosal splenium (Semrud-Clikeman *et al.*, 1994; Lyoo *et al.*, 1996), genu (Hynd *et al.*, 1991), and rostrum (Giedd *et al.*, 1994; Baumgardner *et al.*, 1996)

Down's Syndrome   Reduced frontal lobe volume (Jernigan *et al.*, 1993) and reduced anterior callosal area (Wang *et al.*, 1992); reduced volume in cerebellar hemispheres and hippocampus (Raz *et al.*, 1995)

Fetal Alcohol Syndrome   Striatal callosal, and cerebellar abnormalities (Roebuck *et al.*, 1998); complete or partial agenesis of the corpus callosum (Riley *et al.*, 1995; Johnson *et al.*, 1996); midline anomalies (callosal hypoplasia, cavum septi pellucidi and cavum vergae) associated with increased number of facial anomalies (Swayze *et al.*, 1997); reduced size of basal ganglia (Mattson *et al.*, 1996) and vermian lobules I–V (Sowell *et al.*, 1996)

Obsessive Compulsive Disorder (OCD)   Smaller striatal volumes in pediatric OCD, inversely correlated with symptom severity (Rosenberg *et al.*, 1997; see Aylward *et al.*, 1996, for a contrary view); larger third ventricle volumes, but no differences in prefrontal cortical, lateral ventricular, or intracranial volumes (Rosenberg *et al.*, 1997); white matter reductions also reported (Jenike *et al.*, 1996; $N = 20$)

Sydenham's chorea   Increased sizes of the caudate, putamen, and globus pallidus, but not total cerebral, prefrontal, or midfrontal volumes (Giedd *et al.*, 1996)

Multiple Sclerosis[a]   Recurrent inflammatory lesions throughout the white matter; T2-hyperintense lesions most prominent in the corpus callosum, internal capsule, and around the lateral ventricular body and occipital horns (Lee *et al.*, 1999)

Demographic Factors

Age   Cerebral volume declines by 2% per decade after age 50; corpus callosum area shows greater age-related decline in males (Burke *et al.*, 1993, $N = 97$); large-scale morphometric studies of brain development include Giedd *et al.*, 1996 ($N = 99, 104$); and Paus *et al.*, 1999 ($N = 111$); adolescent brain volume increases are specific to dorsal cortices (Jernigan *et al.*, 1991; Pfefferbaum *et al.*, 1994)

Gender   9% greater mean forebrain volume in adult men that women ($N = 71$ men/49 women; Jäncke *et al.*, 1997); controversy over whether splenium of corpus collosum is larger in women (DeLacoste-Utamsing and Holloway, 1982); in women, splenium may also be more bulbous (Clarke *et al.*, 1989) or shaped differently (Davatzikos, 1996; Bookstein, 1997); for a review of this controversy, see Bishop and Wahlsten, 1997; Beaton, 1997; and Thompson *et al.*, 1999

Handedness   Normal left > right volume asymmetry of the planum temporale (PT) reduced in left-handers, without significant gender effects; same result in monozygotic twins discordant for handedness (Steinmetz, 1996); hand preference associated with increased connectivity (increased neuropil in left area 4) and an increased intrasulcal surface of the precentral gyrus in the dominant hemisphere (Amunts et al., 1996)

[a] Unlike the above disorders, lesions alter image intensity rather than morphology; stereotaxic maps have revealed regional biases in lesion deposition (Narayanan *et al.*, 1997; Lee *et al.*, 1998).

evant clinical information and case histories. Other collections focus on postmortem data. The *On-line Neuropathology Atlas* developed by the University of Debrecen Medical School (Hegedüs and Molnár, 1996) includes labeled images of the normal brain, with an extensive collection of pathological images from patients with cerebrovascular disease, neoplasms, and inflammatory and degenerative disorders.

Perhaps surprisingly, few atlases of neuropathology use a standardized three-dimensional coordinate system to integrate data across patients, techniques, and acquisitions. To suggest why this can be advantageous, we review recent developments in the brain atlasing field that have created a framework for interlaboratory communication. Digital templates placed in a well-defined coordinate space (Evans *et al.*, 1992; Friston *et al.*, 1995; Drury and Van Essen, 1997), together with algorithms to align data with them (Toga, 1998), have enabled the pooling of brain mapping data from multiple subjects and sources, including large patient populations. As we shall see, standardized coordinate systems also allow parameterized, anatomical models to be statistically combined (Thompson *et al.*, 1996a,b). By combining anatomical models, the results of morphometric studies can be leveraged to create disease-specific brain templates. Automated algorithms can then capitalize on atlas descriptions of anatomical variance to guide image segmentation (Le Goualher *et al.*, 1999; Pitiot *et al.*, 1999), tissue classification (Zijdenbos and Dawant, 1994), functional image analysis (Dinov *et al.*, 1999), and pathology detection (Thompson *et al.*, 1997).

Before describing atlases specialized to represent disease, we first review developments in informatics that led to the creation of deformable anatomic templates. These templates are a flexible type of brain atlas that can be customized to represent a given individual and then subsequently to represent a population.

B. Early Anatomic Templates

Research on brain atlases was originally based on the premise that brain structure and function imaged in any modality can be better localized by correlation with higher resolution anatomic data placed in an appropriate spatial coordinate system. Three-dimensional brain templates also provide reference information for surgical planning and have been used to guide stereotaxic radiosurgery and electrode implantations (Talairach and Szikla, 1967; Kikinis *et al.*, 1996). Because of their detailed characterization of anatomy, most early brain atlases were derived from one, or a few, postmortem specimens (Brodmann, 1909; Schaltenbrand and Bailey, 1959; Van Buren and Maccubin, 1962; Talairach and Szikla, 1967; Van Buren and Borke, 1972; Schaltenbrand and Wahren, 1977;

Matsui and Hirano, 1978; Talairach and Tournoux, 1988; Ono *et al.*, 1990). These anatomical references typically represent a particular feature of the brain, such as a neurochemical distribution (Mansour *et al.*, 1995), myelination patterns (Smith, 1907; Mai *et al.*, 1997), or the cellular architecture of the cortex (Brodmann, 1909).

C. Digital Manipulations

Atlas templates became considerably easier to manipulate with the transition from paper to digital format. Widely used brain atlases were converted to electronic form, and image registration algorithms made it feasible to overlay digital atlas data onto volumetric radiologic scans. In an endeavor to fuse data from multiple atlases, Nowinski *et al.* (1997) merged the thalamic and brainstem anatomical maps of Schaltenbrand and Wahren (1977) and the Talairach atlases (1988) with cortical and sulcal atlases employed in radiology (Brodmann, 1909; Ono *et al.*, 1990). Recent neurosurgical systems (e.g., the CASS system; Computer Assisted Stereotaxic Surgery, Midco Corp., San Diego, California) now support the digital overlay of the Schaltenbrand, Talairach, and Brodmann atlas data onto individual patient MR scans to create composite maps and simulation displays for surgical planning (Hardy, 1994; St.-Jean *et al.*, 1998).

D. Multimodality Atlases

Because of the superior anatomic resolution, several digital atlases have been created using cryosection imaging. This technique allows the serial collection of photographic images from a cryoplaned specimen blockface (Bohm *et al.*, 1983; Greitz *et al.*, 1991; Toga *et al.*, 1994). Using 1024^2, 24 bits/pixel digital color cameras, cryosection imaging offers a spatial resolution as high as 100 μm/voxel for whole human head cadaver preparations, or higher for isolated brain regions (Toga *et al.*, 1997). Unlike paper atlases, digital cryosection volumes are amenable to a variety of resampling and repositioning schemes. Structures can therefore be rendered and visualized from any angle. In the *Visible Human Project* (Spitzer *et al.*, 1996), two (male and female) cadavers were cryoplaned and imaged at 1.0-mm intervals (0.33 mm for the female data), and the entire bodies were also reconstructed via 5,000 postmortem CT and MRI images. The resulting digital data sets consist of over 15 GB of image data. While not an atlas per se, the *Visible Human Project* data have served as the foundation for developing related atlases of regions of the cerebral cortex (Drury and Van Essen, 1997) and high-quality brain models and visualizations (Schiemann *et al.*, 1996; Stewart *et al.*, 1996). Using multimodality data from a patient with a localized pathology, and more recently the *Visible Human Project* data, Höhne and co-workers developed a commercially available brain atlas designed for teaching neuroanatomy (VOXEL-MAN; Höhne *et al.*, 1990, 1992; Tiede *et al.*, 1993; Pommert *et al.*, 1994).

E. Postmortem Data Fusion

Fusion of metabolic and functional images acquired *in vivo* with postmortem biochemical maps provides a unique view of the relationship between brain function and pathology. Mega *et al.* (1997) scanned Alzheimer's patients in the terminal stages of their disease using both MRI and PET. Using elastic registration techniques (Thompson *et al.*, 1996), we combined these data with postmortem histologic images showing the gross anatomy (Toga *et al.*, 1994), a Gallyas stain of neurofibrillary tangles, and a variety of spatially indexed biochemical assays (Fig. 2). The resulting multimodality maps of the Alzheimer's disease brain relate the anatomic and histopathologic underpinnings of the disease in a standardized coordinate space. These data are further correlated with *in vivo* metabolic and perfusion maps of this disease. The resulting maps are key components of a growing disease-specific atlas (Mega *et al.*, see Chapter 9).

III. Analyzing Brain Data

One of the driving forces that made anatomical templates important in brain imaging was the need to perform brain-to-brain comparisons. Anatomic variations severely hamper the integration and comparison of data across subjects and can lead to misleading results (Meltzer and Frost, 1994; Woods, 1996). Motivated by the need to standardize and pool data across subjects and compare results across laboratories, investigators have developed several registration methods to align brain mapping data with an atlas. The simplest registration techniques are linear, removing global differences in brain size. Nonlinear approaches, however, can eliminate even the most local size and shape differences that distinguish one brain from another. Transforming individual data sets into the shape of a single reference anatomy, or onto a 3D digital brain atlas, allows subsequent comparison of brain function across individuals (Christensen *et al.*, 1993; Ashburner *et al.*, 1997). Interestingly, the transformations required to remove individual differences in anatomy are themselves a rich source of morphometric data (Thompson and Toga, 1999b). As we shall see later (Section X), these data can be used to create disease-specific atlases.

A. Spatial Normalization

In the earliest brain atlases, spatial normalization systems were proposed to transform new data to match the space occupied by the atlas. In the Talairach stereotaxic system (Talairach and Tournoux, 1988), piecewise affine transformations are applied to 12 rectangular regions of brain to reposition the subject's brain in a defined space. The Talairach stereotaxic system rapidly became an international reporting standard for functional activation sites in PET and fMRI studies (Fox *et al.*, 1985, 1988; Friston *et al.*, 1989, 1991).

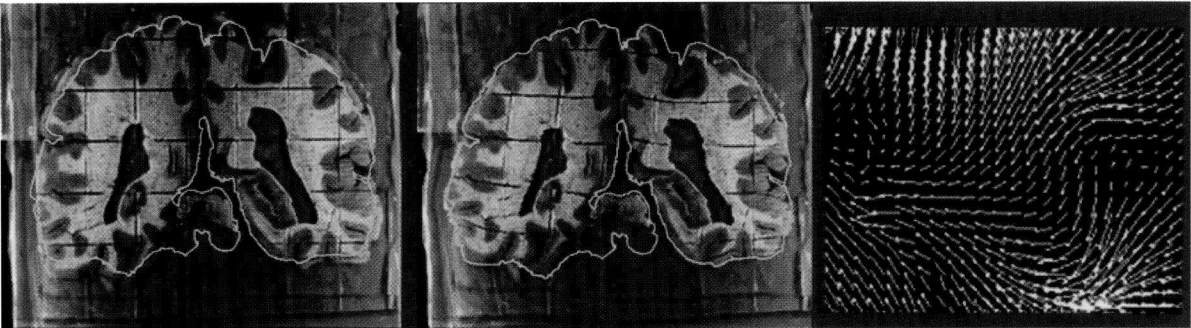

Figure 2 Elastic registration of brain maps and molecular assays. Postmortem tissue sections, from patients with Alzheimer's disease, are gridded (left) to produce tissue elements for biochemical assays. These assays provide detailed quantitative measures of the major hallmarks of AD, including β-amyloid and synaptophysin density. To pool these data in a common coordinate space, tissue elements are elastically warped back into their original configuration in the cryosection blockface (middle panel). Image data acquired from the same patient *in vivo* can then be correlated with regional biochemistry (Mega et al., 1997). When tissue sections are warped to the blockface, continuum-mechanical models are used to make the deformations reflect how real physical tissues deform. The complexity of the required deformation vector field in a small tissue region (magnified vector map, right) demonstrates that very flexible, high-dimensional transformations are essential (Christensen *et al.*, 1996; Schormann *et al.*, 1996; Thompson and Toga, 1996). These deformation vector fields project histologic and biochemical data back into their *in vivo* configuration, populating a growing Alzheimer's disease atlas with maps of molecular content and histology.

B. Use of Digital Templates

While stereotaxic methods provide a common coordinate system to pool activation data for multisubject comparisons, the accuracy and utility of the underlying atlas are equally dependent on the anatomical template itself (Roland and Zilles, 1994). Clearly, the success of any brain atlas depends on how well the anatomies of individual subjects match the representation of anatomy in the atlas. The Talairach templates, for example, are based on postmortem brain sections from a 60-year-old female subject. They therefore poorly reflect the *in vivo* anatomy of subjects in activation studies and are even less representative of brains undergoing degenerative or developmental change. To address some of these limitations, a composite MRI data set was constructed from several hundred young normal subjects (239 males, 66 females; age: 23.4 ± 4.1 years; Evans *et al.*, 1992, 1994). These subjects' scans were individually mapped into the Talairach system by linear transformation, intensity normalized, and averaged on a voxel-by-voxel basis. Although the resulting average brain has regions where individual structures are blurred out due to spatial variability in the population (Evans *et al.*, 1992, 1994; see Section X), the effect of anatomical variability in different brain areas is illustrated qualitatively by this image template. The average-intensity template is part of the widely used Statistical Parametric Mapping package (SPM; Friston *et al.*, 1995). Automated methods were subsequently developed to map new MRI and PET data into a common space. Because the composite MR target atlas was digital, algorithms could align new MR data with the template by maximizing a measure of intensity similarity, such as 3D cross-correlation (Collins *et al.*, 1994, 1995), ratio image uniformity (Woods *et al.*, 1992), or mutual information (Viola and Wells, 1995; Wells *et al.*, 1997). Any alignment transformation defined for one modality, such as MRI, can be identically applied to another modality, such as PET, if a previous cross-modality intrasubject registration has been performed (Woods *et al.*, 1993). For the first time, then, PET data could be mapped into stereotaxic space via a correlated MR data set (Woods *et al.*, 1993; Evans *et al.*, 1994). Registration algorithms therefore made it feasible to automatically map data from a variety of modalities into an atlas coordinate space based directly on the Talairach reference system.

IV. Individualized Brain Atlases

A. Anatomic Variations

No two brains are the same, which presents a challenge in creating standardized atlases. Even without pathology, brain structures vary between individuals in every metric: shape, size, position, and orientation relative to each other (Steinmetz *et al.*, 1989, 1990). Due to the obvious limitations of a fixed atlas, new algorithms were developed to elastically reshape an atlas to the anatomy of new individuals. The resulting *deformable brain atlases* more accurately project atlas data into new scans. Their uses include surgical planning (St.-Jean *et al.*, 1998; Warfield *et al.*, 1998), anatomical labeling (Iosifescu *et al.*, 1997), and shape measurement (Haller *et al.*, 1997; Subsol *et al.*, 1997; Thompson *et al.*, 1997; Csernansky *et al.*, 1998). The shape of the digital atlas is adapted using local warping transformations (dilations, contractions, and shearing), producing an *individualized* brain atlas (Evans *et al.*, 1991; Christensen *et al.*, 1993; Miller *et al.*, 1993; Sandor and Leahy, 1994, 1995; Rizzo *et al.*, 1995). Pioneered by Bajcsy and colleagues at the University of Pennsylvania (Broit, 1981; Bajcsy and Kovacic, 1989; Gee *et al.*, 1993, 1995, 1998), this approach was adopted by the *Karolinska Brain Atlas* Program (Seitz *et al.*, 1990; Thurfjell *et al.*, 1993; Ingvar *et al.*, 1994). Warping algorithms can transfer maps of cytoarchitecture, biochemistry, and functional and vascular territories into the coordinate system of different subjects (see Toga, 1998, for a review). Intricate patterns of structural variation in anatomy can be accommodated. These transformations must allow any segment of the atlas anatomy, however small, to grow, shrink, twist, and even rotate to produce a transformation that encodes local differences in topography from one individual to another.

Nonlinear mapping of raster volumes or 3D geometric atlases onto individual data sets has empowered many studies of disease. These include brain structure labeling for hippocampal morphometry in dementia (Haller *et al.*, 1997), analysis of subcortical structure volumes in schizophrenia (Iosifescu *et al.*, 1997; Csernansky *et al.*, 1998), estimation of structural variation in normal and diseased populations (Collins *et al.*, 1994b; Thompson *et al.*, 1997), and segmentation and classification of multiple sclerosis lesions (Warfield *et al.*, 1995). Digital anatomic models can also be projected into PET data to define regions of interest for quantitative calculations of regional cerebral blood flow (Ingvar *et al.*, 1994; Dinov *et al.*, 1999). These template-driven segmentations require extensive validation relative to more labor-intensive manual delineation of structures but show considerable promise in studies of disease.

B. Analyzing Brain Data with an Atlas

The ability to relate atlas data to a new subject's brain images also operates in reverse. By inverting the deformation field that reconfigures an atlas to match an individual, an individual's data can be nonlinearly registered with the atlas, removing subject-specific anatomical differences. Functional data can then be compared and integrated across subjects, with confounding anatomical effects factored out. Since they transfer multisubject data more accurately into a stereotaxic framework, nonlinear registration algorithms are now increasingly used in functional image analysis packages (Seitz *et al.*, 1990; Friston *et al.*, 1995; Ashburner *et al.*, 1997).

Because variations in structure and function are so great, and both are altered in disease, nonlinear registration approaches become relevant in creating disease-specific templates. These algorithms eliminate the anatomic component of functional variation and are required to separate variations in structure and function. They are also vital in creating deformable atlases, which offer a means to represent, and measure, variations in structure.

C. Intensity-Driven and Model-Driven Algorithms

Elastic registration, or *warping*, algorithms that drive a deformable atlas can be classified into two basic types (see Thompson and Toga, 1998a,b, for a review). We briefly review these, because of their fundamental role in transferring data between scans and atlas templates, in either direction.

1. Intensity-Driven Approaches

These algorithms measure the similarity between the deforming atlas and the patient's scan. They then optimize the measure by tuning parameters of the deformation field. The widely used Automated Image Registration (AIR; Woods *et al.*, 1998) and Statistical Parametric Mapping algorithms (SPM; Ashburner *et al.*, 1997) are examples of this approach. As the cost

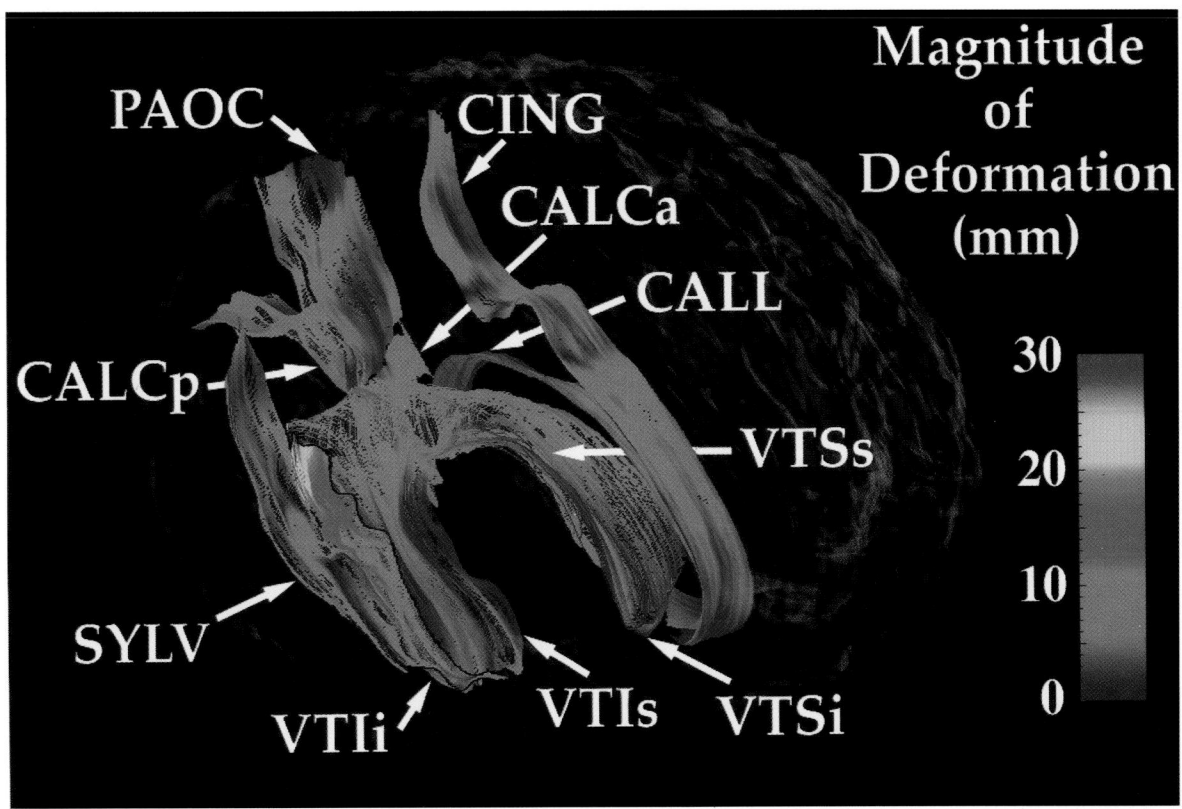

Figure 3 Models of deep sulcal and ventricular surfaces can measure anatomic differences across subjects. Parametric surface models of deep anatomic structures can guide the mapping of one brain to another (data from Thompson and Toga, 1996). This mapping measures anatomic differences between two subjects. Deep sulcal surfaces include the anterior and posterior calcarine (CALCa/p), cingulate (CING), parieto-occipital (PAOC) and callosal (CALL) sulci, and sylvian fissure (SYLV). Also shown are the superior and inferior surfaces of the rostral horn (VTSs/i) and inferior horn (VTIs/i) of the right lateral ventricle. Ventricles and deep sulci are represented by connected systems of rectangularly parameterized surface meshes, whereas the external surface has a spherical parameterization (Section VIII). Connections are introduced between elementary mesh surfaces at known tissue-type and cytoarchitectural field boundaries and at complex anatomical junctions (such as the PAOC/CALCa/CALCp junction shown here). Color-coded profiles show the magnitude of the 3D deformation maps warping these surface components (in the right hemisphere of a 3D T_1-weighted SPGR MRI scan of an Alzheimer's patient) onto their counterparts in an age-matched normal subject.

function is optimized, increasingly complex warping fields are expressed in terms of a 3D cosine basis (SPM) or by tuning parameters of 3D polynomials (AIR). Since they use purely mathematical criteria, intensity-driven approaches bypass information on the topological organization of the brain. Accurate anatomic correspondences, especially at the cortex, are sometimes difficult to establish. Nonetheless, their speed and automation make them ideal for many applications.

2. Model-Driven Approaches

These algorithms establish biologically constrained object-to-object correspondences in the brain. They first build large systems of anatomical models in the atlas and patient's scan (Thompson *et al.*, 1996). This topological model of the brain's anatomy is used to derive a transformation matching anatomical surface boundaries and cortical features exactly, guiding the volumetric mapping of the atlas onto the patient (Fig. 3).

When extended to accommodate more subjects, deformations that match an atlas with each patient in a population can be used to create statistical maps of anatomy, revealing patterns of variability, asymmetry, or abnormality in a group (Thompson *et al.*, 1996b, 1997). With a model-driven approach, graphical surface models represent each major anatomic system, so a comprehensive geometric atlas can be built. Average representations can be created for each anatomical element, along with statistical maps that can be visualized directly or used to guide subsequent image analysis.

D. Continuum-Mechanical Atlases

Many brain atlases have been developed to deform according to the principles of continuum mechanics (Broit, 1981; Bajcsy and Kovacic, 1989; Christensen *et al.*, 1993, 1996; Davatzikos, 1996; Gee and Bajcsy, 1998; Thompson and Toga, 1998a). This feature is relevant to understanding how variations in structure can be encoded. In modeling the atlas deformations, differential equations are used to make the deforming atlas conform to the behavior of elastic or fluid materials. An advantage of this approach is that the well-understood mathematics enforces several desirable characteristics in the mappings. For instance, atlas-to-patient mappings should be one-to-one (i.e., the deformed atlas should not tear or self-intersect). This is surprisingly difficult to guarantee, unless continuum-mechanical or variational methods are applied (Christensen *et al.*, 1995b; Dupuis *et al.*, 1998; see Footnote 2). The continuum-mechanical operators that govern the atlas deformations also have a spectral (or eigenfunction) representation that helps

calculate the mappings rapidly (Miller *et al.*, 1993; Ashburner *et al.*, 1997).

E. Statistical Templates

The deformable template framework has also been widely tested in computer vision applications where shape variability needs to be accommodated, such as written digit identification or face recognition. This makes it easier to build a statistical theory of shape for encoding brain variation, using Gaussian fields (Davatzikos *et al.*, 1996b; Thompson *et al.*, 1996a,b, 1997a,b; Ashburner *et al.*, 1997; Dupuis *et al.*, 1998; Gee and

---

[2]Applying Large Deformations to Brain Atlases. Christensen *et al.* (1995b) observed that most warping methods that are used to adapt deformable brain atlases do not maintain the topological integrity of the deforming template if large deformations are required. In 2D, for example, a specification of correspondences at point landmarks is usually extended to produce a deformation field by minimizing a specific regularizing functional such as the *thin-plate spline energy* (Bookstein, 1997):

$$J_{\text{thin-plate}}(\mathbf{u}) = \int_{\mathbb{R}^2}[(\partial_{11}\mathbf{u})^2 + 2(\partial_{12}\mathbf{u})^2 + (\partial_{22}\mathbf{u})^2]dx_1dx_2,$$

where $\partial_{ij}\mathbf{u} = \partial^2\mathbf{u}/\partial x_i\partial x_j$, or the *membrane spline* (Amit *et al.*, 1991; Gee *et al.*, 1993; Ashburner *et al.*, 1997) or *elastic body energies* (Miller *et al.*, 1993):

$$J_{\text{memb}}(\mathbf{u}) = \int[(\partial_1 u_1)^2 + (\partial_1 u_2)^2 + (\partial_2 u_1)^2 + (\partial_2 u_2)^2]dx_1dx_2$$

$$J_{\text{elas}}(\mathbf{u}) = \int \Sigma\Sigma[(\lambda/2)(\partial_i u_i)(\partial_j u_j) + (\mu/4)((\partial_i u_j) + (\partial_j u_i))^2]dx_1dx_2.$$

Since the transformations that minimize these energies do not guarantee that the deforming template remains intact after large landmark deformations, Christensen *et al.* (1995b) suggested a way to avoid this problem. By forcing the warping field to arise by Euler integration of a continuously differentiable velocity field $\mathbf{v}(\mathbf{x},t) \in \Omega\times[0,1] \rightarrow \mathbf{v}(\mathbf{x},t) \in R^3$, the resulting transform,

$$\mathbf{u}(\mathbf{x},t) = \mathbf{u}(\mathbf{x},0) + \int \mathbf{v}(\mathbf{u}(\mathbf{x},s),s)ds,$$

is a unique diffeomorphism. Dupuis *et al.* (1998) further suggested that the Dirichlet problem for image matching (Joshi et al., 1995) could be reformulated by forcing the optimal velocity field to minimize quadratic energetics on the space–time element $\Omega\times[0,1] = [0,1]^4 \subset R^4$, governed by a matrix constant coefficient differential operator: $\mathbf{E}(\mathbf{v}) = \int_{\Omega\times[0,1]}\|L\mathbf{v}(\mathbf{u}(\mathbf{x},t),t)\|^2dxdt$. For matching a system of N point landmarks with this type of flow, the optimal solution was then shown (Joshi, 1998) to be derivable from the paths of the N landmarks directly, resulting in the parameterization $\hat{\mathbf{u}}(\mathbf{x},t) = \Sigma_i c_i(t)GG^\dagger(\mathbf{x},\hat{\mathbf{u}}(\mathbf{x}_i,t))$. Here the coefficients $c_i(t)$ are R^3-valued functions on [0,1], and $G(\mathbf{x},\mathbf{y})$ is the matrix Green's operator corresponding to L. The resulting framework for constructing diffeomorphic (smooth one-to-one) brain maps allows arbitrarily complex atlas transformations while guaranteeing that the digital template remains intact and connected under the transformation.

Bajcsy, 1998; Thirion *et al.*, 1998; Cao and Worsley, 1999; Le Goualher *et al.*, 1999) or Riemannian shape manifolds (Bookstein, 1997). Probabilistic transformations can then be applied to deformable anatomic templates to create a type of probabilistic atlas that measures variability and detects pathology (Thompson *et al.*, 1997, 1998a; Grenander and Miller, 1998; Joshi *et al.*, 1998).

F. Individualizing an Atlas

To understand how deformable atlases work, consider the deforming atlas to be embedded in a 3D elastic or fluid medium (Fig. 4). The medium is subjected to distributed internal forces, which reconfigure it and lead the image to match the target. In elastic media, the displacement vector field $\mathbf{u}(\mathbf{x})$ resulting from internal driving forces $\mathbf{F}(\mathbf{x})$ (called *body forces*) obeys the Navier–Stokes equilibrium equations for linear elasticity:

$$\mu\nabla^2\mathbf{u} + (\lambda + \mu)\nabla(\nabla^T \cdot \mathbf{u}(\mathbf{x})) + \mathbf{F}(\mathbf{x}) = 0, \forall\mathbf{x} \in R. \quad (1)$$

Here R is a discrete lattice representation of the atlas, $\nabla^T \cdot \mathbf{u}(\mathbf{x}) = \sum \partial u_j/\partial x_j$ is the cubical dilation of the medium, ∇^2 is the Laplacian operator, and Lamé's coefficients λ and μ refer to the elastic properties of the medium. Body forces, $\mathbf{F}(\mathbf{x})$, drive the atlas into correspondence with similar regions in the patient. In a model-based approach, these forces are computed from a large set of anatomical surface models (Thompson and Toga, 1998a), so that the deformations of structure boundaries are propagated to the image volume. In an intensity-based approach, the body forces are derived from the gradient of a local cost function. The cost function is a metric designed to capture how well the atlas is aligned with the patient's scan. Common measures include squared differences in normalized pixel intensities (Christensen *et al.*, 1993; Woods *et al.*, 1993, 1998; Ashburner *et al.*, 1997), ratio image uniformity (Woods *et al.*, 1992), regional correlation (Bajcsy and Kovacic, 1989; Collins *et al.*, 1995), or mutual information (Kim *et al.*, 1997; Kjems *et al.*, 1999). If an elastic model is used, the equilibrium equations are solved numerically by finite difference, finite element, or spectral methods, and the three-dimensional deformation field $\mathbf{u}(\mathbf{x})$ warps the atlas into register with the target scan.

G. Massively Parallel Implementations

Recently, Christensen *et al.* (1993, 1995a, 1996) proposed a viscous-fluid-based warping transform to individualize a labeled atlas and segment hippocampal anatomy in new patient scans (cf. Haller *et al.*, 1997;

Csernansky *et al.*, 1998). Manual tag points first roughly align the atlas and patient's anatomy. An intensity-based force (2) then drives a large-distance, nonlinear fluid evolution of the neuroanatomic template. With the introduction of concepts such as deformation velocity and an Eulerian reference frame, the energetics of the deformed medium are hypothesized to be relaxed in a highly viscous fluid:

$$\mathbf{F}((\mathbf{x},\mathbf{u}(\mathbf{x},t)) = -(T(\mathbf{x} - \mathbf{u}(\mathbf{x},t)) - S(\mathbf{x}))\nabla T|_{\mathbf{x}-\mathbf{u}(\mathbf{x},t)}, \quad (2)$$

$$\mu\nabla^2\mathbf{v}(\mathbf{x},t) + (\lambda + \mu)\nabla(\nabla^T \cdot \mathbf{v}(\mathbf{x},t)) + \mathbf{F}(\mathbf{x},\mathbf{u}(\mathbf{x},t)) = \mathbf{0}, \quad (3)$$

$$\partial\mathbf{u}(\mathbf{x},t)/\partial t = \mathbf{v}(\mathbf{x},t) - \nabla\mathbf{u}(\mathbf{x},t)\mathbf{v}(\mathbf{x},t). \quad (4)$$

The deformation velocity of the atlas (3) is governed by the creeping flow momentum equation for a Newtonian fluid, and the conventional displacement field in a Lagrangian reference system (4) is connected to an Eulerian velocity field by the relation of material differentiation. Experimental results were excellent (Christensen *et al.*, 1996). Nonetheless, both elastic and fluid algorithms contain core systems of up to 0.1 billion simultaneous partial differential equations ((1) or (2)–(4)). In early implementations, deformable registration of a 128^3 MRI atlas to a patient took 9.5 and 13 h for elastic and fluid transforms, respectively, on a 128 × 64 DECmpp1200Sx/Model 200 MASPAR (Massively Parallel Mesh-Connected Supercomputer). This spurred work to modify the algorithm to individualize atlases on standard single-processor workstations (Thirion, 1995; Bro-Nielsen and Gramkow, 1996; Freeborough and Fox, 1998).

Bro-Nielsen and Gramkow (1996) used the eigenfunctions of the Navier–Stokes differential operator $L = \mu\nabla^2 + (\lambda + \mu)\nabla(\nabla^T\cdot)$, which governs the atlas deformations, to derive a Green's function solution $\mathbf{u}^*(\mathbf{x}) = \mathbf{G}(\mathbf{x})$ to the impulse response equation $L\mathbf{u}^*(\mathbf{x}) = \delta(\mathbf{x} - \mathbf{x_0})$. This speeds up the core registration step by a factor of 1000. The solution to the full PDE $L\mathbf{u}(\mathbf{x}) = -\mathbf{F}(\mathbf{x})$ was approximated as a rapid filtering operation on the 3D arrays representing body force components:

$$\mathbf{u}(\mathbf{x}) = -\int_\Omega \mathbf{G}(\mathbf{x} - \mathbf{r}) \cdot \mathbf{F}(\mathbf{r})d\mathbf{r} = -(\mathbf{G}*\mathbf{F})(\mathbf{x}), \quad (5)$$

where $\mathbf{G}*$ represents convolution with the impulse response filter. As noted by Gramkow and Bro-Nielsen (1997), a recent fast, "demons-based" warping algorithm (Thirion, 1995) calculates the atlas flow velocity by regularizing the force field driving the template with a Gaussian filter (cf. Collins *et al.*, 1994b). Since this filter may be regarded as a separable approximation to the continuum-mechanical filters derived above, interest has focused on deriving additional

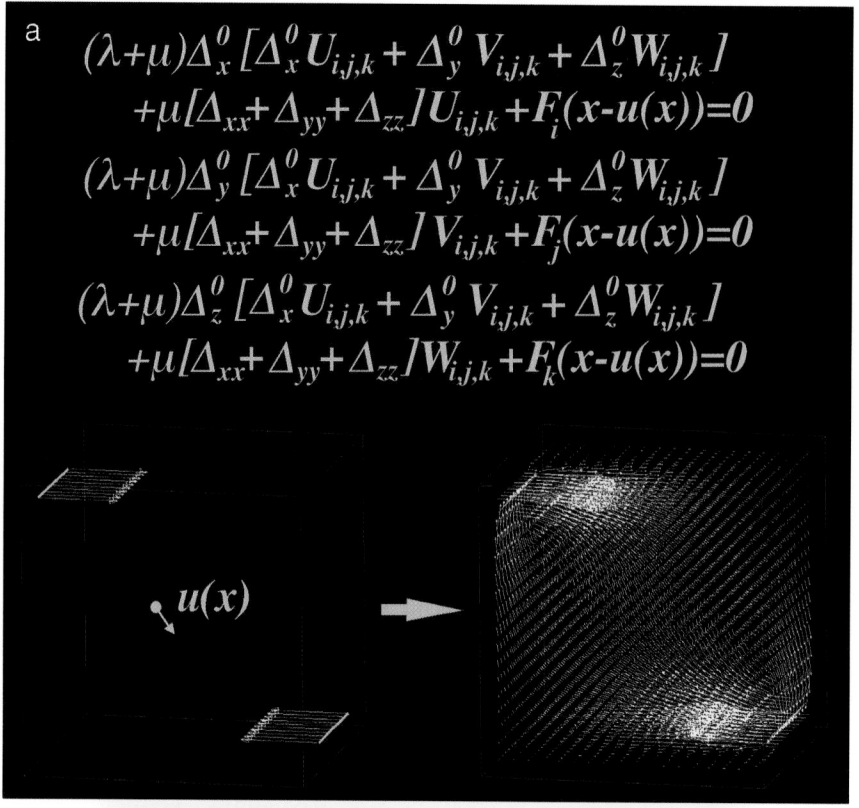

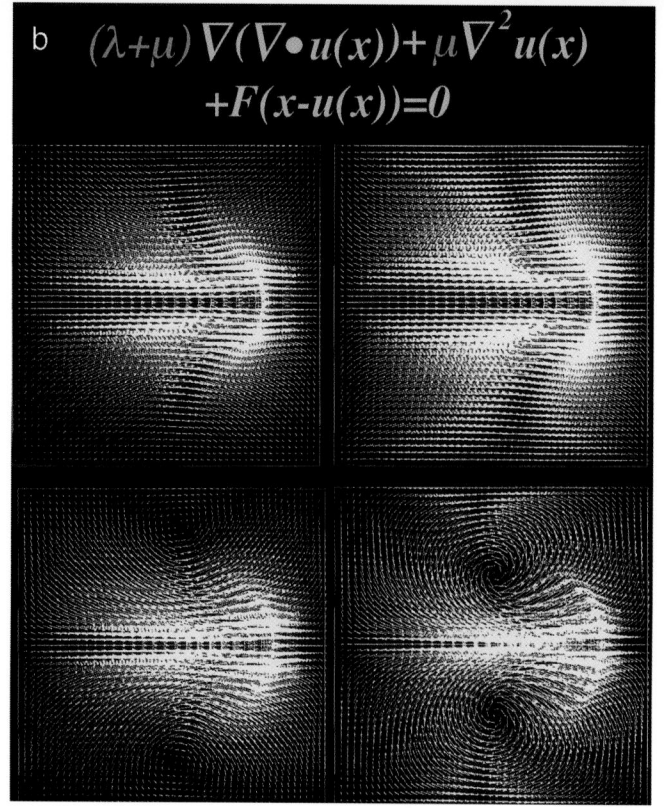

separable (and therefore computationally fast) filters to create subject-specific brain atlases and rapidly label new images (Gramkow, 1996; Lester *et al.,* 1999). Fast multigrid solvers have also accelerated systems for atlas-based segmentation and labeling (Dengler and Schmidt, 1988; Bajcsy and Kovacic, 1989; Gee *et al.,* 1993, 1995; Collins *et al.,* 1994b, 1995; Schormann *et al.,* 1996). Some of these now have sufficient speed for real-time surgical guidance applications (Warfield *et al.,* 1998).

V. Model-Driven Deformable Atlases

The extreme difficulty of finding structures in new patients based on intensity criteria alone has led several groups to develop model-driven deformable atlases (Thompson and Toga, 1997; Toga and Thompson, 1997, 1999b). Anatomic models provide an explicit geometry for individual structures in each scan, such as landmark points, curves, or surfaces. Because the digital models reside in the same stereotaxic space as the atlas data, surface and volume models stored as lists of vector coordinates are amenable to digital transformation, as well as geometric and statistical measurement (Thompson *et al.,* 1996a). The underlying 3D coordinate system is central to all atlas systems, since it supports the linkage of structure models and associated image data with spatially indexed neuroanatomic labels, preserving spatial information and adding anatomical knowledge.

In Sections VI–X, we show how anatomical models can create probabilistic atlases and disease-specific templates. Statistical averaging of models provides a means to analyze brain structure in morphometric projects, localizing disease-specific differences with statistical and visual power. We first describe how models can drive deformable atlases, measuring patient-specific differences in considerable detail.

When an atlas is deformed to match a patient's anatomy, mesh-based models of anatomic systems help guide the mapping of one brain to another. Anatomi-

cally driven algorithms guarantee biological as well as computational validity, generating meaningful object-to-object correspondences, especially at the cortex. Ultimately, accurate warping of brain data requires

1. Matching entire systems of anatomic surface boundaries, both external and internal, and
2. Matching relevant curved and point landmarks, including ones within the surfaces being matched (e.g., primary sulci at the cortex and tissue type boundaries at the ventricular surface).

In our own model-driven warping algorithm (Thompson and Toga, 1996, 1997, 1998), systems of model surfaces are first extracted from each data set, to guide the volumetric mapping. The model surfaces include many functional, cytoarchitectonic, and lobar boundaries in three dimensions. Both the surfaces and the landmark curves within them are reconfigured to match their counterparts in the target data sets exactly. We will discuss this approach in some detail.

A. Anatomical Models

Since much of the functional territory of the human cortex is buried in cortical folds, or *sulci,* a generic structure is built to model them (Figs. 3 and 5; Thompson and Toga, 1996). The underlying data structure is a connected system of surface meshes, in which the individual meshes are parametric. These surfaces are 3D sheets that divide and join at curved junctions to form a connected network of models. With the help of these meshes, each patient's anatomy is modeled in sufficient detail to be sensitive to subtle differences in disease. Separate surfaces model the deep internal trajectories of features such as the parieto-occipital sulcus, the anterior and posterior calcarine sulcus, the sylvian fissure, and the cingulate, marginal, and supracallosal sulci in both hemispheres. Additional gyral boundaries are represented by parameterized curves lying in the cortical surface. The ventricular system is modeled as a closed system of 14 connected surface elements whose junctions reflect cytoarchitectonic boundaries of the adjacent tissue (Fig. 6; Thompson and Toga, 1998b). In-

Figure 4 Continuum-mechanical warping. (a) The complex transformation required to reconfigure one brain into the shape of another can be determined using continuum-mechanical models, which describe how real physical materials deform. In this illustration, two line elements embedded in a linearly elastic 3D block (lower left) are slightly perturbed (arrows). The goal is to find how the rest of the material deforms in response to this displacement. The Cauchy–Navier equations (shown in discrete form, top) are solved to determine the values of the displacement field vectors, $\mathbf{u}(\mathbf{x})$, throughout the 3D volume. (b) Lamé elasticity coefficients. Different choices of elasticity coefficients, λ and μ, in the Cauchy–Navier equations (shown in continuous form, top) result in different deformations, even if the applied internal displacements are the same. In histologic applications where an elastic tissue deformation is estimated, values of the elasticity coefficients can be chosen that limit the amount of curl (lower right) in the deformation field. Stiffer material models (top left) may better reflect the deformational behavior of tissue during histologic staining procedures. Note: To emphasize differences, the displacement vector fields shown in this figure have been multiplied by a factor of 10. The Cauchy–Navier equations, derived using an assumption of small displacements, are valid only when the magnitude of the deformation field is small.

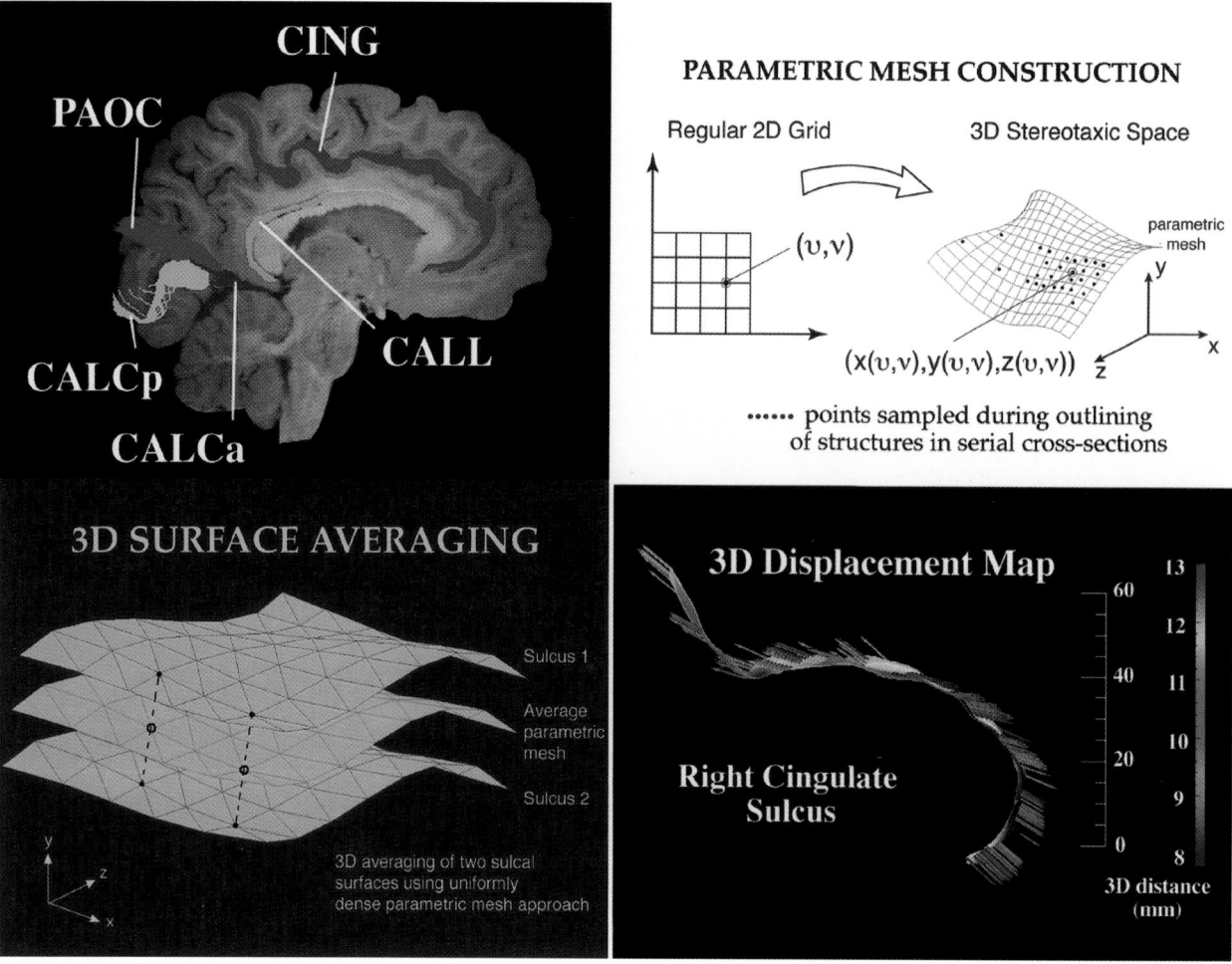

Figure 5 Mesh construction and averaging. The derivation of a standard surface representation for each structure makes it easier to compare anatomical models from multiple subjects. An algorithm converts a set of digitized points on an anatomical structure boundary (e.g., deep sulci, top left) into a parametric grid of uniformly spaced points in a regular rectangular mesh stretched over the surface (Thompson *et al.*, 1996). By averaging nodes with the same grid coordinates across subject (bottom left), an average surface can be produced for the group. However, information on each subject's individual differences is stored as a vector-valued displacement map (bottom right). This map indicates how that subject deviates locally from the average anatomy. These maps can be stored to measure variability and detect abnormalities in different anatomic systems..

formation on the meshes' spatial relations, including their surface topology (*closed* or *open*), anatomical names, mutual connections, directions of parameterization, and common 3D junctions and boundaries, is stored in a hierarchical graph structure. This ensures the continuity of displacement vector fields defined at mesh junctions.

B. Surface Parameterization

After an identical regular grid structure is imposed on anatomic surfaces from different subjects (Fig. 5), the explicit geometry can be exploited to drive and constrain correspondence maps that associate anatomic points in different subjects. Structures that can be ex-

tracted automatically in parametric form include the external cortical surface (discussed in Section VIII), ventricular surfaces, and several deep sulcal surfaces. Recent success of sulcal extraction approaches based on deformable surfaces (Vaillant *et al.*, 1997) led us to combine a 3D skeletonization algorithm with deformable curve and surface governing equations to automatically produce parameterized models of cingulate, parieto-occipital, and calcarine sulci, without manual initialization (Zhou *et al.*, 1999). Additional, manually segmented surfaces can also be given a uniform rectilinear parameterization using algorithms described in Thompson *et al.* (1996a,b) and used to drive the warping algorithm. Each resultant surface mesh is analogous in form to a uniform rectangular grid, drawn on a rubber sheet,

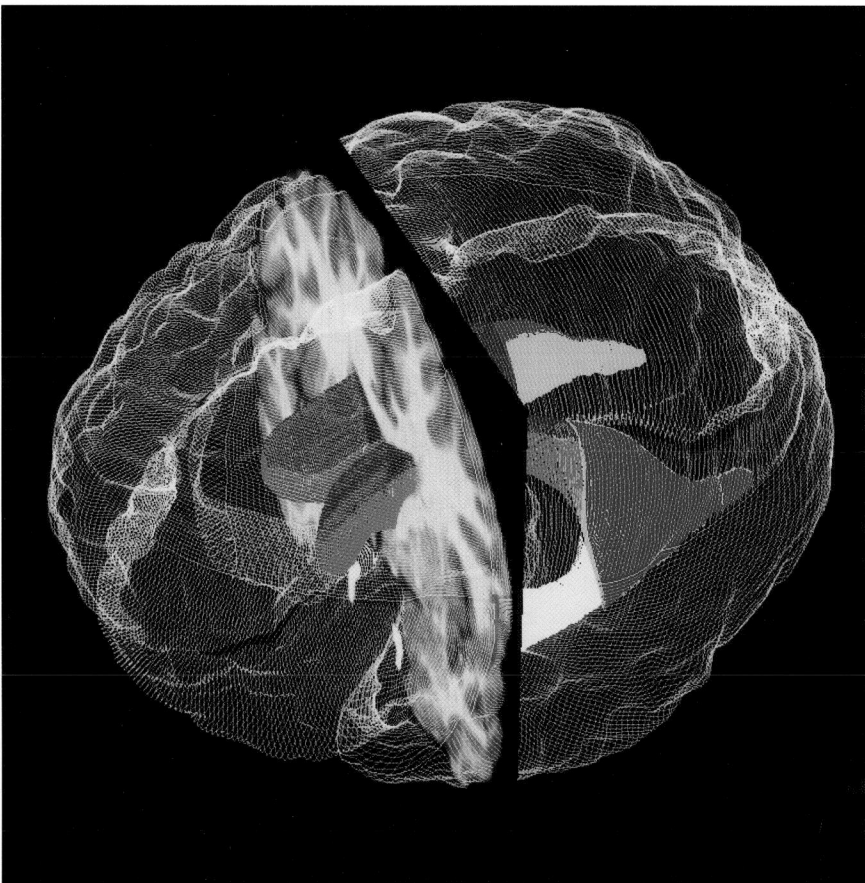

Figure 6 Partitioning the ventricles into 3D surface elements. A model of the lateral ventricles is shown, in the context of a coronal anatomic image and a smoothed cortical surface mesh. The ventricles are partitioned into 14 connected surface elements, whose junctions reflect tissue type boundaries at the ventricular surface. For example, a combination of caudate, thalamic, and septal tissues (as well as callosal fibers) surrounds the superior ventricular horn, and each exhibits different patterns of variation and asymmetry (Fig. 8). To ensure that these parameters are not confounded, each ventricular element is modeled separately.

which is subsequently stretched to match all data points. Association of points on each surface with the same mesh coordinate produces a dense correspondence vector field between surface points in different subjects. This procedure is carried out under stringent conditions (see Footnote 3) to ensure that landmark curves and points known to the anatomist appear in corresponding locations in each parametric grid.

C. Displacement Maps

For each surface mesh $\mathbf{M}_l{}^p$ in a pair of scans $\mathbf{A}_p$ and $\mathbf{A}_q$, we define a 3D displacement field

$$\mathbf{W}_l{}^{pq}[\mathbf{r}_l{}^p(u,v)] = \mathbf{r}_l{}^q(u,v) - \mathbf{r}_l{}^p(u,v), \qquad (6)$$

carrying each surface point $\mathbf{r}_l{}^p(u,v)$ in $\mathbf{A}_p$ into structural correspondence with $\mathbf{r}_l{}^q(u,v)$, the point in the target mesh parameterized by rectangular coordinates (u,v). This family of high-resolution transformations, applied to individual meshes in a connected system deep inside the brain, elastically transforms elements of the surface system in one 3D image to their counterparts in the target scan. Weighted linear combinations of radial functions,

---

[3]For example, the calcarine sulcus (see Fig. 5) is partitioned into two meshes (CALCa and CALCp). This ensures that the complex 3D curve forming their junction with the parieto-occipital sulcus is accurately mapped under both the surface displacement and 3D volumetric maps reconfiguring one anatomy into the shape of another. Figure 5b illustrates this procedure, in a case where three surface meshes in one brain are matched with their counterparts in a target brain. Section VIII describes a separate approach, which is needed to match systems of curves lying *within* a surface, such as the cortex, with their counterparts in a target brain.

describing the influence of deforming surfaces on points in their vicinity, extend the surface-based deformation to the whole brain volume (see Fig. 7). Recent extensions of the core algorithm to include continuum-mechanical and other filter-based models of deformation (Thompson and Toga, 1998a; cf. Joshi *et al.*, 1995; Davatzikos, 1996; Schiemann and Höhne, 1997; Gabrani and Tretiak, 1999) have yielded similar encouraging results. Figure 7 shows how the algorithm performs on cryosection data.

VI. Probabilistic Brain Atlases

A. Encoding Anatomic Variability

Many morphometric studies focus on identifying systematic alterations in anatomy in a variety of diseases (Table I). These studies are complicated by the substantial overlap between measures of normal and diseased anatomy. Normal anatomic complexity makes group-specific patterns hard to discern. However, disease-specific variants may be easier to localize by creating average models of anatomy rather than deriving volumetric descriptors.

In response to these challenges, *probabilistic atlases* are research tools that retain information on anatomic and functional variability (Mazziotta *et al.*, 1995; Thompson *et al.*, 1997). A probabilistic atlas solves many of the limitations of a fixed atlas in representing highly variable anatomy. As the subject database increases in size and content, the digital form of the atlas allows efficient statistical comparisons of individuals or groups. In addition, the population that an atlas represents can be stratified into subpopulations to represent specific disease types, and subsequently by age, gender, handedness, or genetic factors.

B. Parametric Mesh Modeling

Parametric meshes (Thompson *et al.*, 1996a,b) offer a means to create average models of anatomy. Once anatomical data are transformed to a standardized co-

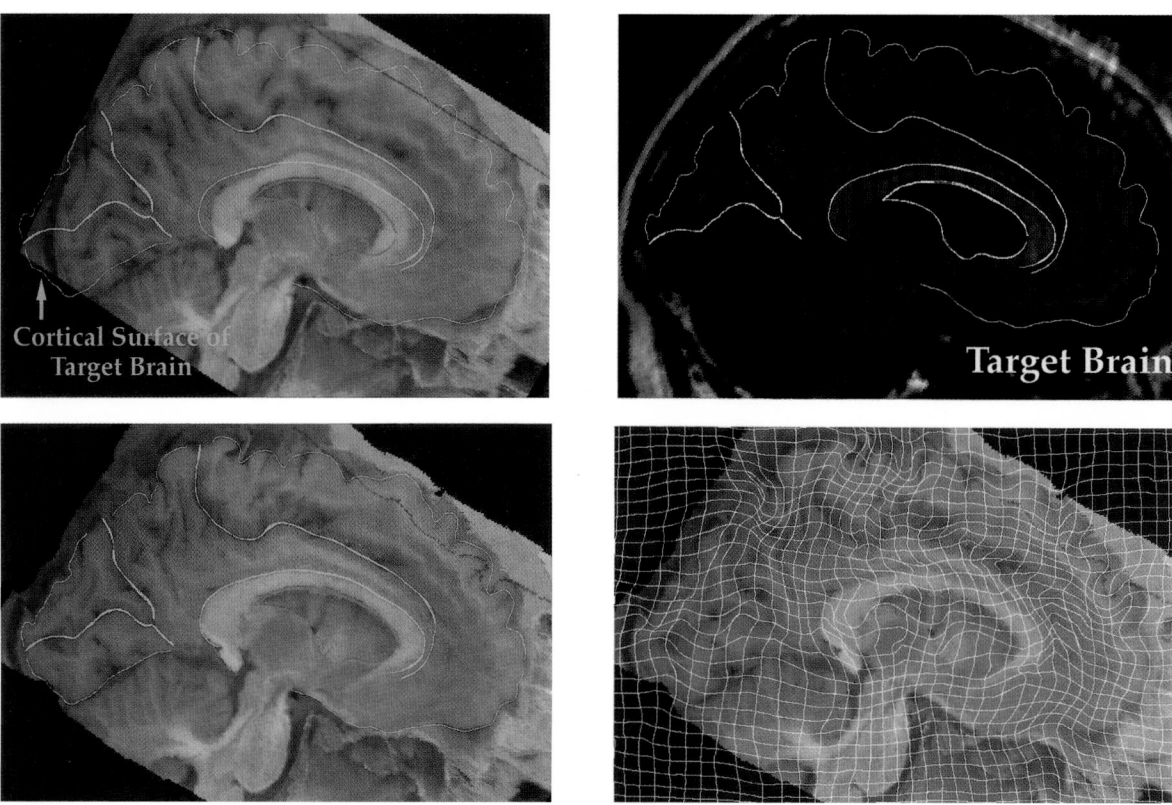

Figure 7 A deformable cryosection atlas measures anatomic differences. Structure boundaries from a patient with Alzheimer's disease (top left) are overlaid on a cryosection atlas (top right), which has been registered to it using a simple linear transformation. A surface-based image warping algorithm is then applied to drive the atlas into the configuration of the patient's anatomy (bottom left). Histologic and neurochemical maps, accessible only postmortem, can be transferred onto the living subject's scan (Mega *et al.*, 1997). The amount of deformation required is displayed as a tensor map (only two components of the fully 3D transformation are shown). Tensor maps, and derived vector or scalar fields, can be analyzed in a statistical setting to examine anatomic variation, detect pathology, or track structural changes over time.

ordinate space, such as the Talairach space, a computational grid structure can be imposed on anatomical surface boundaries. These mesh models represent boundary point locations in stereotaxic coordinates (Fig. 5). Averaging of corresponding grid points across subjects results in an average surface model for each structure. At the same time, knowledge of each subject's deviations from the group average anatomy can be retained as a vector displacement map (Fig. 5). After storing these maps from large numbers of subjects, local biases in the magnitude and direction of anatomic variability can be displayed as a map. Variability maps for deep sulcal surfaces are shown in Fig. 8. In these maps, the color shows the root-mean-square (rms) magnitude of the displacement vectors that map individuals to the group mean. Separate maps are displayed for elderly normals (mean age: 72.9 ± 5.6 years; all 10 right-handed) and demographically matched Alzheimer's patients (age: 71.9 ± 10.7 years; all 10 right-handed; mean Mini-Mental State Exam score: 19.7 ± 5.7, out of 30). As expected, there is an extraordinary increase in anatomical variability from deep structures (0–5 mm at the corpus callosum) to peak rms values of 12–13 mm at the posterior sylvian fissures (Thompson et al., 1998a). In AD, however, sylvian fissure variability rose extremely sharply from an SD of 6.0 mm rostrally on the left to 19.6 mm caudally. Underlying atrophy and possible left greater than right degeneration of perisylvian gyri (Loewenstein et al., 1989; Siegel et al., 1996) may widen the sylvian fissure, superimposing additional individual variation and asymmetry on that seen in normal aging.

C. Brain Asymmetry

A third feature observable from the average anatomical models (Fig. 8) is that consistent patterns of brain asymmetry can be mapped, despite wide variations in asymmetry in individual subjects. In dementia, the increased cortical asymmetry probably reflects asymmetric progression of the disease. Figure 9 shows average maps of the lateral ventricles, again from Alzheimer's disease and matched elderly normal populations. As expected, the ventricles are significantly enlarged in dementia. Notice, however, that a pronounced asymmetry is observed in both groups (left volume larger than right, $p < 0.05$). This is an example of an effect that becomes clear after group averaging of anatomy and is not universally apparent in individual subjects. It is, however, consistent with prior volumetric measurements (Shenton et al., 1992; Aso et al., 1995). Anatomical averaging can also be cross-validated with a traditional volumetric approach. Occipital horns were on average 17.1% larger on the left in the normal group (4070.1 ± 479.9 mm^3) than on the right (3475.3 ± 334.0 mm^3; $p <$

0.05), but no significant asymmetry was found for the superior horns (left: 8658.0 ± 976.7 mm^3; right: 8086.4 ± 1068.2 mm^3; $p > 0.19$) or for the inferior horns (left: 620.6 ± 102.6 mm^3; right: 573.7 ± 85.2 mm^3; $p > 0.37$). The asymmetry is clearly localized in the 3D group average anatomic representations. In particular, the occipital horn extends (on average) 5.1 mm more posteriorly on the left than the right. The capacity to resolve asymmetries in a group atlas can assist in studies of disease-specific cortical organization (Thompson et al., 1997; Mega et al., 1998; Narr et al., 1998, 1999; Zoumalan et al., 1999).

D. Corpus Callosum Differences

We also tested the ability of anatomical averaging to identify disease-specific patterns in clinical populations. First, the approach was used to detect preclinical hippocampal atrophy in patients with minimal cognitive impairment (Kwong et al., 1999; Mega et al., see Chapter 9). To identify more focal effects, we attempted to identify regionally selective patterns of callosal change in patient groups with Alzheimer's disease and schizophrenia (Thompson et al., 1998; Narr et al., 1999). The midsagittal callosum was first partitioned into five sectors (Fig. 10; Duara et al., 1991; Larsen et al., 1992). This roughly segregates callosal fibers from distinct cortical regions. In AD, focal fiber loss was expected at the callosal isthmus (sector 2), whose fibers selectively innervate the temporoparietal regions with early neuronal loss and perfusion deficits (Brun and Englund, 1981; Clarke and Zaidel, 1994). Consistent with this hypothesis, a significant area reduction at the isthmus was found, reflecting a dramatic 24.5% decrease from 98.0 ± 8.6 mm^2 in controls to 74.0 ± 5.3 mm^2 in AD ($p < 0.025$). Terminal sectors (1 and 5) were not significantly atrophied, and the central midbody sector showed only a trend toward significance (16.6% mean area loss; $p < 0.1$), due to substantial intergroup overlap. Average boundary representations, however, localized these findings directly. At the isthmus, average models in AD showed a pronounced shape inflection at stereotaxic location (0.0, −25.0, 19.0) (see Fig. 10).

E. Gender in Schizophrenia

Different shape alterations were observed in schizophrenia (Narr et al., 1999; Fig. 11). A significant bowing effect was observed, reflecting enlargement of the underlying third ventricle. By creating separate average models for male and female patients, significant gender effects also emerged (Figs. 11a–d). The greater bowing effect in male than female patients was confirmed by

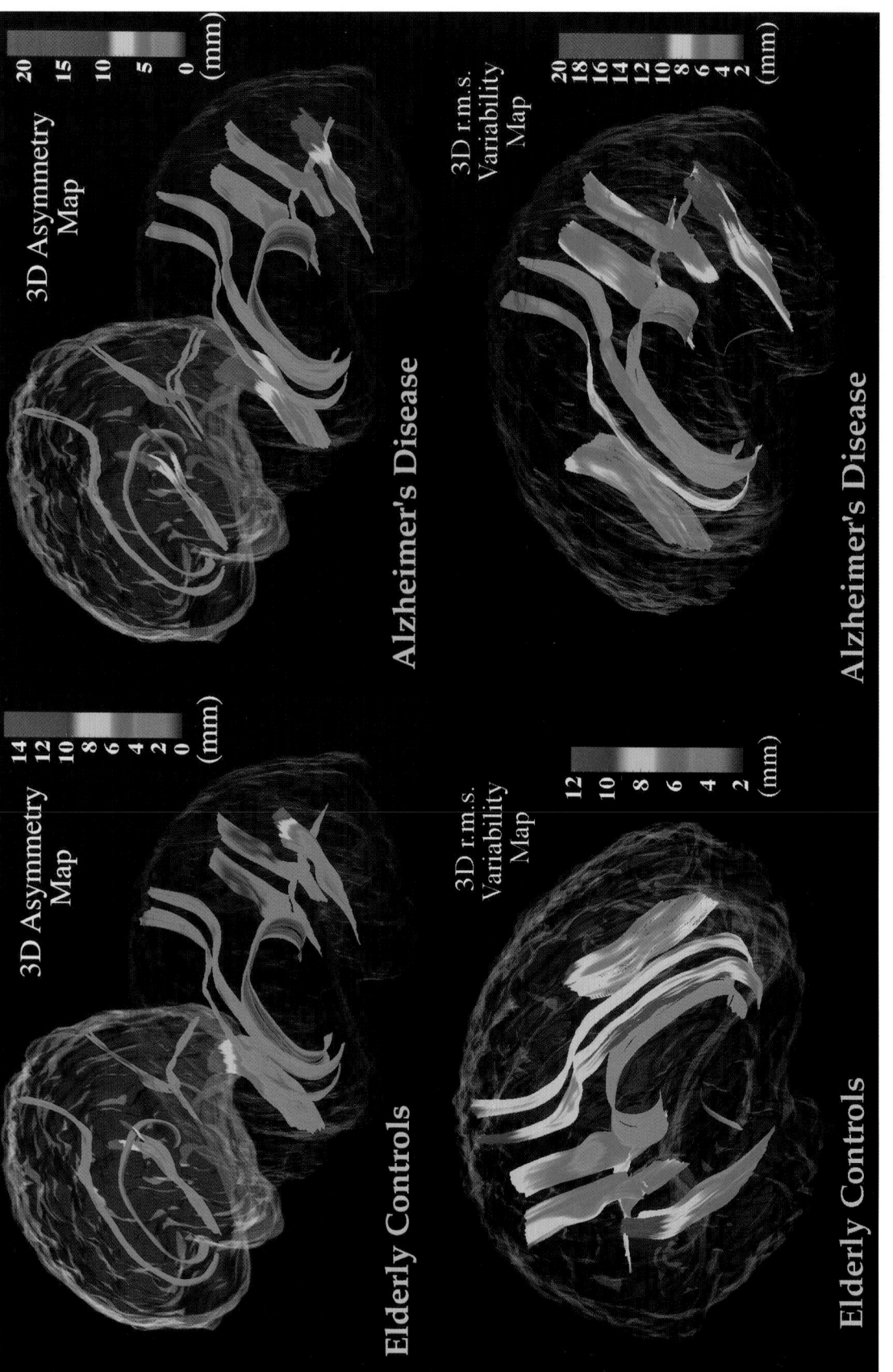

Figure 8 Population-based maps of 3D structural variation and asymmetry. Statistics of 3D deformation maps help define confidence limits on normal anatomic variation. 3D maps of anatomic variability and asymmetry are shown for 10 subjects with Alzheimer's disease (AD: age: 71.9 ± 10.9 years) and 10 normal elderly subjects matched for age (72.9 ± 5.6 years), gender, handedness, and educational level (Thompson *et al.*, 1998). Normal sylvian fissure asymmetries (right higher than left; $p < 0.0005$) were significantly greater in AD than in controls ($p < 0.0002$; top panels). In the 3D variability maps derived for each group (lower panels), the color encodes the root-mean-square magnitude of the displacement vectors that map surfaces from each of the 10 patients' brains onto the average. 3D cortical variability (lower right panel) increased in AD from 2–4 mm at the corpus callosum to a peak standard deviation of 19.6 mm at the posterior left sylvian fissure. From Thompson *et al.* (1998). Cortical variability and asymmetry in normal aging and Alzheimer Disease. *Cerebral Cortex* **8**(6), 492–509, with permission.

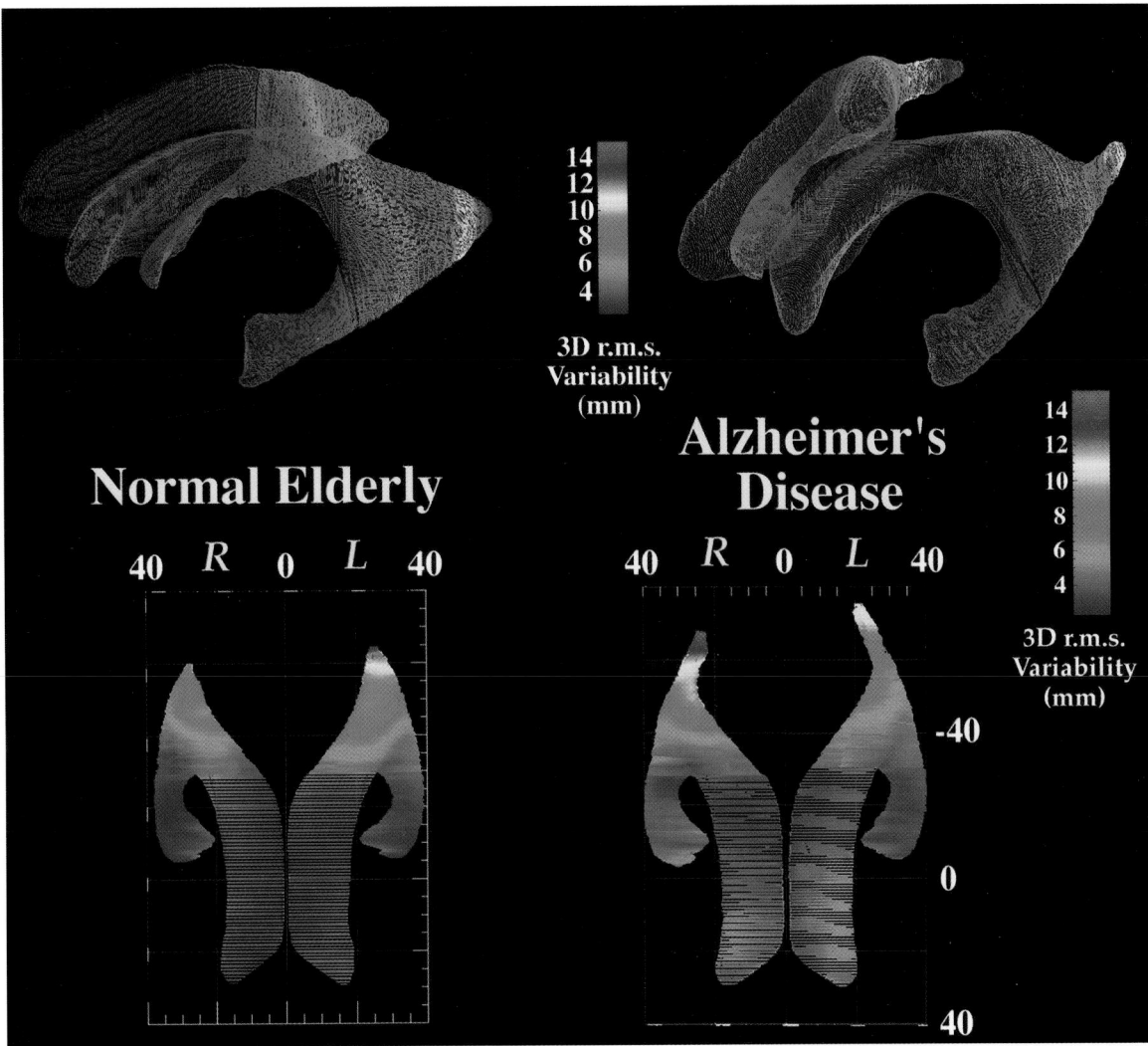

Figure 9 Population-based maps of ventricular anatomy in normal aging and Alzheimer's disease. 3D parametric surface meshes (Thompson *et al.*, 1996) were used to model the 14 ventricular elements, shown in Fig. 6, in 3D MRI scans of 10 Alzheimer's patients (age: 71.9 ± 10.9 years) and 10 matched controls (72.9 ± 5.6 years; Thompson *et al.*, 1998). 3D meshes representing each surface element were averaged by hemisphere in each group. (Top) The color map shows a 3D rms measure of group anatomic variability pointwise on an average surface representation for each group in Talairach stereotaxic space. Oblique side views reveal enlarged occipital horns in the Alzheimer's patients and high stereotaxic variability in both groups. (Bottom) A top view of these averaged surface meshes reveals localized asymmetry, variability, and displacement within and between groups. Asymmetries at the ventricles and sylvian fissure emerge only after averaging of anatomical maps in large groups of subjects. These patterns can be encoded probabilistically to detect structural anomalies in individual patients or groups (Thompson *et al.*, 1997).

multivariate analysis of variance and is highlighted in the average anatomic templates. As emphasized by this example, even if no sex difference is present in normal callosal morphology (see Thompson *et al.*, 1999a, for a review of this controversy), this does not preclude sex effects from interacting with morphometric abnormalities in diseased populations. In schizophrenia, there is typically a later age of onset in female schizophrenics,

and hereditary factors may be unevenly distributed between the sexes (De Lisi *et al.*, 1989; Colombo *et al.*, 1993; Waddington, 1993). Stratification of probabilistic atlases by gender and other genetic factors provides a computationally fast way to visualize these effects and relate them to epidemiologic data (Mazziotta *et al.*, 1995; Mega *et al.*, 1998; Blanton *et al.*, 1999; Le Goualher *et al.*, 1999; Zoumalan *et al.*, 1999).

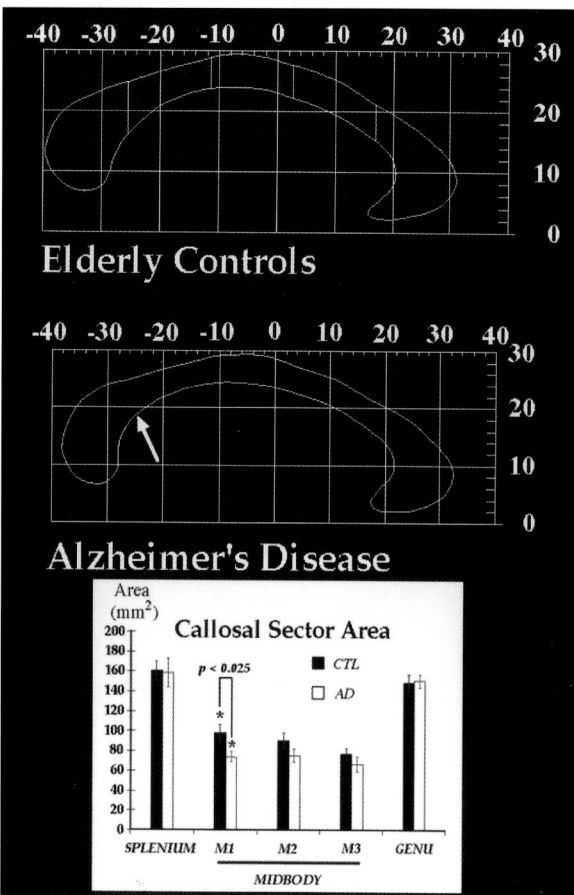

Figure 10 Corpus callosum in Alzheimer's disease. Midsagittal corpus callosum boundaries were averaged from patients with Alzheimer's disease and from elderly controls matched for age, educational level, gender, and handedness. The average representations show a focal shape inflection in the Alzheimer's patients relative to normal elderly subjects of the same age. A statistically significant tissue loss is also found at the isthmus (second sector, when the structure is partitioned into fifths). The isthmus connects regions of temporoparietal cortex that exhibit early neuronal loss and perfusion deficits in AD. From Thompson *et al.* (1998). Cortical variability and asymmetry in normal aging and Alzheimer Disease. *Cerebral Cortex* **8**(6), 492–509, with permission.

VII. Atlas-Based Pathology Detection

A. Deformable Probabilistic Atlases

As noted earlier, *warping* algorithms create deformation maps (Fig. 7) that indicate 3D patterns of anatomic differences between any pair of subjects. By defining probability distributions on the space of deformation transformations that drive the anatomy of different subjects into correspondence (Grenander, 1976; Amit *et al.,* 1991; Grenander and Miller, 1994; Thompson and Toga, 1997; Thompson *et al.,* 1997), statistical parameters of

these distributions can be estimated from databased anatomic data to determine the magnitude and directional biases of anatomic variation. Encoding of local variation can then be used to assess the severity of structural variants outside of the normal range, which, in brain data, may be a sign of disease (Thompson *et al.,* 1997).

B. Encoding Brain Variation

To see if disease-specific features could be detected in individual patients, we developed a *random vector field* approach to construct a population-based brain atlas (Thompson and Toga, 1997). Briefly, given a 3D MR image of a new subject, a warping algorithm calculates a set of high-dimensional volumetric maps, elastically matching this image with other scans from an anatomic image database. Target scans are selected from subjects matched for age, handedness, gender, and other demographic factors (Thompson *et al.,* 1997, 1998a). The resulting family of volumetric warps provides empirical information on local variability patterns. A probability space of random transformations, based on the theory of anisotropic Gaussian random fields (Thompson *et al.,* 1997), is then used to encode the variations. For the cortex, specialized approaches are needed to represent variations in gyral patterns (Thompson *et al.,* 1997; Thompson and Toga, 1998b; Section VIII). Confidence limits in stereotaxic space are determined for points in the new subject's brain, enabling the creation of color-coded probability maps to highlight and quantify regional patterns of deformity (Fig. 12).

C. Comparing a Subject with an Atlas

In one validation experiment (Thompson *et al.,* 1997; Fig. 12), probability maps were created to highlight abnormal deviations in the callosal and midline anatomy of a tumor patient. The two regions of metastatic tissue induced marked distortions in the normal architecture of the brain. After variations in deep surface anatomy were stored as a spatially adaptive covariance tensor field (Thompson *et al.,* 1997), probability maps were generated for the tumor patient. In the tumor patient, the herniation effects apparent in the blockface imagery (Fig. 12a) were detected in the probability maps of structures near the lesion sites (Fig. 12b).

In this approach, the tensor field stores information on the preferred directions of normal variability. For example, it retains information that lateral displacements of the callosum are more unlikely than vertical displacements in a normal group. Local computation of the variance components also means that confidence limits for abnormal structure are appropriately relaxed in re-

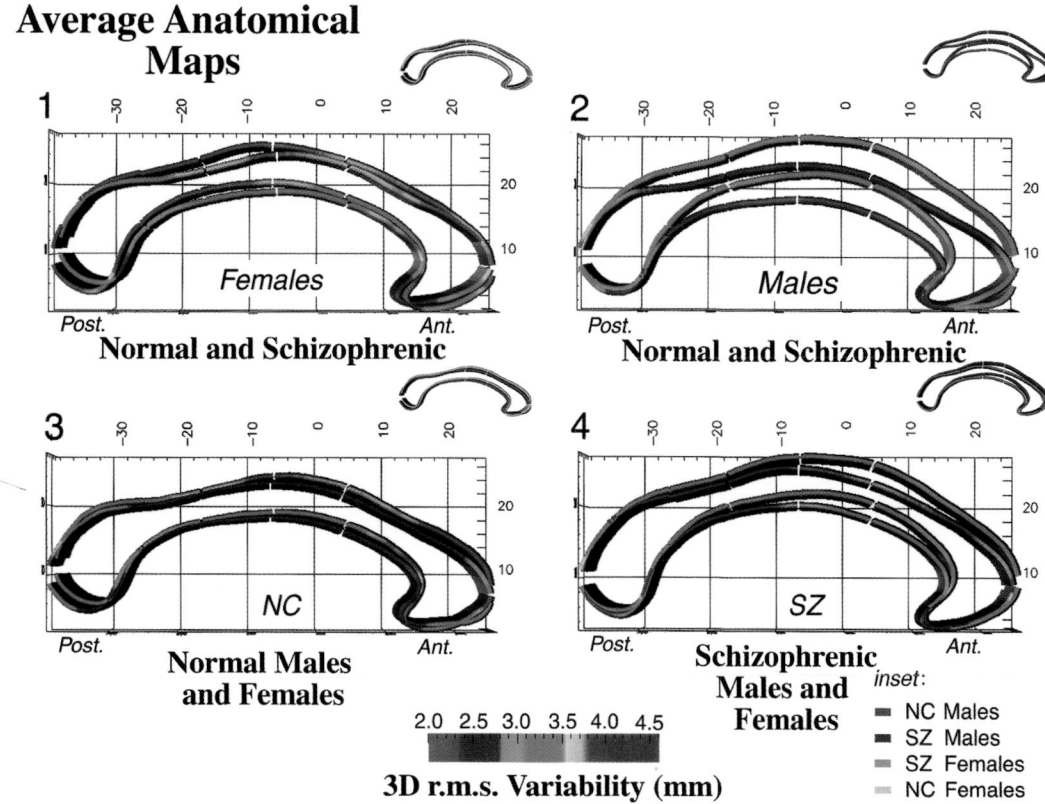

Figure 11 Corpus callosum in schizophrenia (data from Narr *et al.,* 1999). Midsagittal corpus callosum boundaries were averaged from 25 patients with chronic schizophrenia (DSM-III-R criteria; 15 males, 10 females; age: 31.1 ± 5.6 years) and from 28 control subjects matched for age (30.5 ± 8.7 years), gender (15 males, 13 females), and handedness (1 left-handed subject per group). Profiles of anatomic variability around the group averages are also shown (in color) as an rms deviation from the mean. Anatomical averaging reveals a pronounced and significant bowing effect in the schizophrenic patients relative to normal controls. Male patients show a significant increase in curvature for superior and inferior callosal boundaries ($p < 0.001$), with a highly significant Sex × Diagnosis interaction ($p < 0.004$). The sample was stratified by Sex and Diagnosis and separate group averages show that the disease induces less bowing in females (panel 1) than in males (panel 2). While gender differences are not apparent in controls (panel 3), a clear gender difference is seen in the schizophrenic patients (panel 4). Abnormalities localized in a disease-specific atlas can therefore be analyzed to reveal interactions between disease and demographic parameters.

gions of high anatomic variability, so normal differences are not signaled as deficits.

D. Anisotropic Gaussian Fields

In a probabilistic atlas, well-defined statistical criteria are required to identify significant differences in brain structure. These criteria can be formulated in different ways, depending on the attribute whose statistical variation is being modeled. One approach is to use the theory of Gaussian random fields, a modeling technique used widely in functional image analysis (e.g., SPM; Friston *et al.,* 1995). By contrast with functional signals, which are generally treated as random *scalar* fields, the deformation maps that quantify structural differences are treated as random *vector* fields. Instead of a field of variance values, the variability of the deformation vectors, and their directional tendencies, are stored using a covariance tensor at each anatomical point (Thompson *et al.,* 1996; Cao and Worsley, 1999).

In one study (Thompson *et al.,* 1997; cf. Cao and Worsley, 1999), we developed an approach to detect brain structure differences between two groups or between an individual subject and a database of demographically matched subjects. Suppose $\mathbf{W}_{ij}(\mathbf{x})$ is the deformation vector required to match the structure at position $\mathbf{x}$ in an atlas template with its counterpart in subject i of group j. (If surface models are being analyzed rather than full brain volumes (Thompson and Toga, 1998a), $\mathbf{W}_{ij}(\mathbf{x})$ is the deformation vector matching parametric mesh node $\mathbf{x}(u,v)$ with its counterpart

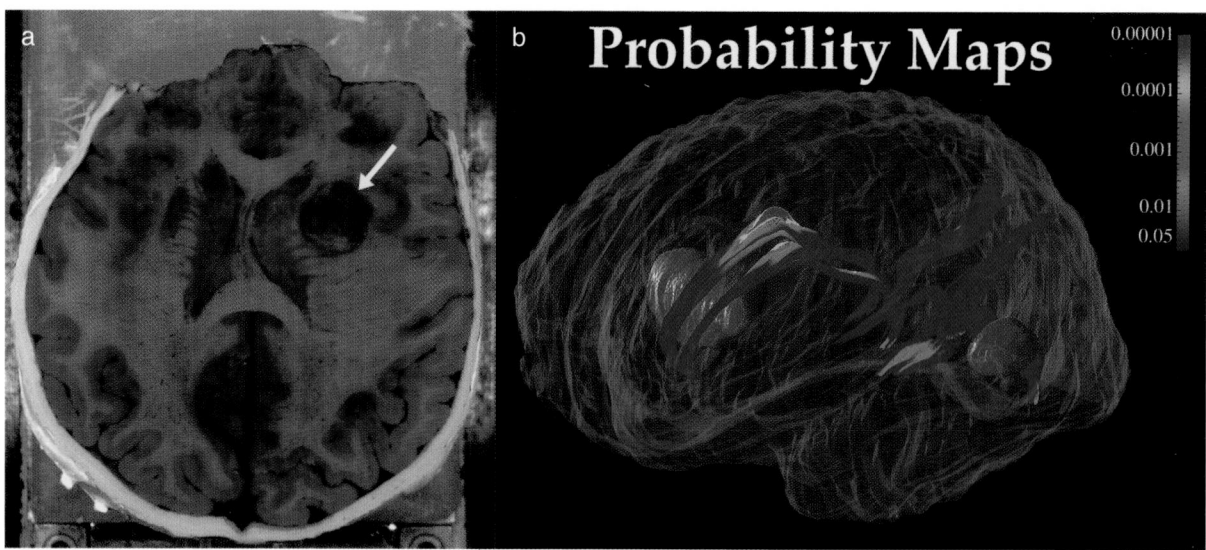

Figure 12 Distortions in brain architecture induced by tumor tissue: probability maps for ventral callosum and major sulci. Color-coded probability maps (b) quantify the impact of two focal metastatic tumors (illustrated in red; see cryosection blockface (a)) on the ventral callosal boundary as well as the parieto-occipital and anterior and posterior calcarine sulci in both brain hemispheres.

in subject i of group j.) We then model the deformations as

$$\mathbf{W}_{ij}(\mathbf{x}) = \mu_j(\mathbf{x}) + \Sigma(\mathbf{x})^{1/2}\epsilon_{ij}(\mathbf{x}). \quad (7)$$

Here $\mu_j(\mathbf{x})$ is the mean deformation for group j, $\Sigma(\mathbf{x})$ is a nonstationary, anisotropic covariance tensor field (Section IX), which relaxes the confidence threshold for detecting abnormal structure in regions where normal variability is extreme, $\Sigma(\mathbf{x})^{1/2}$ is the upper triangular Cholesky factor tensor field, and $\epsilon_{ij}(\mathbf{x})$ is a trivariate random vector field whose components are independent stationary Gaussian random fields.

E. Deformation-Based Morphometry

A T^2 or F statistic that indicates evidence of significant difference in deformations between the groups is calculated at each lattice location in a 3D image or parameterized 3D surface to form a statistic image (Thompson et al., 1997). Specialized algorithms, using corrections for the metric tensor of the underlying surface, are required to calculate these fields at the cortex (see next section). Under the null hypothesis of no abnormal deformations, the statistic image is approximated by a T^2 random field. Specifically, the significance of a difference in brain structure between two subject groups (e.g., patients and controls) of N_1 and N_2 subjects is assessed by calculating the sam-

ple mean and variance of the deformation fields ($j = 1, 2$),

$$\mathbf{W}_{\mu j}(\mathbf{x}) = \sum\mathbf{W}_{ij}(\mathbf{x})/N_j,$$

$$\Psi(\mathbf{x}) = (1/(N_1 + N_2 - 2)) \{\sum\sum[\mathbf{W}_{ij}(\mathbf{x}) - \mathbf{W}_{\mu j}(\mathbf{x})][\mathbf{W}_{ij}(\mathbf{x}) - \mathbf{W}_{\mu j}(\mathbf{x})]^\mathrm{T}\},$$

and computing the following statistical map (Thompson et al, 1997; Cao and Worsley, 1999):

$$T^2(\mathbf{x}) = \{N_1N_2/(N_1 + N_2)(N_1 + N_2 - 2)\} [\mathbf{W}_{\mu 2}(\mathbf{x}) - \mathbf{W}_{\mu 1}(\mathbf{x})]^\mathrm{T}[\Psi(\mathbf{x})]^{-1}[\mathbf{W}_{\mu 2}(\mathbf{x}) - \mathbf{W}_{\mu 1}(\mathbf{x})],$$

Under the null hypothesis, $(N_1 + N_2 - 2)T^2(\mathbf{x})$ is a stationary Hotelling's T^2-distributed random field. At each point, if we let $\nu = (N_1 + N_2 - 2)$ and we let the dimension of the search space be $d = 3$, then

$$F(\mathbf{x}) = ((\nu - d + 1)/d)T^2(\mathbf{x}) \sim F_{d,(\nu-d+1)}.$$

In other words, the field can be transformed pointwise to a Fisher–Snedecor F distribution (Thompson et al., 1997). To obtain a p value for the effect that is adjusted for the multiple comparisons involved in assessing a whole field of statistics, Cao and Worsley (1999) examined the distribution of the global maximum $T^2_{\max}$ of the resulting T^2-distributed random field under the null hypothesis. The resulting probability that $T^2(\mathbf{x})$ ever exceeds a fixed high threshold $T^2_{\max}$ is approximated by the expected Euler characteristic $E[\chi(A(T^2_{\max}))]$ of the excursion sets of the Hotelling's T^2-distributed random field above the threshold $T^2_{\max}$. Then $p[T^2_{\max} \geq t]$ is approximated by $\Sigma R_n \rho_n(t)$, where the number of n-dimensional resolution elements $R_n = V_n/(\text{FWHM})^n$ depends on the effective full width at half-maximum (FWHM) of the component Gaussian images $\varepsilon_{ij}(\mathbf{x})$ and on the Euler characteristic (V_0), caliper diameter ($V_1/2$), surface area ($2V_2$), and volume (V_3) of the search region. The n-dimensional EC densities are given by Cao and Worsley (1999):

$$\rho_0(t) = \int [\Gamma((\nu+1)/2)/(\nu\pi)^{1/2}\Gamma(\nu/2)]$$
$$[1 + (u^2/\nu)]^{-1/2(\nu+1)}du,$$

$$\rho_1(t) = ((4 \ln 2)^{1/2}/2\pi)[1 + (t^2/\nu)]^{-1/2(\nu-1)},$$

$$\rho_2(t) = ((4 \ln 2)/(2\pi)^{3/2})[\Gamma((\nu + 1)/2)/(\nu/2)^{1/2}\Gamma(\nu/2)]$$
$$[t[1 + (t^2/\nu)]^{-1/2(\nu-1)}],$$

$$\rho_3(t) = ((4 \ln 2)^{3/2}/(2\pi)^2)[((\nu - 1)/\nu)t^2 - 1]$$
$$[1 + (t^2/\nu)]^{-1/2(\nu-1)}.$$

The global maximum of the random deformation field, or derived tensor fields (Thompson et al., 1998a,b), can be used to test the hypothesis of no structural change in disease (Worsley, 1994a,b; Cao and Worsley, 1999). Similar multivariate linear models can be used to test for the effect of explanatory variables (e.g., age, gender, and clinical test scores) on a set of deformation field images (Ashburner et al., 1998; Gaser et al., 1998). This can help explore linkages between atlas descriptions of variance and behavioral or cognitive parameters (Fuh et al., 1997; Mega et al., 1998b; Zoumalan et al., 1999).

F. Pattern-Theoretic Approaches

In a related approach based on pattern theory (Grenander and Miller, 1998), a spectral approach to representing anatomic variation is developed. Deformation maps expressing variations in normal anatomies are calculated with a nonlinear registration procedure based on continuum mechanics (Christensen et al., 1993; Miller et al., 1993). Each deformation map is expanded in terms of the eigenfunctions of the governing operator that controls the transformations (such as the Laplacian ∇^2 (Ashburner et al., 1997) or Cauchy–Navier operator $(\lambda + \mu)\nabla(\nabla^{\mathrm{T}}\bullet) + \mu\nabla^2$ (Christensen et al., 1996)). Gaussian probability measures are defined on the resulting sequences of expansion coefficients (Amit et al., 1991; Grenander and Miller, 1998). Essentially, this spectral formulation is a model of anatomic variability. Once the model parameters σ_k are learned (see Footnote 4), every subject's anatomy can be represented by a feature vector $(z_1, \ldots, z_n)$, whose elements are just the

coefficients of the deformation field required to match their particular anatomy with a mean anatomical template. If the parameters of anatomical variation are altered in disease, a pattern classifier can classify new subjects according to their statistical distance from the diseased group mean relative to the normal group mean (Thompson et al., 1997; Joshi et al., 1998). From a validation standpoint, the operating characteristics of such a system can be investigated (i.e., false positives versus false negatives; Thompson et al., 1997; Joshi et al., 1998). Currently being tested as a framework to encode anatomic variation, these deformable atlas systems show considerable promise in identifying disease-specific differences (Haller et al., 1997; Joshi et al., 1998).

G. Bayesian Pattern Recognition

Pattern recognition algorithms for automated identification of brain structures can benefit greatly from encoded information on anatomic variability (Ashburner et al., 1998; Gee and Bajcsy, 1998; Vaillant and Davatzikos, 1999). We recently developed a Bayesian approach to identify the corpus callosum in each image in an MRI database (Pitiot et al., 1999). The shape of a deformable curve (Fig. 13, panel 7) is progressively tuned to optimize a mathematical criterion measuring how likely it is that it has found the corpus callosum. The measure includes terms that reward contours based on their agreement with a diffused edge map (panels 7–9), their geometric regularity, and their statistical abnormality when compared with a distribution of normal shapes. As we have seen, by averaging contours derived

---

[4]In Grenander's formalism, the distribution of the random deformation fields $\mathbf{u}(\mathbf{x})$ is assumed to satisfy the stochastic differential equation

$$L(\mathbf{u}(\mathbf{x})) = \mathbf{e}(\mathbf{x}). \tag{8}$$

Here L is the operator governing the deformation and $\mathbf{e}(\mathbf{x})$ is a 3×1 random noise vector field, whose coefficients in L's eigenbasis are zero-mean independent Gaussian variables with variances σ_k^2. If the differential operator L has eigenbasis $\{\varphi_k\}$ with eigenvalues $\{\lambda_k\}$, a probability density can be defined directly on the deformation field's expansion coefficients $(z_1, \ldots, z_n)$. If

$$\mathbf{u}(\mathbf{x}) = \Sigma z_k \varphi_k(\mathbf{x}), \tag{9}$$

then

$$p(z_1, \ldots, z_n) = \exp[-(1/2)(\Sigma \log\{2\pi\sigma_k^2/\lambda_k^2\} + \Sigma\{|\lambda_k z_k|^2/\sigma_k^2\})], \tag{10}$$

manually from an image database, structural abnormalities associated with Alzheimer's disease and schizophrenia were identified (Figs. 10 and 11; Thompson *et* *al.,* 1998a; Narr *et al.,* 1999). Automated parameterization of structures will accelerate the identification and analysis of disease-specific structural patterns.

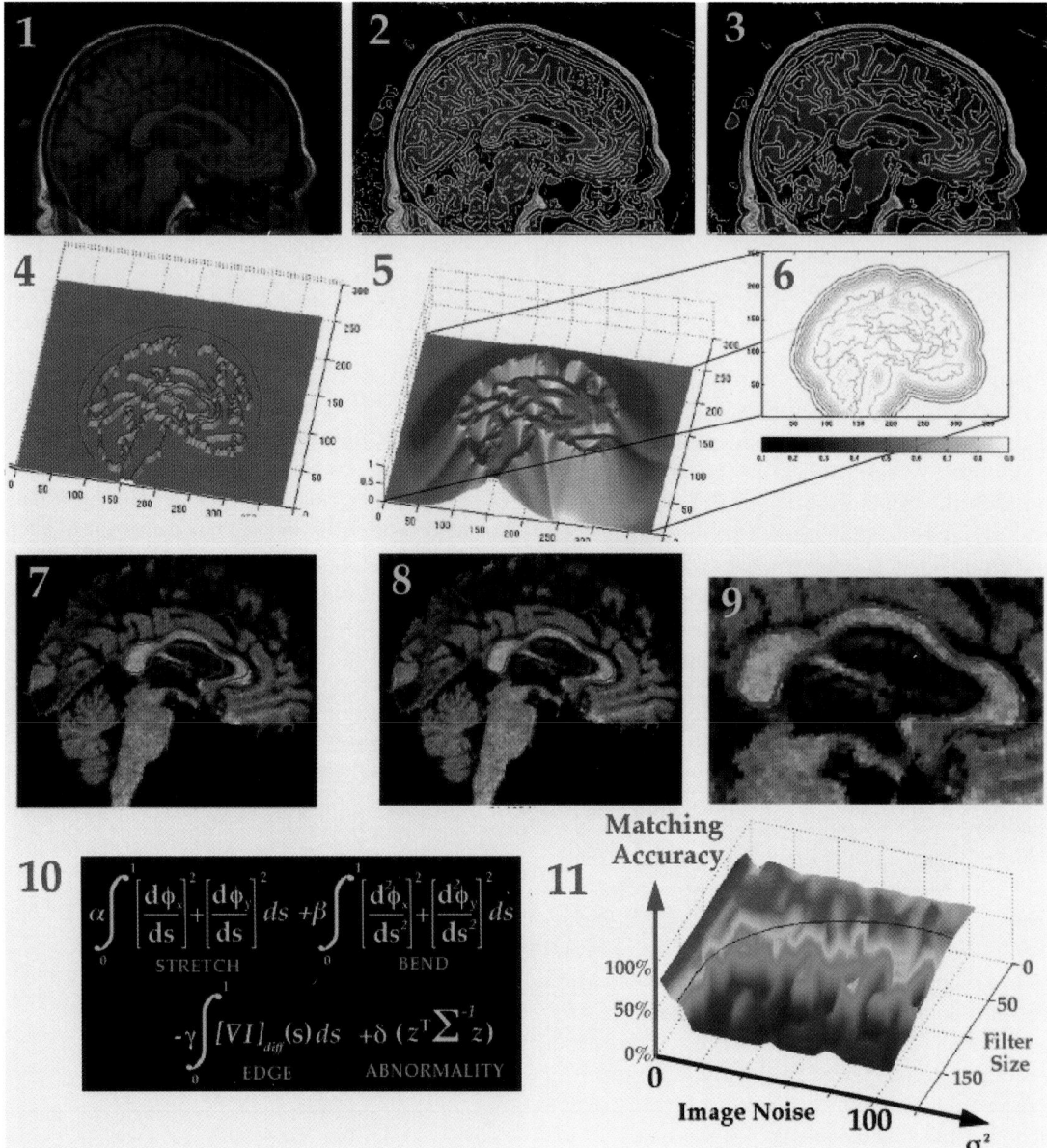

Figure 13 Probabilistic labeling of structures in image databases. An atlas storing information on anatomic variability is used to guide an algorithm in finding the corpus callosum boundary (panel 9) in each image in an anatomic database (*N* = 104; Pitiot *et al.,* 1999). The output of an edge detector (panel 2) is run through a connectivity filter that suppresses the smallest connected sets of edge pixels. The filtered edge image is then diffused over time (panels 4–6) and a deformable curve (panel 7) is adapted to optimize a matching measure (panel 10). This measure penalizes curve shapes (1) that are too bent or stretched, (2) that fail to overlap the diffused edge image, or (3) that are unlikely based on a statistical distribution of normal corpus callosum shapes. Given an image database, algorithm parameters (such as the size of the connectivity filter; panel 11) can be tuned based on their overall performance on an image database. Their optimal values differ depending on how noisy the images are. Boundaries can then be averaged from patients with dementia or schizophrenia to detect anomalies. With separate sets of training and test images, these algorithms can both invoke and generate information on structural variation and pathology.

VIII. Cortical Modeling

Cortical morphology is notoriously complex and presents unique challenges in anatomic modeling. The cortex is also severely affected in disorders such as Alzheimer's disease, Pick's disease, and other dementias, by tumor growth, and in cases of epilepsy, cortical dysplasias, and schizophrenia.

A major challenge in investigations of disease is to determine (1) whether cortical organization is altered, and if so, which cortical systems are implicated, and (2) whether normal features of cortical organization are lost, such as sulcal pattern asymmetries. This requires methods to create (1) a well-resolved average model of the cortex specific for a diseased group and (2) a statistical framework to compare these average models with normative data.

A. Averaging Images or Averaging Geometry

In an atlasing context, it would be ideal to create a disease-specific template for a clinical group with well-resolved anatomical features in their mean anatomical configuration. Unfortunately, this cannot be achieved by averaging together structural images in the traditional way, after a simple linear transformation to a standard space (Evans *et al.*, 1994). If images are averaged in this way, cortical patterns are washed away due to wide variations in gyral organization. We describe a way to avoid this. First, an average cortical surface model is created with well-resolved gyral features in the group mean configuration. Continuum-mechanical mappings are then used to bring each subject's gyral pattern into correspondence with the average cortex. Maps of cortical variation are created as a byproduct. Finally, a high-dimensional mapping (driven by 84 structures per brain) elastically deforms each brain into the group mean geometric configuration. Once elastically reconfigured, the scan intensities are averaged on a voxel-by-voxel basis to produce a group-specific atlas template with a well-resolved cortex. Disease-specific templates in Alzheimer's disease and schizophrenia will be used to illustrate this method.

B. Cortical Matching

Cortical anatomy can be compared, between a pair of subjects, by computing the warped mapping that elastically transforms one cortex into the shape of the other. These transformations can also match large networks of gyral and sulcal features with their counterparts in the target brain (Davatzikos, 1996; Thompson and Toga,

1996, 1997; Van Essen *et al.*, 1997; Fischl *et al.*, 1999). Differences in cortical organization prevent exact gyrus-by-gyrus matching of one cortex with another. Nonetheless, an important intermediate goal has been to match a comprehensive network of sulcal and gyral elements that are consistent in their incidence and topology across subjects (Ono *et al.*, 1990; Collins *et al.*, 1996; Leonard, 1996; Kennedy *et al.*, 1998; see Footnote 5).

C. Technical Details

Although planar (or spherical) maps serve as proxies for the cortex (Fig. 14), different amounts of local dilation and contraction are required to transform the cortex onto these simpler two-parameter surfaces. In the covariant tensor approach (Thompson and Toga, 1998a),

---

[5]Our method (Thompson and Toga, 1996) is conceptually similar to those of Dale and Sereno (1993), Davatzikos (1996), and Fischl *et al.* (1999). 3D active surfaces (Cohen and Cohen, 1992) extract parametric representations of each subject's cortex by deforming a tiled spherical mesh into the shape of the cortex. On these surface models, corresponding networks of anatomical curves are identified. Specifically, 36 parametric curves are created per subject to represent the major elements of the gyral pattern. These include the superior and inferior frontal, central, postcentral, intraparietal, superior and inferior temporal, collateral, olfactory, and occipito-temporal sulci as well as the sylvian fissures. Additional 3D curves are drawn in each hemisphere to represent gyral limits at the interhemispheric margin (Thompson *et al.*, 1997). Stereotaxic locations of contour points derived from the data volume are then redigitized to produce 36 uniformly parameterized cortical contours per brain, representing the primary gyral pattern of each subject (Thompson *et al.*, 1997). The transformation relating these networks is expressed as a vector flow field in the parameter space of the cortex (Fig. 14). This vector flow field in parameter space indirectly specifies a correspondence field in 3D, which drives one cortical surface into the shape of another. This mapping not only matches overall cortical geometry but matches the entire network of the 36 landmark curves with their counterparts in the target brain, and thus is a valid encoding of cortical variation.

Spherical, Planar Maps of Cortex. Several simpler maps of the cortex are made to help calculate the transformation. Because each subject's cortical model is arrived at by deforming a spherical mesh (MacDonald *et al.*, 1993; Davatzikos *et al.*, 1996; Thompson and Toga, 1996), any point on the cortex maps to exactly one point on the sphere, and a *spherical map* of the cortex is made that indexes sulcal landmarks in the normally folded brain surface. These spherical locations, indexed by two parameters, can also be mapped to a plane (Fig. 14; Thompson *et al.*, 1997; Thompson and Toga, 1998). A flow field is then calculated that elastically warps one flat map onto the other (Fig. 15; or equivalently, one spherical map to the other). On the sphere, the parameter shift function $\mathbf{u}(\mathbf{r}): \Omega \to \Omega$ is given by the solution $F_{pq}: \mathbf{r} \to \mathbf{r} - \mathbf{u}(\mathbf{r})$ to a curve-driven warp in the spherical parametric space $\Omega = [0, 2\pi) \times [0, \pi)$ of the cortex (Thompson *et al.*, 1996, 1998). For points $\mathbf{r} = (r, s)$ in the parameter space, a system of simultaneous partial differential equations can be written for the flow field $\mathbf{u}(\mathbf{r})$:   *continued*

exact information on these metric alterations is stored in the metric tensor of the mapping, $g_{jk}(\mathbf{r})$. In the subsequent matching procedure, correction terms (Christoffel symbols, Γ^{i}_{jk}) make the necessary adjustments for fluctuations in the metric tensor of the mapping procedure. Since metric distortions caused by mappings to spheres or planes can always be encoded as a metric tensor field, a covariant approach supports comparisons of cortical data using *either* flattened or spherical maps. In the partial differential equation formulation (1), we replace L by the covariant differential operator $L^{\ddagger}$. In $L^{\ddagger}$, all L partial derivatives are replaced with covariant derivatives. These covariant derivatives are defined with respect to the metric tensor of the surface domain where calculations are performed (see Footnote 6).

---

[5](continued)

$$L^{\ddagger}(\mathbf{u}(\mathbf{r})) + \mathbf{F}(\mathbf{r} - \mathbf{u}(\mathbf{r})) = \mathbf{0}, \forall \mathbf{r} \; \epsilon \; \Omega,$$

$$\mathbf{u}(\mathbf{r}) = \mathbf{u}_0(\mathbf{r}), \forall \mathbf{r} \; \epsilon \; M_0 \cup M_1. \qquad (11)$$

Here M_0 and M_1 are sets of points and (sulcal or gyral) curves where displacement vectors $\mathbf{u}(\mathbf{r}) = \mathbf{u}_0(\mathbf{r})$ matching corresponding anatomy across subjects are known. The flow behavior is modeled using equations derived from continuum mechanics, and these equations are governed by the Cauchy–Navier differential operator $L = \mu \nabla^2 + (\lambda + \mu) \nabla (\nabla^{\mathrm{T}} \cdot)$ with body force $\mathbf{F}$ (Thompson *et al.*, 1996, 1998; Grenander and Miller, 1998). The only difference is that $L^{\ddagger}$ is the *covariant* form of the differential operator L, for reasons explained in the next section.

Covariant Field Equations. Since the cortex is not a *developable* surface (Davatzikos, 1996), it cannot be given a parameterization whose metric tensor is uniform. As in fluid dynamics or general relativity applications, the intrinsic curvature of the solution domain should be taken into account when computing flow vector fields in the cortical parameter space and mapping one mesh surface onto another; otherwise errors will arise. The result is *a covariant regularization* approach (Einstein, 1914; Thompson and Toga, 1998a; Thompson *et al.*, 1999b). From a practical standpoint, this approach uses a mathematical trick that makes it immaterial whether a spherical or planar map is used to simplify the mathematics of cortical matching. Either a spherical or a planar map can be used. Since the flows defined on these maps are adjusted for variations in the metric tensor of the mapping, the results become independent of the underlying parameterization (spherical or planar). In fact, spherical and planar maps involve different amounts of local dilation or contraction of the surface metric, but this metric tensor field is stored and used later to adjust the flow that maps one cortex on another, so which one is used is immaterial. The covariant approach was introduced by Einstein (1914) to allow the solution of physical field equations defined by elliptic operators on manifolds with intrinsic curvature. Similarly, the problem of deforming one cortex onto another involves solving a similar system of elliptic partial differential equations (Davatzikos, 1996; Drury *et al.*, 1996; Thompson and Toga, 1998a; Bakircioglu *et al.*, 1999), defined on an intrinsically curved computational mesh (in the shape of the cortex). In the covariant formalism, the differential operators governing the mapping of one cortex to another are adaptively modified to reflect changes in the underlying metric tensor of the surface parameterizations (Fig. 14).

With this mathematical adjustment, we eliminate the confounding effects of metric distortions that necessarily occur during the mapping procedure. The resulting cortical matching field is independent of the auxiliary mappings (spherical or planar) used to extract it.

---

[6]The covariant derivative of a (contravariant) vector field, $u^i(\mathbf{x})$, is defined as $u^i_{,k} = \partial u^i / \partial x^k + \Gamma^i_{ik} u^i$ (Thompson and Toga, 1998), where the Γ^i_{ik} are *Christoffel symbols of the second kind* (Einstein, 1914). This expression involves not only the rate of change of the vector field itself, as we move along the cortical model, but also the rate of change of the local basis, which itself varies due to the intrinsic curvature of the cortex. On a surface with no intrinsic curvature, the extra terms (Christoffel symbols) vanish. The Christoffel symbols, expressed in terms of derivatives of the metric tensor components $g_{jk}(\mathbf{x})$, are calculated from the cortical model: $\Gamma^i_{jk} = (1/2) g^{il} (\partial g_{lj} / \partial x^k + \partial g_{lk} / \partial x^j - \partial g_{jk} / \partial x^i)$. Scalar, vector, and tensor quantities, in addition to the Christoffel symbols required to implement the diffusion operators on a curved manifold, are evaluated by finite differences. These correction terms are then used in the solution of the Dirichlet problem (Joshi *et al.*, 1995) for matching one cortex with another. A final complication is that different metric tensors $g_{jk}(\mathbf{r}_p)$ and $h_{jk}(\mathbf{r}_q)$ relate (1) the physical domain of the *input* data to the computation mesh (via mapping D_p^{-1}) and (2) the solution on the computation mesh to the *output* data domain (via mapping D_q). To address this problem, two different approaches are possible, using either (1) simultaneous covariant regularization or (2) Polyakov actions and Beltrami flows (concepts from high-energy physics). In the first approach (Fig. 14), the PDE $L^{\ddagger\ddagger}\mathbf{u}(\mathbf{r}_q) = -\mathbf{F}$ is solved first to find a flow field $T_q{:}\mathbf{r} \rightarrow \mathbf{r} - \mathbf{u}(\mathbf{r})$ on the target spherical map with anatomically driven boundary conditions $\mathbf{u}(\mathbf{r}_q) = \mathbf{u}_0(\mathbf{r}_q), \forall \mathbf{r}_q \; \epsilon \; M_0 \cup M_1$. Here $L^{\ddagger\ddagger}$ is the covariant adjustment of the differential operator L with respect to the tensor field $h_{jk}(\mathbf{r}_q)$ induced by D_q. Next, the PDE $L^{\ddagger}\mathbf{u}(\mathbf{r}_p) = -\mathbf{F}$ is solved to find a reparameterization $T_p{:}\mathbf{r} \rightarrow \mathbf{r} - \mathbf{u}(\mathbf{r})$ of the initial spherical map with boundary conditions $\mathbf{u}(\mathbf{r}_p) = \mathbf{0}, \forall \mathbf{r}_p \; \epsilon \; M_0 \cup M_1$. Here $L^{\ddagger}$ is the covariant adjustment of L with respect to the tensor field $g_{jk}(\mathbf{r}_p)$ induced by D_p. The full cortical matching field (Fig. 14, top right) is then defined as $\mathbf{x} \rightarrow D_q(F_{pq}(D_p^{-1}(\mathbf{x})))$, with $F_{pq} = (T_q)^{-1}\mathrm{o}(T_p)^{-1}$. A second (conceptually related) approach uses Beltrami flows to establish a p-harmonic map from one surface to the other. If P and Q are cortical surfaces with metric tensors $g_{jk}(u^i)$ and $h_{jk}(\xi^\alpha)$ defined in local coordinates u^i and ξ^α ($i, \alpha = 1, 2$), the energy density of a map between the surfaces $\xi(u){:}(P,g) \rightarrow (Q,h)$ is the functional $e(\xi){:}P \rightarrow R$ defined in local coordinates as

$$e(\xi)(u) = g^{ij}(u) \; \partial \xi^\alpha(u)/\partial u^i \; \partial \xi^\beta(u)/\partial u^j \; h_{\alpha\beta}(\xi(u))$$

The Dirichlet energy of the mapping $\xi(u)$ (i.e., the generalization of the Hilbert space norm to curved spaces) is defined as $E(\xi) = \int_P e(\xi)(u) dP$, where $dP = (\sqrt{\det[g_{ij}]}) du^1 du^2$. The Euler equations, whose solution $\xi^\alpha(u)$ minimizes the mapping energy, are

$$0 = L(\xi^i) = \Sigma \partial / \partial u^m [(\sqrt{\det[g^{ru}]}) \Sigma g^{ml}_{ur} \partial \xi^i / \partial u^l] \; (i = 1, 2)$$

(Liseikin, 1991). These equations can be discretized to produce a Beltrami flow (Sochen *et al.*, 1998) or a quasi-linear elliptic system (Liseikin, 1991) whose solution is a harmonic map from one surface to the other. This harmonic map (1) minimizes the change in metric from one surface to the other and (2) is again independent of the parameterizations (spherical or planar) used for each surface.

D. Retention of 3D Cortical Information

To ensure that each subject's spherical map can be converted back into a 3D cortical model, cortical surface point position vectors in 3D stereotaxic space are represented on the sphere with a color code (at 16 bits per channel). This forms an image of the parameter space in RGB color image format (Fig. 14; Thompson and Toga, 1997). To find good matches between cortical regions in different subjects, we first derive a color image map for each respective surface model and transfer the entire network of sulcal curves back onto it. After the matching process is performed using a flow in the parametric space (Figs. 14 and 15), the corresponding 3D mapping is recovered, carrying one cortex onto another.

IX. Cortical Averaging

The warping field deforming one cortex into gyral correspondence with another can also be used to create an *average* cortex. To do this, all 36 gyral curves for all subjects are first transferred to the spherical parameter space. Next, each curve is uniformly reparameterized to produce a regular curve of 100 points on the sphere whose corresponding 3D locations are uniformly spaced. A set of 36 average gyral curves for the group is created by vector averaging all point locations on each curve. This *average curve template* (curves in Fig. 15b) serves as the target for alignment of individual cortical patterns (cf. Fischl *et al.*, 1999). Each individual cortical pattern is transformed into the average curve configuration using a flow field within the spherical map (Figs. 15a and 15b). By carrying a color code (that indexes 3D locations) along with the vector flow that aligns each individual with the average folding pattern, information can be recovered at a particular location in the average folding pattern (Fig. 15d) specifying the 3D cortical points mapping each subject to the average. This produces a new coordinate grid on a given subject's cortex (Fig. 15f) in which particular grid points appear in the same location across subjects relative to the mean gyral pattern. Averaging these 3D positions across subjects allows an average 3D cortical model to be constructed for the group (Fig. 16, bottom row). The resulting mapping is guaranteed to average together all points falling on the same cortical locations across the set of brains and ensures that corresponding features are averaged together.

A. Cortical Variability

By using the color code (Fig. 15d) to identify original cortical locations in 3D space (Fig. 15f), displacement fields are recovered mapping each patient into gyrus-by-gyrus correspondence with the average cortex (Fig. 17). Anatomic variability is then defined at each point on the average cortex as the root-mean-square (rms) magnitude of the 3D displacement vectors, assigned to each point, in the surface maps driving individuals onto the group average (Thompson *et al.*, 1996a,b, 1997, 1999b). This variability pattern is visualized as a color-coded map (Fig. 18).

Overall, variability values rise sharply (Fig. 18) from 4–5 mm in primary motor cortex to localized peaks of maximal variability in posterior perisylvian zones and superior frontal association cortex (12–14 mm). Primary sensory and motor areas show a dramatic, localized invariance (2–5 mm), but variability rises sharply with the transition anteriorly from motor area 4 to prefrontal heteromodal association cortex. Intermediate variability values (6–10 mm) over the inferior prefrontal convexity fall sharply with the transition to archicortical orbitofrontal cortex, where the gyral pattern is highly conserved across subjects (2- to 5-mm variability). More laterally, the posterior frontal cortex, including territory occupied by Broca's area, displays intermediate variability (6–10 mm). Temporal lobe variability rose from 2–3 mm deep in the sylvian fissure to 14 mm at the posterior limit of the inferior temporal sulcus in both brain hemispheres, extending into the posterior heteromodal association cortex of the parietal lobe (12–14 mm). When a variety of widely used registration systems were examined in addition to the Talairach system (Thompson *et al.*, 1999b), 3D variability in these higher order processing areas was consistently a factor of 10 greater than the variability in the most highly conserved areas of cortex. The region of maximal variability, in temporal cortex, is tightly linked with the location of human visual area MT (or V5; Watson *et al.*, 1993). This suggests that extreme caution is necessary when referring to activation foci in this important area using stereotaxic coordinates. The overall patterns of variation corroborate recent volumetric findings based on a fine-scale parcellation of the cortex (Kennedy *et al.*, 1998). These studies also suggest a greater morphologic individuality in cortical regions that are phylogenetically more recent.

B. Tensor Maps of Directional Variation

Structures do not vary to the same degree in every coordinate direction (Thompson *et al.*, 1996a), and even these directional biases vary by cortical system. The principal directions of anatomic variability in a group

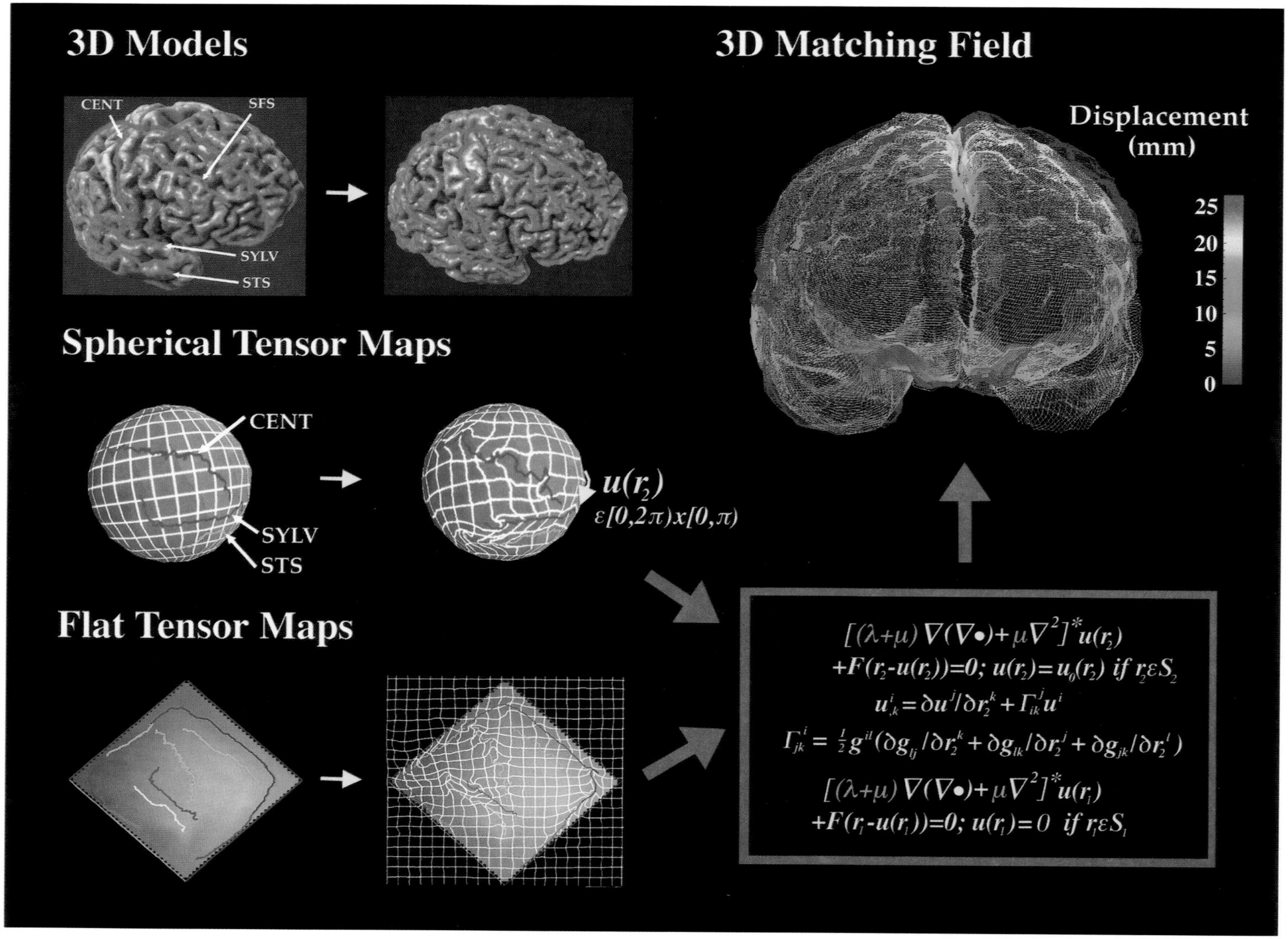

3D Matching Field

Displacement (mm)

25
20
15
10
5
0

$$[(\lambda+\mu)\,\nabla(\nabla\bullet)+\mu\nabla^2]^*\,u(r_2)$$
$$+F(r_2-u(r_2))=0;\ u(r_2)=u_0(r_2)\ if\ r_2\varepsilon S_2$$
$$u^i_{,k}=\partial u^i/\partial r_2^k+\Gamma^i_{lk}u^l$$
$$\Gamma^i_{jk}=\tfrac{1}{2}g^{il}(\partial g_{lj}/\partial r_2^k+\partial g_{lk}/\partial r_2^j+\partial g_{jk}/\partial r_2^l)$$
$$[(\lambda+\mu)\,\nabla(\nabla\bullet)+\mu\nabla^2]^*\,u(r_1)$$
$$+F(r_1-u(r_1))=0;\ u(r_1)=0\ if\ r_1\varepsilon S_1$$

3D Models

CENT
SFS
SYLV
STS

Spherical Tensor Maps

$u(r_2)$
$\varepsilon[0,2\pi]x[0,\pi]$

CENT
SYLV
STS

Flat Tensor Maps

can be shown in a *tensor map* (Figs. 19 and 20). The maps have two uses. First, they make it easier to detect anomalies, which may be small in magnitude but in an unusual direction. Second, they significantly increase the information content of Bayesian priors used for automated structure extraction and identification (Mangin *et al.*, 1994; Gee *et al.*, 1995; Royackkers *et al.*, 1996; Pitiot *et al.*, 1999).

Figure 19 shows a tensor map of variability for normal subjects after mapping 20 elderly subjects' data into Talairach space (all right-handed, 10 males, 10 females). Rectangular glyphs indicate the principal directions of variation—they are most elongated along directions where anatomic variation is greatest across subjects. Each glyph represents the covariance tensor of the vector fields that map individual subjects onto their group average. Because gyral patterns constrain the mappings, the fields reflect variations in cortical organization at a more local level than can be achieved by only matching global cortical geometry. Note the elongated glyphs in anterior temporal cortex and the very low variability (in any direction) in entorhinal and inferior frontal areas. By better defining the parameters of allowable normal variations, the resulting information can be leveraged to distinguish normal from abnormal anatomical variants.

C. Emerging Asymmetries

There is a substantial literature on sylvian fissure cortical surface asymmetries (Eberstaller, 1884; Cunningham, 1892; Geschwind and Levitsky, 1968; Davidson and Hugdahl, 1994). These asymmetries have been related to functional lateralization (Strauss *et al.*, 1983), handedness (Witelson, 1989), language function (Burke *et al.*, 1993; Davidson and Hugdahl, 1994), and asymmetries of associated cytoarchitectonic fields (Galaburda and Geschwind, 1981) and their thalamic projection ar-

eas (Eidelberg and Galaburda, 1982). After group averaging of anatomy, asymmetric features emerge that are not observed in individual anatomies due to their considerable variability. As shown in Fig. 18 (sagittal projection), the marked anatomic asymmetry in posterior perisylvian cortex (Geschwind and Levitsky, 1968) actually extends rostrally into postcentral cortex, with the posterior bank of the postcentral gyrus thrust forward by 8–9 mm on the right compared to the left. The asymmetry also extends caudally across the lateral convexity into superior and inferior temporal cortex. As shown earlier by averaging ventricular models (Fig. 9), this asymmetry penetrates subcortically into the occipital ventricular horn, but not into adjacent parieto-occipital and calcarine cortex.

The improved ability to localize asymmetry and encode its variability in a disease-specific atlas has interesting applications in schizophrenia (Narr *et al.*, 1999). Schizophrenic patients have anatomic alterations in several brain regions, including the superior temporal gyrus (e.g., Nestor *et al.*, 1993). This gyrus is the approximate site of Wernicke's language area and marks the planum temporale on its superior banks (Geschwind and Galaburda, 1985; Galaburda *et al.*, 1990). To focus on language-related cortex in schizophrenic patients, many investigators have measured the asymmetry of the *planum temporale* and sylvian fissure in MR images. Nonetheless, studies of these important regions have not always agreed in their findings, with some reporting a lack of asymmetry in schizophrenic patients (Hoff *et al.*, 1992; Kikinis *et al.*, 1994; Petty *et al.*, 1995) and others not (De Lisi *et al.*, 1994; Kleinschmidt *et al.*, 1994; Frangou *et al.*, 1997).

To see if cortical asymmetries were lost in schizophrenia, we made average cortical representations for schizophrenic patients (15 males, all right-handed) and matched controls (also 15 males, right-handed). As de-

---

Figure 14 Maps of the human cerebral cortex: flat maps, spherical maps, and tensor maps. Extreme variations in cortical anatomy (3D models; top left) present challenges in brain mapping, because of the need to compare cortically derived brain maps from many subjects. Comparisons of cortical geometry can be based on the warped mapping of one subject's cortex onto another (top right; Thompson *et al.*, 1997). These warps can also transfer functional maps from one subject to another or onto a common template for comparison. Current approaches for deforming one cortex into the shape of another typically simplify the problem by first representing cortical features on a 2D plane, sphere, or ellipsoid, where the matching procedure (i.e., finding $\mathbf{u}(\mathbf{r}_2)$, above) is subsequently performed (Davatzikos *et al.*, 1996; Drury *et al.*, 1996; Thompson and Toga, 1996; Van Essen *et al.*, 1997; Bakircioglu *et al.*, 1999; Vaillant and Davatzikos, 1999). In one approach (Thompson *et al.*, 1997), active surface extraction of the cortex provides a continuous inverse mapping from the cortex of each subject to the spherical template used to extract it. These inverse maps are applied to connected networks of curved sulci in each subject. This transforms the problem into one of computing an angular flow vector field $\mathbf{u}(\mathbf{r}_2)$, in spherical coordinates, which drives the network elements into register on the sphere (middle panel; Thompson and Toga, 1996). The full mapping (top right) can be recovered in 3D space as a displacement vector field which drives cortical points and regions in one brain into register with their counterparts in the other brain. *Tensor maps* (middle and lower left): Although these simple two-parameter surfaces can serve as proxies for the cortex, different amounts of local dilation and contraction (encoded in the metric tensor in the mapping, $g_{jk}(\mathbf{r})$) are required to transform the cortex into a simpler two-parameter surface. These variations complicate the direct application of 2D regularization equations for matching their features. A covariant tensor approach is introduced in Thompson and Toga (1998a,b; see red box) to address this difficulty. The regularization operator L is replaced by its covariant form L^*, in which correction terms (Christoffel symbols, Γ^i_{jk}) compensate for fluctuations in the metric tensor of the flattening procedure. A covariant tensor approach (Thompson and Toga, 1998a,b) allows either flat or spherical maps to support cross-subject comparisons and registrations of cortical data by eliminating the confounding effects of metric distortions that necessarily occur in the flattening procedure.

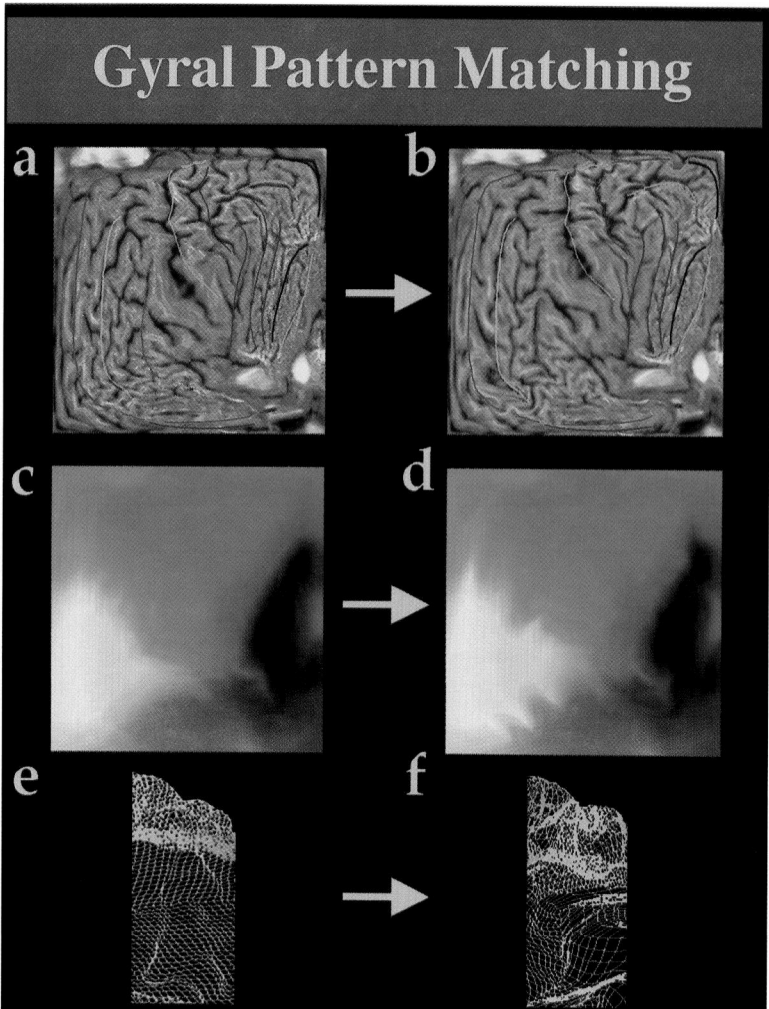

Figure 15 Gyral pattern matching. Gyral patterns can be matched in a group of subjects to create average cortical surfaces. Panel (a) shows a cortical flat map for the left hemisphere of one subject, with the average cortical pattern for the group overlaid (colored lines). Panel (b) shows the result of warping the individual's sulcal pattern into the average configuration for the group, using the covariant field equations (Section VIII). The individual cortex (a) is reconfigured (b) to match the average set of cortical curves. The 3D cortical regions that map to these average locations are then recovered in each individual subject as follows. A color code (c) representing 3D cortical point locations (e) in this subject is convected along with the flow that drives the sulcal pattern into the average configuration for the group (d). Once this is done in all subjects, points on each individual's cortex are recovered (f) that have the same relative location to the primary folding pattern in all subjects. Averaging of these corresponding points results in a crisp average cortex. These transformation fields are stored and used to measure regional variability.

scribed in Section VIII, 36 major sulcal curves were used to drive each subject's gyral pattern into a group mean configuration (Fig. 21). The magnitude of anatomic variation in each brain region was also computed from the deformation vector fields and shown in color as a variability map (Fig. 21, colors). Perhaps surprisingly, asymmetry was not attenuated in the patient group. This can be seen immediately in the sagittal pro-

jections of average anatomy for each group. Significant asymmetries were confirmed by calculating curvature and extent measures from the parametric mesh models (Narr *et al.,* 1999). In frontal cortex, the patients also displayed greater variability than controls. Since relatively subtle asymmetries emerge clearly in a group atlas, population-based atlases may be advantageous for investigating a variety of alterations in cortical organi-

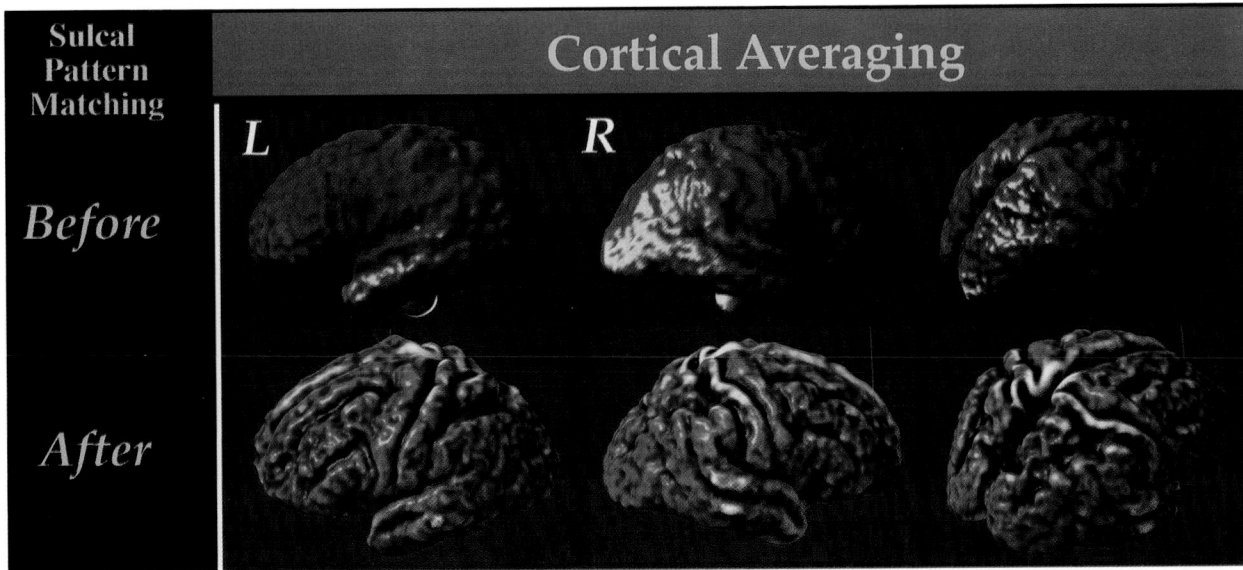

Figure 16 Average cortex in Alzheimer's disease. The average cortical surface for a group of subjects ($N = 9$, Alzheimer's patients) is shown as a graphically rendered surface model. If sulcal position vectors are averaged without aligning the intervening gyral patterns (top), sulcal features are not reinforced across subjects, and a smooth average cortex is produced. By matching gyral patterns across subjects before averaging, a crisper average cortex is produced (bottom row). Sulcal features that consistently occur across all subjects appear in their average geometric configuration.

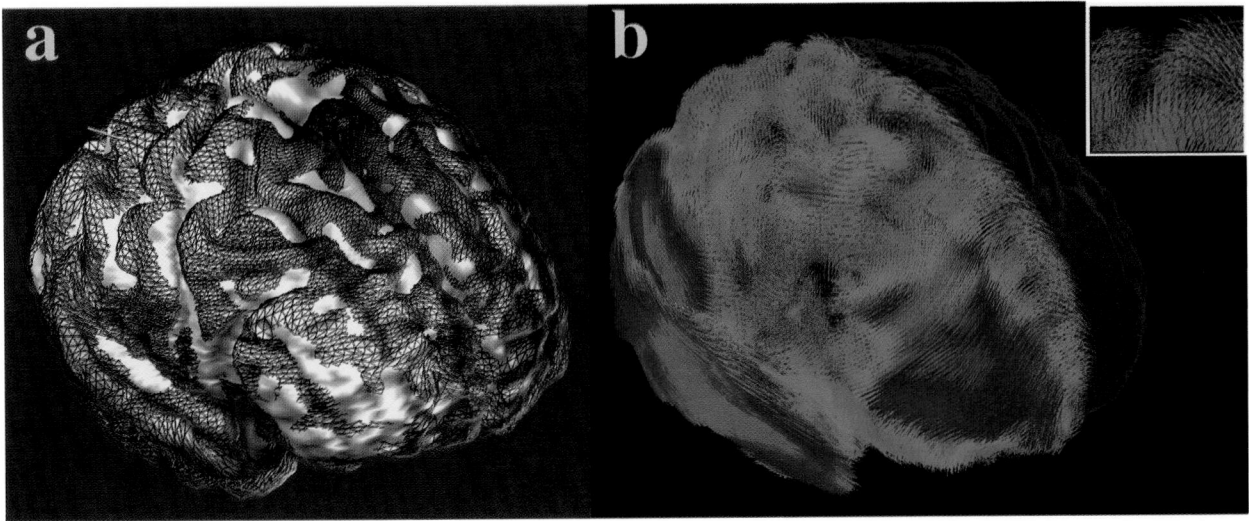

Figure 17 Matching an individual's cortex to the average cortex. 3D variability patterns across the cortex are measured by driving individual cortical patterns into local correspondence with the average cortical model. Panel (a) shows how the anatomy of one subject (brown surface mesh) deviates from the average cortex (white), after affine alignment of the individual data. Panel (b) shows the deformation vector field required to reconfigure the gyral pattern of the subject into the exact configuration of the average cortex. The transformation is shown as a flow field that takes the individual's anatomy onto the right hemisphere of the average cortex (blue surface mesh). The largest amount of deformation is required in the temporal and parietal cortex (pink colors, large deformation). Details of the 3D vector deformation field (b, inset) show the local complexity of the mapping.

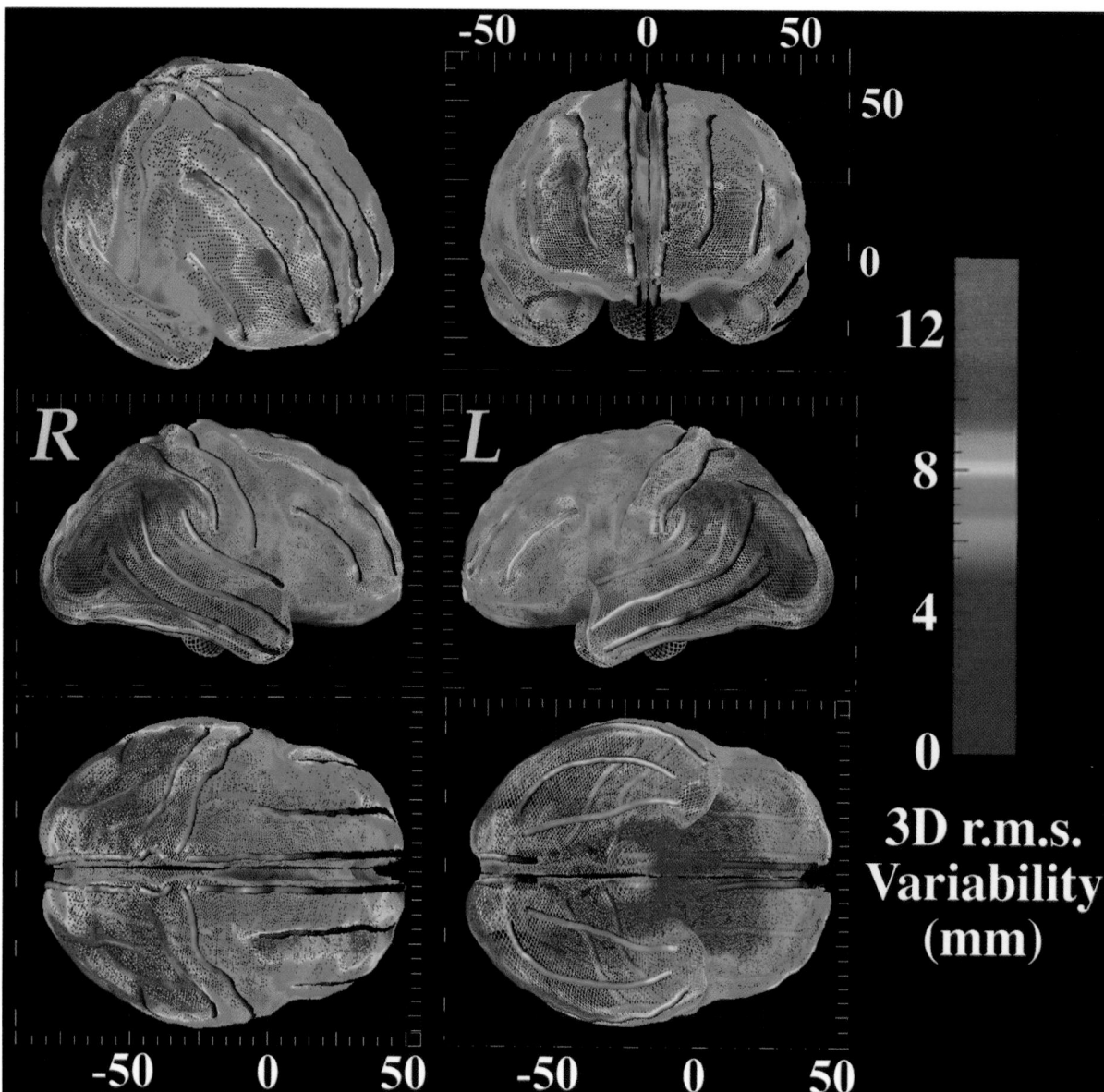

Figure 18 3D cortical variability in Talairach stereotaxic space. (a) The profile of variability across the cortex is shown ($N = 26$ Alzheimer's patients), after differences in brain orientation and size are removed by transforming individual data into Talairach stereotaxic space. The following views are shown: oblique frontal, frontal, right, left, top, bottom. Extreme variability in posterior perisylvian zones and superior frontal association cortex (12–14 mm; red colors) contrasts sharply with the comparative invariance of primary sensory, motor, and orbitofrontal cortex (2–5 mm, blue colors).

Figure 19 Tensor maps reveal directional biases of cortical variation. Tensor maps can be used to visualize these complex patterns of gyral pattern variation at the cortex. The maps are based on a group of 20 elderly normal subjects. Color distinguishes regions of high variability (pink colors) from areas of low variability (blue). Rectangular glyphs indicate the principal directions of variation—they are most elongated along directions where there is greatest anatomic variation across subjects. Each glyph represents the covariance tensor of the vector fields that map individual subjects onto their group average anatomic representation. The resulting information can be leveraged to distinguish normal from abnormal anatomical variants using random field algorithms and can define statistical distributions for feature labeling at the cortex (cf. Le Goualher *et al.,* 1999; Vaillant and Davatzikos, 1999).

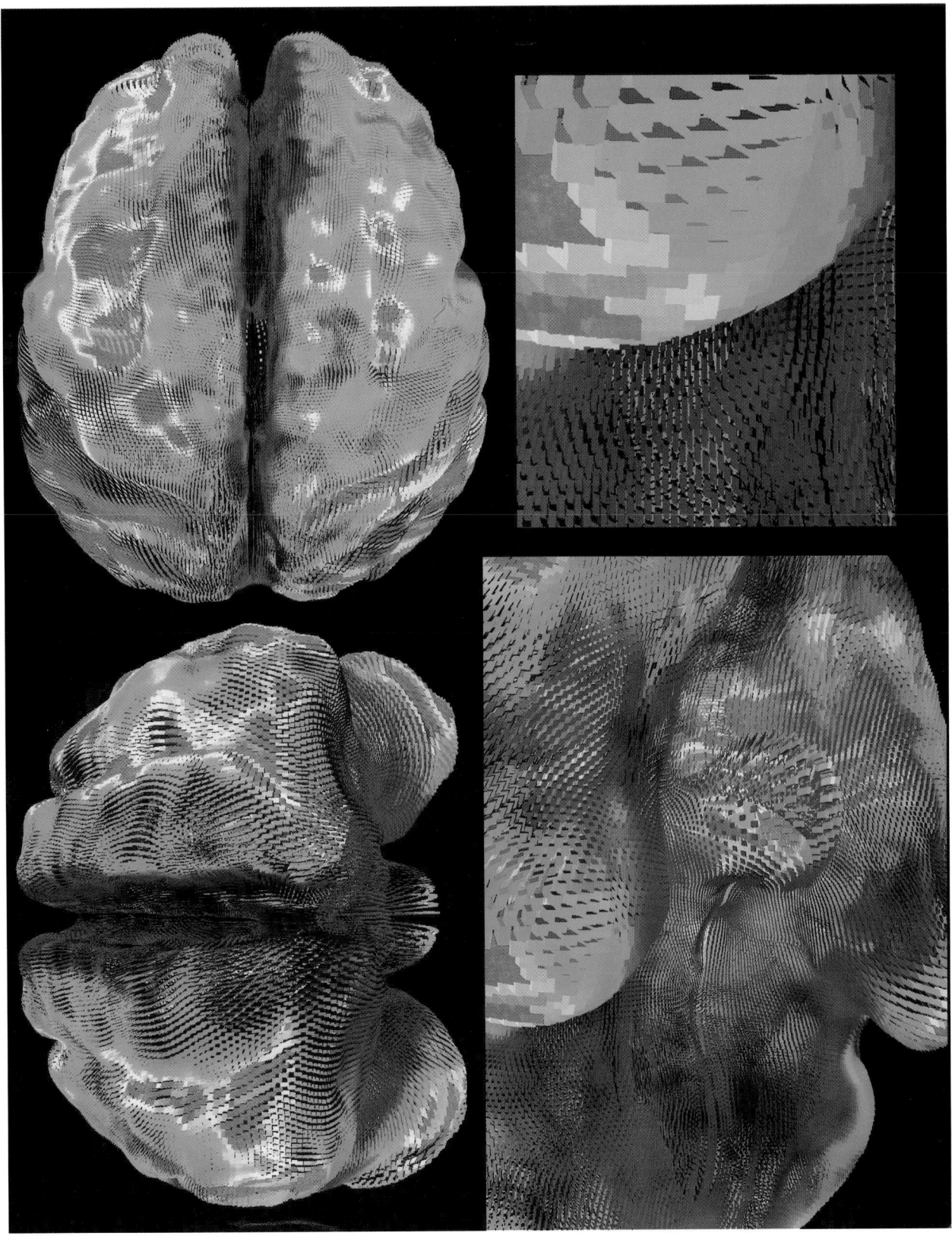

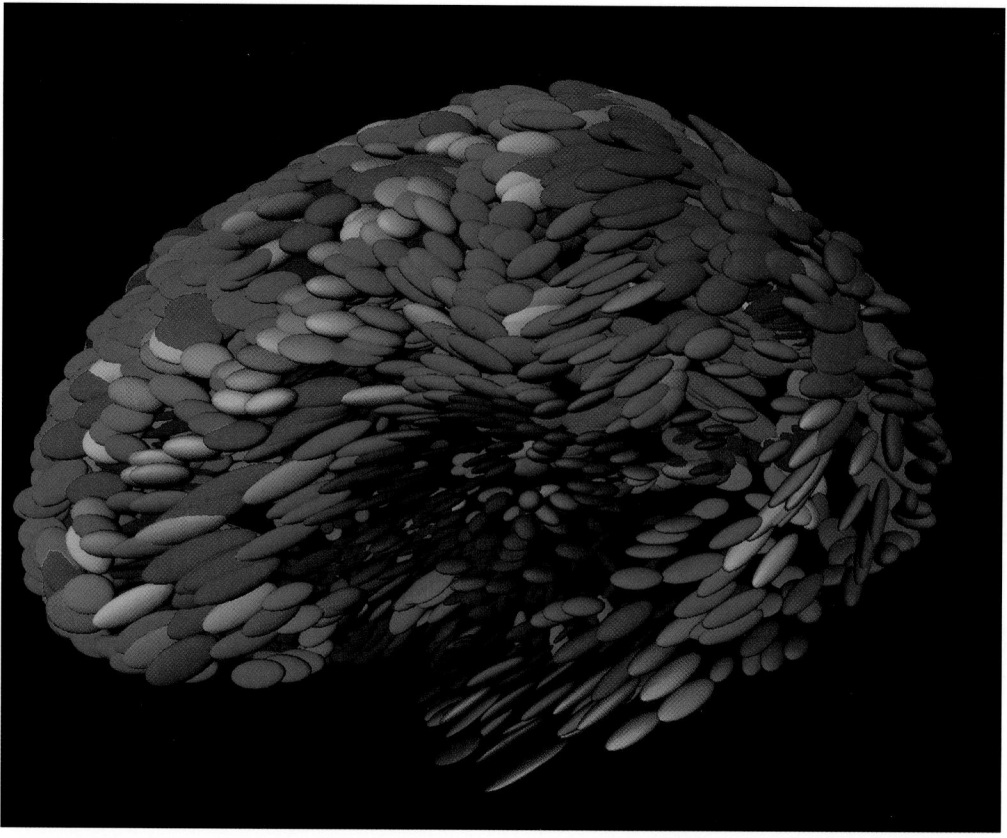

Figure 20 Confidence limits on normal anatomic variation: tensor field representation. Again, a tensor map reveals the preferred directions of cortical variation, after sulcal pattern correspondences are taken into account. Variability is greatest in temporoparietal cortex. If cortical variations are modeled as vector field displacements of an average cortical model, ellipsoids of constant probability density can be computed for positions of cortical regions (relative to the average cortex). These ellipsoids are shown, colored by the determinant of the covariance tensor. Fields of tumbling ellipses have also been used to visualize multidirectional parameters in diffusion imaging data and offer a means to represent cortical variability for anomaly detection and Bayesian image labeling.

zation or lateralization and their dependencies on genetic parameters (Kikinis *et al.,* 1994; cf. Csernansky *et al.,* 1998; Le Goualher *et al.,* 1999).

D. Abnormal Asymmetry

In an interesting development, Thirion *et al.* (1998) applied a warping algorithm to a range of subjects' scans, in each case matching each brain hemisphere with a reflected version of the opposite hemisphere. The resulting asymmetry fields were treated as observations from a spatially parameterized random vector field, and deviations due to lesion growth or ventricular enlargement were detected using the theory developed in Thompson *et al.* (1997). Due to the asymmetric progression of many degenerative disorders (Thompson *et al.,* 1998a), abnormal asymmetry may prove to be a sensitive index of pathology in individual subjects or groups. From a more practical standpoint, asymmetry fields are smaller in magnitude than subject-to-subject deforma-

tion maps. This makes the fields easier to estimate with automated nonlinear registration algorithms (Section IV). When the estimated deformation is small, it is easier to avoid false, nonglobal minima of the matching measure being optimized.

X. Brain Averaging

A. Average Image Templates

So far, we have described a scheme to create average anatomical models for specific patient groups. By assembling these average models for a wide range of systems (cortex, hippocampus, ventricles, deep sulci, and basal ganglia), an annotated atlas of structures can be built. Nonetheless, before new data can be pooled into the atlas, an average intensity image template is also required that reflects the unique morphology of the diseased population. This makes it easier for automated,

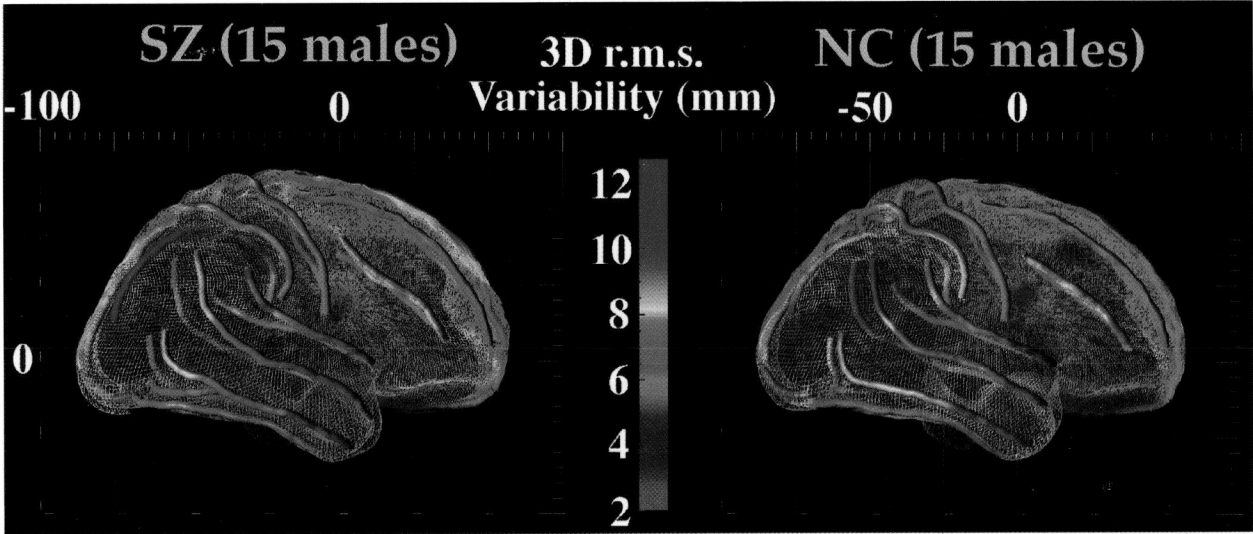

Figure 21 Anatomical variability of the cerebral cortex (N = 15, schizophrenia patients, and N = 15, matched controls; all males) [data from Narr *et al.* (1999)]. 3D maps of variability are shown on average surface representations of the cortex derived from a schizophrenia (left) and normal control population (right). In frontal association cortex (red colors), schizophrenic patients exhibit greater individual variations in gyral patterns. Contrary to several recent research reports, the marked brain asymmetry in temporoparietal cortex is clearly apparent in the schizophrenic patients as well as in controls. Again, variability is calculated based on 3D displacement maps, which locally encode the amount of deformation required to drive each subject's gyral pattern into exact correspondence with the average cortex for the group (see Section VIII).

intensity-based registration algorithms to align new data with the atlas.

To create a mean image template for a group, several approaches are possible. Which one is used depends on the application objectives. We describe a particular approach that guarantees that the average template has (1) well-resolved cortical features (Thompson *et al.*, 1999b) and (2) the average size and shape for a subject group (Woods *et al.*, 1998). To create an atlas template that is consistent with an average set of anatomical models, high-dimensional model-based registration is required. If scans are mutually aligned with only a linear transformation, the resulting average brain is blurred in the more variable anatomical regions. The resulting average brain also tends to exceed the average dimensions of the component brain images.

By averaging geometric and intensity features separately (cf. Ge *et al.*, 1995; Bookstein, 1997; Grenander and Miller, 1998; Thompson *et al.*, 1999), a template can be made with the mean intensity and geometry for a patient population. We illustrate this approach by using the cortical transformations defined above (Section IX) to create a well-resolved disease-specific image template for an Alzheimer's disease population.

First, a group of well-characterized Alzheimer's patients was selected, for whom a range of anatomical surface models (84 per brain) had been created in prior

morphometric projects (Thompson *et al.*, 1998). An initial image template for the group was constructed by (1) using automated linear transformations (Woods *et al.*, 1993) to align the MRI data with a randomly selected image, (2) intensity-averaging the aligned scans, and then (3) recursively reregistering the scans to the resulting average affine image. The resulting average image was adjusted to have the mean affine shape for the group (Woods *et al.*, 1998). Images and surface models were then linearly aligned to this template, and an average surface set was created for the group (Thompson *et al.*, 1997). Displacement maps (Fig. 22) driving the surface anatomy of each subject into correspondence with the average surface set were then computed and were extended to the full volume with surface-based elastic warping (see Fig. 7; Thompson and Toga, 1996, 1998a). These warping fields reconfigured each subject's 3D image into the average anatomic configuration for the group. By averaging the reconfigured images (after intensity normalization), a crisp image template was created to represent the group (Fig. 23). Note the better resolved cortical features in the average images after high-dimensional cortical registration. If desired, this AD-specific atlas can retain the coordinate matrix of the Talairach system (with the AC at (0, 0, 0)) while refining the gyral map of the Talairach atlas to encode the unique anatomy of the AD population. By explicitly computing matching

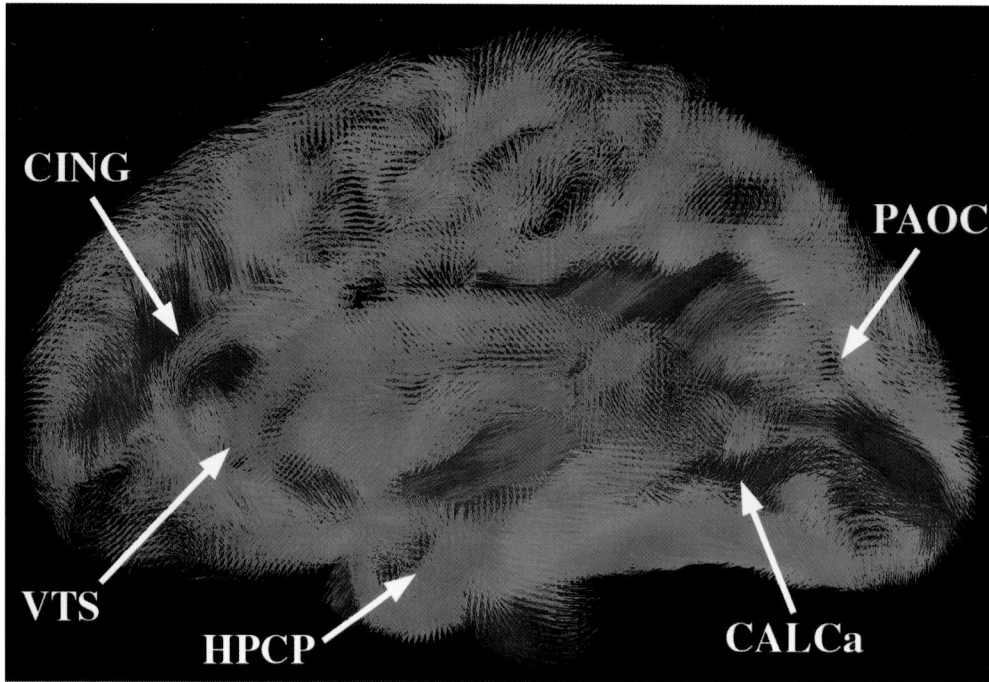

Figure 22 Mapping a patient into the group average configuration. Instead of matching just the cortex, this figure shows the complex transformation required to match 84 different surface models in a given patient, after affine alignment, into the configuration of an average surface set derived for the group (see Thompson *et al.*, 1999, for details). The effects of several anatomic surfaces driving the transformation are indicated, including the cingulate sulcus (CING), hippocampal surface (HPCP), superior ventricular horn (VTS), parieto-occipital sulcus (PAOC), and the anterior calcarine fissure (CALCa). This surface-based vector field is extended to a full volumetric transformation field (0.1 billion degrees of freedom) which reconfigures the anatomy of the patient into correspondence with the average configuration for the group. Storage of these mappings allows quantification of local anatomic variability.

fields that relate gyral patterns across subjects, a well-resolved and spatially consistent set of probabilistic anatomical models and average images can be made to represent the average anatomy and its variation in a subpopulation.

B. Disease Progression

The templates so far described, for the dementia and schizophrenia populations, have been based on homogeneous patient groups, matched for age, gender, handedness, and educational level. Since AD, in particular, is a progressive disease (see next section), the initial atlas template was created to reflect a particular stage in the disease (MMSE score: 19.3 ± 2.0). At this stage, patients often present for initial evaluation, and MR, PET, and SPECT scans have maximal diagnostic value. Nonetheless, by expanding the underlying patient database, atlases are under construction to represent the more advanced stages of Alzheimer's disease. By stratifying the population according to different criteria, different at-

lases can be synthesized to represent other clinically defined groups.

C. Image Distortion and Registration Accuracy

Since the anatomy of a dementia population is poorly reflected by current imaging templates, substantially less distortion will be applied by mapping multimodality brain data into an atlas that reflects AD morphology (Mega *et al.*, 1997; Thompson *et al.*, 1998a; Woods *et al.*, 1998). Incoming subjects deviate least from the mean template in terms of both image intensity and anatomy. Registration of their imaging data to this template therefore requires minimal image distortion. Since the template has the average affine shape for the group (Woods *et al.*, 1998), least distortion is applied when either linear or nonlinear approaches are used. Interestingly, automated registration approaches were able to reduce anatomic variability to a greater degree if a specially prepared image template was used as a

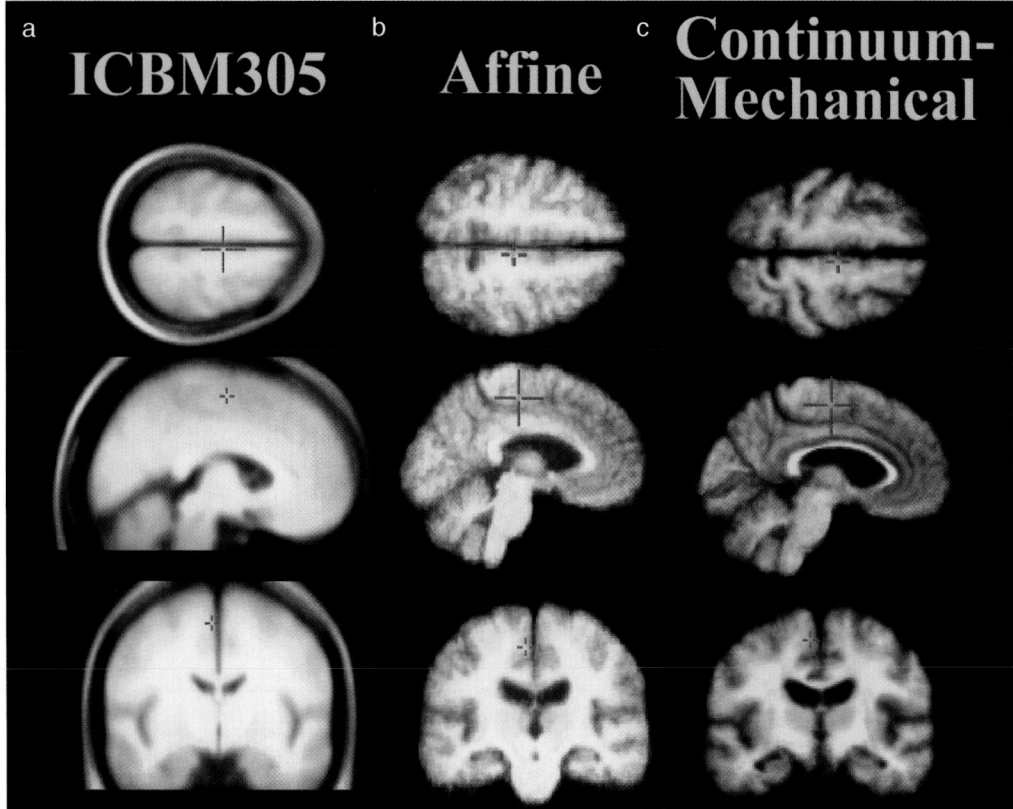

Figure 23 Average brain templates. Axial, sagittal, and coronal images are shown from a variety of population-based brain image templates. For comparison purposes, (a) shows a widely used average-intensity data set (ICBM305) based on 305 young normal subjects, created by the International Consortium for Brain Mapping (Evans *et al.,* 1994); by contrast, templates (b) and (c) are average brain templates created from high-resolution 3D MRI scans of Alzheimer's disease patients. (b) Affine brain template, constructed by averaging normalized MR intensities on a voxel-by-voxel basis after automated affine registration. (c) Continuum-mechanical brain template, based on intensity averaging after continuum-mechanical transformation. By using spatial transformations of increasing complexity, each patient's anatomy can increasingly be reconfigured into the average anatomical configuration for the group. After intensity correction and normalization, the reconfigured scans are then averaged on a pixel-by-pixel basis to produce a group image template with the average geometry and average image intensity for the group. Anatomical features are highly resolved, even at the cortex (c). Transformations of extremely high spatial dimension are required to match cortical features with sufficient accuracy to resolve them after scans are averaged together.

registration target (Woods *et al.,* 1998; Thompson *et al.,* 1999b).

D. Other Average Templates

Several approaches are under active development to create average brain templates. Many of them are based on high-dimensional image transformations. Average templates have been made for the *Macaque* brain (Grenander and Miller, 1998), for individual structures such as the corpus callosum (Davatzikos, 1996; Gee and Bajcsy, 1998), central sulcus (Manceaux-Demiau *et al.,* 1998), cingulate and paracingulate sulci (Paus *et al.,* 1996; Thompson *et al.,* 1997), and hippocampus (Haller *et al.,* 1997; Csernansky *et al.,* 1998; Joshi *et al.,* 1998; Thompson *et al.,* 1999), and for transformed representations of the human and *Macaque* cortex (Drury and Van Essen, 1997; Grenander and Miller, 1998; Fischl *et al.,* 1999; Thompson *et al.,* 1999b). Under various metrics, incoming subjects deviate least from these mean brain templates in terms of both image intensity and anatomy. Registration of new data to these templates not only requires minimal image distortion but also allows faster algorithm convergence. This is because with smaller deformations, nonglobal minima of the registration measure may be avoided, as the parameter space is searched for an optimal match. For these reasons, templates that reflect the mean geometry and intensity of a group are a topic of active research (Grenander and Miller, 1998; Woods *et al.*, 1998; Thompson *et al.*, 1999b).

XI. Dynamic (4D) Brain Atlases

A. 4D Coordinate Systems

Atlasing of data from the developing or degenerating brain presents unique challenges (Toga *et al.*, 1996). However, warping algorithms can be applied to serial scan data to track disease and growth processes in their full spatial and temporal complexity. Maps of anatomical change can be generated by warping scans acquired from the same subject over time (Thirion and Calmon, 1997; Thompson *et al.*, 1998b). Serial scanning of human subjects (Fox *et al.*, 1996; Freeborough *et al.*, 1996, 1998; Subsol *et al.*, 1997; Thompson *et al.*, 1998b) or experimental animals (Jacobs and Fraser, 1994) in a dynamic state of disease or development offers the potential to create 4D models of brain structure. These models incorporate dynamic descriptors of how the brain changes during maturation or disease. They are therefore of interest for investigating and staging brain development. In an atlas setting, these four-dimensional maps can act as normative data to define aberrant growth rates and their modulation by therapy (Haney *et al.*, 1999).

In our initial human studies (Thompson *et al.*, 1998b, 1999), we developed several algorithms to create 4D quantitative maps of growth patterns in the developing human brain. Time series of high-resolution pediatric MRI scans were analyzed. The resulting tensor maps of growth provided spatially detailed information on local growth patterns, quantifying rates of tissue maturation, atrophy, shearing, and dilation in the dynamically changing brain architecture. Pairs of scans were selected to determine patterns of structural change across the interscan interval. Deformation processes recovered by a high-dimensional warping algorithm were then analyzed using vector field operators to produce a variety of tensor maps (Fig. 24). These maps were designed to reflect the magnitude and principal directions of dilation or contraction, the rate of strain, and the local curl, divergence, and gradient of flow fields representing the growth processes recovered by the transformation.

The growth maps obtained in these studies exhibit several striking characteristics. First, foci of rapid growth at the callosal isthmus appeared consistently across puberty. These rates appeared to attenuate as subjects progressed into adolescence (Fig. 24). Rapid

rates of tissue loss were also revealed at the head of the caudate, in an earlier phase of development (Fig. 25).

In the near future, 4D atlases will be able to map growth and degeneration in their full spatial and temporal complexity. Despite the logistic and technical challenges, these mapping approaches hold tremendous promise in analyzing the dynamics of degenerative or neoplastic diseases. They will ultimately play a role in detecting how different therapeutic approaches modulate the course of disease.

XII. Conclusion

Encoding patterns of anatomical variation in disease presents significant challenges. By describing an atlasing scheme that treats intensity and geometric variation separately, we described the creation of well-resolved image templates and probabilistic models of anatomy that reflect the average morphology of a group. The continual refinement of anatomic templates is likely to be leveraged by algorithms for deformation-based morphometry in large image databases (Thompson *et al.*, 1997; Ashburner *et al.*, 1998) and by next-generation probabilistic atlases. Atlas data on anatomic variability can also act as Bayesian prior information to guide algorithms for automated image registration and labeling (Ashburner *et al.*, 1998; Gee and Bajcsy, 1998; Pitiot *et al.*, 1999). The resulting atlases are expandable in every respect and may be stratified into subpopulations according to clinical, demographic, or genetic criteria.

We also described approaches for creating and averaging brain models. These techniques produce statistical maps of group differences, abnormalities, and patterns of variation and asymmetry. These maps and models are key components of disease-specific brain atlases. We also described registration algorithms that transfer postmortem maps into an atlas, to correlate them with functional and metabolic data. The result is a multimodality atlas that relates cognitive and functional measures with the cellular and pathologic hallmarks of the disease (Fig. 1; Mega *et al.*, 1997).

As well as disease-specific atlases reflecting brain structure in dementia and schizophrenia (Narr *et al.*, 1999; Thompson *et al.*, 1999a,b), research is underway to build dynamic brain atlases that retain probabilistic information on growth rates in development and degeneration (Thompson *et al.*, 1999a,b). Refinement of these atlas systems to support dynamic and disease-specific data should generate an exciting framework to investigate variations in brain structure and function in large human populations.

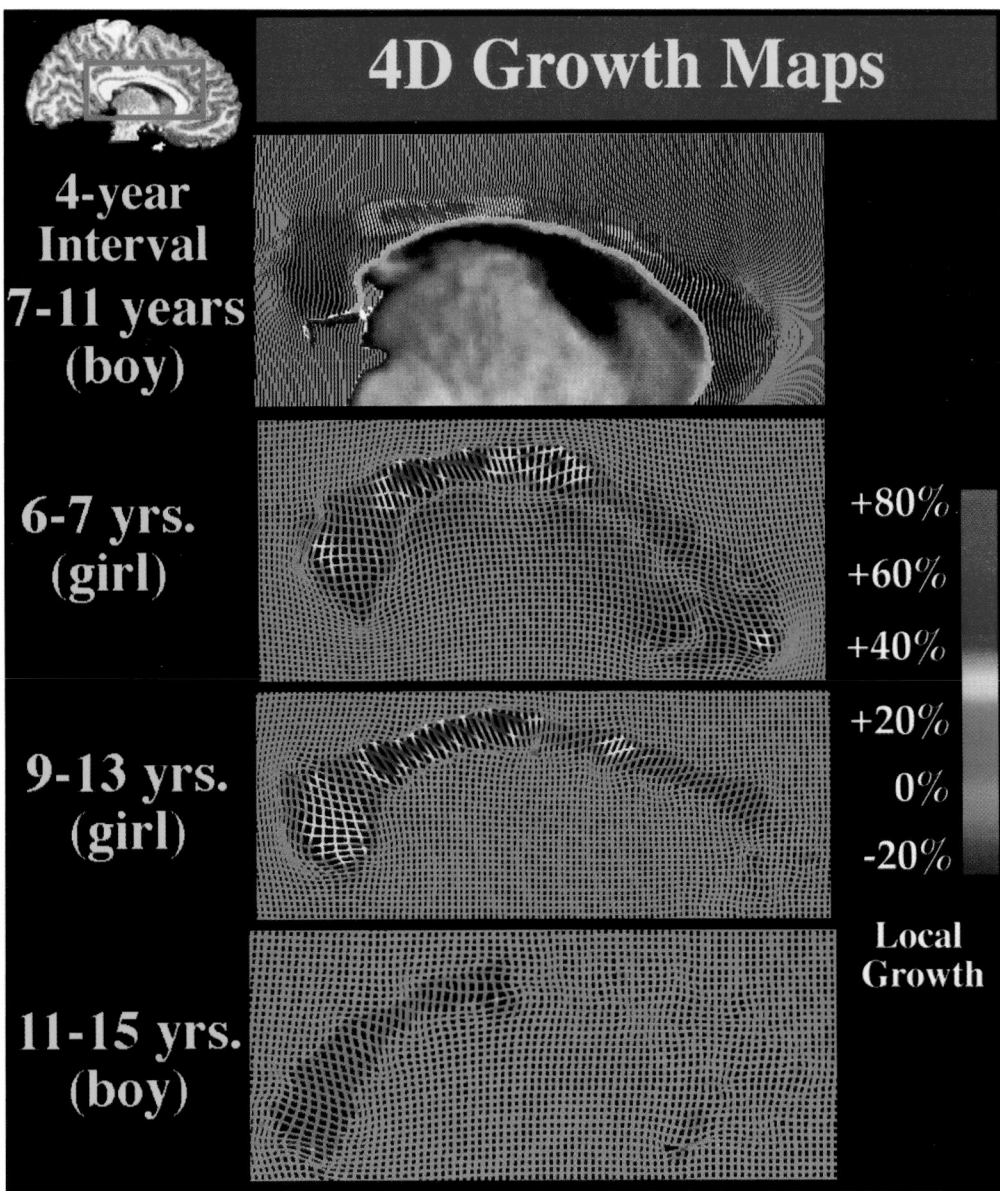

Figure 24 Tensor maps of growth. (Top panel) A complex pattern of growth is detected in the corpus callosum of a young normal male subject in the 4-year period from 7 to 11 years of age. Vector field operators emphasize patterns of contractions and dilations, emphasizing their regional character. The color code shows values of the local Jacobian of the warping field, which indicates local volume loss or gain. The effects of the transformation are shown on a regular grid ruled over the reference anatomy and passively carried along in the transformation that matches it with the later anatomy. Despite minimal changes in overall cerebral volume, callosal growth is dramatic, with peak values throughout the isthmus and posterior midbody (top panel). Rapid heterogeneous growth, with a similar topographic pattern, is also observed in two young normal females (next two panels), during a 1-year period from the age of 6 to 7 and during a 4-year period spanning puberty, from 9 to 13 years of age. The final panel shows growth mapped in a male subject across a period of 4 years from 11 to 15 years of age. Growth rates have clearly attenuated, but have a pattern similar to that of the younger subjects.

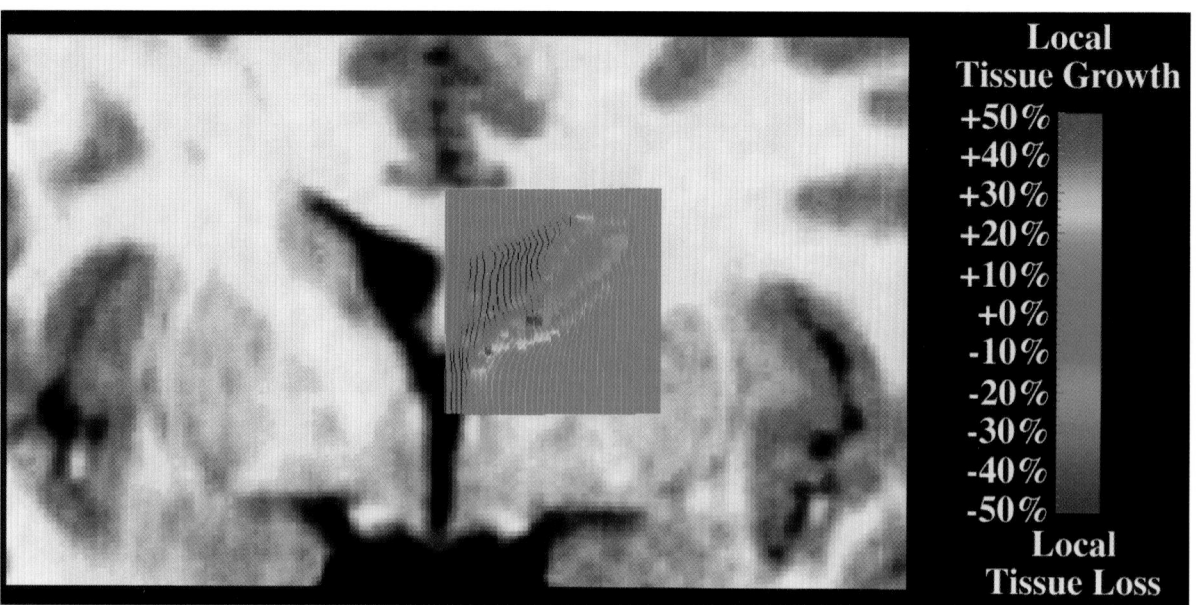

Figure 25 3D patterns of deep nuclear tissue loss. In the 4-year period from 7 to 11 years of age, a young normal male subject displays a local 50% tissue loss at the caudate head as well as a 20–30% growth of the internal capsule and a 5–10% dilation of the superior ventricular horn (Thompson *et al.*, 1999). Graphical visualizations of growth rates indicate the regional complexity of the growth processes between the two scans.

Acknowledgments

This work was supported by the National Center for Research Resources (P41 RR13642 and RR05956), the National Institute of Neurological Disorders and Stroke and the National Institute of Mental Health (NINDS/NIMH NS38753), and a Human Brain Project grant to the International Consortium for Brain Mapping, funded jointly by NIMH and NIDA (P20 MH/DA52176). Additional support was provided by the United States Information Agency (Grant G-1-00001), the Howard Hughes Medical Institute, and the U.S.–U.K. Fulbright Commission and by the National Library of Medicine (LM/MH05639) and National Science Foundation (BIR 93-22434). Special thanks go to our colleagues Roger Woods, Colin Holmes, Jay Giedd, David MacDonald, Alan Evans, and John Mazziotta, whose advice and support have been invaluable in these investigations.

References

Amit, Y., Grenander, U., and Piccioni, M. (1991). Structural image restoration through deformable templates. *J. Am. Stat. Assoc.* **86**(414), 376–386.

Amunts, K., Schlaug, G., Schleicher, A., Steinmetz, H., Dabringhaus, A., Roland, P. E., and Zilles, K. (1996). Asymmetry in the human motor cortex and handedness. *NeuroImage* **4**(3, Part 1), 216–222.

Andreasen, N. C., Arndt, S., Swayze, V., Cizadlo, T., Flaum, M., O'Leary, D., Ehrhardt, J. C., and Yuh, W. T. C. (1994). Thalamic abnormalities in schizophrenia visualized through magnetic resonance image averaging. *Science* **266**, 294–298.

Arnold, S. E., Hyman, B. T., Flory, J., Damasio, A. R., and Van Hoesen, G. W. (1991). The topographical and neuroanatomical distribution of neurofibrillary tangles and neuritic plaques in the cerebral cortex of patients with Alzheimer's disease. *Cereb. Cortex* **1**, 103–116.

Ashburner, J., Neelin, P., Collins, D. L., Evans, A. C., and Friston, K. J. (1997). Incorporating prior knowledge into image registration. *NeuroImage* **6**(4), 344–352.

Ashburner, J., Hutton, C., Frackowiak, R., Johnsrude, I., Price, C., and Friston, K. (1998). Identifying global anatomical differences: Deformation-based morphometry. *Hum. Brain Mapp.* **6**(5–6), 348–357.

Aso, M., Kurachi, M., Suzuki, M., Yuasa, S., Matsui, M., and Saitoh, O. (1995). Asymmetry of the ventricle and age at the onset of schizophrenia. *Eur. Arch. Psychiatry Clin. Neurosci.* **245**(3), 142–144.

Aylward, E. H., Harris, G. J., Hoehn-Saric, R., Barta, P. E., Machlin, S. R., and Pearlson, G. D. (1996). Normal caudate nucleus in obsessive-compulsive disorder assessed by quantitative neuroimaging. *Arch. Gen. Psychiatry* **53**(7), 577–584.

Bajcsy, R., and Kovacic, S. (1989). Multiresolution elastic matching. Comput. *Vision, Graph. Image Process.* **46**, 1–21.

Bakircioglu, M., Joshi, S. C., and Miller, M. I. (1999). Landmark matching on brain surfaces via large deformation diffeomorphisms on the sphere. *Proc. SPIE Med. Imaging.*

Baumgardner, T. L., Singer, H. S., Denckla, M. D., Rubin, B. S., Abrams, B. A., Colli, M. J., and Reiss, A. L. (1996). Corpus callosum morphology in children with Tourette syndrome and attention deficit hyperactivity disorder. *Neurology* **47**, 477–482.

Beaton, A. A. (1997). The relation of planum temporale asymmetry and morphology of the corpus callosum to handedness, gender and dyslexia: A review of the evidence. *Brain Lang.* **60**, 255–322.

Berg, L., McKeel, D. W., Jr., Miller, J. P., Baty, J., and Morris, J. C. (1993). Neuropathological indexes of Alzheimer's disease in demented and nondemented persons aged 80 years and older. *Arch. Neurol.* **50**(4), 349–358.

Bilder, R. M., Wu, H., Chakos, M. H., Bogerts, B., Pollack, S., Aronowitz, J., Ashtari, M., Degreef, G., Kane, J. M., and Lieberman, J. A. (1994). Cerebral morphometry and clozapine treatment in schizophrenia. *J. Clin. Psychiatry* **55** (Suppl. B), 53–56.

Bishop, K. M., and Wahlsten, D. (1997). Sex differences in the human corpus callosum: Myth or reality? *Neurosci. Biobehav. Rev.* **21**, 581–601.

Blanton, R. E., Levitt, J., Thompson, P. M., Badrtalei, S., Capetillo-Cunliffe, L., and Toga, A. W. (1999). Average 3-Dimensional Caudate Surface Representations in a Juvenile-Onset Schizophrenia and Normal Pediatric Population, Presentation #621, 5th International Conference on Functional Mapping of the Human Brain, Dusseldorf, Germany, June 1999.

Bohm, C., Greitz, T., Kingsley, D., Berggren, B. M., and Olsson, L. (1983). Adjustable computerized brain atlas for transmission and emission tomography. *Am. J. Neuroradiol.* **4,** 731–733.

Bookstein, F. L. (1997). Landmark methods for forms without landmarks: Morphometrics of group differences in outline shape. *Med. Image Anal.* **1**(3), 225–243.

Braak, H., and Braak, E. (1991). Neuropathological staging of Alzheimer-related changes. *Acta Neuropathol.* **82,** 239–259.

Brodmann, K. (1909). Vergleichende Lokalisationslehre der Grosshirnrinde in ihren Prinzipien dargestellt auf Grund des Zellenbaues, Barth, Leipzig [translated as: On the comparative localization of the cortex. *In* "Some Papers on the Cerebral Cortex," pp. 201–230. Thomas, Springfield, IL, 1960].

Broit, C. (1981). "Optimal Registration of Deformed Images," Ph.D. Dissertation, Univ. of Pennsylvania.

Bro-Nielsen, M., and Gramkow, C. (1996). Fast fluid registration of medical images. *In* "Visualization in Biomedical Computing" (K. H. Höhne and R. Kikinis, Eds.). Springer-Verlag, Berlin/New York. *Lect. Notes Comput. Sci.* **1131,** 267–276.

Brun, A., and Englund, E. (1981). Regional pattern of degeneration in Alzheimer's disease: Neuronal loss and histopathologic grading. *Histopathology* **5,** 549–564.

Burke, H. L., Yeo, R. A., Delaney, H. D., and Conner, L. (1993). CT scan cerebral hemispheric asymmetries: Predictors of recovery from aphasia. *J. Clin. Exp. Neuropsychol.* **15**(2), 191–204.

Cao, J., and Worsley, K. J. (1999). The geometry of the Hotelling's T-squared random field with applications to the detection of shape changes. *Ann. Stat.* (in press).

Carpenter, W. T., Buchanan, R. W., Kirkpatrick, B., Tamminga, C., and Wood, F. (1993). Strong inference, theory testing, and the neuroanatomy of schizophrenia. *Arch. Gen. Psychiatry* **50,** 825–831.

Casanova, M. F., Sanders, R. D., Goldberg, T. E., Bigelow, L. B., Christison, G., Torrey, E. F., and Weinberger, D. R. (1990). Morphometry of the corpus callosum in monozygotic twins discordant for schizophrenia: A magnetic resonance imaging study. *J. Neurol. Neurosurg. Psychiatry* **53,** 416–421.

Castellanos, F. X., Giedd, J. N., Marsh, W. L., Hamburger, S. D., Vaituzis, A. C., Dickstein, D. P., Sarfatti, S. E., Vauss, Y. C., Snell, J. W., Lange, N., Kaysen, D., Krain, A. L., Ritchie, G. F., Rajapakse, J. C., and Rapoport, J. L. (1996). Quantitative brain magnetic resonance imaging in attention-deficit hyperactivity disorder. *Arch. Gen. Psychiatry* **53**(7), 607–616.

Chakos, M. H., Lieberman, J. A., Bilder, R. M., Borenstein, M., Lerner, G., Bogerts, B., Wu, H., Kinon, B., and Ashtari, M. (1994). Increase in caudate nuclei volumes of first-episode schizophrenic patients taking antipsychotic drugs. *Am. J. Psychiatry* **151**(10), 1430–1436.

Christensen, G. E., Rabbitt, R. D., and Miller, M. I. (1993). A deformable neuroanatomy textbook based on viscous fluid mechanics. *27th Annu. Conf. Inf. Sci. Syst.* 211–216.

Christensen, G. E., Miller, M. I., Marsh, J. L., and Vannier, M. W. (1995a). Automatic analysis of medical images using a deformable textbook. *Proc. Comput. Assist. Radiol.* 152–157.

Christensen, G. E., Rabbitt, R. D., Miller, M. I., Joshi, S. C., Grenander, U., Coogan, T. A., and Van Essen, D. C. (1995b). Topological properties of smooth anatomic maps. *In* "Information Processing in Medical Imaging" (Y. Bizais, C. Barillot, and R. Di Paola, Eds.), pp. 101–112. Kluwer.

Christensen, G. E., Rabbitt, R. D., and Miller, M. I. (1996). Deformable templates using large deformation kinematics. *IEEE Trans. Image Process.* **5**(10), 1435–1447.

Clarke, J. M., and Zaidel, E. (1994). Anatomical–behavioral relationships: Corpus callosum morphometry and hemispheric specialization. *Behav. Brain Res.* **64,** 185–202.

Clarke, S., Kraftsik, R., Van der Loos, H., and Innocenti, G. M. (1989). Forms and measures of adult and developing human corpus callosum. *J. Neuropathol. Exp. Neurol.* **280,** 213–230.

Cohen, L. D., and Cohen, I. (1992). Deformable models for 3D medical images using finite elements and balloons. *In* "Proceedings: 1992 IEEE Computer Society Conference on Computer Vision and Pattern Recognition" (Cat. No. 92CH3168-2), pp. 592–598. IEEE Comput. Soc., Los Alamitos, CA.

Collins, D. L., Neelin, P., Peters, T. M., and Evans, A. C. (1994a). Automatic 3D intersubject registration of MR volumetric data into standardized Talairach space. *J. Comput. Assist. Tomogr.* **18**(2), 192–205.

Collins, D. L., Peters, T. M., and Evans, A. C. (1994b). An automated 3D non-linear image deformation procedure for determination of gross morphometric variability in the human brain. *Proc. Vis. Biomed. Comput. (SPIE)* **3,** 180–190.

Collins, D. L., Holmes, C. J., Peters, T. M., and Evans, A. C. (1995). Automatic 3D model-based neuroanatomical segmentation. *Hum. Brain Mapp.* **3,** 190–208.

Collins, D. L., Le Goualher, G., Venugopal, R., Caramanos, A., Evans, A. C., and Barillot, C. (1996). Cortical constraints for non-linear cortical registration. *In* "Visualization in Biomedical Computing" (K. H. Höhne and R. Kikinis, Eds.). Springer-Verlag, Berlin/New York. *Lect. Notes Comput. Sci.* **1131,** 307–316.

Colombo, C., Bonfanti, A., Livian, S., Abbruzzese, M., and Scarone, S. (1993). Size of the corpus callosum and auditory comprehension in schizophrenics and normal controls. *Schizophr. Res.* **11**(1), 63–70.

Cook, M. J., Free, S. L., Fish, D. R., Shorvon, S. D., Straughan, K., and Stevens, J. M. (1994). Analysis of cortical patterns. *In* "Magnetic Resonance Scanning and Epilepsy" (S. D. Shorvon, Ed.), pp. 263–274. Plenum, New York.

Courchesne, E. (1997). Brainstem, cerebellar and limbic neuroanatomical abnormalities in autism. *Curr. Opin. Neurobiol.* **7**(2), 269–278.

Courchesne, E., Saitoh, O., Yeung-Courchesne, R., Press, G. A., Lincoln, A. J., Haas, R. H., and Schreibman, L. (1994). Abnormality of cerebellar vermian lobules VI and VII in patients with infantile autism: Identification of hypoplastic and hyperplastic subgroups with MR imaging. *AJR, Am. J. Roentgenol.* **162**(1), 123–130.

Csernansky, J. G., Joshi, S., Wang, L., Haller, J. W., Gado, M., Miller, J. P., Grenander, U., and Miller, M. I. (1998). Hippocampal morphometry in schizophrenia by high dimensional brain mapping. *Proc. Natl. Acad. Sci. U.S.A.* **95**(19), 11406–11411.

Cuénod, C. A., Denys, A., Michot, J. L., Jehenson, P., Forette, F., Kaplan, D., Syrota, A., and Boller, F. (1993). Amygdala atrophy in Alzheimer's disease: An in vivo magnetic resonance imaging study. *Arch. Neurol.* **50,** 941–945.

Cunningham, D. J. (1892). Contribution to the surface anatomy of the cerebral hemispheres. *Cunningham Mem. (R. Irish Acad.)* **7,** 372.

Dale, A. M., and Sereno, M. I. (1993). Improved localization of cortical activity by combining EEG and MEG with MRI cortical surface reconstruction—A linear approach. *J. Cog. Neuro.* **5**(2), 162–176.

Davatzikos, C. (1996). Spatial normalization of 3D brain images using deformable models. *J. Comput. Assist. Tomogr.* **20**(4), 656–665.

Davatzikos, C., Vaillant, M., Resnick, S. M., Prince, J. L., Letovsky, S., and Bryan, R. N. (1996). A computerized approach for morphological analysis of the corpus callosum. *J. Comput. Assist. Tomogr.* **20**(1), 88–97.

Davidson, R. J., and Hugdahl, K. (1994). "Brain Asymmetry." MIT Press, Cambridge, MA.

DeLacoste-Utamsing, M. C., and Holloway, R. L. (1982). Sexual dimorphism in the human corpus callosum. *Science* **216,** 1431–1432.

DeLisi, L. E., Dauphinais, I. D., and Hauser, P. (1989). Gender differences in the brain: Are they relevant to the pathogenesis of schizophrenia? *Comp. Psychiatry* **30,** 197–208.

DeLisi, L. E., Hoff, A. L., Neale, C., and Kushner, M. (1994). Asymmetries in the superior temporal lobe in male and female first-episode schizophrenic patients: Measures of the planum temporale and superior temporal gyrus by MRI. *Schizophr. Res.* **12,** 19–28.

Dengler, J., and Schmidt, M. (1988). The dynamic pyramid—A model for motion analysis with controlled continuity. *Int. J. Patt. Recog. Artif. Intell.* **2**(2), 275–286.

DeQuardo, J. R., Bookstein, F. L., Green, W. D., Brunberg, J. A., and Tandon, R. (1996). Spatial relationships of neuroanatomic landmarks in schizophrenia. *Psychiatry Res.* **67**(1), 81–95.

Dinov, I. D., Thompson, P. M., Woods, R. P., Mega, M. S., Holmes, C. J., Sumners, D., Saxena, S., and Toga, A. W. (1999). Probabilistic subvolume partitioning techniques for determining the statistically significant regions of activation in stereotaxic functional data. *J. Comput. Assist. Tomogr.* (in press).

Drury, H. A., Van Essen, D. C., Joshi, S. C., and Miller, M. I. (1996). Analysis and Comparison of Areal Partitioning Schemes Using Two-Dimensional Fluid Deformations, Poster Presentation, 2nd International Conference on Functional Mapping of the Human Brain, Boston, June 17–21, 1996, *NeuroImage* **3,** S130.

Drury, H. A., and Van Essen, D. C. (1997). Analysis of functional specialization in human cerebral cortex using the visible man surface based atlas. *Hum. Brain Mapp.* **5,** 233–237.

Duara, R., Kushch, A., Gross-Glenn, K., Barker, W. W., Jallad, B., Pascal, S., Loewenstein, D. A., Sheldon, J., Rabin, M., and Levin, B. (1991). Neuroanatomic differences between dyslexic and normal readers on magnetic resonance imaging scans. *Arch. Neurol.* **48,** 410–416.

Dupuis, P., Grenander, U., and Miller, M. I. (1998). Variational problems on flows of diffeomorphisms for image matching. *Q. Appl. Math.* **56**(3), 587–600.

Eberstaller, O. (1884). Zür Oberflachen Anatomie der Grosshirn Hemisphaeren. *Wien Med. Bl.* **7,** 479, 642, 644.

Eidelberg, D., and Galaburda, A. M. (1982). Symmetry and asymmetry in the human posterior thalamus: I. Cytoarchitectonic analysis in normal persons. *Arch. Neurol.* **39**(6), 325–332.

Einstein, A. (1914). Covariance properties of the field equations of the theory of gravitation based on the generalized theory of relativity [published as: Kovarianzeigenschaften der Feldgleichungen der auf die verallgemeinerte Relativitätstheorie gegründeten Gravitationstheorie. *Z. Math. Phys.* **63,** 215–225].

Erkinjuntti, T., Lee, D. H., Gao, F., Steenhuis, R., Eliasziw, M., Fry, R., Merskey, H., and Hachinski, V. C. (1993). Temporal lobe atrophy on magnetic resonance imaging in the diagnosis of early Alzheimer's disease. *Arch. Neurol.* **50**(3), 305–310.

Evans, A. C., Dai, W., Collins, D. L., Neelin, P., and Marrett, S. (1991). Warping of a computerized 3D atlas to match brain image volumes for quantitative neuroanatomical and functional analysis. *SPIE Med. Imaging* **1445,** 236–247.

Evans, A. C., Collins, D. L., and Milner, B. (1992). An MRI-based stereotactic brain atlas from 300 young normal subjects. *In* "Proceedings of the 22nd Symposium of the Society for Neuroscience, Anaheim," Paper 408.

Evans, A. C., Collins, D. L., Neelin, P., MacDonald, D., Kamber, M., and Marrett, T. S. (1994). Three-dimensional correlative imaging: Applications in human brain mapping. *In* "Functional Neuroimaging: Technical Foundations" (R. W. Thatcher, M. Hallett, T. Zeffiro, E. R. John, and M. Huerta, Eds.), pp. 145–162, Academic Press, San Diego.

Falkai, P., Bogerts, B., Greve, B., Pfeiffer, U., Machus, B., Folsch-Reetz, B., Majtenyi, C., and Ovary, I. (1992). Loss of sylvian fissure asymmetry in schizophrenia. A quantitative post mortem study. *Schizophr. Res.* **7**(1), 23–32.

Filipek, P. A. (1995). Quantitative magnetic resonance imaging in autism: The cerebellar vermis. *Curr. Opin. Neurol.* **8**(2), 134–138.

Filipek, P. A. (1996). Brief report: Neuroimaging in autism–The state of the science. *J. Autism Dev. Disord.* **26**(2), 211–215.

Fischl, B., Sereno, M. I., Tootell, R. B. H., and Dale, A. M. (1999). High-resolution inter-subject averaging and a coordinate system for the cortical surface. *Human Brain Mapping* **8**(4), 272–284.

Fox, N. C., Freeborough, P. A., and Rossor, M. N. (1996). Visualization and quantification of rates of cerebral atrophy in Alzheimer's disease. *Lancet* **348**(9020), 94–97.

Fox, P. T., Perlmutter, J. S., and Raichle, M. (1985). A stereotactic method of localization for positron emission tomography. *J. Comput. Assist. Tomogr.* **9**(1), 141–153.

Fox, P. T., Mintun, M. A., Reiman, E. M., and Raichle, M. E. (1988). Enhanced detection of focal brain responses using inter-subject averaging and change distribution analysis of subtracted PET images. *J. Cereb. Blood Flow Metab.* **8,** 642–653.

Fox, P. T., Mikiten, S., Davis, G., and Lancaster, J. L. (1994). BrainMap: A database of human functional brain mapping. *In* "Functional Neuroimaging: Technical Foundations" (R. W. Thatcher, M. Hallett, T. Zeffiro, E. R. John, and M. Huerta, Eds.), pp. 95–106. Academic Press, San Diego.

Frangou, S., Sharma, T., Sigmudsson, T., Barta, P., Pearlson, G., and Murray, R. M. (1997). The Maudsley Family study. 4. Normal planum temporale asymmetry in familial schizophrenia: A volumetric MRI study. *Br. J. Psychiatry* **170,** 230–233.

Freeborough, P. A., and Fox, N. C. (1998). Modeling brain deformations in Alzheimer disease by fluid registration of serial 3D MR images. *J. Comput. Assist. Tomogr.* **22**(5), 838–843.

Freeborough, P. A., Woods, R. P., and Fox, N. C. (1996). Accurate registration of serial 3D MR brain images and its application to visualizing change in neurodegenerative disorders. *J. Comput. Assist. Tomogr.* **20**(6), 1012–1022.

Friedland, R. P., and Luxenberg, J. (1988). Neuroimaging and dementia. *In* "Clinical Neuroimaging: Frontiers in Clinical Neuroscience" (W. H. Theodore, Ed.), Vol. 4, pp. 139–163. A. R. Liss, New York.

Friston, K. J., Passingham, R. E., Nutt, J. G., Heather, J. D., Sawle, G. V., and Frackowiak, R. S. J. (1989). Localization in PET images: Direct fitting of the intercommissural (AC–PC) line. *J. Cereb. Blood Flow Metab.* **9,** 690–695.

Friston, K. J., Frith, C. D., Liddle, P. F., and Frackowiak, R. S. J. (1991). Plastic transformation of PET images. *J. Comput. Assist. Tomogr.* **9**(1), 141–153.

Friston, K. J., Holmes, A. P., Worsley, K. J., Poline, J. P., Frith, C. D., and Frackowiak, R. S. J. (1995). Statistical parametric maps in functional imaging: A general linear approach. *Hum. Brain Mapp.* **2,** 189–210.

Fuh, J. L., Mega, M. S., Thompson, P. M., Cummings, J. L., and Toga, A. W. (1997). Cortical complexity maps and cognition in Alzheimer's disease. *In* "Proceedings of the 122nd Annual Meeting of the American Neurological Association, 1997."

Gabrani, M., and Tretiak, O. J. (1999). Surface-based matching using elastic transformations. *Patt. Recog.* **32**(1), 87–97.

Galaburda, A. M., and Geschwind, N. (1981). Anatomical asymmetries in the adult and developing brain and their implications for function. *Adv. Pediatr.* **28,** 271–292.

Galaburda, A. M., Rosen, G. D., and Sherman, G. F. (1990). Individual variability in cortical organization: Its relationship to brain laterality and implications to function. *Neuropsychologica* **28,** 529–546.

Gaser, C., Kiebel, S., Riehemann, S., Volz, H.-P., and Sauer, H. (1998). Statistical Parametric Mapping of Structural Changes in Brain—Application to Schizophrenia Research, Poster 718, 4th International Conference on Functional Mapping of the Human Brain, Montreal, 1998.

Ge, Y., Fitzpatrick, J. M., Kessler, R. M., and Jeske-Janicka, M. (1995). Intersubject brain image registration using both cortical and subcortical landmarks. *SPIE Image Processing* **2434,** 81–95.

Gee, J. C., and Bajcsy, R. K. (1998). Elastic matching: Continuum-mechanical and probabilistic analysis. In "Brain Warping" (A. W. Toga, Ed.). Academic Press, San Diego.

Gee, J. C., Reivich, M., and Bajcsy, R. (1993). Elastically deforming an atlas to match anatomical brain images. *J. Comput. Assist. Tomogr.* **17**(2), 225–236.

Gee, J. C., LeBriquer, L., Barillot, C., Haynor, D. R., and Bajcsy, R. (1995). "Bayesian Approach to the Brain Image Matching Problem," Institute for Research in Cognitive Science Technical Report 95-08, April 1995.

Geschwind, N., and Levitsky, W. (1968). Human brain: Left–right asymmetries in temporal speech region. *Science* **161**, 186.

Geschwind, N., and Galaburda, A. M. (1985). Cerebral lateralization. Biological mechanisms, associations and pathology. *Arch. Neurol.* **42**(5):428–459.

Giedd, J. N., Castellanos, F. X., Casey, B. J., Kozuch, P., King, A. C., Hamburger, S. D., and Rapoport, J. L. (1994). Quantitative morphology of the corpus callosum in attention deficit hyperactivity disorder. *Am. J. Psychiatry* **151**, 665–669.

Giedd, J. N., Rumsey, J. M., Castellanos, F. X., Rajapakse, J. C., Kaysen, D., Vaituzis, A. C., Vauss, A. C., Hamburger, S. D., and Rapoport, J. L. (1996). A quantitative MRI study of the corpus callosum in children and adolescents. *Dev. Brain Res.* **91**, 274–280.

Gramkow, C. (1996). ``Registration of 2D and 3D Medical Images," M.Sc. Thesis, Denmark Technical University.

Gramkow, C., and Bro-Nielsen, M. (1997). Comparison of three filters in the solution of the Navier–Stokes equation in registration. *Proc. Scand. Conf. Image Anal. (SCIA'97)* 795–802.

Greitz, T., Bohm, C., Holte, S., and Eriksson, L. (1991). A computerized brain atlas: Construction, anatomical content and application. *J. Comput. Assist. Tomogr.* **15**(1), 26–38.

Grenander, U. (1976). "Pattern Synthesis: Lectures in Pattern Theory." Springer-Verlag, Berlin/New York. *Appl. Math. Sci.* **13**.

Grenander, U., and Miller, M. I. (1994). Representations of knowledge in complex systems. *J. R. Stat. Soc. B* **56**(4), 549–603.

Grenander, U., and Miller, M. I. (1998). ``Computational Anatomy: An Emerging Discipline," Technical Report, Department of Mathematics, Brown University.

Haller, J. W., Banerjee, A., Christensen, G. E., Gado, M., Joshi, S., Miller, M. I., Sheline, Y., Vannier, M. W., and Csernansky, J. G. (1997). Three-dimensional hippocampal MR morphometry with high-dimensional transformation of a neuroanatomic atlas. *Radiology* **202**(2), 504–510.

Haney, S., Thompson, P. M., Alger, J. R., Cloughesy, T. F., and Toga, A. W. (1999). Tracking Tumor Growth Rates in Patients with Malignant Gliomas, 4th Annual Meeting of the Society for Neuro-Oncology, Scottsdale, AZ, November 1999.

Hardy, T. L. (1994). Computerized atlas for functional stereotaxis, robotics and radiosurgery. *SPIE* **2359**, 447–456.

Harrison, P. J. (1999). The neuropathology of schizophrenia. A critical review of the data and their interpretation. *Brain* **122** (Part 4), 593–624.

Hegedüs, K., and Molnár, P. (1996). Neuroanatomy and Neuropathology on the Internet, Department of Neurology, University Medical School, Debrecen, Hungary, http://www.neuropat.dote.hu/.

Hoff, A. L., Roidan, H., O'Donnell, D., Stritzke, P., Neale, C., Boccio, A., Anand, A. K., and DeLisi, L. E. (1992). Anomalous lateral sulcus asymmetry and cognitive function in first-episode schizophrenia. *Schizophr. Bull.* **18**(2), 257–270.

Höhne, K. H., Bomans, M., Pommert, A., Riemer, M., Schiers, C., Tiede, U., and Wiebecke, G. (1990). 3D visualization of tomographic volume data using the generalized voxel model. *Visual Comput.* **6**, 28–36.

Höhne, K. H., Bomans, M., Riemer, M., Schubert, R., Tiede, U., and Lierse, W. (1992). A 3D anatomical atlas based on a volume model. *IEEE Comput. Graph. Appl.* **12**, 72–78.

Holman, L. B., Chandak, P. K., and Garada, B. M. (1994). Atlas of Brain Perfusion SPECT, http://brighamrad.harvard.edu/education/online/BrainSPECT/BrSPECT.html.

Hynd, G. W., Semrud-Clikeman, M., Lorys, A. R., Novey, E. S., Eliopulos, D., and Lyytinen, J. (1991). Corpus callosum morphology in attention deficit hyperactivity disorder: Morphometric analysis of MRI. *J. Learn. Disabil.* **24**, 141–146.

Ingvar, M., Eriksson, L., Greitz, T., Stoneelander, S., et al. (1994). Methodological aspects of brain activation studies—Cerebral blood flow determined with [O-15]-butanol and positron emission tomography. *J. Cereb. Blood Flow Metab.* **14**(4), 628–638.

Iosifescu, D. V., Shenton, M. E., Warfield, S. K., Kikinis, R., Dengler, J., Jolesz, F. A., and McCarley, R. W. (1997). An automated registration algorithm for measuring MRI subcortical brain structures. *NeuroImage* **6**(1), 13–25.

Jack, C. R., Jr. (1994). MRI-based hippocampal volume measurements in epilepsy. *Epilepsia* **35** (Suppl. 6), S21–S29.

Jacobs, R. E., and Fraser, S. E. (1994). Magnetic resonance microscopy of embryonic cell lineages and movements. *Science* **263**(5147), 681–684.

Jäncke, L., Staiger, J. F., Schlaug, G., Huang, Y., and Steinmetz, H. (1997). The relationship between corpus callosum size and forebrain volume. *Cereb. Cortex* **7**, 1047–3211.

Jenike, M. A., Breiter, H. C., Baer, L., Kennedy, D. N., Savage, C. R., Olivares, M. J., O'Sullivan, R. L., Shera, D. M., Rauch, S. L., Keuthen, N., Rosen, B. R., Caviness, V. S., and Filipek, P. A. (1996). Cerebral structural abnormalities in obsessive-compulsive disorder. A quantitative morphometric magnetic resonance imaging study. *Arch. Gen. Psychiatry* **53**(7), 625–632.

Jernigan, T. L., Salmon, D. P., Butters, N., and Hesselink, J. R. (1991). Cerebral structure on MRI: Part I. Localization of age-related changes. *Biol. Psychiatry* **29**, 55–67.

Jernigan, T. L., Bellugi, U., Sowell, E., Doherty, S., and Hesselink, J. R. (1993). Cerebral morphological distinctions between Williams and Down syndromes. *Arch. Neurol.* **50**, 186–191.

Johnson, K. A. (1996). The Harvard Whole Brain Atlas, http://www.med.harvard.edu/AANLIB/home.html.

Johnson, V. P., Swayze, V. W., Sato, Y., and Andreason, N. C. (1996). Fetal alcohol syndrome: Craniofacial and central nervous system manifestations. *Am. J. Med. Genet.* **61**, 329–339.

Joshi, S. (1998). "Large Deformation Diffeomorphisms and Gaussian Random Fields for Statistical Characterization of Brain Sub-Manifolds," Doctoral Dissertation, Washington University, St. Louis, MO.

Joshi, S. C., Miller, M. I., Christensen, G. E., Banerjee, A., Coogan, T. A., and Grenander, U. (1995). Hierarchical brain mapping via a generalized Dirichlet solution for mapping brain manifolds. In "Vision Geometry IV, Proceedings of the SPIE Conference on Optical Science, Engineering and Instrumentation, San Diego, CA, August 1995," Paper 2573, pp. 278–289.

Joshi, S., Miller, M. I., and Grenander, U. (1998). On the geometry and shape of brain sub-manifolds. *Int. J. Patt. Recog. Artif. Intell.* (in press).

Kennedy, D. N., Lange, N., Makris, N., Bates, J., Meyer, J., and Caviness, V. S., Jr. (1998). Gyri of the human neocortex: An MRI-based analysis of volume and variance. *Cereb. Cortex* **8**(4), 372–384.

Kido, D. K., Caine, E. D., LeMay, M., Ekholm, S., Booth, H., and Panzer, R. (1989). Temporal lobe atrophy in patients with Alzheimer disease: A CT study. *AJNR, Am. J. Neuroradiol.* **7**, 551–555.

Kikinis, R., Shenton, M. E., Gerig, G., Hokama, H., Haimson, J., O'Donnell, B. F., Wible, C. G., McCarley, R. W., and Jolesz, F. A. (1994). Temporal lobe sulco-gyral pattern anomalies in schizophrenia: An in vivo MR three-dimensional surface rendering study. *Neurosci. Lett.* **182**, 7–12.

Kikinis, R., Shenton, M. E., Iosifescu, D. V., McCarley, R. W., Saivi-roonporn, P., Hokama, H. H., Robatino, A., Metcalf, D., Wible, C. G., Portas, C. M., Donnino, R., and Jolesz, F. (1996). A digital brain atlas for surgical planning, model-driven segmentation, and teaching. *IEEE Trans. Vis. Comput. Graph.* **2**(3), 232–241.

Killiany, R. J., Moss, M. B., Albert, M. S., Sandor, T., Tieman, J., and Jolesz, F. (1993). Temporal lobe regions on magnetic resonance imaging identify patients with early Alzheimer's disease. *Arch. Neurol.* **50**, 949–954.

Kim, B., Boes, J. L., Frey, K. A., and Meyer, C. R. (1997). Mutual information for automated unwarping of rat brain autoradiographs. *NeuroImage* **5**(1), 31–40.

Kjems, U., Strother, S. C., Anderson, J., Law, I., and Hansen, L. K. (1999). Enhancing the multivariate signal of [^{15}O]water PET studies with a new nonlinear neuroanatomical registration algorithm. *IEEE Trans. Med. Imaging* **18**(4), 306–319.

Kleinschmidt, A., Falkai, P., Huang, Y., Schneider, T., Furst, G., and Steinmetz, H. (1994). In vivo morphometry of planum temporale asymmetry in first-episode schizophrenia. *Schizophr. Res.* **12**, 9–18.

Kwong, E. M., Mega, M. S., Thompson, P. M., Xu, L. Q., Ercoli, L. M., Cummings, J. L., Small, G. W., Felix, J., and Toga, A. W. (1999). Three-Dimensional Hippocampal Maps in Normal Aging, Older Persons with Mild Cognitive Impairment and Patients with Alzheimer's Disease, Annual Conference of the American Academy of Neurology, 1999.

Larsen, J. P., Høien, T., and ödegaard, H. (1992). Magnetic resonance imaging of the corpus callosum in developmental dyslexia. *Cog. Neuropsychol.* **9**, 123–134.

Lawrie, S. M., and Abukmeil, S. S. (1998). Brain abnormality in schizophrenia. *Br. J. Psychiatry* **172**, 110–120.

Lee, M. A., Smith, S., Palace, J., and Matthews, P. M. (1998). Defining multiple sclerosis disease activity using MRI T2-weighted difference imaging. *Brain* **121**(Part 11), 2095–2102.

Lee, M. A., Smith, S., Palace, J., Narayanan, S., Silver, N., Minicucci, L., Filippi, M., Miller, D. H., Arnold, D. L., and Matthews, P. M. (1999). Spatial mapping of T2 and gadolinium-enhancing T1 lesion volumes in multiple sclerosis: Evidence for distinct mechanisms of lesion genesis? *Brain* **122**(Part 7), 1261–1270.

Le Goualher, G., Procyk, E., Collins, D. L., Venugopal, R., Barillot, C., and Evans, A. C. (1999). Automated extraction and variability analysis of sulcal neuroanatomy. *IEEE Trans. Med. Imaging* **18**(3), 206–217.

Leonard, C. M. (1996). Structural variation in the developing and mature cerebral cortex: Noise or signal? *In* "Developmental Neuroimaging: Mapping the Development of Brain and Behavior" (R. W. Thatcher, G. Reid Lyon, J. Rumsey, and N. Krasnegor, Eds.), pp. 207–231. Academic Press, San Diego.

Lester, H., Arridge, S. R., Jansons, K. M., Lemieux, L., Hajnal, J. V., and Oatridge, A. (1999). Non-linear registration with the variable viscosity fluid algorithm. *16th Int. Conf. Inf. Proc. Med. Imaging, IPMI '99* (in press).

Liseikin, V. D. (1991). On a variational method for generating adaptive grids on *N*-dimensional surfaces. *Dokl. Akad. Nauk CCCP* **319**(3), :546–549.

Loewenstein, D. A., Barker, W. W., Chang, J. Y., Apicella, A., Yoshii, F., Kothari, P., Levin, B., and Duara, R. (1989). Predominant left hemisphere metabolic dysfunction in dementia. *Arch. Neurol.* **46**(2), 146–152.

Lyoo, I. K., Noam, G. G., Chang, K. L., Ho, K. L., Kennedy, B. P., and Renshaw, P. F. (1996). The corpus callosum and lateral ventricles in children with attention deficit hyperactivity disorder: A brain magnetic resonance imaging study. *Biol. Psychiatry* **40**, 1060–1063.

MacDonald, D., Avis, D., and Evans, A. C. (1993). Automatic parameterization of human cortical surfaces. *Annu. Symp. Inf. Proc. Med. Imaging (IPMI)*.

Mai, J., Assheuer, J., and Paxinos, G. (1997). "Atlas of the Human Brain." Academic Press, San Diego.

Manceaux-Demiau, A., Bryan, R. N., and Davatzikos, C. (1998). A probabilistic ribbon model for shape analysis of the cerebral sulci: Application to the central sulcus. *J. Comput. Assist. Tomogr.* **22**(6), 962–971.

Mangin, J.-F., Frouin, V., Bloch, I., Regis, J., and Lopez-Krahe, J. (1994). Automatic construction of an attributed relational graph representing the cortex topography using homotopic transformations. *SPIE* **2299**, 110–121.

Mansour, A., Fox, C. A., Burke, S., Akil, H., and Watson, S. J. (1995). Immunohistochemical localization of the cloned mu opioid receptor in the rat CNS. *J. Chem. Neuroanat.* **8**(4), 283–305.

Matsui, T., and Hirano, A. (1978). "An Atlas of the Human Brain for Computerized Tomography." Igaku Shoin, Tokyo.

Mattson, S. N., Riley, E. P., Sowell, E. R., Jernigan, T. L., Sobel, D. F., and Jones, K. L. (1996). A decrease in the size of the basal ganglia in children with fetal alcohol syndrome. *Alcohol Clin. Exp. Res.* **20**(6), 1088–1093.

Mazziotta, J. C., Toga, A. W., Evans, A. C., Fox, P., and Lancaster, J. (1995). A probabilistic atlas of the human brain: Theory and rationale for its development. *NeuroImage* **2**, 89–101.

Mega, M. S., Chen, S., Thompson, P. M., Woods, R. P., Karaca, T. J., Tiwari, A., Vinters, H., Small, G. W., and Toga, A. W. (1997). Mapping pathology to metabolism: Coregistration of stained whole brain sections to PET in Alzheimer's disease. *NeuroImage* **5**, 147–153.

Mega, M. S., Dinov, I. D., Lee, L., Woods, R. P., Thompson, P. M., Holmes, C. J., Back, C. L., Collins, D. L., Evans, A. C., and Toga, A. W. (1998a). Dissecting neural networks underlying the retrieval deficit from the amnestic memory disorder using [^{99}mTc]-HM-PAO-SPECT. *Proc. Am. Behav. Neurol. Soc.*, February 1998.

Mega, M. S., Thompson, P. M., Cummings, J. L., Back, C. L., Xu, L. Q., Zohoori, S., Goldkorn, A., Moussai, J., Fairbanks, L., Small, G. W., and Toga, A. W. (1998b). Sulcal variability in the Alzheimer's brain: Correlations with cognition. *Neurology* **50**, 145–151.

Mega, M. S., Woods, R. P., Thompson, P. M., Dinov, I. D., Lee, L., Aron, J., Zoumalan, C. I., Cummings, J. L., and Toga, A. W. (1998c). Detecting metabolic patterns associated with minimal cognitive impairment using FDG-PET analysis within a probabilistic brain atlas based upon continuum mechanics. *Proc. Soc. Neurosci.*

Mega, M. S., Thompson, P. M., Toga, A. W., and Cummings, J. L. (2000). Brain mapping in dementia. *In* "Brain Mapping: The Disorders" (J. C. Mazziotta, A. W. Toga, and R. S. J. Frackowiak, Eds.), Chap. 9. Academic Press, San Diego.

Meltzer, C. C., and Frost, J. J. (1994). Partial volume correction in emission-computed tomography: Focus on Alzheimer disease. *In* "Functional Neuroimaging: Technical Foundations" (R. W. Thatcher, M. Hallett, T. Zeffiro, E. R. John, and M. Huerta, Eds.), pp. 163–170. Academic Press, San Diego.

Miller, M. I., Christensen, G. E., Amit, Y., and Grenander, U. (1993). Mathematical textbook of deformable neuroanatomies. *Proc. Natl. Acad. Sci. U.S.A.* **90**, 11944–11948.

Minoshima, S., Koeppe, R. A., Frey, K. A., Ishihara, M., and Kuhl, D. E. (1994). Stereotactic PET atlas of the human brain: Aid for visual interpretation of functional brain images. *J. Nucl. Med.* **35**, 949–954.

Narayanan, S., Fu, L., Pioro, E., De Stefano, N., Collins, D. L., Francis, G. S., Antel, J. P., Matthews, P. M., and Arnold, D. L. (1997). Imaging of axonal damage in multiple sclerosis: Spatial distribution of magnetic resonance imaging lesions. *Ann. Neurol.* **41**(3), 385–391.

Narr, K. L., Cannestra, A. F., Thompson, P. M., Sharma, T., and Toga, A. W. (1998). Morphological Variability Maps of the Corpus Callosum and Fornix in Schizophrenia. 4th International Conference on Human Brain Mapping. *NeuroImage* **7**(4), S506.

Narr, K. L., Thompson, P. M., Sharma, T., Moussai, J., Cannestra, A. F., and Toga, A. W. (1999). Mapping corpus callosum morphology in schizophrenia. *Cereb. Cortex* (in press).

Nestor, P. G., Shenton, M. E., McCarley, R. W., Haimson, J., Smith, R. S., O'Donnell, B., Kimble, M., Kikinis, R., and Jolesz, F. A. (1993). Neuropsychological correlates of MRI temporal lobe abnormalities in schizophrenia. *Am. J. Psychiatry* **150**(12), 1849–1855.

Nowinski, W. L., Fang, A., Nguyen, B. T., Raphel, J. K., Jagannathan, L., Raghavan, R., Bryan, R. N., and Miller, G. A. (1997). Multiple brain atlas database and atlas-based neuroimaging system. *Comput. Aided Surg.* **2**(1), 42–66.

Ono, M., Kubik, S., and Abernathey, C. D. (1990). "Atlas of the Cerebral Sulci." Thieme, Stuttgart.

Paus, T., Tomaioulo, F., Otaky, N., MacDonald, D., Petrides, M., Atlas, J., Morris, R., and Evans, A. C. (1996). Human cingulate and paracingulate sulci: Pattern, variability, asymmetry and probabilistic map. *Cereb. Cortex* **6**, 207–214.

Paus, T., Zijdenbos, A., Worsley, K., Collins, D. L., Blumenthal, J., Giedd, J. N., Rapoport, J. L., and Evans, A. C. (1999). Structural maturation of neural pathways in children and adolescents: *In vivo* study. *Science* **283**(5409), 1908–1911.

Petty, R. G., Barta, P. E., Pearlson, G. D., McGilchrist, I. K., Lewis, R. W., Tien, A. Y., Pulver, A., Vaughn, D. D., Casanova, M. F., and Powers, R. E. (1995). Reversal of asymmetry of the planum temporale in schizophrenia. *Am. J. Psychiatry* **152**(5), 715–721.

Pfefferbaum, A., Mathalon, D. H., Sullivan, E. V., Rawles, J. M., Zipursky, R. B., and Lim, K. O. (1994). A quantitative magnetic resonance imaging study of changes in brain morphology from infancy to late adulthood. *Arch. Neurol.* **51**(9), 874–887.

Pitiot, A., Thompson, P. M., and Toga, A. W. (1999). Spatially and temporally adaptive elastic template matching. *IEEE Trans. Patt. Anal. Machine Intell.*

Piven, J., Bailey, J., Ranson, B. J., and Arndt, S. (1997). An MRI study of the corpus callosum in autism. *Am. J. Psychiatry* **154**(8), 1051–1056.

Pommert, A., Schubert, R., Riemer, M., Schiemann, T., Tiede, U., and Höhne, K. H. (1994). Symbolic modeling of human anatomy for visualization and simulation. *IEEE Vis. Biomed. Comput.* **2359**, 412–423.

Raz, N., Torres, I. J., and Acker, J. D. (1995). Age, gender, and hemispheric differences in human striatum: A quantitative review and new data from in vivo MRI morphometry. *Neurobiol. Learn. Mem.* **63**(2), 133–142.

Riley, E. P., Mattson, S. N., Sowell, E. R., Jernigan, T. L., Sobel, D. F., and Jones, K. L. (1995). Abnormalities of the corpus callosum in children prenatally exposed to alcohol. *Alcoholism Clin. Exp. Res.* **19**(5), 1198–1202.

Rizzo, G., Gilardi, M. C., Prinster, A., Grassi, F., Scotti, G., Cerutti, S., and Fazio, F. (1995). An elastic computerized brain atlas for the analysis of clinical PET/SPET data. *Eur. J. Nucl. Med.* **22**(11), 1313–1318.

Roebuck, T. M., Mattson, S. N., and Riley, E. P. (1998). A review of the neuroanatomical findings in children with fetal alcohol syndrome or prenatal exposure to alcohol. *Alcohol Clin. Exp. Res.* **22**(2), 339–344.

Roland, P. E., and Zilles, K. (1994). Brain atlases—A new research tool. *Trends Neurosci.* **17**(11), 458–467.

Rosenberg, D. R., Keshavan, M. S., Dick, E. L., Bagwell, W. W., MacMaster, F. P., and Birmaher, B. (1997). Corpus callosal morphology in treatment-naive pediatric obsessive compulsive disorder. *Prog. Neuropsychopharmacol. Biol. Psychiatry* **21**(8), 1269–1283.

Royackkers, N., Desvignes, M., and Revenu, M. (1996). Construction automatique d'un atlas adaptatif des sillons corticaux, ORASIS 96, Clermont-Ferrand, pp. 187–192.

Sandor, S. R., and Leahy, R. M. (1994). Matching deformable atlas models to pre-processed magnetic resonance brain images. *Proc. IEEE Conf. Image Processing* **3**, 686–690.

Sandor, S. R., and Leahy, R. M. (1995). Towards automated labeling of the cerebral cortex using a deformable atlas. *In* "Information Processing in Medical Imaging" (Y. Bizais, C. Barillot, and R. Di Paola, Eds.), pp. 127–138. Kluwer.

Schaltenbrand, G., and Bailey, P. (1959). "Introduction to Stereotaxis with an Atlas of the Human Brain." Thieme, Stuttgart.

Schaltenbrand, G., and Wahren, W. (1977). "Atlas for Stereotaxy of the Human Brain," 2nd ed. Thieme, Stuttgart.

Schiemann, T., and Höhne, K. H. (1997). Definition of volume transformations for volume interaction. *In* "IPMI '97" (J. Duncan and G. Gindi, Eds.). Springer-Verlag, Heidelberg.

Schiemann, T., Nuthmann, J., Tiede, U., and Höhne, K. H. (1996). Segmentation of the visible human for high-quality volume-based visualization. *Vis. Biomed. Comput.* **4**, 13–22.

Schmidt, R. (1992). Comparison of magnetic resonance imaging in Alzheimer's disease, vascular dementia and normal aging. *Eur. Neurol.* **32**(3), 164–169.

Schormann, T., Henn, S., and Zilles, K. (1996). A new approach to fast elastic alignment with applications to human brains. *Proc. Vis. Biomed. Comput.* **4**, 337–342.

Seitz, R. J., Bohm, C., Greitz, T., Roland, P. E., Eriksson, L., Blomqvist, G., Rosenqvist, G., and Nordell, B. (1990). Accuracy and precision of the Computerized Brain Atlas Programme for Localization and Quantification in Positron Emission Tomography. *J. Cereb. Blood Flow. Metab.* **10**, 443–457.

Semrud-Clikeman, M., Filipek, P. A., Biederman, J., Steingard, R., Kennedy, D., Renshaw, P., and Bekken, K. (1994). Attention-deficit hyperactivity disorder: Magnetic resonance imaging morphometric analysis of the corpus callosum. *J. Am. Acad. Child Adolesc. Psychiatry* **33**(6), 875–881.

Shenton, M. E., Kikinis, R., Jolesz, F. A., Pollack, S. D., LeMay, M., Wible, C. G., Hokama, H., Martin, J., Metcalf, D., Coleman, M., and McCarley, R. (1992). Abnormalities of the left temporal lobe and thought disorder in schizophrenia. *N. Engl. J. Med.* **327**(9), 604–612.

Siegel, B. V., Jr., Shihabuddin, L., Buchsbaum, M. S., Starr, A., Haier, R. J., and Valladares Neto, D. C. (1996). Gender differences in cortical glucose metabolism in Alzheimer's disease and normal aging. *J. Neuropsychiatry Clin. Neurosci.* **8**(2), 211–214.

Smith, G. E. (1907). A new topographical survey of the human cerebral cortex, being an account of the distribution of the anatomically distinct cortical areas and their relationship to the cerebral sulci. *J. Anat.* **41**, 237–254.

Sobire, G., Goutieres, F., Tardieu, M., Landrieu, P., and Aicardi, J. (1995). Extensive macrogyri or no visible gyri: Distinct clinical, electroencephalographic, and genetic features according to different imaging patterns. *Neurology* **45**, 1105–1111.

Sochen, N., Kimmel, R., and Malladi, R. (1998). A general framework for low level vision. *IEEE Trans. Image Processing* **7**(3), 310–318.

Sowell, E. R., Jernigan, T. L., Mattson, S. N., Riley, E. P., Sobel, D. F., and Jones, K. L. (1996). Abnormal development of the cerebellar vermis in children prenatally exposed to alcohol: Size reduction in lobules I–V. *Alcohol Clin. Exp. Res.* **20**(1), 31–34.

Spitzer, V., Ackerman, M. J., Scherzinger, A. L., and Whitlock, D. (1996). The visible human male: A technical report. *J. Am. Med. Inf. Assoc.* **3**(2), 118–130. http://www.nlm.nih.gov/extramural_research.dir/visible human.html.

Steinmetz, H., Furst, G., and Freund, H.-J. (1989). Cerebral cortical localization: Application and validation of the proportional grid system in MR imaging. *J. Comput. Assist. Tomogr.* **13**(1), 10–19.

Steinmetz, H., Furst, G., and Freund, H.-J. (1990). Variation of perisylvian and calcarine anatomic landmarks within stereotaxic proportional coordinates. *Am. J. Neuroradiol.* **11**(6), 1123–1130.

Steinmetz, H., Staiger, J. F., Schlaug, G., Huang, Y., and Jancke, L. (1995). Corpus callosum and brain volume in women and men. *NeuroReport* **6**(7), 1002–1004.

Stewart, J. E., Broaddus, W. C., and Johnson, J. H. (1996). Rebuilding the Visible Man. *Vis. Biomed. Comput.* **4**, 81–86.

St.-Jean, P., Sadikot, A. F., Collins, L., Clonda, D., Kasrai, R., Evans, A. C., and Peters, T. M. (1998). Automated atlas integration and interactive three-dimensional visualization tools for planning and guidance in functional neurosurgery. *IEEE Trans. Med. Imaging* **17**(5), 672–680.

Strauss, E., Kosaka, B., and Wada, J. (1983). The neurobiological basis of lateralized cerebral function. A review. *Hum. Neurobiol.* **2**(3), 115–127.

Subsol, G., Roberts, N., Doran, M., Thirion, J. P., and Whitehouse, G. H. (1997). Automatic analysis of cerebral atrophy. *Magn. Reson. Imaging* **15**(8), 917–927.

Swayze, V. W., Johnson, V. P., Hanson, J. W., Piven, J., Sato, Y., Giedd, J. N., Mosnik, D., and Andreasen, N. C. (1997). Magnetic resonance imaging of brain anomalies in fetal alcohol syndrome. *Pediatrics* **99**(2), 232–240.

Talairach, J., and Szikla, G. (1967). "Atlas d'Anatomie Stereotaxique du Telencephale: Etudes Anatomo-Radiologiques." Masson & Cie, Paris.

Talairach, J., and Tournoux, P. (1988). "Co-planar Stereotaxic Atlas of the Human Brain." Thieme, Stuttgart.

Thirion, J.-P. (1995). "Fast Non-Rigid Matching of Medical Images," INRIA Internal Report 2547, Projet Epidaure, INRIA, France.

Thirion, J.-P., and Calmon, G. (1997). "Deformation Analysis to Detect and Quantify Active Lesions in 3D Medical Image Sequences," INRIA Technical Report 3101, INRIA, France.

Thirion, J.-P., Prima, S., and Subsol, S. (1998). Statistical analysis of dissymmetry in volumetric medical images. *Med. Image Anal.* (in press).

Thompson, P. M., and Toga, A. W. (1996). A surface-based technique for warping 3-dimensional images of the brain. *IEEE Trans. Med. Imaging* **15**(4), 1–16.

Thompson, P. M., and Toga, A. W. (1997). Detection, visualization and animation of abnormal anatomic structure with a deformable probabilistic brain atlas based on random vector field transformations. *Med. Image Anal.* **1**(4), 271–294.

Thompson, P. M., and Toga, A. W. (1998a). Anatomically-driven strategies for high-dimensional brain image warping and pathology detection. *In* "Brain Warping" (A. W. Toga, Ed.), pp. 311–336. Academic Press, San Diego.

Thompson, P. M., and Toga, A. W. (1998b). Mathematical/Computational Strategies for Creating a Probabilistic Atlas of the Human Brain, Workshop on Statistics of Brain Mapping, Centre de Recherches Mathématiques, McGill University, Canada, June 13–14, 1998.

Thompson, P. M., and Toga, A. W. (1999). Elastic image registration and pathology detection. *In* "Handbook of Medical Image Processing" (I. Bankman, R. Rangayyan, A. C. Evans, R. P. Woods, E. Fishman, and H. K. Huang, Eds.). Academic Press, San Diego.

Thompson, P. M., Schwartz, C., and Toga, A. W. (1996a). High-resolution random mesh algorithms for creating a probabilistic 3D surface atlas of the human brain. *NeuroImage* **3**, 19–34.

Thompson, P. M., Schwartz, C., Lin, R. T., Khan, A. A., and Toga, A. W. (1996b). 3D statistical analysis of sulcal variability in the human brain. *J. Neurosci.* **16**(13), 4261–4274.

Thompson, P. M., MacDonald, D., Mega, M. S., Holmes, C. J., Evans, A. C., and Toga, A. W. (1997). Detection and mapping of abnormal brain structure with a probabilistic atlas of cortical surfaces. *J. Comput. Assist. Tomogr.* **21**(4), 567–581.

Thompson, P. M., Moussai, J., Khan, A. A., Zohoori, S., Goldkorn, A., Mega, M. S., Small, G. W., Cummings, J. L., and Toga, A. W. (1998). Cortical variability and asymmetry in normal aging and Alzheimer's disease. *Cereb. Cortex* **8**(6), 492–509.

Thompson, P. M., Giedd, J. N., Blanton, R. E., Lindshield, C., Woods, R. P., MacDonald, D., Evans, A. C., and Toga, A. W. (1998). Growth Patterns in the Developing Human Brain Detected Using Continuum-Mechanical Tensor Maps and Serial MRI, 5th International Conference on Human Brain Mapping, Montreal, Canada.

Thompson, P. M., Narr, K. L., Blanton, R. E., and Toga, A. W. (1999a). Mapping structural alterations of the corpus callosum during brain development and degeneration. *In* "The Corpus Callosum." Kluwer Academic, Dordrecht/Norwell, MA (in press).

Thompson, P. M., Woods, R. P., Mega, M. S., and Toga, A. W. (1999b). Mathematical/computational challenges in creating population-based brain atlases. *Hum. Brain Mapp.* **8**(2).

Thurfjell, L., Bohm, C., Greitz, T., and Eriksson, L. (1993). Transformations and algorithms in a computerized brain atlas. *IEEE Trans. Nucl. Sci.* **40**(4, Part 1), 1167–1191.

Tiede, U., Bomans, M., Höhne, K. H., Pommert, A., Riemer, M., Schiemann, T., Schubert, R., and Lierse, W. (1993). A computerized 3D atlas of the human skull and brain. *Am. J. Neuroradiol.* **14**, 551–559.

Toga, A. W. (1994). Visualization and warping of multimodality brain imagery. *In* "Functional Neuroimaging: Technical Foundations" (R. W. Thatcher, M. Hallett, T. Zeffiro, E. R. John, and M. Huerta, Eds.), pp. 171–180. Kluwer.

Toga, A. W. (1998). "Brain Warping." Academic Press, San Diego.

Toga, A. W., and Mazziotta, J. C. (1996). "Brain Mapping: The Methods." Academic Press, San Diego.

Toga, A. W., and Thompson, P. M. (1997). Measuring, Mapping, and Modeling Brain Structure and Function, SPIE Medical Imaging Symposium, February 1997, Newport Beach, CA. *SPIE Lect. Notes* **3033.**

Toga, A. W., and Thompson, P. M. (1998a). An introduction to brain warping. *In* "Brain Warping" (A. W. Toga, Ed.). Academic Press, San Diego.

Toga, A. W., and Thompson, P. M. (1998b). Multimodal brain atlases. *In* "Medical Image Databases" (S. T. C. Wong, Ed.), pp. 53–88. Kluwer Academic, Dordrecht/Norwell, MA.

Toga, A. W., and Thompson, P. M. (1999a). An introduction to maps and atlases of the brain. *In* "Brain Mapping: The Applications" (A. W. Toga and J. C. Mazziotta, Eds.). Academic Press, San Diego.

Toga, A. W., and Thompson, P. M. (1999b). Brain atlases and image registration. *In* "Handbook of Medical Image Processing" (I. Bankman, R. Rangayyan, A. C. Evans, R. P. Woods, E. Fishman, and H. K. Huang, Eds.). Academic Press, San Diego.

Toga, A. W., Ambach, K., Quinn, B., Hutchin, M., and Burton, J. S. (1994). Postmortem anatomy from cryosectioned whole human brain. *J. Neurosci. Methods* **54**(2), 239–252.

Toga, A. W., Thompson, P. M., and Payne, B. A. (1996). Modeling morphometric changes of the brain during development. *In* "Developmental Neuroimaging: Mapping the Development of Brain and Behavior" (R. W. Thatcher, G. R. Lyon, J. Rumsey, and N. Krasnegor, Eds.). Academic Press, San Diego.

Toga, A. W., Goldkorn, A., Ambach, K., Chao, K., Quinn, B. C., and Yao, P. (1997). Postmortem cryosectioning as an anatomic reference for human brain mapping. *Comput. Med. Imaging Graph.* **21**(2), 131–141.

Vaillant, M., and Davatzikos, C. (1999). Hierarchical Matching of Cortical Features for Deformable Brain Image Registration, Proceedings, Information Processing in Medical Imaging, Budapest, June 1999.

Vaillant, M., Davatzikos, C., Taylor, R. H., and Bryan, R. N. (1997). A path-planning algorithm for image-guided neurosurgery. *Proc. CVRMed/MRCAS 1997* 467–476.

Van Buren, J. M., and Borke, R. C. (1972). "Variations and Connections of the Human Thalamus," Vols. 1 and 2. Springer-Verlag, Berlin/New York.

Van Buren, J. M., and Maccubin, D. (1962). An outline atlas of human basal ganglia and estimation of anatomic variants. *J. Neurosurg.* **19**, 811–839.

Van Essen, D. C., Drury, H. A., Joshi, S. C., and Miller, M. I. (1997). Comparisons between Human and Macaque Using Shape-Based Deformation Algorithms Applied to Cortical Flat Maps, 3rd International Conference on Functional Mapping of the Human Brain, Copenhagen, May 19–23, 1997, *NeuroImage* 5(**4**), S41.

Viola, P. A., and Wells, W. M. (1995). Alignment by maximization of mutual information. *In* "5th IEEE International Conference on Computer Vision, Cambridge, MA," pp. 16–23.

Waddington, J. L. (1993). Neurodynamics of abnormalities in cerebral metabolism and structure in schizophrenia. *Schizophr. Bull.* **19**, 55–58.

Wang, P. P., Doherty, S., Hesselink, J. R., and Bellugi, U. (1992). Callosal morphology concurs with neurobehavioral and neuropathological findings in two neurodevelopmental disorders. *Arch. Neurol.* **49**(4), 407–411.

Warfield, S., Dengler, J., Zaers, J., Guttmann, C. R. G., Wells, W. M., Ettinger, G. J., Hiller, J., and Kikinis, R. (1995). Automatic identification of gray matter structures from MRI to improve the segmentation of white matter lesions. *Proc. Med. Robotics Comput. Assist. Surg. (MRCAS),* Nov. 4–7, 1995, 55–62.

Warfield, S., Robatino, A., Dengler, J., Jolesz, F., and Kikinis, R. (1998). Nonlinear registration and template driven segmentation. *In* "Brain Warping" (A. W. Toga, Ed.), Chap. 4, pp. 67–84. Academic Press, San Diego.

Watson, J. D. G., Myers, R., Frackowiak, R. S. J., Hajnal, J. V., Woods, R. P., Mazziotta, J. C., Shipp, S., and Zeki, S. (1993). Area V5 of the human brain: Evidence from a combined study using positron emission tomography and magnetic resonance imaging. *Cereb. Cortex* **3**, 79–94.

Wells, W. M., Viola, P., Atsumi, H., Nakajima, S., and Kikinis, R. (1997). Multi-modal volume registration by maximization of mutual information. *Med. Image Anal.* **1**(1), 35–51.

West, M. J., Coleman, P. D., Flood, D. G., and Troncoso, J. C. (1994). Differences in the pattern of hippocampal neuronal loss in normal aging and Alzheimer's disease. *Lancet* **344**, 769–772.

Whitehouse, P. J., Price, D. L., Clark, A. W., Coyle, J. T., and DeLong, M. R. (1981). Alzheimer's disease: Evidence for selective loss of cholinergic neurons in the nucleus basalis. *Ann. Neurol.* **10**, 122–126.

Witelson, S. F. (1989). Hand and sex differences in the isthmus and genu of the human corpus callosum. A postmortem morphological study. *Brain* **112**, 799–835.

Woods, R. P. (1996). Modeling for intergroup comparisons of imaging data. *NeuroImage* **4**(3), 84–94.

Woods, R. P., Cherry, S. R., and Mazziotta, J. C. (1992). Rapid automated algorithm for aligning and reslicing PET images. *J. Comput. Assist. Tomogr.* **16**, 620–633.

Woods, R. P., Mazziotta, J. C., and Cherry, S. R. (1993). MRI-PET registration with automated algorithm. *J. Comput. Assist. Tomogr.* **17**, 536–546.

Woods, R. P., Grafton, S. T., Watson, J. D. G., Sicotte, N. L., and Mazziotta, J. C. (1998). Automated image registration: II. Intersubject validation of linear and nonlinear models. *J. Comput. Assist. Tomogr.* **22**(1), 153–165.

Worsley, K. J. (1994a). "Quadratic Tests for Local Changes in Random Fields with Applications to Medical Images," Technical Report 94-08, Department of Mathematics and Statistics, McGill University.

Worsley, K. J. (1994b). Local maxima and the expected Euler characteristic of excursion sets of chi-squared, *F* and *t* fields. *Adv. Appl. Prob.* **26**, 13–42.

Zhou, Y., Thompson, P. M., and Toga, A. W. (1999). Automatic extraction and parametric representations of cortical sulci. *Comput. Graph. Appl.* **19**(3), 49–55.

Zijdenbos, A. P., and Dawant, B. M. (1994). Brain segmentation and white matter lesion detection in MR images. *Crit. Rev. Biomed. Eng.* **22**(5–6), 401–465.

Zoumalan, C. I., Mega, M. S., Thompson, P. M., Fuh, J. L., Lindshield, C., and Toga, A. W. (1999). Mapping 3D Patterns of Cortical Variability in Normal Aging and Alzheimer's Disease, Annual Conference of the American Academy of Neurology.

II

Neurological Disorders

7

Functional Imaging Studies of Aphasia

Cathy J. Price

Wellcome Department of Cognitive Neurology,
Institute of Neurology, London WC1N 3BG, United Kingdom

I. Neurological and Neuropsychological Models of Language

II. Neuroimaging Studies of Aphasic Patients

III. Examples of How Neuroimaging Experiments on Aphasic Patients Have Contributed to Normal and Abnormal Models of Language

IV. Conclusions

References

This chapter considers how functional neuroimaging studies of aphasic patients can contribute to normal and abnormal models of language processing. The first section discusses the classic neurological and cognitive models of language. The second section discusses the expectations and limitations of functional neuroimaging with aphasic patients. The third reviews some examples of differing types of neuroimaging investigation and the fourth discusses how these studies can contribute to normal and abnormal models of language.

I. Neurological and Neuropsychological Models of Language

A. The Lesion-Deficit Model

Neurological studies of aphasic patients date back more than a century and are based on the application of the lesion-deficit model to human subjects. This approach equates selective brain lesions to selective cognitive deficits. Some classic examples of the lesion-deficit model, as applied to aphasic patients, were first documented by the nineteenth century neurologists. For instance, postmortem studies demonstrated that a patient who had been impaired at articulating language had damage encompassing the third frontal convolution (Broca, 1861) and a patient with a deficit in speech comprehension had damage to the left posterior temporal cortex (Wernicke, 1874). By deduction, Broca's area was associated with speech production and Wernicke's area was associated with speech comprehension. Wernicke developed the model further to predict that patients could have intact speech comprehension and production but a deficit integrating these regions in order to repeat what was heard. This type of disconnection syndrome, referred to as "conduction aphasia," was demonstrated by Lichtheim (1885) in a patient who had damage to the white matter tract that connects Broca's area with Wernicke's area (the arcuate fasciculus). By clinical descriptions and localization of lesions, Wernicke and Lichtheim theorized that disorders of language either arose from damage to the "centers of memory images" or were a result of disconnections between these centers, their sensory input, or their motor output. Figure 1a illustrates a simple version of Lichtheim's theoretical model of language. There are three language centers and five possible interruption points that result in (1) Broca's aphasia (damage to the motor images of speech), (2) Wernicke's aphasia (damage to the auditory images of speech), (3) conduction aphasia (disconnection between Broca's and Wernicke's areas), (4) transcortical sensory aphasia (disconnection between Wernicke's area and the concept center), and (5) transcortical motor aphasia (disconnection between Broca's area and the concept center).

With respect to reading, the first major contribution came from Dejerine (1891; 1892), who distinguished two main alexic syndromes: "alexia with agraphia" and "alexia without agraphia." Alexia with agraphia was thought to arise from damage to the left angular gyrus, proposed to be the center for visual word forms. In contrast, alexia without agraphia (which is associated with lesions to the left occipital lobe and the splenium of the corpus callosum) was thought to arise from a disconnection of the left angular gyrus from visual input. In Fig. 1a, the reading center is added to Lichtheim's model, with connections between the angular gyrus and Wernicke's area (see Geschwind, 1979). The numerals 6 and 7 indicate the disruption sites associated with alexia without agraphia (6) and alexia with agraphia (7).

Although the neurologically based information processing models have been highly influential and bear many similarities to modern-day diagrams (see Fig. 1b), they have endured a range of criticisms that include a lack of anatomical precision and the vagueness of the psychological constructs (see Shallice, 1988, for a review). A contrasting neurologically based view of language is summarized by Mesulam (1990): "There are no centres dedicated to comprehension, articulation or grammar but a distributed network in which nodal foci of relative specialisation work in concert. Each behaviour is represented in multiple sites and each site subserves multiple behaviours, leading to a distributed and interactive system with a one to many and many to one mapping of anatomical substrate onto neural computation and computation onto behaviour."

B. Limitations of the Lesion-Deficit Model

As noted above, the shortcomings of the lesion-deficit model are well recognized. It is very difficult to ascribe a function to a particular region that has been damaged for a number of reasons. Perhaps the most obvious of these is that pathological (as opposed to experimental) lesions seldom conform to functionally homogeneous neuroanatomical systems. Furthermore, the neuropsychological profile is usually complicated, involving more than one functional deficit, and these can be obscured by the compensatory measures adopted by the patient to overcome the deficits. Any reasonable relationship between the functional deficit and the brain systems involved is therefore usually impossible to establish. Another problem with the lesion-deficit model that has a long history dating from the nineteenth century (Goltz, 1881) is that the results of lesion studies are only properly interpreted by referring to the connections between cortical areas: Damage to a selected area may impair nearby connections and therefore the responsiveness of other undamaged areas. Indeed it is impossible to distinguish between the impact of a lesion due to the loss of neuronal infrastructure per se and the more pervasive dysfunction of distributed systems of which the lesioned area is only a component. These considerations mean that all that can be concluded from a lesion-deficit study is that the neuronal systems intrinsic to the lesioned area, or the connections passing through this area, were necessary for the cognitive function. One cannot say that this region was either sufficient for, or uniquely identifiable with, the function in question.

C. Cognitive Neuropsychological Studies of Aphasic Patients

Cognitive neuropsychology is the study of patients with functional deficits where the emphasis is on developing cognitive rather than neurological models of language. Essentially, models of cognition are modified by neuropsychological studies when patients demonstrate a double dissociation in the impairment of selective functions. For instance, differing patterns of reading in dyslexic patients point to a double dissociation in the reading processes. This was demonstrated by Newcombe and Marshall (1973), who observed that some patients retain the ability to read words with regular spelling to sound correspondence but fail to read words that do not comply to spelling rules. In contrast, other patients suffer the reverse dissociation. This particular double dissociation suggests that different routes are available to read words: a lexical route and a grapheme-phoneme conversion route (see Coltheart, 1980). In the past two decades, neuropsy-

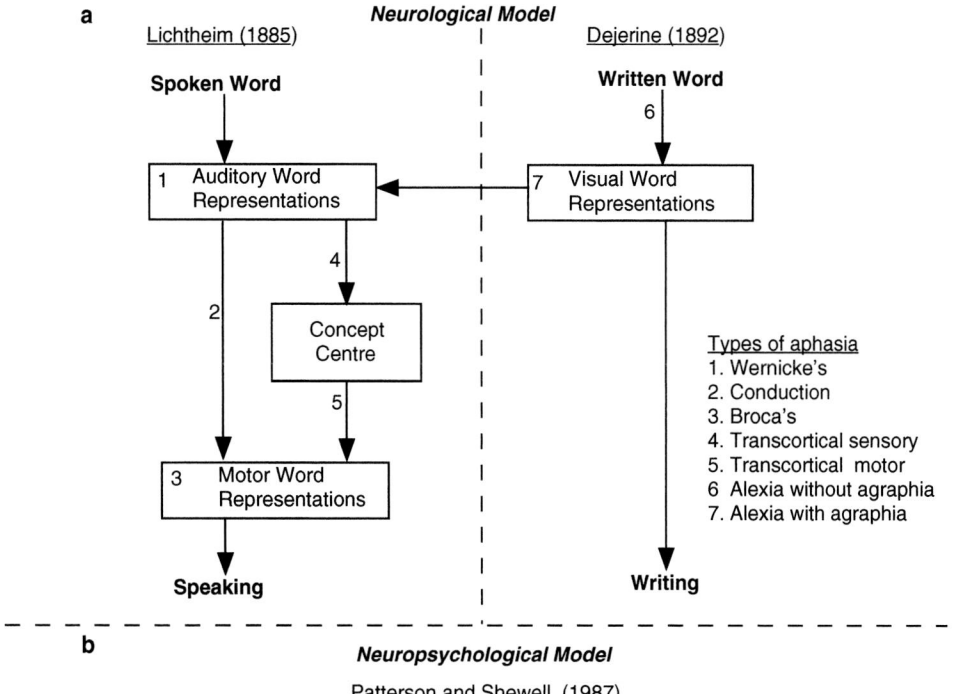

Figure 1 Nineteenth and twentieth century models of word processing. (a) Simplified version of nineteenth century models of language. The numbers refer to the type of aphasia that results from damage to either the language centers or the connections between these centers. (b) Cognitive model proposed by Patterson and Shewell (1987).

chological studies have decomposed the normal language system into many interacting subcomponents. A classic example is the cognitive model proposed by Patterson and Shewell (1987). The many subcomponents involve, at a minimum, auditory and phonological analysis of heard words, visual and orthographic analysis of seen words, a semantic system, phonological output processing, and articulation (see Fig. 1b).

D. Limitations of Cognitive Neuropsychology

The assumption behind traditional neuropsychological studies is that dissociations between cognitive functions indicate that the neural mechanisms underlying the processes are also modular. However, unlike the early neurological models, neuropsychological models are generally not concerned with how the various subcomponents are implemented in the brain. For instance, a particular subcomponent might correspond to a specialized module of cortical tissue or it might be implemented by the connections between a more limited number of specialized modules. In contrast, connectionist models of cognitive processing have demonstrated that many of the cognitive properties that traditional models are based on emerge from a system that has a limited number of highly interactive features, or "nodes." In connectionist models, or parallel distributed processing (PDP) models, representations need not correspond to specific nodes in the system but may be distributed over many nodes, with each node participating in many different operations. Note the similarity here with the neurologically based model proposed by Mesulam (see above).

A popular example of a connectionist model of reading is given by McClelland and Rumelhart (1981). In this model, there is no distinction between the routes taken for known and novel words. Both types of word are processed by a three-level interactive activation model, where the visual features in letters activate corresponding letter nodes and inhibit incongruent letter nodes. Similarly, the activated letter nodes excite and inhibit word nodes, which, in turn, have excitatory and inhibitory feedback connections to the letters and visual features. From positive feedback loops, an excitation pattern closest to the word or novel word emerges and disruptions within this interactive system can result in selective deficits in recognizing known or novel words (Hinton and Shallice, 1991; Plaut and Shallice, 1993).

Despite the emphasis that connectionist models place on distributed rather than modular processing, they are still not constrained by physiological validity. It is in this role that functional neuroimaging has the potential to contribute to normal and abnormal models of language processing. In particular, it can assess whether there is a specialized neural system for a particular process or whether the implementation of that process is governed by a particular pattern of distributed activity in neural systems that are shared by other functions.

II. Neuroimaging Studies of Aphasic Patients

A. Expectations of Neuroimaging Studies of Aphasic Patients

With the advent of functional neuroimaging, it is hoped that some of the incompleteness of the lesion-deficit model can be remedied by studying brain activity in normal and brain-damaged subjects. Functional imaging offers several fundamental advantages over the lesion-deficit model. The most obvious is that, unlike the lesion-deficit model, functional imaging is not limited to a particular region of the brain that has been damaged; rather, the system of distributed cortical areas that sustain sensory, motor, or cognitive tasks can be identified. In particular, we can examine the effect that a lesion has on the responses in undamaged regions. Another advantage of the systems-level approach is that it does not assume that cognitive processes or operations are localized in discrete anatomical modules (functional segregation) but allows for functional specialization that is embodied in the interactions among two or more distributed regions (functional integration).

As hypothesized by Wernicke and Lichtheim (see Section I), some aphasic patients suffer from abnormal functional segregation (i.e., the function of a discrete cortical region such as Broca's area is abnormal) and some patients suffer from abnormal functional integration (i.e., abnormal connections among different brain regions). Abnormalities in functional integration can be demonstrated when the responsiveness of an undamaged region is determined by task-dependent activity in other connected regions. In patients, a nondamaged region might perform normally in some tasks and abnormally in other tasks, depending on whether input from a disconnected region is required. Studies of functional integration are ideally suited for testing the neuronal dysfunction expressed by patients with disconnection syndromes. For example, as shown in Fig. 1, Wernicke predicted that patients with "conduction aphasia" should be able to comprehend speech, indicating intact functioning of Wernicke's area. They should also be able to articulate speech, indicating intact functioning of Broca's area, but they are unable to repeat back what is said to them. The hypothesis is that the deficit results from a disconnection of the fibers that connect Broca's area with Wernicke's area (see Section I). This hypothesis could be tested directly with neuroimaging by demonstrating that Broca's area performs normally during self-generated speech tasks but abnormally during repetition. Similarly, Wernicke's area should respond normally during comprehension tasks but not during repetition. In other words, abnormal regional responses would be observed only when Broca's and Wernicke's areas interact directly.

In the example of classical conduction aphasia, the dysfunctional integration is a direct result of damage to the arcuate fasciculus-the white matter that connects Broca's and Wernicke's regions. This corresponds to an anatomical disconnection. In other cases, there may be no detectable pathological damage but a failure to integrate activity during particular tasks. This is referred to as a functional disconnection. An example of a functional disconnection was recently demonstrated by Paulesu *et al.* (1996) in a group of developmental dyslexics without detectable pathological damage. These patients were able to activate Broca's area during rhyming and the left temporoparietal cortex during a short-term memory task but, in contrast to normals, these regions were never activated in concert. In other words, no single component of the system is malfunctioning but the system as a whole is malfunctioning because the components do not work in concert.

In summary, by assessing the responsiveness of undamaged regions during different tasks, functional neuroimaging provides a potential means to test directly the disconnection syndromes first described by the nineteenth century neurologists.

B. Types of Neuroimaging Experiment That Can Be Performed on Aphasic Patients

Most neuroimaging studies assume that different tasks will be associated with a different set of cortical areas, and experiments aim to identify the areas where there are changes in regional cerebral activity in response to changes in task or pathology. In relation to patients, three types of activation pattern can be distinguished: normal, underactivity, and overactivity. Normal activation is usually only informative when it is unpredicted: for instance, when structural imaging (with CT or MRI) indicates damage to a neuronal system thought to be important for the task and functional imaging indicates that the area is still responsive (see Section IIIF). Underactivity is only interpretable when patients make normal behavioral responses because if a patient cannot perform the task, the corresponding neuronal system cannot be activated and it is not possible to tell whether the abnormal neural processing is a consequence of the performance deficit or whether the performance deficit is a consequence of the abnormal neural processing (see Fletcher *et al.*, 1998). In the context of normal task performance, a failure to activate a component of the normal system implies that the region is not necessary for task performance (but see Section IIIB).

Overactivity indicates either cognitive or neuronal reorganization. Cognitive reorganization is indicated when a patient uses a different set of cognitive processes to implement the same task (e.g., when patients learn a particular strategy to recover the ability to perform a lost function). Neuronal reorganization takes place when patients use a different neuronal system to implement the same set of cognitive processes. Since neuronal systems, and their connections, are fully wired in the mature brain, evidence for a different neuronal system in a patient (with damage postdevelopment) does not represent a rewiring of the system but changes in preexisting connections. Such changes can occur slowly through learning-related plasticity, for example, when a patient learns a new cognitive strategy, or they may come about immediately following deafferentation of inputs (Buonomano and Merzenich, 1998). In the latter case, overactivity could indicate a disinhibitory phenomenon. For instance, neuronal systems for a particular cognitive process may be duplicated in the normal brain, with the dominant system inhibiting the others. When the prepotent system is damaged, the less dominant systems are able to respond. The duplication of functionality in neuronal systems has been referred to as "degeneracy" (see Edelman, 1989). This property renders a function immune from the effects of focal damage. Furthermore, because functionality is preserved, the effect of damage may never be detected by neuropsychological assessment. Functional imaging studies of patients therefore have an important role to play in the identification of degeneracy—the multiplicity of sufficient brain systems for a cognitive function.

In summary, functional imaging studies on patients reveal either normal or abnormal activity. Abnormal activation can occur locally at the site of a lesion or in undamaged tissue distant from the lesion. Failure to activate a component of the normal system despite normal behavioral responses suggests that region was not necessary for task performance. Overactivity reflects either cognitive reorganization (when alternative cognitive strategies and their corresponding brain systems are adopted) or neuronal reorganization (when preexisting cognitive strategies are reinstantiated using duplicated neuronal systems).

C. Limitations of Neuroimaging Studies on Cognitively Impaired Patients

Neuroimaging studies of patients usually investigate how a neuropsychological deficit is characterized in terms of abnormal brain function. The expectation is that the alteration of neuronal responses in the damaged brain will shed some light on the physiological underpinnings of a cognitive deficit. There are fundamental limitations when using neuroimaging in this way. The most critical point relates to task performance. By definition, patients will have reduced performance relative to normals on tasks that reveal a cognitive impairment. In severe cases, the patients may not be able to perform a task at all. In this case, it is meaningless to perform neuroimaging experiments that attempt to compare normal and abnormal brain activity because failure to activate could be due to either a loss of neuronal responsiveness or a failure to perform the task. Neuroimaging studies of patients are therefore only interpretable relative to normal subjects if the patients retain the ability to perform the task. Thus, whereas neuropsychological assessments focus on functional impairments, neuroimaging assessments focus on functional preservation in the context of brain damage.

The second point that should be noted regards the interpretation of abnormal activation when a patient activates a region that is not usually observed in normals. The distinction of whether an abnormal activation reflects cognitive reorganization or neuronal reorganization is not straightforward. Neuronal reorganization is only indicated if the patient performs the task using the same set of cognitive operations as normal subjects; i.e., changes in cognitive reorganization are excluded. This is not possible from functional imaging data, which are derived at a neurophysiological level, not at the cognitive level. The cognitive components of a task (the task analysis) need to be specified from behavioral data. Even here, a task analysis is never as refined or as comprehensive as one would like because there are certain attributes of cognitive processing (e.g., attention, implicit processing, and effects of time) that are not always amenable to measurement and these subtle cognitive differences may result in changes in neuronal implementation. The point being made here is that the cognitive level of description of a task needs to be as detailed as possible to conclude that (i) the patient is using the same cognitive system as normals and (ii) any changes at the neuronal level are due to neuronal reorganization rather than cognitive reorganization. The more detailed the task analysis, the more valid the inferences about changes in neuronal implementation.

Just as inferences regarding neuronal reorganization depend on excluding an explanation in terms of cognitive reorganization, inferences about cognitive reorganization depend on excluding explanations in terms of neuronal reorganization. This might be achieved with neuroimaging if the same neuronal systems activate when normal subjects adopt the same cognitive strategy as the patient. For example, to confirm that the differential pattern of activation seen in a dyslexic patient relates to a letter-by-letter reading strategy, the equivalent

activation pattern needs to be elicited in normals when they also read letter by letter. As our understanding of normal neuronal systems expands, we should be able to identify a patient's strategy from the recorded neuronal responses. This would entail associating the regions activated by a patient to the cognitive processes that produce the same neuronal response in normal subjects.

In summary, changes in neuronal responses may result from cognitive or neuronal reorganization. It is only possible to make inferences about one level of reorganization when the other can be shown to be unaffected. This emphasizes the crucial relationship between behavioral and imaging data and a dependence on a detailed task analysis prior to the imaging experiment. The following section provides some examples of neuroimaging experiments with aphasic patients who can perform the tested task.

III. Examples of How Neuroimaging Experiments on Aphasic Patients Have Contributed to Normal and Abnormal Models of Language

This section is divided into six parts. The first (A) presents a functional imaging study of aphasics with damage to Broca's area and illustrates the effect of the lesion on distant undamaged regions. The second (B) investigates how patients retain the ability to perform a task despite damage to cortical regions known to be activated in normals. These studies demonstrate how functional imaging studies of patients can be used to delineate the necessary and sufficient brain systems. The third (C) investigates task-specific effects of lesions on regions distant to the site of damage. The fourth (D) discusses cognitive and neuronal abnormalities in developmental dyslexics who have no detectable pathological damage. The fifth (E) considers whether there is a duplicated neuronal system for reading in the right hemisphere by imaging aphasic patients with deep dyslexia. Finally, the sixth part (F) describes an investigation of the mechanisms underlying the recovery of language.

A. The Effect of a Lesion on Distant Undamaged Regions

It has been known for more than a century that damage to Broca's area results in speech production deficits but the effect of such lesions on undamaged regions has only been investigated recently (Price *et al.*, submitted). Four aphasic patients who all had damage to the left posterior inferior frontal cortex, classically known as Broca's area, participated in the functional imaging experiment (see Fig. 2). Each patient had speech output deficits but relatively preserved comprehension for single words-spoken and written. During the functional imaging study, the patients were engaged in a simple nonlinguistic task that they could perform normally. This involved a key press response when a visually presented letter string contained a letter with an ascending visual feature (e.g., b, d, f, h, k, l, or t). Attention to the stimuli was therefore maintained and task performance was monitored. While the task remained constant in all conditions, the stimuli presented were either words or consonant letter strings (see Fig. 2). Activations detected for words, relative to letters, are attributed to implicit word processing when subjects are engaged in an

irrelevant task (Price *et al.,* 1996; Büchel *et al.,* 1998; Brunswick *et al.,* 1999). Implicit word processing is an established psychological phenomenon that is classically illustrated by the Stroop effect (Macleod, 1991). The functional imaging experiment investigated how implicit word processing was affected by left frontal lesions incorporating Broca's area.

In 15 normal controls, word relative to consonant letter string conditions activated the right cerebellum and the following regions in the left hemisphere: posterior inferior frontal, inferior parietal, posterior middle and inferior temporal, and medial superior frontal (see Fig. 2). Each patient showed normal activation of the left posterior middle temporal cortex, which has been asso-

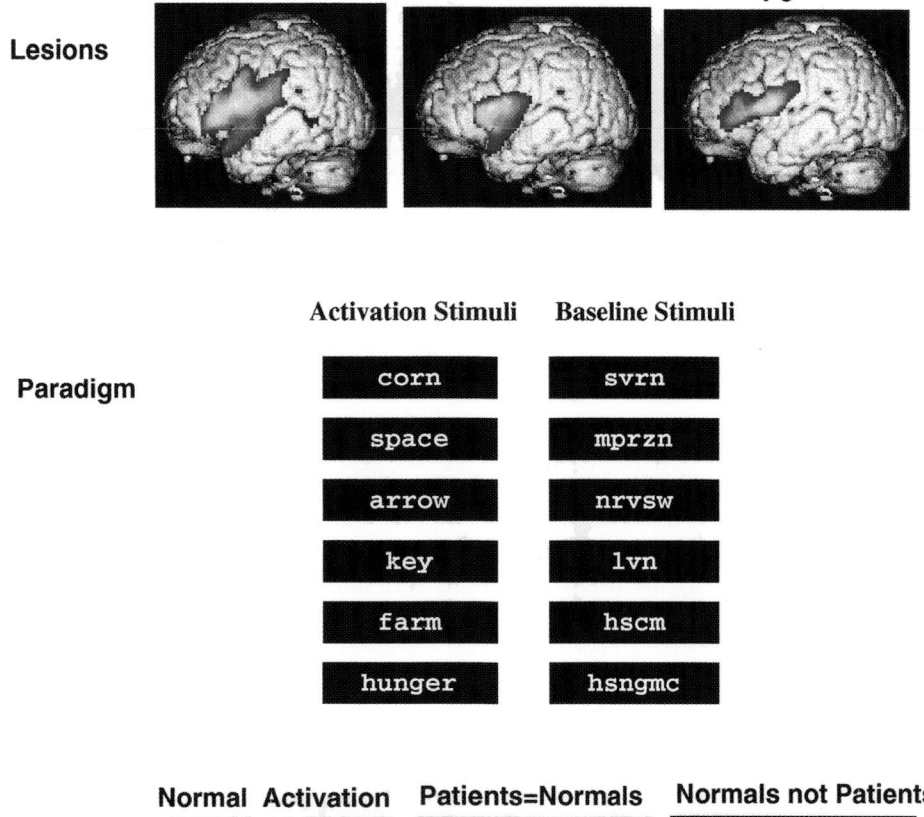

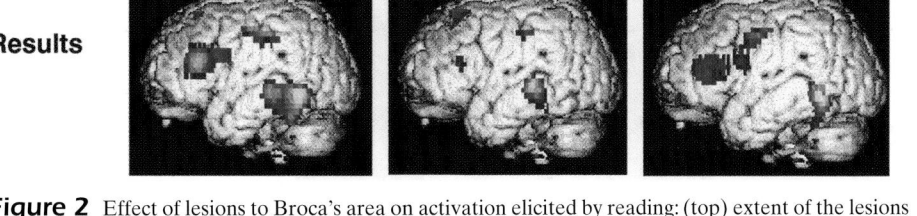

Figure 2 Effect of lesions to Broca's area on activation elicited by reading: (top) extent of the lesions in three of the patients (P1, P2, and P3) suffering from Broca's aphasia; (middle) examples of the activation and baseline stimuli used in the functional imaging paradigm; (bottom) the pattern of normal activation (left), where the patients activated like the normals (middle), and where the patients failed to activate like normals (right).

ciated with semantic processing (Vandenberghe *et al.,* 1996). However, none of the patients activated the left posterior inferior frontal cortex (damaged by the stroke), the left posterior inferior temporal region (undamaged by the stroke), or any other regions that were not activated by the controls (see Fig. 2).

Underactivity in the left posterior inferior frontal cortex was predicted given the extent of the lesion and the associated cognitive deficit. However, abnormal responses in the left posterior inferior temporal cortex were not predicted and occurred even though this undamaged region (i) lies adjacent and posterior to a region of the left middle temporal cortex that activated normally and (ii) is thought to be involved in an earlier stage of word processing than the damaged left inferior frontal cortex (Foundas *et al.,* 1998). The left posterior inferior temporal area is not one of the classic language regions because it is supplied by both the posterior and middle cerebral arteries and therefore protected from the most common types of cerebral vascular events. However, in the past decade, it has been named the "basal temporal language area" because electrical stimulation disrupts naming (Lüders *et al.,* 1986; Burnstine, 1990). There have also been two recent studies showing that isolated lesions to the posterior basal temporal language area result in naming deficits, with comprehension relatively preserved (Raymer *et al.,* 1996; Foundas *et al.,* 1998). Similarly, functional imaging experiments have shown that the left posterior inferior temporal area and the left anterior insula activate together in a wide variety of word generation tasks (Warburton *et al.,* 1996; Price and Friston, 1997; Büchel *et al.,* 1998).

The results of this functional imaging study of aphasic patients with left inferior frontal damage suggest that during the implicit reading task, responses in the left basal temporal language area rely on afferent inputs from the left posterior inferior frontal cortex. Thus neuroimaging has demonstrated the effect of a lesion on the functionality of distant undamaged regions, thereby illustrating the integrative nature of neuronal systems.

B. Necessary and Sufficient Brain Systems

The patients with lesions to Broca's area had impaired speech output but preserved comprehension for single words. By deduction, the damaged area must have been necessary for speech production but not speech comprehension. The application of the lesion-deficit model, however, does not inform us of the full set of regions required for speech production. In contrast, functional imaging studies on normal subjects reveal the full set of regions required for one task relative to another but do not indicate the necessity of the different subcomponents—some activated regions might be

superfluous to task requirements (see Price *et al.,* 1996). Hence, functional imaging studies of normal subjects identify the full set of regions for a task and the lesion-deficit model provides information on whether a particular region is necessary for task performance. An example of how functional imaging can motivate behavioral studies of patients is described by Feiz *et al.* (1992), who report a patient with a right cerebellar infarct following the observation that the right cerebellum is activated during verbal fluency (Petersen *et al.,* 1988). On non-motor tasks, the patient showed deficits completing and learning a word generation task but had normal or above normal behavior when performing standardized language tasks. In this instance, a neuroimaging study of word generation motivated a neuropsychological assessment of word generation in a patient with right cerebellar damage. The neuropsychological study allows inferences to be made regarding the necessity of this subcomponent of the language system following identification by neuroimaging.

An inference designating a region as "not necessary" is not so straightforward. Patients might retain the ability to perform a task despite damage to necessary cortical regions because the function of the damaged region was either (i) carried out by a different brain system (e.g., involving the homolog region in the contralateral hemisphere (Weiller *et al.,* 1995; Buckner *et al.,* 1996) or cognitive reorganization) or (ii) maintained by residual functionality in an area that appears damaged with routine structural imaging (see Section IIIF). To discount these possibilities and conclude that a region is not necessary, functional imaging is required of patients who retain the ability to perform a task despite damage to a region known to be activated when normals perform the task. In summary, the combination of functional neuroimaging experiments on normals and patients can reveal necessary and sufficient brain systems.

This approach is illustrated in Figs. 3–5 with two types of patients who retained the ability to perform semantic decision tasks despite damage to components of the system activated by normal subjects performing the task. The semantic task used involved semantic similarity judgments on words and pictures of objects and the baseline used was perceptual (visual size) judgments on the same stimuli (see Figs. 3 and 4). This paradigm is associated with activation of frontal and temporal regions in the left hemisphere, including the anterior inferior frontal, anterior and posterior inferior temporal, and middle temporal cortices, the bilateral posterior inferior parietal cortices, the anterior cingulate cortex, and the right cerebellum (Mummery *et al.,* 1999). The first patient investigated (SW) was able to perform the paradigm, despite extensive damage to the left frontal and temporoparietal cortex, including the left anterior inferior frontal region activated by normal subjects. The

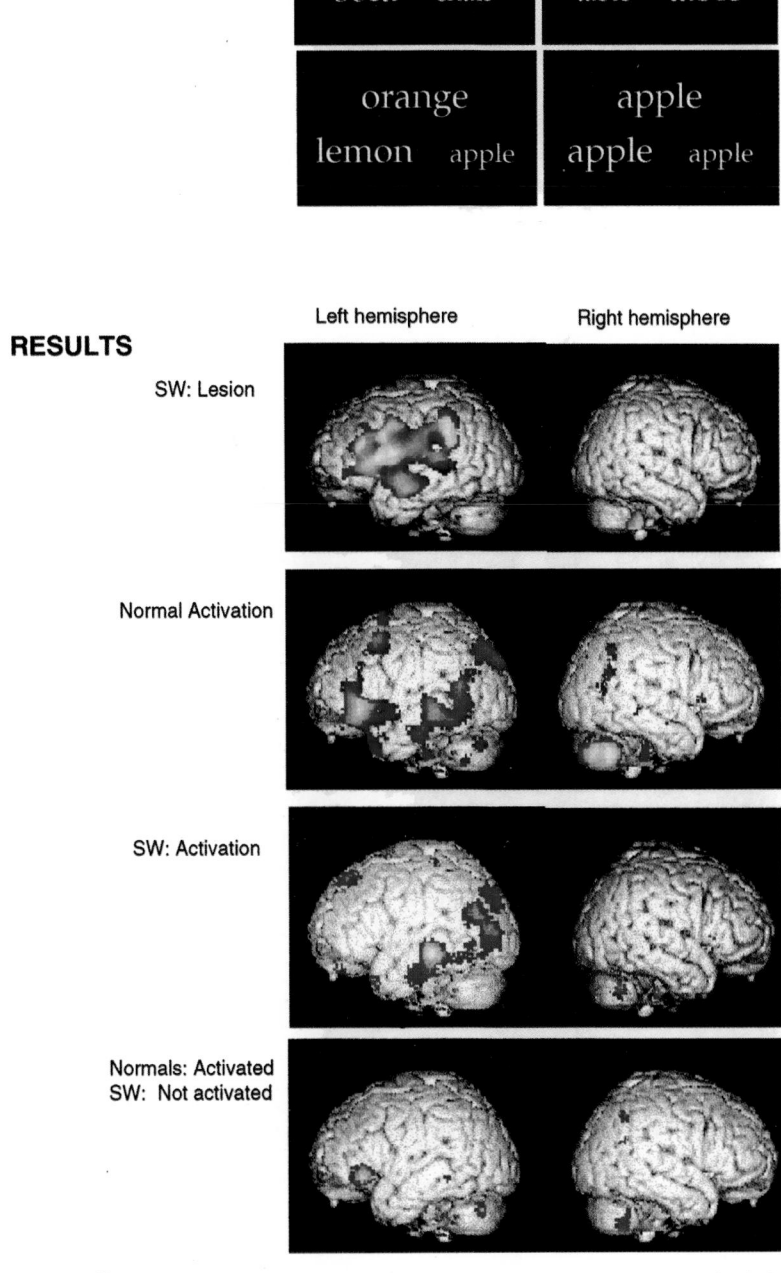

Figure 3 Effect of frontoparietal damage on activation elicited by semantic decisions: (top) examples of the activation and baseline stimuli used for semantic and size decision tasks; (bottom) the extent of patient SW's lesion and the resulting pattern of activation when semantic decision is contrasted to size decision for normals, SW, and normals but not SW.

functional imaging experiment revealed that SW performed the semantic decision task by activating the left temporal and parietal regions (see Fig. 3) but he did not activate the left anterior inferior frontal cortex ob-

served in all normal controls and there was no compensatory activity in the right inferior frontal cortex. The conclusion is that the left anterior frontal cortex activated by normal subjects performing the semantic

PARADIGM

Activation task	Baseline task
Semantic decision	Size decision

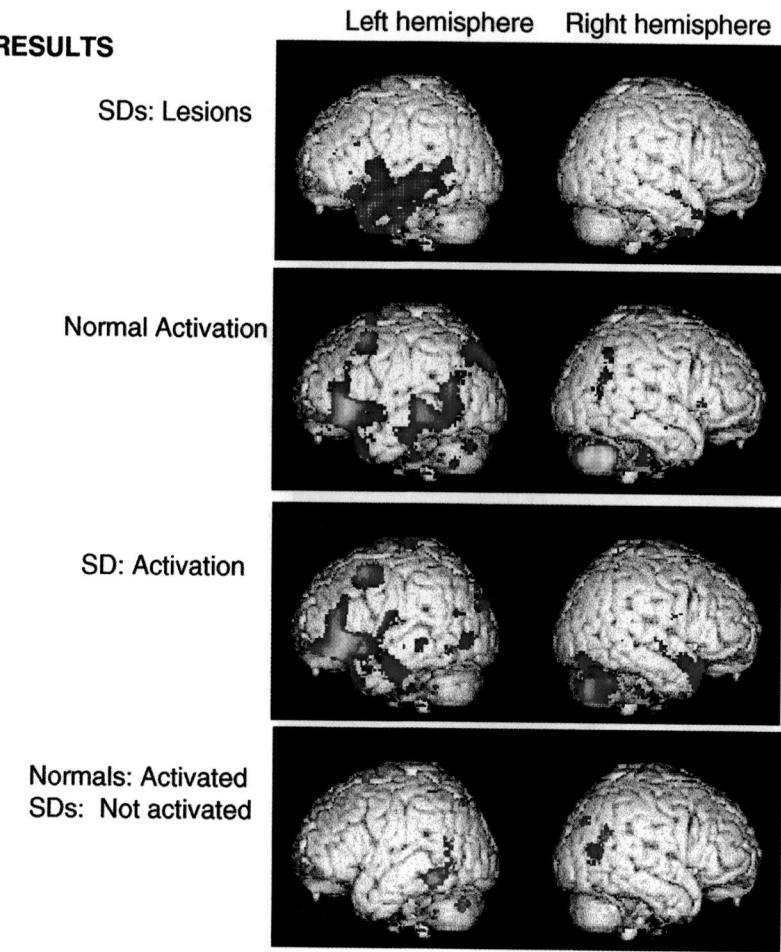

RESULTS

SDs: Lesions

Normal Activation

SD: Activation

Normals: Activated
SDs: Not activated

Figure 4 Effect of anterior temporal lobe damage on activation elicited by semantic decisions: (top) examples of the activation and baseline stimuli used for semantic and size decision tasks; (bottom) the area where all four patients with semantic dementia (SDs) had reduced gray matter and the resulting pattern of activation when semantic decision is contrasted to size decision for normals, patients with semantic dementia (SD), and normals but not SDs.

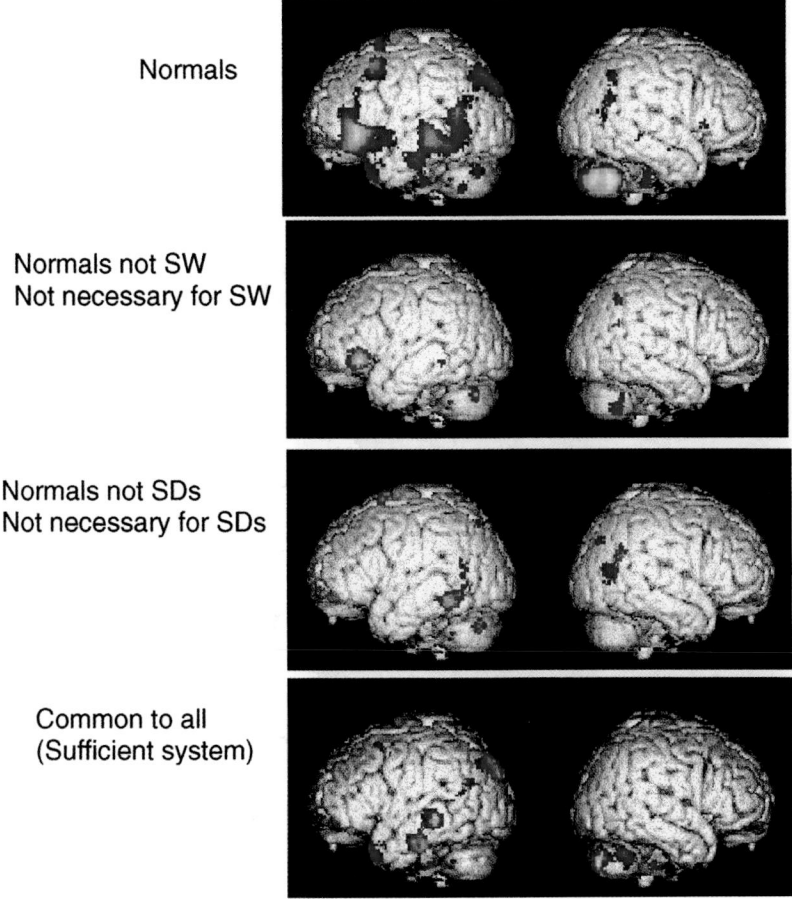

Normals

Normals not SW
Not necessary for SW

Normals not SDs
Not necessary for SDs

Common to all
(Sufficient system)

Figure 5 Delineating the necessary and sufficient semantic system: Activation for semantic relative to size decisions (see Figs. 3 and 4) for, respectively, normals, normals but not SW, and normals but not patients with semantic dementia (SDs), and those areas activated by normals, SW, and SDs.

similarity task is not necessary for SW to perform the same task (Price *et al.*, 1999).

The second type of patient scanned suffered from a progressive deterioration in semantic knowledge and name retrieval, with other cognitive and language functions remaining relatively intact (Sasanuma and Monoi, 1975; Warrington, 1975; Hodges *et al.*, 1992). The anatomical correlates of this condition, known as semantic dementia (SD), are the bilateral temporal lobes, with cortical atrophy commencing in the anterior temporal poles and then spreading posteriorly as the disease progresses. According to the lesion-deficit model, regions within the anterior temporal lobes appear to be the site of the impaired semantic and naming processes. Six such patients who were only mildly impaired on the semantic similarity task were investigated with functional neuroimaging on the paradigm discussed above (see Fig. 4). Only four of the patients were able to perform the task well in the scanner. In order to match per-

formance in the patients and the normal controls, scans where patients had impaired performance relative to the normals were removed. The resulting activations involved the left inferior frontal, left temporoparietal, and left anterior middle temporal cortices, the anterior cingulate cortex, and the right cerebellum. Unlike the normals, the patients did not activate the left posterior inferior temporal cortex or the right temporoparietal junction. Neither of these regions demonstrated structural damage but remarkably, the regions where there *was* structural damage, i.e., the anterior temporal cortices, were more active in the patients than the normals. This study therefore demonstrates (i) activation at the site of damage, (ii) dysfunction distant to the site of damage, and (iii) dysfunctional integration between damaged anterior and undamaged posterior inferior temporal cortices. It also illustrates that the left posterior inferior temporal area that failed to activate in each patient was not necessary for intact task performance.

In this case, however, there was compensatory activation in the left inferior frontal cortex and the left anterior middle temporal cortex, suggesting that there was cognitive and/or neuronal reorganization.

The combined results of SW (Fig. 3) and the SDs (Fig. 4) are illustrated in Fig. 5. This shows that during performance of the semantic similarity judgments, SW failed to activate the left anterior inferior frontal cortex activated by all normal subjects and the SDs failed to activate the left posterior inferior temporal cortex. Together these results suggest that activation in the left inferior frontal and the left posterior inferior temporal cortex was not necessary for the task. The bottom panel in Fig. 5 illustrates the areas that were activated by all the patients and normals. This delineated semantic system indicates the importance of the left middle temporal, left anterior inferior temporal, and left posterior inferior parietal cortices and the right cerebellum. Further studies are required on patients with lesions to these areas. Theoretically, one might hypothesize that intact task performance could follow damage to any subcomponent of a system. An analogy might be that it does not matter which finger you chop off; the remaining fingers are able to adapt to some task requirements. This would be an example of degeneracy (see Section IIIE). However, in the example given above, lesion studies indicate that impairments on semantic tasks do result in patients with extensive damage to the anterior temporal cortices (Hodges *et al.,* 1992) and in patients with transcortical motor aphasia who have damage to the left inferior temporal and posterior inferior parietal cortices, the left thalamus, and the white matter connecting these regions (Alexander *et al.,* 1989). The regions illustrated in Fig. 5 may therefore indicate the necessary and sufficient semantic system.

C. Task-Dependent Lesions

An area of underactivity in a patient can be either at the site of lesion, as in SW (the patient with the damaged left frontal cortex), or distant to the lesion (as with the semantic dementia patients). In the latter case, the dysfunctional, undamaged region must have been dependent on input from the damaged anterior temporal regions. In other words, the posterior temporal region appears to have been functionally disconnected from the semantic system. A disconnected undamaged region is either (i) rendered abnormally responsive in all contexts or (ii) may function normally when it does not rely on inputs from the damaged region. The occurrence of normal activation in some contexts but dysfunctional responses when integration with the damaged region is involved has recently been referred to as "dynamic di-

aschisis" and illustrated in a patient with an extensive left frontal lesion (Price *et al.,* submitted). In the reading paradigm illustrated in Fig. 2, SW failed to activate the damaged frontal cortex and showed abnormal activation in the undamaged left posterior inferior temporal cortex (decreased activity rather than increased activity). The inference is (as discussed in Section IIIA) that activation in the left posterior inferior temporal cortex normally relies on inputs from the frontal cortex. Indeed, when the same patient (SW) performed a semantic paradigm that activates a different set of regions, the undamaged left posterior inferior temporal cortex activated normally. The results of the reading paradigm and the semantic paradigm combined are illustrated in Fig. 6, which shows the areas that were activated in normal subjects and SW and the areas that were activated by normals but not SW. The bottom panel then highlights the area that was activated by SW during semantic decisions but not during reading. The conclusion is that the left posterior inferior temporal cortex responds abnormally only when it depends on input from the damaged frontal region (as in the reading paradigm).

D. Cognitive and Neuronal Abnormalities in Developmental Dyslexics

In the case of patients with known cortical damage, dysfunction at a distant site can be attributed to impaired connections with the damaged area. This can be described as an anatomical disconnection. Context-dependent dysfunction can also occur in the absence of anatomical damage, as, for instance, when no single component of the system is malfunctioning but the system as a whole is malfunctioning because the components do not work in concert. This is referred to as a functional disconnection and has been demonstrated in a group of developmental dyslexics who were able to activate Broca's area during rhyming and the left temporoparietal cortex during a short-term memory task but who failed to activate both regions in concert, despite no detectable pathological damage (Paulesu *et al.,* 1996).

Other functional imaging studies have confirmed abnormalities in the left perisylvian and extrasylvian temporal cortex (Rumsey *et al.,* 1992, 1997; Shaywitz *et al.,* 1998; Brunswick *et al.,* 1999), suggesting that the deficits developmental dyslexics have with reading, short-term memory, phonological awareness, rapid object naming, and speech production may relate to abnormal perisylvian function. However, reading difficulty has also been shown to correlate with abnormalities in the magnocellular component of the visual system (Eden *et al.,* 1996; Demb *et al.,* 1997). Differences between studies may reflect a malfunctioning system where the individual re-

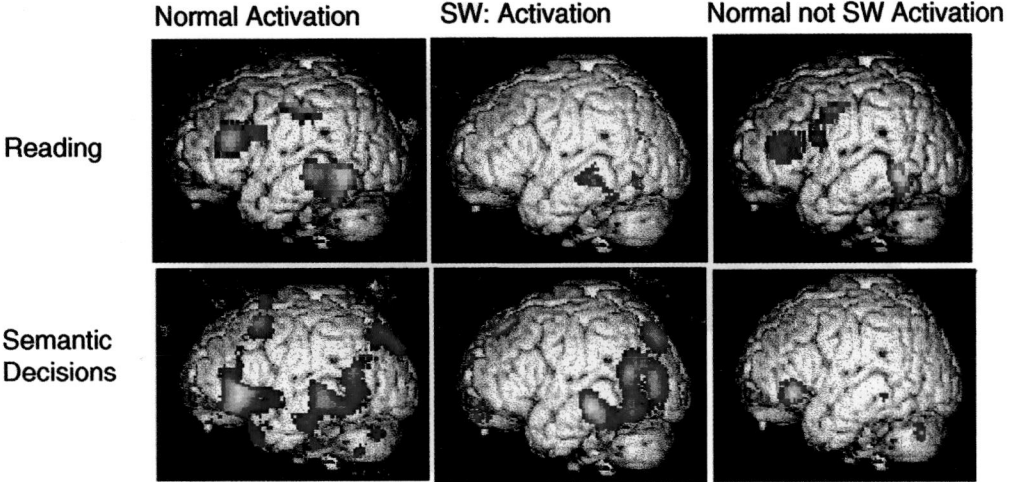

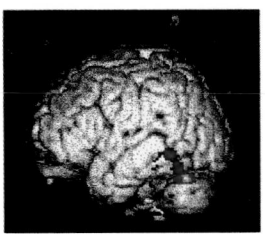

Activated by SW during semantic decisions
Not activated by SW during reading

Figure 6 An example of dynamic diaschisis: Activation patterns for (top row) reading words relative to consonants (see Fig. 2) and (second row) semantic relative to size decisions (see Fig. 4). First column shows normal activation, second column shows activation by SW, and third column shows areas activated by normals but not SW. The regions illustrating dynamic diaschisis are shown at the bottom. Here, the voxels depicted showed abnormal responses during reading but normal responses during semantic decision (see text).

gions perform normally in some tasks and abnormally in others. However, there may also be differences between studies in the degree of reading impairment and the subject's ability to perform the different tasks.

Figure 7 illustrates the results of a functional imaging study of developmental dyslexia by Brunswick *et al.* (1999) that identified the neural system associated with reading by contrasting brain activity during reading familiar words aloud and resting with eyes closed. As can be seen, activations in the visual, motor, and temporal regions were very similar in developmental dyslexics and normal readers. However, there was reduced activation in the dyslexics in the posterior inferior temporal cortex and increased activation in a left premotor region. The interpretation of these results is that the left premotor signal is compensating for reduced activity in the left posterior inferior temporal region associated with modality-independent naming. Other studies are required to determine whether these regions perform normally with different tasks and to determine the underlying cause of the abnormal neuronal responses.

E. A Duplicated Neuronal System for Reading in the Right Hemisphere of Deep Dyslexic Patients?

As discussed in Section II, neuronal systems for a particular cognitive process may be duplicated in the normal brain, with the dominant system inhibiting the others. When the prepotent system is damaged, the less dominant systems are then able to respond. Since functionality is preserved following disinhibition, damage to one system may never be detected in neuropsychological assessments. In contrast, functional imaging studies of patients who have preserved function despite damage to regions of the brain known to be important for a task can reveal multiple, sufficient brain systems for a cognitive function. For instance, functional neuroimaging can determine whether right hemisphere systems can maintain functionality of some language tasks following left hemisphere damage.

The hypothesis that the right hemisphere is maintaining functionality following extensive left hemi-

Reading Aloud Words versus Rest

Left hemisphere   Right hemisphere

Normals: activate
Dyslexics: activate

Normals activate
more than Dyslexics

Dyslexics activate
more than normals

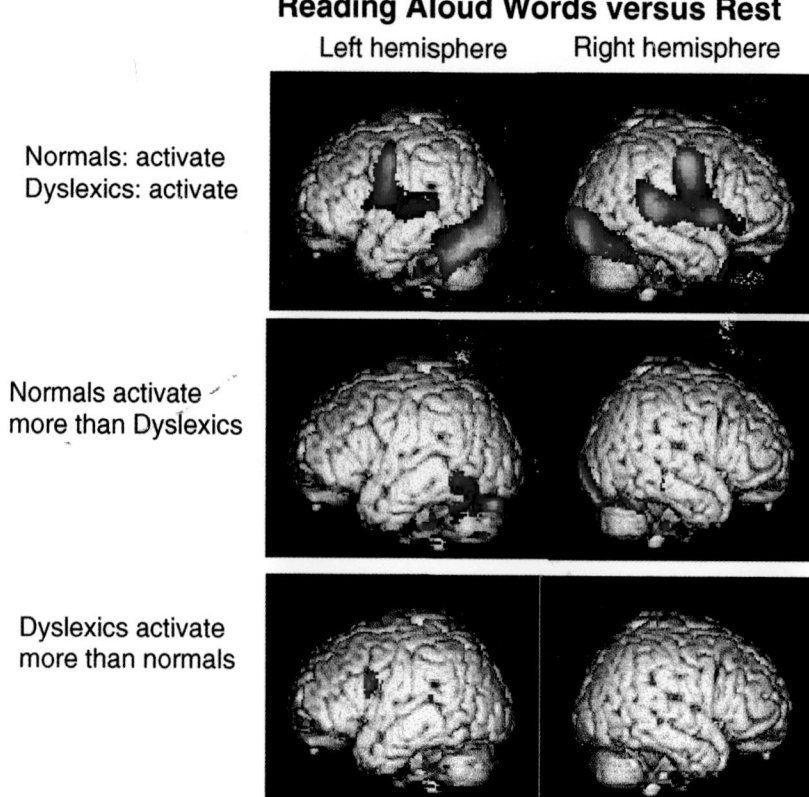

Figure 7 Patterns of reading activity in developmental dyslexia: Activation for reading aloud relative to rest for (top) normals and dyslexics, (middle) normals more than dyslexics, and (bottom) dyslexics more than normals.

sphere damage has been attributed to patients with deep dyslexia. Deep dyslexia is characterized by aphasia and an inability to read aloud (i) the simplest of nonwords (e.g., GAM, DAKE) and (ii) words that lack concrete referents such as function words (OF, THAT) and abstract content words (PROCESS, INVOLVE). Despite this profound inability to generate pronunciations for letter strings on the basis of knowledge of spelling to sound correspondences, the ability to read words with concrete referents (such as object names) is relatively preserved. The most intriguing symptom is that words are sometimes articulated as a semantically related but orthographically and phonologically dissimilar word (e.g., reading SPIRIT as "whisky," YACHT as "ship," and BUSH as "tree"). Clearly, in these cases, the deep dyslexics have comprehended the words to some degree but have failed to produce the appropriate sounds.

Two main classes of explanation for this syndrome have been proposed. One is that deep dyslexia results from a residual left hemisphere reading system that has lost the normal ability to pronounce a printed word without reference to meaning (Morton and Patterson, 1980; Newcombe and Marshall, 1980; Shallice and Warrington, 1980; Patterson and Besner, 1984). The second

hypothesis is that due to extensive left hemisphere damage, the patients rely on a right hemisphere reading system to comprehend words (Coltheart, 1980; Saffran *et al.*, 1980). These neuropsychological predictions can be tested explicitly with functional neuroimaging. If deep dyslexic reading relies on a left hemisphere system that has lost the ability to read words without meaning, then deep dyslexics should activate left hemisphere word processing areas for words they are able to read (i.e., words depicting objects). If, on the other hand, deep dyslexics translate print to meaning in the right hemisphere and then output speech in the left hemisphere, we would expect to see activation in right (but not left) hemisphere regions associated with semantics (as well as activation of left hemisphere regions associated with speech output). These predictions were tested with two deep dyslexic patients (JG and CJ), reported by Price *et al.* (1998) and illustrated in Fig. 8.

Both patients showed activation of semantic and phonological systems in spared regions of the left hemisphere, precluding an explanation of deep dyslexia in terms of purely right hemisphere processing. However, there was also some enhanced activity in the right hemisphere (relative to normals). In one patient (JG) there

Reading Aloud Words versus Rest

Left hemisphere   Right hemisphere

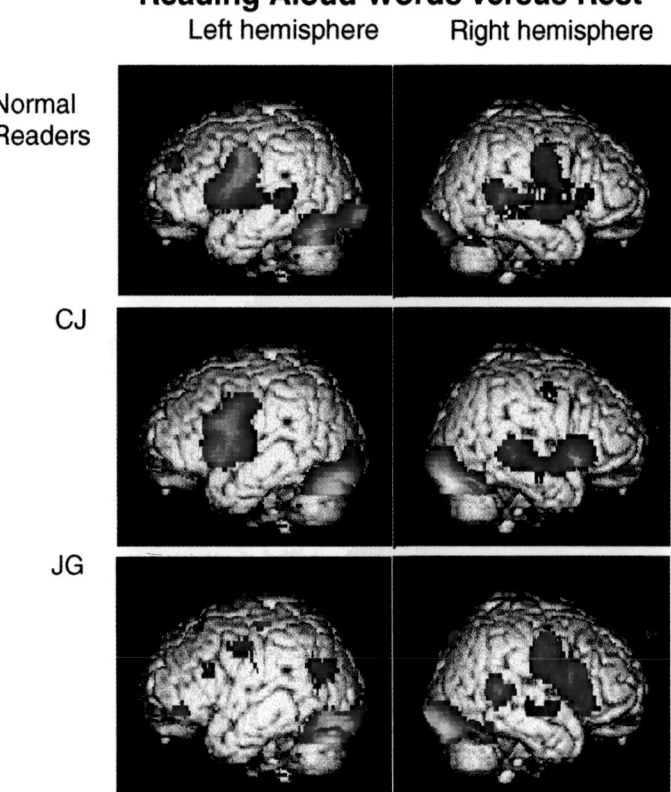

Normal
Readers

CJ

JG

Figure 8 Patterns or reading activity in patients with deep dyslexia: Activation for reading aloud relative to rest for (top) normals, (middle) patient 1 (CJ), and (bottom) patient 2 (JG).

was right hemisphere activation in the frontal operculum and the posterior inferior frontal cortex but no corresponding activation in the left hemisphere. The frontal operculum and Broca's area in the left hemisphere are associated with articulation and speech output. It therefore appears that JG's right hemisphere can articulate speech following damage to the left hemisphere. Since semantic regions were only activated in the left hemisphere, there was no evidence of a duplicated right hemisphere semantic system. In contrast, CJ showed enhanced activations in the right anterior inferior temporal cortex as well as the right frontal operculum. The homolog of the left anterior inferior temporal cortex is associated with semantic processing (Vandenberghe *et al.*, 1996); therefore the right hemisphere may have been contributing to semantic processing in CJ. In summary, the functional imaging studies of these deep dyslexic patients indicate overactivity in right hemisphere language areas but the areas implemented are not consistent across CJ and JG. CJ showed overactivity in a right hemisphere region associated with semantics and both patients showed right hemisphere activity in an area associated with articulation. Figure 7 illustrates the differing activation profiles that led to the same di-

agnosis of deep dyslexia. Since these patients were investigated several years following damage, we do not know whether the overactivity reflects disinhibition or long-term changes in potentiation.

F. Recovery of Language Following Aphasia

How patients might recover a lost function is one of the most crucial questions that needs to be addressed by imaging studies of brain-damaged patients. There are three possible mechanisms of recovery that can be investigated with functional neuroimaging: (i) peri-infarct activation, where viable tissue surrounding the lesion is able to mediate a function that was previously maintained by many more cells in the region; (ii) neuronal reorganization, when the cognitive architecture required for a task is implemented by a different neuronal architecture; and (iii) cognitive reorganization, when a patient compensates for a lost function by using a different cognitive architecture (and its neuronal architecture) to implement the same task.

Structural indices of lesions (from conventional use of CT and MRI scanners) do not necessarily imply a complete loss of function and it is sometimes surprising when a patient with a large lesion makes an unexpectedly good recovery. Functional indices of lesions, in contrast, can detect areas where a degree of functional responsiveness has been maintained even in areas that appear damaged on structural images. Typically, these areas are around the region of insult (e.g., peri-infarct tissue) and sometimes within the lesion. Recovery of a lost function results either by the reactivation of tissue that was initially inactivated (e.g., due to a reduction in edema) or by increases in the synaptic effacity of viable tissue until it can support a function that was originally executed by many more cells. The important role that peri-infarct activity may play in functional recovery has probably been grossly underestimated in imaging studies to date. This is because most previous functional imaging studies have only been able to make inferences by pooling data from different patients into one group and then comparing the patient group to a group of normal subjects (e.g., Weiller *et al.* 1995). Since peri-infarct activity inevitably varies from patient to patient, depending on the size and location of the lesion, it will not be detected in group-to-group comparisons. The demonstration of peri-infarct activity therefore relies on studies where each subject is analyzed individually.

In a recent study by Warburton *et al.* (1999), six patients with large left temporoparietal regions who had lost, and then recovered, the ability to generate words were scanned six times during a word generation task and six times during rest. Data from each patient were analyzed independently and compared to those of a group of nine control subjects. In normal subjects, the word generation task (relative to rest) consistently activates a widely distributed system of language regions in the left hemisphere (in particular, the left prefrontal and posterior temporal cortices). By analyzing data from each subject individually, it was possible to ascertain that half the normal subjects also activate the same set of regions in the right hemisphere (see Figure 9). All six recovered aphasics also evidenced activation in the left prefrontal regions, all but the most impaired patient activated the damaged left temporal lobe, and half activated the right prefrontal and temporal cortices (see Fig. 9). The peri-infarct left temporal activity was not detected when the patients were pooled together for a group analysis.

In summary, the activations associated with cued word retrieval in the recovered aphasics were indistinguishable from the normal controls, except that in the presence of a lesion the activations were perilesional. Similar results have been obtained in a single-case study of a patient who had recovered from auditory agnosia (Engelein *et al.*, 1995). Heiss *et al.* (1997) have also demonstrated in a longitudinal study that the recovery of aphasia is related to the reactivation of left hemi-

spheric speech areas surrounding the area of infarction.

Other functional neuroimaging studies of aphasia (e.g., Weiller *et al.*, 1995) have suggested that recovery occurs following a laterality shift, with homologous regions in the contralateral cortex assuming the functions of the damaged region. A laterality shift could result from disinhibition of a duplicated right hemisphere language system or an increase in the responsiveness of a system that already exists. However, Weiller *et al.*'s (1995) conclusions were based on findings that the right hemisphere language areas are more active in a group of recovered aphasics than a group of normal controls. Another explanation that needs to be excluded is that the patients who make a good recovery from language are those who activated the right hemisphere language areas before damage (see Fig. 9). A group of recovered aphasics would then be biased to encompass individuals who have bilateral representation of language. This will result in greater activation of right hemisphere language areas in relation to a normal sample of the population where less than half activate bilaterally. Analysis of individual subjects is therefore required to determine whether recovered aphasics activate the right hemisphere more than normal subjects with bilateral language representation.

Finally, as discussed in Section II, in order to demonstrate neuronal reorganization has occurred, it is necessary to show that the cognitive implementation of the task is the same in both patients and normal subjects. This may not be possible because some attributes of cognitive processing are not amenable to measurement, particularly attentional set, implicit cognitive processing, and performance level. Indeed, many of the changes attributed to neuronal reorganization may simply reflect different task performance between the patients and the normals. In summary, the mechanisms involved in language recovery have not as yet been elucidated in neuroimaging studies. Further studies are required to distinguish between cognitive and neuronal changes. Nevertheless, functional imaging has made it clear (see Fig. 7) that some patients recover language performance using the same neuronal systems as normals by virtue of residual function in the damaged cortex.

IV. Conclusions

A. Implications for Normal Models of Language Processing

The joint and complementary use of neuroimaging and neuropsychology offers several fundamental advantages over either technique in isolation. The first relates to defining necessary and sufficient normal sys-

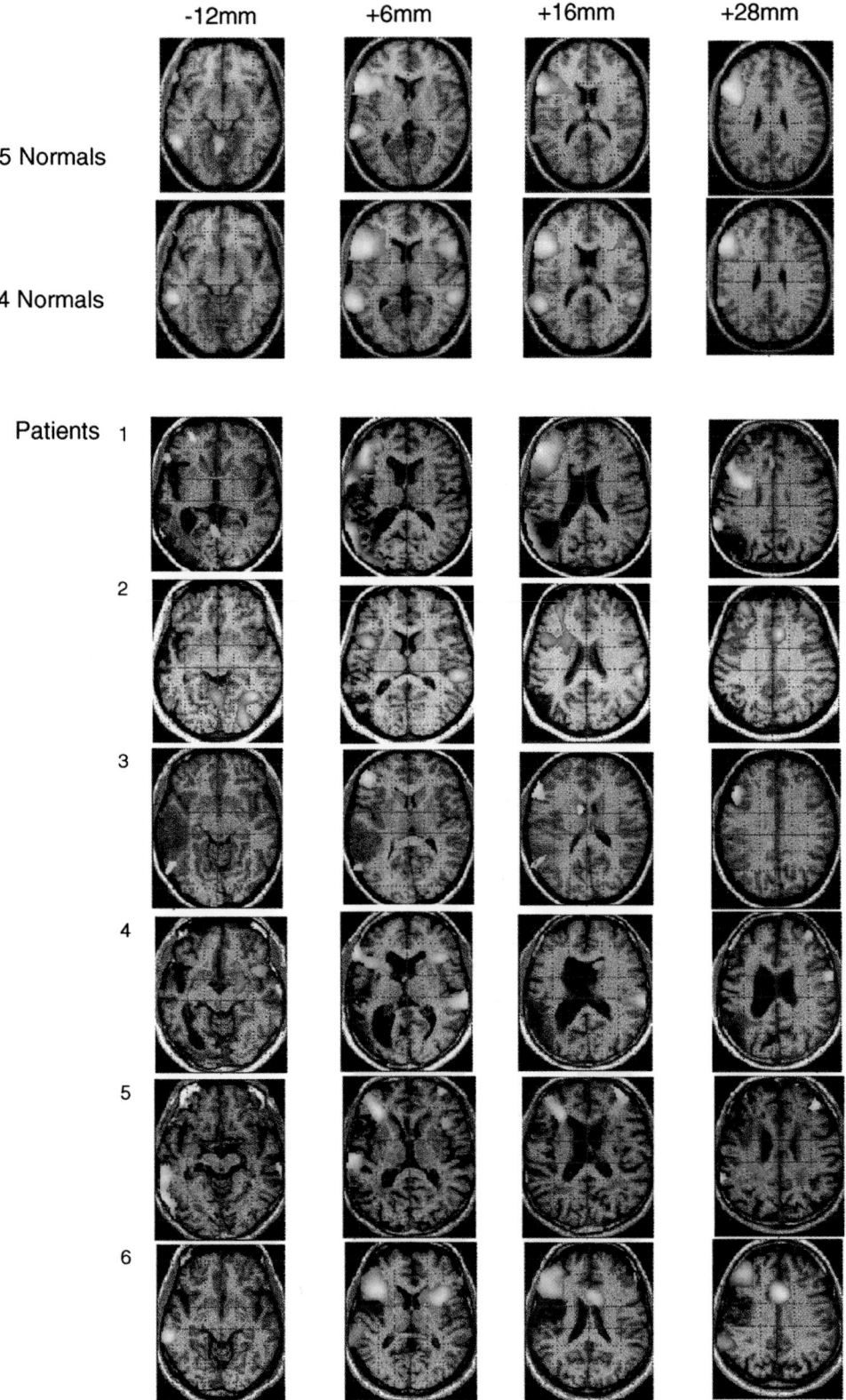

Figure 9 Recovery of verbal fluency following neurological infarct. Four coronal slices showing activation for verbal fluency relative to rest in five normals who show left lateralized responses, four normals who show bilateral responses, and six patients who show peri-damage activation in the left hemisphere and varying degrees of activation in the right hemisphere (like normals). From Warburton *et al.*, 1999.

tems. Neuroimaging in normal subjects defines the sufficient set of regions (the neural architecture) for performing one task relative to another. Neuropsychology establishes the necessity of cortical components by investigating the performance of subjects with damage to the region of interest, and critically, neuroimaging of the patient is required to establish whether the function of a region is superfluous to task performance. By designating each region in the sufficient system as necessary or not necessary, the critical normal system can be identified (see Section IIIB).

The second advantage of imaging neuropsychologically impaired patients relates to the potential identification of duplicated linguistic functions that are normally inhibited by the dominant system. This applies most strongly to the role of the right hemisphere in language processing. If, in some patients, the right hemisphere preserves functionality following left hemisphere damage, changes in neuronal implementation may not be detected in neuropsychological assessment. In contrast, functional imaging of the patient can reveal when a different system is being used (see Section IIIE).

The third advantage of imaging aphasic patients relates to normal functional integration. For example, the finding that patients with lesions to Broca's area failed to activate the left posterior inferior temporal cortex during an implicit reading paradigm (Section IIIA, Fig. 2) indicates that during this task, the left posterior inferior temporal cortex relied on input from Broca's area. Similarly, the finding that patients with anterior temporal lobe damage fail to activate the left posterior inferior temporal cortex during a semantic task (Section IIIB, Fig. 4) indicates that during the semantic task, the left posterior inferior temporal cortex relied on input from the anterior temporal cortex. Finally, the contrasting effects of different lesions on the function of the left inferior temporal lobe indicate that the input to this region depends on the task. This is an illustration of how functional specialization can be specified by differential patterns of activity in the same cortical regions.

B. Implications for Abnormal Models of Language Processing

Functional imaging of aphasic patients provides information about neuronal and cognitive performance that is not available from either structural or behavioral assessments. First, functional imaging can determine whether a region that appears damaged on structural scans maintains any residual function in or around the lesion (Section IIIF). Second, it can identify the effects of a lesion on undamaged tissue (Section IIIA; e.g., lesions to Broca's area disrupt function in the left poste-

rior inferior temporal lobe). Third, dysfunction in undamaged cortex can also be revealed in patients who have no detectable damage on structural scans (Section IIID). Fourth, functional imaging of the patient provides a means of testing directly the disconnection models first proposed by the nineteenth century neurologists. This involves demonstrating that a cortical area responds normally during some tasks but abnormally during others, depending on the required connections (Section IIIC).

In addition, the most important contribution that functional imaging of aphasic patients can make to models of abnormal language processing is an understanding of the mechanisms involved in recovery. Such studies are only in preliminary stages and rely on both technological and analytical advances. The critical questions relate to whether recovery from aphasia results from changes in cognitive or neuronal implementation of a task. As discussed, there are fundamental difficulties related to distinguishing between functional and neuronal implementation. Nevertheless, we might predict that as the anatomical and functional definition of a neurological insult becomes increasingly refined, it might be possible to predict which functions might recover and those that might be lost. Such information could then be used to devise speech or motor therapies that focus on the functions for which viable tissue remains. Even more circumspect is the option that functional imaging might have predictive validity in relation to particular therapeutic drugs. The optimistic aim is therefore that functional imaging studies of language in both normals and aphasic patients will facilitate the recovery of aphasia following brain damage.

References

Alexander, M. P., Hiltbrunner, B., and Fischer, R. S. (1989). Distributed anatomy of transcortical sensory aphasia. *Arch. Neurol.* **46,** 885–892.

Broca, P. (1861). Remarques sur le siege de la faculte du langage articule suivie d'une observation d'aphemie. *Bull. Soc. Anat., Paris* **6,** 330 [translated in Herrnstein, R., and Boring, E. G. (1965). "A Source Book in the History of Psychology." Harvard Univ. Press, Cambridge, MA].

Brunswick, N., McCrory, E., Price, C. J., Frith, C. D., and Frith, U. (1999). Explicit and implicit processing of words and pseudowords by adult developmental dyslexics: A search for Wernicke's Wortschatz. *Brain* **122,** 1901–1917.

Büchel, C., Price, C. J., and Friston, K. J. (1998). A multimodal language area in the ventral visual pathway. *Nature* **394,** 274–277.

Buckner, R. L., Corbetta, M., Schatz, J., Raichle, M. E., and Petersen, S. E. (1996). Preserved speech abilities and compensation following prefrontal damage. *Proc. Natl. Acad. Sci. U.S.A.* **93,** 1249–1253.

Buonomano, D. V., and Merzenich, M. M. (1998). Cortical plasticity: From synapses to maps. *Annu. Rev. Neurosci.* **21,** 149–186.

Burnstine, T. H., Lesser, R. P., Hart, J., Jr., *et al.* (1990). Characterization of the basal temporal language area in patients with left temporal lobe epilepsy. *Neurology* **40**, 966–970.

Coltheart, M. (1980). Deep dyslexia: A right hemisphere hypothesis. *In* "Deep Dyslexia" (M. Coltheart, K. E. Patterson, and J. C. Marshall, Eds.), pp. 326–386. Routledge, London.

Dejerine, J. (1891). Sur un cas de cecite verbale avec agraphie, suivi d'autopsie. *Mem. Soc. Biol.* **3,** 197–201.

Dejerine, J. (1892). Contribution a l'etude anatomoclinique et clinique des differentes varietes de cecite verbal. CR Hebdomadaire des Sceances et Memories de la Societe de Biologie **4,** 61–90.

Demb, J. B., Boynton, G. M., and Heeger, D. J. (1997). Brain activity in visual cortex predicts individual differences in reading performance. *Proc. Nat. Acad. Sci. U.S.A.* **94,** 13363–13366.

Edelman, G. M. (1989). "The Remembered Present. A Biological Theory of Consciousness." Basic Books, New York.

Eden, G. F., VanMeter, J. W., Rumsey, J. M., Maisog, J. M., Woods, R. P., and Zeffiro, T. A. (1996). *Nature* **382,** 66–69.

Engelein, A., Silbersweig, D., Stern, E., Huber, W., Doring, W., Frith, C. D., and Frackowiak, R. S. J. (1995). The functional anatomy of recovery from auditory agnosia. *Brain* **118,** 1395–1409.

Feiz, J. A., Petersen, S. E., Cheney, M. K., and Raichle, M. E. (1992). Impaired nonmotor learning and error detection associated with cerebellar damage. *Brain* **115,** 155–178.

Fletcher, P. C., McKenna, P. J., Frith, C. D., Grasby, P. M., Friston, K. J., and Dolan, R. J. (1998). Brain activations in schizophrenia during a graded memory task studied with functional neuroimaging. *Arch. Gen. Psychiatry* **55,** 1001–1008.

Foundas, A., Daniels, S. K., and Vasterling, J. J. (1998). Anomia: Case studies with lesion localisation. *Neurocase* **4,** 35–43.

Geschwind, N. (1979). Specialisations of the human brain. *In* "The Brain." W. H. Freeman, San Francisco.

Goltz, F. (1881). In "Transactions of the 7th International Medical Congress" (W. MacCormac, Ed.), Vol. 1, pp. 218–228. Kolkmann, London.

Heiss, W. D., Karber, H., Weber-Luxenburger, G., Herholz, K., Kessler, J., Pietrzyk, U., and Pawlik, G. (1997). Speech-induced cerebral metabolic activation reflects recovery from aphasia. *J. Neurol. Sci.* **145**(2), 213–217.

Hinton, G. E., and Shallice, T. (1991). Lesioning an attractor network: Investigations of acquired dyslexia. *Psychol. Rev.* **98,** 74–95.

Hodges, J. R., Patterson, K., Oxbury, S., and Funnell, E. (1992). Semantic dementia. Progressive fluent aphasia with temporal lobe atrophy. *Brain* **115,** 1783–1806.

Lichtheim, L. (1885). On aphasia. *Brain* 7, 433–484.

Lüders, H., Lesser, R. P., Hahn, J., *et al.* (1986). Basal temporal language area demonstrated by electrical stimulation. *Neurology* **36,** 505–510.

Macleod, C. M. (1991). Half a century of research on the Stoop effect: An integrative review. *Psychol. Bull.* **109,** 163–203.

Marshall, J. C., and Newcombe, F. (1973). Patterns of paralexia: A psycholinguistic approach. *J. Psycholing. Res.* **2,** 175–199.

McClelland, J. L., and Rumelhart, D. E. (1981). An interactive activation model of context effects in letter perception. 1. An account of basic findings. *Psychol. Rev.* **88,** 375–407.

Mesulam, M. M. (1990). Large scale neurocognitive networks and distributed processing for attention, language and memory. *Ann. Neurol.* **28,** 597–613.

Morton, J., and Patterson, K. E. (1980). A new attempt at an interpretation, or, an attempt at a new interpretation. *In* "Deep Dyslexia" (M. Coltheart, K. E. Patterson, and J. C. Marshall, Eds.), pp. 160–175. Routledge, London.

Mummery, C. J., Patterson, K., Wise, R., Price, C. J., and Hodges, J. (1999). Disrupted temporal lobe connections in semantic dementia. *Brain* **122,** 61–73.

Patterson, K., and Besner, D. (1984). Is the right hemisphere literate? *Cog. Neuropsychol.* **1,** 315–341.

Patterson, K., and Shewell, C. (1987). Speak and spell: Dissociations and word class effects. *In* "The Cognitive Neuropsychology of Language" (M. Coltheart, G. Sartori, and R. Job, Eds.), pp. 273–294. Erlbaum, London.

Paulesu, E., Frith, U., Snowling, M., Gallagher, A., Morton, J., Frackowiak, R. S. J., and Frith, C. D. (1996). Is developmental dyslexia a disconnection syndrome? Evidence from PET scanning. *Brain* **119,** 143–157.

Petersen, S. E., Fox, P. T., Posner, M. I., Mintum, M., and Raichle, M. E. (1988). Positron emission tomographic studies of the cortical anatomy of single word processing. *Nature* **331,** 585–589.

Plaut, D., and Shallice, T. (1993). Deep dyslexia: A case study of connectionist neuropsychology. *Cog. Neuropsychol.* **10,** 377–500.

Price, C. J., and Friston, K. J. (1997). Cognitive conjunctions: A new approach to brain activation experiments. *NeuroImage* **5,** 261–270.

Price, C. J., Wise, R. J. S., and Frackowiak, R. S. J. (1996). Demonstrating the implicit processing of visually presented words and pseudowords. *Cereb. Cortex* **6,** 62–70.

Price, C. J., Howard, D., Patterson, K., Warburton, E., Moore, C. J., Friston, K., and Frackowiak, R. S. J. (1998). A functional neuroimaging description of two deep dyslexic patients. *J. Cog. Neurosci.* **10**(3), 303–315.

Price, C. J., Mummery, C. J., Moore, C. J., Frackowiak, R. S. J., and Friston, K. J. (1999). Delineating necessary and sufficient neural systems with functional imaging studies of neuropsychological patients. *J. Cog. Neurosci.* **11**(4), 371–382.

Raymer, A. M., Foundas, A. L., Maher, L. M., *et al.* (1997). Cognitive neuropsychological analysis and neuroanatomic correlates in a case of acute anomia. *Brain Lang.* **58,** 137–156.

Rumsey, J. M., Andreason, P., Zametkin, A. J., Aquino, T., King, C., Hamburger, S. D., Pikus, A., Rapoport, J. L., and Cohen, R. (1992). Failure to activate the left temporal cortex in dyslexia: An oxygen-15 positron emission tomographic study. *Arch. Neurol.* **49,** 527–534.

Rumsey, J. M., Nace, K., Donahue, B., Wise, D., Maisog, M., and Andreason, P. (1997). A positron emission tomographic study of impaired word recognition and phonological processing in dyslexic men. *Arch. Neurol.* **54,** 562–573.

Saffran, E. M., Bogyo, L. C., Schwarz, M. F., and Marin, O. S. M. (1980). Does deep dyslexia reflect right hemisphere reading? *In* "Deep Dyslexia" (M. Coltheart, K. E. Patterson, and J. C. Marshall, Eds.), pp. 381–406. Routledge, London.

Sasanuma, S., and Monoi, H. (1975). The syndrome of Gogi (word-meaning) aphasia. *Neurology* **25,** 627–632.

Shallice, T. (1988). "From Neuropsychology to Mental Structure." Cambridge Univ. Press, Cambridge, U.K.

Shallice, T., and Warrington, E. (1980). Single and multiple component central dyslexic syndromes. *In* "Deep Dyslexia" (M. Coltheart, K. E. Patterson, and J. C. Marshall, Eds.), pp. 119–145. Routledge, London.

Shaywitz, S., Shaywitz, B., Pugh, K., Fulbright, R., Constable, R., Mencl, W., Shankweiler, D., Liberman, A., Skudlarski, P., Fletcher, J., Katz, L., Marchione, K., Lacadie, C., Gatenby, C., and Gore, J. (1998). Functional disruption in the organization of the brain for reading in dyslexia. *Proc. Natl. Acad. Sci. U.S.A.* **95,** 2636–2641.

Vandenberghe, R., Price, C. J., Wise, R., Josephs, O., and Frackowiak, R. S. J. (1996). Semantic system(s) for words or pictures: Functional anatomy. *Nature* **383,** 254–256.

Warburton, E., Wise, R. J. S., Price, C. J., *et al.* (1996). Studies with positron emission tomography of noun and verb retrieval in normal subjects. *Brain* **119,** 159–180.

Warburton, E. A., Price, C. J., Swinburn, K., and Wise, R. J. S. (1999). Mechanisms of recovery from aphasia: Evidence from positron emission tomography studies. *J. Neurol., Neurosurg. Psychiatry* **66,** 155–161.

Warrington, E. K. (1975). Selective impairment of semantic memory. *Q. J. Exp. Psychol.* **27,** 635–657.

Weiller, C., Insensee, C., Rijntjes, M., Huber, W., Muller, S., Bier, D., *et al.* (1995). Recovery from Wernicke's aphasia: A positron emission tomography study. *Ann. Neurol.* **37,** 723–732.

Wernicke, C. (1874). "Der Aphasiche Symptomenkomplex." Cohen and Weigert, Breslau, Poland [translation of title: "The aphasias"].

8

The Functional Neuroimaging of Memory Disorders

P. C. Fletcher

*Research Department of Psychiatry, Cambridge University, Addenbrooke's Hospital, Cambridge CB2 2QQ,
United Kingdom*

I. Functional and Structural Studies: Major
Conceptual Differences

II. Exploring Memory Impairment with PET and
fMRI: Conceptual Difficulties

III. Overcoming the Difficulties: Making
Functional Neuroimaging Useful to the Study
of Memory Impairment

IV. Summary
References

This chapter will explore the major conceptual issues surrounding the application of the relatively new functional neuroimaging techniques positron emission tomography (PET) and functional magnetic resonance imaging (fMRI) to the understanding of memory disorders. It is not intended to provide a comprehensive account of the functional anatomy of memory nor does its scope include an exhaustive account of existing reports of the application of these techniques to memory-impaired patients. Rather, its goal is to show that the use of PET and fMRI in the study of memory dysfunction (and, indeed, many other aspects of brain impairment)

brings to the fore many fundamental theoretical issues that must be addressed at the outset. These issues encompass the characteristics of the functional neuroimaging techniques, the questions that they may address, and the nature of the information that they may provide. Discussing them will draw attention to difficulties in designing and analyzing experiments for the investigation of the subject who experiences and exhibits memory impairment. Addressing these difficulties will necessitate, in many cases, experimental modifications such that the functional neuroimaging of memory disorders will differ considerably, practically and conceptually, from neuroimaging studies of normal memory function and from the more established neuropsychological approaches to the study of memory impairment.

The chapter comprises three main sections. In the first, the major conceptual differences between functional neuroimaging techniques and the more traditional approaches to the study of memory disorders will be considered. The second section will draw attention to some of the inherent difficulties, both practical and conceptual, in carrying out functional neuroimaging studies on the subject who is memory impaired. In the third section, a framework for such studies will be suggested. Within this framework will be included suggestions for the optimization of the design, analysis, and interpretation. The chapter will concern itself principally with PET and fMRI. It will touch, only incidentally, upon existing structural imaging studies of memory impairments and, in most cases, they will be alluded to only to contrast them with the more recent techniques.

I. Functional and Structural Studies: Major Conceptual Differences

The broadest endeavor of functional neuroimaging is, of course, to map psychological function onto brain structure as precisely as possible. The extent to which such an endeavor may be successful will be dependent, in part, on the degree to which the psychological systems and classifications used are valid. The application of a flawed psychological model, one that may be consistent with the known behavioral data but that is incomplete or incorrect at the level of real brain organization, will ultimately produce inconsistencies. Memory, like most higher cognitive functions, is only partially understood and functional neuroimaging studies are unlikely to provide a complete mapping of normal memory function. Fundamental gaps in our understanding of the psychological structure of memory, while they remain unfilled, will be a bar to the understanding of functional neuroimaging data, no matter how technically sophisticated the scanning procedures become. The results of PET and fMRI studies can only be interpreted in light of a thorough understanding of the psychological processes whose effects they are measuring. This constraint must apply, likewise, to the mapping of memory dysfunction.

Crucial to an understanding of the relationship between brain structure and memory dysfunction is the capacity to observe correlations between measured brain abnormalities and impairments in precisely defined and described subprocesses that come under the broad heading of memory. These subprocesses are not precisely defined, and, at present, therefore, functional neuroimaging studies of memory must be interpreted cautiously. However, as has been pointed out (Shallice, 1988), it would be a little overcautious to withold a search for the neuronal implementation of a functional architecture even though that functional architecture may be incomplete. With this caveat in mind, I use the existing taxonomy, which, if neither complete nor completely correct, is of heuristic value. This generally accepted taxonomy will not be described in detail here. It is sufficient to say that there is strong evidence that memory can be fractionated into several distinct systems (Schacter and Tulving, 1994). Initially, it can be subdivided into working and long-term memory, the

former referring to a limited capacity short-term store and the latter to a store of practically unlimited capacity enduring over hours, days, months, or years. Long-term memory can be further subdivided into that which is accessible to consciousness (so-called explicit or declarative memory) and that which is manifest unconsciously (implicit memory such as priming, skill-learning, or classical conditioning). A further subdivision is between episodic and semantic memory, the former referring to memories for specific events or episodes and the latter to a non-autobiographical knowledge. Although contention exists over many of the subdivisions in this taxonomy [even at the broader level of, for example, the distinction between implicit and explicit memory (Green and Shanks, 1993)], it is clear that memory is not a unitary concept and may be subdivided, on the basis of neuropsychological and psychophysical observations, into a number of subcomponents. Furthermore, memory disorders may present in a variety of ways, with varying degrees of impairment across the different components of the memory taxonomy. Thus, the classically described case of the amnesic patient HM (Scoville and Milner, 1957) points to an impairment in the ability to retain and recollect events (which would fall under the heading of *episodic memory* as shown in Fig. 1) but with a preserved ability to learn and use new skills (*implicit memory*). Thus, such a patient might be given a series of training sessions in which they learned the difficult skill of writing while viewing their writing hand in a mirror. On subsequent occasions they would retain this skill while having no conscious memory of ever having learned it or of any aspects of the previous training sessions or even of the experimenter who had been with them throughout the training period. Neuropsychological observations such as these have played the greatest part in linking structural abnormalities to memory dysfunction. They are conceptually simple and based upon the notion that the observation of a brain lesion in association with a functional deficit provides evidence that this brain region would normally support, at least in part, the function that is found to be impaired. Of course, there are a number of pitfalls to such a simplistic interpretation (Shallice, 1988).

▲ Such an observation may indicate that the region under consideration is necessary to carry out the psychological task but not that it is sufficient. That is, the damaged region may be only one part of a distributed system necessary for carrying out the psychological task.

▲ Any inferences are highly dependent upon the "purity" of the psychological task. That is, in the context of a memory task, it may be the case that other cognitive capabilities are required for successful performance. If

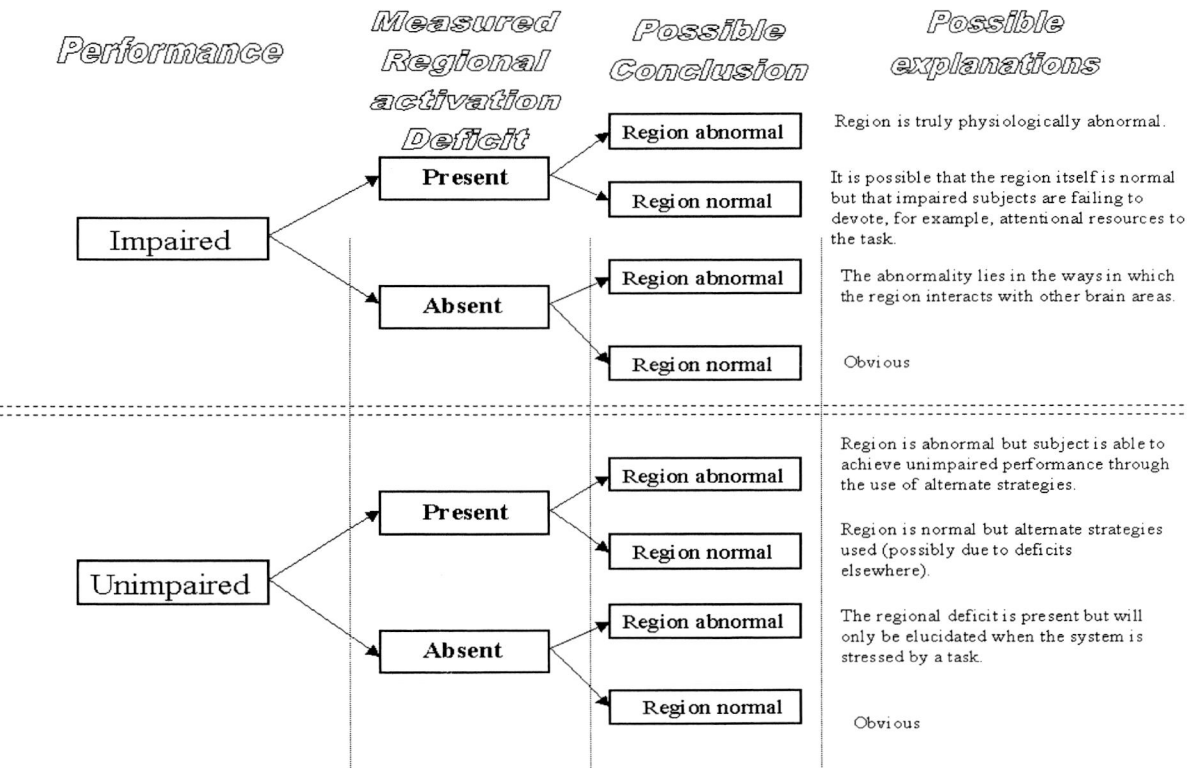

Figure 1 A diagrammatic representation of the possible interpretations of findings from cognitive task activation studies. The figure is described with reference to the practical example in the text.

the lesion affects a brain region required for the other aspects of the task, it is conceivable that the impaired performance could be misinterpreted.

▲ An additional caution is that care must be taken in interpreting negative findings in such studies, particularly with regard to what such a finding may tell us about normal brain function. The observation that a patient with a lesion to a particular region is able to perform a task does not provide the grounds for concluding that a healthy subject, when performing the same task, will not recruit that brain region. A simple illustration of this comes from the observation, across a series of functional neuroimaging studies of recognition memory, that prefrontal cortical activation is a consistent finding in healthy subjects (Tulving *et al.*, 1994; Kapur *et al.*, 1995; Haxby *et al.*, 1996; Rugg *et al.*, 1996, 1997). This observation would seem hard to reconcile with the neuropsychological data, suggesting that patients, even with large areas of prefrontal damage, are unimpaired on such tasks (Stuss *et al.*, 1994). One obvious possibility is that the healthy, and usually high-functioning, subjects who take part in the functional imaging experiments are carrying out the recognition memory task in a different way from the brain-damaged patients. Perhaps they respond to the recognized test items with a rich recollection of having seen the item during the test phase

whereas the patients have no explicit recall of the encoding event but are, nevertheless, able to provide a correct response on the basis of a feeling of familiarity engendered by reexposure to the previously studied items. Such an explanation would reconcile apparently disparate findings from neuropsychology and functional neuroimaging and, additionally, brings to mind the importance of a thorough understanding of the ways in which both patients and control subjects pass and fail the psychological tasks with which they are confronted.

The problems alluded to above are well known and have produced intricate and, at times, entertaining arguments (reviewed in Shallice, 1988) concerning the ultimate validity of the neuropsychological approach. Some of them can, to an extent, be overcome by refining the psychological tasks used; others should simply be borne in mind so that observations are not overinterpreted. The debate will not be rehearsed here. Rather, the constraints are referred to in order to highlight some crucial differences in the experimental approach that, largely, drives neuropsychology and the approach adopted in functional neuroimaging. PET and fMRI apply, to a greater extent, the methods of experimental psychology. Put simply, in a standard psychology experiment, the relationship between two types of variable is measured. The first, the *independent* or *explana-*

tory variable, is manipulated and the extent to which the *dependent* variable alters as a function of this manipulation is assessed. In so doing, the characteristics of the relationship, if there is a relationship, between the two variables may be explored. In both neuropsychology and functional neuroimaging, the experimental psychology experiments are of, initially, background importance, serving to help parcellate function and to delineate the areas of study. However, a difference between the structural and the functional neuroimaging experiment is that the latter takes place within the context of a psychological experiment. A psychological manipulation is made and the subject's behavior is observed at the same time as an estimate of regional cerebral activation is acquired. In structural studies, the "brain measurement" is acquired independently of the task and of the associated behavior. Indeed, it may be acquired after the patient's death. This is a crucial difference. It means that the brain observations made in association with the structural (neuropsychology) experiment, though it may be more limited in terms of the information that it contains, can be used to make inferences that are more clear-cut and beset by fewer difficulties than those associated with functional neuroimaging. In applying the functional neuroimaging techniques, we are confronted by a new set of difficulties in making inferences about the relationship between structure and function.

▲ In trying to characterize the brain region or regions associated with a cognitive component, it is necessary to be able to isolate and to manipulate that function across a series of tasks. Regions of the brain whose activity varies as a function of the presence, absence, or level of the component are assumed to be intimately linked with it. A functional neuroimaging experiment is therefore heavily dependent upon the extent to which components of psychological function can truly be dissected out. There are a number of problems with this assumption, particularly when the standard "cognitive subtraction" approach is used (Friston *et al.,* 1996; Jennings *et al.,* 1997).

▲ Related to the above point, the observation of a task-related functional neuroimaging activation does not provide hard evidence that the activation is necessary in order to be able to carry out the task. This point is discussed in greater detail in Price *et al.* (1999).

▲ Finally, as suggested above, it is clear that, unlike structural imaging, which is the staple of the neuropsychological approach, the pattern of observations in functional imaging cannot be interpreted in isolation from what the subject is actually doing during scanning. In a memory task, the pattern of brain activation associated with that task must have a different implication

when the task is being performed successfully than when it is being performed unsuccessfully. This simple fact has broad implications for the application of functional neuroimaging to memory disorders. These implications are addressed more fully in the next section of the chapter.

To summarize, there are crucial conceptual differences between the technique that has become the standard approach to the characterization of memory disorders, that of neuropsychology, and the more recent and evolving techniques of PET and fMRI. While the former is reliant upon the observation of structural abnormalities which, though, of course, related to the functional deficits, may be observed in isolation from them, the latter represents brain measurement in terms of function. If the functional neuroimaging techniques are to have anything useful to say about the functional anatomy of memory disorders, this must be borne in mind: that the observation of a PET or fMRI abnormality is conceptually different from the observation of a structural abnormality. While, of course, it must be the case that the two are intimately related and, further, that the relationship between them is a worthy field of study, it is, nevertheless, important not to bring a structural approach to a functional experiment. It is clear that the application of functional neuroimaging to memory disorders will be useful only insofar as it is recognized that the information that it can provide is qualitatively different from that produced by structural studies. While PET and fMRI bring to the study of memory disorders a set of new capabilities, they are nevertheless dogged by a set of peculiar difficulties. The questions posed by such studies should be carefully framed. The rest of this chapter will explore ways in which these techniques can produce data that are interpretable and useful to the understanding of memory disorders. This consideration will begin in the next section with a consideration of the problems peculiar to the functional neuroimaging approach.

II. Exploring Memory Impairment with PET and fMRI: Conceptual Difficulties

A major problem in evaluating and characterizing memory disorders and other cognitive impairments using fMRI and PET is that these techniques explore a mysterious relationship, that between brain and behavior. The estimation of regional brain activity across changing behavioral contexts will be informative only to the extent that the observed behavior is a consistent indicator of the psychological state. In memory, as in many aspects of cognitive function, strategies may vary markedly across subjects and yet produce behavioral results that may appear identical. Thus, an initial problem lies in ensuring that the tasks chosen as paradigms for the cognitive components under study are performed in truly similar ways across subjects. The mere observation that measured performance does not differ significantly across subjects is not necessarily an indication that the subjects are setting about the tasks in similar ways. When differences in patterns of brain response are characterized across different groups of subjects, the problems are magnified since, a priori, we may expect the groups to differ behaviorally and it becomes crucial to the understanding of functional imaging differences that the behavioral differences are understood. These differences may be measurable in terms of performance or they may be more subtle, being manifest in differing ways in which the assigned tasks are carried out. Just as it does not follow that two groups of subjects achieving similar levels of performance are necessarily using similar strategies and approaches, it is also important to remember that a failure to achieve good levels of performance does not mean that subjects are failing to engage the cognitive processes of interest. Indeed, it is conceivable that some cognitive processes will, through necessity, be engaged to a greater degree when performance is impaired than when it is successful.

The above observations will be, in the main, obvious and they are highlighted here as a reminder that the technological capabilities in no way absolve the experimenter of the need to have a clear understanding of the cognitive characteristics of the chosen task, of the relationship between these characteristics and the performance measures used, and, crucially, when the tasks are used in clinical studies, of the ways in which the patients are failing or succeeding in these tasks. Without this understanding, it is suggested that the array of possible interpretations of any functional neuroimaging observation will be so broad as to render it largely uninformative. This is illustrated in Fig. 1, which is designed to show that virtually any possible finding, in the absence of an understanding of what the subject is actually doing, is interpretable in ways that differ markedly. It is illustrated with reference to a study of memory impairment in the following practical example.

A. The Difficulties in Interpreting Patterns of Brain Response in Memory Impairment: A Practical Example

Schizophrenia is a common and disabling illness that is frequently associated with memory impairment (Tamlyn *et al.*, 1992). Using PET, we explored the brain

activations associated with increasing memory load in groups of schizophrenic and control subjects (Fletcher *et al.*, 1998). Each subject underwent 12 PET observations, each of which was associated with immediate paced verbal recall of word lists. Across the 12 observations, the memory load (as defined by the length of word list memorized and recalled) was systematically varied (from 1 to 12 words inclusive). Thus, for control and schizophrenic subjects, it was possible to characterize the brain response to the increasing memory demands of the task and to perform a direct comparison of memory-related activations across the two groups. Moreover, it was possible to divide the scans into those in which both groups of subjects were performing at ceiling and those in which performance fell below 100%. The findings with respect to left prefrontal cortex (PFC) will be discussed here but the discussion would be equally relevant to any brain area. The pattern of activity in left PFC is summarized in Fig. 2. Essentially, the response to increasing memory load was as follows: increasing memory load occurring across those scans in which performance was 100% was associated with an increase in left PFC activity, which occurred in both groups (schizophrenics and controls) and was statistically indistinguishable between the two groups. As memory load increased, however, and as performance fell below 100%, the left PFC response differed across the groups. In controls, increasing memory load continued to be associated with increasing lev-

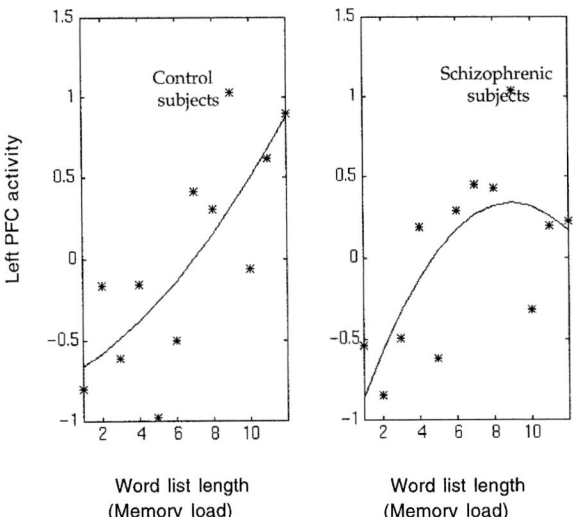

Figure 2 Plots showing the relationship between increasing memory load and regional cerebral blood flow in left PFC for control and for schizophrenic subjects. It can be seen that increasing load across the shorter word lists (1 to 6 items) is associated with a similar response. As the lists become longer, however, control subjects continue to show a left prefrontal increase whereas the response in left PFC in schizophrenic subjects falls away.

els of left PFC activity. In the schizophrenic subjects this was no longer the case, with left PFC activity falling away during the scans in which memory load was maximized. Within this single experiment are illustrated the difficulties in interpreting regional differences (or lack of differences) across memory-impaired subjects. These difficulties, set out in Fig. 1, arise at several stages. The first concerns the question of whether or not performance measured during the scanning procedure is impaired or not. Clearly, one will be interested in the sort of task with which the patient has difficulties. However, the implications of functional neuroimaging abnormalities found in the presence of abnormal task performance are far from clear. Consider the study described above in which, when the performance of the schizophrenic subjects fell below that of controls, there was an associated abnormality in the pattern of left PFC response to the increasing memory load. It is possible that this pattern may be associated with a true disease-related physiological abnormality of left PFC or it may in fact reflect some other phenomenon in which left PFC is itself functionally normal but, due to deficits elsewhere, the region is not recruited. Suppose, for example, that the impaired subjects are insufficiently motivated during increasing task demands to devote the requisite attentional resources to the task and, as such, do not recruit the brain areas that would normally subserve task performance. Alternatively, it may be so that the subjects (through, for example, distraction caused by other symptoms) do not adopt the strategies that are associated with left PFC activation and that would normally be used during performance of the more demanding task. Conversely, if it had been the case that, in the presence of abnormal task performance, the pattern of left PFC response had been measurably normal, this would not necessarily mean that there was no functional imaging abnormality in that region. It could be the case that the region, while showing normal patterns of task-related activation, was, nonetheless, manifesting an abnormality in its interactions with other brain regions: an abnormality that would not be observed using the standard approach to functional imaging analysis.

Thus, functional imaging abnormalities (and lack of abnormalities) are difficult to interpret in the presence of impaired performance. Turning to cases where performance is balanced across groups of patients and control subjects, it might be supposed that an abnormality found in the presence of normal performance could be unequivocally interpreted as a true regional abnormality (with the suggestion further that the subject is achieving normal levels of performance through adopting alternate strategies). However, a priori, it is

equally compelling to suggest that the region itself is normal and that the failure of activation also reflects the adoption of alternative strategies by the patients, possibly due to the presence of regional abnormalities elsewhere. If, on the other hand, as was the case in the experiment described, the left PFC showed a normal pattern of response in the context of measurably normal task performance, it simply does not follow that this region is functionally normal. As we saw, when the task becomes more difficult and performance is impaired, left PFC shows an abnormal pattern of response. It is thus entirely plausible that we may be exploring a region or system that will be normally responsive to milder task demands but that will, under the stress of greater demands, fail to respond. Whatever the explanation for the findings shown here, or for the gamut of possible findings in any functional neuroimaging study of memory impairment, clearly results cannot be interpreted in any simplistic way. Moreover, a further complexity is added in cases where structural abnormalities are detectable since it is unclear how the structural damage affects measurements in PET and fMRI. The usefulness of interpreting a functional deficit in a brain region that is known to be damaged may therefore be debatable. Although this may not be an issue in "functionally" abnormal conditions such as schizophrenia, the structural abnormalities that are present in many conditions of memory impairment cannot be ignored.

Having highlighted difficulties associated with applying the functional neuroimaging techniques to memory, and other cognitive, impairments, we explore in the next section ways in which the techniques may be applied to overcome some of the difficulties raised.

III. Overcoming the Difficulties: Making Functional Neuroimaging Useful to the Study of Memory Impairment

The difficulties described above generally apply to gaining an understanding of the ways in which patterns of brain activity reflect the subprocesses of memory and, further, the ways in which these subprocesses may be impaired. While these difficulties are not completely insurmountable, a failure to address them will undermine the informativeness of functional studies. In the following subsections will be described a number of approaches that seek to address these difficulties or if not to address them, then to modify the scope of the conclusions that may be drawn.

A. Functional Neuroimaging as a Diagnostic Tool in Memory Impairment

Most simply, the functional neuroimaging techniques may provide an extra diagnostic tool for establishing the presence of brain abnormalities in the face of memory impairment. Hitherto, the majority of studies have used them in this way. For example, a relatively early study by Freedman *et al.* (1991) showed structural MRI and PET evidence for atrophy and hypometabolism, respectively, in a male whose cognitive impairment included amnesia. It may be so, in such cases, when the techniques are used simply as a diagnostic tool, that the psychological task performed will be of less crucial importance than when the techniques are used to gain a fuller understanding of the psychological processes that are affected by the pathological process. That is, studies that concern themselves less with the functional significance of patterns of hypo- and hyperactivation in memory-impaired patients and more with the diagnostic significance of those findings will be freer from the constraints described above. In many (but not all) ways a purely diagnostic approach relies less heavily on a clear understanding of the precise psychological state of the patient. This is only true insofar as such studies are not used to draw functional inferences with regard to the patients studied but only to ascertain whether a person with a certain group of signs and symptoms belongs to a particular designated diagnostic group. For example, a PET study by Kuwert *et al.* showed that six out of a group of seven patients with posthypoxic amnesia showed reduced regional cerebral glucose metabolism compared to control subjects (Kuwert *et al.,* 1993). Although there was substantial intragroup variability in the location of the abnormality, group comparisons seemed to indicate consistent thalamic and medial temporal differences between the groups. Such a finding, where a particular group of patients with a pattern of amnesia symptoms, when scanned at rest, is found to have hypoperfusion in a given brain region, can be of diagnostic value in that it would serve to separate groups of patients from controls (and from other patient groups with memory disorders). It would not, in itself, be of any further value unless the difficulties of interpretation, described above, were addressed. It would not even allow us to say for certain that the hypoperfused area was physiologically abnormal since it is feasible that the state of rest in the patient group was different from that in the control group and this could well explain the differences. (Indeed, if one takes the view that the resting state comprises, to a great extent, freeform thoughts that are largely retrieved memories (Andreasen *et al.,* 1995), then it would seem plausible that the resting state in the memory-impaired patient would be different from that in the control subject.)

With this constraint in mind, however, it is worth noting that a number of PET studies of resting blood flow have produced results showing differences between memory-impaired patients and controls. Of particular interest are those that have indicated different patterns of blood flow when structural abnormalities were not detectable. Thus, in a recent study of a patient showing a retrograde amnesia (diagnosed as "psychogenic," that is, of psychological rather than structural origin), standard structural analyses (CT and MRI) showed no abnormalities, but right temporal and frontal perfusion was shown to be reduced on single photon emission computerized tomography (SPECT) (Markowitsch *et al.,* 1997). Similarly, Mattioli *et al.* report the case of a woman with severe retrograde amnesia, again with no evidence of structural abnormalities on CT and MRI investigation, in whom a PET measurement indicated reduced metabolism in the hippocampus and anterior cingulate cortex (Mattioli *et al.,* 1996). It is noteworthy that a number of studies in which functional imaging has been useful in this way (that is, elucidating abnormalities when none are visible with structural imaging) have examined atypical memory loss, particularly retrograde amnesia (Levine *et al.,* 1998), which is thought to have a psychological ("functional") basis rather than a structural one. However, it is by no means true that the functional imaging findings can add to the structural observations only in such cases. In a patient with dense amnesia following alcohol abuse, thalamic abnormalities were observed with PET measurement but not with structural MRI, which showed only right temporal (including hippocampal) damage (Welch *et al.,* 1996). The temporal abnormality was also seen with PET. A further case was reported by Luccheli *et al.* in which a developing amnesic syndrome was associated with PET but not structural MRI abnormalities (Lucchelli *et al.,* 1994). It seems, therefore, that functional imaging has a diagnostic potential in the so-called organic and the functional memory impairments and that this potential lies in establishing the existence of group differences (note, not necessarily abnormalities) that may not be visible in brain structure as assessed using currently available techniques. A further potential, of course, lies in the possibility of carrying out serial studies on subjects and correlating any changes in functional imaging findings with deterioration (or improvement) in memory ability. This approach was used in a PET study of a subject at risk of Alzheimer's disease and showing early memory impairment (Pietrini *et al.,* 1993). Interestingly, with a standard method of comparison, the subject showed no initial functional imaging abnormalities but, a year later, in association with a deterioration in cognitive function, parietal abnormalities were found.

Alternatively, the functional imaging findings may not necessarily add to the structural findings but may provide valuable confirmatory evidence in the face of an unclear pattern of symptoms. Thus, Kopelman *et al.* reported the case of a patient in whom the clinical history was strongly redolent of a psychogenic amnesia but in whom the structural MRI and PET findings both showed evidence of a number of infarctions (including MTL) (Kopelman *et al.*, 1994).

Briefly then, functional neuroimaging may well have a use as an aid to diagnosis and, with such usage, the constraints discussed in the section above may be, to some extent, irrelevant. This is only so, however, when the interpretation is limited to a simple statement of whether or not a patient belongs to a given diagnostic group. Any further statements concerning the functional significance of the findings are not justified without taking these constraints into consideration and making attempts to overcome them. Nevertheless, it is clear that the functional imaging approach adds an important conceptual dimension to the anatomical description of memory disorders. This is well-illustrated by the findings and discussion in a recent study of a group of patients with structural (MRI) evidence of medial temporal lobe lesions following hypoxic episodes (Reed and others, 1999). Patients showed reduced FDG-uptake in spatially remote areas (thalamus and retrosplenial cortex) that were not structurally damaged. The findings (shown in Fig. 3) were interpreted in terms of metabolic changes existing distally to the structural damage within an interconnected system and the authors made the point that it can be easy, but over-hasty, to interpret a functional deficit purely in terms of the observed structural deficit and to ignore the system of which that structure forms a part. Clearly this is an extremely important point with regard to the possible benefits that PET and fMRI may confer on future research into memory impairment. In the following section, a more refined variant of this approach to diagnostic grouping will be discussed. In this approach, the difficulties of a functional interpretation confounding a diagnostic interpretation are, to a small extent, obviated by the inclusion of separate pathological groups in whom similar memory deficits are associated with different functional neuroimaging findings.

B. Comparisons of Functional Neuroimaging Findings across Differing Groups of Memory-Impaired Patients

A major problem with the simple diagnostic studies in which the results of a functional imaging procedure

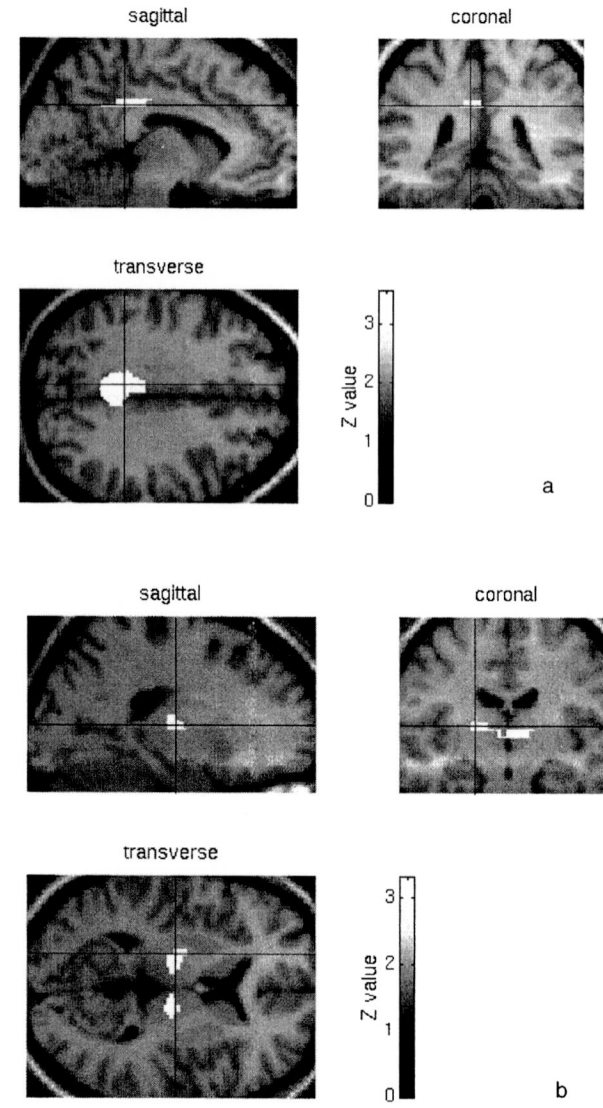

Figure 3 SPM(Z) maximal intensity projections produced as in Reed *et al.* (1999) showing reductions in [18]FDG uptake in thalamus bilaterally (a) and retrosplenial posterior cingulate cortex (b). From L. J. Reed, P. Marsden, Lasserson *et al.* (1999). Memory, FDG-PET Analysis and Findings in Amnesia Resulting from Hypoxia. Reprinted by permission of Psychology Press Ltd.

in one or more patients with impaired memory are compared with results from unimpaired control subjects lies in the performance differences. This has been discussed in detail above but, in essence, the question is simply this: are differences in findings between the groups explicable merely in terms of different behaviors occurring during the scanning (different strategies, different performance, etc.)? Even if the subjects are scanned at rest, although performance may not be an issue, the psychological processes that are being scanned may, nevertheless, differ appreciably between groups and may wholly or partly explain differences in observations.

One possible, though by no means infallible, way of removing the confound of performance effects lies in exploring two patient groups, both of whom show memory impairment. Once again, this approach, at its simplest, will be useful as a diagnostic aid rather than to attribute detailed functional significance to findings. An example of such a study is that carried out by Ouchi et al., who compared PET measurements of glucose metabolism in patients with pure amnesia, in patients with Alzheimer's disease (AD), and in control subjects (Ouchi et al., 1998). They found wider areas of hypometabolism in the AD group, including medial temporal and parietotemporal areas. Hypometabolism in the pure amnesic group was limited to the head of the hippocampus. In a different study, Perani et al. used PET to compare patients with global amnesia to those with AD (Perani et al., 1993). They found that thalamic hypometabolism was specific to the former group and more widespread neocortical hypometabolism was found in the latter group. Although these results may be interpretable in a number of different ways, nevertheless, the study is indicative of an approach that may have two benefits. First, they make it possible to assign functionally similar memory-impaired patients to different diagnostic groups on the basis of the functional neuroimaging findings. Second, they highlight the possibility of interpreting a functional neuroimaging finding in isolation from the behavioral deficit. That is, it may be possible to assess whether a functional neuroimaging finding is specific to a particular disease if it is found in association with one patient group but not with a second group in whom performance is impaired in a similar way.

The remainder of this section will discuss those approaches to the functional neuroimaging of memory disorders in which the intention is to provide the basis for a broader interpretation of the results rather than to allocate subjects to a diagnostic group. This section will be largely theoretical since very few studies of memory-impaired patients have actually been carried out where the intention is to isolate and examine specific cognitive components of memory. Broadly speaking, the approaches come under two headings: those in which tasks are carefully designed to address the problems of interpretation described in detail above and those in which a new approach to the analysis and exploration of data is used—specifically when function (or impairment) is described in terms of brain systems rather than separate brain regions.

C. Using Task-Related Activations to Explore Memory Impairment

If PET and fMRI are to be used as more than a diagnostic resource, then an obvious approach involves the application of cognitive activation studies to the memory-impaired population. Hopefully, the problems inherent in this approach are clearly drawn out in the first two sections. It is crucial that all studies exploring the functional abnormalities associated with memory disorders use paradigmatic tasks that are defined and understood, in terms as precise as possible, with regard to the cognitive subprocesses that they engage. Further, for the reasons stated above (and summarized in Fig. 1), it is of the utmost importance that there is a clear understanding of how patients are addressing these tasks—ways in which they are able to succeed at the tasks as well as ways in which they fail them. Do they engage the optimizing strategies that controls do when faced with the same tasks? If not, is it through an inability to carry out such strategies or does it rather reflect a failure to, for example, allocate appropriate attentional resources to the task? These issues have been drawn out in detail above with respect to the practical example.

One crucial part of the successful application of PET and fMRI to a functional anatomical description of memory impairment will depend heavily upon a thorough understanding of the normal cerebral response to memory tasks. Such an understanding cannot yet be said to exist but a number of memory paradigms have been explored and have produced results that are consistent across several studies (Fletcher et al., 1997). One such task, having been well established in healthy controls (Schacter et al., 1996a-c), has also been applied to a memory-impaired group (subjects with schizophrenia) (Heckers et al., 1998). In this study, the basic task design allowed a differentiation of brain regions associated with the attempt to retrieve words from those associated with successful retrieval of the words. This was done by dint of a simple manipulation of the way in which subjects processed words at study (prior to scanning). Some words were processed according to their meaning, others according to the letters that they contained. When subjects attempted (during scanning) to retrieve the words that had been processed according to meaning, this was done with a greater degree of success than with the other words. Thus, the retrieval attempt was common to both retrieval conditions but success was a more prominent feature of one of them. Previous work on healthy volunteers had suggested that PFC was activated in association with the attempt to retrieve words whereas hippocampal activity was increased during successful retrieval of words. In the case of the schizophrenic subjects, an interesting pattern was observed: a relative failure of hippocampal activation (see Fig. 4) occurred during the successful retrieval condition whereas PFC showed a greater degree of activation, in the same subjects, in association with this condition. By thus manipulating two aspects of explicit memory re-

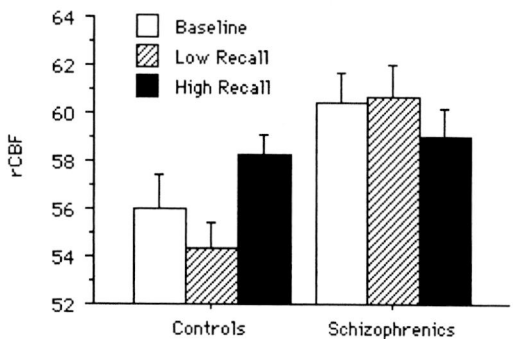

Figure 4 Plots of hippocampal activity for controls and subjects with schizophrenia across the three tasks of the memory-retrieval experiment described Heckers *et al.* (1998).

trieval, Heckers *et al.* have grounds for an interpretation of the observed abnormalities that is free from some of the constraints described in the opening sections of this chapter. Thus, for example, it is unlikely that the relative failure of hippocampal activation in the schizophrenic subjects reflects, in any direct way, a tendency to give up on a task that is too demanding. Clearly, if this were so, we would expect that a hippocampal abnormality would have been apparent in the other condition (that is, the condition that maximized attempt while minimizing success). Rather, the more obvious interpretation is that the schizophrenic subjects are in fact having to make an increased effort, as evidenced by increased right prefrontal activation (right PFC activation being a prominent feature in both control and schizophrenic subjects in association with the high attempt/low recall condition). Of course, further questions must be begged concerning the failure of recruitment of hippocampus in schizophrenic subjects performing the recall task. Clearly, right hippocampal activity is related, in some way, to retrieval success (which was lower in the schizophrenic group during the high-recall condition and, interestingly, higher in this group during the low-recall condition). The question that arises is how, precisely, the hippocampal activity relates to success and whether this relationship is abnormal in the schizophrenic subjects. More specifically, does the hippocampus activate as a result of the successful retrieval of an item? Or is it the case that the activation in the hippocampus is a necessary step toward successful retrieval? Once again, the latter is made unlikely by the absence of hippocampal activation in the low-recall condition (that is, if hippocampal activation is related to retrieval attempt, it would be at least as high in this condition). The question that remains, therefore, concerns the meaning of a failure of hippocampal activation in association with a lower level of performance. Whether this reflects an abnormality of hippocampal function or not can only be

speculated upon. An obvious way to address the question would be to compare successful and unsuccessful retrieval in the schizophrenic subjects, with the prediction that the hippocampal response would be abnormal in schizophrenic subjects even in the face of successful retrieval, if pathology truly lies here. This may be possible with the development of event-related measurements in fMRI, which will be discussed more fully below.

It is also possible to use a cognitive task activation in a single subject using the subject himself as a control. A study carried out by Costello *et al.*, explored brain activation associated with both successful and unsuccessful recognition in a man with a circumscribed retrograde amnesia spanning 19 years (Costello *et al.*, 1998). In brief, the study used PET to explore brain activity associated with three conditions: The first involved recognizing photographs of family occasions (at which he had been present) occurring outside the 19 years spanned by his amnesia. The second involved attempting to recognize photographs of occasions that had occurred during the period for which he was now amnesic. The third served as a baseline in which he was presented with photographs of occasions at which he had not been present and asked to say whether or not he remembered them. Thus for the first condition, activations associated with autobiographical memories (induced by photographs) were elicited. For the second and third conditions, activations associated with the failed attempt to retrieve autobiographical memories were elicited. A crucial difference between the second and third conditions, however, was the fact that for the second condition, the failure to recollect the occasion was associated with the pathological condition whereas for the third condition, this represented the true state of affairs (that is, he had not actually been present at the event and thus his denial of a memory of it was correct). It was found that the right prefrontal activation found frequently in memory-retrieval studies (Fletcher *et al.*, 1997) was greater in the two conditions in which the retrograde amnesia was not a factor whereas an activation in the medial parietal cortex (also a frequent observation in functional imaging studies of memory retrieval) was maximal in association with the failed attempt to retrieve episodic memories from the crucial 19-year period. The authors interpreted this as evidence for a dissociation between brain regions concerned with a "first-pass" retrieval (more implicit and subserved by medial parietal cortex) and a more explicit retrieval attempt resulting in the conscious experience of an episodic memory (subserved by PFC). The phenomenon observed in association with the failed retrieval condition (that is, an overactivation of medial parietal cortex and an underactivation of prefrontal cortex) was interpreted in terms of a suppressive effect produced by the emergence of memories from the period for which the patient was amnesic.

D. Brain Systems Rather Than Brain Regions

So far, the application and interpretation of PET and fMRI have been discussed predominantly in terms of the activations (and failures of activation) of specific regions in association with conditions or tasks. As such, they describe brain function in a spatially segregated way, attributing different functions to discrete regions. Likewise, they have, in the main, attempted to describe disorders in terms of regional dysfunctions. This is not a complete view of brain function (or dysfunction) since clearly, even the simplest of tasks used in PET and fMRI studies will be dependent on the function of complex systems in the brain. With respect to memory disorder, it may be the case that, in some instances, impairment is caused not by a specific regional dysfunction but rather by a failure of the segregated regions to act as an integrated system. This phenomenon may be overlooked if data are analyzed only in terms of regional responses. It is suggested also, with reference to Fig. 1, that some of the potential for misinterpretation of normal and abnormal regional patterns may be obviated by a full analysis and understanding of how distributed brain systems are behaving. Thus, for example, it is feasible that a regional response to a memory task in a patient may be apparently normal but that, nevertheless, that region is functioning abnormally in terms of its interaction with other brain regions. If data are analyzed and interpreted in a way that is solely concerned with isolated regional responses, such a dysfunction will be overlooked.

PET and fMRI allow a number of ways of exploring memory function and dysfunction in terms of integrated brain systems. Most simply, it is a matter of adopting a stance toward interpreting data that allows for the possibility that regionally discrete patterns of activation and deactivation are made mutually informative. Thus, it may be found that a relative failure of activation in a memory-impaired patient is accompanied by a simultaneous overactivation in another region. Although both of these findings are, in isolation, of interest, it may nevertheless be true that a fuller understanding of the disorder will only come from an attempt to integrate them into a broader explanation. Thus, in the study described above (Heckers et al., 1998), the prefrontal overactivation was interpreted (in light of previous findings in healthy volunteers) as an effort to compensate for failed recruitment of the hippocampus. That is, the spatially distributed activations were considered explicitly

in terms of a functional system subserving recognition memory. Interestingly, in a PET study of isolated retrograde amnesia (Levine et al., 1998), MRI evidence of damage to an area including the uncinate fasciculus (a white matter tract connecting frontal and temporal cortices) was associated with the opposite pattern, that is, hypoactivation in right frontal and hyperactivation in left medial temporal cortex when the patient (ML) performed a cued recall task. These results are shown in Fig. 5. The authors observed that ML, on behavioral testing, seemed to show preserved recollection on the basis of nonepisodic memory processes, with a deficiency in the episodic or autobiographical contents of memory. Like Heckers et al., they discussed the cued retrieval activations in terms of a distributed memory system. However, their interpretation of the frontal and temporal activations differed from that of Heckers et al. Rather than an attempt-success system mediated by frontal and hippocampal cortex, respectively, they suggest that the frontal cortex mediates *autonoietic* awareness (Wheeler et al., 1997) (that is, "awareness of oneself as a continuous entity across time") whereas the medial temporal cortex subserves a more impersonal form of memory. Thus, the presence of an isolated retrograde amnesia (affecting episodic or autobiographical memories) while anterograde memory is preserved may be explicable in terms of the frontotemporal disconnection.

A further example of this approach, adopted in a relatively early PET-FDG study (Fazio et al., 1992), was the attempt to characterize amnesia in terms of a dysfunctional network comprising medial temporal, diencephalic, and frontal structures. The results of this study, showing an abnormal functional network, were differentiated from structural findings and represent an attempt to exploit the much broader possibilites of these techniques in exploring integration (and disintegration) of brain function. Such an approach, obvious though it is, has been used in surprisingly few functional neuroimaging studies of memory dysfunction, with many researchers tending to concentrate on localized functional abnormalities. In part, of course, this stems from the fact that a lot of PET and fMRI studies are guided by neuropsychological findings, which are, most frequently, concerned with function of regions rather than systems. Nevertheless, since there is a growing body of data from functional imaging experiments exploring healthy memory function, it seems likely that, in more cases, abnormal functional imaging findings will be interpreted and expressed in terms of memory systems.

The studies above are exemplary of an approach to interpreting neuroimaging data from the perspective of functional integration. They are not dependent upon novel methods of data analysis. However, new methods

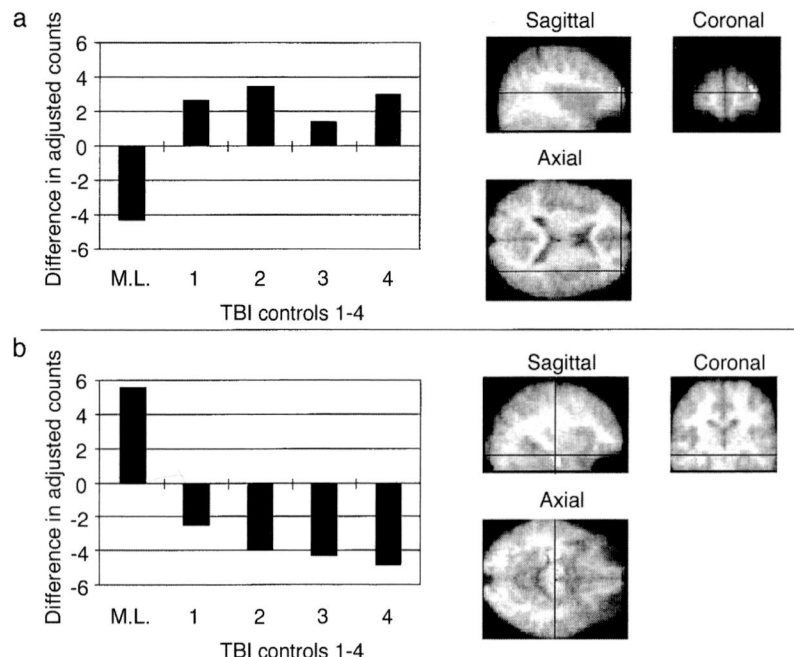

Figure 5 Right hippocampal (a) and left frontal (b) regions in which ML showed a greater (retrieval versus encoding) difference than controls. Data were acquired and analyzed as in Levine *et al.* (1998). Plots of relative activations are shown on the left and their spatial locations shown on the MRI images on the right. From Levine *et al.* (1998), *Brain*, by permission of Oxford University Press.

are being more frequently applied in order to explore human memory and learning in terms of functionally integrated systems (Mcintosh *et al.*, 1996; Nyberg *et al.*, 1996; Kohler *et al.*, 1998; Fletcher *et al.*, 1999). The most common of these uses path analyses to estimate connection strengths between separate regions in a predefined system (Friston *et al.*, 1993; Mcintosh and Gonzales-Lima, 1994; Buchel and Friston, 1997, 1998). These connection strengths are taken to imply interregional influences that may be estimated across tasks and groups. This, of course, raises the possibility that memory disorders may be characterized with respect to changes in connectivity between regions with disorders produced not by dysfunctional regions but by disintegrated systems. An example of this comes from the aforementioned study carried out on memory impairment in schizophrenia, a condition whose name implies a splitting or disintegration of function. We carried out a further analysis of the PET in which we explored a functional system comprising anterior cingulate, left prefrontal, and lateral temporal cortices (Fletcher *et al.*, 1999). On the basis of a number of previous observations, we predicted that frontotemporal connectivity was modulated by anterior cingulate cortical inputs and that this system would be disrupted in schizophrenia. In this analysis, the product of regional cerebral blood flow estimated values in prefrontal and anterior cingulate

cortices was treated as an independent variable. It was predicted that when this product was high (that is, when higher levels of prefrontal activity occurred in the presence of high levels of cingulate activity), then, in control subjects, this would predict lower levels of activity in the temporal neocortex. This was indeed found and we interpreted it as a frontal suppression of temporal activation, with this suppression being positively modulated by the anterior cingulate cortex. The same was not found for the schizophrenic group, in which the results suggested that there was a lack of the normal cingulate modulation of frontotemporal connectivity. One interesting facet of this study is that we were able to explore the brain responses in a way that was, to an extent, isolated from cognitive task manipulation. This task (that is, the varying memory load) was treated as a factor of no explanatory interest for the purposes of this analysis. We could thus be confident that the observed phenomena, including the abnormalities in the patient group, did not relate, in a direct way, to differential responses to changing task demands on the part of the patient group. This makes it possible that such an approach may overcome problems associated with interpreting task-related regional differences. This possibility is raised with caution, however, since it may well be the case that the observed cingulate-frontal-temporal disintegration is, indeed, related, albeit more indirectly, to the memory task, which, after all, provided the context in which the regional interactions were observed. To be more confident that this reflects a core deficit in schizophrenia (rather than a nonspecifically abnormal response to task demands), it would be necessary to observe the same pattern of interregional disconnection across a variety of tasks and levels of performance.[1]

E. New Possibilities

1. Imaging of Single Events

It has been repeatedly observed in this chapter that a pitfall to interpreting the PET and fMRI observations in memory disorders is that the tasks that are of most interest in such patients tend to be the ones that they perform less well than the control group. As such, it becomes impossible to say whether functional neuroimaging abnormalities are specific to the disease or whether they rather reflect simply nonperformance of task. Recent developments in the design and analysis of fMRI experiments, however, have made it possible to characterize brain responses to single events in studies of memory (Schacter et al., 1997; Buckner, 1998; Buckner et al., 1998a,b; Friston et al., 1998). That is, the hemodynamic response occurring following a single type of stimulus or response may be plotted and compared with the hemodynamic response to other events. This is, initially, beneficial in two major ways. First, it is no longer the case that functional neuroimaging experiments must be designed in a blocked way, that is, with groups of similar events being grouped together and the average brain response to this block of events being considered as representative of the typical response to a single event. Such an assumption is not justified since it may be the case that the presentation of similar events in blocks produces an added effect that would not be seen were such events isolated. This, of course, could be a potential confound in studies of memory disorder since a block of events in which a subject is continually showing impaired performance may produce psychological effects that do not occur in unimpaired control subjects or in the same patients when the events are not blocked together. In the case where responses to single events are separated, the experimenter can present event types unblocked and in any order.

A further advantage of event-related measurements derives from the potential to group responses, post hoc, on the basis of subjects' performances. Thus, for example, it has become possible to measure postscan recollection following the acquisition of fMRI data during a series of memory-encoding events. These performance data can be used to group the encoding events according to whether they were followed by successful retrieval or not. Such an approach has been used in healthy volunteers, producing informative compatible results across two studies (Brewer et al., 1998; Wagner et al., 1998) and the potential for its application to memory-impaired patients is exciting. Consider, for example, a series of memory-retrieval events in which the proportion of successful to unsuccessful retrieval is greater in a control than a patient group, as was the case in the study by Heckers et al. described above. With a blocked design, in which blocks of predominantly successful and predominantly unsuccessful retrieval were recorded separately, differences in brain response may be difficult to interpret. The possibility of taking an average brain response to all successful retrieval events in both groups and comparing it to the average response to all unsuccessful events becomes an important step forward since there is a major difference to interpreting abnormal activations occurring in the face of impaired performance to those that occur when performance is successful.

An additional possibility with event-related fMRI lies in a more accurate analysis of the temporal relationships between event-related responses in different brain areas. Thus, for example, it is feasible that an abnormality may be manifest in an abnormal time lag between event-related responses in one brain area and one or more others. Although the full potential of the new approach has yet to be addressed, it seems clear that there are a number of exciting possibilities with respect to the characterization of memory disorders.

2. Larger Data Sets. Multiple Sessions

Unlike PET, fMRI does not involve the use of ionizing radiation. The limitations on the number of measurements that may be made during a scanning session and the number of sessions that a subject may undergo are, therefore, practical ones, depending upon subjects' comfort rather than medico-legal ones. Additionally, with fMRI, single scans may be acquired in a matter of seconds rather than minutes so that more observations may be acquired in unit time than with PET. There are, thus, a number of advantages to the use of fMRI as a tool for the investigation of memory disorders. First, the possibility of acquiring larger data sets during single scanning sessions means that it is more feasible to chart the progress of brain changes occurring as a function of

---

[1] The procedure described does not constitute a full analysis of path strengths, using only a very incomplete model and consigning the majority of influences on a given region to the error term (see Friston et al. (1997) for full description). There are, however, more complex and sophisticated analytical techniques that allow assertions to be made regarding interregional influences with greater confidence and that carry great potential for characterizing disorders of the memory system.

learning, for example. Since memory impairment may be manifest as altered learning patterns, such a possibility becomes highly relevant to the functional neuroanatomical description of the patient group. Additionally, with respect to the application of the path analysis techniques to assess and compare functional integration, collecting large data time series is crucial.

In many cases, the degree or severity of memory impairment is not constant over time. In, for example, the dementias, it may worsen (in a progressive or stepwise manner). In some conditions, it may recover (as, for example, in transient global amnesia). In yet other cases, it may follow a relapsing and remitting course, such as when it is associated with a mental illness such as depression. Furthermore, even in patients in whom the underlying structural deficit shows no sign of recovery, it is possible that alternative strategies and/or brain systems may be engaged. Thus, in all of these cases, the possibility of carrying out a series of measurements over an extended time period can become important.

IV. Summary

In conclusion, it seems likely that the new functional neuroimaging techniques, including PET and fMRI, have a role to play in broadening understanding of memory function and in characterizing the ways in which memory may be impaired. To make full of use, however, of the new possibilities, it is suggested that experiments must be designed in ways that are conceptually different from standard neuropsychological studies. While a structural abnormality may be confidently assumed to represent a genuine regional abnormality irrespective of the patient's behavior at the time of its observation, the same cannot be said of an apparent abnormality observed using PET or fMRI. A functional measurement is only interpretable to the extent that the subject's psychological state, at the time of measurement, is understood. In addition, although it may be the case that some functional imaging disorders are measurable across a variety of memory tasks, irrespective of the degree to which the patient is impaired, it seems highly likely that others will be observed only in cases where the system is stressed and performance falls below normal. In such cases, it becomes important that attempts are made to differentiate between group differences that reflect true physiological abnormalities and those that reflect performance differences between patients and their controls. Although fMRI brings increased spatial and temporal resolution, together with the possibility of collecting larger data sets on multiple occasions, these increased capabilities do not obviate the need for closer understanding of the cognitive states

of patients and controls. Only with such a background of understanding will the techniques, no matter how sophisticated they become, generate results that are meaningful. The techniques, therefore, do not stand alone but rely upon the insights gained from the broad range of approaches that make up the cognitive neuroscience of memory.

References

Andreasen, N. C., O'Leary, D. S., Cizadlo, T., Arndt, S., Rezai, K., Watkins, G. L., Ponto, L. L., and Hichwa, R. D. (1995). Remembering the past: Two facets of episodic memory explored with positron emission tomography. *Am. J. Psychiatry* **152**, 1576–1585.

Brewer, J. B., Zhao, Z., Desmond, J. E., Glover, G. H., and Gabrieli, J. D. (1998). Making memories: Brain activity that predicts how well visual experience will be remembered [see comments]. *Science* **281**, 1185–1187.

Buchel, C., and Friston, K. J. (1997). Modulation of connectivity in visual pathways by attention: Cortical interactions evaluated with structural equation modelling and fMRI. *Cereb. Cortex* **7**, 768–778.

Buchel, C., and Friston, K. J. (1998). Dynamic changes in effective connectivity characterized by variable parameter regression and Kalman filtering. *Hum. Brain Mapp.* **6**, 403–408.

Buckner, R. L. (1998). Event-related fMRI and the hemodynamic response. *Hum. Brain Mapp.* **6**, 373–377.

Buckner, R. L., Goodman, J., Burock, M., Rotte, M., Koutstaal, W., Schacter, D., Rosen, B., and Dale, A. M. (1998a). Functional-anatomic correlates of object priming in humans revealed by rapid presentation event-related fMRI. *Neuron* **20**, 285–296.

Buckner, R. L., Koutstaal, W., Schacter, D. L., Dale, A. M., Rotte, M., and Rosen, B. R. (1998b). Functional-anatomic study of episodic retrieval. II. Selective averaging of event-related fMRI trials to test the retrieval success hypothesis. *NeuroImage* **7**, 163–175.

Costello, A., Fletcher, P. C., Dolan, R. J., Frith, C. D., and Shallice, T. (1998). The origins of forgetting in a case of isolated retrograde amnesia following a haemorrhage: Evidence from functional imaging. *Neurocase* **4**, 437–446.

Fazio, F., Perani, D., Gilardi, M. C., Colombo, F., Cappa, S. F., Vallar, G., Bettinardi, V., Paulesu, E., Alberoni, M., Bressi, S., *et al.* (1992). Metabolic impairment in human amnesia: A PET study of memory networks. *J. Cereb. Blood Flow Metab.* **12**, 353–358.

Fletcher, P. C., Frith, C. D., and Rugg, M. D. (1997). The functional neuroanatomy of episodic memory [see comments]. *Trends Neurosci.* **20**, 213–218.

Fletcher, P. C., McKenna, P. J., Frith, C. D., Grasby, P. M., Friston, K. J., and Dolan, R. J. (1998). Brain activations in schizophrenia during a graded memory task studied with functional neuroimaging. *Arch. Gen. Psychiatry* **55**, 1001–1008.

Fletcher, P. C., Büchel, C., Josephs, O., Friston, K. J., and Dolan, R. J. (1999a). Learning-related neuronal responses in prefrontal cortex studied with functional neuroimaging. *Cereb. Cortex* **9**, 168–178.

Fletcher, P. C., McKenna, P. J., Friston, K. J., Frith, C. D., and Dolan, R. J. (1999b). Abnormal cingulate modulation of fronto-temporal connectivity in schizophrenia. *NeuroImage* **9**, 337–342.

Friston, K. J., Frith, C. D., and Frackowiak, R. S. J. (1993). Time-dependent changes in effective connectivity measured with PET. *Hum. Brain Mapp.* **1**, 69–79.

Friston, K. J., Price, C. J., Fletcher, P., Moore, C., Frackowiak, R. S., and

Dolan, R. J. (1996). The trouble with cognitive subtraction. *NeuroImage* **4,** 97–104.

Friston, K. J., Buechel, C., Fink, G. R., Morris, J., Rolls, E., and Dolan, R. J. (1997). Psychophysiological and modulatory interactions in neuroimaging. *NeuroImage* **6,** 218–229.

Friston, K. J., Fletcher, P., Josephs, O., Holmes, A., Rugg, M. D., and Turner, R. (1998). Event-related fMRI: Characterizing differential responses. *NeuroImage* 7, 30–40.

Green, R. E., and Shanks, D. R. (1993). On the existence of independent explicit and implicit learning systems: An examination of some evidence: *Mem Cog.* **21,** 304–317.

Haxby, J. V., Ungerleider, L. G., Horwitz, B., Maisog, J. M., Rapoport, S. I., and Grady, C. L. (1996). Face encoding and recognition in the human brain. *Proc. Natl. Acad. Sci. U.S.A.* **93,** 922–927.

Heckers, S., Rauch, S. L., Goff, D., Savage, C. R., Schacter, D. L., Fischmann, A., and Alpert, N. M. (1998). Impaired recruitment of the hippocampus during conscious recollection in schizoprehnia. *Nat. Neurosci.* **1,** 318–323.

Heiss, W. D., Pawlik, G., Holthoff, V., Kessler, J., and Szelies, B. (1992). PET correlates of normal and impaired memory functions. *Cerebrovasc. Brain Metab. Rev.* **4,** 1–27.

Jennings, J. M., Mcintosh, A. R., Kapur, S., Tulving, E., and Houle, S. (1997). Cognitive subtractions may not add up: The interaction between semantic processing and response mode. *NeuroImage* **5,** 229–239.

Kapur, S., Craik, F. I., Jones, C., Brown, G. M., Houle, S., and Tulving, E. (1995). Functional role of the prefrontal cortex in retrieval of memories: A PET study. *NeuroReport* **6,** 1880–1884.

Kohler, S., Mcintosh, A. R., Moscovitch, M., and Winocur, G. (1998). Functional interactions between the medial temporal lobes and posterior neocortex related to episodic memory retrieval. *Cereb. Cortex* **8,** 451–461.

Kopelman, M. D., Green, R. E., Guinan, E. M., Lewis, P. D., and Stanhope, N. (1994). The case of the amnesic intelligence officer. *Psychol. Med.* **24,** 1037–1045.

Kuwert, T., Homberg, V., Steinmetz, H., Unverhau, S., Langen, K. J., Herzog, H., and Feinendegen, L. E. (1993). Posthypoxic amnesia: Regional cerebral glucose consumption measured by positron emission tomography. *J. Neurol. Sci.* **118,** 10–16.

Levine, B., Black, S. E., Cabeza, R., Sinden, M., Mcintosh, A. R., Toth, J. P., Tulving, E., and Stuss, D. T. (1998). Episodic memory and the self in a case of isolated retrograde amnesia. *Brain* **121,** 1951–1973.

Lucchelli, F., De Renzi, E., Perani, D., and Fazio, F. (1994). Primary amnesia of insidious onset with subsequent stabilisation. *J. Neurol. Neurosurg. Psychiatry* **57,** 1366–1370.

Markowitsch, H. J., Calabrese, P., Fink, G. R., Durwen, H. F., Kessler, J., Harting, C., Konig, M., Mirzaian, E. B., Heiss, W. D., Heuser, L., and Gehlen, W. (1997). Impaired episodic memory retrieval in a case of probable psychogenic amnesia. *Psychiatry Res.* **74,** 119–126.

Mattioli, F., Grassi, F., Perani, D., Cappa, S. F., Miozzo, A., and Fazio, F. (1996). Persistent post-traumatic retrograde amnesia: A neuropsychological and (^{18}F)FDG PET study. *Cortex* **32,** 121–129.

Mcintosh, A. R., and Gonzales-Lima, F. (1994). Structural equation modelling and its application to network analysis in functional brain imaging. *Hum. Brain Mapp.* **2,** 2–22.

Mcintosh, A. R., Grady, C. L., Haxby, J. V., Ungerleider, L. G., and Horwitz, B. (1996). Changes in limbic and prefrontal functional interactions in a working memory task for faces. *Cereb. Cortex* **6,** 571–584.

Nyberg, L., Mcintosh, A. R., Cabeza, R., Nilsson, L. G., Houle, S., Habib, R., and Tulving, E. (1996). Network analysis of positron emission tomography regional cerebral blood flow data: Ensemble inhibition during episodic memory retrieval. *J. Neurosci.* **16,** 3753–3759.

Ouchi, Y., Nobezawa, S., Okada, H., Yoshikawa, E., Futatsubashi, M.,

and Kaneko, M. (1998). Altered glucose metabolism in the hippocampal head in memory impairment. *Neurology* **51,** 136–142.

Perani, D., Bressi, S., Cappa, S. F., Vallar, G., Alberoni, M., Grassi, F., Caltagirone, C., Cipolotti, L., Franceschi, M., Lenzi, G. L., *et al.* (1993). Evidence of multiple memory systems in the human brain. A [^{18}F]FDG PET metabolic study. *Brain* **116,** 903–919.

Pietrini, P., Azari, N. P., Grady, C. L., Salerno, J. A., Gonzales Aviles, A., Heston, L. L., Pettigrew, K. D., Horwitz, B., Haxby, J. V., and Schapiro, M. B. (1993). Pattern of cerebral metabolic interactions in a subject with isolated amnesia at risk for Alzheimer's disease: A longitudinal evaluation. *Dementia* **4,** 94–101.

Price, C. J., Mummery, C., Moore, C., Frackowiak, R. S. J., and Friston, K. J. (1999). Delineating necessary and sufficient neural systems: Functional imaging studies of neuropsychological patients. *J. Cog. Neurosci.* (in press).

Rugg, M. D., Fletcher, P. C., Frith, C. D., Frackowiak, R. S., and Dolan, R. J. (1996). Differential activation of the prefrontal cortex in successful and unsuccessful memory retrieval. *Brain* **119,** 2073–2083.

Rugg, M. D., Fletcher, P. C., Frith, C. D., Frackowiak, R. S., and Dolan, R. J. (1997). Brain regions supporting intentional and incidental memory: A PET study. *NeuroReport* **8,** 1283–1287.

Schacter, D. L., and Tulving, E. (1994). "Memory Systems." MIT Press, Massachusetts.

Schacter, D. L., Alpert, N. M., Savage, C. R., Rauch, S. L., and Albert, M. S. (1996a). Conscious recollection and the human hippocampal formation: Evidence from positron emission tomography. *Proc. Natl. Acad. Sci. U.S.A.* **93,** 321–325.

Schacter, D. L., Reiman, E., Curran, T., Yun, L. S., Bandy, D., McDermott, K. B., and Roediger, H. L. (1996b). Neuroanatomical correlates of veridical and illusory recognition memory: Evidence from positron emission tomography [see comments]. *Neuron* **17,** 267–274.

Schacter, D. L., Savage, C. R., Alpert, N. M., Rauch, S. L., and Albert, M. S. (1996c). The role of hippocampus and frontal cortex in age-related memory changes: A PET study. *NeuroReport* **7,** 1165–1169.

Schacter, D. L., Buckner, R. L., Koustaal, W., Dale, A. M., and Rosen, B. R. (1997). Late onset of anterior prefrontal activity during true and false recognition: An event-related fMRI study. *NeuroImage* **6,** 259–269.

Scoville, W. B., and Milner, B. (1957). Loss of recent memory after bilateral hippocampal lesions. *J. Neuro. Neurosurg. Psych.* **20,** 11–21.

Shallice, T. (1988). "From Neuropsychology to Mental Structure." Cambridge Univ. Press, Cambridge, UK.

Stuss, D. T., Alexander, M. P., Palumbo, C. L., Buckle, L., Sayer, L., and Pogue, J. (1994). Organisational strategies of patients with unilateral or bilateral frontal lobe injury in word list learning tasks. *Neuropsych.* **8,** 355–373.

Tamlyn, D., McKenna, P. J., Mortimer, A. M., Lund, C. E., Hammond, S., and Baddeley, A. D. (1992). Memory impairment in schizophrenia: Its extent, affiliations and neuropsychological character. *Psychol. Med.* **22,** 101–115.

Tulving, E., Kapur, S., Markowitsch, H. J., Craik, F. I., Habib, R., and Houle, S. (1994). Neuroanatomical correlates of retrieval in episodic memory: Auditory sentence recognition [see comments]. *Proc. Natl. Acad. Sci. U.S.A.* **91,** 2012–2015.

Wagner, A. D., Schacter, D. L., Rotte, M., Koutstaal, W., Maril, A., Dale, A. M., Rosen, B. R., and Buckner, R. L. (1998). Building memories: Remembering and forgetting of verbal experiences as predicted by brain activity [see comments]. *Science* **281,** 1188–1191.

Welch, L. W., Nimmerrichter, A., Kessler, R., King, D., Hoehn, R., Margolin, R., and Martin, P. R. (1996). Severe global amnesia presenting as Wernicke-Korsakoff syndrome but resulting from atypical lesions. *Psychol. Med.* **26,** 421–425.

Wheeler, M. A., Stuss, D. T., and Tulving, E. (1997). Toward a theory of episodic memory: The frontal lobes and autonoetic consciousness. *Psychol. Bull.* **121,** 331–354.

9

Brain Mapping in Dementia

Michael S. Mega,*,†,1 Paul M. Thompson,* Arthur W. Toga,* and Jeffrey L. Cummings†,‡

*Laboratory of Neuro Imaging, Division of Brain Mapping, Department of Neurology,
†Alzheimer's Disease Center, and
‡Department of Psychiatry and Biobehavioral Sciences, UCLA School of Medicine, Los Angeles, California 90095

I. Structural Imaging
II. Functional Imaging
III. Summary
   References

The application of brain mapping techniques in elderly and demented populations presents unique challenges but offers exciting breakthroughs in early detection and the monitoring of treatment efficacy. Utilizing brain mapping techniques in combination with a profile of cognitive performance and risk factors is currently the most powerful predictor of incipient Alzheimer's disease (AD) in elderly individuals with mild cognitive impairment (MCI). In the absence of a biochemical marker, functional and structural neuroimaging is now the best biological marker for AD. Longitudinal prospective studies of individuals presenting to memory disorder clinics who later develop AD, by clinical (Johnson et al., 1998) or pathologic criteria (Jobst et al., 1998), show that these individuals have greater functional defects in parietal and posterior cingulate regions than those individuals who do not develop AD. Similarly,

longitudinal studies of patients with isolated memory impairment reveal that within a 4-year period 80% who also have medial temporal atrophy at baseline develop AD (de Leon et al., 1989, 1993). The combination of medial temporal atrophy and functional defects in parietal or medial temporal/posterior cingulate cortices has a higher sensitivity and specificity of correctly identifying pathologically confirmed AD than a clinician's application of National Institute of Neurological and Communicative Disorders and Stroke/Alzheimer's Disease and Related Disorders Association (NINCDS/ADRDA) criteria (Jobst et al., 1998).

Once a *valid and reliable* methodology is established to identify regional atrophy and functional defects predicting incipient AD, then modern brain mapping techniques will have significantly advanced health care delivery. This chapter reviews the advances in brain mapping techniques as applied to aging and dementia, covering both structural and functional neuroimaging. We focus on efforts to uncover the pattern of regional atrophy and functional defects specific for AD. An emphasis on the application of modern brain mapping techniques will be provided together with the challenges in their application in the elderly population, where cortical atrophy adds to the normal anatomic variability of one person's brain with a population's average brain.

---

[1] To whom correspondence should be addressed.

I. Structural Imaging

Structural studies based on computerized tomography (CT) and magnetic resonance imaging (MRI) provided evidence that the cortex in AD was significantly reduced compared to that in age-matched controls. In the middle stage of AD, atrophy is most pronounced in the temporal lobes (Brun and Englund, 1981). The greatest regional loss in cortical volume is found in the medial temporal allocortex and heteromodal frontal and parietal association cortices. The degree of atrophy varies among patients (Miller *et al.*, 1980), yet cortical neuronal loss is two to three times greater in AD than in age-matched normals (Shefer, 1972). The identification of regional gray matter (GM) loss in AD compared to normal aging indicated that generalized atrophy did not occur in AD but rather specific brain regions are targeted by the pathophysiologic process. A regional predilection of the atrophic process in AD agreed with the emerging microscopic pathologic pattern found for neurofibrillary tangle (NFT) deposition. Seminal work (Braak and Braak, 1991) showed that the pathologic severity in AD could be staged based upon regional NFT density, beginning in medial temporal transentorhinal cortex and then spreading through heteromodal association cortices in frontal and parietal lobes to eventually affect sensory and motor cortices in the late stages of the disease. Similar to the regional progression of NFT burden in postmortem cases, longitudinal imaging studies also showed a greater loss of GM volume with disease progression in medial temporal and frontal/parietal cortices. An agreement among the pathological and imaging data supported a specific pattern of atrophy in the AD brain.

A. Segmentation

Early studies evaluating brain tissue changes in aging were based on serially sectioned pathological specimens outlined to derive planimetric GM and white matter (WM) volumes throughout formalin-fixed specimens (Anton, 1903; Jaeger, 1914) and documented a GM:WM ratio change of 2.26-2.38 at the age of three to 1.05-1.4 in the fifth and sixth decades, later confirmed with more modern pathological assessment (Miller *et al.*, 1980). These initial studies revealed volumetric loss throughout aging and in dementia but they were subject to error due to shrinkage and dehydration during the fixation process. The first *in vivo* imaging attempts delineating brain tissue volumes in aging and dementia were conducted with ^{133}Xe carotid injection (Høedt-Rasmussen and Skinhøj, 1966) and noted a progressive loss of GM: 48.8% in normals, 40.2% in mild dementia, and

33.4% in moderate dementia. The first volumetric CT studies in aging and AD respectively were by Yamaura *et al.* (1980) and Gado *et al.* (1982). Because of the low slice frequency interval, approaching 5 mm to 1 cm, these studies lacked precision in documenting subtle regional changes due to dementia versus normal aging. With the advent of MRI, trinary segmentation was initiated (Seab *et al.*, 1988; Filipek *et al.*, 1989; Jernigan *et al.*, 1990, 1991; Rusinek *et al.*, 1991) utilizing the differential intensities of GM, WM, and cerebral spinal fluid (CSF).

The application of brain mapping techniques in segmentation analysis is challenged by the simultaneous assessment of disease-specific tissue change, compared to normals, while controlling for normal anatomical variation. After segmentation of MRIs into GM, WM, and CSF, automated regions of interest (ROIs) analyses with current brain mapping techniques can be applied. Techniques have used nonlinear registration of a segmented, single-subject atlas to individual structural scans, with a 6% error (Gee *et al.*, 1993). By using nonlinear warping with a probabilistic atlas, which incorporates the spatial distribution of a population's anatomy (Mazziotta *et al.*, 1995; Evans *et al.*, 1996), control over anatomic confounds with preservation of age-related effects can be accomplished in an unbiased manner. We hypothesized significant frontal GM loss in elderly nondemented normals compared to young normals. To assess the GM quantity in lobar ROIs, high-resolution 3D MRI scans of 24 persons (12 normal right-handed males, mean age 30.0, SD = 7.7 years; 12 normal older right-handed males, mean age 70.9, SD = 7.0 years), after inhomogeneity correction (Sled *et al.*, 1998), were processed with a minimum-distance classification algorithm (Kollokian, 1996) to segment the MRI data into WM, GM, and CSF voxels (Fig. 1).

To localize the segmented anatomy from the probabilistically partitioned atlas, a fifth-order intensity-based polynomial warp (Woods *et al.*, 1993) drove the partitioned atlas space into the subject's native data, with subsequent correction of tissue counts for head size (Fig. 2). As hypothesized, there was a significant decrease in frontal GM ($p < 0.02$) and an increase in frontal CSF ($p < 0.001$) in elderly compared to young normals. Figure 3 graphically illustrates the decline in GM and increase in CSF with respect to age. In addition, we also found a significant decrease in occipital GM ($p < 0.05$) and a significant increase in CSF for the aged subjects in the occipital ($p < 0.02$), parietal ($p < 0.001$), and temporal lobes ($p < 0.02$).

This brain mapping technique demonstrates how high-resolution volumetric tissue analysis can be corrected for anatomic differences across subjects. While still preserving group differences using nonlinear warp-

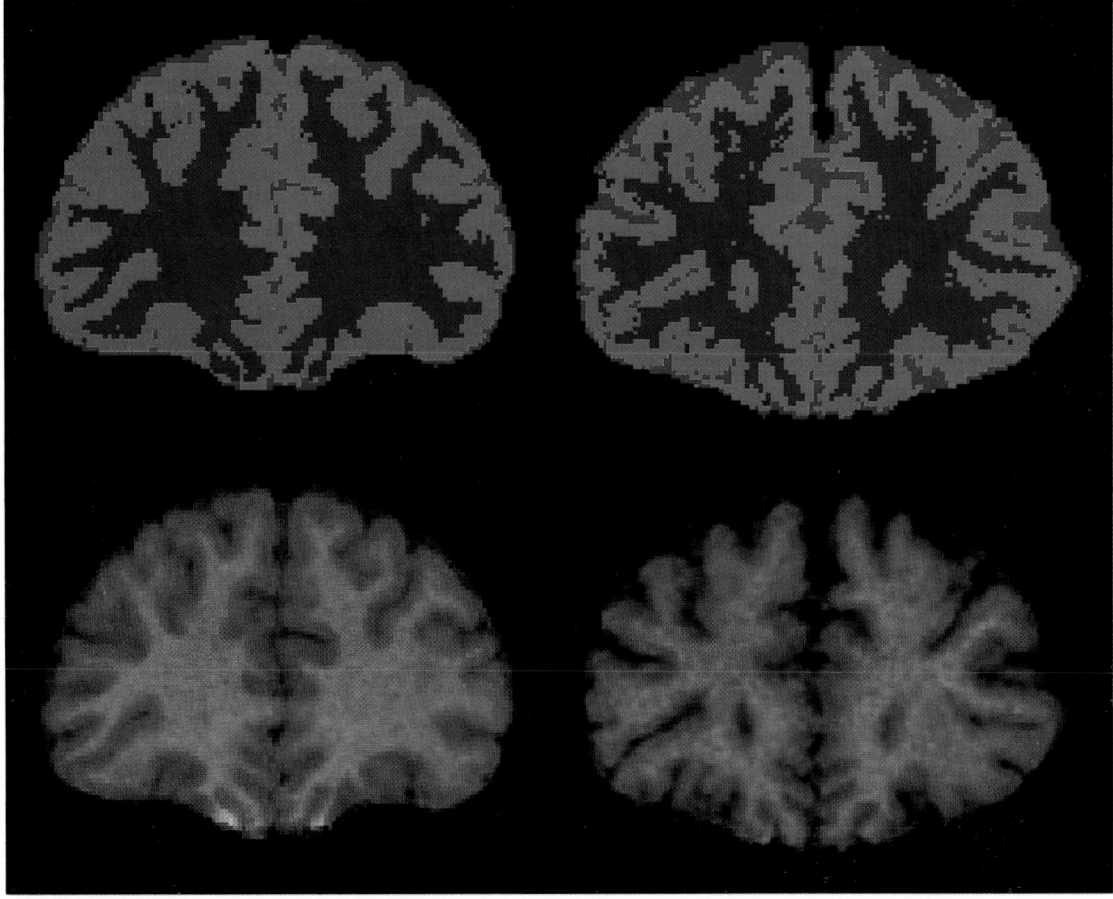

Figure 1 Segmentation of young (left) and old (right) normal brains into gray matter (green), white matter (blue), and cerebral spinal fluid (red) using a minimum-distance classification algorithm (Kollokian, 1996) after inhomogeneity correction (Sled *et al.*, 1998). Once segmented, a template, with probabilistically defined anatomic regions (Mazziotta *et al.*, 1995; Evans *et al.*, 1996), can be registered to each subject's brain scan for subvolume analysis.

ing algorithms within a probabilistic atlas, an unbiased automated analysis is possible. Our findings support previous findings that as we age, we lose GM in frontal lobes and other regions.

B. Volumetrics

From earlier segmentation studies using both pathological data and *in vivo* imaging analysis, a pattern of atrophy associated with AD has emerged. The concentration of the atrophic process early in the course of the disease begins in medial temporal structures. Given the mediation of declarative memory by medial temporal structures, as well as the pathologic concentration of markers in hippocampal and entorhinal cortices, the search for early imaging changes in this region began in the early 1980s. Volumetric assessment of the hippocampus has produced the clearest distinction between AD and normal aging compared to any other re-

gion studied. Attempts also have been made to evaluate amygdalar volume; however, a clear distinction of anatomic landmarks demarcating the amygdala is difficult to achieve with *in vivo* imaging.

Given the wealth of evidence accumulated across studies and institutions showing significant hippocampal loss in AD compared to elderly controls, there is now no longer a need for any future studies to confirm that the hippocampus in AD is reduced in volume compared to normal aged individuals. The present challenge for hippocampal volumetry is to identify the earliest point at which the hippocampus becomes atrophic in patients with MCI and at what rate this change occurs over time. Identifying the first manifestation of hippocampal atrophy, or its rate of decline, makes the early diagnosis of AD possible and the same assessment may serve as a biological marker to evaluate disease-modifying pharmacological treatments.

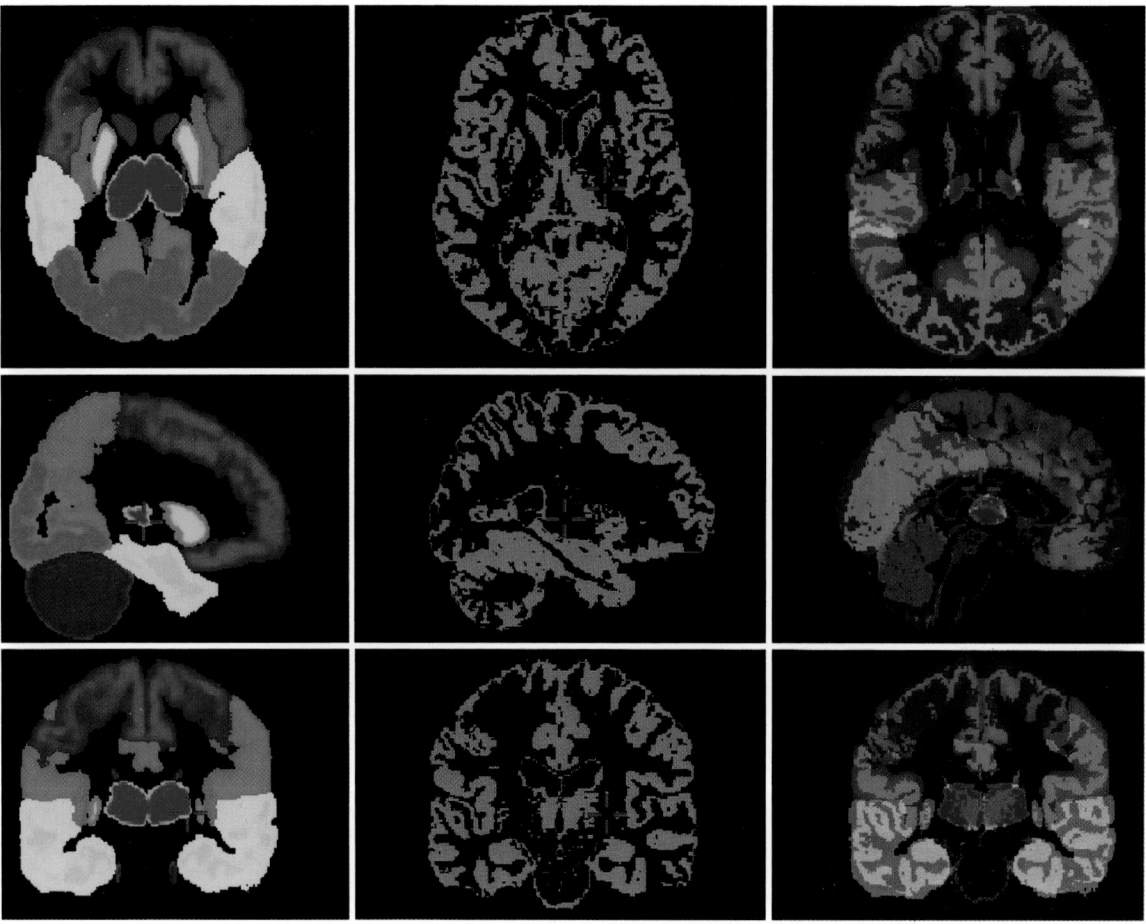

Figure 2 The nine International Consortium of Human Brain Mapping (ICBM) probabilistic ROIs (Evans *et al.,* 1996) are displayed on the left. Each ROI is labeled with a different color. The middle column represents a gray matter map of an individual young normal. To correct for anatomic variability in an unbiased manner, a fifth-order nonlinear warp (Woods *et al.,* 1998) was used to overlay the ICBM ROIs onto individual scans. The third column demonstrates the registration accuracy of the two images.

1. Hippocampus and Amygdala

Hippocampal structures in AD exhibit a mean volume loss of between 20 and 52% compared with age-matched controls (Table I). The challenge presented to neuroimaging in aging and dementia is centered on patients with MCI, those at risk for AD. Qualitative (de Leon *et al.,* 1989) and quantitative (Convit *et al.,* 1993, 1995; de Leon *et al.,* 1993; Soininen *et al.,* 1994; Parnetti *et al.,* 1996) studies of persons with age-associated memory impairment (AAMI) (Blackford and La Rue, 1989; Cook *et al.,* 1992), or those who eventually develop AD but did not meet criteria for the disease at the time of initial evaluation (Kaye *et al.,* 1997), have demonstrated significant hippocampal atrophy compared to normal age-related losses. Hippocampal volume loss due to normal aging may approach 46 mm^3 per year over the age of 65, with a near-linear decline (Jack *et al.,* 1997). The hippocampus of the AD patient in the earliest stage

of the disease is already 1.75 standard deviations beyond the normal age expected loss (Jack *et al.,* 1997) and is correlated with a patient's memory impairment (Scheltens *et al.,* 1992; Golomb *et al.,* 1994, 1996; Deweer *et al.,* 1995; Laakso *et al.,* 1995; Cahn *et al.,* 1998; Kohler *et al.,* 1998). The entorhinal cortex may be the focus of the atrophic process in early AD, with a volume loss of 40% in patients compared to controls (Juottonen *et al.,* 1998).

A normal asymmetry of medial temporal volumes, right larger than left, has been confirmed in morphological studies based on *in vivo* imaging analysis (Table I); this normal asymmetry averages 6.7% across all studies. Longitudinal analysis reveals a trend for the reversal of this normal, right-larger-than-left hippocampal asymmetry as an early morphologic change in persons who later go on to develop AD (Kaye *et al.,* 1997). Although only one study explicitly tested this reversal of normal asymmetry in persons with AAMI (Soininen *et al.,*

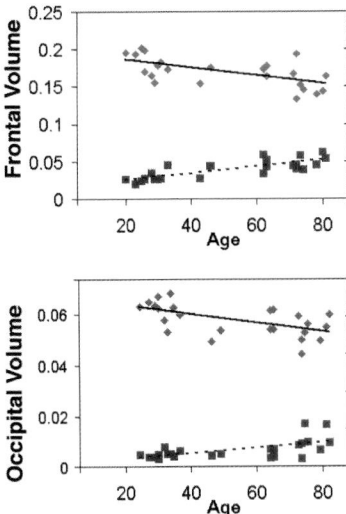

Figure 3 Graphic representation of significantly decreased gray matter (green) and increased CSF (red) as a function of age for the frontal lobe (top) and occipital lobe (bottom).

1994), all studies of these mildly impaired persons reflect a reversal of the normal hippocampal asymmetry when bilateral volumes were reported (Soininen *et al.*, 1994; Parnetti *et al.*, 1996; Kaye *et al.*, 1997). Persons with AAMI who are homozygous for the apolipoprotein E . . . 4 (ApoE-4) allele may have a greater reversal of the normal hippocampal asymmetry than those without the ApoE-4 allele (Soininen *et al.*, 1995); yet the ApoE-4 allele is not associated with global atrophy in AD (Yasuda *et al.*, 1998). As the disease advances, the ApoE-4 allele may be associated with imaging markers of medial temporal atrophy, compared to AD patients without the ApoE-4 allele (Tanaka *et al.*, 1998), but hippocampal volume might not be the cause of this general temporal atrophy (Jack *et al.*, 1998).

The current challenge in applying brain mapping techniques to hippocampal volumetry is to detect subtle early changes in persons with preclinical AD using a standardized technique that controls for normal anatomic variability and allows easy application across centers. Because high-resolution imaging is able to detect subtle early morphological changes, we used high-resolution imaging to test the hypothesis that patients with MCI have a significant reversal of the normal right-larger-than-left hippocampal asymmetry compared to elderly individuals with normal memory function. We also sought to determine which subregions of the hippocampus demonstrate significant atrophy in MCI since a previous study has implicated the hippocampal head as being the most severely affected in AD (Jack *et al.*, 1997). In this example, we applied 1-mm contiguous volumetry within the Talairach coordinate space to evaluate the subtle early changes hypothesized

to occur in the hippocampus of patients with MCI. The use of a common brain space for hippocampal analysis fulfills the brain mapping objective of standardization in methodology and ease of application across centers. All hippocampal volumes were lower in the MCI group compared to the cognitively intact elderly, with the right hippocampal head and total volume showing a significant loss ($p = 0.02$). The mean hippocampal volume results for the two groups are shown in Table II.

Although both hippocampi were reduced, we found a significant ($p = 0.02$) volume loss in the right anterior half of the hippocampus in patients with mild memory impairment. This finding agrees with the location of hippocampal atrophy seen in early AD patients (Jack *et al.*, 1997) and suggests that the pathologic process responsible for this volume loss predates the clinical manifestations by several years.

The cause of the greater right hippocampal volume loss during the preclinical stage of AD, or in MCI, is unclear. The early reversal of the normal medial temporal asymmetry may reflect an asymmetry in the pathophysiology of early AD. There may be greater synaptic pruning, or neuronal loss, in the right hippocampus in patients with MCI or incipient AD that could contribute to volume loss first in this region. Conversely, individuals with developmental or acquired reductions in the neuronal reserve of the right hippocampus, reflected by smaller volumes, may simply be at an increased risk for mild cognitive decline. The cause for the early asymmetric change in MCI is unknown. All of our patients were right-handed; thus the influence of handedness could not be assessed. A trend for reversal of the normal hippocampal asymmetry has been documented to occur in normal elderly subjects nearly 4 years prior to their development of AD (Kaye *et al.*, 1997), with equal rates of change for both hippocampi (-14 to -46 mm^3/year) (Jack *et al.*, 1997; Kaye *et al.*, 1997). These prior data suggest that the right-greater-than-left atrophic process predates the development of AD by more than 4 years. Long-term prospective studies that employ high-resolution imaging, or 3D mapping, will be needed to determine at what point prior to dementia onset hippocampal atrophy occurs.

A further objective in brain mapping is the development of automated techniques. With a semiautomated 3D mapping approach (Haller *et al.*, 1996, 1997; Csernansky *et al.*, 1998a), utilizing a hippocampal MRI template of a healthy control, nonlinear intensity-based warping of the template onto each subject's MRI scan extracts the hippocampal boundaries from new subjects. Digital extraction occurs in two steps. The template is first coarsely aligned to each target scan by landmarks manually placed on the template and target scans. Next the target's local anatomy is determined by a high-dimensional, au-

Table I   Morphologic Studies of Hippocampus (Hippo) and Amygdala (Amyg) in Patients with Age-Related Cognitive Decline (ARCD), Alzheimer's Disease (AD), and Elderly Controls

Reference	Study modality	Contiguous volumetry[a]	Slice spacing	n ARCD	n AD	n Controls	Percentage R > L Hippo	Percentage R > L Amyg	Percentage decline Hippo	Percentage decline Amyg
Herzog and Kemper, 1980	Autopsy	Yes	700 μm	—	4	4			—	24
LeMay et al., 1986	CT	No, qualitative	NA	—	24	22		14	86% accuracy	
Murphy et al., 1987	Autopsy	Yes	350 μm	—	—	19	—	6[h]	—	—
Seab et al., 1988	MRI	No, planimetric	NA	—	10	7			40	—
de Leon et al., 1988	CT	No qualitative	NA	—	9	8			100% accuracy	
de la Monte, 1989	Autopsy	No, planimetric	NA	—	16	14			10	24
de Leon et al., 1989	CT	No, qualitative	NA	27	76	72			70/87/22[b]	—
George et al., 1990	CT	No, qualitative	NA	—	34	20			80% accuracy	
Kesslak et al., 1991	MRI	No, 1-mm gap	5 mm	—	8	7			49	—
Scott et al., 1991	Autopsy	Yes	2.5 mm	—	5	5			—	45
Dahlbeck et al., 1991	MRI	No, linear	NA	—	10	10			41[c]	—
Jack et al., 1992	MRI	Yes	4 mm	—	20	22			40	—
Jobst et al., 1992a	CT	No, linear	NA	—	51	18			41[d]	—
Jobst et al., 1992b	CT	No, linear	NA	—	44	75			44[d] controls 1% loss/year	—
Scheltens et al., 1992	MRI	No, linear	NA	—	21	21	5	—	15	—
Pearlson et al., 1992	MRI	No, two slices	3 mm	—	15	16	3	—	61	66
Coffey et al., 1992	MRI	Yes	5 mm	—	—	76			0.3/year[g]	—
Doraiswamy et al., 1993	MRI	No, linear	NA	—	—	75			21[c] (>50 years)	—
Howieson et al., 1993	MRI	No, linear	NA	—	10	10			30[c]	—
Early et al., 1993	MRI	No, linear	NA	—	12	17			A/H vs IUD; $r = -0.38, p = 0.13$	
Erkinjuntti et al., 1993	MRI	No, planimetric	3–5 mm	—	34	39			35	—
Killiany et al., 1993	MRI	Yes	1.5 mm	—	8	7	3	14	23	26
Cuénod et al., 1993	MRI	Yes	5 mm	—	11	5			—	33
Convit et al., 1993	MRI	No, 1-mm gap	4 mm	17	15	18			12e/20	—
de Leon et al., 1993	MRI/CT	Yes[i]	4 mm	32	25	54	18	—	91 sen/89 spec	
Huesgen et al., 1993	MRI	No, planimetric	NA	—	13	9			31	—
Lehéricy et al., 1994	MRI	Yes	5 mm	—	18	8	4	4	30	37
Ikeda et al., 1994	MRI	No, planimetric	NA	6 (pAD)	8	8			23/29	—
Soininen et al., 1994	MRI	Yes	2 mm	16[j]	—	16	10	−9	11[f]	13[f]
Lehtovirta et al., 1995	MRI	Yes	2 mm	—	26	[k]			40	20
Laakso et al., 1995	MRI	Yes	2 mm	—	32	[k]			38	16
Convit et al., 1995	MRI	Unknown	4 mm	22	—	27			14[e]	—
Parnetti et al., 1996	MRI	No, 1-mm gap	4 mm	6[j]	6	6	3	—	29e/26	—
Maunoury et al., 1996	MRI	Yes	5 mm	—	12	15			—	41
Lehtovirta et al., 1996	MRI	Yes	2 mm	—	58	34	7	−8	30	13
Fox et al., 1996	MRI	Yes		—	7	—			8% loss/year	—
Foundas et al., 1997	MRI	Yes (sagittal)	1.25 mm	—	8	8			52	—
Kaye et al., 1997	MRI	Yes	4 mm	12[j]	—	18	3	—	13 (at entry)	
Jack et al., 1997	MRI	Yes	1.6 mm	—	94	126			(CDR 0.5) decrease of 1.75 SD	

continues

Table I *continued*

Reference	Study modality	Contiguous volumetry[a]	Slice spacing	ARCD	AD	Controls	Hippo	Amyg	Hippo	Amyg
					n		Percentage R > L		Percentage decline	
Schuff *et al.*, 1997	MRI	Yes	1.4 mm	—	12	17	0		21	—
Mauri *et al.*, 1998	MRI	Yes	1.5 mm	—	31	22	7		—	1.5
Pucci *et al.*, 1998	MRI	No, linear	NA	—	39	33			10	—
Krasuski *et al.*, 1998	MRI	Yes	5.0 mm	—	13	21	5	7	19	33
Smith *et al.*, 1999	MRI	Yes		—	20	20			20	33
Jack *et al.*, 1999	MRI	Yes	1.5 mm	80	—	—			26% develop AD in 3 years with a volume in 0–0.6 percentile of controls	
Bobinski *et al.*, 1999	MRI/ Autopsy	Yes	3.0 mm	—	11	4			30	—

[a]Planimetric, an area assessment; linear, a line length. [b]Percentage with qualitative atrophy present: ARCD/AD/control. [c]Interuncal distance (IUD); A/H = amygdala and hippocampus. [d]Medial temporal thickness. [e]Subjects with ARCD compared to controls. [f]Subgroup analysis only. [g]A/H age-related volume loss. [h]The lateral nucleus shows the greatest asymmetry. [i]Volume of parahippocampal fissure (after 4 years, 2 controls and 23 of the ARCD subjects developed AD). [j]reversal of normal R > L asymmetry. Sen, sensitivity; spec, specificity; pAD, possible AD; CDR, Clinical Dementia Rating. [k]Same data used in Soininen (1994).

tomated, continuum-mechanical fluid transformation (Miller *et al.*, 1993, 1997). To determine hippocampal shape as well as volume characteristics, a triangulated graph of points is superimposed onto the surface of the hippocampus in the template and then carried along the warp as the template is transformed onto the target scans. Computing the transformation vector fields from this graphical surface generates a 3D model of each subject's hippocampus. A pooled, within-group covariance matrix computed from the transformation vector fields can be used to compare the shape characteristics of the hippocampus in the two subject groups. This covariance matrix is reduced in its dimensionality by computing a complete orthonormal set of eigenvectors specific to the shape of the hippocampus. The first 15 eigenvectors, in decreasing order of power, are chosen a priori as adequately representing structural shapes, and a linear dis-

criminant function is computed by sequentially using an optimal combination of eigenvectors selected through a stepwise procedure. Logarithms of the likelihood ratios are then calculated as shape metric values for each subject according to these optimal solutions, and the statistical significance of the group difference in shape is tested using Wilk's Lambda. Displacement maps of the hippocampal surface that discriminated 18 very mild AD subjects', Clinical Dementia Rating (Hughes *et al.*, 1982; Berg, 1988) of 0.5, and 18 controls matched for age and gender are shown in Fig. 4 (Csernansky *et al.*, 1998b).

Another 3D mapping technique produces group-average 3D hippocampal surface maps from manual outlines. Understanding the average hippocampal shape for a population, within a common coordinate space, fulfills the brain mapping objective of incorporating normal anatomic variability in the assessment

Table II Mean Hippocampal Volumes (SEM) for Normal Cognitively Intact Elderly and Subjects with Mild Cognitive Impairment (MCI) Registered to the Talairach Coordinate Space via a Seven-Parameter Transformation

	Total hippocampal volume (mm3)		Anterior hippocampal volume (mm3)		Posterior hippocampal volume (mm3)	
	Left	Right	Left	Right	Left	Right
Normal	2087 (114)	2239 (168)	756 (74)	918 (97)	1331 (82)	1320 (111)
MCI	1887 (96)	1814[a] (83)	599 (61)	658[a] (53)	1287 (74)	1156 (66)

[a]*p* = 0.02 compared to right side in normals.

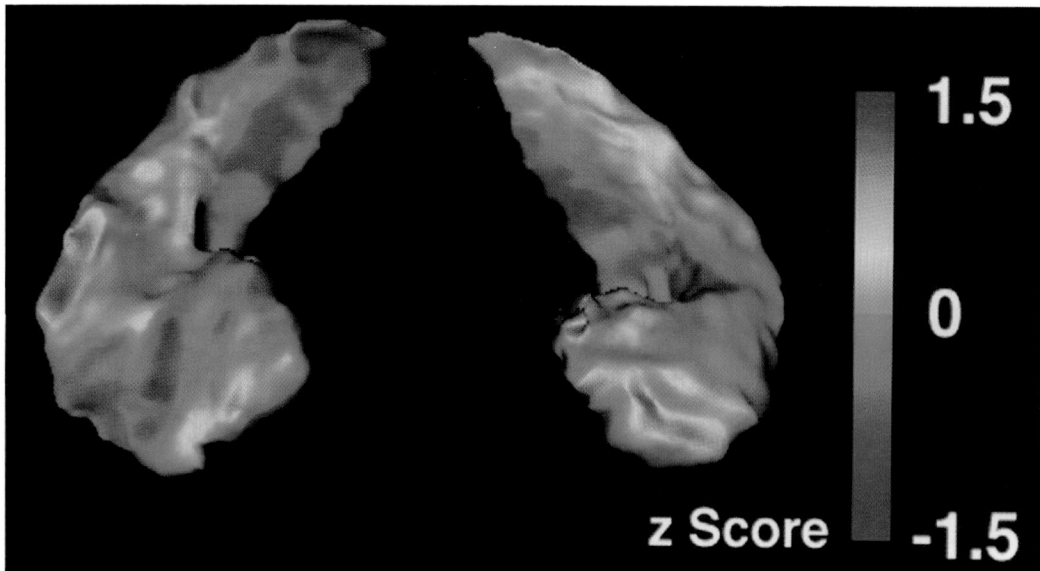

Figure 4 Hippocampal volumes and shapes were derived from the high-dimensional mapping of MRI scans from healthy controls and AD patients in a pilot study (Csernansky *et al.*, 1998b). The volume comparison obtained from high-dimensional brain mapping indicated a statistically significant loss of hippocampal volume in the AD subjects. A map of *z*-score values is shown to examine the shape deformities in very mild AD subjects while accounting for general variability in hippocampal shape. In contrast to the normal pattern of shape variability, shape deformities found in the AD subjects were highly localized. The degree of displacement of hippocampal surfaces (in millimeters) perpendicular to the plane of the structure and relative to the control composite is indicated by a hot body lookup table. Collapse of the hippocampal surface in AD was observed both in hippocampal heads and along the lateral aspect of the hippocampal body on the right side.

method. Creation of these surface maps is based on a parametric mesh construction (Thompson *et al.*, 1996). Hippocampal contours are traced using a track ball on each relevant slice. The points making up a traced contour, after smoothing the effects of irregular hand movements, are connected across slices to create regularly ordered 3D meshes warping the hippocampal surface. To quantify a population's anatomic variability, the shape of the average hippocampus is first resolved. This allows the computation of 3D displacement vectors—the movement needed from each point on a individual's hippocampal surface to that corresponding point on the average hippocampus. Hippocampal variability, expressed as a 3D distance in the Talairach coordinate space, is then computed by taking the square of the mean of the square roots of the vector displacements necessary to align each node of the population's hippocampi onto the average hippocampus. This computation of the root mean square (rms) allows the production of a population variability map shown in Fig. 5 for elderly controls. Understanding the normal anatomic variability in a population is crucial for the application of modern brain mapping techniques in disease states.

C. Cortical Morphometry

Controlling cortical morphology within a population presents one of the greatest challenges to modern brain

mapping (Vannier *et al.*, 1991; Christensen *et al.*, 1997). The importance of mapping gyral morphology within a population is relevant not only to structural imaging in dementia, for the pursuit of a global pattern of atrophic change specific to AD, but also in functional imaging assessments when functional data are averaged across a group of individuals, each having unique gyral morphology. Attempts to control the variability of cortical anatomy have relied mainly on (1) the transformation of structural and functional imaging data into a common coordinate space (typically the Talairach space) and (2) the blurring of functional imaging data with a Gaussian smoothing kernel of varying size (ranging from 8 to 12 mm). Sulcal variability within the Talairach coordinate space for a population of elderly normal individuals can reach 14 mm about mean locations for temporal or parietal sulci. Using a Gaussian smoothing kernel to equalize such extreme cortical variability within a population blurs the functional imaging signal and thereby decreases the sensitivity of localizing subtle functional changes.

As demonstrated in the hippocampal mapping above, the first step in controlling for cortical morphologic variability is to measure it. The starting point for mapping variability in cortical patterns is to pick a common coordinate space in which the expression of variability can be documented. Once again, the Talairach space serves as a convenient initial reference

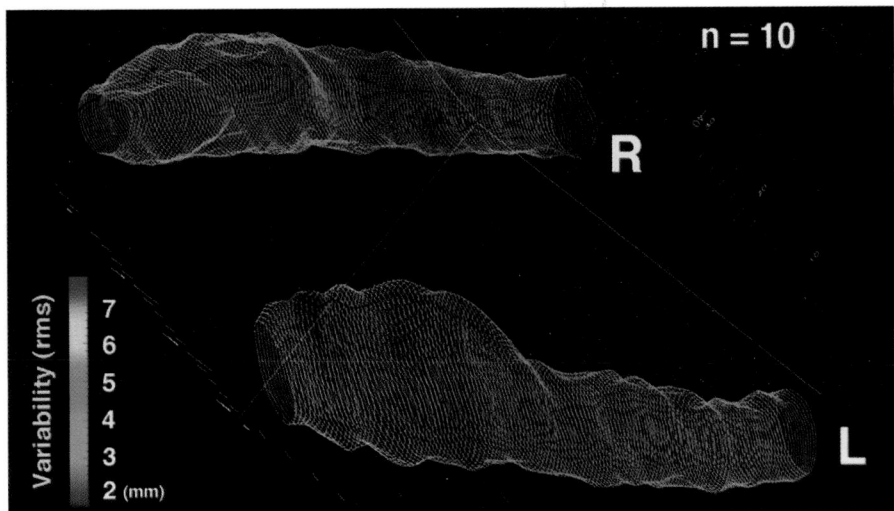

Figure 5 Variability maps can resolve the 3D variability of the hippocampus, in a common coordinate space, for any population as exemplified in this elderly control group.

space for the representation of variability. After high-resolution 3D MRI scans of 19 AD patients (mean age 75.8, SD = 8.7 years) and 20 controls (mean age 72.7, SD = 6.1 years) of similar gender, educational level, and handedness were edited to eliminate nonbrain tissue, a 3D surface model was constructed. Major and minor sulci on both the cortical surface and midline regions are manually outlined and resolved into 3D sulcal meshes. Sulcal averaging (Thompson *et al.*, 1996, 1997) across the individuals within the two populations enabled the mean location of each sulcus to be computed for a population as described above. The mean location of major and minor sulci, and their variability about that mean, were mapped to provide 3D information on cortical morphology within the populations (Thompson *et al.*, 1998). The anatomic variability within the AD population is tightly correlated with neuropsychological test performance for regions traditionally associated with both visual spatial and frontal-executive performance (Mega *et al.*, 1998). Two factors contribute to variability within a population: normal individual gyrification and, in a disease population such as AD, the atrophic process.

After the variability in a normal elderly population is mapped, sulcal displacement from normal locations can be determined for an AD population. Displacement values in the *x, y,* and *z* axes of AD patients from the average control location suggest a specific pattern of 3D atrophy within AD (Zoumalan *et al.*, 1999). Sulcal displacement across all individuals, referenced to the normal location, shows a significant correlation between the right midline frontal region and the Ruff figural fluency test ($r = 0.48, p < 0.01$). The left temporal region and naming ability showed a significant correlation (the left superior temporal sulcus with verbal

fluency $r = 0.41, p < 0.05$). The temporal-occipital region and visuospatial ability also were linked as revealed by displacement of the right inferior temporal sulcus that significantly correlated to the Benton visual retention test ($r = 0.40$ and -0.45 for correct and errors, respectively; $p < 0.05$). The left collateral sulcus and verbal memory were linked as reflected in the Buschke-Fuld recall, storage, and retrieval scores ($r = -0.53, -0.63,$ and $-0.59; < 0.01$). Thus, sulcal displacements in the Talairach space are related to cognitive performance and implicate a specific cortical pattern associated with AD. Once a specific cortical pattern is established for the AD population, brain mapping techniques will discern, from high-resolution 3D MRIs, whether individuals with cognitive complaints have incipient AD.

Another technique for determining brain changes in disease states from normal individuals (Fox *et al.*, 1996a) or within a disease population across time (Fox *et al.*, 1999) is the brain boundary shift integral (BBSI) (Fox and Freeborough, 1997; Freeborough and Fox, 1998). The basis for BBSI determining morphological changes is to average the structural imaging studies of a group and then create a subtraction of that average compared to a control or baseline average. This subtraction produces areas of brain that are dissimilar between the two groups, thus revealing the location of volumetric loss throughout the 3D volume. Rates of total brain volume loss in AD compared to controls are significant (5-20 vs <2 ml/year) (Fox *et al.*, 1996b; Rossor *et al.*, 1997) and are related to cognitive decline in AD (Fox *et al.*, 1999). The averaging of individuals in the two groups is conducted using linear registration. Once the average volume for the two groups is determined, simple subtraction can provide a difference image that

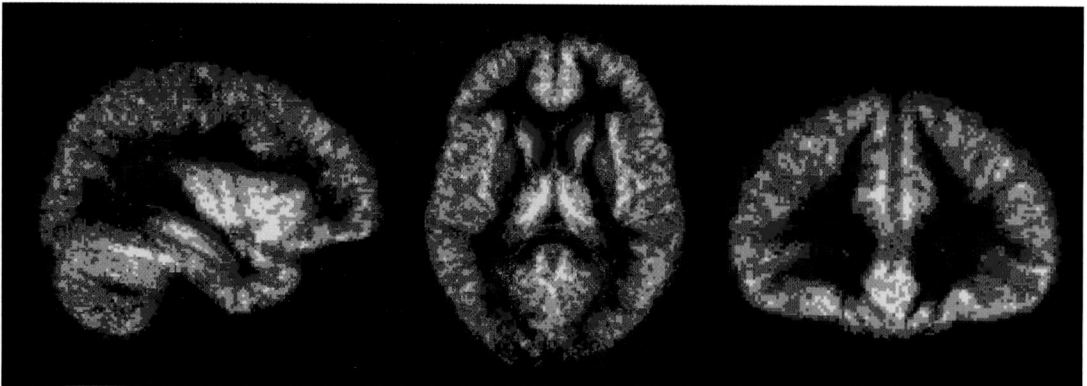

Figure 6 Gray matter difference map between younger normals and older normals. The red pixels represent the difference displayed over the young group's average gray matter. These differences were visualized by linear averaging of the gray matter in each group and subtracting the older normals from the younger. Such a linear approach is not able to correct the normal anatomic variability within each population.

is thought to reflect atrophy. This technique is demonstrated in GM maps of young and old normals (Fig. 6). Linear registration techniques do not control for anatomic variability and thus subtractions of averages based on linear registration will be confounded by registration error and normal morphologic variability that occurs from differences in group anatomy rather than entirely from disease-specific changes. Nonlinear intensity-based registration algorithms are superior in controlling the anatomic variability within a population.

Normal anatomic variability can be controlled with warping algorithms of increasing degrees of freedom from linear to nonlinear. Perfect correspondence between an individual and template, or a group of subjects, can be achieved by using surface-based high-dimensional continuum-mechanical warping. Warping of structural data to their average anatomy according to the constraints of fluid or elastic mechanics has been accomplished for an AD population as demonstrated in Fig. 7. The warping field that governs the transformation of any given subject to the template contains information that is specific to both the individual's anatomy and the atrophic changes imparted by disease. Evaluating the warping field and resolving it into tensors that define the 3D vectors displacing any given point in a reference volume to the corresponding target volume will enable the production of tensor maps. Tensor mapping can then be used to reflect the 3D displacements that result from both the atrophic process and normal anatomic variability. Using principal component analysis to determine which eigenvectors associated with a tensor field are more responsible for normal anatomic variability and which may be most responsible for disease-specific, atrophic morphologic change is the goal of modern research in morphologic brain mapping within dementia and aging.

II. Functional Imaging

The earliest functional imaging studies in aged and demented individuals (Freyham *et al.*, 1951; Kety, 1956; Lassen *et al.*, 1957) were conducted using the nitrous oxide method of Kety and Schmidt (1948). These studies showed significant declines of cerebral perfusion in demented subjects compared to controls but were not capable of regional subhemispheric analysis. Nonetheless, a specific pattern of functional defects in dementia was tentatively suggested in an early review: "From these studies emerges the somewhat surprising suggestion that perhaps even dementia is to some extent a regional cerebral disease, namely, a disease of the dominant hemisphere." [p. 66] (Lassen and Ingvar, 1963). With the advent of [^{18}F]fluorodeoxyglucose positron emission tomography (FDG-PET) to measure *in vivo* metabolism, regional cerebral metabolic defects were established in dementia (Benson, 1982, 1983; Foster *et al.*, 1983), stroke patients (Benson *et al.*, 1983; Metter *et al.*, 1983; Benson, 1984), and subcortical degenerative disorders (Mazziotta *et al.*, 1987; Grafton *et al.*, 1992). In the mid-1980s, the patterns of functional defects involving structurally intact brain tissue emerged across dementing disorders. Functional deactivation of brain regions disconnected from subcortical activating projections was found in patients with thalamic or striatal lesions; this remote deactivation is termed *diaschisis* (Meyer *et al.*, 1993). Thus, stroke patients who present with isolated subcortical lesions but demonstrate classical cortical language abnormalities were found to have *functional* defects in structurally intact cortical language regions. Similarly, in degenerative extrapyramidal disorders such as progressive supranuclear palsy, or Parkinson's disease, individuals who were clinically demented showed functional imaging abnormalities in frontal cor-

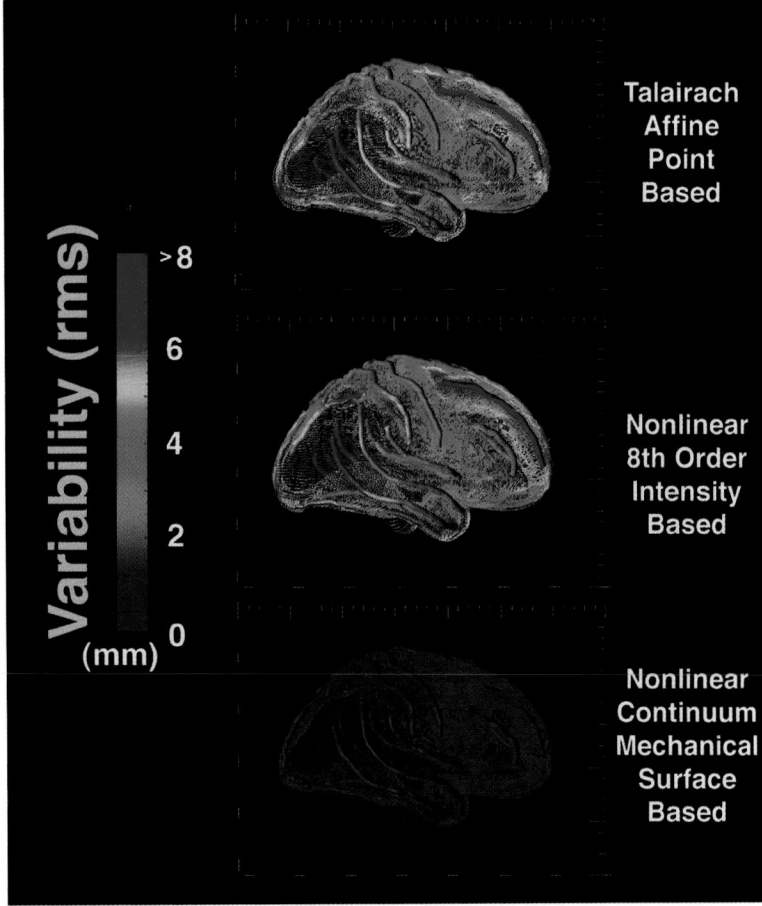

Figure 7 Cortical variability in the Alzhiemer's disease population (n = 19) is demonstrated across three examples of registration techniques. Linear registration to the Talairach space yields cortical variability greater than three times the FWHM of current high-resolution PET scanners. Nonlinear automated intensity-based warping aligns the least variable regions (primary motor and sensory cortex) but heteromodal association cortex escapes the control of automated techniques. Manually assisted continuum-mechanical surface-based warping brings all regions of the cortex into near-perfect registration within a population.

tical regions that were relatively spared from pathological changes. Consequently, any functional imaging assessment in dementia must consider the distributed neuroanatomy of a deactivated region and search for a disconnection of cortical or subcortical projections. In AD deafferentation may result from NFT burden in the entorhinal region, thereby disconnecting projections to frontal, parietal, and temporal association cortices. Such deafferentation of ascending cholinergic input due to cell loss in the nucleus Basalis of Meynert has been suggested. However, studies with lesions in baboons of the nucleus Basalis of Meynert have failed to replicate (Le Mestric *et al.,* 1998) the earlier finding of hypometabolism in frontotemporal cortex (Kiyosawa *et al.,* 1989).

A. Methodological Considerations

1. Partial Volume Correction

In addition to deafferentation, partial volume error also contributes to the loss of signal in functional imaging studies in aging and dementia. Given the resolution of most current PET (3.5-mm full width at half-maximum) or SPECT (6-mm full width at half-maximum) scanners, partial volume error will occur when the tissues of GM, WM, or CSF are averaged in the relatively large functional imaging voxel (Mazziotta *et al.,* 1981). In the elderly and demented brain, this averaging of the three tissue compartments is particularly severe given the degree of atrophy occurring in normal aging and dementia. Atrophy is the strongest correlate to focal hypometabolism (Jamieson *et al.*, 1987; Fazekas *et al.*, 1989). Thus, the interpretation of functional imaging studies in aging and dementia must take into account the partial volume error in assessing the functional viability of brain regions. Conversely, partial volume error may assist the researcher in identifying individuals with AD compared to normal controls in that the effects of atrophy will result in greater hypoperfusion, or hypometabolism, in AD compared to a control cohort. However, if the research question centers on the actual functional nature, or ligand concentration, within brain regions, then partial volume correction (PVC) must be performed.

The increasing power of brain imaging and the development of novel techniques for brain modeling allow incorporating high-resolution structural data into PET analysis. A variety of methods for PVC of PET data, using the kinetic properties of specific tissue components, have been proposed. Metabolic rates of GM are 3-4 times that of WM, while CSF is presumed to have no metabolic activity. Binary segmented MRI scans (total brain tissue separated from CSF) convolved, or reduced in resolution, to the in-plane resolution of the PET images have been done (Meltzer *et al.*, 1990). A limitation of binary correction is that it does not address partial volume averaging of GM with WM. This is crucial in AD, which has a significant reduction in GM volume compared to normal controls (Høedt-Rasmussen and Skinhøj, 1966; Creasey *et al.*, 1986; Prohovnik *et al.*, 1989; Rusinek *et al.*, 1991). Atrophy-corrected PET, using a binary segmentation method, in AD patients and elderly controls eliminated apparent significant reductions of metabolism in many cortical regions within the AD group's uncorrected PET data (Meltzer and Frost, 1994; Meltzer *et al.*, 1996). A trinary correction method has been applied to PET (Müller-Gärtner *et al.*, 1992; Ibanez *et al.*, 1998). Trinary correction (separating GM, WM, and CSF) will eliminate partial volume error and allow more accurate measurement of the metabolic activity from the remaining synapses present in AD patients.

We employed a pixel-based PVC approach to a group of normal elderly control and AD individuals to test the hypothesis that temporal lobe hypometabolism in AD, relative to controls, will be reduced in PVC-PET compared to uncorrected statistical difference maps. The accuracy of PVC depends upon accuracy of the segmentation of the MRI scan, accuracy of the registration of the segmented MRI scans with the PET data to be corrected, and the inhomogeneity of the PET point spread function. Previous PVC methods either have registered a binary segmented MRI image to the PET data set, after convolution to the resolution of the PET, for correction of the GM/WM, and CSF signal or have

used automated MRI image segmentation based on the probability maps within the International Consortium of Human Brain Mapping (ICBM) probabilistic atlas (Evans *et al.*, 1996) to reduce any thresholding bias. We evaluated PVC and uncorrected FDG-PET studies from five AD patients and five healthy controls within the ICBM 305 probabilistic atlas. The two groups were matched for age, gender, and level of education. A semi-automated minimum-distance classification algorithm (Kollokian, 1996) was used to segment the individual native MRI data into WM, GM, CSF, and background voxels after inhomogeneity correction was accomplished (Sled *et al.*, 1998). Native MRIs and tissue maps were registered via point-based affine transforms with the FDG-PET data; these transforms were then concatenated with intensity-based affine registrations (Woods *et al.*, 1993) of the MRIs into the ICBM 305 space. A ratio of 4/1/0 (GM/WM/CSF) was used to recompute native PET voxel values, based upon the segmented high-resolution MRI data; both corrected and uncorrected PET data sets were then normalized to each group's global mean. In this rest-state subtraction paradigm, a subvolume thresholding (SVT) was applied (Dinov *et al.*, 2000) to the uncorrected and corrected PET data to assess change in differences among normals and the AD group. The SVT approach uses the anatomic probabilistic maps associated with the ICBM 305 atlas (Collins and Evans, 1997) to evaluate the ROI between-group variance and estimate both the regional and voxel statistical group differences. PVC was performed prior to SVT.

Figure 8 shows the total brain voxel difference between the partial volume corrected and uncorrected tissue maps. Figure 9 illustrates individual slices in the statistical maps, comparing the PVC technique with uncorrected data. Substantial reduction in the volume of significant voxels was seen in the perisylvian and medial temporal regions with PVC. Regardless of the correction technique, all studies show a similar underestimation of functional signal within AD brains due to partial volume error. Our preliminary findings demonstrate the effect of atrophy correction within a probabilistic brain atlas and suggest that atrophy accounts for a large percentage of the measured hypometabolism in AD FDG-PET data. PVC should not be done when high sensitivity is desired in discriminating between demented and normal elderly.

2. ROI vs. Group Analysis

Evaluating functional imaging data beyond qualitative assessment can be done either with ROIs drawn on functional imaging slices, with ROIs drawn on the same subject's MRI registered to that subject's functional imaging study, or from averaging groups of functional

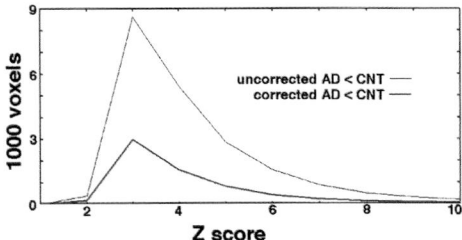

Figure 8 Volume difference in total brain z-scores from statistical maps of hypometabolism on FDG-PET between AD patients and elderly controls (CNT). Partial volume corrected FDG-PET identifies fewer significant voxels compared to uncorrected PET maps.

imaging data sets into a common space for statistical analysis on a voxel-by-voxel basis. There are difficulties with each technique. The ROI drawn on a functional imaging slice is prone to error because of the poor anatomic delineation of the low-resolution functional data set. ROIs drawn on MRI and then coregistered to functional data are time-consuming and limited in the amount of information that can be gained. Any ROI analysis will collapse the pattern of functional signal within an ROI into a single number that is then analyzed for statistical significance. This single value may not adequately reflect the subtle disease-specific variability within the subregions of an ROI. Group analysis of functional mapping on a voxel-by-voxel basis avoids the drawbacks of ROI analysis by permitting all data within a functional scan to participate in a statistical analysis. The advantage of voxel-by-voxel assessments is constrained by the difficulty in controlling for anatomic variability of individuals within a population

and avoiding Type I errors occurring from multiple voxelwise testing. There are solutions to solving these drawbacks.

To control for anatomic variability many statistical mapping approaches apply a Gaussian smoothing kernel to the functional data as described previously; however, this decreases the resolution of the functional data set. Another approach employs a probabilistic atlas to control for anatomic variability within a population by weighting functional imaging voxels with more or less emphasis, depending upon where they fall within an anatomical probability cloud. This type of analysis has been performed on functional imaging data shown in the example using SVT above. The voxel-by-voxel analysis of functional imaging data can be performed without the use of a Gaussian smoothing kernel, thus better localizing functional differences.

Another approach in controlling for anatomic variability utilizes surface-based, high-dimensional, continuum-mechanical registration of a selected ROI. Constrained warping of data, depending upon the hypothesis to be tested, is less processor dependent when high-dimensional warps are used. We apply this technique to the hippocampus, which has a great deal of variability in the Talairach space (Fig. 10), in AD, MCI, and controls. The variability of PET signal decreases when the hippocampal surfaces pull the PET data into perfect correspondence (Fig. 11). Here the variability of the PET signal in the subtraction of the MCI from the control group's FDG-PET is significantly reduced compared to the same cases subtracted with affine registration alone.

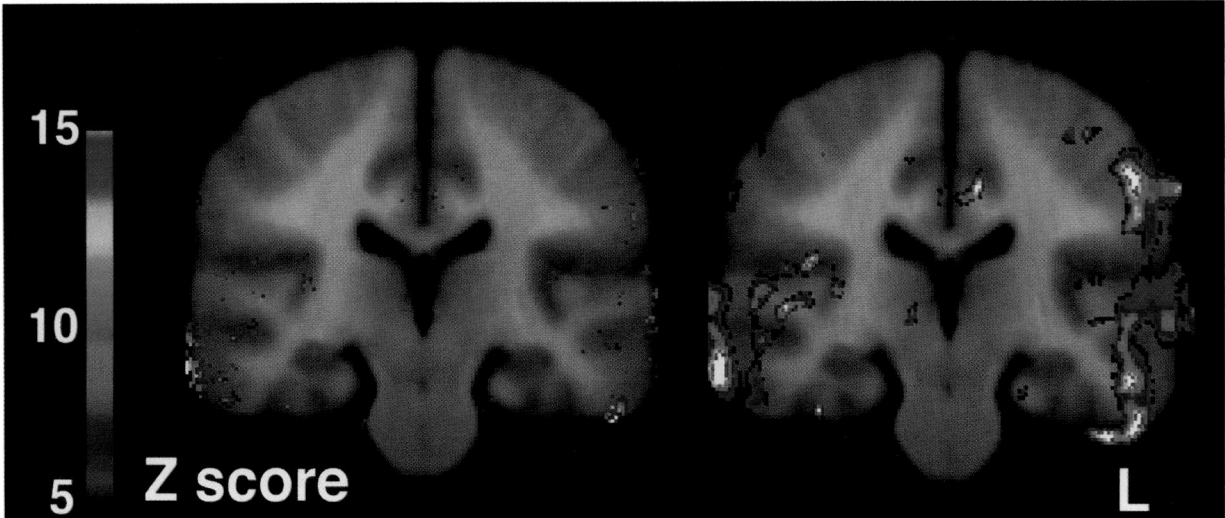

Figure 9 Statistical maps of hypometabolism as measured by FDG-PET in AD patients compared to elderly controls. Atrophy contributes to much of the hypometabolism seen in the AD FDG-PET data as revealed by the uncorrected data (right) compared to the partial volume corrected data (left).

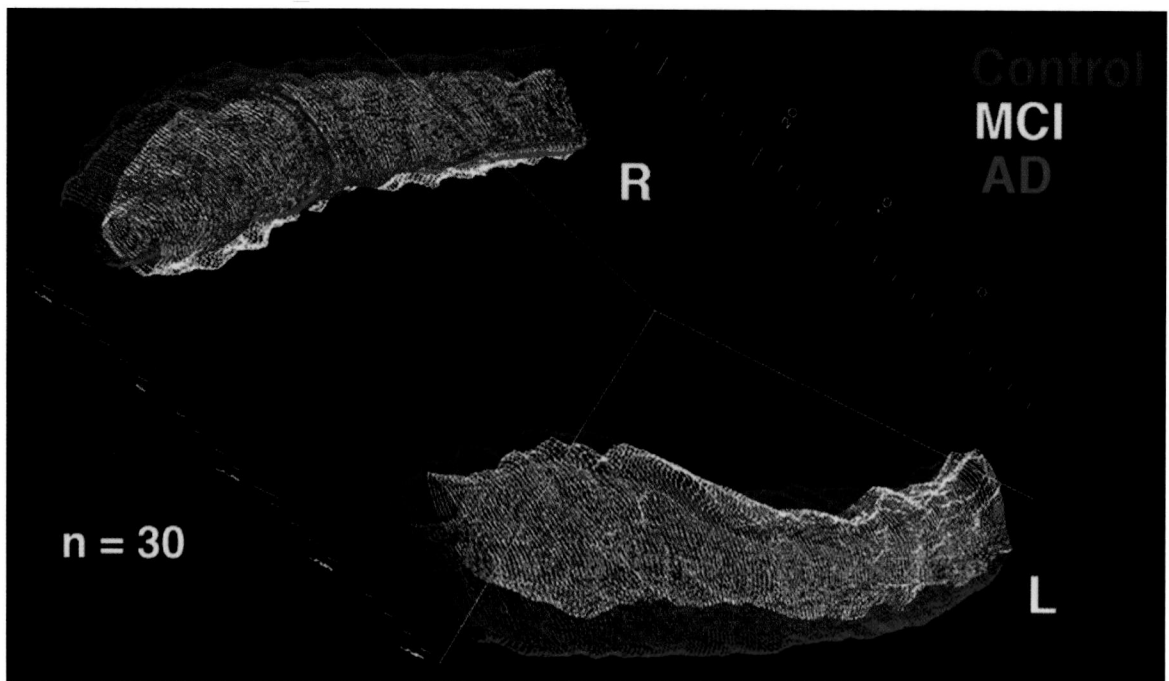

Figure 10 The average 3D surface models of 10 control, 10 MCI, and 10 AD patients' hippocampi derived from manual outlines. Note the mismatch of these averages in the Talairach coordinate space.

B. Resting Studies

1. SPECT and PET

SPECT evaluation in AD compared to elderly controls consistently shows decreased perfusion in temporal and parietal association cortices. The sensitivity/specificity for identifying clinically diagnosed AD patients within a general dementia population by qualitative analysis using HMPAO-SPECT has a broad range, 76-88%/43-87% (Holman, 1991; Masterman *et al.*, 1997), and is helped by qualitative medial temporal atrophy assessment, 77%/93% (Lavenu *et al.*, 1997). Longitudinal evaluation of patients who present to memory disorder clinics but are not frankly demented has shown that the posterior cingulate, left thalamic, and left medial temporal regions are significantly hypoperfused in a group that later goes on to develop AD within 2 years of examination (Johnson *et al.*, 1998), but replication of these results by visual analysis of individual patients may not be possible (McKelvey *et al.*, 1999). Evaluation of a single, questionably demented patient's functional imaging data, with reference to known variability of a normal elderly group, is just beginning (Signorini *et al.*, 1999).

FDG-PET in AD is a window on the metabolic activity of neurons not destroyed by the disease. The earliest abnormalities on PET are found in the posterior parieto-occipito-temporal areas before clinical symptoms emerge; as the disease progresses, the hypometabolism moves anteriorly to affect the temporal and finally frontal cortices (Benson, 1982, 1983; Benson *et al.*, 1983; de Leon *et al.*, 1983; Ferris *et al.*, 1983; Foster *et al.*, 1983; Friedland *et al.*, 1983; Duara *et al.*, 1986; Grady *et al.*, 1986, 1987; Salmon and Franck, 1989; Haxby *et al.*, 1990). These changes may be loosely correlated with data from cross-sectional, postmortem studies showing the progressive regional deposition of amyloid deposits and NFTs (Braak and Braak, 1991). Presumed preclinical AD patients evaluated with FDG-PET have also shown biparietal metabolic abnormalities before clinical criteria for probable AD are met. The use of genetic markers to identify a population at risk for the disease, in combination with functional imaging studies demonstrating the existence of brain pathology, may provide very early detection of AD (Small *et al.*, 1995; Reiman *et al.*, 1996). Functional imaging abnormalities may predate medial temporal volume loss in genetically at-risk persons (Reiman *et al.*, 1998) and preclinical carriers of the Swedish Alzheimer amyloid protein mutation (Julin *et al.*, 1998). Alzheimer's disease patients with the ApoE-4 allele (Lehtovirta *et al.*, 1998) or the α-antichymotrypsin type A allele (Higuchi *et al.*, 1997) have greater regional defects on functional imaging than those without these genetic burdens.

In addition to partial volume error and deafferentation, described above, other possible causes for hypometabolism on FDG-PET include synapse loss and dysfunctional energy metabolism or glucose transport.

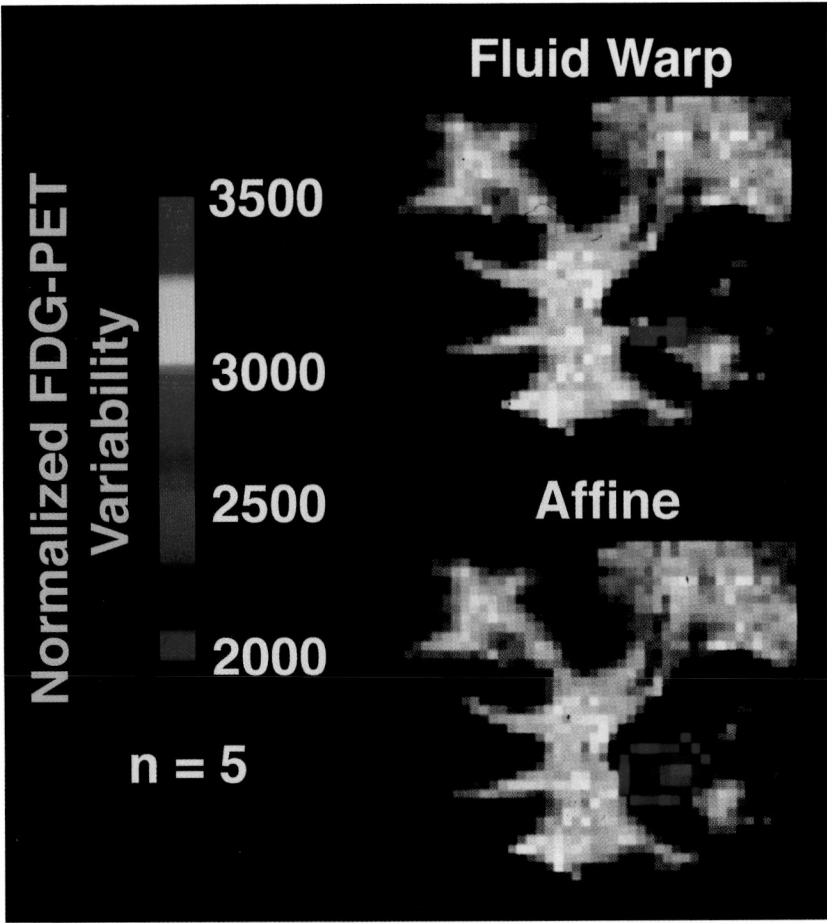

Figure 11 The normalized variability of the FDG-PET signal in a subtraction of five AD patients from five normal controls in the medial temporal lobe with affine (Woods *et al.*, 1993) registration to the Talairach space (bottom). By drawing anchoring contours around the hippocampi on subjects' MRI scans, surface-based fluid warping (top) can bring all hippocampi into perfect alignment. Then statistical mapping of the functional imaging data can be conducted with the anatomic variability that exists in the Talairach space eliminated.

The activity observed on FDG-PET, assuming that blood flow and glucose utilization remain coupled, reflects the metabolism of active synapses as they restore their resting ionic gradients via Na/K-ATPase (Hand *et al.*, 1979; Mata *et al.*, 1980; Kadekaro *et al.*, 1985; Ginsberg *et al.*, 1987). Metabolism of glucose and the production of ATP from the electron transport chain in mitochondria of AD may be normal or dysfunctional. Synapse loss is the best cellular correlate with the degree of cognitive impairment in AD patients (Terry *et al.*, 1991), with the greatest regional loss seen in frontal areas across pathological studies (Davies *et al.*, 1987; Hamos *et al.*, 1989; Samuel *et al.*, 1993). Given their high metabolism, synapse dropout will result in a lower glucose metabolism on FDG-PET studies. Loss of synapses will also result in atrophy. In a pathological study conducted 16 months after an AD patient was evaluated with FDG-PET, regional hypometabolism correlated

with cell loss, gliosis, and amyloid plaques (APs) (McGeer *et al.*, 1986).

The remaining synapses in AD patients may have primary defects in energy metabolism. Five glucose transporters (Glut1-5) have been identified (Bell *et al.*, 1990), and two are present in the brain: Glut1 and Glut3. Glut1 is present in the endothelial cells of the blood-brain barrier and is the glucose transporter in glial cells; Glut3 is found in neurons. Using quantitative immunohistochemistry, Harr *et al.* (1995) found a 49.5% decrease in Glut3 immunoreactivity in the outer portion of the molecular layer of the dentate gyrus in AD brains. This area is the termination zone of the perforant pathway whose cells of origin are the layer II pyramidal neurons of the entorhinal cortex; cells that suffer the earliest burden of NFT in AD. Since glucose uptake influences metabolism, deafferented cells might downregulate Glut3 in AD, contributing to the hypometabolism seen on FDG-PET. An estimation of the kinetic parameters describing the forward and reverse FDG transport (K_1^* and k_2^*, respectively) and the phosphorylation of FDG by the enzyme hexokinase (k_3^*) has been done in AD compared to controls (Friedland *et al.*, 1989; Jagust *et al.*, 1991). A significant decrease of ~20% in K_1^* in the frontal and temporal cortices in AD compared to controls was found (Jagust *et al.*, 1991). Transport of FDG across endothelial cells (Glut1 transporter) through the interstitial space and into neurons (Glut3 transporter) is reflected by K_1^*. Abnormalities in Glut3 could be responsible for the reduction in K_1^*. However, a 20% reduction in glucose delivery to cells will not significantly reduce energy metabolism (Lund-Anderson, 1979)—some other metabolic defect must be present in AD.

Amyloid is deposited in the posterior parietal lobe early in AD (Arnold *et al.*, 1991), an area first affected on FDG-PET in at-risk patients (Small *et al.*, 1995; Reiman *et al.*, 1996). Hippocampal structures are also affected but are not well visualized by PET. The parietal abnormality supports amyloid as contributing to the hypometabolism on FDG-PET in AD. Although plaques do not correlate with the degree of cognitive loss, and presumably neuronal dysfunction, seen in AD (Terry *et al.*, 1991), altered synaptic function is associated with the accumulation of amyloid β-protein (Aβ) (Terry *et al.*, 1994). Accumulation of reactive oxygen species (ROS) may be the mechanism of Aβ neurotoxicity. Aβ induces lipid peroxidation in synaptosomes (Butterfield *et al.*, 1994) and cultured cortical cells (Behl *et al.*, 1994). Injury could result from free radical production in Aβ itself (Hensley *et al.*, 1994) or secondarily from calcium influx (Mark *et al.*, 1995). If Aβ contributes to the hypometabolism on FDG-PET in AD by initiating the production of ROS, oxidative damage is a likely destructive mechanism.

Modern brain mapping techniques allow the mapping of Aβ and other markers such as NFTs with the metabolic changes in AD to provide a unique inquiry into pathologic relationships with functional imaging abnormalities in AD. Figure 12 provides an overview of this methodology in AD patients who undergo functional imaging with FDG-PET close to death, permitting functional-histological-biochemical integration within a common AD atlas (Mega *et al.*, 1997; Mega *et al.*, 1999). Future work is necessary to understand the pathological basis of the functional defects in the AD population.

C. Activation Studies

1. PET and fMRI

Activation protocols in aging and dementia are exploratory and methodologically difficult with cogni-

tively impaired individuals who are easily distractible. Coupled with the methodological problems of activation protocols with demented persons is the challenge of controlling for spatial variability in an elderly and AD population when data are averaged in a common space. The first attempted activation study comparing AD to elderly controls used FDG-PET and a verbal memory task but failed to show any functional increase (Miller *et al.*, 1987). Similar global metabolic increases were found with picture preference and reading memory tasks for both AD and controls (Duara *et al.*, 1990a,b, 1992). Using a continuous visual recognition task with an ROI analysis of FDG-PET, Kessler *et al.* (1991) found a mean, task-related, global metabolic increase for elderly controls that was 15% greater than the increase in the AD group. With the appropriate cognitive challenge, AD patients appeared to activate greater volumes of brain tissue than controls (Deutsh and Halsey, 1990; Grady *et al.*, 1993). This compensatory recruitment of an increased volume of brain activated by cognitive tasks in AD, compared to normal elderly controls, has been found with varying paradigms (Becker *et al.*, 1996; Woodard *et al.*, 1998; Bäckman *et al.*, 1999). Preliminary results suggest that an activation protocol may be a better predictor of clinical disease severity, and perhaps diagnosis, than resting studies (Pietrini *et al.*, 1999). Combining high-resolution MRI with functional MRI in a hippocampal ROI analysis of elderly controls, elderly with isolated memory dysfunction, and AD patients using a face-encoding task, Small *et al.* (1999) observed a progressive reduction in hippocampal activation across the three groups, with the AD patients showing the least activation. Future studies should determine which task activates the most eloquent region in predicting incipient AD and what the variability of a given response is for normal and demented populations. With data on the distribution of possible responses in a given anatomical region, single subjects can then be compared to this distribution to determine their risk of ensuing dementia so that therapies may be initiated in the preclinical disease stage. Population data sets such as this will require control over the anatomic variability across the aged and demented brain that modern brain mapping techniques now offer.

III. Summary

The application of brain mapping techniques in dementia is challenged by the disease-related anatomic changes superimposed over the normal morphological variability of the human brain. The structural variability, both normal and disease related, impairs the ability of

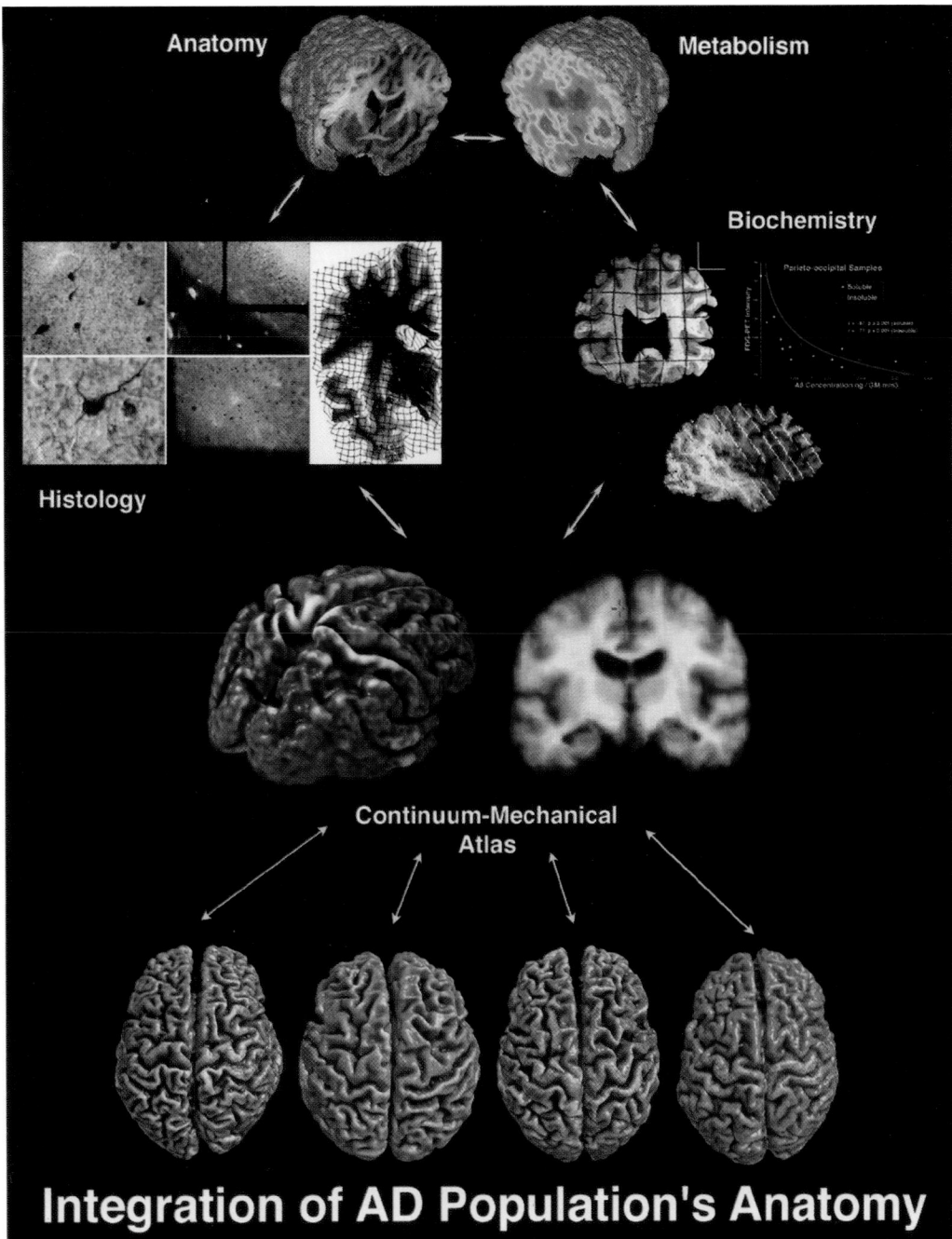

Figure 12 Integration of functional imaging, morphology, histology, and biochemistry into a population-based AD Atlas. Nonlinear surfaced-based warping allows the registration of histological and biochemical samples into each patient's high-resolution cryosliced brain volume (Mega *et al.*, 1997; Mega *et al.*, 1999). MRIs acquired during life are the target for registration of the postmortem cryoimage data and are also used to create a continuum-mechanical brain atlas (see Chapter 6) that serves as the space to integrate the multimodal data.

functional imaging studies that combine individual's data to discern the subtle changes of incipient AD form normal aging. Using nonlinear surface and intensity based warping techniques, combined with novel atlasing strategies, control over this variability is possible. The future automation and dissemination of current brain mapping techniques promise preclinical detection of AD-specific structural and functional abnormalities on an individual basis. With these powerful population-derived tools, early treatment with disease-modifying agents will significantly decrease the prevalence of AD in the next century.

Acknowledgments

Support for this work was provided by an NIA career development award (K08AG100784) to M.S.M. and by an NIA Alzheimer's Disease Research Center grant (P50 AG16570), an Alzheimer's Disease Research Center of California grant, the Sidell-Kagan Foundation, and the Human Brain Project: NIMH/NIDA (P20MH/DA 52176), NSF (BIR9322434), and NCRR (RR05956, RR13642, and NS38753).

References

Anton, G. (1903). Gehirnvermessung mittelst des Kompensations-Polar-Planimeters. *Wien Klin. Wochenschr.* **16**, 1263–1267.

Arnold, S. E., Hyman, B. T., Flory, J., Damasio, A. R., and Van Hoesen, G. W. (1991). The topographical and neuroanatomical distribution of neurofibrillary tangles and neuritic plaques in the cerebral cortex of patients with Alzheimer's disease. *Cerebr. Cortex* **1**, 103–116.

Bäckman, L., Andersson, J. L. R., Nyberg, L., Winblad, B., Nordberg, A., and Almkvist, O. (1999). Brain regions associated with episodic retrieval in normal aging and Alzheimer's disease. *Neurology* **52**, 1861–1870.

Becker, J. T., Mintun, M. A., Aleva, K., Wiseman, M. B., Nichols, T., and DeKosky, S. T. (1996). Compensatory reallocation of brain resources supporting verbal episodic memory in Alzheimer's disease. *Neurology* **46**, 692–700.

Behl, C., Davis, J. B., Lesley, R., and Schubert, D. (1994). Hydrogen peroxide mediates amyloid ß protein toxicity. *Cell* **77**, 817–827.

Bell, G. I., Kayano, T., Buse, J. B., *et al.* (1990). Molecular biology of mammalian glucose transporters. *Diabetes Care* **13**, 198–208.

Benson, D. F. (1982). Cerebral metabolism in aging and dementia. *Neurol. Neurosurg. Update Ser.* **3**(31), 1–8.

Benson, D. F. (1983). Alterations in glucose metabolism in Alzheimer's disease. "Biological Aspects of Alzheimer's Disease" (Banbury Report 15), pp. 309–315. Cold Spring Harbor Laboratory Press, Cold Spring Harbor, New York.

Benson, D. F. (1984). Positron emission tomography in aphasia. *Sem. Neurol.* **4**, 169–173.

Benson, D. F., Kuhl, D. E., Hawkins, R. A., Phelps, M. E., Cummings, J. L., and Tsai, S. Y. (1983). The fluorodeoxyglucose ^{18}F scan in Alzheimer's disease and multi-infarct dementia. *Arch. Neurol.* **40**, 711–714.

Berg, L. (1988). Clinical dementia rating. *Psychopharmacol. Bull.* **24**, 637–639.

Blackford, R. C., and La Rue, A. (1989). Criteria for diagnosing age-associated memory impairment: Proposed improvements from the field. *Dev. Neuropsychol.* **5**, 295–306.

Bobinski, M., de Leon, M. J., Wegiel, J., Desanti, S., Convit, A., Saint Louis, L. A., Rusinek, H., and Wisniewski, H. M. (1999). The histological validation of post mortem magnetic resonance imaging-determined hippocampal volume in Alzheimer's disease. *Neuroscience* **95**, 721–725.

Braak, H., and Braak, E. (1991). Neuropathological staging of Alzheimer-related changes. *Acta Neuropathol.* **82**, 239–259.

Brun, A., and Englund, E. (1981). Regional pattern of degeneration in Alzheimer disease: Neuronal loss and histopathological grading. *Histopathology* **5**, 549–564.

Butterfield, D. A., Hensley, K., Harris, M., Mattson, M. P., and Carney, J. (1994). ß-Amyloid peptide free radical fragments initiate synaptosomal lipoperoxidation in a sequence-specific fashion: Implications to Alzheimer's disease. *Biochem. Biophys. Res. Commun.* **200**, 710–715.

Cahn, D. A., Sullivan, E. V., Shear, P. K., Marsh, L., Fama, R., Lim,

K. O., Yesavage, J. A., Tinklenberg, J. R., and Pfefferbaum, A. (1998). Structural MRI correlates of recognition memory in Alzheimer's disease. *J. Int. Neuropsychol.* **4**, 106–114.

Christensen, G. E., Joshi, S. C., and Miller, M. I. (1997). Volumetric transformation of brain anatomy. *IEEE Trans. Med. Imaging* **16**, 864–877.

Coffey, C. E., Wilkinson, W. E., Parashos, I. A., Soady, S. A. R., Sullivan, R. J., Patterson, L. J., Figiel, G. S., Webb, M. C., Spritzer, C. E., and Djang, W. T. (1992). Quantitative cerebral anatomy of the aging human brain: A cross-sectional study using magnetic resonance imaging. *Neurology* **42**, 527–536.

Collins, D. L., and Evans, A. C. (1997). Animal: Validation and applications of non-linear registration-based segmentation. *Int. J. Patt. Recog. Artif. Intell.* **11**, 1271–1294.

Convit, A., de Leon, M. J., Golomb, J., George, A. E., Tarshish, C. Y., Bobinski, M., Tsui, W., De Santi, S., Wegiel, J., and Wisniewski, H. (1993). Hippocampal atrophy in early Alzheimer's disease: Anatomic specificity and validation. *Psychiatr. Q.* **64**, 371–387.

Convit, A., de Leon, M. J., Tarshish, C. Y., De Santi, S., Kluger, A., Rusinek, H., and George, A. E. (1995). Hippocampal volume losses in minimally impaired elderly. *Lancet* **345**, 266.

Cook, M. J., Fish, D. R., Shorvon, S. D., Straughan, K., and Stevens, J. M. (1992). Hippocampal volumetric and morphometric studies in frontal and temporal lobe epilepsy. *Brain* **115**, 1001–1015.

Creasey, H., Schwartz, M., Frederickson, H., Haxby, J. V., and Rapoport, S. I. (1986). Quantitative computed tomography in dementia of the Alzheimer type. *Neurology* **36**, 1563–1568.

Csernansky, J. G., Joshi, S., Wang, L., Haller, J. W., Gado, M., Miller, J. P., Grenander, U., and Miller, M. I. (1998a). Hippocampal morphology in schizophrenia by high dimensional brain mapping. *Proc. Natl. Acad. Sci. U.S.A.* **95**, 11406–11411.

Csernansky, J. G., Wang, L., Kido, D., Gado, M., Morris, J. C., and Miller, M. I. (1998b). Hippocampal morphometry in dementia of the Alzheimer type by computer algorithm. *Soc. Neurosci. Abstr.* **24**, 264.

Cuénod, C. A., Denys, A., Michot, J. L., Jehenson, P., Forette, F., Kaplan, D., Syrota, A., and Boller, F. (1993). Amygdala atrophy in Alzheimer's disease an *in vivo* magnetic resonance imaging study. *Arch. Neurol.* **50**, 941–945

Dahlbeck, S. W., McCluney, K. W., Yeakley, J. W., Fenstermacher, M. J., Bonmati, C., van Horn, G., and Aldag, J. (1991). The interuncal distance: a new MR measurement for the hippocampal atrophy of Alzheimer disease. *AJNR* **12**, 931–932.

Davies, C. A., Mann, D. M. A., Sumpter, P. Q., and Yates, P. O. (1987). A quantitative morphometric analysis of the neuronal and synaptic content of the frontal and temporal cortex in patients with Alzheimer's disease. *J. Neurol. Sci.* **78**, 151–164.

de la Monte, S. (1989). Quantitation of cerebral atrophy in preclinical and end-stage Alzheimer's disease. *Ann. Neurol.* **25**, 450–459.

de Leon, M. J., George, A. E., Ferris, S. H., *et al.* (1983). Regional correlation of PET and CT in senile dementia of the Alzheimer type. *Am. J. Neuroradiol.* **4**, 553–556.

de Leon, M. J., McRae, T., Tsai, J. R., George, A. E., Marcus, D. L., Freedman, M., Wolf, A. P., and McEwen, B. (1988). Abnormal cortisol response in Alzheimer's disease linked to hippocampal atrophy. *Lancet* August 13, 391–392.

de Leon, M. J., George, A. E., Stylopoulos, L. A., Smith, G., and Miller, D. C. (1989). Early marker for Alzheimer's disease: The atrophic hippocampus. *Lancet* **2**, 672–673.

de Leon, M. J., Golomb, J., George, A. E., Convit, A., Tarshish, C. Y., McRae, T., De Santi, S., Smith, G., Ferris, S. H., Noz, M., and Rusinek, H. (1993). The radiologic prediction of Alzheimer disease: The atrophic hippocampal formation. *AJNR, Am. J. Neuroradiol* **14**, 897–906.

Deutsh, G., and Halsey, J. H. (1990). Cortical blood flow effects of mental rotation in older subjects and Alzheimer patients. *J. Clin. Exp. Neuropsychol.* **12,** 31.

Deweer, B., Lehéricy, S., Pillon, B., Baulac, M., Chiras, J., Marsault, C., Agid, Y., and Dubois, B. (1995). Memory disorders in probable Alzheimer's disease: The role of hippocampal atrophy as shown with MRI. *J. Neurol. Neurosurg. Psychiatry* **58,** 590–597.

Dinov, I. D., Mega, M. S., Thompson, P. M., Lee, L., Woods, R. P., Holmes, C. J., Sumners, D., and Toga, A. W. (2000). Analyzing functional brain images in a probabilistic atlas: A validation of subvolume thresholding. *J. Comput. Assist. Tomogr.* (in press).

Doraiswamy, P. M., McDonald, W. M., Patterson, L., Husain, M. M., Figiel, G. S., Boyko, O. B., and Krishnan, K. R. (1993). Interuncal distance as a measure of hippocampal atrophy: Normative data on axial MR imaging. *AJNR, Am. J. Neuroradiol.* **14,** 141–143.

Duara, R., Grady, C. L., Haxby, J. V., Sundaram, M., Cutler, N. R., Heston, L., Moore, A., Schlageter, N., Larson, S., and Rapoport, S. I. (1986). Positron emission tomography in Alzheimer's disease. *Neurology* **36,** 879–887.

Duara, R., Barker, W. W., Pascal, S., Loewenstein, D. A., and Boothe, T. (1990a). Behavioral activation PET studies in normal aging and Alzheimer's disease. *J. Nucl. Med.* **31,** 730.

Duara, R., Loewenstein, D. A., and Barker, W. W. (1990b). Utilization of behavioral activation paradigms for positron emission tomography studies in normal young elderly subjects and in dementia. "Positron Emission Tomography in Dementia," pp. 131–148. Wiley-Liss, New York.

Duara, R., Barker, W. W., Chang, J., Yoshii, F., Loewenstein, D. A., and Pascal, S. (1992). Viability of neocortical function shown in behavioral activation state PET studies in Alzheimer disease. *J. Cereb. Blood Flow Metab.* **12,** 927–934.

Early, B., Escalona, P. R., Boyko, O. B., Dorasiwamy, P. M., Axelson, D. A., Patterson, L., McDonald, W. M., and Rama Krishman, K. R. (1993). Interuncal measurements in healthy volunteers and in patients with Alzheimer disease. *AJNR* **14,** 907–910.

Erkinjuntti, T., Lee, D. H., Gao, F., Steenhuis, R., Eliasziw, M., Fry, R., Merskey, H., and Hachinski, V. C. (1993). Temporal lobe atrophy on magnetic resonance imaging in the diagnosis of early Alzheimer's disease. *Arch. Neurol.* **50,** 305–310.

Evans, A. C., Collins, D. L., and Holmes, C. J. (1996). Automated 3D regional MRI segmentation and statistical probabilistic anatomical maps. *In* "Quantification of Brain Function Using PET" (R. Myers, V. J. Cunningham, D. L. Bailey, and T. Jones, Eds.), pp. 123–130. Academic Press, New York.

Fazekas, F., Alavi, A., Chawluk, J. B., Zimmerman, R. A., Hackney, D., Bilaniuk, L., Rosen, M., Alves, W. M., Hurtig, H. I., Jamieson, D. G., Kushner, M. J., and Reivich, M. (1989). Comparison of CT, MR, and PET in Alzheimer dementia and normal aging. *J. Nucl. Med.* **30,** 1607–1615.

Ferris, S. H., De Leon, M. J., Wolf, A. P., *et al.* (1983). Positron emission tomography in dementia. *Adv. Neurol.* **38,** 123–129.

Filipek, P. A., Kennedy, D. N., Caviness, V. S., Rossnick, S. L., Spraggins, T. A., and Starewicz, P. M. (1989). Magnetic resonance imaging-based brain morphometry: Development and application to normal subjects. *Ann. Neurol.* **25,** 61–67.

Foster, N. L., Chase, T. N., Fedio, P., Patronas, N. J., Brooks, R. A., and Di Chiro, G. (1983). Alzheimer's disease: Focal cortical changes shown by positron emission tomography. *Neurology* **33,** 961–965.

Foundas, A. L., Leonard, C. M., Mahoney, M., Agee, O. F., and Heilman, K. M. (1997). Atrophy of the hippocampus, parietal cortex, and insula in Alzheimer's disease: A volumetric magnetic resonance imaging study. *Neurol., Neuropsychol., Behav. Neurol.* **10,** 81–89.

Fox, N. C., and Freeborough, P. A. (1997). Brain atrophy progression measured from registered serial MRI: Validation and application to Alzheimer's disease. *J. Magn. Reson. Imaging* **7,** 1069–1075.

Fox, N. C., Freeborough, P. A., and Rossor, M. N. (1996a). Visualisation and quantification of rates of atrophy in Alzheimer's disease. *Lancet* **348,** 94–97.

Fox, N. C., Warrington, E. K., Freeborough, P. A., Hartikainen, P., Kennedy, A. M., Stevens, J. M., and Rossor, M. N. (1996b). Presymptomatic hippocampal atrophy in Alzheimer's disease. A longitudinal MRI study. *Brain* **119,** 2001–2007.

Fox, N. C., Scahill, R. I., Crum, W. R., and Rossor, M. N. (1999). Correlation between rates of brain atrophy and cognitive decline in AD. *Neurology* **52,** 1687–1689.

Freeborough, P. A., and Fox, N. C. (1998). MR image texture analysis applied to the diagnosis and tracking of Alzheimer's disease. *IEEE Trans. Med. Imaging* **17,** 475–479.

Freyham, F. A., Woodford, R. B., and Kety, S. S. (1951). Cerebral blood flow and metabolism in psychosis of senility. *J. Nerv. Ment. Dis.* **113,** 445–456.

Friedland, R. P., Budinger, T. F., Ganz, E., Yano, Y., Mathis, C. A., Koss, B., Ober, B. A., Huesman, R. H., and Derenzo, S. E. (1983). Regional cerebral metabolic alterations in dementia of the Alzheimer type: Positron emission tomography with ^{18}F-fluorodeoxyglucose. *J. Comput. Assist. Tomogr.* **7,** 590–598.

Friedland, R. P., Jagust, W. J., Huesman, R. H., Koss, E., Knittel, B., Mathis, C. A., Ober, B. A., Mazoyer, B. M., and Budinger, T. F. (1989). Regional cerebral glucose transport and utilization in Alzheimer's disease. Neurology 39, 1427–1434.

Gado, M., Hughes, C. P., Warren, D., Chi, D., Jost, G., and Berg, L. (1982). Volumetric measurements of the cerebrospinal fluid spaces in demented subjects and controls. *Radiology* **144,** 535–538.

Gee, J. C., Reivich, M., and Bajcsy, R. (1993). Elastically deforming 3D atlas to match anatomical brain images. *J. Comput. Assist. Tomogr.* **17,** 225–236.

George, A. E., de Leon, M. J., Stylopoulos, L. A., Miller, J., Kluger, A., Smith, G., and Miller, D. C. (1990). CT diagnostic features of Alzheimer disease: Importance of the choroidal/hippocampal fissure complex. *AJNR, Am. J. Neuroradiol.* **11,** 101–107.

Ginsberg, M. D., Dietrich, W. D., and Busto, R. (1987). Coupled forebrain increases of local cerebral glucose utilization and blood flow during physiologic stimulation of a somatosensory pathway in the rat: Demonstration by double-label autoradiography. *Neurology* **37,** 11–19.

Golomb, J., Kluger, A., de Leon, M. J., Ferris, S. H., Convit, A., Mittelman, M. S., Cohen, J., Rusinek, H., De Santi, S., and George, A. E. (1994). Hippocampal formation size in normal human aging: A correlate of delayed secondary memory performance. *Learn. Mem.* **1,** 45–54.

Golomb, J., Kluger, A., de Leon, M. J., Ferris, S. H., Mittelman, M., Cohen, J., and George, A. E. (1996). Hippocampal formation size predicts declining memory performance in normal aging. *Neurology* **47,** 810–813.

Grady, C. L., Haxby, J. V., Schlageter, N. L., et al. (1986). Stability of metabolic and neuropsychological asymmetries in dementia of the Alzheimer type. *Neurology* **36,** 1390–1392.

Grady, C. L., Haxby, J. V., Horwitz, B., et al. (1987). Neuropsychological and cerebral metabolic function in early vs late onset dementia of the Alzheimer type. *Neuropsychologia* **25,** 807–816.

Grady, C. L., Haxby, J. V., Horwitz, B., Gillette, J., Salerno, J. A., Gonzalez-Aviles, A., Carson, R. E., Herscovitch, P., Schapiro, M. B., and Rapoport, S. I. (1993). Activation of cerebral blood flow during a visuoperceptual task in patients with Alzheimer-type dementia. *Neurobiol. Aging* **14,** 35–44.

Grafton, S. T., Mazziotta, J. C., Pahl, J. J., St. George-Hyslop, P., Haines, J. L., Gusella, J., Hoffman, J. M., Baxter, L. R., and Phelps, M. E. (1992). Serial changes of cerebral glucose metabolism and caudate size in persons at risk for Huntington's disease. *Arch. Neurol.* **49,** 1161–1167.

Haller, J. W., Christensen, G. E., Joshi, S. C., Newcomer, J. W., Miller,

M. I., Csernansky, J. G., and Vannier, M. W. (1996). Hippocampal MR imaging morphometry by means of general pattern matching. *Radiology* **199**, 787–791.

Haller, J. W., Banerjee, A., Christensen, G. E., Gado, M., Joshi, S., Miller, M. I., Sheline, Y., Vannier, M. W., and Csernansky, J. G. (1997). Three-dimensional hippocampal MR morphometry with high-dimensional transformation of a neuroanatomic atlas. *Radiology* **202**, 504–510.

Hamos, J. E., DeGennaro, L. J., and Drachman, D. A. (1989). Synaptic loss in Alzheimer's disease and other dementias. *Neurology* **39**, 355–361.

Hand, P., Greenberg, J., Goochee, C., Sylvestro, A., Weller, L., and Reivich, M. (1979). A normal and developmentally-altered cortical column: A laminar analysis of local glucose utilization with natural stimulation of a single receptor organ. *Acta Neurol. Scand.* **60** (Suppl. 72), 46–47.

Harr, S. D., Simonian, N. A., and Hyman, B. T. (1995). Functional activity in Alzheimer's disease: Decreased glucose transporter 3 immunoreactivity in the perforant pathway terminal zone. *J. Neuropathol. Exp. Neurol.* **54**, 38–41.

Haxby, J. V., Grady, C. L., Koss, E., Horwitz, B., Heston, L., Schapiro, M., Friedland, R. P., and Rapoport, S. I. (1990). Longitudinal study of cerebral metabolic asymmetries and associated neuropsychological patterns in early dementia of the Alzheimer type. *Arch. Neurol.* **47**, 753–760.

Hensley, K., Carney, J. M., Mattson, M. P., Aksenova, M., Harris, M., Wu, J. F., Floyd, R. A., and Butterfield, D. A. (1994). A model for β-amyloid aggregation and neurotoxicity based on free radical generation by the peptide: Relevance to Alzheimer's disease. *Proc. Natl. Acad. Sci. U.S.A.* **91**, 3270–3274.

Herzog, A. G., and Kemper, T. L. (1980). Amygdaloid changes in aging and dementia. *Arch. Neurol.* **37**, 625–629.

Higuchi, M., Arai, H., Nakagawa, T., Higuchi, S., Muramatsu, T., Matsushita, S., Kosaka, Y., Itoh, M., and Sasaki, H. (1997). Regional cerebral glucose utilization is modulated by the dosage of apolipoprotein E type 4 allele and α-antichymotrypsin type A allele in Alzheimer's disease. *NeuroReport* **8**, 2639–2643.

Høedt-Rasmussen, K., and Skinhøj, E. (1966). *In vivo* measurements of the relative weights of gray and white matter in the human brain. *Neurology* **16**, 515–521.

Holman, B. L. (1991). Imaging dementia with SPECT. *Ann. N.Y. Acad. Sci.* **620**, 165–174.

Howieson, J., Kaye, J. A., Holm, L., and Howieson, D. (1993). Interuncal distance: Marker of aging and Alzheimer disease. *AJNR, Am. J. Neuroradiol.* **14**, 647–650.

Huesgen, C. T., Burger, P. C., Crain, B. J., and Johnson, G. A. (1993). *In vitro* MR microscopy of the hippocampus in Alzheimer's disease. *Neurology* **43**, 145–152.

Hughes, C. P., Berg, L., Danziger, W. L., Coben, L. A., and Martin, R. L. (1982). A new clinical scale for the staging of dementia. *Br. J. Psychiatry* **140**, 566–572.

Ibanez, V., Pietrini, P., Alexander, G. E., Furey, M. L., Teichberg, D., Rajapakse, J. C., Rapoport, S. I., Schapiro, M. B., and Horwitz, B. (1998). Regional glucose metabolic abnormalities are not the result of atrophy in Alzheimer's disease. *Neurology* **50**, 1585–1593.

Ikeda, M., Tanabe, H., Nakagawa, Y., Kazui, H., Oi, H., Yamazaki, H., Harada, K., and Nishimura, T. (1994). MRI-based quantitative assessment of the hippocampal region in very mild to moderate Alzheimer's disease. *Neuroradiology* **36**, 7–10.

Jack, C. R., Petersen, R. C., O'Brien, P. C., and Tangalos, E. G. (1992). MR-based hippocampal volumetry in the diagnosis of Alzheimer's disease. *Neurology* **42**, 183–188.

Jack, C. R., Petersen, R. C., Xu, Y. C., Waring, S. C., O'Brien, P. C., Tangalos, E. G., Smith, G. E., Ivnik, R. J., and Kokmen, E. (1997). Me-

dial temporal atrophy on MRI in normal aging and very mild Alzheimer's disease. *Neurology* **49**, 786–794.

Jack, C. R., Petersen, R. C., Xu, Y. C., O'Brien, P. C., Waring, S. C., Tangalos, E. G., Smith, G. E., Ivnik, R. J., Thibodeau, S. N., and Kokmen, E. (1998). Hippocampal atrophy and apolipoprotein E genotype are independently associated with Alzheimer's disease. *Ann. Neurol.* **43**, 303–310.

Jack, C. R., Petersen, R. C., Xu, Y. C., O'Brien, P. C., Smith, G. E., Ivnik, R. J., Boeve, B. F., Waring, S. C., Tangalos, E. G., and Kokmen, E. (1999). Prediction of AD with MRI-based hippocampal volume in mild cognitive impairment. *Neurology* **52**, 1397–1403.

Jaeger, R. (1914). Inhaltsberechnungen der rinden- und mark-substanz des grosshirns durch planimetrische messungen. *Arch. Psychiatr. Nervenkrankh.* **54**, 261–272.

Jagust, W. J., Seab, J. P., Huesman, R. H., Valk, P. E., Mathis, C. A., Reed, B. R., Coxson, P. G., and Budinger, T. F. (1991). Diminished glucose transport in Alzheimer's disease: Dynamic PET studies. *J. Cereb. Blood Flow Metab.* **11**, 323–330.

Jamieson, D. G., Chawluk, J. B., Alavi, A., Hurtig, H. I., Rosen, M., Bais, S., Dann, R., Kushner, M., and Reivich, M. (1987). The effect of disease severity on local cerebral glucose metabolism in Alzheimer disease. *J. Cereb. Blood Flow Metab.* **7** (Suppl. 1), S410.

Jernigan, T. L., Press, G. A., and Hesselink, J. R. (1990). Methods for measuring morphologic features on magnetic resonance images. *Arch. Neurol.* **47**, 27–32.

Jernigan, T. L., Archibald, S. L., Berhow, M. T., Sowell, E. R., Foster, D. S., and Hesselink, J. R. (1991). Cerebral structure on MRI, Part I: Localization of age-related changes. *Biol. Psychiatry* **29**, 55–67.

Jobst, K. A., Smith, A. D., Barker, C. S., Wear, A., King, E. M., Smith, A., Anslow, P. A., Molyneux, A. J., Shepstone, B. J., Soper, N., Holmes, K. A., Robinson, J. R., Hope, R. A., Oppenheimer, C., Brockbank, K., and McDonald, B. (1992a). Association of atrophy of the medial temporal lobe with reduced blood flow in the posterior parietotemporal cortex in patients with a clinical and pathological diagnosis of Alzheimer's disease. *J. Neurol. Neurosurg. Psychiatry* **55**, 190–194.

Jobst, K. A., Smith, A. D., Szatmari, M., Molyneux, A. J., Esiri, M. E., King, E., Smith, A., Jaskowski, A., McDonald, B., and Wald, N. (1992b). Detection in life of confirmed Alzheimer's disease using a simple measurement of medial temporal lobe atrophy by computed tomography. *Lancet* **340**, 1179–1183.

Jobst, K. A., Barnetson, L. P. D., Shepstone, B. J., and members of OPTIMA (1998). Accurate prediction of histologically confirmed Alzheimer's disease and the differential diagnosis of dementia: The use of NINCDS-ADRDA and DSM-III-R criteria, SPECT, X-ray CT, and Apo E4 in medial temporal lobe dementias. *Int. Psychogeriatr.* **10**, 271–302.

Johnson, K. A., Jones, K., Holman, B. L., Becker, J. A., Spiers, P. A., Satlin, A., and Albert, M. S. (1998). Preclinical prediction of Alzheimer's disease using SPECT. *Neurology* **50**, 1563–1571.

Julin, P., Almkvist, O., Basun, H., Lannfelt, L., Svensson, L., Winblad, B., and Wahlund, L. O. (1998). Brain volume and regional cerebral blood flow in carriers of the Swedish Alzheimer amyloid protein mutation. *Alzheimer Dis. Assoc. Disord.* **12**, 49–53.

Juottonen, K., Laakso, M. P., Insausti, R., Lehtovirta, M., Pitkanen, A., Partanen, K., and Soininen, H. (1998). Volumes of the entorhinal and perirhinal cortices in Alzheimer's disease. *Neurobiol. Aging* **19**, 15–22.

Kadekaro, M., Crane, A. M., and Sokoloff, L. (1985). Differential effects of electrical stimulation of sciatic nerve on metabolic activity in spinal cord and dorsal root ganglion in the rat. *Proc. Natl. Acad. Sci. U.S.A.* **82**, 6010–6013.

Kaye, J. A., Swihart, T., Howieson, D., Dame, A., Moore, M. M., Karnos, T., Camicioli, R., Ball, M., Oken, B., and Sexton, G. (1997). Volume loss of the hippocampus and temporal lobe in healthy elderly persons destined to develop dementia. *Neurology* **48**, 1297–1304.

Kesslak, J. P., Nalcioglu, O., and Cotman, C. W. (1991). Quantification of magnetic resonance scans for hippocampal and parahippocampal atrophy in Alzheimer's disease. *Neurology* **41**, 51–54.

Kessler, J., Herholz, M., Grond, M., and Heiss, W. D. (1991). Impaired metabolic activation in Alzheimer's disease: A PET study during continuous visual recognition. *Neuropsychologia* **29**, 229–243.

Kety, S. S. (1956). Human cerebral blood flow and oxygen consumption as related to ageing. *Assoc. Res. Nerve Dis. Proc.* **35**, 31–45.

Kety, S. S., and Schmidt, C. F. (1948). The nitrous oxide method for the quantitative determination of cerebral blood flow in man: Theory, procedure and normal values. *J. Clin. Invest.* **27**, 476–483.

Killiany, R. J., Moss, M. B., Albert, M. S., Sandor, T., Tieman, J., and Jolesz, F. (1993). Temporal lobe regions on magnetic resonance imaging identify patients with early Alzheimer's disease. *Arch. Neurol.* **50**, 949–954.

Kiyosawa, M., Baron, J. C., Hamel, E., Pappata, S., Duverger, D., Riche, D., Mazoyer, B., Naquet, R., and MacKenzie, E. T. (1989). Time course of effects of unilateral lesions of the nucleus basalis of Meynert on glucose utilization by the cerebral cortex. Positron tomography in baboons. *Brain* **112**, 435–455.

Kohler, S., Black, S. E., Sinden, M., Szekely, C., Kidron, D., Parker, J. L., Foster, J. K., Moscovitch, M., WIncour, G., Szalai, J. P., and Bronskill, M. J. (1998). Memory impairments associated with hippocampal versus parahippocampal-gyrus atrophy: An MR volumetry study in Alzheimer's disease. *Neuropsychologia* **36**, 901–914.

Kollokian, V. (1996). Performance analysis of automatic techniques for tissue classification in magnetic resonance images of the human brain. Department of Computer Science, Concordia University, Montreal, Quebec, Canada.

Krasuski, J. S., Alexander, G. E., Horwitz, B., Daly, E. M., Murphy, D. G., Rapoport, S. I., and Schapiro, M. B. (1998). Volumes of medial temporal lobe structures in patients with Alzheimer's disease and mild cognitive impairment (and in healthy controls). *Biol. Psychiatry* **43**, 60–68.

Laakso, M. P., Soininen, H., Partanen, K., Helkala, E.-L., Hartikainen, P., Vainio, P., Hallikainen, M., Hänninen, T., and Riekkinen, P. J. (1995). Volumes of hippocampus, amygdala, and frontal lobes in MRI based diagnosis of early Alzheimer's disease: Correlations with memory functions. *J. Neural Transm. [P-D Sect.]* **9**, 73–86.

Lassen, N. A., and Ingvar, D. H. (1963). Regional cerebral blood flow measurement in man. *Arch. Neurol.* **9**, 65–72.

Lassen, N. A., Munck, O., and Tottery, E. R. (1957). Mental function and cerebral oxygen consumption in organic dementia. *Arch. Neurol. Psychiatry* **77**, 126–133.

Lavenu, I., Pasquier, F., Lebert, F., Jacob, B., and Petit, H. (1997). Association between medial temporal lobe atrophy on CT and parietotemporal uptake decrease on SPECT in Alzheimer's disease. *J. Neurol. Neurosurg. Psychiatry* **63**, 441–445.

LeMay, M., Stafford, J. L., Sandor, T., Albert, M., Haykal, H., and Zamani, A. (1986). Statistical assessment of perceptual CT scan ratings in patients with Alzheimer type dementia. *JCAT* **10**, 802–809.

Le Mestric, C., Chavoix, C., Chapon, F., Mezenge, F., Epelbaum, J., and Baron, J. C. (1998). Effects of damage to the basal forebrain on brain glucose utilization: A reevaluation using positron emission tomography in baboons with extensive unilateral excitotoxic lesion. *J. Cereb. Blood Flow Metab.* **18**, 476–490.

Lehéricy, S., Baulac, M., Chiras, J., Piérot, L., Martin, N., Pillon, B., Deweer, B., Dubois, B., and Marsault, C. (1994). Amygdalohippocam-

pal MR volume measurements in the early stages of Alzheimer disease. *AJNR, Am. J. Neuroradiol.* **15**, 927–937.

Lehtovirta, M., Laakso, M. P., Soininen, H., Helisalmi, S., Mannermaa, A., Helkala, E.-L., Partanen, K., Ryyänen, M., Vainio, P., Hartikainen, P., and Riekkinen, P. J. (1995). Volumes of hippocampus, amygdala and frontal lobe in Alzheimer patients with different apolipoprotein E genotypes. *Neuroscience* **67**, 65–72.

Lehtovirta, M., Soininen, H., Laakso, M. P., Partanen, K., Helisalmi, S., Mannermaa, A., Ryyänen, M., Kuikka, J., Hartikainen, P., and Riekkinen, P. J. (1996). SPECT and MRI analysis in Alzheimer's disease: Relation to apolipoprotein E . . . 4 allele. *J. Neurol. Neurosurg. Psychiatry* **60**, 644–649.

Lehtovirta, M., Kuikka, J., Helisalmi, S., Hartikainen, P., Mannermaa, A., Ryynanen, M., Riekkinen, P., Sr., and Soininen, H. (1998). Longitudinal SPECT study in Alzheimer's disease: Relation to apolipoprotein E polymorphism. *J. Neurol. Neurosurg. Psychiatry* **64**, 724–746.

Lund-Anderson, H. (1979). Transport of glucose from blood to brain. *Physiol. Rev.* **59**, 305–352.

Mark, R. J., Hensley, K., Butterfield, A., and Mattson, M. P. (1995). Amyloid β-peptide impairs ion-motive ATPase activities: Evidence for a role in loss of neuronal Ca^{2+} homeostasis and cell death. *J. Neurosci.* **15**, 6239–6249.

Masterman, D. L., Mendez, M. F., Fairbanks, L. A., and Cummings, J. L. (1997). Sensitivity, specificity, and positive predictive value of technetium 99–HMPAO SPECT in discriminating Alzheimer's disease from other dementias. *J. Geriatr. Psychiatry Neurol.* **10**, 15–21.

Mata, M., Fink, D. J., Gainer, H., et al. (1980). Activity-dependent energy metabolism in rat posterior pituitary primarily reflects sodium pump activity. *J. Neurochem.* **34**, 213–215.

Maunoury, C., Michot, J.-L., Caillet, H., Parlato, V., Leroy-Willig, A., Jehenson, P., Syrota, A., and Boller, F. (1996). Specificity of temporal amygdala atrophy in Alzheimer's disease: Quantitative assessment with magnetic resonance imaging. *Dementia* **7**, 10–14.

Mauri, M., Sibilla, L., Bono, G., Carlesimo, G. A., Sinforiani, E., and Martelli, A. (1998). The role of morpho-volumetric and memory correlations in the diagnosis of early Alzheimer dementia. *J. Neurol.* **245**, 525–530.

Mazziotta, J. C., Phelps, M. E., Plummer, D., and Kuhl, D. E. (1981). Quantitation in positron emission tomography. 5. Physical-anatomical effects. *J. Comput. Assist. Tomogr.* **5**, 734–743.

Mazziotta, J. C., Phelps, M. E., Pahl, J. J., Huang, S. C., Baxter, L. R., Riege, W. H., Hoffman, J. M., Kuhl, D. E., Lanto, A. B., Wapenski, J. A., et al. (1987). Reduced cerebral glucose metabolism in asymptomatic subjects at risk for Huntingon's disease. *N. Engl. J. Med.* **316**, 357–362.

Mazziotta, J. C., Toga, A. W., Evans, A. C., Fox, P., and Lancaster, J. (1995). A probabilistic atlas of the human brain: Theory and rationale for its development. *Neuroimage* **2**, 89–101.

McGeer, P. L., Kamo, H., Harrop, R., McGeer, E. G., Martin, W. R. W., Pate, B. D., and Li, D. K. B. (1986). Comparison of PET, MRI, and CT with pathology in a proven case of Alzheimer disease. *Neurology* **36**, 1569–1574.

McKelvey, R., Bergman, H., Stern, J., Rush, C., Zahirney, G., and Chertkow, H. (1999). Lack of prognostic significance of SPECT abnormalities in non-demented elderly subjects with memory loss. *Can. J. Neurol. Sci.* **26**, 23–28.

Mega, M. S., Chen, S. S., Thompson, P. M., Woods, R. P., Karaca, T. J., Tiwari, A., Vinters, H. V., Small, G. W., and Toga, A. W. (1997). Mapping histology to metabolism: Coregistration of stained whole brain sections to premortem PET using 3D cryomacrotome reconstruction as a reference. *NeuroImage* **5**, 147–153.

Mega, M. S., Thompson, P. M., Cummings, J. L., Back, C. L., Xu, L. M.,

Zohoori, S., Goldkorn, A., Moussai, J., Fairbanks, L., Small, G. W., and Toga, A. W. (1998). Sulcal variability in the Alzheimer's brain: Correlations with cognition. *Neurology* **50,** 145–151.

Mega, M. S., Chu, T., Mazziotta, J. C., Trivedi, K. H., Thompson, P. M., Shah, A. K., Aron, J., Frautsch, S. A., Cole, G. M., and Toga, A. W. (1999). Mapping biochemistry to metabolism: FDG-PET and amyloid burden in Alzheimer's disease. *NeuroReport* **10,** 2911–2917.

Meltzer, C. C., and Frost, J. J. (1994). Partial volume correction in emission-computed tomography: Focus on Alzheimer disease. "Functional Neuroimaging," pp. 163–170. Academic Press, San Diego.

Meltzer, C. C., Leal, J. P., Mayberg, H. S., Wagner, H. N., and Frost, J. J. (1990). Correction of PET data for partial volume effects in human cerebral cortex by MR imaging. *J. Comput. Assist. Tomogr.* **14,** 561–570.

Meltzer, C. C., Zubieta, J. K., Brandt, J., Tune, L. E., Mayberg, H. S., and Frost, J. J. (1996). Regional hypometabolism in Alzheimer's disease as measured by positron emission tomography after correction for effects of partial volume averaging. *Neurology* **47,** 454–461.

Metter, E. J., Riege, W. H., Hanson, W. R., Kuhl, D. E., Phelps, M. E., Squire, L. R., Wasterlain, C. G., and Benson, D. F. (1983). Comparison of metabolic rates, language, and memory in subcortical aphasias. *Brain Lang.* **19,** 33–47.

Meyer, J. S., Obara, K., and Muramatsu, K. (1993). Diaschisis. *Neurol. Res.* **15,** 362–366.

Miller, A. K. H., Alston, R. L., and Corsellis, J. A. N. (1980). Variation with age in the volumes of grey and white matter in the cerebral hemispheres of man: Measurements with an image analyzer. *Neuropathol. Appl. Neurobiol.* **6,** 119–132.

Miller, J. D., de Leon, M. J., Ferris, S. H., et al. (1987). Abnormal temporal lobe response in Alzheimer's disease during cognitive processing as measured by ^{11}C-2–deoxy-D-glucose and PET. *J. Cereb. Blood Flow Metab.* **7,** 248–251.

Miller, M. I., Christensen, G. E., Amit, Y., and Grenander, U. (1993). Mathematical textbook of deformable neuroanatomies. *Proc. Natl. Acad. Sci. U.S.A.* **90,** 11944–11948.

Miller, M. I., Banerjee, A., et al. (1997). Statistical methods in computational anatomy. *Stat. Methods Med. Res.* **6,** 267–299.

Müller-Gärtner, H. W., Links, J. M., Prince, J. L., Bryan, R. N., McVeigh, E., Leal, J. P., Davatzikos, C., and Frost, J. J. (1992). Measurement of radiotracer concentration in brain gray matter using positron emission tomography: MRI-based correction for partial volume effects. *J. Cereb. Blood Flow Metab.* **12,** 571–583.

Murphy, G. M., Inger, P., Mark, K., Lin, J., Morrice, W., Gee, C., Gan, S., and Korp, B. (1987). Volumetric asymmetry in the human amygdaloid complex. *J. Hirnforsch.* **28,** 281–289.

Parnetti, L., Lowenthal, D. T., Presciutti, O., Pelliccioli, G., Palumbo, R., Gobbi, G., Chiarini, P., Palumbo, B., Tarducci, R., and Senin, U. (1996). ^{1}H-MRS, MRI-based hippocampal volumetry, and ^{99m}Tc-HMPAO-SPECT in normal aging, age-associated memory impairment, and probable Alzheimer's disease. *J. Am. Geriatr. Soc.* **44,** 133–138.

Pearlson, G. D., Harris, G. J., Powers, R. E., Barta, P. E., Camargo, E. E., Chase, G. A., Noga, J. T., and Tune, L. E. (1992). Quantitative changes in mesial temporal volume, regional cerebral blood flow, and cognition in Alzheimer's disease. *Arch. Gen. Psychiatry.* **49,** 402–408.

Pietrini, P., Furey, M. L., Alexander, G. E., Mentis, M. J., Dani, A., Guazzelli, M., Rapoport, S. I., and Schapiro, M. B. (1999). Association between brain functional failure and dementia severity in Alzheimer's disease: Resting versus stimulation PET study. *Am. J. Psychiatry* **156,** 470–473.

Prohovnik, I., Smith, G., Sackeim, H. A., Mayeux, R., and Stern, Y. (1989). Gray-matter degeneration in presenile Alzheimer disease. *Ann. Neurol.* **25,** 117–124.

Pucci, E., Belardinelli, N., Regnicolo, L., Nolfe, G., Signorine, M., Salvolini, U., and Angeleri, F. (1998). Hippocampus and parahippocampal gyrus linear measurements based on magnetic resonance in Alzheimer's disease. *Eur. Neurol.* **39,** 16–25.

Reiman, E. M., Caselli, R. J., Yun, L. S., Chen, K., Bandy, D., Minoshima, S., Thibodeau, S. N., and Osborne, D. (1996). Preclinical evidence of Alzheimer's disease in persons homozygous for the epsilon 4 allele for apolipoprotein *E. N. Engl. J. Med.* **334,** 752–758.

Reiman, E. M., Uecker, A., Caselli, R. J., Lewis, S., Bandy, D., de Leon, M. J., De Santi, S., Convit, A., Osborne, D., Weaver, A., and Thibodeau, S. N. (1998). Hippocampal volumes in cognitively normal persons at genetic risk for Alzheimer's disease. *Ann. Neurol.* **44,** 288–291.

Rossor, M. N., Fox, N. C., Freeborough, P. A., and Roques, P. K. (1997). Slowing the progression of Alzheimer's disease: Monitoring progression. *Alzheimer Dis. Assoc. Disord.* **11** (Suppl. 5), S6–S9.

Rusinek, H., de Leon, M. J., George, A. E., Stylopoulos, L. A., Chandra, R., Smith, G., Rand, T., Mourino, M., and Kowalski, H. (1991). Alzheimer disease: Measuring loss of cerebral gray matter with MR imaging. *Radiology* **178,** 109–114.

Salmon, E., and Franck, G. (1989). Positron emission tomographic study in Alzheimer's disease and Pick's disease. *Arch. Gerontol. Geriatr.* **1,** 241–247.

Samuel, W., Terry, R. D., DeTeresa, R., Butters, N., and Masliah, E. (1993). Clinical correlates of cortical and nucleus basalis pathology in Alzheimer's disease. *Arch. Neurol.* **51,** 772–778.

Scheltens, P., Leys, D., Barkhof, F., Huglo, D., Weinstein, H. C., Vermersch, P., Kuiper, M., Steinling, M., Wolters, E. C., and Valk, J. (1992). Atrophy of medial temporal lobes on MRI in 'probable' Alzheimer's disease and normal ageing: Diagnostic value and neuropsychological correlates. *J. Neurol. Neurosurg. Psychiatry* **55,** 967–972.

Schuff, N., Amend, D. L., Ezekiel, F., Steinman, S. K., Tanabe, J. L., Norman, D., Jagust, W., Kramer, J. H., Mastrianni, J. A., Fein, G., and Weiner, M. W. (1997). Changes of hippocampal *N*-acetyl aspartate and volume in Alzheimer's disease. A proton MR spectroscopic imaging and MRI study. *Neurology* **49,** 1513–1521.

Scott, S. A., DeKosky, S. T., and Scheff, S. W. (1991). Volumetric atrophy of the amygdala in Alzheimer's disease: Quantitative serial reconstruction. *Neurology* **41,** 351–356.

Seab, J. P., Jagust, W. J., Wong, S. T. S., Roos, M. S., Reed, B. R., and Budinger, T. F. (1988). Quantitative NMR measurements of hippocampal atrophy in Alzheimer disease. *Magn. Reson. Med.* **8,** 200–208.

Shefer, V. F. (1972). Absolute numbers of neurons and thickness of the cerebral cortex during aging, senile and vascular dementia, and Pick's and Alzheimer diseases. *Neurosci. Behav. Physiol.* **6,** 319–324.

Signorini, M., Paulesu, E., Friston, K., Perani, D., Colleluori, A., Lucignani, G., Grassi, F., Bettinardi, V., Frackowiak, R. S. J., and Fazio, F. (1999). Rapid assessment of regional cerebral metabolic abnormalities in single subjects with quantitative and nonquantitative [^{18}F]FDG PET: A clinical validation of statistical parametric mapping. *NeuroImage* **9,** 63–80.

Sled, J. G., Zijdenbos, A. P., and Evans, A. C. (1998). A nonparametric method for automatic correction of intensity nonuniformity in MRI data. *IEEE Trans. Med. Imaging* **17,** 87–97.

Small, G. S., Mazziotta, J. C., Collins, M. T., Baxter, L. R., Phelps, M. E., Mandelkern, M. A., Kaplan, A., La Rue, A., Adamson, C. F., Chang, L., Guze, B. H., Corder, E. H., Saunders, A. M., Haines, J. L., Pericak-Vance, M. A., and Roses, A. D. (1995). Apolipoprotein E type 4 allele and cerebral glucose metabolism in relatives at risk for familial Alzheimer disease. *JAMA* **273,** 942–947.

Small, S. A., Perera, G. M., DeLaPaz, R., Mayeux, R., and Stern, Y. (1999). Differential regional dysfunction of the hippocampal formation among elderly with memory decline and Alzheimer's disease. *Ann. Neurol.* **45,** 466–472.

Smith, C. D., Malcein, M., Meurer, K., Schmitt, F. A., Markesbery, W. R., and Pettigrew, L. C. (1999). MRI temporal lobe volume measures and neuropsychologic function in Alzheimer's disease. *J. Neuroimaging* **9,** 2–9.

Soininen, H. S., Partanen, K., Pitkänen, A., Vainio, P., Hänninen, T., Hallikainen, M., Koivisto, K., and Riekkinen, P. J. (1994). Volumetric MRI analysis of the amygdala and the hippocampus in subjects with age-associated memory impairment. *Neurology* **44,** 1660–1668.

Soininen, H., Partanen, K., Pitkänen, A., Hallikainen, M., Hänninen, T., Helisalmi, S., and Riekkinen, P. (1995). Decreased hippocampal volume asymmetry on MRIs in nondemented elderly subjects carrying the apolipoprotein E . . . 4 allele. *Neurology* **45,** 391–392.

Tanaka, S., Kawamata, J., Shimohama, S., Akaki, H., Akiguchi, I., Kimura, J., and Ueda, K. (1998). Inferior temporal lobe atrophy and APOE genotypes in Alzheimer's disease. X-ray computed tomography, magnetic resonance imaging and Xe-133 SPECT studies. *Dement. Geriatr. Cog. Disord.* **9,** 90–98.

Terry, R. D., Masliah, E., Salmon, D. P., Butters, N., DeTeresa, R., Hill, R., Hansen, L. A., and Katzman, R. (1991). Physical basis of cognitive alterations in Alzheimer's disease: Synapse loss is the major correlate of cognitive impairment. *Ann. Neurol.* **30,** 572–580.

Terry, R. D., Masliah, E., and Hansen, L. A. (1994). Structural basis of the cognitive alterations in Alzheimer disease. "Alzheimer Disease, pp. 179–196. Raven, New York.

Thompson, P. M., and Toga, A. W. (1996). A surface-based technique for warping three-dimensional images of the brain. *IEEE Trans. Med. Imaging* **15,** 1–16.

Thompson, P. M., Schwartz, C., Lin, R. T., Khan, A. A., and Toga, A. W. (1996). Three-dimensional statistical analysis of sulcal variability in the human brain. *J. Neurosci.* **16,** 4261–4274.

Thompson, P. M., MacDonald, D., Mega, M. S., Holmes, C. J., Evans, A. C., and Toga, A. W. (1997). Detection and mapping of abnormal brain structure with a probabilistic atlas of cortical surfaces. *J. Comput. Assit. Tech.* **21,** 567–581.

Thompson, P. M., Moussai, J., Zohoori, S., Goldkorn, A., Khan, A. A., Mega, M. S., Small, G. W., Cummings, J. L., and Toga, A. W. (1998). Cortical variability and asymmetry in normal aging and Alzheimer's disease. *Cerebr. Cortex* **8,** 492–509.

Vannier, M. W., Brunsden, B. S., Hildebolt, C. F., Falk, D., Chevarud, J. M., Figiel, G. S., Perman, W. H., Kohn, L. A., Robb, R. A., Yoffie, R. L., and Bresina, S. J. (1991). Brain surface cortical sulcal lengths: Quantification with three-dimensional MR imaging. *Radiology* **180,** 479–484.

Woodard, J. L., Grafton, S. T., Votaw, J. R., Green, R. C., Dobraski, M. E., and Hoffman, J. M. (1998). Compensatory recruitment of neural resources during overt rehearsal of word lists in Alzheimer's disease. *Neuropsychology* **12,** 491–504.

Woods, R. P., Mazziotta, J. C., and Cherry, S. R. (1993). MRI-PET registration with automated algorithm. *J. Comput. Assist. Tomogr.* **17,** 536–546.

Woods, R. P., Grafton, S. T., Watson, J. D. G., Sicotte, N. L., and Mazziotta, J. C. (1998). Automated image registration: II. Intersubject validation of linear and nonlinear models. *J. Comput. Assist. Tomogr.* **22,** 153–165.

Yamaura, H., Ito, M., Kubota, K., and Matsuzawa, T. (1980). Brain atrophy during aging: A quantitative study with computed tomography. *J. Gerontol.* **35,** 492–498.

Yasuda, M., Mori, E., Kitagaki, H., Hirono, N., Shimada, K., Maeda, K., and Tanaka, C. (1998). Apolipoprotein E epsilon 4 allele and whole brain atrophy in late-onset Alzheimer's disease. *Am. J. Psychiatry* **155,** 779–784.

Zoumalan, C., Mega, M. S., Thompson, P. M., Grigorians, A., Nguyen, A., and Toga, A. W. (1999). Three-dimensional sulcal deformations of the brain in aging and dementia are related to cognitive performance. *NeuroImage* **9,** S602.

10

Movement Disorders: Parkinson's Disease

D. Eidelberg*,†,1 C. Edwards,* M. Mentis,* V. Dhawan,* and J. R. Moeller‡

*Functional Brain Imaging Laboratory, Department of Neurology, North Shore University Hospital, Manhasset, New York 11030
†Department of Neurology, The NY Hospital–Cornell Medical Center, New York, New York 10021
‡Department of Biological Psychiatry, New York State Psychiatric Institute, New York, New York 10032

I. Introduction

II. Dopamine System Imaging in Parkinsonism

III. Functional Brain Imaging in the Resting State

IV. Brain Activation Studies: Motor Execution and Learning

V. Functional Brain Imaging in the Assessment of Therapeutic Interventions

VI. Conclusion

References

I. Introduction

Functional brain imaging with positron emission tomography (PET) has provided novel insights into the pathophysiology of Parkinson's disease (PD) and other movement disorders. In PD, PET has been applied in the investigation of nigrostriatal presynaptic dopaminergic nerve terminals to measure dopa decarboxylase activity using [18F]fluorodopa (FDOPA) (Brooks et al., 1990b; Eidelberg et al., 1990; Antonini et al., 1995; Ishikawa et al., 1996a) or ligands to quantify the dopamine transporter (DAT) and the vesicular monoamine transporter (VMAT) using cocaine derivative tracers (Chaly et al., 1996; Assenbaum et al., 1997) and the labeled tetrabenazines (Frey et al., 1996). The postsynaptic dopamine receptor systems can be quantified using [11C]raclopride (Antonini et al., 1993, 1995) or [11C]methylspiperone (Stoessl and Ruth, 1998) for D_2 neuroreceptors and [11C]SCH23390 (Rinne et al., 1990; Shinotoh et al., 1993b) for D_1 neuroreceptor binding. PET has also been used to study regional neuronal activity by quantifying cerebral blood flow (rCBF) activation responses with [15O]H_2O (Jahanshahi et al., 1995; Catalan et al., 1999) as well as resting-state glucose metabolism (rCMRGlc) with [18F]fluorodeoxyglucose (FDG) (Eidelberg et al., 1990, 1994, 1995a; Antonini et al., 1995). Particularly, studies of brain metabolism and blood flow have contributed considerably to our understanding of abnormal neuronal circuitry underlying the pathophysiology of PD.

---

[1] To whom correspondence should be addressed.

In this chapter, we will emphasize the recent advances in PET studies of PD as well as the relative advantages and shortcomings of these techniques in clinical diagnosis and management. We will additionally focus on the potential application of PET in the selection of suitable candidates for surgical interventions, including implantation of fetal mesencephalic tissue, pallidotomy, thalamotomy, and deep-brain stimulation, as well as in the assessment of their outcome after surgery.

II. Dopamine System Imaging in Parkinsonism

A. Presynaptic Dopaminergic Function

1. Dopa Decarboxylase Activity

FDOPA is probably the most commonly applied radiotracer for the study of striatal dopaminergic nerve terminals in parkinsonism. PET studies with this tracer measure the rate of decarboxylation of [18F]fluorodopa to [18F]fluorodopamine by the enzyme dopa decarboxylase (DDC) and its subsequent storage in the striatal dopaminergic nerve terminals. In the plasma, FDOPA is metabolized by catechol O-methyl-transferase (COMT) to 3-O-methyl-fluorodopa (3OMFD) as well as by DDC (Dhawan et al., 1996; Ishikawa et al., 1996b,c). In PET experiments, peripheral DDC can be blocked by the administration of the DDC inhibitor carbidopa before tracer administration. FDOPA transport across the blood–brain barrier (BBB) follows the same channel as the large neutral amino acid (LNAA) (Banos et al., 1978).

FDOPA/PET data can be analyzed either by using simple noninvasively derived target-to-background ratios, obtained by dividing striatal count rates by those in a neutral brain region such as the occipital cortex (SOR) (Brooks et al., 1990a; Eidelberg et al., 1990; Ishikawa et al., 1996c), or, alternatively, by multiple time graphical analysis (MTGA) (Patlak and Blasberg, 1985; Ishikawa et al., 1996a) utilizing plasma or brain input functions. In MTGA, the gradient of the linear regression of the data, described as the net influx constant (K_i), reflects the rate of FDOPA decarboxylation and storage. Compartmental models have also been developed to estimate the specific kinetic rate constant for striatal DDC activity (k_3) (Gjedde et al., 1991; Dhawan et al., 1996).

A number of prior studies have shown that the assessment of nigrostriatal dopaminergic function using FDOPA/PET yields quantitative parameters that correlate with independent disease severity measures (Brooks et al., 1990b; Eidelberg et al., 1990, 1995b; An-

tonini et al., 1995) and can discriminate early-stage PD patients from normal control subjects (Leenders et al., 1990; Antonini et al., 1995). More importantly, it has been shown that in vivo striatal FDOPA measurements correlate with dopamine cell counts measured in postmortem specimens (Pate et al., 1993; Snow et al., 1993).

Differential diagnosis is possible using PET. Atypical parkinsonian syndromes such as striatonigral degeneration (SND), progressive supranuclear palsy (PSP), hemiparkinson-hemiatrophy syndrome (HPHA), and corticobasal ganglionic degeneration (CBGD), which are often difficult to distinguish from PD on clinical grounds alone, have differential imaging signatures. In patients with early-stage idiopathic PD, FDOPA uptake is relatively preserved in the caudate and anterior putamen. By contrast, in patients with atypical parkinsonian syndromes such as multiple-system atrophy (MSA), equivalent impairment of FDOPA uptake can be observed in the caudate and the putamen (Brooks et al., 1990b).However, this differential dopaminergic topography is often insufficient to discriminate idiopathic PD from MSA at early clinical stages (Eidelberg et al., 1995a). Indeed, striatal FDOPA uptake can be reduced in other parkinsonian movement disorders such as SND, PSP, Wilson's disease (Snow et al., 1991), Guamanian amyotrophic lateral sclerosis (ALS-PD complex; Snow et al., 1990), and X-linked Filippino dystonia parkinsonism (Eidelberg et al., 1993b).

Asymmetrical parkinsonian syndromes such as HPHA and CBGD show relative reductions in basal ganglia FDOPA uptake contralateral to the affected side (Eidelberg et al., 1991; Przedborski et al., 1994). HPHA is a syndrome featuring early onset and slow progression associated with atrophy of the predominantly affected body side. CBGD is a progressive, asymmetrical, levodopa-resistant parkinsonism associated with lateralized cortical abnormalities in the form of apraxia, reflex myoclonus, parietal sensory disturbance, or pyramidal dysfunction.

2. Dopamine Transporter (DAT)

The development of radiotracers that bind to the striatal DAT has led to another means for directly imaging the nigrostriatal dopaminergic system with PET or single photon emission computed tomography (SPECT). The most extensively studied agents in this category are the cocaine analogs, such as 2β-carbomethyl-3β-(4-iodophenyl)tropane (βCIT) and its fluoroalkyl esters (Chaly et al., 1996). DAT is expressed on dopaminergic nigral terminals, and quantification of striatal DAT appears to be directly related to the extent of nigral cell degeneration (Wilson et al., 1996). Imaging of DAT may have advantages over FDOPA for several reasons:

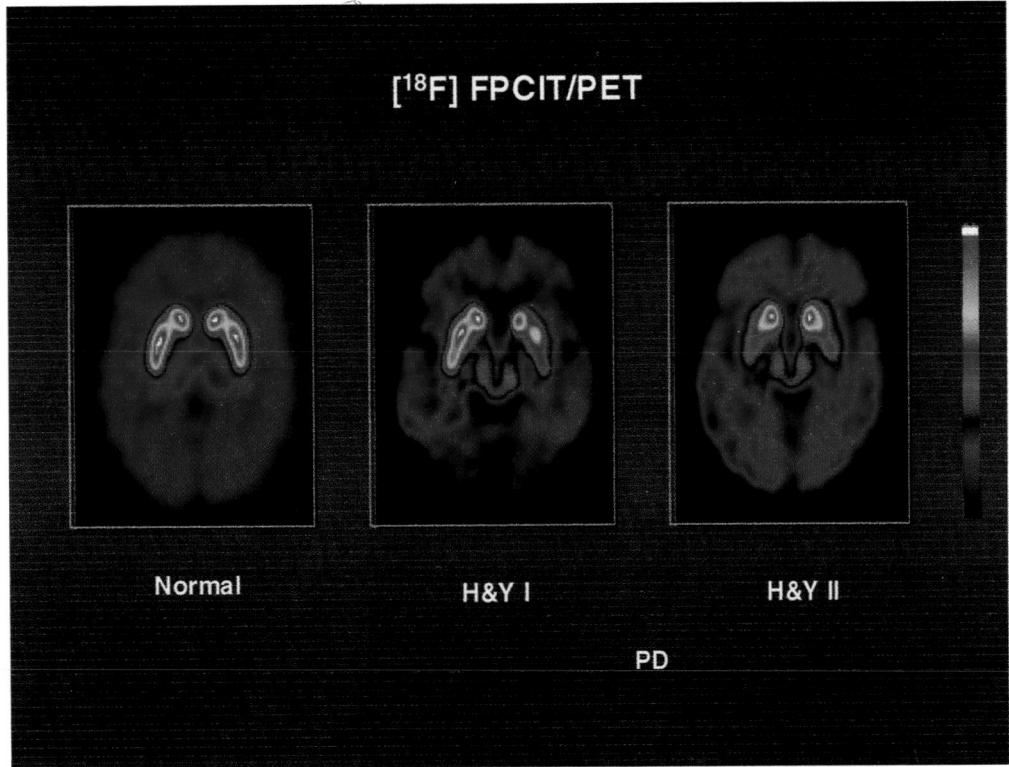

Figure 1 PET images obtained with [^{18}F]FP-β CIT in a normal volunteer (left), in a patient with Hoehn and Yahr Stage I PD (middle), and in a patient with Hoehn and Yahr Stage II PD (right).

(1) As dopaminergic neurons decline in normal senescence, FDOPA uptake may be maintained at a constant level by the upregulation of DDC activity, thus potentially making FDOPA/PET insensitive to age-related decrements in presynaptic dopaminergic function. Since DAT may not be as subject to upregulation as DDC (Kish *et al.*, 1995), the former may be a more sensitive marker for nigrostriatal cell loss in parkinsonism and normal aging (Ishikawa *et al.*, 1996c). (2) Transport of 3OMFD across the BBB may affect quantification of DDC activity, thus confounding results achieved using FDOPA/PET (Ishikawa *et al.*, 1996c). (3) Signal to noise is potentially higher for DAT imaging than for FDOPA/PET (Ishikawa *et al.*, 1996a,c). A number of recent SPECT studies have demonstrated the utility of DAT binding ligands as effective markers of nigrostriatal dopaminergic degeneration in parkinsonism (van Dyck *et al.*, 1995; Ishikawa *et al.*, 1996c; Assenbaum *et al.*, 1997). SPECT imaging with these tracers reliably differentiates PD subjects from normal volunteers, and the degree of striatal binding correlates with clinical measures of PD severity. In contrast to FDOPA/PET, both PET and SPECT measures of DAT binding decline with normal aging (van Dyck *et al.*, 1995; Ishikawa *et al.*, 1996c). Thus the DAT binding agents may have utility in

quantifying the attrition in nigrostriatal dopamine function that occurs in the course of the normal aging process. Nevertheless, this sensitivity may require the introduction of age corrections in longitudinal studies of disease progression in PD. Our laboratory has recently reported that PET imaging with N-3-fluoropropyl 2-β-carboxymethoxy-3-β-(4-iodophenyl) nortropane ([^{18}F]FP-βCIT) can provide images with higher resolution and better quantification than SPECT images acquired with the same ligand (Chaly *et al.*, 1996) (Fig. 1).

3. Vesicular Monoamine Transporter (VMAT)

The presynaptic VMAT is involved in the packaging and transport of monoamines to storage vesicles located in nerve terminals. Radioactive ligands that bind to VMAT sites such as [^{11}C]dihydrotetrabenazine can be used as a reliable measure of monoaminergic and nerve terminal density. Frey *et al.* (1996) reported a reduction in VMAT binding in the putamen of seven PD patients. An age-related decline of 7.7% per year was also observed in normal controls (Frey *et al.*, 1996).

VMAT binding appears not to be affected by antiparkinsonian dopaminergic medications such as levodopa, giving it potential advantage over FDOPA and DAT. However, this method has comparably low signal to noise and may not be specific for dopaminergic terminals (Stoessl and Ruth, 1998).

B. Postsynaptic Dopaminergic Function

Dopamine receptor bearing neurons constitute approximately 80% of the neuronal population in the striatum (Parent and Hazrati, 1995a,b). PET studies using ligands that bind selectively to D_1 and D_2 receptors in the striatum offer a quantitative approach to the measurement of dopamine receptor density in relation to progression of PD and pharmacological treatment with antiparkinsonian medication.

1. D_2 Ligands

Radioligands such as [^{11}C]raclopride (RAC) and [^{11}C] N-methylspiperone (CNMSP) can provide sensitive measures of local D_2 receptor density. In normal subjects RAC/PET studies have demonstrated a decrement of dopamine D_2 receptor binding of approximately 0.6% per year, suggesting that the striatal projection neurons may also progressively decline with normal aging (Antonini et al., 1997). By contrast, in PD the postsynaptic response to nigrostriatal deafferentation is likely to differ from that of normal aging. It has been suggested that loss of dopaminergic nerve terminals in association with changes in postsynaptic dopamine receptors may underlie motor complications occurring in the course of treatment of PD (Brooks et al., 1990b). A relative increase of striatal dopamine D_2 receptor binding has been reported in early untreated parkinsonian patients, particularly in the putamen contralateral to the more affected body side (Antonini et al., 1994). Relative dopamine D_2 receptor upregulation has also been demonstrated in the striatum of subjects with 1-methyl-4-phenyl-1,2,6-tetrahydropyridine (MPTP)-induced parkinsonism (Perlmutter et al., 1987). However, the initial dopamine D_2 receptor upregulation in the putamen may reverse with increasing disease severity and binding values in the range of control subjects or lower may be encountered in advanced PD patients (Brooks et al., 1992; Antonini et al., 1997). Because RAC and FDOPA changes are associated throughout the disease course, it is likely that dopamine D_2 receptor changes result from the decline in presynaptic dopaminergic drive (Antonini et al., 1995).

Studies using CNMSP/PET show conflicting results. Shinotoh et al. (1993a) showed no difference in striatal D_2 receptor binding in untreated and treated PD subjects, whereas in earlier studies, Rutgers et al. (1987) found striatal D_2 binding to be significantly higher bilaterally (close to normal range) in PD patients undergoing dopaminergic treatment than in untreated patients. RAC, although a widely used D_2 tracer, is easily

displaced by endogenous dopamine whereas CNMSP has a higher affinity for D_2 receptors and is not as easily displaced. Stoessl and Ruth (1998) raise the possibility that an increase or decrease in RAC binding may reflect endogenous dopamine levels instead of D_2 receptor density. Nevertheless, this attribute may allow RAC to be utilized with pharmacological and behavioral activation as a means of assessing endogenous dopamine levels. This novel application may provide important new information regarding the dynamic functions of the intact nigrostriatal dopamine terminals in PD and other movement disorders (Leenders et al., 1985).

2. D_1 Ligands

PET studies using [^{11}C]SCH23390, [^{11}C]SCH39166, and [^{11}C]NNC756 show no change in striatal D_1 receptor binding in PD. Rinne et al., (1990) studied D_1 receptor density in untreated early-stage PD patients using [^{11}C]SCH23390/PET. Results showed normal striatal binding (Rinne et al., 1990). Shinotoh et al. (1993b) found similar results using SCH23390 in 10 PD patients, half of whom were chronically treated. Both the treated and untreated PD groups showed normal D_1 binding in the caudate and putamen; 5 SND patients displayed a significant reduction in the posterior putamen (Shinotoh et al., 1993b).

[^{11}C]SCH39166 and [^{11}C]NNC756 are relatively new tracers. In a PET study by Laihinen et al. (1994), both of these ligands were found to be useful in measuring D_1 receptor binding. In their study of four normal subjects and eight early-stage PD subjects, no significant differences in D_1 binding were found in either group, supporting the findings in [^{11}C]SCH23390 studies.

In a PET study measuring D_1 receptor density in PD patients treated with levodopa, both caudate and putamen uptake was reduced. A greater (although nonsignificant) reduction was seen in patients who had developed dyskinesias (Rinne et al., 1990). The ultimate utility of D_1 neuroreceptor quantification with PET in the study of PD remains unknown. It is likely that a better understanding of the mechanism of the levodopa response, including the potentiation of dyskinesias, will emanate from the quantification of both D_1 and D_2 striatal neuroreceptors in individual patients.

In summary, D_1 binding appears to be normal in the caudate and putamen of PD patients undergoing dopaminergic therapy, and may be reduced in the caudate and putamen in PD patients with treatment. Results of D_2 binding studies show normal to slightly increased striatal binding levels in untreated PD patients. With treatment, putamenal levels appear to return to normal while caudate binding decreases (Shinotoh et al., 1993a).

III. Functional Brain Imaging in the Resting State

In addition to quantifying neurochemical abnormalities in PD, PET can be used to identify alterations of regional cerebral glucose metabolism (rCMRGlc) and blood flow (rCBF). [^{18}F]Fluorodeoxyglucose (FDG) and [^{15}O]H$_2$O/PET studies provide unique information regarding the topography of widespread functional changes in the brains of PD patients, which is unavailable from studies with dopamine uptake markers or D$_2$ ligands.

A. Metabolism and Blood Flow Studies

FDG/PET may be useful for quantifying nigrostriatal function as well as for assessing the consequences of nigrostriatal dopamine loss on the functional organization of the basal ganglia. Although the primary pathological abnormality in PD is confined to the substantia nigra, the degeneration of dopaminergic projection neurons from the substantia nigra to the striatum results in widespread alterations in the functional activity of the basal ganglia (Alexander et al., 1986, 1990; Parent and Hazrati, 1995b). Specifically, the functional organization of the basal ganglia predicts that the loss of inhibitory dopaminergic input to the striatum results in increased inhibitory output from the putamen to the external globus pallidus (GPe), diminished inhibitory output from the GPe to the subthalamic nucleus (STN), and functional overactivity of the STN and internal globus pallidus (GPi), resulting in decreased output from the ventrolateral thalamus to the cortex. These functional alterations in basal ganglia activity are accompanied by alterations in regional cerebral glucose metabolism and blood flow.

Increased tonic neuronal activity has been shown in the medial pallidum and STN, and corresponding metabolic alterations are found in the same structures in animal models of parkinsonism (Filion et al., 1991; Bergman et al., 1994). Over the past decade, FDG/PET has been employed to identify alterations in regional brain function in living patients with PD analogous to those discerned in experimental animal models (Takikawa et al., 1994). Nonetheless, differences in technical and data analytical approaches between authors have given rise to inconsistent results. Martin et al. (1987) described contralateral increases in pallidal rCMRGlc in 2/4 hemiparkinsonian patients, although anatomical localization was not demonstrated conclusively. Bilaterally affected patients were found to have globally reduced perfusion with preserved oxygen metabolism (CMRO$_2$) (Eidelberg et al., 1997), whereas in another study, basal ganglionic rCMRGlc was elevated

(Rougemont et al., 1984). In experimental animal models of PD, unilateral nigrostriatal lesions appear to result in raised rCMRGlc in the ipsilateral globus pallidus, as measured with [^{14}C]2-deoxyglucose autoradiography (Mitchell et al., 1994). Blesa et al. (1991) reported that pallidal metabolism in patients with PD may be reversed by administration of levodopa. This suggests that lentiform hypermetabolism in PD may be in part reversible and that these findings in FDG/PET scans may be an indication for neuroprotective agents or surgery.

FDG/PET is a powerful tool to differentiate idiopathic PD from atypical parkinsonian syndromes (Antonini et al., 1995). Subjects with atypical parkinsonism such as SND, PSP, HPHA, and CBGD show significantly reduced rCMRGlc in the striatum as compared with normals as well as duration- and severity-matched PD patients (Eidelberg et al., 1993a). FDG/PET studies in PSP consistently demonstrated extensive hypometabolism most dominantly in the frontal cortex together with reduced striatal metabolism (Foster et al., 1988). In HPHA patients, we found unilateral hypometabolism in the basal ganglia or frontal cortex contralateral to the affected body side, while rCMRGlc measures ipsilateral to the affected body side were normal (Przedborski et al., 1994). In CBGD, FDG/PET demonstrated reduced global metabolic rate and significant metabolic asymmetries in the thalamus and inferior parietal lobes as compared with both normals and hemi-PD controls (Eidelberg et al., 1991). Because of its sensitivity in differential diagnosis, FDG/PET is increasingly utilized in selecting candidates for surgical procedures (Kazumata et al., 1997; Antonini et al., 1998a).

[^{15}O]H$_2$O/PET has been used to measure rCBF in PD patients at rest. Perlmutter and Raichle (1985) used [^{15}O] H$_2$O to examine rCBF in hemiparkinsonian patients before and after administration of levodopa. rCBF measurements before levodopa showed an abnormal asymmetry of pallidal blood flow in four such patients; rCBF was contralaterally reduced in three patients. After levodopa administration, pallidal blood flow returned to normal levels (Perlmutter and Raichle, 1985). Wolfson et al. (1985) reported an asymmetry in rCBF and rCMRO$_2$ in unilateral PD patients. Although within normal range, there was a 10% coupled increase in rCBF and rCMRO$_2$ in the basal ganglia contralateral to the symptomatic side of hemiparkinsonian patients (Wolfson et al., 1985).

B. Metabolic Network Analyses

In most degenerative neurologic conditions, rCBF and rCMRGlc are coupled processes and relate to synaptic activity within a given brain region. The measurements of local rates of metabolism or regional activation responses may not fully describe the complexities of neural systems involved in a neurodegenerative process and their modulation with treatment. These networks may be represented as patterns of covariation among spatially distributed brain regions and can be altered by behavioral activation or the presence of disease. Principal component analysis (PCA) has been used in the analysis of PET data to contrast groups in the same resting state (Moeller and Strother, 1991) and, more recently, in brain activation paradigms (Friston *et al.*, 1990, 1995; Moeller and Strother, 1991).

We have developed a statistical modeling approach for the detection and quantification of regional functional interactions in disease states (Moeller *et al.*, 1987; Moeller and Strother, 1991). This approach, known as the Scaled Subprofile Model (SSM; Moeller and Strother, 1991; Alexander and Moeller, 1994; Moeller *et al.*, 1996), is a general form of the two-way factor analysis of variance model (FANOVA) (Yochmowitz, 1982). [The mathematical relationship of SSM to other statistical approaches such as ANOVA has been described elsewhere (Friston *et al.*, 1990; Alexander and Moeller, 1994).] In SSM modeling, PCA is employed to identify regional metabolic covariance patterns from rCMRGlc data sets obtained from combined samples of patients and normals. This form of analysis is blind to subject class designation and utilizes the variance across the entire population (normals *and* patients) to identify specific patterns associated with the disease state. These patterns reflect covarying regional increases or decreases in brain function relative to a baseline defined by the normal population.

The topographies of these covariance patterns correspond closely to specific physiological and anatomical regional networks known to be involved in disease processes (Alexander and Moeller, 1994; Eidelberg *et al.*, 1995a, 1996, 1997) and are highly reproducible across patient populations and tomographs (Eidelberg *et al.*, 1994; Moeller *et al.*, 1996, 1999). SSM also allows for the quantification of intersubject differences in covariance pattern expression. Subject scores (PCA scalars) for regional covariance patterns can be computed on a prospective individual case basis from functional brain images and can be correlated with individual differences in independently measured clinical or physiological indices (Eidelberg *et al.*, 1995b; Moeller *et al.*, 1996). Additionally, in contrast to neurochemical dopaminergic markers such as FDOPA or βCIT, SSM subject scores have the attribute of increasing signal to noise with disease progression. This property allows for potentially greater sensitivity in the detection of temporal changes in brain network expression. Thus, SSM is ideally suited for the study of intersubject variability in

brain organization as might occur in neurodegenerative processes (Ford, 1995).

A reproducible pattern of regional metabolism characterized by increased lentiform and thalamic metabolism is associated with reduced metabolism in the lateral frontal, paracentral, and parieto-occipital areas that covary in PD (Moeller *et al.*, 1999; Fig. 2). In addition, the degree of individual expression of this PD-related covariance pattern (PDRP) correlates with the disease severity and independent PET measurement of striatal FDOPA uptake in PD (Eidelberg *et al.*, 1990, 1994, 1995a,b). These results are in keeping with the hypothesis that specific elements of cortico–striato–pallido–thalamo–cortical (CSPTC) motor circuits (Alexander *et al.*, 1990; DeLong, 1990) are functionally abnormal in PD. A reduction of nigrostriatal dopaminergic activity results in increased functional activity in the putamen and STN. Increased subthalamic activity promotes the action of the inhibitory pallidothalamic pathway.

Considerable efforts have been made to validate and implement network imaging methods for the diagnosis and evaluation of patients with parkinsonism and related movement disorders. The application of SSM/PCA to PET data reveals a significant pattern of regional metabolic covariation characterized by lentiform and thalamic hypermetabolism associated with regional metabolic decreases in the lateral frontal and paracentral cortical areas [corresponding to the lateral premotor cortex and the supplementary motor area (SMA) (Eidelberg *et al.*, 1994)]. Subject scores for this pattern are abnormally elevated in PD patients (Eidelberg *et al.*, 1990, 1994, 1995b). These PET measures correlate positively with standardized clinical bradykinesia ratings and negatively with striatal FDOPA uptake rate constants. This pattern of regional metabolic covariation was confirmed (Eidelberg *et al.*, 1994) in a totally different age-matched cohort of 22 PD patients and 20 normals with a higher resolution PET instrument (transaxial resolution 6.5-mm FWHM). Again, individual subject scores for this profile are significantly elevated in PD patients as compared with normals and correlate significantly with bradykinesia and rigidity ratings. Recently, the reproducibility of the PDRP in three additional independent groups of PD patients has been reconfirmed, each studied at separate institutions with different tomographs of higher resolution (Moeller *et al.*, 1999).

Importantly, the topography of the PDRP identified in these resting-state studies is consistent with experimental models of parkinsonism (Eidelberg *et al.*, 1994). Specifically, the reproducible findings of relative lentiform-thalamic hypermetabolism in PD, associated with motor cortical hypometabolism, support the hypothesis of excessive pallidothalamic inhibition as the main functional substrate of parkinsonian bradykinesia

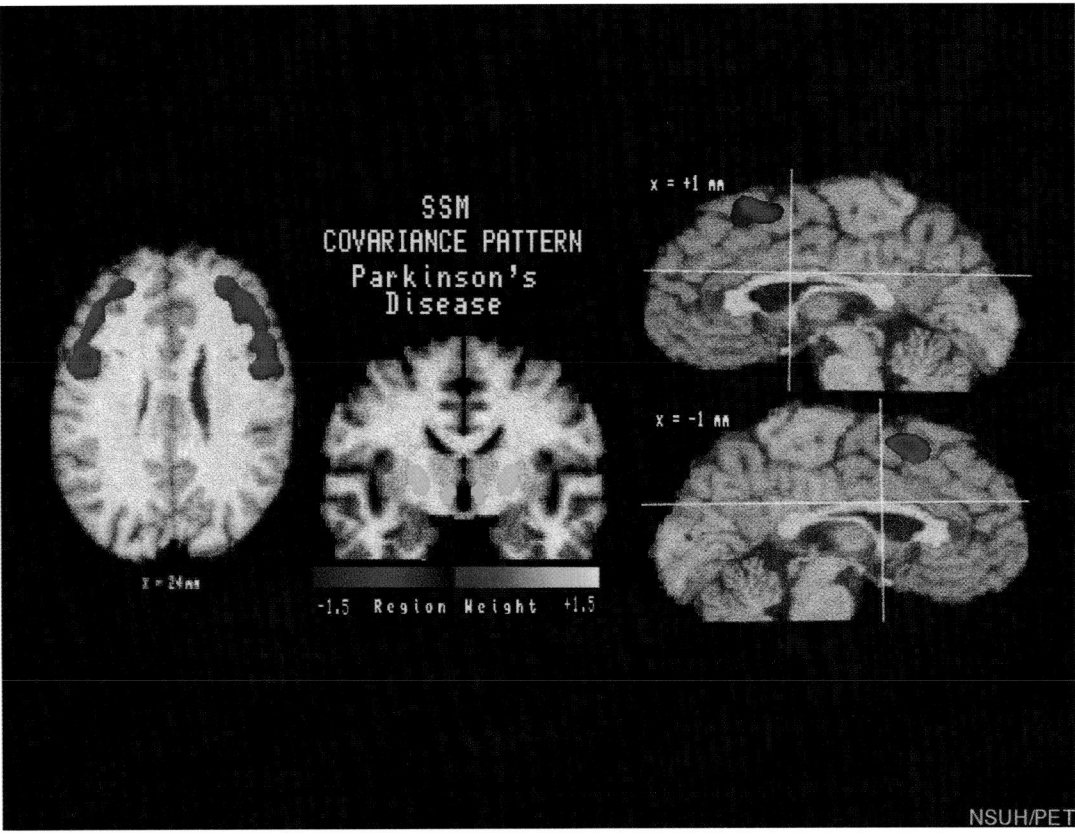

Figure 2 Display of the regional metabolic covariance pattern associated with PD identified using the SSM. Region weights for this disease-related pattern have been overlaid on standard Talairach-transformed MRI sections. The pattern is characterized by relative lentiform and thalamic hypermetabolism (yellow) covarying negatively with bilateral metabolic reductions (blue) in motor and premotor regions and in the supplementary motor area (SMA).

(DeLong, 1990). Subject scores for this pattern correlate with objective disease severity ratings and with independent FDOPA/PET measures of nigrostriatal dopamine function.

We developed a modification of SSM/PCA to compute subject scores for a predetermined topographic profile from individual PD patient rCMRGlc data on a prospective case-by-case basis (Eidelberg *et al.*, 1995a,b; Moeller *et al.*, 1996). This computational algorithm, referred to as Topographic Profile Rating (TPR), is critical to the clinical application of network analysis in disease severity assessment and differential diagnosis. TPR has been used to compute subject scores for the PDRP in four independent populations of PD patients and normals scanned on different tomographs (Moeller *et al.*, 1999). Results showed that subject scores for this pattern accurately discriminated PD patients from normals in all four populations, with a sensitivity of 75–85% ($p < 0.001$). Thus, in addition to being reproducible across populations, this unique pattern can be used as an accurate marker for the differential diagnosis of parkinsonism.

SSM network imaging has also been applied to study the utility of FDG/PET in the diagnosis of PD at its earliest clinical stages (Hoehn and Yahr Stage I). Early-stage PD (EPD) is associated with a distinctive covariance pattern of regional metabolic right–left asymmetries that is also reproducible across study populations and scanning instruments (Eidelberg *et al.*, 1990, 1994, 1995b). The topography of this pattern is characterized by covariate lentiform and thalamic metabolic asymmetries. To assess the clinical utility of this pattern in early differential diagnosis, TPR was used to compute individual subject scores in a prospective cohort of 14 successive patients with early-stage parkinsonism. TPR computations were performed in an automated blinded fashion. Categorization into normal or disease groups by subject score was compared with discrimination based upon independent FDOPA/PET measurements obtained in the same subjects. SSM subject scores were found to accurately discriminate EPD patients from normals as well as atypical early-stage patients who were levodopa resistant (Eidelberg *et al.*, 1995a). Comparable discrimination of EPD patients from normals was achieved in the same subjects with FDOPA/PET, although levodopa-responsive and -resistant early-stage parkinsonians could only be distinguished by SSM network expression, and not by caudate and/or putamen FDOPA uptake.

SSM/PCA has also been used to identify the network correlates of specific features of PD. Antonini *et al.* (1998b) studied a cohort of 16 PD patients consisting of individuals with and without appreciable tremor. Tremulous and atremulous patients were matched for age and disease severity. SSM analysis revealed that the tremor-predominant patients expressed a specific brain network (Fig. 3). This tremor pattern was characterized by thalamocortical covariance, which was topographically different from the PDRP described above. The latter pattern, associated mainly with pallidothalamic output, was expressed equally in tremulous and atremulous patients (Antonini *et al.*, 1998b). Recently, FDG/PET studies using SSM analysis have revealed specific network topographies associated with cognitive abnormalities in PD patients (Mentis *et al.*, 1999).

In SSM analyses of metabolic data from normal volunteers (Moeller *et al.*, 1996), we have identified several covariance topographies associated with normal aging. Specifically, we found that the age-related SSM pattern identified in a cohort of 20 subjects predicted chronological age in 130 other volunteers scanned at another

laboratory as well as in 15 subsequent normal subjects scanned on a second, newer PET camera. Moreover, subject scores for this pattern were also found to be highly reproducible in 22 volunteers who were restudied (Moeller *et al.*, 1996). These results demonstrate that metabolic networks identified by SSM analysis can serve as useful imaging markers in PET studies of the normal aging process. Because the normal aging patterns and the PDRP are topographically discrete (Moeller *et al.*, 1996), we sought to use these patterns as independent functional network markers to determine the role of the normal aging process in the evolution of PD. Having identified and validated a quantifiable aging-related pattern in normals, we prospectively calculated the expression of this marker in 37 PD patients and 20 normals. We used network analysis to compute the metabolic age for each subject, i.e., the age predicted solely by the expression of the aging pattern in the FDG/PET data of that individual. We found that in the normals, real age and metabolic age were similar and that the difference between the two values (defined as Δ) approximated zero. By contrast, we noted a significant underestimation of age by FDG/PET in the PD cohort, suggesting a disruption of the normal age–metabolism relationship. Additionally, we found that in PD the magnitude of Δ correlated with disease dura-

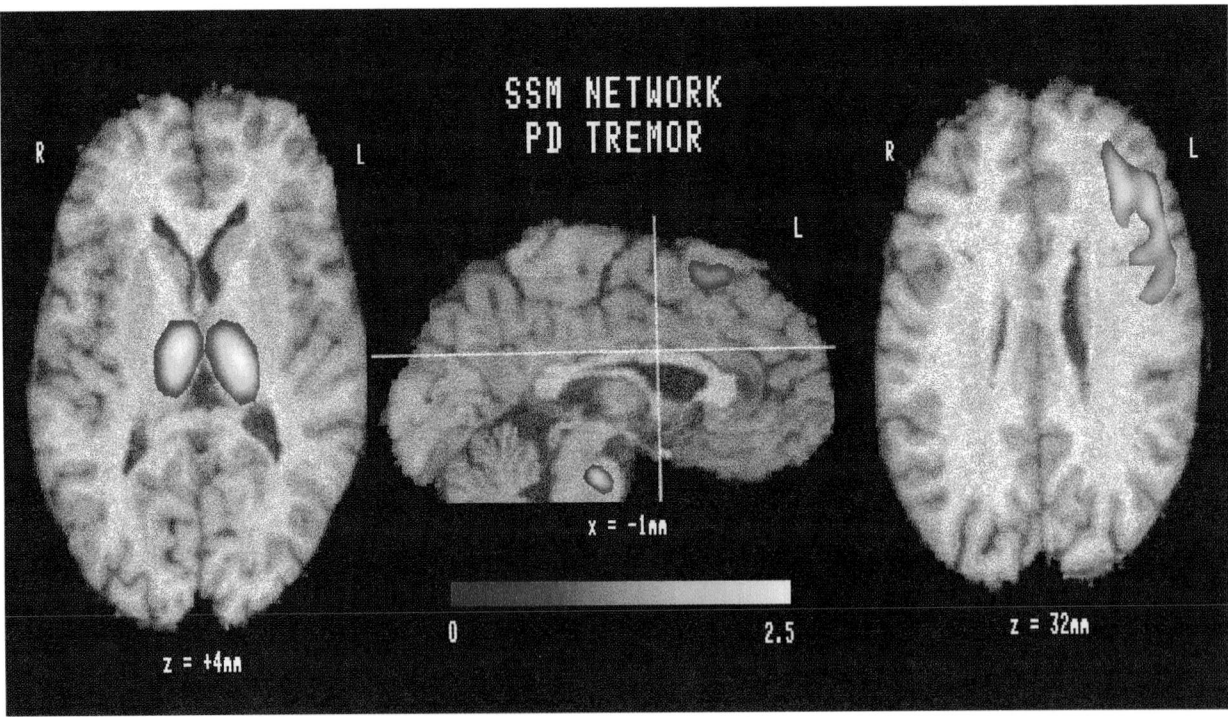

Figure 3 Display of the region weights of the SSM topography associated with parkinsonian tremor, overlaid on a normal MRI scan. Relative hypermetabolism (scale) of the thalamus (transverse, left) is associated with covariate metabolic increases in the pons and in the left paracentral region (sagittal) and in the left lateral frontal cortex (transverse, right).

tion. Extrapolation of this linear relationship to the normal Δ value of zero yielded an estimate of the preclinical period of approximately 4.5 years (Fig. 4). Indeed, this cross-sectional estimate of the parkinsonian preclinical period is in close agreement with other estimates derived through longitudinal dopaminergic PET measurements (Moeller *et al.*, 1996). These findings demonstrate how metabolic network imaging allows for the development of comprehensive models of disease progression that may be useful in understanding the natural history of neurodegenerative processes.

IV. Brain Activation Studies: Motor Execution and Learning

[^{15}O]H$_2$O/PET activation studies have also been used to investigate the role of volitional motor planning and motor execution (Goldman-Rakic, 1987; Thaler and Passingham, 1989; Mushiake *et al.*, 1990; Goldberg, 1995). Playford *et al.* (1992) studied rCBF in 6 PD and 6 normal controls who were scanned while performing different motor tasks using a joystick. The PD subjects showed reduced rCBF in the contralateral putamen, anterior cingulate, SMA, and dorsolateral prefrontal cortex (DLPFC; Playford *et al.*, 1992). Jahanshahi *et al.*

(1995) showed similar results while PD subjects performed a self-initiated finger extension task. When compared to normals, the PD group showed lower activation in the SMA, anterior cingulate, left putamen, left insular cortex, right DLPFC, and right parietal area 40. Hypoperfusion in these areas may play a part in the initiation of movement that many PD patients find difficult (Playford *et al.*, 1992).

Catalan *et al.* (1999) examined motor sequence learning in 13 PD patients and 13 normal control subjects. Subjects performed simple repetitive movements, increased complexity of sequential finger movements, and self-selected movements. Results showed similar activation patterns in sequential finger movements in both groups; however, the PD group showed increased rCBF in the precuneus, premotor, and parietal areas. With increasing complexity of sequential movements, PD patients and normals showed greater increases in precuneus, premotor, and parietal areas, with additional increases in the anterior SMA and cingulate shown in the PD group. Lastly, in the self-selected condition, increased rCBF in the anterior SMA and cingulate was only shown in the normal controls. The authors conclude that in PD patients, more cortical areas are recruited to perform sequential finger movements, resulting from increasing corticocortical activity to compensate for striatal dysfunction (Catalan *et al.*, 1999).

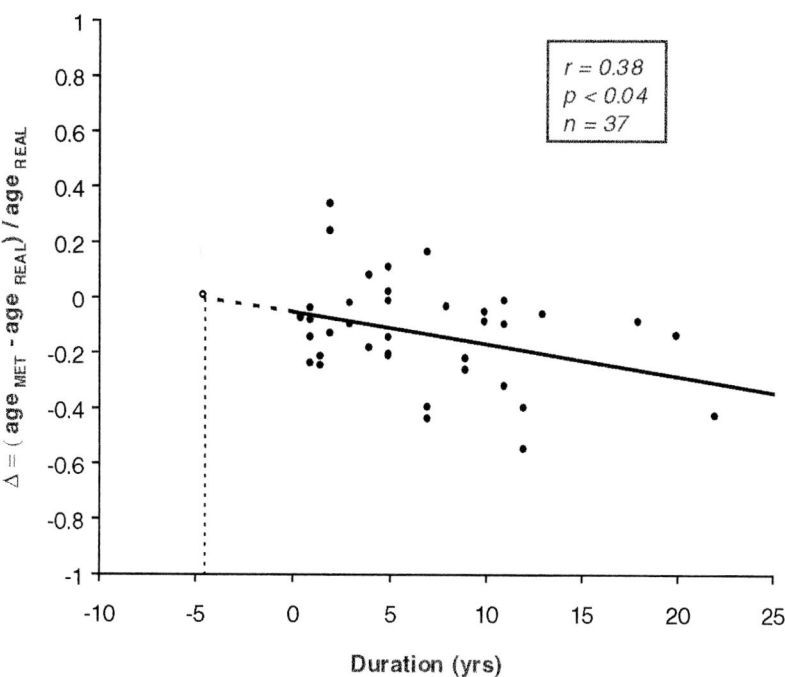

Figure 4 Correlation between Δ and disease duration in 37 PD patients (closed circles). A significant negative correlation ($r = -0.38, p < 0.04$) was found between these variables. Linear extrapolation (dashed line) to the expected normal mean value (Δ = 0, open circle) yielded a mean estimate of the metabolic preclinical period of 4.5 years.

V. Functional Brain Imaging in the Assessment of Therapeutic Interventions

A. Symptomatic Pharmacology: Effects of Dopaminergic Therapy on Brain Function

The pharmacotherapy of PD has been aimed at improving dopaminergic neurotransmission (Lang and Lozano, 1998) through the administration of levodopa, dopamine receptor agonists, or both. Recent advances in the understanding of the underlying pathophysiology of these neurodegenerative processes have led to the development of new therapeutic strategies aimed at slowing the progression of the disease. To assess these new therapies, reliable and accurate markers of disease progression are needed.

The mechanism of action of dopaminergic pharmacotherapy is not fully known. PET has been applied to the measurement of regional functional changes associated with successful drug therapy and its complication. The effect of levodopa on brain metabolism and cerebral blood flow has been of considerable interest to researchers. Rougemont *et al.* (1984) used FDG/PET to measure rCMRGlc in four early-onset PD patients with akinesia. The patients were scanned twice—once without treatment and once with levodopa treatment. Results showed marked clinical improvement in symptoms but no change in rCMRGlc between the two conditions (Rougemont *et al.*, 1984). Broussolle *et al.* (1993) had similar results when measuring rCMRGlc after using a dopamine agonist (apomorphine) to relieve akinesia in 14 advanced-stage PD patients with severe on-off fluctuations. Results showed an improvement in clinical symptoms but no change in rCMRGlc (Broussolle *et al.*, 1993). Leenders *et al.* (1985) examined the effect of an acute administration of levodopa on rCBF in 18 PD patients compared to 6 normal controls. Results showed a small increase in rCBF in the basal ganglia but no change in oxygen metabolism, indicating that levodopa does not alter brain metabolism. After several weeks of levodopa treatment, clinical improvement remained even though rCBF returned to baseline levels (Leenders *et al.*, 1985).

Perlmutter and Raichle (1985) used PET to measure global and regional blood flow in 11 patients with hemiparkinsonism and in 26 normal subjects. The PD group was scanned before and after an acute oral dose of levodopa. Results before levodopa administration showed no significant difference in global blood flow between the PD and normal groups. However, without medication, a reduction in mesocortical blood flow contralateral to the patients' symptoms was seen in the PD group; pallidal blood flow was less tightly coupled in pa-

tients than controls. Results after levodopa administration showed the mesocortical blood flow remained below normal levels in the PD group while no difference was seen in pallidal blood (Perlmutter and Raichle, 1985).

Hershey *et al.* (1998) studied rCBF response to levodopa with PET in patients with dopa-induced dyskinesias (DID) due to chronic levodopa exposure. Results showed a greater response in the ventrolateral thalamus of the DID group as compared to 10 chronically treated PD patients without DID, 17 dopa-naïve PD patients, and 11 normals. The authors suggest that these findings are consistent with increased inhibitory output from the internal segment of the globus pallidus (GPi) to the thalamus after levodopa administration (Hershey *et al.*, 1998). In an interesting study of DID with ^{133}Xe/SPECT, Rascol *et al.* (1998) investigated rCBF changes in 15 PD patients with DID, 23 PD patients without DID, and 14 normals who executed a finger-to-thumb opposition task during scanning. The authors found that the PD patients with DID overactivated the SMA and primary motor areas (Rascol *et al.*, 1998). In contrast to the above, these findings suggest that DID may be related to decreased inhibitory output of the GPi, resulting in concomitant functional activation of the primary and associated motor cortex (Hershey *et al.*, 1998).

B. Stereotaxic Surgical Therapies

1. Pallidotomy/Thalamotomy

Reliable *in vivo* markers of neuronal activity are needed to assess surgical outcome. Currently available clinical scales are relatively insensitive and inherently variable, and may not accurately reflect the extent of neuropathological change. By contrast, quantitative functional brain imaging markers may be suitable as outcome measures for the surgical treatment of PD. Indeed, we have found that PET may serve as a useful tool in predicting optimal candidates for certain surgical interventions. Posterolateral ventral pallidotomy has been shown to significantly improve akinetic symptoms in PD as well as to relieve dyskinesia associated with levodopa administration (Laitinen *et al.*, 1992; Lozano *et al.*, 1995). The pathophysiology of pallidal ablation for the relief of parkinsonism is not completely understood. The ameliorative effects of pallidotomy have been attributed to the reduction of excessive inhibitory outflow from the GPi (Marsden and Obeso, 1994).

We originally reported eight PD patients undergoing pallidotomy who were scanned with FDG/PET preoperatively and 6 months postoperatively (Eidelberg *et al.*, 1997). We found that pallidotomy resulted in a metabolic decline in the thalamus, which occurred in conjunction with a metabolic increase in primary and associative motor cortical regions. Indeed, the improvement in limb performance at 6 months following surgery was significantly correlated with the operative metabolic decline in the thalamus and with the accompanying increases in lateral premotor cortex. To quantify potential modulations in the expression of motor networks by pallidotomy, we applied SSM/PCA to operative differences in regional glucose metabolism. We found that the topography identified in this analysis closely resembled the PDRP identified previously (Eidelberg *et al.*, 1990, 1995a) and was characterized by a postoperative decline in the lentiform and thalamic metabolism ipsilateral to the surgical side associated with bilateral increases in SMA metabolism. The individual expression of this pattern of metabolic operative change correlated significantly with improvements in both contralateral and ipsilateral limb CAPIT scores. These findings indicate that metabolic brain networks comprising functionally and anatomically interconnected brain regions remote from the lesion site may be modulated by pallidotomy, including motor cortical regions of the hemisphere contralateral to the surgery.

Because local rates of glucose are a reflection of net afferent synaptic activity (Mata *et al.*, 1980), it is reasonable to expect a decline in thalamic metabolism subsequent to surgical interference with pallidofugal inhibitory projections to the thalamus. Similarly, the metabolic increases in motor cortical areas occurring with pallidotomy may be related to enhanced cortical afferent synaptic activity from the ventral thalamus following surgical reduction in pallidothalamic inhibitory output (Alexander *et al.*, 1990; DeLong, 1990; Mitchell *et al.*, 1994; Parent and Hazrati, 1995b). Indeed, we have recently shown that spontaneous GPi single-unit activity recorded intraoperatively during pallidotomy correlated significantly with preoperative measures of thalamic glucose utilization obtained in the same patients under comparable behavioral conditions (Eidelberg *et al.*, 1997). This physiological–metabolic relationship was reproduced in the subgroups of patients scanned on different PET cameras. Moreover, we found that GPi firing rates were also significantly correlated with the expression of an SSM network related to the pallidum and its major efferent projections (Fig. 5). It is therefore likely that pallidal ablation may exert its primary metabolic effect in spatially distributed projection fields lying in the ventral tier and intralaminar thalamic nuclei as well as the brainstem.

PET activation studies using [¹⁵O]H₂O have also supported the notion of pallidotomy-induced modulation of the CSPTC motor circuit. Subtraction of resting-state data from task-specific data allows isolation of changes in rCBF relating to various motor tasks (Friston *et al.*, 1995). Grafton *et al.* (1995) reported postpallidotomy increases in rCBF in the ipsilateral premotor and in the SMA with movement. In another study employing network analysis of motor system connectivity, Grafton and colleagues (1994) found significant postoperative reductions in the strength of interactions between the globus pallidus and thalamus and the thalamus and mesial frontal motor area. These findings are consistent with the notion that pallidal ablation induces alterations in the normal functioning of CSPTC motor networks (DeLong, 1990; Bathia and Marsden, 1994; Marsden and Obeso, 1994).

In addition to adding to our understanding of the network modulations occurring with pallidotomy, FDG/PET may be used to select optimal candidates for this procedure. In our original study of 10 patients undergoing pallidotomy, we found that preoperative FDG/PET measurements of lentiform metabolism in the off state correlated with clinical outcome up to 6 months following surgery (Eidelberg *et al.*, 1996). We subsequently studied an additional cohort of 22 PD patients to assess the usefulness of preoperative FDG/PET, quantitative motor performance indices as well as MRI measurements of lesion size and location as potential predictors of surgical outcome (Kazumata *et al.*, 1997). We found that the pallidotomy lesions were comparable in location and size in all patients, and therefore did not correlate with individual differences in surgical outcome. Nonetheless, in this series we confirmed that preoperative measures of lentiform glucose metabolism offered an accurate prediction of ultimate surgical improvement.

The finding of a significant correlation between levodopa response and clinical outcome following surgery suggests that the dynamic range of the CSPTC motor loop with pallidal suppression may be clinically estimated using a levodopa challenge. These findings have been supported by FDG/PET experiments in PD patients showing reduction of pallidal hypermetabolism during levodopa infusion (Blesa *et al.*, 1991). To the extent that the severity of clinical manifestations is correlated with the expression of abnormal CSPTC metabolic networks, the measured clinical benefit with levodopa administration may provide a simple indicator of the dynamic range of network modulation that can occur with either pharmacological or surgical pallidal suppression. In this vein, we found that preoperative measurements of lentiform metabolism with FDG/PET reproducibly predict approximately 50% of operative improvement in off-state CAPIT scores. The preoperative levodopa responsiveness index, though simpler to

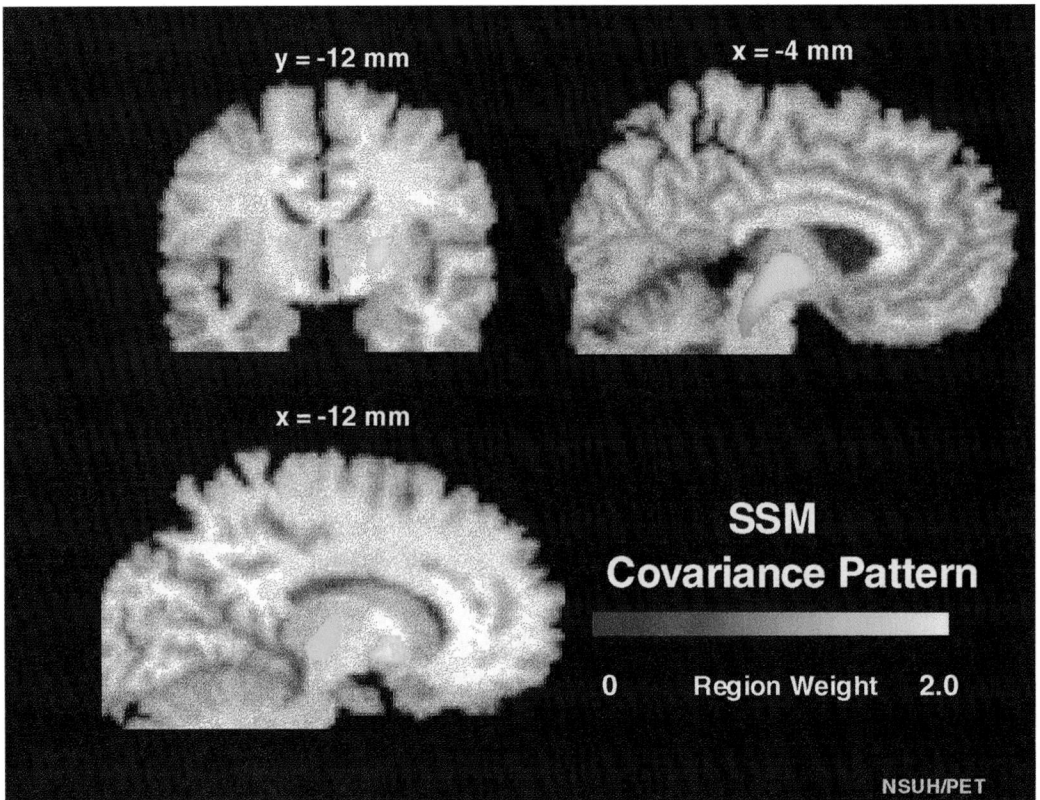

Figure 5 Display of the regional covariance pattern identified by the SSM in 18 PD patients. Region weights have been mapped on standardized Talairach MRI sections. This pattern was characterized by a network of regional metabolic covariation involving the putamen, pallidum, and thalamus (left) as well as the midbrain and pons (right). Subject scores for this network were highly correlated with intraoperative recordings of spontaneous internal pallidal firing rate.

measure, predicted approximately 36% of individual differences in outcome. Together, however, both preoperative measures predicted approximately 70% of the pallidotomy response. The combination of preoperative FDG/PET measurements of lentiform metabolism together with clinical–pharmacological estimates of the patient's individual capacity for network modulation can provide useful and complementary criteria for patient selection for pallidotomy.

Stereotactic thalamotomy has been shown to significantly improve drug-resistant tremor in Parkinson's disease. The thalamic ventralis intermedius nucleus (Vim) is targeted in thalamotomy as it relays cerebellar and proprioceptive output to sensorimotor cortical areas. Using $[^{15}O]H_2O$/PET, Boecker *et al.* (1997) observed significant operative declines in sensorimotor, premotor, and parietal rCBF both at rest and during motor activation in two PD patients who underwent thalamotomy for tremor. This supports the notion that parkinsonian tremor arises through the overaction of ventral thalamic projections to cortical motor regions

(Antonini *et al.*, 1998b). These findings indicate that in relieving PD tremor, thalamotomy may also alter functional input from the surgical target to sensorimotor cortical regions. Nevertheless, a comprehensive network modeling approach to define the mechanism of surgical improvement using imaging awaits future investigation.

2. Deep-Brain Stimulation

Deep-brain high-frequency stimulation (DBS) has the advantage of avoiding permanent side effects due to the ablative lesion, therefore inducing the possibility of a reversible amelioration of parkinsonian symptoms. In addition, this technique is adaptable as the stimulation frequency can be increased or decreased. Inhibitory DBS of the ventral portion of the thalamus, usually Vim, was originally employed for the control of parkinsonian resting tremor. More recently, DBS has been applied to the GPi to improve rigidity and bradykinesia in PD (Sigfried and Lippitz, 1994) and to the STN to improve rigidity, akinesia, and tremor (Krack *et al.*, 1998). Parker *et al.* (1992) investigated the effect of thalamic stimulation. They observed that rCBF increased in the sensorimotor, premotor, supplementary motor area, caudate nucleus, and cerebellar vermis and hemisphere during the tremor period compared with the period of tremor arrest. Deiber *et al.* (1993) observed that suppression of tremor induced by sufficient stimulation specifically associated with the reduction of cerebellar rCBF, whereas the incomplete arrest of tremor induced by insufficient stimulation only reduced rCBF in frontal cortex. They suggest that tremor suppression is mainly associated with a decrease of synaptic activity in the cerebellum (Deiber *et al.*, 1993).

Davis *et al.* (1997) examined rCBF and parkinsonian symptoms in patients who had Vim DBS implants. Using [^{15}O]H$_2$O/PET , patients were scanned at rest under three stimulation conditions ("off," "suboptimal," and "optimal" stimulation). They found a decrease in rCBF in the SMA, contralateral cerebellum, cingulate motor area, and the ipsilateral sensory association areas during thalamic stimulation. An increase in rCBF was seen in bilateral frontal and ipsilateral occipital regions and in the DLPFC during Vim stimulation (Davis *et al.*, 1997).

[^{15}O]H$_2$O/PET has also been used to study the mechanism of pallidal stimulation. In their study, Davis *et al.* (1997) also measured the effects of GPi DBS on cerebral blood flow. They found an increase in rCBF in ipsilateral premotor areas during GPi stimulation, which clinically improved rigidity and bradykinesia (Davis *et al.*, 1997). Comparable rCBF changes with GPi DBS were not evident in the joystick activation study of Limousin *et al.* (1997). Recently, we studied six patients with advanced PD undergoing pallidal stimulation (Eidelberg *et al.*, 1999). GPi stimulation during a kinematically controlled motor execution task resulted in significant rCBF increases in contralateral primary motor cortex and in premotor regions as well as in the cerebellar hemisphere ipsilateral to the moving hand (Fig. 6). These findings are similar to our previous FDG/PET studies in pallidotomy and suggest that disrupting the excessive inhibitory GPi output to the thalamus reverses the symptoms of parkinsonism by activating areas involved in the initiation of movement.

STN is thought to be more effective than GPi stimulation in that it can influence more than one area of basal ganglia inhibitory output, i.e., both the GPi and the substantia nigra reticulata (SNr). Limousin *et al.* (1997) investigated the effect of STN stimulation on rCBF during an activation task. Using [^{15}O]H$_2$O/PET,

six PD patients with STN DBS implants were scanned while performing self-directed movements with a joystick. Results showed significant rCBF increases in SMA, cingulate cortex, and DLPFC during effective STN stimulation. This suggests that STN DBS may play a role in potentiating nonprimary motor cortical areas, especially the DLPFC, which showed greater activation than in GPi during effective stimulation (Limousin *et al.*, 1997).

In summary, high-frequency DBS has the potential benefit of avoiding permanent side effects due to an ablative lesion. In addition, the technique offers the opportunity of inducing a reversible alteration in functional brain circuitry. It is therefore a useful experimental method to assess the modulation of structure-functional relationships during the successful treatment of PD. Nonetheless, the role of preoperative PET in patient selection for DBS in parkinsonism remains a topic of future investigation.

C. Neurorestorative Therapies: Dopamine Cell Implantation Procedures

The implantation of fetal mesencephalon may prove to be useful in the treatment of PD (Bjorklund, 1979; Fahn, 1992). *In vivo* assessment of presynaptic nigrostriatal dopaminergic function with FDOPA/PET may be used to monitor graft survival and development after transplantation. To date, only a limited number of patients have been studied with PET. One patient exhibited sustained postoperative increases in FDOPA uptake in the grafted putamen 33 months after transplantation (Freed *et al.*, 1992). Sawle *et al.* (1992) subsequently reported two PD patients who underwent unilateral transplantation into the putamen and had FDOPA/PET imaging before and 12 and 13 months following surgery. These patients had substantial increases in FDOPA uptake in the grafted putamen and decreases in the nongrafted striatum, presumably associated with disease progression. In one of these patients, clinical improvement continued 3 years after transplantation and was accompanied by a further increase in FDOPA uptake in the grafted putamen. In the other patient, there was no further increase in FDOPA uptake or improvement in clinical symptoms between 1 and 3 years after transplantation (Sawle *et al.*, 1992). Remy *et al.* (1995) reported increases in FDOPA uptake in five patients with unilateral implantation in the putamen; changes in FDOPA uptake rate constant were significantly correlated with clinical improvement measures. Hauser *et al.* (1999) reported six patients with advanced-stage PD treated with bilateral implantation in the putamen. These patients showed significant clinical

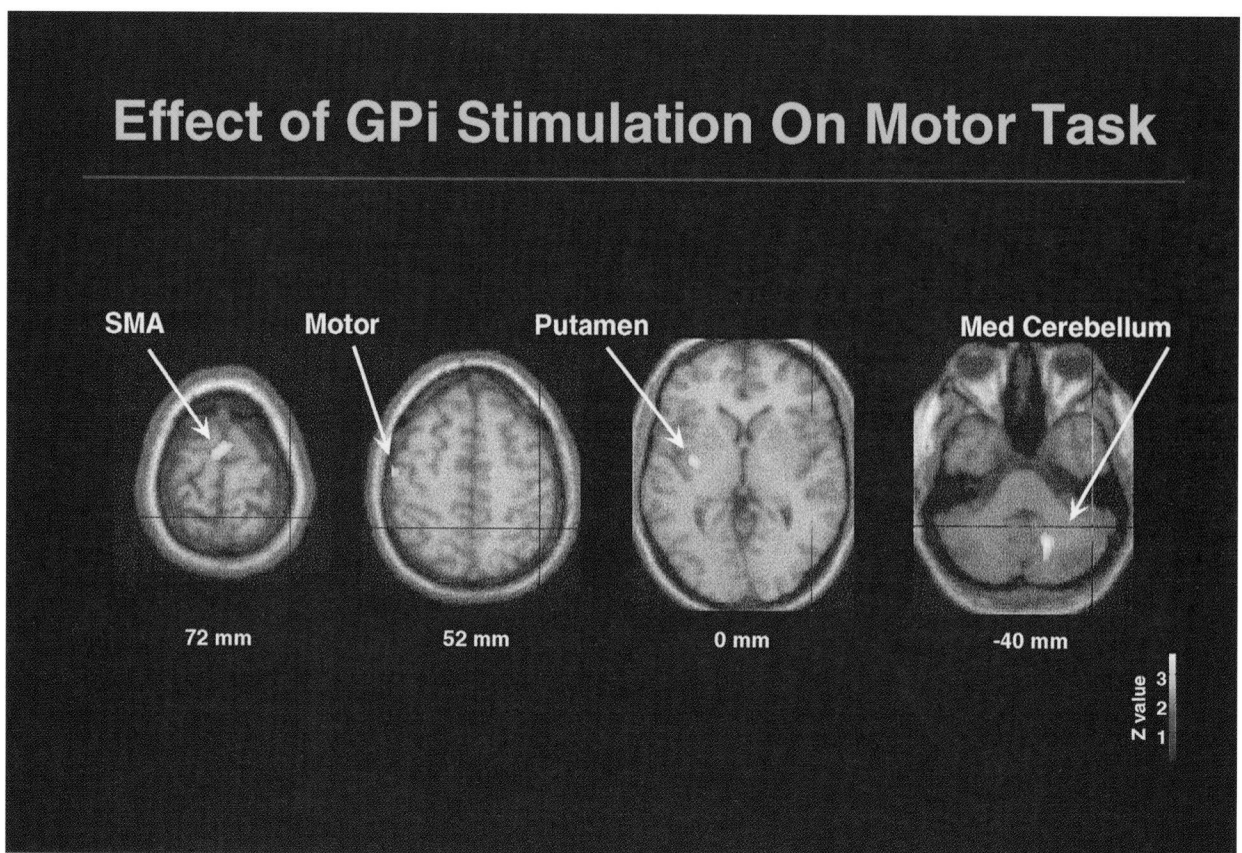

Figure 6 $H_2^{15}O$/PET images of changes in rCBF in six advanced PD patients undergoing unilateral DBS of the internal pallidum. The patients were scanned in the DBS ON and OFF conditions while performing a kinematically controlled motor execution task using the hand opposite the pallidal stimulator. (The rate and amplitude of movement were controlled across DBS conditions.) During task execution, pallidal DBS resulted in a significant increase in rCBF in the putamen and primary motor cortex (BA 4, hand region) ipsilateral to stimulation, bilaterally in the SMA, and in the cerebellum contralateral to the stimulator.

improvement 1 year after transplantation as well as appreciable bilateral increases in striatal FDOPA uptake (Hauser *et al.*, 1999). Similar results have been reported by us recently (Dhawan *et al.*, 1999) in a cohort of 19 advanced PD patients undergoing fetal nigral implantation as part of a randomized blinded comparison with sham-operated controls (Fig. 7).

A number of caveats must be considered in the interpretation of these findings. FDOPA crosses the BBB in competition with other neutral amino acids (Leenders *et al.*, 1990; Ishikawa *et al.*, 1996b). Thus, any alteration of plasma amino acid content (including changes in peripheral levodopa levels) may alter FDOPA transport and can confound FDOPA/PET quantifiers of dopaminergic function. In addition, the assessment of graft survival and growth using FDOPA/PET may potentially be overestimated by postoperative increases in the net transport of FDOPA across the BBB (Martin *et al.*, 1987; Sawle and Myers, 1993). Indeed, evidence in-

dicates that both permeability and surface area can increase after tissue transplantation. Fluorescence histochemical measurements of monoamine uptake into central neurons suggest that the BBB to monoamines may not be fully developed in infant rats (Loizou, 1970). The influx rate of amino acids is greater during the first weeks of life and falls subsequently to the level found in adult rats at 8–10 weeks of age (Banos *et al.*, 1978). Guttman *et al.* (1989) reported five patients who underwent adrenal medulla implantation and who had pre- and postoperative FDOPA/PET scans. In two patients, increased permeability of the BBB was shown at the implanted sites using gallium ([^{68}Ga]EDTA) and PET (Guttman *et al.*, 1989). Furthermore, Dusart *et al.* (1989) demonstrated two different types of angiogenesis in the thalamus after transplantation of dissociated fetal tissue. Although these authors observed mature angiogenesis in the grafted site, reactive angiogenesis may also induce significant increases in local FDOPA uptake

FDOPA / PET

Sham surgery

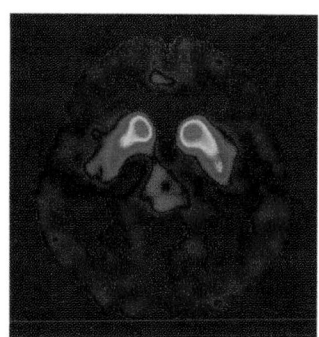 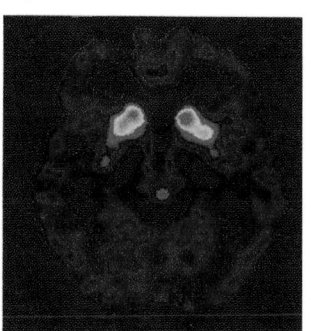

Preop.   Postop.

Fetal mesencephalic cell implant

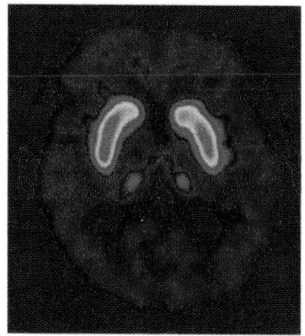

Normal

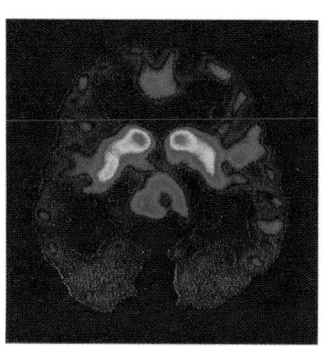 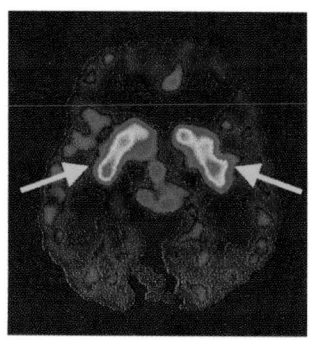

Preop.   Postop.

Figure 7 FDOPA/PET images from a normal volunteer (left) and from two patients studied in the course of a blinded sham-controlled study of fetal nigral dopamine cell implantation for advanced PD (right). The upper tier represents baseline and 15-month follow-up FDOPA/PET scans from a PD patient randomized to sham surgery. The lower tier represents baseline and 15-month postoperative scans from another patient who was randomized to bilateral implantation in the putamen. At baseline, both patients demonstrated typical significant reductions in FDOPA uptake involving predominantly the posterior putamen. This decline was comparatively greater in the sham patient at follow-up. By contrast, the patient undergoing implantation demonstrated bilateral increases in putamen FDOPA uptake compatible with engraftment (arrows).

without necessarily implicating reinnervation from the grafted tissues (Dusart *et al.*, 1989).

Postmortem results from two patients who underwent fetal mesencephalic implantation and postoperative FDOPA/PET have been reported. Kordower and colleagues (1995) reported one patient who showed substantial clinical improvement as well as an increase in FDOPA uptake in the grafted putamen (Kordower *et al.*, 1995). PET examination 18 months following surgery (8 months before death) disclosed increased radiotracer uptake in the pericaudate area and in the putamen sites where implantation had been performed. Necropsy revealed only gliosis and inflamma-

tory response in these areas (Folkerth and Durso, 1996).

Further technological development is required to optimize fetal implantation surgery as a therapeutic option in advanced parkinsonism. The safety and efficacy of these procedures are currently unknown and are a topic of several controlled clinical studies, including investigations of porcine nigral cells (Deacon *et al.*, 1997). Whether or not changes in striatal FDOPA uptake accurately reflect graft survival is also the subject of ongoing investigation. Alternative dopaminergic imaging methods may afford a complementary tool for the assessment of outcome following fetal implantation.

Specifically, adjunctive PET imaging of striatal DAT and VMAT activity before and after implantation procedures, as well as the application of functional mapping methods with FDG and $[^{15}O]H_2O$/PET, may lead to a broader understanding of the effects of neural transplantation on the nigrostriatal dopaminergic system.

D. Neuroprotective Therapies

1. Native Progression Studies: Longitudinal PET

Clinical trials to assess the efficacy of new therapies for PD usually utilize clinical scales such as the Unified Parkinson's Disease Rating Scale (UPDRS) (Fahn et al., 1987). This scale assesses four clinical domains, including mentation and behavior, activities of daily living, motor examination, and complications due to therapy. The UPDRS has been used extensively to assess symptomatic effects of new medications for PD, and its reliability and validity have been verified (Richards et al., 1994; van Hilten et al., 1994). In addition, the UPDRS has been used in clinical trials to assess potential "neuroprotective" therapies (Parkinson Study Group, 1989a,b), but this has been problematic for several reasons. First, since the UPDRS is sensitive to symptomatic therapies, potential neuroprotective therapies that have even very small symptomatic effects may be perceived as neuroprotective if washout periods at the end of clinical trials are inadequate. Second, clinical scales such as the UPDRS may be more sensitive to changes early in the course of PD and may therefore falsely give the impression of a nonlinear rate of progression that may not accurately reflect the rate of neuropathological change. The actual rate of nigrostriatal dopaminergic degeneration in PD cannot be determined, as neuropathological data are by definition cross-sectional. Indeed, longitudinal studies using reliable, quantitative in vivo markers of cell loss are needed to provide valid estimates of the temporal progression of PD and other neurodegenerative disorders.

FDOPA/PET qualifies as one such marker as it directly relates to the amount of nigral dopaminergic neurons and correlates to UPDRS scores in PD subjects (Snow et al., 1993). In an early longitudinal PD study, Bhatt et al. (1991) examined disease progression in nine PD patients and seven normal controls using FDOPA/PET. All subjects were scanned twice during a 3- to 4-year interval. Results showed a similar 5% decline in the striatal/temporoparietal ratio in both groups, which the authors interpreted as a slow progression (Bhatt et al., 1991). In a follow-up study, Vingerhoets et al. (1994) found an annual SOR decrement of 7.8% in 16 PD patients compared to 3% in 10

normal controls. This suggests a more rapid rate of progression in the PD group. Indeed, Morrish et al. (1996) showed an even faster annual rate of decline (14%) in a study of 17 PD patients when measuring putamenal FDOPA uptake. There was no significant change in FDOPA uptake in a control group of 10 normals (Morrish et al., 1996).

Several issues confound the interpretation of these varying results. It appears that the rate of progression is dependent on demographic factors such as the age of onset of symptoms as well as disease duration at the time of PET imaging (Eidelberg et al., 1995b). Importantly, the rate of progression is likely to be nonlinear and may vary with the phase of disease analyzed, i.e., may be faster during the clinical and early symptomatic phases. Disparities in the estimates of rate of progression may also evolve through differences in PET instrumentation and image quantification procedures. Lastly, these estimates may depend in part on the choice of dopaminergic PET tracers. Different radioligands for dopaminergic imaging may vary in their sensitivity to ongoing neurodegenerative processes. Longitudinal PET studies employing multiple imaging probes quantified at several time points throughout the natural history of disease will be needed before a comprehensive mathematical model of progression becomes available using PET (Morrish et al., 1998).

In addition to using FDOPA/PET in progression studies, RAC/PET is also being used to measure longitudinal changes in dopamine receptor density. Rinne et al. (1993) and Antonini et al. (1994) studied D_2 receptor changes in early-stage PD patients and normal controls with RAC/PET. In the earlier of the two PET studies, the unmedicated PD group showed an increase in D_2 binding in the striatum contralateral to the more symptomatic side of the body as compared with the ipsilateral striatum, which was still present after repeated scans 1 year later (Rinne et al., 1993).

Antonini (1994) studied the effects of dopaminergic therapy on D_2 receptors in 18 de novo PD patients, 9 of whom had repeated scans 3–5 years (Antonini et al., 1994, 1997) after being treated with dopaminergic therapy. Baseline results showed an increase in the putamen of the PD patients before the treatment condition began. After 3–4 months of pharmacotherapy, no specific change in caudate or putamen D_2 uptake was noted in the patients treated with levodopa. However, analysis of repeat scans 3-5 years later demonstrated a significant reduction in the putamen and caudate compared with the baseline scans. This suggests that there is a downregulation of D_2 receptor binding in PD that may be induced by long-term dopaminergic therapy or is a result of an adaptive response to nigrostriatal denervation.

PET is likely to play an important role in the development and evaluation of neuroprotective therapies for PD. To date, the validation of such therapies has been hampered by the absence of biological markers that reliably parallel the extent or pace of nigral degeneration. PET holds promise of providing biologic imaging markers that can be applied in controlled clinical trials of potential neuroprotective agents in PD.

VI. Conclusion

Considerable attention has been dedicated to the development of novel data analytical methods for the characterization and quantification of neural networks in functional brain imaging data. Such metabolic covariance patterns may serve as clinically useful markers of disease severity as well as aid in the differential diagnosis of parkinsonism. Moreover, new radiotracer techniques have been developed to quantify neurochemical deficits associated with neurodegenerative processes. A major contribution of this research has been a combined approach utilizing both network analytical strategies and *in vivo* neurochemical measurements to investigate the relationships between localized neuronal attrition and the expression of widely distributed functional brain networks. These complementary PET techniques may greatly advance the understanding of the pathophysiology of PD and the functional changes that occur with successful therapy.

Acknowledgments

We acknowledge support by the National Institues of Health (NIH NS RO1 32368, 35069, and 37564) as well as generous grants from the National Parkinson Foundation, the Parkinson Disease Foundation, and the American Parkinson's Disease Association.

References

Alexander, G., and Moeller, J. (1994). Application of the scaled subprofile model to functional imaging in neuropsychiatric disorders: A prinicipal component approach to modeling brain function in disease. *Hum. Brain Mapp.* **2,** 1–16.

Alexander, G. E., DeLong, M. R., and Strick, P. L. (1986). Parallel organization of functionally segregated circuits linking basal ganglia and cortex. *Annu. Rev. Neurosci.* **9,** 357–381.

Alexander, G. E., Crutcher, M. D., and DeLong, M. R. (1990). Basal ganglia-thalamocortical circuits: Parallel substrates for motor, oculomotor, "prefrontal" and "limbic" functions. *Prog. Brain Res.* **85,** 119–146.

Antonini, A., Leenders, K., Reist, H., Thomann, R., Beer, H., and Locher, J. (1993). Effect of age on D$_2$ dopamine receptors in normal human brain measured by positron emission tomography and ^{11}C-raclopride. *Arch. Neurol.* **50** (5), 474–480.

Antonini, A., Schwarz, J., Oertel, W. H., Beer, H. F., Madeja, U. D., and Leenders, K. L. (1994). [^{11}C]Raclopride and positron emission tomography in previously untreated patients with Parkinson's disease: Influence of L-dopa and lisuride therapy on striatal dopamine D$_2$-receptors. *Neurology* **44** (7), 1325–1329.

Antonini, A., Vontobel, P., Psylla, M., Gunther, I., Maguire, P., Missimer, J., and Leenders, K. (1995). Complementary positron emission tomographic studies of the striatal dopaminergic system in Parkinson's disease. *Arch. Neurol.* **52** (12), 1183–1190.

Antonini, A., Schwarz, J., Oertel, W., Pogarell, O., and Leenders, K. (1997). Long-term changes of striatal dopamine D$_2$ receptors in patients with Parkinson's disease: A study with positron emission tomography and [^{11}C]Raclopride. *Mov. Disord.* **12,** 33–38.

Antonini, A., Kazumata, K., Feigin, A., Mandel, F., Dhawan, V., Margouleff, C., and Eidelberg, D. (1998a). Differential diagnosis of parkinsonism with [^{18}F]fluorodeoxyglucose and PET. *Mov. Disord.* **13** (2), 268–74.

Antonini, A., Moeller, J. R., Nakamura, T., Spetsieris, P., Dhawan, V., and Eidelberg, D. (1998b). The metabolic anatomy of tremor in Parkinson's disease. *Neurology* **51** (3), 803–810.

Assenbaum, S., Brucke, T., Pirker, W., Podreka, I., Angelberger, P., Wenger, S., Wober, C., Muller, C., and Deecke, L. (1997). Imaging of dopamine transporters with iodine-123-β-CIT and SPECT in Parkinson's disease. *J. Nucl. Med.* **38,** 1–6.

Banos, G., Daniel, P. M., and Pratt, O. E. (1978). The effect of age upon the entry of some amino acids into the brain, and their incorporation into cerebral protein. *Dev. Med. Child Neurol.* **20** (3), 335–346.

Bathia, K ., and Marsden, C. (1994). The behavioral and motor consequences of focal lesions of the basal ganglia in man. *Brain* **117,** 859–876.

Bergman, H., Wichmann, T., Karmon, B., and DeLong, M. R. (1994). The primate subthalamic nucleus. II. Neuronal activity in the MPTP model of parkinsonism. *J. Neurophysiol.* **72** (2), 507–520.

Bhatt, M. H., Snow, B. J., Martin, W. R., Pate, B. D., Ruth, T. J., and Calne, D. B. (1991). Positron emission tomography suggests that the rate of progression of idiopathic parkinsonism is slow. *Ann. Neurol.* **29** (6), 673–677.

Bjorklund, A. (1979). Reconstruction of the nigrostriatal dopamine pathway by intracerebral nigral transplants. *Brain Res.* **177** (3), 555–560.

Blesa, R., Blin, J., and Miletich, R. (1991). Levodopa-reduced glucose metabolism in striatopallido-thalamocortical circuit in Parkinson's disease. *Neurology* **41** (S1), 359.

Boecker, H., Wills, A. J., Ceballos-Baumann, A., Samuel, M., Thomas, D. G., Marsden, C. D., and Brooks, D. J. (1997). Stereotactic thalamotomy in tremor-dominant Parkinson's disease: An H$_2$(15)O PET motor activation study. *Ann. Neurol.* **41** (1), 108–111.

Brooks, D. J., Ibanez, V., Sawle, G. V., Quinn, N., Lees, A. J., Mathias, C. J., Bannister, R., Marsden, C. D., and Frackowiak, R. S. (1990a). Differing patterns of striatal ^{18}F-dopa uptake in Parkinson's disease, multiple system atrophy, and progressive supranuclear palsy [see comments]. *Ann. Neurol.* **28** (4), 547–555.

Brooks, D., Salmon, E., Mathias, C., Quinn, N., Leenders, K., Bannister, R., Marsden, C., and Frackowiak, R. (1990b). The relationship between locomotor disability, autonomic dysfunction, and the integrity of the striatal dopaminergic system in patients with multiple system atrophy, pure autonomic failure, and Parkinson's disease, studied with PET. *Brain* **113,** 1539–1552.

Brooks, D., Ibanez, V., Sawle, G., Playford, E., Quinn, N., Mathias, C., Lees, A., Marsden, C., Bannister, R., and Frackowiak, R. (1992). Striatal D$_2$ receptor status in patients with Parkinson's disease, striatonigral degeneration, and progressive supranuclear palsy, measured with ^{11}C-raclopride and positron emission tomography. *Ann. Neurol.* **31** (2), 184–192.

Broussolle, E., Cinotti, L., Pollak, P., Landais, P., Le Bars, D., Galy, G., Lavenne, F., Khalfallah, Y., Chazot, G., and Mauguiere, F. (1993).

Relief of akinesia by apomorphine and cerebral metabolic changes in Parkinson's disease. *Mov. Disord.* **8** (4), 459–462.

Catalan, M. J., Ishii, K., Honda, M., Samii, A., and Hallett, M. (1999). A PET study of sequential finger movements of varying length in patients with Parkinson's disease. *Brain* **122** (Part 3), 483–495.

Chaly, T., Dhawan, V., Kazumata, K., Antonini, A., Margouleff, C., Dahl, J. R., Belakhlef, A., Margouleff, D., Yee, A., Wang, S., Tamagnan, G., Neumeyer, J. L., and Eidelberg, D. (1996). Radiosynthesis of [^{18}F]*N*-3-fluoropropyl-2β-carbomethoxy-3β-(4-iodophenyl) nortropane and the first human study with positron emission tomography. *Nucl. Med. Biol.* **23** (8), 999–1004.

Davis, K. D., Taub, E., Houle, S., Lang, A. E., Dostrovsky, J. O., Tasker, R. R., and Lozano, A. M. (1997). Globus pallidus stimulation activates the cortical motor system during alleviation of parkinsonian symptoms [see comments]. *Nat. Med.* **3** (6), 671–674.

Deacon, T., Schumacher, J., Dinsmore, J., Thomas, C., Palmer, P., Kott, S., Edge, A., Penney, D., Kassissieh, S., Dempsey, P., and Isacson, O. (1997). Histological evidence of fetal pig neural cell survival after transplantation into a patient with Parkinson's disease. *Nat. Med.* **3** (3), 350–353.

Deiber, M. P., Pollak, P., Passingham, R., Landais, P., Gervason, C., Cinotti, L., Friston, K., Frackowiak, R., Mauguiere, F., and Benabid, A. L. (1993). Thalamic stimulation and suppression of parkinsonian tremor. Evidence of a cerebellar deactivation using positron emission tomography. *Brain* **116** (Part 1), 267–279.

DeLong, M. R. (1990). Primate models of movement disorders of basal ganglia origin. *Trends Neurosci.* **13** (7), 281–285.

Dhawan, V., Ishikawa, T., Patlak, C., Chaly, T., Robeson, W., Belakhlef, A., Margouleff, C., Mandel, F., and Eidelberg, D. (1996). Combined FDOPA and 3OMFD PET studies in Parkinson's disease. *J. Nucl. Med.* **37** (2), 209–216.

Dhawan, V., Nakamura, T., Margouleff, C., Freed, C., Breeze, R., Fahn, S., Greene, P., Tsai, W., Kao, R., and Eidelberg, D. (1999). Double-blind controlled trial of human embryonic dopaminergic tissue transplants in advanced Parkinson's disease: Fluorodopa PET imaging. *Neurology* **52** (6), A405.

Dusart, I., Nothias, F., Roudier, F., Besson, J., and Peschanski, M. (1989). Vascularization of fetal cell suspension grafts in the excitotoxically lesioned adult rat thalamus. *Brain Res. Dev. Brain Res.* **48** (2), 215–228.

Eidelberg, D., Moeller, J. R., Dhawan, V., Sidtis, J. J., Ginos, J. Z., Strother, S. C., Cedarbaum, J., Greene, P., Fahn, S., and Rottenberg, D. A. (1990). The metabolic anatomy of Parkinson's disease: Complementary [^{18}F]fluorodeoxyglucose and [^{18}F]fluorodopa positron emission tomographic studies. *Mov. Disord.* **5** (3), 203–213.

Eidelberg, D., Dhawan, V., Moeller, J. R., Sidtis, J. J., Ginos, J. Z., Strother, S. C., Cedarbaum, J., Greene, P., Fahn, S., Powers, J. M., *et al.* (1991). The metabolic landscape of cortico-basal ganglionic degeneration: Regional asymmetries studied with positron emission tomography. *J. Neurol. Neurosurg. Psychiatry* **54** (10), 856–862.

Eidelberg, D., Takikawa, S., Moeller, J. R., Dhawan, V., Redington, K., Chaly, T., Robeson, W., Dahl, J. R., Margouleff, D., Fazzini, E. ., *et al.* (1993a). Striatal hypometabolism distinguishes striatonigral degeneration from Parkinson's disease. *Ann. Neurol.* **33** (5), 518–527.

Eidelberg, D., Takikawa, S., Wilhelmsen, K., Dhawan, V., Chaly, T., Robeson, W., Dahl, R., Margouleff, D., Greene, P., Hunt, A., *et al.* (1993b). Positron emission tomographic findings in Filipino X-linked dystonia-parkinsonism. *Ann. Neurol.* **34** (2), 185–191.

Eidelberg, D., Moeller, J. R., Dhawan, V., Spetsieris, P., Takikawa, S., Ishikawa, T., Chaly, T., Robeson, W., Margouleff, D., Przedborski, S., *et al.* (1994). The metabolic topography of parkinsonism. *J. Cereb. Blood Flow Metab.* **14** (5), 783–801.

Eidelberg, D., Moeller, J. R., Ishikawa, T., Dhawan, V., Spetsieris, P., Chaly, T., Belakhlef, A., Mandel, F., Przedborski, S., and Fahn, S. (1995a). Early differential diagnosis of Parkinson's disease with

^{18}F-fluorodeoxyglucose and positron emission tomography. *Neurology* **45** (11), 1995–2004.

Eidelberg, D., Moeller, J. R., Ishikawa, T., Dhawan, V., Spetsieris, P., Chaly, T., Robeson, W., Dahl, J. R., and Margouleff, D. (1995b). Assessment of disease severity in parkinsonism with fluorine-18-fluorodeoxyglucose and PET. *J. Nucl. Med.* **36** (3), 378–383.

Eidelberg, D., Moeller, J. R., Ishikawa, T., Dhawan, V., Spetsieris, P., Silbersweig, D., Stern, E., Woods, R. P., Fazzini, E., Dogali, M., and Beric, A. (1996). Regional metabolic correlates of surgical outcome following unilateral pallidotomy for Parkinson's disease. *Ann. Neurol.* **39** (4), 450–459.

Eidelberg, D., Moeller, J. R., Kazumata, K., Antonini, A., Sterio, D., Dhawan, V., Spetsieris, P., Alterman, R., Kelly, P. J., Dogali, M., Fazzini, E., and Beric, A. (1997). Metabolic correlates of pallidal neuronal activity in Parkinson's disease. *Brain* **120** (Part 8), 1315–1324.

Eidelberg, D., Nakamura, T., Mentis, M., Antonini, A., Dhawan, V., Ghilardi, M. F., Ghez, C., Hammerstad, J. P., and Koller, W. C. (1999). Brain activation responses with internal pallidal stimulation in Parkinson's disease. *Neurology* **52** (6), A176.

Fahn, S. (Editorial). (1992). Fetal-tissue transplants in Parkinson's disease [comment]. *N. Engl. J. Med.* **327** (22), 1589–1590.

Fahn, S., and Elton, R. (1987). Unified Parkinson's disease rating scale. *In* "Recent Developments in Parkinson's Disease" (S. Fahn *et al.*, Eds)., pp. 293–304. Macmillan, New York.

Filion, M., Tremblay, L., and Bedard, P. J. (1991). Effects of dopamine agonists on the spontaneous activity of globus pallidus neurons in monkeys with MPTP-induced parkinsonism. *Brain Res.* **547** (1), 152–161.

Folkerth, R. D ., and Durso, R. (1996). Survival and proliferation of nonneural tissues, with obstruction of cerebral ventricles, in a parkinsonian patient treated with fetal allografts [see comments]. *Neurology* **46** (5), 1219–1225.

Ford, I. (1995). Commentary and opinion: III. Some nonontological and functionally unconnected views on current issues in the analysis of PET datasets. *J. Cereb. Blood Flow Metab.* **15** (3), 371–377.

Foster, N. L., Gilman, S., Berent, S., Morin, E. M., Brown, M. B., and Koeppe, R. A. (1988). Cerebral hypometabolism in progressive supranuclear palsy studied with positron emission tomography. *Ann. Neurol.* **24** (3), 399–406.

Freed, C., Breeze, R., Rosenberg, N., Schneck, S., Kriek, E., Qi, J., Lone, T., Zhang, Y., Snyder, J., Wells, T., *et al.* (1992). Survival of implanted fetal dopamine cells and neurologic improvement 12 to 46 months after transplantation for Parkinson's disease. *N. Engl. J. Med.* **327** (22), 1549–1555.

Frey, K., Koeppe, R., Kilbourn, M., Vander Borght, T., Albin, R., Gilman, S., and Kuhl, D. (1996). Presynaptic monoaminergic vesicles in Parkinson's disease and normal aging. *Ann. Neurol.* **40** (6), 873–874.

Friston, K. J., Frith, C. D., Liddle, P. F., Dolan, R. J., Lammertsma, A. A., and Frackowiak, R. S. (1990). The relationship between global and local changes in PET scans. *J. Cereb. Blood Flow Metab.* **10** (4), 458–466.

Friston, K., Holmes, A., Worsley, K., Poline, J.-P., Frith, C., and Frackowiak, R. (1995). Statistical parametric maps in functional imaging: A general linear approach. *Hum. Brain Mapp.* **2**, 189–210.

Gjedde, A., Reith, J., Dyve, S., Leger, G., Guttman, M., Diksic, M., Evans, A., and Kuwabara, H. (1991). Dopa decarboxylase activity of the living human brain. *Proc. Natl. Acad. Sci. U.S.A.* **88** (7), 2721–2725.

Goldberg, G. (1995). Supplementary motor area structure and function: Review and hypothesis. *Brain Behav. Sci.* **8**, 567–616.

Goldman-Rakic, P. (1987). Circuitry of primate prefrontal cortex and regulation of behavior by representational memory. *In* "The Nervous System: Higher Functions of the Brain" (F. Plum, Ed.), pp. 373–417. American Physiology Society, Bethesda, MD.

Grafton, S., Sutton, J., Couldwell, W., Lew, M., and Waters, C. (1994). Network analysis of motor system connectivity in Parkinson's Disease: Modulation of thalamocortical interactions after pallidotomy. *Hum. Brain Mapp.* **2,** 45–55.

Grafton, S. T., Waters, C., Sutton, J., Lew, M. F., and Couldwell, W. (1995). Pallidotomy increases activity of motor association cortex in Parkinson's disease: A positron emission tomographic study. *Ann. Neurol.* **37** (6), 776–783.

Guttman, M., Burns, R. S., Martin, W. R., Peppard, R. F., Adam, M. J., Ruth, T. J., Allen, G., Parker, R. A., Tulipan, N. B., and Calne, D. B. (1989). PET studies of parkinsonian patients treated with autologous adrenal implants. *Can. J. Neurol. Sci.* **16** (3), 305–309.

Hauser, R. A., Freeman, T. B., Snow, B. J., Nauert, M., Gauger, L., Kordower, J. H., and Olanow, C. W. (1999). Long-term evaluation of bilateral fetal nigral transplantation in Parkinson disease. *Arch. Neurol.* **56** (2), 179–187.

Hershey, T., Black, K. J., Stambuk, M. K., Carl, J. L., McGee-Minnich, L. A., and Perlmutter, J. S. (1998). Altered thalamic response to levodopa in Parkinson's patients with dopa-induced dyskinesias. *Proc. Natl. Acad. Sci. U.S.A.* **95** (20), 12016–12021.

Ishikawa, T., Dhawan, V., Chaly, T., Margouleff, C., Robeson, W., Dahl, J. R., Mandel, F., Spetsieris, P., and Eidelberg, D. (1996a). Clinical significance of striatal DOPA decarboxylase activity in Parkinson's disease. *J. Nucl. Med.* **37** (2), 216–222.

Ishikawa, T., Dhawan, V., Chaly, T., Robeson, W., Belakhlef, A., Mandel, F., Dahl, R., Margouleff, C., and Eidelberg, D. (1996b). Fluorodopa positron emission tomography with an inhibitor of catechol-*O*-methyltransferase: Effect of the plasma 3-*O*-methyldopa fraction on data analysis. *J. Cereb. Blood Flow Metab.* **16** (5), 854–863.

Ishikawa, T., Dhawan, V., Kazumata, K., Chaly, T., Mandel, F., Neumeyer, J., Margouleff, C., Babchyck, B., Zanzi, I., and Eidelberg, D. (1996c). Comparative nigrostriatal dopaminergic imaging with iodine-123-βCIT- FP/SPECT and fluorine-18-FDOPA/PET. *J. Nucl. Med.* **37** (11), 1760–1765.

Jahanshahi, M., Jenkins, I. H., Brown, R. G., Marsden, C. D., Passingham, R. E., and Brooks, D. J. (1995). Self-initiated versus externally triggered movements. I. An investigation using measurement of regional cerebral blood flow with PET and movement-related potentials in normal and Parkinson's disease subjects. *Brain* **118** (Part 4), 913–933.

Kazumata, K., Antonini, A., Dhawan, V., Moeller, J. R., Alterman, R. L., Kelly, P., Sterio, D., Fazzini, E., Beric, A., and Eidelberg, D. (1997). Preoperative indicators of clinical outcome following stereotaxic pallidotomy. *Neurology* **49** (4), 1083–1090.

Kish, S. J., Zhong, X. H., Hornykiewicz, O., and Haycock, J. W. (1995). Striatal 3,4-dihydroxyphenylalanine decarboxylase in aging: Disparity between postmortem and positron emission tomography studies? *Ann. Neurol.* **38** (2), 260–264.

Kordower, J., Freeman, T., Snow, B., Vingerhoets, F., Mufson, E., Sanberg, P., Hauser, R., Smith, D., Nauert, G., Perl, D., *et al.* (1995). Neuropathological evidence of graft survival and striatal reinnervation after the transplantation of fetal mesencephalic tissue in a patient with Parkinson's disease. *N. Engl. J. Med.* **332** (17), 1118–1124.

Krack, P., Pollak, P., Limousin, P., Hoffmann, D., Xie, J., Benazzouz, A., and Benabid, A. L. (1998). Subthalamic nucleus or internal pallidal stimulation in young onset Parkinson's disease. *Brain* **121** (Part 3), 451–457.

Laihinen, A. O., Rinne, J. O., Ruottinen, H. M., Nagren, K. A., Lehikoinen, P. K., Oikonen, V. J., Ruotsalainen, U. H., and Rinne, U. K. (1994). PET studies on dopamine D_1 receptors in the human brain with carbon-11–SCH 39166 and carbon-11–NNC 756. *J. Nucl. Med.* **35** (12), 1916–1920.

Laitinen, L. V., Bergenheim, A. T., and Hariz, M. I. (1992). Leksell's posteroventral pallidotomy in the treatment of Parkinson's disease [see comments]. *J. Neurosurg.* **76** (1), 53–61.

Lang, A. E ., and Lozano, A. M. (1998). Parkinson's disease. Second of two parts. *N. Engl. J. Med.* **339** (16), 1130–1143.

Leenders, K. L., Wolfson, L., Gibbs, J. M., Wise, R. J., Causon, R., Jones, T., and Legg, N. J. (1985). The effects of L-DOPA on regional cerebral blood flow and oxygen metabolism in patients with Parkinson's disease. *Brain* **108** (Part 1), 171–191.

Leenders, K., Salmon, E., Tyrrell, P., Perani, D., Brooks, D., Sager, H., Jones, T., Marsden, C., and Frackowiak, R. (1990). The nigrostriatal dopaminergic system assessed *in vivo* by positron emission tomography in healthy volunteer subjects and patients with Parkinson's disease. *Arch. Neurol.* **47** (12), 1290–1298.

Limousin, P., Greene, J., Pollak, P., Rothwell, J., Benabid, A. L., and Frackowiak, R. (1997). Changes in cerebral activity pattern due to subthalamic nucleus or internal pallidum stimulation in Parkinson's disease. *Ann. Neurol.* **42** (3), 283–291.

Loizou, L. A. (1970). Uptake of monoamines into central neurones and the blood-brain barrier in the infant rat. *Br. J. Pharmacol.* **40** (4), 800–813.

Lozano, A. M., Lang, A. E., Galvez-Jimenez, N., Miyasaki, J., Duff, J., Hutchinson, W. D., and Dostrovsky, J. O. (1995). Effect of GP_i pallidotomy on motor function in Parkinson's disease. *Lancet* **346** (8987), 1383–1387.

Marsden, C ., and Obeso, J. (1994). The functions of the basal ganglia and the paradox of stereotaxic surgery in Parkinson's disease. *Brain* **117,** 877–897.

Martin, W. R ., *et al.* (1987). Positron emission tomography in Parkinson's disease: Glucose and DOPA metabolism. *Adv. Neurol.* **45,** 95–98.

Mata, M., Fink, D., and Gainer, H. (1980). Activity-dependent energy metabolism in rat posterior pituitary primarily reflects sodium pump activity. *J. Neurochem.* **34,** 213–215.

Mentis, M., Edwards, C., Krch, D., Perrine, K., Beric, A., Mattis, P., Nakamura, T., Moeller, J., and Eidelberg, D. (1999). Metabolic substrate underlying cognitive dysfunction in Parkinson's disease. *J. Nucl. Med.* **40** (5), 1184.

Mitchell, I., Boyce, S., Sambrook, M., *et al.* (1994). A 2-deoxyglucose study of the effects of dopamine agonists on the parkinsonian primate brain. *Brain* **115,** 809–824.

Moeller, J. R., and Strother, S. C. (1991). A regional covariance approach to the analysis of functional patterns in positron emission tomographic data. *J. Cereb. Blood Flow Metab.* **11** (2), A121–A135.

Moeller, J. R., Strother, S. C., Sidtis, J. J., and Rottenberg, D. A. (1987). Scaled subprofile model: A statistical approach to the analysis of functional patterns in positron emission tomographic data. *J. Cereb. Blood Flow Metab.* **7** (5), 649–658.

Moeller, J. R., Ishikawa, T., Dhawan, V., Spetsieris, P., Mandel, F., Alexander, G. E., Grady, C., Pietrini, P., and Eidelberg, D. (1996). The metabolic topography of normal aging. *J. Cereb. Blood Flow Metab.* **16** (3), 385–398.

Moeller, J. R., Nakamura, T., Mentis, M. J., Dhawan, V., Spetsieres, P., Antonini, A., Missimer, J., Leenders, K. L., and Eidelberg, D. (1999). Reproducibility of regional metabolic covariance patterns: Comparison of four populations. *J. Nucl. Med.* **40** (8), 1264–1269.

Morrish, P. K., Sawle, G. V., and Brooks, D. J. (1996). An [18F]dopa-PET and clinical study of the rate of progression in Parkinson's disease. *Brain* **119** (Part 2), 585–591.

Morrish, P. K., Rakshi, J. S., Bailey, D. L., Sawle, G. V., and Brooks, D. J. (1998). Measuring the rate of progression and estimating the preclinical period of Parkinson's disease with [^{18}F]dopa PET. *J. Neurol. Neurosurg. Psychiatry* **64** (3), 314–319.

Mushiake, H., Inase, M., and Tanji, J. (1990). Selective coding of motor sequence in the supplementary motor area of the monkey cerebral cortex. *Exp. Brain Res.* **82** (1), 208–210.

Parent, A., and Hazrati, L. (1995a). Functional anatomy of the basal ganglia. II. The place of subthalamic nucleus and external pallidum in basal ganglia circuitry. *Brain Res. Brain Res. Rev.* **20** (1), 128–154.

Parent, A., and Hazrati, L. N. (1995b). Functional anatomy of the basal ganglia. I. The cortico–basal ganglia–thalamo-cortical loop. *Brain Res. Brain Res. Rev.* **20** (1), 91–127.

Parker, F., Tzourio, N., Blond, S., Petit, H., and Mazoyer, B. (1992). Evidence for a common network of brain structures involved in parkinsonian tremor and voluntary repetitive movement. *Brain Res.* **584** (1–2), 11–17.

Parkinson Study Group. (1989a). DATATOP: A multicenter controlled clinical trial in early Parkinson's disease. *Arch. Neurol.* **46** (10), 1052–1060.

Parkinson Study Group. (1989b). Effect of deprenyl on the progression of disability in early Parkinson's disease. *N. Engl. J. Med.* **321** (20), 1364–1371.

Pate, B. D., Kawamata, T., Yamada, T., McGeer, E. G., Hewitt, K. A., Snow, B. J., Ruth, T. J., and Calne, D. B. (1993). Correlation of striatal fluorodopa uptake in the MPTP monkey with dopaminergic indices. *Ann. Neurol.* **34** (3), 331–338.

Patlak, C., and Blasberg, R. (1985). Graphical evaluation of blood-to-brain transfer constants from multiple-time uptake data. Generalizations. *J. Cereb. Blood Flow Metab.* **5**, 584–590.

Perlmutter, J. S., and Raichle, M. E. (1985). Regional blood flow in hemiparkinsonism. *Neurology* **35** (8), 1127–1134.

Perlmutter, J., Kilbourn, M., Raichle, M., and Welch, M. (1987). MPTP-induced up-regulation of *in vivo* dopaminergic radioligand-receptor binding in humans. *Neurology* **37**, 1575–1579.

Playford, E. D., Jenkins, I. H., Passingham, R. E., Nutt, J., Frackowiak, R. S., and Brooks, D. J. (1992). Impaired mesial frontal and putamen activation in Parkinson's disease: A positron emission tomography study. *Ann. Neurol.* **32** (2), 151–161.

Przedborski, S., Giladi, N., Takikawa, S., Ishikawa, T., Dhawan, V., Spetsieris, P., Chaly, T., Fahn, S., and Eidelberg, D. (1994). Metabolic topography of the hemiparkinsonism-hemiatrophy syndrome. *Neurology* **44** (9), 1622–1628.

Rascol, O., Sabatini, U., Brefel, C., Fabre, N., Rai, S., Senard, J. M., Celsis, P., Viallard, G., Montastruc, J. L., and Chollet, F. (1998). Cortical motor overactivation in parkinsonian patients with L-dopa-induced peak-dose dyskinesia. *Brain* **121** (Part 3), 527–533.

Remy, P., Samson, Y., Hantraye, P., Fontaine, A., Defer, G., Mangin, J. F., Fenelon, G., Geny, C., Ricolfi, F., Frouin, V., *et al.* (1995). Clinical correlates of [^{18}F]fluorodopa uptake in five grafted parkinsonian patients. *Ann. Neurol.* **38** (4), 580–588.

Richards, M., Marder, K., Cote, L., and Mayeux, R. (1994). Interrater reliability of the Unified Parkinson's Disease Rating Scale motor examination. *Mov. Disord.* **9** (1), 89–91.

Rinne, J., Laihinen, A., Nagren, K., Bergman, J., Solin, O., Haaparanta, M., Ruotsalainen, U., and Rinne, U. K. (1990). PET demonstrates different behaviour of striatal dopamine D-1 and D-2 receptors in early Parkinson's disease. *J. Neurosci. Res.* **27** (4), 494–499.

Rinne, J. O., Laihinen, A., Rinne, U. K., Nagren, K., Bergman, J., and Ruotsalainen, U. (1993). PET study on striatal dopamine D$_2$ receptor changes during the progression of early Parkinson's disease. *Mov. Disord.* **8** (2), 134–138.

Rougemont, D., Baron, J. C., Collard, P., Bustany, P., Comar, D., and Agid, Y. (1984). Local cerebral glucose utilisation in treated and untreated patients with Parkinson's disease. *J. Neurol. Neurosurg. Psychiatry* **47** (8), 824–830.

Rutgers, A. W., Lakke, J. P., Paans, A. M., Vaalburg, W., and Korf, J. (1987). Tracing of dopamine receptors in hemiparkinsonism with positron emission tomography (PET). *J. Neurol. Sci.* **80** (2-3), 237–248.

Sawle, G., Bloomfield, P., Bjorklund, A., Brooks, D., Brundin, P., Leenders, K., Lindvall, O., Marsden, C., Rehncrona, S., and Widner, H. (1992). Transplantation of fetal dopamine neurons in Parkinson's disease: PET [^{18}F]6-L-fluorodopa studies in two patients with putaminal implants. *Ann. Neurol.* **31** (2), 166–173.

Sawle, G. V., and Myers, R. (1993). The role of positron emission tomography in the assessment of human neurotransplantation. *Trends Neurosci.* **16** (5), 172–176.

Shinotoh, H., Hirayama, K., and Tateno, Y. (1993a). Dopamine D$_1$ and D$_2$ receptors in Parkinson's disease and striatonigral degeneration determined by PET. *Adv. Neurol.* **60**, 488–493.

Shinotoh, H., Inoue, O., Hirayama, K., Aotsuka, A., Asahina, M., Suhara, T., Yamazaki, T., and Tateno, Y. (1993b). Dopamine D$_1$ receptors in Parkinson's disease and striatonigral degeneration: A positron emission tomography study. *J. Neurol. Neurosurg. Psychiatry* **56** (5), 467–472.

Sigfried, J., and Lippitz, B. (1994). Bilateral chronic electrostimulation of ventroposterolateral pallidum: A new therapeutic approach for alleviating all parkinsonian symptoms. *Neurosurgery* **35**, 1126–1130.

Snow, B. J., Peppard, R. F., Guttman, M., Okada, J., Martin, W. R., Steele, J., Eisen, A., Carr, G., Schoenberg, B., and Calne, D. (1990). Positron emission tomographic scanning demonstrates a presynaptic dopaminergic lesion in Lytico-Bodig. The amyotrophic lateral sclerosis-parkinsonism-dementia complex of Guam. *Arch. Neurol.* **47** (8), 870–874.

Snow, B. J., Bhatt, M., Martin, W. R., Li, D., and Calne, D. B. (1991). The nigrostriatal dopaminergic pathway in Wilson's disease studied with positron emission tomography. *J. Neurol. Neurosurg. Psychiatry* **54** (1), 12–17.

Snow, B., Tooyama, I., McGeer, E., Yamada, T., Calne, D., Takahashi, H., and Kimura, H. (1993). Human positron emission tomographic [^{18}F]fluorodopa studies correlate with dopamine cell counts and levels. *Ann. Neurol.* **34** (3), 324–330.

Stoessl, A., and Ruth, T. (1998). Neuroreceptor imaging: New developments in PET and SPECT imaging of neuroreceptive binding (including dopamine transporters, vesicle transporters, and post-synaptic receptor sites). *Curr. Opin. Neurol.* **11** (4), 327–333.

Takikawa, S., and Eidelberg, D. (1994). Movement disorders. In "Functional Neuroimaging" (R. Kelley, Ed.), pp. 247–262. Futura, Armonk, NY.

Thaler, D., and Passingham, R. (1989). The supplementary motor cortex and internally directed movement. *In* "Neural Mechanisms in Disorders of Movement" (A. Crossman and M. Sambrook, Eds.), pp. 175–181. Libby, England.

van Dyck, C. H., Seibyl, J. P., Malison, R. T., Laruelle, M., Wallace, E., Zoghbi, S. S., Zea-Ponce, Y., Baldwin, R. M., Charney, D. S., and Hoffer, P. B. (1995). Age-related decline in striatal dopamine transporter binding with iodine-123-β-CIT SPECT [see comments]. *J. Nucl. Med.* **36** (7), 1175–1181.

van Hilten, J. J., van der Zwan, A. D., Zwinderman, A. H., and Roos, R. A. (1994). Rating impairment and disability in Parkinson's disease: Evaluation of the Unified Parkinson's Disease Rating Scale. *Mov. Disord.* **9** (1), 84–88.

Vingerhoets, F. J., Snow, B. J., Lee, C. S., Schulzer, M., Mak, E., and Calne, D. B. (1994). Longitudinal fluorodopa positron emission tomographic studies of the evolution of idiopathic parkinsonism. *Ann. Neurol.* **36** (5), 759–764.

Wilson, J. M., Levey, A. I., Rajput, A., Ang, L., Guttman, M., Shannak, K., Niznik, H. B., Hornykiewicz, O., Pifl, C., and Kish, S. J. (1996). Differential changes in neurochemical markers of striatal dopamine nerve terminals in idiopathic Parkinson's disease. *Neurology* **47** (3), 718–726.

Wolfson, L. I., Leenders, K. L., Brown, L. L., and Jones, T. (1985). Alterations of regional cerebral blood flow and oxygen metabolism in Parkinson's disease. *Neurology* **35** (10), 1399–1405.

Yochmowitz, M. (1982). Factor Analysis of Variance Model (FANOVA). In "Encyclopedia of Statistical Sciences" (S. Kotz and N. Johnson, Eds.), pp. 8–13. Wiley: New York.

11

Movement Disorders: Other Hypokinetic Disorders

David J. Brooks

MRC Cyclotron Unit, Hammersmith Hospital, London TW7 4QN, United Kingdom

I. Multiple-System Atrophy

II. Progressive Supranuclear Palsy

III. Corticobasal Degeneration

IV. Dopa-Responsive Dystonia

V. Akinetic-Rigid Huntington's Disease

VI. Conclusions

    References

In this chapter the relative roles of magnetic resonance imaging (MRI), proton magnetic resonance spectroscopy (MRS), positron emission tomography (PET), and single photon emission computed tomography (SPECT) for understanding the patterns of structural and functional abnormalities underlying atypical parkinsonian disorders will be compared and contrasted. The relative diagnostic values of these various approaches will also be assessed. The atypical parkinsonian disorders that are reviewed include multiple-system atrophy/striatonigral degeneration, progressive supranuclear palsy, corticobasal degeneration, and the akinetic-rigid variants of dopa-responsive dystonia and Huntington's disease. Some of the more common tracers in use for studying these disorders are detailed in Table I.

I. Multiple-System Atrophy

Multiple-system atrophy (MSA), also known as Shy–Drager syndrome, includes striatonigral degeneration (SND), olivopontocerebellar atrophy (OPCA), and pure autonomic failure (PAF) in its spectrum (see Fig. 1). In its early stages, MSA most frequently presents as an isolated akinetic-rigid syndrome (MSA-P), when it is often termed SND, but can also present as autonomic failure or as progressive ataxia (MSA-C). Fifty percent of patients with the akinetic-rigid variant show a good sustained response to levodopa, which makes them difficult to distinguish from idiopathic Parkinson's disease (PD) on clinical criteria alone (Fearnley and Lees, 1990). About 10% of cases thought to have PD during life are found to have MSA at autopsy (Hughes *et al.*, 1992).

The pathology of MSA is quite distinct from that of PD and is characterized by neuronal loss from the nigra and striatum, brainstem and cerebellar nuclei, and intermediolateral columns of the cord, with argyrophilic neuronal and glial rather than Lewy body inclusions (Papp and Lantos, 1994). Whereas in PD the ventrolateral nigral dopaminergic projections to the dorsal putamen are particularly targeted, the nigral involvement tends to be more extensive in MSA (Spokes *et al.*, 1979; Kish *et al.*, 1988; German *et al.*, 1989; Goto *et al.*, 1989;

Table I   PET and SPECT Tracers in Common Use for Studying Movement Disorders

Biological application	Tracer
Blood flow	$C^{15}O_2$, $H_2{}^{15}O$, $[^{99m-}Tc]$ HMPAO, ^{133}Xe
Glucose metabolism	$[^{18}F]$-2-Fluoro-2-deoxyglucose (^{18}FDG)
Dopamine storage	$[^{18}F]$-6-Fluorodopa ($[^{18}F]$ dopa)
Dopamine transporters	$[^{11}C]$Nomifensime, $[^{11}C]$CFT, $[^{11}C]$RTI 32, $[123I]\beta$-CIT, $[^{123}I]$-FP-CIT,$[^{123}]$IPT
Dopamine vesicle transport	$[^{11}C]$ dihydrotetrabenazine
Dopamine D_1 sites	$[^{11}C]$ SCH23390/39166
Dopamine D_2/D_3 sites	$[^{11}C]$Raclopride, $[^{11}C]$ methylspiperone, $[^{18}F]$ Fluorospiperone, $[^{76}Br]$ bromospiperone, $[^{18}F]$ fluoroethylspiperone (FSESP), $[^{123}I]$ iodebenzamide (IBZM), $[^{123}I]$ epidepride
MAOB activity	$[^{11}C]$ Deprenyl
Opioid binding	$[^{11}C]$ Diprenorphine
Benzodiazepine binding	$[^{11}C]$ Flumazenil
Muscarinic binding	$[^{11}C]$ N-Methyl-4-piperidylbenylate (NMPB)
Acetylcholinesterase	$[^{11}C]$ Physostigmine

Fearnley and Lees, 1990, 1991). At one time it was felt that SND and OPCA were distinct syndromes but at postmortem the majority of patients clinically diagnosed as SND are found to have subclinical cerebellar degeneration with argyrophilic inclusions while those diagnosed as OPCA show subclinical striatonigral degeneration (Papp and Lantos, 1994). As a consequence, SND and OPCA have now been reclassified as MSA-P and MSA-C.

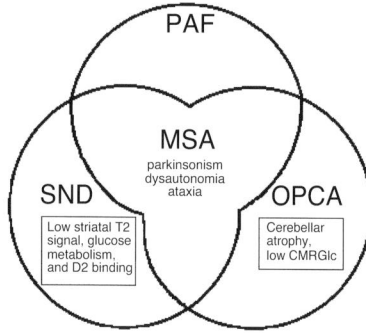

Figure 1 A Venn diagram showing the relationship between multiple-system atrophy (MSA), striatonigral degeneration (SND), olivo-pontocerebellar atrophy (OPCA), and pure autonomic failure (PAF).

A. Magnetic Resonance Imaging

MRI findings are often normal in idiopathic PD though occasional patients show increased signal from the substantia nigra, probably representing gliosis, on T_2-weighted sequences (Duguid *et al.*, 1986; Rutledge *et al.*, 1987). Formal measurement of T_2 with high-field MRI using a multiple-echo sequence can, however, demonstrate reduced nigral, caudate, and putamen relaxation times in 50% of PD cases (Antonini *et al.*, 1993).

Patients with SND have been reported to show reduced putamen signal, thought to represent excess iron deposition, on high-field T_2-weighted sequences (Drayer *et al.*, 1986; Pastakia *et al.*, 1986; Rutledge *et al.*, 1987; Savoiardo *et al.*, 1994). This low signal is often accompanied by a slitlike hyperintensity tracking the lateral edge of the putamen, matching the pattern of gliosis observed at autopsy (Konagaya *et al.*, 1994; Savoiardo *et al.*, 1994; Yagashita and Oda, 1996) (Fig. 2). SND patients may also show increased pontine tegmentum signal due to degeneration of pontine nuclei (Yagashita and Oda, 1996) (Fig. 3) and if patients have the full spectrum of multiple-system atrophy with concomitant cerebellar ataxia, then evidence of brainstem and/or cerebellar atrophy may also be present (Rutledge *et al.*, 1987; Savoiardo *et al.*, 1994; Yagashita and Oda, 1996). Although these MRI features are of diagnostic value, they are frequently absent in cases with early disease. Low putamen T_2 signal has also been occasionally reported in elderly normals (Rutledge *et al.*, 1987).

B. Metabolic Studies

Functional imaging provides a more sensitive means of detecting the presence of regional brain abnormalities in MSA. ^{18}FDG PET measurements of regional cerebral glucose metabolism primarily reflect the metabolism of nerve terminal synaptic vesicles while oxygen metabolism studies with $^{15}O_2$ PET also reflect somatic metabolism. Levels of glucose metabolism in the basal ganglia, therefore, reflect the synaptic activity of interneurons and afferent projections to those nuclei rather than that of basal ganglia efferent projections. It should be noted that an increase in synaptic or somatic metabolism when detected with functional imaging can represent either excitatory or inhibitory activity. Currently, PET and SPECT are unable to distinguish between these two possibilities.

Monkeys made hemiparkinsonian by administering the nigral toxin 1-methyl-4-phenyltetrahydropyridine (MPTP) show an increase in external, but not internal, pallidal glucose utilization (Crossman, 1990). This increased lateral pallidal metabolism following nigral de-

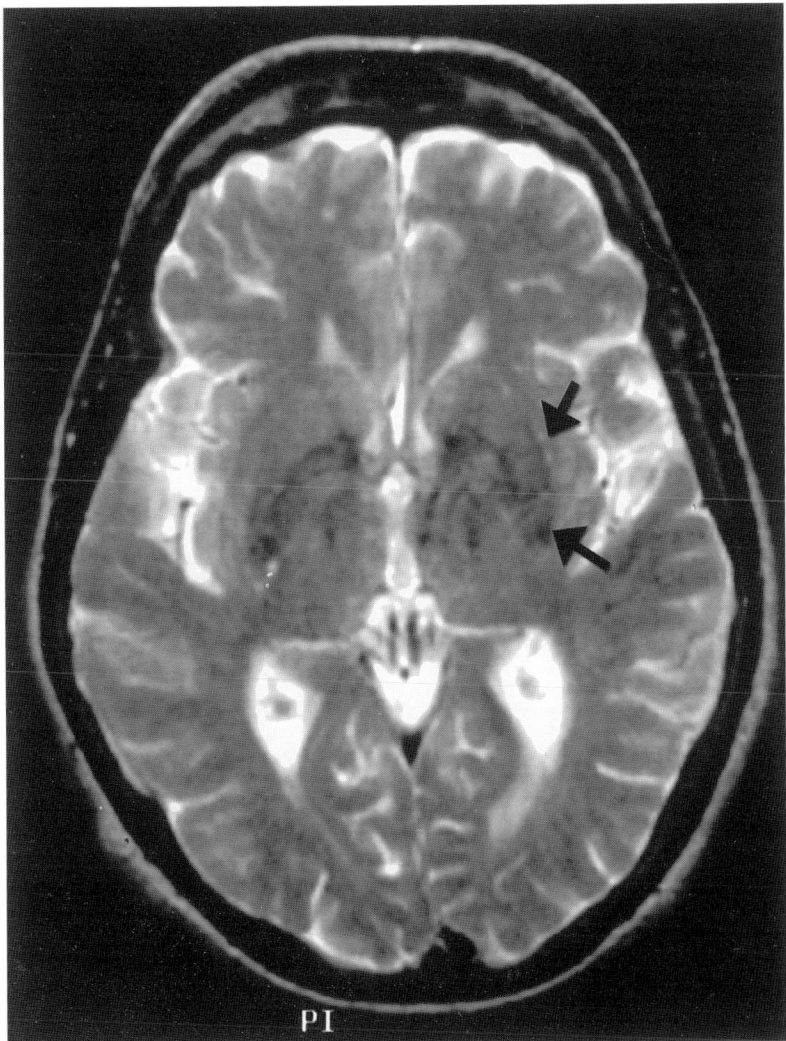

Figure 2 A T_2-weighted MRI showing posterior hypointensity and lateral slitlike hyperintensity in the left putamen of a case of multiple-system atrophy.

struction suggests that the nigrostriatal dopaminergic system normally has an inhibitory action on the GABAergic striatal projections to the external pallidum.

With PET, relatively increased oxygen and glucose metabolism can be demonstrated in the lentiform nucleus contralateral to the affected limbs in patients with early idiopathic PD (Wolfson *et al.,* 1985; Miletich *et al.,* 1988; Dethy *et al.,* 1998). When treated subjects with bilateral PD are studied, metabolism of the lentiform nuclei lies in the normal range, though raised relative to that of the caudate nucleus (Antonini *et al.,* 1997). More sophisticated covariance analysis can sensitively demonstrate an abnormal profile where resting lentiform nucleus metabolism is relatively raised and frontal metabolism reduced in PD patients; expression of this profile correlates with clinical disease severity (Eidelberg *et al.,* 1994).

[18]FDG PET studies in patients with the full syndrome of sporadic multiple-system atrophy have reported reduced levels of striatal, cerebellar, and brainstem glucose metabolism (rCMRGlc) (De Volder *et al.,* 1989; Gilman *et al.,* 1994; Otsuka *et al.,* 1996). In contrast to PD, lentiform rCMRGlc is relatively lower than that of caudate in MSA (Antonini *et al.,* 1997). Levels of lentiform nucleus glucose metabolism have also been reported to be reduced in patients with clinically probable SND (defined as levodopa-resistant parkinsonism without autonomic failure or ataxia) (De Volder *et al.,* 1989; Otsuka *et al.,* 1991; Eidelberg *et al.,* 1993; Dethy *et al.,* 1998). In one series (Eidelberg *et al.,* 1993), eight out of ten clinically probable SND cases showed reduced lentiform nucleus metabolism whereas it was normal or raised in 20 levodopa-responsive PD cases. Levels of lentiform glucose metabolism in the SND cases correlated with their locomotor score. In a series of five cases

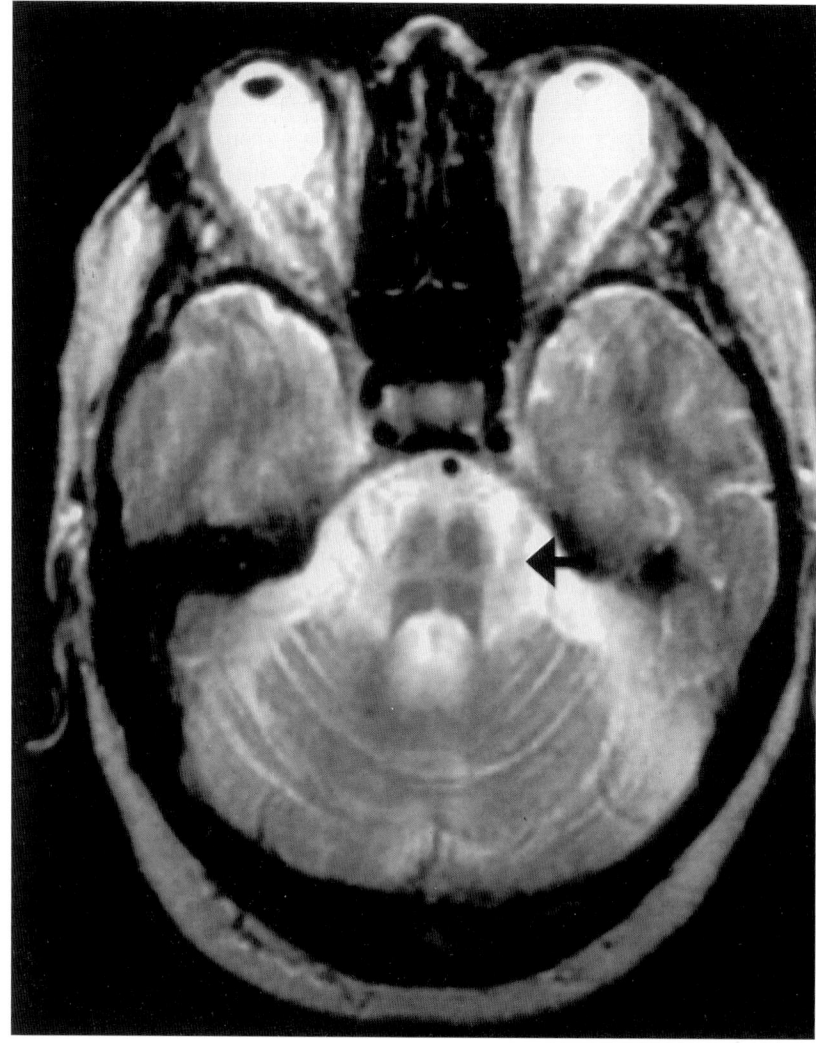

Figure 3 A T_2-weighted MRI showing raised pontine tegmental signal leading to a "hot cross bun" effect in a case of MSA.

of probable striatonigral degeneration and two cases of MSA, reduced levels of putamen and caudate glucose metabolism were found in all seven subjects while the two patients with ataxia showed additional reductions in cerebellar metabolism (De Volder *et al.*, 1989). Otsuka and co-workers (1991) reported significantly reduced striatal glucose metabolism in eight cases of probable SND while striatal glucose metabolism was preserved in their eight PD patients. On the basis of these reports, [18]FDG PET appears able to detect the presence of lentiform nucleus hypometabolism in around 80% of patients clinically suspected of having MSA or SND.

Antonini and colleagues (1998) have recently reviewed the initial [18]FDG PET findings of 104 parkinsonian patients subsequently assigned to clinically probable PD (56) or SND (48) categories based on

preservation or loss of their response to levodopa or development of severe dysautonomia. Discriminant analysis assigned 75% of the probable SND cases to an atypical parkinsonian group based on the presence of relatively reduced caudate, putamen, and thalamic rCMRGlc. All of the probable PD cases were assigned to a typical parkinsonian group. Interestingly, the presence of rest tremor did not prove to be a useful discriminator between the PD and SND cohorts in this series.

C. Proton Magnetic Resonance Spectroscopy

Proton MRS may also be helpful for discriminating SND from PD. *N*-Acetylaspartate (NAA) is a marker of the size of the amino acid pool in neurons and, therefore, a marker of neuronal viability. Davie and colleagues (1995) were able to demonstrate a reduced NAA:creatine signal ratio from the lentiform nucleus in six out of seven clinically probable SND cases while eight out of their nine PD cases had normal ratios.

D. Dopaminergic Dysfunction

Striatal $[^{18}F]$-6-fluorodopa ($[^{18}F]$dopa) uptake provides a marker of the ability of the putamen and caudate to decarboxylate exogenous levodopa and store the resultant dopamine (Firnau et al., 1987). Striatal binding of $[^{11}C]$nomifensine and tropane derivatives, such as $[^{123}I]\beta$-CIT, reflects dopamine transporter density (Hunt et al., 1974; Seibyl et al., 1995). In patients with clinically probable SND, the function of both the pre- and postsynaptic dopaminergic systems is affected. Specific putamen $[^{18}F]$dopa, $[^{11}C]$nomifensine, and $[^{123}I]\beta$-CIT uptake is reduced to 40-50% of normal levels in established SND (Brooks et al., 1990a,b; Salmon et al., 1990; Antonini et al., 1997; Brucke et al., 1997) and individual levels of putamen $[^{18}F]$dopa uptake have been shown to correlate with locomotor function (Brooks et al., 1990a,b; Antonini et al., 1997).

Pathological observations have suggested that the nigra is more uniformly involved by the pathology of MSA than by that of PD, the latter targeting ventral areas (Spokes et al., 1979; Goto et al., 1989; Fearnley and Lees, 1990). If this is so, patients with MSA would be expected to show significantly greater loss of head of caudate $[^{18}F]$dopa, $[^{11}C]$nomifensine, and $[^{123}I]\beta$-CIT uptake compared with PD. This has indeed been reported in some series (Brooks et al., 1990b; Otsuka et al., 1991; Brucke et al., 1997) but two other series found that levels of caudate $[^{18}F]$dopa or $[^{123}I]\beta$-CIT uptake were of little value in discriminating MSA from PD (Antonini et al., 1997; Brucke et al., 1997). Using formal discriminant analysis, Burn and co-workers reported that relative levels of caudate and putamen $[^{18}F]$dopa uptake discriminated clinically probable SND from PD in 70% of cases (Burn et al., 1994). ^{18}FDG PET would, therefore, seem to provide a more sensitive tool than $[^{18}F]$dopa PET or $[^{123}I]\beta$-CIT SPECT for the differential diagnosis of SND from PD.

Dopamine receptors broadly fall into D_1 type (D_1 and D_5), which are adenyl cyclase dependent, and D_2 type (D_2, D_3, and D_4), which are not. $[^{11}C]$SCH23390 PET is a marker of D_1-type receptor binding and studies with this approach have suggested that striatal D_1 binding is normal in untreated PD but 20% reduced in cases chronically exposed to exogenous levodopa (Rinne et al., 1990; Shinotoh et al., 1993; Turjanski et al., 1997). There are now numerous PET and SPECT ligands available for assessing D_2-type receptor binding.

Their findings combined suggest that putamen D_2 binding is normal or mildly upregulated in untreated PD whereas caudate D_2 binding remains normal (Playford and Brooks, 1992). In PD patients with established disease who have been chronically exposed to levodopa, putamen D_2 binding is normal whereas caudate binding becomes 20% decreased (Antonini et al., 1997; Turjanski et al., 1997) whether dyskinesias are present or absent.

Striatal dopamine D_1 binding has also been studied with $[^{11}C]$SCH23390 PET in SND (Shinotoh et al., 1993). A significant reduction in putamen D_1 binding, most evident in posterior putamen, was reported for the SND cases though individual levels of D_1 binding overlapped with both normal and PD ranges. Dopamine D_2 binding has been studied with both $[^{11}C]$raclopride and $[^{11}C]$methylspiperone PET in patients with clinically probable SND (Shinotoh et al., 1990; Brooks et al., 1992). Mild to moderate significant reductions in mean putamen $[^{11}C]$raclopride and striatal $[^{11}C]$methylspiperone binding have been reported but again an overlap between SND, normal, and PD ranges was noted. By contrast, Antonini and co-workers found that patients with the full clinical syndrome of MSA could be reliably discriminated from PD and normal cohorts on the basis of their reduced levels of caudate and putamen $[^{11}C]$raclopride binding (Antonini et al., 1997). In this series, individual levels of putamen $[^{11}C]$raclopride uptake correlated with locomotor scores.

It would seem, therefore, that although striatal D_2 binding can reliably discriminate patients with the full clinical syndrome of MSA from PD, striatal D_1 and D_2 binding are a less reliable discriminator of clinically probable SND (MSA-P) from PD. In support of this viewpoint, Schwarz et al. (1992) found reduced striatal D_2 binding with $[^{123}I]$IBZM SPECT in only 8 of their 12 de novo parkinsonian patients with a negative apomorphine response. As a significant number of parkinsonian patients who respond poorly to levodopa retain normal levels of striatal D_2 binding, it seems likely that degeneration of downstream pallidal and brainstem rather than striatal projections is responsible for their poor response to levodopa.

E. Opioid Dysfunction

Striatal projections to the external pallidum contain the opioid peptide enkephalin whereas those to internal pallidum express dynorphin. The basal ganglia are also rich in μ, κ, and δ opioid receptor subtypes. Autopsy studies have reported that putamen met-enkephalin levels are preserved in PD but reduced in SND (Goto et al., 1990). $[^{11}C]$Diprenorphine is a nonspecific opioid antagonist that binds with similar affinity to μ, κ, and δ sites. In nondyskinetic PD patients with a sustained response to levodopa, caudate and putamen [^{11}C]diprenorphine binding is preserved whereas PD patients with dyskinesias show a uniform reduction in caudate and putamen [^{11}C]diprenorphine binding (Burn *et al.*, 1995; Piccini *et al.*, 1997). In contrast, patients with SND show a selective reduction in putamen tracer uptake. It is likely that in SND the selective loss of putamen opioid binding reflects receptor degeneration whereas in dyskinetic PD increased site occupancy by raised levels of endogenous enkephalin and dynorphin reduces receptor availability to [^{11}C]diprenorphine in both caudate and putamen (Taylor *et al.*, 1992).

F. The Overlap between SND, OPCA, and PAF

A number of functional imaging studies have been designed to determine the overlap between SND, OPCA, and PAF as part of the spectrum of MSA. Seven out of a series of ten sporadic OPCA patients with autonomic failure but no parkinsonism (MSA-C) individually showed reduced putamen [^{18}F]dopa uptake while four had reduced putamen [^{11}C]diprenorphine binding indicative of the presence of subclinical SND (Rinne *et al.*, 1995). A further six cases of sporadic OPCA without parkinsonism showed a mean 30% reduction in putamen [^{18}F]dopa uptake (Otsuka *et al.*, 1994) while Gilman and colleagues (1996) reported mean 26 and 24% reductions in caudate and putamen binding of [^{11}C]dihydrotetrabenazine, a marker of dopamine vesicular transporters, in their eight sporadic OPCA cases. Striatal glucose hypometabolism (Gilman *et al.*, 1994) and reduced lentiform NAA:creatine signal ratio (Davie *et al.*, 1995) have also been reported in other series of sporadic OPCA cases. It would seem, therefore, that, in practice, the majority of sporadic OPCA cases show functional imaging evidence of subclinical striatonigral dysfunction.

One series of seven pure autonomic failure patients has been studied with [^{18}F]dopa PET. Putamen [^{18}F]dopa uptake was found to be abnormal in two of these PAF cases, suggesting that subclinical nigral dysfunction was present (Brooks *et al.*, 1990b). One of these patients subsequently developed MSA.

G. Conclusions

In summary, patients with the full clinical spectrum of MSA show reduced lentiform nucleus, cerebellar, and brainstem glucose metabolism, striatal dopamine terminal function, and putamen dopamine and opioid receptor binding. The presence of reduced putamen glucose metabolism and [^{11}C]raclopride binding reliably discriminates patients with MSA from those with PD and levels of putamen [^{18}F]dopa uptake and D$_2$ binding both correlate with locomotor function.

The presence of reduced lentiform nucleus glucose metabolism discriminates clinically probable SND (MSA-P) from PD in around 80% of cases. Reduced levels of putamen D$_1$ and D$_2$ binding appear to be a less sensitive marker of striatal degeneration. Relative levels of putamen:caudate [^{18}F]dopa or [^{123}I]β-CIT uptake discriminate up to 70% of SND cases from PD.

Sporadic OPCA (MSA-C) patients all show reduced cerebellar and brainstem glucose metabolism and around 70% also have evidence of subclinical nigral and striatal dysfunction as evidenced by reduced striatal [^{18}F]dopa, [^{11}C]dihydrotetrabenazine, and [^{11}C]diprenorphine uptake, lentiform nucleus glucose metabolism, and striatal NAA:creatine signal. These findings confirm that isolated OPCA (MSA-C) and SND (MSA-P) are truly part of an MSA spectrum. The majority of PAF patients have an intact nigrostriatal dopaminergic system, arguing against this condition being a variant of PD or MSA despite some pathological overlap. PET is capable, however, of detecting subclinical nigrostriatal dysfunction in occasional PAF patients when this is present.

II. Progressive Supranuclear Palsy

Progressive supranuclear palsy (PSP) is conventionally taken to refer to the syndrome first described by Steele, Richardson, and Olszewski in 1964. Affected patients have an associated akinetic-rigid syndrome, axial dystonia, bulbar palsy, and dementia of frontal type. At postmortem neuronal loss with neurofibrillary tangle (NFT) inclusions and gliosis are characteristically found in the basal ganglia, brainstem, and diencephalic and cerebellar nuclei, though cerebral cortex may also be affected (Jellinger *et al.*, 1980; Lees, 1986; Maher and Lees, 1986). The NFTs comprise 15-nm straight filaments, rather than the paired helical filaments characteristic of Alzheimer's disease, postencephalitic parkinsonism, and Lytico-Bodig disease. Unlike MSA, the basal ganglia targeted along with the substantia nigra in PSP are the subthalamus and globus pallidus, with a lesser involvement of the neostriatum. Unlike PD, where ventrolateral nigra compacta shows the most severe degeneration, the nigra is uniformly involved in PSP (Jellinger *et al.*, 1980; Fearnley and Lees, 1991). Brainstem areas particularly involved include the pretectal area, midbrain and pontine tegmentum, periaqueductal gray matter, corpora quadrigemina, cranial nerve nuclei controlling eye and tongue movement, pontine, cuneate, gracile, and dentate nuclei.

Although the pathology of PSP is distinctive, this condition can pose diagnostic problems in life when clinical criteria alone are applied. The supranuclear ophthalmoplegia may be absent (Dubas *et al.*, 1983) or occur late into the disease (Perkin *et al.*, 1978). On presentation, the parkinsonism of PSP has been estimated to be levodopa responsive in up to 50% of cases, though this response is rarely sustained (Jackson *et al.*, 1983). Up to 40% of PSP patients have been reported to have tremor (Jellinger *et al.*, 1980). It has been estimated that 3-6% of cases diagnosed as PD in life will turn out to have PSP at autopsy (Jackson *et al.*, 1983; Rajput *et al.*, 1991; Hughes *et al.*, 1992). The supranuclear gaze palsy that characterizes PSP has also been described in association with a number of other neurodegenerative disorders, including diffuse Lewy body disease (Fearnley *et al.*, 1991), corticobasal degeneration (Gibb *et al.*, 1989), progressive subcortical gliosis (Will *et al.*, 1988), olivo-pontocerebellar atrophy (Koeppen and Hans, 1976), and Creutzfeldt-Jacob disease (Ross-Russell, 1980).

A. Magnetic Resonance Imaging

Although an initial report suggested that, like MSA, PSP is associated with low putamen signal on T_2-weighted MR sequences (Drayer *et al.*, 1986), such a finding would run counter to the known distribution of pathology that tends to spare the putamen. Other MRI series have not reproduced this observation (Rutledge *et al.*, 1987; Savoiardo *et al.*, 1994; Yagashita and Oda, 1996). In contrast to MSA, the majority of cases of clinically probable PSP show midbrain atrophy (Fig. 4) and third ventricular dilatation (Fig. 5) while the cerebellum is spared (Giminez-Roldan *et al.*, 1994; Savoiardo *et al.*, 1994; Yagashita and Oda, 1996). Additional MRI features reported in established PSP on T_2-weighted se-

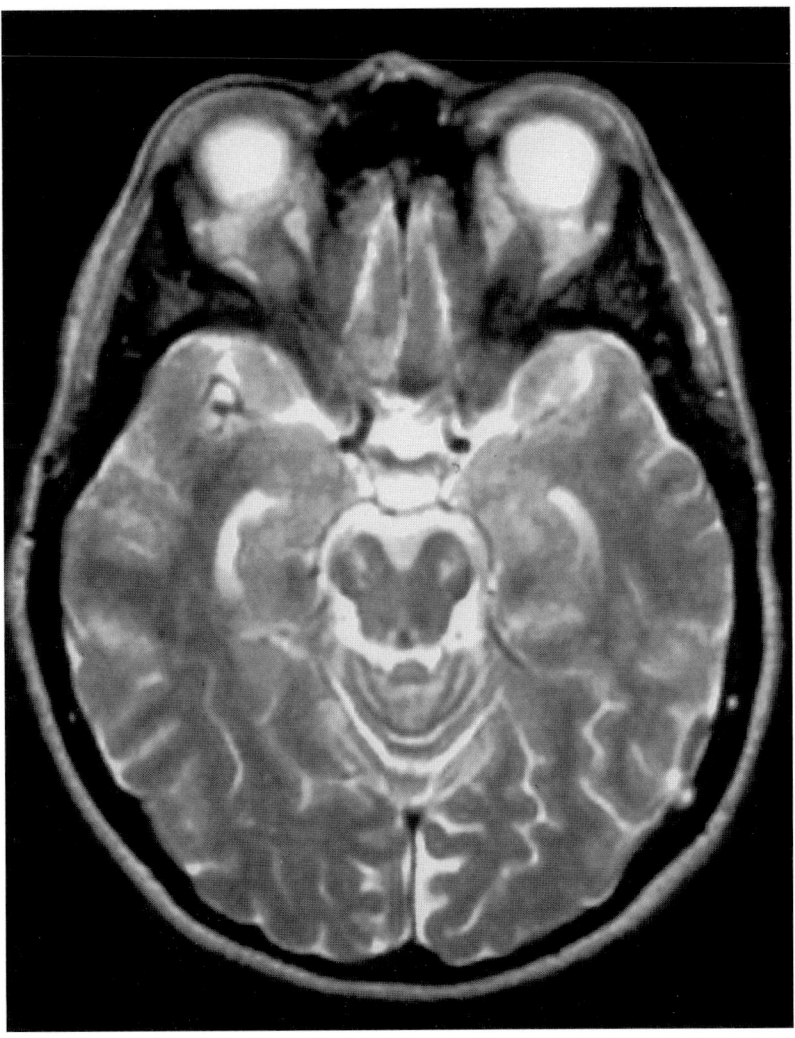

Figure 4 An MRI showing midbrain atrophy in a case of progressive supranuclear palsy.

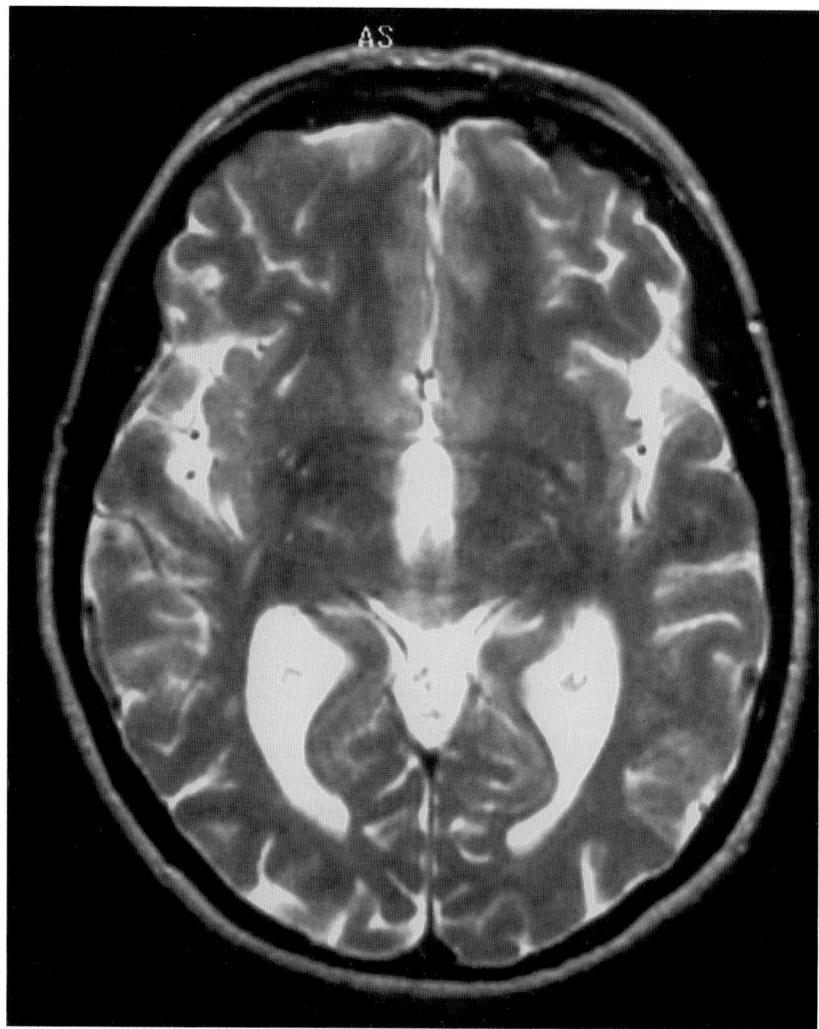

Figure 5 An MRI showing third ventricle dilatation in a case of progressive supranuclear palsy.

quences include diffuse high signal from the midbrain and pontine tegmentum and midbrain tectum (Yagashita and Oda, 1996), increased inferior olivary signal (Yagashita and Oda, 1996), and low superior colliculus signal (Savoiardo *et al.*, 1994).

B. Metabolic Studies

D'Antona and co-workers (1985) were the first to report ^{18}FDG PET findings in PSP. They compared regional cerebral glucose metabolism (rCMRGlc) in six cases with that of eight age-matched controls. Only four of their six patients had voluntary downgaze problems and one of these four had mild parkinsonism so their PSP cohort strictly comprised three clinically probable and three clinically possible cases. Four of the six patients had frontal behavioral abnormalities. Cortical glucose utilization was globally reduced to 83% of nor-

mal levels and medial rCMRGlc and lateral frontal rCMRGlc were particularly targeted (71 and 75% of normal). In this early study, performed with a single-slice tomograph, striatal and brainstem metabolism was not reported.

Leenders and colleagues (1988) reported regional cerebral blood flow (rCBF) and oxygen metabolism (rCMRO$_2$) for four cases of clinically probable PSP; the diagnosis was subsequently confirmed at autopsy in one case. Their findings were compared with those of five age-matched controls. These workers also found that cortical function in PSP was globally affected. Frontal rCMRO$_2$ and striatal rCMRO$_2$ were both reduced to 78% of normal levels; frontal rCMRO$_2$ levels correlated inversely with disease duration.

Foster and co-workers (1988) scanned 14 clinically probable cases of PSP with ^{18}FDG PET. As above, they found global cerebral hypometabolism, with relative targeting of frontal cortex. In contrast to MSA, where caudate function is relatively spared, caudate rCMRGlc and putamen rCMRGlc were uniformly reduced in their PSP cohort to 78 and 79% of normal levels. Thalamic rCMRGlc and brainstem rCMRGlc were also impaired. Three of these 14 PSP cases have subsequently come to autopsy (Foster *et al.*, 1992). Two had the diagnosis of PSP confirmed but the third was found to have progressive subcortical gliosis.

Goffinet and colleagues (1989) compared levels of rCMRGlc in nine cases of clinically probable PSP with those of 10 age-matched controls. The diagnosis in one of their PSP patients was later confirmed at autopsy. Seven of their PSP patients showed hypofrontality, one case had entirely normal cerebral metabolism, and one case showed diffusely reduced rCMRGlc. These workers reported that premotor cortex was marginally more affected than prefrontal cortex, rCMRGlc being reduced to 73% compared with 77% of normal. Striatal, thalamus, and cerebellar rCMRGlc was also significantly reduced. The authors found no correlation between levels of rCMRGlc and cognitive performance.

The largest PET series reported for PSP is that of Blin and colleagues (1990), who extended D'Antona's original study, reporting PET findings on a total of 41 PSP cases (25 clinically probable and 16 clinically possible). Two different PET cameras were used during this series and some patients had glucose while others had oxygen metabolic studies. To facilitate comparisons between PSP and controls, regional cerebral metabolic data for individual patients were described as a fraction of the mean control rCMRGlc or rCMRO$_2$ values for that brain region with the particular PET camera used. As in other series, cerebral metabolism was found to be diffusely reduced in PSP, particularly targeting the frontal cortex. Caudate, putamen, thalamic, and cerebellar metabolism was also significantly impaired. Normalized levels of frontal metabolism were inversely related to disease duration and correlated with intelli-

gence scores derived from subtests of the Wechsler Adult Intelligence Scale (WAIS) and performance on a battery of tests thought to be sensitive to frontal lobe function (verbal fluency, card sorting). Locomotor function, rated with a modified Columbia scale, correlated with caudate and thalamic, but surprisingly not putamen, metabolism.

Karbe and co-workers (1992) reported [18]FDG findings in nine cases of probable PSP. Like other authors, these workers found reduced frontal and striatal metabolism in PSP but in their series striatal function was marginally more affected than frontal function. Otsuka and colleagues (1989) reported a combination of [18]FDG, H$_2$[15]O, and [15]O$_2$ PET findings for four PSP cases. Frontal cortex and striatum were again targeted, with the caudate generally more affected than the lentiform nucleus. Regional cerebral oxygen metabolism and blood flow were reduced in a coupled fashion throughout the cerebrum. Johnson *et al.* (1992) have examined regional cerebral blood flow with [^{123}I]IMP-SPECT in 11 PSP cases and were also able to demonstrate frontal and striatal hypofunction in PSP.

Table II summarizes the regional reductions in cerebral metabolism reported by various workers in PSP. All studies have shown a global reduction in cerebral metabolism in PSP, frontal cortex, striatum, thalamus, and cerebellum being particularly targeted. Figure 6 shows a typical regional cerebral blood flow scan for a PSP patient. Two studies have suggested that superior frontal cortex may be more affected than prefrontal cortex in PSP, though the differences in the relative degrees of involvement have not been statistically significant. One study has reported a significant correlation between frontal metabolism and performance on tests of fluency and sorting ability and between striatal metabolism and locomotor function.

The sensitivity of PET for detecting regional cerebral hypometabolism in PSP has only been formally addressed by one study (Goffinet *et al.*, 1989). These au-

Table II   Regional Cerebral Metabolism in PSP (Percent of Normal)

Source	Number	Whole cortex	Frontal	Occipital	Striatal	Thalamic	Cerebellar
D'Antona *et al.* (1985)	6	83	73	93	—	—	—
Leenders *et al.* (1988)	4	—	78	89	78	—	—
Foster *et al.* (1988)	14	81	75	—	79	84	91
Goffinet *et al.* (1989)	9	85	75	86	79	69	76
Blin *et al.* (1990)	41	—	78	83	81	73	79
Karbe *et al.* (1992)	9		79	87	76	77	81
Otsuka *et al.* (1989)	4	75	72	86	72	73	86
Johnson *et al.* (1992)	11	—	78	—	21		—

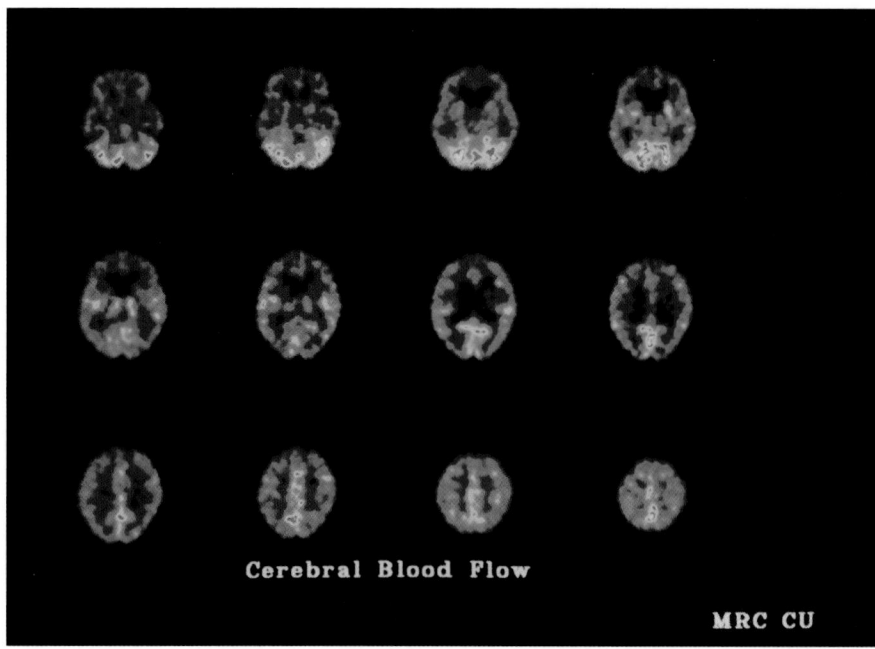

Figure 6 $H_2{}^{15}O$ PET images showing symmetrically reduced regional cerebral blood flow in frontotemporal, basal ganglia, and thalamic areas of a patient with probable progressive supranuclear palsy.

thors found that significant hypofrontality was present in seven out of nine PSP patients; inspection of their data suggests that six also had significant striatal hypometabolism and that altogether eight out of the nine showed significant abnormalities of regional cerebral metabolism compared with controls.

The mechanism of the hypofrontality that features in PSP is still under debate. Although there is widespread subcortical pathology in this condition, cortical involvement is relatively minor (Steele *et al.*, 1964). Dopaminergic, noradrenergic, and serotonergic projections all degenerate in PSP but loss of these fibers in Parkinson's disease and animal models is not necessarily associated with cortical hypometabolism (McCulloch *et al.*, 1984; Savaki *et al.*, 1984; Eidelberg *et al.*, 1995). Acute lesions of the nucleus basalis resulting in loss of cholinergic projections are associated with transient diffuse cortical hypometabolism in monkeys (Kiyosawa *et al.*, 1989) whereas thalamic infarcts cause sustained diffuse hemisphere dysfunction (Pappata *et al.*, 1990). As there is loss of both nucleus basalis and thalamic neurons in PSP, this may contribute toward the diffuse cortical hypofunction present. The most likely explanation for the hypofrontality in PSP, however, is pallidal degeneration. Internal pallidal neurons are known to project via ventrolateral thalamus to premotor and prefrontal areas (Alexander *et al.*, 1986). Toxic and vascular pallidal lesions have been shown to cause selective decreases in

frontal metabolism associated with obsessive behavior (Laplane *et al.*, 1989). As the internal pallidum is targeted in PSP, damage to this structure is likely to contribute to the frontal cognitive deficits found in PSP patients.

There has recently been some debate concerning whether patients with isolated progressive akinesia may have a variant of PSP. Taniwaki *et al.* (1992) have studied three such patients and, as in PSP, found reduced striatal and frontal glucose hypometabolism, in support of this hypothesis. Their akinesia patients all showed pretectal and pontine atrophy on MRI. A case of pure progressive akinesia due to Lewy body disease has, however, been described (Quinn *et al.*, 1989), suggesting that this syndrome may have multiple pathologies.

C. Proton Magnetic Resonance

Three series have examined basal ganglia metabolism with proton MRS in PSP. Davie and co-workers (1997) reported reduced lentiform nucleus NAA:Cr ratios in seven out of their nine patients with clinically probable PSP. Tedeschi and colleagues (1997) noted a nonsignificant 23% reduction in lentiform nucleus NAA:Cr ratio in their 12 PSP cases though the NAA:choline ratio was significantly reduced. The NAA:Cr ratio was, however, significantly reduced in frontal and parietal cortex, centrum semiovale, and the brainstem. No functional abnormalities were detected in 10 PD cases studied but a direct comparison between PSP and PD groups failed to reveal significant regional metabolic differences. Federico and co-workers (1997) reported a significant mean 29% reduction in basal ganglia NAA:Cr ratio for five PSP patients compared with nine controls. No functional abnormalities were noted in eight PD cases. It would appear, therefore, that while proton MRS is capable of detecting basal ganglia dysfunction in a majority of individuals with PSP, its value for discriminating PSP from PD is still unclear.

D. The Dopaminergic System

At autopsy, a uniform loss of nigrostriatal dopaminergic projections has been reported in PSP, caudate and putamen dopamine content being equivalently reduced to 9-25% of normal levels (Bokobza et al., 1984; Kish et al., 1985; Ruberg et al., 1985). This distinguishes PSP from PD, where caudate putamen levels are relatively spared (Kish et al., 1988). Leenders and co-workers (1988) were the first to measure striatal [^{18}F]dopa signals in PSP and found that the mean striatal:temporal cortex uptake ratio was reduced to 87% of normal in their five patients. The 1.7-cm resolution of their camera was unable to separate caudate from putamen signal. Striatal [^{18}F]dopa uptake correlated with disease duration but not with locomotor disability.

Other workers have subsequently reported reduced striatal [^{18}F]dopa uptake both in PSP (Brooks et al., 1990a; Bhatt et al., 1991; Taniwaki et al., 1992) and in pure progressive akinesia (Taniwaki et al., 1992). Brooks and colleagues (1990a) measured putamen and caudate [^{18}F]dopa influx constants (K_i) in 10 patients with clinically probable PSP. Loss of striatal activity in the PSP cases was uniform, mean putamen and caudate tracer uptake being reduced to 37 and 44% of normal. Mean caudate tracer uptake was significantly more impaired in PSP than in equivalently disabled PD cases (47 versus 74% of normal). Ninety percent of individual PSP cases showed reduced caudate [^{18}F]dopa influx compared with only 25% of PD patients. This differential involvement of caudate [^{18}F]dopa uptake subsequently allowed 90% of probable PSP patients to be separated from PD by formal discriminant analysis (Burn et al., 1994). In contrast to PD, neither locomotor function nor disease duration of individual PSP patients was found to correlate with putamen or caudate [^{18}F]dopa uptake. It would appear that, whereas in PD locomotor disability is primarily determined by the extent of loss of nigrostriatal dopaminergic projections, in PSP it is determined by the extent of loss of other basal ganglia and brainstem connections.

There have been two [^{123}I]β-CIT SPECT reports concerning striatal dopamine transporter binding in PSP. In the first series (Brucke et al., 1997), where 13 MSA (9) and PSP (4) patients were considered as a combined group, it was noted that their mean putamen:cerebellar [^{123}I]β-CIT uptake ratio was reduced to 36% of normal, comparable to the 37 and 46% reductions seen for two groups of PD patients. Caudate [^{123}I]β-CIT binding was most severely reduced in the MSA/PSP cohort (44% of normal compared with 60 and 49% for the two groups of PD patients) but these reductions were not found to be significantly different. In the second series (Messa et al., 1998), 13 PD and 5 PSP cases were studied. Again, putamen [^{123}I]β-CIT binding was severely reduced in the PD (30% of normal) and PSP (21% of normal) groups but in this series caudate [^{123}I]β-CIT binding was significantly more affected in the clinically probable PSP (28% of normal) compared with PD (53% of normal) patients. On the basis of their relative levels of caudate and putamen [^{123}I]β-CIT uptake, 10 of the 13 PD cases could individually be discriminated from the PSP patients.

Baron and co-workers (1986) were the first to use PET to measure striatal D$_2$ binding in vivo in PSP. They scanned seven PSP patients with the reversibly bound D$_2$ antagonist [^{76}Br]bromospiperone (BSP) and found a significant mean 24% fall in the equilibrium striatum:cerebellum tracer uptake ratio. Two of their seven PSP patients, however, had striatal:cerebellar BSP uptake ratios within the normal range and only three PSP patients individually showed a significant fall in striatal D$_2$ binding. Six of the seven PSP patients were taking dopaminergic replacement medication at the time of PET. The effects of such medication on striatal D$_2$ binding remains unclear and it is possible that the medication could have contributed to the observed reduction in striatal BSP uptake.

Wienhard and colleagues (1990) scanned two PSP patients using the irreversibly bound D$_2$ antagonist [^{18}F]fluoroethylspiperone. They found a mean 17% fall in caudate [^{18}F]FESP binding potential (B_{max}/K_d) but did not report levels of putamen binding. Brooks et al. (1992) studied nine PSP patients with [^{11}C]raclopride PET. Four of their nine patients were taking dopaminergic medication. The PSP group showed significant 24 and 9% falls in equilibrium caudate:cerebellum and putamen:cerebellum tracer uptake ratios. Individually, five of the nine PSP patients had caudate, and three putamen, [^{11}C]raclopride binding that was significantly reduced. Figure 7 shows the [^{11}C]raclopride PET scan for one of their patients. There was no correlation between striatal [^{11}C]raclopride uptake and either disease

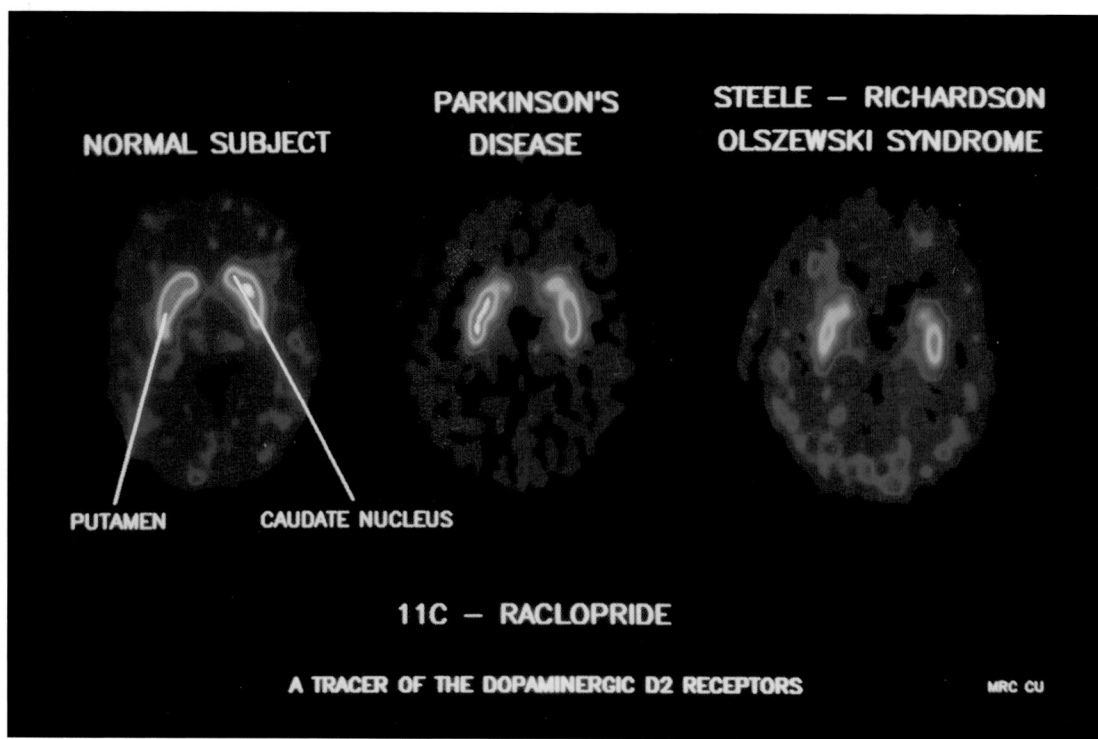

Figure 7 PET images of striatal [^{11}C]raclopride uptake in a normal subject and in PD and progressive supranuclear palsy patients. It can be seen that striatal dopamine D$_2$ binding is reduced in progressive supranuclear palsy.

duration or the presence or absence of dopaminergic treatment. Schwarz and colleagues (1997) have reported reduced striatal uptake of [^{123}I]IBZM in a single case of PSP.

The above PET findings of moderate reductions in striatal D$_2$ receptor binding in clinically probable PSP cases are in broad agreement with postmortem reports. Using [^{3}H]spiperone, Bokobza *et al.* (1984) found that five PSP patients showed mean 37 and 30% reductions in putamen and caudate dopamine D$_2$ receptor density. In a follow-up autopsy series of seven PSP patients, Ruberg *et al.* (1985) reported mean 42 and 48% losses of putamen and caudate D$_2$ sites whereas Pierot *et al.* (1988) found mean 28 and 36% losses of putamen and caudate D$_2$ sites in their 11 PSP patients. Given these relatively minor reductions in striatal D$_2$ binding at autopsy, it is likely that degeneration of downstream pallidal and brainstem projections is in part responsible for the poor L-dopa response in PSP.

E. Opioid and Cholinergic Binding

One study has examined striatal opioid binding in PSP (Burn *et al.*, 1995). Striatal:occipital [^{11}C]diprenorphine uptake ratios were reduced in all six PSP cases studied compared with nondyskinetic PD cases where opioid binding was normal. Caudate and putamen opi-

oid binding were equally affected in PSP; this contrasted with SND, where caudate function was spared in all patients.

Pappata and colleagues (1997) have measured regional cerebral acetylcholinesterase (AChE) activity with [^{11}C]physostigmine ([^{11}C]PHY) PET in eight patients with probable PSP. They found significant reductions in AChE activity in the striatum and a significant correlation between levels of striatal [^{11}C]PHY uptake and locomotor score. This suggests that loss of striatal interneuronal cholinergic function may play a greater role than loss of dopaminergic dysfunction in causing disability in PSP.

With [^{11}C] *N*-methyl-4-piperidylbenzylate ([^{11}C]NMPB) PET, Asahina and colleagues (1998) have compared regional cerebral muscarinic binding in patients with probable PSP and PD. The PSP cases all had impaired mini-mental test scores but showed no significant changes in either cortical or striatal muscarinic binding. In contrast, the nondemented PD patients showed raised levels of frontal muscarinic binding, possibly reflecting lower occupancy by endogenous acetylcholine due to degeneration of cholinergic projections from the nucleus basalis of Meynert. The authors concluded that loss of cortical cholinergic function was unlikely to explain the dementia characteristically present in PSP.

F. Conclusions

In summary, mean striatal [^{18}F]dopa and [^{123}I]β-CIT uptake is reduced to 20–40% of normal in PSP. In contrast to PD, where caudate dopamine terminal function is spared, caudate and putamen are similarly affected in PSP. Locomotor disability in PSP does not correlate with loss of striatal [^{18}F]dopa storage but does correlate with impairment of striatal glucose metabolism and acetylcholineesterase activity, suggesting that it is influenced more by striatal than nigral degeneration. Whereas mean striatal D$_2$ receptor binding is moderately reduced in PSP, 50% of individual patients show normal levels of striatal D$_2$ receptor function. As these patients have L-dopa-resistant akinetic-rigid syndromes, their lack of therapeutic response is likely to reflect degeneration of striatal interneurons and pallidal projections rather than loss of striatal dopamine receptors.

III. Corticobasal Degeneration

Corticobasal degeneration (CBD) has also been labeled corticobasal ganglionic degeneration, corticodentatonigral degeneration, and neuronal achromasia (Rebeiz et al., 1968; Riley et al., 1990). It is a late-onset, slowly progressive degenerative disorder characterized clinically by asymmetrical limb rigidity, dystonia, apraxia, cortical sensory loss and myoclonus, supranuclear gaze problems, pseudobulbar dysfunction, and instability of gait (Gibb et al., 1989; Riley et al., 1990; Litvan et al., 1997). Less commonly, dysphasia, frontal behavior, and frank dementia are features. Tremor, when present, tends to be postural or action rather than resting in character. The limb myoclonus is typically stimulus sensitive, suggestive of a cortical origin (Thompson et al., 1994). The limb that is initially affected may become functionally useless by end-stage and exhibit "alien" features; that is, it performs involuntary but apparently purposeful movements. The akinesis and rigidity associated with CBD rarely show a useful response to dopaminergic agents.

At postmortem swollen ubiquitin- and τ-positive achromatic neurons in the absence of argyrophilic Pick bodies are asymmetrically distributed in posterior frontal, inferior parietal, and superior temporal cortical areas and in the thalamus, striatum, and substantia nigra (Gibb et al., 1989; Halliday et al., 1995). Basophilic τ-positive but ubiquitin-negative inclusions are occasionally present in the substantia nigra (Gibb et al., 1989) and τ-positive astrocytic plaques are also a feature of this disorder (Feaney and Dickson, 1995). In Parkinson's disease the ventrolateral substantia nigra

compacta is particularly targeted, leading to greatest dopamine depletion in the putamen (Kish et al., 1988; Fearnley and Lees, 1991). In contast, in CBD the nigra is uniformly involved, resulting in similar levels of dopamine loss in both caudate and putamen (Riley et al., 1990).

While the pathology and distribution of CBD are distinctive, neuronal achromasia is also a feature of Pick's disease and other rarer neurodegenerative disorders (Riley et al., 1990), and the basophilic inclusions seen in CBD can mimic the neurofibrillary tangles found in progressive supranuclear palsy (Litvan et al., 1997). CBD can pose considerable diagnostic problems in life when clinical criteria alone are applied and the full spectrum of this disorder is still not evident. Combinations of ideomotor apraxia, parkinsonism, dystonia, and myoclonus may be features of Pick's and Alzheimer's diseases, prion disorders, diffuse Lewy body disease, striatonigral degeneration, and hemiatrophy-hemiparkinsonism syndrome whereas supranuclear gaze problems, pseudobulbar dysfunction, parkinsonism, and gait instability are all features of PSP. It has been estimated that only one-third of CBD cases are correctly diagnosed on clinical criteria at first visit and only one-half by end-stage (Litvan et al., 1997).

A. Magnetic Resonance Imaging

There have been a number of reports concerning MRI/CT findings in CBD, though, to date, none have employed a formal volumetric approach. Riley and co-workers (1990) reported structural radiological findings for 15 cases of clinically probable CBD (two later confirmed at autopsy). Eight (53%) of these patients had asymmetrical cortical atrophy contralateral to the more affected limbs while another six (40%) showed symmetrical brain atrophy. In a review of 36 probable cases of CBD (seven confirmed at autopsy), Rinne and co-workers (1994) found evidence of asymmetrical cortical atrophy in 10 (28%) of their subjects while a further 17 (47%) showed symmetrical loss of brain volume. Five out of eight probable CBD cases showed asymmetrical cortical atrophy in Hauser's series (Hauser et al., 1996) but this was also true of two out of their eight Parkinson's disease patients. Grisoli and co-workers (1995) found evidence of asymmetrical cortical atrophy in all 10 of their probable CBD cases when MRIs were reviewed retrospectively but these workers remarked that this asymmetry had not been noted initially. It would seem, therefore, that about half of the patients suspected of having CBD will show CT or MR evidence of asymmetrical cortical volume loss in life.

Loss of putamen signal on high-field T_2-weighted MRI is said to be a feature of striatal degeneration. One of Grisoli's two CBD cases examined with 1.5-T MRI showed this phenomenon and in Hauser's series it was seen in all eight of their CBD patients; however, it was also noted in two out of their six PD cases. In contrast to PSP, midbrain atrophy is not a feature of CBD (Giminez-Roldan *et al.*, 1994; Savoiardo *et al.*, 1994).

To summarize, structural brain imaging can be of diagnostic help in CBD if asymmetric frontoparietal atrophy is evident. In practice, however, only around 50% of suspected cases show such atrophy and it can be fairly subtle when present. The presence of midbrain atrophy may help to distinguish PSP from CBD where diagnostic doubt exists.

B. Metabolic Studies

Sawle and co-workers (1991a) studied resting levels of regional cerebral oxygen metabolism (rCMRO$_2$) with $^{15}O_2$ PET in six patients with probable CBD, one of whom subsequently had the diagnosis confirmed on biopsy. All had an asymmetrical non-levodopa-responsive akinetic-rigid syndrome with limb apraxia while four had a supranuclear gaze palsy, three had limb myoclonus, and three exhibited alien limb phenomena. None of the six were demented at the time of PET but

four had frontal lobe release signs and one had expressive and receptive dysphasia. Five out of the six showed generalized cerebral atrophy on CT while no atrophy was evident in the sixth.

The CBD patients showed a global reduction in cortical oxygen metabolism (rCMRO$_2$). When regional brain rCMRO$_2$ data were normalized to values measured for primary visual cortex, an area spared by CBD pathology, relative reductions in oxygen metabolism were found in superior prefrontal cortex (17%), in lateral (16%) and mesial premotor (19%) areas, and in sensorimotor (16%), inferior parietal (16%), and superior temporal (21%) cortices. The reductions in cortical metabolism were strikingly asymmetrical in these CBD cases (see Fig. 8) and, in contrast to Pick's disease, inferior frontal and temporal cortex was relatively spared. The only significantly affected subcortical area in this series was the thalamus.

Eidelberg and co-workers (1991) measured side-to-side asymmetry of regional cerebral glucose metabolism (rCMRGlc) in five cases of clinically probable CBD, one of whom had the diagnosis subsequently confirmed at necropsy. All five patients had progressively worsening, levodopa-resistant unilateral limb rigidity or tremor associated with apraxia (4), cortical sensory im-

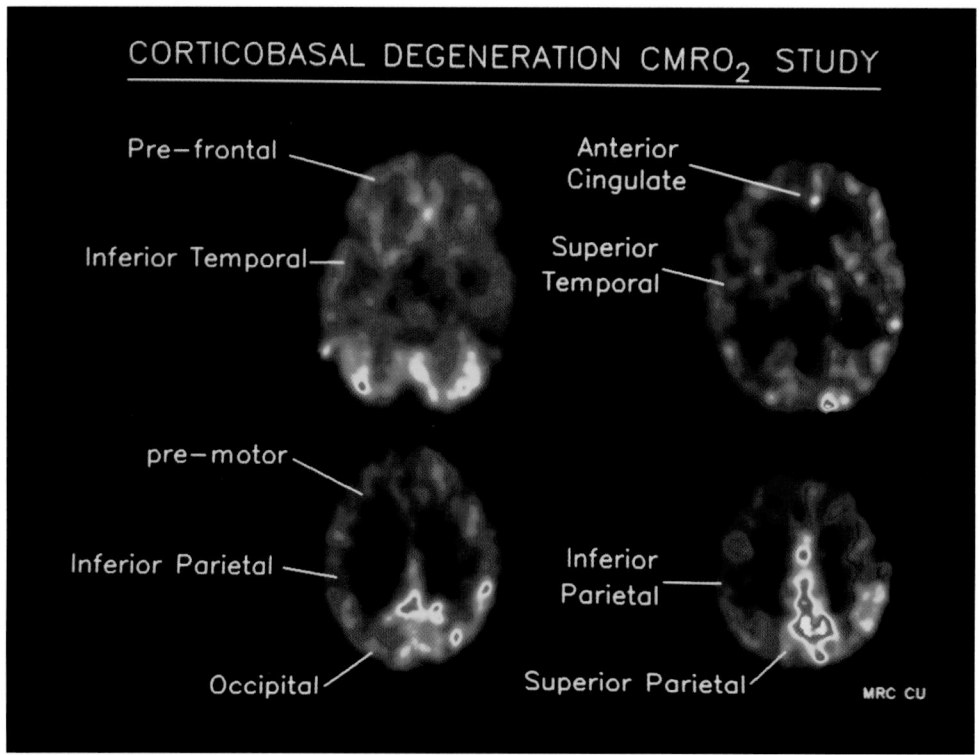

Figure 8 PET images of $^{15}O_2$ uptake showing asymmetrically reduced oxygen metabolism in superior frontal, superior temporal, inferior parietal, and thalamic areas of a patient with probable corticobasal degeneration (picture courtesy of G. V. Sawle).

pairment (4), and pyramidal signs (5). Three of the five had alien limb phenomena and one was dysphasic. No intellectual decline or supranuclear gaze difficulties were evident. Two (40%) patients had asymmetrical cortical atrophy, one had symmetrical cortical atrophy, and the other two had normal structural imaging.

Brain glucose metabolism was globally reduced by 38% in CBD and lateral temporal, inferior parietal, and frontal opercular regions were targeted. Left-right metabolic asymmetries were significantly increased in inferior parietal, medial temporal, and thalamic regions compared with both normal and PD cohorts. Individually, all of the five CBD patients showed a greater than 5% side-to-side asymmetry in parietal rCMRGlc. In contrast, none of the nine PD patients studied showed a greater than 5% side-to-side rCMRGlc asymmetry in any brain region although all these subjects had asymmetrical signs.

Blin and co-workers (1992) reported ^{18}FDG PET findings for five CBD patients. All their patients had a progressive non-levodopa-responsive asymmetric akinetic-rigid syndrome associated with apraxia. Four patients had a dystonic limb and two exhibited alien limb phenomena though none had a supranuclear gaze paresis. Three of the five showed "frontal" behavior though none of the subjects were demented. Four of these five subjects had normal MRI findings while the fifth showed mild left hemisphere atrophy. Mean resting levels of glucose metabolism were significantly reduced in frontal, temporal, sensorimotor, and parietal areas. Subcortically, caudate, lenticular, and thalamic metabolism was also impaired. There was a marked asymmetry in these metabolic reductions, the hemisphere contralateral to the clinically more affected side being worse affected.

Nagasawa and colleagues (1996) studied six cases of clinically probable CBD with ^{18}FDG PET. All six had progressive limb rigidity and apraxia and three exhibited alien limb phenomena. One was demented, another was aphasic, and none responded to levodopa therapy. MRI showed mild symmetrical cortical atrophy in five and right frontotemporal atrophy in the sixth. These workers reported asymmetrically reduced levels of rCMRGlc in parietal and sensorimotor cortex, thalamus, caudate nucleus, and putamen contralateral to the dominantly affected limbs. Occipital, inferior frontal, and temporal cortices were relatively spared, as previously noted.

Markus and co-workers (1995) compared [^{99m}Tc] HMPAO SPECT findings for eight clinically probable CBD patients with those obtained for 12 age-matched PD patients and 12 normal controls. The eight CBD cases all had asymmetric, non-levodopa-responsive limb rigidity, while seven had apraxia, four exhibited

alien limb phenomena, four had a cortical sensory deficit, and three had limb myoclonus. Six out of the eight cases showed cerebral atrophy on MRI but this was felt to be asymmetrical in only one patient. The rCBF images of the CBD patients all showed striking asymmetry and there were significant reductions of rCBF in the more affected hemisphere, targeting posterior frontal cortex (12%), the parietal lobe (13%), caudate (9%), putamen (10%), and thalamus (9%). Seven of the eight CBD cases individually showed significant reductions in contralateral parietal rCBF. Okuda and co-workers (1995) have reported [^{123}I]IMP SPECT findings for two cases of possible CBD. Both patients had increased limb rigidity with apraxia, which was felt to be limb-kinetic in one case and constructional in character in the other. SPECT showed reduced levels of resting rCBF in perirolandic areas and posterior parietal cortex.

Nagahama and colleagues (1997) compared the pattern of rCMRGlc in eight patients with clinically probable CBD and eight patients thought to have PSP. The CBD cases showed the characteristic pattern of metabolic derangement compared with a normal control group. When compared with the PSP cases, the CBD cohort showed relatively reduced levels of inferior parietal, striatal, and thalamic glucose metabolism in their worst affected hemispheres. This was true even when the comparison was run against the five most asymmetric PSP cases.

C. Proton Magnetic Resonance

Regional cerebral function has also been studied with proton MRS in nine patients with probable CBD (Tedeschi *et al.*, 1997). These cases all had a progressive asymmetrical parkinsonian syndrome that was unresponsive to dopaminergic medication with associated limb dystonia, myoclonus, apraxia, or cortical sensory loss. Eight of the nine cases had cortical atrophy. The NAA:Cr ratio was reduced in the centrum semiovale of CBD cases but not in cortical regions or the basal ganglia. The authors concluded that levels of lenticular metabolism could not be reliably used to discriminate CBD from PD.

D. The Dopaminergic System

Sawle and co-workers (1991a) reported that both caudate and putamen [^{18}F]dopa influx constants (K_i) were equivalently and asymmetrically reduced in their six CBD cases, reductions being greatest in the structures contralateral to the clinically more affected limbs (see Fig. 9). Mesial frontal [^{18}F]dopa uptake was also reduced in their patients, suggesting that the function of

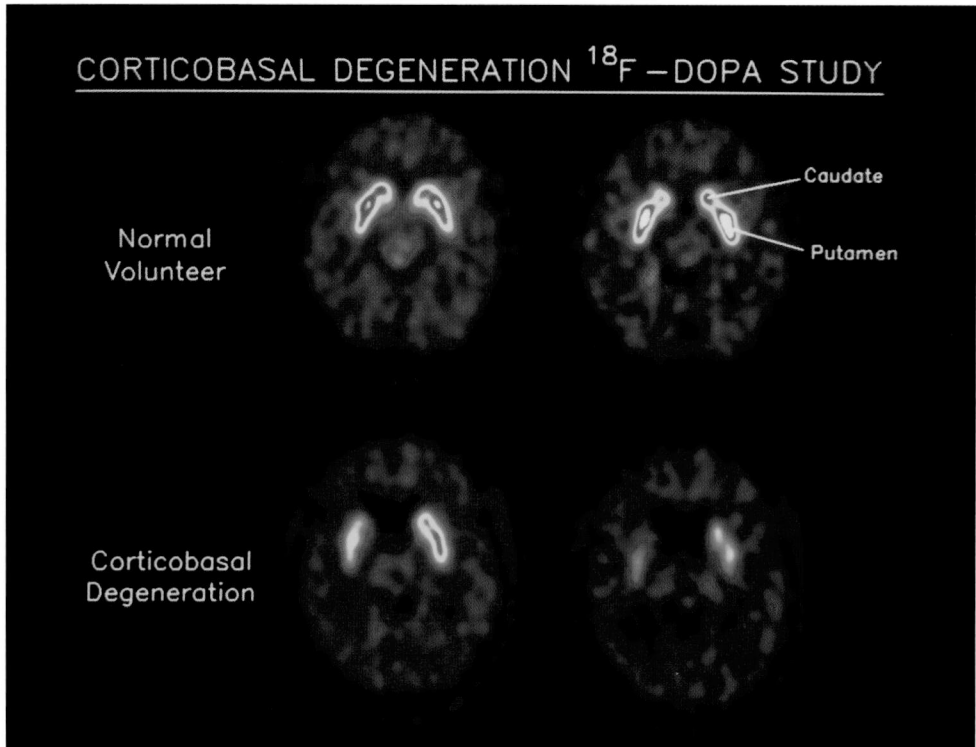

Figure 9 PET images of striatal [^{18}F]dopa uptake in a normal subject and in PD and corticobasal degeneration patients. It can be seen that while striatal dopaminergic function is asymmetrically reduced in both PD and CBD, caudate and putamen are equivalently affected in the latter (picture courtesy of G. V. Sawle).

both mesofrontal and nigrostriatal dopaminergic projections is affected in this condition. Eidelberg and co-workers (1991) also reported that striatal [^{18}F]dopa K_i was asymmetrically reduced in a single case of CBD. Nagasawa *et al.* (1996) studied four of their six CBD cases with [^{18}F]dopa PET. In agreement with Sawle and co-workers, these workers found an asymmetric reduction in striatal K_i. Caudate [^{18}F]dopa uptake and putamen [^{18}F]dopa uptake were equivalently affected and reduced to around 25% of normal in the most affected hemisphere.

There have been two [^{123}I]IBZM SPECT case reports on striatal dopamine D$_2$ receptor binding in CBD. Frisoni *et al.* (1995) scanned a 44-year-old male with an apraxic, akinetic-rigid left arm, sensory inattention, and reflex myoclonus. The MRI showed right-sided cortical atrophy and basal ganglia [^{123}I]IBZM uptake was asymmetric, being 14% lower in the right striatum relative to the left. Schwarz and colleagues (1998) noted that a single case of CBD had reduced striatal IBZM binding.

E. Conclusions

In summary, PET and SPECT metabolic studies on patients with clinically probable CBD have all shown strikingly asymmetrical reductions in resting levels of brain function. The hypometabolism is most evident in the brain hemisphere contralateral to the more affected limbs and particularly targets posterior frontal, inferior parietal, and superior temporal regions along with thalamus and striatum. Proton MRS has demonstrated reduced NAA:Cho ratios in the parietal cortex and lentiform nucleus in CBD.

Striatal [^{18}F]dopa uptake is also asymmetrically reduced in CBD and, in contrast to PD, putamen and caudate are similarly affected. Mean striatal dopamine D$_2$ receptor binding is moderately reduced in CBD, which may in part explain the levodopa-resistant akinetic-rigid syndromes that these patients manifest.

IV. Dopa-Responsive Dystonia

Dominantly inherited dopa-responsive dystonia (DRD) is related to GTP-cyclohydrolase 1 deficiency in the majority of cases, the genetic defect being located on chromosome 14 (Ichinose *et al.*, 1994). This enzyme constitutes part of the tetrahydrobiopterin synthetic pathway, the cofactor for tyrosine hydroxylase. Patients are unable to manufacture dopa, and hence dopamine, from endogenous tyrosine but can still convert exogenous levodopa to dopamine. DRD cases generally present in childhood with diurnally fluctuating dystonia and later develop background parkinsonism (Nygaard *et al.*, 1990). When DRD presents in adulthood, it is usually as pure parkinsonism.

[^{18}F]Dopa PET findings are normal in the majority of DRD patients (Sawle *et al.*, 1991b; Snow *et al.*, 1993), which distinguishes this condition from early-onset dystonia-parkinsonism, where severely reduced putamen [^{18}F]dopa uptake is found (Snow *et al.*, 1993; Turjanski *et al.*, 1993). One might predict that striatal D_2 binding would be raised in dopa-naive DRD cases as an adaptive response to the chronically low levels of striatal dopamine present; this has been reported to be the case (Kishore *et al.*, 1995). Treated DRD cases show normal striatal D_2 binding.

V. Akinetic-Rigid Huntington's Disease

Huntington's disease (HD) is an autosomal dominantly transmitted disorder associated with an excess of CAG triplet repeats (>38) in the IT15 gene on chromosome 4 (The Huntington's Disease Collaborative Research Group, 1993). The function of this gene is still uncertain but the pathology of HD targets medium spiny projection neurons in the striatum. Patients with predominant chorea show a selective loss of striatolateral pallidal projections which express GABA and enkephalin whereas those with the young-onset akinetic-rigid syndrome show additional severe loss of striatomedial pallidal fibers containing GABA and dynorphine (Albin *et al.*, 1990).

In contrast to PD, akinetic-rigid HD patients show severely reduced levels of glucose and oxygen metabolism of both caudate and lentiform nuclei, the levels of resting putamen metabolism correlating with severity of rigidity (Young *et al.*, 1986). Striatal [^{18}F]dopa uptake remains preserved (Leenders *et al.*, 1986). The medium spiny striatal neurons that degenerate in HD express either D_1 or D_2 receptors. Using [^{11}C]-SCH23390 and [^{11}C]raclopride PET, Turjanski and colleagues (1995) reported a parallel reduction in striatal D_1 and D_2 binding in HD, which was most severe in akinetic-rigid cases. Figure 10 shows PET [^{11}C]raclopride and [^{18}F]dopa scans for one of their cases. These findings contrast with PD, where putamen D_2 binding is normal or raised and [^{18}F]dopa uptake is severely reduced.

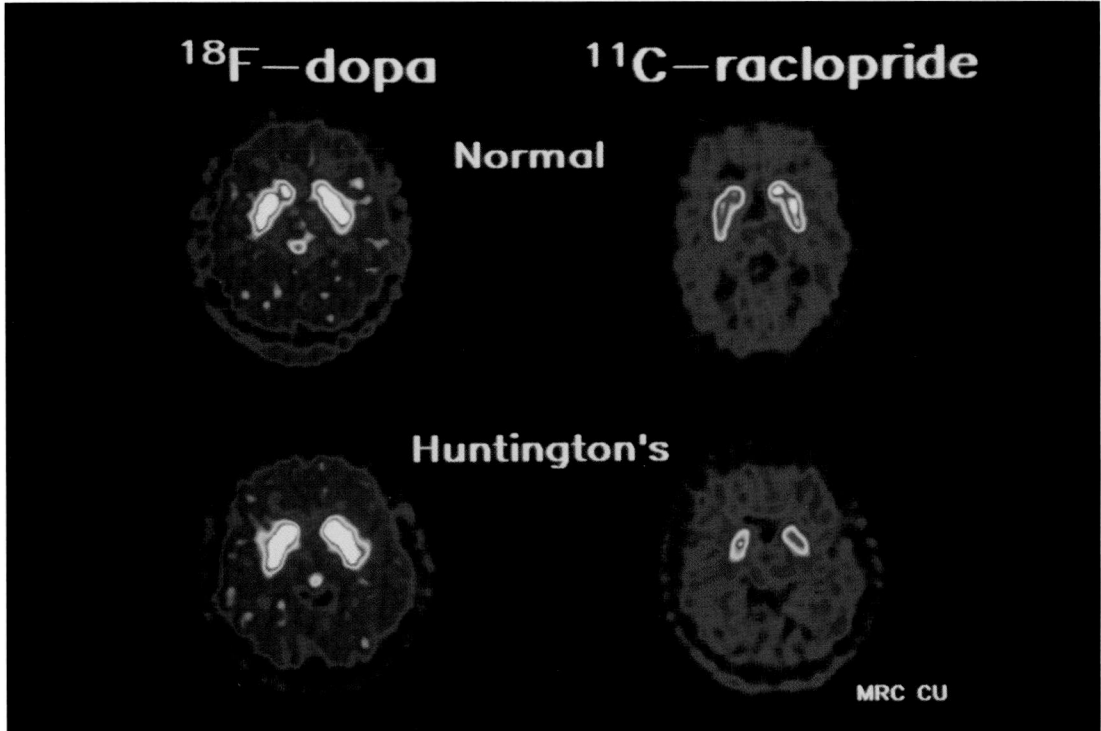

Figure 10 PET images of [^{18}F]dopa and [^{11}C]raclopride uptake in a normal subject and a patient with Huntington's disease. It can be seen that presynaptic dopamine terminal function is normal while a profound loss of dopamine D_2 binding has occurred in HD (picture courtesy of N. Turjanski).

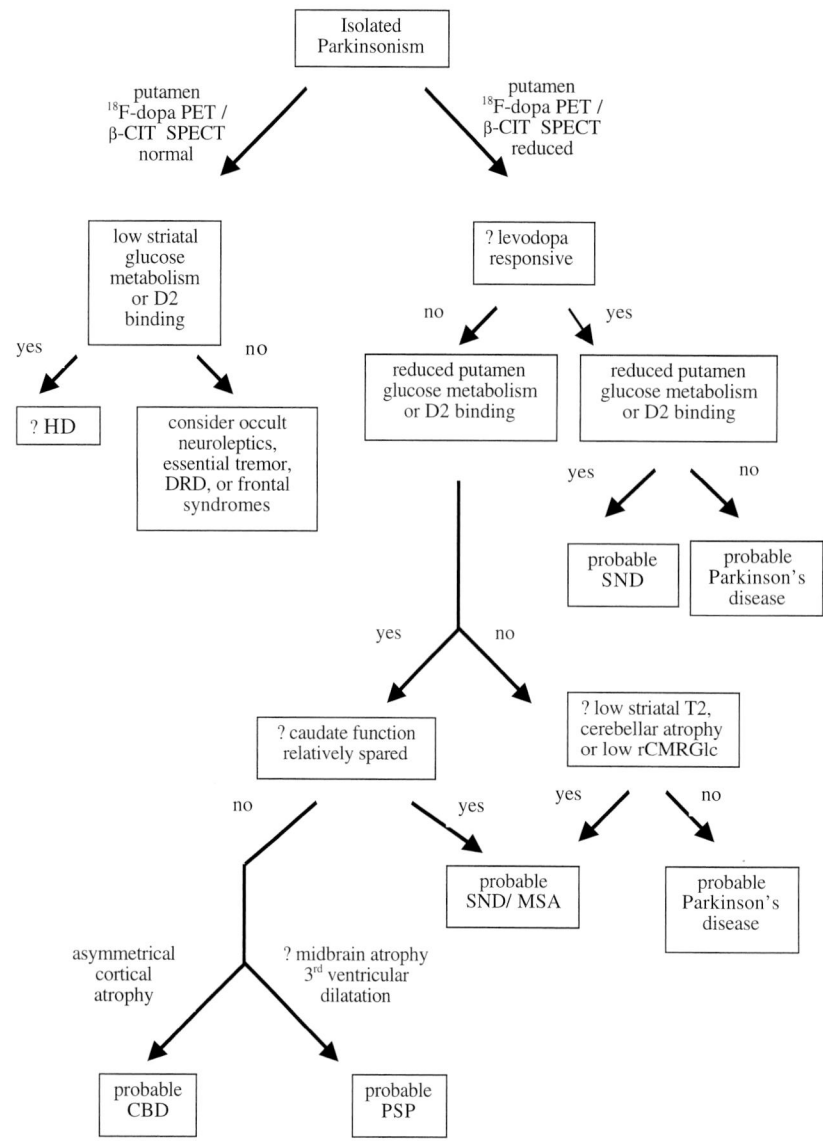

Figure 11 Flow chart showing the diagnosis of isolated parkinsonism with neuroimaging.

Table III PET Findings in Hypokinetic Syndromes

Imaging modality	PD	SND	PSP	CBD	DRD	HD
MRI	Normal or raised nigral T_2	Low putamen and raised pons T_2	Midbrain atrophy	Hemisphere atrophy	Normal	Striatal atrophy, raised T_2
FDG	Normal or raised lentiform	Low putamen and frontal	Low putamen caudate, and frontal	Low thalamic/striatal inferior parietal/ posterior frontal	—	Low striatal
F-Dopa/β-CIT	Low putamen ± caudate	Low putamen ± caudate	Low putamen, low caudate	Low putamen, low caudate	Normal	Normal FD
Putamen D_2 binding	Raised (untreated) or normal (treated)	Normal or low	Normal or low	Low	Normal or raised	Low
Opioid binding	Normal (early) or low striatal if dyskinetic	Low putamen, normal caudate	Low putamen, low caudate	?	?	Low putamen, low caudate
Muscarinic binding	Raised frontal	?	Normal	?	?	?

VI. Conclusions

Figure 11 is a flow chart showing the diagnosis of isolated parkinsonism with neuroimaging, and Table III details the differential radiological findings in the various hypokinetic disorders reviewed.

References

Albin, R. L., Reiner, A., Anderson, K. D., Penney, J. B., and Young, A. B. (1990). Striatal and nigral neuron subpopulations in rigid Huntington's disease: Implications for the functional anatomy of chorea and rigidity-akinesia. *Ann. Neurol.* **27,** 357–365.

Alexander, G. E., Delong, M. R., and Strick, P. L. (1986). Parallel organization of functionally segregated circuits linking basal ganglia and cortex. *Annu. Rev. Neurosci.* **9,** 357–381.

Antonini, A., Leenders, K. L., Meier, D., Oertel, W. H., Boesiger, P., and Anliker, M. (1993). T_2 relaxation-time in patients with Parkinson's disease. *Neurology* **43,** 697–700.

Antonini, A., Leenders, K. L., Vontobel, P., Maguire, R. P., Missimer, J., Psylla, M., and Gunther, I. (1997). Complementary PET studies of striatal neuronal function in the differential diagnosis between multiple system atrophy and Parkinson's disease. *Brain* **120,** 2187–2195.

Antonini, A., Kazumata, K., Feigin, A., Mandel, F., Dhawan, V., Margouleff, C., and Eidelberg, D. (1998). Differential diagnosis of parkinsonism with [^{18}F]fluorodeoxyglucose and PET. *Mov. Disord.* **13,** 268–274.

Asahina, M., Suhara, T., Shinotoh, H., Inoue, O., Suzuki, K., and Hattori, T. (1998). Brain muscarinic receptors in progressive supranuclear palsy and Parkinson's disease: A positron emission tomographic study. *J. Neurol. Neurosurg. Psychiatry* **65,** 155–163.

Baron, J. C., Maziere, B., Loc'h, C., Cambon, H., Sgouropoulos, P., Bonnet, M., and Agid, Y. (1986). Loss of striatal (^{76}Br)bromospiperone binding sites demonstrated by positron tomography in progressive supranuclear palsy. *J. Cereb. Blood Flow Metab.* **6,** 131–136.

Bhatt, M. H., Snow, B. J., Martin, W. R. W., Peppard, R., and Calne, D. B. (1991). Positron emission tomography in progressive supranuclear palsy. *Arch. Neurol.* **48,** 389–391.

Blin, J., Baron, J. C., Dubois, P., Pillon, B., Cambon, H., Cambier, J., and Agid, Y. (1990). Positron emission tomography study in progressive supranuclear palsy. *Arch. Neurol.* **47,** 747–752.

Blin, J., Vidhailhet, M.-J., Pillon, B., Dubois, B., Feve, J.-R., and Agid, Y. (1992). Corticobasal degeneration: Decreased and asymmetrical glucose consumption as studied by PET. *Mov. Disord.* **7,** 348–354.

Bokobza, B., Ruberg, M., Scatton, B., Javoy-Agid, F., and Agid, Y. (1984). ^{3}H-Spiperone binding, dopamine and HVA concentrations in Parkinson's disease and supranuclear palsy. *Eur. J. Pharmacol.* **99,** 167–175.

Brooks, D. J., Ibañez, V., Sawle, G. V., Quinn, N., Lees, A. J., Mathias, C. J., Bannister, R., Marsden, C. D., and Frackowiak, R. S. J. (1990a). Differing patterns of striatal ^{18}F-dopa uptake in Parkinson's disease, multiple system atrophy and progressive supranuclear palsy. *Ann. Neurol.* **28,** 547–555.

Brooks, D. J., Salmon, E. P., Mathias, C. J., Quinn, N., Leenders, K. L., Bannister, R., Marsden, C. D., and Frackowiak, R. S. J. (1990b). The relationship between locomotor disability, autonomic dysfunction, and the integrity of the striatal dopaminergic system, in patients with multiple system atrophy, pure autonomic failure, and Parkinson's disease, studied with PET. *Brain* **113,** 1539–1552.

Brooks, D. J., Ibanez, V., Sawle, G. V., Playford, E. D., Quinn, N., Mathias, C. J., Lees, A. J., Marsden, C. D., Bannister, R., and Frackowiak,

R. S. J. (1992). Striatal D_2 receptor status in Parkinson's disease, striatonigral degeneration, and progressive supranuclear palsy, measured with ^{11}C-raclopride and PET. *Ann. Neurol.* **31,** 184–192.

Brucke, T., Asenbaum, A., Pirker, W., Djamshidian, S., Wenger, S., Wober, Ch., Muller, Ch., and Podreka, I. (1997). Measurement of dopaminergic degeneration in Parkinson's disease with [^{123}I] β-CIT and SPECT. *J. Neural Transm.* **50** (Suppl.), 9–24.

Burn, D. J., Sawle, G. V., and Brooks, D. J. (1994). The differential diagnosis of Parkinson's disease, multiple system atrophy, and Steele-Richardson-Olszewski syndrome: Discriminant analysis of striatal ^{18}F-dopa PET data. *J. Neurol. Neurosurg. Psychiatry* **57,** 278–284.

Burn, D. J., Rinne, J. O., Quinn, N. P., Lees, A. J., Marsden, C. D., and Brooks, D. J. (1995). Striatal opioid receptor binding in Parkinson's disease, striatonigral degeneration, and Steele-Richardson-Olszewski syndrome: An ^{11}C-diprenorphine PET study. *Brain* **118,** 951–958.

Crossman, A. R. (1990). A hypothesis on the pathophysiological mechanisms that underlie levodopa- or dopamine agonist-induced dyskinesia in Parkinson's disease: Implications for future strategies in treatment. *Mov. Disord.* **5,** 100–108.

D'Antona, R., Baron, J. C., Samson, Y., Serdaru, M., Viader, F., Agid, Y., and Cambier, J. (1985). Subcortical dementia: Frontal cortex hypometabolism detected by positron tomography in patients with progressive supranuclear palsy. *Brain* **108,** 785–800.

Davie, C. A., Wenning, G. K., Barker, G. J., Tofts, P. S., Phil, D., Kendall, B. E., Quinn, N., McDonald, W. I., Marsden, C. D., and Miller, D. H. (1995). Differentiation of multiple system atrophy from idiopathic Parkinson's disease using proton magnetic resonance spectroscopy. *Ann. Neurol.* **37,** 204–210.

Davie, C. A., Barker, G. J., Machado, C., Miller, D. H., and Lees, A. J. (1997). Proton magnetic resonance spectroscopy in Steele-Richardson-Olszewski syndrome. *Mov. Disord.* **12,** 767–771.

Dethy, S., Van Blercom, N., Damhaut, P., *et al.* (1998). Asymmetry of basal ganglia glucose metabolism and dopa responsiveness in parkinsonism. *Mov. Disord.* **13,** 275–280.

De Volder, A. G., Francard, J., Laterre, C., Dooms, G., Bol, A., Michel, C., and Goffinet, A. M. (1989). Decreased glucose utilisation in the striatum and frontal lobe in probable striatonigral degeneration. *Ann. Neurol.* **26,** 239–247.

Drayer, B. P., Olanow, W., Burger, P., Johnson, G. A., Herfkens, R., and Riederer, S. (1986). Parkinson Plus syndrome: Diagnosis using high field MR imaging of brain iron. *Radiology* **159,** 493–498.

Dubas, F., Gray, F., and Escourolle, R. (1983). Maladie de Steele-Richardson-Olszewski sans ophthalmologie. *Rev. Neurol.* **139,** 407–416.

Duguid, J. R., De La Paz, R., and DeGroot, J. (1986). Magnetic resonance imaging of the midbrain in Parkinson's disease. *Ann. Neurol.* **20,** 744–747.

Eidelberg, D., Dhawan, V., Moeller, J. R., Sidtis, J. J., Ginos, J. Z., Strother, S. C., Cederbaum, J., Greene, P., Fahn, S., Powers, J. M., and Rottenberg, D. A. (1991). The metabolic landscape of cortico-basal ganglionic degeneration: Regional asymmetries studied with positron emission tomography. *J. Neurol. Neurosurg. Psychiatry* **54,** 856–862.

Eidelberg, D., Takikawa, S., Moeller, J. R., Dhawan, V., Redington, K., Chaly, T., Robeson, W., Dahl, J. R., Margouleff, D., Fazzini, E., Przedborski, S., and Fahn, S. (1993). Striatal hypometabolism distinguishes striatonigral degeneration from Parkinson's disease. *Ann. Neurol.* **33,** 518–527.

Eidelberg, D., Moeller, J. R., Dhawan, V., Spetsieris, P., Takikawa, S., Ishikawa, T., Chaly, T., Robeson, T., Margouleff, D., Przedborski, S., and Fahn, S. (1994). The metabolic topography of parkinsonism. *J. Cereb. Blood Flow Metab.* **14,** 783–801.

Eidelberg, D., Moeller, J. R., Ishikawa, T., Dhawan, V., Spetsieris, P., Chaly, T., Belakhlef, A., Mandel, F., Przedborski, S., and Fahn, S. (1995). Early differential diagnosis of Parkinson's disease with F-

18 fluorodeoxyglucose and positron emission tomography. *Neurology* **45,** 1995–2004.

Feaney, M. B., and Dickson, D. W. (1995). Widespread cytoskeletal pathology characterizes corticobasal degeneration. *Am. J. Pathol.* **146,** 1388–1396.

Fearnley, J. M., and Lees, A. J. (1990). Striatonigral degeneration: A clinicopathological study. *Brain* **113,** 1823–1842.

Fearnley, J. M., and Lees, A. J. (1991). Ageing and Parkinson's disease: Substantia nigra regional selectivity. *Brain* **114,** 2283–2301.

Fearnley, J. M., Revesz, T., Brooks, D. J., Frackowiak, R. S. J., and Lees, A. J. (1991). Diffuse Lewy body disease presenting with a supranuclear downgaze palsy. *J. Neurol. Neurosurg. Psychiatry* **54,** 159–161.

Federico, F., Simone, I. L., Lucivero, V., DeMari, M., Giannini, P., Iliceto, G., Mezzapesa, D. M., and Lamberti, P. (1997). Proton magnetic resonance spectroscopy in Parkinson's disease and progressive supranuclear palsy. *J. Neurol. Neurosurg. Psychiatry* **62,** 239–242.

Firnau, G., Sood, S., Chirakal, R., Nahmias, C., and Garnett, E. S. (1987). Cerebral metabolism of 6–[^{18}F]fluoro-L-3,4–dihydroxyphenylalanine in the primate. *J. Neurochem.* **48,** 1077–1082.

Foster, N. L., Gilman, S., Berent, S., Morin, E. M., Brown, M. B., and Koeppe, R. A. (1988). Cerebral hypometabolism in progressive supranuclear palsy studied with positron emission tomography. *Ann. Neurol.* **24,** 399–406.

Foster, N. L., Gilman, S., Berent, S., Sima, A. A. F., D'Amato, C., Koeppe, R. A., and Hicks, S. P. (1992). Progressive subcortical gliosis and progressive supranuclear palsy can have similar clinical and PET abnormalities. *J. Neurol. Neurosurg. Psychiatry* **55,** 707–713.

Frisoni, G. B., Pizzolato, G., Zanetti, O., Bianchetti, A., Chierichetti, F., and Trabucchi, M. (1995). Corticobasal degeneration: Neuropsychological assessment and dopamine D-2 receptor SPECT analysis. *Eur. Neurol.* **35,** 50–54.

German, D. C., Manaye, K., Smith, W. K., Woodward, D. J., and Saper, C. B. (1989). Midbrain dopaminergic cell loss in Parkinson's disease: Computer visualization. *Ann. Neurol.* **26,** 507–514.

Gibb, W. R. G., Luthert, P., and Marsden, C. D. (1989). Corticobasal degeneration. *Brain* **112,** 1171–1192.

Gilman, S., Koeppe, R. A., Junck, L., Kluin, K. J., Lohman, M., and St. Laurent, R. T. (1994). Patterns of cerebral glucose metabolism detected with positron emission tomography differ in multiple system atrophy and olivopontocerebellar atrophy. *Ann. Neurol.* **36,** 166–175.

Gilman, S., Frey, K. A., Koeppe, R. A., Junck, L., Little, R., Vander Borght, T. M., Lohman, M., Martorello, S., Lee, L. C., Jewett, D. M., and Kilbourn, M. R. (1996). Decreased striatal monoaminergic terminals in olivopontocerebellar atrophy and multiple system atrophy demonstrated with positron emission tomography. *Ann. Neurol.* **40,** 885–892.

Giminez-Roldan, S., Mateo, D., Benito, C., Grandas, F., and Perez-Gilabert, Y. (1994). Progressive supranuclear palsy and corticobasal ganglionic degeneration: Differentiation by clinical features and neuroimaging techniques. *J. Neural Transm.* **42** (Suppl.), 79–90.

Goffinet, A. M., De Volder, A. G., Gillain, C., Rectem, D., Bol, A., Michel, C., Cogneau, M., Labar, D., and Laterre, C. (1989). Positron tomography demonstrates frontal lobe hypometabolism in progressive supranuclear palsy. *Ann. Neurol.* **25,** 131–139.

Goto, S., Hirano, A., and Matsumoto, S. (1989). Subdivisional involvement of nigrostriatal loop in idiopathic Parkinson's disease and striatonigral degeneration. *Ann. Neurol.* **26,** 766–770.

Goto, S., Hirano, A., and Matsumoto, S. (1990). Met-enkephalin immunoreactivity in the basal ganglia in Parkinson's disease and striatonigral degeneration. *Neurology* **40,** 1051–1056.

Grisoli, M., Fetoni, V., Savoiardo, M., Girotti, F., and Bruzzone, M. G. (1995). MRI in corticobasal degeneration. *Eur. J. Neurol.* **2,** 547–552.

Halliday, G. M., Davies, L., McRitchie, D. A., Cartwright, H., Pamphlett, R., and Morris, J. G. (1995). Ubiquitin positive achromatic neurons in corticobasal degeneration. *Acta Neuropathol. (Berl.)* **90,** 68–75.

Hauser, R. A., Murtagh, F. R., Akhter, K., Gold, M., and Olanow, C. W. (1996). Magnetic resonance imaging of corticobasal degeneration. *J. Neuroimaging* **6,** 222–226.

Hughes, A. J., Daniel, S. E., Kilford, L., and Lees, A. J. (1992). The accuracy of the clinical diagnosis of Parkinson's disease: A clinico-pathological study of 100 cases. *J. Neurol. Neurosurg. Psychiatry* **55,** 181–184.

Hunt, P., Kannengiesser, M.-H., and Raynaud, J.-P. (1974). Nomifensine: A new potent inhibitor of dopamine uptake into synaptosomes from rat brain corpus striatum. *J. Pharm. Pharmacol.* **26,** 370–371.

Ichinose, H., Ohye, T., Takahashi, E., Seki, N., Hori, T., Segawa, M., Nomura, Y., Endo, K., Tanaka, H., Tsuji, S., Fujita, K., and Nagatsu, T. (1994). Hereditary progressive dystonia with marked diurnal fluctuation caused by mutations in the GTP-cyclohydrolase-1 gene. *Nat. Genet.* **8,** 236–242.

Jackson, J. A., Jankovic, J., and Ford, J. (1983). Progressive supranuclear palsy: Clinical features and response to treatment in 16 patients. *Ann. Neurol.* **13,** 273–278.

Jellinger, K., Riederer, P., and Tomananga, M. (1980). Progressive supranuclear palsy: Clinico-pathological and biochemical studies. *J. Neural Transm.* **16** (Suppl.), 111–128.

Johnson, K. A., Sperling, R. A., Holman, B. L., Nagel, J. S., and Growdon, J. H. (1992). Cerebral perfusion in progressive supranuclear palsy. *J. Nucl. Med.* **33,** 704–709.

Karbe, H., Grond, M., Huber, M., Herholz, K., Kessler, J., and Heiss, W. D. (1992). Subcortical damage and cortical dysfunction in progressive supranuclear palsy demonstrated by positron emission tomography. *J. Neurol.* **239,** 98–102.

Kish, S. J., Chang, L. J., Mirchandani, L. J., Shannak, K., and Hornykiewicz, O. (1985). Progressive supranuclear palsy: Relationship between extrapyramidal disturbances, dementia, and brain neurotransmitter markers. *Ann. Neurol.* **18,** 530–536.

Kish, S. J., Shannak, K., and Hornykiewicz, O. (1988). Uneven pattern of dopamine loss in the striatum of patients with idiopathic Parkinson's disease. *N. Engl. J. Med.* **318,** 876–880.

Kishore, A., Snow, B. J., Naini, A. B., Przedborski, S., Vingerhoets, F. J. G., and Nygaard, T. G. (1995). Analysis of dopaminergic function in asymptomatic carriers of the DRD gene: CSF dopaminergic metabolites and PET with fluorodopa and raclopride. *Neurology* **45** (Suppl. 4), A187.

Kiyosawa, M., Baron, J. C., Hamel, E., Pappata, S., Duverger, D., Riche, D., Mazoyer, B., Naquet, R., and MacKenzie, E. T. (1989). Time course of effects of unilateral lesions of the Nucleus Basalis of Meynert on glucose utilisation by the cerebral cortex. Positron emission tomography in baboons. *Brain* **112,** 435–455.

Koeppen, A. H., and Hans, M. B. (1976). Supranuclear ophthalmoplegia in olivopontocerebellar degeneration. *Neurology* **26,** 764–768.

Konagaya, M., Konagaya, Y., and Iida, M. (1994). Clinical and magnetic resonance imaging study of extrapyramidal symptoms in multiple system atrophy. *J. Neurol. Neurosurg. Psychiatry* **57,** 1528–1531.

Laplane, D., Levasseur, M., Pillon, B., Dubois, B., Baulac, M., Mazoyer, B., Tran Dinh, S., Sette, G., Danze, F., and Baron, J. C. (1989). Obsessive-compulsive and other behavioural changes with bilateral basal ganglia lesions. *Brain* **112,** 699–725.

Leenders, K. L., Frackowiak, R. S. J., Quinn, N., and Marsden, C. D. (1986). Brain energy metabolism and dopaminergic function in Huntington's disease measured *in vivo* using positron emission tomography. *Mov. Disord.* **1,** 69–77.

Leenders, K. L., Frackowiak, R. S., and Lees, A. J. (1988). Steele-

Richardson-Olszewski syndrome: Brain energy metabolism, blood flow and fluorodopa uptake measured by positron emission tomography. *Brain* **111,** 615–630.

Lees, A. J. (1986). The Steele-Richardson-Olszewski syndrome (progressive supranuclear palsy). *In* "Movement Disorders 2" (C. D. Marsden and S. Fahn, Eds.), pp. 273–287. Butterworths, London.

Litvan, I., Agid, Y., Goetz, C., Jankovic, J., Wenning, G. K., Brandel, J. P., Lai, E. C., Verny, M., Chaudhuri, K. R., McKee, A., Jellinger, K., Pearce, R. K. B., and Bartko, J. J. (1997). Accuracy of the clinical diagnosis of corticobasal degeneration: A clinicopathological study. *Neurology* **48,** 119–125.

Maher, E. R., and Lees, A. J. (1986). The clinical features and natural history of the Steele-Richardson-Olszewski syndrome (progressive supranuclear palsy). *Neurology* **36,** 1005–1008.

Markus, H. S., Lees, A. J., Lennox, G., Marsden, C. D., and Costa, D. C. (1995). Patterns of regional cerebral blood flow in corticobasal degeneration studied using HMPAO SPECT: Comparison with Parkinson's disease and normal controls. *Mov. Disord.* **10,** 179–187.

McCulloch, J., MacKenzie, E. T., Cudennec, A., Duverger, D., Degueurce, A., and Scatton, B. (1984). Influences of the raphe nuclei on brain glucose utilisation. *Soc. Neurosci. Abstr.* **10,** 218.

Messa, C., Volonte, M. A., Fazio, F., Zito, F., Carpinelli, A., d'Amico, A., Rizzo, G., Moresco, R. M., Paulesu, E., Franchesci, M., and Lucignani, G. (1998). Differential distribution of striatal $[^{123}I]\beta$-CIT in Parkinson's disease and progressive supranuclear palsy, evaluated with single-photon emission tomography. *Eur. J. Nucl. Med.* **25,** 1270–1276.

Miletich, R. S., Chan, T., Gillespie, M., Di Chiro, G., and Stein, S. (1988). Contralateral basal ganglia metabolism is abnormal in hemiparkinsonian patients. An FDG-PET study. *Neurology* **38,** S260.

Nagahama, Y., Fukuyama, H., Turjanski, N., Kennedy, A., Yamauchi, H., Ouchi, Y., Kimura, J., Brooks, D. J., and Shibasaki, H. (1997). Cerebral glucose metabolism in corticobasal degeneration: Comparison with progressive supranuclear palsy and normal controls. *Mov. Disord.* **12,** 691–696.

Nagasawa, H., Tanji, H., Nomura, H., Saito, H., Itoyama, Y., Kimura, I., Tuji, S., Fujiwara, T., Iwata, R., Itoh, M., and Ido, T. (1996). PET study of cerebral glucose metabolism and fluorodopa uptake in patients with corticobasal degeneration. *J. Neurol. Sci.* **39,** 210–217.

Nygaard, T. G., Trugman, J. M., de Yebenes, J. G., and Fahn, S. (1990). Dopa-responsive dystonia: The spectrum of clinical manifestations in a large North American family. *Neurology* **40,** 66–69.

Okuda, B., Tachibana, H., Takeda, M., Kawabata, K., Sugita, M., and Fukuchi, M. (1995). Focal cortical hypoperfusion in corticobasal degeneration demonstrated by three-dimensional surface display with ^{123}I-IMP: A possible cause of apraxia. *Neuroradiology* **37,** 642–644.

Otsuka, M., Ichiya, Y., Kuwabara, Y., Miyake, Y., Tahara, T., Masuda, K., Hosokawa, S., Goto, I., Kato, M., Ichimiya, A., and Suetsugu, M. (1989). Cerebral blood flow, oxygen and glucose metabolism with PET in progressive supranuclear palsy. *Ann. Nucl. Med.* **3,** 111–118.

Otsuka, M., Ichiya, Y., Hosokawa, S., Kuwabara, Y., Tahara, T., Fukumura, T., Kato, M., Masuda, K., and Goto, I. (1991). Striatal blood flow, glucose metabolism, and ^{18}F-dopa uptake: Difference in Parkinson's disease and atypical parkinsonism. *J. Neurol. Neurosurg. Psychiatry* **54,** 898–904.

Otsuka, M., Ichiya, Y., Kuwabara, Y., Hosokawa, S., Akashi, Y., Yoshida, T., Fukumura, T., Masuda, K., Goto, I., and Kato, M. (1994). Striatal ^{18}F-Dopa uptake and brain glucose metabolism by PET in patients with syndrome of progressive ataxia. *J. Neurol. Sci.* **124,** 198–203.

Otsuka, M., Ichiya, Y., Kuwabara, Y., Hosokawa, S., Sasaki, M., Yoshida, T., Fukumura, T., Kato, M., and Masuda, K. (1996). Glucose metabolism in the cortical and subcortical brain structures in

multiple system atrophy and Parkinson's disease: A positron emission tomographic study. *J. Neurol. Sci.* **144,** 77–83.

Papp, M. I., and Lantos, P. L. (1994). The distribution of oligodendroglial inclusions in multiple system atrophy and its relevance to clinical symptomatology. *Brain* **117,** 235–243.

Pappata, S., Mazoyer, B., Tran Dinh, S., Cambon, H., Levasseur, M., and Baron, J. C. (1990). Effects of capsular or thalamic stroke on metabolism in the cortex and cerebellum: A positron tomography study. *Stroke* **21,** 519–524.

Pappata, S., Traykov, L., Tavitian, B., Damier, P., Dubois, B., Jobert, A., Crouzel, C., and DiGiamberardino, L. (1997). Striatal reduction of acetylcholinesterase in patients with progressive supranuclear palsy (PSP) as measured *in vivo* by PET and ^{11}C-physostigmine (^{11}C-PHY). *J. Cereb. Blood Flow Metab.* **17** (Suppl1.), S687.

Pastakia, B., Polinsky, R., Di Chiro, G., Simmons, J. T., Brown, R., and Wener, L. (1986). Multiple system atrophy (Shy-Drager syndrome): MR imaging. *Radiology* **159,** 499–502.

Perkin, G. D., Lees, A. J., Stern, G. M., and Kocen, R. S. (1978). Problems in the diagnosis of progressive supranuclear palsy (Steele-Richardson-Olszewski syndrome). *Can. J. Neurol. Sci.* **6,** 167–173.

Piccini, P., Weeks, R. A., and Brooks, D. J. (1997). Opioid receptor binding in Parkinson's patients with and without levodopa-induced dyskinesias. *Ann. Neurol.* **42,** 720–726.

Pierot, L., Desnos, C., Blin, J., Raisman, R., Scherman, D., Javoy-Agid, F., Ruberg, M., and Agid, Y. (1988). D_1 and D_2-type dopamine receptors in patients with Parkinson's disease and progressive supranuclear palsy. *J. Neurol. Sci.* **86,** 291–306.

Playford, E. D., and Brooks, D. J. (1992). *In vivo* and *in vitro* studies of the dopaminergic system in movement disorders. *Cerebrovasc. Brain Metab. Rev.* **4,** 144–171.

Quinn, N. P., Luthert, P., Honavar, M., and Marsden, C. D. (1989). Pure akinesia due to Lewy body Parkinson's disease: A case with pathology. *Mov. Disord.* **4,** 85–89.

Rajput, A. H., Rozdilsky, B., and Rajput, A. (1991). Accuracy of clinical diagnosis in Parkinsonism: A prospective study. *Can. J. Neurol. Sci.* **18,** 275–278.

Rebeiz, J. J., Kolodny, E. H., and Richardson, E. P. (1968). Corticodentatonigral degeneration with neuronal achromasia. *Arch. Neurol.* **18,** 20–33.

Riley, D. E., Lang, A. E., Lewis, A., Resch, L., Ashby, P., Hornykiewicz, O., and Black, S. (1990). Cortical-basal ganglionic degeneration. *Neurology* **40,** 1203–1212.

Rinne, J. O., Laihinen, A., Nagren, K., Bergman, J., Solin, O., Haapparanta, M., Ruotsalainen, U., and Rinne, U. K. (1990). PET demonstrates different behaviour of striatal dopamine D_1 and D_2 receptors in early Parkinson's disease. *J. Neurosci. Res.* **27,** 494–499.

Rinne, J. O., Lee, M. S., Thompson, P. D., and Marsden, C. D. (1994). Corticobasal degeneration: A clinical study of 36 cases. *Brain* **117,** 1183–1196.

Rinne, J. O., Burn, D. J., Mathias, C. J., Quinn, N. P., Marsden, C. D., and Brooks, D. J. (1995). PET studies on the dopaminergic system and striatal opioid binding in the olivopontocerebellar atrophy variant of multiple system atrophy. *Ann. Neurol.* **37,** 568–573.

Ross-Russell, R. (1980). Supranuclear palsy of eyelid closure. *Brain* **103,** 71–82.

Ruberg, M., Javoy-Agid, F., Hirsch, E., Scatton, B., L'Heureux, R., Hauw, J.-J., Duyckaerts, C., Gray, F., Morel-Maroger, A., Rascol, A., Serdaru, M., and Agid, Y. (1985). Dopaminergic and cholinergic lesions in progressive supranuclear palsy. *Ann. Neurol.* **18,** 523–529.

Rutledge, J. N., Hilal, S. K., Silver, A. J., Defendini, R., and Fahn, S. (1987). Study of movement disorders and brain iron by MR. *A.J.R., Am. J. Roentgenol.* **149,** 365–379.

Salmon, E. P., Brooks, D. J., Leenders, K. L., Turton, D. R., Hume, S. P., Cremer, J. E., Jones, T., and Frackowiak, R. S. J. (1990). A two-compartment description and kinetic procedure for measuring regional cerebral [^{11}C]nomifensine uptake using positron emission tomography. *J. Cereb. Blood Flow Metab.* **10**, 307–316.

Savaki, H. E., Graham, D. I., Grome, J. J., and McCulloch, J. (1984). Functional consequences of unilateral lesion of the locus coeruleus: A quantitative [^{14}C]2–deoxyglucose investigation. *Brain Res.* **292**, 239–249.

Savoiardo, M., Girotti, F., Strada, L., and Ciceri, E. (1994). Magnetic resonance imaging in progressive supranuclear palsy and other parkinsonian disorders. *J. Neural Transm.* **42** (Suppl.), 93–110.

Sawle, G. V., Brooks, D. J., Marsden, C. D., and Frackowiak, R. S. J. (1991a). Corticobasal degeneration: A unique pattern of regional cortical oxygen metabolism and striatal fluorodopa uptake demonstrated by positron emission tomography. *Brain* **114**, 541–556.

Sawle, G. V., Leenders, K. L., Brooks, D. J., Harwood, G., Lees, A. J., Frackowiak, R. S. J., and Marsden, C. D. (1991b). Dopa-responsive dystonia: [^{18}F]Dopa positron emission tomography. *Ann. Neurol.* **30**, 24–30.

Schwarz, J., Tatsch, K., Arnold, G., Gasser, T., Trenkwalder, C., Kirsch, C. M., and Oertel, W. (1992). ^{123}I-Iodobenzamide-SPECT predicts dopaminergic responsiveness in patients with *de novo* parkinsonism. *Neurology* **42**, 556–561.

Schwarz, J., Tatsch, K., Linke, R., Pogarell, O., Mozley, D., and Kung, H. F. (1997). Measuring the decline of dopamine transporter binding in patients with Parkinson's disease using ^{123}I-IPT and SPECT. *Neurology* **48** (Suppl. 2), A208.

Schwarz, J., Tatsch, K., Gasser, T., Arnold, G., Pogarell, O., Kunig, G., and Oertel, W. H. (1998). ^{123}I-IBZM binding compared with long-term clinical follow up in patients with *de novo* parkinsonism. *Mov. Disord.* **13**, 16–19.

Seibyl, J. P., Marek, K. L., Quinlan, D., Sheff, K., Zoghbi, S., Zea-Ponce, Y., Baldwin, R. M., Fussell, B., Smith, E. O., Charney, D. S., Hoffer, P. B., and Innis, R. B. (1995). Decreased single-photon emission computed tomographic [^{123}I]β-CIT striatal uptake correlates with symptom severity in Parkinson's disease. *Ann. Neurol.* **38**, 589–598.

Shinotoh, H., Aotsuka, A., Yonezawa, H., Fukuda, H., Inoue, O., Yamasaki, T., Tateno, Y., and Hirayama, K. (1990). Striatal dopamine D$_2$ receptors in Parkinson's disease and striato-nigral degeneration determined by positron emission tomography. *In* "Basic, Clinical, and Therapeutic Advances of Alzheimer's and Parkinson's Diseases" (T. Nagatsu, A. Fisher, and M. Yoshida, Eds.), Vol. 2, pp. 107–110. Plenum, New York.

Shinotoh, H., Inoue, O., Hirayama, K., Aotsuka, A., Asahina, M., Suhara, T., Yamazaki, T., and Tateno, Y. (1993). Dopamine D$_1$ receptors in Parkinson's disease and striatonigral degeneration: A positron emission tomography study. *J. Neurol. Neurosurg. Psychiatry* **56**, 467–472.

Snow, B. J., Nygaard, T. G., Takahashi, H., and Calne, D. B. (1993). Positron emission tomography studies of dopa-responsive dystonia and early-onset idiopathic parkinsonism. *Ann. Neurol.* **34**, 733–738.

Spokes, E. G. S., Bannister, R., and Oppenheimer, D. R. (1979). Multiple system atrophy with autonomic failure. Clinical, histological, and neurochemical observations on four cases. *J. Neurol. Sci.* **43**, 59–62.

Steele, J. C., Richardson, J. C., and Olszewski, J. (1964). Progressive supranuclear palsy. A heterogeneous degeneration involving the brain stem, basal ganglia, and cerebellum, with vertical gaze and pseudobulbar palsy. *Arch. Neurol.* **10**, 333–359.

Taniwaki, T., Hosokawa, S., Goto, I., Fujii, N., Otsuka, M., Kuwabara, Y., Ichiya, Y., Hasuo, K., and Kato, M. (1992). Positron emission tomography (PET) in "pure akinesia." *J. Neurol. Sci.* **107**, 34–39.

Taylor, M. D., De Ceballos, M. L., Rose, S., *et al.* (1992). Effects of unilateral 6–hydroxydopamine lesion and prolonged L-3,4–dihydroxyphenylalanine treatment on peptidergic systems in rat basal ganglia. *Eur. J. Pharmacol.* **219**, 183–192.

Tedeschi, G., Litvan, I., Bonavita, S., Bertolino, A., Lundbom, N., Patronas, N. J., and Hallett, M. (1997). Proton magnetic resonance spectroscopic imaging in progressive supranuclear palsy, Parkinson's disease and corticobasal degeneration. *Brain* **120**, 1541–1552.

The Huntington's Disease Collaborative Research Group. (1993). A novel gene containing a trinucleotide repeat that is expanded and unstable on Huntington's disease chromosomes. *Cell* **72**, 971–983.

Thompson, P. D., Day, B. L., Rothwell, J. C., Brown, P., Britton, T. C., and Marsden, C. D. (1994). The myoclonus in corticobasal degeneration: Evidence for two forms of cortical reflex myoclonus. *Brain* **117**, 1197–1207.

Turjanski, N., Bhatia, K., Burn, D. J., Sawle, G. V., Marsden, C. D., and Brooks, D. J. (1993). Comparison of striatal ^{18}F-dopa uptake in adult-onset dystonia-parkinsonism, Parkinson's disease, and dopa-responsive dystonia. *Neurology* **43**, 1563–1568.

Turjanski, N., Weeks, R., Dolan, R., Harding, A. E., and Brooks, D. J. (1995). Striatal D$_1$ and D$_2$ receptor binding in patients with Huntington's disease and other choreas: A PET study. *Brain* **118**, 689–696.

Turjanski, N., Lees, A. J., and Brooks, D. J. (1997). PET studies on striatal dopaminergic receptor binding in drug naive and L-dopa treated Parkinson's disease patients with and without dyskinesia. *Neurology* **49**, 717–723.

Wienhard, K., Coenen, H. H., Pawlik, G., Rudolf, J., Laufer, P., Jovkar, S., Stocklin, G., and Heiss, W. D. (1990). PET studies of dopamine receptor distribution using [^{18}F]fluoroethylspiperone: Findings in disorders related to the dopaminergic system. *J. Neural Transm.* **81**, 195–213.

Will, R. G., Lees, A. J., Gibb, W., and Barnard, R. O. (1988). A case of progressive subcortical gliosis presenting clinically as Steele-Richardson-Olszewski syndrome. *J. Neurol. Neurosurg. Psychiatry* **51**, 1224–1227.

Wolfson, L. I., Leenders, K. L., Brown, L. L., and Jones, T. (1985). Alterations of regional cerebral blood flow and oxygen metabolism in Parkinson's disease. *Neurology* **35**, 1399–1405.

Yagashita, A., and Oda, M. (1996). Progressive supranuclear palsy: MRI and pathological findings. *Neuroradiology* **38**, S60–S66.

Young, A. B., Penney, J. B., Starosta-Rubinstein, S., Markel, D. S., Berent, S., Giordani, B., and Ehrenkaufer, R. (1986). PET scan investigations of Huntington's disease: Cerebral metabolic correlates of neurological features and functional decline. *Ann. Neurol.* **20**, 296–303.

12

Functional Imaging in Hyperkinetic Disorders

Guy Sawle

Department of Neurology, Queens Medical Centre, Nottingham NG7 2UH, United Kingdom

I. Tremor

II. Dystonia

III. Tics

IV. Chorea

V. Tardive Dyskinesia

VI. Restless Legs Syndrome

References

A broad range of functional imaging modalities have been employed in the study of hyperkinetic movement disorders, including tremor, dystonia, and chorea. In most cases, the principal focus of attention has been to explain the pathophysiology.

The classical resting tremor of Parkinson's disease has been shown to have a poor correlation with the nigrostriatal dopaminergic defect, but to correlate with expression of an abnormal metabolic network involving the thalamus, pons, and premotor cortex. Essential tremor is instead characterized functionally by a bilateral increase in cerebellar activity at rest, with further increases in cerebellar activity and red nucleus activity on posture and striatal, thalamic, and sensorimotor activation on voluntary movement.

In dystonia, the main findings have been overactivity in the premotor cortex and lentiform nuclei and relative underactivity in the sensorimotor cortex. The premotor activity may drive the dystonic movements, it may be part of a network trying to stop them, or it may simply be a correlate of the heightened striatal output. The underactivity in sensorimotor cortex may represent a reduction in inhibitory activity.

Patients with dopa-responsive dystonia may be distinguished from those with young-onset Parkinson's disease on the basis of preserved fluorodopa uptake.

In chorea, patients have either decreased striatal metabolism (as in Huntington's disease) or, less commonly, increased metabolism (as in Sydenham's chorea). This division is most probably relevant to the mechanism of involuntary movement generation in these two groups of disorders.

Patients with the restless legs syndrome have impaired fluorodopa uptake, confirming the dopamine system pathology long suspected in this condition on the basis of clinical response to dopaminergic medication.

I. Tremor

Tremor is an involuntary movement that is approximately rhythmic and roughly sinusoidal. Some kinds of tremor are present only at rest (such as the pill rolling tremor of Parkinson's disease) whereas others are most marked on posture (such as essential tremor) or during particular motor activities (such as primary writing tremor). In most tremor disorders the pathophysiological cause of the tremor is unknown and it is unclear whether the tremor origin is entirely central (thalamic or olivary oscillation), entirely peripheral (oscillating feedback loops), or a mixture of both. A variety of functional imaging modalities have been used in an effort to better understand these diverse conditions

A. Parkinson's Disease and Other Resting Tremors

A variety of PET and SPECT studies have shown correlation between the known derangement of the nigrostriatal dopamine system and both bradykinesia and rigidity. Correlations with rest tremor have been much less clear, however, with reports of either a weak relationship (Morrish et al., 1995) or else none at all between tremor severity and dopamine function (Eidelberg et al., 1990). It is known from primate studies that a nigrostriatal dopamine lesion alone will not induce a resting tremor and it is assumed that nigrostriatal dopamine deficiency is probably necessary but not sufficient to cause resting tremor. In patients with an isolated resting tremor, even those who do not have other signs of parkinsonism have impaired striatal fluorodopa uptake (Brooks et al., 1992).

Network analysis of resting fluorodeoxyglucose data from patients with tremulous Parkinson's disease compared with data from patients with Parkinson's disease but no tremor has shown an increased expression of a metabolic network comprising the thalamus, pons, and premotor cortical regions (Antonini et al., 1998); since both the patient groups expressed a separate Parkinson's disease related pattern (which the authors have called PDRP), it is argued that parkinsonian tremor requires not only abnormal expression of the motor network associated with bradykinesia but also abnormal expression of a second functionally distinct network. This finding lends support to the hypothesis that tremor generation in Parkinson's disease depends upon a central mechanism.

It has been suggested that parkinsonian tremor may be due to abnormal activation of a cerebral circuit that is designed for voluntary repetitive movement. To investigate this hypothesis, patients undergoing stereotactic thalamic stimulation for severe parkinsonian tremor have undergone blood flow PET studies (using oxygen-15-labeled water) with tremor present (stimulator off) and with tremor absent (stimulator on). When the tremor stops, there is reduced blood flow in the con-

tralateral sensorimotor cortex, supplementary motor area, and cortico-cerebellar pathways (Parker et al., 1992).

Tremors resulting from lesions of the cerebellar outflow, or else sited close to the red nucleus (hence the alternative name "rubral tremor"), typically have a resting component, which worsens on trying to maintain a steady posture and is then further and dramatically enhanced on attempting controlled movement. Fluorodopa PET scans in such patients show severe dopamine denervation, at least as severe as seen in Parkinson's disease (Remy et al., 1995), indicating involvement of the nigrostriatal dopamine projection.

In a patient who developed a resting left-sided tremor several months after rupture of a midbrain arteriovenous malformation, a fluorodopa PET study showed reduced tracer uptake in the right striatum (Defer et al., 1994).

B. Essential Tremor

Essential tremor is a monosymptomatic disorder comprising a postural tremor that is usually unaffected by movement and is absent at rest. The most typical phenotype is a mild symmetrical postural tremor of the arms. In a study of hereditary essential tremor, tremor of the legs, head, facial muscles, voice, jaw, and tongue were all seen, but never in isolation. Many patients are responsive to alcohol.

In an early FDG-PET study of resting patients, hypermetabolism was noted in the medulla and thalami but not in the cerebellum (Hallett and Dubinsky, 1993). In a subsequent blood flow PET study, control subjects were scanned at rest, during arm extension without tremor, and during voluntary wrist oscillation. Patients with essential tremor were also scanned at rest, during postural tremor, and during passive wrist oscillation. The essential tremor patients had increased resting cerebellar blood flow, which increased further involving also the contralateral striatal, thalamic, and sensorimotor cortex when postural tremor was present. All the other tasks (in patients and controls) were associated with ipsilateral cerebellar activation only (Jenkins et al., 1993). In a further blood flow study (using radioactive water PET), resting cerebellar blood flow was bilaterally increased by 30–40% (Wills et al., 1994). Arm elevation resulted in postural tremor and a further increase in cerebellar and red nuclear blood flow (Wills et al., 1994). When control subjects were scanned during voluntary wrist oscillation, this led to ipsilateral cerebellar activation only.

fMRI studies of patients with postural tremor reveal activation in contralateral primary motor and sensory cortex, globus pallidus, and thalamus, together with bilateral activation of the dentate nucleus, cerebellar hemispheres, and red nucleus (Bucher *et al.*, 1997a). The authors identified medullary activity in only two of twelve patients. Unilateral passive wrist oscillation in the same patients led to unilateral activation of the dentate, red nucleus, and cerebellar hemisphere. The time resolution of this study did not enable the authors to decide which of the changes they observed might be of primary relevance to tremor generation and which may be secondary effects.

Many patients with essential tremor report improvement in their tremor after drinking alcohol. Measures of blood flow in control subjects and essential tremor patients given two to three units of alcohol have shown a bilateral decrease in cerebellar blood flow in patients and controls. In the patients, blood flow in the inferior olivary region was increased after alcohol ingestion (Boecker *et al.*, 1996), suggesting that the mechanism of symptom relief is via reduced cerebellar overactivity, resulting in an increased afferent input to the inferior olive.

In an electrophysiological study, when patients with essential tremor, patients with parkinsonian postural tremor, or control subjects imitating a postural tremor were given transcranial magnetic brain shocks, the result was a modulation of the tremor. In all these tremors, when the tremor returned, its reappearance was time-locked to the stimulus. In the case of the parkinsonian postural tremor (but not the essential tremor or mimicked tremor), the time to reappearance of rhythmic EMG activity following magnetic brain stimulation varied with the period of the tremor. This suggests that there are different central pathophysiological mechanisms underlying essential tremor and parkinsonian postural tremor (Britton *et al.*, 1993).

C. Other Tremors

Some patients experience tremor only when writing; it is thought clinically that this may be a variant of essential tremor or dystonia. A PET study designed to contrast the activation changes associated with writing tremor and essential tremor showed bilateral cerebellar activation in both subject groups. This suggests that writing tremor may indeed be a variant of essential tremor, but does not exclude other causes (Wills *et al.*, 1995).

Primary orthostatic tremor is characterized by a fast (14–16 Hz) lower limb tremor caused unsteadiness on walking. The tremor is too fast to see with the naked eye, but the patellae may "judder" and it may be possible to hear a helicopter-like sound with a stethoscope held over the muscles (Brown, 1995). Some patients also have a postural upper limb tremor of similar fre-

quency. PET activation scans show bilateral cerebellar activation together with contralateral activation of thalamus and lentiform nuclei (Wills *et al.*, 1996) (though, of course, scans in the upright position have not, to date, been possible). Magnetic stimulation over the motor cortex in patients with primary orthostatic tremor leads to resetting of the subsequent tremor bursts, suggesting involvement of the cortex in the networks underlying tremor timing in this condition (Tsai *et al.*, 1998).

Palatal "myoclonus" is more properly termed palatal tremor even though many authors continue to use the original term. The condition is associated with hypertrophic degeneration of the inferior olivary nucleus and indicates a lesion in the pathway from the dentate to the contralateral olive via the superior cerebellar peduncle, red nucleus, and the central tegmental tract. The inferior olive is presumed to be the pacemaker in this condition, a proposition strengthened by the observation of hypermetabolism at this site identified by fluorodeoxyglucose PET (Dubinsky *et al.*, 1991).

II. Dystonia

Dystonia is a syndrome dominated by sustained muscle contractions, frequently causing twisting and repetitive movements or abnormal posture. It may begin in childhood or adult life, may be focal or generalized, or can be present at rest or only on action or even only during execution of a particular task such as writing. The commonest form, idiopathic generalized dystonia, is a genetic condition with very variable manifestations among affected family members and gene carriers.

A. Activation Studies in Idiopathic Dystonia

The earliest activation studies in (predominantly unilateral) idiopathic dystonia used vibrotactile stimulation to induce dystonic spasms and reported decreased activation of the primary sensorimotor cortex whether the affected or apparently normal hand was stimulated (Tempel and Perlmutter, 1990, 1993). In a subsequent PET study of patients with generalized dystonia performing paced joystick movements in freely chosen directions, the authors reported significant overactivity in contralateral lateral premotor cortex, rostral (pre)supplementary motor area (SMA), Brodmann area 8, anterior cingulate area 32, ipsilateral dorsolateral prefrontal cortex, and bilateral lentiform nucleus. Significant underactivity was found in caudal SMA (proper), bilateral sensorimotor cortex, posterior cingulate, and mesial parietal cortex (Ceballos-Baumann *et al.*, 1995c). A broadly similar study of patients with idiopathic torsion dystonia studied during the same task showed increased activity in contralateral premotor cortex, SMA, anterior cingulate, and dorsolateral prefrontal cortex. Subcortical increases were seen in the putamen and cerebellum and there was reduced activation in the contralateral sensorimotor cortex (Playford *et al.,* 1998).

In a further activation study, patients with writer's cramp were instructed to write the word "dog" every 4 s. This paradigm was chosen to minimize "thinking time" and to maximize activity in the motor output circuits. The patients showed impaired activation of contralateral primary motor cortex, caudal SMA, mesial prefrontal cortex, anterior cingulate, mesial parietal cortex, and thalamus in comparison to control subjects (Ceballos-Baumann *et al.,* 1997). The location of the impaired activation of primary motor cortex was over the hand area of the precentral gyrus and extended toward the premotor cortex rather than posteriorly. When the same patients were treated with botulinum toxin, their writing improved. Repeat scanning revealed enhanced activation of parietal cortex and motor accessory areas, particularly caudal SMA, but failed to normalize the impaired activation of primary motor cortex. When control subjects were scanned while imitating writer's cramp, this increased ipsilateral parietal activation but did not reduce contralateral primary motor cortex activity (Ceballos-Baumann *et al.,* 1995b).

Using intravenous bolus injections of [^{15}O]butanol to measure cerebral blood flow, Odergren and colleagues studied patients with writer's cramp when drawing horizontal lines or writing a prelearned text (Odergren *et al.,* 1998). Correlating with the duration of writing, these patients had progressively increased activity in the left primary sensorimotor and premotor cortices, left thalamus, and right cerebellum as well as a progressive decrease in blood flow in the left supramarginal and angular gyri. The control subjects also showed increased activity in left sensorimotor cortex and right cerebellum during writing, but these changes did not become progressively greater as the control subjects continued to write. The extent to which the progressive increases in regional blood flow are due to sensory feedback as the writing task becomes progressively more difficult with time is not known.

Patients with idiopathic torsion dystonia scanned while imagining hand movements showed similar activity to controls, namely, activation in occipital cortex (area A18), left lentiform nucleus, left posterior thalamus, bilateral insula, bilateral parietal cortex, bilateral dorsolateral prefrontal cortex (areas 9 and 46), rostral SMA, and lateral premotor cortex, with sparing of sensorimotor cortex. On this basis the authors argued that patients' abilities to plan and prepare movements were likely to be normal (Ceballos-Baumann and Brooks, 1998). Dystonia appears to be a disorder of the motor executive system rather than a problem with motor planning.

B. Fluorodeoxyglucose Studies in Idiopathic Dystonia

Using fluorodeoxyglucose PET, Eidelberg and colleagues scanned a group of patients with predominantly right-sided idiopathic torsion dystonia at rest. Although global and regional metabolic rates were normal in comparison to a control group, when the scaled subprofile model was applied to these data, the authors were able to identify relative bilateral increases in the metabolic activity of the lateral prefrontal and paracentral cortices, associated with relative covariate hypermetabolism in the contralateral lentiform nucleus, pons, and midbrain, but not the thalamus (Eidelberg *et al.,* 1995). Magyar-Lehmann and colleagues also studied resting glucose metabolism in dystonia; their patients had spasmodic torticollis and the findings were of bilateral lentiform hypermetabolism (Magyar-Lehmann *et al.,* 1997). These patients had undergone botulinum toxin treatment and had only mild to moderate symptoms at the time of scanning. Galardi and colleagues also studied patients with cervical dystonia and reported more extensive areas of hypermetabolism, including the basal ganglia, thalamus, premotor-motor cortex, and cerebellum (Galardi *et al.,* 1996). These authors applied a discriminant function analysis and showed that by this means all but one (of 10 patients and 15 controls) could be correctly assigned to their diagnostic group on the basis of the imaging data.

To avoid possible confounding effects of dystonic movements occurring during resting scans, Eidelberg and colleagues subsequently studied nonmanifesting carriers of the DYT1 gene. Using scaled subprofile model analysis again, the authors identified a significant covariance pattern characterized by bilateral lentiform hypermetabolism and thalamo-lentiform dissociation associated with covarying resting-state increases in SMA metabolism. Scores for this pattern were abnormally elevated in the gene carriers despite their lack of clinical symptoms (Eidelberg, 1998). These data suggest that abnormal expression of this network pattern may be a necessary but not sufficient criterion for the development of symptoms.

C. Ligand Studies in Idiopathic Dystonia

Small numbers of patients were studied in the 1980s and reported to have reduced striatal fluorodopa uptake (Lang *et al.,* 1988; Leenders *et al.,* 1988). A subsequent larger study of eleven patients with familial idiopathic generalized dystonia reported eight with uptake in the lower half of the normal range and three whose fluorodopa uptake fell more than two standard deviations below the normal mean (Playford *et al.,* 1993). The authors suggested that an abnormality in the nigrostriatal dopamine pathway was unlikely to be the primary determinant of the nature and severity of the condition. In contrast, a single publication has described increased putamen and caudate uptake in eight patients with focal, segmental, and generalized idiopathic dystonia (not all of whom had a positive family history) (Otsuka et al., 1991). The cause of the increased uptake in these subjects is unclear.

Naumann reported normal [^{123}I]β-CIT binding in the striatum, whereas [^{123}I]epidepride binding, a marker for D_2 receptors, was bilaterally lower in patients than in controls (Naumann et al., 1998). Using spiperone (which binds to D_2-like (about 74%) and serotoninergic S_2-like sites (about 26%)), Perlmutter et al. reported decreased uptake in a group of 14 patients with adult-onset cranial dystonia and 7 with adult-onset dystonic hand cramp (Perlmutter et al., 1998); this is the first demonstration of a receptor abnormality in idiopathic dystonia. The authors suggest that this finding accords with a preferential reduction in D_2-mediated inhibition in the indirect striatopallidal pathway (via the external pallidum and subthalamic nucleus) as a cause of dystonic movements.

D. Acquired Hemidystonia

Patients with hemidystonia due to structural lesions in the contralateral basal ganglia or posterior thalamus have also been studied during joystick movements in freely chosen directions. Using their dystonic arm, these patients showed increased activity in both primary and accessory motor areas; similar but less marked changes were shown when the patients used their clinically unaffected arm (Ceballos-Baumann et al., 1995a). This contrasts with the decreased sensorimotor cortical activation reported in idiopathic dystonia (Ceballos-Baumann et al., 1995c). While accepting that this could in part be explained by a lesser quantum of physical movement in the idiopathic group, the authors postulated that the higher level of cortical activation in the group with acquired dystonia was a reflection of normal cortical response to subcortical pathology, whereas in the idiopathic group the disease process may affect the primary sensorimotor cortex also.

E. Dopa-Responsive Dystonia

Around 5–10% of patients with inherited dystonia have dopa-responsive dystonia (Fig. 1). This condition is inherited as an autosomal dominant condition and it has been shown to be due to a variety of defects in the GTP cyclohydrolase-I gene (Ichinose et al., 1994; Bandmann et al., 1996). The ratio of mutant/normal mRNA encoding GTP cyclohydrolase-I may vary among members of a single family who share the same genetic mutation, perhaps explaining their different clinical characteristics (Hirano et al., 1996). The two principal differential diagnoses are idiopathic generalized dystonia and juvenile parkinsonism. Patients with juvenile parkinsonism rarely present before the age of 8, but in other respects the differential diagnosis may be difficult in the early stages and since both patient groups respond well to levodopa, it may not be until response fluctuations and other complications of levodopa treatment emerge in those with juvenile parkinsonism that the diagnosis becomes clear. Fluorodopa scans in patients with dopa-responsive dystonia have normal or near-normal tracer uptake, whereas patients with juvenile parkinsonism have scans similar to those in adult Parkinson's disease (Sawle et al., 1991; Snow et al., 1993). This is one case where fluorodopa scans have important diagnostic use, since patients with juvenile Parkinson's disease are best not exposed to levodopa too soon, whereas patients with dopa-responsive dystonia may be given levodopa without fear of treatment-related complications.

F. Wilson's Disease

When Wilson's disease presents as a movement disorder, it most commonly leads to parkinsonism. A minority of patients present instead with a dystonia and, of course, the clinical picture may be mixed. It is usual for speech and hand function to be the most severely affected. As expected in a predominantly parkinsonian disorder, the principal findings of functional imaging have involved the nigrostriatal system (Fig. 2), including an impairment of striatal fluorodopa uptake (Snow et al., 1991), reduced binding to the dopamine transporter (Jeon et al., 1998), and reduced binding to striatal D_2 receptors (Oertel et al., 1992).

G. Paroxysmal Dystonias

Rarely, patients with dystonia have symptoms present only on exercise. In two affected patients, Kluge and colleagues injected [^{99m}Tc]ECD (ethyl cysteinate dimer) at the peak of a dystonic attack. They reported a 7–19% reduction in frontal perfusion in both subjects and a reduction in basal ganglia perfusion in one (Kluge et al., 1998). Cerebellar perfusion was increased during the dystonic episode in both subjects. The lack of frontal cortical activity contrasts with the heightened cortical flow seen ictally in focal epilepsy studies.

In patients with paroxysmal kinesogenic choreoathetosis, interictal and postictal SPECT HMPAO studies showed changes matched to focal EEG changes and

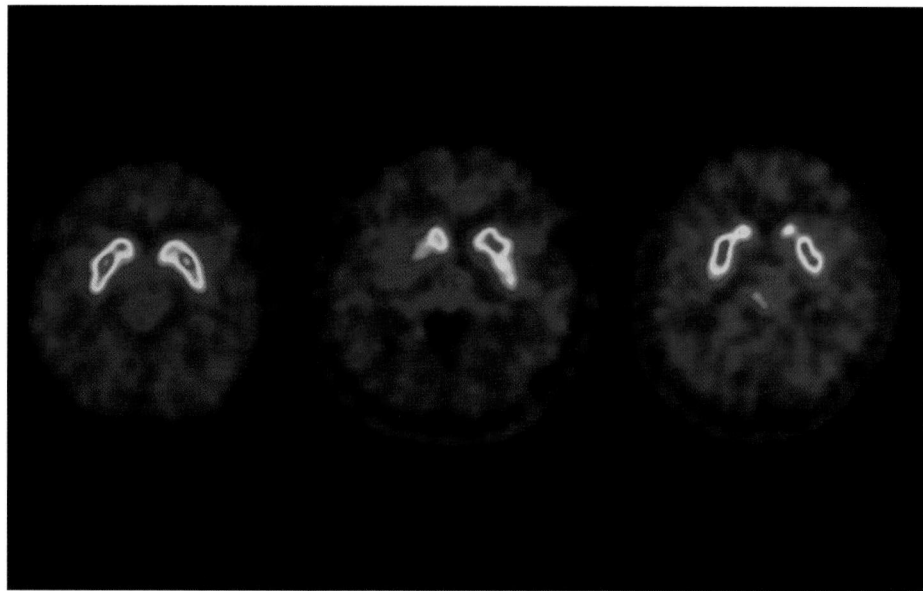

Figure 1 Fluorodopa uptake in a healthy control (left), a patient with Parkinson's disease (middle), and a patient with dopa-responsive dystonia (right). The apparently normal uptake in the DRD patient contrasts with the reduced striatal uptake in the Parkinson's disease patient.

suggestive of focal epilepsy (Lombroso, 1995). In familial paroxysmal dystonic choreoathetosis, PET using $(+)$-α-$[^{11}C]$dihydrotetrabenazine as ligand failed to show any alteration of striatal binding, suggesting that the nigrostriatal nerve terminal density is normal (Bohnen *et al.*, 1999).

III. Tics

Tics are the most variable of the movement disorders. Motor tics range from simple brief rapid movements of a single body part to complex sequences of coordinated movements. Vocal tics vary from brief elementary noises to complete words and phrases. Complex tics may look like normal movements, but are irresistible, repetitive, purposeless, and unwanted. Most patients can temporarily suppress the tics but this is usually accompanied by mounting inner tension and followed by a flurry of noises or movements as soon as the effort is relaxed.

SPECT studies in Tourette syndrome have been reported as showing reduction in blood flow through either right (Klieger *et al.*, 1997) or left (Riddle *et al.*, 1992; Sieg *et al.*, 1993) basal ganglia. Other studies have shown no such asymmetry of blood flow but have shown changes in dopaminergic markers. For example, Wolf and colleagues (1996) studied five pairs of monozygotic twins who were discordant for their severity of Tourette's syndrome using iodobenzamide (IBZM) to measure D_2 receptor binding. They found greater IBZM binding in the caudate nuclei of the more severely affected twins.

PET ligand studies have not shown any consistent changes in dopaminergic systems. Turjanski and colleagues reported normal fluorodopa and raclopride (D_2) binding in Tourette patients, some of whom were taking neuroleptic medication but some of whom were drug free (Turjanski *et al.*, 1994). Singer and colleagues reported a greater variability in their patients' D_2 receptor density, but no absolute difference between patient and control mean values (Singer *et al.*, 1992). Metabolic PET studies using fluorodeoxyglucose have not revealed any absolute change in regional cerebral metabolism, but covariance analysis of fluorodeoxyglucose scans suggests a relative increase in glucose metabolism in lateral premotor, supplementary motor cortex, and midbrain (reflecting a nonspecific increase in cortical motor activity) and a reduction in metabolism in caudate, lentiform, thalamic, and hypothalamic nuclei (which perhaps reflects underactivity of an inhibitory network) (Eidelberg *et al.*, 1997). The involvement of the midbrain in these circuits is supported by a case report in which tics remitted after the dorsal midbrain was damaged by Wernicke's encephalopathy due to hyperemesis (Pantoni *et al.*, 1997). In a SPECT study using $[^{123}I]\beta$-CIT, Heinz and colleagues reported a negative correlation between tic severity and β-CIT binding (a

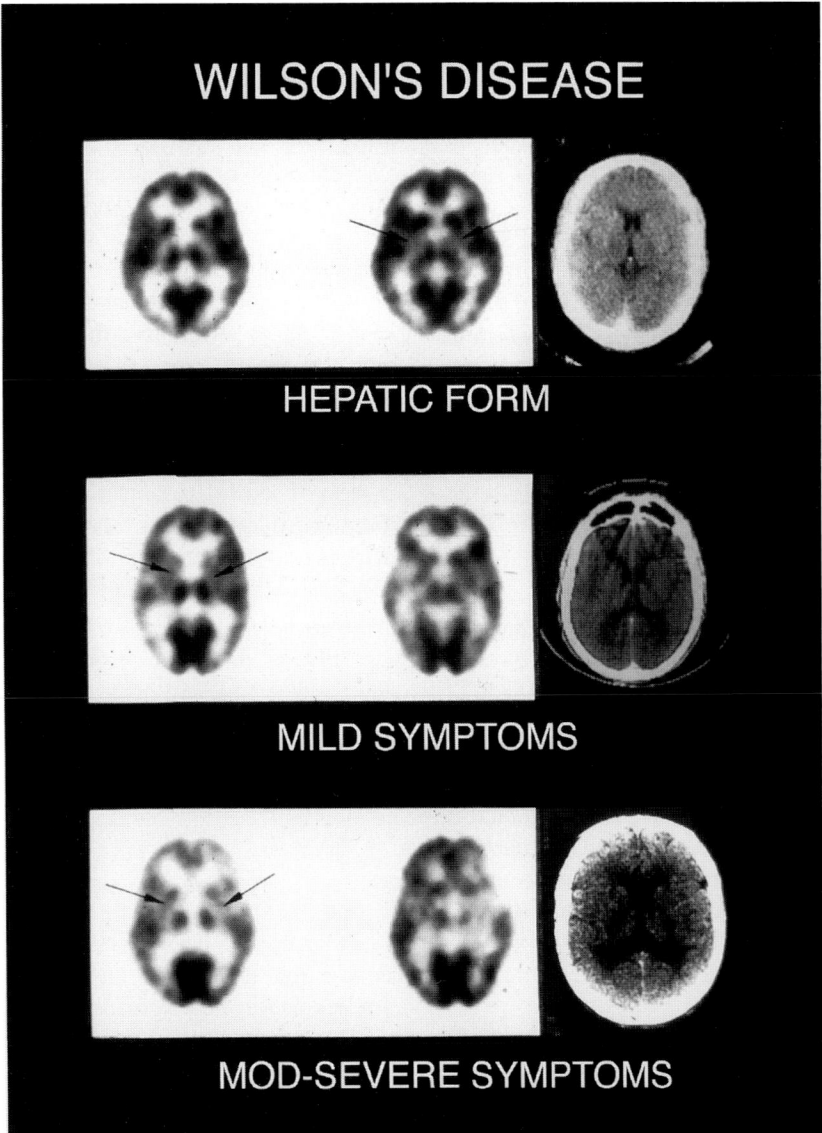

Figure 2 Wilson's disease studied with PET to determine cerebral glucose metabolism. In Wilson's disease, abnormalities of the brain are most concentrated in the lenticular nuclei as well as cortical regions. Three different patients are illustrated in this figure, demonstrating glucose metabolism as compared with X-ray CT images. In the patient with the "hepatic form" of the disease, there is mild hypometabolism of the posterior portion of the lenticular nuclei and no corresponding abnormalities for the structural imaging study. This patient was neurologically asymptomatic. In the patient with mild neurological symptoms (middle row), more profound hypometabolism of the lenticular nuclei is seen in the glucose metabolic image but again the structural study is normal. The patient with moderately severe symptoms (bottom row) has both lenticular and cortical abnormalities as evidenced by hypometabolism of the posterior putamen and globus pallidus as well as right frontotemporal hypometabolism. Atrophy is evidenced by dilation of the lateral ventricles seen on X-ray CT. Courtesy of Hawkins, Mazziotta, and Phelps (*Neurology* **37,** 1707–1711 (1987)).

measure of presynaptic monoaminergic transporters) in the midbrain and thalamus (Heinz *et al.*, 1998).

Functional magnetic resonance imaging techniques have recently been applied to this condition. Patients were asked to suppress their tics during scanning, and this led to significant changes in signal intensity in the basal ganglia and anatomically connected cortical regions believed to subserve attention-demanding tasks (Peterson *et al.*, 1998).

IV. Chorea

Chorea is Greek for "dance," and patients with chorea have jerky, restless, purposeless movements that give a fidgety appearance. The movements typically flit from one body part to another. Patients may try to hide choreic movements by continuing the movement into an apparently purposeful action, such as adjusting their clothing or hair. Chorea may be caused by a variety of hereditary disorders, of which the commonest is Huntington's disease, and there are also metabolic, infectious, and structural causes, the latter comprising both vascular and mass lesions. Chorea can also occur as a side effect of certain drugs.

A. Huntington's Disease

The cardinal clinical features of Huntington's disease are the movement disorder, dementia, and psychiatric features, usually in the context of a positive family history. While chorea usually dominates the clinical picture in the earlier stages of the disease, patients with Huntington's disease also have other abnormal movements, including myoclonus, parkinsonism, and dystonia, which become more obvious later. All patients have abnormal eye movements, usually from very early on in the clinical phase of their illness.

It is now recognized that Huntington's disease is caused by an unstable expanded DNA trinucleotide (cytosine-adenosine-guanine, or CAG) repeat within the coding region for a protein called huntingtin, located on the short arm of chromosome 4. Before the gene had been identified, a variety of approaches had been tried in an effort to predict which at-risk patients (for example, children of affected parents) would develop the disorder in the future. PET was explored as a tool to identify subclinical disease (Fig. 3; also see Chapter 1, Fig. 8). Now that the diagnosis can be made by a blood test instead, attention has focused on an improved understanding of the pathophysiology of this disorder.

In resting patients with mild symptoms, fluorodeoxyglucose PET has shown increased thalamic tracer uptake (Clark *et al.*, 1986; Grafton *et al.*, 1992), which may be functionally related to the choreic movements. Measuring cerebral blood flow using oxygen-15-labeled water in Huntington's disease patients performing an externally triggered finger opposition task,

Bartenstein and colleagues reported impaired activity in the striatum and its frontal projection areas (rostral SMA, anterior cingulate, and premotor cortex) as well as enhanced activity in the parietal lobe (Bartenstein *et al.*, 1997). The activity in the rostral SMA was closely correlated with the patient's ability to perform the set motor task accurately. The authors suggested that the pathology of Huntington's disease caused impairment of the output limb of the basal ganglia–thalamocortical motor circuit, leading in turn to a compensatory recruitment of alternative motor pathways involving the parietal cortex.

Oxygen-15-labeled water PET scans in patients with Huntington's disease performing freely chosen (but time-paced) joystick movements showed impaired activation of contralateral primary motor, medial premotor, bilateral parietal, and bilateral prefrontal areas, along with increased activation of the insular region bilaterally (Weeks *et al.*, 1997). The lowered sensorimotor cortical activity may reflect either decreased afferent or interneuronal firing or else direct involvement of the cortex in the disease process. Since, of course, the lowered sensorimotor activation may be a loss of inhibition, this could contribute to the genesis of involuntary movements. It has been assumed that the bilateral increase in insular activity represents a compensatory mechanism.

B. Other Causes of Chorea

Patients with chorea resulting from a variety of other conditions have also undergone functional imaging studies. Fluorodeoxyglucose studies have been the most plentiful. In most cases, the principal finding has been a decrease in striatal metabolism. This has been demonstrated in small numbers of patients with neuroacanthocytosis (Hosokawa *et al.*, 1987) (in whom striatal blood flow and oxygen metabolism are reduced also (Tanaka *et al.*, 1998)), the pseudo-Huntington form of dentato-rubro-pallido-luysian atrophy (Hosokawa *et al.*, 1987), Lesch Nyhan syndrome (Palella *et al.*, 1981), and vascular (heminballismus) (Heiss *et al.*, 1989) origins for their movement disorder. In benign hereditary chorea, there have been reports of striatal glucose hypometabolism (Suchowersky *et al.*, 1986) but also of normal metabolism (Kuwert *et al.*, 1990).

These conditions together with Huntington's disease contrast with those in which the findings have chiefly been striatal hypermetabolism. In the primary antiphospholipid syndrome, an increase in striatal glucose metabolism has been reported (Furie *et al.*, 1994), with return to normal metabolic levels after clinical improvement (Sundén-Cullberg *et al.*, 1998). Increased striatal glucose metabolism has also been reported in

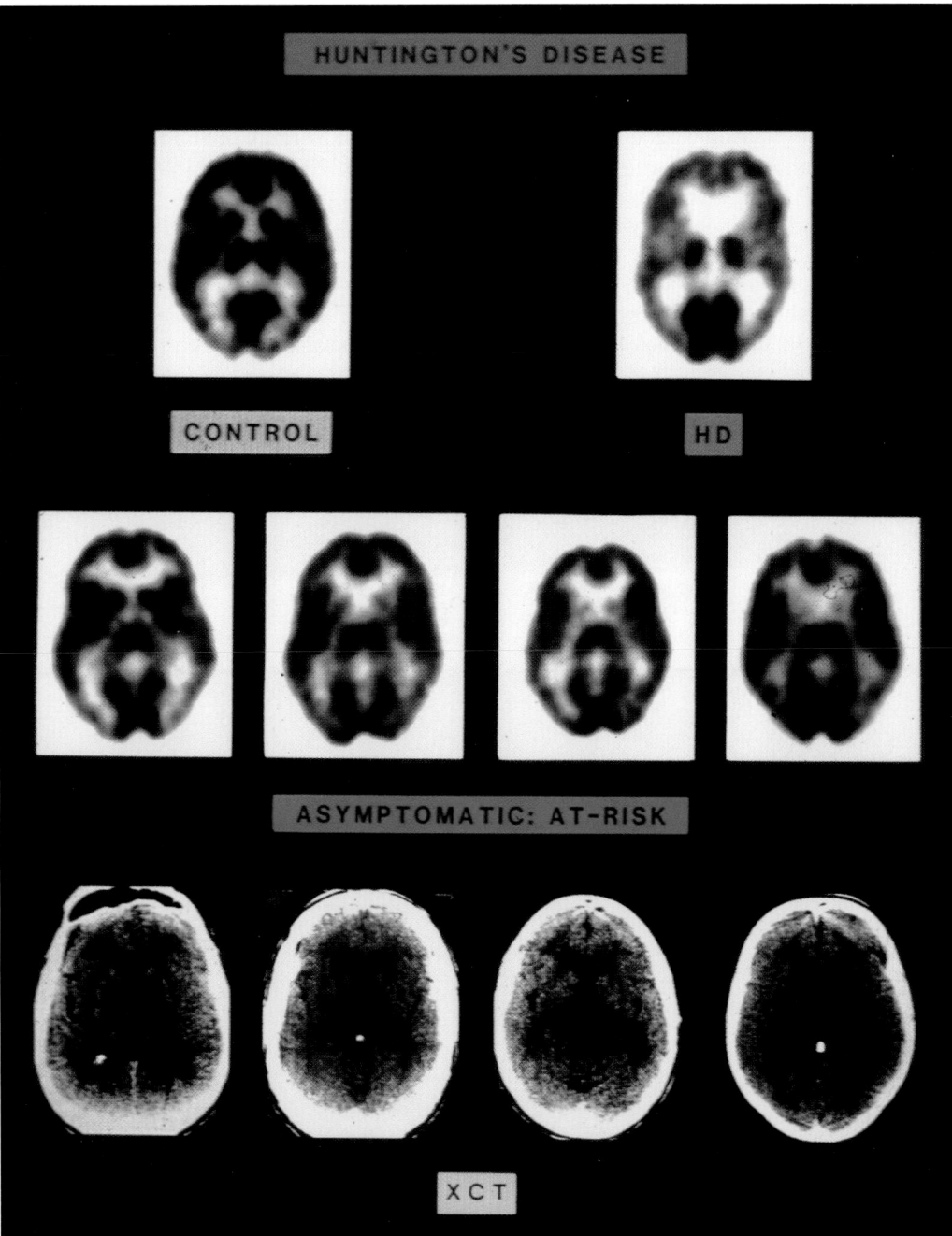

Figure 3 Huntington's disease studied with PET to determine cerebral glucose metabolism. Top left image shows a typical glucose metabolic scan through the striatum and thalamus of a normal control subject. Upper right image shows a patient with symptomatic Huntington's disease, demonstrating near-complete absence of the striata (only minimal activity remains in the posterior putamen) as well as evidence of atrophy (increased ventricular size). The middle and bottom rows demonstrate glucose metabolic images and X-ray CT images, respectively, of four different asymptomatic subjects, each of whom has a parent with clinically diagnosed Huntington's disease. In each case, the structural images are normal. Nevertheless, the glucose metabolic images demonstrate a spectrum of change from normal (left image, middle row) to profound hypometabolism (right image, middle row) of the caudates even in this presymptomatic group (also see Chapter 1, Fig. 8). Studies such as these have demonstrated that in patients who are genetically presymptomatic for Huntington's disease (i.e., have expanded triplet repeat segments for the huntingtin gene), reduced metabolism of the caudate nuclei and striata occurs 5–7 years before the disorder can be confirmed clinically. Courtesy of Mazziotta and colleagues, UCLA School of Medicine, Los Angeles, California [*N. Engl. J. Med.* **316,** 357–362 (1987). Copyright © 1987 Massachusetts Medical Society. All rights reserved.].

Sydenham's chorea. Scans in young symptomatic patients have shown increased caudate and lentiform metabolism in the acute phase, returning to normal or near normal as symptoms settle (Goldman *et al.,* 1993; Weindl *et al.,* 1993). In a single patient aged 74 who was scanned after 8 years hemichorea and who had previously suffered with chorea in childhood after rheumatic fever, the contralateral lentiform nucleus was again hypermetabolic (Weindl *et al.,* 1993). Various suggestions have been put forward to explain why some choreic patients have striatal hypometabolism while others are hypermetabolic. It may be that these immunological conditions are associated with antibody binding to striatal interneurons, leading to "hypermetabolic dysfunction" of these cells. Some authors have considered a vascular explanation unlikely on the basis that patients may not have lesions visible on MRI scans and they may respond quickly to corticosteroids (Sundén-Cullberg *et al.,* 1998).

V. Tardive Dyskinesia

This, the most troublesome drug-induced movement disorder, usually follows chronic exposure to dopamine

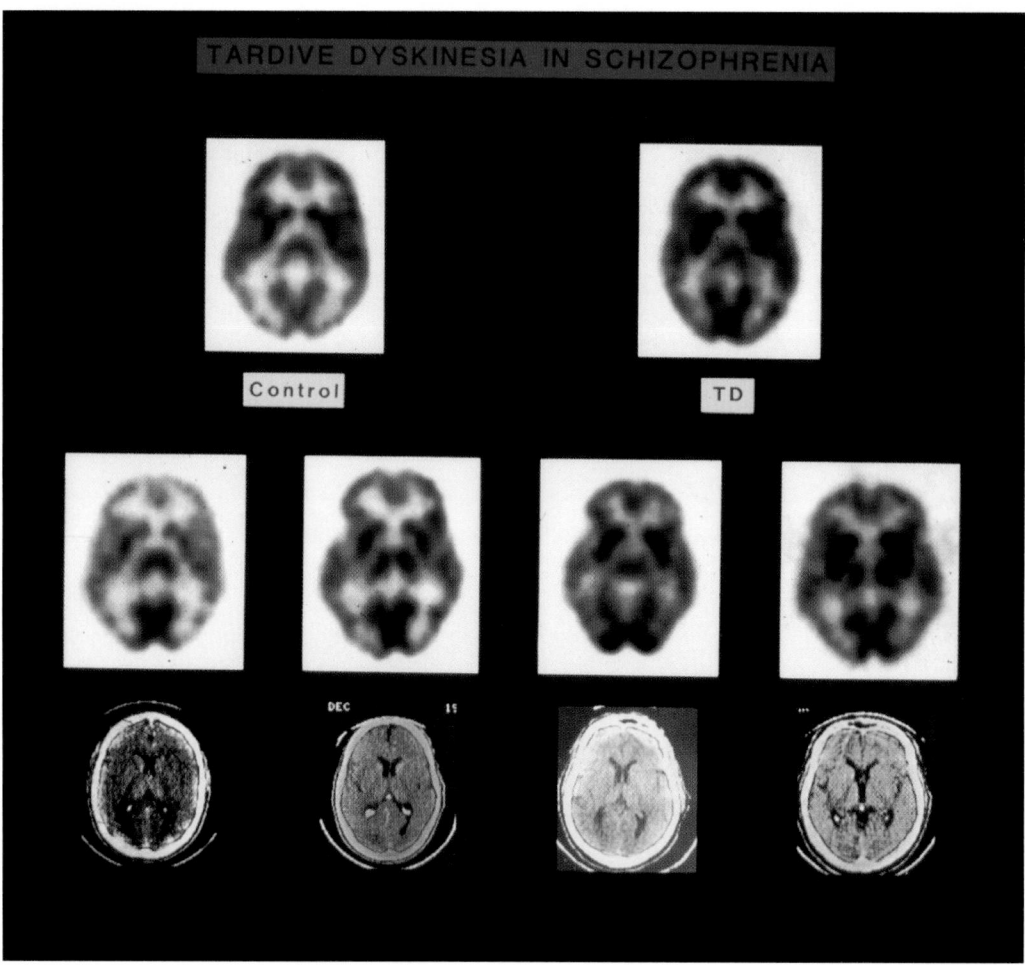

Figure 4 Glucose metabolic imaging in patients with tardive dyskinesia. The upper left image shows a glucose metabolic image obtained with PET in a normal control through the striatum and midthalamic region. This is compared to the study of a patient with moderately severe tardive dyskinesia in the upper right image (TD). In the tardive dyskinesia patient, there is hypermetabolism of the striatum, particularly in the lenticular nuclei. The middle and bottom rows demonstrate four patients with schizophrenia and tardive dyskinesia. Their symptoms range from mild (left column) to severe (right column). The bottom row displays X-ray CT structural imaging while the middle row contains glucose metabolic images from PET. Note that as the patients' symptoms increase from mild to severe, so too does the relative hypermetabolism of the striata and, in particular, the lenticular nuclei. Courtesy of Pahl, Mazziotta, and colleagues, UCLA School of Medicine, Los Angeles, California. (*J. Neuropsychiatry Clin. Neurosci.* **7,** 457–465 (1995)).

receptor blocking drugs. The movements encompass almost all movements seen in other conditions, including (in decreasing order of frequency) stereotypy, dystonia, akathisia, tremor, chorea, and myoclonus.

Dopamine D_2 receptors have been reported to be either in the normal range, but yet with receptor density correlating with the severity of orofacial movements (Blin *et al.*, 1989), or else entirely normal (Andersson *et al.*, 1990). Fluorodeoxyglucose estimates of glucose metabolism in patients with schizophrenia and established tardive dyskinesia show increased metabolism in the globus pallidus and primary motor cortex, suggesting a role for increased pallidal synaptic activity due to either altered striato-pallidal input or increased pallidal interneuronal firing (Pahl *et al.*, 1995) (Fig. 4). In a prospective fluorodeoxyglucose study of schizophrenic patients without tardive dyskinesia, relative hypermetabolism in the temporolimbic, brainstem, and cerebellar regions as well as reduced activity in the parietal and cingulate gyrus was found in those patients who went on to develop tardive dyskinesia over the following 3 years (Szymanski *et al.*, 1996).

VI. Restless Legs Syndrome

This condition has a prevalence around 5% in the general population. Because some patients report benefit from levodopa or dopamine agonist medication and because the condition is common in patients with Parkinson's disease, it has been assumed that central dopaminergic dysfunction underpins this disorder. Staedt and colleagues used SPECT to study patients with restless legs and periodic movements of sleep and reported a reduction in striatal dopamine D_2 receptor binding (Staedt *et al.*, 1995). Turjanski and colleagues performed fluorodopa and raclopride PET studies in affected subjects and reported significantly impaired putamen fluorodopa uptake and reduced raclopride binding in both the caudate and putamen (Turjanski *et al.*, 1999). Both of these studies indicate an abnormality of dopaminergic function in this disorder.

Using fMRI, Bucher and colleagues reported bilateral cerebellar and contralateral thalamic activity during sensory leg discomfort. When periodic leg movements accompanied the discomfort, there was additional activation of the red nuclei and brainstem close to the reticular formation; the authors suggested that these brainstem structures were involved as movement generators (Bucher *et al.*, 1997b) as there was no recorded cortical activation. None of the published functional imaging studies to date have shown any indication of a cortical origin for these involuntary movements.

References

Andersson, U., Eckernas, S. A., Hartvig, P., Ulin, J., Langstrom, B., and Haggstrom, J. E. (1990). Striatal binding of ^{11}C-NMSP studied with positron emission tomography in patients with persistent tardive dyskinesia: No evidence for altered dopamine D_2-receptor binding. *J. Neural Trans.* **79**, 215–226.

Antonini, A., Moeller, J. R., Nakamura, T., Spetsieris, P., Dhawan, V., and Eidelberg, D. (1998). The metabolic anatomy of tremor in Parkinson's disease. *Neurology* **51**, 803–810.

Bandmann, O., Nygaard, T. G., Surtees, R., Marsden, C. D., Wood, N. W., and Harding, A. E. (1996). Dopa-responsive dystonia in British patients: New mutations of the GTP-cyclohydrolase I gene and evidence for genetic heterogeneity. *Hum. Mol. Genet.* **5**, 403–406.

Bartenstein, P., Weindl, A., Spiegel, S., Boecker, H., Wenzel, R., Ceballos-Baumann, A. O., *et al.* (1997). Central motor processing in Huntington's disease: A PET study. *Brain* **120**, 1553–1567.

Blin, J., Baron, J. C., Cambon, H., Bonnet, A. M., Dubois, B. V., Loc'h, C., Maziere, B., and Agid, Y. (1989). Striatal dopamine D_2 receptors in tardive dyskinesia: A PET study. *J. Neurol. Neurosurg. Psychiatry* **52**, 1248–1252.

Boecker, H., Wills, A. J., Ceballos-Baumann, A., Samuel, M., Thompson, P. D., Findley, L. J., *et al.* (1996). The effect of alcohol on alcohol-responsive essential tremor: A positron emission tomographic study. *Ann. Neurol.* **39**, 650–658.

Bohnen, N. I., Albin, R. L., Frey, K. A., and Fink, J. K. (1999). (+)-α-[^{11}C]Dihydrotetrabenazine PET imaging in familial paroxysmal dystonic choreoathetosis. *Neurology* **52**, 1067–1069.

Britton, T. C., Thompson, P. D., Day, B. L., Rothwell, J. C., Findley, L. J., and Marsden, C. D. (1993). Modulation of postural wrist tremors by magnetic stimulation of the motor cortices in patients with Parkinson's disease or essential tremor and in normal subjects mimicking tremor. *Ann. Neurol.* **33**, 473–479.

Brooks, D. J., Playford, E. D., Ibanez, V., Sawle, G. V., Thompson, P. D., Findley, L. J., *et al.* (1992). Isolated tremor and disruption of the nigrostriatal dopaminergic system. *Neurology* **42**, 1554–1560.

Brown, P. (1995). New clinical sign for orthostatic tremor. *Lancet* **346**, 306–307.

Bucher, S. F., Seelos, K. C., Dodel, R. C., Reiser, M., and Oertel, W. H. (1997a). Activation mapping in essential tremor with functional magnetic resonance imaging. *Ann. Neurol.* **41**, 32–40.

Bucher, S. F., Seelos, K. C., Oertel, W. H., Reiser, M., and Trenkwalder, C. (1997b). Cerebral generators involved in the pathogenesis of the restless legs syndrome. *Ann. Neurol.* **41**, 639–645.

Ceballos-Baumann, A. O., and Brooks, D. J. (1998). Activation positron emission tomography scanning in dystonia. *Adv. Neurol.* **78**, 135–152.

Ceballos-Baumann, A. O., Passingham, R. E., Marsden, C. D., and Brooks, D. J. (1995a). Motor reorganization in acquired hemidystonia. *Ann. Neurol.* **37**, 746–757.

Ceballos-Baumann, A. O., Passingham, R. E., Marsden, C. D., and Brooks, D. J. (1995b). Differential brain activation in idiopathic and simulated writer's cramp. *J. Neurol.* **242** (Suppl. 2), S47.

Ceballos-Baumann, A. O., Passingham, R. E., Warner, T., Playford, E. D., Marsden, C. D., and Brooks, D. J. (1995c). Overactive prefrontal and underactive motor cortical areas in idiopathic dystonia. *Ann. Neurol.* **37**, 363–372.

Ceballos-Baumann, A. O., Sheean, G., Passingham, R. E., Marsden, C. D., and Brooks, D. J. (1997). Botulinum toxin does not reverse the cortical dysfunction associated with writer's cramp. A PET study. *Brain* **120**, 571–582.

Clark, C. M., Hayden, M. R., Stoessl, A. J., and Martin, W. R. (1986). Regression model for predicting dissociations of regional cerebral glucose metabolism in individuals at risk for Huntington's disease. *J. Cereb. Blood Flow Metab.* **6**, 756–762.

Defer, G. L., Remy, P., Malapert, D., Ricolfi, F., Samson, Y., and Degos, J. D. (1994). Rest tremor and extrapyramidal symptoms after midbrain haemorrhage: Clinical and ^{18}F-dopa PET evaluation. *J. Neurol. Neurosurg. Psychiatry* **57,** 987–989.

Dubinsky, R. M., Hallett, M., Di Chiro, G., Fulham, M., and Schwankhaus, J. (1991). Increased glucose metabolism in the medulla of patients with palatal myoclonus. *Neurology* **41,** 557–562.

Eidelberg, D. (1998). Abnormal brain networks in DYT1 dystonia. *Adv. Neurol.* **78,** 127–133.

Eidelberg, D., Moeller, J. R., Dhawan, V., Sidtis, J. J., Ginos, J. Z., Strother, S. C., *et al.* (1990). The metabolic anatomy of Parkinson's disease: Complementary [(18)F][fluorodeoxyglucose and [(18)F]fluorodopa positron emission tomographic studies. *Mov. Disord.* **5,** 203–213.

Eidelberg, D., Moeller, J. R., Ishikawa, T., Dhawan, V., Spetsieris, P., Przedborski, S., *et al.* (1995). The metabolic topography of idiopathic torsion dystonia. *Brain* **118,** 1473–1484.

Eidelberg, D., Moeller, J. R., Antonini, A., *et al.* (1997). The metabolic anatomy of Tourette's syndrome. *Neurology* **48,** 927–934.

Furie, R., Ishikawa, T., Dhawan, V., and Eidelberg, D. (1994). Alternating hemichorea in primary antiphospholipid syndrome: Evidence for contralateral striatal hypermetabolism. *Neurology* **44,** 2197–2199.

Galardi, G., Perani, D., Grassi, F., Bressi, S., Amadio, S., Antoni, M., *et al.* (1996). Basal ganglia and thalamo-cortical hypermetabolism in patients with spasmodic torticollis. *Acta Neurol. Scand.* **94,** 172–176.

Goldman, S., Amron, D., Szliwowski, H. B., Detemmerman, D., Bidaut, L. M., Stanus, E., *et al.* (1993). Reversible striatal hypermetabolism in a case of Sydenham's chorea. *Mov. Disord.* **8,** 355–358.

Grafton, S. T., Mazziotta, J. C., Pahl, J. J., St. George-Hyslop, P., Haines, J. L., Gusella, J., *et al.* (1992). Serial changes of cerebral glucose metabolism and caudate size in persons at risk for Huntington's disease. *Arch. Neurol.* **49,** 1161–1167.

Hallett, M., and Dubinsky, R. M. (1993). Glucose metabolism in the brain of patients with essential tremor. *J. Neurol. Sci.* **114,** 45–48.

Heinz, A., Knable, M. B., Wolf, S. S., Jones, D. W., Gorey, J. G., Hyde, T. M., *et al.* (1998). Tourette's syndrome: [I-123]β-CIT SPECT correlates of vocal tic severity. *Neurology* **51,** 1069–1074.

Heiss, W., Pawlik, G., Herholz, K., Szelies, B., and Wienhard, K. (1989). Positron emission tomography as an imaging tool in psychiatric disorders. *Psychiatry Res.* **29,** 351–352.

Hirano, M., Tamaru, Y., Ito, H., Matsumoto, S., Imai, T., and Ueno, S. (1996). Mutant GTP cyclohydrolase I mRNA levels contribute to dopa-responsive dystonia onset. *Ann. Neurol.* **40,** 796–798.

Hosokawa, S., Ichiya, Y., Kuwabara, Y., *et al.* (1987). Positron emission tomography in cases of chorea with different underlying causes. *J. Neurol. Neurosurg. Psychiatry* **50,** 1284–1287.

Ichinose, H., Ohye, T., Takahashi, E. I., Seki, N., Hori, T. A., Segawa, M., *et al.* (1994). Hereditary progressive dystonia with marked diurnal fluctuation caused by mutations in the GTP cyclohydrolase I gene. *Nat. Genet.* **8,** 236–242.

Jenkins, I. H., Bain, P. G., Colebatch, J. G., Thompson, P. D., Findley, L. J., Frackowiak, R. S. J., *et al.* (1993). A positron emission tomography study of essential tremor: Evidence for overactivity of cerebellar connections. *Ann. Neurol.* **34,** 82–90.

Jeon, B., Kim, J. M., Jeong, J. M., *et al.* (1998). Dopamine transporter imaging with [^{123}I]-β-CIT demonstrates presynaptic nigrostriatal dopaminergic damage in Wilson's disease. *J. Neurol. Neurosurg. Psychiatry* **65,** 60–64.

Klieger, P. S., Fett, K. A., Dimitsopulos, T., and Kurlan, R. (1997). Asymmetry of basal ganglia perfusion in Tourette's syndrome shown by technetium-99m-HMPAO SPECT. *J. Nucl. Med.* **38,** 188–191.

Kluge, A., Kettner, B., Zschenderlein, R., Sandrock, D., Munz, D. L., Hesse, S., *et al.* (1998). Changes in perfusion pattern using ECD-SPECT indicate frontal lobe and cerebellar involvement in exercise-induced paroxysmal dystonia *Mov. Disord.* **13,** 125–134.

Kuwert, T., Lange, H. W., Langen, K., Herzog, H., Hefter, H., Aulich, A., *et al.* (1990). Normal striatal glucose consumption in two patients with benign hereditary chorea as measured by positron emission tomography. *J. Neurol.* **237,** 80–84.

Lang, A. E., Garnett, E. S., Firnau, G., Nahmias, C., and Talalla, A. (1988). Positron tomography in dystonia. *Adv. Neurol.* **50,** 249–253.

Leenders, K. L., Quinn, N., Frackowiak, R. S. J., and Marsden, C. D. (1988). Brain dopaminergic system studied in patients with dystonia using positron emission tomography. *Adv. Neurol.* **50,** 243–247.

Lombroso, C. T. (1995). Paroxysmal choreoathetosis: An epileptic or non-epileptic disorder? *Ital. J. Neurol. Sci.* **16,** 271–277.

Magyar-Lehmann, S., Antonini, A., Roelcke, U., Maguire, R. P., Missimer, J., Meyer, M., *et al.* (1997). Cerebral glucose metabolism in patients with spasmodic torticollis. *Mov. Disord.* **12,** 704–708.

Morrish, P. K., Sawle, G. V., and Brooks, D. J. (1995). Clinical and [^{18}F]dopa PET findings in early Parkinson's disease. *J. Neurol. Neurosurg. Psychiatry* **59,** 597–600.

Naumann, M., Pirker, W., Reiners, K., Lange, K. W., Becker, G., and Brücke, T. (1998). Imaging the pre- and post-synaptic side of striatal dopaminergic synapses in idiopathic cervical dystonia: A SPECT study using [^{123}I]epidepride and [^{123}I]β-CIT. *Mov. Disord.* **13,** 319–323.

Odergren, T., Stone-Elander, S., and Ingvar, M. (1998). Cerebral and cerebellar activation in correlation to the action-induced dystonia in writer's cramp. *Mov. Disord.* **13,** 497–508.

Oertel, W. H., Tatsch, K., Schwarz, J., *et al.* (1992). Decrease of D_2 receptors indicated by ^{123}I-iodobenzamide single photon emission computed tomography relates to neurological deficit in Wilson's disease. *Ann. Neurol.* **32,** 743–748.

Otsuka, M., Ichiya, Y., Kuwabara, Y., Yoshikai, T., Fukumura, T., Masuda, K., *et al.* (1991). Increased striatal dopamine uptake and normal glucose metabolism in patients with idiopathic dystonia. *J. Nucl. Med.* **32,** 1015.

Pahl, J. J., Mazziotta, J. C., Bartzokis, G., Cummings, J., Altschuler, L., Mintz, J., Marder, S. R., and Phalps, M. E. (1995). Positron emission tomography in tardive dyskinesia. *J. Neurol. Clin. Neurosci.* **7,** 457–465.

Palella, T. D., Hichwa, R. D., Ehrenkaufer, R. C., *et al.* (1981). ^{18}F-Fluorodeoxyglucose PET scanning in HPRT deficiency. *Am. J. Hum. Genet.* **37,** A70.

Pantoni, L., Poggesi, I., Repice, A., and Inzitari, D. (1997). Disappearance of motor tics after Wernicke's encephalopathy in a patient with Tourette syndrome. *Neurology* **48,** 381–383.

Parker, F., Tzourio, N., Blond, S., Petit, H., and Mazoyer, B. (1992). Evidence for a common network of brain structures involved in parkinsonian tremor and voluntary repetitive movement. *Brain Res.* **584,** 11–17.

Perlmutter, J. S., Stambuk, M. K., Markham, J., Black, K. J., McGee-Minnich, L., Jankovic, J., *et al.* (1998). Decreased [^{18}F]spiperone binding in putamen in dystonia. *Adv. Neurol.* **78,** 161–168.

Peterson, B. S., Skudlarski, P., Anderson, A. W., *et al.* (1998). A functional magnetic resonance imaging study of tic suppression in Tourette syndrome. *Arch. Gen. Psychiatry* **54,** 326–333.

Playford, E. D., Fletcher, N. A., Sawle, G. V., Marsden, C. D., and Brooks, D. J. (1993). Striatal [^{18}F]dopa uptake in familial idiopathic dystonia. *Brain* **116,** 1191–1199.

Playford, E. D., Passingham, R. E., Marsden, C. D., and Brooks, D. J. (1998). Increased activation of frontal areas during arm movements in idiopathic torsion dystonia. *Mov. Disord.* 13, 309–318.

Remy, P., De Recondo, A., Defer, G., Loch, C., Amarenco, P., Plante-Bordeneuve, V., *et al.* (1995). Peduncular 'rubral' tremor and dopaminergic denervation: A PET study. *Neurology* **45,** 472–477.

Riddle, M., Rasmussen, A., Woods, S., and Hoffer, P. B. (1992). SPECT imaging of cerebral blood flow in Tourette's syndrome. *Adv. Neurol.* **58,** 207–211.

Sawle, G. V., Leenders, K. L., Brooks, D. J., Harwood, G., Lees, A. J., Frackowiak, R. S. J., *et al.* (1991). Dopa-responsive dystonia: [^{18}F]Dopa positron emission tomography. *Ann. Neurol.* **30,** 24–30.

Sieg, K., Buckingham, D., Gaffney, G., Preston, D. F., and Sieg, K. G. (1993). Technetium-99m-HMPAO SPECT imaging of Gilles de la Tourette's syndrome. *Clin. Nucl. Med.* **18,** 255.

Singer, H. S., Wong, D. F., Brown, J. E., *et al.* (1992). Positron emission tomography evaluation of dopamine-D_2 receptors in adults with Tourette syndrome. *Adv. Neurol.* **58,** 233–239.

Snow, B. J., Bhatt, M., Martin, W. R., Li, D., and Calne, D. B. (1991). The nigrostriatal dopaminergic pathway in Wilson's disease studied with positron emission tomography. *J. Neurol. Neurosurg. Psychiatry* **54,** 12–17.

Snow, B. J., Nygaard, T. G., Takahashi, H., and Calne, D. B. (1993). Positron emission tomographic studies of dopa-responsive dystonia and early-onset idiopathic parkinsonism. *Ann. Neurol.* **34,** 733–738.

Staedt, J., Stoppe, G., Kogler, A., *et al.* (1995). Nocturnal myoclonus syndrome (periodic movements in sleep) related to central dopamine D_2-receptor alteration. *Eur. Arch. Psychiatry Clin. Neurosci.* **245,** 8–10.

Suchowersky, O., Hayden, M. R., Martin, W. R. W., Stoessl, A. J., Hildebrand, A. M., and Pate, B. D. (1986). Cerebral metabolism of glucose in benign hereditary chorea. *Mov. Disord.* **1,** 33–44.

Sundén-Cullberg, J., Tedroff, J., and Aquilonius, S.-M . (1998). Reversible chorea in primary antiphospholipid syndrome. *Mov. Disord.* **13,** 147–149.

Szymanski, S., Gur, R. C., Gallacher, F., Mozley, L. H., and Gur, R. E. (1996). Vulnerability to tardive dyskinesia: An FDG-PET study of cerebral metabolism. *Neuropsychopharmacology* **15,** 567–575.

Tanaka, M., Hirai, S., Kondo, S., Sun, X., Nakagawa, T., Tanaka, S., *et al.* (1998). Cerebral hypoperfusion and hypometabolism with altered striatal signal intensity in chorea-acanthocytosis: A combined PET and MRI study. *Mov. Disord.* **13,** 100–107.

Tempel, L. W., and Perlmutter, J. S. (1990). Abnormal vibration-induced cerebral blood flow responses in idiopathic dystonia. *Brain* **113,** 691–707.

Tempel, L. W., and Perlmutter, J. S. (1993). Abnormal cortical responses in patients with writer's cramp. *Neurology* **43,** 2252–2257.

Tsai, C. H., Semmler, J. G., Kimber, T. E., Thickbroom, G., Stell, R., Mastaglia, F. L., *et al.* (1998). Modulation of primary orthostatic tremor by magnetic stimulation over the motor cortex. *J. Neurol. Neurosurg. Psychiatry* **64,** 33–36.

Turjanski, N., Sawle, G. V., Playford, E. D., Weeks, R., Lammertsma, A. A., Lees, A. J., *et al.* (1994). PET studies of the presynaptic and postsynaptic dopaminergic system in Tourette's syndrome. *J. Neurol. Neurosurg. Psychiatry* **57,** 688–692.

Turjanski, N., Lees, A. J., and Brooks, D. J. (1999). Striatal dopaminergic function in restless legs syndrome. ^{18}F-Dopa and ^{11}C-Raclopride PET studies. *Neurology* **52,** 932–937.

Weeks, R. A., Ceballos-Baumann, A., Piccini, P., Boecker, H., Harding, A. E., and Brooks, D. J. (1997). Cortical control of movement in Huntington's disease. *Brain* **120,** 1569–1578.

Weindl, A., Kuwert, T., Leenders, K. L., Poremba, M., Graffin von Einsiedel, H., Antonini, A., *et al.* (1993). Increased striatal glucose consumption in Sydenham's chorea. *Mov. Disord.* **8,** 437–444.

Wills, A. J., Jenkins, I. H., Thompson, P. D., Findley, L. J., and Brooks, D. J. (1994). Red nuclear and cerebellar but no olivary activation associated with essential tremor: A positron emission tomography study. *Ann. Neurol.* **36,** 636–642.

Wills, A. J., Jenkins, I. H., Thompson, P. D., Findley, L. J., and Brooks, D. J. (1995). A positron emission tomography study of cerebral activation associated with essential and writing tremor. *Arch. Neurol.* **52,** 299–305.

Wills, A. J., Thompson, P. D., Findley, L. J., and Brooks, D. J. (1996). A positron emission tomography study of primary orthostatic tremor. *Neurology* **46,** 747–752.

Wolf, S. S., Jines, D. W., Knable, M. B., *et al.* (1996). Tourette syndrome: Prediction of phenotypic variation in monozygotic twins by caudate nucleus D_2 receptor binding. *Science* **273,** 1225–1227.

13

Functional Imaging in Vascular Disorders

J. C. Baron and G. Marchal

INSERM U320, CYCERON, University of Caen, Caen 14074, France

I. Brief Overview of Methods Employed

II. Normal Physiology and Basic Pathophysiology of Brain Perfusion and Metabolism

III. Long-Standing Arterial Obstruction: Mapping Hemodynamic Failure

IV. Acute Ischemic Stroke: Mapping the Core, the Penumbra, and the Reperfused Tissue

V. Remote Metabolic Effects of Stroke

VI. Receptor Studies in Vascular Disorders

References

The application of functional imaging in cerebrovascular disorders started in the late sixties in the group of the late Niels Lassen in Copenhagen with intracarotid ^{133}xenon cerebral blood flow studies. However, it was with the advent of tomographic brain mapping techniques and especially of positron imaging of brain perfusion and metabolism in the late seventies and early eighties that major pathophysiologic insights were gained in the understanding of acute cerebral ischemia, the hemodynamic and metabolic effects of carotid-artery obstruction, and the neurobiological mechanisms underlying the clinical expression of stroke. Further

studies addressed the clinical correlates of such physiologic changes in terms of patient classification, clinical prognosis, and mechanisms of recovery. More recently, the concepts that issued from PET investigations have been applied to larger patient samples with more clinically accessible techniques such as single photon emission tomography (SPECT) and stable xenon CT (Xe-CT) as well as with newer MR-based techniques such as proton and phosphorus spectroscopic imaging, diffusion-weighted imaging, and perfusion imaging. In this chapter, we review these findings, with techniques other than PET being cited whenever they substantially added to the understanding already afforded by PET. Studies focusing on the neuronal reorganization following stroke and their relationships with recovery, as assessed by activation studies, are the topic of Chapter 24.

I. Brief Overview of Methods Employed

A. Brain Perfusion and Hemodynamics

Table I shows the main hemodynamic variables assessable by functional imaging in man, together with their commonly used abbreviations and their typical values in predominantly gray matter regions of interests (ROIs).

One hemodynamic variable not shown in Table I is

299

Table I Physiologic Variables

Physiologic	Abbreviation	Normal values (gray matter)
Cerebral blood flow	CBF	60 ml/100 ml/min
Cerebral blood volume	CBV	4 ml/100 ml
Mean transit time	t	0.08 min
CBF/CBV ratio	CBF/CBV	12 min^{-1}
Local tissue hematocrit	tHt	0.28
Cerebral metabolic rate of oxygen	CMRO$_2$[a]	4 ml/100 ml/min
Cerebral metabolic rate of glucose	CMRGlc[a]	8 mg/100 ml/min
Tissular pH	pHt	7.04
Oxygen extraction fraction	OEF	0.40
Glucose extraction fraction	GEF	0.12
Tissue partial O$_2$ tension	PtO$_2$	31.2 mmHg

[a] The conversion factors to obtain CMRO$_2$ and CMRGlc in micromolar units are 44.6 and 5.56 μmol/ml, respectively.

the "hemodynamic reserve," which expresses the vasodilatory capacity of the cerebrovascular bed, assessed with a vasodilatation challenge such as inhalation of 5% CO$_2$ in air (Levine *et al.*, 1991) or systemic administration of acetazolamide (Vorstrup *et al.*, 1986) (Fig. 1). It is expressed as the absolute or percentage increase in tissue perfusion from the resting state (per mmHg of PaCO$_2$ change in the case of CO$_2$ inhalation).

Activation studies are dealt with elsewhere in this book. However, it is important to note here that, since their principle is to exploit the hemodynamic response coupled with neuronal activation, interference from reduced hemodynamic reserve is a theoretical concern. Although the literature available on this topic has been conflicting, it appears from recent work that this concern may not necessarily be well founded (Inao *et al.*, 1998).

B. Brain Metabolism

1. Terminology

Table I also shows the main physiological parameters related to brain metabolism that can be quantitated with functional imaging. Not shown in the table is the CMRGlc/CMRO$_2$ ratio, which assesses the stoichiometry of glucose use.

2. Techniques

The measurement of CBF and OEF with PET allows the generation of quantitative maps of oxygen consumption (CMRO$_2$). It must be recalled that the use of ^{18}FDG to measure CMRGlc in acute stroke may be complicated by unpredictable changes in the "lumped constant" (Baron *et al.*, 1989). PET also allows one to

measure the intracellular pH (Syrota *et al.*, 1985; Senda *et al.*, 1989).

MR spectroscopy (MRS) allows assessment of important cellular biochemicals, such as *N*-acetylaspartate (NAA, a specific neuronal marker). Proton MRS can be used to determine lactate, and ATP and phosphocreatine can be assessed with ^{31}P MRS. For technical reasons, most applications so far have been with single-volume ROIs, but actual imaging with acceptable resolution is feasible, especially with high-field new-generation magnets (Gillard *et al.*, 1996; Kamada *et al.*, 1997).

Diffusion-weighted MRI assesses the compartmentation of water in brain tissues and is able to document reduced water diffusion within minutes of onset of cerebral ischemia, reflecting intracellular cytotoxic edema and thus failure of cell energy metabolism (Sorensen *et al.*, 1996; Baird and Warach, 1998).

Electromagnetic techniques such as EEG, MEG, and transcranial magnetic stimulation provide essential information about neuronal function not obtainable by other techniques and as such are a useful complement in certain clinical situations (Kamada *et al.*, 1997). However, their main interest thus far has been in the understanding of the changes in large-scale neural networks following focal stroke in relation to clinical recovery (Traversa *et al.*, 1997).

C. Specific Radiotracers

In addition to the above physiologic variables, PET also allows one to investigate specific binding sites/receptors using, for instance, [^{11}C]flumazenil and [^{11}C]PK11195, which serve as markers of neuronal death and glial proliferation, respectively (Ramsay *et al.*, 1992; Sette *et al.*, 1993). Corresponding single photon emission computed tomography (SPECT) tracers have been developed for clinical use.

Recently, ^{18}F-labeled fluoromisonidazole has been developed as a tracer of hypoxic brain tissue and is being tested as a potential marker of the ischemic penumbra in acute stroke (Read *et al.*, 1998).

II. Normal Physiology and Basic Pathophysiology of Brain Perfusion and Metabolism

A. Normal Brain

In physiological conditions, there exists a matching of local values of CBF, CMRGlc, CMRO$_2$, and CBV according to linearly proportional relationships (Baron *et al.*, 1984; Sette *et al.*, 1989). This reflects the metabolic regulation of the cerebral circulation and explains why in physiological conditions the distribution of CBF is superimposable on that of CMRO$_2$ and CMRGlc.

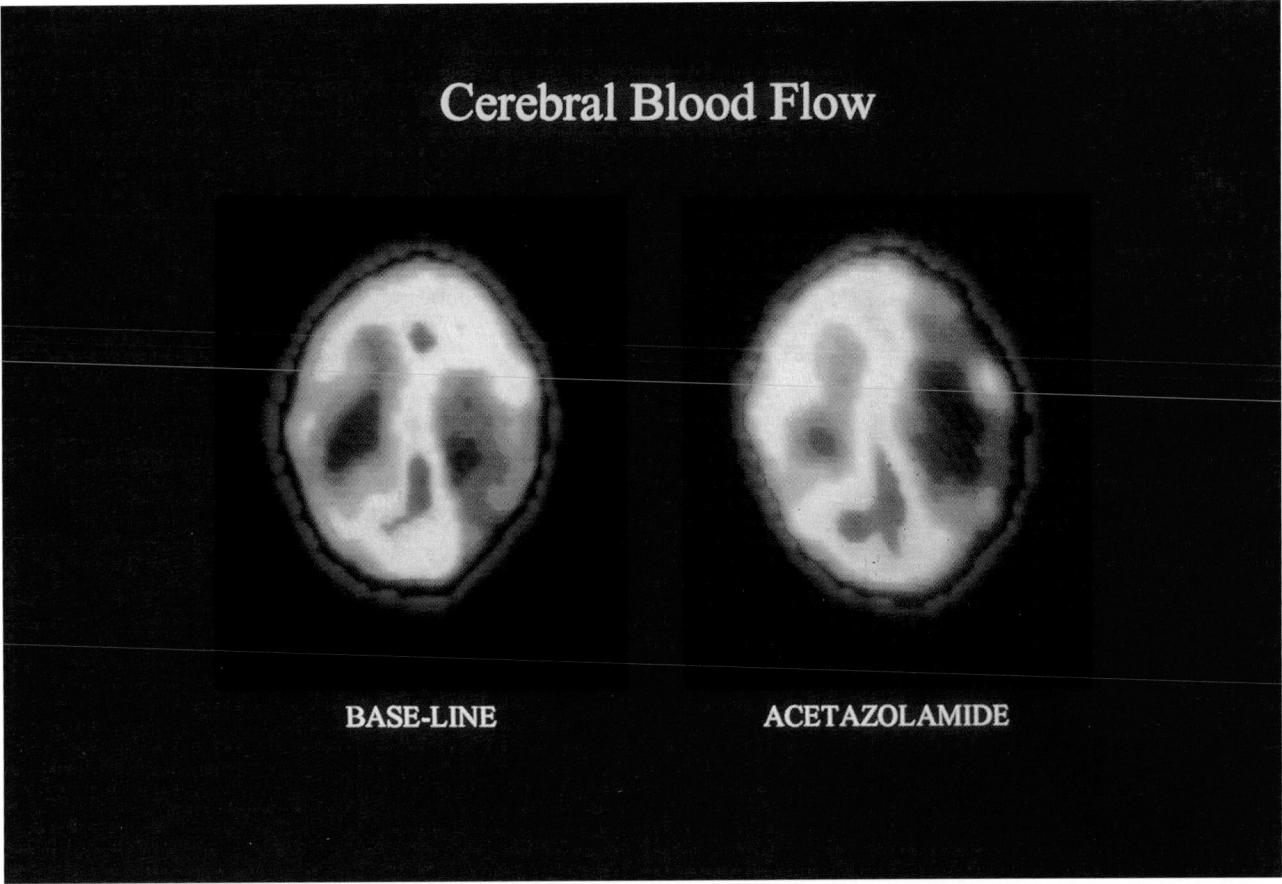

Figure 1 Impaired hemodynamic reserve. Resting and postacetazolamide PET scans of CBF in a patient with repeated transient ischemic attacks (TIAs) and left ICA occlusion. Although the resting scan (left) showed little or no alteration of CBF in the affected hemisphere, the vasodilatation challenge with acetazolamide induced markedly asymmetric CBF (right). Clearly visible is a marked increase of CBF in the nonoccluded side but a lack of increase of CBF on the occluded side with a decrease in the posterior and anterior parts of the left carotid territory, respectively, suggesting hemodynamic steal.

B. Autoregulation and Hemodynamic Failure

Shown in Table II are the main hemodynamic and metabolic changes that occur in response to a fall in the cerebral perfusion pressure (CPP) distal to an arterial obstruction, subdivided into four stages of increasing severity. The marked increase in CBV during the phase

of autoregulation reflects the physiologic vasodilatation of resistance vessels, i.e., the "hemodynamic reserve." As soon as the CPP falls below the lower threshold of autoregulation, the CBF starts to decline but the $CMRO_2$ at first remains unaltered. This flow–metabolism uncoupling, which translates as a focal increase in the OEF up to the theoretical maximum of 1.00, has been termed "misery perfusion" (Baron *et al.*, 1981a)

Table II Four Stages of Brain Hemodynamic and Metabolic Impairment as a Function of Severity in CPP Drop[a]

Stage	CPP	CBV	CBF	OEF	$CMRO_2$
1. Hemodynamic reserve (autoregulation)	60–100%	↗	N	N	N
2. Perfusion reserve (oligemia)	40–60%	↗↗	↘	↗	N
3. Metabolic reserve (ischemic penumbra)	20–40%	↗	↘↘	↗↗	↘
4. Irreversible damage (ischemic necrosis)	<40%	↘	↘↘↘	↗ to ↘	↘↘

[a] CPP, cerebral perfusion pressure; CBV, cerebral blood volume; CBF, cerebral blood flow; OEF, oxygen extraction fraction; $CMRO_2$, cerebral metabolic rate of oxygen. N, normal range.

(Fig. 2). In the moderate stage of misery perfusion, the brain is able to maintain its $CMRO_2$ despite reduced CBF, although at the expense of tissue hypoxia. This phase, which calls on the perfusion reserve, is designated "oligemia." If the CPP drops further, neuronal function becomes impaired and the $CMRO_2$ falls despite maximally increased OEF, characterizing true ischemia, which comprises a reversible stage (the ischemic penumbra) and an irreversible stage (Baron *et al.*, 1981b; Frackowiak, 1985; Powers *et al.*, 1985).

Under certain conditions, the CBF/CBV ratio reliably reflects the CPP (Schumann *et al.*, 1998). The use of this ratio has allowed a clear demonstration of the successive stages of declining CPP (Gibbs *et al.*, 1984; Sette *et al.*, 1989). As already stated, another way to assess the vasodilatory capacity is with 5% CO_2 or acetazolamide. Whereas these agents physiologically induce marked increases in CBF, a blunted or even absent response characterizes brain regions with an already vasodilated vas-

cular bed, for instance when perfusion is maintained thanks to the mechanism of autoregulation; paradoxical decreases in CBF may even occur in areas with an exhausted reserve ("hemodynamic steal") (Fig. 1).

C. Luxury Perfusion

Luxury perfusion is characterized by an oxygen supply in excess of demand (Lassen, 1966), and its hallmark is a focal reduction of the OEF (Baron *et al.*, 1981b). It indicates full or partial reestablishment of perfusion within an ischemic or already irreversibly damaged tissue. In luxury perfusion, the CBF may be increased (hyperperfusion), normal, or even decreased (relative luxury perfusion), although, by definition, in excess of prevailing $CMRO_2$, which itself may be normal, increased, or reduced. In hyperperfused areas, the CBV is increased (true hyperemia) (Marchal *et al.*, 1996a), documenting abnormal vasodilatation ("vasoparalysis").

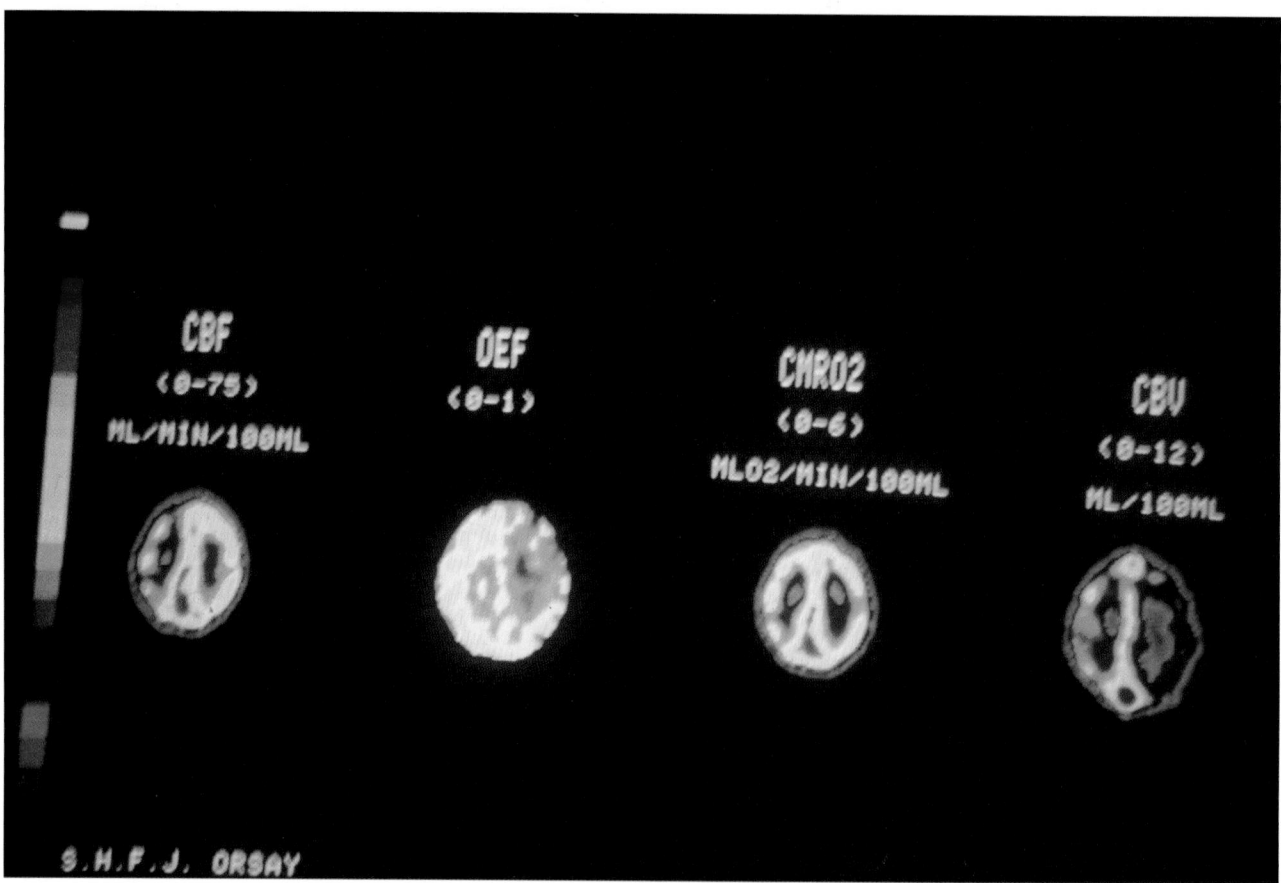

Figure 2 Misery perfusion: Stage 2 of hemodynamic failure. Occlusion of the right internal carotid artery was diagnosed when the patient presented following an ipsilateral transient ischemic attack. Repeated TIAs followed despite closely supervised antiplatelet and then anticoagulant treatment. Some of the TIAs were triggered on standing. PET performed several weeks later showed moderate hypoperfusion in the territory of the right carotid artery, with a completely normal $CMRO_2$. As a result, the OEF was increased, reflecting "misery perfusion." These data suggest inability of the collateral circulation to compensate fully for occlusion of the carotid artery and that the pressure of the blood supply to the brain downstream of the circle of Willis is insufficient to maintain CBF (i.e., the local autoregulation mechanism has been overcome). This interpretation is supported by the observation of a marked increase in the cerebral blood volume on the side of the occlusion.

III. Long-Standing Arterial Obstruction: Mapping Hemodynamic Failure

A. Hemodynamic Reserve and Misery Perfusion

The original observation of misery perfusion was in a patient with carotid-artery occlusion and continuing reversible ischemic attacks, some triggered by standing up (Baron et al., 1981a). Since then, both PET and SPECT studies have repeatedly documented that internal carotid-artery disease may have hemodynamic consequences on the distal cerebral vascular bed (Leblanc et al., 1987a; Powers et al., 1987; Herold et al., 1988; Yamauchi et al., 1990; Levine et al., 1991; Derlon et al., 1992). Overall, these investigations have shown that the severity of such consequences is related both to the degree of obstruction (i.e., only <50% stenosis or occlusion may have measurable effects) and to the compensation afforded by the circle of Willis (with the most marked effects seen when compensation is essentially or exclusively via the ipsilateral ophthalmic artery). Similar effects have been reported in patients with long-standing MCA stem stenosis or occlusion (Sgouropoulos et al., 1985). The hemodynamic effects observed, which reflect the extent to which the CPP is reduced, range from simply autoregulated (i.e., tapping the hemodynamic reserve, with normal resting CBF but increased CBV, reduced CBF/CBV ratio, and reduced vasodilatory capacity to CO_2 or acetazolamide stress) to true oligemia (i.e., tapping the perfusion reserve, with reduced resting CBF, increased OEF, increased CBV, markedly reduced CBF/CBV ratio, and abolished vasodilatory capacity with occasional hemodynamic steal) (Gibbs et al., 1984; Vorstrup et al., 1984, 1985, 1986) (Fig. 1). Whatever their severity, these changes generally predominate in watershed (border zone) territories, as expected (Samson et al., 1985; Leblanc et al., 1987b; Carpenter et al., 1990). Furthermore, focal chronic misery perfusion has been documented as a forerunner of the development of watershed infarction in occasional patients with tight carotid-artery stenosis or occlusion (Itoh et al., 1988; Yamauchi et al., 1992).

B. Clinical Correlates

Although the clinical correlates of these hemodynamic changes can be straightforward, as in the rare instances of orthostatic transient ischemic attacks (Baron et al., 1981a), in many instances they are difficult to ascertain. However, although hemodynamic abnormalities can be found in asymptomatic subjects or in the asymptomatic hemisphere of symptomatic subjects, there exists a relationship between the presence of high OEF and the occurrence of ipsilateral ischemic symptoms (Derdeyn et al., 1998). In the early 1980s, the documentation of a clear-cut compromise of brain circulation was considered the only rational basis for extracranial–intracranial arterial bypass (EIAB) (Baron et al., 1981a). However, this surgical procedure has now been largely abandoned due to lack of demonstrated clinical benefit. Consistent with this, the retrospective nonrandomized studies of Powers et al. (1989a,b) suggested that the finding of hemodynamic compromise accurately predicted neither poor outcome if medical therapy was elected nor good outcome if EIAB was performed. However, several recent studies, including a prospective one by Powers et al. (1998), now suggest that, if cerebrovascular reactivity is severely impaired, there is significantly increased risk of ipsilateral stroke despite the best medical treatment (Derlon et al., 1992; Kleiser and Widder, 1992; Kuroda et al., 1993; Yonas et al., 1993; Webster et al., 1995; Yamauchi et al., 1996).

C. Implications for Therapy

It has been amply documented that successful cerebral revascularization by means of either carotid endarterectomy or EIAB at least partially reverses the preoperatively observed hemodynamic compromise, documenting the latter's reversibility (Baron et al., 1981a; Powers et al., 1984; Samson et al., 1985; Gibbs et al., 1987; Leblanc et al., 1987a; Derlon et al., 1992; Muraishi et al., 1993). However, it has also been suggested that patients with the most compromised cerebrovascular physiology may also be those most at risk of perioperative complications such as low-pressure breakthrough of autoregulation (presumably as a result of long-term dysregulation of the cerebral circulation) (Derlon et al., 1992). Thus, the results from functional imaging of each candidate for revascularization surgery need to be weighed carefully in relation to other clinical and instrumental data to assess the risk/benefit ratio of the surgical procedure under consideration.

Apart from the issue of surgical management, impaired brain hemodynamics in a patient with carotid-artery disease should be considered in planning medical management. For instance, systemic hypotension (as a result of drug therapy or any surgical procedure, especially cardiac) should be carefully avoided. Furthermore, because embolic events may have more serious tissue consequences in a dysregulated vascular bed than they would in normal brain, medical measures to prevent embolism should always be considered.

IV. Acute Ischemic Stroke: Mapping the Core, the Penumbra, and the Reperfused Tissue

By convention, the acute stage of stroke will be defined in the following as the first 24 h after onset of clinical symptoms, because available evidence suggests this is the maximal time window within which salvageable tissue may still be present.

A. PET Studies

Three findings in acute MCA territory stroke have been investigated in detail with respect to both their time course and their prognostic value for tissue and clinical outcome: the core of *irreversibly damaged tissue*, the *penumbral tissue*, and the *hyperperfused tissue*.

1. Irreversibly Damaged Tissue

Irreversibly damaged tissue has been defined and validated with PET as having a $CMRO_2$ below a threshold of about 1.4 ml/100 ml/min for gray matter (Baron *et al.*, 1984; Powers *et al.*, 1985), or, more conservatively, ~0.9 ml/100 ml/min for any voxel in brain tissue (Marchal *et al.*, 1999a). In a large proportion of patients, irreversible damage affects the striato-capsular area very early, associated in most instances with cortical misery perfusion (Wise *et al.*, 1983; Marchal *et al.*, 1993) (Fig. 3). Presumably because of poor collaterals, and unlike the cerebral cortex, the lenticulo-striate territory consti-

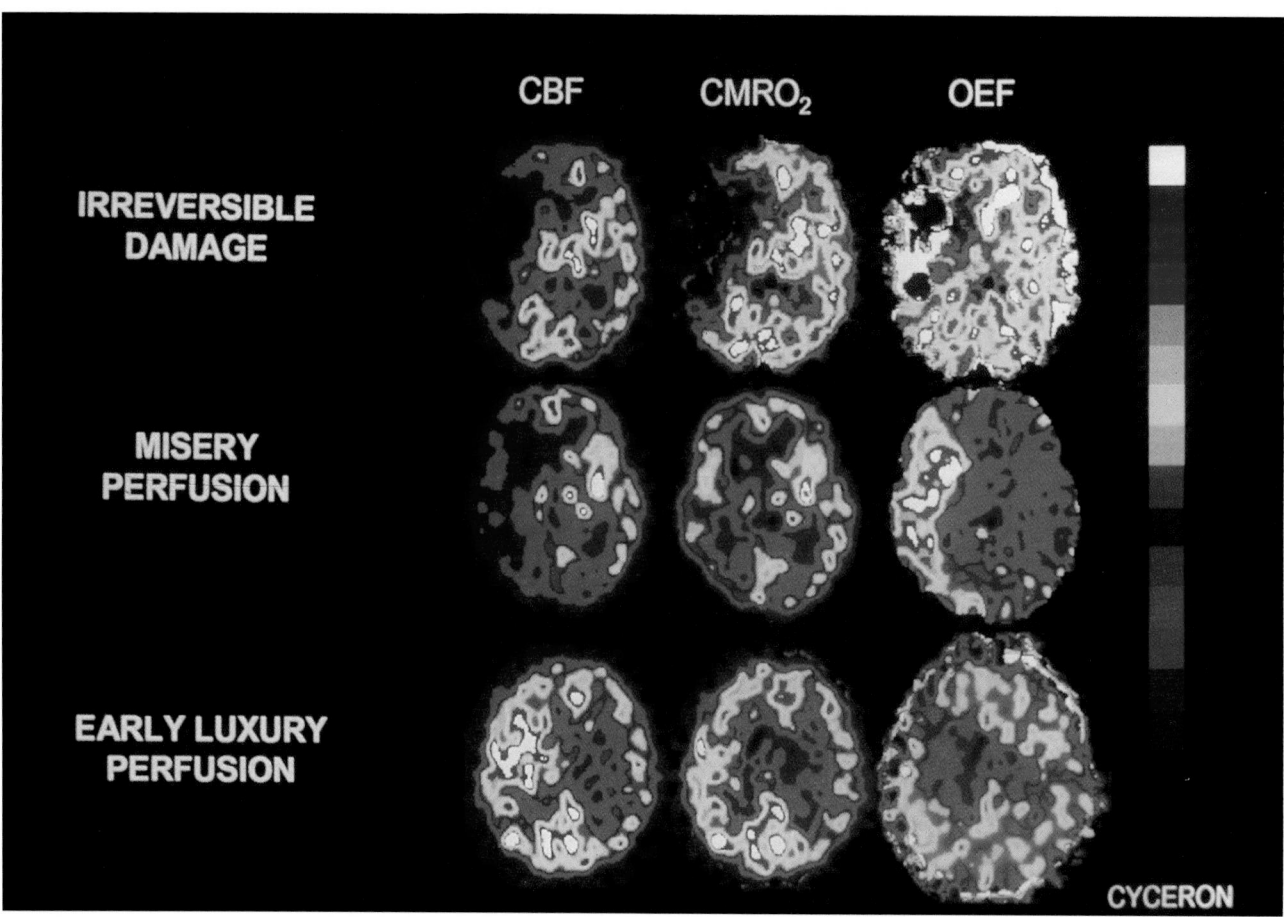

Figure 3 PET patterns in acute MCA territory stroke. This figure illustrates the three PET patterns of CBF and $CMRO_2$ changes observed within 18 h of onset of MCA territory stroke. The top row shows an example of early extensive irreversible damage in a patient with right-sided MCA territory stroke studied with PET 17 h after symptom onset. There was a near-zero CBF and $CMRO_2$ in the whole right MCA territory (Pattern 1), together with patchy OEF (black pixels represent unmeasurable OEF). The patient survived but outcome was poor and the whole MCA territory was infarcted at follow-up CT. Misery perfusion (middle row): in this patient with acute-stage right MCA territory stroke, the PET study performed 12 h after onset revealed a markedly reduced CBF in the whole right MCA territory, associated with relatively preserved $CMRO_2$ (except in the lenticulo-striate area) and extremely elevated OEF. This example corresponds to a typical Pattern 2 of PET changes. This patient died 3 days after the PET study. The bottom row images illustrate an example of early luxury perfusion (Pattern 3) in a patient studied with PET 13 h after onset of right-sided MCA territory stroke. There is markedly increased CBF in the central right MCA territory, associated with normal or slightly increased $CMRO_2$ and decreased OEF. This patient made a full recovery and the follow-up CT scan showed a small periventricular infarct.

tutes the core of ischemia and rapidly suffers irreversible damage. However, in a subset of patients, the area of irreversible damage affects extensive parts of the cortical territory only hours after stroke (Marchal *et al.*, 1993), suggesting inadequate pial collaterals (Fig. 3).

In the vast majority of cases, the profoundly hypometabolic areas express similarly reduced CBF (Fig. 3). However, in cases with early recanalization and extensive cortical hyperperfusion, the associated small deep infarcts may occasionally express only mildly reduced or even nearly normal CBF (Marchal *et al.*, 1996b).

Marchal *et al.* (1999a) found that the volume of profoundly hypometabolic tissue as assessed with PET 5–18 h postonset of stroke was highly linearly correlated to final infarct volume, as measured by CT scan about 1 month later; the former, however, underestimated the latter by a factor of two, because of subsequent metabolic deterioration of the surrounding penumbra (see below). Thus, mapping the profoundly hypometabolic tissue in the acute stage of stroke may provide an early assessment of already established damage and predict a *minimum* volume of final infarction.

2. Penumbral Tissue

One major finding from PET studies both in man and in the baboon with MCA occlusion has been the demonstration, hours into an episode, of wide zones of cortex with persistently ischemic tissue (Ackerman *et al.*, 1981; Baron *et al.*, 1981b; Wise *et al.*, 1983; Pappata *et al.*, 1993) (Fig. 3). This tissue is characterized by reduced CBF (below the penumbral threshold of ~20 ml/100 g/min), massively increased OEF (usually above 0.80), and mildly to moderately reduced $CMRO_2$ (i.e., above the threshold for irreversibility described earlier) (Fig. 3). These alterations are consistent with at-risk but recuperable (i.e., penumbral) tissue. The finding of acidosis in the affected tissue (Hakim *et al.*, 1987; Senda *et al.*, 1989) is also consistent with the concept of a penumbra. A substantial cortical penumbra, reflecting efficient pial collaterals, has been reported in over 50% of patients studied within 9 h of onset, in up to 25% of cases at 24 h, and occasionally up till 30 h postonset (Baron *et al.*, 1983; Wise *et al.*, 1983; Marchal *et al.*, 1995, 1996b), suggesting a protracted window for therapeutic opportunity in a fraction of acute stroke cases. In one study, it was found that up to 52 % of the ultimately infarcted tissue still exhibited physiological characteristics compatible with penumbra as late as 16 h after symptom onset, suggesting even delayed neuroprotection might have significantly altered the functional outcome in such cases (Marchal *et al.*, 1996b).

Transition of such "penumbral" areas toward infarction has been documented within hours to days and is documented by a decline in $CMRO_2$ regardless of the local CBF, which may decline in parallel, remain stable, or even at times increase (Wise *et al.*, 1983; Heiss *et al.*, 1992; Marchal *et al.*, 1996a). With time, perfusion increases progressively in necrotic tissue, representing neovascularization, before falling again in the mature gliotic and/or cavitated area (Baron *et al.*, 1981b; Lenzi *et al.*, 1982). This whole process is strikingly illustrated by the associated dramatic fall in OEF, from initially very high to increasingly low values, signaling the exhaustion of the tissue's oxygen needs (Wise *et al.*, 1983). Such a deleterious course of events occasionally does not take place, however, and all or part of the penumbral tissue eventually escapes infarction (Heiss *et al.*, 1992, 1998; Furlan *et al.*, 1996), consistent with the potential for reversibility that characterizes the penumbra. In this event, one hypothesis is that some favorable event (e.g., partial reperfusion) occurred after the PET study to save part or all of this tissue, a hypothesis supported by PET investigations in baboons (Young *et al.*, 1996; Touzani *et al.*, 1997).

3. Early Spontaneous Hyperperfusion

Early hyperperfusion, which suggests recanalization of the occluded artery, has been observed in up to one-third of cases studied between the fifth and the eighteenth hour after stroke onset (Marchal *et al.*, 1993, 1995) (Fig. 3). In the majority of cases, hyperperfusion is not associated with reduced metabolism, but instead with a mildly increased $CMRO_2$, suggesting postischemic rebound of cellular energy-dependent processes (Marchal *et al.*, 1996a). However, the OEF is significantly reduced and the CBV significantly increased in these areas, indicating luxury perfusion with abnormal vasodilatation. In a sample of 10 such patients, no instance of MCA stem occlusion was recorded at transcranial doppler, and the hyperperfused areas consistently exhibited intact morphology on chronic-stage CT (Marchal *et al.*, 1996a). The interpretation is that these changes represent early spontaneous recanalization of the occluded MCA artery occurring at some undefined time point before the PET study, with resulting efficient reperfusion of the previously ischemic and dysregulated, but still viable, tissue. Thus, at odds with the concept that sudden tissue reoxygenation might exacerbate ischemic damage, these findings in humans suggest that, consistent with some evidence from the animal literature, postischemic hyperperfusion is not detrimental (see Marchal *et al.*, 1999b, for review). This is, in turn, consistent with the well-established notion that infarct size is reduced by early recanalization.

4. Clinical Correlates

Marchal *et al.* (1993, 1995) conducted a prospective study that assessed the relationship between acute-stage

PET findings and clinical outcome. They studied 30 patients with first-ever MCA territory stroke and compared the changes in CBF and $CMRO_2$, observed 5–18 h postonset of stroke, against the subsequent spontaneous neurological course over 2 months, quantitated with validated stroke scales. Each patient was classified into one of the three above-mentioned patterns of PET changes (Fig. 3), namely, (i) *Pattern 1,* characterized by a large subcortico-cortical area of extensive necrosis; (ii)) *Pattern 2,* characterized by the presence of presumably penumbral tissue without associated irreversible damage, except possibly in the lenticulo-striate area; and (iii)) *Pattern 3,* characterized by hyperperfusion without associated irreversible damage, except again possibly in a small, distinct area. There was a statistically highly significant relationship between PET patterns and subsequent neurological course. Thus, all patients classified as Pattern 1 did poorly (early death from massive infarct or poor outcome), whereas all patients classified as Pattern 3 did well (complete or nearly complete recovery in all); patients classified as Pattern 2 had a very variable course, ranging from death to full recovery. These findings from functional imaging are consistent with evidence from clinical studies that early recanalization is associated with rapid recovery whereas persistence of MCA trunk occlusion is a risk factor for poor outcome and massive brain swelling. This 100% predictive value of PET Patterns 1 and 3 remained statistically significant even when initial neurological scores were taken into account in the model, such that PET predicted outcome better than neurological scores in the group with intermediate admission scores (Marchal *et al.,* 1995). Consistent findings regarding hyperperfusion have been reported by Heiss *et al.* (1997), who showed in a few cases studied both pre- and post-early intravenous thrombolysis that the occurrence of hyperperfusion, unlike severe and persisting hypoperfusion, was associated with good clinical and tissue outcome.

5. Alleviation of Penumbra: Its Role in Clinical Recovery

Alleviation of penumbra has long been hypothesized as one major mechanism underlying early recovery from ischemic stroke. The above-mentioned finding that early hyperperfusion is invariably associated with good outcome is further, though indirect, evidence for this hypothesis. However, this mechanism was only recently directly documented in a quantitative manner by Furlan *et al.* (1996). In 11 patients with MCA territory stroke evaluated by PET within 18 h of onset, these authors found that the degree of neurological recovery over the subsequent 2 months was positively correlated with the individual volume of acute-stage penumbral tissue that escaped infarction, as assessed with chronic-stage CT (Fig. 4). Somewhat unexpectedly, the best correlation was observed with 2-month recovery scores, which suggests that survival of the penumbra influences not only early, but also late, recovery. The hypothesis given to explain this finding was that survival of the penumbra not only allows for early return of function in peri-infarct tissue but also provides an important opportunity for subsequent neural reorganization processes in a synergistic rather than simply cumulative way. Thus, survival of the penumbra would appear to be an important early mechanism for subsequent functional recovery.

Confirmatory results were reported subsequently by Heiss *et al.* (1998a). These authors found a significant positive relationship between the reduction in volume of critically hypoperfused tissue between prethrombolysis and postthrombolysis PET and the change in neurological scores between admission and 3 weeks.

B. Perfusion Studies with SPECT and Xe-CT

Because of the logistics involved, ^{133}Xe SPECT has been very little applied in acute stroke. Thus, essentially all studies have employed semiquantitative perfusion radiotracers, essentially [^{99m}Tc]HMPAO and, more recently, [^{99m}Tc]ECD. Overall, the findings have been consistent with the above-described results from PET studies, although with less accuracy due to poor spatial resolution, problems with reliability of the tracers as markers of perfusion, and lack of metabolic data (Marchal *et al.,* 1999c). In close to 100% of patients with MCA territory stroke, acute-stage SPECT reveals a focally reduced tracer uptake, the extent of which is proportional to the severity of the contemporaneous neurological deficit. Severe and extensive tracer hypofixation is almost invariably predictive of persistent MCA occlusion with subsequent large or malignant infarction (Giubilei *et al.,* 1990; Berrouschot *et al.,* 1998b), whereas mild or moderate hypoperfusion carries a variable outcome but is only exceptionally associated with subsequent death (Giubilei *et al.,* 1990; Limburg *et al.,* 1990; Davis *et al.,* 1993; Hanson *et al.,* 1993). Profound hypoperfusion is also associated with increased risk of massive hemorrhagic transformation (Ueda *et al.,* 1994) or no reflow (Herderschee *et al.,* 1991) following therapeutic thrombolysis. Conversely, normal uptake (with or without mild hyperfixation) is invariably associated with reversible neurological deficits (Giubilei *et al.,* 1990; Hanson *et al.,* 1993; Berrouschot *et al.,* 1998a; Marchal *et al.,* 1999c). Well-demarcated areas of massively increased tracer uptake appear to predict subsequent infarction (Shimosegawa *et al.,* 1994), but such "hot spots" may not represent true hyperperfusion, but rather abnormal penetration of

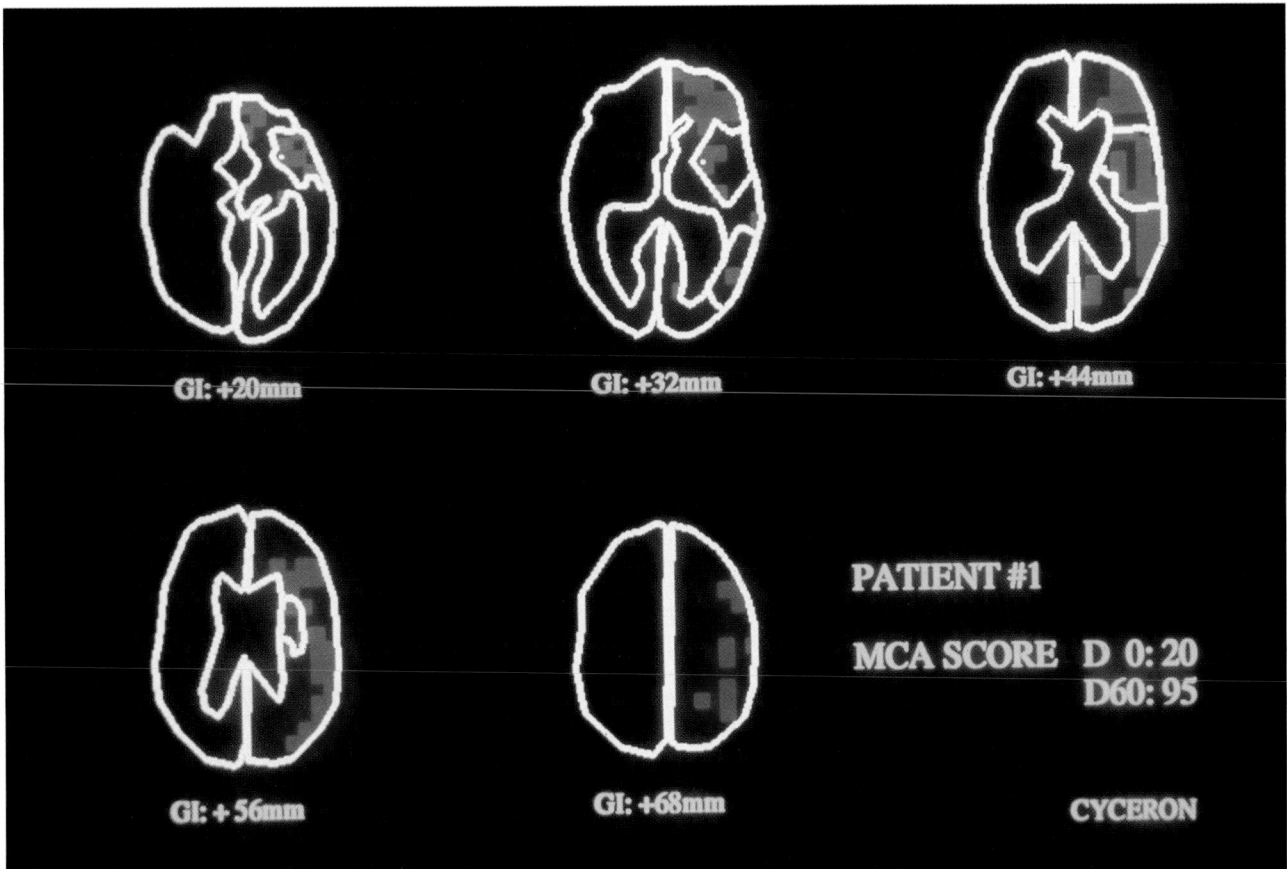

GI: +20mm   GI: +32mm   GI: +44mm

GI: +56mm   GI: +68mm

PATIENT #1

MCA SCORE D 0: 20
                    D60: 95

CYCERON

Figure 4 The ischemic penumbra. This diagram shows, superimposed on the contours of brain tissue, ventricles and hypodense lesions on CT scanning obtained in the chronic stage, the voxels identified as the ischemic penumbra (red dots) according to the method described by Furlan *et al.* (1996), in a patient studied with PET 8 h after onset of middle cerebral artery territory stroke. Five CT scan levels parallel to the glabella-inion (GI) plane are illustrated. Note that penumbral voxels were detected both within and outside the final infarct. The very good spontaneous neurological recovery in this patient can be related to the fact that extensive areas of penumbra ultimately escaped infarction.

HMPAO into brain parenchyma due to an altered blood–brain barrier in already severely damaged tissue (Sperling and Lassen, 1993).

Either spontaneous or thrombolysis-induced early reperfusion of previously hypoperfused tissue is associated with good outcome (Herderschee *et al.,* 1991; Barber *et al.,* 1998), and spectacular recovery is associated with complete, extensive reperfusion (Baird *et al.,* 1995). However, when the repeat SPECT is done >36–48 h after onset, reperfusion may not always indicate good outcome because it may have occurred too late, into already necrotic tissue (Jorgensen *et al.,* 1994).

Studies with Xe-CT in acute MCA territory stroke are entirely consistent with the above findings from PET and SPECT. Thus, patients with normal or near-normal CBF had spontaneous resolution of their deficits, whereas those with considerably reduced CBF in affected territories developed severe brain edema and herniation (Firlik *et al.,* 1998a,b).

C. Implications for Therapy

1. Implications of High OEF and Low OEF for Clinical Management

Demonstration of high OEF in the setting of acute stroke implies that the autoregulation of CBF has been overcome in the affected territory. This finding is especially important in view of the frequent occurrence of reactive hypertension in this clinical setting. Thus, any lowering of the systemic arterial pressure (SAP) is likely to further reduce the cerebral perfusion pressure and, in turn, the CBF in the affected tissue, effects that can have potentially damaging effects. This may explain why reductions in SAP in acute stroke have frequently been associated with poorer outcome. Conversely, if low OEF with hyperperfusion is found, management of arterial hypertension may be warranted, particularly if early edema is demonstrated by CT or MRI, since some experimental studies suggest that hyperperfusion in necrotic tissue may promote the development of malignant brain swelling.

2. Implications for Trials of Neuroprotection

The available functional imaging data in acute stroke reviewed above suggest that pathophysiologically blind inclusion of acute stroke patients into trials may blur any beneficial effects of the agent being tested, because of underlying pathophysiological heterogeneity documented by imaging studies (i.e., the three patterns referred to above). Thus, tentative guidelines with respect to patient screening into trials of neuroprotective agents would be as follows: (i) The finding of a normal scan or hyperperfusion earlier than 18 h postonset suggests spontaneous recanalization has already occurred and almost invariably predicts good outcome. Thus, these patients need not be entered into a neuroprotection trial, although whether antioxidants may further reduce tissue damage cannot be entirely ruled out. (ii) The finding of a large area of near-zero perfusion and/or metabolism suggests poor outcome with considerable risk of massive brain swelling and early death. Patients with this profile should also be excluded from classic neuroprotection trials. However, because vasogenic edema in itself could cause further tissue damage to the penumbra, these patients might benefit from early preventive antiedema therapy with or without surgical brain decompression. (iii) The remaining patients have substantial penumbral tissue and an unpredictable spontaneous outcome. Their potential for recovery depends on the extent of penumbra, and thus they constitute the best candidates for a trial of neuroprotective agent. Furthermore, as this pattern has been observed up to 16 h after stroke onset in some patients, the therapeutic window needs to be considered in each individual case. In such patients, however, neuroprotective agents that tend to reduce arterial blood pressure should be avoided for reasons explained above.

Functional imaging can also be of use in therapeutic trials in two additional ways. First, as severe hypoperfusion/hypometabolism reliably predicts the extent of irretrievably damaged tissue, the effectiveness of a given therapeutic agent at arresting infarction should be easily testable. Second, the effects of therapy on tissue physiology can be monitored, providing essential information about the mechanisms by which an agent produces its beneficial effects, if any (Grotta and Alexandrov, 1998; Heiss et al., 1998a).

V. Remote Metabolic Effects of Stroke

Remote metabolic depression is characterized by coupled reductions in perfusion and metabolism in brain structures remote from, but connected with, the area damaged by a stroke. This effect is widely explained as depressed synaptic activity as a result of disconnection (either direct or transneural). Thus, remote effects allow mapping of the disruption in distributed networks as a result of focal infarction. Although they are often referred to collectively as "diaschisis" (Feeney and Baron, 1986), this term conceals a variety of cellular derangements, from reversible hypofunction to evolving Wallerian or transsynaptic degeneration, which all have the same PET expression. Importantly, some of these effects reflect purely functional, potentially recoverable synaptic abnormalities which therefore may participate in both the acute clinical expression of stroke and the recovery therefrom.

A. Crossed Cerebellar Diaschisis

This phenomenon (Baron et al., 1980), also known as crossed cerebellar hypometabolism (CCH), consists of an up to 50% matched reduction in both perfusion and metabolism in the cerebellar hemisphere contralateral to supratentorial stroke. It occurs in about 50% of patients with either cortical or subcortical stroke but is more frequent and severe with large hemispheric infarcts or with capsular stroke (Martin and Raichle, 1983; Kushner et al., 1984; Pantano et al., 1986). Evidence indicates CCH results from damage to the glutamatergic cortico–ponto–cerebellar system (CPCS), inducing transneuronal functional depression, although in rare instances it may result from retrograde cerebello-cortical effects (Pappata et al., 1990). Although CCH is correlated with both the presence and severity of hemiparesis (Pantano et al., 1986), this association is not systematic and presumably simply reflects the anatomical relationships between the pyramidal and the corticopontine fibers. Observations of CCH after a unilateral brainstem lesion at the level of the crus cerebri or basis pontis further support the CPCS mechanism.

In the vast majority of patients, CCH exhibits no tendency toward recovery (Lenzi et al., 1982; Pantano et al., 1986), and this chronicity suggests CCH might evolve into transneuronal degeneration in the long run. However, the fact that CCH may develop within the first hours of stroke and subsequently disappear within a few days (Serrati et al., 1994) indicates that it can also be a transient manifestation of deafferentation. This is further documented by the fact that CCH can transiently manifest in instances of reversible functional depression of the cerebral cortex, such as transient ischemic attacks, unilateral carotid infusion of barbiturates in epileptic patients (Kurthen et al., 1990), or balloon occlusion of the internal carotid artery (Brunberg et al., 1994). Finally, even in cases with chronic CCH from MCA stroke developing in the adult, crossed cerebellar atrophy is not clearly demonstrated by standard MRI, even though ipsilateral atrophy of the cerebral peduncle is occasionally seen, which documents CPCS damage (Pappata et al., 1987).

It has been shown that the lack of CCH in the acute stage of MCA territory stroke predicts a good neurological outcome, whereas its presence has little predictive value in individual patients (Serrati et al., 1994). A relationship between CCH and ipsilateral ataxia would seem straightforward and has been anecdotally reported (Tanaka et al., 1992). However, other studies indicate a lack of one-to-one association between ataxia and CCH (Pappata et al., 1990). Apart from ataxia, a significant relationship between CCH and ipsilateral flaccidity has been reported (Pantano et al., 1995), but again this was not an invariable association.

B. Contralateral Cerebral Effects

Although contralateral cerebral effects have long been thought to underlie "diffuse" symptoms of acute supratentorial stroke (such as agitation, confusion, and coma) and to exacerbate focal deficits, they have been difficult to document either in man or in animal models, because of many confounding factors, above all the lack of adequate controls and the frequent use of CBF as the indicator variable despite this parameter's intrinsic variability. Thus PET studies of this effect in heterogeneous samples of carotid-territory stroke patients are inconsistent (Lenzi et al., 1982; Wise et al., 1986). A recent study showed no relationship between changes in contralateral hemisphere $CMRO_2$ and early changes in neurological deficit (Iglesias et al., 1996). Evidence is, however, accumulating that contralateral hemisphere hypometabolism develops in the subacute stage of MCA stroke, dissociated from clinical recovery that takes place at this stage, presumably reflecting the degeneration of severed transcallosal fibers (Heiss et al., 1993; Iglesias et al., 1996). Some data, however, support the idea that, in the chronic stage, slow recovery from contralateral hemisphere hypometabolism takes place and underlies late improvements in cognitive deficits (Baron et al., 1986; Heiss et al., 1993; Mimura et al., 1998).

C. Subcortico-cortical Effects

1. Subcortical Aphasia

Earlier reports of mildly reduced cortical CMRGlc in small, deep infarcts (Kuhl et al., 1980; Metter et al., 1981) suggested that some aspects of language impairment of subcortical origin could be related to this remote effect. Baron et al. (1986) subsequently reported a statistically significant hypometabolism of the whole left cortical mantle in patients with verbal impairment from left thal-

amic or thalamo-capsular stroke, a finding confirmed by Perani et al. (1987) (Fig. 5). In an extensive study, Metter et al. (1988) documented that although the subcortical lesion itself did have a direct relationship with some of the aphasia measures, left frontal and temporal hypometabolism indirectly played a role in verbal fluency and comprehension tasks, respectively. Supporting these findings, Karbe et al. (1990) also found a positive correlation between impairment in several aphasia items (oral and written comprehension, naming, and repetition) and left parieto-temporal hypometabolism; remarkably, there was no consistent correlation between these aphasia measures and left prefrontal or left basal ganglia metabolism.

2. Subcortical Neglect

Marked ipsilateral cortical hypometabolism has been consistently reported after right-sided subcortical infarcts with left hemineglect (Perani et al., 1987; Bogousslavsky et al., 1988; Fiorelli et al., 1991). Predominance of these effects over the frontal and parietal cortices, especially in the case of prominent motor hemineglect (Fiorelli et al., 1991), suggests involvement of the subcortico-cortical network for directed attention, which involves parieto-frontal interactions. Consistent with this interpretation, motor neglect is characterized by sparing of the primary motor circuit (striatum, cerebellum, and motor strip) but hypometabolism of higher order motor areas (i.e., premotor, prefrontal, cingulate, and parietal cortices) (Von Giesen et al., 1994).

3. Hemianopia

Damage to the optic radiations induces a significant reduction in glucose utilization in the disconnected part of the ipsilateral primary visual cortex (Bosley et al., 1985), sometimes spreading to the visual association areas and even to the contralateral visual cortex (Kiyosawa et al., 1990).

4. Thalamo-cortical Diaschisis

As already stated, even small unilateral infarcts in the anterior, medial, or lateral thalamus almost invariably induce a metabolic depression of the entire ipsilateral cortical mantle, with lesser effects contralaterally (Baron et al., 1986). Accentuation of this diffuse effect in the projection area corresponding to the apparent nuclear topography of the thalamic infarct (Kuwert et al., 1991; Szelies et al., 1991) suggests involvement of the thalamo-cortical excitatory projections from the intralaminar, anterior, ventral, and mediodorsal nuclei. A relation between cognitive impairment and cortical hypometabolism after thalamic stroke is supported by the findings that, following ventrolateral thalamotomy, the magnitude of ipsilateral neocortical hypometabolism significantly correlates with the severity of global cog-

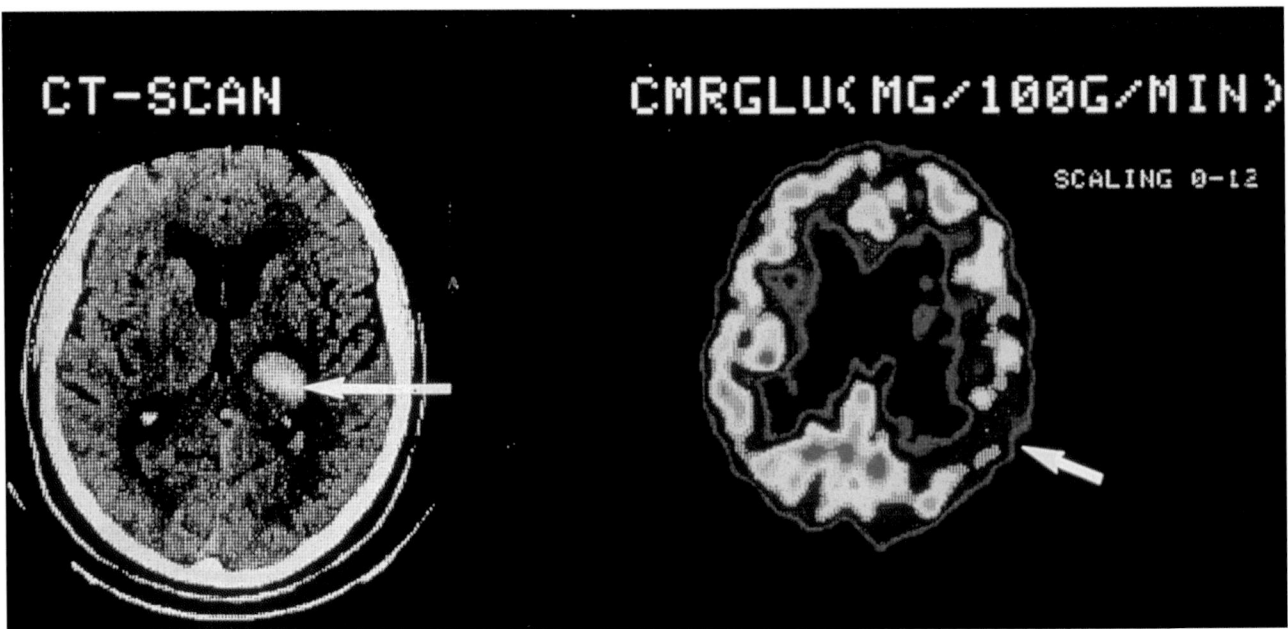

Figure 5 Thalamo-cortical diaschisis. In this chronically hypertensive patient who presented with sudden-onset right hemiparesis and speech difficulties the CT scan (right-hand side) shows a highly dense area in the left thalamo-capsular region (arrows) due to hemorrhage. The PET scan to investigate cerebral glucose uptake, performed 2 weeks later, shows hypometabolism in the left parietal and temporal cortex (arrows, in the axial plane at the level of the corona radiata). This phenomenon provides an explanation for the speech difficulties of the patient by documenting dysfunction of the thalamo-cortical loop.

nitive impairment (Baron *et al.*, 1986), and accordingly postero-lateral thalamic infarcts with pure sensorimotor stroke do not exhibit significant cortical hypometabolism (Chabriat *et al.*, 1992). As noted above, there are also significant relationships between the pattern of cortical hypometabolism and the aphasia profile or hemi-neglect after thalamic stroke (Baron *et al.*, 1986) (Fig. 5). Furthermore, preferential frontal cortex hypometabolism has been associated with frontal-like syndromes and global amnesia after right- or left-sided thalamic infarction (Kuwert *et al.*, 1991; Szelies *et al.*, 1991). In the bilateral paramedial thalamic infarction syndrome, which includes severe permanent amnesia and apathy, marked neocortical hypometabolism has been reported, consistent with the idea that thalamo-cortical deafferentation underlies "thalamic dementia" (Levasseur *et al.*, 1992). Bogousslavsky *et al.* (1991) described isolated apathy in two patients with this syndrome, both of whom exhibited predominantly bifrontal hypoperfusion, consistent with the notion that the prefrontal–striatal–pallidal–thalamic loops play an important role in the control of behavior.

D. Ipsilateral Effects

Striatal and the thalamic hypometabolism ipsilateral to cortico-subcortical stroke is a frequent finding (Kuhl *et al.*, 1980; Baron *et al.*, 1981; Pappata *et al.*, 1987;

De Reuck *et al.*, 1995). Thalamic hypometabolism develops a few days after the stroke and presumably represents active retrograde degeneration of the damaged thalamo-cortical neurons, whereas striatal hypometabolism probably reflects loss of glutamatergic input from the cortex. Left caudate and thalamic hypometabolism are significantly associated with Broca's (i.e., nonfluent) aphasia, as compared to Wernicke's, or conduction, aphasia (Metter *et al.*, 1989). Thalamic hypometabolism has been associated with poor recovery of hand function after ischemic stroke (Binkofski *et al.*, 1996).

Ipsilateral cortical effects have been documented not only in cases of subcortical stroke, as described above, but also after cortico-subcortical infarcts, and the mechanism involved is disconnection via cortico-cortical pathways (Pantano *et al.*, 1996; Iglesias *et al.*, 1997).

E. Role of Cortical Hypometabolism in Behavioral Recovery from Stroke

Following unilateral thalamic stroke, both the bilateral metabolic depression and the cortical asymmetry are more profound early on and tend to recover monoexponentially over the ensuing months, in parallel with cognitive recovery (Baron *et al.*, 1986, 1992). This suggests that, following early trans-synaptic depression, some mechanism of synaptic reorganization is slowly established, which underlies recovery from cognitive impairment. More specifically, and regardless of the topography of the stroke, recovery from subcortical aphasia is associated with a return toward normality of cortical hypometabolism/hypoperfusion (Metter *et al.*, 1992), although exceptions at the individual level do occur. Karbe *et al.* (1995) found that patients with lesser defects in resting CMRGlc around Wernicke's and Broca's areas in the subacute stage of stroke had better outcome in terms of language comprehension and verbal fluency, respectively. Furthermore, performing the PET scan during language activation improved this predictive value for subsequent recovery from aphasia (Heiss *et al.*, 1993). Recently, Mimura *et al.* (1998) reported the results of an extensive longitudinal investigation of the relationship between cortical hypoperfusion and language performance in a large cohort of aphasic stroke patients (both cortical and subcortical). They concluded that language recovery within the first year of stroke is linked primarily to functional recovery in the dominant hemisphere, where an increase in CBF was observed at 9 months postonset. Increased perfusion adjacent to a lesion may be crucial for recovery of aphasia in the early months, consistent with above-described findings of Furlan *et al.* (1996), inasmuch as these authors reported that neurological recovery at 2 months was related to the amount of acute-stage penumbra that escaped infarction. Subsequent language recovery and long-term recovery from aphasia are related to slow and gradual compensatory mechanisms involving the contralateral hemisphere, specifically in the homotopic frontal and thalamic areas (Cappa *et al.*, 1997). Taken together, therefore, the results concur to suggest that recovery of cortical metabolism, both ipsilateral and contralateral, subtends functional recovery after stroke and is one expression of neuronal reorganization after damage to the language network. The relationship between these changes in resting metabolism and alterations in large-scale neural networks occurring following focal brain damage remains to be established.

F. White Matter Stroke and Subcortical Dementia

There is converging evidence that after isolated white matter stroke, there is an association between hypometabolism of the overlying cortex and the presence of cognitive impairment (Pappata *et al.*, 1990), e.g., the frontal lobe syndrome after capsular genu infarction (Tatemichi *et al.*, 1992) and neglect or aphasia after partial anterior choroidal artery stroke (Bogousslavsky *et al.*, 1988). Cortical hypometabolism after white matter stroke results from disconnection (Herholz *et al.*, 1990).

Similarly, leukoaraiosis, even when extensive, does not seem to greatly affect cortical metabolism unless cognitive impairment, albeit subtle, is present (Delpla *et al.*, 1990; Meguro *et al.*, 1990; De Carli *et al.*, 1995). Accordingly, in Binswanger's dementia, neocortical metabolism is diffusely depressed, with frequent accentuation in the frontal lobes (Yao *et al.*, 1990). This finding also applies to another type of subcortical dementia, that due to bilateral paramedian thalamic infarction, as already discussed above (Levasseur *et al.*, 1992), further highlighting the importance of strategic damage to subcortico-cortical networks in the development of cognitive impairment of vascular origin.

VI. Receptor Studies in Vascular Disorders

A. Neuronal Marker

[^{11}C]Flumazenil is an ideal potential *in vivo* marker of neuronal loss, because it binds to the central benzodiazepine receptor, which is expressed only by neurons and is part of the GABA-A complex, which has widespread brain distribution (Sette *et al.*, 1989). In both experimental and clinical stroke, loss of [^{11}C]flumazenil binding occurs in the ischemic core hours after arterial occlusion (Sette *et al.*, 1993; Heiss *et al.*, 1998b). Several studies in the chronic stage have reported decreases in cortical binding of [^{11}C]flumazenil or its SPECT analog, [^{123}I]iomazenil, including areas without signs of frank infarction on structural imaging (Nakagawara *et al.*, 1997). This finding has been taken as evidence for selective neuronal loss in overlying cortex, which may, in turn, explain neuropsychological deficits after subcortical infarction, e.g., language impairment. Thus, benzodiazepine receptor imaging may allow a distinction between cortical dysfunction induced by diaschisis from that due to occult cortical damage. However, the finding of reduced tracer uptake in the cerebral cortex after striato-capsular infarction is not universal (Takahashi *et al.*, 1997), and in addition, histopathologic confirmation of neuronal loss is lacking. A recent experimental investigation in the baboon suggests that reduced [^{11}C]flumazenil uptake in cortical zones overlying striato-capsular infarction may not necessarily correspond to cell loss (Watanabe *et al.*, 1999), but may reflect other processes such as functional changes due to deafferentation.

B. Glial Marker

[^{11}C]PK11195 is a ligand of the peripheral benzodiazepine receptor, which is expressed on microglia and macrophages and as such may be a good marker of glial proliferation after stroke. This hypothesis has been confirmed by studies in man and nonhuman primates, showing progressively increasing [^{11}C]PK11195 uptake within infarcts, peaking at around 10–15 days. (Sette *et al.*, 1993; Ramsay *et al.*, 1992).

References

Ackerman, R. H., Correia, J. A., Alpert, N. M., Baron, J. C., Gouliamos, A., Grotta, J. C., Brownell, G. L., and Taveras, J. M. (1981). Positron imaging in ischemic stroke disease using compounds labeled with oxygen-15. *Arch. Neurol.* **38,** 537–543.

Baird, A. E., and Warach, S. (1998). Magnetic resonance imaging of acute stroke. *J. Cereb. Blood Flow Metab.* **18,** 583–609.

Baird, A. E., Donnan, G. A., Austin, M. C., and MacKay, W. J. (1995). Early reperfusion in the "spectacular shrinking deficit" demonstrated by single-photon emission computed tomography. *Neurology* **45,** 1335–1339.

Barber, P. A., Davis, S. M., Infeld, B., Baird, A. E., Donnan, G. A., Jolley, D., and Lichtenstein, M. (1998). Spontaneous reperfusion after ischemic stroke is associated with improved outcome. *Stroke* **29,** 2522–2528.

Baron, J. C., Bousser, M. G., Comar, D., and Castaigne, P. (1980). "Crossed cerebellar diaschisis" in human supratentorial brain infarction. *Trans. Am. Neurol. Assoc.* **105,** 459–461.

Baron, J. C., Bousser, M. G., Rey, A., Guillard, A., Comar, D., and Castaigne, P. (1981a). Reversal of focal "misery-perfusion syndrome" by extra-intracranial arterial bypass in hemodynamic cerebral ischemia: A case study with ^{15}O positron tomography. *Stroke* **12,** 454–459.

Baron, J. C., Bousser, M. G., Comar, D., Soussaline, F., and Castaigne, P. (1981b). Noninvasive tomographic study of cerebral blood flow and oxygen metabolism *in vivo:* Potentials, limitations and clinical applications in cerebral ischemic disorders. *Eur. Neurol.* **20,** 273–284.

Baron, J. C., Rougemont, D., Bousser, M. G., Lebrun-Grandie, P., Iba-Zizen, M. T., and Chiras, J. (1983). Local CBF, oxygen extraction fraction (OEF) and CMRO$_2$: Prognostic value in recent supratentorial infarction in humans. *J. Cereb. Blood Flow Metab.* **3**(Suppl. 1), S1–S2.

Baron, J. C., Rougemont, D., Soussaline, F., Bustany, P., Crouzel, C., Bousser, M. G., and Comar, D. (1984). Local interrelationship of cerebral oxygen consumption and glucose utilization in normal subjects and in ischemic stroke patients: A positron tomography study. *J. Cereb. Blood Flow Metab.* **4,** 140–149.

Baron, J. C., D'Antona, R., Pantano, P., Serdaru, M., Samson, Y., and Bousser, M. G. (1986). Effects of thalamic stroke on energy metabolism of the cerebral cortex. *Brain* **109,** 1243–1259.

Baron, J. C., Frackowiak, R. S. J., Herholz, K., Jones, T., Lammertsma, A. A., Mazoyer, B. M., and Wienhardt, K. (1989). Use of positron emission tomography in the investigation of cerebral hemodynamics and energy metabolism in cerebrovascular disease. *J. Cereb. Blood Flow Metab.* **9,** 723–742.

Baron, J. C., Levasseur, M., Mazoyer, B., Legault Demare, F., Mauguiere, F., Pappata, S., Jedynak, P., Derome, P., Cambier, J., Tran Dinh, S., and Cambon, H. (1992). Thalamo-cortical diaschisis: PET study in humans. *J. Neurol., Neurosurg., Psychiatry* **55,** 935–942.

Berrouschot, J., Barthel, H., Hesse, S., Köster, J., Knapp, W. H., and Schneider, D. (1998a). Differentiation between transient ischemic attack and ischemic stroke within the first six hours after onset of symptoms by using ^{99m}Tc-ECD-SPECT. *J. Cereb. Blood Flow Metab.* **18,** 921–929.

Berrouschot, J., Barthel, H., von Kummer, R., Knapp, W. H., Hesse, S., and Schneider, D. (1998b). ^{99m}Technetium-ethyl-cysteinate-dimer single-photon emission CT can predict fatal ischemic brain edema. *Stroke* **12,** 2556–2562.

Binkofski, F., Seitz, R. J., Arnold, S., Classen, J., Benecke, R., and Freund, H. J. (1996). Thalamic metabolism and corticospinal tract integrity determine motor recovery in stroke. *Ann. Neurol.* **39,** 460–470.

Bogousslavsky, J., Miklossy, J., Regli, F., Deruaz, J. P., Assal, G., and Delaloye, G. (1988). Subcortical neglect: Neuropsychological, SPECT, and neuropathological correlations with anterior choroidal artery territory infarction. *Ann. Neurol.* **23,** 448–452.

Bogousslavsky, J., Regli, F., Delaloye, G., *et al.* (1991). Loss of psychic self-activation with bithalamic infarction. *Acta Neurol. Scand.* **83,** 309–316.

Bosley, T., Rosenquist, A. C., Kushner, M., Burke, A., Stein, A., Dann, R., Cobbs, W., Savino, P. J., Schatz, N. J., Alavi, A., and Reivich, M. (1985). Ischemic lesions of the occipital cortex and optic radiations: Positron emission tomography. *Neurology* **35,** 470–484.

Brunberg, J. A., Frey, K. A., Horton, J. A., Deveikis, J. P., Ross, D. A., and Koeppe, R. A. (1994). [^{15}O]H$_2$O positron emission tomography determination of cerebral blood flow during balloon test occlusion of the internal carotid artery. *AJNR, Am. J. Neuroradiol.* **15,** 725–732.

Cappa, S. F., Perani, D., Grassi, F., Bressi, S., Alberoni, M., Franceschi, M., Bettinardi, V., Todde, S., and Fazio, F. (1997). A PET follow-up study of recovery after stroke in acute aphasics. *Brain Lang.* **56,** 55–67.

Carpenter, D. A., Grubb, R. L., and Powers, W. J. (1990). Borderzone hemodynamics in cerebrovascular disease. *Neurology* **40,** 1587–1592.

Chabriat, H., Levasseur, M., Pappata, S., Fiorelli, M., and Baron, J. C. (1992). Cortical metabolism in postero-lateral thalamic stroke: A PET study. *Acta Neurol. Scand.* **86,** 285–290.

Davis, S., Chua, M. G., Lichtenstein, M., Rossiter, S., Binns, D., and Hopper, J. L. (1993). Cerebral hypoperfusion in stroke prognosis and brain recovery. *Stroke* **24,** 1691–1696.

De Carli, C., Murphy, D. G., Tranh, M., Grady, C. L., Haxby, J. V., Gillette, J. A., Salerno, J. A., Gonzales-Aviles, A., Horwitz, B., and Rapoport, S. I. (1995). The effect of white matter hyperintensity volume on brain structure, cognitive performance, and cerebral metabolism of glucose in 51 healthy adults. *Neurology* **45,** 2077–2084.

Delpla, P. A., Zatorre, R., Meyer, E., Geraud, G., Bes, A., Etheir, R., and Hakim, A. M. (1990). Leucoaraïose et dysfonctionnement frontal précoce chez le sujet âgé non dément: Approche neuropsychologique et par la caméra à positons. *In* "Cerveau et hypertension artérielle" (A. Bes and G. Géraud, Eds.), pp. 123–139. Libbey, Paris.

Derdeyn, C. P., Yundt, K. D., Videen, T. O., Carpenter, D. A., Grubb, R. L., and Powers, W. J. (1998). Increased oxygen extraction fraction is associated with prior ischemic events in patients with carotid occlusion. *Stroke* **29,** 754–758.

De Reuck, J., Decoo, D., Lemahieu, I., Strijckmans, K., Goethals, P., and Van Maele, G. (1995). Ipsilateral thalamic diaschisis after middle cerebral artery infarction. *J. Neurol. Sci.* **134,** 130–135.

Derlon, J. M., Bouvard, G., Viader, F., Petit, M. C., Dupuy, B., Khoury, S., Thenint, J. P., and Houtteville, J. P. (1992). Impaired cerebral hemodynamics in internal carotid occlusion. *Cerebrovasc. Dis.* **2,** 72–81.

Feeney, D., and Baron, J. C. (1986). Diaschisis. *Stroke* **17,** 817–830.

Fiorelli, M., Blin, J., Bakchine, S., Laplane, D., and Baron, J. C. (1991). PET studies of cortical diaschisis in patients with motor hemineglect. *J. Neurol. Sci.* **104,** 135–142.

Firlik, A. D., Rubin, G., Yonas, H., and Wechsler, L. R. (1998a). Relation between cerebral blood flow and neurologic deficit resolution in acute ischemic stroke. *Neurology* **51,** 177–182.

Firlik, A. D., Yonas, H., Kaufmann, A. M., Wechsler, L. R., Jungreis, C. A., Fukui, M. B., and Williams, R. L. (1998b). Relationship between cerebral blood flow and the development of swelling and life-threatening herniation in acute ischemic stroke. *J. Neurosurg.* **89,** 243–249.

Frackowiak, R. S. J. (1985). The pathophysiology of human cerebral ischaemia: A new perspective obtained with positron tomography. *Q. J. Med.* **223,** 713–727.

Furlan, M., Marchal, G., Viader, F., Derlon, J.-M., and Baron, J.-C. (1996). Spontaneous neurological recovery after stroke and the fate of the ischemic penumbra. *Ann. Neurol.* **40,** 216–226.

Gibbs, J. M., Wise, R. J. S., Leenders, K. L., and Jones, T. (1984). Evaluation of cerebral perfusion reserve in patients with carotid-artery occlusion. *Lancet* **1,** 310–314.

Gibbs, J. M., Wise, R. J. S., Thomas, D. J., Mansfield, A. U., and Ross Russel, R. W. (1987). Cerebral haemodynamic changes after extracranial-intracranial bypass surgery. *J. Neurol. Neurosurg. Psychiatry* **50,** 140–150.

Gillard, J. H., Barker, P. B., Van Zijl, P. C., Bryan, R. N., and Oppenheimer, S. M. (1996). Proton MR spectroscopy in acute middle cerebral artery stroke. *Am. J. Neuroradiol.* **17,** 873–886.

Giubilei, F., Lenzi, G. L., Di Piero, V., Pozzilli, C., Pantano, P., Bastianello, S., Argentino, C., and Fieschi, C. (1990). Predictive value of brain perfusion single-photon emission computed tomography in acute ischemic stroke. *Stroke* **21,** 895–900.

Grotta, J. C., and Alexandrov, A. V. (1998). tPA-associated reperfusion after acute stroke demonstrated by SPECT. *Stroke* **29,** 428–432.

Hakim, A. M., Pokrupa, R. P., Villanueva, J., Diksic, M., Evans, A. C., Thompson, C. J., *et al.* (1987). The effects of spontaneous reperfusion on metabolic function in early human cerebral infarcts. *Ann. Neurol.* **21,** 279–289.

Hanson, S. K., Grotta, J. C., Rhoades, H., Tran, H. D., Lamki, L. M., Barron, B. J., and Taylor, W. J. (1993). Value of single-photon emission-computed tomography in acute stroke therapeutic trials. *Stroke* **24,** 1322–1329.

Heiss, W. D., Huber, M., Fink, G. R., Herholz, K., Pietryk, U., Wagner, R., and Wienhard, K. (1992). Progressive derangement of peri-infarct viable tissue in ischemic stroke. *J. Cereb. Blood Flow Metab.* **12,** 193–203.

Heiss, W. D., Kessler, J., Karbe, H., Fink, G. R., and Pawlik, G. (1993). Cerebral glucose metabolism as a predictor of recovery from aphasia in ischemic stroke. *Arch. Neurol.* **50,** 958–964.

Heiss, W. D., Graf, R., Löttgen, J., Ohta, K., Fujita, T., Wagner, R., Grond, M., and Wienhard, K. (1997). Repeat positron emission tomographic studies in transient middle cerebral artery occlusion in cats: Residual perfusion and efficacy of postischemic reperfusion. *J. Cereb. Blood Flow Metab.* **17,** 388–400.

Heiss, W. D., Grond, M., Thiel, A., Von Stockhausen, H. M., Rudolf, J., Ghaemi, M., Löttgen, J., Stenzel, C., and Pawlik, G. (1998a). Tissue at risk of infarction rescued by early reperfusion: A positron emission tomography study in systemic recombinant tissue plasminogen activator thrombolysis of acute stroke. *J. Cereb. Blood Flow Metab.* **18,** 1298–1307.

Heiss, W. D., Grond, M., Thiel, A., Ghaemi, M., Sobesky, J., Rudolf, J., Bauer, B., and Wienhard, K. (1998b). Permanent cortical damage detected by flumazenil positron emission tomography in acute stroke. *Stroke* **29,** 454–461.

Herderschee, D., Limburg, M., van Royen, E. A., Hijdra, K., Buller, H. R., and Koster, P. A. (1991). Thrombolysis with recombinant tissue plasminogen activator in acute ischemic stroke: Evaluation with rCBF-SPECT. *Acta Neurol. Scand.* **83,** 317–322.

Herholz, K., Heindel, W., Rackl, A., Neubauer, I., Steinbrich, W., Pietrzyk, U., Erasmi-Körber, H., and Heiss, W.-D. (1990). Regional cerebral blood flow in patients with leuko-araiosis and atherosclerotic carotid artery disease. *Arch. Neurol.* **47,** 392–396.

Herold, S., Brown, M. M., Frackowiak, R. S. J., Mansfield, A. O., Thomas, D. J., and Marshall, J. (1988). Assessment of cerebral haemodynamic reserve: Correlation between PET parameters and CO_2 reactivity measured by the intravenous [133]xenon injection technique. *J. Neurol., Neurosurg., Psychiatry* **51,** 1045–1050.

Iglesias, S., Marchal, G., Rioux, P., Beaudouin, V., Hauttement, J. L., De La Sayette, V., Le Doze, F., Derlon, J. M., Viader, F., and Baron, J. C. (1996). Do changes in oxygen metabolism in the unaffected cerebral hemisphere underlie early neurological recovery after stroke? A positron emission tomography study. *Stroke* **27,** 1192–1199.

Iglesias, S., Marchal, G., Furlan, M., Viader, F., Derlon, J. M., and Baron, J. C. (1997). Affected hemisphere metabolism and early recovery (ER) after acute MCA stroke. *Eur. J. Neurol.* **4** (Suppl. 1), S13.

Inao, S., Tadokoro, M., Nishino, M., Mizutani, N., Terada, K., Bundo, M., Kuchiwaki, H., and Yoshida, J. (1998). Neural activation of the brain with hemodynamic insufficiency. *J. Cereb. Blood Flow Metab.* **18,** 960–967.

Itoh, M., Hatazawa, J., Pozzilli, C., Matsuzawa, T., Abe, Y., Fukuda, H., Fujiwara, T., Watanuki, S., and Ido, T. (1988). Positron CT imaging of an impending stroke. *Neuroradiology* **30,** 276–279.

Jorgensen, H. S., Sperling, B., Nakayama, H., Raaschou, H. O., and Olsen, T. S. (1994). Spontaneous reperfusion of cerebral infarcts in patients with acute stroke. *Arch. Neurol.* **51,** 865–873.

Kamada, K., Saguer, M., Möller, M., Wicklow, K., Katenhäuser, M., Kober, H., and Vieth, J. (1997). Functional and metabolic analysis of cerebral ischemia using magnetoencephalography and proton magnetic resonance spectroscopy. *Ann. Neurol.* **42,** 554–563.

Karbe, H., Szelies, B., Herholz, K., and Heiss, W. D. (1990). Impairment of language is related to left parieto-temporal glucose metabolism in aphasic stroke patients. *J. Neurol.* **237,** 19–23.

Karbe, H., Kessler, J., Herholz, K., Fink, G. R., and Heiss, W. D. (1995). Long-term prognosis of poststroke aphasia studied with positron emission tomography. *Arch. Neurol.* **52,** 186–190.

Kiyosawa, M., Bosley, T. M., Kushner, M., Jamieson, D., Alavi, A., and Reivich, M. (1990). Middle cerebral artery strokes causing homonymous hemianopia: Positron emission tomography. *Ann. Neurol.* **28,** 180–183.

Kleiser, B., and Widder, B. (1992). Course of carotid artery occlusions with impaired cerebrovascular reactivity. *Stroke* **23,** 171–174.

Kuhl, D. E., Phelps, M. E., Kowell, A. P., Metter, E. J., Selin, C., and Winter, J. (1980). Effects of stroke on local cerebral metabolism and perfusion. Mapping by emission computed tomography of [18]FDG and [13]NH_3. *Ann. Neurol.* **8,** 47–60.

Kuroda, S., Kamiyama, H., Abe, H., Houkin, K., Isobe, M., and Mitsumori, K. (1993). Acetazolamide test in detecting reduced cerebral perfusion reserve and predicting long-term prognosis in patients with internal carotid artery occlusion. *Neurosurgery* **32,** 912–919.

Kurthen, M., Reichman, K., Linke, D. B., Biersack, H. J., Reuter, B. M., Durwen, J. F., and Grunmann, F. (1990). Crossed cerebellar diaschisis in intracarotid sodium amytal procedures: A SPECT study. *Acta Neurol. Scand.* **81,** 416–422.

Kushner, M., Alair, A., Reivich, M., Dann, R., Burke, A., and Robinson, G. (1984). Contralateral cerebellar hypometabolism following cerebral insult: A positron emission tomographic study. *Ann. Neurol.* **15,** 425–434.

Kuwert, T., Hennerici, M., Langen, K. L., Aulich, A., Herzog, H., Sitzer, M., and Feinendegen, L. E. (1991). Regional cerebral glucose consumption measured by positron emission tomography in patients with unilateral thalamic infarction. *Cerebrovasc. Dis.* **1,** 327–336.

Lassen, N. A. (1966). The luxury perfusion syndrome and its possible relation to acute metabolic acidosis localised within the brain. *Lancet* **2,** 1113–1115.

Leblanc, R., Tyler, J. L., Mohr, G., Meyer, E., Diksic, M., Yamamoto, L., Taylor, L., Gauthier, S., and Hakim, A. (1987a). Hemodynamic and metabolic effects of cerebral revascularization. *J. Neurosurg.* **66**, 529–535.

Leblanc, R., Yamamoto, Y. L., Tyler, J. L., Diksic, M., and Hakim, A. (1987b). Borderzone ischemia. *Ann. Neurol.* **22**, 707–713.

Lenzi, G. L., Frackowiak, R. S. J., and Jones, T. (1982). Cerebral oxygen metabolism and blood flow in human cerebral ischemic infarction. *J. Cereb. Blood Flow Metab.* **2**, 231–235.

Levasseur, M., Baron, J. C., Sette, G., Legault-Demare, F., Pappata, S., Mauguiere, F., Benoit, N., Tran-Dinh, S., Degos, J. D., Laplane, D., and Mazoyer, B. (1992). Brain energy metabolism in bilateral paramedian thalamic infarcts: A positron emission tomography study. *Brain* **115**, 795–807.

Levine, R. L., Dobkin, J. A., Rozental, J. M., Satter, M. R., and Nickles, R. J. (1991). Blood flow reactivity to hypercapnia in strictly unilateral carotid disease: Preliminary results. *J. Neurol., Neurosurg., Psychiatry* **54**, 204–209.

Limburg, M., Van Royen, E. A., Hijdra, A., De Bruïne, J. F., and Verbeeten, B. W. J. (1990). Single-photon emission computed tomography and early death in acute ischemic stroke. *Stroke* **21**, 1150–1155.

Marchal, G., Serrati, C., Rioux, P., Petit-Taboue, M. C., Viader, F., De La Sayette, V., Le Doze, F., Lochon, P., Derlon, J. M., Orgogozo, J. M., and Baron, J. C. (1993). PET imaging of cerebral perfusion and oxygen consumption in acute ischaemic stroke: Relation to outcome. *Lancet* **341**, 925–927.

Marchal, G., Rioux, P., Serrati, C., Furlan, M., Derlon, J. M., Viader, F., and Baron, J. C. (1995). Value of acute-stage PET in predicting neurological outcome after ischemic stroke: Further assessment. *Stroke* **26**, 524–525.

Marchal, G., Furlan, M., Beaudouin, V., *et al.* (1996a). Early spontaneous hyperperfusion after stroke: A marker of favorable tissue outcome. *Brain* **119**, 409–419.

Marchal, G., Beaudouin, V., Rioux, P., de la Sayette, V., Le Doze, F., Viader, F., Derlon, J.-M., and Baron, J.-C. (1996b). Prolonged persistence of substantial volumes of potentially viable brain tissue after stroke: A correlative PET-CT study with voxel-based data analysis. *Stroke* **27**, 599–606.

Marchal, G., Benali, K., Iglesias, S., Viader, F., Derlon, J. M., and Baron, J. C. (1999a). Voxel-based mapping of irreversible tissue damage by PET in the acute stage of ischemic stroke. *Brain* **123**, 2387–2400.

Marchal, G., Young, A. R., and Baron, J. C. (1999b). Early postischaemic hyperperfusion: Pathophysiological insights from positron emission tomography. *J. Cereb. Blood Flow Metab.* **19**, 467–482.

Marchal, G., Bouvard, G., Iglesias, S., Sebastien, B., Benali, K., Defer, G., Viader, F., and Baron, J. C. (1999c). Predictive value of ^{99m}Tc-HMPAO for neurological outcome/recovery in the acute stage of stroke. *Cerebrovasc. Dis.* (in press).

Martin, W. R., and Raichle, M. E. (1983). Cerebellar blood flow and metabolism in cerebral hemisphere infarction. *Ann. Neurol.* **14**, 168–176.

Meguro, K., Hatazawa, J., Yamaguchi, T., *et al.* (1990). Cerebral circulation and oxygen metabolism associated with subclinical periventricular hyperintensity as shown by magnetic resonance imaging. *Ann. Neurol.* **28**, 378–383.

Metter, E. J., Wasterlain, C. G., Kuhl, D. E., Hanson, W. R., and Phelps, M. E. (1981). FDG positron emission computed tomography in a study of aphasia. *Ann. Neurol.* **10**, 173–183.

Metter, E. J., Riege, W. H., Hanson, W. R., Cambras, L. R., Phelps, M. E., and Kuhl, D. E. (1984). Correlations of glucose metabolism and structural damage to language function in aphasia. *Brain Lang.* **21**, 187–207.

Metter, E. J., Riege, W. H., Hanson, W. R., *et al.* (1988). Subcortical structures in aphasia. *Arch. Neurol.* **45**, 1229–1234.

Metter, E. J., Kempler, D., Jackson, C., *et al.* (1989). Cerebral glucose metabolism in Wernicke's, Broca's, and conduction aphasia. *Arch. Neurol.* **46**, 27–34.

Metter, E. J., Jackson, C. A., Kempler, D., and Hanson, W. R. (1992). Temporoparietal cortex and the recovery of language comprehension in aphasia. *Aphasiology* **6**, 349–358.

Mimura, M., Kato, M., Kato, M., Sano, Y., Kojima, T., Naeser, M., and Kashima, H. (1998). Prospective and retrospective studies of recovery in aphasia. Changes in cerebral blood flow and language functions. *Brain* **121**, 2083–2094.

Miura, H., Nagata, K., Hirata, Y., Satoh, Y., Watahiki, Y., and Hatazawa, J. (1994). Evolution of crossed cerebellar diaschisis in middle cerebral artery infarction. *J. Neuroimaging* **4**, 91–96.

Muraishi, K., Kameyama, M., Sato, K., Sirane, R., Ogawa, A., Yoshimoto, T., Hatazawa, J., and Itoh, M. (1993). Cerebral circulatory and metabolic changes following EC/IC bypass surgery in cerebral occlusive diseases. *Neurol. Res.* **15**, 97–103.

Nakagawara, J., Sperling, B., and Lassen, N. A. (1997). Incomplete brain infarction of reperfused cortex may be quantitated with iomazenil. *Stroke* **28**, 124–132.

Pantano, P., Baron, J. C., Samson, Y., Bousser, M. G., Derouesne, C., and Comar, D. (1986). Crossed cerebellar diaschisis: Further studies. *Brain* **109**, 677–694.

Pantano, P., Formisano, R., Ricci, M., Di Piero, V., Sabatini, U., Barbanti, P., Fiorelli, M., Bozzao, L., and Lenzi, G. L. (1995). Prolonged muscular flaccidity after stroke. Morphological and functional brain alterations. *Brain* **118**, 1329–1338.

Pantano, P., Formisano, R., Ricci, M., Di Piero, V., Sabatini, U., Di Pofi, B., Rossi, R., Bozzao, L., and Lenzi, G. L. (1996). Motor recovery after stroke. Morphological and functional brain alterations. *Brain* **119**, 1849–1857.

Pappata, S., Tran Dinh, S., Baron, J. C., Cambon, H., and Syrota, A. (1987). Remote metabolic effects of cerebrovascular lesions: Magnetic resonance and positron tomography imaging. *Neuroradiology* **29**, 1–6.

Pappata, S., Mazoyer, B., Tran-Dinh, S., Cambon, H., Levasseur, M., and Baron, J. C. (1990). Cortical and cerebellar hypometabolic effects of capsular, thalamo-capsular, and thalamic stroke: A positron tomography study. *Stroke* **21**, 519–524.

Pappata, S., Fiorelli, M., Rommel, T., Hartmann, A., Dettmers, C., Yamaguchi, T., Chabriat, H., Poline, J. B., Crouzel, C., Di Giamberardino, L., and Baron, J. C. (1993). PET study of changes in local brain hemodynamics and oxygen metabolism after unilateral middle cerebral artery occlusion in baboons. *J. Cereb. Blood Flow Metab.* **13**, 416–424.

Perani, D., Vallar, G., Cappa, S., Messa, C., and Fazio, F. (1987). Aphasia and neglect after subcortical stroke. A clinical/cerebral perfusion correlation study. *Brain* **110**, 1211–1229.

Powers, W. J., Martin, W. R. W., Herscovitch, P., Raichle, M. E., and Grubb, R. L. (1984). Extracranial–intracranial bypass surgery: Hemodynamic and metabolic effects. *Neurology* **34**, 1168–1174.

Powers, W. J., Grubb, R. L., Jr., Darriet, D., and Raichle, M. E. (1985). Cerebral blood flow and cerebral metabolic rate of oxygen requirements for cerebral function and viability in humans. *J. Cereb. Blood Flow Metab.* **5**, 600–608.

Powers, W. J., Press, G. A., Grubb, R. L., Gado, M., and Raichle, M. E. (1987). The effect of hemodynamically significant carotid artery disease on the hemodynamic status of the cerebral circulation. *Ann. Int. Med.* **106**, 27–35.

Powers, W. J., Grubb, R. L., and Raichle, M. (1989a). Clinical results of extracranial-intracranial bypass surgery in patients with hemodynamic cerebrovascular disease. *J. Neurosurg.* **70**, 61–67.

Powers, W. J., Templel, L. W., and Grubb, R. L. (1989b). Influence of cerebral hemodynamics on stroke risk: One-year follow-up of 30 medically treated patients. *Ann. Neurol.* **25**, 325–330.

Powers, W. J., Derdeyn, C. P., Yundt, K. D., Carpenter, D. A., Videen, T. O., Fritsch, S. M., Spitznagel, E. L., and Grubb, R. L. (1998). PET predicts subsequent stroke in patients with symptomatic carotid occlusion. *Neurology* **50,** A195.

Ramsay, S. C., Weiller, C., Myers, R., Cremer, J. E., Luthra, S. K., Lammertsma, A. A., and Frackowiak, R. S. J. (1992). Monitoring by PET of macrophage accumulation in brain after ischaemic stroke. *Lancet* **239,** 1054–1055.

Read, S. J., Hirano, T., Abbott, D. F., Sachinidis, J. I., Tochon-Danguy, H. J., Chan, J. G., Egan, G. F., Scott, A. M., Bladin, C. F., McKay, W. J., and Donnan, G. A. (1998). Identifying hypoxic tissue after acute ischemic stroke using PET and ^{18}F-fluoromisonidazole. *Neurology* **51,** 1617–1621.

Samson, Y., Baron, J. C., Bousser, M. G., Rey, A., Derlon, J. M., David, P., and Comoy, J. (1985). Effects of extra-intracranial arterial bypass on cerebral blood flow and oxygen metabolism in humans. *Stroke* **16,** 609–616.

Schumann, P., Touzani, O., Young, A. R., Morello, R., Baron, J. C., and MacKenzie, E. T. (1998). Evaluation of the ratio of cerebral blood flow to cerebral blood volume as an index of local cerebral perfusion pressure. *Brain* **121,** 1369–1379.

Senda, M., Alpert, N. M., Mackay, B. C., Buxton, R. B., Correia, J. A., Weise, S. B., Ackerman, R. H., Dorer, D., and Buonanno, F. S. (1989). Evaluation of the $^{11}CO_2$ positron emission tomographic method for measuring brain pH. II. Quantitative pH mapping in patients with ischemic cerebrovascular diseases. *J. Cereb. Blood Flow Metab.* **9,** 859–873.

Serrati, C., Marchal, G., Rioux, P., Viader, F., Petit-Taboue, M. C., Lochon, P., Luet, D., Derlon, J. M., and Baron, J. C. (1994). Contralateral cerebellar hypometabolism: A predictor for stroke outcome? *J. Neurol., Neurosurg., Psychiatry* **57,** 174–179.

Sette, G., Baron, J. C., Mazoyer, B., Levasseur, M., Pappata, S., and Crouzel, C. (1989). Local brain hemodynamics and oxygen metabolism in cerebrovascular disease: Positron emission tomography. *Brain* **112,** 931–951.

Sette, G., Baron, J. C., Young, A. R., Miyazawa, H., Tillet, I., Barré, L., Travère, J. M., Derlon, J. M., and MacKenzie, E. T. (1993). *In vivo* mapping of brain benzodiazepine receptor changes by positron emission tomography after focal ischemia in the anesthetized baboon. *Stroke* **24,** 2046–2058.

Sgouropoulos, P., Baron, J. C., Samson, Y., Bousser, M. G., Comar, D., and Castaigne, P. (1985). Sténoses et occlusions persistantes de l'artère cérébrale moyenne: Conséquences hémodynamiques et métaboliques étudiées par tomographie à positons. *Rev. Neurol. (Paris)* **141,** 698–705.

Shimosegawa, E., Hatazawa, J., Inugami, A., Fujita, H., Ogawa, T., Aizawa, Y., Kanno, I., Okudera, T., and Uemura, K. (1994). Cerebral infarction within six hours of onset: Prediction of completed infarction with technetium-99m-HMPAO SPECT. *J. Nucl. Med.* **35,** 1097–1103.

Sorensen, A. G., Buonanno, F. S., Gonzalez, R. G., Schwamm, L. H., Lev, M. H., Huang-Hellinger, F. R., Reese, T. G., Wisskoff, R. M., Davis, T. L., Suwanwela, N., Can, U., Moreira, J. A., Copen, W. A., Look, R. B., Finklestein, S. P., Rosen, B. R., and Koroshetz, W. J. (1996). Hyperacute stroke: Evaluation with combined multisection diffusion-weighted and hemodynamically weighted echo-planar MR imaging. *Radiology* **199,** 391–401.

Sperling, B., and Lassen, N. A. (1993). Hyperfixation of HMPAO in subacute ischemic stroke leading to spuriously high estimates of cerebral blood flow by SPECT. *Stroke* **24,** 193–194.

Syrota, A., Samson, Y., Boullais, C., Wajnberg, P., Loc'h, C., Crouzel, C., Maziere, B., Soussaline, F., and Baron, J. C. (1985). Tomographic mapping of brain intracellular pH and extracellular water space in stroke patients. *J. Cereb. Blood Flow Metab.* **5,** 358–385.

Szelies, B., Herholz, K., Pawlik, G., Karbe, H., Hebold, I., and Heiss, W. D. (1991). Widespread functional effects of discrete thalamic infarction. *Arch. Neurol.* **48,** 178–182.

Takahashi, W., Ohnuki, Y., Ohta, T., Hamano, H., Yamamoto, M., and Shinohara, Y. (1997). Mechanism of reduction of cortical blood flow in striatocapsular infarction: Studies using [^{123}I]iomazenil SPECT. *NeuroImage* **6,** 75–80.

Tanaka, M., Kondo, S., Hirai, S., Ishiguro, K., Ishihara, T., and Morimatsu, M. (1992). Crossed cerebellar diaschisis accompanied by hemiataxia: A PET study. *J. Neurol., Neurosurg., Psychiatry* **55,** 121–125.

Tatemichi, T. K., Desmond, D. W., Prohovnik, I., Cross, D. T., Gropen, T. I., Mohr, J. P., and Stern, Y. (1992). Confusion and memory loss from capsular genu infarction: A thalamocortical disconnection syndrome? *Neurology* **42,** 1966–1979.

Touzani, O., Young, A. R., Derlon, J. M., Baron, J. C., and MacKenzie, E. T. (1997). Progressive impairment of brain oxidative metabolism reversed by reperfusion following middle cerebral artery occlusion in anaesthetized baboons. *Brain Res.* **767,** 17–25.

Traversa, R., Cicinelli, P., Bassi, A., Rossini, P. M., and Bernardi, G. (1997). Mapping of motor cortical reorganization after stroke. A brain stimulation study with focal magnetic pulses. *Stroke* **28,** 110–117.

Ueda, T., Hatakeyama, T., Kumon, Y., Sakaki, S., and Uraoka, T. (1994). Evaluation of risk of hemorrhagic transformation in local intra-arterial thrombolysis in acute ischemic stroke by initial SPECT. *Stroke* **25,** 298–303.

Von Giesen, H. J., Schlaug, G., Steinmetz, H., Benecke, R., Freund, H. J., and Seitz, R. J. (1994). Cerebral network underlying unilateral motor neglect: Evidence from positron emission tomography. *J. Neurol. Sci.* **125,** 29–38.

Vorstrup, S., Engell, H. C., Lindewald, H., and Lassen, N. A. (1984). Hemodynamically significant stenosis of the internal carotid artery treated with endarterectomy. *J. Neurosurg.* **60,** 1070–1075.

Vorstrup, S., Lassen, N. A., Henriksen, L., Haase, J., Lindewald, H., Boysen, G., and Paulson, O. B. (1985). CBF before and after extracranial–intracranial bypass surgery in patients with ischemic cerebrovascular disease studied with ^{133}Xe-inhalation tomography. *Stroke* **16,** 616–626.

Vorstrup, S., Brun, B., and Lassen, N. A. (1986). Evaluation of the cerebral vasodilatory capacity by the acetazolamide test before EC–IC bypass surgery in patients with occlusion of the internal carotid artery. *Stroke* **17,** 1291–1298.

Watanabe, N., Young, A. R., Garcia, J. H., Liu, K.-F., Mézenge, F., Derlon, J.-M., Lassen, N. A., and Baron, J.-C. (1999). Focal cerebral ischaemia and chronic-stage reduced flumazenil uptake in anaesthetized baboons: A PET study with combined histological analysis. *J. Cereb. Blood Flow Metab.* **19**(1), 316.

Webster, M. W., Makaroun, M. S., Steed, D. L., Smith, H. A., Johnson, D. W., and Yonas, H. (1995). Compromised cerebral blood flow reactivity is a predictor of stroke in patients with symptomatic carotid artery occlusive disease. *J. Vasc. Surg.* **21,** 338–345.

Wise, R. J. S., Bernardi, S., Frackowiak, R. S. J., Legg, N. J., and Jones, T. (1983). Serial observations on the pathophysiology of acute stroke. The transition from ischaemia to infarction as reflected in regional oxygen extraction. *Brain* **106,** 197–222.

Wise, R. J. S., Gibbs, J., Frackowiak, R. S. J., Marshall, J., and Jones, T. (1986). No evidence for transhemispheric diaschisis after human cerebral infarction. *Stroke* **17,** 853–860.

Yamauchi, H., Fukuyama, H., Kimura, J., Konishi, J., and Kameyama, M. (1990). Hemodynamics in internal carotid artery occlusion examined by positron emission tomography. *Stroke* **21,** 1400–1406.

Yamauchi, H., Fukuyama, H., Fujimoto, N., Nabatame, H., and Kimura, J. (1992). Significance of low perfusion with increased oxygen extraction fraction in a case of internal carotid artery stenosis. *Stroke* **23,** 431–432.

Yamauchi, H., Fukuyama, H., Nagahama, Y., Nabatame, H., Nakamura, K., Yamamoto, Y., Yonekura, Y., Konishi, J., and Kimura, J. (1996). Evidence of misery perfusion and risk for recurrent stroke in major cerebral arterial occlusive diseases from PET. *J. Neurol., Neurosurg., Psychiatry* **61,** 18–25.

Yao, H., Sadoshima, S., Kuwabara, Y., Ichiya, Y., and Fujishima, M. (1990). Cerebral blood flow and oxygen metabolism in patients with vascular dementia of the Binswanger type. *Stroke* **21,** 1694–1699.

Yonas, H., Smith, H. A., Durham, S. R., Pentheny, S. L., Johnson, P. A. C., and Gur, D. (1993). Increased stroke risk predicted by compromised cerebral blood flow reactivity. *J Neurosurg.* **79,** 483–489.

Young, A. R., Sette, G., Touzani, O., Rioux, P., Derlon, J. M., MacKenzie, E. T., and Baron, J. C. (1996). Relationships between high oxygen extraction fraction in the acute stage and final infarction in reversible middle cerebral artery occlusion. An investigation in anaesthetized baboons with positron emission tomography. *J. Cereb. Blood Flow Metab.* **16,** 1176–1188.

14

The Epilepsies

Institute of Neurology, London WC1N 3BG, United Kingdom

II. Positron Emission Tomography Studies of
Cerebral Blood Flow and
Glucose Metabolism

III. Positron Emission Tomography Studies of
Specific Ligands

IV. Single Photon Emission
Computerized Tomography

V. Functional MRI

VI. Diffusion-Weighted MRI

VII. Magnetic Resonance Spectroscopy

VIII. Electrophysiological Imaging

References

I. Introduction

Magnetic resonance imaging (MRI) has been applied to the investigation of epilepsy for 15 years. The principal role of MRI is in the definition of structural abnormalities that underlie seizure disorders (Duncan, 1997). Hippocampal sclerosis (HS) may be reliably identified, and quantitative studies are useful for re-search and, in equivocal cases, for clinical purposes. A range of malformations of cortical development (MCD) may be determined. In patients with refractory partial seizures who are candidates for surgical treatment, a relevant abnormality is identifiable using MRI in 85% of cases; it is likely that subtle MCD or gliosis accounts for the majority of the remainder. The proportion of cryptogenic cases will decrease with improvements in MRI hardware, new and refined acquisition sequences, and postprocessing of data. A consequence of the advances in MRI is that the role of functional imaging in the clinical evaluation of patients with epilepsy has had to be redefined.

High-quality structural imaging with MRI is necessary for the understanding of the structural basis of functional disturbances and for the correlation of structural and functional data at voxel-based and region-based levels (Woods *et al.,* 1993; Duncan and Fish, 1998).

This review will consider positron emission tomography (PET) and then single photon emission computerized tomography (SPECT). This is followed by the MR techniques, MR spectroscopy (MRS), functional MRI (fMRI), and diffusion-weighted MRI. Finally, the electrophysiological imaging methods are considered: electroencephalograph (EEG) source localization and magnetoencephalography (MEG).

Brain Mapping: The Disorders

317

Copyright © 2000 by Academic Press.
All rights of reproduction in any form reserved.

II. Positron Emission Tomography Studies of Cerebral Blood Flow and Glucose Metabolism

A. Introduction

Positron emission tomography may be used to map cerebral blood flow, using ^{15}O-labeled water, and regional cerebral glucose metabolism, using [^{18}F]fluorodeoxyglucose (^{18}FDG). PET is costly and scarce but produces quantitative data with spatial resolution superior to that of SPECT. The radiation exposure from an ^{18}FDG PET scan is approximately 5 mSv, which is similar to that received from a series of [^{15}O]H$_2$O scans to determine areas of brain involved in a cerebral task. These exposures limit the comparative studies that may be carried out on patients and normal volunteers.

Positron emission tomography has been available for longer than MRI. MRI has made great strides in the past 10 years and this has had two major consequences for the application of PET techniques to epilepsy. First, the role of PET in the clinical investigation of patients needs to be re-evaluated, as contemporary MRI may result in PET data being superfluous. Second, clinical and research PET data should always be interpreted in the light of high-quality anatomical MRI, providing a structural-functional correlation. The development of computer programs to coregister MRI and PET data sets on a pixel-by-pixel basis has been fundamental to making these correlations, as demonstrated initially by Woods *et al.* (1993). Statistical parametric mapping has been shown to be useful in the evaluation of ^{18}FDG PET scans for clinical purposes, with the advantage of allowing a rapid and objective evaluation (Wong *et al.*, 1996; Signorini *et al.*, 1999).

B. Localization-Related Epilepsies

An epileptogenic focus, studied interictally, is associated with an area of reduced glucose metabolism, and reduced blood flow, that is usually considerably larger than the pathological abnormality (Engel *et al.*, 1982a; Franck *et al.*, 1986). Analysis of dynamic ^{18}FDG PET scans has shown that the hypometabolic epileptogenic area is also a region of reduced blood-brain barrier glucose transport (Cornford *et al.*, 1998). ^{18}FDG PET scans provide superior resolution and greater reliability for identifying a focal deficit than do PET scans using [^{15}O]water or SPECT scans of cerebral blood flow (Leiderman *et al.*, 1992; Ryvlin *et al.*, 1992a; Theodore *et al.*, 1994). There may be focal uncoupling of glucose metabolism and blood-brain barrier transport or blood flow in mesial temporal lobe epilepsy (Gaillard *et al.*, 1995c; Fink *et al.*, 1996). The degree of uncoupling appears to

increase with longer duration of epilepsy (Breier *et al.*, 1997).

The most likely reason for the large region of reduced blood flow and metabolism around an epileptogenic focus is inhibition or deafferentation of neurons. Comparison of ^{18}FDG PET scans with [^{11}C]flumazenil scans (Henry *et al.*, 1993a; Savic *et al.*, 1993) indicates that neuronal loss is confined to a more restricted area than the region of reduced metabolism.

Partial seizures are associated with an increase in regional cerebral glucose metabolism and blood flow in the region of the epileptogenic focus and often a suppression elsewhere (Engel *et al.*, 1983; Chugani *et al.*, 1994). In general, ictal PET scans can only be obtained fortuitously, because of the 2-min half-life of ^{15}O and the fact that cerebral uptake of ^{18}FDG occurs over 40 min after injection, so that cerebral glucose utilization data will reflect an amalgam of the ictal and postictal conditions. Hypometabolism is accentuated after a seizure and may not return to the interictal state for more than 24 h (Leiderman *et al.*, 1994). In a patient with fixation-off sensitivity, interictal spiking produced focal increases in cerebral blood flow and glucose metabolism (Bittar *et al.*, 1999), but this was not found in patients with benign childhood epilepsy with centrotemporal spikes (Van Bogaert *et al.*, 1998a).

Attempts have been made to obtain ictal H$_2$^{15}O PET scans using intravenous pentylenetetrazole to induce seizure activity. A high proportion of seizures was associated with bilateral increases in cerebral blood flow (CBF) and the results were not clear-cut (Theodore *et al.*, 1996). This does not appear to be a promising technique to develop further.

Over the past decade, some epilepsy surgery programs have relied extensively on ^{18}FDG PET as a tool for localizing the epileptic focus. The current place of this technique needs to be reevaluated in the light of developments in MRI as the finding of a definite focal abnormality with the latter technique, such as HS, may render an ^{18}FDG PET scan superfluous (Heinz *et al.*, 1994; Gaillard *et al.*, 1995a). Comparative studies of two imaging techniques, such as MRI and PET, are difficult to interpret, however, because of the likelihood that the methodologies and equipment will not both be developed to the same degree in any one center.

1. Temporal Lobe Epilepsy

Several studies of ^{18}FDG PET have found a 60–90% incidence of hypometabolism in the temporal lobe interictally in adults and children with temporal lobe epilepsy (TLE) (Kuhl *et al.*, 1980; Engel *et al.*, 1982a,b; Theodore *et al.*, 1983; Franck *et al.*, 1986; Abou-Khalil *et al.*, 1987; Stefan *et al.*, 1987a; Ryvlin *et al.*, 1991; Sadzot *et al.*, 1992; Gaillard *et al.*, 1995b). In studies carried out before the recent developments in assessment of the hippocampus using MRI, [18]FDG PET was thought to be a highly useful clinical tool, particularly in patients in whom scalp EEG recordings were nonlocalizing (Engel et al., 1990; Theodore et al., 1992a). The results of comparative studies depend critically on the relative sophistication of the techniques used. In a recent comparative study of patients with temporal lobe epilepsy, it could be concluded that [18]FDG PET data did not provide clinically useful data if the MRI findings were definite, but had some additional sensitivity (Gaillard et al., 1995a). The same authors concluded that FDG PET may reduce the numbers of patients requiring invasive EEG studies (Theodore et al., 1997). MRI is critical in such a decision, and in our current practice the absence of a demonstrated structural pathology would mandate invasive EEG studies. In a comparative study of patients with TLE, FDG PET found lateralized abnormalities but did not predict good postoperative outcome in the absence of asymmetric hippocampal volumes (Knowlton et al., 1997a).

Visual assessment of hypometabolism is less accurate than quantitation. The asymmetry of lateral, but not mesial temporal, glucose metabolism was greater in patients who became seizure free following surgery. Patients with at least 15% hypometabolism were more likely to become seizure free (Theodore et al., 1992a). The degree and extent of temporal lobe hypometabolism have been strongly correlated with the seizure outcome following temporal lobectomy (Radtke et al., 1993; Delbeke et al., 1996). In a similar study, the degree of hypometabolism in the medial temporal lobe, but not the lateral temporal lobe, correlated with the chances of being rendered seizure free by anterior temporal lobe resection (Manno et al., 1994). Absence of unilateral temporal hypometabolism, however, does not preclude a good result from surgery (Chee et al., 1993). Not surprisingly, bilateral temporal hypometabolism was associated with a poor prognosis for seizure remission after surgery (Blum et al., 1998).

Interictal temporal slow waves on the EEG have been shown to be associated with lateral temporal hypometabolism, both being nonspecific signs of neuronal dysfunction (Koutroumanidis et al., 1998). The area of hypometabolism often extends beyond the temporal lobe. In a series of 27 patients, with no specific abnormality evident on the MRI available at the time, this involved the thalamus in 63%, basal ganglia in 41%, and frontal lobe in 30% (Henry et al., 1993b). Similar results were found in other series (Sperling et al., 1990). In patients with medial temporal lobe foci, the hypometabolism has frequently been found to be more pronounced in the lateral neocortex (Sackellares et al., 1990; Sadzot et al., 1992). This finding may reflect the limited spatial resolution of the cameras used; but, in consequence, [18]FDG PET studies have been thought less reliable for answering the question of precise localization of seizure onset (e.g., inferior frontal vs medial or lateral temporal lobe) than for answering the question of lateralization (e.g., left or right temporal lobe). In an evaluation of [18]FDG PET in patients with temporal lobe epilepsy and different pathologies, those with HS had the lowest glucose metabolism in the whole temporal lobe, followed by patients whose seizures arose laterally. Patients with mesiobasal tumors generally had only a slight reduction of glucose uptake in the temporal lobe. The metabolic pattern was different between patients with mesial and lateral temporal seizure onset, but there was not a clear correlation between the location of the defined epileptogenic focus defined with EEG and the degree of hypometabolism (Hajek et al., 1993). Studies such as this are of interest in understanding the metabolic consequences of epilepsy of different etiologies, but from a clinical perspective have been rendered less useful by modern MRI.

In one study, hippocampal neuron loss was not correlated with the severity of temporal lobe hypometabolism (Henry et al., 1994). In contrast, a comparison of hippocampal atrophy and glucose metabolism in patients with temporal lobe epilepsy found that there was most marked hypometabolism in the hippocampus and temporal pole. The degree of hippocampal atrophy correlated with the degree of hypometabolism in these structures, but not elsewhere in the temporal lobe (Semah et al., 1995). In a separate series, there was a significant correlation between hippocampal volume and inferior mesial and lateral temporal lobe cerebral metabolic rate of glucose, suggesting that hypometabolism may reflect hippocampal atrophy (Gaillard et al., 1995a). In another series, however, temporal hypometabolism and hippocampal atrophy were not correlated (O'Brien et al., 1997). Thus there is not a consensus on this point.

In the presurgical evaluation of temporal lobe epilepsy, [18]FDG PET scans provide superior resolution and greater reliability for identifying a focal deficit than do PET scans using $H_2^{15}O$ (Leiderman et al., 1992; Theodore et al., 1994).

False lateralizations may occur with [18]FDG PET in patients with temporal lobe epilepsy. The risk of false lateralization may be reduced by undertaking quantitative rather than just qualitative assessments of scans. An area of reduced glucose metabolism indicates a focal deficit in function and this does not necessarily imply epileptogenicity, which must still be verified neurophysiologically. Another important potential source of error may be seizure activity during the scan that is unrecognized clinically or on scalp EEG recordings (Sperling et al., 1995).

The degree of left hemisphere hypometabolism has been correlated with impairment of verbal IQ, and lateral left temporal lobe hypometabolism with impairment of verbal memory, in patients with mesial temporal lobe epilepsy (Rausch et al., 1994). Patients with left temporal lobe epilepsy and secondarily generalized seizures were more likely to have prefrontal metabolic asymmetry, implying dysfunction in this area (Jokeit et al., 1997). Ictal dystonia in mesial TLE is associated with hypometabolism in the basal ganglia, supporting involvement of the latter structures in this feature (Dupont et al., 1998).

After selective amygdalohippocampectomy in patients with temporal lobe epilepsy, there was an increase of regional cerebral glucose metabolism in the ipsilateral and also the contralateral hemisphere in patients with mesiobasal temporal lobe epilepsy and HS. There was also a trend toward a normalization of cerebral glucose metabolism in the ipsilateral temporal neocortex 12 months after surgery in such patients, implying a consequence of the removal of the epileptic focus (Hajek et al., 1994).

Cavernomas may be associated with an area of hypometabolism around the lesion, particularly in overlying temporal neocortex if connections from limbic structures are disrupted. This finding did not correlate with the size of the abnormality, or it being associated with epilepsy, implying that the hypometabolism was a consequence of deafferentation (Ryvlin et al., 1995).

In patients with cryptogenic temporal lobe epilepsy who had never received antiepileptic drugs, significantly increased glucose metabolism was found in the temporal lobes, thalami, basal ganglia, and cingulate cortex and also in the frontal, mesial temporal, and cerebellar cortices (Franceschi et al., 1995). Hypermetabolic areas have also been found in children with partial seizures who were not having overt seizures. In some, this was associated with focal spike-wave activity, indicating the likelihood of increased neuronal activity underlying the increased glucose metabolism and the need to have EEG recordings during [18]FDG PET studies (Chugani et al., 1993). An alternative explanation is that the increased glucose metabolism reflects increased neuronal numbers, as may occur in a focal MCD, resulting in thickening of the neocortical ribbon (Richardson et al., 1997a).

[18]FDG PET scans have also been used to investigate the metabolic consequences of partial epilepsy and its treatment. The administration of the $GABA_A$ receptor agonist THIP to patients with temporal lobe epilepsy was followed by an increase in glucose metabolism, measured with [18]FDG PET, in areas with reduced interictal glucose metabolism that was greater than the increase seen in the brain as a whole (Peyron et al., 1994a,b).

These data suggested that activation of GABAergic neurons results in increased metabolic demands and that $GABA_A$ receptors are not lost in the neocortex in temporal lobe epilepsy. Regions of interest were placed in medial and lateral temporal lobe as the spatial resolution of the study was not sufficient to identify the hippocampus separately, but the results were consistent with the results of studies of ^{11}C]flumazenil binding in HS (Koepp et al., 1996a, 1997b). In patients with complex partial seizures, interictal cerebellar metabolism was reduced bilaterally, and this did not appear to be entirely attributable to the usage of phenytoin (Theodore et al., 1987a).

2. Frontal Lobe Epilepsy

[18]FDG PET shows hypometabolism in about 60% of patients with frontal lobe epilepsy. In 90% of those with a hypometabolic area, structural imaging shows a relevant underlying abnormality. In common with temporal lobe epilepsy, the area of reduced metabolism in frontal lobe epilepsy may be much larger than the pathological abnormality. In contrast, however, the hypometabolic area may be restricted to the underlying lesion (Henry et al., 1991; Engel et al., 1995).

There have been three main patterns of hypometabolism described in patients with frontal lobe epilepsy:

- ▲ No abnormality
- ▲ A discrete focal area of hypometabolism
- ▲ Diffuse widespread hypometabolism. This may be multilobar and involve subcortical structures. There is often an area of particularly reduced metabolism in one part of the frontal lobe.

An uncommon finding has been the patient with a multilobar or extensive structural lesion and a small area of hypometabolism (Swartz et al., 1989). It has been the general experience that the epileptic focus is contained within the hypometabolic area, if one exists (Theodore et al., 1986; Swartz et al., 1989; Henry et al., 1991; Sadzot et al., 1992). Quantitative analysis of [18]FDG PET scans was found to enhance sensitivity and accuracy of determination over visual assessment (Swartz et al., 1995). Analysis of groups of patients with neocortical seizures showed that those with unilateral clonic seizures had principally contralateral perirolandic hypometabolism and a lesser degree of contralateral frontomesial hypometabolism. Those with focal tonic seizures had hypometabolism within the frontomesial and perirolandic regions that was unilateral in all patients with lateralized tonic seizures. Patients with versive seizures had mainly contralateral hypometabolism without a consistent pattern. Patients in all groups had bilateral hypometabolism of the thalamus and cerebellum (Schlaug et al., 1997).

These patterns imply the cerebral areas involved in particular features of seizures and seizure spread.

Overall, published clinical series indicate that [18]FDG PET does not appear to provide additional clinically useful information in the majority of patients with frontal lobe epilepsy. Future research will determine whether studies with specific ligands may have a greater ability to identify focal abnormalities in patients in whom structural imaging has not revealed an etiology for the seizure disorder.

Status epilepticus, arising from the motor cortex, has been associated with hypermetabolism and also with hypometabolism (Engel et al., 1983; Franck et al., 1986; Hajek et al., 1991). Scans performed with $H_2^{15}O$ have shown increased blood flow and oxygen consumption and reduced oxygen extraction fraction in the frontal lobe in patients with epilepsia partialis continua (Franck et al., 1986). A rare finding, in patients with frontal epilepsy of early childhood onset, has been an interictal focal increase in metabolism (Chugani et al., 1993).

Dysfunctional working memory in patients with frontal lobe epilepsy has been associated with impaired task-related metabolic response (Swartz et al., 1996a). In view of the limitations of PET activation studies, however, functional MRI investigations will become the method of choice to resolve the functional anatomy of cognitive deficits.

3. Malformations of Cortical Development

Focal cortical hypometabolism is commonly seen in patients with tuberous sclerosis (Szelies et al., 1983; Pawlik et al., 1990). Glucose metabolism has been detected using [18]FDG PET in the layers of ectopic neurons in band heterotopia (Miura et al., 1993; De Volder et al., 1994) and in heterotopic nodules and displaced gray matter (Bairamian et al., 1985; Falconer et al., 1990), implying synaptic activity.

In 15 of 17 patients with MCD, both MRI and [18]FDG PET identified ectopic gray matter and hypometabolism concurred with MRI findings of abnormal cortex. Analysis was by visual inspection and [18]FDG PET did not identify abnormalities that were not evident on MRI, although in some cases the area of hypometabolism was more extensive than the MRI lesion (Lee et al., 1994). Abnormalities of glucose metabolism, identified with [18]FDG PET, in the contralateral hemisphere in patients with hemimegalencephaly have been associated with a less good prognosis following surgery (Rintahaka et al., 1993). In perisylvian dysgenesis, there was a heterogeneous pattern, with areas of relatively normal metabolism and of reduction in areas of polymicrogyria and in nearby cortex that appeared normal on MRI (Van Bogaert et al., 1998b).

4. Other Syndromes

Some patients with severe secondarily generalized epilepsies and syndromes such as West's syndrome, Lennox-Gastaut syndrome, Landau-Kleffner syndrome, and other pediatric epilepsy syndromes appear to be of cryptogenic etiology on MRI. There may, however, be single or multiple cortical areas of hypometabolism on [18]FDG PET scans which indicate areas that may be epileptogenic, the pathological correlate of which is MCD (Chugani, 1994). Patients with the Lennox-Gastaut syndrome have shown a variety of cerebral metabolic patterns: normal, focal, and diffuse (unilateral and bilateral) hypometabolism (Chugani et al., 1987; Theodore et al., 1987b), with a predominance of temporal lobe abnormalities (da Silva et al., 1997). This heterogeneity is not surprising and reflects the diverse etiologies of the syndrome. Operative electrocorticography studies have found epileptic activity in areas shown to be hypometabolic on [18]FDG PET scans (Olson et al., 1990). In children with apparently cryptogenic epileptic encephalopathies, [18]FDG PET scans have shown focal metabolic abnormalities in a third and also diffuse cortical and thalamic abnormalities (Ferrie et al., 1997), indicating the widespread nature of the disturbance, and the patterns were stable over time (Parker et al., 1998).

Continuous spike-wave discharges may be associated with a focal increase in glucose metabolism (Park et al., 1994). In one study of patients with the Landau-Kleffner syndrome, the metabolic pattern was variable but abnormalities were most evident over the temporal lobes (Maquet et al., 1990). In another, widespread bilateral hypometabolism or a mild increase in glucose metabolism was noted in the left inferior temporal cortex in the awake state. In scans performed during slow-wave sleep, there was a marked bilateral increase in glucose metabolism in the temporal cortex, generating the hypothesis that the increase in glucose metabolism reflected areas of epileptic activity (Rintahaka et al., 1995).

5. Studies of the Effect of Antiepileptic Drugs

The effects of antiepileptic drugs on cerebral glucose metabolism have also been studied with [18]FDG PET. Carbamazepine was associated with a diffuse 12% reduction, valproate a 22% reduction, phenytoin a nonuniform, mean 13% reduction, and phenobarbitone a 37% reduction in cerebral glucose metabolism (Theodore, 1988; Theodore et al., 1989; Leiderman et al., 1991). It was also reported that valproate was associated with a reduced blood flow, but no reduction of glucose metabolism in the thalamus (Gaillard et al., 1996).

6. Activation Studies

Activation studies using ^{18}FDG PET have been been reported in attempts to highlight hypometabolic areas. They may improve the delineation of dysfunctional areas of brain (Pawlik *et al.*, 1990, 1994; Bromfield *et al.*, 1991). These studies have the severe limitation of cerebral uptake of ^{18}FDG continuing over at least 15 min.

Activation studies performed with $H_2^{15}O$ have been used for some years to delineate the functional anatomy of a variety of cognitive and other cerebral tasks. These may be used in the surgical planning in patients with epileptogenic areas close to eloquent areas (Hojo *et al.*, 1995), particularly the language cortex (Bookheimer *et al.*, 1997). Deficits may also be demonstrated. Impaired naming in patients with left temporal lobe epilepsy has been associated with impaired activation of the left fusiform gyrus on PET studies (Henry *et al.*, 1998a). Following right functional hemispherectomy, motor activation studies using the weak hand have shown involvement of the motor areas of the retained hemisphere and the cerebellum bilaterally (Muller *et al.*, 1998a)

$H_2^{15}O$ PET studies of cognitive activation tasks involving motor learning, visual attention, and other tasks in patients with MCD have shown that malformed cortex may participate in higher cerebral functions. The studies also showed widespread atypical cortical organization, indicating that there may be extensive disorganization of normal structure—function correlates in these patients that would have implications for the planning of surgical resections (Muller *et al.*, 1998b; Richardson *et al.*, 1998a).

As in healthy control subjects and patients with other pathologies, cognitive activation tasks in those with epilepsy will increasingly be carried out using fMRI rather than $H_2^{15}O$ PET.

7. Vagal Nerve Stimulation

Repetitive electrical stimulation of the left vagus nerve by means of an implanted stimulator is a palliative procedure for refractory epilepsy. PET studies of cerebral blood flow have shown that the stimulation was associated with increases in CBF in the rostral, dorsal-central medulla, the right postcentral gyrus, the bilateral hypothalami, thalami, and insular cortices, and the inferior cerebellar hemispheres and decreases in the bilateral hippocampi, amygdalae, and posterior cingulate gyri (Ko *et al.*, 1996; Henry *et al.*, 1998b). This suggested that vagal nerve stimulation increases synaptic activity in structures directly innervated by central vagal structures and those that process left-sided somatosensory information, but also affects synaptic activity in multiple bilateral limbic structures, which may reflect sites of therapeutic actions of vagal nerve stimulation, although the mechanisms remain unclear.

C. Generalized Epilepsy

There have been three studies of cerebral glucose metabolism using ^{18}FDG and PET in patients with idiopathic generalized epilepsy (IGE). Interictal studies have been unremarkable; studies carried out when frequent absences were occurring have shown a diffuse increase in cerebral glucose metabolism of 30–300%, with greater increases being seen in children. Absence status, however, was associated with a reduction in cerebral glucose metabolism. There were no focal abnormalities and the rate of metabolism did not correlate with the amount of spike-wave activity (Engel *et al.*, 1985; Theodore *et al.*, 1985; Ochs *et al.*, 1987). ^{18}FDG PET studies have poor temporal resolution, 70–80% of cerebral uptake occurring over 15 min following intravenous injection, with the consequence that in a patient with frequent absences, there is an amalgam of preictal, ictal, and postictal periods contributing to one scan.

In an autoradiographic study of cerebral glucose utilization in a rat model of spontaneous generalized absences, there was a widespread increase in cerebral glucose metabolism, compared with controls. The relationship between spike-wave activity and cerebral glucose utilization is not straightforward, however, as a dose of ethosuximide that suppressed all spike-wave activity was not associated with a reduction of glucose metabolism (Nehlig *et al.*, 1991, 1992).

Regional cerebral blood flow (rCBF) has been assessed using PET and bolus injections of ^{15}O-labeled water, achieving a temporal resolution of <30 s and a spatial resolution of $8 \times 8 \times 4.3$ mm. Typical absences were precipitated with hyperventilation and the distribution of rCBF was compared when absences did and did not occur. During absences there was a global 15% increase in CBF and, in addition, a focal 4–8% increase in blood flow in the thalamus. There were no focal increases of CBF in the cortex and no focal decreases (Prevett *et al.*, 1995a). Spike-wave activity in typical absences oscillates in thalamocortical circuits and the site of primary abnormality remains uncertain. The preferential increase in thalamic blood flow is evidence to the key role of this structure in the pathophysiology of absences in man but does not clarify whether this is the result of converging activated thalamocortical pathways or reflects a primary thalamic process. No significant increases in thalamic blood flow were detected in the 30 s prior to generalized spike-wave activity on the EEG. A focal increase in blood flow developing 5 s before generalized spike-wave activity appeared would be most unlikely to be detected with this technique. Functional MRI studies have superior temporal resolution but have the similar limitations of the temporal resolution of the hemodynamic response.

Patients with juvenile myoclonic epilepsy (JME) did not show increased frontal metabolism on visual working memory tasks (Swartz *et al.*, 1996b). fMRI will be a more potent tool to investigate issues of frontal lobe function in patients with JME.

D. Conclusion

Studies with ^{18}FDG PET have defined the major cerebral metabolic associations and consequences of epilepsy but the data are nonspecific with regard to etiology and abnormalities are more widespread than the pathological lesions. The role of ^{18}FDG PET in the clinical evaluation of patients has been reduced by the advances made in MRI over the past 5 years. Activation studies with $H_2{}^{15}O$ are useful for determining the functional anatomy of cerebral processes in both healthy and pathological brains, but these studies are now increasingly performed with functional MRI.

III. Positron Emission Tomography Studies of Specific Ligands

A. Introduction

Positron emission tomography may be used to demonstrate the binding of specific ligands, for example, [^{11}C]flumazenil to the central benzodiazepine-GABA$_A$ receptor complex, [^{11}C]diprenorphine and [^{11}C]carfentanil to opiate receptors, and [^{11}C]deprenyl to MAO-B receptors. The technique is costly and scarce but gives quantitative data with spatial resolution superior to that of single photon emission computerized tomography.

B. Localization-Related Epilepsies

The main findings in localization-related epilepsies are summarized in Table I.

1. Central Benzodiazepine Receptors

The most important inhibitory transmitter in the central nervous system, γ-aminobutyric acid (GABA), acts

Table I Principal Findings of PET Receptor Studies in Localization-Related Epilepsies

Ligand	Labels	Epilepsy	Findings
[^{11}C]Flumazenil	Central benzodiazepine receptor	Temporal lobe epilepsy–hippocampal sclerosis	Binding decreased over and above atrophy, loss of neurons; may be bilateral, with unilateral MRI evidence of HS
		TLE, MRI-negative	Decreased binding in 30%, occasional increases
		Acquired lesions	Decreased binding; may exceed MRI-visible lesion
		Malformation of cortical development	Decreases and increases of binding
		Extratemporal lobe, MRI–negative	Decreases and increases in 70%
[^{11}C]Diprenorphine	Opioid receptors (mu, kappa, delta)	Temporal lobe epilepsy	No asymmetry
[^{18}F]Cyclofoxy	Opioid receptors (mu, kappa)	Temporal lobe epilepsy	No asymmetry
[^{11}C]Carfentanil	Opioid receptors (mu)	Temporal lobe epilepsy	Increases in lateral temporal neocortex
[^{11}C]Methylnaltrindole	Opioid receptors (delta)	Temporal lobe epilepsy	Increases in mid- and superior temporal neocortex
[^{11}C]Deprenyl	Monoamine oxidase B	Temporal lobe epilepsy	Increases in temporal lobe
[^{11}C]Doxepin	Histamine H_1 receptors		Increases in region of focus
[^{11}C]PK11195	Peripheral benzodiazepine receptors	Rasmussen's encephalitis	Increases in affected hemisphere

at the $GABA_A$-central benzodiazepine receptor (cBZR) complex to increase chloride conductance and thereby to hyperpolarize the resting membrane potential (Meldrum et al., 1989). Flumazenil is a specific, reversibly bound antagonist at the α-subunit types 1, 2, 3, and 5 of the cBZR (Olsen et al., 1990) and [^{11}C]flumazenil is a positron emission tomography ligand that acts as a useful marker of the $GABA_A$ cBZR complex in vivo (Maziere et al., 1984; Samson et al., 1985).

The binding of [^{11}C]flumazenil to central benzodiazepine receptors in epileptogenic foci was initially found to be reduced by an average of 30%, with no change in the affinity of the ligand for the receptor (Savic et al., 1988). Other groups have also found this and comparative studies with ^{18}FDG PET scans have shown the area of reduced [^{11}C]flumazenil binding to be more restricted than is the area of reduced glucose metabolism in temporal lobe epilepsy (Henry et al., 1993a; Savic et al., 1993; Szelies et al., 1996; Debets et al., 1997). Using the pixel-based method of statistical parametric mapping (SPM) (Friston et al., 1995) applied to parametric images of cerebral [^{11}C]flumazenil binding coregistered with MRI, Koepp et al. (1996a) found a reduction of cBZR confined to the sclerotic hippocampus, with no significant abnormalities elsewhere, in patients with unilateral hippocampal sclerosis.

Although a pixel-based analysis identifies areas of brain where there is a focal increase or decrease of receptor binding, the method does not quantify the changes and does not differentiate between abnormalities that are the result of changes in the amount of gray matter or of receptor density. Quantification of PET data requires a volume of interest based approach and, in view of the size of the hippocampus and the limited spatial resolution of PET, correction for partial volume effect (Muller-Gartner et al., 1992; Meltzer et al., 1996; Labbe et al., 1998).

Correction of partial volume effect resulted in increased sensitivity of [^{11}C]flumazenil PET in detection of unilateral hippocampal sclerosis (Koepp et al., 1997b) and also identified bilateral hippocampal abnormalities of cBZR in one-third of patients who appeared to have unilateral HS on MRI (Koepp et al., 1997c). Furthermore, after correction of partial volume effect, [^{11}C]flumazenil binding in vivo in hippocampal sclerosis in patients having temporal lobe resections was reduced by a mean of 38%, indicating that the loss of binding was not simply due to hippocampal atrophy (Koepp et al., 1997b).

The underlying pathological basis of reduced [^{11}C]flumazenil binding in hippocampal sclerosis has been extensively investigated. Autoradiographic and histopathological studies of surgically removed sclerotic hippocampi have shown reduced neuron counts and cBZR densities, with a further reduction of density of BZR per remaining neuron (Johnson et al., 1992). Burdette found a statistically highly significant correlation coefficient between neuron loss and reduced cBZR binding on autoradiographic analysis of hippocampi from patients treated surgically for hippocampal sclerosis and concluded that neuron loss was the basis of reduced [^{11}C]flumazenil binding in vivo (Burdette et al., 1995). This analysis did not exclude an additional reduction of cBZR per neuron. More recently, quantitative autoradiographic and quantitative neuropathological studies of resected HS showed that cBZR density (B_{max}) was reduced in the CA1 subregion of the hippocampus, over and above the loss of receptors that was attributable to neuron loss. In other hippocampal subregions, loss of receptors paralleled loss of neurons and increases in affinity were noted in the subiculum, hilus, and dentate gyrus (Hand et al., 1997). A direct comparison of quantitative in vivo hippocampal [^{11}C]flumazenil binding and ex vivo quantitative [^{3}H]flumazenil autoradiographic studies showed a mean 42% reduction of the two measures in 10 patients with hippocampal sclerosis, compared with control material, and a good correlation between the in vivo and ex vivo measures in individual patients (Fig. 1). This study demonstrated that reduction of available cBZR on remaining neurons in HS can be reliably detected in vivo using [^{11}C]-flumazenil PET, after correction for partial volume effect (Koepp et al., 1998a).

It seems most likely that cBZR changes reflect localized neuronal and synaptic loss in the epileptogenic zone and that the more extensive hypometabolism is a result of diaschisis (Feeney and Baron, 1986). In clinical terms, [^{11}C]flumazenil PET may be superior to ^{18}FDG for the localization of the source of the seizure. It is not certain whether these data confer additional clinically useful information in patients with clear-cut MRI findings of unilateral hippocampal sclerosis. In a clinical series of 100 patients with partial seizures having presurgical evaluation, 94% of those with temporal lobe epilepsy had an abnormality of [^{11}C]flumazenil PET detected, as did 50% of those with other forms of partial epilepsy. Eighty-one percent of abnormalities found using [^{11}C]flumazenil PET were concordant with abnormalities on MRI. [^{11}C]Flumazenil PET was useful in the identification of bilateral temporal lobe pathology (Ryvlin et al., 1998).

Malformations of cortical development commonly underlie partial seizures, are of heterogenous appearance, and may not be detectable on MRI (Desbiens et al., 1993). Epilepsy surgery in patients with MCD is less successful than with discrete lesions, most likely because the anatomical and functional abnormality is more widespread than a feasible resection. An SPM

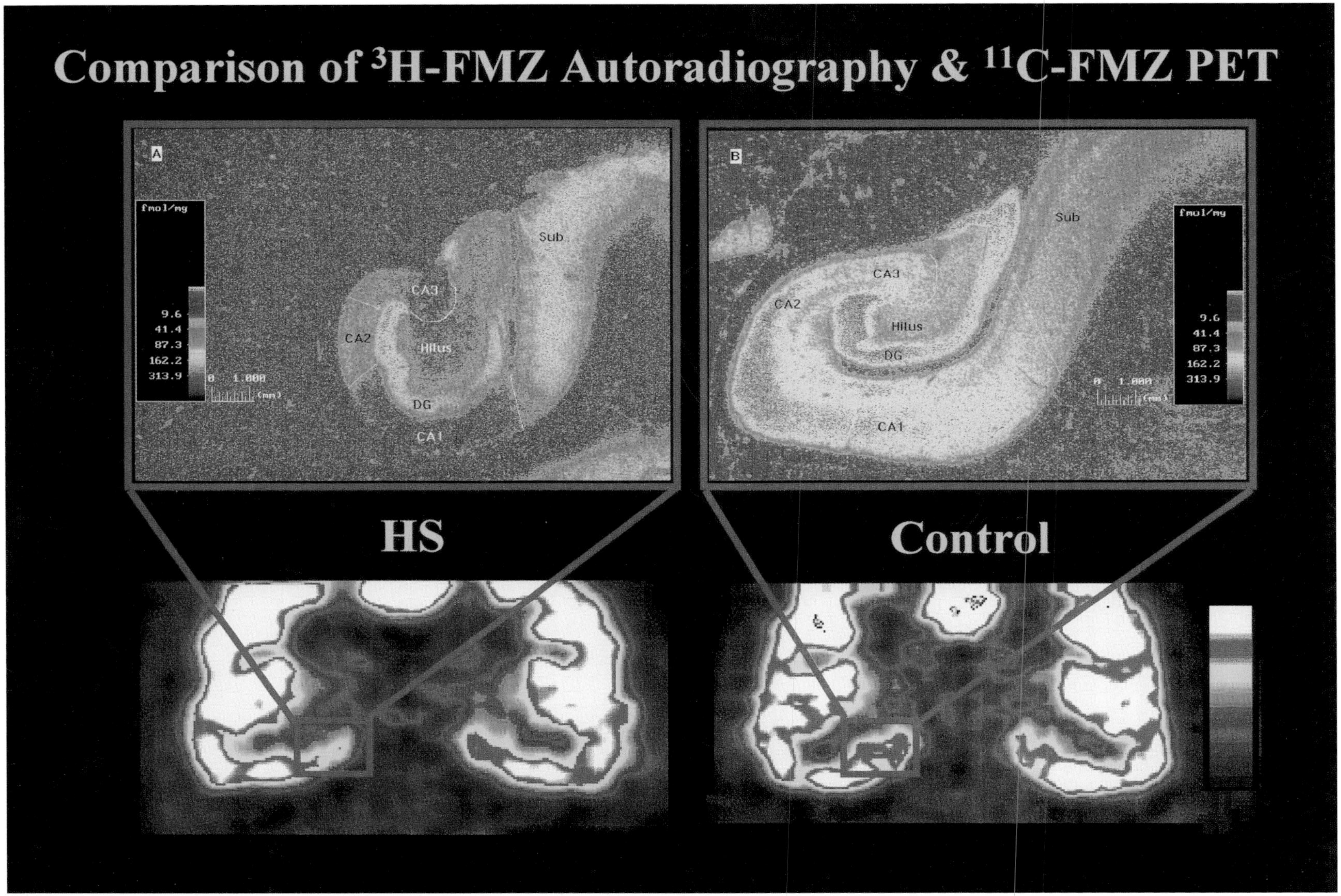

Comparison of ³H-FMZ Autoradiography & ¹¹C-FMZ PET

Figure 1 Comparison of cross-sectional views of the hippocampal body with [³H]flumazenil ([³H]FMZ) autoradiography (top row) and coronal views through the hippocampal body with [¹¹C]FMZ positron emission tomography ([¹¹C]FMZ PET; bottom row) from a patient with hippocampal sclerosis (HS; left) and normal control material (right). The color scale in the upper row reflects [³H]FMZ B_{max}, and in the lower row [¹¹C]FMZ V_d. B_{max} = receptor availability; V_d = volume of distribution. Reprinted with permission from Koepp et al. (1998a). *In vivo* ¹¹C-flumazenil PET correlates with *ex vivo* ³H-flumazenil autoradiography in hippocampal sclerosis. *Ann. Neurol.* **43**, 618–626.

analysis of cBZR visualized with [^{11}C]flumazenil PET and coregistered with high-quality MRI in 12 patients with partial seizures and MCD found areas of abnormal cerebral cBZR binding in 10. The abnormal regions were frequently more extensive than the abnormality seen with MRI and were also noted in distant sites at which the cortex appeared unremarkable on MRI (Richardson *et al.*, 1996). In contrast to studies with ^{18}FDG, in which reduced metabolism may be the result of diaschisis, reduced [^{11}C]flumazenil binding implies neuronal deficits. A further, novel finding of this investigation was of areas of increased binding to cBZR in many patients with MCD, which has not been found in patients with epilepsy caused by other pathologies. Possible explanations of this finding include increased neuron density, the presence of ectopic neurons bearing cBZR, and an increased number of available receptors, which may reflect abnormal neurons or a response to the abnormal circuitry implicit in MCD, with a change in available receptor numbers and/or affinity.

A subsequent voxel-based comparison of the binding of flumazenil to cBZR with the distribution of cortical gray matter in 10 patients with MCD showed that some regions with abnormal [^{11}C]flumazenil binding were accounted for by abnormalities of the cortical gray matter volume. In other areas, there was disproportionate [^{11}C]flumazenil binding, compared to the local gray matter, including areas where analysis of the PET data alone did not reveal an abnormality, implying abnormal receptor density per neuron or a change in affinity (Richardson *et al.*, 1997a). A complementary region-based analysis of partial volume corrected [^{11}C]flumazenil binding to determine the cBZR binding per unit volume of dysgenetic cortex has not yet been performed. These results underline the importance of interpreting PET data in the light of high-resolution structural imaging.

Elucidation of the neurobiological basis of these findings awaits correlative pathological and *in vitro* receptor studies. These findings may also be of clinical importance in the evaluation of patients with MCD for possible surgical treatment. Further studies need to be done to determine whether the finding of widespread abnormalities of cBZR is an adverse prognostic factor for surgical outcome in patients in whom MRI and other investigatory data implicate a single restricted focus, although these data will not be easy to acquire.

The utility of [^{11}C]flumazenil PET in the clinical investigation of patients with refractory partial seizures and unremarkable high-quality MRI is currently under investigation. In six patients with frontal lobe epilepsy, reduction in BZR binding was demonstrated with [^{11}C]flumazenil PET, which was consistent with clinical and EEG data (Savic *et al.*, 1995). In two of four patients who also had ^{18}FDG scans, the focus was charac-

terized by a region of reduced metabolism that was more extensive than the reduction in cBZR. The MRI was unremarkable in five, implying superior sensitivity of the PET technique over MRI. Comparisons of this nature, however, clearly depend on the relative sophistication of the instruments and methodologies used.

In 18 patients with refractory extratemporal partial seizures and unremarkable high-resolution MRI, [^{11}C]flumazenil PET showed focal decreases in 6 and focal increases in binding in 10 (Richardson *et al.*, 1998b). In patients with temporal lobe epilepsy and normal MRI, [^{11}C]flumazenil PET showed focal decreases and focal increases in binding in 50% (Koepp *et al.*, 1998b). The clinical significance of these findings is not yet clear.

In six patients with partial seizures as a result of acquired lesions, however, only decreases in flumazenil binding were seen (Richardson *et al.*, 1998b). The implication of these data is that focally increased [^{11}C]flumazenil binding is a marker of MCD and may indicate occult MCD in patients who are MRI-negative. At the present, there is no correlative neuropathological material available to confirm or refute this.

Analysis of groups of patients who have similar forms of epilepsy is also of interest. In a group of 10 patients with frontal lobe epilepsy, there was an increase in flumazenil binding in the putamen, particularly ipsilateral to the seizure focus, and in related motor areas. Furthermore, the extent of the increase was inversely related to the seizure frequency, suggesting the possibility that increased inhibition in the basal ganglia may modulate neocortical seizure threshold (Richardson *et al.*, 1997b). In this study, there was not a correlation between extent of abnormality of neocortical flumazenil binding and seizure frequency. In contrast, Savic *et al.* (1996) reported in 19 patients with partial seizures and normal MRI that the severity of cBZR reduction, in regions of interest that were placed visually on the PET images, correlated with seizure frequency. Furthermore, in those patients with daily seizures, cBZR was also reduced in the primary projection areas of the focus. The same authors (Savic *et al.*, 1998) have reported that, in four patients who had anterior temporal lobe resections for refractory epilepsy, cBZR was reduced in primary projection areas preoperatively and was normalized months postoperation. This raised the possibility of dynamic changes in cBZR in relation to seizures, the basis of which is not clear.

In a series of 10 patients with MRI-negative temporal lobe epilepsy, one-third of individuals had focal abnormalities of flumazenil binding in the hippocampus using SPM and, when analyzed as a group, reduced flumazenil binding in the hippocampus on the side of the seizure focus (Koepp *et al.*, 1996b). Preliminary data suggest that an analysis using a template of regions with partial volume correction will give additional useful information in the evaluation of MRI-negative patients.

In a clinical series of 100 patients with partial seizures having presurgical evaluation, 94% of those with temporal lobe epilepsy had an abnormality on [^{11}C]flumazenil PET detected, as did 50% of those with other forms of partial seizures. Eighty-one percent of abnormalities on [^{11}C]flumazenil PET scans were concordant with abnormalities on MRI. [^{11}C]flumazenil PET was useful in the identification of bilateral temporal lobe pathology and in patients with cryptogenic frontal lobe epilepsy, in which the technique gave evidence of lateralization and localization of seizure onset in 55% (Ryvlin et al., 1998).

Studies of drug action in vivo in patients with epilepsy may also be carried out using [^{11}C]flumazenil PET. It was shown that 1.5 mg intravenous flumazenil occupied 55% of BZR, whereas 15 mg occupied nearly all receptors (Savic et al., 1991).

In conclusion, [^{11}C]flumazenil PET scan data are good markers for neuronal integrity in the hippocampus and neocortex and may also identify ectopic neurons in MCD. There are suggestions that there may also be abnormalities of cBZR availability on neurons, but definitive correlative in vitro studies have not yet been completed. The future clinical role of [^{11}C]flumazenil PET is likely to be in the presurgical evaluation of those patients in whom MRI is not definitive and in those with evidence of MCD.

2. Opioid Receptors

Endogenous opioids are released following partial and generalized tonic-clonic seizures and contribute to the postictal rise in seizure threshold (Bajorek et al., 1986). Investigations of opioid receptors in patients with temporal lobe epilepsy have shown an increase of the binding of the specific mu agonist [^{11}C]carfentanil to mu receptors in lateral temporal neocortex in areas that also showed reduced glucose metabolism, reflecting an increase in number of available receptors or increased affinity (Frost et al., 1988). It has been speculated that an increase in mu-opioid receptors in the temporal neocortex may be a manifestation of a tonic antiepileptic system that serves to limit the spread of electrical activity from other temporal lobe structures. In a second study, the increase in [^{11}C]carfentanil binding to lateral temporal neocortex was confirmed and reduced binding to the amygdala was noted (Mayberg et al., 1991). The finding of reduced signal from such a small structure may be, at least in part, due to partial volume effects. Methods to correct for this have recently been developed and implemented (Muller-Gartner et al., 1992; Meltzer et al., 1996; Labbe et al., 1998). The same patients were also studied with [^{11}C]diprenorphine, which binds with similar affinities

in vivo to the mu, kappa, and delta subtypes of opioid receptors. There were no overall asymmetries of binding of [^{11}C]diprenorphine in the temporal lobe or elsewhere (Mayberg et al., 1991), an observation confirmed by others (Bartenstein et al., 1994). Similarly, there was no overall asymmetry of binding of the PET tracer [^{18}F]cyclofoxy, which binds to mu- and kappa-opioid receptors, in patients with temporal lobe epilepsy, although some patients had higher binding of the ligand in the temporal lobe ipsilateral to the EEG focus (Theodore et al., 1992b). The precise explanation of these findings is not clear. Hypotheses include an upregulation of mu receptors and reduction of number or affinity of kappa receptors, or upregulation of mu receptors and occupation of kappa receptors by an endogenous opioid ligand. Further investigations with a kappa-specific opioid PET tracer may clarify this issue. Delta-opioid receptors, identified using the delta-specific antagonist [^{11}C]methylnaltrindole, were found to be increased in the midinferior and anterior part of the middle and superior temporal neocortex in patients with temporal lobe epilepsy (Madar et al., 1997). This contrasted with the increase in mu receptors, identified with [^{11}C]carfentanil, that was confined to the middle part of the inferior temporal neocortex. The suggestion has been that the different opioid receptor subtypes have specific roles in the neocortical response to an epileptic focus and may have an antiepileptic role.

These studies have all been carried out in patients with temporal lobe epilepsy. No investigations of opioid receptors in vivo using PET have yet been reported in patients with extratemporal seizure disorders or MCD.

Dynamic ictal studies of opioid receptors have been carried out in patients with reading epilepsy, using [^{11}C]diprenorphine. To localize dynamic changes of opioid neurotransmission associated with partial seizures and higher cognitive function, release of endogenous opioids in patients with reading epilepsy was compared with that in healthy volunteers (Koepp et al., 1998c). Reading-induced seizures were associated with reduced [^{11}C]diprenorphine binding to opioid receptors in the left parieto-temporo-occipital cortex and to a lesser extent the left middle temporal gyrus and the posterior parieto-occipital junction (Fig. 2). These data gave evidence for localized endogenous opioid peptide release during seizures induced by reading and demonstrate the potential of PET to image release of specific neurotransmitters in response to brain activity in specific cerebral areas in vivo.

3. Monoamine Oxidase Type B (MAO-B)

Deprenyl binds with high specificity and affinity to monoamine oxidase type B (MAO-B), which is mainly located in astrocytes. In patients with temporal lobe epilepsy, uptake of [^{11}C]deprenyl was increased in the

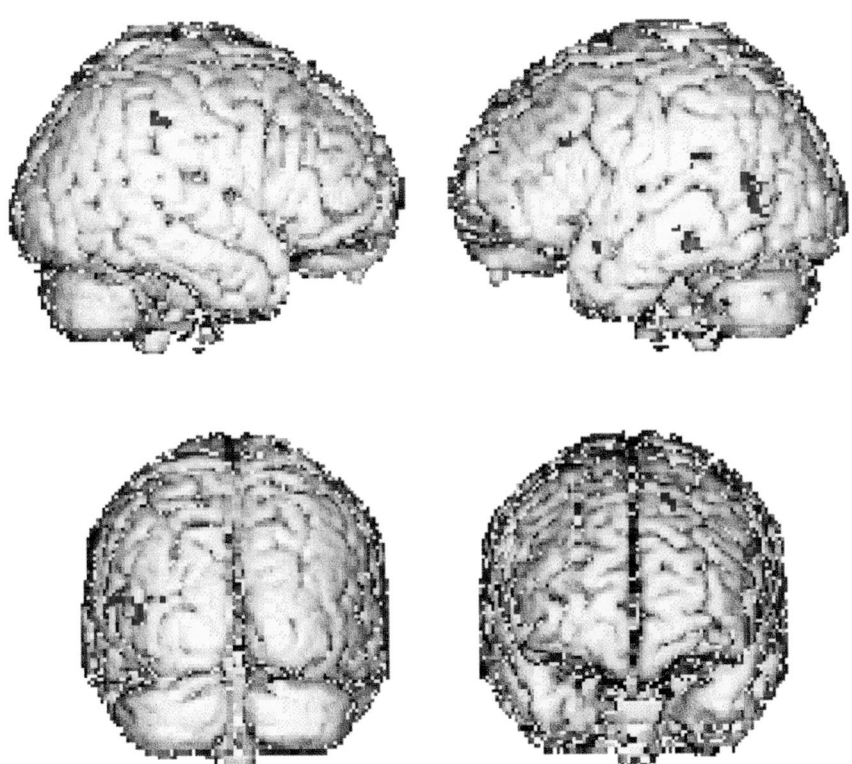

Figure 2 Left lateral view of cerebral hemisphere, demonstrating areas in patients with reading epilepsy in which there was ictal reduction of [^{11}C]- to diprenorphine-binding, implying endogenous opioid release. Reprinted with permission from M. J. Koepp, M. P. Richardson, D. J. Brooks, and J. S. Duncan (1998c). Focal cortical release of endogenous opioids during reading induced seizures. *Lancet* **352,** 952–955. © by The Lancet Ltd.

epileptogenic temporal lobe, possibly reflecting gliosis (Kumlein *et al.,* 1995; Bergstrom *et al.,* 1998). This technique does not appear to be likely to contribute to precise localization of an epileptic focus within the temporal lobe, but the anatomical discrimination of medial temporal lobe structures was rather limited. It is possible that higher resolution images may produce more refined findings, and the technique has not yet been evaluated in extratemporal seizure disorders.

4. NMDA Receptor

[^{11}C](*S*)-*N*-Methylketamine binds to the NMDA receptor and is thus of interest in studies of epilepsy. In eight patients with medial temporal lobe epilepsy, there was a reduction in tracer binding potential that paralleled hypometabolism. It is not clear, however, whether the reduction was due to reduced perfusion, loss of tissue, or reduction of receptor binding (Kumlein *et al.,* 1999) and further work is needed to clarify this.

5. Serotoninergic Neurons

Increased concentrations of serotonin and serotonin immunoreactivity have been reported in resected human epileptic cortex. α-[^{11}C]Methyl-L-tryptophan ([^{11}C]-AMT) is a marker for serotonin synthesis. In children with tuberous sclerosis, uptake was increased in some tubers that appeared to be the sites of seizure onset. Other tubers showed decreased uptake. In contrast, FDG PET showed hypometabolism in all tubers. This study suggests that [^{11}C]AMT PET may be useful to detect epileptogenic foci in patients with tuberous sclerosis, and possibly other forms of cerebral malformation (Chugani *et al.,* 1998). Confirmation of this finding and further evaluation of the basis of these results are required.

6. Histamine Receptors

An increase of H$_1$ receptors, visualized with PET and [^{11}C]doxepin in epileptic foci, that also show reduced interictal glucose metabolism has been reported (Iinuma *et al.,* 1993; Itoh *et al.,* 1995). It has been suggested that this finding is compatible with an increase of mu-opioid receptors. It is not certain, however, how specific this tracer is for H$_1$ receptors and this finding has not been replicated by other groups.

7. Peripheral Benzodiazepine Receptors

The peripheral benzodiazepine receptor ligand [^{11}C]PK11195 labels macophages and activated microglia and as such is a marker of inflammatory responses in the brain. Increased binding has been demonstrated in Rasmussen's encephalitis, reflecting the inflammatory nature of the condition, and is in contrast to hippocampal sclerosis, in which there is no increased binding (Banati *et al.,* 1999).

C. Idiopathic Generalized Epilepsy

1. Central Benzodiazepine Receptors

Savic *et al.* (1990) reported a slight reduction in the cortical binding of [^{11}C]flumazenil to central benzodiazepine receptors in a heterogeneous group of patients with generalized seizures, compared with the "nonfocus" cortical areas of patients with partial seizures. It was subsequently reported that, compared with normal subjects, patients with primary generalized tonic-clonic seizures had an increased BZ receptor density in the cerebellar nuclei and decreased density in the thalamus (Savic *et al.,* 1994). Reliable identification of cerebellar nuclei is not easy on [^{11}C]flumazenil PET images and these results have not been replicated. Prevett *et al.* (1995b) found no significant difference in [^{11}C]flumazenil binding to cerebral cortex, thalamus, or cerebellum between patients with childhood and juvenile absence epilepsy, not taking valproate, and control subjects. The volume of distribution of [^{11}C]flumazenil, however, was 9% less in patients receiving valproate, with a 20% reduction of receptor density, suggesting that this drug may result in reduced number of available benzodiazepine receptors. In a further study, however, comparing 10 patients with idiopathic generalized epilepsy, before and after introduction of valproate, with 20 control subjects, patients with idiopathic generalized epilepsy had an 11% higher binding to cBZR in neocortex and a 14% increase in the thalamus and cerebellar cortex. The introduction of valproate was not associated with a significant change in cBZR (Koepp *et al.,* 1997a). These data provide evidence for the structural and functional basis of idiopathic generalized epilepsy. It is not certain whether this finding is the result of an increased number of neurons or a functional change in available receptors. Correlative neuropathological studies will be needed to answer this question. It does appear, however, that valproate does not affect the numbers of available cBZR, but this does not exclude the possibility of the drug having a modulatory effect on receptor function.

Acutely, the cerebral binding of [^{11}C]flumazenil was not affected by flurries of absences in any area of neocortex or the thalamus, implying that binding to this part of the GABA$_A$-BZR is not involved in the pathophysiology of absences (Prevett *et al.,* 1995c).

2. Opioid Receptors

Systemic administration of opioids tends to cause an increase in generalized spike-wave activity, in contrast to the anticonvulsant effect of endogenous opioids on generalized tonic-clonic seizures, suggesting that opioid transmission may have a role in the pathogenesis of absences (Frey and Voits, 1991). Diprenorphine is a weak opiate receptor partial agonist with similar *in vivo* affinities for the mu, kappa, and delta receptor subtypes. There was no significant difference in [^{11}C]diprenorphine binding between control subjects and patients with childhood and juvenile absence epilepsy, suggesting there is no overall abnormality of opioid receptors in this condition (Prevett *et al.,* 1994).

In a dynamic study, however, it was found that serial absences were associated with an acute 15–41% reduction in [^{11}C]diprenorphine binding to association areas of neocortex, with no effect on binding to thalamus, basal ganglia, or cerebellum. The results implied release of endogenous opioids in the neocortex that may have a role in the pathophysiology of typical absences (Bartenstein *et al.,* 1993). Further studies with subtype-specific ligands may provide information about the role of different opioid receptor subtypes in typical absences.

Investigations with ligands to study GABA$_B$ and excitatory amino acid receptors would give useful information on the involvement of these systems in the pathogenesis of typical absences and idiopathic generalized epilepsy, but suitable PET tracers for these receptors are not yet available.

D. Conclusion

Studies with PET are useful for investigating the neurochemical abnormalities associated with the epilepsies, both static interictal derangements and dynamic changes in ligand-receptor interaction that may occur at the time of seizures. The development of further ligands in the coming years, particularly tracers that are specific for excitatory amino acid receptors, subtypes of the opioid receptors, and the GABA$_B$ receptor, is necessary to further understand the processes that give rise to and respond to the various forms of the epilepsies. All functional data need to be interpreted in the light of the structure of the brain. Coregistration with high-quality MRI is now readily achievable and essential. It will also be important to carry out parallel studies with *in vitro* autoradiography and quantitative neuropathological studies on surgical specimens and postmortem material.

IV. Single Photon Emission Computerized Tomography

Single photon emission computerized tomography is principally used, in the investigation of the epilepsies, to image the distribution of cerebral blood flow. In addition, there have been a few studies of specific receptors in the brain.

A. Cerebral Blood Flow Tracers

1. Introduction

The most commonly used SPECT tracer for imaging CBF is [^{99m}Tc]hexamethylpropylenamine oxime ([^{99m}Tc]HMPAO). The superiority of this tracer over radioxenon and iodoamphetamine was established over a decade ago (Longostrevi, 1986). Other CBF tracers in current use include [^{123}I]*N*-isopropyl-*p*-iodoamphetamine and [^{99m}Tc]ethyl cysteinate dimer (ECD, bicisate).

[^{99m}Tc]HMPAO is given by vein and 70% brain uptake occurs in 1 min. The subsequent image is stable for 6 h, as after crossing the blood-brain barrier [^{99m}Tc]HMPAO reacts with intracellular glutathione, becoming hydrophilic and so is much less able to recross the blood-brain barrier. A limitation is that the tracer is chemically unstable 30 min after preparation. This is less of a problem for interictal than ictal studies. Derivatives, stabilized with cobalt chloride, with a longer shelf life have recently been developed. Radiolabeled ECD is stable for 6 h, easing study of brief ictal events (Grunwald *et al.*, 1994; Runge *et al.*, 1997). ECD does have a cerebral distribution pattern different from that of HMPAO, which needs consideration in the interpretation of scans (Oku *et al.*, 1997).

2. Hardware

The spatial resolution depends on the imaging equipment used, being 14–17 mm (full width half-maximum) for single-head rotating γ cameras, 8–10 mm for three-headed instruments, and 7–8 mm for ring-type cameras. Unlike PET, the data remain, at best, semiquantitative, with the use of a reference region as an internal standard. The correct orientation of imaging planes is important to visualize the structures optimally. The commonly used orbito-meatal line is about 40° off the long axis of the temporal lobe. In consequence, there are inevitable partial volume effects that give rise to difficulty in discriminating between inferior frontal and superior temporal lobe, with loss of potentially useful localizing information. If thin slices are acquired, this deficit can, to some extent, be corrected for by reformatting. This is of limited utility, however, if slices are 8 mm thick and it is much more satisfactory to acquire data in the optimum plane.

3. Analysis

As with PET data, coregistration with MRI allows for a structure-function correlation that enhances the interpretation of both individual data sets and the consensus diagnosis (Woods *et al.*, 1993; Zubal *et al.*, 1995).

4. Interictal SPECT Studies

It was established in the 1980s that the marker of an epileptic focus studied interictally in adults and children with SPECT is a region of reduced CBF. It was soon noted that the results were not always reliable (Podreka *et al.*, 1987; Stefan *et al.*, 1987b; Denays *et al.*, 1988; Chiron *et al.*, 1989; Andersen *et al.*, 1990, 1994; Cordes *et al.*, 1990; Duncan *et al.*, 1990; Iivanainen *et al.*, 1990; Vles *et al.*, 1990; Dietrich *et al.*, 1991; Ryvlin *et al.*, 1992b; Andersen *et al.*, 1994; Schmitz *et al.*, 1995). Focal abnormalities of regional CBF have been visualized with SPECT in patients with infantile spasms (Chiron *et al.*, 1993), Landau-Kleffner syndrome (O'Tuama *et al.*, 1992), continuous spike waves during slow wave (Gaggero *et al.*, 1995), Lennox-Gastaut syndrome (Heiskala *et al.*, 1993), and tuberous sclerosis (Tamaki *et al.*, 1991). Reduced CBF has been found in agyria-pachygyria (Chiron *et al.*, 1996) and normal gray matter and increased cerebral perfusion has been found in laminar heterotopia (Matsuda *et al.*, 1995). In several of the above studies, the sensitivity of SPECT imaging appeared to be superior to the MRI available at the time. With the advances in MRI made in the 1990s, the same may not currently be the case.

In [^{99m}Tc]HMPAO SPECT studies of patients with temporal lobe epilepsy, a significant asymmetry of cerebral blood flow has been noted in about 50% of cases, ranging from 11 to 80% in different studies. Concordance with interictal EEG lateralization has been noted in 65% of cases. Lobar localization (e.g., frontal vs temporal) has been more difficult, with, in one large representative series, correct localization in 38% in interictal studies of patients with unilateral temporal lobe EEG focus (Rowe *et al.*, 1991a,b). In children with TLE, there was reasonable concordance between interictal SPECT and proton MRS findings (Cross *et al.*, 1997a).

Localization with interictal SPECT is more difficult in patients with extratemporal epilepsy (Dasheiff, 1992; Marks *et al.*, 1992). In a blinded comparative study, interictal SPECT was less effective at lateralizing the focus of temporal lobe epilepsy than MRI, with correct lateralization in 45% compared to 86%. Furthermore, agreement of MRI and EEG data was a good predictor of a satisfactory result from surgical treatment, whereas SPECT was not and was prone to give an incorrect result in patients whose MRI was not lateralizing (Jack *et al.*, 1994a). In consequence to this poor sensitivity and specificity, interictal SPECT studies have little place in the routine investigation of patients with epilepsy.

Interictal HMPAO SPECT has been shown to have inferior resolution and reliability for identifying a focal deficit than [18]FDG PET in the evaluation of patients with partial seizures (Leiderman *et al.*, 1992; Ryvlin *et al.*, 1992a; Coubes *et al.*, 1993; Theodore *et al.*, 1994; Nagata *et al.*, 1995; Mastin *et al.*, 1996).

Interictal SPECT, however, may delineate the "irritative zone" of interictal epileptiform activity. The distribution of interictal epileptiform activity correlates with the area of reduced perfusion. In patients with TLE studied with depth EEG, reduced perfusion of the medial temporal lobe was associated with spikes being confined to the mesial structures. In contrast, more widespread temporal spikes were associated with reduced perfusion throughout and beyond the temporal lobe (Guillon *et al.*, 1998).

5. Ictal and Postictal SPECT Studies

It was established over a decade ago that the increase in CBF associated with a seizure may be detected using SPECT (Bonte *et al.*, 1983; Lee *et al.*, 1986). This may provide useful localizing information in patients with partial seizures. It has also been suggested that a focal increase of blood flow may occur before seizure activity detected on a scalp EEG (Baumgartner *et al.*, 1998). An injection of [^{99m}Tc]HMPAO at the time of a seizure results in an image of the distribution of CBF 1–2 min after tracer administration. The distribution is then stable for several hours so that the patient may be imaged when the seizure is over. The general pattern is of localized ictal hyperperfusion, with surrounding hypoperfusion, that is followed by accentuated hypoperfusion in the region of the focus, which gradually returns to the interictal state. Combined data from interictal and ictal SPECT scans give a lot more data than interictal scans alone and may be useful in the evaluation of both temporal and extratemporal epilepsy. In complex partial seizure disorders, the epileptic focus has been identified in 69–93% of ictal SPECT studies (Lee *et al.*, 1988; Shen *et al.*, 1990; Stefan *et al.*, 1990; Rowe *et al.*, 1991a; Marks *et al.*, 1992; R. Duncan *et al.*, 1993; Harvey *et al.*, 1993a; Markand *et al.*, 1994; Lee *et al.*, 1997). A recent meta-analysis of published data showed that in patients with temporal lobe seizures, the sensitivities of SPECT relative to diagnostic evaluation were 0.44 (interictal), 0.75 (postictal), and 0.97 (ictal) (Devous *et al.*, 1998).

Ictal SPECT investigations may be carried out successfully in children as well as adults (Cross *et al.*, 1995, 1997b; O'Brien *et al.*, 1998a). In one large series, ictal SPECT achieved 97% correct localization in unilateral temporal lobe epilepsy, compared with 71% for postictal SPECT and 48% for interictal scans. In extratemporal seizures, ictal SPECT studies localized the focus in

92%, compared to 46% for postictal studies, and interictal SPECT was of little value (Newton *et al.*, 1995).

In temporal lobe seizures, the occurrence of contralateral dystonic posturing was associated with an ictal increase in CBF in the basal ganglia ipsilateral to the focus (Newton *et al.*, 1992a). A characteristic feature of temporal lobe seizures is an initial hyperperfusion of the temporal lobe, followed by medial temporal hyperperfusion and lateral temporal hypoperfusion (Fig. 3) (Newton *et al.*, 1992b). In patients with TLE, extensive areas of ictal perfusion did not predict poor outcome, indicating that this pattern of blood flow response probably reflected seizure propagation rather than intrinsically epileptogenic tissue. In contrast, atypical patterns of hyperperfusion were associated with an absence of pathology in the resected specimen and poor outcome and may indicate diffuse or extratemporal epileptogenicity (Ho *et al.*, 1997). Particular ictal patterns of cerebral perfusion have been identified in subtypes of TLE. In those with hippocampal sclerosis or medial temporal lobe lesions, increased blood flow was in the medial and lateral ipsilateral temporal lobe. In those with lateral temporal lesions, blood flow increased bilaterally, but particularly in relation to the lesion. In those with no identifiable pathology, blood flow increased in the anterior-medial part of the temporal lobe. These findings suggest the pathways of seizure spread in TLE and indicate that lateral-onset seizures spread bilaterally (Ho *et al.*, 1996).

Ictal [^{99m}Tc]HMPAO scans may be useful in the evaluation of patients with extratemporal seizures and unremarkable MRI (Marks *et al.*, 1992; Harvey *et al.*, 1993b; Duncan *et al.*, 1997a). Asymmetric tonic posturing, contralateral head and eye deviation, and unilateral clonic jerking were associated with an ictal increase in CBF in the frontocentral, medial frontal, or dorsolateral areas (Harvey *et al.*, 1993b). Varying patterns have been seen in patients with autosomal-dominant frontal lobe epilepsy (Hayman *et al.*, 1997).

Ictal SPECT with injection 2–5 s after seizure onset showed two patterns in supplementary sensorimotor area (SSMA) seizures: first, involvement of the ipsilateral SSMA, dorsal premotor, and motor cortex; second, bilateral asymmetric medial frontal increases in blood flow, reflecting propagation of seizure activity (Laich *et al.*, 1997). Injection of tracer at the onset of the seizure is important in the evaluation of brief extratemporal seizures. In another series of SSMA seizures, in which tracer was injected up to 30 s after seizure onset, findings were less clear-cut (Ebner *et al.*, 1996). Focal ictal increases in CBF have also been found in patients with parietal lobe epilepsy, both with and without demonstrable parietal lesions and corresponding with the latter when present (Ho *et al.*, 1994). Distinct patterns of

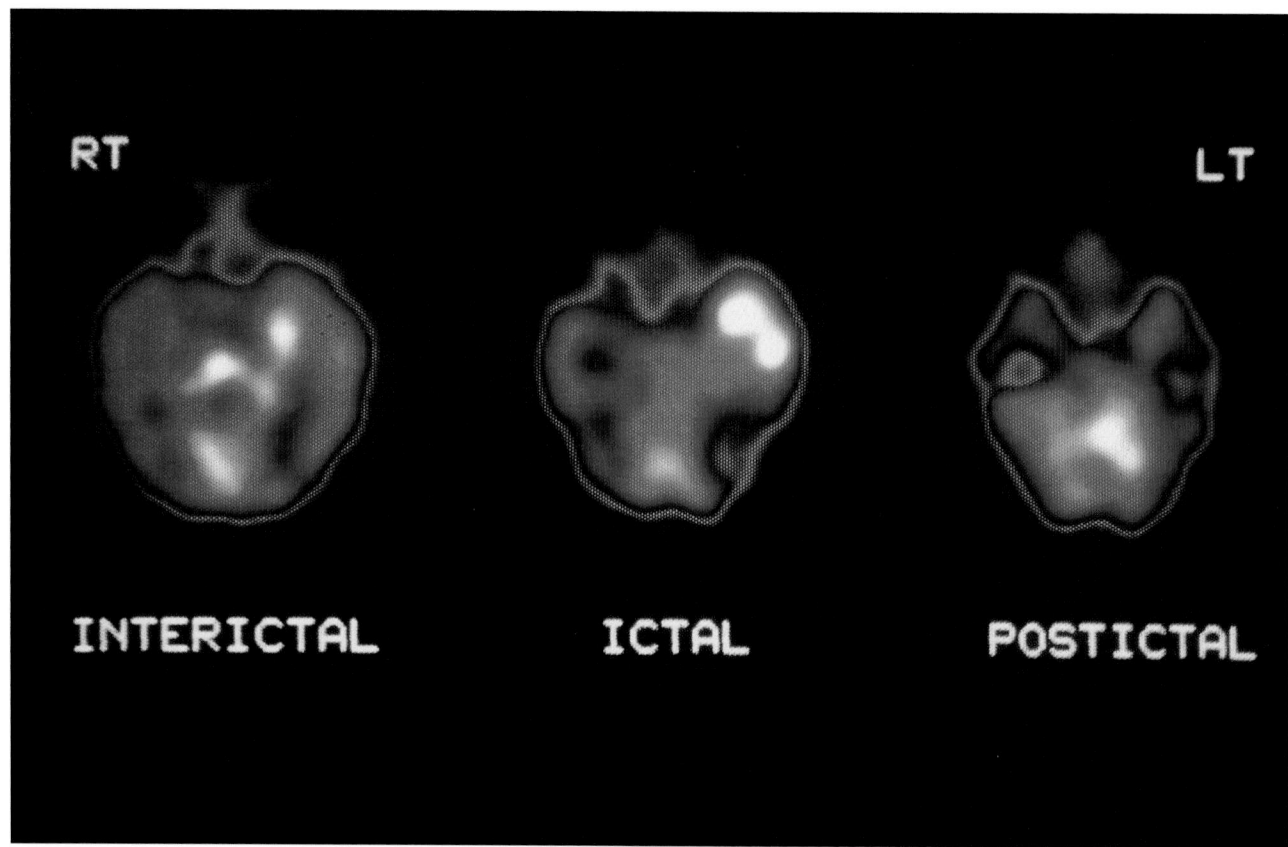

Figure 3 Transverse interictal, ictal, and postictal HMPAO SPECT scans of a patient with left temporal lobe seizures. The color scale from white to red to blue indicates a spectrum of high to low counts. Interictally, the right mesial temporal lobe appears hypoperfused. The ictal study shows marked left temporal hyperperfusion. The postictal scan shows lateral temporal hypoperfusion with residual mesial hyperperfusion on the left. Figure courtesy of Prof. S. Berkovic.

ictal CBF have been identified in occipital seizures (Duncan *et al.*, 1997b).

Postictal SPECT has shown a focal increase of CBF in areas of focal cortical dysplasia (Otsubo *et al.*, 1993). A focal ictal CBF rise has been demonstrated in patients with malformations of cortical development and nonlocalizing ictal scalp EEG and used to identify surgically resectable epileptic tissue (Kuzniecky *et al.*, 1993c). A focal increase in CBF may be seen in epilepsia partialis continua, even when the EEG does not show focal epileptic activity (Katz *et al.*, 1990).

The coregistration of postictal SPECT images with a patient's MRI improves anatomical determination of abnormalities of CBF (Hogan *et al.*, 1997). A greater advance, however, has been the coregistration of interictal with ictal or postictal SPECT images, to result in an "ictal difference image" that may be coregistered with an individual's MRI. This technique enhances objectivity and the accuracy of data interpretation (Zubal *et al.*, 1995; O'Brien *et al.*, 1998b, 1999).

Ictal [^{99m}Tc]HMPAO scans must always be interpreted with caution. Simultaneous video–EEG is essen-

tial to determine the relationship between the onset of a seizure and tracer delivery; without this there is the risk of confusing ictal and postictal data. A further problem is that spread to other areas of the brain, such as the contralateral temporal lobe, may occur within seconds of seizure onset and so an image of cerebral blood flow distribution 1–2 min after the onset of a seizure may indicate other than the site of onset. Ictal and postictal SPECT studies carried out in patients with intracranial EEG have shown that the former generally accurately reflect the site of seizure activity (Spanaki *et al.*, 1999).

Until recently, [^{99m}Tc]HMPAO had to be constituted immediately prior to injection, resulting in a delay of up to 1 min. A preparation has now been developed that is stabilized with cobalt chloride. This allows the labeled tracer to be prepared in advance and injected into a patient at any time over the subsequent 6 h, without the need for further preparatory work. The advantage of this development is that the interval between seizure onset and tracer delivery to the brain can be significantly reduced. An alternative is to use ready-constituted [^{99m}Tc]ECD, or bicisate, which is stable for several hours, may be injected within 2–20 s of seizure onset, and demonstrates a focal increase in CBF (Grunwald *et al.*, 1994; Packard *et al.*, 1996; Lancman *et al.*, 1997). The interval between seizure onset and injection may also be shortened by the use of an automated injection device that may be activated by the patient when they detect the beginning of a seizure (Sepkuty *et al.*, 1998). Extratemporal seizures may be very brief, increasing the need for injection of blood flow tracer as soon as possible after the start of a seizure. With the inevitable interval between injection and fixation of the tracer in the brain, however, it may not be possible to obtain true ictal studies.

Although the technique has been reported as being of use (Biersack *et al.*, 1988), ictal [^{99m}Tc]HMPAO studies do not reliably differentiate between epileptic and nonepileptic attacks. Cases of nonepileptic attacks that were not associated with a focal increase in cerebral blood flow have been reported, but a focal increase in cerebral blood flow may occur, particularly if there is prominent motor activity.

In patients with temporal lobe epilepsy, ictal SPECT data were of superior lateralizing ability than were interictal ^{18}FDG PET data, particularly when MRI did not show a structural lesion, but the investigations had complementary roles when localization was difficult (Ho *et al.*, 1995; Markand *et al.*, 1997). When compared against EEG data, ictal SPECT and interictal PET had lower sensitivity and higher specificity for extratemporal than temporal seizures (Spencer, 1994).

In conclusion, interictal SPECT imaging of CBF is only moderately sensitive; ictal SPECT markedly improves the yield. Positron emission tomography imaging of interictal cerebral glucose metabolism is more sensitive than measurement of CBF in temporal lobe epilepsy.

6. Carotid Amytal Testing and SPECT

Intravenous HMPAO and SPECT have been used to demonstrate a unilateral hemispheric reduction of CBF of 50–90% following intracarotid injection of sodium amytal to assess language lateralization and memory function (Biersack *et al.*, 1987). The distribution of hypoperfusion was variable between patients and did not always involve the medial temporal lobe, and the degree of hypoperfusion correlated with the duration of hemiplegia and of drug-induced delta activity on the EEG (Coubes *et al.*, 1995; McMackin *et al.*, 1997). A different approach has been to inject the tracer and the sodium amytal together into the carotid artery. This has shown that the distribution of the tracer varied from patient to patient and did not always include the medial temporal lobe (Jeffery *et al.*, 1991; Hart *et al.*, 1993). Al-

though of interest and potential importance, it is not yet evident that such investigations are of clinical benefit.

B. Specific Ligand Tracers

1. Iomazenil

[^{123}I]Iomazenil is a derivative of the central benzodiazepine receptor antagonist flumazenil. Early studies with [^{123}I]iomazenil showed a reduction in binding in region of the epileptic focus (Van Huffelen *et al.*, 1990; Bartenstein *et al.*, 1991), with generally concordant results with ictal EEG recordings. ^{18}FDG PET concurred with ictal EEG in all cases. Analysis of the former was by visual assessment, whereas the ^{18}FDG data were quantified, resulting in greater precision (Van Huffelen *et al.*, 1990). In the latter investigation, 10 of the 12 patients had reduced blood flow demonstrable with [^{99m}Tc]HMPAO and it was suggested that [^{123}I]iomazenil did not confer additional benefits (Bartenstein *et al.*, 1991). Similar results have been reported elsewhere (Cordes *et al.*, 1992; Haldemann *et al.*, 1992; S. Duncan *et al.*, 1993). Studies with higher resolution cameras and optimal scan orientation, however, have suggested that the area of reduced specific binding of [^{123}I]iomazenil is more restricted than the defect of blood flow and of greater sensitivity for the localization of an epileptogenic focus (Johnson *et al.*, 1992; Venz *et al.*, 1994; Sjoholm *et al.*, 1995; Tanaka *et al.*, 1997). Reduced binding of [^{123}I]iomazenil has been found in a focal area of MCD, in which reduced CBF was not detectable, implying that the former may have greater sensitivity for detecting areas of cortical abnormality (Bartenstein *et al.*, 1992). In this case, however, there was a very evident abnormality on MRI, which appeared to be as extensive as the abnormality of BZR. In a comparison with flumazenil and FDG PET, [^{123}I]iomazenil SPECT was innacurate at localizing epileptic foci (Debets *et al.*, 1997; Lamusuo *et al.*, 1997).

2. Other Tracers

Increased uptake of the MAO-B SPECT tracer [^{123}I]RO43-0463 has been reported in the mesial temporal lobe of patients with TLE (Buck *et al.*, 1998). This is concordant with PET studies using [^{11}C]deprenyl (Kumlein *et al.*, 1995; Bergstrom *et al.*, 1998). As with the PET studies, it is not clear whether this finding reflects more than the presence of gliosis.

Another tracer that may prove to be of clinical utility is [^{123}I]iododexetimide, which labels muscarinic acetylcholine receptors. Reduced binding of this tracer has been found at epileptic foci; at present, it appears that this represents loss of neurons (Muller-Gartner *et al.*, 1993).

C. Conclusion

An advantage of SPECT over PET is that the former is much less expensive and the equipment is more widespread. A further advantage is the ability to obtain images representative of CBF at the time of seizures. These data need careful and cautious interpretation and are nonquantitative. If further SPECT tracers that probe the integrity of specific receptors and neurotransmitters are developed, the technique may have further applications in the future in the investigation and management of the epilepsies.

V. Functional MRI

The first functional MRI study applied to epilepsy was in 1988: abnormal perfusion was demonstrated using MRI with a phase-mapping technique in a patient with epilepsia partialis continua (Fish *et al.*, 1988). Considerable technical advances have been made since that time. Increased perfusion in temporoparietal cortex in a patient with partial status epilepticus was demonstrated using a susceptibility-weighted sequence and dynamic contrast enhancement with gadolinium (Warach *et al.*, 1994).

A. Ictal and Interictal Epileptiform Activity

Functional MRI can detect ictal changes in cerebral blood flow (Jackson *et al.*, 1994; Detre *et al.*, 1995;

Warach *et al.*, 1996). Activation of the thalamus coupled with activation of the motor cortex has been shown in a patient with frequent partial seizure, indicating the ictal activation of a neural network (Detre *et al.*, 1996). Limitations of the method include the effects of movement-induced artifact over a series of scans, although this may be compensated for by image realignment and coregistration, and the fact that it is impracticable for a patient, even with very frequent seizures, to lie for hours in an MRI scanner awaiting the onset of an attack.

The development of safe and reliable EEG recording, with removal of the pulse artifact, from subjects having fMRI studies has been a major step forward in the fMRI of interictal epileptiform activity (Ives *et al.*, 1993; Lemieux *et al.*, 1997; Allen *et al.*, 1998). Focal increases in cerebral blood delivery have been identified in patients with frequent interictal spikes (Krakow *et al.*, 1999; Symms *et al.*, 1999) (Fig. 4). The time course of the increase was similar to that seen in cognitive activation tasks, with maxima at 4–6 s. In these studies, results were obtained by averaging 40 spikes and 40 rest periods, and fMRI acquisition was triggered manually after identification of a spike on the EEG trace. Rest periods were defined as those in which there was no interictal epileptiform activity in the preceding 10 s. At present, EEG cannot be recorded during the actual fMRI acquisition. The next advances will be the ability to do this and to record continuously, so that the temporal and spatial evolution of fMRI changes in relation to interictal activity may be determined. Improvements in signal/noise

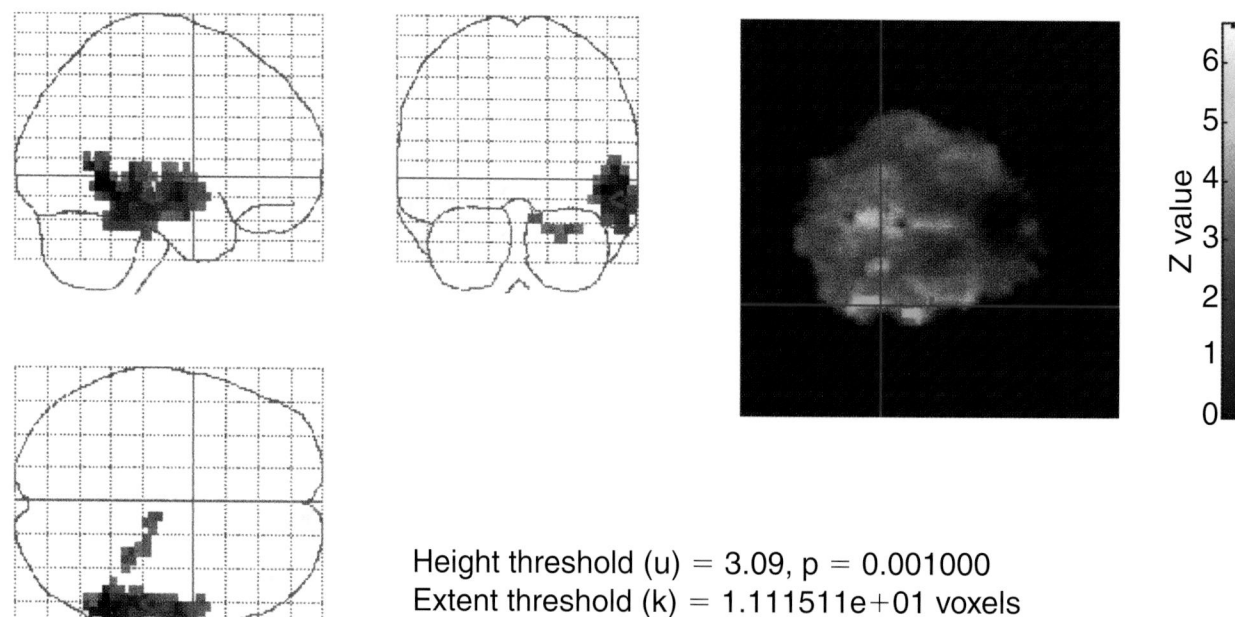

Height threshold (u) = 3.09, p = 0.001000
Extent threshold (k) = 1.111511e+01 voxels

Figure 4 SPM projection of fMRI of interictal epileptiform activity obtained at 1.5 T after 45 acquisitions at rest and 45 acquisitions following interictal spikes, showing activation in the left temporal lobe.

will also allow fewer events having to be recorded and so will make the technique applicable to many more patients. fMRI has demonstrated focal activation of sensorimotor cortex in a patient with benign rolandic epilepsy who had central spkes in the left central area in response to sensory stimulation of the contralatral hand (Manganotti *et al.*, 1998). This demonstrates the potential to map areas of hyperexcitable cortex and that the method may be applied to reflex seizure disorders.

EEG-triggered fMRI and 3D EEG source localization have been used together. In a patient with frontal lobe seizures fMRI showed multiple areas of activation. Three-dimensional EEG source localization identified the same areas and suggested a left frontal onset (Seeck *et al.*, 1998). A potential development will be to use the fMRI data to constrain the solutions of dipole modeling of the source of interictal epileptiform activity and to mitigate the "inverse problem."

Clinically, these methods will aid EEG interpretation and understanding of the pathophysiological basis of epileptic activity. Their application, utility, and limitations in defining the irritative zone of the cortex (which generates interictal spikes) and its relationship with the epileptogenic zone (which gives rise to seizures) in patients in whom surgical treatment is being considered will need careful and critical evaluation. The next step will be the comparison of the areas of cortex identified using fMRI with the areas of spiking that are identified using electrocorticography.

B. Localization and Lateralization of Cognitive Functions

A further important use of fMRI in patients with epilepsy is to delineate areas of brain that are responsible for specific functions, such as the primary sensory and motor cortex, and to identify their anatomical relation to areas of planned neurosurgical resection (Hammeke *et al.*, 1994; Jack *et al.*, 1994b; Morris *et al.*, 1994; Puce *et al.*, 1995; Rao *et al.*, 1995). In patients with cerebral lesions, the localization of cognitive activation may differ from the pattern in normal subjects (Alsop *et al.*, 1996). These data may be helpful in the planning of neocortical resections of epileptic foci, in order to minimize the risk of causing a fixed deficit.

Lateralization of language function may also be accomplished using functional MRI (Binder *et al.*, 1995; Desmond *et al.*, 1995). There was a strong correlation between language lateralization measured with the carotid amytal test and that using fMRI with a single-word semantic decision task (Binder *et al.*, 1996) and other fMRI language studies have generally concurred with carotid amytal testing (Desmond *et al.*, 1995; Benson *et al.*, 1999). This technique has also been accomplished in children and adolescents (Hertz-Pannier *et al.*, 1997). Comparison of fMRI and the results of carotid amytal testing has shown that the duration of speech arrest during the latter is not a reliable measure of language dominance (Benbadis *et al.*, 1998). fMRI results, however, do not always accord with the carotid amytal data (Worthington *et al.*, 1997). There is discrepancy in approximately 1 in 10 persons. Artifacts and technical difficulties may adversely affect both methods. Sedation may complicate amytal testing, and poor understanding of instructions may confound fMRI. Furthermore, identification of the areas of brain involved in language is not the same as determining if someone can speak when half of the brain is anesthetized.

Activation of language areas in the frontal and parietal lobes, using a semantic decision task alternating with an auditory perceptual task, was similar in groups of patients with left and with right TLE. The right TLE group had a stronger activation of the left medial temporal lobe than did the left TLE group (Bellgowan *et al.*, 1998a). It is possible that this method may be useful for predicting lateralization of seizure onset and postoperative deficits.

The choice of language activation paradigm affects results. At present, various approaches are under evaluation in different centers, and a universal consensus has not been established. Activation patterns related to verb generation, object naming, and reading have been compared, and only verb generation gave reliable lateralization compared with carotid amytal and cortical stimulation testing (Benson *et al.*, 1999). A limitation of this study, however, was that the task was carried out silently, so compliance was not assessed, and the only control state was visual fixation. Studies in which word generation is compared with the rest state usually show activation of anterior language areas. Comparisons of semantic tasks with nonlinguistic control tasks show more widespread activation in the dominant hemisphere (Binder *et al.*, 1997).

As well as predicting the lateralization of language function, fMRI may localize cerebral areas involved in language. In the future, these data may assist in planning surgical resections in the language-dominant hemisphere. There are, however, important caveats. Absence of activation on one language task does not guarantee that that part of the brain is inert. Conversely, an area that is activated may have only a peripheral and nonessential role in verbal communication.

At present, many candidates for anterior temporal lobe resection require carotid amytal tests to determine whether they would be at risk of severe memory impairment following the proposed surgery. The test is crude and invasive and serves to determine whether the contralateral temporal lobe is able to subserve basic short-term memory functions.

Until recently, little research had demonstrated hippocampal function in the normal brain during memory processing or had been applied to memory mapping in epilepsy (Bookheimer, 1996). Functional imaging visualization of the medial temporal lobe and its role in memory is difficult because of the small size of the hippocampus, which may be reduced further if there is hippocampal sclerosis, and because of MRI artifacts in this area of the brain. Several fMRI studies have shown activation of bilateral posterior medial temporal lobe, particularly posterior parahippocampal gyri, during encoding of novel verbal and spatial material (Stern *et al.,* 1996; Gabrieli *et al.,* 1997; Fernandez *et al.,* 1998), but it is not clear whether these activations reflect memory processing per se. A complex visual scene encoding task that activates mesial temporal structures has been used during fMRI. Activation patterns were nearly symmetric in normal subjects, but asymmetries were noted in patients with TLE. In nine patients the asymmetry agreed with the findings of the carotid amytal test (Detre *et al.,* 1998). Thus, fMRI can identify asymmetries in memory activation in patients with TLE. These data suggest that fMRI may have a role in preoperative assessment of memory function.

Memory tasks that activate the anterior medial temporal lobe would be expected to be of most value in predicting the consequences of anterior temporal lobe resection. Recent studies using visual encoding tasks that had differing novelty or spatial scrambling found that the former gave rise to mainly posterior parahippocampal activation. The latter, however, resulted in bilateral anterior medial temporal lobe activation, suggesting that this may be an appropriate paradigm to use in the evaluation of patients with temporal lobe epilepsy (Bellgowan *et al.,* 1998b).

Clinical application of fMRI studies to patients with epilepsy in whom surgical treatment is being considered needs to be cautious. While interpretation of activation studies of primary motor and somatosensory cortex may be straightforward, the integration of cognitive activation data into surgical decision making needs to be circumspect, particularly for more sophisticated paradigms. There are two principal caveats: First, if a cerebral area does not activate on a specific task, this does not imply that it may be removed with impunity. Second, the activation of a particular cerebral area with a specific task does not necessarily imply that surgery to that area would cause a clinically significant fixed deficit.

VI. Diffusion-Weighted MRI

Diffusion-weighted imaging reflects the apparent diffusion coefficient (ADC) of water, which is reduced in injured areas of brain, including excitotoxic injury. ADC fell by 20% during status epilepticus induced by biculline (Zhong *et al.,* 1993). Similar, but smaller, effects were seen using hexafluorodiethyl, with reversal by barbiturates (Zhong *et al.,* 1995). Decreased ADC was observed in rat piriform cortex 12 h after kainate administration and paralleled decrease in NAA and increase in lactate at a time when there was little change in T_2-weighted MRI or histology (Ebisu *et al.,* 1996). Reductions in ADC of a mean 2.4% have also been noted after electrical afterdischarges lasting less than 5 s, with a time course of approximately 1 min (Prichard *et al.,* 1995).

Diffusion-weighted imaging obtained during focal motor status epilepticus showed reduced ADC in the motor cortex and increased ADC in the underlying white matter. The changes resolved after seizure activity stopped (Wieshmann *et al.,* 1997). Interictally, the hippocampi of patients with epilepsy showed increased ADC that correlated with the severity of hippocampal sclerosis, assessed by hippocampal volumetrics and T_2 relaxation time measurement, suggesting that diffusion changes were a reflection of structural changes (Wieshmann *et al.,* 1999).

Diffusion-weighted imaging obtained after seizures or time-locked to the occurrence of interictal epileptiform activity may show focal abnormalities that illuminate the processes of epileptic activity and that may also be of utility as a localizing tool.

VII. Magnetic Resonance Spectroscopy

A. Introduction

As with all functional imaging studies, accurate definition of brain anatomy and the identification of structural abnormalities using MRI are necessary for the interpretation of magnetic resonance spectroscopy data. These techniques may be carried out with MRI in a single session. *In vivo* MRS investigations in epilepsy have examined the nuclei ^{31}P and ^{1}H and have mainly been performed in patients with temporal lobe epilepsy who are candidates for surgical treatment.

B. Proton Spectroscopy

Proton (^{1}H) spectroscopy has a greater signal-to-noise ratio and better spatial resolution than ^{31}P MRS and is more easily carried out in a single examination, with MRI (Connelly and Duncan, 1995). The metabolites that are detectable using ^{1}H MRS depend on the conditions used for the acquisition. Some molecules, for example, GABA, glutamate, glutamine, and lactate, give rise to MR signals that exhibit spin-spin coupling, which results in the signals changing with time. As a result, the detection of their resonance is dependent on the echo time used. In epilepsy studies *in vivo*, the principal signals of interest have been those from *N*-acetylaspartate (NAA), creatine +phosphocreatine (Cr), choline-containing compounds (Cho), and lactate (Lac). There is evidence that NAA is located primarily within neurons and precursor cells (Urenjak *et al.*, 1992, 1993) and a reduction of NAA signal is usually regarded as indicating loss or dysfunction of neurons. Cr and Cho are found both in neurons and in glial cells, and cell studies suggest that they are present at much higher concentrations in glia than neurons (Urenjak *et al.*, 1992, 1993).

1. Proton MRS Investigation in the Temporal Lobe

A structural etiology is not identified with optimal MRI in 15–25% of patients with refractory temporal lobe epilepsy (Duncan, 1997; Van Paesschen *et al.*, 1997).

Reduced NAA was found in the temporal lobes of two patients with Rasmussen's encephalitis, and focally increased lactate was found in one patient who had epilepsia partialis continua (Matthews *et al.*, 1990). In temporal lobe epilepsy caused by hippocampal sclerosis, MRS at an echo time of 135 ms showed reduction of NAA and increases of Cho and Cr, reflecting neuronal loss or dysfunction and astrocytosis (Connelly *et al.*, 1994; Gadian *et al.*, 1994). The mean NAA/(Cho +Cr) ratios were significantly less in the patients with epilepsy than in the control subjects, both ipsilateral and contralateral to the focus, with the ipsilateral side being more affected. Analysis of individual patients showed a reduced NAA/(Cho +Cr) ratio on the side of the focus in 88%, with 40% having bilateral abnormalities. In 6 of the 10 patients with bilateral abnormalities, the abnormality was asymmetric, and when lateralization was possible with MRS, this concurred with MRI findings in all but one case. Two patients with no abnormality shown on MRI, however, had abnormal MRS and three patients with normal MRS on both sides had HS. All 15 of the patients who had been followed for more than 1 year after temporal lobectomy had a good seizure outcome. The number of patients was not sufficient to ascribe any prognostic significance to the data. Similar results have been reported, using the same technique, in children with temporal lobe epilepsy (Cross *et al.*, 1996). Abnormalities of the NAA/(Cho +Cr) ratio were found in 75%, with bilateral abnormalities in 45%, and correct lateralization of the seizure focus was made in 55%. Bilateral increases in choline and creatine were noted, suggesting gliosis. Other studies confirm these findings (Duc *et al.*, 1998).

The implication from these data was that there is neuronal loss or dysfunction and astrocytosis in the temporal lobes of patients with temporal lobe epilepsy. The magnitude of the reduction of NAA was such that the abnormality could not be confined to the hippocampus, as that structure occupies only a small proportion of the 8-ml voxel used. This finding is consistent with PET data on cerebral glucose metabolism, in which there is commonly an area of hypometabolism that is larger than the anatomically defined focus (Engel *et al.*, 1982a; Sackellares *et al.*, 1990). The basis of the regional area of NAA reduction is not certain. Comparative PET studies with [11C]flumazenil and [18F]fluorodeoxyglucose in patients with temporal lobe epilepsy have suggested that diaschisis is a more likely explanation than is neuronal loss (Henry *et al.*, 1993a; Savic *et al.*, 1993). Comparisons of quantitative MRS and [18]FDG PET indicate that reduction of the NAA/(Cho +Cr) ratio detected on MRS reflects the degree of hypometabolism in TLE (Lu *et al.*, 1997). Abnormalities demonstrated by ^{1}H MRS correlated well with hypoperfusion seen on interictal SPECT (Cross *et al.*, 1997a). In a further comparative study, both [18]FDG PET and ^{1}H MRSI lateralized patients with normal MRI, but only relative unilateral hippocampal atrophy predicted a seizure-free outcome. Bilaterally abnormal MRI and ^{1}H MRSI, however, did not preclude a good result (Knowlton *et al.*, 1997a). Results of magnetic resonance spectra in a heterogeneous group of children with epileptic encephalopathies were not uniform, with relative decreases and increases of NAA/choline in hypometabolic areas (Parker *et al.*, 1998). The difficulty with the interpretation of this study is that the underlying pathologies were not certain.

The cellular mechanisms that underlie the reduction of NAA and the elevation of Cho and Cr need to be clarified with correlative neuropathological studies. There are dynamic changes in the concentrations of NAA. After successful temporal lobe resection, there is a tendency for the NAA in the contralateral medial temporal lobe to normalize (Hugg *et al.*, 1996; Cendes *et al.*, 1997a). Other metabolites in relation to the occurrence of seizures also need further evaluation.

Magnetic resonance spectroscopy imaging has the advantages of giving information on the regional distribution of metabolites and identifying areas of maximal abnormality but is more susceptible to artifacts. It is technically more demanding than single-voxel MRS. There are problems with magnetic field homogeneity (shimming), water suppression, and leakage of signal from subcutaneous fat into voxels other than just those adjacent to the scalp. This is a particular problem with studies of the temporal lobes, because of the proximity of the petrous temporal bones. MRSI data must be interpreted with caution (Sauter *et al.*, 1991).

^{1}H MRSI correctly localized the seizure focus in eight patients with TLE cases, by reduced NAA, with no significant changes in Cho or Cr (Hugg *et al.*, 1993). Xue *et al.* (1993) reported problems with suboptimal shimming when performing CSI in a large region including both temporal lobes and in consequence adopted the strategy of acquiring CSI volumes from each temporal lobe separately. The same group subsequently reported localized reductions of NAA/Cho in 53 patients with temporal lobe epilepsy, unilateral in 34 and bilateral in 19 (Xue *et al.*, 1994). In 10 patients with temporal lobe epilepsy and five controls, the left-right asymmetry of NAA/Cr ratios was found to be significantly different from that of controls in all cases (Cendes *et al.*, 1994). This ratio was low in the midtemporal lobe in five cases and in the posterior temporal lobes in eight patients. The asymmetry was maximal in the midtemporal region in three patients and in the posterior temporal region in six patients. The use of an asymmetry index alone precludes the detection of bilateral abnormalities. Comparison of NAA/Cr ratios in patients and controls, however, indicated that two patients had a bilateral reduction in the NAA/Cr ratio in the posterior temporal region and that the greatest reduction was ipsilateral to the maximum EEG disturbance. One of the 10 patients had no MRI evidence of hippocampal atrophy but had a decrease of the NAA/Cr ratio in the midposterior temporal lobe and the resected specimen revealed mild mesial temporal sclerosis. In large series of patients with TLE, EEG, MRSI, and hippocampal volumetry were highly concordant. The combination of the two latter methods lateralized TLE accurately and noninvasively in most patients (Cendes *et al.*, 1997b; Kuzniecky *et al.*, 1998a), without a close correlation between the degree of the volumetric and metabolic abnormalities.

Using high-resolution single-slice MRSI obtained at 4.1 T and mapping MRS data onto the equivalent MRI slice, Hetherington *et al.* found a reduced NAA/Cr ratio in the epileptogenic hippocampus in all of 10 patients with temporal lobe epilepsy. Four of the 10 patients also had abnormalities in the NAA/Cr ratio in the contralateral hippocampus and in two of these four, invasive EEG recordings demonstrated seizure onset from both hippocampi. Significant reductions in the NAA/Cho ratio were found in eight of the 10 patients. In this population, MRI showed hippocampal atrophy in seven patients and was normal in three. In the three patients with normal MRI, the pathological specimen showed gliosis and minimal neuronal loss, suggesting that this technique may be useful for the *in vivo* identification of subtle pathology (Hetherington *et al.*, 1995).

Abnormalities of metabolite profiles may be found in the temporal lobes, with no abnormality evident on MRI (Knowlton *et al.*, 1997a; Connelly *et al.*, 1998). Bilateral abnormalities have been noted in up to 50% of patients with apparently unilateral structural abnormality (Ende *et al.*, 1997), indicating that MRS may be more sensitive in detecting pathology. Further correlative studies are needed to evaluate the clinical utility and reliability of these data. NAA was not reduced in the hippocampi of patients with neocortical epilepsy, either ipsilateral or contralateral to the focus (Vermathen *et al.*, 1997), suggesting that hippocampal dysfunction is not a feature of neocortical epilepsy and that the technique may have utility in defining this.

Temporal lobe MRS data correlate with cognitive function. In a postoperative study of 48 children who had had temporal lobe resections, normal or abnormal ^{1}H MRS on the unoperated side was not a prognostic factor for seizure outcome. Patients who had right temporal resections and who had abnormalities of MRS in the left temporal lobe had some verbal memory deficits, suggesting that MRS data may be a useful indicator of the functional integrity of this part of the brain (Incisa della Rocchetta *et al.*, 1995). This was corroborated by the finding that abnormalities of ^{1}H MRS in the left temporal lobe were associated with a loss of verbal cognitive functions and abnormalities on the right were associated with impaired nonverbal functions (Gadian *et al.*, 1996).

2. ^{1}H MRS Investigations in Extratemporal Epilepsies

A ^{1}H MRSI study reported reduced NAA in frontal lobes ipsilateral to frontal lobe epileptic foci in eight patients (Garcia *et al.*, 1993). In frontal lobe epilepsy, NAA/Cr was reduced by a mean 27% in the lobe of seizure onset (Garcia *et al.*, 1995) and the decrease in NAA was inversely related to seizure frequency in patients with frontal lobe epilepsy, suggesting that a higher seizure frequency is associated with more neuronal dysfunction or loss (Garcia *et al.*, 1997). In patients with extratemporal lobe epilepsy, NAA/Cho and Cr were reduced over a wide area, particularly at the site of seizure onset (Stanley *et al.*, 1998).

3. Malformations of Cortical Development

Reduced NAA/Cho and NAA/Cr have been shown in focal cortical dysplasia (Kuzniecky *et al.*, 1997). In two children with hemimegalencephaly, the white matter of the affected hemisphere had markedly reduced concentrations of NAA and glutamate, with mild abnormalities in the contralateral hemisphere and less marked changes in the gray matter. One child had increased myoinositol and choline-containing compounds in gray matter, suggesting gliosis (Hanefield *et al.*, 1996). NAA/Cr has been reported to be reduced in hamartomas in comparison with normal hypothalami (Tasch et al., 1998). In a patient with focal cortical dysplasia studied with single-voxel MRS, no abnormality was detected (Lee et al., 1998). In an MRSI study of cerebral malformations, there was a decrease of NAA in heterotopia, and abnormalities were also noted in brain that appeared normal on MRI (Li et al., 1998). Interpretation of these studies, however, is difficult without correction being made for partial volume effects.

At present, ^{1}H MRS appears to be a sensitive method for detecting regional neuronal integrity and may identify areas of gliosis. In the future, the technique may contribute in clinical practice to the lateralization and localization of the epileptic focus and the identification of bilateral abnormalities and may further reduce dependence on invasive EEG studies. It will be difficult to firmly establish this role as presurgical evaluation depends on establishing a consensus between different strands of data. Examination of a larger number of patients is needed to determine the strength of the associations between ^{1}H MRS and other investigatory data and the clinical significance of discrepancies.

4. Lactate

An increase of lactate was shown by ^{1}H MRS in the brains of rabbits after bicuculline-induced status epilepticus (Petroff et al., 1986). In vivo microdialysis has shown an increase of lactate for 60–90 min after isolated complex partial seizures. Interictally, extracellular lactate concentrations correlated with interictal spike frequency (During et al., 1994). Decrease in NAA and increase in lactate were noted in rat piriform cortex 12 h after kainate administration at a time when there was little change in T_2-weighted MRI or histology (Ebisu et al., 1996), illustrating the potential for MRS to identify subtle functional changes. Kainic acid induced seizures in rat hippocampi were associated with elevation of lactate/Cr for up to 24 h, a transient increase of NAA/Cr, and a subsequent fall of NAA/Cr (Najm et al., 1997, 1998).

Using in vivo ^{1}H/^{13}C magnetic resonance spectroscopy, Petroff et al. showed that cortical electroshock caused a prolonged rise of brain lactate levels without significant change in intracellular pH or high-energy phosphorylated compounds. The brain lactate approached equilibrium with blood glucose within 1 h, with nearly complete turnover of the raised brain lactate pool (Petroff et al., 1992).

Elevated lactate was reported in patients with focal motor seizures and Rasmussen's encephalitis (Matthews et al., 1990; Cendes et al., 1995). A postictal rise in lactate has been shown using ^{1}H MRSI in the ipsilateral temporal lobe in patients with unilateral temporal lobe epilepsy, and also confined to one side in patients who

appeared to have bilateral temporal lobe epilepsy. The elevation in lactate persisted for up to 6.5 h and this may be a useful technique for lateralizing seizure foci (Ng et al., 1994). An elevation of cerebral lactate has been noted during and for a few hours after complex partial seizures, with no change in NAA. In contrast, there were no changes following absences (Cendes et al., 1997c).

5. Functional MRS and Lactate

Dynamic changes in the concentration of lactate in the visual cortex have been demonstrated following photic stimulation, which correlate with EPI BOLD fMRI time courses (Menon and Gati, 1997). These increases in lactate are generally interpreted as a transient excess of glycolysis over respiration, perhaps as a result of stimulation of astrocytic glycolysis by glutamate uptake (Pellerin and Magistretti, 1994).

C. ^{1}H MRS Investigations of Neurotransmitters

1. GABA

GABA is the principal inhibitory neurotransmitter in the brain, acting at up to 40% of synapses, with a resting concentration of 1–2 mmol/l and a major role in regulation of seizure activity (Loscher and Schmidt, 1994). GABA is concentrated in GABAergic neurons, and its concentration in vesicles is dependent on cytosol GABA concentrations. Decreased GABAergic interneurons have been noted in epileptic cortex (Marco et al., 1996; Spreafico et al., 1998). Microdialysis studies have shown low GABA levels in epileptic hippocampi (During and Spencer, 1993). ^{1}H MRS, using spectral editing, can identify cerebral GABA in vivo and estimate the rise in cerebral GABA concentrations that occurs after administration of vigabatrin (Rothman et al., 1993; Preece et al., 1994; Petroff et al., 1995a), gabapentin (Petroff et al., 1996a), and topiramate. Following a single oral dose of vigabatrin, the rate of GABA synthesis in human brain was estimated to account for 17% of the Krebs cycle rate (Petroff and Rothman, 1998). In healthy control subjects, MRS showed that topiramate resulted in elevation of brain GABA concentrations by 72% at 3 h (Kuzniecky et al., 1998b). Low GABA concentrations have been associated with continued seizure activity (Petroff et al., 1996b).

Homocarnosine is a dipeptide of GABA and histidine in the brain, most likely contained in a subclass of GABAergic neurons (Rothman et al., 1997). Patients whose seizure control improved with vigabatrin had a higher cerebral homocarnosine concentration than those who did not improve (Petroff et al., 1998), implying a role of this compound in seizure suppression. Topiramate has also been associated with increases in homocarnosine and pyrrolidone, which is the internal lactam of GABA (Petroff *et al.*, 1999).

An alternative MRS technique for quantitation of cerebral GABA concentrations, which enables editing to be achieved in a single shot, is the multiple-quantum method. This produces better suppression of the creatine signal at 3.03 ppm, which obscures the GABA resonance, and may allow detection of lower GABA concentrations than is achievable with difference methods (Keltner *et al.*, 1997; Shen *et al.*, 1999). This promises to be most useful in studies of conditions in which GABA concentrations may be reduced.

2. Glutamate and Glutamine

Glutamate is the principal excitatory neurotransmitter in the brain and is responsible for mediating excitotoxicity and initiating epileptic activity (Loscher, 1998). Glutamate is also an intermediary metabolite and is present at a concentration of 8-12 mmol/l. Aspartate, also an excitatory transmitter, is present at a concentration of 1-3 mmol/l. High proportions of neocortical neurons are glutamatergic. Neocortical glutamate concentration reflects the amounts in glutamatergic neurons. Neuronal glutamate enters the synaptic cleft during transmitter release and is taken up by glia and converted into glutamine, which is taken back into neurons and reconverted into glutamate prior to storage in vesicles. Increased excitatory neurons (Spreafico *et al.*, 1998) and increased glutamate concentrations have been noted in epileptogenic cortex (Petroff *et al.*, 1995b). Microdialysis studies have shown increased levels of glutamate in epileptogenic hippocampi, with further increases of glutamate preceding seizures (During and Spencer, 1993). Increased glutamate levels have been noted in the CSF of newly diagnosed patients with epilepsy and to be predictive of a good response to γ-vinyl-GABA (Kalviainen *et al.*, 1993), which decreased CSF glutamate and glutamine (Ben-Menachem, 1989). There are anecdotal reports of high concentrations of glutamate and aspartate, measured using MRS, in cysts adjoining epileptic brain tissue, raising the possibility that the chemical abnormality was the basis of the epileptogenesis (Hajek *et al.*, 1997).

Glutamate is not completely visible to MRS, at least at the long echo times (TE > 100 ms) and high fields used in animal studies (De Graaf *et al.*, 1990; Kauppinen *et al.*, 1991) and only a proportion is in the neurotransmitter pool. Nevertheless, cytosolic glutamate concentrations have a crucial role for maintaining the vesicular glutamate concentrations (Fyske and Fonnum, 1996).

Discrimination between glutamate and glutamine *in vivo* on clinical scanners requires spectral modeling, because of the large number of coupled overlapping peaks and the limited achievable spectral resolution. A primitive approach of modeling several signal intensities between 2 and 3 ppm as a linear combination of glutamate, glutamine, GABA, and NAA has been reported to obtain separate estimates of glutamate and glutamine from 17-ms TE ISIS spectra at 2 T (Petroff *et al.*, 1995a). More recently, separate concentrations of glutamate and glutamine have been reported in the temporal lobe using LCModel (Provencher *et al.*, 1993) on short TE (20 ms) STEAM spectra at 2 T (Choi and Frahm, 1999).

MRS using short echo times (30 ms), voxels tailored to individual hippocampi, and quantitative assessment with the LCModel (Provencher *et al.*, 1993) has shown reduced NAA and increased myoinositol (reflecting gliosis) in epileptogenic sclerotic hippocampi and similar, but less severe, abnormalities contralaterally (Woermann *et al.*, 1999). In patients with TLE and normal MRI, there was a different MRS profile, characterized by elevation of glutamate and glutamine. This suggested a different pathogenetic process in those patients, that there may be a neurochemical abnormality without evident structural abnormalities, and that this may predict the site of onset of epileptic activity (Woermann *et al.*, 1999). An increased concentration of combined glutamate +glutamine was noted following focal status epilepticus, with resolution at 3 months, but persistence of low levels of NAA (Fazekas *et al.*, 1995). The scene is now set for the regional measurement of GABA, glutamate, and glutamine in specific epilepsy syndromes. Research in this area should lead to the neurochemical classification of epilepsies, clarification of the effects of pharmacological, surgical, and other therapies, subsequent prediction of treatments that are most likely to be efficacious, and identification of new pharmacological targets.

D. ^{31}P Spectroscopy

Cerebral metabolites detectable with ^{31}P MRS in epilepsy include compounds related to high-energy phosphate and phospholipid metabolism such as adenosine triphosphate (ATP), phosphomonoesters (PME), phosphodiesters (PDE), phosphocreatine (PCr), and inorganic phosphate (P_i). At neutral pH, P_i exists principally as HPO_4 and H_2PO_4. The chemical shift of ^{31}P in these two molecules differs by approximately 2.4 ppm, but rapid exchange between the two forms results in only a single MR spectral peak being detected. The resonance frequency of the peak is determined by the proportion of the two species present and as the equilibrium is dependent on the pH of the tissue, this is reflected in the effective chemical shift of P_i and is measurable *in vivo*.

In animal models, epileptic seizures can produce marked changes in energy metabolism and tissue pH (Siesjo, 1978). ^{31}P MRS has been used to investigate the metabolic changes associated with partial seizure foci.

Laxer et al. (1992) studied the anterior temporal lobes of eight patients with complex partial seizures and HS and found no significant asymmetries between ipsilateral and contralateral temporal lobes of ATP, PCr, or PDE concentrations. In seven of the eight patients, the temporal lobe ipsilateral to the focus had increased pH (mean in all eight patients of 7.25 vs 7.08) and, in all eight, increased P_i (mean 1.9 vs 1.1 mM). Concentrations of PME were less on the side of the focus, although this was not statistically significant. No significant side-to-side asymmetries were noted in eight normal subjects. There was no apparent relationship between the pH and P_i levels and severity of abnormality shown by MRI. The spatial resolution of this study was very limited and tissue heterogeneity may have confounded interpretation of the data. The same group subsequently investigated eight patients with partial seizures (seven temporal and one frontal) using ^{31}P MRSI, with an effective voxel size of 25 cm^3, enabling more precise delineation of regions of interest (Hugg et al., 1992). The same lateralizing abnormalities were found, that is, increased P_i, decreased PME, and increased pH (7.17 ipsilateral vs 7.06 contralateral). The side-to-side asymmetry of all metabolite intensities in a control group was less than 10%. Using ^{31}P CSI, Laxer et al. (1993) found increased pH (7.13 ipsilateral vs 6.97 contralateral), higher P_i, and reduced PME in the anterior hippocampus of 11 patients with seizures arising from the anterior hippocampus. In the temporal lobe apart from the hippocampus, PME and P_i showed similar asymmetries, but pH did not. In eight patients with frontal lobe epilepsy, increased pH in all eight and decreased PME in seven patients were found in the epileptogenic frontal lobes, but no alteration in P_i levels was detected (Garcia et al., 1994). In patients with TLE, P_i and pH were higher and PME was lower throughout the temporal lobe ipsilateral to the epileptic focus, compared with the contralateral side, and there were no significant asymmetries outside the temporal lobe. On the contralateral side, there were changes in P_i, pH, and PME in persons with epilepsy as compared with controls that were opposite to changes on the ipsilateral side (Van der Grond et al., 1998).

There was no statistically significant difference in the pH in either temporal lobe in seven patients with temporal lobe epilepsy (Kuzniecky et al., 1992). These were 7.11 (ipsilateral) and 7.05 (contralateral), with the ipsilateral value being closer to that measured in controls (control pH 7.12). No change in PME levels was reported. The PCr/P_i ratios were lower in the ipsilateral

than the contralateral temporal lobe of the patients, and both sides were lower than the control data. This appears consistent with the studies of Laxer et al. (1993) and Hugg et al. (1992) in that each of these two studies reported an increase in P_i with no change in PCr (giving a decrease in PCr/P_i). It should be noted, however, that Kuzniecky et al. interpreted this change in terms of a decrease in PCr, and not an increase in P_i, but did not report intensity levels for the individual metabolites.

The pathophysiological significance of these data is not certain at present. It has been suggested that a decrease in PME may reflect altered metabolism associated with neuronal cell loss and glial proliferation (Hugg et al., 1992), both of which are associated with HS and cortical lesions, although the data do not provide direct evidence of this. All studies found a higher pH on the ipsilateral side compared to the contralateral side. In one series, this difference was not found to be significant (Kuzniecky et al., 1992), whereas the other three studies reported a higher ipsilateral pH that was statistically significant. Kuzniecky et al. (1992) found that the higher ipsilateral pH values were similar to control measurements, whereas Garcia et al. reported that the contralateral pH in the frontal lobe was similar to control values. In conclusion, the evidence for an increase in pH in the region of a seizure focus is not yet definitive and further work is needed on larger groups of both patients and control subjects.

If confirmed, the neurobiological significance of an increase in pH associated with a seizure focus is not clear. Seizures have been shown to produce acidosis. Hugg et al. (1992) postulated that an increase in pH might be the consequence of an adaptation in brain buffering in response to repeated acidotic episodes associated with seizures, but there is no direct evidence for this.

In several of the above studies, ^{31}P MRS appeared to be superior to MRI in determining the lateralization of the seizure focus. The MRI methods and analysis used were not described and it is not clear whether ^{31}P MRS provides useful lateralizing and localizing data over and above that available from optimal MRI, and comparative studies using suboptimal MRI should be treated with caution. Even with this caveat, ^{31}P MRS yields potentially very useful data which will be enhanced by technological developments such as the use of improved radiofrequency coils, particularly double-tuned coils, which would allow ^{1}H imaging and ^{31}P spectroscopy to be carried out in the same examination without changing coils. This would simplify investigations and the anatomical localization of the spectra. Future studies will need to compare ^{31}P MRS data with optimal MRI and address the question of whether MRSI can make a useful contribution to the precise localization of epileptic foci, in addition to lateralization.

E. Carbon Spectroscopy

Carbon-13 is only present in 1% of naturally occurring compounds. Compounds labeled with ^{13}C need to be administered for studies to be carried out. Dynamic investigations allow measurement of glucose transport, oxidation, glutamate turnover, and glutamine synthesis in brain *in vivo* (Gruetter *et al.,* 1992; Rothman *et al.,* 1992; Mason *et al.,* 1995; Van Zijl and Rothman, 1995). Studies in patients with epilepsy are only just beginning but promise important data on the metabolic abnormalities associated with epileptogenesis.

F. Conclusion

Over the past decade MRS has advanced as a noninvasive tool for investigating cerebral metabolism. Implementation of the rapid developments now being made in MR hardware and software may enable parametric imaging of the cerebral concentrations of these compounds, and this may have important consequences for the noninvasive investigation and the medical and surgical treatment of patients with epilepsy.

VIII. Electrophysiological Imaging

A. Electroencephalograph Source Localization

The digital recording of the EEG allows sophisticated postacquisition processing of the data. The equivalent current dipole (ECD) is the model used of the source of the brain's electrical activity. The distribution of the potential at the surface of the scalp, modeled as though arising from a single source, is affected by the shape of the skull and the head. The most useful model of the head is a series of concentric spheres of varying conductivity, representing the brain, skull, and scalp. If several sources are active at the same time, however, the activation detected at the scalp will be a smeared amalgam of the individual sources. There is also the "inverse problem" to contend with when attempting to localize an ECD. This is that an infinite number of source configurations can give rise to the same distribution of potentials at the surface of the scalp. The distribution of the potential of a single ECD depends on six parameters: three for the location, two for the orientation, and one for the amplitude. Iterative algorithms are used to find the most convincing biologically valid solution despite the inverse. During millisecond-by-millisecond analysis of the EEG, the distribution of potentials is not stable over time. In general, analysis over a period of time suggests that a dipole moves within a larger vol-

ume. An alternative explanation is that there are more than one dipole source that are not activated in synchrony. These issues are particularly relevant to the analysis of interictal epileptiform activity, the location and extent of which are unknown (Scherg and Ebersole, 1993).

Studies of interictal epileptiform activity have concluded that most such activity recorded on scalp electrodes arises from underlying neocortex. Delays of up to 220 ms between interictal paroxysms at different recording sites show that interictal epileptiform activity can propagate over several milliseconds to other areas of cortex. Analysis may also indicate spurious deep dipoles (Alarcon *et al.,* 1994). In temporal lobe epilepsy, spike modeling demonstrated dipolar sources in both medial and lateral temporal cortex that were in the hypometabolic area demonstrated using ^{18}FDG PET. Hypometabolism was not more or less marked at the sites of the dipoles, suggesting the presence of epileptogenic networks in the temporal lobe (Merlet *et al.,* 1996). Dipole mapping may localize epileptogenic areas in patients with medial TLE. In patients with frontal lobe epilepsy, the accuracy has been regarded to be at a lobar level (Shindo *et al.,* 1998). Previous surgery and skull defects result in breach rhythms that complicate the analysis

Furthermore, the site of interictal epileptiform activity (the irritative zone) does not necessarily colocate with the site of seizure onset (the epileptogenic zone). Thus, the results of mapping dipoles of interictal epileptiform activity must be interpreted with great caution in the clinical context. It is possible that more sophisticated scalp EEG recording systems, for example, with 124 electrodes, will allow spatial deconvolution and more precise modeling (Gevins *et al.,* 1994). A persisting problem, however, will be to identify and localize activities in the limbic or deep areas of the brain, which are obscured by the activity in the superficial neocortex.

B. Magnetoencephalography

Magnetoencephalographic (MEG) dipole source localization is potentially more accurate than EEG localization techniques because magnetic fields are not attenuated or distorted by the skull and scalp, with the consequence that cerebral sources may be modeled more simply. MEG sources related to interictal epileptiform activity are coregistered with the patient's MRI as an anatomical correlate and for clinical interpretation. Several studies have shown concordance of MEG localization of foci with lesions, intracranial EEG recordings, and the results of surgery (Ebersole, 1997). Sources of epileptic activity may be mapped on the same MRI data set as somatosensory and other evoked cortical responses, which is particularly useful if surgical resection in the vicinity of eloquent cortex is planned (Paetau *et al.*, 1994). In seven of eight patients with tumors having surgery for epilepsy, the 3D magnetic source localization was within 1 cm of the edge of the neoplasm. In eight of nine patients with a temporal/hippocampal atrophy, dipoles of epileptiform activity were located within the atrophic lobe (Stefan *et al.*, 1994). MEG dipole sources gave more consistent results than EEG source analyses (Nakasato *et al.*, 1994).

An advantage of MEG recordings is that in a patient who appears to have more than one focus, the order of activation can be resolved in millisecond scale and the neuronal ciruitry involved potentially elucidated (Forss, 1997). EEG and MEG data are the result of different neuronal activations. The MEG and EEG peaks of focal interictal spikes may be simultaneous, or the MEG peak occurs first, by 9–40 ms (Merlet *et al.*, 1997). An application of the temporal resolution of MEG has been in the Landau-Kleffner syndrome. Secondary spikes occurred within 10–60 ms in ipsilateral perisylvian, temporo-occipital, and parieto-occipital areas. In some patients, a single intrasylvian pacemaker initiated all epileptic activity; in others there were independent left- and right-hemisphere circuits or focal spikes (Paetau *et al.*, 1999).

As with the analysis of ECDs, a major limitation of interpreting MEG data is the "inverse problem," whereby there may be more than a single dipole solution, and more than a single dipole, that is concordant with the MEG data recorded from the scalp. An increased number of recording channels and improved signal to noise help the situation but do not resolve it. Prior knowledge, such as the location of a lesion that is likely to be the cause of seizures, which allows the model to be constrained, aids the situation. Spherical head models have been used in the past for MEG analysis, but as the site and direction of an epileptic dipole source are unknown, it is not possible to estimate the error of localizations except by using models of individual head volumes. The accuracy of magnetic source localization depends on the distance between the sensors and the source, the characteristics of the source, the signal-to-noise ratio, and the adequacy of the source model. Sources in the superficial neocortex may be located within a few millimeters, but those deep in the brain are not likely to be localized with an accuracy of less than 10–15 mm (Rose *et al.*, 1991). Constraints derived from fMRI studies of interictal epileptiform activity may be useful in the analysis of ECDs and of magnetic dipoles.

It is patients who have no anatomical abnormality demonstrated on MRI and no clear localizing EEG data that present the most difficulties in analysis. In a recent study, for example, using a 37-channel neuromagnetometer, source locations within or close to the presurgically defined primary epileptogenic area were evident in only 3 of 15 patients and it was noted that spikes arising from the medial temporal lobe may not be detected with MEG (Brockhaus *et al.*, 1997). Patients need to have an adequate amount of interictal epileptiform activity for a successful study, which may last 2–3 h. The yield is greater with neocortical than with medial temporal lobe epilepsy. In a recent study, MEG spike sources colocalized with the lesions in 5 of 6 patients with focal epileptogenic pathology. Furthermore, results were encouraging in the difficult group of patients with cryptogenic neocortical epilepsies. MEG spike sources were identified in the area of the epileptogenic zone, as determined by clinical and EEG data (including intracranial EEG) in 11 of 12 patients with no definitive MRI abnormality (Knowlton *et al.*, 1997b).

In an evaluation of 17 patients who had not become seizure free following epilepsy surgery, source analysis of interictal epileptic MEG discharges revealed dipole localizations adjacent to the borders of previous resections in 10 patients. Scalp EEG recordings had suggested a wide area of epileptogenicity in the region of the previous resection in all patients, without precise localization. In five patients who subsequently had intracranial EEG studies, the MEG source was confirmed and three became seizure free with further surgery (Kirchberger *et al.*, 1998). Thus, MEG may be useful in the diagnostic reevaluation of patients who have had surgery and in whom breaches in the skull and dura distort the EEG.

In addition to the localization of interictal epileptiform activity, MEG may have a role in the cortical localization of the somatosensory and motor cortices (Smith *et al.*, 1994; Gallen *et al.*, 1995) when surgical resections are planned near these areas. In practice, however, it is likely that this role will be primarily undertaken by fMRI as it is the anatomical localization rather than the analysis of the time course of activations that is the principal concern in this situation.

Acknowledgments

I am grateful to the National Society for Epilepsy, the Medical Research Council, The Wellcome Trust, Action Research, and the National Lottery Charity Board for support.

References

Abou-Khalil, B. W., Siegel, G. J., Sackellares, J. C., Gilman, S., Hichwa, R., and Marshall, R. (1987). Positron emission tomography studies of cerebral glucose metabolism in chronic partial epilepsy. *Ann. Neurol.* **22**, 480–486.

Alarcon, G., Guy, C. N., Binnie, C. D., Walker, S. R., Elwes, R. D., and Polkey, C. E. (1994). Intracerebral propagation of interictal activity in partial epilepsy: Implications for source localisation. *J. Neurol. Neurosurg. Psychiatry* **57**, 435–449.

Allen, P. J., Polizzi, G., Krakow, K., Fish, D. R., and Lemieux, L. (1998). Identification of EEG events in the MR scanner: The problem of pulse artifact and a method for its subtraction. *NeuroImage* **8**, 229–239.

Alsop, D. C., Detre, J. A., D'Esposito, M., Howard, R. S., Maldjian, J. A., Grossman, M., Listerud, J., Flamm, E. S., Judy, K. D., and Atlas, S. W. (1996). Functional activation during an auditory comprehension task in patients with temporal lobe lesions. *NeuroImage* **4**, 55–59.

Andersen, A. R., Waldemar, G., Dam, M., Fuglsang-Frederiksen, A., Herning, M., and Kruse-Larsen, C. (1990). SPECT in the presurgical evaluation of patients with temporal lobe epilepsy: A preliminary report. *Acta Neurochir. Suppl. (Wien)* **50**, 80–83.

Andersen, A. R., Rogvi-Hansen, B., and Dam, M. (1994). Utility of interictal SPECT of rCBF for focal diagnosis of the epileptogenic zone(s). *Acta Neurol. Scand. Suppl.* **152**, 129–134.

Bairamian, D., Di Chiro, G., Theodore, W. H., *et al.* (1985). MR imaging and positron emission tomography of cortical heterotopia. *Comput. Assist. Tomogr.* **9**, 1137–1139.

Bajorek, J. G., Lee, R. L., and Lomax, P. (1986). Neuropeptides: Anticonvulsant and convulsant mechanisms in epileptic model systems and in humans. *Adv. Neurol.* **44**, 489–500.

Banati, R. B., Goerres, G. W., Myers, R., Gunn, R. N., Turkheimer, F. E., Kreutzberg, G. W., Brooks, D. J., Jones, T., and Duncan, J. S. (1999). [^{11}C](R)-PK1119S positron emission tomography imaging of activated microglia *in vivo* in Rasmussen's encephalitis. *Neurology* **53**, 2199–2203.

Bartenstein, P., Ludolph, A., Schober, O., *et al.* (1991). Benzodiazepine receptors and cerebral blood flow in partial epilepsy. *Eur. J. Nucl. Med.* **18**, 111–118.

Bartenstein, P., Lehmenkuhler, C., Sciuk, J., and Schuierer, G. (1992). Cortical dysplasia as an epileptogenic focus: Reduced binding of ^{123}I-iomezanil with barely perceptible ^{99m}Tc-HMPAO SPECT. *Nuklearmedizin* **31**, 142–144.

Bartenstein, P. A., Duncan, J. S., Prevett, M. C., Cunningham, V. J., Fish, D. R., Jones, A. K., Luthra, S. K., Sawle, G. V., and Brooks, D. J. (1993). Investigation of the opioid system in absence seizures with positron emission tomography. *J. Neurol. Neurosurg. Psychiatry* **56**, 1295–1302.

Bartenstein, P. A., Prevett, M. C., Duncan, J. S., Hajek, M., and Wieser, H. G. (1994). Quantification of opiate receptors in two patients with mesiobasal temporal lobe epilepsy, before and after selective amygdalohippocampectomy, using positron emission tomography. *Epilepsy Res.* **18**, 119–125.

Baumgartner, C., Serles, W., Leutmezer, F., Pataraia, E., Aull, S., Czech, T., Pietrzyk, U., Relic, A., and Podreka, I. (1998). Preictal SPECT in temporal lobe epilepsy: Regional cerebral blood flow is increased prior to electroencephalography-seizure onset. *J. Nucl. Med.* **39**, 978–982.

Bellgowan, P. S., Binder, J. R., Swanson, S. J., Hammeke, T. A., Springer, J. A., Frost, J. A., Mueller, W. M., and Morris, G. L. (1998a). Side of seizure focus predicts left medial temporal lobe activation during verbal encoding. *Neurology* **51**, 479–484.

Bellgowan, P. S. F., Binder, J. R., Hammeke, T. A., Frost, J. A., and Possing, E. T. (1998b). Configural associations and novelty are necessary for hippocampal-dependent encoding. *NeuroImage* **7**, S51.

Benbadis, S. R., Binder, J. R., Swanson, S. J., Fischer, M., Hammeke, T. A., Morris, G. L., Frost, J. A., and Springer, J. A. (1998). Is speech arrest during Wada testing a valid method for determining hemispheric representation of language? *Brain Lang.* **65**, 441–446.

Ben-Menachem, E. (1989). Pharmacokinetic effects of vigabatrin on cerebrospinal fluid amino acids in humans. *Epilepsia* **20** (Suppl. 3), S12–S14.

Benson, R. R., Fitzgerald, D. B., LeSueur, L. L., Kennedy, D. N., Kwong, K. K., Buchbinder, B. R., *et al.* (1999). Language dominance determined by whole brain functional MRI in patients with brain lesions. *Neurology* **52**, 798–809.

Bergstrom, M., Kumlein, E., Lilja, A., Tyrefors, N., Westerberg, G., and Langstrom, B. (1998). Temporal lobe epilepsy visualized with PET with ^{11}C-L-deuterium-deprenyl: Analysis of kinetic data. *Acta Neurol. Scand.* **98** (4), 224–231.

Biersack, H. J., Linke, D., Brassel, F., Reichmann, K., Kurthen, M., Durwen, H. F., Reuter, B. M., Wappenschmidt, J., and Stefan, H. (1987). Technetium-99m HM-PAO brain SPECT in epileptic patients before and during unilateral hemispheric anesthesia (Wada test): Report of three cases. *J. Nucl. Med.* **28**, 1763–1767.

Biersack, H. J., Elger, C. E., Grunwald, F., Reichmann, K., and Durwen, H. F. (1988). Brain in epilepsy. *Adv. Func. Neuroimaging* **1**, 4–9.

Binder, J. R., Rao, S. M., Hammeke, T. A., Frost, J. A., Bandettini, P. A., Jesmanowicz, A., and Hyde, J. S. (1995). Lateralized human brain language systems demonstrated by task subtraction functional magnetic resonance imaging. *Arch. Neurol.* **52**, 593–601.

Binder, J. R., Swanson, S. J., Hammeke, T. A., Morris, G. L., Mueller, W. M., Fischer, M., Benbadis, S., Frost, J. A., Rao, S. M., and Haughton, V. M. (1996). Determination of language dominance using functional MRI: A comparison with the Wada test. *Neurology* **46**, 978–984.

Binder, J. R., Frost, J. A., Hammeke, T. A., Cox, R. W., Rao, S. M., and Prieto, S. M. (1997). Human brain language areas identified by functional magnetic resonance imaging. *J. Neurosci.* **17**, 353–362.

Bittar, R. G., Andermann, F., Olivier, A., Dubeau, F., Dumoulin, S. O., Pike, G. B., and Reutens, D. C. (1999). Interictal spikes increase cerebral glucose metabolism and blood flow: A PET study. *Epilepsia* **40**, 170–178.

Blum, D. E., Ehsan, T., Dungan, D., Karis, J. P., and Fisher, R. S. (1998). Bilateral temporal hypometabolism in epilepsy. *Epilepsia* **39**, 651–659.

Bonte, F. J., Stokely, E. M., Devous, M. D., Sr., and Homan, R. W. (1983). Single-photon tomographic study of regional cerebral blood flow in epilepsy. A preliminary report. *Arch. Neurol.* **40**, 267–270.

Bookheimer, S. Y. (1996). Functional MRI applications in clinical epilepsy. *NeuroImage* **4**, S139–S146.

Bookheimer, S. Y., Zeffiro, T. A., Blaxton, T., Malow, B. A., Gaillard, W. D., Sato, S., Kufta, C., Fedio, P., and Theodore, W. H. (1997). A direct comparison of PET activation and electrocortical stimulation mapping for language localization. *Neurology* **48**, 1056–1065.

Breier, J. I., Mullani, N. A., Thomas, A. B., Wheless, J. W., Plenger, P. M., Gould, K. L., Papanicolaou, A., and Willmore, L. J. (1997). Effects of duration of epilepsy on the uncoupling of metabolism and blood flow in complex partial seizures. *Neurology* **48**, 1047–1053.

Brockhaus, A., Lehnertz, K., Wienbruch, C., Kowalik, A., Burr, W., Elbert, T., Hoke, M., and Elger, C. E. (1997). Possibilities and limitations of magnetic source imaging of methohexital-induced epileptiform patterns in temporal lobe epilepsy patients. *Electroencephalogr. Clin. Neurophysiol.* **102**, 423–436.

Bromfield, E. B., Ludlow, C. L., Sedory, S., Leiderman, D. B., and Theodore, W. H. (1991). Cerebral activation during speech discrimination in temporal lobe epilepsy. *Epilepsy Res.* **9**, 49–58.

Buck, A., Frey, L. D., Blauenstein, P., Kramer, G., Siegel, A., Weber, B., Schubiger, P. A., and Wieser, H. G. (1998). Monoamine oxidase B single-photon emission tomography with [^{123}I]RO 43–0463: Imaging in volunteers and patients with temporal lobe epilepsy. *Eur J Nucl Med* **25**, 464–470.

Burdette, D. E., Sakurai, S. Y., Henry, T. R., Ross, D. A., Pennell, P. B., Frey, K. A., Sackellares, J. C., and Albin, R. L. (1995). Temporal lobe central benzodiazepine binding in unilateral mesial temporal lobe epilepsy. *Neurology* **45**, 934–941.

Cendes, F., Andermann, F., Preul, M. C., and Arnold, D. L. (1994). Lateralization of temporal lobe epilepsy based on regional metabolic abnormalities in proton magnetic resonance spectroscopic images. *Ann. Neurol.* **35**, 211–216.

Cendes, F., Andermann, F., Silver, K., and Arnold, D. L. (1995). Imaging of axonal damage *in vivo* in Rasmussen's syndrome. *Brain* **118**, 753–758.

Cendes, F., Andermann, F., Dubeau, F., Matthews, P. M., and Arnold, D. L. (1997a). Normalization of neuronal metabolic dysfunction after surgery for temporal lobe epilepsy. Evidence from proton MR spectroscopic imaging. *Neurology* **49**, 1525–1533.

Cendes, F., Caramanos, Z., Andermann, F., Dubeau, F., and Arnold, D. L. (1997b). Proton magnetic resonance spectroscopic imaging and magnetic resonance imaging volumetry in the lateralization of temporal lobe epilepsy: A series of 100 patients. *Ann. Neurol.* **42**, 737–746.

Cendes, F., Stanley, J. A., Dubeau, F., Andermann, F., and Arnold, D. L. (1997c). Proton magnetic resonance spectroscopic imaging for discrimination of absence and complex partial seizures. *Ann. Neurol.* **41**, 74–81.

Chee, M. W., Morris, H. H., 3rd, Antar, M. A., Van Ness, P. C., Dinner, D. S., Rehm, P., and Salanova, V. (1993). Presurgical evaluation of temporal lobe epilepsy using interictal temporal spikes and positron emission tomography. *Arch. Neurol.* **50**, 45–48.

Chiron, C., Raynaud, C., Dulac, O., Tzourio, N., Plouin, P., and Tran-Dinh, S. (1989). Study of the cerebral blood flow in partial epilepsy of childhood using the SPECT method. *J. Neuroradiol.* **16**, 317–324.

Chiron, C., Dulac, O., Bulteau, C., Nuttin, C., Depas, G., Raynaud, C., and Syrota, A. (1993). Study of regional cerebral blood flow in West syndrome. *Epilepsia* **34**, 707–715.

Chiron, C., Nabbout, R., Pinton, F., Nuttin, C., Dulac, O., and Syrota, A. (1996). Brain functional imaging SPECT in agyria-pachygyria. *Epilepsy Res.* **24**, 109–117.

Choi, C. G., and Frahm, J. (1999). Localized proton MRS of the human hippocampus: Metabolite concentrations and relaxation times. *Magn. Reson. Med.* **41**, 204–207.

Chugani, D. C., Chugani, H. T., Muzik, O., Shah, J. R., Shah, A. K., Canady, A., Mangner, T. J., and Chakraborty, P. K. (1998). Imaging epileptogenic tubers in children with tuberous sclerosis complex using α-[^{11}C]methyl-L-tryptophan positron emission tomography. *Ann. Neurol.* **44**, 858–866.

Chugani, H. T. (1994). The role of PET in childhood epilepsy. *J. Child Neurol.* **9** (Suppl. 1), S82–S88.

Chugani, H. T., Mazziotta, J. C., Engel, J., Jr., and Phelps, M. E. (1987). The Lennox-Gastaut syndrome: Metabolic subtypes determined by 2–deoxy-2[^{18}F]fluoro-D-glucose positron emission tomography. *Ann. Neurol.* **21**, 4–13.

Chugani, H. T., Shewmon, D. A., Khanna, S., and Phelps, M. E. (1993). Interictal and postictal focal hypermetabolism on positron emission tomography. *Pediatr. Neurol.* **9**, 10–15.

Chugani, H. T., Rintahaka, P. J., and Shewmon, D. A. (1994). Ictal patterns of cerebral glucose utilization in children with epilepsy. *Epilepsia* **35**, 813–822.

Connelly, A., Jackson, G. D., Duncan, J. S., King, M. D., and Gadian, D. G. (1994). Magnetic resonance spectroscopy in temporal lobe epilepsy. *Neurology* **44**, 1411–1417.

Connelly, A., Van Paesschen, W., Porter, D. A., Johnson, C. L., Duncan, J. S., and Gadian, D. G. (1998). Proton magnetic resonance spectroscopy in MRI-negative temporal lobe epilepsy. *Neurology* **51**, 61–66.

Connelly, A., and Duncan, J. S. (1995). Magnetic resonance spectroscopy in epilepsy. *In* "Recent Advances in Epilepsy" (T. A. Pedley and B. S. Meldrum, Eds.), Vol. 6, pp. 23–40. Churchill Livingstone, Edinburgh.

Cordes, M., Christe, W., Henkes, H., Delavier, U., Eichstadt, H., Schorner, W., Langer, R., and Felix, R. (1990). Focal epilepsies: HM-PAO SPECT compared with CT, MR, and EEG. *J. Comput. Assist. Tomogr.* **14**, 402–409.

Cordes, M., Henkes, H., Ferstl, F., Schmitz, B., Hierholzer, J., Schmidt, D., and Felix, R. (1992). Evaluation of focal epilepsy: A SPECT scanning comparison of ^{123}I-iomazenil versus HM PAO. *Am. J. Neuroradiol.* **13**, 249–253.

Cornford, E. M., Gee, M. N., Swartz, B. E., Mandelkern, M. A., Blahd, W. H., Landaw, E. M., and Delgado-Escueta, A. V. (1998). Dynamic [^{18}F]fluorodeoxyglucose positron emission tomography and hypometabolic zones in seizures: Reduced capillary influx. *Ann. Neurol.* **43**, 801–808.

Coubes, P., Awad, I. A., Antar, M., Magdinec, M., and Sufka, B. (1993). Comparison and spacial correlation of interictal HMPAO-SPECT and FDG-PET in intractable temporal lobe epilepsy. *Neurol. Res.* **15**, 160–168.

Coubes, P., Baldy-Moulinier, M., Zanca, M., Boire, J. Y., Child, R., Bourbotte, G., and Frerebeau, P. (1995). Monitoring sodium methohexital distribution with [^{99m}Tc]HMPAO with single photon emission computed tomography during Wada test. *Epilepsia* **36**, 1041–1049.

Cross, J. H., Gordon, I., Jackson, G. D., Boyd, S. G., Todd-Pokropek, A., Anderson, P. J., and Neville, B. G. (1995). Children with intractable focal epilepsy: Ictal and interictal 99mTc HMPAO single photon emission computed tomography. *Dev. Med. Child Neurol.* **37**, 673–681.

Cross, J. H., Connelly, A., Jackson, G. D., Johnson, C. L., Neville, B. G. R., and Gadian, D. G. (1996). Proton magnetic resonance spectroscopy in children with temporal lobe epilepsy. *Ann. Neurol.* **39**, 107–113.

Cross, J. H., Gordon, I., Connelly, A., Jackson, G. D., Johnson, C. L., Neville, B. G., and Gadian, D. G. (1997a). Interictal ^{99}Tc(m) HMPAO SPECT and ^{1}H MRS in children with temporal lobe epilepsy. *Epilepsia* **38**, 338–345.

Cross, J. H., Boyd, S. G., Gordon, I., Harper, A., and Neville, B. G. (1997b). Ictal cerebral perfusion related to EEG in drug resistant focal epilepsy of childhood. *J. Neurol. Neurosurg. Psychiatry* **62**, 377–384.

Dasheiff, R. M. (1992). A review of interictal cerebral blood flow in the evaluation of patients for epilepsy surgery. *Seizure* **1**, 117–125.

da Silva, E. A., Chugani, D. C., Muzik, O., and Chugani, H. T. (1997). Landau-Kleffner syndrome: Metabolic abnormalities in temporal lobe are a common feature. *J. Child Neurol.* **12**, 489–495.

Debets, R. M., Sadzot, B., van Isselt, J. W., Brekelmans, G. J., Meiners, L. C., van Huffelen, A. O., Franck, G., and van Veelen, C. W. (1997). Is ^{11}C-flumazenil PET superior to ^{18}FDG PET and ^{123}I-iomazenil SPECT in presurgical evaluation of temporal lobe epilepsy? *J. Neurol. Neurosurg. Psychiatry* **62**, 141–150.

De Graaf, A. A., and Bovee, W. M. M. (1990). Improved quantification of *in vivo* ^{1}H NMR spectra by optimization of signal acquisition and processing and by incorporation of prior knowledge into spectra fitting. *Magn. Reson. Med.* **15**, 305–319.

Delbeke, D., Lawrence, S. K., Abou-Khalil, B. W., Blumenkopf, B., and Kessler, R. M. (1996). Postsurgical outcome of patients with uncontrolled complex partial seizures and temporal lobe hypometabolism on ^{18}FDG-positron emission tomography. *Invest. Radiol.* **31**, 261–266.

Denays, R., Rubinstein, M., Ham, H., Piepsz, A., and Noel, P. (1988). Single photon emission computed tomography in seizure disorders. *Arch. Dis. Child.* **63**, 1184–1188.

Desbiens, R., Berkovic, S. F., Dubeau, F., Andermann, F., Laxer, K. D., Harvey, S., Leproux, F., Melanson, D., Robitaille, Y., Kalnins, R., *et al.* (1993). Life threatening focal status epilepticus due to occult cortical dysplasia. *Arch. Neurol.* **50**, 695–700.

Desmond, J. E., Sum, J. M., Wagner, A. D., *et al.* (1995). Functional MRI measurement of language lateralization in Wada-tested patients. *Brain* **118**, 1411–1419.

Detre, J. A., Sirven, J. I., Alsop, D. C., O'Connor, M. J., and French, J. A. (1995). Localization of subclinical ictal activity by functional magnetic resonance imaging: Correlation with invasive monitoring. *Ann. Neurol.* **38**, 618–624.

Detre, J. A., Alsop, D. C., Aguirre, G. K., and Sperling, M. R. (1996). Coupling of cortical and thalamic ictal activity in human partial epilepsy: Demonstration by functional magnetic resonance imaging. *Epilepsia* **37**, 657–661.

Detre, J. A., Maccotta, L., King, D., Alsop, D. C., Glosser, G., D'Esposito, M., Zarahn, E., Aguirre, G. K., and French, J. A. (1998). Functional MRI lateralization of memory in temporal lobe epilepsy. *Neurology* **50**, 926–932.

De Volder, A. G., Gadisseux, J. F., Michel, C. J., Maloteaux, J. M., Bol, A. C., Grandin, C. B., Duprez, T. P., and Evrard, P. (1994). Brain glucose utilization in band heterotopia: Synaptic activity of "double cortex." *Pediatr. Neurol.* **11**, 290–294.

Devous, M. D., Sr., Thisted, R. A., Morgan, G. F., Leroy, R. F., and Rowe, C. C. (1998). SPECT brain imaging in epilepsy: A meta-analysis. *J. Nucl. Med.* **39**, 285–293.

Dietrich, M. E., Bergen, D., Smith, M. C., Fariello, R., and Ali, A. (1991). Correlation of abnormalities of interictal *n*-isopropyl-*p*-iodoamphetamine single-photon emission tomography with focus of seizure onset in complex partial seizure disorders. *Epilepsia* **32**, 187–194.

Duc, C. O., Trabesinger, A. H., Weber, O. M., Meier, D., Walder, M., Wieser, H. G., and Boesiger, P. (1998). Quantitative ^{1}H MRS in the evaluation of mesial temporal lobe epilepsy *in vivo. Magn. Reson. Imaging* **16**, 969–979.

Duncan, J. S. (1997). Imaging and epilepsy. *Brain* **120**, 339–378.

Duncan, J. S., and Fish, D. R. (1998). Integration of structural and functional data. *Curr. Opinion Neurol.* **11**(2), 119–122.

Duncan, R., Patterson, J., Hadley, D. M., Macpherson, P., Brodie, M. J., Bone, I., McGeorge, A. P., and Wyper, D. J. (1990). CT, MR and SPECT imaging in temporal lobe epilepsy. *J. Neurol. Neurosurg. Psychiatry* **53**, 11–15.

Duncan, R., Patterson, J., Roberts, R., Hadley, D. M., and Bone, I. (1993). Ictal/postictal SPECT in the pre-surgical localisation of complex partial seizures. *J. Neurol. Neurosurg. Psychiatry* **56**, 141–148.

Duncan, R., Patterson, J., Hadley, D., and Roberts, R. (1997a). Ictal regional cerebral blood flow in frontal lobe seizures. *Seizure* **6**, 393–401.

Duncan, R., Biraben, A., Patterson, J., Hadley, D., Bernard, A. M., Lecloirec, J., Vignal, J. P., and Chauvel, P. (1997b). Ictal single photon emission computed tomography in occipital lobe seizures. *Epilepsia* **38**, 839–843.

Duncan, S., Gillen, G. J., and Brodie, M. J. (1993). Lack of effect of concomitant clobazam on interictal ^{123}I-iomazenil SPECT. *Epilepsy Res.* **15**, 61–66.

Dupont, S., Semah, F., Baulac, M., and Samson, Y. (1998). The underlying pathophysiology of ictal dystonia in temporal lobe epilepsy: An FDG-PET study. *Neurology* **51**(5), 1289–1292.

During, M. J., and Spencer, D. D. (1993). Extracellular hippocampal glutamate and spontaneous seizures in the conscious human brain. *Lancet* **341**, 1607–1610.

During, M. J., Fried, I., Leone, P., Katz, A., and Spencer, D. D. (1994). Direct measurement of extracellular lactate in the human hippocampus during spontaneous seizures. *J. Neurochem.* **62**, 2356–2361.

Ebersole, J. S. (1997). Magnetoencephalography/magnetic source imaging in the assessment of patients with epilepsy. *Epilepsia.* **38** (Suppl. 4), S1–S5.

Ebisu, T., Rooney, W. D., Graham, S. H., Mancuso, A., Weiner, M. W., and Maudsley, A. A. (1996). MR spectroscopic imaging and diffusion-weighted MRI for early detection of kainate-induced status epilepticus in the rat. *Magn. Reson. Med.* **36**, 821–828.

Ebner, A., Buschsieweke, U., Tuxhorn, I., Witte, O. W., and Seitz, R. J. (1996). Supplementary sensorimotor area seizure and ictal single-photon emission tomography. *Adv. Neurol.* **70**, 363–368.

Ende, G. R., Laxer, K. D., Knowlton, R. C., Matson, G. B., Schuff, N., Fein, G., and Weiner, M. W. (1997). Temporal lobe epilepsy: Bilateral hippocampal metabolite changes revealed at proton MR spectroscopic imaging. *Radiology* **202**, 809–817.

Engel, J., Kuhl, D. E., Phelps, M. E., and Mazziotta, J. C. (1982a). Interictal cerebral glucose metabolism in partial epilepsy and its relation to EEG changes. *Ann. Neurol.* **12**, 510–517.

Engel, J., Brown, W. G., Kuhl, D. E., Phelps, M. E., Mazziota, J. C., and Crandall, P. H. (1982b). Pathological findings underlying focal temporal hypometabolism in partial epilepsy. *Ann. Neurol.* **12**, 518–528.

Engel, J., Kuhl, D. E., Phelps, M. E., Rausch, R., and Nuwer, M. (1983). Local cerebral metabolism during partial seizures. *Neurology* **33**, 400–413.

Engel, J., Jr., Lubens, P., Kuhl, D. E., and Phelps, M. E. (1985). Local cerebral metabolic rate for glucose during petit mal absences. *Ann. Neurol.* **17**, 121–128.

Engel, J., Jr., Henry, T. R., Risinger, M. W., Mazziotta, J. C., Sutherling, W. W., Levesque, M. F., and Phelps, M. E. (1990). Presurgical evaluation for partial epilepsy: Relative contributions of chronic depth-electrode recordings versus FDG-PET and scalp-sphenoidal ictal EEG. *Neurology* **40**, 1670–1677.

Engel, J., Henry, T. R., and Swartz, B. E. (1995). Positron emission tomography in frontal lobe epilepsy. *In* "epilepsy and the Functional Anatomy of the Frontal Lobe" (H. H. Jasper, S. Riggio, and P. S. Goldman-Rakic, Eds.), pp. 223–238. Raven Press, New York.

Falconer, J., Wada, J. A., Martin, W., and Li, D. (1990). PET, CT, and MRI imaging of neuronal migration anomalies in epileptic patients. *Can. J. Neurol. Sci.* **17**, 35–39.

Fazekas, F., Kapeller, P., Schmidt, R., *et al.* (1995). Magnetic resonance imaging and spectroscopy findings after focal status epilepticus. *Epilepsia* **36**, 946–949.

Feeney, D. M., and Baron, J. C. (1986). Diaschisis. *Stroke* **17**, 817–830.

Fernandez, G., Weyerts, H., Schrader-Bolsche, M., *et al.* (1998). Successful verbal encoding into episodic memory engages the posterior hippocampus: A parametrically analyzed functional magnetic resonance imaging study. *J. Neurosci.* **18**, 1841–1847.

Ferrie, C. D., Marsden, P. K., Maisey, M. N., and Robinson, R. O. (1997). Cortical and subcortical glucose metabolism in childhood epileptic encephalopathies. *J. Neurol. Neurosurg. Psychiatry* **63**, 181–187.

Fink, G. R., Pawlik, G., Stefan, H., Pietrzyk, U., Wienhard, K., and Heiss, W. D. (1996). Temporal lobe epilepsy: Evidence for interictal uncoupling of blood flow and glucose metabolism in temporomesial structures. *J. Neurol. Sci.* **137**, 28–34.

Fish, D. R., Brooks, D. J., Young, I. R., and Bydder, G. M. (1988). Use of magnetic resonance imaging to identify changes in cerebral blood flow in epilepsia partialis continua. *Magn. Reson. Med.* **8**, 238–240.

Forss, N. (1997). Magnetoencephalography (MEG) in epilepsy surgery. *Acta Neurochir. Suppl. (Wien)* **68**, 81–84.

Franceschi, M., Lucignani, G., Del Sole, A., Grana, C., Bressi, S., Minicucci, F., Messa, C., Canevini, M. P., and Fazio, F. (1995). Increased interictal cerebral glucose metabolism in a cortical-subcortical network in drug naive patients with cryptogenic temporal lobe epilepsy. *J. Neurol. Neurosurg. Psychiatry* **59**, 427–431.

Franck, G., Sadzot, B., Salmon, E., Depresseux, J. C., Grisar, T., Peters, J. M., Guillaume, M., Quaglia, L., Delfiore, G., and Lamotte, D. (1986). Regional cerebral blood flow and metabolic rates in human focal epilepsy and status epilepticus. *In* "Advances in Neurology" (A. V. Delgado-Escueta, A. A. Ward, D. M. Woodbury, and R. J. Porter, Eds.), Vol. 44, pp. 935–948. Raven Press, New York.

Frey, H. H., and Voits, M. (1991). Effect of psychotropic agents on a model of absence epilepsy in rats. *Neuropharmacology* **30,** 651–656.

Friston, K. J., Holmes, A. P., Worsley, K. J., Poline, J. P., Frith, C. D., and Frackowiak, R. S. J. (1995). Statistical parametric maps in functional imaging: A general linear approach. *Hum. Brain Mapp.* **2,** 189–210.

Frost, J. J., Mayberg, H. S., Fisher, R. S., Douglass, K. H., Dannals, R. F., Links, J. M., Wilson, A. A., Ravert, H. T., Rosenbaum, A. E., Snyder, S. H., *et al.* (1988). Mu opiate receptors measured by positron emission tomography are increased in temporal lobe epilepsy. *Ann. Neurol.* **23,** 231–237.

Fyske, E. M., and Fonnum, F. (1996). Amino acid neurotransmission: Dynamics of vesicular uptake. *Neurochem. Res.* **21,** 1053–1060.

Gabrieli, J. D. E., Brewer, J. B., Desmond, J. E., and Glover, G. H. (1997). Separate neural bases of two fundamental memory processes in human medial temporal lobe. *Science* **276,** 264–266.

Gadian, D. G., Connelly, A., Duncan, J. S., Cross, J. H., Kirkham, F. J., Johnson, C. L., Vargha-Khadem, F., Neville, B. G. R., and Jackson, G. D. (1994). ^{1}H magnetic resonance spectroscopy in the investigation of intractable epilepsy. *Acta Neurol. Scand.* **89** (Suppl. 152), 116–122.

Gadian, D. G., Isaacs, E. B., Cross, J. H., Connelly, A., Jackson, G. D., King, M. D., Neville, B. G., and Vargha-Khadem, F. (1996). Lateralization of brain function in childhood revealed by magnetic resonance spectroscopy. *Neurology* **46,** 974–977.

Gaggero, R., Caputo, M., Fiorio, P., Pessagno, A., Baglietto, M. G., Muttini, P., and De Negri, M. (1995). SPECT and epilepsy with continuous spike waves during slow-wave sleep. *Childs Nerv. Syst.* **11,** 154–160.

Gaillard, W. D., Bhatia, S., Bookheimer, S. Y., Fazilat, S., Sato, S., and Theodore, W. H. (1995a). FDG-PET and volumetric MRI in the evaluation of patients with partial epilepsy. *Neurology* **45,** 123–126.

Gaillard, W. D., White, S., Malow, B., Flamini, R., Weinstein, S., Sato, S., Kufta, C., Schiff, S., Devinsky, O., Fazilat, S., *et al.* (1995b). FDG-PET in children and adolescents with partial seizures: Role in epilepsy surgery evaluation. *Epilepsy Res.* **20,** 77–84.

Gaillard, W. D., Fazilat, S., White, S., Malow, B., Sato, S., Reeves, P., Herscovitch, P., and Theodore, W. H. (1995c). Interictal metabolism and blood flow are uncoupled in temporal lobe cortex of patients with complex partial epilepsy. *Neurology* **45,** 1841–1847.

Gaillard, W. D., Zeffiro, T., Fazilat, S., DeCarli, C., and Theodore, W. H. (1996). Effect of valproate on cerebral metabolism and blood flow: An ^{18}F-2–deoxyglucose and ^{15}O-water positron emission tomography study. *Epilepsia* **37,** 515–521.

Gallen, C. C., Schwartz, B. J., Bucholz, R. D., *et al.* (1995). Presurgical localization of functional cortex using magnetic source imaging. *J. Neurosurg.* **82,** 988–994.

Garcia, P. A., Laxer, K. D., Van der Grond, J., Hugg, J. W., Matson, G. B., and Weiner, M. W. (1993). ^{1}H magnetic resonance spectroscopic imaging in patients with frontal lobe epilepsy. *Epilepsia* **34** (Suppl. 6), 122.

Garcia, P. A., Laxer, K. D., Van der Grond, J., Hugg, J. W., Matson, G. B., and Weiner, M. W. (1994). Phosphorus magnetic resonance spectroscopic imaging in patients with frontal lobe epilepsy. *Ann. Neurol.* **35,** 217–221.

Garcia, P. A., Laxer, K. D., Van der Grond, J., Hugg, J. W., Matson, G. B., and Weiner, M. W. (1995). Proton magnetic resonance spectroscopic imaging in patients with frontal lobe epilepsy. *Ann. Neurol.* **37,** 279–281.

Garcia, P. A., Laxer, K. D., Van der Grond, J., Hugg, J. W., Matson, G. B., and Weiner, M. W. (1997). Correlation of seizure frequency with *N*-acetyl-aspartate levels determined by ^{1}H magnetic resonance spectroscopic imaging. *Magn. Reson. Imaging* **15,** 475–478.

Gevins, A., Le, J., Martin, N. K., Brickett, P., Desmond, J., and Reutter, B. (1994). High resolution EEG: 124–channel recording, spatial deblurring and MRI integration methods. *Electroencephlogr. Clin. Neurophysiol.* **90,** 337–358.

Gruetter, R., Novotny, E. J., Boulware, S. D., Rothman, D. L., Mason, G. F., Shulman, G. I., Shulman, R. G., and Tamborlane, W. T. (1992). Direct measurements of brain glucose concentrations in humans by ^{13}C NMR spectroscopy. *Proc. Natl. Acad. Sci. U.S.A.* **89,** 1109–1112.

Grunwald, F., Menzel, C., Pavics, L., Bauer, J., Hufnagel, A., Reichmann, K., Sakowski, R., Elger, C. E., and Biersack, H. J. (1994). Ictal and interictal brain SPECT imaging in epilepsy using technetium-99m-ECD. *J. Nucl. Med.* **35,** 1896–1901.

Guillon, B., Duncan, R., Biraben, A., Bernard, A. M., Vignal, J. P., and Chauvel, P. (1998). Correlation between interictal regional cerebral blood flow and depth-recorded interictal spiking in temporal lobe epilepsy. *Epilepsia.* **39,** 67–76.

Hajek, M., Antonini, A., Leenders, K. L., and Wieser, H. G. (1991). Epilepsia partialis continua studied by PET. *Epilepsy Res.* **9,** 44–48.

Hajek, M., Antonini, A., Leenders, K. L., and Wieser, H. G. (1993). Mesiobasal versus lateral temporal lobe epilepsy: Metabolic differences in the temporal lobe shown by interictal ^{18}F-FDG positron emission tomography. *Neurology* **43,** 79–86.

Hajek, M., Wieser, H. G., Khan, N., Antonini, A., Schrott, P. R., Maguire, P., Beer, H. F., and Leenders, K. L. (1994). Preoperative and postoperative glucose consumption in mesiobasal and lateral temporal lobe epilepsy. *Neurology* **44,** 2125–2132.

Hajek, M., Do, K. Q., Duc, C., Boesiger, P., and Wieser, H. G. (1997). Increased excitatory amino acid levels in brain cysts of epileptic patients. *Epilepsy Res.* **28,** 245–254.

Haldemann, R. C., Bicik, I., Pfeiffer, A., Wieser, H. G., Hasler, P. H., Schubiger, P., and von Schulthess, G. K. (1992). ^{123}I-Iomazenil: A quantitative study of the central benzodiazepine receptor distribution. *Nuklearmedizin* **31,** 91–97.

Hammeke, T. A., Yetkin, F. Z., Mueller, W. M., Morris, G. L., Haughton, V. M., Rao, S. M., and Binder, J. R. (1994). Functional magnetic resonance imaging of somatosensory stimulation. *Neurosurgery* **35,** 677–681.

Hand, K. S. P., Baird, V. H., Van Paesschen, W., Koepp, M. J., Revesz, T., Thom, M., Harkness, W. F. J., Duncan, J. S., and Bowery, N. G. (1997). Central benzodiazepine receptor autoradiography in hippocampal sclerosis. *Br. J. Pharmacol.* **122,** 358–364.

Hanefield, F., Kruse, B., Holzbach, U., *et al.* (1996). Hemimegalencephaly: Localized proton magnetic resonance spectroscopy *in vivo. Epilepsia* **36,** 1215–1224.

Hart, J., Jr., Lewis, P. J., Lesser, R. P., Fisher, R. S., Monsein, L. H., Schwerdt, P., Bandeen-Roche, K., and Gordon, B. (1993). Anatomic correlates of memory from intracarotid amobarbital injections with technetium Tc 99m hexamethylpropyleneamine oxime SPECT. *Arch Neurol.* **50,** 745–750.

Harvey, A. S., Bowe, J. M., Hopkins, I. J., *et al.* (1993a). Ictal ^{99m}Tc-HM-PAO single photon emission computed tomography in children with temporal lobe epilepsy. *Epilepsia* **34,** 869–877.

Harvey, A. S., Hopkins, I. J., Bowe, J. M., Cook, D. J., Shield, L. K., and Berkovic, S. F. (1993b). Frontal lobe epilepsy: Clinical seizure characteristics and localization with ictal ^{99m}Tc-HMPAO SPECT. *Neurology* **43,** 1966–1980.

Hayman, M., Scheffer, I. E., Chinvarun, Y., Berlangieri, S. U., and Berkovic, S. F. (1997). Autosomal dominant nocturnal frontal lobe epilepsy: Demonstration of focal frontal onset and intrafamilial variation. *Neurology* **49,** 969–975.

Heinz, R., Ferris, N., Lee, E. K., Radtke, R., Crain, B., Hoffman, J. M., Hanson, M., Paine, S., and Friedman, A. (1994). MR and positron emission tomography in the diagnosis of surgically correctable temporal lobe epilepsy. *Am. J. Neuroradiol.* **15,** 1341–1348.

Heiskala, H., Launes, J., Pihko, H., Nikkinen, P., and Santavuori, P. (1993). Brain perfusion SPECT in children with frequent fits. *Brain Dev.* **15,** 214–218.

Henry, T. R., Sutherling, W. W., Engel, J., Jr., Risinger, M. W., Levesque, M. F., Mazziotta, J. C., and Phelps, M. E. (1991). Interictal cerebral metabolism in partial epilepsies of neocortical origin. *Epilepsy Res.* **10**, 174–182.

Henry, T. R., Frey, K. A., Sackellares, J. C., Gilman, S., Koeppe, R. A., Brunberg, J. A., Ross, D. A., Berent, S., Young, A. B., and Kuhl, D. E. (1993a). *In vivo* cerebral metabolism and central benzodiazepine-receptor binding in temporal lobe epilepsy. *Neurology* **43**, 1998–2006.

Henry, T. R., Mazziotta, J. C., and Engel, J., Jr. (1993b). Interictal metabolic anatomy of mesial temporal lobe epilepsy. *Arch. Neurol.* **50**, 582–589.

Henry, T. R., Babb, T. L., Engel, J., Jr., Mazziotta, J. C., Phelps, M. E., and Crandall, P. H. (1994). Hippocampal neuronal loss and regional hypometabolism in temporal lobe epilepsy. *Ann. Neurol.* **36**, 925–927.

Henry, T. R., Buchtel, H. A., Koeppe, R. A., Pennell, P. B., Kluin, K. J., and Minoshima, S. (1998a). Absence of normal activation of the left anterior fusiform gyrus during naming in left temporal lobe epilepsy. *Neurology* **50**, 787–790.

Henry, T. R., Bakay, R. A., Votaw, J. R., Pennell, P. B., Epstein, C. M., Faber, T. L., Grafton, S. T., and Hoffman, J. M. (1998b). Brain blood flow alterations induced by therapeutic vagus nerve stimulation in partial epilepsy: I. Acute effects at high and low levels of stimulation. *Epilepsia* **39**, 983–990.

Hertz-Pannier, L., Gaillard, W. D., Mott, S. H., Cuenod, C. A., Bookheimer, S. Y., Weinstein, S., Conry, J., Papero, P. H., Schiff, S. J., Le Bihan, D., and Theodore, W. H. (1997). Noninvasive assessment of language dominance in children and adolescents with functional MRI: A preliminary study. *Neurology* **48**, 1003–1012.

Hetherington, H., Kuzniecky, R., Pan, J., Mason, G., Morawetz, R., Harris, C., Faught, E., Vaughan, T., and Pohost, G. (1995). Proton nuclear magnetic resonance spectroscopic imaging of human temporal lobe epilepsy at 4.1 T. *Ann. Neurol.* **38**, 396–404.

Ho, S. S., Berkovic, S. F., Newton, M. R., Austin, M. C., McKay, W. J., and Bladin, P. F. (1994). Parietal lobe epilepsy: Clinical features and seizure localization by ictal SPECT. *Neurology* **44**, 2277–2284.

Ho, S. S., Berkovic, S. F., Berlangieri, S. U., Newton, M. R., Egan, G. F., Tochon-Danguy, H. J., and McKay, W. J. (1995). Comparison of ictal SPECT and interictal PET in the presurgical evaluation of temporal lobe epilepsy. *Ann. Neurol.* **37**, 738–745.

Ho, S. S., Berkovic, S. F., McKay, W. J., Kalnins, R. M., and Bladin, P. F. (1996). Temporal lobe epilepsy subtypes: Differential patterns of cerebral perfusion on ictal SPECT. *Epilepsia* **37**, 788–795.

Ho, S. S., Newton, M. R., McIntosh, A. M., Kalnins, R. M., Fabinyi, G. C., Brazenor, G. A., McKay, W. J., Bladin, P. F., and Berkovic, S. F. (1997). Perfusion patterns during temporal lobe seizures: Relationship to surgical outcome. *Brain* **120**, 1921–1928.

Hogan, R. E., Cook, M. J., Binns, D. W., Desmond, P. M., Kilpatrick, C. J., Murrie, V. L., and Morris, K. F. (1997). Perfusion patterns in postictal ^{99m}Tc-HMPAO SPECT after coregistration with MRI in patients with mesial temporal lobe epilepsy. *J. Neurol. Neurosurg. Psychiatry* **63**, 235–239.

Hojo, M., Miyamoto, S., Nakahara, I., Kikuchi, H., Ishikawa, M., Taki, W., Nagata, I., Yamamoto, K., Yonekura, Y., Nishizawa, S., *et al.* (1995). A case of arteriovenous malformation successfully treated with functional mapping of the language area by PET activation study. *No Shinkei Geka* **23**, 537–541.

Hugg, J. W., Laxer, K. D., Matson, G. B., Maudsley, A. A., Husted, C. A., and Weiner, M. W. (1992). Lateralization of human focal epilepsy by ^{31}P magnetic resonance imaging spectroscopy. *Neurology* **42**, 2011–2018.

Hugg, J. W., Laxer, K. D., Matson, G. B., Maudsley, A. A., and Weiner, M. W. (1993). Neuron loss localises human focal epilepsy by *in-vivo* proton MR spectroscopic imaging. *Ann. Neurol.* **34**, 788–794.

Hugg, J. W., Kuzniecky, R. I., Gilliam, F. G., Morawetz, R. B., Fraught, R. E., and Hetherington, H. P. (1996). Normalization of contralateral metabolic function following temporal lobectomy demonstrated by ^{1}H magnetic resonance spectroscopic imaging. *Ann Neurol.* **40**, 236–239.

Iinuma, K., Yokoyama, H., Otsuki, T., Yanai, K., Watanabe, T., Ido, T., and Itoh, M. (1993). Histamine H_1 receptors in complex partial seizures. *Lancet* **341**, 238.

Iivanainen, M., Launes, J., Pihko, H., Nikkinen, P., and Lindroth, L. (1990). Single-photon emission computed tomography of brain perfusion: Analysis of 60 paediatric cases. *Dev. Med. Child Neurol.* **32**, 63–68.

Incisa della Rocchetta, A., Gadian, D. G., Connelly, A., Polkey, C. E., Jackson, G. D., Watkins, K. E., Johnson, C. L., Mishkin, M., and Vargha Khadem, F. (1995). Verbal memory impairment after right temporal lobe surgery: Role of contralateral damage as revealed by ^{1}H magnetic resonance spectroscopy and T_2 relaxometry. *Neurology* **45**, 797–802.

Itoh, M., Yanai, K., Yamaguchi, S., Fujiwara, T., Nagasawa, H., Yokoyama, H., Iinuma, K., and Ido, T. (1995). *In vivo* visualization of neurotransmitter function in the human brain by PET. *No To Hattatsu* **27**, 146–152.

Ives, J. R., Warach, S., Schmitt, F., Edelman, R. R., and Schomer, D. L. (1993). Monitoring the patient's EEG during echoplanar MRI. *Electroencephalogr. Clin. Neurophysiol.* **87**, 417–420.

Jack, C. R., Jr., Mullan, B. P., Sharbrough, F. W., Cascino, G. D., Hauser, M. F., Krecke, K. N., Luetmer, P. H., Trenerry, M. R., O'Brien, P. C., and Parisi, J. E. (1994a). Intractable nonlesional epilepsy of temporal lobe origin: Lateralization by interictal SPECT versus MRI. *Neurology* **44**, 829–836.

Jack, C. R., Jr., Thompson, R. M., Butts, R. K., Sharbrough, F. W., Kelly, P. J., Hanson, D. P., Riederer, S. J., Ehman, R. L., Hangiandreou, N. J., and Cascino, G. D. (1994b). Sensory motor cortex: Correlation of presurgical mapping with functional MR imaging and invasive cortical mapping. *Radiology* **190**, 85–92.

Jackson, G. D., Connelly, A., Cross, J. H., Gordon, I., and Gadian, D. G. (1994). Functional magnetic resonance imaging of focal seizures. *Neurology* **44**, 850–856.

Jeffery, P. J., Monsein, L. H., Szabo, Z., Hart, J., Fisher, R. S., Lesser, R. P., Debrun, G. M., Gordon, B., Wagner, H. N., Jr., and Camargo, E. E. (1991). Mapping the distribution of amobarbital sodium in the intracarotid Wada test by use of Tc-99m HMPAO with SPECT. *Radiology* **178**, 847–850.

Johnson, E. W., de Lanerolle, N. C., Kim, J. H., Sundaresan, S., Spencer, D. D., Mattson, R. H., Zoghbi, S. S., Baldwin, R. M., Hoffer, P. B., Seibyl, J. P., *et al.* (1992). "Central" and "peripheral" benzodiazepine receptors: Opposite changes in human epileptogenic tissue. *Neurology* **42**, 811–815.

Jokeit, H., Seitz, R. J., Markowitsch, H. J., Neumann, N., Witte, O. W., and Ebner, A. (1997). Prefrontal asymmetric interictal glucose hypometabolism and cognitive impairment in patients with temporal lobe epilepsy. *Brain* **120**, 2283–2294.

Kalviainen, R., Halonen, T., Pitkanen, A., and Riekkinen, P. J. (1993). Amino acid levels in the cerebrospinal fluid of newly diagnosed epileptic patients: Effect of vigabatrin and carbamazepine monotherapies. *J. Neurochem.* **60**, 1244–1250.

Katz, A., Bose, A., Lind, S. J., and Spencer, S. S. (1990). SPECT in patients with epilepsia partialis continua. *Neurology* **40**, 1848–1850.

Kauppinen, R. A., and Williams, S. R. (1991). Nondestructive detection of glutamate by ^{1}H nuclear magnetic resonance spectroscopy in cortical brain slices from the guinea pig: Evidence for changes in detectability during severe anoxic insults. *J. Neurochem.* **57**, 1136–1144.

Keltner, J. R., Wald, L., De Frederick, B., and Renshaw, P. F. (1997). *In vivo* detection of GABA in human brain using a localized double quantum filter technique. *Magn. Reson. Med.* **37**, 366–371.

Kirchberger, K., Hummel, C., and Stefan, H. (1998). Postoperative multichannel magnetoencephalography in patients with recurrent seizures after epilepsy surgery. *Acta Neurol. Scand.* **98**, 1–7.

Knowlton, R. C., Laxer, K. D., Ende, G., Hawkins, R. A., Wong, S. T., Matson, G. B., Rowley, H. A., Fein, G., and Weiner, M. W. (1997a). Presurgical multimodality neuroimaging in electroencephalographic lateralized temporal lobe epilepsy. *Ann. Neurol.* **42**, 829–837.

Knowlton, R. C., Laxer, K. D., Aminoff, M. J., Roberts, T. P., Wong, S. T., and Rowley, H. A. (1997b). Magnetoencephalography in partial epilepsy: Clinical yield and localization accuracy. *Ann. Neurol.* **42**, 622–631.

Ko, D., Heck, C., Grafton, S., Apuzzo, M. L., Couldwell, W. T., Chen, T., Day, J. D., Zelman, V., Smith, T., and DeGiorgio, C. M. (1996). Vagus nerve stimulation activates central nervous system structures in epileptic patients during PET H_2 ^{15}O blood flow imaging. *Neurosurgery* **39**, 426–430.

Koepp, M. J., Richardson, M. P., Brooks, D. J., Poline, J. B., Friston, K. J., Cunningham, V. J., and Duncan, J. S. (1996a). Central benzodiazepine receptors in hippocampal sclerosis: An objective *in vivo* analysis. *Brain* **119**, 1677–1687.

Koepp, M. J., Richardson, M. P., Brooks, D. J., Van Paesschen, W., and Duncan, J. S. (1996b). [^{11}C]-Flumazenil PET in temporal lobe epilepsy with normal or non-diagnostic MRI. *Epilepsia* **37** (Suppl. 4), 153.

Koepp, M. J., Richardson, M. P., Brooks, D. J., Cunningham, V. J., and Duncan, J. S. (1997a). Central benzodiazepine-GABAA receptors in idiopathic generalized epilepsy: An [^{11}C]-flumazenil positron emission tomography study. *Epilepsia* **38**, 1089–1097.

Koepp, M. J., Richardson, M. P., Labbe, C., Cunningham, V. J., Ashburner, J., Van Paesschen, W., Revesz, T., Brooks, D. J., and Duncan, J. S. (1997b). ^{11}C-Flumazenil PET, volumetric MRI and quantitative pathology in mesial temporal lobe epilepsy. *Neurology* **49**, 764–773.

Koepp, M. J., Labbe, C., Richardson, M. P., Brooks, D. J., Van Paesschen, W., Cunningham, V. J., and Duncan, J. S. (1997c). Regional hippocampal [^{11}C]-flumazenil PET in temporal lobe epilepsy with unilateral and bilateral hippocampal sclerosis. *Brain* **120**, 1865–1876.

Koepp, M. J., Hand, K. S. P., Labbe, C., Richardson, M. P., Van Paesschen, W., Baird, V. H., Bowery, N. G., Brooks, D. J., and Duncan, J. S. (1998a). *In vivo* ^{11}C-flumazenil PET correlates with *ex vivo* ^{3}H-flumazenil autoradiography in hippocampal sclerosis. *Ann. Neurol.* **43**, 618–626.

Koepp, M. J., Hammers, A., Richardson, M. P., Labbe, C., Brooks, D. J., Woermann, F. G., and Duncan, J. S. (1998b). Partial volume effect corrected [^{11}C]flumazenil PET reveals abnormalities in temporal lobe epilepsy with normal quantitative MRI. *Epilepsia* **39** (Suppl. 6), 141.

Koepp, M. J., Richardson, M. P., Brooks, D. J., and Duncan, J. S. (1998c). Focal cortical release of endogenous opioids during reading induced seizures. *Lancet* **352**, 952–955.

Koutroumanidis, M., Binnie, C. D., Elwes, R. D., Polkey, C. E., Seed, P., Alarcon, G., Cox, T., Barrington, S., Marsden, P., Maisey, M. N., and Panayiotopoulos, C. P. (1998). Interictal regional slow activity in temporal lobe epilepsy correlates with lateral temporal hypometabolism as imaged with ^{18}FDG PET: Neurophysiological and metabolic implications. *J. Neurol. Neurosurg. Psychiatry* **65**, 170–176.

Krakow, K., Woermann, F. G., Symms, M. R., Allen, P. J., Lemieux, L., Barker, G. J., Duncan, J. S., and Fish, D. R. (1999). EEG-triggered functional MRI of interictal epileptiform activity in patients with partial seizures. *Brain* **122**, 1679–1688.

Kuhl, D. E., Engel, J., Phelps, M. E., and Selin, C. (1980). Epileptic patterns of local cerebral metabolism and perfusion in humans deter-

mined by emission computed tomography of ^{18}FDG and $^{13}NH_3$. *Ann. Neurol.* **8**, 348–360.

Kumlein, E., Bergstrom, M., Lilja, A., Andersson, J., Szekeres, V., Westerberg, C. E., Westerberg, G., Antoni, G., and Langstrom, B. (1995). Positron emission tomography with [^{11}C]deuterium deprenyl in temporal lobe epilepsy. *Epilepsia* **36**, 712–721.

Kumlein, E., Hartvig, P., Valind, S., Oye, I., Tedroff, J., and Langstrom, B. (1999). NMDA-receptor activity visualized with (S)-N-[methyl-^{11}C]ketamine and positron emission tomography in patients with medial temporal lobe epilepsy. *Epilepsia.* **40**, 30–37.

Kuznicky, R., Elgavish, G. A., Hetherington, H. P., Evanochko, W. T., and Pohost, G. M. (1992). *In vivo* ^{31}P nuclear magnetic resonance spectroscopy of human temporal lobe epilepsy. *Neurology* **42**, 1586–1590.

Kuznicky, R., Mountz, J. M., Wheatley, G., and Morawetz, R. (1993). Ictal single-photon emission computed tomography demonstrates localized epileptogenesis in cortical dysplasia. *Ann. Neurol.* **34**, 627–631.

Kuznicky, R., Hetherington, H., Pan, J., *et al.* (1997). Proton spectroscopic imaging at 4.1 tesla in patients with malformations of cortical development and epilepsy. *Neurology* **48**, 1018–1024.

Kuznicky, R., Hugg, J. W., Hetherington, H., Butterworth, E., Bilir, E., Faught, E., and Gilliam, F. (1998). Relative utility of ^{1}H spectroscopic imaging and hippocampal volumetry in the lateralization of mesial temporal lobe epilepsy. *Neurology* **51**, 66–71.

Kuznicky, R., Hetherington, H., Ho, S., Pan, J., Martin, R., Gilliam, F., Hugg, J., and Faught, E. (1998). Topiramate increases cerebral GABA in healthy humans. *Neurology* **51**, 627–629.

Labbe, C., Koepp, M. J., Ashburner, J., Spinks, T., Richardson, M. P., Duncan, J. S., and Cunningham, V. J. (1998) Absolute PET quantification with correction for partial volume effects within cerebral structures. *In* "Quantitative Functional Brain Imaging with Positron Emission Tomography" (C. Carson, M. Daube-Witherspoon, and P. Herscovitch, Eds.), pp. 59–66. Academic Press, San Diego.

Laich, E., Kuznicky, R., Mountz, J., Liu, H. G., Gilliam, F., Bebin, M., Faught, E., and Morawetz, R. (1997). Supplementary sensorimotor area epilepsy. Seizure localization, cortical propagation and subcortical activation pathways using ictal SPECT. *Brain* **120**, 855–864.

Lamusuo, S., Ruottinen, H. M., Knuuti, J., Harkonen, R., Ruotsalainen, U., Bergman, J., Haaparanta, M., Solin, O., Mervaala, E., Nousiainen, U., Jaaskelainen, S., Ylinen, A., Kalviainen, R., Rinne, J. K., Vapalahti, M., and Rinne, J. O. (1997). Comparison of [^{18}F]FDG-PET, [^{99m}Tc]-HMPAO-SPECT, and [^{123}I]-iomazenil-SPECT in localising the epileptogenic cortex. *J. Neurol. Neurosurg. Psychiatry* **63**, 743–748.

Lancman, M. E., Morris, H. H., 3rd, Raja, S., Sullivan, M. J., Saha, G., and Go, R. (1997). Usefulness of ictal and interictal ^{99m}Tc ethyl cysteinate dimer single photon emission computed tomography in patients with refractory partial epilepsy. *Epilepsia* **38**, 466–471.

Laxer, K. D., Hubesch, B., Sappey-Marinier, D., and Weiner, M. W. (1992). Increased pH and inorganic phosphate in temporal seizure foci demonstrated by [^{31}P]MRS. *Epilepsia* **33**, 618–623.

Laxer, K. D., van der Grond, J., Gerson, J. R., Hugg, J. W., Matson, G. B., and Weiner, M. W. (1993). Temporal lobe epilepsy localization by ^{31}P magnetic resonance spectroscopic imaging. *Epilepsia* **34** (Suppl. 6), 122.

Lee, B. C., Schmidt, R. E., Hatfield, G. A., Bourgeois, B., and Park, T. S. (1998). MRI of focal cortical dysplasia. *Neuroradiology* **40**, 675–683.

Lee, B. I., Markand, O. N., Siddiqui, A. R., Park, H. M., Mock, B., Wellman, H. H., Worth, R. M., and Edwards, M. K. (1986). Single photon emission computed tomography (SPECT) brain imaging using

N,N,N' trimethyl-*N'* (2   hydroxy-3–methyl-5–¹²³*I*-iodobenzyl)-1,3–propanediamine 2HCl (HIPDM): Intractable complex partial seizures. *Neurology* **36**, 1471–1477.

Lee, B. I., Markand, O. N., Wellman, H. N., Siddiqui, A. R., Park, H. M., Mock, B., Worth, R. M., Edwards, M. K., and Krepshaw, J. (1988). HIPDM-SPECT in patients with medically intractable complex partial seizures. Ictal study. *Arch. Neurol.* **45**, 397–402.

Lee, B. I., Lee, J. D., Kim, J. Y., Ryu, Y. H., Kim, W. J., Lee, J. H., Lee, S. J., and Park, S. C. (1997). Single photon emission computed tomography-EEG relations in temporal lobe epilepsy. *Neurology* **49**, 981–991.

Lee, N., Radtke, R. A., Gray, L., Burger, P. C., Montine, T. J., DeLong, G. R., Lewis, D. V., Oakes, W. J., Friedman, A. H., and Hoffman, J. M. (1994). Neuronal migration disorders: Positron emission tomography correlations. *Ann. Neurol.* **35**, 290–297.

Leiderman, D. B., Balish, M., Bromfield, E. B., and Theodore, W. H. (1991). Effect of valproate on human cerebral glucose metabolism. *Epilepsia* **32**, 417–422.

Leiderman, D. B., Balish, M., Sato, S., Kufta, C., Reeves, P., Gaillard, W. D., and Theodore, W. H. (1992). Comparison of PET measurements of cerebral blood flow and glucose metabolism for the localization of human epileptic foci. *Epilepsy Res.* **13**, 153–157.

Leiderman, D. B., Albert, P., Balish, M., Bromfield, E., and Theodore, W. H. (1994). The dynamics of metabolic change following seizures as measured by positron emission tomography with fluorodeoxyglucose F18. *Arch. Neurol.* **51**, 932–936.

Lemieux, L., Allen, P. J., Franconi, F., Symms, J. M. R., and Fish, D. R. (1997). Recording of EEG during fMRI experiments: Patient safety. *Magn. Reson. Med.* **38**, 943–952.

Li, L. M., Cendes, F., Bastos, A. C., Andermann, F., Dubeau, F., and Arnold, D. L. (1998). Neuronal metabolic dysfunction in patients with cortical developmental malformations: A proton magnetic resonance spectroscopic imaging study. *Neurology* **50**, 755–759.

Longostrevi, G. P. (1986). Diagnostic usefulness of SPECT with Tc99m HM-PAO in cerebral pathology in outpatient practice. *Minerva Med.* **77**, 1777–1788.

Loscher, W. (1998). Pharmacology of glutamate receptor antagonists in the kindling model of epilepsy. *Prog. Neurobiol.* **54**, 721–741.

Loscher, W., and Schmidt, D. (1994). Strategies in antiepileptic drug development: Is rational drug design superior to random screening and structural variation? *Epilepsy Res.* **17**, 95–134.

Lu, D., Margouleff, C., Rubin, E., Labar, D., Schaul, N., Ishikawa, T., Kazumata, K., Antonini, A., Dhawan, V., Hyman, R. A., and Eidelberg, D. (1997). Temporal lobe epilepsy: Correlation of proton magnetic resonance spectroscopy and ¹⁸F-fluorodeoxyglucose positron emission tomography. *Magn. Reson. Med.* **37**, 18–23.

Madar, I., Lesser, R. P., Krauss, G., Zubieta, J. K., Lever, J. R., Kinter, C. M., Ravert, H. T., Musachio, J. L., Mathews, W. B., Dannals, R. F., and Frost, J. J. (1997). Imaging of delta- and mu-opioid receptors in temporal lobe epilepsy by positron emission tomography. *Ann. Neurol.* **41**, 358–367.

Manganotti, P., Zanette, G., Beltramello, A., Puppini, G., Miniussi, C., Maravita, A., Santorum, E., Marzi, C. A., Fiaschi, A., and Dalla-Bernardina, B. (1998). Spike topography and functional magnetic resonance imaging (fMRI) in benign rolandic epilepsy with spikes evoked by tapping stimulation. *Electroencephalogr. Clin. Neurophysiol.* **107**, 88–92.

Manno, E. M., Sperling, M. R., Ding, X., Jaggi, J., Alavi, A., O'Connor, M. J., and Reivich, M. (1994). Predictors of outcome after anterior temporal lobectomy: Positron emission tomography. *Neurology* **44**, 2331–2336.

Maquet, P., Hirsch, E., Dive, D., Salmon, E., Marescaux, C., and Franck, G. (1990). Cerebral glucose utilization during sleep in Landau-Kleffner syndrome: A PET study. *Epilepsia* **31**, 778–783.

Marco, P., Sola, R. G., Pulido, M. T., *et al.* (1996). Inhibitory neurons in the human epileptogenic temporal neocortex: An immunocytochemical study. *Brain* **119**, 1327–1347.

Markand, O. N., Salanova, V., Worth, R. M., Park, H. M., and Wellman, H. H. (1994). Ictal brain imaging in presurgical evaluation of patients with medically intractable complex partial seizures. *Acta Neurol. Scand. Suppl.* **152**, 137–144.

Markand, O. N., Salanova, V., Worth, R., Park, H. M., and Wellman, H. N. (1997). Comparative study of interictal PET and ictal SPECT in complex partial seizures. *Acta Neurol. Scand.* **95**, 129–136.

Marks, D. A., Katz, A., Hoffer, P., and Spencer, S. S. (1992). Localization of extratemporal epileptic foci during ictal single photon emission computed tomography. *Ann. Neurol.* **31**, 250–255.

Mason, G., Gruetter, R., Rothman, D. L., Behar, K. L., Shulman, R. G., and Novotny, E. J. (1995). Simultaneous determination of the rates of the TCA cycle, glucose determination, and α-ketoglutarate/glutamate exchange and glutamine synthesis in human brain by NMR. *J. Cereb. Blood Flow Metab.* **15**, 12–25.

Mastin, S. T., Drane, W. E., Gilmore, R. L., Helveston, W. R., Quisling, R. G., Roper, S. N., Eikman, E. A., and Browd, S. R. (1996). Prospective localization of epileptogenic foci: Comparison of PET and SPECT with site of surgery and clinical outcome. *Radiology* **199**, 375–380.

Matsuda, H., Onuma, T., and Yagishita, A. (1995). Brain SPECT imaging for laminar heterotopia. *J. Nucl. Med.* **36**, 238–240.

Matthews, P. M., Andermann, F., and Arnold, D. L. (1990). A proton magnetic resonance spectroscopy study of focal epilepsy in humans. *Neurology* **40**, 985–989.

Mayberg, H. S., Sadzot, B., Meltzer, C. C., Fisher, R. S., Lesser, R. P., Dannals, R. F., Lever, J. R., Wilson, A. A., Ravert, H. T., Wagner, H. N., Jr., *et al.* (1991). Quantification of mu and non-mu opiate receptors in temporal lobe epilepsy using positron emission tomography. *Ann. Neurol.* **30**, 3–11.

Maziere, M., Hantraye, P., Prenant, C., Sastre, J., and Comar, D. (1984). Synthesis of ethyl 8–fluoro-5,6–dihydro-5–[¹¹C]methyl-6–oxo-4*H*-imidazol[1,5–*a*][1,4]benzodiazepine-3–carboxylate (RO15,1788–¹¹C): A specific radioligand for the *in vivo* study of central benzodiazepine receptors by positron emission tomography. *Int. J. Appl. Radiat. Isot.* **35**, 973–976.

McMackin, D., Dubeau, F., Jones-Gotman, M., Gotman, J., Lukban, A., Dean, G., Evans, A., Tampieri, D., and Lisbona, R. (1997). Assessment of the functional effect of the intracarotid sodium amobarbital procedure using co-registered MRI/HMPAO-SPECT and SEEG. *Brain Cog.* **33**, 50–70.

Meldrum, B. S. (1989). GABAergic mechanisms in the pathogenesis and treatment of epilepsy. *Br. J. Clin. Pharmacol.* **27** (Suppl. 1), 3S–11S.

Meltzer, C. C., Zubieta, J. K., Links, J. M., Brakeman, P., Stumpf, M. J., and Frost, J. J. (1996). MR-based correction for brain PET measurements for heterogenous gray matter radioactivity distribution. *J. Cereb. Blood Flow Metab.* **16**, 650–658.

Menon, R. S., and Gati, J. S. (1997). Two second temporal resolution measurements of lactate correlate with EPI BOLD fMRI time courses during photic stimulation. *Proc. 5th ISMRM* 152.

Merlet, I., Garcia-Larrea, L., Gregoire, M. C., Lavenne, F., and Mauguiere, F. (1996). Source propagation of interictal spikes in temporal lobe epilepsy. Correlations between spike dipole modelling and [¹⁸F]fluorodeoxyglucose PET data. *Brain* 119, 377–392.

Merlet, I., Paetau, R., Garcia-Larrea, L., Uutela, K., Granstrom, M. L., and Mauguiere, F. (1997). Apparent asynchrony between interictal electric and magnetic spikes. *NeuroReport* **8**, 1071–1076.

Miura, K., Watanabe, K., Maeda, N., Matsumoto, A., Kumagai, T., Ito, K., and Kato, T. (1993). Magnetic resonance imaging and positron emission tomography of band heterotopia. *Brain Dev.* **15**, 288–290.

Morris, G. L., 3rd, Mueller, W. M., Yetkin, F. Z., Haughton, V. M., Hammeke, T. A., Swanson, S., Rao, S. M., Jesmanowicz, A., Estkowski, L. D., Bandettini, P. A., *et al.* (1994). Functional magnetic resonance imaging in partial epilepsy. *Epilepsia* **35,** 1194–1198.

Muller, R. A., Chugani, H. T., Muzik, O., and Mangner, T. J. (1998a). Brain organization of motor and language functions following hemispherectomy: A [(15)O]-water positron emission tomography study. *Child Neurol.* **13,** 16–22.

Muller, R. A., Behen, M. E., Muzik, O., Rothermel, R. D., Downey, R. A., Mangner, T. J., and Chugani, H. T. (1998b). Task-related activations in heterotopic brain malformations: A PET study. *NeuroReport* **9,** 2527–2533.

Muller-Gartner, H. W., Links, J. M., Prince, J. L., Bryan, R. N., McVeigh, E., Leal, J. P., Davatzikos, C., and Frost, J. J. (1992). Measurement of radiotracer concentration in brain gray matter using positron emission tomography: MRI-based correction for partial volume effects. *J. Cereb. Blood Flow Metab.* **12,** 571–583.

Muller-Gartner, H. W., Mayberg, H. S., Fisher, R. S., *et al.* (1993). Decreased hippocampal muscarinic cholinergic receptor binding measured by ^{123}I-iododexetimide and single-photon emission computed tomography in epilepsy. *Ann. Neurol.* **34,** 235–238.

Nagata, T., Tanaka, F., Yonekura, Y., Ikeda, A., Nishizawa, S., Ishizu, K., Okazawa, H., Terada, K., Mikuni, N., Yamamoto, I., *et al.* (1995). Limited value of interictal brain perfusion SPECT for detection of epileptic foci: High resolution SPECT studies in comparison with FDG-PET. *Ann. Nucl. Med.* **9,** 59–63.

Najm, I. M., Wang, Y., Hong, S. C., Luders, H. O., Ng, T. C., and Comair, Y. G. (1997). Temporal changes in proton MRS metabolites after kainic acid-induced seizures in rat brain. *Epilepsia* **38,** 87–94.

Najm, I. M., Wang, Y., Shedid, D., Luders, H. O., Ng, T. C., and Comair, Y. G. (1998). MRS metabolic markers of seizures and seizure-induced neuronal damage. *Epilepsia* **39,** 244–250.

Nakasato, N., Levesque, M. F., Barth, D. S., Baumgartner, C., Rogers, R. L., and Sutherling, W. W. (1994). Comparisons of MEG, EEG, and ECoG source localization in neocortical partial epilepsy in humans. *Electroencephalogr. Clin. Neurophysiol.* **91,** 171–178.

Nehlig, A., Vergnes, M., Marescaux, C., Boyet, S., and Lannes, B. (1991). Local cerebral glucose utilization in rats with petit mal-like seizures. *Ann. Neurol.* **29,** 72–77.

Nehlig, A., Vergnes, M., Marescaux, C., and Boyet, S. (1992). Mapping of cerebral energy metabolism in rats with genetic generalized nonconvulsive epilepsy. *J. Neural Transm. Suppl.* **35,** 141–153.

Newton, M. R., Berkovic, S. F., Austin, M. C., Reutens, D. C., McKay, W. J., and Bladin, P. F. (1992a). Dystonia, clinical lateralization, and regional blood flow changes in temporal lobe seizures. *Neurology* **42,** 371–377.

Newton, M. R., Berkovic, S. F., Austin, M. C., Rowe, C. C., McKay, W. J., and Bladin, P. F. (1992b). A postictal switch in blood flow distribution and temporal lobe seizures. *J. Neurol. Neurosurg. Psychiatry* **55,** 891–894.

Newton, M. R., Berkovic, S. F., Austin, M. C., Rowe, C. C., McKay, W. J., and Bladin, P. F. (1995). SPECT in the localisation of extratemporal and temporal seizure foci. *J. Neurol. Neurosurg. Psychiatry* **59,** 26–30.

Ng, T., Comair, Y. G., Xue, M., *et al.* (1994). Temporal lobe epilepsy: Presurgical localization with proton chemical shift imaging. *Radiology* **193,** 465–472.

O'Brien, T. J., Newton, M. R., Cook, M. J., Berlangieri, S. U., Kilpatrick, C., Morris, K., and Berkovic, S. F. (1997). Hippocampal atrophy is not a major determinant of regional hypometabolism in temporal lobe epilepsy. *Epilepsia* **38,** 74–80.

O'Brien, T. J., Zupanc, M. L., Mullan, B. P., O'Connor, M. K., Brinkmann, B. H., Cicora, K. M., and So, E. L. (1998a). The practical utility of performing peri-ictal SPECT in the evaluation of children with partial epilepsy. *Pediatr. Neurol.* **19,** 15–22.

O'Brien, T. J., So, E. L., Mullan, B. P., Hauser, M. F., Brinkmann, B. H., Bohnen, N. I., Hanson, D., Cascino, G. D., Jack, C. R., Jr., and Sharbrough, F. W. (1998b). Subtraction ictal SPECT co-registered to MRI improves clinical usefulness of SPECT in localizing the surgical seizure focus. *Neurology* **50,** 445–454.

O'Brien, T. J., So, E. L., Mullan, B. P., Hauser, M. F., Brinkmann, B. H., Jack, C. R., Jr., Cascino, G. D., Meyer, F. B., and Sharbrough, F. W. (1999). Subtraction SPECT co-registered to MRI improves postictal SPECT localization of seizure foci. *Neurology* **52,** 137–146.

Ochs, R. F., Gloor, P., Tyler, J. L., Wolfson, T., Worsley, K., Andermann, F., Diksic, M., Meyer, E., and Evans, A. (1987). Effect of generalized spike-and-wave discharge on glucose metabolism measured by positron emission tomography. *Ann. Neurol.* **21,** 458–464.

Oku, N., Matsumoto, M., Hashikawa, K., Moriwaki, H., Ishida, M., Seike, Y., Terakawa, H., Watanabe, Y., Uehara, T., and Nishimura, T. (1997). Intra-individual differences between technetium-99m-HM-PAO and technetium-99m-ECD in the normal medial temporal lobe. *J. Nucl. Med.* **38,** 1109–1111.

Olsen, R. W., McCabe, R. T., and Wamsley, J. K. (1990). GABAA receptor subtypes: Autoradiographic comparison of GABA, benzodiazepine, and convulsant binding sites in the rat central nervous system. *J. Chem. Neuroanat.* **3,** 59–76.

Olson, D. M., Chugani, H. T., Shewmon, D. A., Phelps, M. E., and Peacock, W. J. (1990). Electrocorticographic confirmation of focal positron emission tomographic abnormalities in children with intractable epilepsy. *Epilepsia* **31,** 731–739.

Otsubo, H., Hwang, P. A., Jay, V., Becker, L. E., Hoffman, H. J., Gilday, D., and Blaser, S. (1993). Focal cortical dysplasia in children with localization-related epilepsy: EEG, MRI, and SPECT findings. *Pediatr. Neurol.* **9,** 101–107.

O'Tuama, L. A., Urion, D. K., Janicek, M. J., Treves, S. T., Bjornson, B., and Moriarty, J. M. (1992). Regional cerebral perfusion in Landau-Kleffner syndrome and related childhood aphasias. *J. Nucl. Med.* **33,** 1758–1765.

Packard, A. B., Roach, P. J., Davis, R. T., Carmant, L., Davis, R., Riviello, J., Holmes, G., Barnes, P. D., O'Tuama, L. A., Bjornson, B., and Treves, S. T. (1996). Ictal and interictal technetium-99m-bicisate brain SPECT in children with refractory epilepsy. *J. Nucl. Med.* **37,** 1101–1106.

Paetau, R., Hamalainen, M., Hari, R., Kajola, M., Karhu, J., Larsen, T. A., Lindahl, E., and Salonen, O. (1994). Magnetoencephalographic evaluation of children and adolescents with intractable epilepsy. *Epilepsia* **35,** 275–284.

Paetau, R., Granstrom, M. L., Blomstedt, G., Jousmaki, V., Korkman, M., and Liukkonen, E. (1999). Magnetoencephalography in presurgical evaluation of children with the Landau-Kleffner syndrome. *Epilepsia* **40,** 326–335.

Park, Y. D., Hoffman, J. M., Radtke, R. A., and DeLong, G. R. (1994). Focal cerebral metabolic abnormality in a patient with continuous spike waves during slow-wave sleep. *J. Child Neurol.* **9,** 139–143.

Parker, A. P., Ferrie, C. D., Keevil, S., Newbold, M., Cox, T., Maisey, M., and Robinson, R. O. (1998). Neuroimaging and spectroscopy in children with epileptic encephalopathies. *Arch. Dis. Child.* **79,** 39–43.

Pawlik, G., Holthoff, V. A., Kessler, J., Rudolf, J., Hebold, I. R., Lottgen, J., and Heiss, W. D. (1990). Positron emission tomography findings relevant to neurosurgery for epilepsy. *Acta Neurochir. Suppl. (Wien)* **50,** 84–87.

Pawlik, G., Fink, G. R., Kessler, J., and Heiss, W. D. (1994). PET and functional testing in temporal lobe epilepsy. *Acta Neurol. Scand. Suppl.* **152,** 150–156.

Pellerin, L., and Magistretti, P. J. (1994). Glutamate uptake into astrocytes stimulates aerobic glycolysis: A mechanism coupling neuronal activity to glucose utilization. *Proc. Natl. Acad. Sci. U.S.A.* **91,** 10625–10629.

Petroff, O. A., and Rothman, D. L. (1998). Measuring human brain GABA *in vivo*: Effects of GABA-transaminase inhibition with vigabatrin. *Mol. Neurobiol.* **16,** 97–121.

Petroff, O. A. C., Prichard, J. W., Ogino, T., Avison, M. J., Alger, J. R., and Shulman, R. G. (1986). Combined ¹H and ³¹P NMR studies of bicuculline-induced seizures *in vivo. Ann. Neurol.* **20,** 185–193.

Petroff, O. A. C., Novotny, E. J., Avison, M., Rothman, D. L., Alger, J. R., Ogino, T., Shulman, G. I., and Prichard, J. W. (1992). Cerebral lactate turnover after electroshock: *In vivo* measurements by ¹H/¹³C magnetic resonance spectroscopy. *J. Cereb. Blood Flow Metab.* **12,** 1022–1029.

Petroff, O. A. C., Rothman, D. L., Behar, K. L., and Mattson, R. H. (1995a). Initial observations on the effect of vigabatrin on *in vivo* ¹H spectroscopic measurements of GABA, glutamate and glutamine in human brain. *Epilepsia* **36,** 457–464.

Petroff, O., Pleban, L. A., and Spencer, D. D. (1995b). Symbiosis between *in vivo* and *in vitro* NMR spectroscopy: The creatine, *N*-acetylaspartate, glutamate and GABA content of the epileptic human brain. *Magn. Reson. Imaging* **13,** 1197–1211.

Petroff, O. A. C., Rothman, D. L., Behar, K. L., and Mattson, R. H. (1996a). The effect of gabapentin on brain γ-aminobutyric acid in patients with epilepsy. *Ann. Neurol.* **39,** 95–99.

Petroff, O. A. C., Rothman, D. L., Behar, K. L., and Mattson, R. H. (1996b). Low brain GABA level is associated with poor seizure control. *Ann. Neurol.* **40,** 908–911.

Petroff, O. A., Mattson, R. H., Behar, K. L., Hyder, F., and Rothman, D. L. (1998). Vigabatrin increases human brain homocarnosine and improves seizure control. *Ann. Neurol.* **44,** 948–952.

Petroff, O., Hyder, F., Mattson, R. H., and Rothman, D. L. (1999). Topiramate increases brain GABA, homocarnosine and pyrrolidine in patients with epilepsy. *Neurology* **52,** 473–478.

Peyron, R., Le Bars, D., Cinotti, L., Garcia-Larrea, L., Galy, G., Landais, P., Millet, P., Lavenne, F., Froment, J. C., Krogsgaard-Larsen, P., *et al.* (1994a). Effects of GABAA receptors activation on brain glucose metabolism in normal subjects and temporal lobe epilepsy (TLE) patients. A positron emission tomography (PET) study. Part I: Brain glucose metabolism is increased after GABAA receptors activation. *Epilepsy Res.* **19,** 45–54.

Peyron, R., Cinotti, L., Le Bars, D., Garcia-Larrea, L., Galy, G., Landais, P., Millet, P., Lavenne, F., Froment, J. C., Krogsgaard-Larsen, P., *et al.* (1994b). Effects of GABAA receptors activation on brain glucose metabolism in normal subjects and temporal lobe epilepsy (TLE) patients. A positron emission tomography (PET) study. Part II: The focal hypometabolism is reactive to GABAA agonist administration in TLE. *Epilepsy Res.* **19,** 55–62.

Podreka, I., Suess, E., Goldenberg, G., Steiner, M., Brucke, T., Muller, C., Lang, W., Neirinckx, R. D., and Deecke, L. (1987). Initial experience with technetium-99m HM-PAO brain SPECT. *J. Nucl. Med.* **28,** 1657–1666.

Preece, N. E., Jackson, G. D., Williams, S. F., Houseman, J. A., Duncan, J. S., and Williams, S. R. (1994). NMR detection of elevated cortical GABA in the vigabatrin-treated rat *in vivo. Epilepsia* **35,** 431–436.

Prevett, M. C., Cunningham, V. J., Brooks, D. J., Fish, D. R., and Duncan, J. S. (1994). Opiate receptors in idiopathic generalised epilepsy measured with [¹¹C]diprenorphine and positron emission tomography. *Epilepsy Res.* **19,** 71–77.

Prevett, M. C., Duncan, J. S., Jones, T., Fish, D. R., and Brooks, D. J. (1995a). Demonstration of thalamic activation during typical absence seizures using H₂¹⁵O and PET. *Neurology* **45,** 1396–1402.

Prevett, M. C., Lammertsma, A. A., Brooks, D. J., Bartenstein, P. A., Patsalos, P. N., Fish, D. R., and Duncan, J. S. (1995b). Benzodiazepine GABAA receptors in idiopathic generalized epilepsy measured with [¹¹C]flumazenil and positron emission tomography. *Epilepsia* **36,** 113–121.

Prevett, M. C., Lammertsma, A. A., Brooks, D. J., Cunningham, V. J., Fish, D. R., and Duncan, J. S. (1995c) Benzodiazepine-GABAA receptor binding during absence seizures. *Epilepsia* **36,** 592–599.

Prichard, J. W., Zhong, J., Petroff, O. A. C., and Gore, J. C. (1995). Diffusion-weighted NMR imaging changes caused by electrical activation of the brain. *NMR Biomed.* **8,** 359–364.

Provencher, S. W., *et al.* (1993). Estimation of metabolite concentrations from localized *in vivo* proton NMR spectra. *Magn. Reson. Med.* **30,** 672–679.

Puce, A., Constable, R. T., Luby, M. L., McCarthy, G., Nobre, A. C., Spencer, D. D., Gore, J. C., and Allison, T. (1995). Functional magnetic resonance imaging of sensory and motor cortex: Comparison with electrophysiological localization. *J. Neurosurg.* **83,** 262–270.

Radtke, R. A., Hanson, M. W., Hoffman, J. M., Crain, B. J., Walczak, T. S., Lewis, D. V., Beam, C., Coleman, R. E., and Friedman, A. H. (1993). Temporal lobe hypometabolism on PET: Predictor of seizure control after temporal lobectomy. *Neurology* **43,** 1088–1092.

Rao, S. M., Binder, J. R., Hammeke, T. A., Bandettini, P. A., Bobholz, J. A., Frost, J. A., Myklebust, B. M., Jacobson, R. D., and Hyde, J. S. (1995). Somatotopic mapping of the human primary motor cortex with functional magnetic resonance imaging. *Neurology* **45,** 919–924.

Rausch, R., Henry, T. R., Ary, C. M., Engel, J., Jr., and Mazziotta, J. (1994). Asymmetric interictal glucose hypometabolism and cognitive performance in epileptic patients. *Arch. Neurol.* **51,** 139–144.

Richardson, M. P., Koepp, M. J., Brooks, D. J., Fish, D. R., and Duncan, J. S. (1996). Benzodiazepine receptors in focal epilepsy associated with cortical dysgenesis: An ¹¹C-flumazenil PET study. *Ann. Neurol.* **40,** 188–198.

Richardson, M. P., Friston, K. J., Sisodiya, S. M., Koepp, M. J., Ashburner, J., Free, S. L., Brooks, D. J., and Duncan, J. S. (1997a). Benzodiazepine receptors in malformations of cortical development: A voxel-based comparison of structural and functional data in cortical grey matter. *Brain* **120,** 1961–1974.

Richardson, M. P., Koepp, M. J., Brooks, D. J., and Duncan, J. S. (1997b). Extratemporal localization-related epilepsy with normal MRI: Abnormalities of cortical and subcortical [¹¹C]-flumazenil binding. *Neurology* **48** (Suppl. 2), A20–A21.

Richardson, M. P., Koepp, M. J., Brooks, D. J., Coull, J. T., Grasby, P., Fish, D. R., and Duncan, J. S. (1998a). Cerebral activation in malformations of cortical development. *Brain* **121,** 1295–1304.

Richardson, M. P., Koepp, M. J., Brooks, D. J., and Duncan, J. S. (1998b). ¹¹C-Flumazenil PET in neocortical epilepsy. *Neurology* **51,** 485–492.

Rintahaka, P. J., Chugani, H. T., Messa, C., and Phelps, M. E. (1993). Hemimegalencephaly: Evaluation with positron emission tomography. *Pediatr. Neurol.* **9,** 21–28.

Rintahaka, P. J., Chugani, H. T., and Sankar, R. (1995). Landau-Kleffner syndrome with continuous spikes and waves during slow-wave sleep. *J. Child Neurol.* **10,** 127–133.

Rose, D. F., Sato, S., Duda-Soares, E., and Kufta, C. V., (1991). Magnetoencepholographic localization of subdural dipoles in a patient with temporal lobe epilepsy. *Epilepsia* **32,** 635–641.

Rothman, D. L., Novotny, E. J., Shulman, G. I., Howseman, A. M., Petroff, O. A. C., Mason, G., Nixon, T., Hanstock, C. C., Prichard, J. W., and Shulman, R. G. (1992). ¹H-[¹³C] NMR measurements of [4-¹³C]-glutamate turnover in human brain. *Proc. Natl. Acad. Sci. U.S.A.* **89,** 9603–9606.

Rothman, D. L., Petroff, O. A. C., Behar, K. L., and Mattson, R. H. (1993). Localized ¹H NMR measurements of γ-aminobutyric acid in human brain *in vivo. Proc. Natl. Acad. Sci. U.S.A.* **90,** 5662–5666.

Rothman, D. L., Behar, K. L., Prichard, J. W., and Petroff, O. A. (1997). Homocarnosine and the measurement of neuronal pH in patients with epilepsy. *Magn. Reson. Med.* **38,** 924–929.

Rowe, C. C., Berkovic, S. F., Austin, M. C., McKay, W. J., and Bladin, P. F. (1991a). Patterns of postictal cerebral blood flow in temporal lobe epilepsy: Qualitative and quantitative analysis. *Neurology* **41**, 1096–1103.

Rowe, C. C., Berkovic, S. F., Austin, M. C., Saling, M., Kalnins, R. M., McKay, W. J., and Bladin, P. F. (1991b). Visual and quantitative analysis of interictal SPECT with technetium-99m-HMPAO in temporal lobe epilepsy. *J. Nucl. Med.* **32**, 1688–1694.

Runge, U., Kirsch, G., Petersen, B., Kallwellis, G., Gaab, M. R., Piek, J., and Kessler, C. (1997). Ictal and interictal ECD-SPECT for focus localization in epilepsy. *Acta Neurol. Scand.* **96**, 271–276.

Ryvlin, P., Cinotti, L., Froment, J. C., Le Bars, D., Landais, P., Chaze, M., Galy, G., Lavenne, F., Serra, J. P., and Mauguiere, F. (1991). Metabolic patterns associated with non-specific magnetic resonance imaging abnormalities in temporal lobe epilepsy. *Brain* **114**, 2363–2383.

Ryvlin, P., Philippon, B., Cinotti, L., Froment, J. C., Le Bars, D., and Mauguiere, F. (1992a). Functional neuroimaging strategy in temporal lobe epilepsy: A comparative study of ^{18}FDG-PET and ^{99m}Tc-HMPAO-SPECT. *Ann. Neurol.* **31**, 650–656.

Ryvlin, P., Garcia-Larrea, L., Philippon, B., Froment, J. C., Fischer, C., Revol, M., and Mauguiere, F. (1992b). High signal intensity on T_2-weighted MRI correlates with hypoperfusion in temporal lobe epilepsy. *Epilepsia* **33**, 28–35.

Ryvlin, P., Mauguiere, F., Sindou, M., Froment, J. C., and Cinotti, L. (1995). Interictal cerebral metabolism and epilepsy in cavernous angiomas. *Brain* **118**, 677–687.

Ryvlin, P., Bouvard, S., Le Bars, D., De Lamerie, G., Gregoire, M. C., Kahame, P., Froment, J. C., and Mauguiere, F. (1998). Clinical utility of flumazenil-PET versus FDG-PET and MRI in refractory partial epilepsy. A prospective study in 100 patients. *Brain* **121**, 2067–2081.

Sackellares, J. C., Siegel, J. C., Abou-Khalil, B. W., *et al.* (1990). Differences between lateral and mesial temporal metabolism interictally in epilepsy of mesial temporal origin. *Neurology* **40**, 1420–1426.

Sadzot, B., Debets, R. M., Maquet, P., van Veelen, C. W., Salmon, E., van Emde Boas, W., Velis, D. N., van Huffelen, A. C., and Franck, G. (1992). Regional brain glucose metabolism in patients with complex partial seizures investigated by intracranial EEG. *Epilepsy Res.* **12**, 121–129.

Samson, Y., Hantraye, P., Baron, J. C., Soussaline, F., Comar, D., and Maziere, M. (1985). Kinetics and displacement of [^{11}C]RO15,1788, a benzodiazepine antagonist, studied in human brain *in vivo* by positron tomography. *Eur. J. Pharmacol.* **110**, 247–251.

Sauter, R., Schneider, M., Wiclow, K., and Kolem, H. (1991). Localized ^{1}H MRS of the human brain: Single-voxel versus CSI techniques. *J. Magn. Reson. Imaging* **1**, 241.

Savic, I., Persson, A., Roland, P., Pauli, S., Sedvall, G., and Widen, L. (1988). *In-vivo* demonstration of reduced benzodiazepine receptor binding in human epileptic foci. *Lancet* **2**, 863–866.

Savic, I., Widen, L., Thorell, J. O., Blomqvist, G., Ericson, K., and Roland, P. (1990). Cortical benzodiazepine receptor binding in patients with generalized and partial epilepsy. *Epilepsia* **31**, 724–730.

Savic, I., Widen, L., and Stone-Elander, S. (1991). Feasibility of reversing benzodiazepine tolerance with flumazenil. *Lancet* **337**, 133–137.

Savic, I., Ingvar, M., and Stone-Elander, S. (1993). Comparison of [^{11}C]flumazenil and [^{18}F]FDG as PET markers of epileptic foci. *J. Neurol. Neurosurg. Psychiatry* **56**, 615–621.

Savic, I., Pauli, S., Thorell, J. O., and Blomqvist, G. (1994). *In vivo* demonstration of altered benzodiazepine receptor density in patients with generalised epilepsy. *J. Neurol. Neurosurg. Psychiatry* **57**, 797–804.

Savic, I., Thorell, J. O., and Roland, P. (1995). [^{11}C]Flumazenil positron emission tomography visualizes frontal epileptogenic regions. *Epilepsia* **36**, 1225–1232.

Savic, I., Svanborg, E., and Thorell, J. O. (1996). Cortical benzodiazepine receptor changes are related to frequency of partial seizures: A positron emission tomography study. *Epilepsia* **37**, 236–244.

Savic, I., Blomqvist, G., Halldin, C., Litton, J. E., and Gulyas, B. (1998). Regional increases in [^{11}C]flumazenil binding after epilepsy surgery. *Acta Neurol. Scand.* **97**, 279–286.

Scherg, M., and Ebersole, J. S. (1993). Models of brain sources. *Brain Topogr.* **5**, 419–423.

Schlaug, G., Antke, C., Holthausen, H., Arnold, S., Ebner, A., Tuxhorn, I., Jancke, L., Luders, H., Witte, O. W., and Seitz, R. J. (1997). Ictal motor signs and interictal regional cerebral hypometabolism. *Neurology* **49**, 341–350.

Schmitz, E. B., Costa, D. C., Jackson, G. D., Moriarty, J., Duncan, J. S., Trimble, M. R., and Ell, P. J. (1995). Optimised interictal HMPAO-SPECT in the evaluation of partial epilepsies. *Epilepsy Res.* **21**, 159–167.

Seeck, M., Lazeyras, F., Michel, C. M., Blanke, O., Gericke, C. A., Ives, J., Delavelle, J., Golay, X., Haenggeli, C. A., de Tribolet, N., and Landis, T. (1998). Non-invasive epileptic focus localization using EEG-triggered functional MRI and electromagnetic tomography. *Electroencephalogr. Clin. Neurophysiol.* **106**, 508–512.

Semah, F., Baulac, M., Hasboun, D., Frouin, V., Mangin, J. F., Papageorgiou, S., Leroy-Willig, A., Philippon, J., Laplane, D., and Samson, Y. (1995). Is interictal temporal hypometabolism related to mesial temporal sclerosis? A positron emission tomography/magnetic resonance imaging confrontation. *Epilepsia* **36**, 447–456.

Sepkuty, J. P., Lesser, R. P., Civelek, C. A., Cysyk, B., Webber, R., and Shipley, R. (1998). An automated injection system (with patient selection) for SPECT imaging in seizure localization. *Epilepsia* **39**, 1350–1356.

Shen, J., Shungu, D. C., and Rothman, D. L. (1999). Chemical shift imaging of γ-aminobutyric acid in the human brain. *Magn. Reson. Med.* **41**, 35–42.

Shen, W., Lee, B. I., Park, H. M., Siddiqui, A. R., Wellman, H. H., Worth, R. M., and Markand, O. N. (1990). HIPDM-SPECT brain imaging in the presurgical evaluation of patients with intractable seizures. *J. Nucl. Med.* **31**, 1280–1284.

Shindo, K., Ikeda, A., Musha, T., Terada, K., Fukuyama, H., Taki, W., Kimura, J., and Shibasaki, H. (1998). Clinical usefulness of the dipole tracing method for localizing interictal spikes in partial epilepsy. *Epilepsia* **39**, 371–379.

Siesjo, B. K. (1978). "Brain Energy Metabolism," pp. 345–379. Wiley-Interscience, Rochester, NY.

Signorini, M., Paulesu, E., Friston, K., Perani, D., Colleluori, A., Lucignani, G., Grassi, F., Bettinardi, V., Frackowiak, R. S. J., and Fazio, F. (1999). Rapid assessment of regional cerebral metabolic abnormalities in single subjects with quantitative and nonquantitative [^{18}F]FDG PET: A clinical validation of statistical parametric mapping. *NeuroImage* **9**, 63–80.

Sjoholm, H., Rosen, I., and Elmqvist, D. (1995). Role of I-123–iomazenil SPECT imaging in drug resistant epilepsy with complex partial seizures. *Acta Neurol. Scand.* **92**, 41–48.

Smith, J. R., Gallen, C. C., and Schwartz, B. J. (1994). Multichannel magnetoencephalographic mapping of sensorimotor cortex for epilepsy surgery. *Stereotact. Funct. Neurosurg.* **62**, 245–251.

Spanaki, M. V., Zubal, I. G., MacMullan, J., and Spencer, S. S. (1999). Periictal SPECT localization verified by simultaneous intracranial EEG. *Epilepsia* **40**, 267–274.

Spencer, S. S. (1994). The relative contributions of MRI, SPECT, and PET imaging in epilepsy. *Epilepsia* **35** (Suppl. 6), S72–S89.

Sperling, M. R., Gur, R. C., Alavi, A., Gur, R. E., Resnick, S., O'Connor, M. J., and Reivich, M. (1990). Subcortical metabolic alterations in partial epilepsy. *Epilepsia* **31**, 145–155.

Sperling, M. R., Alavi, A., Reivich, M., French, J. A., and O'Connor, M. J. (1995). False lateralization of temporal lobe epilepsy with FDG positron emission tomography. *Epilepsia* **36,** 722–727.

Spreafico, R., Battaglia, G., Arcelli, P., *et al.* (1998). Cortical dysplasia: An immunocytochemical study of three patients. *Neurology* **50,** 27–36.

Stanley, J. A., Cendes, F., Dubeau, F., Andermann, F., and Arnold, D. L. (1998). Proton magnetic resonance spectroscopic imaging in patients with extratemporal epilepsy. *Epilepsia* **39,** 267–273.

Stefan, H., Kuhnen, C., Biersack, H. J., and Reichmann, K. (1987a). Initial experience with ^{99m}Tc-hexamethyl-propylene amine oxime (HM-PAO) single photon emission computed tomography (SPECT) in patients with focal epilepsy. *Epilepsy Res.* **1,** 134–138.

Stefan, H., Pawlik, G., Bocher-Schwartz, H. G., Biersack, H. J., Burr, W., Penin, H., and Heiss, W.-D. (1987b). Functional and morphological abnormalities in temporal lobe epilepsy: A comparison of interictal and ictal EEG, CT, MRI, SPECT and PET. *J. Neurol.* **234,** 377–384 .

Stefan, H., Bauer, J., Feistel, H., *et al.* (1990). Regional cerebral blood flow during focal seizures of temporal and frontocentral onset. *Ann. Neurol.* **27,** 162–166.

Stefan, H., Schuler, P., Abraham-Fuchs, K., Schneider, S., Gebhardt, M., Neubauer, U., Hummel, C., Huk, W. J., and Thierauf, P. (1994). Magnetic source localization and morphological changes in temporal lobe epilepsy: Comparison of MEG/EEG, ECoG and volumetric MRI in presurgical evaluation of operated patients. *Acta Neurol. Scand. Supp.l.* **152,** 83–88.

Stern, C. E., Corkin, S., Gonzalez, R. G., *et al.* (1996). The hippocampal formation participates in novel picture encoding: Evidence from functional magnetic resonance imaging. *Proc. Natl. Acad. Sci. U.S.A.* **93,** 8660–8665.

Swartz, B. E., Halgren, E., Delgado-Escueta, A. V., Mandelkern, M., Gee, M., Quinones, N., Blahd, W. H., and Repchan, J. (1989). Neuroimaging in patients with seizures of probable frontal lobe origin. *Epilepsia* **30,** 547–558.

Swartz, B. E., Khonsari, A., Vrown, C., Mandelkern, M., Simpkins, F., and Krisdakumtorn, T. (1995). Improved sensitivity of ^{18}FDG-positron emission tomography scans in frontal and "frontal plus" epilepsy. *Epilepsia* **36,** 388–395.

Swartz, B. E., Halgren, E., Simpkins, F., Fuster, J., Mandelkern, M., Krisdakumtorn, T., Gee, M., Brown, C., Ropchan, J. R., and Blahd, W. H. (1996a). Primary or working memory in frontal lobe epilepsy: An ^{18}FDG-PET study of dysfunctional zones. *Neurology* **46,** 737–747.

Swartz, B. E., Simpkins, F., Halgren, E., Mandelkern, M., Brown, C., Krisdakumtorn, T., and Gee, M. (1996a). Visual working memory in primary generalized epilepsy: An ^{18}FDG-PET study. *Neurology* **47,** 1203–1212.

Symms, M. R., Allen, P. J., Woermann, F. G., Polizzi, G., Krakow, K., Barker, G. J., Fish, D. R., and Duncan, J. S. (1999). Reproducible localisation of interictal epileptiform activity using functional MRI. *Phys. Med. Biol.* **44,** 161–168.

Szelies, B., Herholz, K., Heiss, W. D., Rackl, A., Pawlik, G., Wagner, R., Ilsen, H. W., and Wienhard, K. (1983). Hypometabolic cortical lesions in tuberous sclerosis with epilepsy: Demonstration by positron emission tomography. *J. Comput. Assist. Tomogr.* **7,** 946–953.

Szelies, B., Weber-Luxenburger, G., Pawlik, G., Kessler, J., Holthoff, V., Mielke, R., Herholz, K., Bauer, B., Wienhard, K., and Heiss, W. D. (1996). MRI-guided flumazenil- and FDG-PET in temporal lobe epilepsy. *NeuroImage* **3,** 109–118.

Tamaki, K., Okuno, T., Iwasaki, Y., Yonekura, Y., Konishi, J., and Mikawa, H. (1991). Regional cerebral blood flow in relation to MRI and EEG findings in tuberous sclerosis. *Brain Dev.* **13,** 420–424.

Tanaka, F., Yonekura, Y., Ikeda, A., Terada, K., Mikuni, N., Nishizawa, S., Ishizu, K., Okazawa, H., Hattori, N., Shibasaki, H., Konishi, J., and Onishi, Y. (1997). Presurgical identification of epileptic foci with iodine-123 iomazenil SPET: Comparison with brain perfusion SPET and FDG PET. *Eur. J. Nucl. Med.* **24,** 27–34.

Tasch, E., Cendes, F., Li, L. M., Dubeau, F., Montes, J., Rosenblatt, B., Andermann, F., and Arnold, D. (1998). Hypothalamic hamartomas and gelastic epilepsy: A spectroscopic study. *Neurology* **51,** 1046–1050.

Theodore, W. H. (1988). Antiepileptic drugs and cerebral glucose metabolism. *Epilepsia* **29** (Suppl. 2), S48–S55.

Theodore, W. H., Newmark, M. E., Sato, S., Brooks, R., Patronas, N., De La Paz, R., Di Chiro, G., Kessler, R. M., Manning, R., Channing, M., and Porter, R. J. (1983). ^{18}F-Fluorodeoxyglucose positron emission tomography in refractory complex partial seizures. *Ann. Neurol.* **14,** 429–437.

Theodore, W. H., Brooks, R., Margolin, R., Patronas, N., Sato, S., Porter, R. J., Mansi, L., Bairamian, D., and Di Chiro, G. (1985). Positron emission tomography in generalized seizures. *Neurology* **35,** 684–690.

Theodore, W. H., Holmes, M. D., Dorwart, R. H., Porter, R. J., Di Chiro, G., Sato, S., and Rose, D. (1986). Complex partial seizures: Cerebral structure and cerebral function. *Epilepsia* **27,** 576–582.

Theodore, W. H., Fishbein, D., Dietz, M., Baldwin, P., and Dietz, M. (1987a). Complex partial seizures: Cerebellar metabolism. *Epilepsia* **28,** 319–323.

Theodore, W. H., Rose, D., Patronas, N., Sato, S., Holmes, M., Bairamian, D., Porter, R. J., Di Chiro, G., Larson, S., and Fishbein, D. (1987b). Cerebral glucose metabolism in the Lennox-Gastaut syndrome. *Ann. Neurol.* **21,** 14–21.

Theodore, W. H., Bromfield, E., and Onorati, L. (1989). The effect of carbamazepine on cerebral glucose metabolism. *Ann. Neurol.* **25,** 516–520.

Theodore, W. H., Sato, S., Kufta, C., Balish, M. B., Bromfield, E. B., and Leiderman, D. B. (1992a). Temporal lobectomy for uncontrolled seizures: The role of positron emission tomography. *Ann. Neurol.* **32,** 789–794.

Theodore, W. H., Carson, R. E., Andreasen, P., Zametkin, A., Blasberg, R., Leiderman, D. B., Rice, K., Newman, A., Channing, M., Dunn, B., *et al.* (1992b). PET imaging of opiate receptor binding in human epilepsy using [^{18}F]cyclofoxy. *Epilepsy Res* **13,** 129–139.

Theodore, W. H., Gaillard, W. D., Sato, S., Kufta, C., and Leiderman, D. (1994). Positron emission tomographic measurement of cerebral blood flow and temporal lobectomy. *Ann. Neurol.* **36,** 241–244.

Theodore, W. H., Balish, M., Leiderman, D., Bromfield, E., Sato, S., and Herscovitch, P. (1996). Effect of seizures on cerebral blood flow measured with ^{15}O-H$_2$O and positron emission tomography. *Epilepsia* **37,** 796–802.

Theodore, W. H., Sato, S., Kufta, C. V., Gaillard, W. D., and Kelley, K. (1997). FDG-positron emission tomography and invasive EEG: Seizure focus detection and surgical outcome. *Epilepsia* **38,** 81–86.

Urenjak, J., Williams, S. R., Gadian, D. G., and Noble, M. (1992). Specific expression of N-acetylaspartate in neurons, oligodendrocyte-type-2–astrocyte progenitors, and immature oligodendrocytes *in vitro. J. Neurochem.* **59,** 55–61.

Urenjak, J., Williams, S. R., Gadian, D. G., and Noble, M. (1993). Proton nuclear magnetic resonance spectroscopy unambiguously identifies different neural cell types. *J. Neurosci.* **13,** 981–989.

Van Bogaert, P., Wikler, D., Damhaut, P., Szliwowski, H. B., and Goldman, S. (1998a). Cerebral glucose metabolism and centrotemporal spikes. *Epilepsy Res.* **29,** 123–127.

Van Bogaert, P., David, P., Gillain, C. A., Wikler, D., Damhaut, P., Scalais, E., Nuttin, C., Wetzburger, C., Szliwowski, H. B., Metens, T., and Goldman, S. (1998b). Perisylvian dysgenesis. Clinical, EEG,

MRI and glucose metabolism features in 10 patients. *Brain* **121**, 2229–2238.

Van der Grond, J., Gerson, J. R., Laxer, K. D., Hugg, J. W., Matson, G. B., and Weiner, M. W. (1998). Regional distribution of interictal ^{31}P metabolic changes in patients with temporal lobe epilepsy. *Epilepsia* **39**, 527–536.

Van Huffelen, A. C., van Isselt, J. W., van Veelen, C. W., van Rijk, P. P., van Bentum, A. M., Dive, D., Maquet, P., Franck, G., Velis, D. N., van Emde Boas, W., *et al.* (1990). Identification of the side of epileptic focus with ^{123}I-iomazenil SPECT. A comparison with ^{18}FDG-PET and ictal EEG findings in patients with medically intractable complex partial seizures. *Acta Neurochir. Suppl. (Wien)* **50**, 95–99.

Van Paesschen, W., Connelly, A., Jackson, G. D., King, M., and Duncan, J. S. (1997). The spectrum of hippocampal sclerosis. A quantitative MRI study. *Ann. Neurol.* **41**, 41–51.

Van Zijl, P. C. M., and Rothman, D. L. (1995). NMR studies of brain ^{13}C-glucose uptake and metabolism: Present status. *Magn. Reson. Imaging* **13**, 1213–1221.

Venz, S., Cordes, M., Straub, H. B., Hierholzer, J., Schroder, R., Richter, W., Schmitz, B., Meencke, H., and Felix, R. (1994). Preoperative evaluation of drug resistant focal epilepsies with ^{123}I-iomazenil SPECT. Comparison with video/EEG monitoring and postoperative results. *Nuklearmedizin* **33**, 189–193.

Vermathen, P., Ende, G., Laxer, K. D., Knowlton, R. C., Matson, G. B., and Weiner, M. W. (1997). Hippocampal *N*-acetylaspartate in neocortical epilepsy and mesial temporal lobe epilepsy. *Ann. Neurol.* **42**, 194–199.

Vles, J. S., Demandt, E., Ceulemans, B., de Roo, M., and Casaer, P. J. (1990). Single photon emission computed tomography (SPECT) in seizure disorders in childhood. *Brain Dev.* **12**, 385–389.

Warach, S., Levin, J. M., Schomer, D. L., Holman, B. L., and Edelman, R. R. (1994). Hyperperfusion of ictal seizure focus demonstrated by MR perfusion imaging. *Am. J. Neuroradiol.* **15**, 965–968.

Warach, S., Ives, J. R., Schlaug, G., Patel, M. R., Darby, D. G., Thangaraj, V., Edelman, R. R., and Schomer, D. L. (1996). EEG-triggered echo-planar functional MRI in epilepsy. *Neurology* **47**, 89–93.

Wieser, H. G., Duc, C., Meier, D., and Boesiger, P. (1995). [^{1}H]MR spectroscopy in the human hippocampus: Its role in the surgical treatment of temporal lobe epilepsy. *Epilepsia* **36** (Suppl. 3), S144.

Wieshmann, U. C., Symms, M. R., and Shorvon, S. D. (1997). Diffusion changes in status epilepticus. *Lancet* **350**, 493–494.

Wieshmann, U. C., Clark, C. A., Symms, M. R., Barker, G. J., Birnie, K. D., and Shorvon, S. D. (1999). Water diffusion in the human hippocampus in epilepsy. *Magn. Reson. Imaging* **17**, 29–36.

Woermann, F. G., Maclean, M. A., Bartlett, P. A., Parker, G., Barker, G. J., and Duncan, J. S. (1999). Short echo time single voxel MRS of hippocampal sclerosis and temporal lobe epilepsy. *Ann. Neurol.* **45**, 369–375.

Wong, C. Y., Geller, E. B., Chen, E. Q., MacIntyre, W. J., Morris, H. H., 3rd., Raja, S., Saha, G. B., Luders, H. O., Cook, S. A., and Go, R. T. (1996). Outcome of temporal lobe epilepsy surgery predicted by statistical parametric PET imaging. *J. Nucl. Med.* **37**, 1094–1100.

Woods, R. P., Mazziotta, J. C., and Cherry, S. R. (1993). MRI-PET registration with automated algorithm. *J. Comput. Assist. Tomogr.* **17**, 536–546.

Worthington, C., Vincent, D. J., Bryant, A. E., Roberts, D. R., Vera, C. L., Ross, D. A., and George, M. S. (1997). Comparison of functional magnetic resonance imaging for language localization and intracarotid speech amytal testing in presurgical evaluation for intractable epilepsy. Preliminary results. *Stereotact. Funct. Neurosurg.* **69**, 197–201.

Xue, M., Ng, T. C., Modic, M., Comair, Y., and Kolem, H. (1993). Oblique angle proton chemical shift imaging for the localization of hippocampal epilepsy. *In* "Book of Abstracts, 12th Annual Scientific Meeting, Society of Magnetic Resonance in Medicine," p. 435.

Xue, M., Ng, T. C., Comair, Y. G., Modic, M., and Luders, H. (1994). Presurgical localization of temporal lobe epilepsy using noninvasive proton chemical shift spectroscopic imaging. *In* "Book of Abstracts, 13th Annual Scientific Meeting, Society of Magnetic Resonance in Medicine," p. 572.

Zhong, J., Petroff, O. A. C., Prichard, J. W., and Gore, J. C. (1993). Changes in water diffusion and relaxation properties of rat cerebrum during status epilepticus. *Magn. Reson. Med.* **30**, 241–246.

Zhong, J., Petroff, O. A. C., Prichard, J. W., and Gore, J. C. (1995). Barbiturate-reversible reduction of water diffusion coefficient in fluorothyl-induced status epilepticus in rats. *Magn. Reson. Med.* **33**, 253–256.

Zubal, I. G., Spencer, S. S., Imam, K., Seibyl, J., Smith, E. O., Wisniewski, G., and Hoffer, P. B. (1995). Difference images calculated from ictal and interictal technetium-99m-HMPAO SPECT scans of epilepsy. *J. Nucl. Med.* **36**, 684–689.

15

MRI in Multiple Sclerosis

Guojun Zhao,[*,1] David K. B. Li,[†] and Donald Paty[*]

[*]Division of Neurology, Vancouver Hospital and Health Sciences Center, University of British Columbia, Vancouver, British Columbia, Canada V6T 2B5
[†]Department of Diagnostic Radiology, Vancouver Hospital and Health Sciences Center, University of British Columbia, Vancouver, British Columbia, Canada V6T 2B5

I. Imaging in the Diagnosis of Multiple Sclerosis

II. MRI in Natural History Studies of Multiple Sclerosis

III. Clinical Correlations with MRI Findings

IV. MRI–Pathological Correlation

V. Application of MRI in Monitoring of Clinical Trials

VI. Evaluation of in Vivo Pathology with Newer MR Techniques

VII. Summary

References

MR techniques have opened a literal "window on the brain" for visualizing and measuring the evolution of pathology in the living patient. Using a combination of proton density (PD), T_2-weighted (T_2W), T_1-unenhanced, and T_1-enhanced imaging, one can visualize the evolution of MS lesions to see the sequence of evolution from blood–brain barrier disruption and inflammation to chronic changes reflecting demyelination and/or gliosis

and axonal loss. Application of these MRI analysis techniques has had a major impact on diagnosis, natural history, and therapeutic studies. Additional MR techniques, such as spectroscopy, magnetization transfer imaging, and T_2 relaxation analysis, will probably help considerably in the specificity of MR measurements for identifying pathological changes.

I. Imaging in the Diagnosis of Multiple Sclerosis

A. Conventional MRI Techniques

Multiple sclerosis (MS) is the most common of the demyelinating diseases. It is an inflammatory and demyelinating disease of the white matter of the central nervous system (CNS) that, in the majority of cases, presents with attacks of neurological dysfunction. The diagnosis is based on the demonstration of acute and chronic white matter lesions disseminated in time and space. It is also a disease of young adults (mostly female) with genetic and environmental background factors.

Magnetic resonance imaging (MRI) of the brain and spinal cord is being used increasingly in the clinical evaluation of demyelinating disease. It reveals the

---

[1]To whom correspondence should be addressed.

lesions dramatically and can be used as the single modality to satisfy the criterion for dissemination in space. Traditionally, the role of radiographic imaging studies in the diagnosis of MS has been predominantly indirect, such as to rule out a space-occupying lesion. Although there are abnormal findings on computed tomography (CT) in patients with MS, the incidence of positive CT findings varies between 9 and 80%. The CT abnormalities were often nonspecific areas of low density, atrophy, and/or contrast enhancement (Fig. 1). Thus CT was used primarily to eliminate the possibility of other intracranial lesions, such as a neoplasm or arteriovenous malformation. Similarly, myelography was used to evaluate patients with spinal cord symptoms to exclude an intrinsic spinal cord mass or compression on the spinal cord.

Before MRI, patients with clinical symptoms suspicious for MS had to undergo CT and/or myelography, a battery of paraclinical studies (including visual, auditory, and somatosensory evoked potentials), and cerebrospinal fluid (CSF) studies (including electrophoresis for detection of oligoclonal bands) to establish a diagnosis. Now, in many cases, with an appropriate clinical history, MRI studies allow the neurologist to make the diagnosis of MS with a high degree of confidence, thereby obviating the need for other ancillary studies.

Since the initial report by Young *et al.* (1981) describing the increased sensitivity of MRI compared

with CT in the diagnosis of MS, many reports have documented the usefulness of MRI (Li *et al.,* 1984; Barnes *et al.,* 1988; Grossman *et al.,* 1988; Miller *et al.,* 1988; Paty, 1988a). MRI is easy to perform and is less invasive than CSF analysis and/or myelography. Furthermore, unlike other paraclinical tests, MRI is able to directly visualize the demyelinating plaques and their anatomical sites of involvement (Fig. 2). The information from MRI studies potentially enables correlation with clinical symptoms, has prognostic significance, and permits MRI to be used for follow-up to determine regression or progression of lesions. This last advantage has proven important in evaluating new therapies (Paty, 1987; Paty and Li, 1993).

Although the MRI appearance may vary slightly from patient to patient or even among lesions in the same patient, certain characteristic findings are common to most MS lesions. MS lesions are generally detected as areas of increased T_1 and T_2 relaxation times relative to white matter. Since MS lesions have a prolonged T_1 relaxation time, they are lower in signal intensity (darker) and are particularly well defined on inversion-recovery (IR) scans. Indeed, in many of the initial MR reports, MS lesions were shown by IR. However, a major disadvantage of conventional IR scanning is its relative insensitivity for detecting lesions that lie immediately adjacent to the ventricles or the subarachnoid spaces. These lesions are very common in MS and

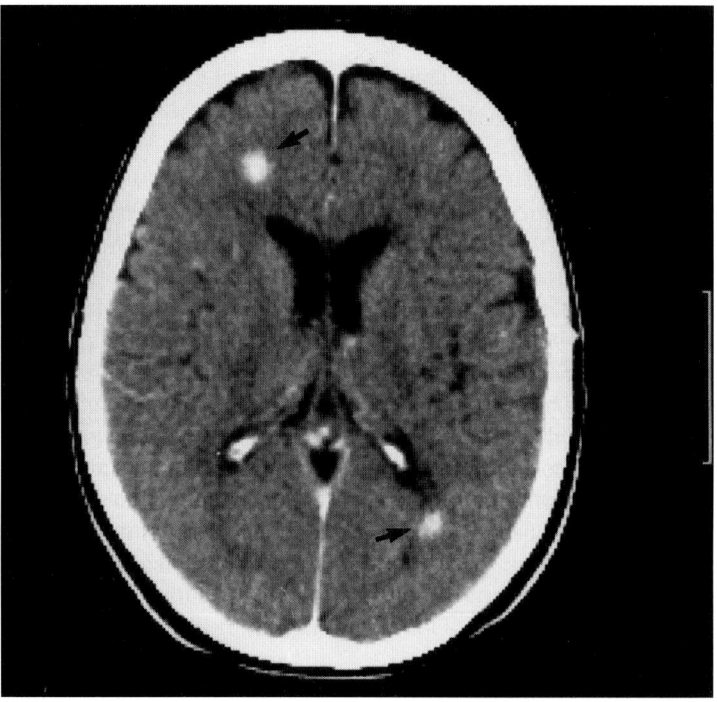

Figure 1 An enhanced CT scan at the middle level of lateral ventricles. Two enhancing lesions can be seen (arrows).

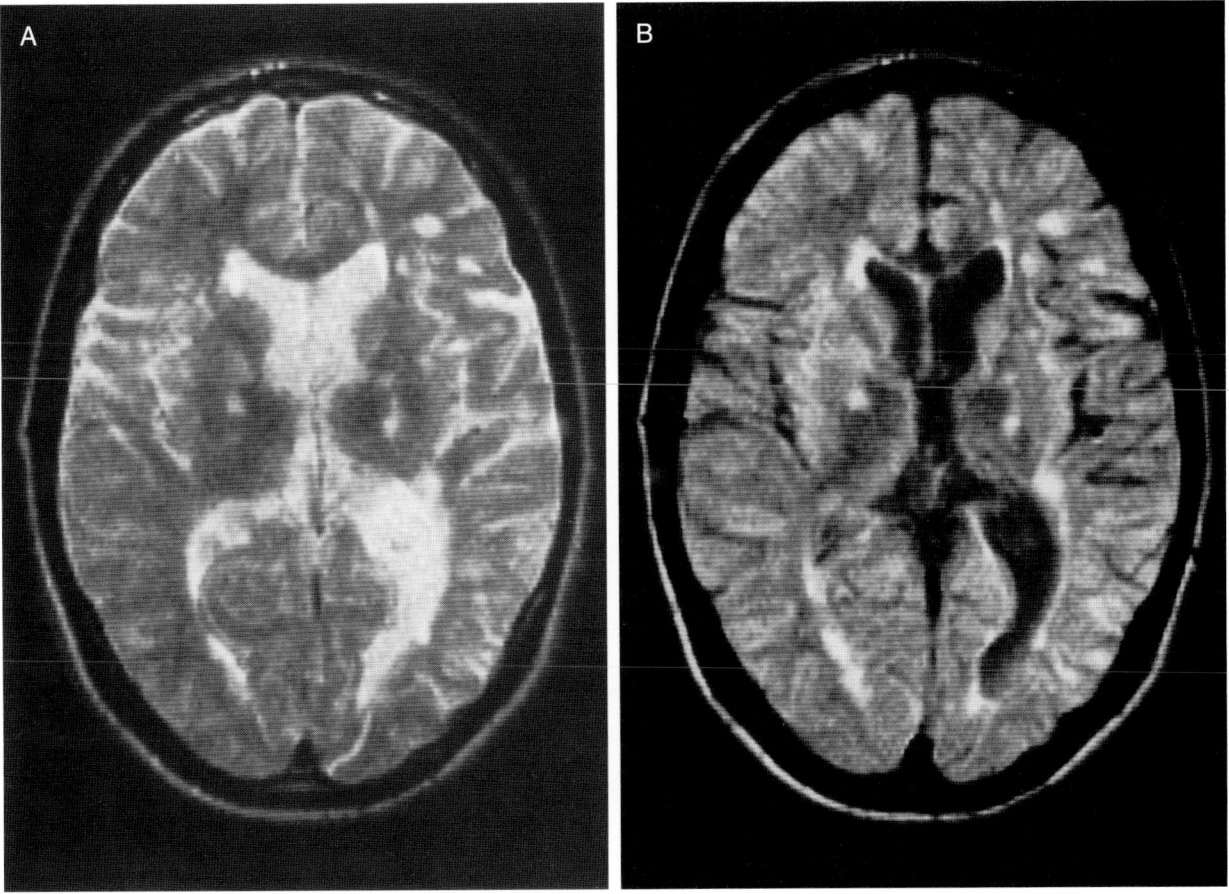

Figure 2 MS lesions in PD and T₂W images. (A) A T₂W image at the level of the ventricles showing a number of lesions. Note that it is hard to define the borders of periventricular lesions. (B) A PD image of the same patient. The periventricular lesions can be easily distinguished from CSF. Therefore, the burden of disease (BOD) measure should always be done on the PD image.

because of partial volume averaging can be confused with the low intensity of the adjacent CSF. Some lesions may also be missed if they are contiguous with the gray matter, which also has a lower intensity than white matter on IR.

The spin–echo (SE) technique is a commonly used method for screening patients with suspected MS (Fig. 2). In contrast to IR, long TR SE sequences reveal MS lesions as areas of high signal intensity (white spots). Spin-echo sequences with a long TR of 2500–3000 ms and dual echos using short and long TEs of 30 and 80 ms are preferred. The MS lesions have moderately increased signal intensity compared with brain or CSF on PD images (TE of 30 ms), with a further increase in signal intensity on the more heavily T₂W images (TE of 80 ms). Partial volume effects on heavily T₂W spin–echo images can obscure periventricular lesions when both CSF and MS lesions are seen as very bright areas (Fig. 2A). However, these periventricular MS lesions can easily be distinguished from the adjacent CSF spaces by comparison with the PD images (Fig.

2B), in which the lesions will have a greater signal intensity than the adjacent CSF. In general, lesions in the periventricular regions are best seen on the first (short) echo images, whereas posterior fossa and deep white matter lesions are often better seen on the second, longer echo images. Dual-echo sequences are particularly useful whenever there are subtle lesions that may be equivocal on one echo, since the images can be compared.

In the past few years, fast spin–echo (FSE) techniques with similar T₂ lesion contrast but obtained in one-quarter to one-third the acquisition times have often replaced conventional SE sequence as a scanning sequence. Although occasionally small lesions may be missed on FSE because of edge blurring, the time advantage and the ability to obtain thinner slices more than compensate for the blurring. FSE sequences have not been used universally in clinical trials because of variation in the implementation of the method among different scanners. Another recent development is the use of fast fluid attenuated inversion–recovery sequences (FLAIR) with T_2 weighting, combining the speed advantage of FSE, T_2 lesion contrast, and the suppression of the partial volume effect of CSF (Hashemi *et al.*, 1995). This sequence is of particular help in demonstrating periventricular and gray–white matter junction lesions. Depending on the parameter chosen, FLAIR may be less sensitive in detecting some brainstem and spinal cord lesions.

Minimum possible slice thickness should be used to decrease partial volume averaging from the surrounding low-intensity white matter. A 5-mm slice with a 0.5- to 1-mm gap is typically used, the gap being primarily determined by cross-slice contamination or "cross talk." In the FSE technique, contiguous 3-mm slices are routinely used. Three-millimeter slices reveal a higher lesion load and greater number of lesions than do 5-mm slices. However, the use of thinner slices does not significantly increase the sensitivity of brain MRI in the detection of lesion changes in MS over time (Rovaris *et al.*, 1998).

Characteristic MRI findings in MS consist of multiple, usually small, lesions within the white matter that have increased proton density and long T_1 and long T_2 relaxation times in comparison with normal white matter (Figs. 3A–3D). The increased T_1 and T_2 relaxation times seen with MS lesions probably relate to the gliosis, loss of axons, and tissue destruction that occur in chronic plaque formation. In acute lesions, edema resulting from the blood–brain barrier (BBB) breakdown results in prolonged relaxation times and may produce an area of abnormality considerably larger than the actual demyelinated area. Demyelination itself, with the breakdown of fatty myelin sheaths within the MS lesion, probably does not contribute significantly to the prolonged T_2 changes. The amount of lipid lost is not great enough to cause the magnitude of change demonstrated on MRI because the lipids related to myelin breakdown have extremely short T_2 relaxation times and are effectively invisible on MRI. The loss of myelin lipid does, however, result in a more hydrophilic environment and this increase in water content leads to the observed increases in proton density, T_1, and T_2 (Ormerod *et al.*, 1989).

Figure 3 shows a number of appearances of lesions. There is a considerably variability in the signal, which must reflect differences in pathology. New, perhaps pathologically more specific MR techniques are discussed at the end of this review.

The individual lesions of MS are usually less than 10 mm in diameter, most often between 1 and 5 mm. Confluent plaque formation from the merging of multiple small individual lesions that are contiguous with each other or secondary to a large acute, actively demyelinating process can occur. In some cases, acute lesions can become quite large, simulating tumors.

The most common location for lesions seen on MRI is in the periventricular region adjacent to the superolateral angles of the lateral ventricles. This distribution corresponds with pathologic descriptions (Stewart *et al.*, 1986). Lesions in the corona radiata often appear oval or elongated, with the long axis of the lesion oriented along the subependymal veins, perpendicular to the walls of the ventricles (Fig. 3C). This orientation corresponds to the perivenular inflammation seen pathologically. It is a highly characteristic appearance on MRI and helps to distinguish MS from other white matter diseases, most notably deep white matter ischemia and infarction in older patients. The sagittal sequence reveals these periventricular lesions to best advantage. MS lesions are also often seen within the body of the corpus callosum, a site rarely affected by microinfarcts, further allowing these two entities to be distinguished. MS lesions also tend to be more focal than deep white matter ischemic changes.

Other common sites of involvement include the walls of ventricles adjacent to the atrial trigones and occipital horns, the white matter of the centrum semiovale, the forceps major and minor, and the temporal lobes.

MS lesions are also often seen on MRI in the brainstem and cerebellum (Fig. 3C), where they often abut the CSF spaces. However, lesions are seen within the cerebellum less commonly than in the brainstem. A lesion distribution study (Wang *et al.*, 1997) was done in 501 patients with clinically definite relapsing–remitting (RR) MS and 108 patients with secondary progressive (SP) MS. The study found that in RRMS 89.4% of all brain lesions were seen in the cerebrum, 7.8% in the brainstem, and 2.9% in the cerebellum, whereas in SPMS 86.6% of the lesions were seen in the cerebrum, 9.0% in the brainstem, and 4.5% in the cerebellum. Brainin *et al.* (1987) reported one or more pontine lesions in 71% of 24 patients with clinically definite MS (CDMS), medullary le-

Figure 3 The variable appearances of lesions on different pulse sequences. (A) An MRI (T_2W) axial image showing a number of MS lesions. Note the large lesion seen in cross section in the right hemisphere. (B) A coronal view (T_2W) showing the large lesion seen in Fig. 3A. Note the leaflike shape that does not seem to go into the gray matter. (C) A parasagittal view (T_2W) showing the same large lesion with a flamelike shape pointing away from the lateral ventricle. Note other lesions along the ventricle and also in the cerebellum. (D) A parasagittal view (T_1W) of the same patient more lateral than Fig. 3C showing a number of black holes (lesions). The large lesion is an acute black hole and the smaller lesions are chronic black holes (arrows).

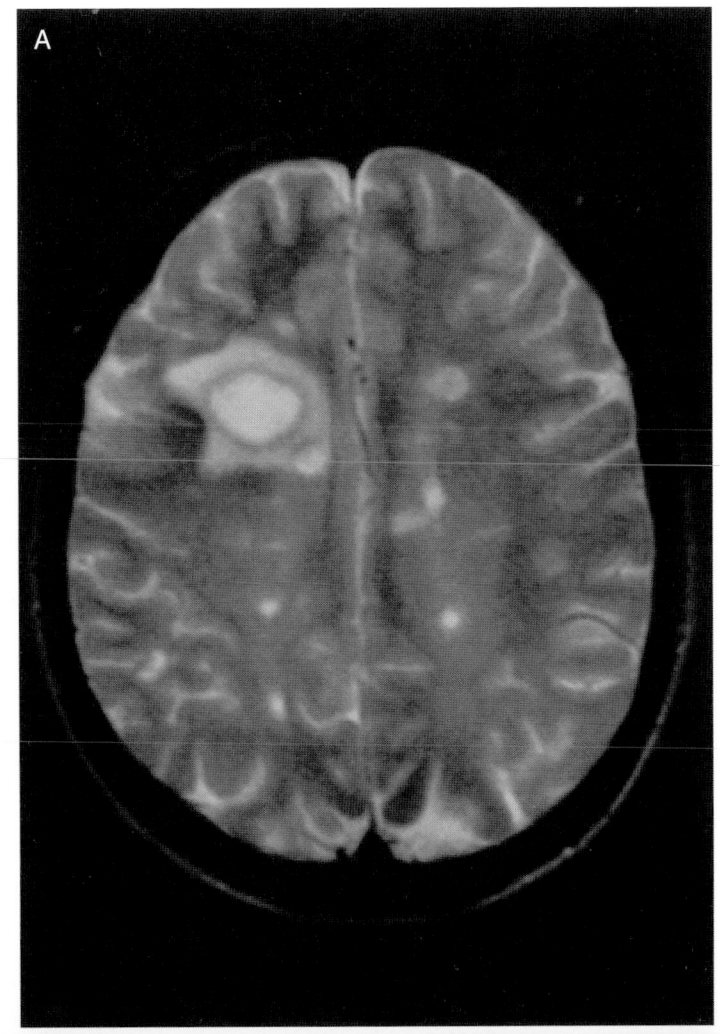

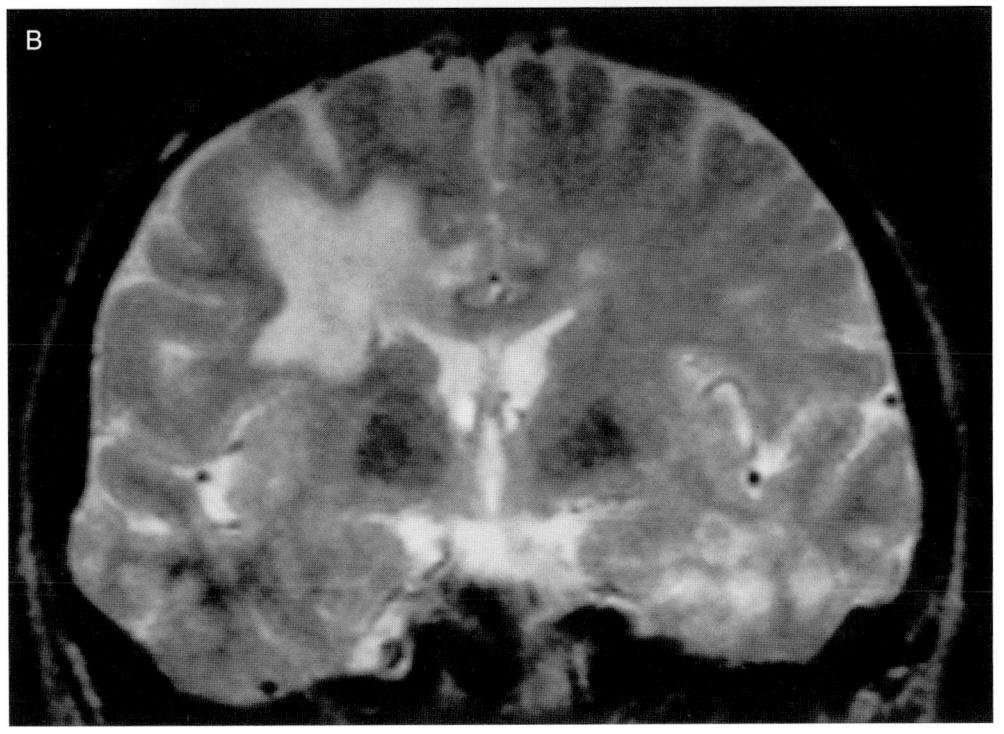

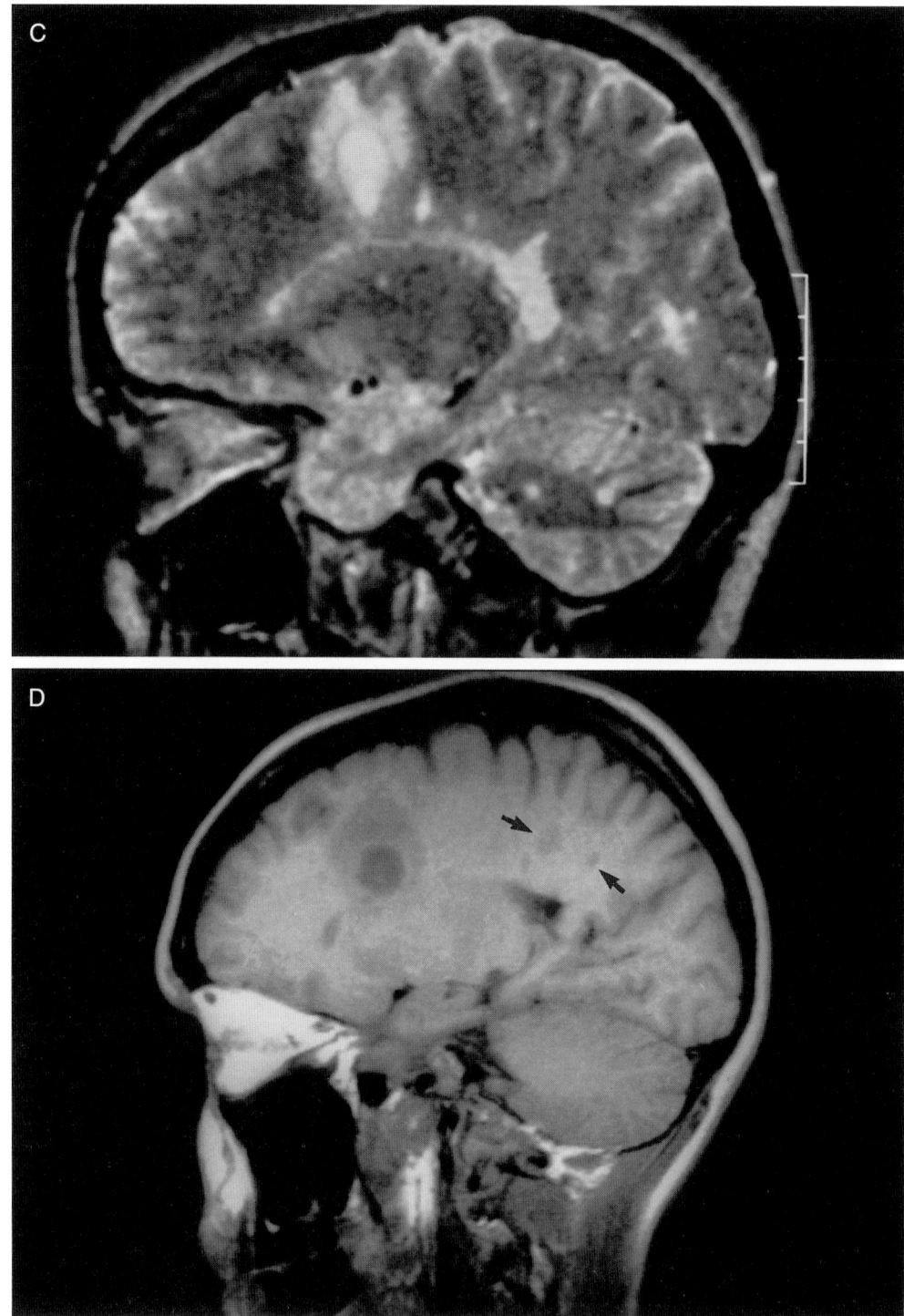

Figure 3 *(continued)*

sions in 50%, and midbrain lesions in 25%. The middle cerebellar peduncles and the white matter of the corpus medullaris are other preferred lesion locations.

Although MS is generally considered a white matter disease, approximately 5–10% of lesions occur in the gray matter.

Some MS patients may have no lesions in the brain and only have lesions in the spinal cord. In some studies, about half of MS patients had only a single spinal cord plaque and more than half of the cord lesions occurred in the cervical area and most frequently involved the posterior and lateral aspects of the cord (Tartaglino *et al.*, 1995; Thielen and Miller, 1996). On MRI, spinal cord lesions are characteristically peripherally located, extend two vertebral segments or less in length, and occupy less than half the cross-sectional area of the cord (Tartaglino *et al.,* 1995). Although most lesions are focal, in some patients more diffuse changes involving the entire cord may occur. Over time, measurement of spinal cord area may show atrophy which correlates with increasing disability (Stevenson *et al.,* 1998). With current MRI technology, imaging the spinal cord provides only modest gains in lesion detection, since brain MRI images alone are able to detect 90% of the active lesions in RRMS. Also, 95% of CDMS patients have abnormal head scans (Thielen and Miller, 1996; Thorpe *et al.,* 1996). However, when the clinical diagnosis is not certain and MR findings in the head are negative, a follow-up head and/or spinal cord study can be helpful.

MRI can be of great value in the diagnostic process, but MS should never be diagnosed without consideration of the clinical features. In conjunction with laboratory tests, MRI can be used to differentiate MS from, for example, degenerative diseases and damage caused by toxins, vasculitis, and tumors. MRI is most useful in patients with a monophasic MS-like presentation, in whom the dissemination of pathology in time and space typical of MS is not clinically evident.

MRI diagnostic criteria for MS have been proposed. Paty *et al.* (1988) suggested that a diagnosis of MS is supported if PD/T_2 scanning reveals four or more lesions, or three or more lesions with one or more bordering a lateral ventricle. To increase the specificity of MRI interpretation in elderly subjects, Fazekas and co-workers (1988) proposed that for a positive MRI diagnosis of MS, there should be at least three PD/T_2 lesions that are greater than or equal to 3 mm, and that two of the following three criteria be present: (a) size greater than or equal to 6 mm, (b) abutting ventricular bodies, and (c) infratentorial location. Using MRI, Barkhof *et al.* (1997) studied 74 patients with clinically isolated neurological symptoms suggestive of MS. Thirty-three of them developed CDMS. They suggested that if PD/T_2 lesions are used to predict CDMS, the balance between sensitivity and specificity is optimized with a cutoff point of nine lesions. If both PD/T_2 and gadolinium-enhancing T_1 lesions are used, the best predictors of CDMS are in the following order: (a) one or more juxtacortical lesions are seen; (b) one or more gadolinium-enhancing lesions are found; (c) one or more infratentorial lesions are present; and (d) three or more periventricular lesions are identified.

A new diagnostic category MRI-supported definite multiple sclerosis was recently suggested by Paty and Li (1999) to be added to the categories of CDMS and laboratory-supported definite MS (LSDMS) for research protocols. To this category, patients should (1) be younger than 45 years, (2) have at least one MS-like clinical episode with appropriate clinical findings (no remission is necessary), and (3) have an abnormal MRI scan as follows: (a) four 3-mm white matter lesions, or three, with one periventricular; (b) at least one of the following specific findings: >6 mm diameter, oval shape (flamelike), located in or above the corpus callosum or infratentorially, one or more—but not all—lesions enhancing or one with an open-ring enhancement.

B. Other Techniques

Neurophysiological techniques are not used as frequently now as in the past with the wide availability of MRI. The EEG findings in MS may be abnormal but are not specific. The most useful physiological studies have been the evoked potentials (EPs). Evoked potentials are used to detect changes in nerve conduction properties, but the exact site of abnormality along the conduction pathway cannot be determined. Heide and co-workers (1998) found that cerebral *N*-acetylaspartate was significantly lower in MS patients with abnormal visual EPs (VEPs) than in normal volunteers with normal VEPs. Another study (Schurmann *et al.,* 1993) showed that auditory EPs may show reduction in amplitude of alpha response in vertex recordings. Different degrees of amplitude reduction can be seen in different frequency channels: alpha (7–12 Hz) components were reduced whereas theta (4–7 Hz) responses were not altered. Schurmann *et al.* suggested changes in alpha responses might be related mostly to primary sensory processing. In contrast, theta response may be related to associative and cognitive processing rather than to primary sensory processing. Somatosensory evoked potentials (SEPs) can also be used to detect subclinical evidence for dissemination in space. A delay in the latency of SEPs can be suggestive of demyelination along somatosensory pathways.

Positron emission tomography (PET) is used to show oxygen utilization and blood flow in both white matter and gray matter. Recent studies (Roelcke *et al.,* 1997; Bakshi *et al.,* 1998) showed a reduction in total brain glucose metabolism in MS patients. This hypometabolism can be seen in cerebral cortex, subcortical nuclei, supratentorial white matter, and infratentorial structures. The most dramatic absolute reductions occurred in the superior mesial frontal cortex, superior dorsolateral frontal cortex, mesial occipital cortex, lateral occipital cortex, deep inferior parietal white matter, and pons. Fatigue was associated with frontal cortex and basal ganglia dysfunction. Memory impairment in MS is associated with bilateral reduction of glucose metabolism in the hippocampus, cingulate gyrus, thalamus, associative occipital cortex, and cerebellum (Paulesu *et al.*, 1996). However, MS lesions seen on MRI do not consistently show hypometabolism. Schiepers *et al.* (1997) found that 10 out of 15 MS lesions showed paradoxically relative hypermetabolism, only two showed hypometabolism.

A small study with 17 MS patients was conducted using MRI and single photon emission computerized tomography (SPECT) (Pozzilli *et al.*, 1992), using [^{99m}Tc]hexamethylpropylenamine oxime ([^{99m}Tc] HMPAO) to correlate with cognitive functions. This study found that activity was significantly reduced in the frontal lobes and in the left temporal lobe of MS patients. A correlation was found between left temporal abnormalities in [^{99m}Tc]HMPAO uptake and deficits in verbal fluency and verbal memory. Another study (Joosten *et al.*, 1995) used SPECT and cobalt-57 (^{57}Co) as a calcium analogue to visualize brain tissue damage in five MS patients but found ^{57}Co-SPECT using a single-headed camera not an appropriate imaging modality for studying MS.

II. MRI in Natural History Studies of Multiple Sclerosis

A. Morphological Changes

Morphological changes (changes in number, size, and appearance of lesions) can be seen on serial MRI study in MS, with PD/T$_2$W and pregadolinium T$_1$W scans. PD/T$_2$W scans are currently most commonly used in monitoring therapeutical trials. PD/T_2 morphologically active lesions can be defined as follows:

▲ *New lesion*—A lesion that has not previously been seen (Figs. 4A and 4B)

▲ *Enlarging lesion*—Enlargement of a previously seen stable lesion (Figs. 5A–5C)

▲ *Recurrent lesion*—A lesion that develops at the same site where a previous lesion had been seen (not shown)

1. Morphologically New Lesions on PD/T$_2$W Scans (Figs. 4A and 4B)

The morphologically new lesions seen on PD/T$_2$W images have a characteristic temporal course. New lesions reach a maximum size in approximately 4 weeks and subsequently regress and decrease in size, leaving a smaller residual lesion characterized by a prolonged relaxation time (Koopmans *et al.*, 1989; Willoughby *et*

al., 1989). A few such lesions have been shown to continuously increase in size with time. The majority of morphologically new MS lesions also enhance. Many new lesions are also clinically asymptomatic so that MRI activity is much greater than is clinical activity. Because MRI is so sensitive, it has proven useful in monitoring the results of therapy (Paty, 1987). It has also been shown that MRI activity is useful for distinguishing between benign and primary chronic progressive forms of MS (Thompson *et al.*, 1990).

2. Morphologically Enlarging Lesions on PD/T$_2$W Scans (Figs. 5A–5C)

Morphologically enlarging lesions are those that increase in size from a previously stable lesion or those that show continued enlargement of a previously seen active lesion. Stable lesions are defined as those that have not changed morphologically over follow-up until the current activity is seen. Enlarging lesions are more similar in size and shape to stable lesions than to new and recurrent ones.

To study the growth pattern of lesions that enlarged from previously stable lesions, we studied 50 patients who were part of an interferon β-1b therapeutic trial. The patients had MRI examinations every 6 weeks over 2 years (Zhao *et al.*, 1993a). Sixty-three enlarging lesions were identified. Of the 24 enlarging periventricular lesions, 11 enlarged predominantly away from ventricle, 3 anteriorly, 2 posteriorly, 2 rostrally, 2 caudally, and 4 equally in all directions. Of the 39 nonperiventricular enlarging lesions, 11 enlarged away from and 8 toward the ventricles, 2 posteriorly, 7 rostrally, 4 caudally, and 7 in all directions. Internal capsule lesions enlarged only rostrally or caudally. All lesions but 4 were visible on the final scan of the series. Seventy-five percent (44/59) of the lesions had become smaller on the subsequent study (6 weeks). Forty-four percent (26/59) of the lesions returned to their original size, whereas 49% (29/59) decreased in size but the residual size was larger than the original size. This study provides evidence that MS lesions enlarge asymmetrically more often than concentrically. The enlargement may be along the course of venules or along the projections of white matter tracts, as was seen in the internal capsules.

Enlarging lesions that originate from previously stable ones probably differ from new lesions. It is unclear what signals a stable lesion to become active. Most enlarging lesions also enhance as a sign of activity (Fig. 5C). Whereas new lesions are often seen in nonperiventricular regions, are round or oval in shape, and are small or medial in size, enlarging lesions are most often seen in the periventricular region, are irregular

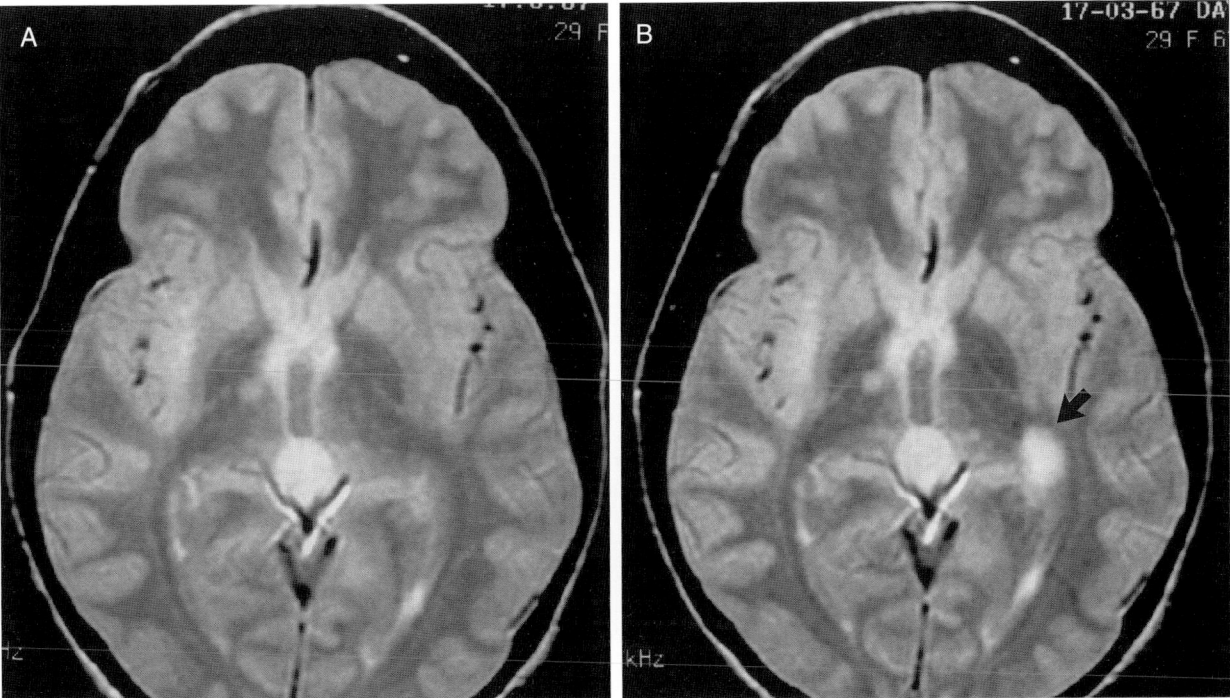

Figure 4 A new lesion. (A) An axial (PD) image showing several small lesions (white spots). A pineal cyst is incidentally noted. (B) The same view as Fig. 4A, 1 month later. Note the new lesion (arrow).

in shape, and tend to be larger in size than new lesions. The factors that control the pattern of growth in MS lesions on MRI have yet to be determined. The process may spread along the course of small veins and venules by direct expansion from the edge of active plaques.

3. Morphologically Recurrent Lesions on PD/T₂W Scans

Recurrent lesions are lesions that reappear at the same site at which an earlier lesion had disappeared. They are relatively uncommon. In the patients in the placebo arm of the interferon β-1a study, the mean number of T_2-active lesions per patient per biannual scan over 2 years was 3.4, with a mean of 2.3 new lesions per patient per year, 1.0 enlarging lesions, and only 0.001 recurrent lesions (PRISMS, unpublished data). It is important not to mistake truly recurrent lesions from apparent recurrences that result from partial volume effects that may obscure a lesion on one scan but not on another. One study (Zhao *et al.*, 1991) showed that the margins of most morphologically active lesions were well-defined. When the margins were ill-defined, the lesions were more commonly new or recurrent lesions rather than stable or enlarging ones. While enlarging and stable lesions were slightly more likely to be found in a periventricular than a nonperiventricular location,

recurrent lesions together with new lesions were more commonly located away from the ventricles. Recurrent lesions were more likely round or oval in shape and the average size of recurrent lesions is smaller than that of enlarging and stable ones.

B. Morphological Change on T₁-Weighted Scans (Figs. 6A–6C)

The lesions seen on the unenhanced T_1W scan have been called black holes. These black holes are not as frequently seen as are the PD/T_2W lesions. New PD/T_2W and enhancing lesions may appear isointense (20%) or hypointense, or acute "black holes" (80%), on unenhanced T_1W images. During 6 months of follow-up, four MR patterns were observed: initially isointense lesions remained isointense (15%); initially isointense lesions became hypointense (5%, most of which reenhanced); initially hypointense lesions became isointense (44%); and initially hypointense lesions remained hypointense (36%) (van Waesberghe *et al.*, 1998). van Waesberghe *et al.* therefore concluded that many acute hypointensities are reversible and probably represent intense inflammation, whereas those lesions that persisted as chronic black holes may have irreversible structural changes such as demyelination, gliosis, and axonal loss (Fig. 6).

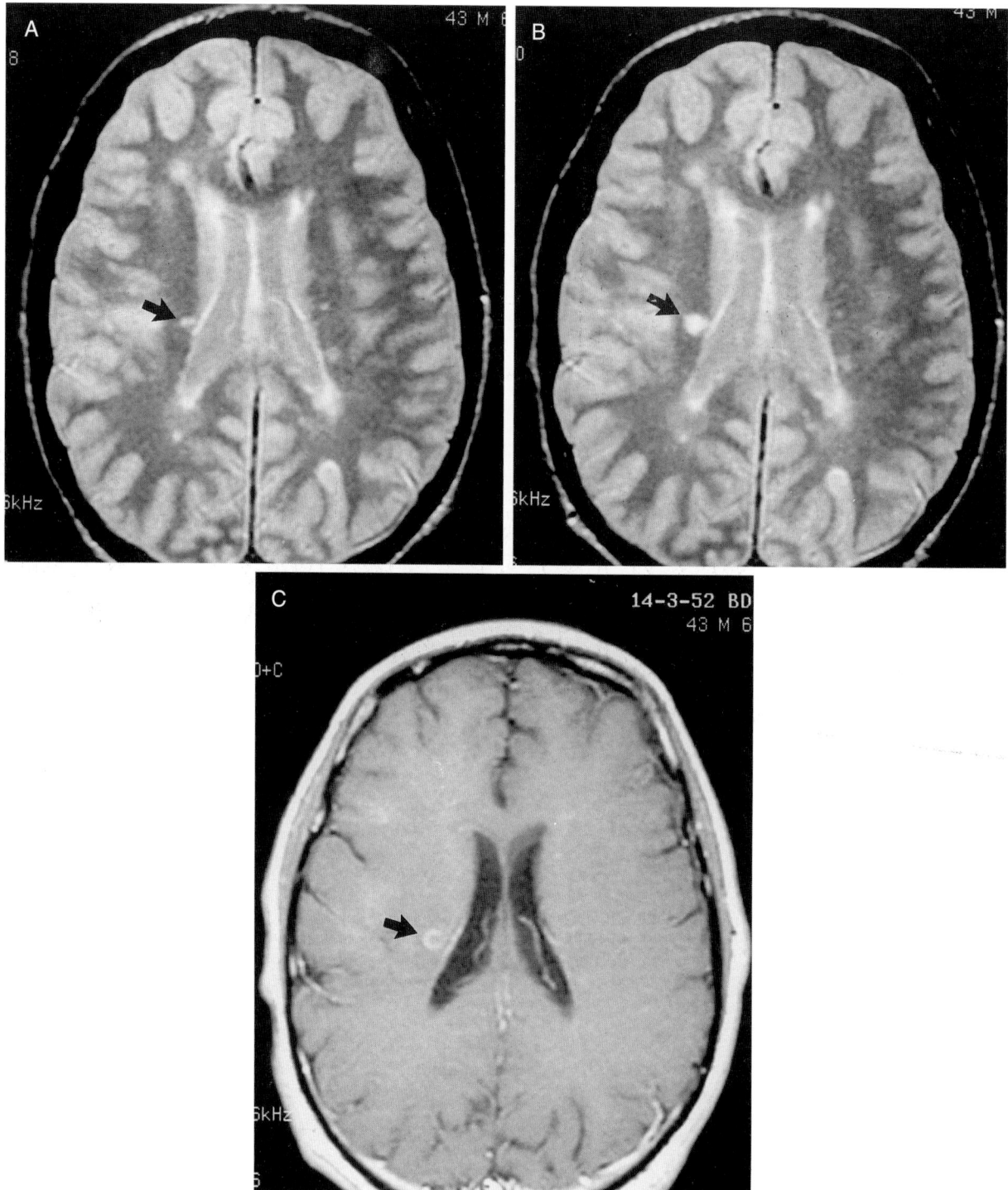

Figure 5 An enlarging lesion with enhancement. (A) A PD axial image showing several small lesions. Note the lesion in the right periventricular white matter (arrow). (B) A PD axial image 1 month later in the same patient. Note that the periventricular lesion has enlarged. (C) A post-Gd T_1-weighted axial image at the same time as Fig. 5B showing that the enlarging lesion also shows ring enhancement (arrow).

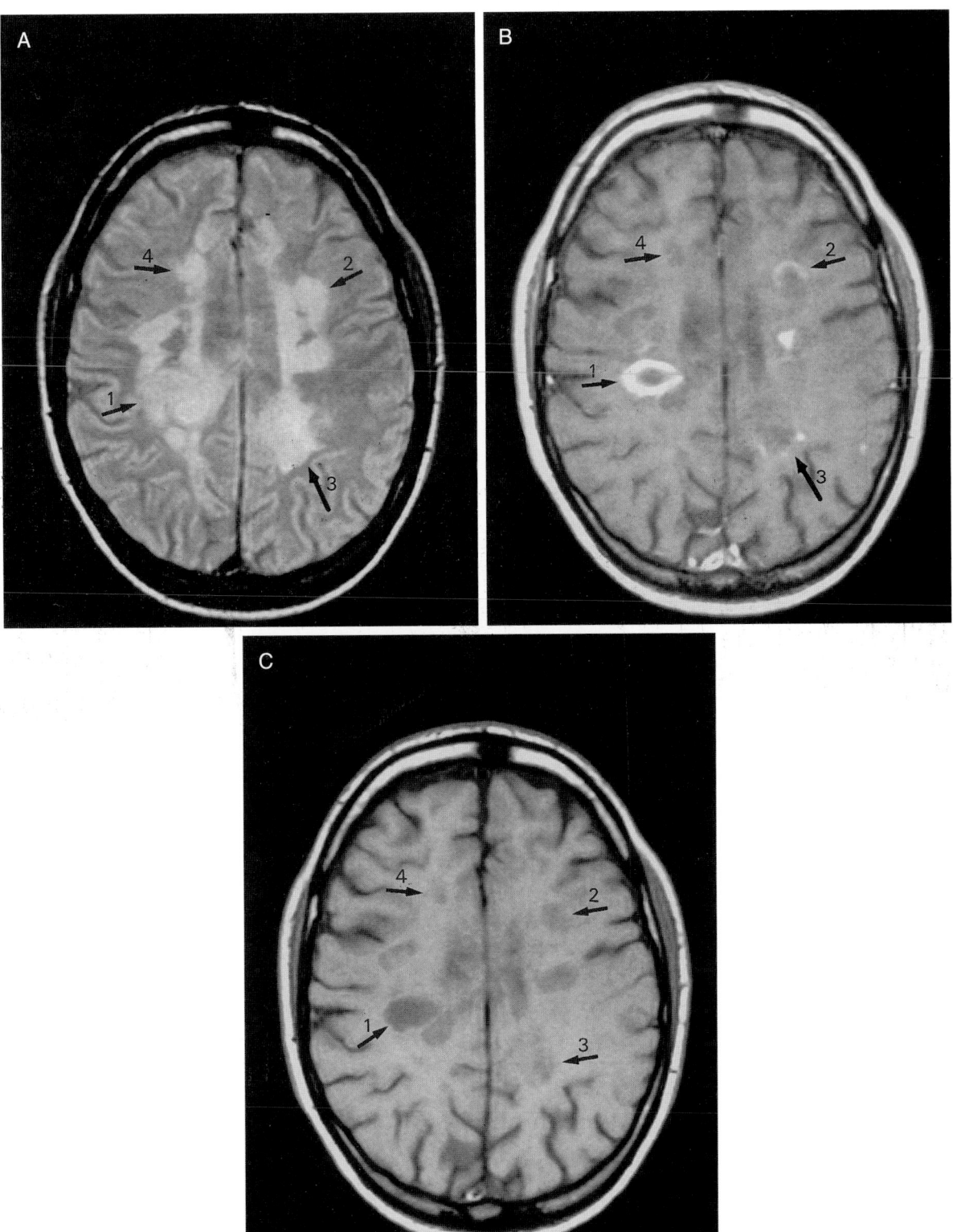

Figure 6 Acute black holes (with enhancement) and chronic black holes (without enhancement). (A) A PD image showing a number of large lesions superior to the lateral ventricles. Note the arrows with numbers. Compared to the previous scan (not shown), lesion 1 is new, lesions 2 and 3 are enlarging, and lesion 4 is stable. (B) Post-Gd T_1W image performed on the same date as Fig. 6A showing a number of enhancing lesions (arrows 1, 2, and 3). Some lesions are not enhancing (arrow 4). One enhancing lesion (arrow 1) is a new lesion on PD and therefore an acute black hole. Lesions 2 and 3 are both enlarging and enhancing as well so the black holes seen are a mixture of acute and chronic pathologies. Lesion 4 is stable on both PD and T_1W images and, therefore, is a chronic black hole (see Fig. 6C). (C) Pre-Gd T_1W image on the same date as Fig. 6B showing both acute (arrows 1, 2, and 3) and chronic black holes (arrow 4).

C. Gadolinium Enhancement (Figs. 5–7)

Gadolinium (Gd) is paramagnetic. Paramagnetic compounds have at least one unpaired electron. That electron has a magnetic moment approximately 1000 times stronger than the magnetic moment of a proton. As a result of motion (diffusion, rotation, etc.), the paramagnetic compounds act at the atomic level to produce a rapid fluctuation in the local magnetic field. This process facilitates energy transfer among the excited protons and also from the protons to their environment, and the magnetic moments of the hydrogen nuclei deflected by the rf pulse return more rapidly to their initial state. Paramagnetic compounds therefore shorten the relaxation times of adjacent protons; they can affect both T_1 and T_2 relaxation times.

Gd, a rare-earth element with seven unpaired electrons, has become widely used as a contrast agent in MRI. Free Gd ion is unsuitable for clinical use because it is toxic. However, when it is chelated (for example, to DTPA, DTPA-BMA, HP-Do3A, and DOTA), it becomes tightly bound and metabolically inert but continues to exhibit its paramagnetic effect. There is no evidence of dissociation of the Gd chelated complexes *in vivo*. They are rapidly excreted. For example, more than 90% of the injected Gd-DTPA is excreted through the kidneys within 24 h (Weinmann *et al.*, 1984). The Gd chelates are well tolerated *in vivo*, and adverse drug reactions occur in only 1.46% of patients (Niendorf *et al.*, 1991). In the majority of cases, adverse events consist of minor side effects, such as nausea, local warmth/pain, headache, and paraesthesia. Severe side effects are very rare.

Although Gd chelates shorten both T_1 and T_2 relaxation *in vivo*, with the usual clinical dose given (0.1–0.3 mmol/kg), the T_1 effect is predominant and the T_2 effect virtually absent (except in areas of high concentrations such as the bladder, where the excreted contrast pools). When given intravenously, contrast enhancement in the central nervous system occurs on the basis of either lesion vascularity or, in the case of active MS lesions, blood–brain (or blood–cord) barrier disruption. Enhancement appears as areas of increased signal intensity (bright areas) on T_1-weighted images.

Gadolinium-enhanced MRI (Gd-MRI) is more sensitive in the detection of active lesions than the clinical examination alone (Grossman *et al.*, 1988; Miller *et al.*, 1988). BBB disruption is a consistent finding in morphologically new lesions detected by MRI. BBB disruption can also develop in older, previously nonenhancing stable lesions without showing any evidence of an increase in size morphologically. Enhancing areas are often smaller than the corresponding area of high signal on PD/T_2-weighted images. The maximum intensity of enhancement occurs from 4 to 120 min after Gd injection,

with the average peaking around 29 min (Kermode *et al.*, 1990a). Lesions may enhance for 20–30 days, although some lesions may enhance for only 1 week whereas others may persist even up to 100 days.

Gd enhancement on MRI is an indicator of MS lesion activity (Figs. 7A–7D). It reflects BBB breakdown and pathological correlation reveals that Gd-enhancing lesions are associated with marked inflammatory infiltration (primarily macrophage) accompanied by edema (Nesbit *et al.*, 1991; Katz *et al.*, 1993). The lesions may or may not be demyelinated. Although Gd enhancement is currently believed to be the earliest MRI evidence of MS lesion activity, neuropathological and immunocytochemical studies show that BBB leakage can be found in varying degrees in every MS lesion. These pathological findings differ from the observation on MRI studies that show enhancement being restricted to active MS lesions. This discrepancy is probably a matter of the sensitivity being lower for MRI in the detection of BBB breakdown compared to neuropathological and immunocytochemical methods. Increasing the sensitivity by using double and triple dose of contrast and magnetization transfer can increase the number of enhancing lesions detected (Filippi *et al.*, 1995b).

Steroids continue to be the principal treatment for acute exacerbations because of their anti-inflammatory effect (van Oosten *et al.*, 1998). Treatment with methylprednisolone (Barkhof *et al.*, 1991, 1994) reduces the duration and severity of clinical relapses in MS and also reduces the number of Gd-enhancing lesions, indicating some restoration of the BBB and suppression of inflammation. The steroid effect and the suppression of MR activity are not seen on PD/T_2 morphological activity, perhaps because of persistent changes such as edema, cellular infiltration, and/or demyelination. A recent study (Gasperini *et al.*, 1998) reported the effect of steroids on Gd-MRI before and during interferon β-1a treatment in patients with RRMS. There was a significant reduction in the mean number and volume of enhancing lesions on the first scan after methylprednisolone injection. There were persistently low levels of enhancement in the follow-up scans of patients treated with interferon β-1a, but a rebound effect (i.e., an increase in the number and volume of gadolinium-enhancing lesions) was observed in those patients who were not on interferon β-1a. These data suggest that interferon β-1a prolongs the therapeutic effect of steroids on the BBB.

D. Combined Studies of Activity

In systematic studies, about twice as many newly enhancing (active) lesions are seen than are morphologically active ones. In a therapeutic trial where

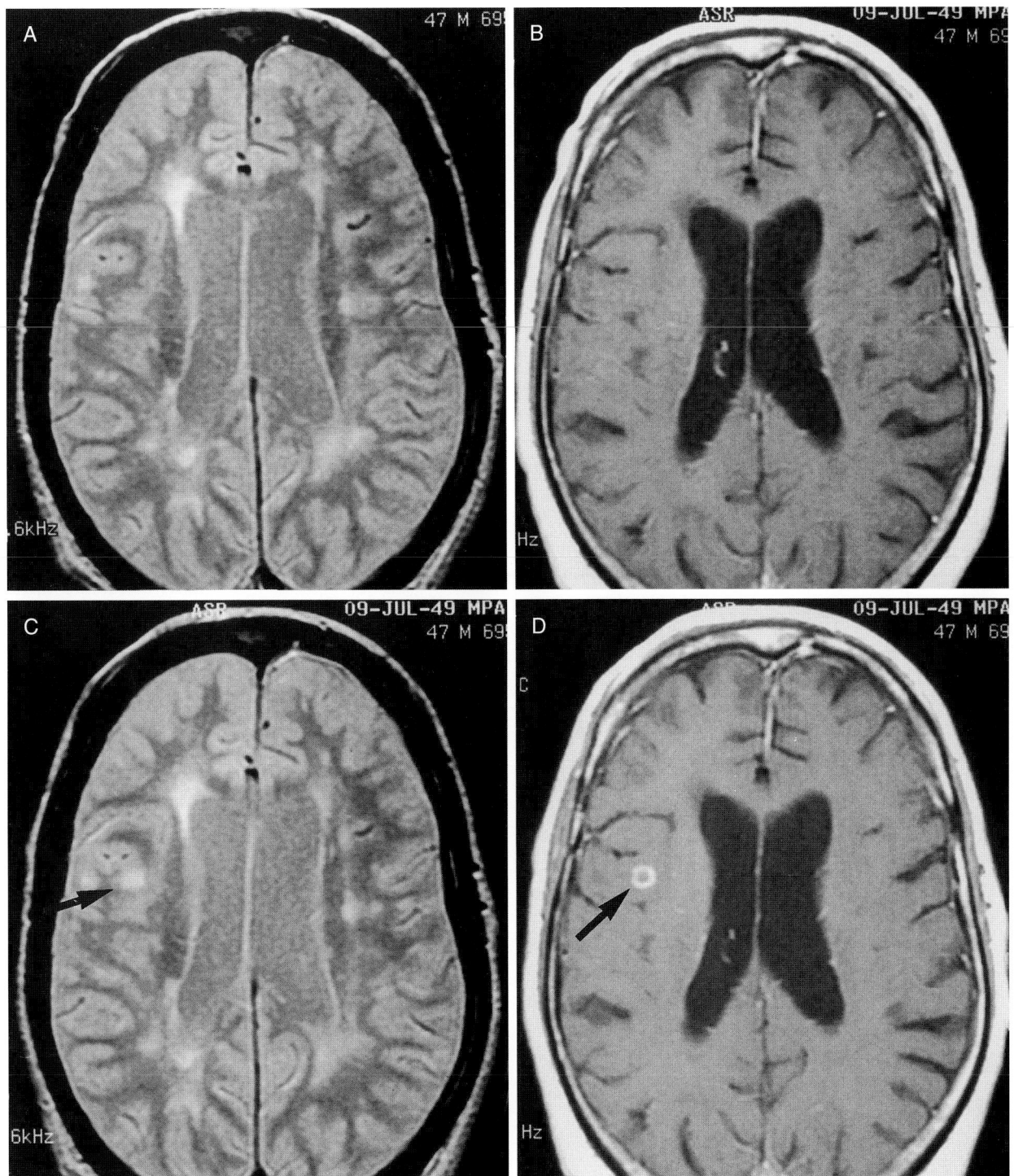

Figure 7 A new lesion that is also enhancing. (A) A PD image showing a number of lesions. (B) Post-Gd T_1W image performed on the same date without any enhancing lesion. (C) One month later, a new lesion is seen in the right insula (arrow) on the PD image. (D) Post-Gd T_1W image showing that the new lesion is also enhancing (arrow).

patients were examined monthly, 77.7% of the morphologically active lesions showed enhancement (Li *et al.*, 1997). When morphological activity is detected without corresponding enhancement, it is usually at-

tributed to the transient nature of the BBB breakdown identifiable on MRI. Enhancement can be missed due to the 4-week scanning interval. Two-thirds of enhancing lesions do not persist for more than 4 weeks, with very few lesions enhancing beyond 8 weeks. Therefore when scans are obtained every 4 weeks, it is possible that the enhancement associated with a newly active lesion, detected by morphological change, could be missed. One study (Li *et al.*, 1997) also showed that when enhancement occurred in morphologically active lesions, most of them enhanced on the same date. Uncommonly (12/522 lesions, or 2.3%), morphologically active lesions were seen prior to the development of enhancement and a few enhancing lesions (37/522, or 7.1%) were seen before they became evident morphologically. Recent ultrastructural studies of stereotactically obtained brain biopsy specimens, which were pathologically consistent with MS, have revealed evidence of myelin degeneration occurring outside areas of maximal inflammation or macrophage infiltration (Rodriguez and Scheithauer, 1994). These findings suggest that demyelination may have preceded inflammation. In addition, a new, enlarging or recurrent lesion associated with demyelination might be seen on PD/T_2, but because there is only minimal inflammation, there may not be sufficient BBB breakdown for gadolinium enhancement to be seen. With further progression and increase in the BBB breakdown, there might be enhancement on follow-up.

To avoid double counting of active lesions with simultaneous detection of Gd enhancement and morphological activity in clinical trials, we have performed a combined unique active lesion analysis (Li *et al.*, 1999). The activity analysis is performed in three phases. Initially, the T_1-weighted scans are evaluated for Gd enhancement. After at least a 4-week delay, the PD/T_2 scans are then assessed for morphological activity. Finally, the combined unique active lesion analysis is performed. When a Gd-enhancing lesion and a morphologically active lesion are identified as being the same lesion, the lesions are linked in the database. Links can involve the current, previous, or subsequent scan in the series. Nonlinked Gd-enhancing and morphologically active lesions and linked active lesions are then combined to give counts of new combined unique active (CU) lesions (sum of new CU and recurring CU lesions) and persistent CU active lesions. Although the three-phase analysis is time-consuming and laborious, it produces data that permit comparison between combined unique activity, morphological activity, and Gd enhancement. In the PRISMS study of interferon β-1a in RRMS, although fewer numbers of morphologically active lesions were identified than on the CU active and Gd-enhanced lesions, it was the morphological activity, and particularly new T_2 lesion activity, that was better able to detect a differential dose effect.

E. Natural History Studies and the Behavior of Placebo Groups in Clinical Trials

In a previous natural history study using morphological analysis only, Isaac *et al.* (1988) found the total MRI activity rate in RR patients (mean EDSS 3.0) was 7.7 MRI activity events per patient per year (scanning interval every 4 weeks over 6 months in seven relapsing MS patients). In a study of less impaired RR (mean EDSS 0.5) patients, Willoughby *et al.* (1989) found the morphological MRI activity was 3.1 MRI activity events per patient per year (scanning interval every 2 weeks for an average of 5 months in nine patients). In more disabled SP patients (mean EDSS 6.1), Koopmans *et al.* (1989) found the morphological activity to be 21.5 MRI activity events per patient per year. Using Gd-enhanced scans, Thompson *et al.* (1990) found that primary progressive patients had the lowest rate of new lesion development, at 3.3 new lesions per patient per year; the next lowest rate was for benign patients, who had 8.8 new lesions per patient per year, whereas typical RR and SP patients had 17.2 and 18.2 new lesions per patient per year, respectively, in their study.

The behavior of placebo patients in clinical trials can provide some information on the natural history of MS. However, because clinical trial patients are highly selected, they do not reflect the behavior of all patients. When comparing different studies, one should exercise caution since the clinical entry criteria of the patients (RR, SP, EDSS, number of relapses preceding entry, etc.), the definition of MRI activity, and the scanning frequency may vary considerably. Table I shows the numbers of active lesions in placebo groups of patients in serial clinical trials. Note that the MRI activity rate in particular can be dramatically affected by the scanning frequency. One study (Koopmans *et al.*, 1993c) suggested that only 21% of morphological activity events were detected when the scanning frequency was reduced from once every 6 weeks to once a year.

The net accumulation of new and enlarging lesions over time results in the increase in total PD/T_2W MRI burden of disease (BOD). Chronic enlargement also plays a part. While month-to-month fluctuations in BOD can occur because of changes in disease activity, patients in the placebo arm of clinical trials have consistently demonstrated a progressive increase in BOD compared to baseline (Table II). In fact, the increase occurred not only in the placebo groups but also in those treated groups where no clinical or MRI benefit of treatment was observed (Koopmans *et al.*, 1993a; Zhao *et al.*, 1997a). We believe this consistent behavior of the placebo group provides an internal validation of the accuracy of the quantitation methodology. Although a previous study of IFN β-1a given by intramuscu-

Table I Lesion Activity in Placebo Patients

Reference	Treatment and type[a] of MS	Placebo patient number	Time (months)	Scanning frequency	Rate of types of lesion activity				
					Total PD/T2 active lesions	Total active lesions (T2 and T1)	New lesions	Enlarging lesions	Enhancing lesions
Multiple Sclerosis Study Group, 1990; Zhao et al., 1997a	Cyclosporine in CP	83	24	Biannually		1.5 (mean)	0.8 (mean)		
Koopmans et al., 1993a	Interferon α- in CP	27	24	Baseline, 6 and 24 months	1.61 at month 6 and 6.11 at month 24 (mean)				
Paty and Li, 1993	Interferon β-1b in RR	110	36	Four scans (baseline and annually)	4.9/year		3.2/year		
IFNB Multiple Sclerosis Group, 1995; Zhao et al., 2000	Interferon β-1b in RR	115	60	Annually	6.44 (mean)		3.57 (mean)	2.65 (mean)	
Jacobs et al., 1996	Interferon β-1a in RR	132	24	Annually					2.32 at baseline; 1.59 at year 1; 1.65 at year 2 (mean)
Karussis et al., 1996	Linomide in SP	15	6	Monthly					0.42/scan (mean)
van Oosten et al., 1998	cM-T412 in RR and SP	36	18	Monthly for the first 9 months and months 12 and 18		15.6 at month 9; 24.7 at month 18 (mean)			
PRISMS Study Group, 1998; Li et al., 1999	Interferon β-1a in RR	187	24	Semiannually	2.25/scan (median)		1.5/scan (median)	0.75/scan (median)	

[a]RR, relapsing–remitting; RP, relapsing progressive; CP, chronic progressive; SP, secondary progressive.

Table II   Percentage Change of BOD in Placebo Patients in Clinical Trials

Reference	Treatment	Placebo patient number	Type of MS	Time (months)	Number of MRI scans	Percentage change of BOD from baseline
Zhao *et al.*, 1997a	Cyclosporine DB, PC	83	CP	24	2 (baseline and 24 months)	7.2 (mean)
Koopmans *et al.*, 1993a	Interferon - α; DB, PC	27	CP	24	3 (baseline, 6 and 24 months)	10 at year 2 (mean)
Paty and Li, 1993	Interferon β-1b; DB, PC	110	RR	36	4 (baseline and annually)	10.9 at year 1, 16.5 at year 2, 15 at year 3 (median)
IFNB Multiple Sclerosis Study Group, 1995	Interferon β-1b; DB, PC	73	RR	60	6 (baseline and annually)	6.7 at year 1, 11.9 at year 2, 21 at year 3, 18.7 at year 4, 30.2 at year 5 (median)
Jacobs *et al.*, 1996	Interferon β-1a; DB, PC	143	RR	24	3 (baseline and annually)	-3.3 at year 1, -6.5 at year 2 (median)
Simon ; 1998a	Interferon β-1a; DB, PC	76	RR	24	3 (baseline and annually)	12.0 at year 1, 19.0 at year 2 (median)
PRISMS Study Group, 1998	Interferon β-1a; DB, PC	187	RR	24	5 (baseline and semiannually)	10.9 over 2 years (median)
European Study Group, 1998	Interferon β-1b; DB, PC	358	SP	24	3 (baseline and annually)	8 (mean)

lar injection (Jacobs *et al.*, 1996) showed an effect of treatment in reducing the percentage change in T_2 lesion volume in the first year ($p = 0.02$), but not in the second year ($p = 0.36$), the placebo group did not demonstrate an increase in lesion volume but instead showed median decreases of 3.3% from baseline after the first year and 6.5% after the second year. This unexpected behavior of the placebo group with a decrease in the BOD over time may have been an artifact of the method used for lesion volume quantitation. In that study, each of the T_2 lesion volume measurements was made independently, without reference to the patient's previous scans. This methodology probably resulted in significant interseries errors. Simon *et al.* (1998a) have recently reported a revision of the BOD analysis of that study. This time, using a manual tracing method, they found an increase of 12% at year two in the placebo group compared to 3.0% at year two for the treated group ($p = 0.057$).

III. Clinical Correlations with MRI Findings

MRI has been proven to be most helpful in MS to aid diagnosis. The reason that MRI is so useful in the diag-

nosis of MS is that it reveals lesions that are clinically asymptomatic, most of which must have occurred in the preclinical phase of MS. Therefore, the lack of a strong correlation between the EDSS and MRI changes in clinical trials should not be a surprise. In fact, MRI studies are revealing pathology as it evolves and clinical symptoms are only a reflection of a small proportion of the pathology that manifests clinically due to the location and severity of the pathological changes (Paty and Moore, 1998). Nevertheless, recent studies have shown that there is significant correlation between the number of new or enlarging lesions and increasing disability in RRMS (Filippi *et al.*, 1995a). In a study of the 115 patients in the placebo arm of the interferon β-1b trial, a moderate Spearman rank correlation coefficient (SRCC) of 0.2–0.3 but statistically significant correlation between clinical status and MRI-active lesions was observed, with the strongest correlation being between changes in EDSS and BOD (IFNB Multiple Sclerosis Study Group, 1995; Zhao *et al.*, 1997b). Mammi *et al.* (1996) also showed a significant correlation ($r = 0.3$, $p = 0.0006$) between total lesion load and EDSS in their 130 patients. Giugni *et al.* (1997) found a weak relationship between disability and BOD on T_1 and T_2 images in RRMS. The presence of cerebellar, brainstem, and mental impairment was significantly associated with a greater BOD on MRI.

The EDSS is the clinical standard for determining neurologic impairment in MS, but it is also highly subjective. As most MRI lesions are neurologically silent, some studies have not demonstrated a strong correlation between MRI lesions and clinical status. Possible reasons for this lack of correlation between clinical and conventional brain MRI findings include the small numbers of patients in the studies and too short a follow-up or MRI scanning intervals too infrequent to allow for the detection of active lesions (for example, to miss the time of BBB breakdown). The location of MS lesions in clinically silent areas of the brain and the varying pathological characteristics and severity of the MS lesion are probably the most likely factors for the lack of neurological correlation.

CNS dysfunction in MS, especially early in the disease, has traditionally been attributed to conduction block secondary to changes in membrane electrical properties resulting from demyelination. More recently, attention has also been directed to the integrity of the axons. Several investigators have found a reduction in the number of axons that traverse the plaque as well as thinning, amputation, and other structural abnormalities of the remaining axons. (Ferguson et al., 1997; Trapp et al., 1998). Although conventional PD/T_2 scans are able to show MS plaques, they do not distinguish the pathology within individual lesions. Compared to T_2 BOD, the T_1 lesion load of hypointense lesions (black holes on Gd-enhanced images) has shown a higher correlation with the EDSS in a cross-sectional analysis (van Walderveen et al., 1995; Truyen et al., 1996). In SP patients, a relative increase in T_1 lesion load after 3 years of follow-up was strongly correlated with disease progression (SRCC 0.81, $p < 0.001$). In a postmortem study (van Walderveen et al., 1998), the degree of hypointensity of T_1 lesions correlated significantly with widening of the extracellular space and axonal loss.

As would be expected, the location of a lesion plays an important role in the patient's symptoms. When lesions are located in clinically eloquent regions, symptoms are likely to be appropriate. Lesions located in the corticospinal tracts have been found to correlate with appropriate contralateral motor deficits (Zhao et al., 1993b). However, quantitative determination of the corticospinal tract lesion area showed only slight improvement in the correlation than the entire lesion load (SRCC 0.67 versus 0.6) (Riahi et al., 1998). Large postchiasmal lesions are likely to cause symptomatic homonymous field defects (Plant et al., 1992). Similarly, trigeminal neuralgia in a MS patient seems to be caused by demyelinating lesions affecting pontine trigeminal pathways (Gass et al., 1997).

Although earlier reports did not find a correlation between spinal cord MRI lesion activity and changes in disability (Kidd et al., 1996), a positive correlation between disability and spinal cord atrophy has been reported (Losseff et al., 1996a; Stevenson et al., 1998). Using a reproducible semiautomated method for outlining the spinal cord area at C2, Losseff et al. observed a significant loss in cord cross-sectional area during 12 months in MS patients but not in control subjects ($p = 0.001$). This reduction in cord size was most marked in the primary progressive patients, who had a mean cord cross-sectional area loss of 3.52 mm^2 (5.2%), and least in the secondary progressive (-0.26 mm^2, -0.7%) and benign patients (-0.41 mm^2, -0.8%). The baseline cord cross-sectional area correlated strongly with the EDSS ($r = -0.52$, $p = 0.005$) and with disease duration ($r = -0.75$, $p < 0.001$). However, there was no significant difference in cord area or change in cord area between patients with a definite increase in EDSS and those without. Nijeholt et al. (1998) found that primary progressive patients more often had diffuse abnormalities in the brain and/or spinal cord than did relapsing-remitting and secondary progressive patients. In their entire study population, EDSS correlated with both brain and spinal cord MRI parameters, which were independent of each other. The urological complaint score correlated only with spinal cord MRI measures.

Longer term follow-up studies have shown the prognostic value of MRI in patients presenting with clinically isolated syndromes of the optic nerve, spinal cord, or brainstem suggestive of MS. Filippi et al. (1994) reported on a 5-year follow-up in 84 patients with clinically isolated syndromes who had MRI abnormalities at baseline. A strong correlation was seen between MRI BOD at baseline and increased BOD and disability at follow-up. O'Riordan et al. (1998) did a 10-year follow-up of 81 patients from the same cohort. Follow-up of the 67% with an initially abnormal brain MRI ($n = 54$) showed progression to clinically definite multiple sclerosis in 45 out of 54 (83%). Eleven (20%) had relapsing-remitting disease (EDSS > 3), 13 (24%) secondary progressive disease, and 21 (39%) benign (relapsing-remitting with EDSS ≦ 3) disease. For those with a normal MRI, progression to clinically definite multiple sclerosis occurred in only 3 out of 27 (11%), all benign at 10 years. There was a significant relationship between the number of lesions at presentation and both EDSS ($r = 0.45$, p 0.001) and the category of disease at follow-up ($p = 0.0001$). Koudriavtseva et al. (1997) studied 68 patients with clinically definite relapsing–remitting MS with monthly Gd-MRI for 6 months. The occurrence of relapses during the follow-up period was predicted by the presence of at least one enhancing lesion on the baseline MRI ($p < 0.05$). The number and volume of enhancing lesions at baseline were significantly associated with both enhancing lesions seen during the follow-up period ($p < 0.0001$) and the accumulation of abnormality on T_2-weighted images ($p < 0.0001$). This study also showed that the presence of three or more enhancing lesions at baseline scan was consistently associated with the development of permanent abnormalities on T_2-weighted images 6 months later. Additional long-term studies are required to confirm the value of MRI activity in predicting future clinical status.

There have been a number of studies correlating MRI and neuropsychological testing. Rao *et al.* (1989) measured three MRI variables: total lesion area, ventricular-brain ratio, and size of the corpus callosum. Using stepwise multiple regression analyses, they found total lesion area was a robust predictor of cognitive dysfunction, particularly for measures of recent memory, abstract/conceptual reasoning, language, and visuospatial problem solving. The size of the corpus callosum predicted test performance on measures of mental processing speed and rapid problem solving, whereas ventricular-brain ratio did not independently predict cognitive test findings. Pozzilli *et al.* (1992) reported that atrophy of the corpus callosum, particularly the anterior portion, affected performance on verbal fluency tasks and correlated with dementia in MS. A study of 41 patients (Gonzalez *et al.*, 1994) with SPMS demonstrated that large or confluent lesions located predominantly in the periventricular area of the parieto-occipital region could be accompanied by significant cognitive dysfunction and severe personality changes. The presence of lesions in the left suprainsular white matter, the region that mainly includes the arcuate fasciculus, was specifically associated with depressive symptoms (Pujol *et al.*, 1997). Foong *et al.* (1998) studied 13 patients with acute relapses using Gd-MRI. They found that MS patients performed significantly worse than controls on most tests of attention and memory during both acute relapse and in remission. At follow-up there was a significant or trend for improvement in performance on some tests of attention in patients in whom the enhanced lesion load had decreased. The improvement correlated significantly with the reduction of acute lesion load. Hohol and co-workers (1997) showed that the MRI BOD and brain to intracranial cavity volume ratio correlated with neuropsychological performance, especially in patients with chronic progressive MS. Worsening MRI BOD correlated with cognitive de-

cline. Proton MRS was performed in a group of patients with MS and matched control subjects to examine the relationship between frontal lobe pathology and performance on tests of executive function (Foong *et al.*, 1999). The *N*-acetylaspartate/creatine ratio (NAA/Cr) was significantly reduced in frontal lesions and/or normal-appearing white matter in the patient group compared with the control group, but choline/creatine ratios did not differ. Although MRS abnormalities and executive deficits were not correlated for MS patients as a group, a few patients with more severe abnormalities of NAA/Cr ratio performed worse than other patients on the spatial working memory test, suggesting that subtle frontal neuropathological abnormalities detected by MRS may contribute to executive deficits (Foong *et al.*, 1999). Van Buchem *et al.* (1998) assessed differences between groups of patients ($n = 44$) who had been classified on the basis of neuropsychological test performance as severely impaired, moderately impaired, and normal with respect to magnetization transfer imaging (MTI). MTI measures corrected for brain volume were found to correlate with disease duration ($p < 0.01$) and showed suggestive correlations with measures of neurologic impairment ($p < 0.05$). Individual neuropsychological tests correlated with MTI measures both corrected for and not corrected for brain volume ($p < 0.001$). A MTI measure not corrected for brain volume differed ($p < 0.05$) between severely impaired, moderately impaired, and normal patients.

IV. MRI–Pathological Correlation

Stewart *et al.* (1984, 1986) reported pathologic correlation in eight postmortem cases and showed that MRI can accurately reveal the extent of chronic demyelination. The data from these eight cases also suggested that longer T_1 values were seen in the more heavily gliotic lesions. An important finding was that MRI studies can be done on both immediate postmortem (*in situ*) and fixed brains. Our current protocol is to scan patients immediately after death before removing the brain. The brain is then scanned in the fixed state (Moore *et al.*, 1999). The distribution and appearance of lesions seen immediately postmortem are the same as in life. Fixing the brain in formalin does not seem to interfere with the ability to visualize demyelinated lesions using SE sequences, though there is obviously a change in the relaxation parameters and some shrinkage of the brain due to fixation.

Nesbit and co-workers (1991) retrospectively reviewed the biopsies of 37 patients and autopsy material in 3 other patients with inflammatory demyelination consistent with multiple sclerosis. They found a wide spectrum of radiologic appearance, including size and enhancement pattern, in the histologically active lesions that were all enhancing on MRI and/or CT. Lesions with a more aggressive radiologic appearance (e.g., ring lesions and infiltrative lesions) showed moderate to marked macrophage infiltration, with variable amounts of gliosis and perivascular lymphocyte infiltration. Less aggressive-appearing lesions typically had less macrophage infiltration and more gliosis. There was no significant enhancement in histologically inactive lesions.

Newcombe *et al.* (1991) examined postmortem unfixed whole brains from 17 MS and 6 control cases by MRI and histology. They found that MS was a more diffuse process than had been previously thought. In five of the brains, extensive MRI abnormalities were observed when only small periventricular plaques were present histologically. These changes were thought to be the result of vascular permeability changes in the normal-appearing white matter surrounding plaques. Lesions in the hindbrain and cerebral gray matter seen pathologically, however, were much less frequently seen on MRI.

Using an animal model of MS, experimental allergic encephalomyelitis (EAE), Hawkins *et al.* (1990) studied the duration and selectivity of BBB breakdown and the nature and mechanism of enhancement with gadolinium. They showed that the duration of enhancement was very short in acute EAE, usually less than 5 days. However, in chronic relapsing EAE, the duration of enhancement ranged from 5 days to greater than 5 weeks. Not all lesions showing enhancement were visible on the T_2W image, particularly if the duration of enhancement was less than 4 weeks. When animals were perfused at the enhancement stage, gadolinium was present in the perivenular spaces and extended for a short distance into the parenchyma. Gd diffusion was associated with perivascular inflammatory cells that were invariably present in enhancing lesions. The transit of Gd through the BBB was also found to be a metabolically active event. The study also provided evidence that gadolinium enhancement reflects active inflammation (with or without demyelination). Serial gadolinium studies in MS patients have shown similar patterns (Kermode *et al.*, 1990a,b). In a preliminary study of five postmortem MS brains (van Walderveen *et al.*, 1998), the histopathologic characteristics of hypointense lesions on T_1W scans (chronic black holes) were studied. Nineteen lesions were selected for analysis. Nearly all lesions had the features of chronic

MS plaques: hypocellularity, absence of myelinated axons, and the presence of reactive astrocytes. The degree of hypointensity on T_1W images did not correlate with degree of demyelination or number of reactive astrocytes but was associated with axonal density. The authors concluded that hypointense lesions seen on T_1W images are associated histopathologically with widening of the extracellular space and more severe tissue destruction, including axonal loss. The histopathologic characteristics of acute black holes, hypointense lesions seen on precontrast scans, which are associated with Gd-enhancing lesions (van Walderveen *et al.*, 1995), have not been studied. Because of their acute nature and the observation that some of these hypointense lesions resolve on follow-up suggest that they are likely inflammatory in nature.

V. Application of MRI in Monitoring of Clinical Trials

Because of the highly variable clinical course of MS, the recognition of treatment effects requires the use of placebo control patients. While clinical endpoints such as the relapse rate and EDSS remain the primary outcome measure for clinical trials, conventional MRI endpoints, despite a limited correlation with disability, have been increasingly used as secondary outcome measures. MRI provides an index of pathological progression. A major advantage of MRI is that, unlike the subjectivity of the clinical examination, it provides an objective measurement that is devoid of examiner bias, which might still be present in spite of precautions to ensure study blinding. In addition, because many new and active lesions seen on MRI are unassociated with new symptoms, MRI is more sensitive than the clinical examination in detecting a treatment effect.

The sensitivity of frequent (e.g., monthly) MRI examinations to detect disease activity has led to the recommendation that MRI be used as a primary outcome measure in preliminary short-term (phase II) trials (Miller *et al.*, 1996). Using relatively small numbers of patients, these studies are designed to provide a quick, initial suggestive of efficacy of new agents to prevent activity on MRI, before embarking on costly pivotal trials to prove clinical efficacy. Several limitations associated with these screening studies have been described. Because of the short-term nature of these studies, a treatment effect may not be apparent if there is a long delay between initiation of the therapy and the therapeutic response. Also, should a treatment effect be observed in a short-term study, it would not be possible to determine if the effect will be sustained in the longer term. The predictive value of short-term MRI activity for long-term clinical disability and impairment remains to be clearly shown. Despite these limitations, several such studies have been conducted to screen new agents such as mitoxantrone, a chimeric monoclonal anti-CD4 antibody and a tumor necrosis factor (TNF) receptor neutralizing immunoglobulin fusion protein.

In the past 15 years, several pivotal therapeutic trials in MS have been carried out using MRI monitoring. These pivotal trials include the use of interferon β-1a and β-1b (IFNB Multiple Sclerosis Study Group, 1993; Jacobs *et al.*, 1996; PRISMS Study Group, 1998) and glatiramer acetate (Johnson *et al.*, 1995) in patients with RRMS and interferon β-1b in SPMS (European Study Group on Interferon β-1b in Secondary Progressive MS, 1998). Most of the recent clinical trials have used MRI as a secondary outcome measure. The two MRI measurements that have been most widely used are counting the number of active lesions (activity analysis) and measuring total lesion area or volume (BOD analysis). Some studies have used both methods, but others have used only one.

A. MRI Activity Acute-Phase Analysis

The rate and severity of relapses are measures of clinical activity. MRI activity includes numbers of new, enlarging, and recurrent lesions seen on PD/T_2W images and the number of enhancing lesions seen on post-Gd T_1W images. Other activity measures such as new hypointense lesions on T_1W images (acute black holes) have not yet been used in clinical trials. Activity can be expressed in terms of the numbers or rates of active lesions or the scan rate.

The first trial in which activity analysis was done was a study involving cyclosporine (Multiple Sclerosis Study Group, 1990). The MRI evaluation, carried out in about 50% of patients, involved an initial scan and a follow-up scan at the end of 2 years (Zhao *et al.*, 1997a). New and enlarging lesions on PD/T_2W scans were identified, but no beneficial effect of cyclosporine was found.

The next trial to be assessed by MRI activity analysis was a study using interferon β-1b (Betaseron) (Paty and Li, 1993). All the patients in the study ($n = 372$) had yearly MRI scans and a subset ($n = 52$) at one site had frequent scans (once every 6 weeks) for 2 years. In this study, as gadolinium was not yet universally available, the analysis was performed only on the PD/T_2W scans. In the frequent-MRI subgroup, PD/T_2 activity was significantly reduced, comparing the placebo and both treatment groups (1.6 and 8 mIU) with regard to the percentage of active scans and the annual rate of active and new lesions. Compared with the placebo group, the high dose treated group had significant reductions of 75–83% in the rate of new ($p = 0.0026$) and active ($p = 0.0089$) lesions and percentage of active scans ($p = 0.0062$). Other studies that followed used enhanced scans for determination of MRI activity. Jacobs *et al.* (1996) and Simon *et al.* (1998a) reported a study with interferon β-1a (Avonex) (30 mIU given intramuscularly once per week). Treated patients had less gadolinium-enhanced lesions than the placebo patients. The proportion of positive scans was 29.9% for treated patients and 42.3% for placebo patients at year 1 ($p = 0.05$). There were also significant reductions in the number of new ($p = 0.006$) and enlarging lesions ($p = 0.024$).

Karussis *et al.* (1996) reported a pilot study with linomide, monitored by monthly MRI in a small number of patients. The number of new enhancing lesions was reduced ($p = 0.0387$) when treated patients were compared with placebo. Most recently, a randomized, placebo-controlled, multicenter trial of interferon β-1b (Betaseron) in treatment of secondary progressive MS was reported (European Study Group on Interferon β-1b in Secondary Progressive MS, 1998). Seven hundred eighteen patients were enrolled in this single-dose (8 mIU) study. All patients had an annual PD/T_2W scan and 125 patients underwent monthly PD/T_2W and enhanced T_1W scans in months 0–6 and 18–24. Patients in the treated group showed a 65% reduction of newly active lesions from months 1 to 6 ($p < 0.0001$) and 78% reduction from months 19 to 24 ($p = 0.0008$) compared with placebo.

Another multicenter, randomized, placebo-controlled study of interferon β-1a (Rebif) in RRMS was reported recently (PRISMS Study Group, 1998). Five hundred sixty patients from 22 centers in nine countries were entered in the 2-year study. Two doses of drug were used at 22 μg (6 mIU) and 44 μg (12 mIU) versus placebo given subcutaneously three times a week. All patients had semiannual PD/T_2W scans. A subgroup of 205 patients had monthly PD/T_2W and enhanced T_1W MRI scans for 1 month prior to the treatment and 9 months posttreatment. There was a 67 and 78% reduction in the median number of PD/T_2W lesions in the low-dose and high-dose groups, respectively, compared to the placebo group ($p < 0.0001$). The decrease in the number of active lesions included reduced numbers of both new PD/T_2W lesions and newly enlarging PD/T_2W lesions for both treatment groups compared with placebo ($p < 0.0001$). The greatest reductions were in the higher dose (44 μg) group ($p < 0.001$). In the frequent-scanning cohort (Li *et al.*, 1999), a three-phase activity analysis (see above) was performed for the first time in a clinical trial to determine separately morphological activity (PD/T_2 lesion activity), BBB activity (Gd enhancement), and a combined unique activity (to avoid double counting of simultaneous morphological and BBB activity in the same lesion). The median number of combined unique (CU) active lesions was reduced 80.7 and 87.5% and the percentage of CU active scans reduced 71.5 and 75% ($p < 0.0001$) for the 22 and 44 μg groups, respectively, when compared to placebo. When the morphological and BBB activities were analyzed separately, a distinct dose effect was observed in the reduction of the median number of new PD/T_2W lesions favoring the 44 μg dose ($p < 0.02$), but not for the median numbers of CU active lesions, new Gd-enhancing lesions, and enlarging PD/T_2Wlesions. Although far fewer numbers of active lesions were identified on the PD/T_2W scans than on the Gd-T_1W-enhanced scans, PD/T_2W activity, and particularly new PD/T_2W lesion activity, appeared more robust in detecting a dose effect. This finding may have important implications for the design of future therapeutic trials. A dose effect favoring the high dose was also seen in a subgroup of 94 patients who were more disabled at baseline with EDSS scores of 3.5 or higher (Paty *et al.*, 1998). In comparison with the placebo group, the number of active lesions was reduced by almost 74% in patients treated with the high dose ($p = 0.0003$) but was not reduced significantly in those treated with the low dose. This study was powered to show a difference in this small number of patients, so the profound effect of the high dose is even more impressive.

In most studies, the treatment effects on MRI lesion activity have been proportionally greater than on any other outcome measure, including the clinical ones. For example, in the 2 years of the PRISMS study frequent-scanning cohort, patients in the higher dose 44 μg group had a 33% reduction in the mean number of clinical relapses, compared with a 67% reduction in active MRI scans, a 78% reduction in the median number of active lesions on biannual scans, a 75% reduction in the active scan rate, and an 87.5% reduction in active lesions. In the OWIMS study (Freedman *et al.*, 1998), 293 patients were enrolled in a randomized, placebo-controlled study where patients were given subcutaneous injections of interferon β-1a (Rebif), either 22 or 44 μg, or placebo once per week. Although there was a consistent trend favoring the higher dose, most of the improvements in the clinical parameters measured were not significant. The MRI activity measures, however, clearly showed a significant reduction (53.5%, $p < 0.01$) in numbers of active lesions for the high-dose group compared to placebo, but not for the low-dose group. The

great sensitivity of MRI activity measures for detecting treatment effects also enhances its ability to detect a dose difference. For example, a significant dose effect favoring the higher dose group was seen in the net accumulation of MRI activity on annual scans in a further analysis of the data from the original interferon β-1b study (Zhao *et al.*, 2000).

B. MRI Burden of Disease Analysis

The total brain lesion load, or MRI burden of disease (BOD), is determined by measuring the area of the lesions seen on each slice and then summing the area slice by slice to obtain a total lesion area. The total lesion volume is obtained by multiplying the area by the slice thickness. Changes in the BOD (either absolute change or percentage from baseline) measured from PD/T_2W images have been used as a secondary endpoint for several clinical trials. In fact, if the results are going to be expressed as percentage change from baseline, conversion from area to volume only introduces some errors and does not make sense. In addition to the actual biological change in BOD over time, other factors may introduce variations. These factors include errors related to data acquisition (MR scanner performance, accuracy of scan repositioning, consistency of pulse sequences, and other imaging parameters) and data analysis (image processing and lesion quantitation, including lesion identification and delineation). The biological variability in the extent of lesions is very great. In fact, this variability is so great that it overwhelms the other factors that can produce errors (Cover *et al.*, 1999). To detect changes on sequential follow-up scans with reliability, accurate repositioning of the patient from scan to scan is essential. An alternative that is being developed is the use of 3D-volumetric acquisition techniques and computer programs to retrospectively coregister slices with each other. To minimize variations in data acquisition, there should not be any changes in the scanner, the pulse sequences, or the imaging parameters as well as the image processing. This standardization can be difficult when a trial extends for 2–3 years because of advances in scanner hardware and frequent software upgrades.

At UBC we have used a computer-assisted manual tracing method to outline the lesion area for BOD on the PD images. We always do the measurement on the PD image because of the confusion between lesion and CSF on the T_2W scans (Figs. 2A and 2B). Trained technicians have been able to achieve intrarater reproducibility of 94%. To minimize interobserver variability, each technician always analyzes the entire series of scans sequentially from the same patient. Although time- and labor-intensive, it is a method that has been validated in several trials. It was the method used in the pivotal interferon β-1b (Betaseron) trial in RRMS, which was the first clinical trial to ever demonstrate a positive clinical and MRI treatment (Paty and Li, 1993). The same BOD method had been previously used in two other clinical trials (α-interferon and cyclosporine), in which no treatment effect was found (Koopmans et al., 1993a; Zhao et al., 1997b). However, in both trials there was an increase in BOD over time in the placebo groups (Table II), which is now an expected result in such studies. In the interferon β-1b study, there was also an increase in the median MRI BOD each year for up to 5 years in the placebo group. In contrast, in the high-dose (8 mIU) treatment arm, there was no demonstrable increase in BOD, either yearly or over the entire 5 years of the study. In the PRISMS study of interferon β-1a, using the same computer-assisted manual tracing method, the placebo group had a median increase of 10.9% at 24 months. Both active treatment groups differed significantly from the placebo group. The low-dose group had a median decrease of 1.2% and the high-dose group an even greater median decrease of 3.8% from baseline.

To improve reproducibility, other groups have used semiautomatic and automatic (by definition, automatic methods should be perfectly reproducible) segmentation methods for lesion detection and delineation. The European Study Group on Interferon β-1b in Secondary Progressive MS (1998), using a semiautomatic contour technique, showed an increase of about 8% in the mean MRI T_2 volume in the placebo group compared to a 5% decrease in the treated group. These investigators had previously shown that the contour technique had better reproducibility than their manual outlining method. Cover et al. (1999) pointed out that an improvement in the reproducibility of the measurement of BOD does not necessarily lead to a corresponding increase in the statistical sensitivity. This assumption would be true if the dominant source of variance in mean lesion load was due to the lack of precision of the measurements. However, it is now recognized that the main source of variability is, in fact, the biological variability (the interpatient difference in both the BOD and the change in BOD), which exceeds the variability in measurements irrespective of the measurement method used, assuming the usual strict training and quality standards are followed. Where improved reproducibility may make a difference might be in future trials where new treatments would be tested against existing treatments rather than against placebo. Also precision would be very important if patients were stratified for entry BOD or if measures other than the

PD/T_2W BOD were used. The biological variations with these other MRI measures may be smaller and under these circumstances improved reproducibility and decreased variability in the measurements may be more important. The greatest advantage of semiautomatic and automatic methods will be the potential for a reduction in the time and effort needed for the human operator to assess a scan. An automatic method could also reduce the time needed for training and experience with the technique.

Improvements in reproducibility and time and labor efficiency must, however, never be achieved at the price of loss of accuracy. Accuracy is difficult to determine. Since there is currently no gold standard against which to measure accuracy, improved reproducibility and efficiency do not necessarily equate to an increase in overall accuracy. The original BOD measurements from the once a week intramuscular interferon β-1a (Avonex) study were performed using a semiautomatic method. Whereas other studies have consistently shown a progressive median increase of 10–20% in the BOD of the placebo group over 2 years, the placebo group in the first report of this study (Jacobs et al., 1996) showed a median decrease of 3.3% from baseline at 1 year and a decrease of 6.5% in the second year. This unexpected decrease in the BOD over time may have been an artifact of the method used for lesion volume quantitation. For this study, each of the T_2 lesion volume measurements was made independently, without reference to the patient's previous scans, which probably resulted in significant interseries errors. Simon et al. (1998a) reported a reanalysis of the BOD in the Avonex study using a manual tracing method with a standard reference to the patient's previous scans and found an increase in BOD of 12% at year one and 19% at year two in the placebo. This finding was consistent with other placebo groups.

Although BOD measurements from PD/T_2 scans have been able to demonstrate a treatment effect in clinical trials, the correlation between changes in BOD and clinical disability as measured with EDSS has been statistically significant but weak (SRCC 0.2–0.3). The pathological heterogeneity of the lesions seen on T_2-weighted scans is obviously a contributing factor to the modest level of correlation. One cannot differentiate between early (edema, inflammation, early demyelination) and chronic lesions (severe demyelination, axonal loss and gliosis). Another factor is that many MRI lesions occur in clinically silent areas of the brain. The use of MRI in the diagnosis of MS is based mostly upon the fact that MRI is sensitive enough to show asymptomatic lesions and is therefore helpful in identifying lesion dissemination in space in suspected MS cases. On the other hand, lesions in the spinal cord will affect clinical disability but are not included in the usual brain BOD measurements. Microscopic pathology in normal-appearing white matter may also contribute to disability but is not visible on conventional PD/T_2 scans. A number of more pathologically specific MR techniques that may have stronger correlations with clinical disability are currently being evaluated. These include MR spectroscopy, magnetization transfer, diffusion imaging, T_2 relaxation measurements, and cerebral and spinal cord atrophy, which will be discussed briefly in the following section.

Of these newer techniques, hypointense lesions (black holes) seen on T_1-weighted spin-echo MR scans have received the most attention. Compared with PD/T_2 BOD, T_1 black hole BOD (measured on Gd-enhanced T_1-weighted images) has shown a higher correlation with the EDSS. In a 3-year follow-up study of patients with SPMS (Truyen et al., 1996), an increase in black hole BOD was strongly correlated with disease progression ($r = 0.81, p < 0.001$). The degree of hypointensity of black holes also correlates significantly with magnetization transfer (MT) ratios, a marker of matrix destruction. A preliminary postmortem study showed the degree of hypointensity correlated significantly with widening of the extracellular space and axonal loss. There are, however, a number of practical issues regarding the quantitative BOD measurement of black holes that need to be addressed. There are gradations of "blackness" for black holes. Some hypointense areas are more "gray holes" than black holes, the result of partial volume averaging as well as possible differences in the severity of tissue destruction. Differentiation of true black hole from CSF and gray matter can also be difficult at times. The practical implication is that black holes are much more difficult to identify and delineate than PD/T_2 lesions. Intra- and interobserver variability is expected to be higher, while the actual extent of abnormality measured is smaller. Experience with PD/T_2W BOD shows that small areas are associated with high measurement error rates. Another potential difficulty is with acute black holes. Eighty percent of new enhancing lesions are acutely hypointense on precontrast scans. Over several months they can either remain hypointense or become isointense. Most studies on black holes have used Gd-enhanced scans so that the acute black holes are masked out by the enhancing lesions. If a therapy produces a reduction in enhancement, these acute black holes may be unmasked and in the short term before they become isointense, there will be an apparent increase in the BOD of black holes in the treated patients. This factor becomes quite confusing because the untreated patients paradoxically may not show an increase in the BOD compared to baseline, since they will continue to have

enhancing lesions that mask the acute black holes. In the longer term, this effect may be less of a problem but until the use of black hole BOD has been validated in a trial, one does not know for certain whether one should use the post-Gd-enhanced scans alone or whether pre-contrast scans or both will be required.

VI. Evaluation of in Vivo Pathology with Newer MR Techniques

A. Magnetic Resonance Spectroscopy

Proton magnetic resonance spectroscopy (MRS) evaluates a number of interesting metabolites, of which N-acetylaspartate (NAA) and mobile lipids have been of most interest in MS. In the normal brain, NAA is found uniquely in neurons and changes in brain NAA reflect neuronal pathology and/or dysfunction. Initial studies showed a variable loss of brain NAA in MS patients, with smaller decreases in patients with lower levels of clinical disability (Arnold et al., 1994). In a 6-year longitudinal study of a patient with RRMS, using MRS from a large central brain volume, De Stefano et al. (1997) found changes in the brain NAA to creatine (NAA/Cr) ratio correlated strongly with clinical disability (Spearman rank coefficient = $-0.73, p\ 0.001$). Because the NAA signal in the large volume of interest originated predominantly from white matter that appeared normal on conventional MRI, these results also suggest that some degree of axonal dysfunction, reflected by the NAA changes, may be widespread in acute, severe relapses. The same group also reported a 30-month longitudinal study of 29 patients with RRMS and SPMS who underwent single-voxel MRS examinations obtained at intervals of 6–8 months with concurrent clinical evaluation (De Stefano et al., 1998). At the onset of the study, the brain NAA/Cr ratio was abnormally low for the whole group of patients (control mean = 2.93 ± 0.2, patient mean = 2.56 ± 0.4, $p < 0.005$), with no significant differences seen between the RRMS and SPMS subgroups. Over the follow-up period, there was a trend toward a decrease (8%) in the brain NAA/Cr ratio for the 11 relapsing patients and a significant ($p = 0.001$) correlation between changes in the brain NAA/Cr ratio and the EDSS for the patients in this group. This correlation was even more evident for the seven patients who had clinically relevant relapses during the 30 months of follow-up. These findings support recent immunohistochemical studies that suggest that axonal transection may begin early, possibly at the time of disease onset (Ferguson et al., 1997; Trapp et al., 1998) and then progressively stepwise increase with time. Although serial studies generally show a progressive decrease in NAA with time, magnetic resonance spectroscopic imaging (MRSI) studies of individual acute lesions show that decreases in NAA may be transient and reversible, indicating that not all the changes in NAA are necessarily irreversible. While the recovery of NAA may be in part related to resolution of edema, it could also reflect reversible metabolic changes on axons rather than irreversible axonal loss.

Other changes that are seen by MRS with the acute lesion are related to myelin breakdown. The lipid in myelin is normally so tightly bound into the molecular structure of the myelin that it has no recognizable signal on MRI. However, as myelin is disrupted, the breakdown products, which contain neutral fat, can be identified on MRS (Wolinsky et al., 1990; Richards et al., 1991; Narayana et al., 1998). Increased choline and inositol, which may also reflect the breakdown of myelin membrane phospholipids, have also been observed with acute lesions (Koopmans et al., 1993b). A recent observation that lipid peaks may occur in the absence of gadolinium enhancement and MRI-defined lesions suggests that demyelination may occur independently of inflammatory changes and that there may be more than one pathophysiological process leading to demyelination (Narayana et al., 1998).

B. Magnetization Transfer Imaging

Magnetization transfer (MT) is a contrast technique based on the principle that protons in macromolecules such as myelin, which are not normally visible on conventional MRI, can be studied indirectly by measuring their effect on visible mobile protons. Normal white matter, because of its dense structure, has a high MT ratio. Initial results indicate that MT is decreased in MS lesions as well as in the adjacent normal appearing white matter (NAWM) (Filippi et al., 1998). Therefore, MT may provide some indication of demyelination and axonal loss. Differences in MT between enhancing and nonenhancing lesions, and between the inner and outer sections of ring-enhancing lesions, suggest that there is an evolutionary pattern to MT changes (Hiehle et al., 1995). New lesions (less than 1 year) had lower MT ratios than did older ones (Tomiak et al., 1994). Serial studies of enhancing lesions have shown a decrease in MT ratio of the NAWM in the region of the enhancing lesion preceding the appearance of the lesion itself (Filippi et al., 1998). This finding was not observed by Silver et al. (1998). The decrease in MT ratio in newly enhancing lesion was noted during the first 2 months. Following the decrease in MT ratio that is seen in newly enhancing lesions during the first 2 months, there is a variable increase in MT ratio, although some le-

sions continue to show a progressive decline in MT ratio (Dousset et al., 1998). A recent study (Filippi et al., 1998) showed the MT ratio changes to be a more accurate predictor of clinical disability in the chronic phase of MS than other measures derived from conventional MRI. Volumetric MT techniques, such as the generation of a whole-brain histogram of calculated MT ratios, provide a global assessment of both visible and microscopic lesions in NAWM and may be of value in monitoring therapy in the future. The decrease of MT ratio can also be seen in enhancing lesions and NAWM (Filippi et al., 1998; Silver et al., 1998). The results from MT ratio studies suggest that MS lesions have heterogeneous pathological substrates, which may change over time.

One paradox in interpreting MT reports is that some reports claim that MT is a sensitive measure of subtle changes in NAWM while at the same time some reports claim that MT measures the most severe of destructive pathologies. If both claims are true, there must be a quantitative difference identifiable that can distinguish between the subtle and severe pathologies. No such quantitative differences have been reported to date.

C. Diffusion Imaging

Diffusion-weighted imaging is based on diffusional motion of water on a molecular scale and measures the structural integrity of the cellular components of which tissue is composed. Diffusion imaging is not yet routinely available on standard MR systems, as it requires powerful field gradients. In addition, images obtained with this technique are prone to artifacts caused by involuntary patient movements. The interest in diffusion imaging lies in its potential to supply information about the inflammatory process and the integrity of the neural structures. The acute brain lesion in nonhuman primate EAE shows a decrease in the diffusion MR signal with the diffusion-sensitizing gradient in all three orthogonal directions (Richards et al., 1995). Chronic lesions showed either a decrease in diffusion signal with the diffusion-sensitizing gradient in the two orthogonal directions perpendicular to the fibers of the internal capsule or no significant change in the diffusion signal with diffusion-sensitizing gradient parallel to the fibers of the internal capsule (Richards et al., 1995). Verhoye and co-workers (1996) studied the rat brain with EAE and found a significant positive correlation between the clinical score and diffusivity values in the external capsule. As the clinical signs became severe, a rise in water diffusion could be measured. Preliminary studies in MS patients suggest that the apparent diffusion coefficient (ADC) is increased in chronic lesions and decreased in the peripheral rims of acute lesions. It has been suggested that the former finding is due to tissue destruction, whereas the latter is related to cellular inflammation. The degree of elevation of the ADC within individual lesions did not relate to the disability of the patients. NAWM in patients with MS has also been reported to have a higher ADC than white matter in healthy control subjects (Horsfield *et al.*,1996).

D. Relaxation Measurements

In vivo measurement of T_1 and T_2 relaxation times of protons can give information about the tissue water environment. Until recently, the literature contained many conflicting and confusing results, mostly due to the use of inappropriate measurement and analysis techniques. T_2 relaxation is usually multicomponent in nature because of the existence of multiple local water environments. Techniques employing only two- or even four-point determinations do not take the multiexponential nature of the T_2 decay curve into account. Instead 32-point or even 48- or 64-point determinations are required. Using a 32-echo technique, MacKay *et al.* (1994) distinguished three water reservoirs in normal white matter: (1) a small component with a short T_2 between 10 and 50 ms due to water compartmentalized in myelin membranes, so-called myelin water; (2) a major component with T_2 between 70 and 95 ms due to water in cytoplasmic and extracellular spaces; (3) CSF with T_2 of 1 s or more. Myelin water depicted by the short-T_2 component can be displayed as a map of brain myelin. Myelin water maps show the expected myelin distribution in normal volunteers, with different white matter areas showing variations between them. However, all white matter structures have significantly higher myelin water percentage than all gray matter structures (Whittall *et al.*, 1997). Preliminary studies of MS lesions showed varying levels of myelin water within lesions, ranging from normal to reduced to absent. NAWM in MS patients also showed reductions in myelin water compared to white matter in normal controls (Vavasour *et al.*, 1998). T_2 relaxation measurements can therefore be used to study *in vivo* demyelination and remyelination. One limitation of the technique is that it can only be applied to one MRI slice due to poor signal to noise in multiple slices.

E. Atrophy

The use of MRI to study the development of atrophy in MS has been stimulated by recent immunohistochemical studies that suggest that axonal transection occurs as a consistent consequence of inflammatory demyelination and, importantly, that this axonal transection may begin at the time of disease onset and be later increasingly accompanied by axonal degeneration (Ferguson *et al.*, 1997; Trapp *et al.*, 1998). Since shrinkage and atrophy follow axonal loss, it has been suggested that atrophy may occur earlier than previously thought and may be a simple global measure that may be linked with worsening disability and clinical progression. Preliminary studies (Losseff *et al.*, 1996a,b; Silver *et al.*, 1997; Simon *et al.*, 1998b) have shown that both spinal cord atrophy and brain atrophy are significantly higher in those patients who sustain deterioration in their EDSS score compared with those who do not. Simon *et al.* (1998b) measured third ventricular width, lateral ventricular width, brain width, and midsagittal corpus callosum thickness on MRI. They observed changes of atrophy occurring in RRMS patients with low disability over 1- and 2-year follow-up. In the future, the measurement of atrophy on MRI may also become an important outcome measure in clinical trials.

VII. Summary

MR techniques have shown themselves to be useful in the diagnosis of MS as well as in understanding its evolving pathology. The application of MR to clinical trials has also allowed an objective and sensitive outcome measure that can be used to shorten the time for screening and evaluating new therapies. Some MR techniques now have the ability to detect and measure evolving specific pathologies. The systematic use of MR has changed the understanding of both clinical and laboratory approaches to MS.

References

Arnold, D. L., Riess, G. T., Matthews, P. M., *et al.* (1994). Use of proton magnetic resonance spectroscopy for monitoring disease progression in multiple sclerosis. *Ann. Neurol.* **36**, 76–82.

Bakshi, R., Miletich, R. S., Kinkel, P. R., *et al.* (1998). High-resolution fluorodeoxyglucose positron emission tomography shows both global and regional cerebral hypometabolism in multiple sclerosis. *J. Neuroimaging* **8**, 228–234.

Barkhof, F., Hommes, O. R., Scheltens, P., and Valk, J. (1991). Quantitative MRI changes in gadolinium-DTPA enhancement after high dose intravenous methylprednisolone in multiple sclerosis. *Neurology* **41**, 1219–1222.

Barkhof, F., Tas, M. W., Frequin, S. T., *et al.* (1994). Limited duration of the effect of methylprednisolone on changes on MRI in multiple sclerosis. *Neuroradiology* **36**, 382–387.

Barkhof, F., Filippi, M., Miller, D. H., *et al.* (1997). Comparison of MRI criteria at first presentation to predict conversion to clinically definite multiple sclerosis. *Brain* **120**, 2059–2069.

Barnes, D., *et al.* (1988). The characterization of experimental gliosis by quantitative nuclear magnetic resonance imaging. *Brain* **111**, 83–94.

Brainin, M., Reisner, T., Beuhold, A., *et al.* (1987). Topological characteristics of brainstem lesions in clinically definite and clinically probable cases of multiple sclerosis: A MRI study. *Neuroradiology* **29,** 530–534.

Cover, K. S., Petkau, J., Li, D. K. B., and Paty, D. W. (1999). Lesion load reproducibility and statistical sensitivity of clinical trials in multiple sclerosis. *Neurology* **52,** 433–434.

De Stefano, N., Matthews, P. M., Narayanan, S., *et al.* (1997). Axonal dysfunction and disability in a relapse of multiple sclerosis: Longitudinal study of a patient. *Neurology* **49,** 1138–1141.

De Stefano, N., Matthews, P. M., Fu, L., *et al.* (1998). Axonal damage correlates with disability in patients with relapsing-remitting multiple sclerosis. Results of a longitudinal magnetic resonance spectroscopy study. *Brain* **121,** 1469–1477.

Dousset, V., Gayou, A., Brochet, B., and Caille, J. M. (1998). Early structural changes in acute MS lesions assessed by serial magnetization transfer studies. *Neurology* **51,** 1150–1155.

European Study Group on Interferon β-1b in Secondary Progressive MS. (1998). Placebo-controlled multicentre randomized trial of interferon β-1b in treatment of secondary progressive multiple sclerosis. *Lancet* **352,** 1491–1497.

Fazekas, F., Offenbacher, H., Fuchs, S., *et al.* (1988). Criteria for an increased specificity of MRI interpretation in elderly subjects with suspected multiple sclerosis. *Neurology* **38,** 1822–1825.

Ferguson, B., Matyszak, M. K., Esiri, M. M., and Perry, V. H. (1997). Axonal damage in acute multiple sclerosis lesions. *Brain* **120,** 393–399.

Filippi, M., Horsfield, M. A., Morrissey, S. P., *et al.* (1994). Quantitative brain MRI lesion load predicts the course of clinically isolated syndromes suggestive of multiple sclerosis. *Neurology* **44,** 635–641.

Filippi, M., Paty, D. W., Kappos, L., *et al.* (1995a). Correlations between changes in disability and T_2-weighted brain MRI activity in multiple sclerosis: A follow-up study. *Neurology* **45,** 255–260.

Filippi, M., Campi, A., Martinelli, V., *et al.* (1995b). Comparison of triple dose versus standard dose gadolinium-DTPA for detection of MRI enhancing lesions in patients with primary progressive multiple sclerosis. *J. Neurol. Neurosurg. Psychiatry* **59,** 540–544.

Filippi, M., Rocca, M. A., Martino, G., *et al.* (1998). Magnetization transfer changes in the normal appearing white matter precede the appearance of enhancing lesions in patients with multiple sclerosis. *Ann. Neurol.* **43,** 809–814.

Foong, J., Rozewicz, L., Quaghebeur, G., *et al.* (1998). Neuropsychological deficits in multiple sclerosis after acute relapse. *J. Neurol. Neurosurg. Psychiatry* **64,** 529–532.

Foong, J., Rozewicz, L., Davie, C. A., *et al.* (1999). Correlates of executive function in multiple sclerosis: The use of magnetic resonance spectroscopy as an index of focal pathology. *J. Neuropsychiatry Clin. Neurosci.* **11,** 45–50.

Freedman, M. S., for the Once Weekly Interferon for Multiple Sclerosis (OWIMS) Study Group. (1998). Dose-dependent clinical and magnetic resonance imaging efficacy of interferon β-1a in multiple sclerosis. *Ann. Neurol.* **44,** 992.

Gasperini, C., Pozzilli, C., Bastianello, S., *et al.* (1998). Effect of steroids on Gd-enhancing lesions before and during recombinant beta interferon 1a treatment in relapsing-remitting multiple sclerosis. *Neurology* **50,** 403–406.

Gass, A., Kitchen, N., MacManus, D. G., *et al.* (1997). Trigeminal neuralgia in patients with multiple sclerosis: Lesion localization with magnetic resonance imaging. *Neurology* **49,** 1142–1144.

Giugni, E., Pozzilli, C., Bastianello, S., *et al.* (1997). MRI measures and their relations with clinical disability in relapsing-remitting and secondary progressive multiple sclerosis. *Mult. Scler.* **3,** 221–225.

Gonzalez, C. F., Swirsky-Sacchetti, T., Mitchell, D., *et al.* (1994). Distributional patterns of multiple sclerosis brain lesions. Magnetic resonance imaging—clinical correlation. *J. Neuroimaging* **4,** 188–195.

Grossman, R. I., Braffman, B. H., Brorson, J. R., *et al.* (1988). Multiple sclerosis: Serial study of gadolinium-enhanced MR imaging. *Radiology* **169,** 117–122.

Hashemi, R. H., Bradley, W. G., Chen, D. Y., Jordan, J. E., Queralt, J. A., Cheng, A. E., and Henrie, J. N. (1995). Suspected multiple sclerosis: MR imaging with a thin-section fast FLAIR pulse sequence. *Radiology* **196,** 505–510.

Hawkins, C. P., Munro, P. M. G., Mackenzie, F., *et al.* (1990). Duration and selectivity of blood–brain barrier breakdown in chronic relapsing experimental allergic encephalomyelitis studied by gadolinium-DTPA and protein markers. *Brain* **113,** 365–378.

Heide, A. C., Kraft, G. H., Slimp, J. C., *et al.* (1998). Cerebral *N*-acetylaspartate is low in patients with multiple sclerosis and abnormal visual evoked potentials. *AJNR, Am. J. Neuroradiol.* **19,** 1047–1054.

Hiehle, J. F., Jr., Grossman, R. I., Ramer, K. N., *et al.* (1995). Magnetization transfer effects in MR-detected multiple sclerosis lesions: Comparison with gadolinium-enhanced spin–echo images and nonenhanced T1-weighted images. *AJNR, Am. J. Neuroradiol.* **16,** 69–77.

Hohol, M. J., Guttmann, C. R., Orav, J., *et al.* (1997). Serial neuropsychological assessment and magnetic resonance imaging analysis in multiple sclerosis. *Arch. Neurol.* **54,** 1018–1025.

Horsfield, M. A., Lai, M., Webb, S. L., *et al.* (1996). Apparent diffusion coefficients in benign and secondary progressive multiple sclerosis by nuclear magnetic resonance. *Magn. Reson. Med.* **36,** 393–400.

IFNB Multiple Sclerosis Study Group. (1993). Interferon β-1b is effective in relapsing-remitting multiple sclerosis. I. Clinical results of a multicenter, randomized, double-blind, placebo-controlled trial. *Neurology* **43,** 655–661.

IFNB Multiple Sclerosis Study Group and the University of British Columbia MS/MRS Analysis Group. (1995). Interferon β-1b in the treatment of multiple sclerosis: Final outcome of the randomised controlled trial. *Neurology* **45,** 1277–1285.

Isaac, C., Li, D. K. B., Genton, M., *et al.* (1988). Multiple sclerosis: A serial study using MRI in relapsing patients. *Neurology* **38,** 1511–1515.

Jacobs, L. D., Cookfair, D. L., Rudick, R. A., *et al.* (1996). Intramuscular interferon beta-1a for disease progression in relapsing multiple sclerosis. *Ann. Neurol.* **39,** 285–294.

Johnson, K. P., Brooks, B. R., Cohen, J. A., *et al.* (1995). Copolymer 1 reduces relapse rate and improves disability in relapsing-remitting multiple sclerosis: Results of a phase III multicenter, double-blind, placebo-controlled trial. *Neurology* **45,** 1268–1276.

Joosten, A. A., Hansen, H. M., Piers, D. A., *et al.* (1995). Cobalt-57 SPET in relapsing-progressive multiple sclerosis: A pilot study. *Nucl. Med. Commun.* **16,** 703–705.

Karussis, D. M., Meiner, Z., Lehmann, D., *et al.* (1996). Treatment of secondary progressive multiple sclerosis with the immunomodulator linomide: A double-blind, placebo-controlled pilot study with monthly magnetic resonance imaging evaluation. *Neurology* **47,** 341–346.

Katz, D., Taubenberger, J. K., Cannella, B., McFarlin, D. E., Raine, C. S., and McFarland, H. F. (1993). Correlation between magnetic resonance imaging findings and lesion development in chronic, active multiple sclerosis. *Ann. Neurol.* **34,** 661–669.

Kermode, A. G., Tifts, P. S., Thompson, A. J., *et al.* (1990a). Heterogeneity of blood–brain barrier changes in multiple sclerosis: A MRI study with gadolinium-DTPA enhancement. *Neurology* **40,** 229–235.

Kermode, A. G., Thompson, A. J., Tofts, P., *et al.* (1990b). Breakdown of the blood–brain barrier precedes symptoms and other MRI signs of new lesions in multiple sclerosis: Pathogenetic and clinical implications. *Brain* **113,** 1477–1489.

Kidd, D., Thorpe, J. W., Kendall, B. E., *et al.* (1996). MRI dynamics of brain and spinal cord in progressive multiple sclerosis. *J. Neurol. Neurosurg. Psychiatry* **60,** 15–19.

Koopmans, R. A., Li, D. K. B., Oger, J. J. F., Kastrukoff, L. F., Jardine, C., Costley, L., Hall, S., Grochowski. E. W., and Paty, D. W. (1989). Chronic progressive multiple sclerosis: Serial magnetic resonance brain imaging over six months. *Ann. Neurol.* **26,** 248–256.

Koopmans, R. A., Li, D. K. B., Redekop, W. K., Zhao, G. J., *et al.* (1993a). The use of magnetic resonance imaging in monitoring interferon therapy of multiple sclerosis. *J. Neuroimaging* **3,** 163–168.

Koopmans, R. A., Li, D. K. B., Zhu, G., *et al.* (1993b). Magnetic resonance spectroscopy of multiple sclerosis: *In vivo* detection of myelin breakdown products. *Lancet* **341**(8845), 631–632.

Koopmans, R. A., Zhao, G. J., Li, D. K. B., Tanton, B., Redekop, W. K., and Paty, D. W. (1993c). MR detection of multiple sclerosis activity: Effects of scanning frequency. *Neurology* **43,** A183.

Koudriavtseva, T., Thompson, A. J., Fiorelli, M., *et al.* (1997). Gadolinium enhanced MRI predicts clinical and MRI disease activity in relapsing-remitting multiple sclerosis. *J. Neurol. Neurosurg. Psychiatry* **62,** 285–287.

Li, D. K. B., Mayo, J., Fache, S., Robertson, W. D., Paty, D. W., and Genton, M. (1984). Early experience in nuclear resonance imaging of multiple sclerosis. *Ann. N.Y. Acad. Sci.* **436,** 483–486.

Li, D. K. B., Zhao, G. J., Koopmans, R. A., *et al.* (1997). Can morphological changes in MRI active lesions occur prior to their blood–brain barrier breakdown in MS. *Neurology* **48** (Suppl. 2), A311.

Li, D. K. B., Paty, D. W., UBC MS/MRI Analysis Research Group and the PRISMS Study Group. (1999). Magnetic resonance imaging results of the PRISMS trial: A randomized, double-blind, placebo-controlled study of interferon beta-1a in relapsing–remitting multiple sclerosis. *Ann. Neurol.* **46,** 197–206.

Losseff, N. A., Webb, S. L., O'Riordan, J. I., *et al.* (1996a). Spinal cord atrophy and disability in multiple sclerosis. A new reproducible and sensitive MRI method with potential to monitor disease progression. *Brain* **119,** 701–708.

Losseff, N. A., Wang, L., Lai, H. M., *et al.* (1996b). Progressive cerebral atrophy in multiple sclerosis. A serial MRI study. *Brain* **119,** 2009–2019.

MacKay, A. L., Whittall, K. P., Adler, J., *et al.* (1994). *In vivo* visualization of myelin water in brain by magnetic resonance. *Magn. Reson. Med.* **31,** 673–677.

Mammi, S., Filippi, M., Martinelli, V., *et al.* (1996). Correlation between brain MRI lesion volume and disability in patients with multiple sclerosis. *Acta Neurol. Scand.* **94,** 93–96.

Miller, D. H., Rudge, P., Johnson, G., *et al.* (1988). Serial gadolinium enhanced magnetic resonance imaging in multiple sclerosis. *Brain* **111,** 927–939.

Miller, D. H., Albert, P. S., Barkhof, F., *et al.* (1996). Guidelines for the use of magnetic resonance techniques in monitoring the treatment of multiple sclerosis. *Ann. Neurol.* **39,** 6–16.

Moore, G. R. W., Leung, E., MacKay, A. L., Vavasour, I. M., Whittall, K. P., Cover, K. S., Li, D. K. B., Hashimoto, S. A., Oger, J., and Paty, D. W. (1999). The short T_2 component is absent in chronically demyelinated plaques in formalin-fixed multiple sclerosis brain. *Neurology* **52** (suppl. 2), A567.

Multiple Sclerosis Study Group. (1990). Efficacy and toxicity of cyclosporine in chronic progressive multiple sclerosis: A randomized, double-blind, placebo-controlled clinical trial. *Ann. Neurol.* **27,** 591–605.

Narayana, P. A., Doyle, T. J., Lai, D., and Wolinsky, J. S. (1998). Serial proton magnetic resonance spectroscopic imaging, contrast-enhanced magnetic resonance imaging, and quantitative lesion volumetry in multiple sclerosis. *Ann. Neurol.* **43,** 56–71.

Nesbit, G. M., Forbes, G. S., Scheithauer, B. W., *et al.* (1991). Multiple sclerosis: Histopathologic and MR and/or CT correlation in 37 cases at biopsy and three cases at autopsy. *Radiology* **180,** 467–474.

Newcombe, J., Hawkins, C. P., Henderson, C. L., *et al.* (1991). Histopathology of multiple sclerosis lesions detected by magnetic resonance imaging in unfixed postmortem central nervous system tissue. *Brain* **114,** 1013–1023.

Niendorf, H. P., Haustein, J., Cornelius, I., *et al.* (1991). Safety of gadolinium-DTPA: Extended clinical experience. *Magn. Reson. Med.* **22,** 222–228.

Nijeholt, G. J., van Walderveen, M. A., Castelihns, J. A., *et al.* (1998). Brain and spinal cord abnormalities in multiple sclerosis. Correlation between MRI parameters, clinical subtypes and symptoms. *Brain* **121,** 687–697.

O'Riordan, J. I., Thompson, A. J., Kingsley, D. P., *et al.* (1998). The prognostic value of brain MRI in clinically isolated syndromes of the CNS. A 10-year follow-up. *Brain* **121,** 495–503.

Ormerod, I. E. C., Miller, D. H., du Boulay, E. P. G. H., and McDonald, W. I. (1989). "Magnetic Resonance Imaging in Multiple Sclerosis." Thieme Medical Publishers, New York.

Paty, D. W. (1987). Multiple sclerosis: Assessment of disease progression and effects of treatment. *Can. J. Neurol. Sci.* **14,** 518–520.

Paty, D. W. (1988a). Magnetic resonance imaging in the assessment of disease activity in multiple sclerosis. *Can. J. Neurol. Sci.* **15,** 266–272.

Paty, D. W. (1988b). Trial measures in multiple sclerosis: The use of magnetic resonance imaging in the evaluation of clinical trials. *Neurology* **38,** 82–83.

Paty, D. W., and Li, D. K. B. (1999). Diagnosis of multiple sclerosis 1998: Do we need new diagnostic criteria? *In* "Frontiers in Multiple Sclerosis" (A. Siva, J. Kesselring, and A. J. Thompson, Eds.), Vol. II, pp. 47–50. Martin Dunitz, London.

Paty, D. W., and Moore, G. R. W. (1998). Magnetic resonance imaging changes as living pathology in multiple sclerosis. *In* "Multiple Sclerosis" (D. W. Paty and G. C. Ebers, Eds.), pp. 328–360. Davis, Philadelphia.

Paty, D. W., Oger, J. J., Kastrukoff, L. F., *et al.* (1988). MRI in the diagnosis of MS: A prospective study with comparison of clinical evaluation, evoked potentials, oligoclonal banding and CT. *Neurology* **38,** 180–185.

Paty, D. W., Li, D. K. B., UBC MS/MRI Study Group, and the IFNB Multiple Sclerosis Study Group. (1993). Interferon beta-1b is effective in relapsing–remitting multiple sclerosis. II. MRI analysis results of a multicenter, randomized, double-blind, placebo-controlled trial. *Neurology* **43,** 662–667.

Paty, D. W., Blumhardt, L. D., *et al.* (1998). High-dose subcutaneous interferon β-1a is efficacious in transitional multiple sclerosis, a group at high risk for progression in disability. *Ann. Neurol.* **44,** 503.

Paulesu, E., Perani, D., Fazio, F., *et al.* (1996). Functional basis of memory impairment in multiple sclerosis: a [^{18}F]FDG PET study. *NeuroImage* **4,** 87–96.

Plant, G. T., Kermode, A. G., Turano, G., *et al.* (1992). Symptomatic retrochiasmal lesions in multiple sclerosis: Clinical features, visual evoked potentials, and magnetic resonance imaging. *Neurology* **42,** 68–76.

Pozzilli, C., Fieschi, C., Perani, D., *et al.* (1992). Relationship between corpus callosum atrophy and cerebral metabolite asymmetries in multiple sclerosis. *J. Neurol. Sci.* **112,** 51–57.

PRISMS Study Group. (1998). Randomised double-blind placebo-controlled study of interferon β-1a in relapsing/remitting multiple sclerosis. *Lancet* **352,** 1498–1504.

Pujol, J., Bello, J., Deus, J., *et al.* (1997). Lesions in the left arcuate fasciculus region and depressive symptoms in multiple sclerosis. *Neurology* **49,** 1105–1110.

Rao, S. M., Leo, G. J., Haughton, V. M., *et al.* (1989). Correlation of magnetic resonance imaging with neuropsychological testing in multiple sclerosis. *Neurology* **39,** 161–166.

Riahi, F., Zijdenbos, A., Narayanan, S., *et al.* (1998). Improved correlation between scores on the expanded disability status scale and cerebral lesion load in relapsing–remitting multiple sclerosis. Results of the application of new imaging methods. *Brain* **121,** 1305–1312.

Richards, T. L. (1991). Proton MR spectroscopy in MS: Value in establishing diagnosis, monitoring progression, and evaluating therapy. *AJR, Am. J. Roentgenol.* **157,** 1073–1078.

Richards, T. L., Alvord, E. C., He, Y., *et al.* (1995). Experimental allergic encephalomyelitis in non-human primates: Diffusion imaging of acute and chronic brain lesions. *Mult. Scler.* **1,** 109–117.

Rodriguez, M., and Scheithauer, B. (1994). Ultrastructure of multiple sclerosis. *Ultrastruct. Pathol.* **18,** 3–13.

Roelcke, U., Kappos, L., Lechner-Scott, J., *et al.* (1997). Reduced glucose metabolism in the frontal cortex and basal ganglia of multiple sclerosis patients with fatigue: A ^{18}F-fluorodeoxyglucose positron emission tomography study. *Neurology* **48,** 1566–1571.

Rovaris, M., Rocca, M. A., Capra, R., Prandini, F., Martinelli, V., Comi, G., and Filippi, M. (1998). A comparison between the sensitivity of 3-mm and 5-mm thick serial brain MRI for detecting lesion volume changes in patients with multiple sclerosis. *J. Neuroimaging* **8,** 144–147.

Schiepers, C., Van Hecke, P., Vandenberghe, R., *et al.* (1997). Positron emission tomography, magnetic resonance imaging and proton NMR spectroscopy of white matter in multiple sclerosis. *Mult. Scler.* **3,** 8–17.

Schurmann, M., Warecka, K., Basar-Eroglu, C., and Basar, E. (1993). Auditory evoked potential in multiple sclerosis: Alpha responses are reduced in amplitude, but theta responses are not altered. *Int. J. Neurosci.* **73,** 259–276.

Silver, N. C., Barker, G. J., Losseff, N. A., *et al.* (1997). Magnetisation transfer ratio measurement in the cervical spinal cord: A preliminary study in multiple sclerosis. *Neuroradiology* **39,** 441–445.

Silver, N. C., Lai, M., Symms, M. R., *et al.* (1998). Serial magnetization transfer imaging to characterize the early evolution of new MS lesions. *Neurology* **51,** 758–764.

Simon, J. H., Jacobs, L. D., Campion M., *et al.* (1998a). Magnetic resonance studies of intramuscular interferon β-1a for relapsing multiple sclerosis. *Ann. Neurol.* **43,** 79–87.

Simon, J. H., Jacobs, L. D., Campion, M., *et al.* (1998b). A longitudinal study of brain atrophy in relapsing MS. *Neurology* **50** (Suppl. 4), A192.]

Stevenson, V. L., Leary, S. M., Losseff, N. A., *et al.* (1998). Spinal cord atrophy and disability in MS: A longitudinal study. *Neurology* **51,** 1234–1238.

Stewart, W. A., Hall, L. D., Berry, K., and Paty, D. W. (1984). Correlation between NMR scan and brain slices: Data in multiple sclerosis. *Lancet* **2,** 412.

Stewart, W. A., Hall, L. D., Berry, K., *et al.* (1986). Magnetic resonance imaging (MRI) in multiple sclerosis (MS): Pathological correlation studies in eight cases. *Neurology* **36** (Suppl. 1), 320.

Tartaglino, L. M., Friedman, D. P., Flanders, A. E., *et al.* (1995). Multiple sclerosis in the spinal cord: MR appearance and correlation with clinical parameters. *Radiology* **195,** 725–732.

Thielen, K. R., and Miller, G. M. (1996). Multiple sclerosis of the spinal cord: Magnetic resonance appearance. *J. Comput. Assist. Tomogr.* **20,** 434–438.

Thompson, A. J., Kermode, A. G., MacManus, D. G., *et al.* (1990). Patterns of disease activity in multiple sclerosis: Clinical and magnetic resonance imaging study. *Br. Med. J.* **300,** 631–634.

Thorpe, J. W., Kidd, D., Moseley, I. F., *et al.* (1996). Serial gadolinium-enhanced MRI of the brain and spinal cord in early relapsing-remitting multiple sclerosis. *Neurology* **46,** 373–378.

Tomiak, M. M., Rosenblum, J. D., Prager, J. M., and Metz, C. E. (1994). Magnetization transfer: A potential method to determine the age

of multiple sclerosis lesions. *AJNR, Am. J. Neuroradiol.* **15,** 1569–1574.

Trapp, B. D., Peterson, J., Ransohoff, R. M., *et al.* (1998). Axonal transection in the lesions of multiple sclerosis. *N. Engl. J. Med.* **338,** 278–285.

Truyen, L., van Waesberghe, J. H., van Walderveen, M. A., *et al.* (1996). Accumulation of hypointense ("black holes") on T1 spin–echo MRI correlates with disease progression in multiple sclerosis. *Neurology* **47,** 1469–1476.

van Buchem, M. A., Grossman, R. I., Armstrong, C., *et al.* (1998). Correlation of volumetric magnetization transfer imaging with clinical data in MS. *Neurology* **50,** 1069–1017.

van Oosten, B. W., Truyen, L., Barkhof, D., and Polman, C. H. (1998). Choosing drug therapy for multiple sclerosis. An update. *Drugs* **56,** 555–569.

van Waesberghe, J. H., *et al.* (1998). Patterns of lesion development in multiple sclerosis: Longitudinal observations with T1-weighted spin–echo and magnetization transfer MR. *AJNR, Am. J. Neuroradiol.* **19,** 675–683.

van Walderveen, M. M. A., Barkhof, F., Hommes, O. R., *et al.* (1995). Correlating MRI and clinical disease activity in multiple sclerosis: Relevance of hypointense lesions on short-TR/short-TE (T_1-weighted) spin–echo images. *Neurology* **45,** 1684–1690.

van Walderveen, M. A., Kamphorst, W., Scheltens, P., *et al.* (1998). Histopathologic correlate of hypointense lesion on T1-weighted spin-echo MRI in multiple sclerosis. *Neurology* **50,** 1282–1288.

Vavasour, I. M., Whittall, K. P., MacKay, A. L., *et al.* (1998). A comparison between magnetization transfer ratio and myelin water percentages in normals and multiple sclerosis patients. *Magn. Reson. Med.* **40,** 763–768.

Verhoye, M. R., Gravenmade, E. J., Raman, E. R., *et al.* (1996). *In vivo* noninvasive determination of abnormal water diffusion in the rat brain studied in an animal model for multiple sclerosis by diffusion-weighted NMR imaging. *Magn. Reson. Imaging* **14,** 521–532.

Wang, X. Y., Zhao, G. J., Li, D. K. B., Paty, D. W., the UBC MS research group, and the PRISMS Study Group. (1997). Comparison of lesion distribution between relapsing-remitting and secondary progressive MS monitored by MRI. *Neurology* **48** (Suppl.), A311–A312.

Weinmann, H. J., Brasch, R. C., Press, W. R., and Wesbey, G. E. (1984). Characteristics of gadolinium-DTPA-complex: A potential NMR contrast agent. *AJR, Am. J. Roentgenol.* **142,** 619–624.

Whittal, K. P., MacKay, A. L., Graeb, D. A., *et al.* (1997). *In vivo* measurement of T2 distribution and water contents in normal human brain. *Magn. Reson. Med.* **37,** 34–43.

Willoughby, E. W., Grochowski, E., Li, D. K. B., Oger, J., Kastrucoff, L. F., and Paty, D. W. (1989). Serial magnetic resonance scanning in multiple sclerosis: A second prospective study in relapsing patients. *Ann. Neurol.* **25,** 43–49.

Wolinsky, J. S., Narayana, P. A., and Fenstermacher, M. J. (1990). Proton magnetic resonance spectroscopy in multiple sclerosis. *Neurology* **40,** 1764–1769.

Young, I. R., Hall, A. S., Pallis, C. A., Legg, N. J., Bydder, G. M., and Steiner, R. E. (1981). Nuclear magnetic resonance imaging of the brain in multiple sclerosis. *Lancet* **2,** 1063–1066.

Zhao, G. J., Li, D. K. B., Tanton, B. L., and Paty, D. W. (1991). Assessment of activity of the individual lesions in relapsing–remitting multiple sclerosis on MRI. *Ann. Neurol.* **30,** 270.

Zhao, G. J., Li, D. K. B., Koopmans, R. A., Tanton, B. L., and Paty, D. W. (1993a). The growth pattern of enlarging lesions in multiple sclerosis: Observation on serial MRI. *Can. J. Neurol. Sci.* Suppl. 4, S219.

Zhao, G. J., Koopmans, R. A., Li, D. K. B., *et al.* (1993b). Corticospinal tract lesions in multiple sclerosis: Relationship between MRI activity and clinical course. *Neurology* **43,** A246.

Zhao, G. J., Li, D. K. B., Koopmans, R. A., *et al.* (1997a). Clinical and magnetic resonance imaging changes correlate in a clinical trial monitoring cyclosporine therapy for multiple sclerosis. *J. Neuroimaging* **7,** 1–7.

Zhao, G. J., Li, D. K. B., Koopmans, R. A., *et al.* (1997b). Correla-

tion of clinical status with MS lesion changes in untreated patients: A 5-year study by yearly MRI. *Neurology* **48** (Suppl.), A361.

Zhao, G. J., Koopmans, R. A., Li, D. K. B., *et al.* (2000). The effect of interferon β-1b on yearly MRI activity in relapsing-remitting multiple sclerosis: Assessment of the annual accumulation of proton density/T2 activity. *Neurology* **54,** 1–7.

16

Structural and Functional Imaging of Cerebral Neoplasia

Jeffry R. Alger[*,1] and Timothy F. Cloughesy[†]

[*]Department of Radiological Sciences, Brain Research Institute, Jonsson Comprehensive Cancer Center, University of California, Los Angeles, California 90095
[†]Department of Neurology, Henry E. Singleton Brain Cancer Research Program, Jonsson Comprehensive Cancer Center, University of California, Los Angeles, California 90095

I. Introduction
II. The Clinical Challenges
III. Recent Neuroimaging Progress Related to Intracranial Neoplasms
IV. Summary and a Look into the Crystal Ball
   References

Structural and functional brain imaging now provides the clinician with heretofore unprecedented insights, at a variety of different levels, into the characteristics and into the evolution of central nervous system neoplastic lesions. Beyond the lesion itself, modern neuroimaging procedures also provide insight into how the neoplastic lesion affects the structure and the function of the surrounding brain tissue.

This chapter reviews the structural and functional brain imaging technologies that can be used to evaluate neoplastic processes within the brain and is intended for two audiences: biomedical imaging scientists interested in pursuing research related to intracranial neoplasia and clinicians interested in understanding how functional and structural neuroimaging methodologies can be used to optimize management of the neuro-oncologic patient. Key background on the clinical "landscape" is provided and related to the more significant of the neuroimaging tools that are presently in use for assessment of intracranial neoplasia, and recent progress in brain tumor neuroimaging in relation to the clinical background is summarized. While the present brain tumor imaging capabilities are often viewed with nothing less than amazement, more progress is clearly needed in terms of managing routine brain tumor cases and establishing more accurate and more efficient clinical trial methodology. Structural and functional neuroimaging has a large role to play in these future developments. It is the authors' hope that this chapter may stimulate increased expenditure of intellectual energy directed toward controlling central nervous system neoplasia.

---

[1]To whom correspondence should be addressed.

I. Introduction

Brain cancer is a significant health problem. Intracranial neoplastic processes are relatively uncommon and present a relatively benign course compared to the entire spectrum of human oncologic disorders. Nevertheless they have a disproportionate clinical importance. Brain cancer is a universally feared disease. Indeed, until the latter part of the twentieth century, a brain tumor was usually viewed as an untreatable condition. This has changed as a result of technological advances that include neuroimaging. Despite the more aggressive attitudes that are now taken, a brain tumor is fatal for many patients. It is estimated that about 17,500 residents of the United States will die each year from cerebral neoplastic syndromes. Primary brain tumors account for 2% of all cancer deaths but are responsible for 7% of the years lost from cancer before the age of 70. Relatively small high-grade primary glial neoplasms inevitably lead to death in a time counted in months rather than years. Moreover, the "survival time" is often complicated by neurologic decline and the accompanying functional disability. While patients with lower grade lesions may survive for years, they can also suffer from irreversible disabilities.

Neuro-oncology is a rapidly evolving multidisciplinary field. In the present day, neuro-oncologists, neurosurgeons, neuroradiologists, and radiation oncologists work together to manage brain tumor cases with the help of a previously undreamed of technological arsenal. Advances in brain imaging, molecular genetics, immunology, neurosurgery, neuropathology, radiation oncology, and pharmacology are all currently being integrated into the fabric of neuro-oncology. The research frontiers of a few years ago are rapidly becoming part of today's routine patient management. The days when the high-grade intracranial glioma or the metastatic brain lesion was viewed as hopeless are past! Very aggressive procedures that make extensive use of functional and structural neuroimaging are becoming very routine.

Given the breadth and expansiveness of neuro-oncology, it is impossible to provide a comprehensive review of (even) the neuroimaging aspects in a relatively short chapter such as the present one. Therefore, scope is limited in a number of ways. Spinal tumors are not specifically discussed, although it is apparent that many of the technological principles used in brain imaging are applicable to the spine provided that spatial resolution limitations and the practical problems associated with body imaging are surmountable. The chapter focuses on only the more common histologic forms of primary glioma (astrocytoma, glioblastoma multiforme, oligodendroglioma, lymphoma, meningioma). The unique problems of pediatric brain tumors are dis-

cussed only in passing. Metastatic tumors in the brain are also discussed only briefly despite their being somewhat more prevalent than primary neoplasms because the management of metastatic disease is somewhat less complicated compared to that of primary neoplasms. Space limitations necessitate a focus on only the more recent and the more significant of the imaging advances. No systematic attempt to survey the relevant neuroimaging literature prior to 1994 is made. Furthermore, no systematic description of the technical underpinnings of the different techniques is provided. The reader is referred to earlier chapters in the present volume and to earlier volumes in the series for this information. Two comprehensive monographs edited by Black and Loeffer (1997) and by Kaye and Laws (1995) are available. These include many chapters covering all aspects of brain tumor science and medicine. These texts were used extensively in preparing this chapter. The reader is thereby referred to them for background material. Any factual material asserted in this chapter but not explicitly referenced is discussed in detail in one or both of these two superbly comprehensive monographs. Finally, an arbitrary decision was made to focus the chapter on clinical neuro-oncologic management and relevant human clinical research. Therefore, animal studies are not discussed except in cases where they indicate a "proof of concept" that is immediately applicable to work in the "human arena."

II. The Clinical Challenges

Structural and functional imaging of intracranial neoplastic disease differs from more esoteric brain mapping endeavors in that the ultimate goal is the pursuit of the appropriate clinical management strategies that lead to a lessening of human suffering combined with the preservation of as much central nervous system function as is possible. For this reason, this chapter is organized from a clinical perspective. Here the concept is that a comprehensive understanding of the present status and likely future progress in brain tumor neuroimaging cannot be had without some minimal background picture of the clinical "landscape." There are many instances—the evolution of histopathologic grading systems for glioma is but one—in which clinical or biological issues have direct influence on the neuro-oncologic application of neuroimaging technologies.

A. Key Histological Features of Brain Tumor

Histological classification schemes for central nervous system (CNS) tumors have been evolving since the time of Harvey Cushing (the 1920s). This is a result of the complex microscopic appearance of these lesions and the large variety of unique histological patterns that can be identified. During the past 20 years, the World Health Organization (WHO) has attempted to formulate and codify the consensus of neuropathological opinion. Table I provides a summary of the WHO classification system that has been in use since 1979. The table aptly illustrates the variety of different tumor types that can be identified histologically. It illustrates that tumors can arise from a number of different cell types present in the brain. It also illustrates that variants (e.g., fibrillary astrocytoma) of many different tumor types are known. The most frequently encountered primary CNS neoplasms are generally referred to as "glioma" because these tumors arise from glial cells. The classification terminology for glioma that is used by many present-day neuropathologists is generally based on a scheme proposed by Kernohan in 1949. The Kernohan scheme identifies five categories of primary brain tumors: astrocytoma, ependymoma, oligodendroglioma, neuroastrocytoma, and medulloblastoma. The Kernohan scheme also recognizes that individual tumors may contain mixtures of these different components—most frequently astrocytoma mixed with oligodendroglioma.

Grading systems for primary brain tumors that distinguish tumors with respect to outcome have been developed over the past 50 years and are in common use. In 1950, Ringherz proposed a three-tier grading system specifically applicable to astrocytoma, the most frequently encountered of the primary brain tumors. This system used the following terminology to denote lesions of increasing malignancy: astrocytoma, astrocytoma with anaplastic foci, and glioblastoma multiforme. In 1988, Daumas-Duport et al. (1988) proposed a simplified grading system for astrocytic tumors based on concepts central to the Ringherz system. This system, now known as the St. Anne-Mayo system is based upon the identification to four key morphological features: nuclear atypia, mitoses, endothelial proliferation, and necrosis. Figure 1 illustrates each of these features. Part of the system's strength lies in the fact that these morphologic features are readily identified with a high degree of between-reader consensus (Daumas-Duport et al. 1988). The system recognizes four grades, with Grade I being the most benign and Grade IV being the most malignant. A tumor having none of the features is assigned to Grade I in the St. Anne-Mayo system. If only one of the features is recognized, the tumor is assigned Grade II. If two features are recognized, the system assigns the tumor to Grade III. If three or more features are identified, the tumor is assigned to Grade IV. In most cases, the presence of necrosis is the defining feature that distinguishes between Grade III and Grade

IV. Another significant strength of the St. Anne-Mayo system is its relative objectivity. It includes no categorization as to the relative frequency at which the individual features are seen in the entire sample. One only need identify a single instance of one of the features anywhere within the sample to establish its presence. In other words, the identification of a single mitosis is sufficient to elevate the grade by one. The St. Anne-Mayo grading system has been demonstrated to be a valid predictor of survival (Daumas-Duport, 1988). However, improvements have been suggested. The relative infrequency of Grade I lesions has been a shortcoming. The most recently proposed WHO classification scheme has recognized this in the way it incorporated the St. Anne-Mayo concepts. For primary astrocytic tumors, the WHO scheme specifies three grades with the following terminology: (1) Astrocytoma is characterized by degrees of cellularity and anaplasia, (2) anaplastic astrocytoma has in addition mitotic figures and endothelial proliferation, and (3) glioblastoma multiforme has necrosis in addition to all the others.

Grading systems for other tumors are not as well defined as is the case for astrocytoma. Pilocytic astrocytoma is an example. These tumors are frequently found on the midline (optic chiasm, third ventricle, and brainstem) in children and young adults. They are not graded although usually they display slow growth and are described as being "low grade." Oligodendroglioma is sometimes graded using the four-tiered Smith system (Smith et al., 1983). However, the WHO system suggests two classifications for oligodendroglioma, which are termed "anaplastic" and "malignant."

It is important for neuroimaging specialists to recognize several facts regarding the discussion of histological features. First, histological classification/grading is subjective and has been in evolution for the past 50 years. Many neuropathologists still use the Kernohan or Ringherz system, although the St. Anne-Mayo system seems to be gaining favor. Therefore, any systematic evaluation of the relationship between histology and the structural or functional imaging features of brain tumor is effectively limited by the evolution of subjectivity inherent in the histological reading. A substantial neuropathological limitation has been differing opinions about what features are present and to what extent they are present. Second, the ultimate purpose of a grading system is to distinguish tumors with respect to outcome as an aid to clinical management. In other words, the systems are used not just for the purpose of biological categorization but, instead, for the categorization of the likely course of the disease. In most cases, the length of survival is used as the primary outcome measure. Third, even when tissue is obtained after some form of therapy (most frequently radiation therapy), it

Table I An Outline of the 1979 WHO Classification System for Central Nervous System Tumors[a]

I. Tumors of neuroepithelial tissue
  A. Astrocytic tumors
    1. Astrocytoma
      a. Fibrillary
      b. Protoplasmic
      c. Gemistocytic
    2. Pilocytic astrocytoma
    3. Subependymal giant cell astrocytoma (ventricular tumor of tuberous sclerosis)
    4. Astroblastoma
    5. Anaplastic (malignant) astrocytoma
  B. Oligodendroglial tumors
    1. Oligodendroglioma
    2. Mixed oligoastrocytoma
    3. Anaplastic (malignant) oligodendroglioma
  C. Ependymal and choroid plexus tumors
    1. Ependymoma
      a. Myxopapillary ependymoma
      b. Papillary ependymoma
      c. Subependymoma
    2. Anaplastic (malignant) ependymoma
    3. Choroid plexus papilloma
    4. Anaplastic (malignant) choroid plexus papilloma
  D. Pineal cell tumors
    1. Pineocytoma (pineal cytoma)
    2. Pineoblastoma (pineal blastoma)
  E. Neuronal tumors
    1. Gangliocytoma
    2. Ganglioglioma
    3. Ganglioneuroblastoma
    4. Anaplastic (malignant) gangliocytoma and ganglioglioma
    5. Neuroblastoma
  F. Poorly differentiated and embryonal tumors
    1. Glioblastoma
      a. Glioblastoma with sarcomatous component (mixed glioblastoma and sarcoma)
      b. Giant cell glioblastoma
    2. Medulloblastoma
      a. Desmoplastic
      b. Medulloepithelioma
    3. Medulloepithelioma
    4. Primitive polar spongioblastoma
    5. Gliomastosis cerebri
II. Tumors of nerve sheath cells
  A. Neurilemoma (schwannoma, neurinoma)
  B. Anaplastic (malignant) neurilemoma (schwannoma, neurinoma)
  C. Neurofibroma
  D. Anaplastic (malignant neurofibroma (neurofibrosarcoma, neurogenic sarcoma)
III. Tumors of meningeal and related tissues
  A. Meningioma
    1. Meningothelioma (endotheliomatous syncytial, archnotheliomatous)
    2. Fibrous (fibroblastic)
    3. Transitional (mixed)
    4. Psammomatous
    5. Angiomatous
    6. Hemangioblastic
    7. Hemangiopericytic
    8. Papillary
    9. Anaplastic (malignant) meningioma

  B. Meningeal sarcomas
    1. Fibrosarcoma
    2. Polymorphic cell sarcoma
    3. Primary meningeal sarcomatosis
  C. Xanthomatous tumors
    1. Fibroxanthoma
    2. Xanthosarcoma (malignant fibroxanthoma)
  D. Primary melanotic tumors
    1. Melanoma
    2. Meningeal melanamatosis
  E. Others
IV. Primary malignant lymphomas
V. Tumors of blood vessel origin
  A. Hemangioblastoma (capillary hemangioblastoma)
  B. Monstrocellular sarcoma
VI. Germ cell tumors
  A. Germinoma
  B. Embryonal carcinoma
  C. Choriccarcinoma
  D. Teratoma
VII. Other malformative tumors and tumorlike lesions
  A. Craniopharyngioma
  B. Rathke's cleft cyst
  C. Epidermoid cyst
  D. Dermoid cyst
  E. Colloid cyst of the third ventricle
  F. Enterogenous cyst
  G. Other cysts
  H. Lipoma
  I. Choristoma (pituicytoma, granular cell "myoblastoma")
  J. Hypothalamic neuronal hamartoma
  K. Nasal glial heterotopia (nasal glioma)
VIII. Vascular malformations
  A. Capillary telangiectasta
  B. Cavernous angioma
  C. Arteriovenous malformation
  D. Venous malformation
  E. Sturge–Weber disease (cerebrofacial or cerebrotrigeminal)
IX. Tumors of the anterior pituitary
  A. Pituitary adenomas
    1. Acidophil
    2. Basophil (mucoid cell)
    3. Mixed acidophil–basophil
    4. Chromophohe
  B. Pituitary adenocarcinoma
X. Local extensions from regional tumors
  A. Glomus jugulare tumor (chemodectoma, paraganglioma
  B. Chordoma
  C. Chondroma
  D. Chondrosarcoma
  E. Olfactory neuroblastoma (esthesioneuroblastoma)
  F. Adenoid cystic carcinoma (cylindroma)
  G. Others
XI. Metastatic tumors
XII. Unclassified tumors

[a] Adapted from Zuelch, 1979.

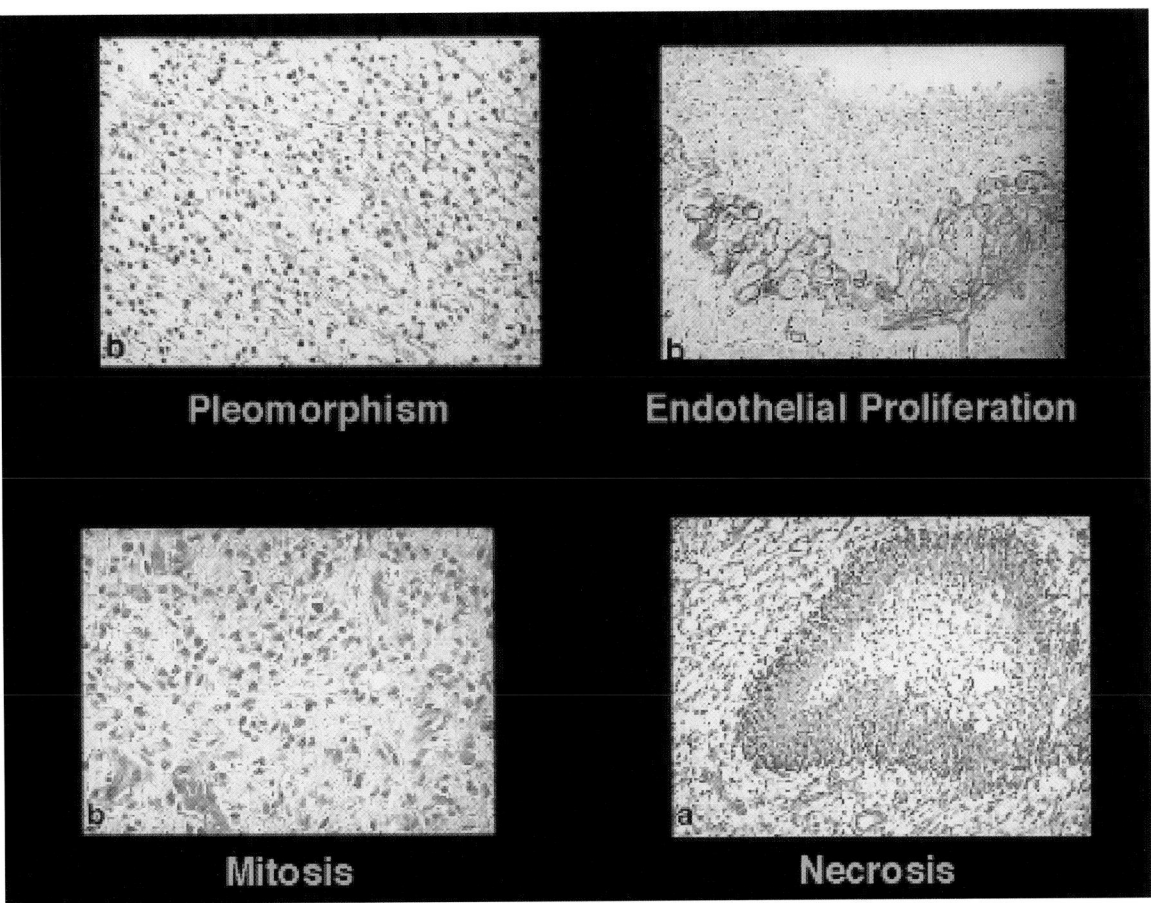

Figure 1 Examples of the four key histological features of astrocytic tumors used in the St. Anne-Mayo grading system. (Adapted from Lantos *et al.*, 1997.)

is problematic to distinguish the necrosis that arose from natural tumor evolution from therapeutically induced necrosis. Therefore the histopathological necrosis feature becomes essentially unusable in the post-therapeutic setting.

Neuroimaging studies can take several "cues" from neuropathology. First, the evolution of neuropathology illustrates that between-reader consensus is crucial. By analogy it is also important that emerging neuroimaging technology evolve in the direction of the highest possible degree of between-reader consensus. The inherently quantitative nature of many neuroimaging findings is likely to be helpful here. Second, the stated purpose of many of the structural and functional imaging studies that will be discussed below was to identify image features that "correlate with histology." In these instances, this was justified in terms of evaluating lesions from which tissue could not be surgically obtained. This line of reasoning misses the point that any measured neuroimaging parameter that could predict outcome would be powerful in its own right and not necessarily need to be correlated with histological grading. Third, given that

most patients have only one preradiation cytoreductive surgery and that histology becomes less powerful at outcome prediction after therapy even if a second resection is performed, clinicians are looking to structural and functional imaging for advances that will predict outcome from that point forward. Therefore in the post-therapy setting, it is more important to develop neuroimaging measures of outcome than it is to establish correlations with histological features.

B. Incidence and Survival

The incidence rate in the United States for primary brain tumors varies between 5 and 14 per 100,000 population per year depending on geographic location. Geographic variability is also seen worldwide and is largely unexplained. Epidemiological studies of brain tumors have, to some extent, been influenced by the evolution of imaging technologies. The majority of the presently available epidemiological data were amassed in the era before X-ray computed tomography (CT) was widely available. Accordingly, the available data probably underestimate the true rate of incidence. An unproven suspicion exists that geographic variability may be related to availability of high-quality cross-sectional neuroimaging technologies. However, this is probably more relevant to the incidence of the more benign forms of disease (e.g., meningioma) because such lesions can be present, but silent, throughout life. Table II summarizes the incidence rates for a few of the most common types of primary brain tumors The table provides estimated incidences for symptomatic and occult lesions detected through incidental neuroimaging and autopsy based on Rochester, Minnesota, data. Meningiomas account for the majority of the occult lesions and are found in 1–2% of all autopsies. There appear to be research opportunities in the utilization of brain mapping techniques to study individuals who carry such small benign lesions for the purpose of understanding how these lesions may influence individual characteristics.

Primary brain tumor syndromes tend to occur in the early and the later decades of life. Primary brain tumors are the second most frequent type of childhood cancer after leukemia. They account for about one-fifth of all childhood cancers. Primary brain tumors diagnosed in the pediatric population tend to be rare and low grade (usually pilocytic astrocytoma), whereas adults are more likely to present with more aggressive lesions—most often glioblastoma multiforme (GBM). Intracranial germ cell tumors typically develop at puberty and are thereby likely related to systemic endrocrinologic alterations. Table III provides a summary of the age dependence of the incidence of primary brain tumors taken from Rochester, Minnesota, data. The table shows that primary brain tumors represent a significant health problem in older individuals. The significance of primary brain tumors in older adults is especially important when one considers that the majority of the lesions diagnosed in these patients are quite aggressive.

Table III Age Dependence of the Incidence of Primary Brain Tumors[a]

Age	Incidence (Case per 100,000 per year)
12	4.0
35	6.5
55	18.0
75	>70

[a] Adapted from Bohnen et al., 1997.

Studies have compared brain tumor incidence rates in the 1950s to those in the 1980s. These studies suggest an increasing incidence of unknown cause. Concern exists that this may be an artifact of improved neuroimaging technology (CT and MRI) and its more widespread availability. It could also be related to increased attention to the health problems of the elderly.

Gender and racial differences in brain tumor incidence are also reported. Females tend to be less frequently diagnosed with primary brain tumors. Averaged across geographic regions, the male to female incidence ratio is about 1.4 (range 0.9–2.6). In a study of incidence in Los Angeles County, Blacks, Asians, and Latinos tended to be diagnosed less frequently than Whites. The possibility exists that these gender and racial differences may be related to the provision of diagnostic technologies such as CT and MRI.

Primary lymphoma of the CNS—sometimes referred to as microglioma—was previously rare. It is now taking on an increased clinical importance as a result of the AIDS (Acquired Immune Deficiency Syndrome) pandemic. It is estimated that between 2 and 5% of all AIDS patients who die have intracranial lymphoma lesions. Structural and functional neuroimaging has a large role to play in the differential diagnoses of lymphoma from a wide variety of other intracranial lesions that occur within the context of AIDS. Most significantly, there is a necessity of distinguishing primary lymphoma from opportunistic infections because of the radically different treatment regimens used for these two conditions.

The development of metastatic disease in the brain is a relatively common complication of systemic cancer. Primary cancers that frequently metastasize include lung, breast, melanoma, gastrointestinal, and renal. Moreover, the frequency of diagnosis of metastatic brain lesions is increasing. This is likely the result of improved availability of sensitive neuroimaging technology and advances in the management of systemic disease.

Brain tumor is second only to stroke in causes of death from neurologic disease. Table IV provides 5-year

Table II Incidence (Cases per 100,000 per year) of Common Forms of Primary Brain Tumor[a]

	Symptomatic	Total[b]
Glioma	5.0	5.7
Meningioma	2.0	7.8
Pituitary Adenoma	2.4	2.8
All	11.8	19.1

[a] Adapted from Bohnen et al., 1997.
[b] Includes occult lesions detected through incidental neuroimaging and autopsy.

Table IV   Five-Year Survival from Malignant CNS Tumors in Victoria, 1982–1991[a]

Tumor	Survival (%)
Glioblastoma multiforme	4
Astrocytoma	43
Medulloblastoma	41
Oligodendroglioma	47
Ependymoma	59
Other glioma	30
Malignant meningioma	50
Other	29
Unspecified	28
No microscopy	21

[a] Adapted from Bohnen *et al.*, 1997.

survival statistics for a few types of primary brain tumors. The table shows the percentage of patients who survive 5 years following their initial diagnosis. The table illustrates considerable variability in survival statistics. GBM patients succumb relatively quickly, whereas many of the other forms of primary tumors are less aggressive and patients have rather prolonged survival compared to GBM or many other forms of human cancer. However, it is important to emphasize that even a small slow-growing brain tumor can be lethal because of its confinement within the cranium.

C. Etiology

The cause of primary brain tumors is not definitively known. In general, primary CNS tumors display a very great deal of phenotypic variety (see above). Molecular genetic studies (see below) have shown a high degree of genotypic variability. This phenotypic and genotypic heterogeneity suggests there is no single pathogenic factor. Genetic, environmental, and biological factors have been suggested as being key causative factors, but none of these has been overwhelmingly supported from an epidemiological point of view. There are likely to be some causes of an inherited nature. A family history of cancer is reported in 19% of patients diagnosed with primary brain tumors, indicating genetic involvement in pathogenesis. Fifteen percent of patients having neurofibromatosis type I—a syndrome resulting from an autosomal dominant defective gene located on chromosome 17 that is responsible for the production of the protein neurofibromin—develop a variety of tumors, including low-grade optic pathway glioma, cerebellar astrocytoma, pilocytic astrocytoma of the third ventricle, and higher grade astrocytoma. Neurofibromatosis type

II patients develop bilateral eighth nerve Schwannomas. Certain chemicals are known to induce brain tumors in animals and there is some evidence that chemical industry workers have higher incidences of brain cancer. It has been proposed that the increased use of aspartame, a chemical compound related to several neurotransmitters, as an artificial sweetener beginning about 20 years ago is, in part, responsible for the increased incidence seen in recent decades. Similarly, there are unsubstantiated inferences that the exposure to electromagnetic fields through the use of cellular telephones is causative and responsible for increased incidence. Certain viruses are also known to induce brain tumors in animals but there has been no clear indication that this is of any significance in human disease.

D. Genetics

The techniques of molecular biology are being used to an increasing extent to evaluate brain tumors. This has been done to improve the understanding of pathogenesis.

Cytogenetic aberrancy in primary brain tumor is now fairly well characterized. While there is some heterogeneity of pattern, relatively consistent patterns of numerical and structural chromosomal alterations have been identified and, furthermore, these are correlated with the progression and grade of disease. This provides substantiation for the hypothesis that the histological progression from astrocytoma to glioblastoma multiforme, for instance, is driven by the accumulation of chromosomal aberrancies as the astrocytoma cells multiply. The following general picture has emerged. Multiple copies of chromosome 7 are often seen in glioma and this occurs in lower grades. Chromosome 7 is the site of two known oncologically important genes encoding for EGFR (epidermal growth factor receptor) and platelet-derived growth factor alpha chain (PDGF-alpha). This suggests that autocrine control factors associated with overexpression of these growth factor genes are a key early feature in disease progression. Similarly, losses of the q-arm of chromosome 13, the site of the tumor suppressor gene that is causative for retininoblastoma, in all grades is involved in glioma pathogenesis. Over 80% of glioblastoma multiforme lesions show a loss of the q-arm of chromosome 10. It is hypothesized that this may be the site of a "glioma suppressor gene." Many gliomas also show aberrancy in the "p53" gene locus, which is implicated in the pathogenesis of other forms of cancer.

E. Clinical Features

Many of the clinical symptoms of brain tumors are the direct result of an expanding tumor mass within a closed cranium. All brain tumors may produce symptoms by one or more of the following mechanisms:

1. Increased intracranial pressure due to the tumor mass, to peritumoral edema, or to obstruction of CSF pathways

2. Local biochemical changes, which may act to depolarize neurons within or adjacent to the tumor, thereby triggering seizure activity

3. Local destruction of brain tissue, resulting in neurologic deficits

4. Stretching, distortion, or compression of surrounding neural structures (cranial nerves), which are themselves not infiltrated by tumors

5. Remote effects on other organ systems due to the alteration of neuroendocrine function or elaboration of active substances by the tumor

6. Stretching of the dura and stretching or distortion of the basal arteries

Intracranial pressure may be elevated as a result of the tumor mass itself or the peritumoral edema. In some cases, the peritumoral edema is more significant compared to the actual tumor mass in this regard. The magnitude of peritumoral edema is particularly great in the more rapidly growing primary CNS neoplasms and in metastatic tumors. Intracranial pressure may also be elevated because the tumor and surrounding edema obstruct the ventricular system. This would most commonly occur at or near the interventricular foramen, at the third ventricle and the aqueduct of Sylvius, or near the outlet foramina of the fourth ventricle. Increased intracranial pressure gives rise to rather nonspecific symptoms and signs, including headache, nausea and vomiting, obscuration of vision, papilledema, and, in young children, spreading of cranial sutures. The irritate action that tumor or edema may exert upon the involved brain can lead to seizures; these may be focal or generalized. In rare instances, a prolonged state of seizure activity may develop, which could be fatal if not treated adequately.

The local destruction of brain tissue may be the result of tumor or peritumoral edema, or both, and can cause the loss of neural function appropriate to the area of brain involved. Accordingly, the resulting clinical picture can be quite variable and may include disturbances of motor, speech, sensory, visual, or intellectual function or personality changes. Cranial nerves often are stretched or distorted by tumor so as to interfere in their function. The clinical picture resulting from such cranial nerve deficit can be of localizing value, although stretching of the most frequently involved nerve—the abducens (VI)—is nonspecific and often does not aid the anatomic localization of the tumor. Intracranial tumors may also exhibit remote systemic effects if they impinge upon or involve neuroendocrine structures such as the hypothalamus or pituitary gland.

The most common cause of death from brain tumors is a herniation syndrome resulting from increased mass effect. In the case of cerebral hemisphere lesions, medial temporal lobe herniation commonly results in the medial displacement of the uncus of the temporal lobe, producing compression and stretching of the brainstem. Posterior fossa tumors may compress the lower brainstem directly or produce herniation of the cerebellar tonsils with medullary compression. In all these situations, coma and respiratory arrest will ensue. Not infrequently, the compressive effect on the brainstem is somewhat more gradual, and the patient first becomes comatose. During this time, there is, of course, a great risk of aspiration and pneumonia. Some patients die as a result of uncontrolled seizure activity. In the setting of hospice care, it is not uncommon for a patient to die from dehydration after the ability to swallow is lost. In some cases, patients also can succumb to a pulmonary embolism.

F. Diagnosis

In the majority of cases, the patient initially presents with new symptoms, commonly seizure, headache, or some neurofunctional or neuropsychiatric deficiency consistent with intracranial mass lesion. The functional representation of symptoms can be highly variable and is dependent on the location of the mass lesion within the brain. Motor, speech, visual, and somatosensory deficiencies are often attributed to mass lesions. At this stage of the clinical assessment, the primary use of neuroimaging is to confirm the presence of an intracranial mass lesion. Noninvasive neuroimaging techniques are rarely able to do more in terms of establishing a specific diagnosis than to confirm the presence of mass lesion. Conventional structural MRI and to a less frequent extent CT are used for this purpose (see below). In some cases, fairly reliable assessments of tumor tissue type can be had through structural and contrast enhancement characteristics together with clinical history. The most obvious example of such a situation is the patient with a known primary tumor elsewhere in the body with new neurologic symptoms and a contrast-enhancing cerebral lesion. Here the clinical information and the neuroimaging lead to a diagnosis of sufficient probability (i.e., metastatic tumor) to proceed with therapy. In cases of primary intracranial lesions or cases where there is no known primary tumor elsewhere in the body, the neuroradiologist may suggest a probable tumor type and grade. However, this information cannot be viewed as being fully conclusive. Some of the newer technological innovations (e.g., nuclear medicine or magnetic resonance spectroscopy) show promise of providing more accurate tumor type and grade diagnoses. These advancements are discussed in the relevant sections (see below).

Magnetic resonance imaging is now widely acknowledged as the method of choice for imaging intracranial mass lesions. It offers modestly higher resolution and better soft-tissue contrast compared to CT. Its ability to depict the anatomic structure of intracranial mass lesions is much superior to that of PET and SPECT. Indeed, in most instances, it is necessary to obtain an MRI for structural correlation in subjects who undergo PET or SPECT studies. A number of inherent properties of MRI are responsible for this evolution. The nonradioactive hydrogen nuclei (which are also known as "protons") in the tissue water and in other relevant endogenous molecules generate the signal used for image production. MRI provides excellent soft-tissue contrast, which can depend on inherent nuclear relaxation properties of the tissue water molecules or on the presence of paramagnetic contrast material. MRI is also capable of imaging planes having an arbitrary orientation relative to standard anatomic directions; in addition to axial imaging, saggittal, coronal, and oblique planar imaging are readily performed. This permits mass lesions to be visualized from a number of unique and arbitrary anatomic perspectives. It is now routine to perform "three-dimensional" MR imaging of intracranial mass lesions for presurgical evaluation. Here image intensities are collected into a three-dimensional array in which the voxel elements are nearly isotropic. This makes possible several important "postacquisition" techniques that aid in presurgical planning, in radiation planning, and in response evaluation. These include three-dimensional tumor volume measurement and the postacquisition reformatting to produce images in any arbitrary viewing plane in an interactive manner.

For the imaging evaluation of intracranial neoplastic lesions, conventional MRI without contrast is performed using T_1-weighting and T_2-weighting. A comprehensive quantitative model-based physical understanding of T_1 and T_2 relaxation times (in tumors or normal tissue) is not available. Therefore, tissue relaxation times are explained in terms of general tissue characteristics; empirical measures of relaxation times are relied upon extensively. In this fashion, the prolonged T_1 and T_2 relaxation times seen in tumors are usually attributed to the presence of a relatively large proportion of extracellular water (edema) resulting from the presence of fields of micronecrosis and the highly disordered cell growth. The tumoral mass and (unfortunately) the surrounding edema are characterized by increased T_2 and tend to appear hyperintense

on T_2-weighted images. In many cases, the more "cellular" tumor volumes appear with a slightly lower signal than do purely edematous volumes. When cystic/necrotic volumes are also present, these produce even higher T_2-weighted signal than does cellular tumor or edema. T_1-weighted imaging results in approximately the opposite appearance. Tumor, edema, and cystic/necrotic volumes appear with low intensity compared to normal gray or white tissue. Cystic necrotic tissue displays the lowest signal. Edema is the next most intense, with cellular tumor being the most intense. Figure 2 provides a few typical MR images of primary brain tumors that illustrate these general features.

At the present, there are four commercially available MRI contrast agents. Three of these (Magnevist, Prohance, and Omniscan) are chelates of paramagnetic gadolinium ion. These are used extensively with MRI as "blood-brain barrier" agents; they are functionally analogous to the contrast agents used in CT. In the instance of intracranial neoplasia, they penetrate the damaged blood-brain barrier when present and "label" the extracellular space. The water within the space that is labeled with these three contrast agents acquires a reduced T_1 and thereby produces stronger signal on T_1-weighted imaging. Blood-brain barrier contrast is always indicated for evaluation of intracranial mass lesions. It is helpful for identifying the location and macroscopic heterogeneity of the mass lesion. However, it is not conclusive as to disease grade or histologic classification because of the variable characteristics of the blood-brain barrier and the blood-tumor barrier. The majority of high-grade astrocytic lesions (e.g., anaplastic astrocytoma, glioblastoma multiforme, and anaplastic oligodendroglioma) and metastatic lesions do contrast enhance, but they do so in a topographically heterogeneous fashion. Portions of GBMs often do not enhance. In rare circumstances, a high-grade astrocytic lesion does not enhance at all. Lower grade lesions (astrocytoma and oligodendroglioma) tend not to enhance but contradictions are not unknown. Figure 2 illustrates these "typical" enhancement patterns. On the other hand, still lower grade lesions (pilocytic astrocytoma and menigioma) do tend to enhance. The fourth commercially available contrast agent (Feridex) is a superparamagnetic iron particle that reduces T_2. It is presently used in liver imaging and has no known utility in the context of intracranial neoplasia.

G. Management: Extending Survival

Prognosis and successful management strategies for the brain tumor patient are, in large part, related to the histopathological diagnosis. Given that present neuroimaging technology is unable to provide a definitive

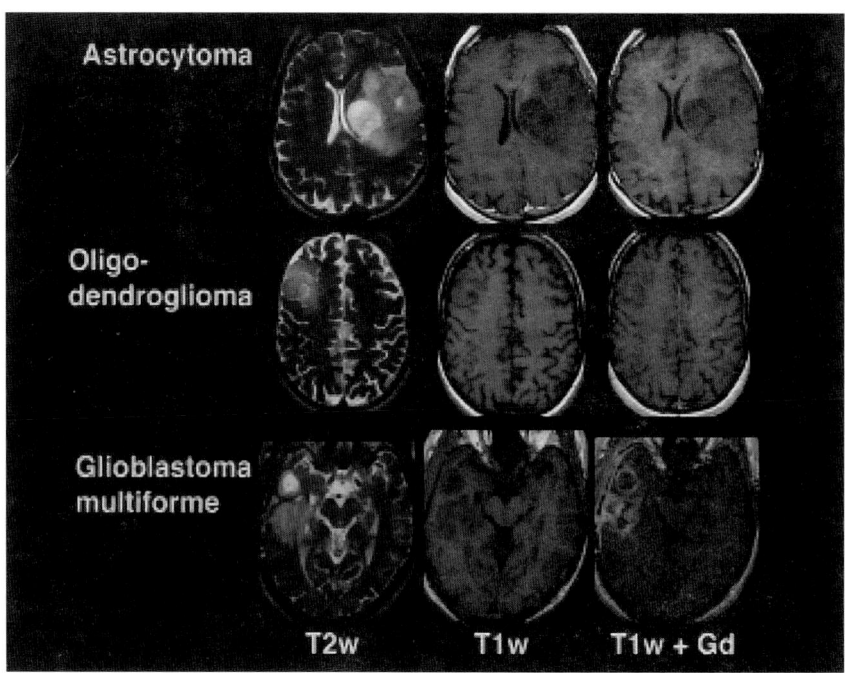

Figure 2 Representative magnetic resonance images of primary brain tumors (gliomas) prior to treatment except steroids. The 1.5-T images were obtained from 3-mm-thick sections using T_2 weighting, T_1 weighting, and T_1 weighting with contrast (left to right). Images from different cases are provided. Top row: left frontal lobe astrocytoma. Middle row: right frontal lobe oligodendroglioma. Bottom row: right temporal lobe glioblastoma multiforme.

histopathological diagnosis, there is usually a critical need for tissue biopsy through a surgical procedure. Most clinicians will not proceed with treatment unless there is a histopathological diagnosis. In cases where the tumor is located in a brain region that is not readily sampled without endangering function, the clinician must proceed only on the basis of the presumptive diagnosis provided by neuroimaging. Such cases are becoming ever more rare due to technological advances in intraoperative neuronavigation and high-resolution three-dimensional neuroimaging.

As is the case with the majority of other forms of human cancer, there is no enduring cure for brain tumor. The managing physician's goal is to control rather than to ablate expansion of the burden of neoplastic cells. Structural and functional neuroimaging has a major role to play here. Because of the relatively inaccessible location of the lesion within the cranial vault and the associated limitations on palpation and biopsy, the clinician looks to neuroimaging to define the lesion in terms of both its macroscopic size and the fraction of active cells that actually reside within it, to provide information on the physiological status of the active tumor cells, and to define the extent to which normal surrounding brain function is affected by the tumor presence. Moreover, this must be done longitudinally as treatments are

applied or as the lesion is allowed to evolve naturally. The neuroimaging specialist (and the patient) are fortunate in the case of aggressive primary glioma in that these lesions do not metastasize systematically. Instead they tend to spread by local infiltration, usually through white matter tracts but sometimes through cerebrospinal fluid. This limits the field of view of neuroimaging vigilance. Unfortunately, in addition to making surgical extirpation difficult, the cellular invasion along white matter tracts presents sensitivity and background problems for neuroimaging. In the extreme, total neuroimaging vigilance would require the ability to detect one transformed cell on the lesion margin.

Conventional therapy for primary glioma involves the application of combinations of surgery, radiation therapy, and chemotherapy. In the majority cases, cytoreductive surgery is the first avenue pursued. The next is radiation therapy, which is followed by chemotherapy. In many cases, neuroimaging plays a key role in defining when therapies subsequent to the initial therapy is given. This is particularly true in the case of patients who have surgical resections that provide diagnoses of low-grade astrocytoma. Often these patients can experience prolonged periods of stability in good health and with good neurological function. Surveillance neuroimaging is used in these instances to assess for early signs of continued growth or progression to higher grades. Cases that show progression on neuroimaging often have a second surgery that leads to a more ominous diagnosis and that prompts radiation therapy. Higher grade lesions usually go directly to radiation therapy following the initial resection. High-grade patients may show prolonged periods of stability on neuroimaging following radiation therapy. Others may show rapid deterioration. Hence neuroimaging plays a large role in defining when and whether chemotherapy should be considered.

Metastatic deposits in the brain are usually conclusively identified through neuroimaging, although one should be cautious when a single brain lesion is found in a patient with systemic cancer since there is a 10% chance that this is something other than a metastatic lesion. These are treated most frequently with combinations of surgery and radiation therapy. Surgical therapy is used if there are a relatively small number of lesions in accessible locations. Newer methods in which high-dose ablative radiation therapy is stereotactically guided (with the aid of three-dimensional neuroimaging) to each isolated metastatic deposit are employed when the lesions are many or are located in surgically inaccessible locations. In recent years, great strides in management of brain metastatic disease have been made as a result of the technological advances in image-guided ablative radiation therapy. In many cases, this technology has extended quality survival in these patients by controlling the growth of brain metastatic deposits and limiting the associated functional loss.

1. Image-Guided Neurosurgery

The goal of brain tumor neurosurgery is to reduce neoplastic cell burden to the greatest extent possible without causing functional deficits. In the case of high-grade primary tumors, there is likely infiltration of isolated glioma cells beyond the macroscopically identified margin. For this reason, it is not thought possible to "completely resect the lesion," because dispersed infiltrating cells are likely present beyond this margin. These infiltrating cells can be controlled through subsequent radiation therapy. Therefore there is little benefit to resecting functional, but possibly infiltrated, tissue. Similarly, low-grade lesions may remain stable for prolonged periods after surgery even if remnants of tumor remain. Loss of function in this setting would also be tragic. Modern-day tumor neurosurgery relies on neuroimaging for preoperative planning. Neuroimaging is used to preoperatively define the tumor location with respect to surrounding anatomy and to define which of surrounding tissues play functional roles. Tumor localization using a variety of contrast mechanisms available in conventional MRI or CT is now well established and

is used as part of sophisticated neuronavigation technologies. In addition to these, less conventional types of imaging are being investigated as means of preoperatively distinguishing tumor from normal functioning tissue (see below). Preoperative functional imaging approaches—mostly fMRI and PET—are being used with increasing frequency to preoperatively define which of the peritumoral brain areas play key functional roles; this represents an area of active current research.

a. Preoperative Identification of Motor and Speech Areas Using Functional Neuroimaging Since the discovery of fMRI in the early 1990s, there has been considerable enthusiasm for using it as a neurosurgical planning tool. Yet there has been relatively little effort to look into the technique's validity in the target population of brain tumor patients. The vast majority of fMRI studies have been performed with normal subjects, with the goals being to study macroscopic neurofunctional organization. A review of the neurosurgical literature reveals a relatively small number of contributions that empirically support the validity of fMRI as a neurosurgical planning tool. For instance, Jack et al. (1994) evaluated two cases, as did Yetkin et al. (1995). Sasahira et al. (1995) examined only one tumor patient. Virtually all of the reported studies involved activation of motor cortex during hand movement. Most focused on the ability of fMRI to detect the central sulcus, with mixed results. For instance, of the seven patients reported by Atlas et al. (1996), only four showed activation (two were lost due to head motion); the remainder showed a reduced magnitude of activation compared to the unaffected hemisphere. Few studies have offered data on the typicality of activation in patients with tumors. Nakayama (1997) examined six brain tumor patients, finding two with no activation and four with displaced activation due to mass effect. Righini et al. (1996) studied 17 patients with motor cortex lesions, using the contralateral activation as a metric of normality: five were lost to head motion. The remainder had normally extended activation, although the relation to tumor and edema was not described. In Roux et al.'s (1997) study of 17 patients, 11 showed "displaced" activation throughout the tumoral region, and four with displaced sulcal boundaries showed activation. Kahn et al. (1996) examined seven patients with fMRI for the purpose of identifying motor cortex as a means to guide thermotherapy procedures; they found three of seven cases with normal activation, four with a scattered pattern throughout the tumor, and one with no activation. Mueller et al. (1996) discuss the preoperative fMRI evaluation of 12 patients with motor cortex involvement and four with language mapping, with intraoperative recording being performed in only three of these subjects. They found evidence for activation in areas of edematous, not tumoral, tissue, although this was not quantified. By far, the most extensive study was performed by Pujol *et al.* (1998), who studied 50 cases with motor cortex associated lesions preoperatively with an image acquisition technique that has quite limited volume coverage. Eighty-two percent showed activation in probable central sulcus, though anatomical confirmation was obtained in 22 cases. Only one study focused on how motor impairment affected activation on fMRI. Yoshiura *et al.* (1997) studied the effect of motor deficits on activation in seven patients, finding more activation in motor cortex contralateral to the lesion in paretic patients. Several other studies emphasized the development of new technologies rather than fMRI results per se. For instance, Ohue *et al.* (1998) used a study of seven patients to justify a conclusion that fMRI can augment 3D-neuronavigation technology and that functional PET imaging is needed for the identification of the central sulcus in cases where it was anatomically displaced. Schulder *et al.* (1998) examined 12 patients with a motor paradigm to describe and test a method of registering preoperative fMRI with a frameless stereotactic surgical navigation device. A similar goal was pursued by Stapleton *et al.* (1997) in a study of 16 children with low-grade glioma.

A second issue that arises from this review is that the majority of the reported studies sought only to evaluate motor function. Language center evaluation has been largely unexplored in the context of fMRI neurosurgical planning, with the exception of a very small number of case studies (Mueller *et al.*, 1996). Unlike motor cortex, where there is very little intersubject variability in functional cortical representations, language areas are highly variable across individuals (Ojemann *et al.*, 1989; Steinmetz *et al.*, 1991). Different language tasks can produce markedly different activation patterns, making it difficult to determine which areas are critical and will lead to aphasia if resected. To reliably identify language-processing areas, a combined paradigm approach can be used. Here language tasks with very different sensory input characteristics and motor output demands are performed; the functional imaging results of each task are then compared to produce a composite map, showing areas that are active during all of the tasks. Figure 3 provides an example of a language fMRI study done as part of the preoperative planning in a brain tumor case. The study used object naming, covert word generation, and sentence comprehension tasks. Object naming is one of the most widely used tasks in electrocorticographic determination of eloquent cortex. Subjects are presented with line drawings and must name them. This task typically produces brain activation in Broca's area, Wernicke's area, SMA, the bilateral

fusiform gyrus, and primary visual cortex. Some patients also show activation in the frontal eyefields and dorsal prefrontal cortex. The covert word generation task has been widely used in functional imaging studies (Frith *et al.*, 1991; McCarthy *et al.*, 1993; Rueckert *et al.*, 1994; Cuenod *et al.*, 1995). In the sentence comprehension task, subjects hear pairs of sentences that are very similar, differing in a small way that may or may not change the meaning of the sentence. Subjects then have to determine whether the two sentences have the same or different meanings. Statistical parametric maps for each task are shown in Fig. 3. All pixels showing activation above a specified threshold for each individual task are shown in the figure. All pixels that are active for all three tasks are then displayed in the composite image (see Fig. 3, right panel). This provides a robust method for identifying critical speech areas.

Positron emission tomography offers an alternative to fMRI for neurosurgical planning. However, even this technology has not been exploited extensively in neurosurgical planning. Recent efforts (Herholz *et al.*, 1997; Thiel *et al.*, 1998), using verb generation and naming tasks to identify the location of language cortex through cerebral blood flow (CBF) increases, involved evaluation of cases having temporal lobe lesions with $H_2^{15}O$ PET. While sensitivity and specificity results were promising, the authors concluded that the $H_2^{15}O$ PET preoperative assessment could not supplant intraoperative recording. These investigators also noted differences between verb generation and naming tasks in the CBF response that have methodological implications. Bookheimer *et al.* (1997) used multiple language tasks to identify language cortex, finding close associations to direct cortical stimulation. Nariai *et al.* (1997) also explored the issue of PET neurosurgical planning. However, the majority of their work involved [^{11}C]methionine PET to identify neoplastic tissue (see below). Preoperative functional assessment using $H_2^{15}O$ PET was preformed only when the mass lesion distorted anatomic features so substantially that the central sulcus could not be identified on three-dimensional structural MRI. In an $H_2^{15}O$ PET study of a single case, Ojemann (1998) demonstrated an exaggerated vascular response in the tumor-infiltrated supplementary motor cortex in response to a motor task. This follows from a series of contributions that have demonstrated, through intraoperative recording techniques, that tissue having abnormal radiological and visual appearance at surgery can retain residual function (Haglund *et al.*, 1996; Ojemann *et al.*, 1996; Skirboll *et al.*, 1996). The fMRI technique is putatively based on sensing the same blood flow increases that are sensed with $H_2^{15}O$ PET. The PET data suggest that, in the absence of further study, caution should be exercised in interpretation of fMRI

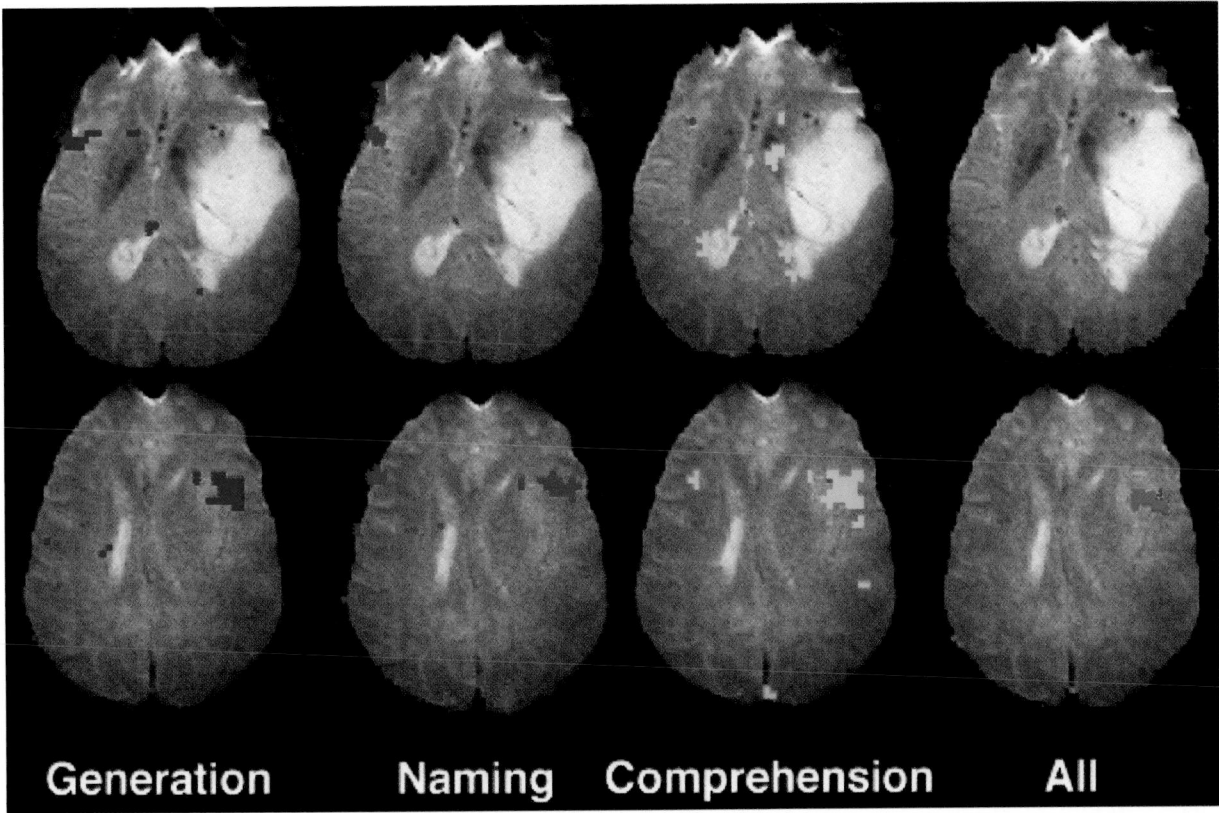

Figure 3 fMRI speech localization in a brain tumor case. The grayscale shows T_2-weighted MRI. The colors demonstrate regions that show physiological activation while the patient performed certain speech tasks (see text). Two slices are shown. (Images provided by Dr. Susan Bookheimer.)

findings in tumor-infiltrated regions. Brain regions that are near or partially infiltrated by glioma may display increased or decreased CBF activation on fMRI compared to normal due to alterations of the microvasculature, to hypothesized disruption of flow—metabolism coupling due to glial mediating factors (Ojemann *et al.*, 1998), or to the possible presence of chemotoxic agents produced by the tumor. Similarly, the presence of tumor or edema in white matter tracts associated with a cortical area may interfere with activity, leading to the absence of an fMRI signal. Therefore, there are conceivably several reasons to exercise caution when interpreting fMRI results from glioma patients.

2. Image-Guided Radiation Therapy

Radiation therapy is, at one time or another, considered for all brain tumor cases. Radiation therapy for malignant glioma, while virtually never achieving long-term tumor growth control, can nevertheless significantly improve short-term endpoints such as median survival. Fractionated external-beam radiation therapy, administered postoperatively, has therefore become the standard of management for patients with malignant glioma.

For instance, for anaplastic astrocytoma, modern irradiation techniques give 5-year survival rates of 15–30%; such figures are rarely achieved with surgery alone. For low-grade CNS lesions, the role of radiation therapy is less well-defined. In some cases it is used and in other cases it is postponed until there is distinct evidence of progression of the disease to a more malignant form.

The proper dose of radiation for control of malignant astrocytomas remains in dispute. The recent trend has been to deliver tumor doses in excess of 50 gray (Gy). Many radiation oncologists now recommend 60 Gy. This dose is usually fractionated over a period of about 7 weeks. The tumor dose is limited by the tolerance of surrounding normal brain parenchyma rather than by any strict relationship between the number of viable tumor cells present and the theoretical dose necessary for tumor control. For those malignant glioma patients with poor prognosis, a shorter course of radiation is sometimes recommended. While this is somewhat less burdensome for the patient, the efficacy of hypofractionation has not been substantiated.

Neuroimaging is used to define anatomic location and extent of the lesion so that the size of an external-beam portal for irradiation can be defined. In general, for malignant glioma the irradiated field must be a few centimeters larger than the macroscopically identifiable tumor defined on neuroimaging. The infiltrative nature of malignant glioma suggests the probability of significant glioma cell burden beyond the neuroimaging-defined lesion border. It is particularly important that the irradiated field extend beyond the margins of the contrast-enhancing tumor as visualized on MRI or on CT because of the probability that adjacent nonenhancing tissues can contain substantial tumor burden. Figure 4 illustrates the use of sophisticated planning systems that permit considerable control in the shaping of the field that is irradiated to conform to anatomic features for individual patients. Whole-brain irradiation may be helpful in certain high-risk patients, namely those having one of the following features: lesions near a ventricle, poorly defined thalamic lesions, "butterfly gliomas," lesions that display irregular enhancement on CT scan in any site, and the rare multicentric astrocytoma. However, in most cases, some form of regional radiation is used. Prior to the advent of modern CT and MRI, employment of inaccurate diagnostic imaging (isotope brain scans, pneumoencephalograms, and angiograms) resulted in ineffective regional radiation (Salazar and Rubin, 1976) and whole-brain radiation was often advocated. Hochberg and Pruitt (1980) also questioned the use of whole-brain radiation. By comparing contrast-enhancing CT scans obtained within 2 months of death with the microscopically defined tumor margin determined at necropsy, they demonstrated that while tumor extended beyond the contrast-enhancing margin, it did not do so to a great extent. In the majority of cases, tumor cells were not found more than 2 cm beyond the contrast-enhancing margin. Accordingly, these authors concluded that, with the aid of modern neuroimaging, there is little justification for routine whole-brain irradiation in the management of malignant astrocytoma. Moreover, it is important to note that no controlled study has incriminated tumor recurrence in brain areas remote from the primary site as being as clinically significant as in-field recurrence. Therefore, there is no *a priori* reason to expect that whole-brain irradiation would yield survival rates superior to limited field radiation.

In addition to aiding in therapeutic targeting, neuroimaging can be used to assess tumor response through longitudinal volume measurement. This can be done by relatively unsophisticated but rapid procedures (measuring a few key linear dimensions with a ruler) or by more sophisticated volume tracing software using three-dimensional data with supplementation by automated image intensity segmentation. Studies have shown that malignant gliomas are only moderately sensitive to radi-

ation therapy. Moreover, conventional thinking holds that fractionated radiation impairs the tumor cell division machinery rather than actually killing cells. Therefore, a protracted response in the tumor volume is to be expected with fractionated radiation therapy. In one study, postradiation tumor volume reduction (defined by comparing pre- with postirradiation tumor volumes) was seen in only 22 of 63 cases and only three cases showed complete resorption (Gaspar *et al.*, 1993). Furthermore, in contradistinction to the conventional thinking, the incidence of protracted radiation response was uncommon. Barker and colleagues reported at least 25% reduction in contrast-enhancing tumor in only 43% of patients (Barker *et al.*, 1996). Tumor volume was stable or increased in the remainder. External-beam radiation is often initiated relatively soon following the tumor resection. Therefore one has to consider the possibility that evolving neuroimaging feature changes associated with the surgery may render the preradiation tumor volume measure inaccurate. Barker *et al.* (1996) cautioned that postoperative, preradiation imaging should be obtained within 4 days of surgery. Postoperative scans performed after that time could be confounded by increased contrast uptake, hyperintense T_2-weighted MRI signal, or hypodense CT related to a postsurgical inflammatory response that would tend to resolve during radiation therapy. These evolving image characteristics due to resolving postsurgical inflammation could easily masquerade as a radiation-induced reduction in tumor volume. Moreover, the managing clinician often uses steroid for intracranial pressure control prior to surgery and tends to taper off this mediation if indicated during the period of radiation therapy. These steroid changes can conceivably be reflected in altered image characteristics—particularly T_2-weighted MRI signal and CT hypodensity. Therefore it is important that possible steroid changes be considered in the neuroimaging assessment of radiation-induced tumor volume.

3. Chemotherapy

Chemotherapy for brain tumor can be considered in a number contexts. Candidate chemotherapeutic agents that can be considered for GBM include nitrosourea-based compounds, cisplatin, procarbazine, vincristine, CPT-11, carboplatin, temozolmide, and etoposide. However, in general, the response rate for malignant glioma is low. A meta-analysis showed benefit in up to 10% of patients with GBM (Fine *et al.*, 1993). Other studies (Walker *et al.*, 1978, 1979) showed a benefit in the last quartile of patients who received chemotherapy and radiation versus those who received only radiation. Thus far, no correlation between chemosensitivity and histopathologic features for astrocytoma has emerged (DeAngelis, 1998). However, anaplastic oligodendro-

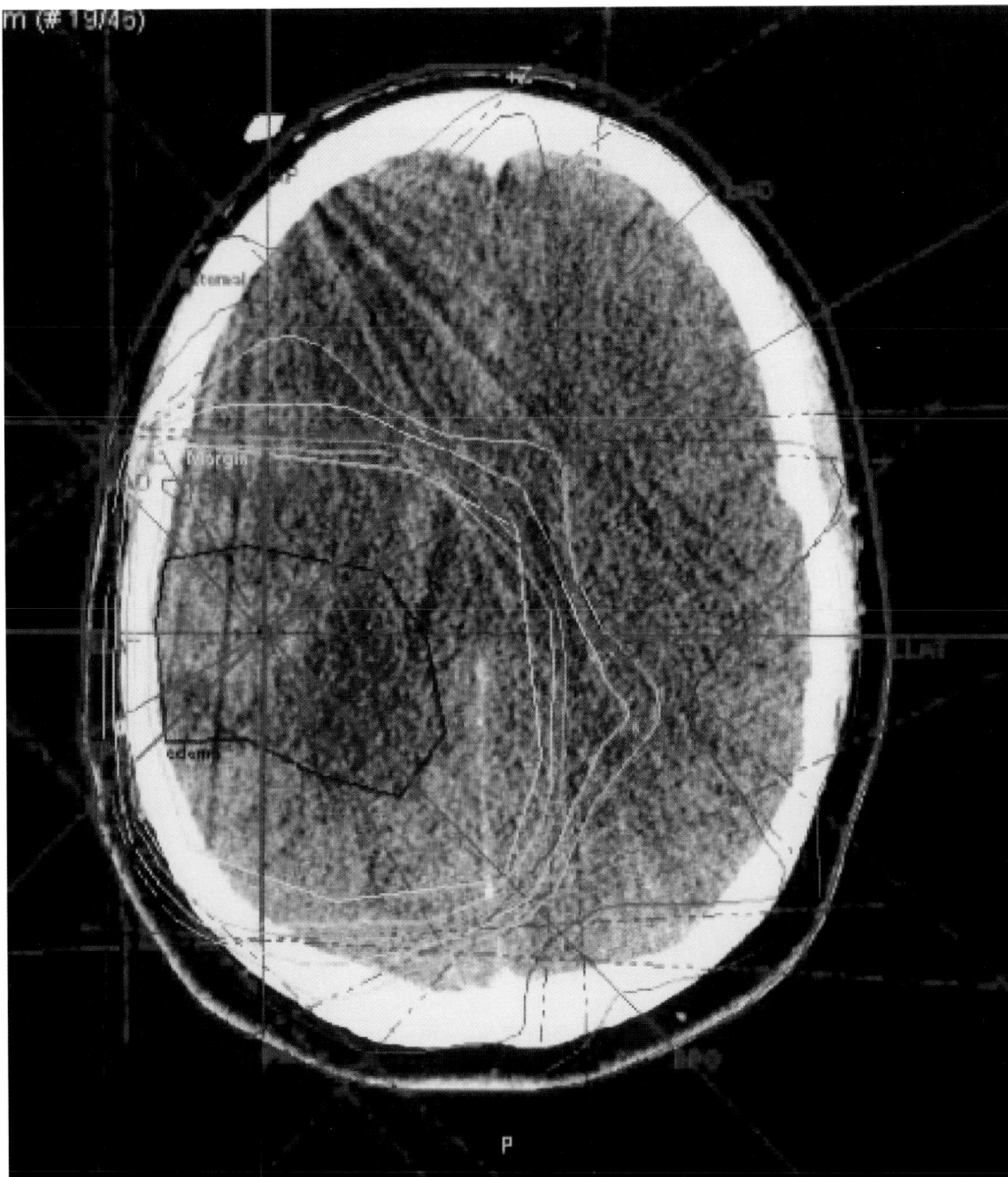

Figure 4 Representation of radiation therapy planning. The tumor and edema are visualized on a CT image (blue contour). These data are used to adjust the radiation field to limit the dose to a shaped area that is limited to a few centimeters beyond the tumor-edema margin. The additional colored lines display the isodose contours. (Data provided by Dr. Timothy Solberg)

glioma and mixed oligodendroglioma/astrocytoma frequently show chemotherapeutic responsiveness (Cairncross and Macdonald, 1988; Macdonald *et al.*, 1990; Cairncross *et al.*, 1994).

Chemotherapies given immediately following radiation can be beneficial in limited instances, but these benefits are balanced by a high rate of side effects. Patients who are younger (<45 years of age), who have better neurological performance status, and who have smaller tumor volumes tend to show benefit. However, one has to consider that even untreated GBM may remain stable for a period of 3-12 months following radiation.

Therefore it has been difficult to separate "chemotherapeutic successes" (which are invariably declared if the tumor volume remains stable during treatment) in patients who receive chemotherapy immediately following radiation from those whose tumor would have remained stable without treatment. For this reason, chemotherapy is often delayed until there is some obvious clinical or neuroimaging sign of disease recurrence or progression. Neuroimaging has a role to play here. There is a need for accurate definition of when to treat and whether treatment is being effective. Neuroimaging definition of when to treat is problematic. Here the long-standing problem has been distinguishing the radiologically similar appearance of radiation necrosis from recurrent tumor. Each of these produce similar characteristics in pre- and postcontrast MRI or CT images. That contrast enhancement may appear at any time following radiation therapy and be confused with tumor recurrence and progression is well known. This is illustrated in Fig. 5 by serial MRI scans obtained from a glioma patient who was followed at UCLA. The patient underwent fractionated radiation therapy for glioma prior to September 1992. Between September and November 1992, MRI scanning demonstrated the development of a ring-enhancing lesion surrounded by hypointense T_1-weighted signal intensity. This is exactly the pattern that recurrent glioma would exhibit. Between November 1992 and May 1993, the patient underwent surgery for the presumed recurrence. The May 1993 scans demonstrate that a large portion of the previously contrast-enhancing region was resected. Histopathological reading of the resected tissue revealed nothing but radiation necrosis. Subsequently, serial MRI scanning demonstrated a new ring-enhancing lesion in March 1994, which then spontaneously regressed without treatment by June 1994. A case such as this underscores the danger of using contrast-enhancement patterns to guide clinical management de-

cisions. In May 1992 or in March 1994, this patient could have been subjected to inappropriate chemotherapeutic management or could have been inappropriately enrolled in a poorly designed clinical trial for recurrent glioma. Had the patient been enrolled in a clinical trial between March and June 1994, the patient would have been inaccurately identified as a positive responder.

4. Investigational Therapies

Beyond radiation and surgical therapies, the best hope for many patients lies in investigational therapies. There are currently many novel therapeutic concepts at various stages of development. These include biological agents designed to interfere with cell growth, sensitizers that make tumor more sensitive to radiation, immunotherapies, and genetic therapies to name a few. These are too many and too complicated to summarize here. However, there is a connection with structural and functional neuroimaging. Clinical trials that support the development of new approaches must have methodology for assessing success and failure. Conventionally, this has been done using survival as an endpoint. This can be problematic because patients are understandably eager to try as many trials as they qualify for. In this way, the effects of independent trials become entangled in each other. The results of any one trial can become dependent on successes or failures of other trials that the patient subsequently participates in. Even if the patient does not seek other investigational trials, conventional therapies, particularly surgery, are being reapplied to an increasing extent. It is not unusual for some patients with high-grade glioma to have two or three resections during the evolution of their disease. In rare cases, stereotactic ablative radiosurgery can also be given in cases where it clinically seems appropriate. These confounding features of clinical management of brain tumor call for the definition of surrogate endpoints that enable success and failure to be defined when the patients leave trials. Simi-

Treatment:

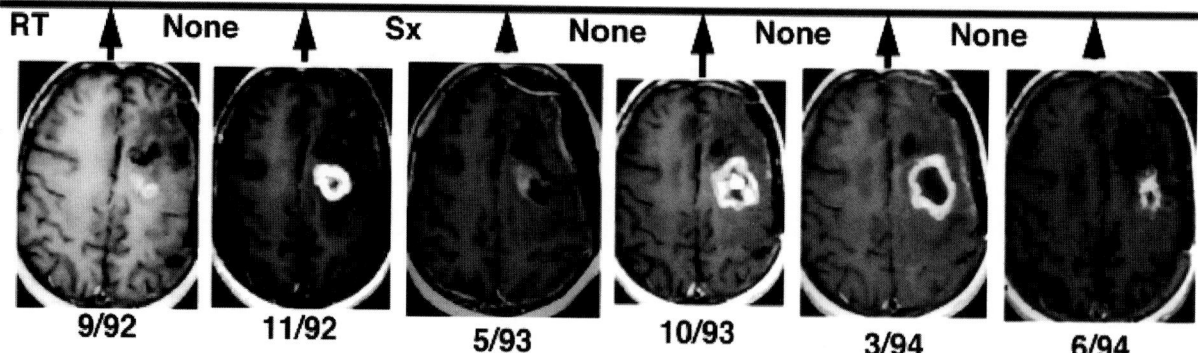

Figure 5 Longitudinal MR images illustrating the complexity of postradiation imaging (see text for further details).

larly, the case illustrated in Fig. 5 demonstrates that entry criteria need to be better defined. Patients having radiation necrosis could inappropriately enter a clinical trial and inappropriately be assigned as "responders" because their tumors failed to grow during the trial. These arguments call for the increased use of neuroimaging in clinical trials for brain tumor. Moreover, it is clear that further refinement of existing structural imaging techniques and the incorporation of functional imaging techniques that measure some aspect of physiological function will be needed to improve the accuracy of neuroimaging and to fully validate neuroimaging tools for establishing entry and endpoint criteria.

III. Recent Neuroimaging Progress Related to Intracranial Neoplasms

A. Radionuclide Scanning

Nuclear scanning encompasses a variety of functional imaging technologies. In general, nuclear scanning uses radioactive tracers to probe some function of intracranial tumor tissue or the normal tissues that surround the tumor. A key concept here is that imaging of functional properties has clinical value. This section will review recent studies that used nuclear scanning to address various clinical challenges that are described in the previous section.

1. Positron Emission Tomography (PET)

a. FDG-PET   PET is a now mature procedure that is used relatively routinely for brain tumor evaluation in many centers. The majority of the clinically relevant studies use fluorodeoxyglucose (FDG) as a tracer of glucose metabolism. FDG-PET was one of the first functional imaging technologies to be used for disease evaluation. The technique has basis in Di Chiro's realization that the imaging of glucose metabolism would have utility in the clinical assessment of brain tumors. Here the reasoning rested on Warburg's principle, that tumors use the relatively inefficient anaerobic portion of glucose metabolic machinery to maintain energy stores and, thereby, are expected to utilize glucose more quickly than they otherwise might. This reasoning suggests that a tumor should appear as a hypermetabolic focus on FDG-PET. In many cases—particularly in untreated high-grade lesions and in recurrent or progressing high-grade lesions—hypermetabolism is indeed seen in the tumor bed. During the 1980s and early 1990s, Di Chiro and colleagues (Di Chiro *et al.*, 1985; Di Chiro, 1987) demonstrated that high-grade and low-grade astrocytic tumors could be distinguished because of their differing glucose metabolism. They also

demonstrated that recurrent lesions appeared as hypermetabolic areas on FDG-PET images.

Figure 6 illustrates a case in which "hot spots" on FDG-PET images can be somewhat subtle relative to the normal cortex. This is a case of a progressing glioma. The patient had a low-grade lesion that had progressed after a long course (69 months) following the initial diagnosis of low-grade astrocytoma. Serial MRI scans obtained during that course demonstrated the transcallosal infiltration of the right frontal lobe from the initial location in the left frontal lobe. The patient had a stereotactic biopsy of the left frontal lesion 5 months prior to the FDG-PET imaging that provided a diagnosis of anaplastic astrocytoma and was not treated in the intervening time. The image demonstrates subtle hypermetabolism within the MRI-defined lesion that is particularly apparent on the left. However, even here the tumor's hypermetabolism is actually lower than the metabolism in the nearby cortex. It is now clear that in many instances tumors demonstrate metabolic rates that are intermediate between those of normal white matter and normal gray matter. Accordingly, sometimes FDG fails to provide adequate contrast to distinguish tumor from surrounding normal tissue (particularly cortex). To some extent, this has been addressed by the development of procedures wherein MRI scans are topographically registered to FDG-PET for improved anatomic reference. There has also been interest in different tracers that may provide more contrast for tumor visualization than FDG does.

FDG-PET offers a number of opportunities for clinical assessments. Deshmukh *et al.* (1996) reviewed the range of circumstances in which FDG-PET was employed at their center over a series of 75 cases. They found that in the majority (87%) of cases, the PET scan was requested to distinguish radiation necrosis from recurrent tumor. The second most frequent use (11% of the cases) was the assessment of malignancy as a substitute for biopsy. Others uses, which included pretherapeutic baseline studies for monitoring the effect of a therapy, mapping of hypermetabolic regions before surgery or biopsy, mapping of hypermetabolic regions before radiotherapy, and postsurgical evaluation for residual tumor, were employed in less than 2% of the cases. This study illustrates the significance of the radiation necrosis versus recurrent tumor dilemma; FDG-PET offers one of the few viable options in this area. There have been a number of recent investigations that support the clinical use of FDG-PET for intracranial tumor evaluation for a variety of purposes beyond radiation necrosis/recurrence evaluation. Barker *et al.* (1998) demonstrated that FDG-PET provides statistically significant prognosis of survival time in patients having

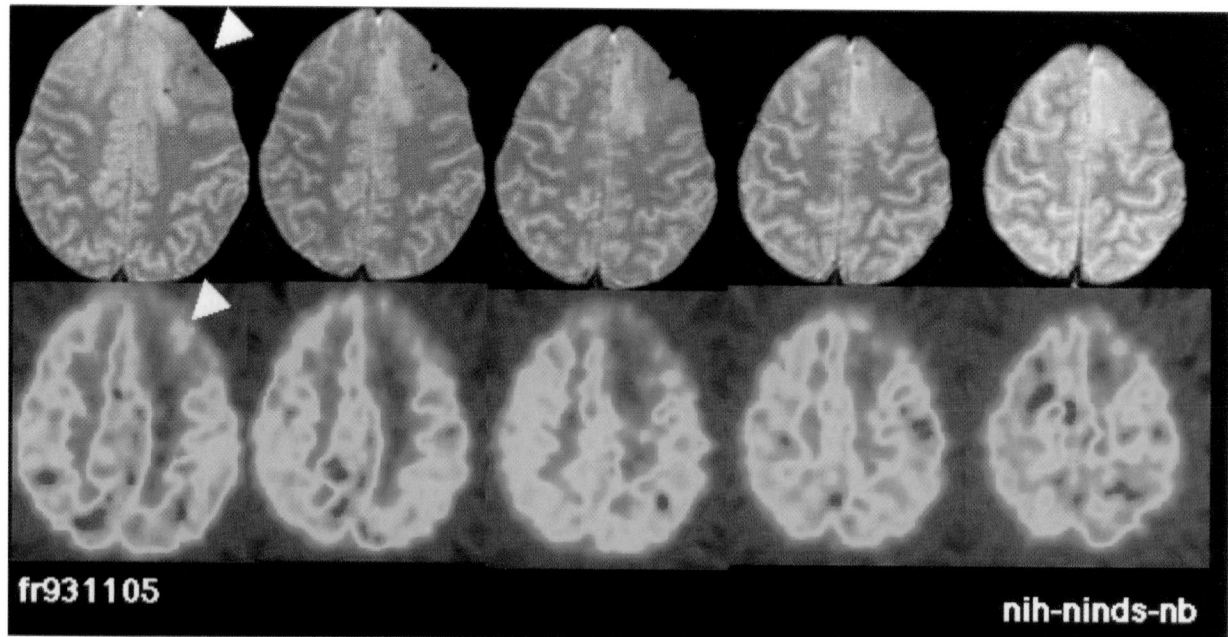

Figure 6 FDG-PET scans with MRI correlation of a case of progressing glioma. The FDG-PET data have been reformatted to register with the MRI slices. The left frontal lesion displays an FDG hot spot (arrows). However, the extent of infiltration of the right frontal area is not as obvious on FDG-PET.

high-grade recurrent tumors after radiation therapy. De Witte *et al.* (1996) supported the role of FDG-PET in predicting progression from low- to high-grade disease. A number of studies (Levivier *et al.*, 1995; Goldman *et al.*, 1996, 1997; Gross, 1998) evaluated spatial heterogeneity and, thereby, advocated the use of FDG-PET for guiding stereotactic biopsy to the most appropriate sampling location or for radiation treatment planning.

Recent research studies have also addressed a number of issues related to quantitation and problems of interpretation. Since the introduction of the FDG-PET procedure for brain tumor evaluation, there has been an unresolved issue regarding what kinetic parameter values are appropriate for accurate quantification of tumor metabolism. Spence *et al.* (1998) addressed this issue by using [^{11}C]glucose and FDG tracers sequentially in a study of 40 patients suffering from malignant glioma. This protocol permitted them to model the metabolism of the true physiologic tracer (glucose) and test assumptions about the nonmetabolizable tracer (FDG) kinetics. The study showed that the glioma FDG "lumped constant" exceeds that of contralateral brain and therefore that quantitative overestimation of the tumor glucose metabolism occurs when the normal brain and tumor lumped constant are assumed to be identical. This study emphasizes that caution must be exercised in interpreting hot spots on FDG-PET; hot spots visualized in FDG-PET studies of gliomas could represent regions where the FDG lumped constant rather than the metabolic rate is elevated. Two groups

(Fulham *et al.*, 1995; Roelcke *et al.*, 1998) have questioned how steroid medication affects FDG-PET readings and showed that steroid does influence FDG uptake, indicating that care should be taken in interpreting serial FDG-PET scans done for treatment evaluation. Marriott *et al.* (1998) reported that there are instances when the hypermetabolic "rim" of FDG accumulation that is usually taken as a sign of active tumor is not necessarily tumor; it can be the result of macrophage infiltrates.

Two groups (Wang *et al.*, 1996; Bruehlmeier *et al.*, 1999) began the process of investigating the functional alterations in surrounding brain that are induced by radiation therapy. Radiation therapy is now administered so that the normal brain tissue is affected to a minimal extent (see above) and alterations of structural imaging measures are usually not observed except in the instance of late-stage (months to years) radiation necrosis. The demonstration by these groups that glucose utilization is altered rather acutely by irradiation illustrates the sensitivity of FDG-PET to subtle physiological changes and suggests that FDG-PET may prove useful for quantitatively and objectively following the functional neurological decline that is sometimes apparent in patients after radiation therapy. The extent to which these subtle radiation-induced functional alterations are predictive of radiation necrosis remains to be studied.

b. Other Novel Tracers Progress on new tracers for brain tumor PET imaging has been fast paced. New tracers that are under investigation include amino acid analogs, "hot" metal ions, and analogs of nucleic acid metabolites. In one instance, MRS studies of brain tumors (see below) prompted the development of a novel PET tracer for choline metabolism and uptake (Hara, 1997; Shinoura, 1997). The latter might be best regarded as a tracer of membrane metabolism. A key justification for this work has been to find tracers that distinguish active tumor tissue from normal brain tissues (whether gray matter or white matter) with higher contrast than is provided by FDG-PET. The need to "trace" a variety of unique physiological functions of the tumor also justifies the investigation of new tracers. With regard to tumor versus brain contrast, many of the tracers, particularly the amino acid analogs, have been quite successful. Figure 7 provides an example. These images from a low-grade (Grade II) astrocytoma illustrate a theme consistent with many studies. Amino acid tracers seem to accumulate in low-grade lesions to a sufficiently high level relative to normal brain to provide high contrast definition of tumor. This is particularly significant since such lesions commonly do not contrast enhance on MRI.

The amino acid tracers used in brain tumor studies during the past few years include analogs of tyrosine (Willemsen et al., 1995; de Wolde et al., 1997; Inoue et al., 1999; Wester et al., 1999), phenylalanine (Ogawa et al., 1996a; Imahori et al., 1998), and methionine (Ogawa et al., 1996b; Wurker et al., 1996; Derlon et al., 1997;

Goldman et al., 1997; Voges et al., 1997; Kaschten et al., 1998). There are also reports indicating that at least one patient with residual GBM was examined with a new tracer, $[^{18}F]$1-amino-3-fluorocyclobutane-1-carboxylic acid (Shoup et al., 1999), and that one patient was evaluated with F-Dopa tracer (Heiss et al., 1996). The general rationale underlying the use of amino acid analogs is that they cross the blood-brain barrier and trace amino acid transport and incorporation into newly synthesized polypeptides. These physiological activities are elevated in CNS tumors relative to normal brain because of the requirements of neoplastic growth. Some of these tracers (Imahori et al., 1998; Inoue et al., 1999; Shoup et al., 1999; Wester et al., 1999) are at an early stage of development, whereas others (see below) are coming into relatively routine use for clinical studies.

Many of the recent clinical studies advocate amino acid tracer PET techniques for some clinical purpose related to CNS neoplasia evaluation. Many studies sought to demonstrate a high degree of tumor to brain contrast that could be used for lesion identification, for tumor margin delineation, for monitoring the effects of treatment, and for detecting recurrence. In some cases (Ogawa et al., 1996b; Derlon et al., 1997; Kaschten et al., 1998), side-by-side comparisons between amino acid tracer and FDG tracer contrast were made to define the relative strengths of the two approaches. In a series of 54 patients studied prior to treatment, Kaschten et al.

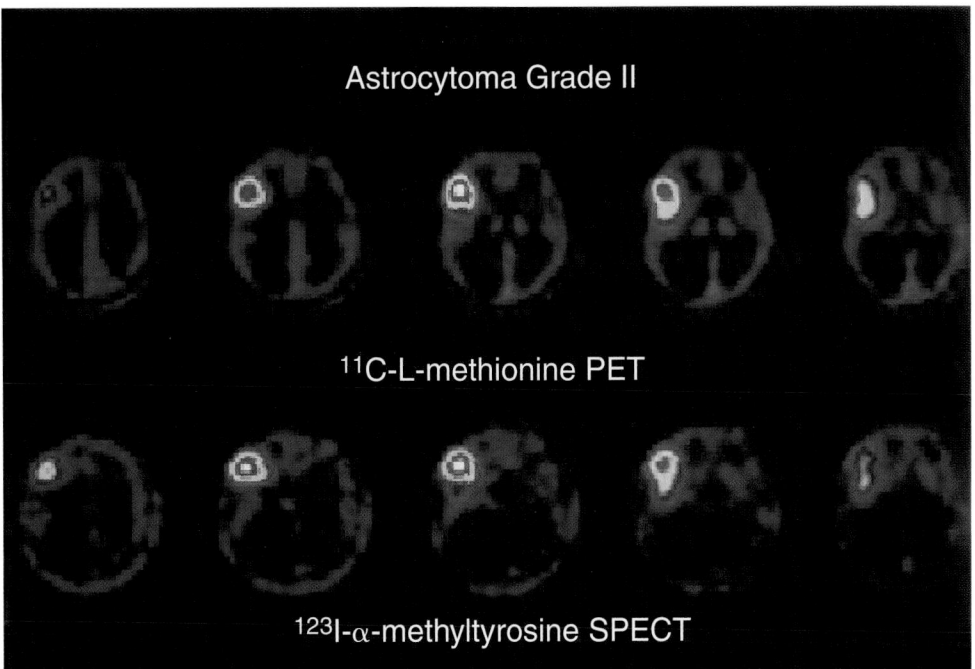

Figure 7 PET and SPECT detection of amino acid tracer labeling of low-grade astrocytoma. [Courtesy of Dr. K. J. Langen (Langen et al., 1997). Reprinted by permission of the Society of Nuclear Medicine from: Langen KJ, et al. 3-[^{123}I]Iodo-α-methyltyrosine and [methyl-^{11}C]-L-methionine uptake in cerebral gliomas. *Journal of Nuclear Medicine*, 1997; 38:517–522.]

(1998) showed that methionine tracer was superior to FDG at delineating low-grade gliomas and was slightly superior to FDG for predicting the histologic grade but that the clinically important differentiation between Grades II and III was not possible. Derlon *et al.* (1997) studied 22 patients with oligodendroglioma and astrocytoma using a methylmethionine tracer. They showed that both types of low-grade tumors exhibited glucose hypometabolism but that they strongly differed in methionine uptake, with oligodendroglioma showing the greater labeling. From a study of 10 patients, Ogawa *et al.* (1996b) suggested that methionine PET labeled tumor tissue more uniformly than FDG, which tended to show a more heterogeneous pattern of hot spots. Wurker *et al.* (1996) followed a group of low-grade glioma patients that had been treated with iodine-125 brachytherapy. They found that after 1 year glucose metabolism was not significantly altered up to a radiation dose of 300 Gy, whereas methionine uptake showed a significant dose-dependent decrease. Higher rates of decline were found in tumors with high basal methionine incorporation activity before therapy. Similarly, Voges *et al.* (1997) suggested that methionine PET may improve tumor delineation and, following brachytherapy, provides more information on the therapeutic effects than does FDG. Goldman *et al.* (1997) concluded that both FDG and methionine help to evaluate the metabolic heterogeneity of human gliomas and that anaplasia is a factor of increased uptake of both tracers but that microscopic necrosis influences their uptake differently. Furthermore, they noted the importance of the nonneoplastic components of necrotic tumors in defining tracer uptake characteristics. De Wolde *et al.* (1997) attempted to correlate the protein synthesis rate measured through tyrosine PET imaging with cell proliferation assays (MIB-1 antigen) of subsequently excised tissues but found little indication of a statistically significant relationship between the two supposedly related measures of neoplastic activity. This provides an indication that neoplastic cell proliferation and protein synthesis may well be unrelated.

Tracers of nucleic acid transport and intracellular incorporation into nucleic acid polymers are also available and have been used in PET studies of brain tumors. These tracers include analogs of thymidine (Vander Borght *et al.*, 1994; Eary *et al.*, 1999) and deoxyuridine (Kameyama *et al.*, 1995). These tracers offer value because of the increased nucleic acid polymer anabolism that characterizes neoplastic states. In a study of 20 patients with untreated and recurrent CNS tumors, Vander Borght *et al.* (1994) compared thymidine tracer to FDG tracer. Normal brain structures showed a low and homogeneous thymidine trace distribution 10 min after injection. Tumors showed a 20% increased uptake rela-

tive to normal cortex across both tumor types (untreated and recurrent). The data indicate the feasibility of brain tumor thymidine tracer imaging and suggest that this tracer could be used for detection of tumor recurrences. Eary *et al.* (1999) used a thymidine PET tracer to image the rate of brain tumor cellular proliferation. Their study included comparative FDG-PET and contrast-enhanced MRI evaluation arms. They found the thymidine tracer produced qualitatively distinct information compared to FDG-PET and MRI in half the cases. They went on to use a kinetic analysis to eliminate the confound introduced by a carbon dioxide label that appears as a result of thymidine catabolism. In a preliminary study, Kameyama *et al.* (1995) showed that deoxyuridine tracer could distinguish high- from low-grade cases.

A number of additional studies have exploited the use of certain metal ion tracers, including cobalt (Jansen *et al.*, 1997), gallium (Black *et al.*, 1997), iron (Roelcke *et al.*, 1996b), and rubidium (Roelcke *et al.*, 1995, 1996b). These studies tended to be developmental or directed at answering relatively specific biological questions as opposed to evaluating utility in large patient groups. Rubidium transport is inhibited by an intact blood-brain barrier. Roelcke *et al.* (1995, 1996a) used it to independently evaluate blood-brain barrier integrity in comparative studies with FDG and methionine PET imaging in patients with glioma and meningioma. The results suggested that methionine uptake in tumors is governed by tracer influx across the blood-brain barrier, whereas FDG uptake is related to tracer metabolism. Accordingly, they advocated FDG as a more appropriate tracer, particularly for the differential diagnosis of contrast-enhancing lesions in operated and irradiated patients. Roelcke *et al.* (1996b) addressed which of blood-brain barrier transport or transferrin-mediated uptake is rate limiting for iron tracer. They concluded that iron accumulation in tumors is governed by tracer uptake at the blood-brain barrier. Black *et al.* (1997) used gallium-EDTA as a tracer to test the ability of the bradykinin analog RMP-7 to open the blood-tumor barrier in nine cases of recurrent malignant gliomas. For each patient, two PET studies (one with and one without RMP-7) were performed. They demonstrated that intracarotid infusion of RMP-7 significantly increased tracer transport into tumor regions and concluded that intracarotid infusion of RMP-7 is a novel technique for selective delivery of antitumor compounds into brain tumors. Jansen *et al.* (1997) used cobalt as a calcium tracer to visualize decaying tumor tissue, based on the fact that Ca influx is essential in both cell death and leukocyte activation. They studied three patients with primary malignant brain tumors. The tracer demonstrated good topographical agreement with the tumor as visualized on CT and MRI. Neither necrotic core nor viable tumor tissue as defined by CT and MRI demonstrated labeling. This demonstrates that the tracer labels the component of the tumor that is progressing toward necrosis and therefore that it may have utility for therapeutic evaluation.

Through multiple tracer studies, PET offers the opportunity to perform a "global" physiology evaluation that includes blood flow, blood volume, oxygen extraction, oxygen metabolism, and glucose metabolism. In a few instances, this has been performed to study basic brain physiology and functional activation in normal subjects and in some disorders such as stroke. However, it is too complex for the routine evaluation of the brain tumor patient. Mineura et al. (1994) offer one of the few examples of such a global physiology evaluation in brain tumor patients. They performed a complete PET study of cerebral circulation and metabolism with long-term follow-up (57 months) in 23 patients. Their principal goal was to evaluate whether the global PET study predicted survival time from the date of the study. As one would expect, they found that histologic grade and performance status were statistically related to survival. They categorized PET parameters as being above or below the median and tested for association between this categorization and survival time. They found a probable relationship between tumor glucose metabolism and survival. Median survival time for patients with tumor metabolic rates greater than the group median of 4.4 mg/100 ml/min was 9 months. This was significantly shorter than the median survival time of 113 months for patients in the lower tumor glucose utilization category. They also found that the "normal" gray matter blood flow, blood volume, and oxygen metabolism were predictors of survival. The patients having higher values of these parameters lived longer. These findings suggest that glucose metabolism is an imaging variable that predicts survival and that additional measures of gray matter glucose metabolism in normal brain may have additional predictive power.

2. Single Photon Emission Computed Tomography

SPECT scanning is also being pursued with great vigor as a means of better evaluating the neuro-oncologic patient. In many instances, this is justified by the lower cost and the more widespread availability of SPECT compared to PET. In some cases, the tracers are analogous to those used in PET amino acid tracer studies (Kuwert et al., 1997, 1998; Langen et al., 1997; Weber et al., 1997; Woesler et al., 1997). Figure 7 illustrates that the SPECT analog of an amino acid tracer delineates a low-grade astrocytoma essentially as well as the similar PET tracer. Beyond the use of SPECT tracers of amino acid metabolism and transport, perfusion tracers (e.g.,

Namba et al., 1996), nucleic acid metabolism tracers (e.g., Tjuvajev et al., 1994), and certain receptor binding tracers (e.g., Maini et al., 1995) have been explored for work with brain tumors. The workhorse tracer of neuro-oncologic SPECT scanning has been thallium-201, a tracer of potassium ion transport. The close parallels between PET and SPECT have led to a number of SPECT studies that parallel the PET studies that are discussed above. Because of this, the SPECT studies will only be briefly reviewed here.

Studies involving thallium-201 tracer have addressed interpretative issues as well as clinical ones. Namba et al. (1996) demonstrated that steroid medication affected the tumor to normal brain tracer uptake ratio and therefore that steroid medication could be a confound in longitudinal studies. Kallen et al. (1996) compared thallium-201 SPECT with conventional CT for grading the malignancy of gliomas. In high-grade gliomas, they observed a significant correlation between CT contrast and thallium-201 uptake. They found 8 of the 31 high-grade gliomas had low thallium-201 uptake that was indistinguishable from low-grade glioma. Two of these were nonenhancing and the other six showed ring enhancement on CT scans. This led to a finding of no significant difference between high- and low-grade gliomas. These findings are consistent with the presence of wide fields of necrosis that would not be labeled with tracer in the high-grade lesions. In a study in which thallium-201 SPECT, methionine PET, and FDG-PET were comparatively evaluated, Sasaki et al. (1998) reported that distinguishing Grades II and III astrocytoma with thallium-201 tracer imaging was problematic. Methionine tracer could detect astrocytoma and differentiate between benign and malignant histologies, but it could also not conclusively predict histological grade. In this series, FDG proved not to be useful either for evaluating the histological grade or for differentiating between benign and malignant lesions. Black et al. (1994) evaluated thallium-201 SPECT and FDG-PET comparatively for the detection of recurrent glial lesions and found the two imaging techniques to have equal efficiency. A growing series of studies have evaluated thallium-201 SPECT as a means of evaluating intracranial lesions in AIDS patients (De La Pena et al., 1998; Kessler et al., 1998; Miller et al., 1998; Antinori et al., 1999). Together, these indicate that thallium-201 SPECT can be quite helpful for distinguishing between intracranial lymphoma and opportunistic infections. Others have investigated the utility of thallium-201 SPECT for therapeutic assessment (Roesdi et al., 1998) and for following longitudinal changes in tumor burden that accompany therapy and progression (Lorberboym et al., 1995; Kline et al., 1996; Schwartz et al., 1998). Zingale et al. (1995) performed a dual-isotope study using

SPECT flow tracer ([^{99m}Tc]HMPAO) and thallium-201 SPECT. They illustrated how the two tracers report on different features of the tumor microphysiology. [^{99m}Tc]HMPAO SPECT images revealed information about both tumoral perfusion and intracellular concentration of mediators converting [^{99m}Tc]HMPAO to hydrophilic derivatives whereas thallium-201 SPECT images defined permeability, extension of tumoral capillary network, and viable tumoral cell presence. They argued that the dual-tracer approach offered more power for the histology prediction.

B. Magnetic Resonance

Magnetic resonance methods are still relatively new and evolving. Present activity involving MRI is directed at using it to measure metabolic and physiologic parameters to augment its role in structural imaging. One example that has already been discussed is the use of fMRI for presurgical evaluation of functional tissue. In regard to these items, it is important to emphasize that the developing functional and physiological MRI procedures use an MRI scanner and can be done in conjunction with a diagnostic (structural) MRI study. These properties of functional and physiological MRI offer examination efficiencies that are not characteristic of PET and SPECT.

1. Magnetic Resonance Spectroscopy

Magnetic resonance spectroscopy (MRS) is an emerging neuroimaging technique that detects magnetic resonance signals from endogenous tissue molecules using the same principles as MRI. MRS measurements are possible because nuclear magnetic resonance frequencies are influenced to a small, but detectable, extent by the chemical structure of molecule in which the signal-producing nucleus is located. This causes otherwise identical nuclei that are situated within different molecular structures to produce oscillatory magnetic dipole resonances at different characteristic frequencies. Knowledge of the aforementioned property predates the discovery of MRI. Chemists have used it for almost 50 years to identify molecular structures in samples of unknown chemicals. The availability of human-size magnets, provided through the development of MRI, allows spectroscopic examination of humans. MRS signals from several nonradioactive nuclei (^{1}H, ^{31}P, ^{13}C, ^{19}F) may be used for human examinations. At the present time, proton (^{1}H) MRS is the most well developed of these and discussion is restricted to it. For technical reasons, only a small subset of the molecules that are present within a tissue produce a detectable MRS signal. For instance, MRS is usually not sensitive enough to detect drugs or tissue metabolites that are present at very low concentration (<1 μmol/ml).

^{1}H MRS examinations can be performed on single geometric tissue volumes of greater than 2 cm^3 located with MRI. This is known as "single-volume spectroscopy." Recently, one MRI manufacturer began marketing a commercial product that does this. This product is known as PROBE (for *Pro*ton *B*rain *E*xamination). Alternately, ^{1}H MRS signals may be sampled from a grid of tissue locations using procedures that closely resemble MRI phase encoding; such examinations are referred to as magnetic resonance spectroscopic imaging (MRSI), spectroscopic imaging (SI), or chemical shift imaging (CSI). Figure 8 provides a typical example of a ^{1}H-MRSI study of a brain tumor patient. MRS signals are not strong enough to permit the MRSI spatial sampling to be as fine as is used in MRI; ^{1}H-MRSI examinations commonly use a volumetric resolution of greater than 0.2 cm^3. The technique used in the study shown in the figure was long echo time (TE = 272 ms) multiple-slice ^{1}H-MRSI (Herholz *et al.*, 1992; Duyn, 1993; Meyerhoff, 1994; Tedeschi *et al.*, 1995). It offers spectra from approximately 1.0-cm^3 volume elements within a limited number of slices. The data may be presented as spectra arising from specific volume elements chosen from an MR image or as tomographic images that show the topographic variation in the strengths of spectroscopic signals. The long echo time signal acquisition provides signals from choline-containing compounds, creatine-containing compounds, *N*-acetylaspartate (NAA), and co-resonant lactate and lipid compounds. The relationship of these signals to brain metabolites is described elsewhere (Miller, 1991; Howe *et al.*, 1993). Systematic studies of the topographic patterns produced by these signals in ^{1}H-MRSI normal adults have appeared (Tedeschi *et al.*, 1995). Spectra arising from intracranial tumors generally show "choline" signals that are elevated with respect to surrounding brain (Bruhn *et al.*, 1989; Gill *et al.*, 1990; Fulham *et al.*, 1992; Ott *et al.*, 1993; Negendank *et al.*, 1996). Elevated choline signals from tumors have been attributed to active cell proliferation. Recent studies (Gillies *et al.*, 1994; Aiken and Gillies, 1996; Aiken *et al.*, 1996) on glioma cell cultures demonstrate the existence of a relationship between the phosphorylcholine (a contributor to the choline signal) level and cell growth. The hypothetical connection has been that choline and phosphorylcholine are precursors of the phospholipids needed for the membrane development and that their levels are high in tumors because biosynthetic pathways related to membrane synthesis are activated in growing cells. Miller *et al.* (1996) and Chang *et al.* (1995) have confirmed that tumor choline signal levels correlate with cellularity and the tissue levels of membrane precursors. Gupta *et al.* (1999) have demonstrated that variation in choline levels is most likely closely related

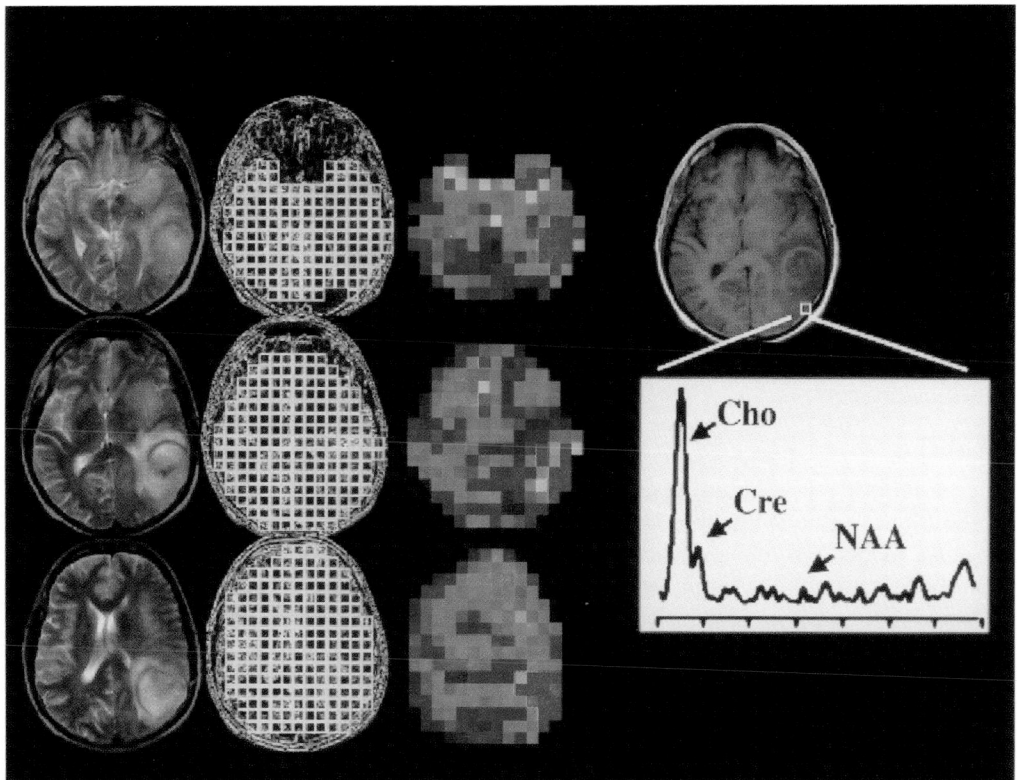

Figure 8 Example of a ¹H-MRSI study of a brain tumor patient. The left panel shows three slices that were examined. The middle panel shows the grid of voxels having volumes of 10 × 10 3 12 mm³ that produced interpretable frequency spectra. One of the frequency spectra chosen from within the tumor is displayed on the far right. This spectrum shows a prominence of the choline signal with little apparent NAA signal. The color images show how the intensity of a particular signal in frequency space (in this case choline) varies with the tomographic coordinate.

to local cellularity. An NAA signal generated by the N-acetylaspartate that is located only within neurons is also routinely detected. Generally, NAA signal is reduced within the tumor-infiltrated region because the tumor either displaces or kills normal neurons in its vicinity. In addition to choline and NAA, a variety of additional signals produced by intracellular metabolites can be detected in human brain studies. The extent to which a number of these are useful for brain tumor evaluation is currently an active research field.

The majority of MRS studies of brain tumors have emphasized the possible clinical uses of MRS. Several of these have been moderately large studies and there have been two organized multicenter efforts (Sijens et al., 1995b; Negendank et al., 1996). The very first MRS study of cerebral neoplasia suggested that the MRS signals served to identify tumor type and grade (Bruhn et al., 1989). Several other studies looked specifically at the relationship between choline signal level and histologic grade of astrocytoma (Gill et al., 1990; Fulham et al., 1992; Ott et al., 1993; Negendank et al., 1996). In general, these showed high-grade glioma tends to display

more substantial choline signal elevation than does low-grade glioma. Preul et al. (1996) demonstrated that pattern recognition analysis of MRS data from intracranial tumors shows a high degree of agreement with histopathological readings. Hence, there is evidence to suggest that MRS can be used as an adjunct to preoperative diagnosis. Beyond preoperative diagnosis, there have been exploratory studies in a number of additional clinically significant areas. Studies have demonstrated that progressive or recurrent glioma manifests itself through pronounced change in the elevation of the choline signal (Fulham et al., 1992; Tedeschi et al., 1997; Wald et al., 1997) and that effective treatment reverses choline signal elevations (Fulham et al., 1992; Bizzi et al., 1995; Sijens et al., 1995a). Taken together, the body of existing MRS investigation of intracranial neoplasia suggests the following: (1) MRS shows promise of accurate diagnosis without tissue sampling. (2) MRS can conceivably be used to guide stereotactic biopsy procedures to improve the probability of representative sampling. (3) Similarly, MRS may be useful for targeting when focally ablative radiation therapy is used. (4)

MRS can be used to longitudinally follow patients who are being treated. (5) MRS can be used for detecting recurrent or progressing disease.

Figure 9 illustrates a comparison between ¹H-MRSI and FDG-PET findings in the same case illustrated in Fig. 6. This figure illustrates the sensitivity of the choline signal to new infiltration. In this case, the right frontal lobe was known to have the newest infiltration. The choline signal shows a pronounced choline signal elevation in both frontal lobes, with the right frontal lobe showing the greatest signal elevation. Compared to FDG-PET, the choline signal displays the newest tumor growth with much higher contrast.

2. Diffusion MRI

The MRI signal intensity is inherently sensitive to the motion of the tissue water molecules. In diffusion MRI, the MRI signal intensity is made sensitive to random diffusion motions made by the tissue water molecules as they undergo Brownian movement on a microscopic scale (Le Bihan, 1991). Diffusion MRI uses a modified spin-echo pulse sequence similar to that used for T_2-weighted MRI. Gradient pulses having a variety of different levels are played out on both sides of the 180° pulse to sensitize the MRI signal to Brownian diffusion. Stronger gradient pulses convey increased sensitivity to diffusion. The gradient sensitization can be performed in any number of unique anatomic directions. The strength of the applied magnetic field gradient conveys a controlled sensitivity to motion. The discovery that newly ischemic brain tissue produces remarkable contrast on diffusion-weighted MRI relative to T_2-weighted imaging (Mintorovitch *et al.*, 1991) has led to a growing enthusiasm for using the method to evaluate stroke (Baird and Warach, 1998); this has been technically feasible for only a handful of years. Diffusion-weighted image intensity does not depend solely on diffusion. There is also a strong influence of T_2 on the image intensity related to the fact that the pulse sequences inherently convey T_2 weighting. Parametric calculation of the apparent diffusion coefficient (ADC) is a frequently used quantitative approach for separating the effects of T_2 and diffusion. In this approach, one acquires a series of diffusion-weighted images using a variety of levels of diffusion weighting. These images can be, then, subjected to a relatively simple calculation that synthesizes an image in which the image intensity conveys the value of the ADC, a measure of the diffusion rate.

What has been learned about diffusion imaging of brain ischemia is potentially applicable to glioma. The brain ischemia studies suggest a model in which the water diffusion coefficients characteristic of intracellular and extracellular space differ to a measurable extent and that the ADC can be used as a measure of cell density through its sensitivity to the relative amounts of intracellular and extracellular water. On the basis of this model, it should be possible to differentiate regions of hypercellularity from regions of hypocellularity in glioma through measurement of the volume-average diffusion properties. This appears to be useful for dealing with the microscopic heterogeneity characteristic of glioma. Clinical studies have illustrated that diffusion imaging enables the identification of the unique histopathologic components of glioma which may not be clearly distinguishable on conventional MRI (Tien *et al.*, 1993; Brunberg *et al.*, 1995; Yanaka *et al.*, 1995; Gupta *et al.*, 1999). Higher ADC readings are suggestive of cyst, necrosis, or edema, whereas lower ADC values suggest cellularity. Figure 10 provides an illustrative example. The figure shows two cases in which contrast enhancement is present following radiation therapy. Com-

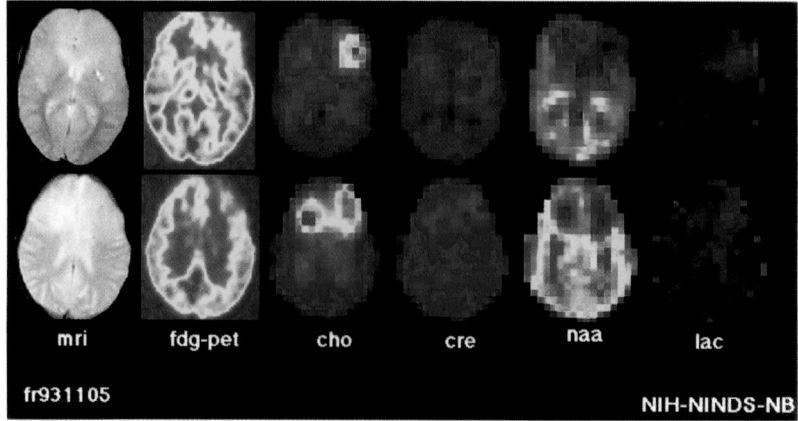

Figure 9 Comparison of FDG-PET and ¹H-MRSI studies of progressing glioma. The same case shown in Fig. 6 is illustrated. Two slices in which all images are registered to the ¹H-MRSI slicing are shown.

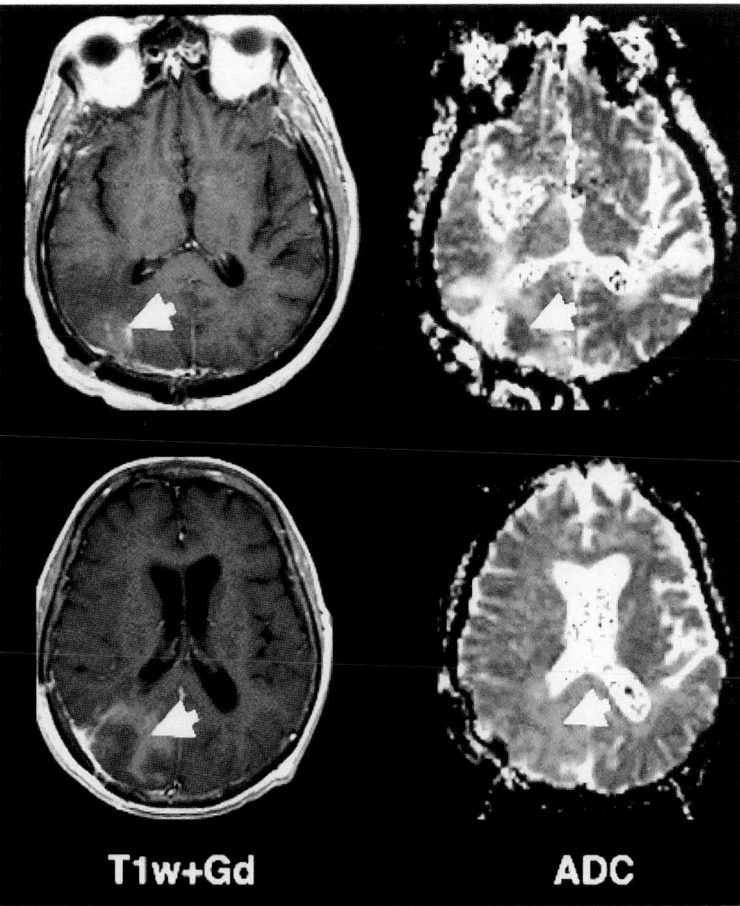

Figure 10 Apparent diffusion coefficient (ADC) contrast-enhanced T_1-weighted images. Two cases of intracranial glioma (top row and bottom row) are shown. Both cases were imaged as part of a routine follow-up following radiation therapy. Each showed contrast enhancement.

parison of the ADC images to the contrast-enhanced images allows the following interpretation. In the pair of images shown in the top row, the contrast-enhancing tissue appears hypointense in the ADC image (arrows). This ADC appearance is consistent with the presence of relatively dense cellularity. On the other hand, the case illustrated in the bottom two images shows the contrast-enhancing tissue (arrows) produces a higher ADC, suggesting a relative hypocellularity. The example illustrates how ADC imaging may help differentiate contrast enhancement resulting from dense tumor cellularity from contrast enhancement that appears as a result of necrosis. Finally, the concept that microscopic necrosis can be detected using ADC imaging may have utility for assessing therapy. It is expected that successful therapy will lead to a reduced cellularity that can be detected with ADC imaging. Preliminary studies of therapy in animals have demonstrated this potential (Zhao *et al.*, 1996; Chenevert *et al.*, 1997; Poptani *et al.*, 1998).

IV. Summary and a Look into the Crystal Ball

Hopefully, the reader has seen that brain tumors represent a significant neurological health problem and that neuroimaging has already contributed much to lessening the suffering of individuals who are unfortunate enough to become afflicted with this condition. The contributions made by neuroimaging thus far have been primarily in the area of better structural depiction of the lesion. MRI and CT imaging now permit the neuroradiologist to, with very few exceptions, definitively diagnose the presence of a neoplastic mass within the cranium. These structural imaging techniques also now provide the neurosurgeon and the radiation oncologist with a previously undreamed of visual representation of the mass and its spatial relationships to other key eloquent brain structures. In the rare case where an open resection is dangerous, structural imaging provides coordinates for an accurate stereotactic biopsy. Metastatic deposits in the brain, even if multiple in number, can be effectively controlled through image-guided minimally invasive neurosurgery or ablative stereotactic radiation therapy so that neurological health does become the limiting factor in patient management. Structural imaging also provides the radiation oncologist and the neuro-oncologist with the ability to detect subtle tumor volume changes in good or bad directions.

Functional neuroimaging is beginning to play a significant additional role. Functional MRI and $H_2^{15}O$ PET scanning each provide a means of accurately identifying cortex that has a high level of functional significance, permitting more informed choices to be made regarding whether to and how to resect lesions in "dangerous" areas such as the frontotemporal junction of the dominant hemisphere. PET, SPECT, MRS, and diffusion MRI are providing useful information related to individual case management. Each of these offers some potential as a preoperative diagnostic aid. Given that almost all patients benefit from cytoreductive surgery and that additional tests (e.g., gene profiling) that require tissue are likely to play a larger role in the future, it is unlikely that biopsy will be abandoned no matter how predictive these functional imaging measures prove to be. However, presurgical prediction of probable histology permits the neurosurgeon and the patients to make more informed decisions about how to proceed. AIDS patients who develop new intracranial masses are likely to be the exception to the assertion made in the previous two sentences. The radiosensitivity of CNS lymphoma and the medical therapy that is applied in the instance of opportunistic infection argue against the use of open surgery in this instance. Functional neuroimaging is also providing many clues regarding a variety of physiological parameters and the ultrastructural composition of the neoplastic tissue. This shows great promise of teaching us much about the behavior of intracranial neoplastic lesions at a variety of different levels of integration.

The situation is "not so bad" for many patients. Cytoreductive surgery is curative for many cases of low-grade primary neoplastic lesions, including meningioma, pilocytic astrocytoma, and astrocytoma. Many forms of oligodendroglioma benefit from cytoreductive surgery and are, additionally, exquisitely chemosensitive. Even patients with anaplastic astrocytoma can experience prolonged periods of good health with proper management. AIDS patients suffering from CNS lymphoma can be managed appropriately as can patients with intracranial metastatic deposits. However, the situation remains dire for other patients. For the large proportion of patients who suffer from glioblastoma multiforme, survival beyond 2 years from diagnosis is an extraordinary occurrence. What is needed to help these patients? New in-

sights into the biological, molecular, physiological, and ultrastructural characteristics (and their evolution) of the neoplastic tissue in the postsurgical, postradiation setting are needed. It is only through these insights that the appropriate combinations of existing treatments can be applied with proper timing. Furthermore, there is the hope that these insights will lead to novel therapeutic strategies. Given that there are no realistic subhuman models of the primary high-grade glioma in the postsurgical, postradiation setting, it is likely that the needed information will have to be gathered through neuroimaging observations of human patients. Structural and functional neuroimaging offer much hope in this regard.

Acknowledgments

The authors acknowledge grant support from the American Cancer Society (EDT-119) and the National Cancer Institute (CA76524).

References

Aiken, N. R., and Gillies, R. J. (1996). Phosphomonoester metabolism as a function of cell proliferative status and exogenous precursors. *Anticancer Res.* **16,** 1393–1397.

Aiken, N. R., Szwergold, E. S., Kappler, F., Stoyanova, R., Kuesel, A. C., Shaller, C., and Brown, T. R. (1996). Metabolism of phosphonium choline by rat-2 fibroblasts: Effects of mitogenic stimulation studied using ^{31}P NMR spectroscopy. *Anticancer Res.* **16,** 1357–1363.

Antinori, A., De Rossi, G., Ammassari, A., Cingolani, A., Murri, R., Di Giuda, D., De Luca, A., Pierconti, F., Tartaglione, T., Scerrati, M., Larocca, L. M., and Ortona, L. (1999). Value of combined approach with thallium-201 single-photon emission computed tomography and Epstein-Barr virus DNA polymerase chain reaction in CSF for the diagnosis of AIDS-related primary CNS lymphoma. *J. Clin. Oncol.* **17,** 554–560.

Atlas, S. W., Howard, R. S., II, Maldjian, J., Alsop, D., Detre, J. A., Listerud, J., D'Esposito, M., Judy, K. D., Zager, E., and Stecker, M. (1996). Functional magnetic resonance imaging of regional brain activity in patients with intracerebral gliomas: Findings and implications for clinical management. *Neurosurgery* **38,** 329–338.

Baird, A. E., and Warach, S. (1998). Magnetic resonance imaging of acute stroke. *J. Cereb. Blood Flow Metab.* **18,** 583–609.

Barker, F. G., 2nd, Prados, M. D., Chang, S. M., Gutin, P. H., Lamborn, K. R., Larson, D. A., Malec, M. K., McDermott, M. W., Sneed, P. K., Wara, W. M., and Wilson, C. B. (1996). Radiation response and survival time in patients with glioblastoma multiforme. *J. Neurosurg.* **84,** 442–448.

Barker, F. G., 2nd, Chang, S. M., Gutin, P. H., Malec, M. K., McDermott, M. W., Prados, M. D., and Wilson, C. B. (1998). Survival and functional status after resection of recurrent glioblastoma multiforme. *Neurosurgery* **42,** 709–720; discussion 720–723.

Bizzi, A., Movsas, B., Tedeschi, G., Phillips, C. L., Okunieff, P., Alger, J. R., and Di Chiro, G. (1995). Early and long term response to therapy monitored by ^{1}H-MR spectroscopic imaging in CNS non-Hodgkin lymphoma. *Radiology* **194,** 271–276.

Black, K. L., Emerick, T., Hoh, C., Hawkins, R. A., Mazziotta, J., and Becker, D. P. (1994). Thallium-201 SPECT and positron emission tomography equal predictors of glioma grade and recurrence. *Neurol. Res.* **16,** 93–96.

Black, K. L., Cloughesy, T., Huang, S. C., Gobin, Y. P., Zhou, Y., Grous, J., Nelson, G., Farahani, K., Hoh, C. K., and Phelps, M. (1997). Intracarotid infusion of RMP-7, a bradykinin analog, and transport of gallium-68 ethylenediamine tetraacetic acid into human gliomas. *J. Neurosurg.* **86,** 603–609.

Black, P. M., and Loeffer, J. S. (Eds.). (1997). "Cancer of the Nervous System." Blackwell Science, Cambridge, MA.

Bohnen, N. I., Radhakrishnan, K., O'Neill, B. P., and Kurland, L. T. (1997). Descriptive and analytic epidemiology of brain tumors. *In* "Cancer of the Nervous System" (P. M. Black and J. S. Loeffer, Eds.), Ch. 1. Blackwell Science, Cambridge, MA.

Bookheimer, S. Y., Zeffiro, T. A., Blaxton, T., Malow, B. A., Gaillard, W. D., Sato, S., Kufta, C., Fedio, P., and Theodore, W. H. (1997). A direct comparison of PET activation and electrocortical stimulation mapping for language localization. *Neurology* **48,** 1056–1065.

Bruehlmeier, M., Roelcke, U., Amsler, B., Schubert, K. H., Hausmann, O., von Ammon, K., Radu, E. W., Gratzl, O., Landmann, C., and Leenders, K. L. (1999). Effect of radiotherapy on brain glucose metabolism in patients operated on for low grade astrocytoma. *J. Neurol. Neurosurg. Psychiatry* **66,** 648–653.

Bruhn, H., Frahm, J., Gyngell, M. L., Merboldt, K. D., Hänicke, W., Sauter, R., and Hamburg, C. (1989). Noninvasive differentiation of tumors with use of localized H-1 MR spectroscopy *in vivo:* Initial experience in patients with cerebral tumors. *Radiology* **172,** 541–548.

Brunberg, J. A., Chenevert, T. L., McKeever, P. E., Ross, D. A., Junck, L. R., Muraszko, K. M., Dauser, R., Pipe, J. G., and Betley, A. T. (1995). *In vivo* MR determination of water diffusion coefficients and diffusion anisotropy: Correlation with structural alteration in gliomas of the cerebral hemispheres. *AJNR, Am. J. Neuroradiol.* **16,** 361–371.

Cairncross, J. G., and Macdonald, D. R. (1988). Successful chemotherapy for recurrent malignant oligodendroglioma. *Ann. Neurol.* **23,** 360–364.

Cairncross, G., Macdonald, D., Ludwin, S., Lee, D., Cascino, T., Buckner, J., Fulton, D., Dropcho, E., Stewart, D., Schold, C., Jr., *et al.* (1994). Chemotherapy for anaplastic oligodendroglioma. National Cancer Institute of Canada Clinical Trials Group. *J. Clin. Oncol.* **12,** 2013–2021.

Chang, L., McBride, D., Miller, B. L., Cornford, M., Booth, R. A., Buchthal, S. D., Ernst, T. M., and Jenden, D. (1995). Localized *in vivo* [1]H magnetic resonance spectroscopy and in vitro analyses of heterogeneous brain tumors. *J. Neuroimaging* **5,** 157–163.

Chenevert, T. L., McKeever, P. E., and Ross, B. D. (1997). Monitoring early response of experimental brain tumors to therapy using diffusion magnetic resonance imaging. *Clin. Cancer Res.* **3,** 1457–1466.

Cuenod, C. A., Bookheimer, S. Y., Hertz-Pannier, L., Zeffiro, T. A., Theodore, W. H., and Le Bihan, D. (1995). Functional MRI during word generation, using conventional equipment: A potential tool for language localization in the clinical environment. *Neurology* **45,** 1821–1827.

Daumas-Duport, C., Scheithauer, B., O'Fallon, J., and Kelly, P. (1988). Grading of astrocytomas. A simple and reproducible method. *Cancer* **62,** 2152–2165.

DeAngelis, L. M., Burger, P. C., Green, S. B., and Cairncross, J. G. (1998). Malignant glioma: Who benefits from adjuvant chemotherapy? *Ann. Neurol.* **44,** 691–695.

De La Pena, R. C., Ketonen, L., and Villanueva-Meyer, J. (1998). Imaging of brain tumors in AIDS patients by means of dual-isotope thallium-201 and technetium-99m sestamibi single-photon emission tomography. *Eur. J. Nucl. Med.* **25,** 1404–1411.

Derlon, J. M., Petit-Taboue, M. C., Chapon, F., Beaudouin, V., Noel, M. H., Creveuil, C., Courtheoux, P., and Houtteville, J. P. (1997). The *in vivo* metabolic pattern of low-grade brain gliomas: A positron emission tomographic study using ^{18}F-fluorodeoxyglucose and [11]C-

L-methylmethionine [see comments]. *Neurosurgery* **40,** 276–287; discussion 287–288.

Deshmukh, A., Scott, J. A., Palmer, E. L., Hochberg, F. H., Gruber, M., and Fischman, A. J. (1996). Impact of fluorodeoxyglucose positron emission tomography on the clinical management of patients with glioma. *Clin. Nucl. Med.* **21,** 720–725.

De Witte, O., Levivier, M., Violon, P., Salmon, I., Damhaut, P., Wikler, D., Jr., Hildebrand, J., Brotchi, J., and Goldman, S. (1996). Prognostic value of positron emission tomography with ^{18}F]fluoro-2-deoxy-D-glucose in the low-grade glioma. *Neurosurgery* **39,** 470–476.

de Wolde, H., Pruim, J., Mastik, M. F., Koudstaal, J., and Molenaar, W. M. (1997). Proliferative activity in human brain tumors: Comparison of histopathology and L-[1-^{11}C]tyrosine PET. *J. Nucl. Med.* **38,** 1369–1374.

Di Chiro, G. (1987). Positron emission tomography using ^{18}F]fluorodeoxyglucose in brain tumors. A powerful and prognostic tool. *Invest. Radiol.* **22,** 360–371.

Di Chiro, G., Brooks, R. A., Bairamaian, D., *et al.* (1985). Diagnostic and prognostic value of positron emission tomography using ^{18}F]fluorodeoxyglucose in brain tumors. *In* "Positron Emission Tomography" (M. Reivitch and A. Alavi, Eds.), pp. 291–309. A. R. Liss, New York.

Duyn, J. H., Gillen, J., Sobering, G., van Zijl, P. C., and Moonen, C. T. (1993). Multisection proton MR spectroscopic imaging of the brain. *Radiology* **188,** 277–282.

Eary, J. F., Mankoff, D. A., Spence, A. M., Berger, M. S., Olshen, A., Link, J. M., O'Sullivan, F., and Krohn, K. A. (1999). 2-[C-11]Thymidine imaging of malignant brain tumors. *Cancer Res.* **59,** 615–621.

Fine, H. A., Dear, K. B., Loeffler, J. S., Black, P. M., and Canellos, G. P. (1993). Meta-analysis of radiation therapy with and without adjuvant chemotherapy for malignant gliomas in adults [see comments]. *Cancer* **71,** 2585–2597.

Frith, C. D., Friston, K., Liddle, P. F., and Frackowiak, R. S. (1991). Willed action and the prefrontal cortex in man: A study with PET. *Proc. R. Soc. London B, Biol. Sci.* **244,** 241–246.

Fulham, M. F., Bizzi, A., Dietz, M. J., Shih, H. H.-L., Raman, R., Sobering, G. S., Frank, J. A., Dwyer, A., Alger, J. R., and Di Chiro, G. (1992). Mapping of brain tumor metabolites with proton MR spectroscopic imaging: Clinical relevance. *Radiology* **185,** 675–686.

Fulham, M. F., Brunetti, A., Aloj, L., Raman, R., Dwyer, A. J., and Di Chiro, G. (1995). Decreased cerebral glucose metabolism in patients with brain tumors: An effect of corticosteroids. *J. Neurosurg.* **83,** 657–664.

Gaspar, L. E., Fisher, B. J., MacDonald, D. R., LeBer, D. V., Halperin, E. C., Schold, S. C., Jr., and Cairncross, J. G. (1993). Malignant glioma: Timing of response to radiation therapy. *Int. J. Radiat. Oncol. Biol. Phys.* **25,** 877–879.

Gill, S. S., Thomas, D. G., Van Bruggen, N., Gadian, D. G., Peden, C. J., Bell, J. D., Cox, I. J., Menon, D. K., Iles, R. A., and Bryant, D. J. (1990). Proton MR spectroscopy of intracranial tumours: *In vivo* and *in vitro* studies. *J. Comp. Assist. Tomogr.* **14,** 497–504.

Gillies, R. J., Barry, J. A., and Ross, B. D. (1994). *In vitro* and *in vivo* ^{13}C and ^{31}P NMR analyses of phosphocholine metabolism in rat glioma cells. *Magn. Reson. Med.* **32,** 310–318.

Goldman, S., Levivier, M., Pirotte, B., Brucher, J. M., Wikler, D., Damhaut, P., Stanus, E., Brotchi, J., and Hildebrand, J. (1996). Regional glucose metabolism and histopathology of gliomas. A study based on positron emission tomography-guided stereotactic biopsy. *Cancer* **78,** 1098–1106.

Goldman, S., Levivier, M., Pirotte, B., Brucher, J. M., Wikler, D., Damhaut, P., Dethy, S., Brotchi, J., and Hildebrand, J. (1997). Regional methionine and glucose uptake in high-grade gliomas: A comparative study on PET-guided stereotactic biopsy. *J. Nucl. Med.* **38,** 1459–1462.

Gross, M. W., Weber, W. A., Feldmann, H. J., Bartenstein, P., Schwaiger, M., and Molls, M. (1998). The value of F-18-fluorodeoxyglucose PET for the 3-D radiation treatment planning of malignant gliomas. *Int. J. Radiat. Oncol. Biol. Phys.* **41**, 989–995.

Gupta, R. K., Sinha, U., Cloughesy, T. F., and Alger, J. R. (1999). Inverse correlation between choline magnetic resonance spectroscopy signal intensity and the apparent diffusion coefficient in human glioma. *Magn. Reson. Med.* **41**, 2–7.

Haglund, M. M., Berger, M. S., and Hochman, D. W. (1996). Enhanced optical imaging of human gliomas and tumor margins. *Neurosurgery* **38**, 308–317.

Hara, T., Kosaka, N., Shinoura, N., and Kondo, T. (1997). PET imaging of brain tumor with [*methyl*-¹¹C]choline. *J. Nucl. Med.* **38**, 842–847.

Heiss, W. D., Wienhard, K., Wagner, R., Lanfermann, H., Thiel, A., Herholz, K., and Pietrzyk, U. (1996). F-Dopa as an amino acid tracer to detect brain tumors. *J. Nucl. Med.* **37**, 1180–1182.

Herholz, K., Heindel, W., Luyten, P. R., den Hollander, J. A., Pietrzyk, U., Voges, J., Kugel, H., Friedmann, G., and Heiss, W. D. (1992). *In vivo* imaging of glucose consumption and lactate concentration in human gliomas. *Ann. Neurol.* **31**, 319–327.

Herholz, K., Reulen, H. J., von Stockhausen, H. M., Thiel, A., Ilmberger, J., Kessler, J., Eisner, W., Yousry, T. A., and Heiss, W. D. (1997). Preoperative activation and intraoperative stimulation of language-related areas in patients with glioma. *Neurosurgery* **41**, 1253–1260; discussion 1260–1262.

Hochberg, F. H., and Pruitt, A. (1980). Assumptions in the radiotherapy of glioblastoma. *Neurology* **30**, 907–911.

Howe, F. A., Maxwell, R. J., Saunders, D. E., Brown, M. M., and Griffiths, J. R. (1993). Proton spectroscopy *in vivo*. *Magn. Reson. Q.* **9**, 31–59.

Imahori, Y., Ueda, S., Ohmori, Y., Kusuki, T., Ono, K., Fujii, R., and Ido, T. (1998). Fluorine-18-labeled fluoroboronophenylalanine PET in patients with glioma. *J. Nucl. Med.* **39**, 325–333.

Inoue, T., Shibasaki, T., Oriuchi, N., Aoyagi, K., Tomiyoshi, K., Amano, S., Mikuni, M., Ida, I., Aoki, J., and Endo, K. (1999). [¹⁸F]α-Methyltyrosine PET studies in patients with brain tumors. *J. Nucl. Med.* **40**, 399–405.

Jack, C. R., Jr., Thompson, R. M., Butts, R. K., Sharbrough, F. W., Kelly, P. J., Hanson, D. P., Riederer, S. J., Ehman, R. L., Hangiandreou, N. J., and Cascino, G. D. (1994). Sensory motor cortex: Correlation of presurgical mapping with functional MR imaging and invasive cortical mapping. *Radiology* **190**, 85–92.

Jansen, H. M., Dierckx, R. A., Hew, J. M., Paans, A. M., Minderhoud, J. M., and Korf, J. (1997). Positron emission tomography in primary brain tumours using cobalt-55. *Nucl. Med. Commun.* **18**, 734–740.

Kahn, T., Schwabe, B., Bettag, M., Harth, T., Ulrich, F., Rassek, M., Schwarzmaier, H. J., and Modder, U. (1996). Mapping of the cortical motor hand area with functional MR imaging and MR imaging-guided laser-induced interstitial thermotherapy of brain tumors. Work in progress. *Radiology* **200**, 149–157.

Kallen, K., Heiling, M., Andersson, A. M., Brun, A., Holtas, S., and Ryding, E. (1996). Preoperative grading of glioma malignancy with thallium-201 single-photon emission CT: Comparison with conventional CT. *AJNR, Am. J. Neuroradiol.* **17**, 925–932.

Kameyama, M., Ishiwata, K., Tsurumi, Y., Itoh, J., Sato, K., Katakura, R., Yoshimoto, T., Hatazawa, J., Ito, M., and Ido, T. (1995). Clinical application of ¹⁸F-FUdR in glioma patients: PET study of nucleic acid metabolism. *J. Neurooncol.* **23**, 53–61.

Kaschten, B., Stevenaert, A., Sadzot, B., Deprez, M., Degueldre, C., Del Fiore, G., Luxen, A., and Reznik, M. (1998). Preoperative evaluation of 54 gliomas by PET with fluorine-18-fluorodeoxyglucose and/or carbon-11-methionine. *J. Nucl. Med.* **39**, 778–785.

Kaye, A. H., and Laws, E. R., Jr. (Eds). (1995). "Brain Tumors." Churchill Livingston, Edinburgh.

Kessler, L. S., Ruiz, A., Donovan Post, M. J., Ganz, W. I., Brandon, A. H., and Foss, J. N. (1998). Thallium-201 brain SPECT of lymphoma in AIDS patients: Pitfalls and technique optimization. *AJNR, Am. J. Neuroradiol.* **19**, 1105–1109.

Kline, J. L., Noto, R. B., and Glantz, M. (1996). Single-photon emission CT in the evaluation of recurrent brain tumor in patients treated with gamma knife radiosurgery or conventional radiation therapy. *AJNR, Am. J. Neuroradiol.* **17**, 1681–1686.

Kuwert, T., Probst-Cousin, S., Woesler, B., Morgenroth, C., Lerch, H., Matheja, P., Palkovic, S., Schafers, M., Wassmann, H., Gullotta, F., and Schober, O. (1997). Iodine-123–alpha-methyltyrosine in gliomas: Correlation with cellular density and proliferative activity. *J. Nucl. Med.* **38**, 1551–1555.

Kuwert, T., Woesler, B., Morgenroth, C., Lerch, H., Schafers, M., Palkovic, S., Matheja, P., Brandau, W., Wassmann, H., and Schober, O. (1998). Diagnosis of recurrent glioma with SPECT and iodine-123-alpha-methyltyrosine [published erratum appears in *J. Nucl. Med.* **39** (3), 574 (1998)]. *J. Nucl. Med.* **39**, 23–27.

Langen, K. J., Ziemons, K., Kiwit, J. C., Herzog, H., Kuwert, T., Bock, W. J., Stocklin, G., Feinendegen, L. E., and Muller-Gartner, H. W. (1997). 3-[¹²³I]Iodo-alpha-methyltyrosine and [*methyl*-¹¹C]-L-methionine uptake in cerebral gliomas: A comparative study using SPECT and PET. *J. Nucl. Med.* **38**, 517–522.

Lantos, P. L., VandenBerg, S. R., Kleihues, P. (1997). Tumours of the central nervous system. *In* "Greenfield's Neuropathology" (P. Lantos and D. I. Graham, eds.), 6th Ed., Vol. II, Ch. 9. Edward Arnold, London.

Le Bihan, D. (1991). Molecular diffusion nuclear magnetic resonance imaging. *Magn. Reson. Q.* **7**, 1–30.

Levivier, M., Goldman, S., Pirotte, B., Brucher, J. M., Baleriaux, D., Luxen, A., Hildebrand, J., and Brotchi, J. (1995). Diagnostic yield of stereotactic brain biopsy guided by positron emission tomography with [¹⁸F]fluorodeoxyglucose. *J. Neurosurg.* **82**, 445–452.

Lorberboym, M., Baram, J., Feibel, M., Hercbergs, A., and Lieberman, L. (1995). A prospective evaluation of thallium-201 single photon emission computerized tomography for brain tumor burden. *Int. J. Radiat. Oncol. Biol. Phys.* **32**, 249–254.

Macdonald, D. R., Gaspar, L. E., and Cairncross, J. G. (1990). Successful chemotherapy for newly diagnosed aggressive oligodendroglioma. *Ann. Neurol.* **27**, 573–574.

Maini, C. L., Sciuto, R., Tofani, A., Ferraironi, A., Carapella, C. M., Occhipinti, E., Mottolese, M., and Crecco, M. (1995). Somatostatin receptor imaging in CNS tumours using ¹¹¹In-octreotide. *Nucl. Med. Commun.* **16**, 756–766.

Marriott, C. J., Thorstad, W., Akabani, G., Brown, M. T., McLendon, R. E., Hanson, M. W., and Coleman, R. E. (1998). Locally increased uptake of fluorine-18-fluorodeoxyglucose after intracavitary administration of iodine-131-abeled antibody for primary brain tumors. *J. Nucl. Med.* **39**, 1376–1380.

McCarthy, G., Blamire, A. M., Rothman, D. L., Gruetter, R., and Shulman, R. G. (1993). Echo-planar magnetic resonance imaging studies of frontal cortex activation during word generation in humans. *Proc. Natl. Acad. Sci. U.S.A.* **90**, 4952–4956.

Meyerhoff, D. J. (1994). Magnetic resonance spectroscopic imaging. *In* "NMR in Physiology and Biomedicine" (R. J. Gillies, Ed.), Chapter 10. Academic Press, San Diego.

Miller, B. L. (1991). A review of chemical issues in ¹H NMR spectroscopy: *N*-Acetyl-L-aspartate, creatine, and choline. *NMR Biomed.* **4**, 47–52.

Miller, B. L., Chang, L., Booth, R., Ernst, T., Cornford, M., Nikas, D., McBride, D., and Jenden, D. J. (1996). *In vivo* ¹H MRS choline: Correlation with *in vitro* chemistry/histology. *Life Sci.* **58**, 1929–1935.

Miller, R. F., Hall-Craggs, M. A., Costa, D. C., Brink, N. S., Scaravilli, F., Lucas, S. B., Wilkinson, I. D., Ell, P. J., Kendall, B. E., and Harrison, M. J. (1998). Magnetic resonance imaging, thallium-201 SPET scanning, and laboratory analyses for discrimination of cerebral lymphoma and toxoplasmosis in AIDS. *Sex. Transm. Infect.* **74**, 258–264.

Mineura, K., Sasajima, T., Kowada, M., Ogawa, T., Hatazawa, J., Shishido, F., and Uemura, K. (1994). Perfusion and metabolism in predicting the survival of patients with cerebral gliomas. *Cancer* **73**, 2386–2394.

Mintorovitch, J., Moseley, M. E., Chileuitt, L., Shimizu, H., Cohen, Y., and Weinstein, P. R. (1991). Comparison of diffusion- and T2-weighted MRI for early detection of cerebral ischemia and reperfusion in rats. *Magn. Reson. Med.* **18,** 39–50.

Mueller, W. M., Yetkin, F. Z., Hammeke, T. A., Morris, G. L., 3rd, Swanson, S. J., Reichert, K., Cox, R., and Haughton, V. M. (1996). Functional magnetic resonance imaging mapping of the motor cortex in patients with cerebral tumors. *Neurosurgery* **39,** 515–520; discussion 520–521.

Nakayama, K. (1997). Localization of the cortical motor area by functional magnetic resonance imaging with gradient echo and echo-planar methods, using clinical 1.5 Tesla MR imaging systems. *Osaka City Med.* **43,** 29–48.

Namba, H., Togawa, T., Yui, N., Yanagisawa, M., Kinoshita, F., Iwadate, Y., Ohsato, K., and Sueyoshi, K. (1996). The effect of steroid on thallium-201 uptake by malignant gliomas. *Eur. J. Nucl. Med.* **23,** 991–992.

Nariai, T., Senda, M., Ishii, K., Maehara, T., Wakabayashi, S., Toyama, H., Ishiwata, K., and Hirakawa, K. (1997). Three-dimensional imaging of cortical structure, function and glioma for tumor resection. *J. Nucl. Med.* **38,** 1563–1568.

Negendank, W. G., Sauter, R., Brown, T. R., Evelhoch, J. L., Falini, A., Gotsis, E. D., Heerschap, A., Kamada, K., Lee, B. C., Mengeot, M. M., *et al.* (1996). Proton magnetic resonance spectroscopy in patients with glial tumors: A multicenter study. *J. Neurosurg.* **84,** 449–458.

Ogawa, T., Miura, S., Murakami, M., Iida, H., Hatazawa, J., Inugami, A., Kanno, I., Yasui, N., Sasajima, T., and Uemura, K. (1996a). Quantitative evaluation of neutral amino acid transport in cerebral gliomas using positron emission tomography and fluorine-18 fluorophenylalanine. *Eur. J. Nucl. Med.* **23,** 889–895.

Ogawa, T., Inugami, A., Hatazawa, J., Kanno, I., Murakami, M., Yasui, N., Mineura, K., and Uemura, K. (1996b). Clinical positron emission tomography for brain tumors: Comparison of fluorodeoxyglucose F-18 and L-methyl-^{11}C-methionine. *AJNR, Am. J. Neuroradiol.* **17,** 345–353.

Ohue, S., Kumon, Y., Kohno, K., Nagato, S., Nakagawa, K., Ohta, S., Sakaki, S., and Kusunoki, K. (1998). [Surgical management for preserving motor function in patients with gliomas near the primary motor cortex: Usefulness of preoperative identification of motor cortex and intraoperative monitoring of motor evoked potentials]. *No Shinkei Geka* **26,** 599–606.

Ojemann, J. G., Ojemann, E., Lettich, E., and Berger, M. (1989). Cortical language localization in left, dominant hemisphere. An electrical stimulation mapping investigation in 117 patients. *J. al.Neurosurg.* **71,** 316–326.

Ojemann, J. G., Miller, J. W., and Silbergeld, D. L. (1996). Preserved function in brain invaded by tumor. *Neurosurgery* **39,** 253–258; discussion 258–259.

Ojemann, J. G., Neil, J. M., MacLeod, A. M., Silbergeld, D. L., Dacey, R. G., Jr., Petersen, S. E., and Raichle, M. E. (1998). Increased functional vascular response in the region of a glioma. *J. Cereb. Blood Flow Metab.* **18,** 148–153.

Ott, D., Hennig, J., and Ernst, T. (1993). Human brain tumors: Assessment with *in vivo* proton MR spectroscopy. *Radiology* **186,** 745–752.

Poptani, H., Puumalainen, A. M., Grohn, O. H., Loimas, S., Kainulainen, R., Yla-Herttuala, S., and Kauppinen, R. A. (1998). Monitoring thymidine kinase and ganciclovir-induced changes in rat malignant glioma *in vivo* by nuclear magnetic resonance imaging. *Cancer Gene Ther.* **5,** 101–109.

Preul, M. C., Caramanos, Z., Collins, D. L., Villemure, J. G., Leblanc, R., Olivier, A., Pokrupa, R., and Arnold, D. L. (1996). Accurate, noninvasive diagnosis of human brain tumors by using proton magnetic resonance spectroscopy. *Nat. Med.* **2,** 323–325.

Pujol, J., Conesa, G., Deus, J., Lopez-Obarrio, L., Isamat, F., and Capdevila, A. (1998). Clinical application of functional magnetic resonance imaging in presurgical identification of the central sulcus. *J. Neurosurg.* **88,** 863–869.

Righini, A., de Divitiis, O., Prinster, A., Spagnoli, D., Appollonio, I., Bello, L., Scifo, P., Tomei, G., Villani, R., Fazio, F., and Leonardi, M. (1996). Functional MRI: Primary motor cortex localization in patients with brain tumors. *J. Comput. Assist. Tomogr.* **20,** 702–708.

Roelcke, U., Radu, E. W., von Ammon, K., Hausmann, O., Maguire, R. P., and Leenders, K. L. (1995). Alteration of blood–brain barrier in human brain tumors: Comparison of [^{18}F]fluorodeoxyglucose, [^{11}C]methionine and rubidium-82 using PET. *J. Neurol. Sci.* **132,** 20–27.

Roelcke, U., Leenders, K. L., von Ammon, K., Radu, E. W., Vontobel, P., Gunther, I., and Psylla, M. (1996a). Brain tumor iron uptake measured with positron emission tomography and [52]Fe-citrate. *J. Neurooncol.* **29,** 157–165.

Roelcke, U., Radu, E., Ametamey, S., Pellikka, R., Steinbrich, W., and Leenders, K. L. (1996b). Association of rubidium and *C*-methionine uptake in brain tumors measured by positron emission tomography. *J. Neurooncol.* **27,** 163–171.

Roelcke, U., Blasberg, R. G., von Ammon, K., Hofer, S., Vontobel, P., Maguire, R. P., Radu, E. W., Herrmann, R., and Leenders, K. L. (1998). Dexamethasone treatment and plasma glucose levels: Relevance for fluorine-18-fluorodeoxyglucose uptake measurements in gliomas. *J. Nucl. Med.* **39,** 879–884.

Roesdi, M. F., Postma, T. J., Hoekstra, O. S., van Groeningen, C. J., Wolbers, J. G., and Heimans, J. J. (1998). Thallium-201 SPECT as response parameter for PCV chemotherapy in recurrent glioma. *J. Neurooncol.* **40,** 251–255.

Roux, F. E., Ranjeva, J. P., Boulanouar, K., Manelfe, C., Sabatier, J., Tremoulet, M., and Berry, I. (1997). Motor functional MRI for presurgical evaluation of cerebral tumors. *Stereotact. Funct. Neurosurg.* **68,** 106–111.

Rueckert, L., Appollonio, I., Grafman, J., Jezzard, P., Johnson, R., Jr., Le Bihan, D., and Turner, R. (1994). Magnetic resonance imaging functional activation of left frontal cortex during covert word production. *J. Neuroimaging* **4,** 67–70.

Salazar, O. M., and Rubin, P. (1976). The spread of glioblastoma multiforme as a determining factor in the radiation treated volume. *Int. J. Radiat. Oncol. Biol. Phys.* **1,** 627–637.

Sasahira, M., Asakura, T., Niiro, M., Haruzono, A., Hirakawa, W., Matsumoto, T., and Fujimoto, T. (1995). Functional magnetic resonance imaging of the human motor cortex. *Neurol. Med. Chir. (Tokyo)* **35,** 277–284.

Sasaki, M., Kuwabara, Y., Yoshida, T., Nakagawa, M., Fukumura, T., Mihara, F., Morioka, T., Fukui, M., and Masuda, K. (1998). A comparative study of thallium-201 SPET, carbon-11 methionine PET and fluorine-18 fluorodeoxyglucose PET for the differentiation of astrocytic tumours. *Eur. J. Nucl. Med.* **25,** 1261–1269.

Schulder, M., Maldjian, J. A., Liu, W. C., Holodny, A. I., Kalnin, A. T., Mun, I. K., and Carmel, P. W. (1998). Functional image-guided surgery of intracranial tumors located in or near the sensorimotor cortex. *J. Neurosurg.* **89,** 412–418.

Schwartz, R. B., Holman, B. L., Polak, J. F., Garada, B. M., Schwartz, M. S., Folkerth, R., Carvalho, P. A., Loeffler, J. S., Shrieve, D. C., Black, P. M., and Alexander, E., 3rd. (1998). Dual-isotope single-photon emission computerized tomography scanning in patients with glioblastoma multiforme: Association with patient survival and histopathological characteristics of tumor after high-dose radiotherapy. *J. Neurosurg.* **89,** 60–68.

Shinoura, N., Nishijima, M., Hara, T., Haisa, T., Yamamoto, H., Fujii, K., Mitsui, I., Kosaka, N., Kondo, T., and Hara, T. (1997). Brain tumors: Detection with C-11 choline PET. *Radiology* **202,** 497–503.

Shoup, T. M., Olson, J., Hoffman, J. M., Votaw, J., Eshima, D., Eshima, L., Camp, V. M., Stabin, M., Votaw, D., and Goodman, M. M. (1999).

Synthesis and evaluation of [^{18}F]1-amino-3-fluorocyclobutane-1-carboxylic acid to image brain tumors. *J. Nucl. Med.* **40**, 331–338.

Sijens, P. E., Vecht, C. J., Levendag, P. C., van Dijk, P., and Oudkerk, M. (1995a). Hydrogen magnetic resonance spectroscopy follow-up after radiation therapy of human brain tumors. *Invest. Radiol.* **30**, 738–744.

Sijens, P. E., Knopp, M. V., Brunetti, A., Wicklow, K., Alfano, B., Bachert, P., Sanders, J. A., Stillman, A. E., Kett, H., Sauter, R., *et al.* (1995b). ^{1}H MR spectroscopy in patients with metastatic brain tumors: A multicenter study. *Magn. Reson. Med.* **33**, 818–826.

Skirboll, S. S., Ojemann, G. A., Berger, M. S., Lettich, E., and Winn, H. R. (1996). Functional cortex and subcortical white matter located within gliomas. *Neurosurgery* **38**, 678–684; discussion 684–685.

Smith, M. T., Ludwig, C. L., Godfrey, A. D., and Armbrustmacher, V. W. (1983). Grading of oligodendrogliomas. *Cancer* **52**, 2107–2114.

Spence, A. M., Muzi, M., Graham, M. M., O'Sullivan, F., Krohn, K. A., Link, J. M., Lewellen, T. K., Lewellen, B., Freeman, S. D., Berger, M. S., and Ojemann, G. A. (1998). Glucose metabolism in human malignant gliomas measured quantitatively with PET, 1-[C-11]glucose and FDG: Analysis of the FDG lumped constant. *J. Nucl. Med.* **39**, 440–448.

Stapleton, S. R., Kiriakopoulos, E., Mikulis, D., Drake, J. M., Hoffman, H. J., Humphreys, R., Hwang, P., Otsubo, H., Holowka, S., Logan, W., and Rutka, J. T. (1997). Combined utility of functional MRI, cortical mapping, and frameless stereotaxy in the resection of lesions in eloquent areas of brain in children. *Pediatr. Neurosurg.* **26**, 68–82.

Steinmetz, H., and Seitz, R. J. (1991). Functional anatomy of language processing: Neuroimaging and the problem of individual variability. *Neuropsychologia* **29**, 1149–1161.

Tedeschi, G., Bertolino, A., Righini, A., Campbell, G., Raman, R., Duyn, J. H., Moonen, C. T., Alger, J. R., and Di Chiro, G. (1995). Brain regional distribution pattern of metabolite signal intensities in young adults by proton magnetic resonance spectroscopic imaging. *Neurology* **45**, 1384–1391.

Tedeschi, G., Lundbom, N., Raman, R., Bonavita, S., Duyn, J. H., Alger, J. R., and Di Chiro, G. (1997). Increased choline signal coincides with malignant degeneration of cerebral gliomas: A serial proton magnetic resonance spectroscopic imaging study. *J. Neurosurg.* **87**, 516–524.

Thapar, K., Fukuyama, K., and Rutka, J. T. (1995). Neurogenetics and the molecular biology of human brain tumors. *In* "Brain Tumors: An Encyclopedic Approach" (A. H. Kaye and E. R. Laws, Jr., eds.), Ch. 5. Churchill-Livingstone, Edinburgh.

Thiel, A., Herholz, K., von Stockhausen, H. M., van Leyen-Pilgram, K., Pietrzyk, U., Kessler, J., Wienhard, K., Klug, N., and Heiss, W. D. (1998). Localization of language-related cortex with ^{15}O-labeled water PET in patients with gliomas. *NeuroImage* **7**, 284–295.

Tien, R. D., Felsberg, G. L., Friedman, H., Brown, M., and MacFall, J. (1993). MR imaging of high grade cerebral gliomas: Value of diffusion weighted echoplanar pulse sequences. *Am. J. al.Roentgenol.* **162**, 671–677.

Tjuvajev, J. G., Macapinlac, H. A., Daghighian, F., Scott, A. M., Ginos, J. Z., Finn, R. D., Kothari, P., Desai, R., Zhang, J., Beattie, B., *et al.* (1994). Imaging of brain tumor proliferative activity with iodine-131–iododeoxyuridine. *J. Nucl. Med.* **35**, 1407–1417.

Vander Borght, T., Pauwels, S., Lambotte, L., Labar, D., De Maeght, S., Stroobandt, G., and Laterre, C. (1994). Brain tumor imaging with PET and 2-[carbon-11]thymidine. *J. Nucl. Med.* **35**, 974–982.

Voges, J., Herholz, K., Holzer, T., Wurker, M., Bauer, B., Pietrzyk, U., Treuer, H., Schroder, R., Sturm, V., and Heiss, W. D. (1997). ^{11}C-Methionine and ^{18}F-2-fluorodeoxyglucose positron emission tomography: A tool for diagnosis of cerebral glioma and monitoring after brachytherapy with ^{125}I seeds. *Stereotact. Funct. Neurosurg.* **69**, 129–135.

Wald, L. L., Nelson, S. J., Day, M. R., Nooworolski, S. E., Henry, R. G., Huhn, S. L., Chang, S., Prados, M. D., Sneed, P. K., Larson, D. A.,

Wara, W. M., McDermott, M., Dillion, W. P., Gutin, P. H., and Vigneron, D. B. (1997). Serial proton magnetic resonance spectroscopy imaging of glioblastoma multiforme after brachytherapy. *J. Neurosurg.* **87**, 525–534.

Walker, M. D., Alexander, E., Jr., Hunt, W. E., MacCarty, C. S., Mahaley, M. S., Jr., Mealey, J., Jr., Norrell, H. A., Owens, G., Ransohoff, J., Wilson, C. B., Gehan, E. A., and Strike, T. A. (1978). Evaluation of BCNU and/or radiotherapy in the treatment of anaplastic gliomas. A cooperative clinical trial. *J. Neurosurg.* **49**, 333–343.

Walker, M. D., Strike, T. A., and Sheline, G. E. (1979). An analysis of dose-effect relationship in the radiotherapy of malignant gliomas. *Int. J. Radiat. Oncol. Biol. Phys.* **5**, 1725–1731.

Wang, G. J., Volkow, N. D., Lau, Y. H., Fowler, J. S., Meek, A. G., Park, T. L., Wong, C., Roque, C. T., Adler, A. J., and Wolf, A. P. (1996). Glucose metabolic changes in nontumoral brain tissue of patients with brain tumor following radiotherapy: A preliminary study. *J. Comput. Assist. Tomogr.* **20**, 709–714.

Weber, W., Bartenstein, P., Gross, M. W., Kinzel, D., Daschner, H., Feldmann, H. J., Reidel, G., Ziegler, S. I., Lumenta, C., Molls, M., and Schwaiger, M. (1997). Fluorine-18-FDG PET and iodine-123–IMT SPECT in the evaluation of brain tumors. *J. Nucl. Med.* **38**, 802–808.

Wester, H. J., Herz, M., Weber, W., Heiss, P., Senekowitsch-Schmidtke, R., Schwaiger, M., and Stocklin, G. (1999). Synthesis and radiopharmacology of O-(2-[^{18}F]fluoroethyl)-L-tyrosine for tumor imaging. *J. Nucl. Med.* **40**, 205–212.

Willemsen, A. T., van Waarde, A., Paans, A. M., Pruim, J., Luurtsema, G., Go, K. G., and Vaalburg, W. (1995). *In vivo* protein synthesis rate determination in primary or recurrent brain tumors using L-[1–^{11}C]-tyrosine and PET. *J. Nucl. Med.* **36**, 411–419.

Woesler, B., Kuwert, T., Morgenroth, C., Matheja, P., Palkovic, S., Schafers, M., Vollet, B., Schafers, K., Lerch, H., Brandau, W., Samnick, S., Wassmann, H., and Schober, O. (1997). Non-invasive grading of primary brain tumours: Results of a comparative study between SPET with ^{123}I-alpha-methyltyrosine and PET with ^{18}F-deoxyglucose. *Eur. J. Nucl. Med.* **24**, 428–434.

Wurker, M., Herholz, K., Voges, J., Pietrzyk, U., Treuer, H., Bauer, B., Sturm, V., and Heiss, W. D. (1996). Glucose consumption and methionine uptake in low-grade gliomas after iodine-125 brachytherapy. *Eur. J. Nucl. Med.* **23**, 583–586.

Yanaka, Y., Shirai, S., Kimura, H., Kamezaki, T., Matsumuna, A., and Nose, T. (1995). Clinical application of diffusion-weighted magnetic resonance imaging to intracranial disorders. *Neurol. Med. Chir. (Tokyo)* **35**, 648–654.

Yetkin, F. Z., Papke, R. A., Mark, L. P., Daniels, D. L., Mueller, W. M., and Haughton, V. M. (1995). Location of the sensorimotor cortex: Functional and conventional MR compared. *AJNR, Am. J. Neuroradiol.* **16**, 2109–2113.

Yoshiura, T., Hasuo, K., Mihara, F., Masuda, K., Morioka, T., and Fukui, M. (1997). Increased activity of the ipsilateral motor cortex during a hand motor task in patients with brain tumor and paresis. *AJNR, Am. J. Neuroradiol.* **18**, 865–869.

Zhao, M., Pipe, J. G., Bonnett, J., and Evelhoch, J. L. (1996). Early detection of treatment response by diffusion-weighted ^{1}H-NMR spectroscopy in a murine tumour *in vivo*. *Br. J. Cancer* **73**, 61–64.

Zingale, A., Musumeci, S., Nicoletti, G., Zingale, R., and Albanese, V. (1995). Thallium-201–SPECT and ^{99}Tc-HM-PAO SPECT imaging to study functionally cerebral supratentorial neoplasms: The biological basis of the functional imaging interpretation. *J. Neurosurg. Sci.* **39**, 227–235.

Zuelch, K. J. (1979). "Histological Typing of Tumours of the Central Nervous System" (*International Histological Classification of Tumours, No. 21*), pp. 19–24. World Health Organization, Geneva.

17

Neurodegenerative Disorders of the Cerebellum

Sid Gilman, Mary Heumann, and Larry Junck

Department of Neurology, University of Michigan Medical Center, Ann Arbor, Michigan 48109

I. Introduction

II. Friedreich's Ataxia

III. The Hereditary Cerebellar Degenerations

IV. The Sporadic Cerebellar Degenerations

References

Imaging provides important diagnostic information about cerebellar neurodegenerative disorders, and functional imaging has begun to elucidate the biochemical abnormalities involved. Friedreich's ataxia is recessively inherited, resulting from a mutation with expansions of an unstable GAA trinucleotide repeat on chromosome 9. Magnetic resonance imaging (MRI) reveals severe atrophy of the cervical spinal cord, with mild atrophy of the brainstem and cerebellum. Positron emission tomography with [^{18}F]fluorodeoxyglucose (PET-FDG) demonstrates glucose hypermetabolism in patients still ambulatory and a decline toward normal in wheelchair-bound patients. Identification of genes responsible for dominantly inherited cerebellar degenerations led to a new classification based upon genotype. Spinocerebellar ataxia (SCA) types 1, 2, 3, 6, and 7 result from mutations consisting of unstable CAG trinucleotide repeat expansions on chromosomes 6, 12, 14,

19, and 3, respectively. SCA8 is due to a CTG trinucleotide repeat expansion on chromosome 13. Mutations responsible for SCA4, -5, and -10 are on chromosomes 16, 11, and 22, respectively, but the types are unknown. Neither the chromosome nor the mutation responsible for SCA9 is published. MRI of SCA1, -2, -3, and -7 reveals cerebellar and brainstem atrophy with fourth ventricular enlargement, and the caudate nucleus and putamen volumes are reduced in SCA3 as well. SCA5 and -6 have cerebellar atrophy with minimal or no brainstem atrophy. PET-FDG reveals cerebellar and brainstem hypometabolism in SCA1, -2, -3, -6, and -7 and additional striatal and thalamic hypometabolism in SCA3. SCA5 has not been tested with functional imaging, and neither structural nor functional imaging findings are reported for SCA4, -8, -9, or -10. In olivopontocerebellar atrophy, the most frequent of the sporadic disorders, MRI reveals cerebellar and brainstem atrophy, and PET-FDG shows hypometabolism in these structures and the cerebral cortex, striatum, and thalamus.

I. Introduction

The neurodegenerative disorders of the central nervous system are characterized by the progressive degeneration and subsequent loss of neurons accompanied by reactive gliosis, degeneration of fibers from the deteriorating neurons, and clinical symptoms reflecting the locations of the lost neurons. Many of these disorders affect specific neuronal groups, with preservation of others, and thus can be viewed as system degenerations. These disorders occur sporadically, usually without known cause, or from genetic disease, with either dominant or recessive inheritance. Many neurodegenerative disorders affect the cerebellum and its pathways, resulting in progressive deterioration of cerebellar function manifested by increasing unsteadiness of gait, incoordination of limb movements with impairment of skilled movements such as handwriting, and a distinctive dysarthria. Other neuronal systems are affected in some of these disorders, notably the corticospinal pathway, basal ganglia, and autonomic nuclei of the brainstem and spinal cord.

This chapter focuses on the most frequently encountered neurodegenerative disorders of the cerebellum and its pathways: Friedreich's ataxia (FA), autosomal dominant cerebellar atrophy (also termed hereditary olivopontocerebellar atrophy and currently termed the spinocerebellar ataxias), sporadic olivopontocerebellar atrophy (sOPCA; also termed idiopathic sporadic cerebellar atrophy and late-onset cerebellar atrophy), sporadic cerebellar cortical atrophy (sCCA), and multiple-system atrophy (MSA).

II. Friedreich's Ataxia

A. Introduction

An autosomal recessive neurodegenerative disorder, FA is one of the most common of the hereditary ataxias. Its prevalence in Europe is 1 in 50,000. The disorder results from a mutation on chromosome 9 consisting of an unstable expansion of a GAA repeat in the first intron with production of an abnormal protein, frataxin.

B. Clinical Presentation

Prior to the identification and cloning of the gene, FA was thought to begin in childhood or early adolescence, at times in late adolescence, and only rarely above age 20. In these "typical" patients, the disorder begins with progressive ataxia of the legs followed by the arms and then progresses to cause leg weakness, leading to paraplegia, ataxic dysarthria, decreased peripheral sensation to all modalities, with particularly severe involvement of vibration sense and position sense, areflexia, and bilateral extensor plantar reflexes. Skeletal deformities are common, particularly scoliosis and pes cavus, and a

cardiomyopathy occurs in most patients late in the course. FA patients have an increased prevalence of impaired glucose tolerance and diabetes mellitus in comparison with the general population.

After the FA gene was cloned and genetic testing became widely available, several studies of large groups of patients with the FA gene demonstrated a much broader clinical spectrum than had been appreciated previously (Dürr *et al.*, 1996a; Schöls *et al.*, 1997b; De Michele *et al.*, 1998). Late onset is not uncommon; in one study, 14% of the patients developed their initial symptoms between 26 and 51 years of age (Dürr *et al.*, 1996a). In this study, the most common features were ataxia of gait (affecting 100% of cases) and ataxia of limb movement (99% of cases), followed by dysarthria (91% of cases). Areflexia of the legs was found in 87% of cases, loss of vibration sense in 78%, and extensor plantar reflexes in 79%, but scoliosis was found in only 60% of cases and pes cavus in only 55% (Dürr *et al.*, 1996a). Axonal neuropathy, defined as abnormal motor conduction velocity in the upper limbs, occurred in 98% of cases (Dürr *et al.*, 1996a).

In another large series of cases, the phenotypic spectrum of FA again was much broader than considered previously, leading to the conclusion that early-onset, areflexia, extensor plantar responses, and reduced vibration sense should no longer be considered the central diagnostic criteria of the disease (Schöls *et al.*, 1997b). As an explanation of the variability of the clinical features, several studies have shown a relationship between the length of the GAA expansion and the clinical presentation. In one investigation, larger GAA expansions were correlated with an earlier age of onset and a more rapid progression to loss of ambulation (Dürr *et al.*, 1996a). A second study confirmed this observation, showing that repeat length correlated inversely with age of symptom onset, age of onset of dysarthria, and rate of progression (Schöls *et al.*, 1997b). In a third report, the size of the smaller allele (GAA1) was inversely correlated with the age of symptom onset (De Michele *et al.*, 1998). A fourth study showed a significant inverse correlation between the GAA expansion size on the shorter expanded allele, the amplitudes of the sensory action potentials at the wrist and medial malleous, and the percentage of myelination fibers found in sural nerve biopsies (Santoro *et al.*, 1999). This fourth study suggests that the severity of the sensory neuropathy is probably genetically determined and not progressive.

C. Neuropathology

The characteristic changes consist of degeneration of sensory nerve fibers, posterior roots, and ganglion cells in the lower segments of the spinal cord, with severe loss of thick myelinated fibers and preservation of large numbers of fine unmyelinated axons. The spinal cord contains shrunken and degenerated posterior columns, with loss of fibers from the gracile and cuneate fasciculi and degeneration of the nuclei of Clarke's column and the posterior spinocerebellar tract. Degeneration of the anterior spinocerebellar tract occurs but is less marked. In some cases, the brainstem shows atrophy. Usually an interstitial myocarditis affects the heart, with enlargement and necrosis of some fibers and hypertrophy of others. Skeletal abnormalities such as kyphoscoliosis and pes cavus occur commonly.

D. Genetics

Current evidence indicates that FA is a mitochondrial disease caused by a mutation in the nuclear genome. The mutation consists of expansions of an unstable GAA trinucleotide repeat in intron 1 on chromosome 9q13 (Campuzano et al., 1996). The normal chromosome contains 7–29 trinucleotide repeats whereas the FA chromosome has 66–1360 repeats (Campuzano et al., 1996; Lamont et al., 1997; Schöls et al., 1997b). In one family with a typical FA phenotype, however, the alleles were not expanded, suggesting that the disease can result from point mutations in the X25 gene on both chromosomes or from locus heterogeneity (Schöls et al., 1997b). The GAA instability in Friedreich's ataxia is a DNA-directed loss-of-function mutation in the gene encoding the mitochondrial protein frataxin (Priller et al., 1997) resulting from improper DNA structure in the repeat region (Gacy et al., 1998). Unlike CAG or CGG repeats, which form hairpins, the GAA repeats form a YRY triple helix that mediates intergenerational instability in 96% of transmissions. The overall result of the mutation is a reduction of frataxin levels, and this may affect mitochondrial function (Campuzano et al., 1997).

Frataxin mRNA is expressed predominantly in tissues rich in mitochondria with a high metabolic rate, including liver, kidney, brown fat, and heart (Koutnikova et al., 1997). Mouse and yeast frataxin homologues contain a potential mitochondrial targeting sequence in their N-terminal domains, and disruption of the yeast gene results in mitochondrial dysfunction (Koutnikova et al., 1997). Frataxin is involved in iron homeostasis in mitochondria (Babcock et al., 1997), and iron accumulation in mitochondria may underlie the pathophysiology of FA; however, serum iron and ferritin concentrations are within normal limits in FA patients (Wilson et al., 1998). Consequently, iron-chelation therapy does not appear advisable.

E. Structural Imaging

Magnetic resonance imaging (MRI) reveals severe atrophy of the cervical spinal cord in patients with both early- and late-onset FA (Wessel et al., 1988; Klockgether et al., 1993). The anterior-posterior diameter of the cervical cord is decreased, and degenerative changes in the posterior and lateral columns result in an abnormal signal (Wessel et al., 1988; Klockgether et al., 1996). The brainstem and cerebellum usually show mild atrophy (Klockgether et al., 1991; Abyad and Kligman, 1995; Spieker et al., 1995), but greater atrophy is found in advanced cases (Wessel et al., 1988; Junck et al., 1994).

F. Functional Imaging

1. Cerebral Metabolism

Local cerebral metabolic rates for glucose (lCMRglc), studied with [^{18}F]fluorodeoxyglucose ([^{18}F]FDG) and positron emission tomography (PET), are significantly *increased* in the cerebral cortex, basal ganglia, thalamus, cerebellum, and brainstem of FA patients who are still ambulatory in comparison to normal control subjects (Fig. 1) (Gilman et al., 1990). In contrast, lCMRglc is increased only in the caudate and lenticular nuclei in nonambulatory patients. These findings suggest that lCMRglc is increased extensively in the central nervous system (CNS) early in the course and, with disease progression, lCMRglc decreases in a regionally specific manner. This is in keeping with the notion that FA is a mitochondrial disorder leading to abnormal storage of iron in the mitochondria.

2. Cerebral Blood Flow

In a study utilizing Tc-HMPAO and single photon emission computed tomography (SPECT), local cerebral blood flow (lCBF) was decreased in the cerebellum of FA patients (Giroud et al., 1994). This was attributed to deafferentation owing to degeneration of the spinocerebellar pathways.

3. Cerebral Neurotransmitter Receptors

A single PET study of γ-aminobutyric acid type A/benzodiazepine (GABA-A/BDZ) receptors with [^{11}C]RO15-178, a specific antagonist of the central type of GABA-A/BDZ receptors, revealed no differences between FA patients and normal control subjects (Chavoix et al., 1990).

G. Magnetic Stimulation of the Cerebral Cortex

In a longitudinal study of 13 FA patients over 9–12 years, peripheral motor and sensory nerve conduction was compared with central motor conduction evoked

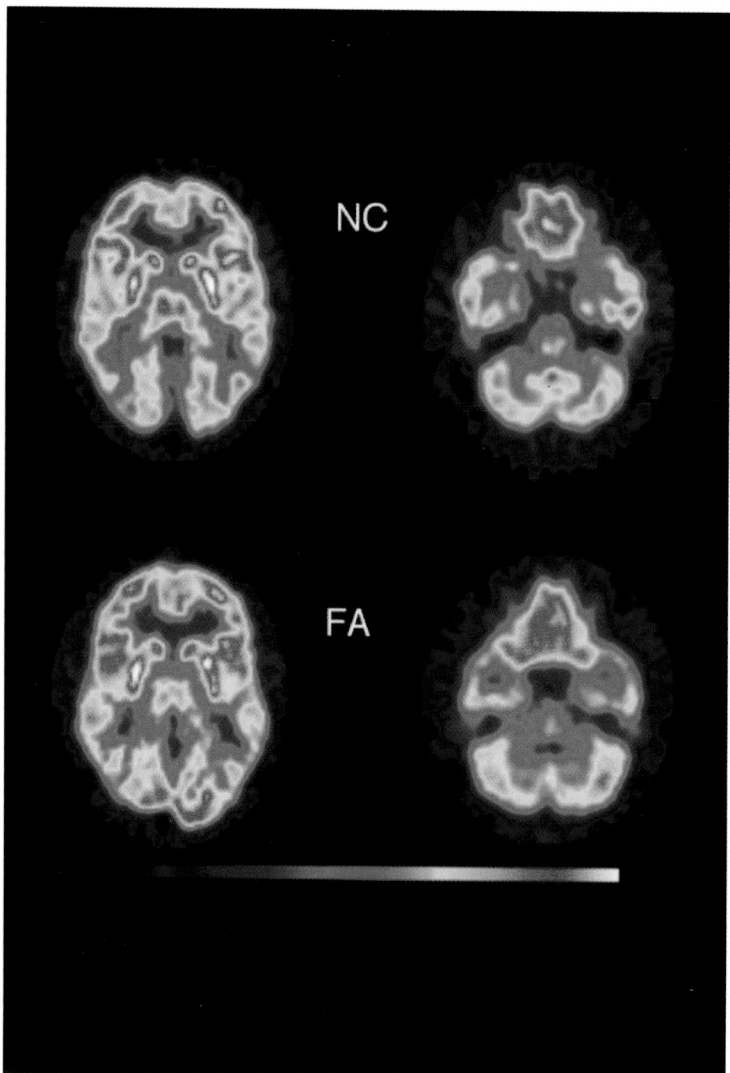

Figure 1 PET scans illustrating lCMRglc as detected with [^{18}F]FDG in a 36-year-old female normal control subject (upper two images, NC) and a 23-year-old female patient with clinically symptomatic Friedreich's ataxia who is still able to walk (lower two images, FA). The left upper and lower scans show horizontal sections at the level of the cerebral cortex, basal ganglia, and thalamus, and the right upper and lower scans show horizontal sections at the level of the cerebellum and brainstem. The scans reveal *increased* lCMRglc in the cerebral cortex, caudate nucleus, putamen, thalamus, cerebellum, and brainstem of the patient as compared to the control subject. The color bar indicates the relative rate of cerebral glucose utilization for all scans illustrated, with values increasing from left to right on the bar.

by magnetic stimulation of the cerebral cortex (Cruz-Martinez and Palau, 1997). The nerve action potential amplitudes were reduced and peripheral conduction velocities were delayed, and these abnormalities remained unchanged over time. In contrast, central motor conduction worsened progressively over time, with significant increases in threshold and significant decreases in amplitude of motor-evoked potentials. The findings suggest that clinical progression results from ongoing degeneration of corticospinal and cerebellar pathways, not from progressive degeneration of peripheral nerve fibers. This finding buttresses another study (Santoro *et al.*, 1999) suggesting that the sensory neuropathy in FA is static and not progressive.

III. The Hereditary Cerebellar Degenerations

A. Introduction

Most of the hereditary cerebellar degenerations are transmitted by autosomal dominant inheritance, although autosomal recessive inheritance has been reported. These disorders are heterogeneous, consisting principally of a slowly progressive ataxia of gait, coordinated limb movements, extraocular movements, and speech. Many of the hereditary cerebellar degenerations include symptoms referable to corticospinal and basal ganglia impairment, and some also lead to ophthalmoparesis, amyotrophy, and peripheral neuropathy. Both clinical and neuropathological classifications of the hereditary cerebellar degenerations have been attempted, but most have been unsuccessful, as the families reported have a large and diverse array of clinical symptoms and signs as well as neuropathological changes that do not conform well to the classifications. Harding (1982) proposed a classification of the autosomal dominant cerebellar atrophies that was widely adopted despite its shortcomings. Now that identification of the responsible genes has become feasible, a new genetically based classification has emerged. The autosomal dominant cerebellar degenerations are now designated as spinocerebellar ataxias, with a sequential number given as each genetic locus is identified. Currently, the classification includes spinocerebellar ataxia type 1 (SCA1), which has a locus on chromosome 6p22–p23 (Zoghbi et al., 1991; Orr et al., 1993); SCA2 on 12q23–q24.1 (Pulst et al., 1996); SCA3/MJD (Machado-Joseph disease) on 14q24.3–q32.1 (Takiyama et al., 1993; Stevanin et al., 1994b); SCA4 on 16q22.1 (Gardner et al., 1994; Flanigan et al., 1996); SCA5 on 11q13 (Ranum et al., 1994a; Zhu and Gerhard, 1998); SCA6 on 19p13 (Ishikawa et al., 1997; Stevanin et al., 1997); SCA7 on 3p12–13 (Gouw et al., 1995; David et al., 1997); SCA8 on 13q21 (Koob et al., 1999); SCA9 on a chromosomal site not identified in the literature yet; and SCA10 on 22q13 (Matsuura et al., 1999; Zu et al., 1999).

B. Spinocerebellar Ataxia Type 1

1. Introduction

The genetic disorder currently recognized as SCA1 was described initially in 1941 in a family that was later studied in detail (Gray and Oliver, 1941; Schut, 1950, 1951, 1991; Schut and Haymaker, 1953; Haines et al., 1984). SCA1 is less common than SCA2 or -3; between 6 and 15% of index cases have SCA1 (Ranum et al., 1994b; Silveira et al., 1996; Schöls et al., 1997a; Klockgether et al., 1998).

2. Clinical Presentation

The disorder begins at a highly variable age, but usually in the third and forth decades with mild ataxia of gait, subtle ataxia of speech, and impaired handwriting. In the Schut kindred, which is the largest known, the average age of onset was 26 years, with a range of 17–35, the disease duration is 12–14 years, and the average age at death is 37 years (Haines et al., 1984). In other large kindreds, the mean age of onset is 36 years, the range is 15–63 years, and the duration is 21 years (Sasaki et al., 1996). The initial clinical findings include slowness in initiating movements, slow and infrequent eye blinking, and abnormal eye movements, including saccadic pursuit movements, hypermetric saccades, and nystagmus on lateral gaze. As the disease progresses, the saccades may become progressively slowed until complete ophthalmoplegia occurs, and this is accompanied by severe, incapacitating ataxia of gait, incoordination of limb movements, and ataxic dysarthria. Impairment of other systems is common, including (a) the corticospinal projections, with spasticity of gait, limb movements, and speech, hyperreflexia with clonus, and extensor plantar responses; (b) the basal ganglia, with parkinsonism, involuntary movements, or dystonia; and (c) the motor nuclei of the medulla, with severe dysarthria, dysphagia, impairment of the cough reflex, and tongue fasciculations. A peripheral neuropathy often occurs, usually late in the course, and dementia has been reported.

3. Neuropathology

Gross examination usually reveals mild to moderate widening of sulci in the frontotemporal areas, marked atrophy of the pons and cerebellum, loss of the inferior olivary bulge in the medulla, shrinkage of the cervical and lumbar enlargements of the spinal cord, and atrophy of the ventral roots (Genis et al., 1995; Robitaille et al., 1995). The basal ganglia appear normal. Microscopic examination shows mild neuronal loss with gliosis in the cerebral cortex, severe loss of Purkinje cells and dentate neurons with atrophy of the dentatofugal pathway, severe neuronal loss with gliosis in the inferior olive and pontine nuclei, and atrophy of the superior, middle, and inferior cerebellar peduncles. Purkinje cell loss is worse in the vermis, less severe in the hemispheres, and moderate in the flocculonodular lobes. Neuronal cell loss also occurs in the nuclei of the third, tenth, and twelfth cranial nerves. The spinal cord shows loss of anterior horn cells and neurons of Clarke's columns, and degeneration of the posterior spinocerebellar tracts, posterior columns, and corticospinal tracts. In one family, the neuropathological changes resembled those seen in multiple-system atrophy, with extensive glial cytoplasmic inclusions accompanying degeneration of neurons in the basal ganglia, cerebellum, brainstem, and spinal autonomic nuclei (Gilman et al., 1996b).

4. Genetics

SCA1 results from a mutation in chromosome 6p consisting of a highly polymorphic, unstable CAG repeat expansion (Zoghbi *et al.*, 1991). The CAG repeat size is inversely related to the age of onset, with the largest allele lengths associated with juvenile onset. Normal alleles contain 6–37 repeats whereas SCA1 alleles contain 40–81 repeats (Orr *et al.*, 1993; Servadio *et al.*, 1995). The CAG repeat codes for a polyglutamine sequence within a novel protein termed ataxin-1 that is found in both the nucleus and the cytoplasm of Purkinje cells (Chong *et al.*, 1995).

5. Structural Imaging

MRI scans reveal atrophy of the cerebellar vermis, cerebellar hemispheres, base of the pons, middle cerebellar peduncles, and medulla oblongata, with enlargement of the fourth ventricle and atrophy of the cervical spinal cord (Fig. 2) (Bürk *et al.*, 1996; Gilman *et al.*, 1996b; Mascalchi *et al.*, 1998).

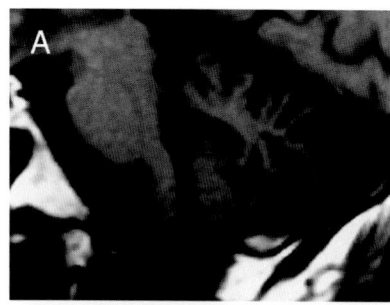
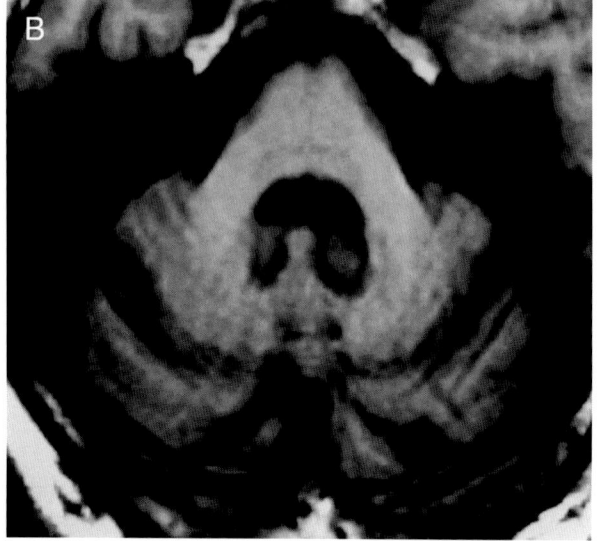

Figure 2 MR images in the sagittal (A) and horizontal (B) planes of a 73-year-old man with clinically symptomatic spinocerebellar ataxia type 1. The images reveal severe atrophy of the cerebellar vermis, cerebellar hemispheres, middle cerebellar peduncles, base of the pons, and the medulla, with enlargement of the fourth ventricle and the basal cistern.

6. Functional Imaging

In three members of a single family with the SCA1 gene, PET with [^{18}F]FDG revealed significant hypometabolism in the entire cerebral cortex, caudate nucleus, putamen, thalamus, cerebellar hemispheres, cerebellar vermis, and brainstem (Gilman *et al.*, 1996b). This pattern is similar to that seen in MSA, a sporadic disease. In the same patients, PET with [^{11}C]flumazenil, [^{11}C]FMZ, revealed normal distribution volumes of GABA-A/BDZ receptors in the cerebral hemispheres and cerebellum. There are no other published reports of functional imaging in SCA1, but we have used PET with [^{18}F]FDG to study seven symptomatic carriers of the SCA1 gene, including the three cases described earlier (Gilman *et al.*, 1996b). The results are shown in Table I in comparison to a group of normal control subjects and five cases of SCA3. Significant differences were found between normal controls and SCA1 cases in the cerebellar vermis, cerebellar hemispheres, and brainstem, and marginally in the thalamus, but not in the caudate nucleus or putamen. These findings verify the previous study showing cerebellar and brainstem hypometabolism in SCA1 (Gilman *et al.*, 1996b), but the cerebral cortical, striatal, and thalamic hypometabolism found in that study was not replicated in this larger group even though data from the three family members reported previously were included in the analysis (Table I, Fig. 3). These results suggest that the family reported previously might be unique among families with SCA1, and the usual finding will be decreased lCMRglc in the cerebellum and brainstem.

Table I Local Cerebral Metabolic Rates for Glucose Normalized to the Cerebral Cortex Studied with [^{18}F]Fluorodeoxyglucose and Positron Emission Tomography in Normal Control Subjects as Compared to Patients with Spinocerebellar Ataxia Type 1 (SCA1) and Type 3 (SCA3)[a]

Region	Normal controls $n = 51$	SCA1 $n = 7$	SCA3 $n = 5$
Caudate nucleus	1.17 ± 0.08	1.15 ± 0.10	1.09 ± 0.03[b]
Putamen	1.26 ± 0.07	1.22 ± 0.08	1.12 ± 0.06[b]
Thalamus	1.25 ± 0.07	1.18 ± 0.09[b]	1.13 ± 0.07[b]
Cerebellar vermis	0.90 ± 0.08	0.73 ± 0.07[b]	0.79 ± 0.04[b]
Cerebellar hemispheres	0.96 ± 0.07	0.76 ± 0.04[b]	0.89 ± 0.08[b,c]
Brainstem	0.86 ± 0.07	0.67 ± 0.12[b]	0.72 ± 0.04[b]

[a] No adjustments have been made for age or gender. The data presented are mean ± standard deviation. In later studies with additional SCA3 patients, the differences found in the caudate nucleus and putamen were nonsignificant.
[b] Significantly different from normal controls at $p < 0.05$.
[c] Significantly different from SCA1 at $p < 0.05$.

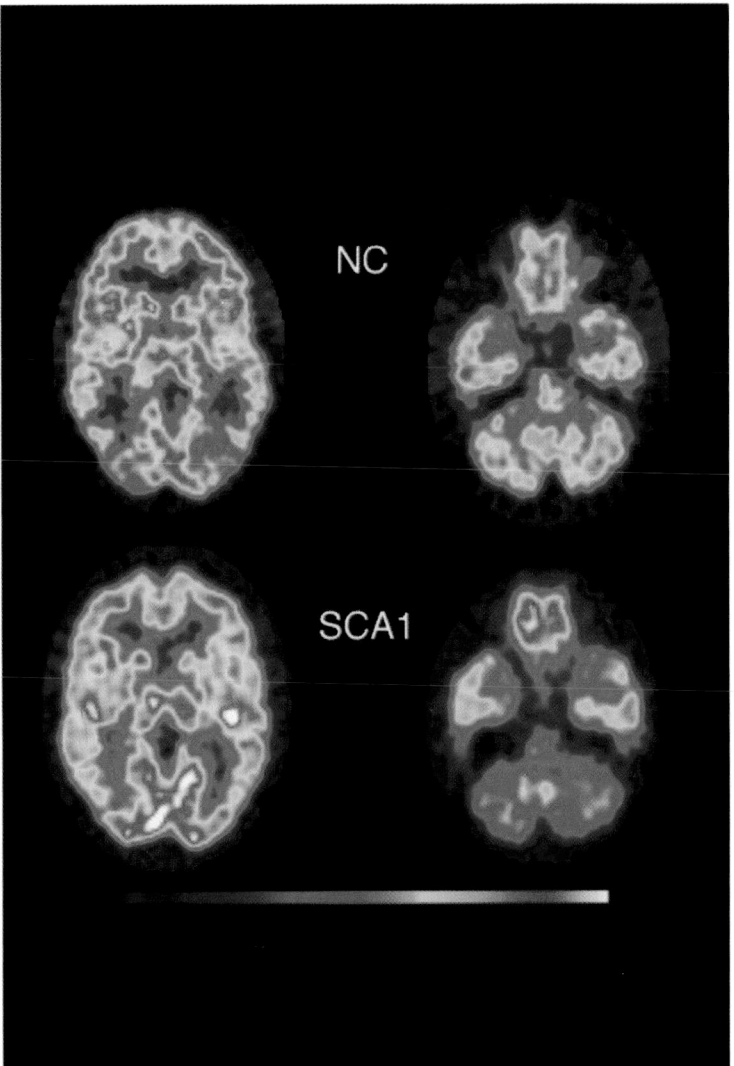

Figure 3 PET scans illustrating lCMRglc as detected with [18F]FDG in a 44-year-old male normal control subject (upper two images, NC) and a 42-year-old male patient with clinically symptomatic spinocerebellar ataxia type 1 (lower two images, SCA1). The left upper and lower scans show horizontal sections at the level of the cerebral cortex, basal ganglia, and thalamus, and the right upper and lower scans show horizontal sections at the level of the cerebellum and brainstem. The scans demonstrate decreased lCMRglc in the cerebellum and brainstem of the patient as compared to the control subject, with no differences between patient and control subject in the cerebral cortex, basal ganglia, or thalamus. The color bar indicates the relative rate of cerebral glucose utilization for all scans illustrated, with values increasing from left to right on the bar.

7. Magnetic Resonance Spectroscopy (MRS)

In a single study, proton MRS revealed a marked decrease of the *N*-acetylaspartate/creatine ratio and the choline/creatine ratio in the basal pons of symptomatic patients and a moderate decrease in asymptomatic carriers (Mascalchi *et al.,* 1998). The authors suggest that reduction of the *N*-acetylaspartate/creatine ratio in the pons might be a useful biochemical marker of SCA1.

8. Electrophysiological Studies

Visual-evoked potentials and motor potentials evoked by transcranial magnetic stimulation were abnormal in most of the SCA1 patients examined in one study (Abele *et al.,* 1997). Somatosensory-evoked potentials were delayed or absent in most, and abnormalities of brainstem auditory-evoked potentials were found in about half. In another study, peripheral nerve conduction was abnormal, reflecting the predominantly demyelinating polyneuropathy (Schöls *et al.,* 1995).

C. Spinocerebellar Ataxia Type 2

1. Introduction

This disorder was originally described in a large kindred in Cuba that was similar in phenotype to SCA1 but was not linked to the SCA1 locus (Orozco *et al.,* 1990; Gispert *et al.,* 1993). Subsequently, the SCA2 locus was mapped to chromosome 12q (Pulst *et al.,* 1996), and several additional kindreds were found with a locus in the same region (Lopes-Cendes *et al.,* 1994; Filla *et al.,* 1995; Pulst *et al.,* 1996).

2. Clinical Presentation

The mean age of onset is in the second to fourth decade, but varies from age 2 to 65 years, with about 40% of cases developing symptoms before the age of 25 years. Early onset is more common in SCA2 than in SCA1 or SCA3. The mean duration of the disease is a little over 10 years. The disorder begins with a progressive ataxia of gait accompanied by an ataxic dysarthria, cerebellar tremor, slow saccades progressing to ophthalmoparesis, and hyporeflexia of the upper limbs. Although unusual, parkinsonism, choreoathetosis, and signs of corticospinal tract disease can be seen (Bürk *et al.,* 1997). Symptomatic carriers of the SCA2 gene are more likely than those with the SCA1 or SCA3 gene to show postural and action tremor, myoclonus, early dementia, slowed ocular movements, and hyporeflexia (Geschwind *et al.,* 1997a; Schöls *et al.,* 1997a). A recent study revealed that 25% of symptomatic carriers of the SCA2 gene have dementia (Bürk *et al.,* 1999).

3. Neuropathology

Gross examination reveals marked atrophy of the brainstem and cerebellum (Orozco *et al.,* 1989; Dürr *et al.,* 1995; Robitaille *et al.,* 1997). The cerebral hemispheres show no abnormality except in the occasional patient with dementia, in which case there is cerebral cortical atrophy and enlargement of the lateral ventricle. Microscopic examination discloses depletion of neurons with reactive gliosis in the basal pons and degeneration of the transverse pontine fibers and the middle and inferior cerebellar peduncles. The inferior olives have severe neuronal loss and gliosis. In the cerebellum, Purkinje cells are markedly decreased and subcortical fibers are degenerated, but the dentate nucleus is less affected. In most cases, the substantia nigra has marked loss of neurons whether parkinsonism was found during life or not. The locus ceruleus remains preserved. The spinal cord shows degeneration of the posterior columns and spinocerebellar tracts, with loss of large

myelinated fibers in the dorsal and ventral roots (Robitaille *et al.,* 1997). Neurons of Clarke's column and anterior horn cells are depleted.

4. Genetics

The SCA2 gene mutation consists of a CAG repeat expansion in chromosome 12q with coding of a novel protein called ataxin-2 (Pulst *et al.,* 1996). Normal alleles contain 14-31 CAG repeats whereas mutated alleles contain 34-59 repeats (Imbert *et al.,* 1996; Sanpei *et al.,* 1996; Cancel *et al.,* 1997; Brice *et al.,* 1997). Anticipation occurs in SCA2, especially with paternal transmission, with long repeat sequences resulting in earlier onset of disease. Ataxin-2 is cytoplasmic in location, but its function is unknown.

5. Structural Imaging

MRI shows marked atrophy of the cerebellar vermis and medulla, with enlargement of the fourth ventricle, and atrophy of the cervical spinal cord. The middle cerebellar peduncles and base of the pons are more severely degenerated in SCA2 than in SCA1 or SCA3 (Bürk *et al.,* 1996), and the cerebellar hemispheres are smaller in SCA2 than in SCA3 (Bürk *et al.,* 1996; Klockgether *et al.,* 1998).

6. Functional Imaging

No studies are reported in the literature. We used PET with [^{18}F]FDG to study lCMRglc in one symptomatic patient with the SCA2 gene, a 52-year-old woman (Fig. 4). The data are presented in Table II. Although the sample is insufficient in size to draw conclusions, the results suggest that there is marked hypometabolism in the cerebellar vermis, cerebellar hemispheres, and brainstem, but not in the caudate nucleus, putamen, or thalamus.

D. Spinocerebellar Ataxia Type 3/ Machado–Joseph Disease (MJD)

1. Introduction

MJD was identified initially as an autosomal dominant disorder common among Portuguese families living in the United States (Nakano *et al.,* 1972; Romanul *et al.,* 1977) and the Azore Islands of Portugal (Coutinho and Andrade, 1978). Later, MJD was identified in many countries around the world, affecting people with diverse ethnic backgrounds. The mutation responsible for MJD, a CAG expansion, was found on chromosome 14q (Takiyama *et al.,* 1993). Independently, the search for a dominantly inherited ataxia with a phenotype different from MJD yielded a locus on 14q (Stevanin *et al.,* 1994a) that was identified as the SCA3 lo-

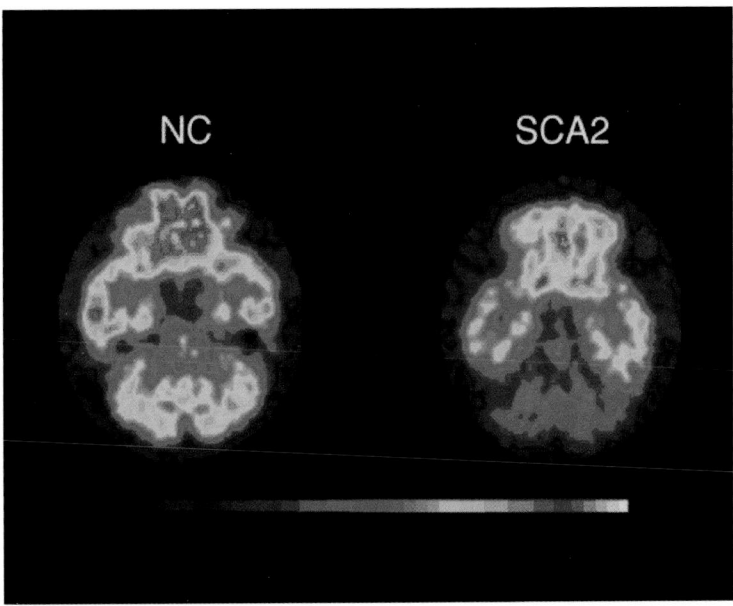

Figure 4 PET scans illustrating lCMRglc as detected with [18F]FDG in a 54-year-old female normal control subject (left image, NC) and a 52-year-old female patient with clinically symptomatic spinocerebellar ataxia type 2 (right image, SCA2). The scans reveal decreased lCMRglc in the cerebellum and brainstem of the patient as compared to the normal control. The color bar indicates the relative rate of cerebral glucose utilization for both scans, with values increasing from left to right on the bar.

cus. Subsequent studies demonstrated that the SCA3 gene mutation is the same as the MJD mutation; both are located at 14q 24.23–q31 (Stevanin *et al.*, 1995). The clinical manifestations of SCA3 and MJD reflect the considerable phenotypic variability of this mutation.

Table II   Local Cerebral Metabolic Rates for Glucose Normalized to the Cerebral Cortex Studied with [18F]Fluorodeoxyglucose and Positron Emission Tomography in Normal Control Subjects as Compared to Patients with Spinocerebellar Ataxia Type 2 (SCA2) and Type 7 (SCA7)

Region	Normal controls $n = 51$	SCA2 $n = 1$	SCA6 $n = 2$	SCA7 $n = 3$
Caudate nucleus	1.17 ± 0.08	1.14	1.16 ± 0.07	1.16 ± 0.06
Putamen	1.26 ± 0.07	1.24	1.27 ± 0.07	1.25 ± 0.09
Thalamus	1.25 ± 0.07	1.23	1.25 ± 0.08	1.18 ± 0.12
Cerebellar vermis	0.90 ± 0.08	0.67	0.85 ± 0.02	0.78 ± 0.05
Cerebellar hemispheres	0.96 ± 0.07	0.62	0.86 ± 0.10	0.83 ± 0.12
Brainstem	0.86 ± 0.07	0.56	0.77 ± 0.00	0.68 ± 0.07

2. Clinical Presentation

Three SCA3 phenotypes are common. Type 1 affects adolescent age groups, causing ataxia and ophthalmoplegia, but these problems can be overshadowed by corticospinal tract disease, with spasticity and hyperreflexia, and basal ganglia disorders, with rigidity, dystonia, and involuntary movements. Type 2 begins in early to middle adulthood, with a combination of ataxia and corticospinal tract signs of spasticity and hyperreflexia along with dystonia. Type 3 usually presents after the age of 40 years, with ataxia and ophthalmoparesis along with peripheral neuropathy or loss of anterior horn cells, resulting in decreased muscle bulk, fasciculations, and weakness. A fourth and much less frequent presentation includes parkinsonism, with tremor, rigidity, and gait disturbances; neuropathy, with oral and facial fasciculations; and bulging eyes (Dürr *et al.*, 1996b).

3. Neuropathology

Degenerative changes affect the substantia nigra, dentate nucleus, pontine nuclei, lower motor neurons of the spinal cord, and neurons of Clarke's column. Both segments of the globus pallidus are affected, but the internal segment is more damaged than the external. The subthalamic nucleus can be markedly involved. Degenerative changes in the inferior olives and Purkinje cells are less severe in SCA3 than in SCA1 or SCA2. Cranial nerve nuclei often are affected, particularly 3, 4, 6, 7, and 11, and the nucleus ambiguus and dorsal nucleus of the vagus nerve can be involved (Rosenberg *et al.*, 1976; Romanul *et al.*, 1977; Takiyama *et al.*, 1994; Dürr *et al.*, 1996b; Robitaille *et al.*, 1997).

4. Genetics

The disease results from a CAG repeat expansion in chromosome 14q24.3-q32.1 (Takiyama *et al.*, 1993). Normal alleles have 13-41 CAG repeats whereas expanded alleles have 56 to more than 80 (Giunti *et al.*, 1995; Maruyama *et al.*, 1995; Higgins *et al.*, 1996; Takiyama *et al.*, 1997). As noted above, MJD and SCA3 are allelic disorders resulting from a single mutation but with marked clinical heterogeneity. There is an inverse correlation between the length of the repeats and the age of onset, and anticipation occurs, especially with paternal transmission (Maruyama *et al.*, 1995). The gene product, ataxin-3, is in the cellular cytoplasm of multiple organs and in many regions of the brain. It accumulates abnormally in the nucleus, forming insoluble inclusions (Paulson, 1997).

5. Structural Imaging

MRI reveals atrophy of the cerebellar vermis, cerebellar hemispheres, base of the pons, middle cerebellar peduncles, and medulla, with enlargement of the fourth ventricle, and atrophy of the cervical spinal cord. The degree of atrophy of the brainstem and cerebellar vermis is correlated with the size of the expanded CAG repeat as well as with the patient's age (Onodera *et al.*, 1998). The cerebellar hemispheres and middle cerebellar peduncles are less atrophied in SCA3 than in SCA2 (Bürk *et al.*, 1996; Abe *et al.*, 1998). The caudate nucleus and putamen are reduced in volume in SCA3, but not in SCA1 or SCA2 (Klockgether *et al.*, 1998). Atrophy of the frontal and temporal lobes and the globus pallidus has been reported in SCA3 (Murata *et al.*, 1998b).

6. Functional Imaging

Studies with [^{18}F]FDG and PET show hypometabolism in the cerebellar hemispheres, cerebellar vermis, and brainstem (Soong *et al.*, 1997; Tanawaki *et al.*, 1997). Hypometabolism has also been found in the striatum and the entire cerebral cortex (Tanawaki *et al.*, 1997) and also focally in the occipital cortex (Soong *et al.*, 1997). A PET study with [^{18}F]FDG in asymptomatic carriers of the SCA3 gene revealed hypometabolism in the cerebellar hemispheres, brainstem, and occipital cortex and hypermetabolism in the parietal and temporal lobes (Soong and Liu, 1998). We studied five symptomatic carriers of the SCA3 gene using [^{18}F]FDG and PET. The results, which are presented in

Table I, show hypometabolism of the striatum, thalamus, cerebellar vermis, cerebellar hemispheres, and brainstem (Fig. 5).

In a single patient with the SCA3 gene, the local cerebral metabolic rate of oxygen (lCMRO$_2$) and lCBF studied with PET were reduced in the cerebral cortex and cerebellum (Yamazaki *et al.*, 1992). In another study with PET, lCBF was decreased in the cerebellum and pons but not in the cerebral cortex of symptomatic SCA3 carriers in comparison to a normal control group (Koshi *et al.*, 1995). In a third study utilizing SPECT and [^{123}I]iodoamphetamine, lCBF was significantly decreased in the cerebellum (Takahashi *et al.*, 1994).

Studies of nigrostriatal projections with [^{18}F]fluorodopa ([^{18}F]FDA) and PET revealed decreased uptake in the putamen with normal levels in the caudate nucleus of patients with the SCA3 gene in comparison to normal controls (Tanawaki *et al.*, 1997). In another study, however, striatal uptake was decreased in two patients of six, but the values were normal in the other four patients (Shinotoh, 1997). A study with SPECT and [^{123}I]iomazenil (RO 16-0154) revealed decreased binding to GABA-A/BDZ receptors in the cerebral cortex, thalamus, striatum, and cerebellum (Ishibashi *et al.*, 1998).

Taken together, these functional imaging studies demonstrate consistently decreased metabolic rate and blood flow in the cerebellum and brainstem of SCA3 patients. The high variability of the reported changes in the cerebral cortex and basal ganglia probably reflects the variable phenotype of the disorder.

E. Spinocerebellar Ataxia Type 4

This disorder was identified in a large kindred in Utah with cerebellar ataxia, axonal with peripheral sensory loss, neuropathy, and symptoms of corticospinal tract disease (Gardner *et al.*, 1994) and was later linked to chromosome 16q22.1 (Flanigan *et al.*, 1996). It is unclear whether SCA4 results from a CAG repeat expansion. The combination of dominantly inherited ataxia with sensory loss in SCA4 may correspond to the findings in a French family with six affected members (Biemond, 1954) and a New Zealand family with three affected siblings who had a late onset of symptoms (Pollack and Kies, 1990). Other cases have been described as well (Bennett *et al.*, 1984; Nachmanoff *et al.*, 1997). The disease begins between ages 19 and 59, but usually in the fourth or fifth decade, with ataxia of gait followed by incoordination of the upper extremities, ataxic dysarthria, loss of the deep tendon reflexes, and decreased peripheral sensation. The neuropathological findings in one case consisted of degeneration of cerebellar Purkinje cells, dorsal root ganglion neurons, and posterior columns (Nachmanoff *et al.*, 1997).

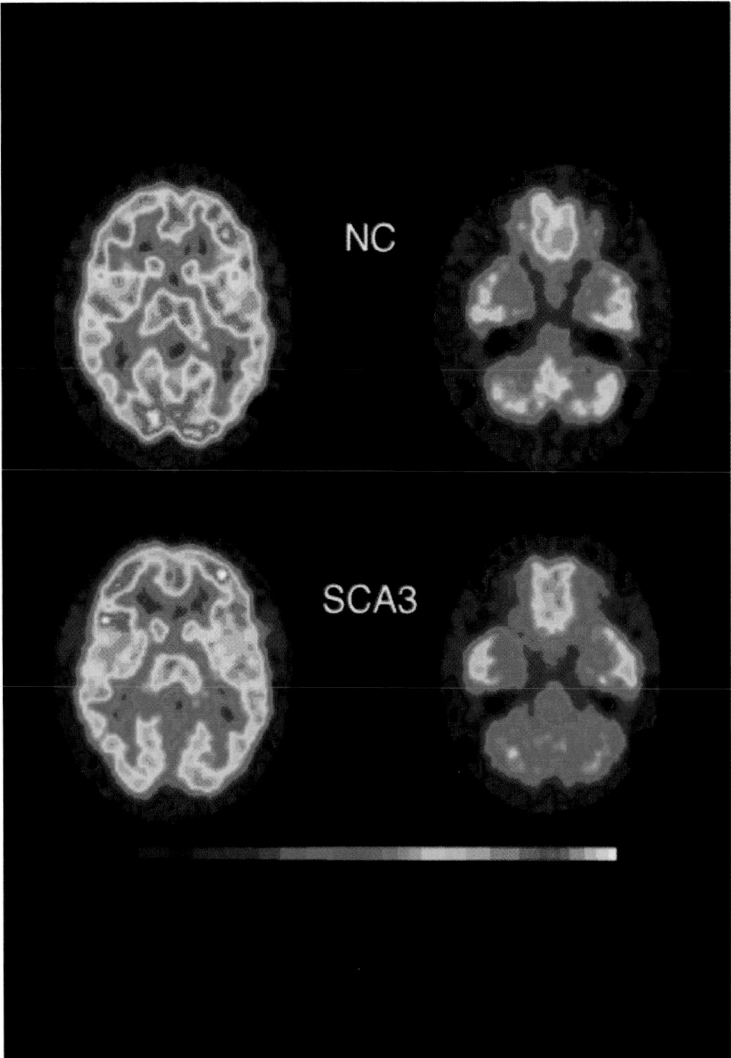

Figure 5 PET scans illustrating lCMRglc as detected with [^{18}F]FDG in a 34-year-old male normal control subject (upper two images, NC) and a 31-year-old male patient with clinically symptomatic spinocerebellar ataxia type 3 (lower two images, SCA3). The left upper and lower scans show horizontal sections at the level of the cerebral cortex, basal ganglia, and thalamus, and the right upper and lower scans show horizontal sections at the level of the cerebellum and brainstem. The scans reveal decreased lCMRglc in the caudate nucleus, putamen, thalamus, cerebellum, and brainstem of the patient as compared to the control subject. The color bar indicates the relative rate of cerebral glucose utilization for all scans illustrated, with values increasing from left to right on the bar.

F. Spinocerebellar Ataxia Type 5

This disorder was identified in a 10-generation kindred with a dominantly inherited ataxia (Ranum *et al.*, 1994a). The family had two major branches, both of which initially were thought to have descended from the paternal grandparents of President Abraham Lincoln. Subsequently, however, this disorder was found to be linked to Abraham Lincoln's name only through a distant connection by marriage, as no person in his nu-clear family or direct parental linkage developed ataxia (Nee and Higgins, 1997). SCA5 results from a mutation on chromosome 11q13 (Kitamura *et al.*, 1997; Zhu and Gerhard, 1998), but it is unclear whether the mutation consists of a CAG repeat expansion. The onset of the disease varies between ages 10 and 68 years, and anticipation is prominent. Usually the symptoms begin in the third or fourth decade, consisting principally of ataxia of gait, limb movements, and speech. The disorder progresses slowly and life span is usually normal in adult-onset cases, but with juvenile onset, life span can be shortened. Structural imaging in adult-onset cases shows cerebellar atrophy without brainstem involvement (Ranum *et al.*, 1994a).

G. Spinocerebellar Ataxia Type 6

1. Clinical Presentation

This recently described autosomal dominant ataxia has protean manifestations, but frequently the onset is later than in most other dominantly inherited ataxias, and the disease usually has an indolent course. The disorder usually presents with a slowly progressive ataxia of gait and limbs, with ataxic dysarthria, nystagmus, slow eye movements, hypoactive deep tendon reflexes, and mildly decreased vibration sense and position sense (Zhuchenko *et al.*, 1997). Other abnormalities have been described, including ophthalmoparesis, dysphagia, spasticity with hyperreflexia and extensor plantar responses, sphincter disturbances, parkinsonism and dystonia, and pes cavus (Geschwind *et al.*, 1997b; Ikeuchi *et al.*, 1997; Matsumura *et al.*, 1997; Schöls *et al.*, 1998). The disorder can present as an episodic ataxia with dysarthria. The onset can be from age 24 to 63 but usually it begins in the fourth or fifth decade (Gomez *et al.*, 1997). In two studies, the age of onset showed an inverse correlation with repeat length (Murata *et al.*, 1998a; Schöls *et al.*, 1998).

Although the disease usually progresses slowly, an indolent course, patients have been described with dramatic anticipation resulting in early onset, rapid disease progression, and severe ataxia with action tremor or action myoclonus (Watanabe *et al.*, 1998).

2. Neuropathology

The neurodegenerative changes are confined to the cerebellar cortex and inferior olivary complex. The cerebellar cortex shows severe loss of Purkinje cells and proliferation of Bergmann glia, worse in the superior parts of the cerebellar vermis and hemispheres than inferior (Gomez *et al.*, 1997; Sasaki *et al.*, 1998; Takahashi *et al.*, 1998).

3. Genetics

The mutation responsible for SCA6 is on chromosome 19p13, where a polymorphic CAG repeat was identified in the human alpha 1a-voltage-dependent calcium channel subunit gene (CACNAIA) (Diriong *et al.*, 1995; Ophoff *et al.*, 1996). Normal alleles have 4–20 repeats and expanded alleles have 21-29 repeats (Ishikawa *et al.*, 1997; Stevanin *et al.*, 1997; Zhuchenko *et al.*, 1997).

4. Structural Imaging

In most reports, MRI revealed cerebellar atrophy without brainstem involvement (Geschwind *et al.*,

1997b; Gomez *et al.*, 1997; Stevanin *et al.*, 1997; Nagai *et al.*, 1998; Satoh *et al.*, 1998; Schöls *et al.*, 1998). One study, however, showed a mild decrease of the anterior–posterior diameter of the pons and the diameter of the middle cerebellar peduncles, with mild shrinkage of the red nucleus (Murata *et al.*, 1998a).

5. Functional Imaging Has Not Been Reported in SCA6

We used PET with [^{18}F]FDG to study lCMRglc in two symptomatic carriers of the SCA6 gene. The data are presented in Table II. Although the sample is insufficient in size to warrant statistical analysis, the results suggest that lCMRglc is decreased in the cerebellar vermis, cerebellar hemispheres, and brainstem (Fig. 6), but within normal limits in the caudate nucleus, putamen, and thalamus.

H. Spinocerebellar Ataxia Type 7

1. Introduction

This is an autosomal dominant cerebellar ataxia with macular degeneration that typically begins with a visual disorder and evolves to include progressive ataxia of gait, limb movements, and speech.

2. Clinical Presentation

The disorder can begin at almost any age between birth and age 65 years (Benomar *et al.*, 1994; Enevoldson *et al.*, 1994; Benton *et al.*, 1998; David *et al.*, 1998; Johansson *et al.*, 1998). Most patients initially develop progressive loss of vision, and this leads to blindness from macular degeneration. As vision worsens, an ataxia develops that initially affects gait and then later speech and limb movements. Supranuclear ophthalmoplegia can occur. In early-onset cases, the deep tendon reflexes may be decreased, and in other cases, the disorder can include dementia, nystagmus, signs of corticospinal tract disease, and peripheral sensory loss. Rarely, signs of basal ganglia disease may occur, principally chorea and dyskinesias of the mouth and face. Auditory hallucinations were described in one patient (Benton *et al.*, 1998). In a severe infantile-onset form, death can occur by 7 months of age (Benton *et al.*, 1998).

Two studies have shown correlations between repeat lengths and clinical presentation. In one, visual impairment was the most common initial symptom in patients with 59 or more repeats, whereas ataxia predominated in patients with fewer than 59 repeats (Johansson *et al.*, 1998). In the other study, a strong negative correlation was found between the age of onset and the length of the CAG repeat expansion (David *et al.*, 1998). Larger expansions were also associated with a severe, rapid clinical course and a high frequency of decreased vision, oph-

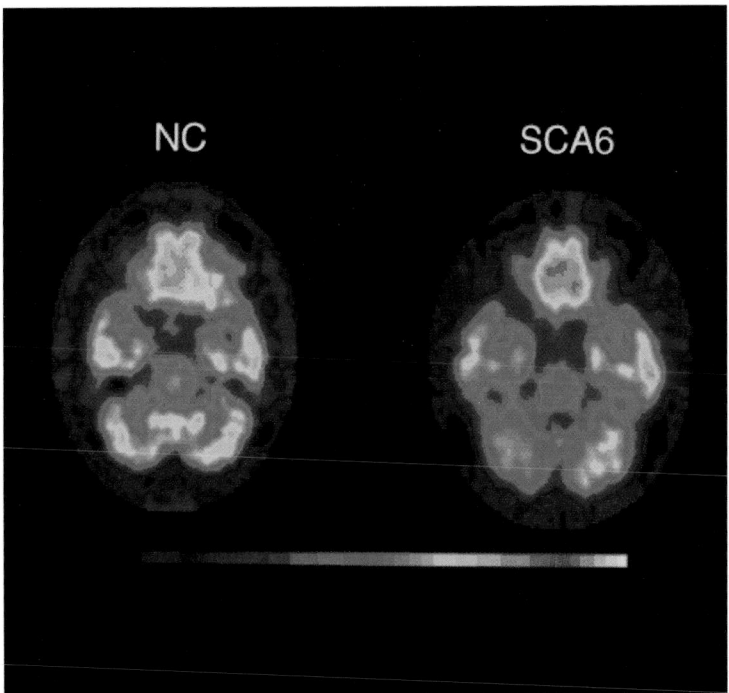

Figure 6 PET scans showing lCMRglc as detected with [^{18}F]FDG in a 68-year-old male normal control subject (left image, NC) and a 49-year-old male patient with clinically symptomatic spinocerebellar ataxia type 6 (right image, SCA6). The scans demonstrate slightly decreased lCMRglc in the cerebellum and brainstem of the patient as compared to the control subject. The color bar indicates the relative rate of cerebral glucose utilization for both scans, with values increasing from left to right on the bar.

thalmoplegia, extensor plantar responses, and scoliosis. The frequency of other disorders such as dysphagia and sphincter disturbances increased with disease duration.

3. Neuropathology

Degenerative changes have been found in the cerebellum, retinal ganglion cells, cerebellum, base of the pons, and inferior olives (Gouw *et al.*, 1994). Neuronal intranuclear inclusions have been found, predominantly in the inferior olivary complex, but also in other brain regions, including the cerebral cortex. Some of the inclusions are ubiquinated (Holmberg *et al.*, 1998).

4. Genetics

SCA7 was mapped initially to a mutation on the short arm of chromosome 3p14–21.1 (Gouw *et al.*, 1995). In other kindreds, however, SCA7 was linked to a mutation on chromosome 3p12–p21.1 (Holmberg *et al.*, 1995; Jobsis *et al.*, 1997). Recently, CAG repeats were identified in chromosome 3q12-13, with a repeat size ranging from 5 to 18 in normal alleles and 37 to over 300 in mutated alleles (David *et al.*, 1997; DelFavero *et al.*, 1998; Johansson *et al.*, 1998). The SCA7 repeat length increases in subsequent generations.

5. Structural Imaging

In the single study available in the literature, MRI scanning showed cerebellar and pontine atrophy (Gouw *et al.*, 1994). We have obtained MRI scans in three symptomatic carriers of the SCA7 gene, and they show mild atrophy of the cerebellar cortex and pons, with modest enlargement of the fourth ventricle (Fig. 7).

6. Functional Imaging

No studies have been reported. We used PET with [^{18}F]FDG to study lCMRglc in the three symptomatic carriers of the SCA7 gene mentioned above. The results are presented in Table II. Although the sample is insufficient in size to warrant statistical analysis, the results suggest that lCMRglc is decreased in the cerebellar vermis, cerebellar hemispheres, and brainstem (Fig. 8). lCMRglc appears to be within normal limits in the caudate nucleus and putamen, but decreased in the thalamus.

I. Spinocerebellar Ataxia Type 8

SCA8 is a newly described dominantly inherited ataxia resulting from a CTG expansion located on chro-

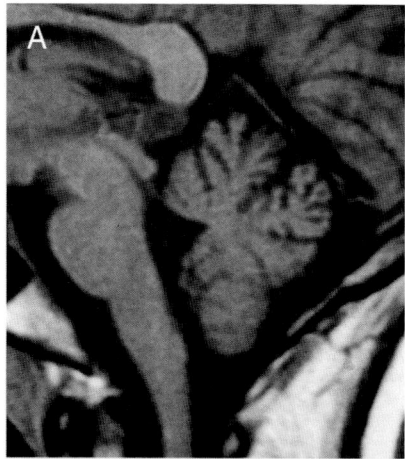

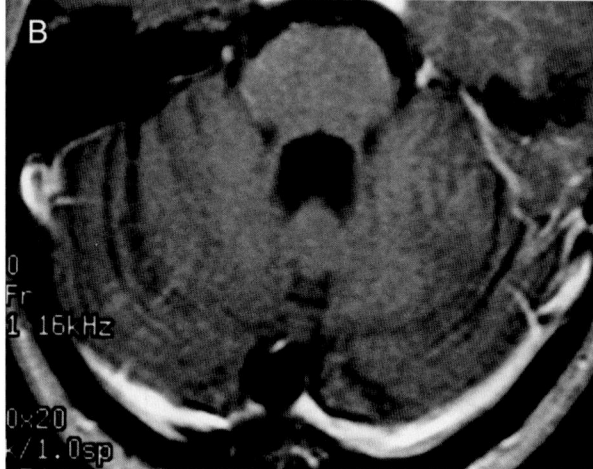

Figure 7 MR images in the sagittal (A) and horizontal (B) planes of a 27-year-old woman with clinically symptomatic spinocerebellar ataxia type 7. The images reveal mild atrophy of the cerebellar vermis, cerebellar hemispheres, and base of the pons, with enlargement of the fourth ventricle.

mosome 13q21 (Koob *et al.*, 1999). This is the first dominantly inherited cerebellar degeneration discovered that does not result from a CAG expansion. Myotonic dystrophy is the only other disease reported to be caused by a CTG expansion. Symptomatic carriers of the SCA8 gene have expansions similar in size (107–127 repeats) to those found in myotonic dystrophy. The clinical and imaging characteristics of the SCA8 mutation have not been published.

J. Spinocerebellar Ataxia Type 9

The clinical and genetic characteristics of SCA9 have not been published yet.

K. Spinocerebellar Ataxia Type 10

SCA10 is a newly described dominantly inherited ataxia affecting Hispanic families of Mexican origin

with a disease locus on chromosome 22q13 (Matsuura *et al.*, 1999; Zu *et al.*, 1999). It is unclear whether SCA10 results from a CAG repeat expansion. In the first of the two families described, the disorder resulted in limb and truncal ataxia, dysarthria, and nystagmus (Zu *et al.*, 1999). In addition, two of the 11 affected members had complex partial seizures. In the second family described, eight of the 12 affected members had a history of generalized motor seizures beginning in the third to fifth decade (Matsuura *et al.*, 1999). In six of the eight, the disease presented with ataxia of gait in the third to fifth decade and the other two developed seizures before the ataxia began. In 10 of the 12 affected members, the average at onset was 36 ± 6 years with a range of 26–45 years, and the age at onset tended to be earlier through successive generations. The principal clinical features are ataxia of gait and limb movements, ataxic dysarthria, and gaze-evoked nystagmus with impaired smooth-pursuit movements. The gene responsible for SCA10 has not been identified, and no information is available on the neuropathologic findings or the results of structural or functional imaging.

L. Dentatorubropallidoluysian Atrophy (DRPLA)

1. Introduction

This is a multiple-system degeneration causing ataxia, a hyperkinetic movement disorder, and dementia that has distinguishing neuropathological findings (Smith *et al.*, 1958). DRPLA has been reported in Europe and North America but appears to occur more frequently in Japan than in other countries (Ikeuchi *et al.*, 1995a,b).

2. Clinical Presentation

The age of onset varies from the first to the seventh decade, and anticipation occurs frequently, particularly with paternal transmission. The principal clinical features include progressive dementia with ataxia, chorea, myoclonus, and epilepsy. Two typical phenotypes have been described in Japan. Type 1 (adult onset) DRPLA has an average age of onset of 44 years and presents with progressive ataxia of gait, limb movements, and speech accompanied by chorea and dementia, and infrequently, epilepsy and myoclonus. The disease progresses over 10–20 years, terminating in death. Type 2 (childhood onset) DRPLA is a rapidly progressive disorder with prominent myoclonus, epilepsy, learning disabilities, mental retardation, and relatively infrequent chorea and ataxia (Komure *et al.*, 1995). The clinical phenotype of DRPLA shows marked heterogeneity both within and between involved kindreds (Iizuka *et al.*, 1984; Iizuka and Hirayama, 1991; Komure *et al.*,

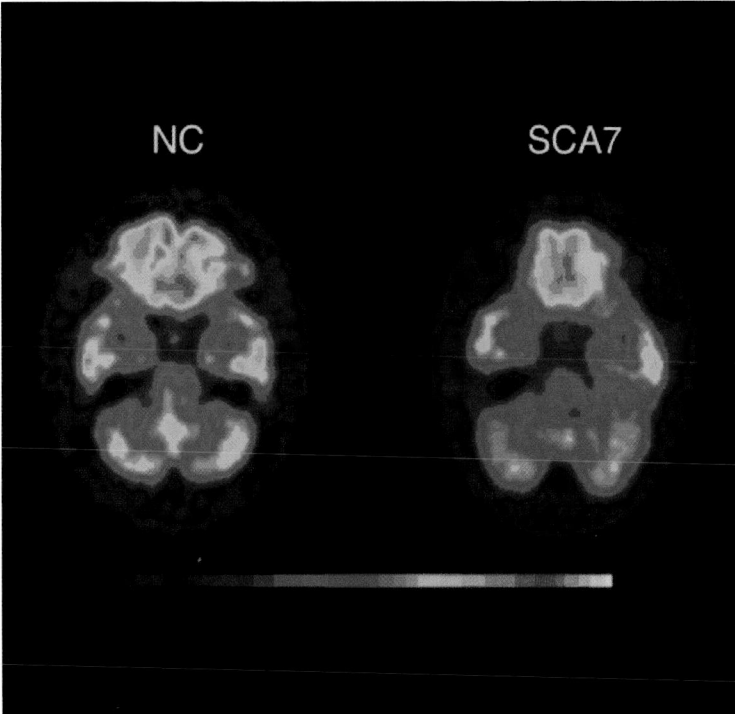

Figure 8 PET scans showing lCMRglc as detected with [^{18}F]FDG in a 32-year-old female normal control subject (left image, NC) and a 27-year-old female patient with clinically symptomatic spinocerebellar ataxia type 7 (right image, SCA7). The scans reveal decreased lCMRglc in the cerebellum and brainstem of the patient as compared to the control subject. The color bar indicates the relative rate of cerebral glucose utilization for both scans, with values increasing from left to right on the bar.

1995), and many sporadic cases have been described in Europe, North America (Smith *et al.*, 1958; Neumann, 1959; Smith, 1975), and Japan (Goto *et al.*, 1982; Naito and Oyanagi, 1982).

A disorder related to DRPLA was described in a large African-American kindred in the United States and identified as the Haw River Syndrome (HRS) (Farmer *et al.*, 1989; Burke *et al.*, 1994). This disorder is associated with mental retardation and psychiatric disturbances and causes progressive ataxia, chorea, dementia, and seizures. Distinctive features have been described in HRS consisting of marked demyelination in the centrum semiovale of the cerebral hemispheres, with atrophy of the dorsal columns in the spinal cord. Although these findings have been described only rarely in DRPLA, HRS results from a CAG repeat expansion in the DRPLA-associated gene (Burke, 1994). Hence DRPLA and HRS are considered variants of the same disorder.

3. Neuropathology

The name dentatorubropallidoluysian atrophy reflects the structures principally affected, the *d*entate nucleus, *r*ed nucleus, globus *p*allidus, and corpus *L*uisi (subthalamic nucleus). Neuronal loss and gliosis of the dentate nucleus are usually severe. In most cases, the cerebellar cortex is relatively spared, although occasional cases show marked involvement (Takahashi *et al.*, 1988). The superior cerebellar peduncle is atrophic, but the middle cerebellar peduncle is relatively unaffected. The olivary nuclei are moderately to severely degenerated. The red nucleus undergoes variable cell loss and gliosis. The tectum and tegmentum of the brainstem, including parts of the reticular formation, are often atrophic (Iizuka and Hirayama, 1991). The substantia nigra usually is spared. The globus pallidus is severely affected and the subthalamic nucleus may be involved (Warner *et al.*, 1994a,b, 1995). The neostriatum may be affected, but the damage is mild in comparison to the globus pallidus. The thalamus may be involved. In many cases, degeneration can be found in the posterior columns, spinocerebellar tracts, corticospinal tracts, and anterior horn cells (Iizuka and Hirayama, 1991). Demyelination of the white matter in the centrum ovale is marked in HRS and occasionally found in DRPLA. Neurons in the dentate nucleus contain ubiquinated filamentous inclusions, and similar neuronal inclusions have been described in the striatum, pontine nuclei, inferior olivary complex, cerebellar cortex, and dentate nucleus (Becher *et al.,* 1998; Hayashi *et al.,* 1998a,b). The neuropathological changes responsible for the dementia in DRPLA have not been identified; in seven autopsied patients with dementia, the typical subcortical features were found, but there was no neuronal loss in the nucleus basalis of Meynert and no change in the cerebral cortex (Hayashi *et al.,* 1998b).

4. Genetics

The mutation consists of a CAG repeat expansion on chromosome 12p13.31 (Goto *et al.,* 1982; Koide *et al.,* 1994; Nagafuchi *et al.,* 1994a; Ikeuchi *et al.,* 1995a,b; Uyama *et al.,* 1995; Takano *et al.,* 1996). The gene product is a protein, atrophin-1, that is localized in neuronal cytoplasm (Yazawa *et al.,* 1995). The CAG repeat in the neuronal cytoplasmic gene product codes for a polyglutamine (Nagafuchi *et al.,* 1994b; Schmitt *et al.,* 1995; Margolis *et al.,* 1996). Normal alleles contain 8–25 CAG repeats, whereas mutated alleles have 54–83 repeats. The age of onset is inversely related to CAG repeat length; juvenile-onset cases have large repeat sizes. These cases are paternally transmitted, present clinically with a rapid progression of myoclonus, epilepsy, and mental retardation, and often are mistaken for cases of progressive myoclonic epilepsy. Late-onset cases have shorter repeat lengths and present clinically with ataxia, choreoathetosis, and dementia, making them difficult to distinguish from Huntington's disease without genetic testing.

5. Structural Imaging

In one study (Arai *et al.,* 1993), CT scanning showed progressive cerebral cortical atrophy, but this has not been reported subsequently. In another study (Mizoi *et al.,* 1994), CT scanning revealed diffuse low-intensity cerebral white matter without cerebral cortical atrophy. Similarly discrepant findings are reported for MRI scans. Progressive cortical atrophy was reported in one study (Arai *et al.,* 1993), but in others the findings consisted of diffuse high-intensity white matter in T_2-weighted images (Mizoi *et al.,* 1994; Potter *et al.,* 1995; Koide *et al.,* 1997), high-intensity signals in the globus pallidus on proton and T_2-weighted images (Imamura *et al.,* 1994), and atrophy of the brainstem and cerebellum (Koide *et al.,* 1997). In a single-case report of DRPLA with childhood onset, MRI revealed generalized cerebral atrophy, with particular involvement of the brainstem tegmentum and cerebellum, and periventricular hyperintensity on T_2-weighted images (Miyazaki *et al.,* 1995). MRI with fluid attenuation inversion-recovery (FLAIR) was carried out in a 60-year-old woman with a family history of DRPLA (Yoshii *et al.,* 1998). The scan revealed atrophy of the cerebral cortex, cerebel-

lum, and pontomesencephalic tegmentum, with prominent high signal intensity of the cerebral white matter and some of the white matter tracts within the brainstem, the latter indicating pathological extension to the white matter.

6. Functional Imaging

There are no reports of PET studies in the literature. In one SPECT study of lCBF with [^{123}I]iodoamphetamine, low perfusion of the cerebral cortex was reported in cases with evidence of cerebral cortical atrophy on structural imaging (Arai *et al.,* 1993). In another SPECT study of cases showing no cerebral cortical atrophy, lCBF was normal (Mizoi *et al.,* 1994). These latter cases had white matter lesions in MRI scans, and the investigators concluded that this study demonstrated that the lesions cannot be attributed to ischemia (Mizoi *et al.,* 1994).

7. Magnetic Resonance Spectroscopy

In three children with DRPLA manifested by progressive myoclonus with epilepsy, proton MRS showed markedly reduced ratios of *N*-acetylaspartate to both choline and creatine in the basal ganglia and parietal cortex (Miyazaki *et al.,* 1996b). The ratio of *N*-acetylaspartate to creatine was correlated with the number of expanded repeats.

8. Electroencephalography

EEGs in DRPLA show epileptiform patterns (Inazuki *et al.,* 1989; Saitoh *et al.,* 1998). Atypical spike-wave complexes, slow-wave bursts, and photosensitivity are found in patients with myoclonus and epilepsy (Inazuki *et al.,* 1989). Brainstem auditory-evoked potentials show prolonged latency and reduced or absent brainstem components (Miyazaki *et al.,* 1996a). Short-latency somatosensory-evoked potentials have prolonged central conduction times, and the cortical components have reduced amplitudes. In two patients with symptom onset in the first decade of life, flash visual-evoked potentials were extremely enlarged and had shortened latency (Miyazaki *et al.,* 1996a).

IV. The Sporadic Cerebellar Degenerations

A. Sporadic Olivopontocerebellar Atrophy

1. Introduction

Selective neuronal degeneration in the inferior olives, pons, and cerebellum can occur not only as an inherited disorder but also as a sporadic disease of unknown cause (Eadie, 1975a,b; Gilman *et al.,* 1981;

Duvoisin, 1984). When sporadic, it is usually termed sporadic olivopontocerebellar atrophy (sOPCA), but it is also called idiopathic late-onset cerebellar degeneration. A related sporadic neurodegenerative disorder of unknown cause is sporadic parenchymatous cerebellar cortical atrophy (sCCA), which consists of neuronal degeneration with reactive gliosis in the cerebellum without concomitant degeneration in the brainstem (Marie et al., 1922; Greenfield, 1954; Harding, 1984). Both sOPCA and sCCA result in progressive ataxia of gait, with ataxic dysarthria and ataxia of limb movements. The symptoms are similar in both disorders, making the diagnosis difficult; however, sCCA progresses more slowly, causing less disability over time, and does not evolve into MSA whereas sOPCA can lead to MSA. Most studies in recent years have utilized MRI to distinguish between sOPCA and sCCA. Cases with atrophy involving both the brainstem and cerebellum are labeled sOPCA and those with atrophy of the cerebellum without brainstem involvement are designated sCCA. Although useful, this distinction presents the problem that in sOPCA, cerebellar atrophy usually can be recognized long before brainstem atrophy can be recognized. Hence diagnosis that relies upon MRI early in the course can result in misdiagnosis.

2. Clinical Presentation

The clinical manifestations of sOPCA usually begin in middle age or later; however, symptoms can begin in children, adolescents, or young adults (Chand et al., 1996). Most patients initially complain of difficulty in walking or loss of balance and describe inability to walk in a straight line, frequent bumping into walls or companions, difficulty maneuvering narrow passageways, and a tendency to trip when going down stairs. Usually a speech disorder is an early complaint and is described as "slurring" or running words together. After this, incoordination of the hands and arms causes loss of gracefulness in handwriting, difficulty in sewing, and loss of dexterity in typing or playing musical instruments. The disease generally progresses steadily, but at a highly variable rate between patients. In some, the symptoms increase slowly over many years, and these patients can continue to ambulate with assistive devices such as walkers for 15–20 years. In others, the disease progresses so rapidly that the patient can become confined to a wheelchair in 3–5 years. These rapidly progressing cases often progress to develop multiple-system atrophy, as shown by the emergence of autonomic failure and parkinsonism features.

Examination reveals ataxia of gait, ataxic dysarthria, abnormal extraocular movements, and difficulty with coordinated limb movements. Soon after the onset of the disease, the gait is only slightly wide based, consisting of irregularly placed steps, some too short and some too long, and at times a step may occur to the side rather than forward. Turning may be difficult, causing unsteadiness with a tendency to fall, and walking in tandem (heel-to-heel) cannot be performed without steps to the side when the patient loses balance. Often patients watch the floor while walking. At this stage, handwriting may be slightly irregular and the finger-nose-finger test of coordination is normal, but the heel-knee-shin test shows a mildly wavering trajectory. With disease progression, stance becomes more wide based, the gait becomes more irregular, and turning may cause sudden loss of balance and falling. Handwriting becomes highly irregular, with some letters too big and others too small, and a tremor may cause wavy strokes. The finger-nose-finger test shows a side-to-side tremor developing from the shoulder as the finger approaches the target. Rapid supination-pronation movements of the hands show dysdiadocokinesis, with irregular, incomplete movements, a slow rate, and disrupted rhythm. The heel-knee-shin test shows marked side-to-side tremor generated at the hip, with the heel repeatedly slipping off the shin.

Many abnormalities of extraocular movements occur, principally square wave jerks, gaze paretic nystagmus, saccadic intrusions into smooth-pursuit movements, ocular dysmetria, and nystagmus in the primary position. Some patients show marked slowness of both saccades and pursuit movements. Speech is usually abnormal by the time the patient seeks medical attention. The abnormalities are principally ataxic, with abnormal modulations of pitch and volume and a pattern of excess and equalized stress, but spasticity can accompany the ataxia, resulting in a strained-strangled quality (Kluin et al., 1988). With disease progression, speech becomes increasingly dysarthric, making patients in late stages difficult to comprehend. Several other neurologic disturbances have been described in sOPCA but are unusual. These include ophthalmoplegia, optic atrophy, pigmentary retinal degeneration, dementia, chorea, athetosis, and amyotrophy of the limbs or tongue (Eadie, 1975a,b; Gilman et al., 1981; Harding, 1984).

Patients with sOPCA may develop signs of corticospinal disease consisting of limb spasticity, usually most markedly in the legs than the arms, with hyperreflexia and extensor plantar responses (Gilman et al., 1988; Rosenthal et al., 1988). Muscle strength is usually preserved. Signs of corticospinal disease in sOPCA do not necessarily indicate that the patient will progress to develop MSA. Patients with sOPCA can also develop the parkinsonian features of bradykinesia, masked face with infrequent blinking, limb rigidity, stooped posture, and shuffling gait, and this indicates probable evolution of the disease into MSA. Parkinsonian features can occur before symptoms of autonomic failure begin, but more commonly appear after the onset of postural hypotension and urinary frequency, urgency, and incontinence.

3. Neuropathological Findings

In sCCA the neuropathological changes are limited to the cerebellar cortex, and in sOPCA the abnormalities included the cerebellar cortex, cerebellar nuclei, and components of the brainstem. In both disorders, neuronal loss with reactive gliosis in the cerebellar cortex affects essentially all elements, including Purkinje cells, granule cells, and interneurons. The deep nuclei show neuronal depletion and reactive gliosis. In the brainstem, the pons is markedly shrunken, with loss of neurons and decreased numbers of myelinated fibers. The inferior olives are decreased in size, showing neuronal loss and reactive gliosis. Degenerative changes may be seen in the corticospinal tracts within the brainstem and spinal cord. Glial cytoplasmic inclusions (GCIs) are a diagnostic feature of MSA and are found in MSA patients with principally cerebellar ataxia as well as those with predominantly parkinsonian features (Lantos, 1998). It is not clear, however, whether patients with sOPCA who do not progress to MSA also have GCIs on neuropathological examination.

4. Structural Imaging

a. CT Scanning CT reveals some degree of cerebellar atrophy in most, but not all, patients with sOPCA, and many show brainstem atrophy as well (Huang and Plaitakis, 1984; Gilman *et al.,* 1988). In both sOPCA and MSA, the degree of atrophy of the brainstem and cerebellum in CT scans is not correlated with either the duration of the disease or the severity of the neurologic signs, including ataxia, ocular movement abnormalities, and the various manifestations of MSA (Staal *et al.,* 1990). sOPCA does not affect children frequently, but on occasion it appears as an idiopathic sporadic disorder. In these patients, CT can be helpful diagnostically in showing atrophy of the brainstem and cerebellum (Kumar et al., 1995).

Several studies have examined the utility of CT and MRI in differentiating sCCA from sOPCA and MSA (Klockgether *et al.,* 1990a,b; Wittkamper *et al.,* 1993). In one study, CT was used to clarify the relationship of atrophy to clinical symptomatology in 35 patients with autosomal dominant or idiopathic cerebellar ataxia (Wittkamper *et al.,* 1993). Thirteen patients with a clinically pure cerebellar disorder were given a diagnosis of sCCA and 22 patients with both cerebellar and noncerebellar signs were given a diagnosis of sOPCA. Thirty percent of those with the clinical diagnosis of sCCA had atrophy of the brainstem in addition to the

cerebellum, and 40% of the group with the clinical diagnosis of sOPCA had atrophy of the cerebellum but not the brainstem. The intensity of the clinical signs was poorly correlated with the degree of atrophy on CT. The authors concluded that CT has limited utility in differentiating sCCA from sOPCA.

Another study evaluated the relationship of infratentorial atrophy assessed by CT or MRI to clinical symptomatology over time by retrospective evaluation of consecutive scans in six patients with idiopathic late-onset cerebellar ataxia (Klockgether *et al.,* 1990a). On the basis of both clinical testing and MRI, four patients were thought to have sOPCA. In these patients, atrophy of the cerebellum and brainstem became visible at the same time and progressed in parallel. In all four patients, brainstem atrophy was visible earlier than the noncerebellar clinical symptoms. In the remaining two patients, the brainstem did not become atrophic over the course of the study, indicating that they probably had sCCA. The investigators suggest that CT scanning at regular intervals may be helpful prognostically in the cerebellar ataxias.

In another study, 28 patients with a clinical diagnosis of idiopathic late-onset cerebellar ataxia were subdivided into those with a pure cerebellar disorder and those with both cerebellar and noncerebellar symptoms and signs (Klockgether *et al.,* 1990b). The patients with pure cerebellar findings had a longer mean life span (20.7 years) than those with additional noncerebellar disorders (7.7 years). MRI or CT revealed cerebellar atrophy without apparent involvement of brainstem in all the patients with a pure cerebellar syndrome, suggesting a diagnosis of sCCA. Most of those with both cerebellar and noncerebellar findings had atrophy of the cerebellum and brainstem, suggesting that they had sOPCA (and probably many had MSA as well). Two of the patients with both cerebellar and noncerebellar findings, however, did not show MRI morphology compatible with the diagnosis of sOPCA. Hence the use of MRI in addition to CT improves the utility of structural imaging in differentiating sCCA from sOPCA.

A recent investigation evaluated the utility of CT or MRI in the diagnosis of MSA among patients with idiopathic cerebellar ataxia (Vuadens *et al.,* 1997). Twenty-one patients with a diagnosis of idiopathic sOPCA were evaluated clinically and by MRI or CT. The cases were divided into probable (48%) and possible (52%) MSA using generally accepted criteria. Multiple system involvement was clear in all patients because of various combinations of cerebellar, basal ganglia, corticospinal, and autonomic symptoms and signs. Both cerebellar atrophy and brainstem atrophy were found in only 43% of cases, and no relationship could be detected between the degree of atrophy and either the duration of the disease or the severity of the clinical features. The authors concluded that structural imaging was not helpful in supporting the diagnosis of MSA among patients with sOPCA.

b. Magnetic Resonance Imaging MRI can be helpful in the diagnosis of sOPCA by demonstrating atrophy of the cerebellar hemispheres and vermis and the pons, with enlargement of the fourth ventricle and basal cistern (Fig. 9). MRI can assist in differentiating sOPCA from sCCA. In a study of 11 patients with sOPCA and 5 with sCCA, atrophy of the components of the pons and cerebellum was found in sOPCA but not in sCCA (Kojima and Hirayama, 1988). Atrophy of the middle cerebellar peduncle observed in the coronal plane in sOPCA usually began in the inferior part of the pons and was more marked than or equal to the degree of atrophy at the base of the pons visualized in the midsagittal plane. In both sOPCA and sCCA, the superior face of the cerebellar vermis was more at-

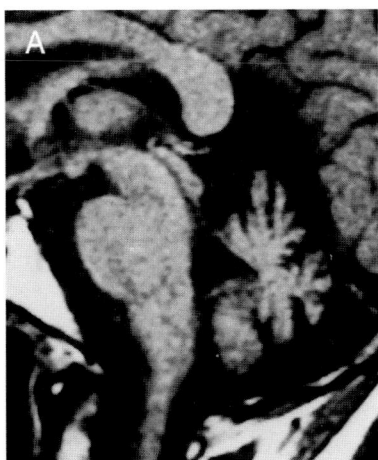

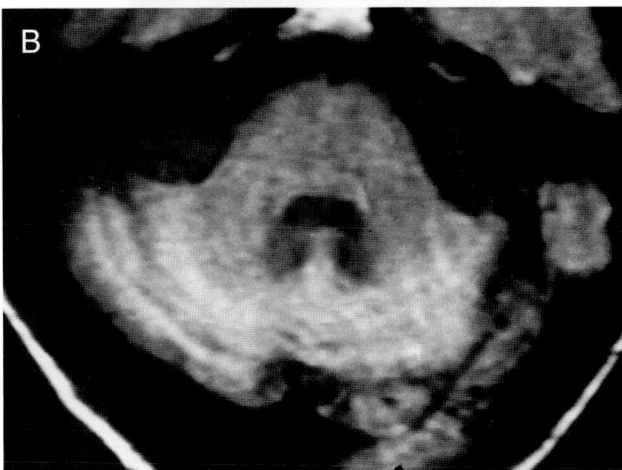

Figure 9 MR images in the sagittal (A) and horizontal (B) planes of a 45-year-old woman with clinically symptomatic sporadic olivopontocerebellar atrophy. The images reveal marked atrophy of the cerebellar vermis and cerebellar hemispheres, with pontine atrophy and enlargement of the fourth ventricle and the basal cistern.

rophic than the inferior, whereas in sOPCA, but not in sCCA, atrophy on the superior face was more prominent in the posterior than the anterior lobe. In another study, MRI was reported to be helpful in the diagnosis of sOPCA by showing atrophy of the olivary eminences of the medulla (Giuliani *et al.*, 1992). This resulted in straightening of the angle found on the ventral interface between the pons and medulla. In another investigation, 23 patients with a clinical diagnosis of sOPCA were evaluated by MRI (Savoiardo *et al.*, 1990). All cases showed atrophy of the cerebellum, pons, and middle cerebellar peduncles. On intermediate and T_2-weighted MR images, abnormal signal intensity was seen in the transverse pontine fibers, middle cerebellar peduncles, and cerebellum. The 23 patients included eight with a clinical diagnosis of MSA, and four of these had abnormal signal intensities in the putamen.

Several investigations have utilized CT or MRI to evaluate the contribution of tissue atrophy to the decrease of tissue metabolic rate in PET studies of glucose and oxygen metabolism. In one study, a significant relationship was found between the degree of atrophy detected with CT and the level of lCMRglc in the cerebellum and brainstem (Gilman *et al.*, 1988). Nevertheless, several patients had minimal atrophy and substantially reduced lCMRglc, suggesting that atrophy does not fully account for the hypometabolism. In another investigation, $[^{15}O]O_2$ was used to examine lCBF and lCMRO$_2$ in sOPCA (Yamaguchi *et al.*, 1994). The volumes of the pons and cerebellar hemispheres were quantified by MRI. Pontine volume was significantly correlated with lCBF and lCMRO$_2$ in sOPCA patients, whose values were decreased in comparison to a group of normal control subjects. In contrast, cerebellar hemispheric volumes were not correlated with either lCBF or lCMRO$_2$, suggesting that atrophy does not fully account for the changes in cerebellar blood flow and metabolism. In a third investigation, PET with $[^{15}O]O_2$ was used to study lCBF and lCMRO$_2$ in patients with sCCA as compared with sOPCA (Sun *et al.*, 1994). In both groups of cases, lCMRO$_2$ and lCBF were decreased in the cerebellar hemispheres and vermis, but markedly decreased lCMRO$_2$ and lCBF and metabolic rate for oxygen were found in the pons only in the group with sOPCA. PET data corrected for the tissue shrinkage on the basis of MRI morphometry showed a net reduction in lCMRO$_2$ and lCBF. Nevertheless, the amount of atrophy in the cerebellum could not fully account for the decreased values observed in the PET studies. Hence, each of these studies has shown that tissue atrophy does not account fully for the finding of hypometabolism and diminished blood flow in the brainstem and cerebellum in sOPCA.

5. Functional Imaging

PET with [^{18}F]FDG has been used to study lCMRglc in both sOPCA and sCCA. In an initial investigation, 30 patients with both sporadic and dominantly inherited (dOPCA) forms of OPCA were studied in comparison to 30 normal control subjects, with a similar distribution of ages and genders between groups (Gilman et al., 1988). This study was conducted before genetic testing became available for dOPCA. Hypometabolism was found in the cerebellar vermis, cerebellar hemispheres, and brainstem, and lCMRglc showed a significant negative correlation with the degree of atrophy in the cerebellum and brainstem. In a subsequent study of sOPCA and dOPCA, the same investigative group found a significant negative correlation between the severity of motor impairment and lCMRglc within the cerebellar vermis, cerebellar hemispheres, and brainstem (Rosenthal et al., 1988). A significant but weaker positive correlation was found between the severity of motor impairment and the degree of tissue atrophy in these regions as assessed by CT scanning. Partial correlation analysis showed that motor dysfunction was correlated more strongly with lCMRglc than with the degree of tissue atrophy. This suggests that the clinical manifestations of both sOPCA and dOPCA are more closely related to the metabolic state of the tissue than to the structural changes in the cerebellum. In a related study by this group, perceptual analysis was used to examine speech disorders in sOPCA and dOPCA patients, and rating scales were devised to quantitate the degree of ataxia and spasticity of speech (Kluin et al., 1988). A significant negative correlation was found between the severity of ataxic dysarthria and lCMRglc in the cerebellar vermis, cerebellar hemispheres, and brainstem. The intensity of spasticity in speech was not correlated with lCMRglc in these or other structures. The findings support the notion that the degree of ataxic dysarthria in both sOPCA and dOPCA is related to the functional activity of the cerebellum and its projections to the brainstem.

In a later study by the same group, PET with [^{18}F]FDG was used to study lCMRglc with MSA patients as compared with sOPCA and dOPCA patients and normal controls (Gilman et al., 1994). In MSA, lCMRglc was significantly decreased not only in the brainstem, cerebellar vermis, and cerebellar hemispheres but also in the putamen, thalamus, and cerebral cortex. In sOPCA, lCMRglc was significantly decreased in the same structures (Fig. 10). In dOPCA, lCMRglc was significantly decreased in the brainstem, cerebellar vermis, and cerebellar hemispheres, but not in the putamen, thalamus, or cerebral hemispheres. The findings demonstrated widespread CNS hypometabolism in MSA and indicated that many of the sOPCA patients had subclinical disease of the basal ganglia, thalamus, and cerebral cortex and were likely to progress into MSA. In contrast, the dOPCA patients had no involvement of forebrain structures and appeared unlikely to develop multiple system involvement.

Several subsequent studies utilizing PET with [^{18}F]FDG and [^{15}O]O$_2$ have confirmed the finding of decreased metabolic rate in the cerebellum and brainstem in sOPCA (Otsuka et al., 1994; Sun et al., 1994; Yamaguchi et al., 1994). In one investigation, lCMRglc was examined in CCA and sOPCA patients as compared with normal controls (Otsuka et al., 1994). In the CCA group, lCMRglc included three with late-onset sCCA and two with the hereditary cerebello-olivary degeneration of Holmes. In the sCCA group, lCMRglc was significantly decreased in the cerebellum but not the brainstem as compared with normal controls. In the sOPCA group, lCMRglc was decreased in both the cerebellar hemispheres and the brainstem as compared with normal controls. In both groups, lCMRglc was unchanged in the cerebral cortex, striatum, and thalamus. In a second study, lCBF and lCMRO$_2$ were examined with PET in sOPCA patients and normal controls (Yamaguchi et al., 1994). The volumes of the pons and the cerebellar hemispheres were quantified by MRI. In sOPCA, both lCBF and lCMRO$_2$ in the cerebellar hemispheres were significantly decreased. Pontine volume was significantly correlated with lCBF and lCMRO$_2$, but cerebellar hemisphere volume was not correlated with these measurements. The authors suggest that disruption of the pontocerebellar pathway contributes to the reduction of both blood flow and oxygen metabolism in the cerebellum of sOPCA patients and that detection of cerebellar circulatory impairment without marked cerebellar atrophy by neuroimaging may suggest the diagnosis of OPCA.

In a third investigation, sCCA and sOPCA patients were compared with each other and with normal controls utilizing quantitative MRI and PET measurements of lCBF and lCMRO$_2$ (Sun et al., 1994). MRI scans showed significant cerebellar and pontine atrophy in the sOPCA group compared with controls and significant cerebellar atrophy but no significant pontine atrophy in the sCCA group compared with controls. lCMRO$_2$ was decreased in the cerebellar hemispheres and vermis in both sCCA and sOPCA groups compared with controls; however, a marked decrease of lCBF and lCMRO$_2$ was found in the pons only in the sOPCA group. In a final study, lCBF (PET-[^{15}O]H$_2$O) and lCMRglc (PET-[^{18}F]FDG) were assessed in 17 sOPCA patients compared with 21 normal control subjects (Gilman et al., 1995b). In the sOPCA group, lCBF was significantly decreased in the cerebellum but not in the cerebral cortex, basal ganglia, thalamus, or brainstem. In

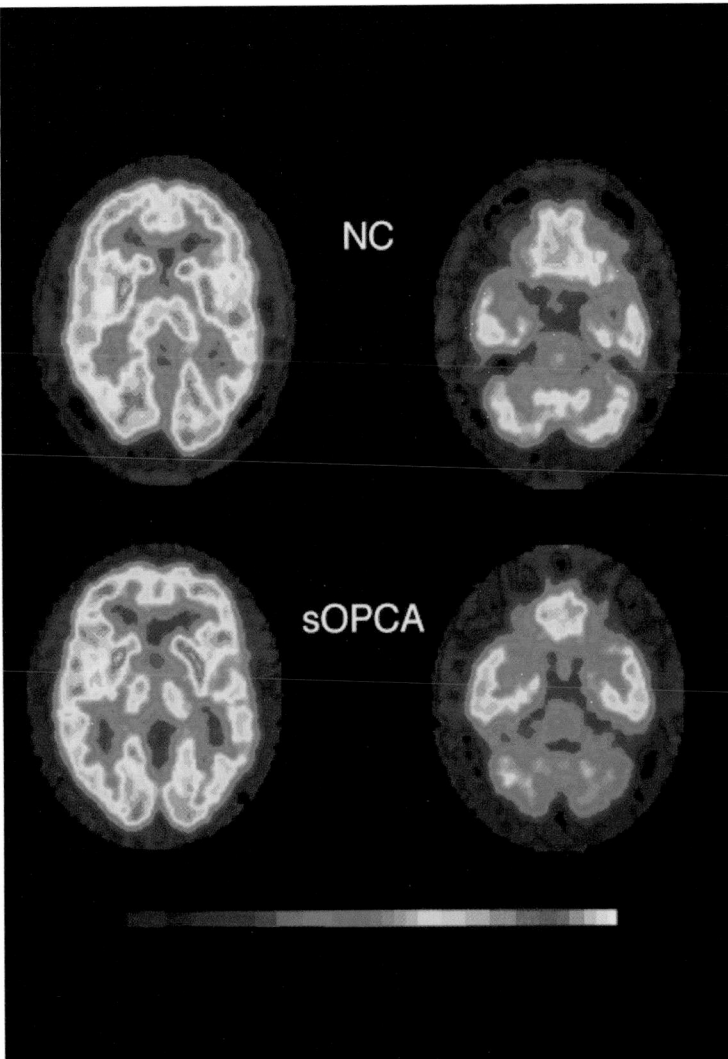

Figure 10 PET scans illustrating lCMRglc as detected with [^{18}F]FDG in a 68-year-old male normal control subject (upper two images, NC) and a 63-year-old male patient with clinically symptomatic sporadic olivopontocerebellar ataxia (lower two images, sOPCA). The left upper and lower scans show horizontal sections at the level of the cerebral cortex, basal ganglia, and thalamus, and the right upper and lower scans show horizontal sections at the level of the cerebellum and brainstem. The scans demonstrate decreased lCMRglc in the caudate nucleus, putamen, thalamus, cerebellum, and brainstem of the patient as compared to the control subject. The color bar indicates the relative rate of cerebral glucose utilization for all scans illustrated, with values increasing from left to right on the bar.

contrast, lCMRglc was significantly decreased in the cerebellum and brainstem, where the largest changes were observed, and also in the cerebral cortex, basal ganglia, and thalamus. The ratio of lCBF to lCMRglc, which indicates the coupling of blood flow to metabolism, was similar in the patient and control groups for all structures except the brainstem, where the ratio was marginally decreased in the sOPCA group. Logistic discriminant analysis was used to assess the ability of lCBF and lCMRglc to differentiate sOPCA patients from normal subjects. It was found that the cross-validated sensitivity of absolute lCMRglc as a predictor of sOPCA was 82%, with a corresponding specificity of 71%. In contrast, the sensitivity of absolute lCBF was 71% and the specificity was 76%. The conclusion was that in sOPCA, lCBF is reduced in the cerebellum, lCBF remains coupled to lCMRglc, and the lCBF pattern is a useful predictor of the diagnosis.

SPECT with N-isopropyl-p-[^{123}I]iodoamphetamine was used to compare lCBF in two groups of patients with cerebellar degeneration, sporadic cerebello-olivary atrophy, and sOPCA (Ohkoshi *et al.*, 1995). Uptake in the cerebellar hemispheres and vermis was decreased in both groups, but uptake in the pons was decreased in sOPCA only. The degree of decrease of uptake in both structures was related to the intensity of the clinical signs.

Several investigations have elucidated the neurochemical changes in sOPCA and MSA. In one study, PET was used with [^{11}C]FMZ to study the density of GABA-A/BDZ receptors (Gilman *et al.*, 1995a). Quantitative analyses were made of binding in 14 MSA patients with predominantly cerebellar features (MSA-C), 5 MSA patients with principally parkinsonism (MSA-P), 18 patients with sOPCA, and 15 patients with dOPCA in comparison to 20 normal control subjects. Mean ligand transport (K_1) was significantly decreased in the cerebellum and brainstem of the MSA-C and sOPCA groups, but not in the MSA-P group. Despite these differences, specific ligand binding to the GABA-A/BDZ receptor was largely preserved in the cerebral hemispheres, basal ganglia, thalamus, cerebellum, and brainstem in all patient groups. The finding of relative preservation of GABA-A/BDZ receptors suggests that these sites may be available for pharmacological therapy in these disorders.

In a pilot study, PET was used with [^{11}C]dihydrotetrabenazine ([^{11}C])DTBZ), a new ligand for the type 2 vesicular monoamine transporter (VMAT2), to study the density of striatal monoaminergic presynaptic terminals in four patients with MSA, eight with sOPCA, and nine normal control subjects (Gilman *et al.*, 1996a). Mean K_1 was significantly decreased in the cerebellum of the sOPCA group and was unchanged in the MSA group. Specific binding was significantly decreased in the striatum of the MSA group and nonsignificantly reduced in the sOPCA group. The study confirmed the expected significant decrease of VMAT2 binding in MSA. The nonsignificant decrease of VMAT2 binding in sOPCA suggests some degree of nigrostriatal pathology, indicating that some may later develop symptomatic parkinsonism and thereby evolve to develop MSA. In a follow-up study, utilization of the (+) enantiomer of [^{11}C]DTBZ to examine striatal VMAT2 provided a stronger signal and additional information about nigrostriatal pathology in sOPCA (Fig. 11) and MSA (Gilman *et al.*, 1999b). The subjects included eight patients with MSA-P, eight with MSA-C, and six with sOPCA compared with seven normal controls. Mean K_1 was significantly reduced in the putamen of all patient groups and in the cerebellar hemispheres of MSA-C and sOPCA, but not in the cerebellar hemispheres of

the MSA-P group. Specific binding was significantly reduced in the putamen of all patient groups, with the greatest decrease found in MSA-P, less in MSA-C, and least in sOPCA. Significant negative correlations were found between the density of striatal binding and the intensity of parkinsonian features and between the level of cerebellar K_1 and the severity of cerebellar ataxia. The results demonstrate differences between MSA-P and MSA-C groups that reflect the relative severity of nigrostriatal and cerebellar system degeneration in these two forms of MSA. The findings also verify earlier studies showing subclinical nigrostriatal dysfunction in sOPCA (Gilman *et al.*, 1994, 1996a).

In another study, binding to dopaminergic terminals with [^{18}F]FDA and to striatal opioid receptors with [^{11}C]diprenorphine ([^{11}C]DPR) was examined with PET in patients described as having sOPCA (Rinne *et al.*, 1995). Ten patients were studied, and all had autonomic failure. The mean caudate : occipital uptake ratio for [^{11}C]DPR was reduced by 88% and the mean putamen : occipital uptake ratio by 85% in comparison to a group of normal controls. Individually, 4 of the 10 patients had significantly reduced opioid binding in the putamen. Mean [^{18}F]FDA uptake in the putamen was decreased by 71% in comparison to the normal controls, and individually, 7 of 10 patients had significantly reduced uptake. A significant positive correlation was found between putamen : occipital uptake ratio for [^{11}C]DPR and putamen uptake of [^{18}F]FDA. The authors concluded that a majority of sOPCA patients have subclinical nigrostriatal dysfunction and that sOPCA therefore is part of the spectrum of MSA. This conclusion is not warranted, as all the sOPCA patients studied had significant autonomic failure, and by definition these patients have MSA-C. Hence this study does not demonstrate that sOPCA can be viewed categorically as part of the spectrum of MSA.

6. Cerebellar Activation

PET with [^{18}F]FDG was used to examine changes in lCMRglc during bipedal walking in seven sOPCA patients compared with seven normal control subjects (Mishina *et al.*, 1995). Glucose metabolism was examined under two conditions, supine resting and 30 min of treadmill walking without support or handrails. Activation ratios were defined as FDG uptake with walking divided by FDG uptake with resting and computed for the cerebellar vermis, cerebellar hemispheres, pons, and thalamus. No significant difference was found between normal controls and sOPCA patients for normalized resting FDG uptake in any region. In the normal controls, only small differences were found in FDG uptake in the cerebellar hemispheres, pons, and thalamus between resting and walking conditions. In the sOPCA pa-

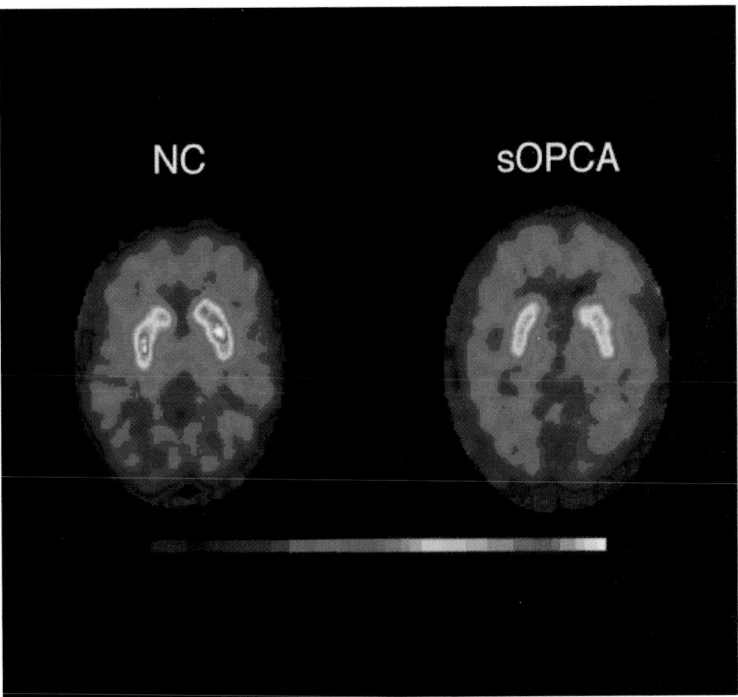

Figure 11 Distribution volume of (+)-[^{11}C]dihydrotetrabenazine detected with PET scans in the striatum of a 55-year-old female normal control subject (left image, NC) and a 63-year-old female patient with clinically symptomatic sporadic olivopontocerebellar atrophy (right image, sOPCA). The scans reveal decreased ligand binding in the caudate nucleus and putamen of the patient as compared to the control subject, indicating decreased numbers of presynaptic monoaminergic terminals. This finding suggests that the patient may progress to develop parkinsonian symptoms and thus her disorder may evolve to multiple-system atrophy. The color bar indicates the degree of ligand binding for both scans, with values increasing from left to right on the bar.

tients, FDG uptake in these regions was moderately increased by walking, and the activation ratio was significantly decreased. The authors suggest that the reduced activation ratio in the cerebellar vermis reflects the tissue dysfunction resulting from degeneration.

7. Magnetic Resonance Spectroscopy

Proton MRS was used with SPECT to compare spectra with lCBF in sOPCA patients and normal control subjects (Ikuta, 1998). MRS was used to examine the distribution and relative signal intensities of *N*-acetyl-aspartate (NAA), creatine (Cre), and choline (Cho). In the cerebellum of sOPCA patients, mean NAA/Cre ratios were significantly decreased and mean NAA/Cho ratios were slightly decreased as compared to controls. The patients with more severe ataxia of gait and speech had slightly lower NAA/Cre ratios than those less affected. Cho/Cre ratios in the cerebellum were not significantly different in the two groups. The SPECT scans showed a decreased cerebellar:occipital lobe ratio in the sOPCA group as compared to the controls. The author concludes

that the MRS and SPECT findings provide further evidence of cerebellar hypofunction in sOPCA.

8. Transcranial Magnetic Stimulation (TMS)

TMS over the motor cortex provides a method of studying central motor conduction time (CMCT) when coupled with electromyographic recording and comparative studies of peripheral nerve conduction time. In the normal human, a single electrical stimulus across the base of the skull, presumably by activating cerebellar circuits, causes a short-lasting (5–6 ms) disfacilitation of the response to TMS over the motor cortex (Ugawa *et al.*, 1991). This effect is thought to occur at the level of the motor cortex. A later (12–15 ms), less specific suppression occurs, probably from interactions at the spinal level. Magnetic stimulation over the back of the head has a similar effect upon the responses to TMS in the normal human (Ugawa *et al.*, 1995). In patients with ataxia from degenerative diseases, the suppression of motor cortical excitability from electrical (Ugawa *et al.*, 1994) or magnetic (Ugawa *et al.*, 1997) stimulation over the cerebellum is reduced or absent. In patients with hemicerebellar lesions, the threshold for evoking a response to magnetic stimulation is higher in the motor cortex contralateral to the cerebellar lesion than ipsilateral, suggesting that cerebellar circuits normally facilitate central motor mechanisms (Di Lazzaro *et al.*, 1994). In a longitudinal study of the long-term effects of isolated infarctions, both the threshold and the CMCT were abnormal when evoked from the motor cortex contralateral to the impaired hemicerebellum, and with time these abnormalities returned toward normal in parallel with the clinical improvement in cerebellar function (Cruz-Martinez and Arpa, 1997). In patients with sOPCA, the responses to TMS over the motor cortex were increased in threshold and CMCT was prolonged (Cruz-Martinez *et al.*, 1995). The finding of prolonged CMCT in the dominantly inherited and idiopathic sporadic cerebellar degenerations was verified in another investigation, which also demonstrated decreased motor cortex excitability, likely from enhancement of inhibitory mechanisms (Wessel *et al.*, 1996).

B. Multiple-System Atrophy

1. Introduction

MSA is a progressive neurodegenerative disease of undetermined cause that occurs sporadically and causes parkinsonian and cerebellar, autonomic, urinary, and corticospinal dysfunction in various combinations (Graham and Oppenheimer, 1969; Beck *et al.*, 1994; Albanese *et al.*, 1995; Consensus Committee, 1996; Quinn, 1996).

2. Clinical Presentation

The median age of onset of MSA is 53 years, with a range of 33–76 (Wenning *et al.*, 1994). The incidence in the 50- to 99-year age group is 3.0 new cases per 100,000 person-years (Bower *et al.*, 1997). The sexes appear to be affected equally. The course is characterized by relentless progression, with a median survival of 9.3 years from the first symptom (Wenning *et al.*, 1994; Klockgether *et al.*, 1998). The parkinsonian features include bradykinesia with rigidity, postural instability, hypokinetic speech, and, at times, tremor. Parkinsonian features occur in at least 90% of patients with MSA (Wenning *et al.*, 1994). Unlike patients with Parkinson's disease (PD), however, MSA patients rarely receive continued benefit from levodopa. Mild beneficial effects can occur, but frequently they are poorly sustained over time. Resting tremor is also uncommon in MSA whereas it affects 60-70% of patients with PD. The cerebellar dysfunction of MSA consists initially of ataxia of gait and

then later speech and limb movements (Gilman, 1989). Cerebellar ataxia is a presenting complaint in a minority of patients but accompanies the parkinsonism and autonomic disorders in a substantial percentage of cases. Autonomic dysfunction in MSA includes orthostatic hypotension often with an inadequate heart rate response to standing, postprandial hypotension, supine hypertension, anhydrosis with thermoregulatory disturbances, constipation, impotence in males, and poor lacrimation and salivation (Sandroni *et al.*, 1991; Mathias and Williams, 1994). A recent consensus conference led to the publication of guidelines for diagnosis (Gilman *et al.*, 1998a,b, 1999b). The guidelines include three levels of diagnostic certainty, possible, probable, and definite. The diagnosis of definite MSA requires neuropathological confirmation of the clinical diagnosis. The diagnosis of probable MSA requires demonstration of the criterion for autonomic failure/urinary dysfunction plus poorly responsive parkinsonism or cerebellar dysfunction. The diagnosis of possible MSA requires features from at least two of the three domains (autonomic dysfunction, parkinsonism, and cerebellar ataxia), though of insufficient severity to qualify for the diagnosis of probable MSA (Gilman *et al.*, 1998a,b, 1999b).

Sleep disorders occur commonly in MSA, notably rapid eye movement sleep behavior disorder (RBD), which affects about 90% of cases as shown by polysomnography (Plazzi *et al.*, 1997, 1998), and sleep apnea (Munschauer *et al.*, 1990). Polysomnography can be used to differentiate MSA from pure autonomic failure (PAF), as RBD is common in MSA and not in PAF (Plazzi *et al.*, 1997). PD is also associated with RBD, but sleep apnea is distinctly unusual. RBD can herald the onset of both MSA and idiopathic PD. Other abnormalities in MSA include emotional liability; corticospinal tract disease causing spasticity, hyperreflexia, and extensor plantar reflexes; supranuclear ophthalmoplegia; antecollis; myoclonus; respiratory stridor; polyneuropathy; and amyotrophy (Quinn *et al.*, 1989; Tison *et al.*, 1996). Dementia is distinctly uncommon, but abnormalities of executive functioning occur frequently.

MRI, MRS, PET and SPECT scanning, autonomic function tests, and external sphincter EMG help to differentiate idiopathic PD from MSA (Wenning and Quinn, 1997). Urodynamic evaluation and sphincter EMG are particularly useful in the differential diagnosis between PD and MSA; urodynamic findings may be abnormal before patients with MSA reach an advanced stage of the disease. Recordings of EMGs from perineal muscles become abnormal as the disease progresses in MSA but not in PD (Stocchi *et al.*, 1997).

3. Neuropathology

The neuropathological findings of MSA include loss of neurons with reactive gliosis in the substantia nigra, caudate nucleus, and putamen; cerebellar cortex; inferior olives; pons; locus ceruleus; dorsal motor nucleus of the vagus; intermediolateral cell column of the spinal cord; and Onuf's nucleus in the spinal cord. These neuropathological changes are accompanied by a high density of glial cytoplasmic inclusions (GCIs) in the sites with degenerative changes (Daniel *et al.*, 1992; Lantos, 1998). GCIs are ubiquitin-, tau-, and alpha-synuclein-positive oligodendroglial inclusions (Lantos, 1998).

4. Structural Imaging

a. Computed Tomography CT scans show cerebellar and pontine atrophy in patients with MSA presenting with cerebellar dysfunction (Abe *et al.*, 1983; Uematsu *et al.*, 1987) but do not show basal ganglia abnormalities.

b. Magnetic Resonance Imaging Numerous studies have shown distinctive findings in MRI scans of MSA. In MSA-C the findings consist of variable degrees of atrophy involving the cerebellar hemispheres, cerebellar vermis, middle cerebellar peduncles, base of the pons, and the medulla, with enlargement of the fourth ventricle and the basal cistern (Figs. 12 and 13). In some MSA-C cases, putaminal hypointensity can be seen in T_2-weighted images, and this can be accompanied by a hyperintense lateral putaminal rim (Fig. 13). In MSA-P the atrophic changes in the posterior fossa can be minimal or absent, but putaminal hypointensity is much more frequent.

An initial investigation of MRI changes in eight patients with a diagnosis of Shy-Drager syndrome (currently recognized as MSA-P) revealed putaminal atrophy and an abnormal decrease in signal intensity of the putamen, particularly along the lateral and posterior portions visualized predominantly in T_2-weighted sequences (Pastakia *et al.*, 1986). A subsequent study of five patients with Shy-Drager syndrome confirmed the finding of putaminal hypointensity in T_2-weighted images at high field strength (1.5 T) (Savoiardo *et al.*, 1989). This study also demonstrated hypointensity in the pallidum at low or intermediate field strength and absence of magnetic susceptibility effects in the putamen, which appeared hyperintense. A later, more complete study of MSA further demonstrated the clinical utility of MRI in the diagnosis of MSA (Testa *et al.*, 1993). The study consisted of 42 patients with probable or possible MSA diagnosed with strict clinical criteria and included 9 with striatonigral degeneration (SND), 13 with sOPCA, and 20 with combined parkinsonism and cerebellar ataxia. All 9 patients with SND had put-

aminal abnormalities, and all 13 with sOPCA had atrophy of the cerebellum and brainstem. The 20 patients with parkinsonism and cerebellar ataxia were classified as having probable MSA, and putaminal abnormalities were found in 7 of the 20, atrophy of the pons and cerebellum in 3, and a combination of both in 10.

Several studies have related MRI findings to clinical presentation. In an investigation of five MSA patients presenting with hemiparkinsonism, asymmetry of parkinsonian features was correlated with asymmetry of MRI findings (Kato *et al.*, 1992). T_2-weighted MR images contained hypointense areas in the posterior lateral putamen that were larger or lower in intensity on the side contralateral than on the side ipsilateral to the hemiparkinsonsim. Another study attempted a correlation between the severity of parkinsonian features and quantitative measures of MRI characteristics in 18 MSA patients as compared with 16 age-matched controls utilizing 1.5-T MRI (Waikai *et al.*, 1994). In MSA the width of the substantia nigra pars compacta was narrowed and the intensity of the putaminal signal was decreased, but the clinical severity of the parkinsonism was not significantly correlated with either of these measurements. Putaminal atrophy was found as well, however, and the degree of atrophy was correlated with the severity of the parkinsonian features.

Several investigations have utilized MRI in attempting to differentiate MSA from PD and normal controls. One study determined the sensitivity, specificity, and positive predictive value of a selection of abnormal findings in the putamen and infratentorial structures on MRI in distinguishing between MSA, idiopathic PD, and age-matched controls (Schrag *et al.*, 1998). Axial T_2-weighted and proton density MRI scans of 44 MSA patients were compared with MRI scans of 47 PD patients and 45 controls. A small subset of patients was scanned with 1.5-T machines, the rest with 0.5-T machines. In scans from both 0.5- and 1.5-T machines, the following items had high specificity but low sensitivity: putaminal atrophy, a hyperintense putaminal rim, and an infratentorial signal change. Demonstration of any infratentorial abnormality gave higher sensitivity but lower specificity. The overall sensitivity was 73% on 0.5-T machines and 88% on 1.5-T machines. The specificity of these findings for MSA compared to PD and compared to controls on 0.5-T machines was 95 and 100%, respectively, and on 1.5-T machines these values were 93 and 91%, respectively. Finding any of the described abnormalities on MRI gave a positive predictive value of 93% on the 0.5-T machines and 85% on 1.5-T machines.

Another investigation used three-dimensional MRI-based volumetry to study atrophy of the caudate nucleus, putamen, brainstem, and cerebellum in 12 MSA-P, 17 MSA-C, and 11 PD patients in comparison to 46

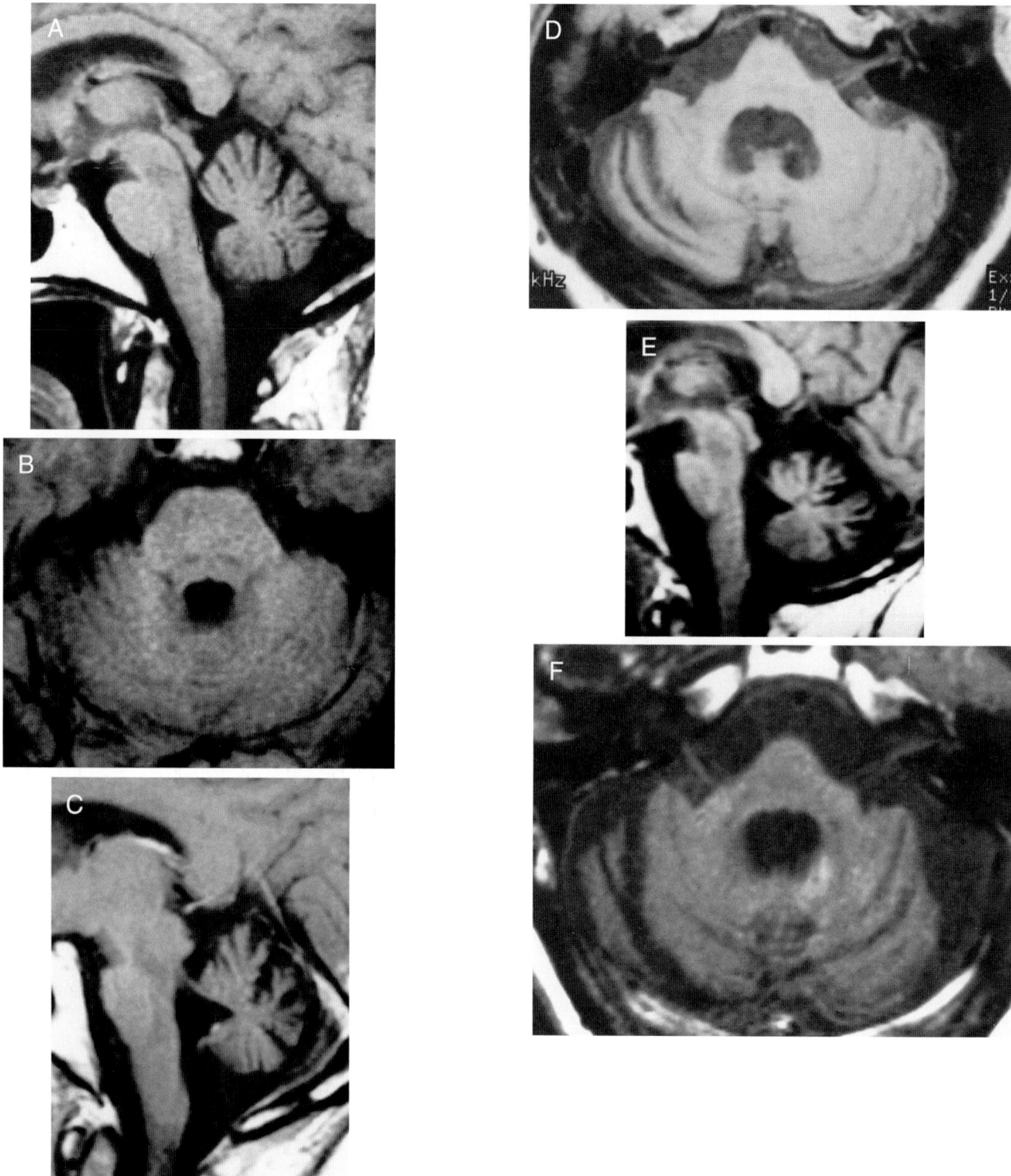

Figure 12 MR images in the sagittal (A, C, E) and horizontal (B, D, F) planes of a 50-year-old woman (A, B), a 52-year-old woman (C, D), and a 68-year-old woman (E, F), all of whom have clinically symptomatic multiple-system atrophy with predominantly cerebellar ataxia and autonomic failure. The images reveal varying degrees of atrophy of the cerebellar vermis, cerebellar hemispheres, middle cerebellar peduncles, base of the pons, and the medulla, with enlargement of the fourth ventricle and the basal cistern.

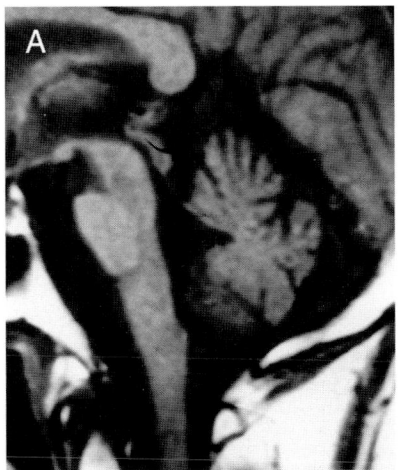

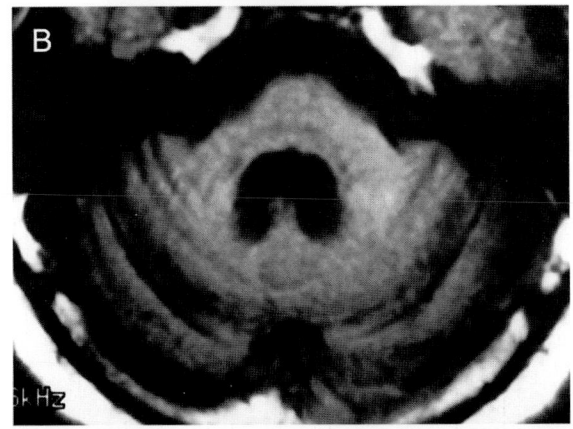

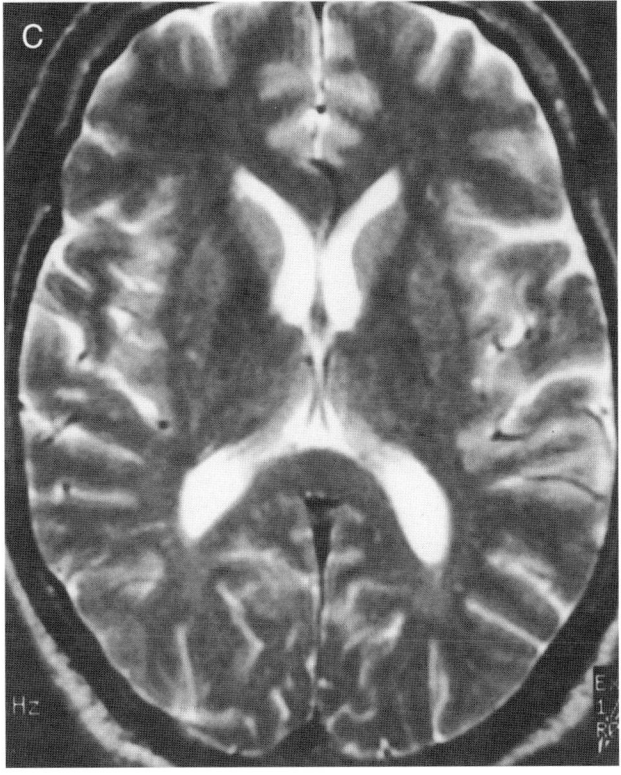

normal controls (Schulz *et al.*, 1999). Mean striatal, cerebellar, and brainstem volumes were normal in the PD group. Significant reductions in mean striatal and brainstem volumes were found in MSA-P and MSA-C, and cerebellar volume was also reduced in both groups. Stepwise discriminant analysis allowed differentiation of all 12 MSA-P patients and 15 of the 17 with MSA-C from the normal and PD cohorts. The PD and control group could not be separated. The study demonstrated that intracranial volume-normalized MRI-based volumetric measurements provide a sensitive marker to discriminate typical PD from MSA.

In another study, the frequency and specificity of hypointense putaminal MRI signal changes were compared with the frequency and specificity of a pattern including a hyperintense putaminal lateral rim signal and a hypointense dorsolateral putaminal signal in T_2-weighted images (Kraft *et al.*, 1999). MRI scans of 65 PD patients were compared with scans of 15 MSA-P patients. The pattern of a hyperintense lateral rim and a hypointense dorsolateral signal was found in 9 of the 15 MSA-P patients and in none in the PD group. Hypointense putaminal signal changes alone were found in 6 PD patients and 5 MSA-P patients. The study suggests that the pattern of hypointense and hyperintense putaminal signal changes in T_2-weighted images is a highly specific MRI sign of MSA whereas hypointensity alone remains a sensitive but nonspecific sign of MSA (Kraft *et al.*, 1999).

Increased iron deposition in the putamen has been described in striatonigral degeneration and MRI can detect iron. A recent investigation utilized a novel imaging method to compare a single patient with SND to previously studied PD patients and age-matched controls (Martin *et al.*, 1998). The study demonstrated an increase in putaminal iron content in the SND patient beyond the 95% confidence limit for inclusion in the PD group, even considering clinical severity.

5. Functional Imaging

Several PET studies of regional cerebral metabolism have shown distinctive findings in MSA that help differentiate it from PD and PSP (Brooks, 1993). In MSA-C, PET studies with [18F]FDG reveal markedly decreased lCMRglc in the cerebellar vermis, cerebellar

Figure 13 MR images in the sagittal (A) and horizontal (B, C) planes of a 56-year-old man with clinically symptomatic multiple-system atrophy with principally cerebellar ataxia and autonomic failure. The images reveal severe atrophy of the cerebellar vermis, cerebellar hemispheres, middle cerebellar peduncles, base of the pons, and the medulla, with enlargement of the fourth ventricle and the basal cistern (A, B), and putaminal hypointensity with a lateral rim of hyperintensity in the putamen in the T_2-weighted image (C).

hemispheres, and brainstem, with a moderate decline of lCMRglc in the cerebral cortex, caudate nucleus, putamen, and thalamus (Fig. 14). In MSA-P, lCMRglc is markedly decreased in the cerebral cortex, caudate nucleus, putamen, and thalamus and only mildly or modestly diminished in the cerebellar vermis, cerebellar hemispheres, and brainstem (Fig. 15).

PET studies of cerebral metabolism and biochemistry have been used to differentiate MSA from PD and other parkinsonian syndromes (Brooks, 1993). In a study utilizing PET with $[^{18}F]$FDG, 11 MSA patients were compared with 12 PD patients (Otsuka *et al.*, 1996). The MSA group included 7 with OPCA (i.e., MSA-C), 2 with striatonigral degeneration (MSA-P),

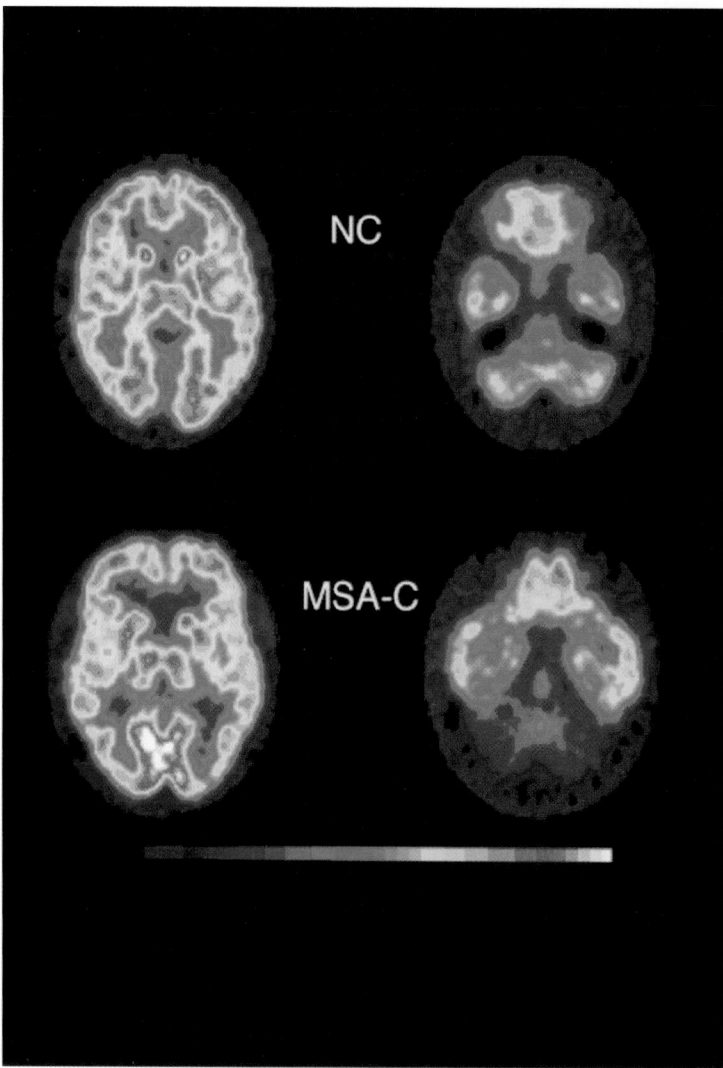

Figure 14 PET scans illustrating lCMRglc as detected with $[^{18}F]$FDG in a 54-year-old female normal control subject (upper two images, NC) and a 51-year-old female patient with clinically symptomatic multiple-system atrophy with severe cerebellar ataxia and autonomic failure (lower two images, MSA-C). The left upper and lower scans show horizontal sections at the level of the cerebral cortex, basal ganglia, and thalamus, and the right upper and lower scans show horizontal sections at the level of the cerebellum and brainstem. The scans reveal moderately decreased lCMRglc in the cerebral cortex, caudate nucleus, putamen, and thalamus and markedly decreased lCMRglc in the cerebellum and brainstem of the patient as compared to the control subject. The color bar indicates the relative rate of cerebral glucose utilization for all scans illustrated, with values increasing from left to right on the bar.

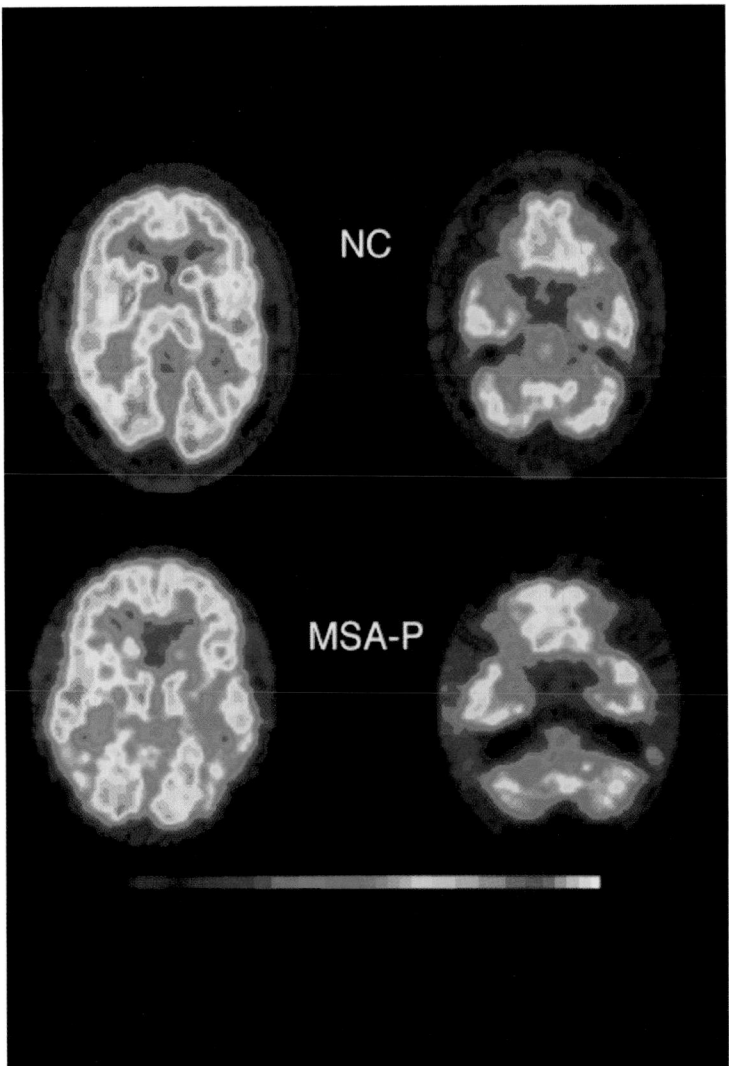

Figure 15 PET scans illustrating lCMRglc as detected with [^{18}F]FDG in a 68-year-old male normal control subject (upper two images, NC) and a 64-year-old male patient with clinically symptomatic multiple-system atrophy characterized by severe parkinsonism and autonomic failure (lower two images, MSA-P). The left upper and lower scans show horizontal sections at the level of the cerebral cortex, basal ganglia, and thalamus, and the right upper and lower scans show horizontal sections at the level of the cerebellum and brainstem. The scans reveal markedly decreased lCMRglc in the cerebral cortex, caudate nucleus, putamen, and thalamus and mildly diminished lCMRglc in the cerebellum and brainstem of the patient as compared to the control subject. The color bar indicates the relative rate of cerebral glucose utilization for all scans illustrated, with values increasing from left to right on the bar.

and 2 with Shy-Drager syndrome (MSA-P). In PD, lCMRglc in multiple regions was no different from controls. In MSA, lCMRglc was different from controls in the frontal, temporal, and parietal lobes of the cerebral cortex and in the caudate nucleus, putamen, cerebellum, and brainstem. MRI scans revealed that cases with greater brainstem and cerebellar atrophy had greater hypometabolism in those structures, although some

cases without detectable atrophy had decreased lCMRglc. Hypometabolism of the cerebellum and brainstem was found in all MSA cases except for the two with the diagnosis of SND.

In another investigation, [^{18}F]FDG was used with PET to differentiate 56 typical PD patients from 48 patients with atypical parkinsonian disorders (Antonini *et al.*, 1998). The group described as having atypical parkinsonism disorders were thought to have MSA because of deteriorating responses to dopaminergic treatment, development of autonomic dysfunction, or both. Discriminant analysis based upon lCMRglc was used to characterize the patients. The results showed that a linear combination of caudate, lentiform, and thalamic values accurately and highly significantly discriminated the atypical parkinsonian patients from the PD group. Significant metabolic abnormalities were found in the striatum and thalamus of 36 of the 48 atypical parkinsonian patients.

PET with [^{18}F]FDG was used to evaluate the symmetry of lCMRglc in five MSA patients who presented with hemiparkinsonism (Kato et al., 1992). Significantly reduced lCMRglc was found in the posterior putamen contralateral to the hemiparkinsonian side. As noted above, T_2-weighted MRI scans showed larger or lower intensity hypointense areas in the putamen contralateral to the hemiparkinsonian side but no significant correlation was found between PET and MRI data.

In three patients with Shy-Drager syndrome, PET with [^{18}F]FDA was used to determine whether the degree of parkinsonism correlated with impaired functional integrity of the presynaptic nigrostriatal pathway (Bhatt et al., 1990). One patient had a short duration of disease and mild parkinsonism, and [^{18}F]FDA uptake was within the normal range. Two patients with a longer duration of disease and more severe parkinsonism had reduced uptake of [^{18}F]FDA, suggesting impaired nigrostriatal dopamine function. In another investigation, PET with a combination of [^{18}F]FDG and [^{18}F]FDA was used to differentiate MSA from PD and normal controls (Otsuka et al., 1997). In the MSA group, lCMRglc was decreased significantly in the frontal and temporal lobes, putamen, cerebellum, and brainstem as compared to controls. [^{18}F]FDA uptake ratios of caudate nucleus/occipital lobe and putamen/occipital lobe were decreased in MSA patients compared to controls. A similar decrease was found in the PD group, but the decrease in the putamen was more prominent in PD than MSA. There was considerable overlap in lCMRglc between patients in the two groups; hence the accuracy of the FDG study for differentiating MSA from PD was lower than that of the FDA study. The conclusion is that FDG is useful for evaluating the regional metabolic activity of the brain, but FDA seems more useful for differentiating MSA from PD.

In another investigation, the utility of PET in differentiating MSA from PD with [^{18}F]FDG and [^{18}F]FDA was augmented by additional scanning with [^{11}C]raclopride ([^{11}C]RAC) to examine dopaminergic D_2 receptor binding (Antonini et al., 1997). In studies of 9 MSA patients, 10 PD patients, and 10 normal controls, striatal FDA binding separated normal controls from MSA and PD groups but could not distinguish between MSA and PD. Both striatal [^{11}C]RAC binding and lCMRglc values discriminated MSA patients from PD patients as well as control subjects. A significant correlation was found between metabolic and receptor binding decrements in the putamen of MSA patients. Stepwise regression analysis revealed that a linear combination of putaminal [^{11}C]RAC and [^{18}F]FDA values accurately predicted the clinical measures of disease severity in the MSA group. The conclusion is that striatal [^{18}F]FDG binding and especially [^{11}C]RAC binding are sensitive and effective measures of striatal function and help characterize patients with MSA. In contrast, [^{18}F]FDA measurements are accurate in detecting abnormalities of the nigrostriatal dopamine system but do not appear to differentiate among the various forms of parkinsonism (Antonini et al., 1997).

PET has been used with other ligands to characterize the neurochemical changes in MSA. PET with [^{11}C]FMZ was used to study GABA-A/BDZ receptor binding in 14 patients with MSA-C, 5 with MSA-P, 18 with sOPCA, and 15 with dOPCA in comparison to 20 normal controls (Gilman et al., 1995a). K_1 was significantly decreased in the cerebellum and brainstem in the MSA-C and sOPCA groups, but not in MSA-P. GABA-A/BDZ binding, however, was largely preserved in the cerebral hemispheres, basal ganglia, thalamus, cerebellum, and brainstem in all patient groups as compared with the normal controls. The finding of relative preserved binding indicates that these sites are available for pharmacologic intervention in these disorders (Gilman et al., 1995a). (S)-[^{11}C]nomifensine is a positron-emitting tracer that binds to both dopaminergic and noradrenergic reuptake sites in the striatum and thalamus (Salmon et al., 1990). A PET study with this ligand demonstrated that the specific compartmentation of this ligand was significantly decreased in both MSA and PD in comparison to age-matched volunteers (Salmon et al., 1990).

Recently, PET with [^{15}O]O_2 was used to examine autoregulation and the coupling between lCBF and lCMRO$_2$ in MSA (Ogawa et al., 1998). Excellent coupling between lCBF and lCMRO$_2$ was found in MSA patients in the resting state. Both elevation of blood pressure induced by leg raising and inhalation of CO_2 increased lCBF. The authors concluded that both autoregulation and coupling of lCBF to lCMRO$_2$ are preserved in MSA.

SPECT has been used to image dopaminergic D_2 receptor density in studies seeking to differentiate idiopathic PD from other movement disorders (Pirker et al., 1997; Hierholzer et al., 1998). In a longitudinal study, 9 patients with PD and 9 with "parkinsonism plus" syndrome (presumably MSA) were followed from 11 to 53 months (Hierholzer et al., 1998). Dopamine D_2 receptor binding was assessed with SPECT and [^{123}I]iodobenzamide both at the beginning of the study and at the end of the follow-up period. Mean specific receptor binding declined during the course of the study in the "parkinsonism plus" group but did not change in the PD group. Another investigation used SPECT with [^{123}I]epidopride, a benzamine derivative with a high affinity for D$_2$ receptors, to compare PD with MSA, progressive supranuclear palsy, and Huntington's disease (Pirker *et al.*, 1997). The results showed reduced binding in MSA and HD compared to controls, but no differences between PD and control groups. The studies were limited by the high specific to nondisplaceable binding ratio of the ligand, and the studies did not discriminate well between the various basal ganglia diseases.

References

Abe, S., Miyasaka, K., Tashiro, K., Takei, H., Isu, T., and Tsuru, M. (1983). Evaluation of the brainstem with high-resolution CT in cerebellar atrophic processes. *AJNR, Am. J. Neuroradiol.* **4,** 446–449.

Abe, Y., Tanaka, F., Matsumoto, M., Doyu, M., Hirayama, M., Kiachi, T., and Sobue, G. (1998). CAG repeat number correlates with the rate of brainstem and cerebellar atrophy in Machado-Joseph disease. *Neurology* **51,** 882–884.

Abele, M., Bürk, K., Andres, F., Topka, H., Laccone, F., Bösch, S., Brice, A., Cancel, G., Dichgans, J., and Klockgether, T. (1997). Autosomal dominant cerebellar ataxia type 1. Nerve conduction and evoked potential studies in families with SCA1, SCA2 and SCA3. *Brain* **120,** 2141–2148.

Abyad, A., and Kligman, E. (1995). Friedreich's ataxia in the elderly. *J. Int. Med. Res.* **23,** 74–84.

Albanese, A., Colosimo, C., Bentivoglio, A. R., Fenici, R., Melillo, G., Colosimo, C., and Tonali, P. (1995). Multiple system atrophy presenting as parkinsonism: Clinical features and diagnostic criteria. *J. Neurol. Neurosurg. Psychiatry* **59,** 144–151.

Antonini, A., Leenders, K. L., Vontobel, P., Maguire, R. P., Missimer, J., Psylla, M., and Gunther, I. (1997). Complementary PET studies of striatal neuronal function in the differential diagnosis between multiple system atrophy and Parkinson's disease. *Brain* **120,** 2187–2195.

Antonini, A., Kazumata, K., Feigin, A., Mandel, F., Dhawan, V., Margouleff, C., and Eidelberg, D. (1998). Differential diagnosis of parkinsonism with [^{18}F]fluorodeoxyglucose and PET. *Mov. Disord.* **13,** 268–274.

Arai, T., Mizukami, K., Matsuzaka, H., Iwakuma, A., Shiraishi, H., and Koizumi, J. (1993). CNS changes in DRPLA with dementia and personality changes: CT, MR and SPECT findings. *Jpn. J. Psychiatry Neurol.* **47,** 105–110.

Babcock, M., deSilva, D., Oaks, R., Davis-Kaplan, S., Jiralerspong, S., Montermini, L., Pandolfo, M., and Kaplan, J. (1997). Regulation of mitochondrial iron accumulation by Yfh1p, a putative homolog of frataxin. *Science* **276,** 1709–1712.

Becher, M. W., Kotzuk, J. A., Sharp, A. H., Davies, S. W., Bates, G. P., Price, D. L., and Ross, C. A. (1998). Intranuclear neuronal inclusions in Huntington's disease and dentatorubral and pallidoluysian atrophy: A correlation between the density of inclusions and 1T15 CAG triplet repeat length. *Neurobiol. Dis.* **4,** 387–397.

Beck, R. O., Betts, C. D., and Fowler, C. K. (1994). Genitourinary dysfunction in mutiple system atrophy: Clinical features and treatment in 62 cases. *J. Urol.* **151,** 1336–1341.

Bennett, R. H., Ludvigson, P., DeLeon, G., and Berry, G. (1984). Large-fiber sensory neuronopathy in autosomal-dominant spinocerebellar degeneration. *Arch. Neurol.* **41,** 175–178.

Benomar, A., Le Guern, E., Dürr, A., Ouhabi, H., Stevanin, G., Yahyaoui, M., Chkili, T., Agid, Y., and Brice, A. (1994). Autosomal dominant cerebellar ataxia with retinal degeneration (ADCA type II) is genetically different from ADCA type I. *Ann. Neurol.* **35,** 439–444.

Benton, C. S., de Silva, R., Rutledge, S. L., Bohlega, S., Ashizawa, T., and Zoghbi, H. Y. (1998). Molecular and clinical studies in SCA-7 define a broad clinical spectrum and the infantile phenotype. *Neurology* **51,** 1081–1086.

Bhatt, M. H., Snow, B. J., Martin, W. R., Cooper, S., and Calne, D. B. (1990). Positron emission tomography in Shy-Drager syndrome. *Ann. Neurol.* **28,** 101–103.

Biemond, A. (1954). La forme radiculo-cordonnale postérieure des dégénérescences spino-cérébelleuses. *Rev. Neurol.* **91,** 2–21.

Bower, J. H., Maraganore, D. M., McDonnell, S. K., and Rocca, W. A. (1997). Incidence of progressive supranuclear palsy and multiple system atrophy in Olmsted County, Minnesota, 1976 to 1990. *Neurology* **49,** 1284–1288.

Brice, A., Cancel, G., Dürr, A., Didierjean, O., Imbert, G., Bürk, K., Saudou, G., Yvert, G., Abada, M., Belal, S., Benomar, A., Klockgether, T., Mandel, J. L., and Agid, Y. (1997). SCA 2 (spinocerebellar ataxia 2): Another unstable CAG expansion. Molecular and clinical analysis of 101 patients. *Neurology* **48,** A210.

Brooks, D. J. (1993). PET studies on the early and differential diagnosis of Parkinson's disease. *Neurology* **43,** S6–S16.

Bürk, K., Abele, M., Fetter, M., Dichgans, J., Skalej, M., Laccone, F., Didierjean, O., Brice, A., and Klockgether, T. (1996). Autosomal dominant cerebellar ataxia type I. Clinical features and MRI in families with SCA1, SCA2 and SCA3. *Brain* **119,** 1497–1505.

Bürk, K., Stevanin, G., Didierjean, O., Cancel, G., Trottier, Y., Skalej, M., Abele, M., Brice, A., Dichgans, J., and Klockgether, T. (1997). Clinical and genetic analysis of three German kindreds with autosomal dominant cerebellar ataxia type I linked to the SCA2 locus. *J. Neurol.* **244,** 256–261.

Bürk, K., Globas, C., Bösch, S., Gräber, S., Abele, M., Brice, A., Dichgans, J., Daum, I., and Klockgether, T. (1999). Cognitive deficits in spinocerebellar ataxia 2. *Brain* **122,** 769–777.

Burke, J. R., Wingfield, M. S., Lewis, K. E., Roses, A. D., Lee, J. E., Hulette, C., Pericak-Vance, M. A., and Vance, J. M. (1994). The Haw River syndrome: Dentatorubral pallidoluysian atrophy (DRPLA) in an African-American family. *Nat. Genet.* **7,** 521–524.

Campuzano, V., Montermini, L., Molto, M. D., Pianese, L., Cossee, M., Cavalcanti, F., Monros, E., *et al.* (1996). Friedreich's ataxia: Autosomal recessive disease caused by an intronic GAA triplet repeat expansion. *Science* **271,** 1423–1427.

Campuzano, V., Montermini, L., Lutz, Y., Cova, L., Hindelang, C., Jiralerspong, S., Trottier, Y., Kish, S. J., Faucheux, B., Trouillas, P., Authier, F. J., Dürr, A., Mandel, J. L., Vescovi, A., Pandolfo, M., and Koenig, M. (1997). Frataxin is reduced in Friedreich ataxia patients and is associated with mitochondrial membrames. *Hum. Mol. Genet.* **6,** 1771–1780.

Cancel, G., Dürr, A., Didierjean, O., Imbert, G., Bürk, K., and Lezin, A. (1997). Molecular and clinical correlations in spinocerebellar ataxia 2: A study of 32 families. *Hum. Mol. Genet.* **6,** 709–715.

Chand, R. P., Tharakan, J. K., Koul, R. L., and Kumar, S. D. (1996). Clinical and radiological features of juvenile onset olivopontocerebellar atrophy. *Clin. Neurol. Neurosurg.* **98,** 152–156.

Chavoix, C., Samson, Y., Pappata, S., Prenant, C., Maziere, M., Seck, A., and Agid, Y. (1990). Positron emission tomography study of brain benzodiazepine receptors in Friedreich's ataxia. *Can. J. Neurol. Sci.* **17,** 404–409.

Chong, S. S., McCall, A. E., Cota, J., Subramony, S. H., Orr, H. T., Hughes, M. R., and Zoghbi, H. Y. (1995). Gametic and somatic tissue-specific heterogeneity of the expanded SCA1 CAG repeat in spinocerebellar ataxia type 1. *Nat. Genet.* **10**, 344–350.

Consensus Committee of the American Autonomic Society and the American Academy of Neurology. (1996). Consensus statement on the definition of orthostatic hypotension, pure autonomic failure, and multiple system atrophy. *Neurology* **46**, 1470.

Coutinho, P., and Andrade, C. (1978). Autosomal dominant system degeneration in Portuguese families of the Azores Islands. A new genetic disorder involving cerebellar, pyramidal, extrapyramidal and spinal cord motor functions. *Neurology* **28**, 703–709.

Cruz-Martinez, A., and Arpa, J. (1997a). Transcranial magnetic stimulation in patients with cerebellar stroke. *Eur. Neurol.* **38**, 82–87.

Cruz-Martinez, A., and Palau, F. (1997b). Central motor conduction by magnetic stimulation of the cortex and peripheral nerve conduction follow-up studies in Friedreich's ataxia. *Electroencephalogr. Clin. Neurophys.* **105**, 458–461.

Cruz-Martinez, A., Arpa, J., Alonso, M., Palomo, F., and Villoslada, C. (1995). Transcranial magnetic stimulation in multiple system and late onset cerebellar atrophies. *Acta Neurol. Scand.* **92**, 218–224.

Daniel, S. E. (1992). The neuropathology and neurochemistry of multiple system atrophy. *In* "Autonomic Failure: A Textbook of Clinical Disorders of the Autonomic Nervous System" (R. Bannister and C. J. Mathias, Eds.), pp. 564–585. Oxford Univ. Press, Oxford.

David, G., Abbas, N., Stevanin, G., Dürr, A., Yvert, G., Cancel, G., Weber, C., Imbert, G., Saudou, F., Antoniou, E., Drabkin, H., Gemmill, R., Giunti, P., Benomar, A., Wood, N., Ruberg, M., Agid, Y., Mandel, J. L., and Brice, A. (1997). Cloning of the SCA7 gene reveals a highly unstable CAG repeat expansion. *Nat. Genet.* **17**, 65–70.

David, G., Dürr, A., Stevanin, G., Cancel, G., Abbas, N., Benomar, A., Belal, S., Lebre, A. S., Abada-Bendib, M., Grid, D., Holmberf, M., Yahyaoui, M., Hentati, F., Chkili, T., Agid, Y., and Brice, A. (1998). Molecular and clinical correlations in autosomal dominant cerebellar ataxia with progressive macular dystrophy (SCA7). *Hum. Mol. Genet.* **7**, 165–170.

De Michele, G., Filla, A., Criscuolo, C., Scarano, V., Cavalcanti, F., Pianese, L., Monticelli, A., and Cocozza, S. (1998). Determinants of onset age in Friedreich's ataxia. *J. Neurol.* **245**, 166–168.

DelFavero, J., Krols, L., Michalik, A., Theuns, J., Lofgren, A., Goossens, D., Wehnert, A., Van den Bossche, D., Van Zand, K., Backhovens, H., van Regenmorter, N., Martin, J. J., and Van Broeckhoven, C. (1998). Molecular genetic analysis of autosomal dominant cerebellar ataxia with retinal degeneration (ADCA type II) caused by CAG triplet repeat expansion. *Hum. Mol. Genet.* **7**, 177–186.

Di Lazzaro, V., Restuccia, D., Molinari, M., Leggio, M. G., Nardone, R., Fogli, D., and Tonali, P. (1994). Excitability of the motor cortex to magnetic stimulation in patients with cerebellar lesions. *J. Neurol. Neurosurg. Psychiatry* **57**, 108–110.

Diriong, S., Lory, P., Williams, M. E., Ellis, S. B., Harpold, M. M., and Taviaux, S. (1995). Chromosomal localization of the human genes for alpha 1A, alpha 1B, and alpha 1E voltage-dependent Ca2$^+$ channel subunits. *Genomics* **30**, 605–609.

Dürr, A., Smadja, D., Cancel, G., Lezin, A., Stevanin, G., Mikol, J., Bellance, R., Buisson, G.G., Chneiwiess, H., Dellanave, J., Agid, Y., Brice, A., and Vernant, J.-C. (1995). Autosomal dominant cerebellar ataxia type I in Martinique (French West Indies). Clinical and neuropathological analysis of 53 patients from three unrelated SCA2 families. *Brain* **118**, 1573–1581.

Dürr, A., Cossee, M., Agid, Y., Campuzano, V., Mignard, C., Penet, C., Mandel, J. L., Brice, A., and Koenig, M. (1996a). Clinical and genetic abnormalities in patients with Friedreich's ataxia. *N. Engl. J. Med.* **335**, 1169–1174.

Dürr, A., Stevanin, G., Cancel, G., Duyckaerts, C., Abbas, N., Didier-

jean, O., Chneiwiess, H., Benomar, A., Lyon-Caen, O., Julien, J., Serdaru, M., Penet, C., Agid, Y., and Brice, A. (1996b). Spinocerebellar ataxia 3 and Machado-Joseph disease: Clinical, molecular, and neuropathological features. *Ann. Neurol.* **39**, 490–499.

Duvoisin, R. C. (1984). An apology and an introduction to the olivopontocerebellar atrophies. *In* "The Olivopontocerebellar Atrophies" (R. C. Duvoisin and A. Plaitakis, Eds.), pp. 5–12. Raven Press, New York.

Eadie, M. J. (1975a). Olivo-ponto-cerebellar atrophy (Dejerine-Thomas type). *In* "Handbook of Clinical *Neurology*" (P. J. Vinken and G. W. Bruyn, Eds.), Vol. 21, pp. 415–431. North-Holland, Amsterdam.

Eadie, M. J. (1975b). Olivo-ponto-cerebellar atrophy (Menzel type). *In* "Handbook of Clinical *Neurology*" (P. J. Vinken and G. W. Bruyn, Eds.), Vol. 21, pp. 433–449. North-Holland, Amsterdam.

Enevoldson, J. P., Sanders, M. D., and Harding, A. E. (1994). Autosomal dominant cerebellar ataxia with pigmentary macular dystrophy. A clinical and genetic study of eight families. *Brain* **117**, 445–460.

Farmer, T. W., Wingfield, M. S., Lynch, S. A., Vogel, F. S., Hulette, C., Katchioff, B., and Jacobson, P. L. (1989). Ataxia, chorea, seizures, and dementia. Pathologic features of a newly defined familial disorder. *Arch. Neurol.* **46**, 774–779.

Filla, A., DeMichele, G., Banfi, S., Santoro, L., Perretti, A., Calvalcanti, F., Pianese, L., Castaldo, I., Barbieri, F., Campanella, G., and Cocozza, S. (1995). Has spinocerebellar ataxia type 2 a distinct phenotype? Genetic and clinical study of an Italian family. *Neurology* **45**, 793–796.

Flanigan, K., Gardner, K., Alderson, K., Galster, B., Otterud, B., Leppert, M. F., Kaplan, C., and Ptacek, L. J. (1996). Autosomal dominant spinocerebellar ataxia with sensory axonal neuropathy (SCA4): Clinical description and genetic localization to chromosome 16q22.1. *Am. J. Hum. Genet.* **59**, 392–399.

Gacy, A. M., Goellner, G. M., Spiro, C., Chen, X., Gupta, G., Bradbury, E. M., Dyer, R. B., Mikesell, M. J., Yao, J. Z., Johnson, A. J., Richter, A., Melancon, S. B., and McMurray, C. T. (1998). GAA instability in Friedreich's ataxia shares a common, DNA-directed and intraallelic mechanism with other trinucleotide diseases. *Mol. Cell* **1**, 583–593.

Gardner, K., Alderson, K., Galster, B., Kaplan, C., Leppert, M., and Ptacek, L. (1994). Autosomal dominant spinocerebellar ataxia: Clinical description of a distinct hereditary ataxia and genetic localization to chromosome 16 (SCA4) in a Utah kindred. *Neurology* **44**, A361.

Genis, D., Matilla, T., Volpini, V., *et al.* (1995). Clinical neuropathologic and genetic studies of a large spinocerebellar ataxia type 1 (SCA1) kindred: (CAG)n expansion and early premonitory signs and symptoms. *Neurology* **45**, 24–30.

Geschwind, D. H., Perlman, S., Figueroa, C. P., Treiman, L. J., and Pulst, S. M. (1997a). The prevalence and wide clinical spectrum of the spinocerebellar ataxia type 2 (SCA2) trinucleotide repeat in patients with autosomal dominant cerebellar ataxia. *Am. J. Hum. Genet.* **60**, 842–850.

Geschwind, D. H., Perlman, S., Figueroa, K. P., Karrim, J., Baloh, R. W., and Pulst, S. M. (1997b). Spinocerebral ataxia type 6. Frequency of the mutation and genotype-phenotype correlations. *Neurology* **49**, 1247–1251.

Gilman, S. (1989). Cerebellar diseases: Studies with positron emission tomography. *Semin. Neurol.* **9**, 370–376.

Gilman, S., Bloedel, J. R., and Lechtenberg, R. (1981). "Disorders of the Cerebellum." Davis, Philadelphia.

Gilman, S., Markel, D. S., Koeppe, R. A., Junck, L., Kluin, K. J., Gebarski, S. S., and Hichwa, R. D. (1988). Cerebellar and brainstem hypometabolism in olivopontocerebellar atrophy detected with positron emission tomography. *Ann. Neurol.* **32**, 223–230.

Gilman, S., Junck, L., Markel, D. S., Koeppe, R. A., and Kluin, K. J. (1990). Cerebral glucose hypermetabolism in Friedreich's ataxia detected with positron emission tomography. *Ann. Neurol.* **28**, 750–757.

Gilman, S., Koeppe, R. A., Junck, L., Kluin, K. L., Lohman, M., and St. Laurent, R. T. (1994). Patterns of cerebral glucose metabolism detected with positron emission tomography differ in multiple system atrophy and olivopontocerebellar atrophy. *Ann. Neurol.* **36**, 166–175.

Gilman, S., Koeppe, R. A., Junck, L., Kluin, K. J., Lohman, M., and St. Laurent, R. T. (1995a). Benzodiazepine receptor binding in cerebellar degenerations studied with positron emission tomography. *Ann. Neurol.* **38**, 176–185.

Gilman, S., St. Laurent, R. T., Koeppe, R. A., Junck, L., Kluin, K. J., and Lohman, M. (1995b). A comparison of cerebral blood flow and glucose metabolism in olivopontocerebellar atrophy using PET. *Neurology* **45**, 1345–1352.

Gilman, S., Frey, K. A., Koeppe, R. A., Junck, L., Little, R., Vander Borght, T. M., Lohman, M., Martorello, S., Lee, L. C., Jewett, D. M., and Kilbourn, M. R. (1996a). Decreased striatal monoaminergic terminals in olivopontocerebellar atrophy and multiple system atrophy demonstrated with positron emission tomography. *Ann. Neurol.* **40**, 885–892.

Gilman, S., Sima, A. A. F., Junck, L., Kluin, K. J., Koeppe, R. A., Lohman, M. E., and Little, R. (1996b). Spinocerebellar ataxia type 1 with multiple system degeneration and glial cytoplasmic inclusions. *Ann. Neurol.* **39**, 241–255.

Gilman, S., Low, P. A., Quinn, N., Albanese, A., Ben-Shlomo, Y., Fowler, C. J., Kaufmann, H., Klockgether, T., Lang, A. E., Lantos, P. L., Litvan, I., Mathias, C. J., Oliver, E., Robertson, D., Schatz, I., and Wenning, G. K. (1998a). Consensus statement on the diagnosis of multiple system atrophy. *J. Auton. Nerv. Syst.* **74**, 189–192.

Gilman, S., Low, P. A., Quinn, N., Albanese, A., Ben-Shlomo, Y., Fowler, C. J., Kaufmann, H., Klockgether, T., Lang, A. E., Lantos, P. L., Litvan, I., Mathias, C. J., Oliver, E., Robertson, D., Schatz, I., and Wenning, G. K. (1998b). Consensus statement on the diagnosis of multiple system atrophy. *Clin. Auton. Res.* **8**, 359–362.

Gilman, S., Koeppe, R. A., Junck, L., Little, R., Kluin, K. J., Heumann, M., Martorello, S., and Johanns, J. (1999a). Decreased striatal monoaminergic terminals in multiple system atrophy detected with PET. *Ann. Neurol.* **45**, 769–777.

Gilman, S., Low, P. A., Quinn, N., Albanese, A., Ben-Shlomo, Y., Fowler, C. J., Kaufmann, H., Klockgether, T., Lang, A. E., Lantos, P. L., Litvan, I., Mathias, C. J., Oliver, E., Robertson, D., Schatz, I., and Wenning, G. K. (1999b). Consensus statement on the diagnosis of multiple system atrophy. *J. Neurol. Sci.* **163**, 94–98.

Giroud, M., Septien, L., Pelletier, J. L., Dueret, N., and Dumas, R. (1994). Decrease in cerebellar blood flow in patients with Friedreich's ataxia: A Tc-HMPAO SPECT study of three cases. *Neurol. Res.* **16**, 342–344.

Gispert, S., Twells, R., Orozco, G., Brice, A., Weber, J., Heredero, L., Scheufler, K., Riley, B., Allotey, R., Nothers, C., Hillermann, R., Lunkes, A., Khati, C., Stevanin, G., Hernandez, A., Magarino, C., Klockgether, T., Dürr, A., Chneieiss, H., Enzman, J., Farrall, M., Beckmann, J., Multan, M., Wernet, P., Agid, Y., Freund, H.-J., Williamson, R., Auburger, G., and Chamberlain, S. (1993). Chromosomal assignment of the second locus for autosomal dominant cerebellar ataxia (SCA2) to chromosome 12q23-24.1. *Nat. Genet.* **4**, 295–299.

Giuliani, G., Chiaramoni, L., Foschi, N., and Terziani, S. (1992). The role of MRI in the diagnosis of olivopontocerebellar atrophy. *Ital. J. Neurol. Sci.* **13**, 151–156.

Giunti, P., Sweeney, M. G., and Hardin, A. E. (1995). Detection of the Machado-Joseph disease/spinocerebellar ataxia three trinculeo-tide repeat expansion in families with autosomal dominant motor disorders, including the Drew family of Walworth. *Brain* **118**, 1077–1085.

Gomez, C. M., Thompson, R. M., Gammack, J. T., Perlman, S. L., Dobyns, W. B., Truwit, C. L., Zee, D. S., Clark, H. B., and Anderson, J. H. (1997). Spinocerebellar ataxia type 6: Gaze-evoked and vertical nystagmus, Purkinje cell degeneration, and variable age of onset. *Ann. Neurol.* **42**, 933–950.

Goto, I., Tobimatzu, S., Ohta, M., Hosokawa, S., Shibasaki, H., and Kuroiwa, Y. (1982). Dentatorubropallidoluysian degeneration: Clinical, neuro-ophthalmologic, biochemical, and pathologic studies on autosomal dominant dorm. *Neurology* **32**, 1395–1399.

Gouw, L. C., Digre, K. B., Harris, C. P., Haines, J. H., and Ptacek, L. J. (1994). Autosomal dominant cerebellar ataxia with retinal degeneration: Clinical, neuropathologic, and genetic analysis of a large kindred. *Neurology* **44**, 1441–1447.

Gouw, L. C., Kaplan, C. D., Haines, J. H., Digre, K. B., Rutledge, S. L., Matilla, A., *et al.* (1995). Retinal degeneration characterizes a spinocerebellar ataxia mapping to chromosome 3p. *Nat. Genet.* **10**, 89–93.

Graham, J. G., and Oppenheimer, D. R. (1969). Orthostatic hypotension and nicotine sensitivity in a case of multiple system atrophy. *J. Neurol. Neurosurg. Psychiatry* **32**, 28–34.

Gray, R. C., and Oliver, C. P. (1941). Marie's hereditary cerebellar ataxia (olivoponto-cerebellar atrophy). *Minn. Med.* **24**, 327–335.

Greenfield, J. G. (1954). "The Spino-cerebellar Degenerations." Blackwell, Oxford.

Haines, J. L., Schut, L. J., Weitkamp, L. R., Thayer, M., and Anderson, V. E. (1984). Spinocerebellar ataxia in a large kindred: Age at onset, reproduction, and genetic linkage studies. *Neurology* **34**, 1542–1548.

Harding, A. E. (1982). The clinical features and classification of late onset autosomal dominant cerebellar ataxias. A study of 11 families, including descendents of "the Drew family of Walwirth." *Brain* **105**, 1–28.

Harding, A. E. (1984). "The Hereditary Ataxias and Related Disorders." Churchill Livingstone, London.

Hayashi, Y., Kakita, A., Yamada, M., Egawa, S., Oyanagi, S., Naito, H., Tsuji, S., and Takahashi, H. (1998a). Hereditary dentatorubralpallidoluysian atrophy: Ubiquitinated filamentous inclusions in the cerebellar dentate nucleus neurons. *Acta Neuropathol.* **95**, 479–482.

Hayashi, Y., Kakita, A., Yamada, M., Koide, R., Igarashi, S., Takano, H., Ikeuchi, T., Wakabayashi, K., Egawa, S., Tsuji, S., and Takahashi, H. (1998b). Hereditary dentatorubral-pallidoluysian atrophy: Detection of widespread ubiquitinated neuronal and glial intranuclear inclusions in the brain. *Acta Neuropathol.* **96**, 547–552.

Hierholzer, J., Cordes, M., Venz, S., Schelosky, L., Harisch, C., Richter, W., Keske, U., Hosten, N., Maurer, J., Poewe, W., and Felix, R. (1998). Loss of dopamine-D$_2$ receptor binding sites in Parkinsonian plus syndromes. *J. Nucl. Med.* **39**, 954–960.

Higgins, J. J., Nee, L. E., Vasconcelos, O., Ide, S. E., Lavedan, C., Goldfarb, L. G., and Polymeropoulus, M. H. (1996). Mutations in American families with spinocerebellar ataxia (SCA) type 3: SCA3 is allelic to Machado–Joseph disease. *Neurology* **46**, 208–213.

Holmberg, M., Johannson, J., Forsgren, L., Heijbel, J., Sandgren, O., and Holmgren, G. (1995). Localization of autosomal dominant cerebellar ataxia associated with retinal degeneration and anticipation to chromosome 3p12-p21.1. *Hum. Mol. Genet.* **4**, 1441–1445.

Holmberg, M., Duyckaerts, C., Dürr, A., Cancel, G., Gourfinkel-An, I., Damkier, P., Faucheux, B., Trottier, Y., Hirsch, E. C., Agid, Y., and Brice, A. (1998). Spinocerebellar ataxia type 7 (SCA7): A neurodegenerative disorder with neuronal intranuclear inclusions. *Hum. Mol. Genet.* **7**, 913–918.

Huang, Y. P., and Plaitakis, A. (1984). Morphological changes of olivo-pontocerebellar atrophy in computed tomography and comments on its pathogenesis. *Adv. Neurol.* **41,** 39–85.

Iizuka, R., and Hirayama, K. (1991). Dentato-rubro-pallidoluysian atrophy. *In* "Handbook of Clinical Neurology. Hereditary Neuropathies and Spinocerebellar Atrophies" (P. Vinken, G. Bruyn, H. Klawans, and J. M. B. V. DeJong, Eds.), Vol. 60, pp. 607–618. Elsevier, Amsterdam/New York.

Iizuka, R., Hirayama, K., and Maehara, K. (1984). Dentatorubropallidoluysian atrophy: A clinico-pathological study. *J. Neurol. Neurosurg. Psychiatry* **47,** 1288–1298.

Ikeuchi, T., Koide, R., Tanaka, H., Onodero, O., Igarishi, S., Takahashi, H., Kondo, R., Ishikawa, A., Tomoda, A., Miike, T., Satao, K., Ihara, Y., Hayabara, T., Isa, F., Tanabe, H., Tokiguchi, S., Hayashi, M., Shimuzu, N., Ikuta, F., Naito, H., and Tsuji, S. (1995a). Dentatorubro-pallidoluysian atrophy: Clinical features are closely related to unstable expansions of trinucleotide (CAG) repeat. *Ann. Neurol.* **37,** 769–775.

Ikeuchi, T., Onodero, O., Oyake, M., Koide, R., Tanaka, H., and Tsuji, S. (1995b). Dentatorubral-pallidoluysian atrophy (DRLPA): Close correlation of CAG repeat expansion with the wide spectrum of clinical presentations and prominent anticipation. *Semin. Cell. Biol.* **6,** 37–44.

Ikeuchi, T., Takano, H., Koide, R., Horikawa, Y., Honma, Y., Onishi, Y., Igarashi, S., Tanaka, H., Nakao, N., Sahashi, K., Tsukagoshi, H., Inoue, K., Takahashi, H., and Tsuji, S. (1997). Spinocerebellar ataxia type 6: CAG repeat expansion in α_{1A} voltage-dependent calcium channel gene and clinical variations in Japanese population. *Ann. Neurol.* **42,** 879–884.

Ikuta, N. (1998). Proton magnetic resonance spectroscopy and single photon emission CT in patients with olivopontocerebellar atrophy. *Clin. Neurol.* **38,** 289–294.

Imamura, A., Ito, R., Tanaka, S., Fukutomi, O., Shimozawa, N., Nishimura, M., Suzuki, Y., Kondo, N., Yamada, M., and Orii, T. (1994). High-intensity proton and T2-weighted MRI signals in the globus pallidus in juvenile-type of dentatorubral and pallidoluysian atrophy. *Neuropediatrics* **25,** 234–237.

Imbert, G., Saudou, F., Yvert, G., Devys, D., Trottier, Y., Garnier, J. M., Weber, C., Mandel, J. L., Cancel, G., Abbas, N., Dürr, A., Didierjean, O., Stevanin, G., Agid, Y., and Brice, A. (1996). Cloning of the gene for spinocerebellar ataxia reveals a locus with high sensitivity to expanded CAG/glutamine repeats. *Nat. Genet.* **14,** 285–291.

Inazuki, G., Baba, K., and Naito, H. (1989). Electroencephalographic findings of hereditary dentatorubral-pallidoluysian atrophy (DRLPA). *Jpn. J. Psychiatry Neurol.* **43,** 213–220.

Ishibashi, M., Sakai, T., Matsuishi, T., Yonekura, Y., Yamashita, Y., Abe, T., Ohnishi, Y., and Hayabuchi, N. (1998). Decreased benzodiazepine receptor binding in Machado-Joseph disease. *J. Nucl. Med.* **39,** 1518–1520.

Ishikawa, K., Tanaka, H., Saito, M., Onkoshi, N., Fujita, T., Yoshizawa, K., Ikeuchi, T., Watanabe, M., Hayashi, A., Takiyama, Y., Nishizawa, M., Nakano, I., Matsubayashi, K., Miwa, M., Shoji, S., Kanazawa, I., Tsuji, S., and Mizusawa, H. (1997). Japanese families with autosomal dominant pure cerebellar ataxia map to chromosome 19p13.1-p13.2 and are strongly associated with mild CAG expansion in the spinocerebellar ataxia type 6 gene in chromosome 19p13.1. *Am. J. Hum. Genet.* **61,** 336–346.

Jobsis, G. J., Weber, J. W., Barth, P. G., Keizers, H., Baas, F., van Schooneveld, M. J., van Hilten, J. J., Troost, D., Geesink, H. H., and Bolhuis, P. A. (1997). Autosomal dominant cerebellar ataxia with retinal degeneration (ADCA II): Clinical and neuropathological findings in two pedigrees and genetic linkage to 3p12-p21.1. *J. Neurol. Neurosurg. Psychiatry* **62,** 367–371.

Johansson, J., Forsgren, L., Sandgren, O., Brice, A., Holmgren, G., and Holmberg, M. (1998). Expanded CAG repeats in Swedish spinocerebellar ataxia type 7 (SCA7) patients: Effect of CAG repeat length on the clinical manifestation. *Hum. Mol. Genet.* **7,** 171–176.

Junck, L., Gilman, S., Gebarski, S. S., Koeppe, R. A., Kluin, K. J., and Markel, D. S. (1994). Structural and functional brain imaging in Friedreich's ataxia. *Arch. Neurol.* **51,** 349–355.

Kato, T., Kume, A., Ito, K., Tadokoro, M., Takahashi, A., and Sakuma, S. (1992). Asymmetrical FDG-PET and MRI findings of striatonigral system in multiple system atrophy with hemiparkinsonism. *Radiat. Med.* **10,** 87–93.

Kitamura, E., Hosoda, F., Fukushima, M., Asakawa, S., Shimizu, N., Imai, T., Soeda, E., and Ohki, M. (1997). A 3-Mb sequence-ready contig map encompassing the multiple disease gene cluster on chromosome 11q13.q13.3. *DNA Res.* **4,** 281–289.

Klockgether, T., Faiss, J., Poremba, M., and Dichgans, J. (1990a). The development of infratentorial atrophy in patients with idiopathic cerebellar ataxia of late onset: A CT study. *J. Neurol.* **237,** 420–423.

Klockgether, T., Schroth, G., Diener, H. C., and Dichgans, J. (1990b). Idiopathic cerebellar ataxia of late onset: Natural history and MRI morphology *J. Neurol. Neurosurg. Psychiatry* **53,** 297–305.

Klockgether, T., Petersen, D., Grodd, W., and Dichgans, J. (1991). Early onset cerebellar ataxia with retained tendon reflexes. Clinical, electrophysiological and MRI observations in comparison with Friedreich's ataxia. *Brain* **114,** 1559–1573.

Klockgether, T., Chamberlain, S., Wullner, U., Fetter, M., Dittman, H., Petersen, D., and Dichgans, J. (1993). Late-onset Friedreich's ataxia. Molecular genetics, clinical neurophysiology, and magnetic resonance imaging. *Arch. Neurol.* **50,** 803–806.

Klockgether, T., Zühlke, C., Schulz, J. B., Bürk, K., Fetter, M., Dittmann, H., Skalej, M., and Dichgans, J. (1996). Friedreich's ataxia with retained tendon reflexes: Molecular genetics, clinical neurophysiology, and magnetic resonance imaging. *Neurology* **46,** 118–121.

Klockgether, T., Lüdtke, R., Kramer, B., Abele, M., Bürk, K., Schöls, L., Riess, O., Laccone, R., Boesch, S., Lopes-Cendes, I., Brice, A., Inzelberg, R., Zilber, N., and Dichgans, J. (1998). The natural history of degenerative ataxia: A retrospective study in 466 patients. *Brain* **121,** 589–600.

Kluin, K. J., Gilman, S., Markel, D. S., Koeppe, R. A., Rosenthal, G., and Junck, L. (1988). Speech disorders in olivopontocerebellar atrophy correlate with positron emission tomography findings. *Ann. Neurol.* **23,** 547–554.

Koide, R., Ikeuchi, T., Onodero, O., Tanaka, H., Igarashi, S., Endo, K., Takahashi, H., Kondo, R., Ishikawa, A., Hayashi, T., Saito, M., Tomoda, A., Miike, T., Naito, H., Ikuta, F., and Tsuji, S. (1994). Unstable expansion of CAG repeat in hereditary dentatorubro-pallidoluysian atrophy (DRLPA). *Nat. Genet.* **6,** 9–13.

Koide, R., Onodera, O., Ikeuchi, T., Kondo, R., Tanaka, H., Tokiguchi, S., Tomoda, A., Miike, T., Isa, F., Beppu, H., Shimizu, N., Watanabe, Y., Horikawa, Y., Shimohata, T., Hirota, K., Ishikawa, A., and Tsuji, S. (1997). Atrophy of the cerebellum and brainstem in dentatorubral pallidoluysian atrophy. Influence of CAG repeat size on MRI findings. *Neurology* **49,** 1605–1612.

Kojima, S., and Hirayama, K. (1988). Magnetic resonance imaging in spinocerebellar degeneration. *Brain Nerve* **40,** 187–193.

Komure, O., Sano, A., Nishino, N., Yamauchi, N., Ueno, S., Kondoh, K., Sano, N., Takahashi, M., Murayama, N., Kondo, I., Nagafuchi, S., Yamada, M., and Kanazawa, I. (1995). DNA analysis in hereditary dentatorubral-pallidoluysian atrophy: Correlation between CAG repeat length and phenotypic variation and the molecular basis of anticipation. *Neurology* **45,** 143–149.

Koob, M. D., Moseley, M. L., Schut, L. J., Benzow, K. A., Bird, T. D.,

Day, J. W., and Ranum, L. P. (1999). An untranslated CTG expansion causes a novel form of spinocerebellar ataxia (SCA8). *Nat. Genet.* **21,** 379–384.

Koshi, Y., Kitamura, S., and Terashi, A. (1995). Relationship between cognitive function and regional CBF in hereditary spinocerebellar degeneration: Comparison between Joseph disese and OPCA of Menzel type. *Clin. Neurol.* **35,** 237–242.

Koutnikova, H., Campuzano, V., Foury, F., Dollé, P., Cazzalini, O., and Koenig, M. (1997). Studies of human, mouse and yeast homologues indicate a mitochondrial function for frataxin. *Nat. Genet.* **16,** 345–351.

Kraft, E., Schwarz, J., Trenkwalkder, C., Vogl, T., Pfluger, T., and Oertel, W. H. (1999). The combination of hypointense and hyperintense signal changes on T_2-weighted magnetic resonance imaging sequences: A specific marker of multiple system atrophy? *Arch. Neurol.* **56,** 225–228.

Kumar, S. D., Chand, R. P., Gururaj, A. K., and Jeans, W. D. (1995). CT features of olivopontocerebellar atrophy in children. *Acta Radiol.* **36,** 593–596.

Lamont, P. J., Davis, M. B., and Wood, N. W. (1997). Identification and sizing of the GAA trinucleotide repeat expansion of Friedreich's ataxia in 56 patients. Clinical and genetic correlations. *Brain* **120,** 673–680.

Lantos, P. L. (1998). The definition of multiple system atrophy: A review of recent developments. *J. Neuropathol. Exp. Neurol.* **57,** 1099–1111.

Lopes-Cendes, I., Andermann, E., Attig, E., Cendes, F., Bosch, S., Wagner, M., Gerstenbrand, F., Andermann, F., and Rouleau, G. (1994). Confirmation of the SCA-2 locus as an alternative locus for dominantly inherited spinocerebellar ataxias and refinement of the candidate region. *Am. J. Hum. Genet.* **54,** 774–781.

Margolis, R. L., Li, S.-H., Young, W. S., Wagster, V., Stine, O. C., Kidwa, A. S., Ashworth, R. G., and Ross, C. A. (1996). DRPLA gene (atrophin-1) sequence and mRNA expression in human brain. *Mol. Brain Res.* **36,** 219–226.

Marie, P., Foix, C., and Alajouanine, T. (1922). De l'atrophie cérébelleus tardive à prédominance corticale. *Rev. Neurol.* **38,** 849–1082.

Martin, W. R., Roberts, T. E., Ye, F. Q., and Allen, P. S. (1998). Increased basal ganglia iron in striatonigral degeneration: *In vivo* estimation with magnetic resonance. *Can. J. Neurol. Sci.* **25,** 44–47.

Maruyama, H., Nakamura, S., Matsuyama, Z., Sakai, T., Doyu, M., Sobue, G., Seto, K. M., Tsujihata, M., Oh-i, T., and Nishio, T. (1995). Molecular features of the CAG repeats and clinical manifestation of Machado–Joseph disease. *Hum. Mol. Genet.* **4,** 807–812.

Mascalchi, M., Tosetti, M., Plasmati, R., Bianchi, M. C., Tessa, C., Salvi, F., Frontali, M., Valzania, F., Bartolozzi, C., and Tassinari, C. A. (1998). Proton magnetic resonance spectroscopy in an Italian family with spinocerebellar ataxia type 1. *Ann. Neurol.* **43,** 244–252.

Mathias, C. J., and Williams, A. C. (1994). The Shy–Drager syndrome (and multiple system atrophy). *In* "Neurodegenerative Diseases" (D. B. Calne, Ed.), pp. 743–768. Saunders, Philadelphia.

Matsumura, R., Futamura, N., Fujimoto, Y., Yanagimoto, S., Horikawa, H., Suzumura, A., and Takayanagi, T. (1997). Spinocerebellar ataxia type 6. Molecular and clinical features of 35 Japanese patients including one homozygous for the CAG repeat expansion. *Neurology* **49,** 1238–1243.

Matsuura, T., Achari, M., Khajavi, M., Bachinski, L. L., Zoghbi, H. Y., and Ashizawa, T. (1999). Mapping of the gene for a novel spinocerebellar ataxia with pure cerebellar signs and epilepsy. *Ann. Neurol.* **45,** 407–411.

Mishina, M., Senda, M., Ohyama, M., Ishii, K., Kitamura, S., and Terashi, A. (1995). Regional cerebral glucose metabolism associated with ataxic gait: An FDG-PET activation study in patients with olivopontocerebellar atrophy. *Clin. Neurol.* **35,** 1199–1204.

Miyazaki, M., Kato, T., Hashimoto, T., Harada, M., Kondo, I., and Kuroda, Y. (1995). MR of childhood-onset dentatorubral-pallidoluysian atrophy. *AJNR, Am. J. Neuroradiol.* **16,** 1834–1836.

Miyazaki, M., Hashimoto, T., Nakagawa, R., Yoneda, Y., Tayama, M., Kawano, N., Muyayama, N., Kondo, I., and Kuroda, Y. (1996a). Characteristic evoked potentials in childhood-onset dentatorubral-pallidoluysian atrophy. *Brain Dev.* **18,** 389–393.

Miyazaki, M., Hashimoto, T., Yoneda, Y., Tayama, M., Harada, M., Miyoshi, H., Kawano, N., Murayama, N., Kondo, I., and Kuroda, Y. (1996b). Proton magnetic resonance spectroscopy in childhood-onset dentatorubral-pallidoluysian atrophy (DRLPA). *Brain Dev.* **18,** 142–146.

Mizoi, Y., Segawa, F., Kamada, K., Sunohara, N., Nakayama, H., and Akashi, T. (1994). Investigation of involvement of cerebral white matter in DRPLA, including MRI perfusion study. *Brain Nerve* **46,** 145–151.

Munschauer, F. E., Loh, L., Bannister, R., and Newsom-Davis, J. (1990). Abnormal respiration and sudden death during sleep in multiple system atrophy with autonomic failure. *Neurology* **40,** 677–679.

Murata, Y., Kawakami, H., Yamaguchi, S., Nishimura, M., Kohriyama, T., Ishizaki, F., Matsuyama, Z., Mimori, Y., and Nakamura, S. (1998a). Characteristic magnetic resonance imaging findings in spinocerebellar ataxia 6. *Arch. Neurol.* **55,** 1348–1352.

Murata, Y., Yamaguchi, S., Kawakami, H., Imon, Y., Maruyama, H., Sakai, T., Kazuta, T., Ohtake, T., Nishimura, M., Saida, T., Chiba, S., Oh-i, T., and Nakamura, S. (1998b). Characteristic magnetic resonance imaging findings in Machado–Joseph disease. *Arch. Neurol.* **55,** 33–37.

Nachmanoff, D. B., Segal, R. A., Dawson, D. M., Brown, R. B., and DeGiolami, U. (1997). Hereditary ataxia with sensory neuronopathy: Biemond's ataxia. *Neurology* **48,** 273–275.

Nagafuchi, S., Yanagisawas, H., Ohsaki, E., Shirayama, T., Tadokoro, K., Inoue, T., and Yamada, M. (1994b). Structure and expression of the gene responsible for the triplet repeat disorder, dentatorubral and pallidoluysian atrophy (DRPLA). *Nat. Genet.* **8,** 177–182.

Nagafuchi, S., Yanagisawas, H., Sato, K., Shirayama, T., Ohsaki, E., Bundo, M., Takeda, T., Tadokoro, K., Kondo, I., Murayama, N., Tanaka, Y., Kikushima, H., Umino, K., Kurosawa, H., Furukawa, T., Nihei, K., Inoue, T., Sano, A., Komure, O., Takahashi, M., Yoshizawa, T., Kanazawa, I., and Yamada, M. (1994a). Dentatorubral and pallidoluysian atrophy expansion of an unstable CAG trinucleotide on chromosome 12p. *Nat. Genet.* **6,** 14–18.

Nagai, Y., Azuma, T., Funauchi, M., Fujita, J., Umi, M., Hirano, M., Matsubara, T., and Ueno, S. (1998). Clinical and molecular genetic study in seven Japanese families with spinocerebellar ataxia type 6. *J. Neurol. Sci.* **157,** 52–59.

Naito, H., and Oyanagi, S. (1982). Familial myoclonus epilepsy and choreoathetosis: Hereditary dentatorubral-pallidoluysian atrophy. *Neurology* **32,** 798–807.

Nakano, K. K., Dawson, D. M., and Spence, A. (1972). Machado disease. A hereditary ataxia in Portuguese emigrants to Massachusetts. *Neurology* **22,** 49–55.

Nee, L. E., and Higgins, J. J. (1997). Should spinocerebellar ataxia type 5 be called Lincoln ataxia? *Neurology* **49,** 298–302.

Neumann, M. A. (1959). Combined degeneration of globus pallidus and dentate nucleus and their projections. *Neurology* **9,** 430–438.

Ogawa, M., Fukuyama, J., Harada, K., and Kimura, J. (1998). Cerebral blood flow and metabolism in multiple system atrophy of the Shy–Drager syndrome type: A PET study. *J. Neurol. Sci.* **158,** 173–179.

Ohkoshi, N., Ishii, A., and Shoji, S. (1995). Single photon emission computed tomography using N-isopropyl-p-[^{123}I]iodoamphetamine in spinocerebellar degeneration. *Eur. Neurol.* **35,** 156–61.

Onodera, O., Idezuka, J., Igarashi, S., Takiyama, Y., Endo, K., Takano, H., Oyake, M., Tanaka, H., Inuzuka, T., Hayashi, T., Yuasa, T., Ito, J., Miyatake, T., and Tsuji, S. (1998). Progressive atrophy of cerebellum and brainstem as a function of age and the size of the expanded CAG repeats in the MJD1 gene in Machado–Joseph disease. *Ann. Neurol.* **43,** 288–296.

Ophoff, R. A., Terwindt, G. M., Vergouwe, M. N., van Eijk, R., Oefner, P. J., Hoffman, S. M., Lamerdin, J. E., Mohrenweiser, H. W., Bulman, D. E., Ferrari, M., Haan, J., Lindouut, D., van Ommen, G. J., Hofker, M. H., Ferrari, M. D., and Frants, R. R. (1996). Familial hemiplegic migraine and episodic ataxia type-2 are caused by mutations in the Ca^{2+} channel gene CANCL1A4. *Cell* **87,** 543–552.

Orozco, G., Estrada, R., Perry, T. L., Arana, J., Fernandez, R., Gonzales-Quevedo, A., Galarraga, J., and Hansen, S. (1989). Dominantly inherited olivopontocerebellar atrophy from eastern Cuba. Clinical, neuropathological and biochemical findings. *J. Neurol. Sci.* **92,** 37–50.

Orozco, D. H., Nodarse, F. A., Cordoves, S. R., and Auburger, G. (1990). Autosomal dominant cerebellar ataxia: Clinical analysis of 263 patients from a homogeneous population in Holgui, Cuba. *Neurology* **40,** 1369–1375.

Orr, H. T., Chung, M.-Y., Banfi, S., Kwiatkowski, T. J., Servadio, A., Beaudet, A. L., McCall, A. E., Duvick, L. A., Ranum, L. P. W., and Zoghbi, H. (1993). Expansion of an unstable trinucleotide CAG repeat in spinocerebellar ataxia type 1. *Nat. Genet.* **4,** 221–226.

Otsuka, M., Ichiya, Y., Kuwabara, Y., Hosokawa, S., Akashi, Y., Yoshida, T., Fukumura, T., Masuda, K., Goto, I., and Kato, M. (1994). Striatal ^{18}F-dopa uptake and brain glucose metabolism by PET in patients with syndrome of progressive ataxia. *J. Neurol. Sci.* **124,** 198–203.

Otsuka, M., Ichiya, Y., Kuwabara, Y., Hosokawa, S., Sasaki, M., Yoshida, T., Fukumura, T., Kato, M., and Masuda, K. (1996). Glucose metabolism in the cortical and subcortical brain structures in multiple system atrophy and Parkinson's disease: A positron emission tomographic study. *J. Neurol. Sci.* **144,** 77–83.

Otsuka, M., Kuwabara, Y., Ichiya, Y., Hosokawa, S., Sasaki, M., Yoshida, T., Fukumura, T., and Kato, M. (1997). Differentiating between multiple system atrophy and Parkinson's disease by positron emission tomography with ^{18}F-dopa and ^{18}F-FDG. *Ann. Nucl. Med.* **11,** 251–257.

Pastakia, B., Polinsky, R., DiChiro, G., Simmons, J. T., Brown, R., and Wener, L. (1986). Multiple system atrophy (Shy–Drager syndrome): MR imaging. *Radiology* **159,** 499–502.

Paulson, H. L., Perez, M. K., Trottier, Y., Trojanowski, J. Q., Subramony, S. H., Dass, S. S., Vig, P., Mandel, J. L., Fischbeck, K. H., and Pittman, R. N. (1997). Intranuclear inclusions of expanded polyglutamine protein in spinocerebellar ataxia type 3. *Neuron* **19,** 333–344.

Pirker, W., Asenbaum, S., Wenger, S., Kornhuber, J., Angelberger, P., Deecke, L., Podreka, I., and Brucke, T. (1997). Iodine-123-epidepride-SPECT: Studies in Parkinson's disease, multiple system atrophy and Huntington's disease. *J. Nucl. Med.* **38,** 1711–1717.

Plazzi, G., Corsini, R., Provini, F., Pierangeli, G., Martinelli, P., Montagna, P., Lugaresi, E., and Cortelli, P. (1997). REM sleep behavior disorders in multiple system atrophy. *Neurology* **48,** 1094–1097.

Plazzi, G., Cortelli, P., Montagna, P., DeMonte, A., Corsini, R., Contin, M., Provini, F., Pierangeli, G., and Lugaresi, E. (1998). REM sleep behaviour disorder differentiates pure autonomic failure from multiple system atrophy with autonomic failure *J. Neurol. Neurosurg. Psychiatry* **64,** 683–685.

Pollack, M., and Kies, B. (1990). Benign hereditary cerebellar ataxia with extensive thermoanalgesia. *Brain* **113,** 857–865.

Potter, N. T., Meyer, M. A., Zimmerman, A. W., Eisenstadt, M. L., and Anderson, I. J. (1995). Molecular and clinical findings in a family with dentatorubral-pallidoluysian atrophy. *Ann. Neurol.* **37,** 273–277.

Priller, J., Scherzer, C. R., Faber, P. W., MacDonald, M. E., and Young, A. B. (1997). Frataxin gene of Friedreich's ataxia is targeted to mitochondria. *Ann. Neurol.* **42,** 265–269.

Pulst, S. M., Nechiporuk, A., Nechiporuk, T., Gispert, S., Chen, X.-N., Lopes-Cendes, I., Pearlman, S., Starkman, S., Orozco, G., Lunkes, A., DeJong, P., Rouleau, G. A., Auburger, G., Korenberg, J. R., Figueroa, C., and Sahba, S. (1996). Moderate expansion of a normally biallelic trinucleotide repeat in spinocerebellar ataxia type 2. *Nat. Genet.* 14, 269–276.

Quinn, N. (1996). Multiple system atrophy. *In:* Movement Disorders 3 (C. D. Marsden, S. Fahn, Eds.) Butterworth: London, pp. 262–281.

Quinn, N., Luthert, P., Honavar, M., and Marsden, C. D. (1989). Pure akinesia due to Lewy body Parkinson's disease: A case with pathology. *Mov. Disord.* **4,** 85–89.

Ranum, L. P., Schut, L. J., Lundgren, J. K., Orr, H. T., and Livingston, D. M. (1994a). Spinocerebellar ataxia type 5 in a family descended from the grandparents of President Lincoln maps to chromosome 11. *Nat. Genet.* **8,** 280–284.

Ranum, L. P., Chung, M.-Y., Banfi, S., Bryer, A., Schut, L. J., Ramesar, R., Duvick, L. A., McCall, A., Subramony, S. H., Goldfarb, L., Gomez, C., Sandkuijl, L. A., Orr, H. T., and Zoghbi, H. Y. (1994b). Molecular and clinical correlations in spinocerebellar ataxia type 1: Evidence for familial effects on the age at onset. *Am. J. Hum. Genet.* **55,** 244–252.

Rinne, J. O., Burn, D. J., Mathias, C. J., Quinn, N. P., Marsden, C. D., and Brooks, D. J. (1995). Positron emission tomography studies on the dopaminergic system and striatal opioid binding in the olivopontocerebellar atrophy variant of multiple system. *Ann. Neurol.* **37,** 568–573.

Robitaille, Y., Schut, L. J., and Kish, S. J. (1995). Structural and immunocytochemical features of olivopontocerebellar atrophy caused by spinocerebellar ataxia type 1 (SCA1) mutation define a unique phenotype. *Acta Neuropathol.* **90,** 572–581.

Robitaille, Y., Lopes-Cendes, I., Becher, M., Rouleau, G., and Clark, A. W. (1997). The neuropathology of CAG repeat diseases: Review and update of genetic and molecular features. *Brain Pathol.* **7,** 901–926.

Romanul, F. C. A., Fowler, H. L., Radvany, J., Feldman, R. G., and Feingold, M. (1977). Azorean disease of the nervous system. *N. Engl. J. Med.* **296,** 1505–1508.

Rosenberg, R. N., Nyhan, W. L., Bay, C., and Shore, P. (1976). Autosomal dominant striatonigral degeneration. A clinical, pathological and biochemical study of a new genetic disorder. *Neurology* **26,** 703–714.

Rosenthal, G., Gilman, S., Koeppe, R. A., Kluin, J., Markel, D. S., Junck, L., and Gebarski, S. S. (1988). Motor dysfunction in olivopontocerebellar atrophy is related to cerebral metabolic rate studied with positron emission tomography. *Ann. Neurol.* **24,** 414–419.

Saitoh, S., Momoi, M. Y., Yamagata, T., Miyao, M., and Suwa, K. (1998). Clinical and electroencephalographic findings in juvenile type DRPLA. *Pediatr. Neurol.* **18,** 265–268.

Salmon, E., Brooks, D. J., Leenders, K. L., Turton, D. R., Hume, S. P., Cremer, J. R., Jones, T., and Frackowiak, R. S. (1990). A two-compartment description and kinetic procedure for measuring regional cerebral [^{11}C]nomifensine uptake using positron emission tomography. *J. Cereb. Blood Flow Metab.* **10,** 307–316.

Sandroni, P., Ahlskog, J. E., Fealey, R. D., and Low, P. A. (1991). Autonomic involvement in extrapyramidal and cerebellar disorders. *Clin. Auton. Res.* **1,** 147–155.

Sanpei, K., Takano, H., Igarashi, S., Sato, T., Oyake, M., Sasaki, H., Wakisaka, A., Tashiro, K., Ishida, Y., Ikeuchi, T., Koide, R., Saito, M., Sata, A., Tanaka, T., Hanyu, S., Takiyama, Y., Nishizawa, M., Shimizu, N., Nomura, Y., Segawa, M., Iwabuchi, K., Eguchi, I., Tanaka, H., Takahashi, H., and Tsuji, S. (1996). Identification of the spinocerebellar ataxia type 2 gene using a direct identification for repeat expansion and cloning technique, DIRECT. *Nat. Genet.* **14,** 277–284.

Santoro, L., De Michele, G., Perretti, A., Crisci, C., Cocozza, S., Cavalcanti, F., Ragno, M., Monticelli, A., Filla, A., and Caruso, G. (1999). Relation between trinucleotide GAA repeat length and sensory neuropathy in Friedreich's ataxia. *J. Neurol. Neurosurg. Psychiatry* **66,** 93–96.

Sasaki, H., Fukazawa, T., Yanagihara, T., Hamada, T., Shima, K., Matsumoto, A., Hashimoto, K., Ito, N., Wakisaka, A., and Tashiro, K. (1996). Clinical features and natural history of spinocerebellar ataxia type 1. *Acta Neurol. Scand.* **93,** 64–71.

Sasaki, H., Kojima, H., Yabe, I., Tashiro, K., Hamada, T., Sawa, H., Hiraga, H., and Nagashima, K. (1998). Neuropathological and molecular studies of spinocerebellar ataxia type 6 (SCA6). *Acta Neuropathol.* **95,** 199–204.

Satoh, J. I., Tokumoto, H., Yukitake, M., Matsui, M., Matsuyama, Z., Kawakami, H., Nakamura, S., and Kuroda, Y. (1998). Spinocerebellar ataxia type 6: MRI of three Japanese patients. *Neuroradiology* **40,** 222–227.

Savoiardo, M., Strada, L., Girotti, F., D'Incerti, L., Sberna, M., Soliveri, P., and Balzarini, L. (1989). MR imaging in progressive supranuclear palsy and Shy–Drager syndrome. *J. Comput. Assist. Tomogr.* **13,** 555–560.

Savoiardo, M., Strada, L., Girotti, F., Zimmerman, R. A., Grisoli, M., Testa, D., and Petrillo, R. (1990). Olivopontocerebellar atrophy: MR diagnosis and relationship to multisystem atrophy. *Radiology* **174,** 693–696.

Schmitt, I., Epplein, J. T., and Reiss, O. (1995). Predominant neuronal expression of the gene responsible for dentatorubral-pallidoluysian atrophy (DRPLA) in rat. *Hum. Mol. Genet.* **4,** 1619–1624.

Schöls, L., Riess, O., Schöls, S., Zeck, S., Amoiridis, G., Langkafel, M., Epplen, J. T., and Przuntek, H. (1995). Spinocerebellar ataxia type 1: Clinical and neurophysiological characteristics in German kindreds. *Acta Neurol. Scand.* **92,** 478–485.

Schöls, L., Amoirdis, G., Büttner, T., Przuntek, H., Epplen, J. T., and Riess, O. (1997a). Autosomal dominant cerebellar ataxia: Phenotypic differences in genetically defined subtypes? *Ann. Neurol.* **42,** 924–932.

Schöls, L., Amoiridis, G., Przuntek, H., Frank, G., Epplen, J. T., and Epplen, C. (1997b). Friedreich's ataxia. Revision of the phenotype according to molecular genetics. *Brain* **120,** 2131–2140.

Schöls, L., Kruger, R., Amoiridis, G., Przuntek, H., Epplen, J. T., and Riess, O. (1998). Spinocerebellar ataxia type 6: Genotype and phenotype in German kindreds *J. Neurol. Neurosurg. Psychiatry* **64,** 67–73.

Schrag, A., Kingsley, D., Phatouros, C., Mathias, C. J., Lees, A. J., Daniel, S. E., and Quinn, N. P. (1998). Clinical usefulness of magnetic resonance imaging in multiple system atrophy *J. Neurol. Neurosurg. Psychiatry* **65,** 65–71.

Schulz, J. B., Skalej, M., Wedekind, D., Luft, A. R., Abele, M., Voigt, K., Dichgans, J., and Klockgether, T. (1999). Magnetic resonance imaging-based volumetry differentiates idiopathic Parkinson's syndrome from multiple system atrophy and progressive supranuclear palsy. *Ann. Neurol.* **45,** 65–74.

Schut, J. W. (1950). Hereditary ataxia: Clinical study through six generations. *Arch. Neurol. Psychiatry* **63,** 535–568.

Schut, J. W. (1951). Hereditary ataxia: A survey of certain clinical, pathologic, and genetic features with linkage data on five additional hereditary factors. *Am. J. Hum. Genet.* **3,** 169–179.

Schut, J. W., and Haymaker, W. (1953). Hereditary ataxia: A pathologic study of five cases of common ancestry. *J. Neuropathol. Clin. Neurol.* **90,** 183–213.

Schut, L. J. (1991). Schut family ataxia. *In* "Handbook of Clinical Neurology. Hereditary Neuropathies and Spinocerebellar Atrophies" (P. Vinken, G. Bruyn, H. Klawans, and J. M. B. V. DeJong, Eds.), Vol. 60, pp. 481–490. Elsevier, Amsterdam/New York.

Servadio, A., Koshy, B., Armstrong, D., Antalfy, B., Orr, H. T., and Zoghbi, H. Y. (1995). Expression analysis of the ataxin-1 protein in tissues from normal and spinocerebellar ataxia type 1 individuals. *Nat. Genet.* **10,** 94–98.

Shinotoh, H., Thiessen, B., Snow, B. J., Hashimoto, S., MacLeod, P., Silveira, I., Rouleau, G. A., Schulzer, M., and Calne, D. B. (1997). Fluorodopa and raclopride PET analysis of patients with Machado–Joseph disease. *Neurology* **49,** 1133–1136.

Silveira, I., Lopes-Cendes, I., Kish, S., Maciel, P., Gaspar, C., Coutinho, P., Botez, M. I., Teive, H., Arruda, W., Steiner, C. E., Pinto-Junior, W., Maciel, J. A., Jain, S., Sack, G., Andermann, E., Sudarsky, L., Rosenberg, R., MacLeod, P., Chitayat, D., Babul, R., Sequeiros, J., and Rouleau, G. A. (1996). Frequency of spinocerebellar ataxia type 1, dentatorubropallidoluysian atrophy, and Machado–Joseph disease mutations in a large group of spinocerebellar ataxia patients. *Neurology* **46,** 214–218.

Smith, J. K. (1975). Dentatorubropallidoluysian atrophy. *In* "Handbook of Clinical Neurology. System Disorders and Atrophies" (P. J. Vinken and G. W. Bruyn, Eds.), Vol. 21, Part I, pp. 519–553. North-Holland, Amsterdam.

Smith, J. K., Gonda, V. E., and Malamud, N. (1958). Unusual form of cerebellar ataxia: Combined dentato-rubral and pallido-luysian degeneration. *Neurology* **8,** 205–209.

Soong, B., and Liu, R. S. (1998). Positron emission tomography in asymptomatic gene carriers of Machado–Joseph disease. *J. Neurol. Neurosurg. Psychiatry* **64,** 499–504.

Soong, B., Cheng, C., Liu, R., and Shan, D. (1997). Machado–Joseph disease: Clinical, molecular, and metabolic characterization in Chinese kindreds. *Ann. Neurol.* **41,** 446–452.

Spieker, S., Schulz, J. B., Petersen, D., Fetter, M., Klockgether, T., and Dichgans, J. (1995). Fixation instability and oculomotor abnormalities in Friedreich's ataxia. *J. Neurol.* **242,** 517–521.

Staal, A., Meerwaldt, J. D., van Dongen, K. J., Mulder, P. G., and Busch, H. F. (1990). Non-familial degenerative disease and atrophy of brainstem and cerebellum. Clinical and CT data in 47 patients. *J. Neurosci.* **95,** 259–269.

Stevanin, G., Le Guern, E., Ravisé, M., Chneiweiss, H., Dürr, A., Cancel, G., Vignal, A., Boch, A. L., Ruberg, M., Penet, C., Pothin, Y., Lagroua, I., Haguenau, M., Rancurel, G., Weissenback, J., Agid, Y., and Brice, A. (1994a). A third locus for autosomal dominant cerebellar ataxia type 1 maps to chromosome 14q24.3-qter: Evidence for the existence of a fourth locus. *Am. J. Hum. Genet.* **54,** 11–20.

Stevanin, G., Sousa, P. S., Cancel, G., Dürr, A., Dubourg, O., Nicholson, G. A., Weissenbach, J., Jardim, E., Agid, Y., Cassa, E., and Brice, A. (1994b). The gene for Machado–Joseph disease maps to the same 3-cM interval as the spinal cerebellar ataxia 3 gene on chromosome 14q. *Neurobiol. Dis.* **1,** 79–82.

Stevanin, G., Cancel, G., Dürr, A., Chneiweiss, H., Dubourg, O., Weissenbach, J., Cann, H. M., Agid, Y., and Brice, A. (1995). The gene for spinal cerebellar ataxia 3 (SCA3) is located in a region of approximately 3 cM on chromosome 14q24.3-132.2. *Am. J. Hum. Genet.* **56,** 193–201.

Stevanin, G., Dürr, A., David, G., Didierjean, O., Cancel, G., Rivaud, S., Tourbah, A., Warter, J. M., Agid, Y., and Brice, A. (1997). Clinical and molecular features of spinocerebellar ataxia type 6. *Neurology* **49,** 1243–1246.

Stocchi, R. F., Carbone, A., Inghilleri, M., Monge, A., Ruggieri, S., Berardelli, A., and Manfredi, M. (1997). Urodynamic and neurophysiological evaluation in Parkinson's disease and multiple system atrophy. *J. Neurol. Neurosurg. Psychiatry* **62,** 507–511.

Sun, X., Tanaka, M., Kondo, S., Hirai, S., and Ishihara, T. (1994). Reduced cerebellar blood flow and oxygen metabolism in spinocerebellar degeneration: A combined PED and MRI study. *J. Neurol.* **241,** 295–300.

Takahashi, H., Ohama E., Naito, H., Takeda, S., Nakashiima, S., Makifuchi, T., and Ikuta, F. (1988). Hereditary dentatorubral-pallidoluysian atrophy: clinical and pathologic variants in a family. *Neurology* **38,** 1065–1070.

Takahashi, H., Odano, I., Nishihara, M., Yuasa, T., and Sakai, K. (1994). Regional cerebral blood flow measured with N-isopropyl-p-[^{123}I]iodoamphetamine single-photon emission tomography in patients with Joseph disease. *Eur. J. Nucl. Med.* **21,** 615–620.

Takahashi, H., Ikeuchi, T., Honma, Y., Hayashi, S., and Tsuji, S. (1998). Autosomal dominant cerebellar ataxia (SCA6): Clinical, genetic and neuropathological study in a family. *Acta Neuropathol.* **95,** 333–337.

Takano, T., Yamanouchi, Y., Nagafuchi, S., and Yamada, M. (1996). Assignment of the dentatorubral and pallidoluysian atrophy (DRPLA) gene to 12p 13.31 by fluorescence *in situ* hybridization. *Genomics* **32,** 171–172.

Takiyama, Y., Nishizawa, M., Tanaka, H., Kawashima, S., Sakamoto, H., Karuba, Y., Shimazaki, H., Soutome, M., Endo, K., Ohta, S., Kagawa, Y., Kanazawa, I., Mizuno, Y., Yoshida, M., Yuasa, T., Horikawa, Y., Oyanagi, K., Nagai, H., Kondo, T., Inuzuka, T., Onodera, O., and Tsuji, S. (1993). The gene for Machado–Joseph disease maps to human chromosome 14q. *Nat. Genet.* **4,** 300–304.

Takiyama, Y., Oyanagi, S., Kawashima, S., Sakamato, H., Saito, K., Yoshida, M., Tsuji, S., Mizuno, Y., and Nishizawa, M. (1994). A clinical and pathologic study of a large Japanese family with Machado–Joseph disease tightly linked to the DNA markers on chromosome 14q. *Neurology* **44,** 1302–1308.

Takiyama, Y., Sakoe, K., Nakano, I., and Nishizawa, M. (1997). Machado–Joseph disease: Cerebellar ataxia and autonomic dysfunction in a patient with the shortest known expanded allele (56 CAG repeat units) of the MJD1 gene. *Neurology* **49,** 604–606.

Tanawaki, T., Sakai, T., Kobayashi, T., Kuwabara, Y., Otsuka, M., Ichiya, Y., Masuda, K., and Goto, I. (1997). Positron emission tomography (PET) in Machado–Joseph disease. *J. Neurol. Sci.* **145,** 63–67.

Testa, D., Savoiardo, M., Fetoni, V., Strada, L., Palazzini, E., Bertulezzi, G., and Girotti, F. (1993). Multiple system atrophy. Clinical and MR observations on 42 cases. *Ital. J. Neurol. Sci.* **14,** 211–216.

Tison, F., Wenning, G. K., Volonte, M. A., Poewe, W. R., Henry, P., and Quinn, N. P. (1996). Pain in multiple system atrophy. *J. Neurol.* **243,** 153–156.

Uematsu, D., Hamada, J., and Gotoh, F. (1987). Brainstem auditory evoked responses and CT findings in multiple system atrophy. *J. Neurol. Sci.* **77,** 161–171.

Ugawa, Y., Day, B. L., Rothwell, J. C., Thompson, P. D., Merton, P. A., and Marsden, C. D. (1991). Modulation of motor cortical excitability by electrical stimulation over the cerebellum in man. *J. Physiol.* **441,** 57–72.

Ugawa, Y., Genba-Shimizu, K., Rothwell, J. C., Iwata, M., and Kanazawa, I. (1994). Suppression of motor cortical excitability by electrical stimulation over the cerebellum in ataxia. *Ann. Neurol.* **36,** 90–96.

Ugawa, Y., Uesaka, Y., Terao, Y., Hanajima, R., and Kanazawa, I. (1995). Magnetic stimulation over the cerebellum in humans. *Ann. Neurol.* **37,** 703–713.

Ugawa, Y., Terao, Y., Hanajima, R., Sakai, K., Furubayashi, T., Machii, K., and Kanazawa, I. (1997). Magnetic stimulation over the cerebellum in patients with ataxia. *Electroencephalogr. Clin. Neurophys.* **104,** 453–458.

Uyama, E., Kondo, I., Uchino, M., Fukushima, T., Murayama, N., Kuwano, A., Inokuchi, N., Ohtani, Y., and Ando, M. (1995). Dentatorubral-pallidoluysian atrophy (DRLPA): Clinical, genetic, and neuroradiologic studies in a family. *J. Neurol. Sci.* **130,** 146–153.

Vuadens, P., Ghika, J., and Regli, F. (1997). Olivo-ponto-cerebellous degeneration. A study of 21 patients defined by the Quinn criteria. *Rev. Neurol.* **153,** 412–416.

Waikai, M., Kume, A., Takahashi, A., Ando, T., and Hashizume, Y. (1994). A study of parkinsonism in multiple system atrophy: Clinical and MRI correlation. *Acta Neurol. Scand.* **90,** 225–231.

Warner, T. T., Williams, L., and Harding, A. E. (1994a). DRPLA in Europe. *Nat. Genet.* **6,** 225.

Warner, T. T., Lennox, G. G., Janota, I., and Harding, A. E. (1994b). Autosomal-dominant dentatorubropallidoluysian atrophy in the United Kingdom. *Mov. Disord.* **9,** 289–296.

Warner, T. T., Williams, L. D., Walker, R. W., Flinter, F., Robb, S. A., Bundey, S. E., Honavar, M., and Harding, A. E. (1995). A clinical and molecular genetic study of dentatorubropallidoluysian atrophy in four European families. *Ann. Neurol.* **37,** 452–459.

Watanabe, H., Tanaka, F., Matsumoto, M., Doyu, M., Ando, T., Mitsuma, T., and Sobue, G. (1998). Frequency analysis of autosomal dominant cerebellar ataxias in Japanese patients and clinical characterization of spinocerebellar ataxia type 6. *Clin. Genet.* **53,** 13–19.

Wenning, G. K., and Quinn, N. P. (1997). Parkinsonism. Multiple system atrophy. *Baillieres Clin. Neurol.* **6,** 187–204.

Wenning, G. K., Ben-Shlomo, Y., Magalhaes, M., Daniel, S. E., and Quinn, N. P. (1994). Clinical features and natural history of multiple system atrophy: An analysis of 100 cases. *Brain* **117,** 835–845.

Wessel, K., Schroth, G., Diener, H. C., Müller-Forell, W., and Dichgans, J. (1989). Significance of MRI-confirmed atrophy of the cranial spinal cord in Friedreich's ataxia. *Eur. Arch. Psychiatry Neurol. Sci.* **238,** 225–230.

Wessel, K., Tegenthoff, M., Vorgerd, M., Otto, V., Nitschke, M. F., and Malin, J. P. (1996). Enhancement of inhibitory mechanisms in the motor cortex of patients with cerebellar degeneration: A study with transcranial magnetic brain stimulation. *Electroencephalogr. Clin. Neurophys.* **101,** 273–280.

Wilson, R. B., Lynch, D. R., and Fischbeck, K. H. (1998). Normal serum iron and ferritin concentrations in patients with Friedreich's ataxia. *Ann. Neurol.* **44,** 132–134.

Wittkamper, A., Wessel, K., and Bruckmann, H. (1993). CT in autosomal dominant and idiopathic cerebellar ataxia. *Neuroradiology* **35,** 520–524.

Yamaguchi, S., Fukuyama, H., Ogawa, M., Yamauchi, H., Harada, K., Nakamura, S., and Kimura, J. (1994). Olivopontocerebellar atrophy studied by positron emission tomography and magnetic resonance imaging. *J. Neurol. Sci.* **125,** 56–61.

Yamazaki, M., Araki, T., Imazu, O., Kitamura, S., and Terashi, A. (1992). A case of Machado–Joseph disease: Cerebral blood flow and cerebral metabolic rate of oxygen. *Clin. Neurol.* **32,** 755–757.

Yazawa, I., Nukina, N., Hashida, H., Goto, J., Yamada, M., and Kanazawa, I. (1995). Abnormal gene product identified in hereditary dentatorubral-pallidoluysian atrophy (DRPLA) brain. *Nat. Genet.* **10,** 99–103.

Yoshii, R. F., Tomiyasu, H., and Shinohara, Y. (1998). Fluid attenuation inversion recovery (FLAIR) images of dentatorubropallidoluysian atrophy: Case report. *J. Neurol. Neurosurg. Psychiatry* **65,** 396–399.

Zhu, S., and Gerhard, D. S. (1998). A transcript map of an 800-kb region on human chromosome 11q13, part of the candidate region for SCA5 and BBS1. *Hum. Genet.* **103,** 674–680.

Zhuchenko, O., Bailey, J., Bonnen, P., Ashizawa, T., Stockton, D. W., Amos, C., Dobyns, W. B., Subramony, S. H., Zoghbi, H. Y., and Lee, C. C. (1997). Autosomal dominant cerebellar ataxia (SCA6) associated with small polyglutamine expansions in the α1A-voltage-dependent calcium channel. *Nat. Genet.* **15,** 62–69.

Zoghbi, H. Y., Jodice, C., Sandkuijl, L. A., Kwiatkowski, T. J., Jr.,

McCall, A. E., Huntoon, S. A., Lulli, P., Spadaro, M., Litt, M., Cann, H. M., Frontali, M., and Terrenato, L. (1991). The gene for autosomal dominant spinocerebellar ataxia (SCA1) maps telomeric to the HLA complex and is closely linked to the D6S89 locus in three large kindreds. *Am. J. Hum. Genet.* **49,** 23–30.

Zu, L., Figueroa, K. P., Grewal, R., and Pulst, S. M. (1999). Mapping of a new autosomal dominant spinocerebellar ataxia to chromosome 22. *Am. J. Hum. Genet.* **64,** 594–599.

III

Pediatric Disorders

18

Dyslexia and Related Learning Disorders: Recent Advances from Brain Imaging Studies

Michel Habib[*,1] and Jean-François Démonet[†]

*Cognitive Neurology Laboratory, Department of Neurology, CHU Timone, 13385 Marseille, France
†INSERM U455, Department of Neurology, CHU Purpan, 31059 Toulouse, France*

I. Brain Mechanisms in Dyslexia: An Overview, with Special Emphasis on Morphological Brain Imaging

II. Event-Related Potentials and Developmental Language Disorders

III. Functional Brain Imaging in Dyslexia

IV. Conclusion

References

During the past few years, dyslexia has been the focus of considerable interest from researchers in different scientific areas, for both theoretical and practical reasons. First, public awareness that this condition, which affects approximately 10% of the population, has a neurobiological basis gave rise to the hope of rational and effective therapy, which stimulated research in such different areas as neuropathology, neuropsychology, linguistics, and the educational sciences. As a consequence,

dyslexia has become a fertile ground for transdisciplinary studies and a model for elucidating biological, educational, and sociocultural factors of brain-cognition interactions and development.

In this special context, the advent of brain imaging methods has provided a fruitful opportunity for testing various hypotheses about the brain mechanisms underlying dyslexia and other related learning difficulties, for understanding brain systems involved in reading, and probably, in the near future, for evaluating remediation methods.

In this chapter, we will first briefly overview available evidence for a neurological basis of dyslexia and related disorders as well as the main postulated mechanisms leading from neurological impairment to learning disorder. The contribution of morphological imaging of the brain to this topic will be summarized, with special emphasis on work suggesting the functional significance of morphological characteristics of the dyslexic brain. Finally, functional imaging studies in dyslexia will be reviewed in three different parts: studies of phonological processing in dyslexia, studies of reading and orthographic mechanisms, and studies of visual impairment in dyslexia.

---

[1] To whom correspondence should be addressed.

I. Brain Mechanisms in Dyslexia: An Overview, with Special Emphasis on Morphological Brain Imaging

Developmental dyslexia is defined as a specific and significant impairment in reading abilities, unexplainable by any kind of deficit in general intelligence, learning opportunity, general motivation, or sensory acuity (WHO, 1993). Children with this condition often have associated deficits in related domains such as oral language (dysphasia), writing abilities (dysgraphia and misspelling), mathematical abilities (dyscalculia), motor coordination (dyspraxia), visuospatial abilities (developmental right hemisphere syndrome), and attentional abilities (hyperactivity and attention deficit disorder) (Weintraub and Mesulam, 1983; Rapin and Allen, 1988; Gross-Tsur *et al.*, 1995, 1996).

Such comorbidity suggests a common origin involving genetic factors, prenatal environmental influences, or both. The basic postulate of current research in this field is that dyslexia and related disorders are fundamentally linked to a constitutional characteristic of the brain, very probably of genetic origin, since they most often occur in families, although genetic transmission is probably complex and nonexclusive (Fisher *et al.*, 1999). The large prevalence of oral or written language deficits among these learning-disordered children suggests a special vulnerability of left hemisphere cortical systems subserving various aspects of language-related abilities to these etiological factors (Geschwind and Galaburda, 1985). Hormonal factors, such as fetal testosterone levels during late pregnancy, may play a crucial role, possibly reflected in the large male predominance in most of these conditions. The overwhelming representation of reading and spelling disorders probably relates to the complex, multifactorial mechanisms underlying the acquisition of a written code, which requires the highly sophisticated integration of various separate cognitive subsystems. Finally, that such deficits usually persist into adulthood is a good argument for a biological conceptualization of these conditions and a justification for studying adults in researching this field (Rumsey, 1996).

In this chapter, we will focus mainly on dyslexia as the most frequent, best defined, and most representative entity. Autism and related disorders, Tourette syndrome, stuttering, and rarer entities sometimes included within the realm of learning disorders will not be considered here, since there are insufficient arguments to suspect common mechanisms in their origin.

A. The Origins of a Neurological Conceptualization of Dyslexia: The Atypical Language Lateralization Theory

Although suspected as early as at the turn of the century (Morgan, 1896; Hinshelwood, 1917), the neurological nature of dyslexia only gained credibility with the clinical studies of S. Orton (1937), who very intuitively suggested that dyslexic children may have incomplete or delayed lateralization of language processes to the left hemisphere. For instance, the high incidence of left-handers and the mirror-writing phenomenon were taken as evidence for abnormal lateralization in these children. However, these early intuitions were not developed until the influential contribution of Galaburda and colleagues.

In eight (five male, three female) dyslexic brains, Galaburda and his collaborators (Galaburda and Kemper, 1979; Galaburda *et al.*, 1985; Humphreys *et al.*, 1990) showed an atypical pattern of symmetry of the planum temporale (PT), in that this region, a posterior component of the language area in the left hemisphere, appeared abnormally symmetric in all the brains examined. This absence of asymmetry, compared to 70% leftward asymmetry of the same region in unselected control specimens (Geschwind and Levitsky, 1968), could be viewed as the neural basis of previously suspected incomplete language lateralization in dyslexia (see, for instance, Obrzut, 1988). Although the developmental mechanisms leading to such atypical symmetry still remain a subject of debate (see, for instance, Steinmetz, 1996), these findings have been generally considered good evidence of maturational deviance being at the origin of the learning difficulties of dyslexics.

1. Morphological Imaging of Cortical Asymmetry in Dyslexia

Several attempts have been made at replicating these findings since, through *in vivo* examination of larger populations of dyslexic individuals using morphological brain MRI. MRI is able to demonstrate the cortical anatomy reliably. Recently, several studies have attempted to review the available data, and most of them concluded that the evidence is not fully convincing (Beaton, 1997; Morgan and Hynd, 1998; Shapleske *et al.*, 1999). Table I summarizes the main characteristics and results of these different studies.

Whereas initial studies seemed to confirm Galaburda's findings statistically, with a larger incidence of reversed (or absent) asymmetry, it is noteworthy that more recent studies (using more refined MRI technology) have failed to confirm such a tendency. For instance, Leonard *et al.*'s (1993) study, one of the most sophisticated and reliable available, reports an atypical pattern of gyrification in right temporal, left temporal, and parietal perisylvian cortices. One intriguing finding in this study is the suggestion that in addition to *inter*hemispheric asymmetry, it would be interesting to consider *intra*hemispheric asymmetries, i.e., the relative importance, within each hemisphere, of the temporal and parietal banks of the posterior sylvian fissure.

Among studies finding significant differences between dyslexics and controls, that of Larsen *et al.* (1990) was the first to suggest that atypical symmetry in dyslexia is specifically linked to phonological impairment. They showed that a subgroup of their dyslexics with impaired performance on a nonword reading task had symmetrical PTs, whereas those with impaired word recognition did not differ in this respect from controls.

As summarized in Table I, although there is a global tendency in the literature to confirm *in vivo* the initial neuropathological observations of Galaburda *et al.*, there are notable exceptions when authors fail to find any significant bias toward symmetry in dyslexics. This is especially so in the most recent studies that measured directly the surface area of the planum temporale. These inconsistencies may relate to the selection of subjects or to the mode of anatomical measurement.

In a preliminary study (Habib and Robichon, 1996) of 16 dyslexic young adults and 14 controls, all students from the same engineering school (to ensure that both groups had a similar intellectual as well as academic level), we found that a parietal area, situated in front of the planum temporale on the other bank of the sylvian fissure (Fig. 1), is less asymmetrical in dyslexics than in controls and that the degree of asymmetry of this area is inversely proportional to the individuals' performances on a phonological categorization task. This finding suggests that parietal rather than temporal asymmetry may be the most relevant morphological characteristic of the dyslexic brain. It is noteworthy, in this context, that recent functional imaging studies have found dissociations between anatomical preponderance of the left planum temporale and functional asymmetry of the temporal cortex during auditory verbal tasks (Karbe *et al.*, 1995). Moreover, it seems that the planum temporale itself is not specifically activated by verbal auditory stimuli, since it responds equally to tones and words during passive listening tasks and more strongly to tones during active listening (Binder *et al.*, 1996; Celsis *et al.*, 1998).

Finally, as the posterior region of the inferior frontal gyrus is classically related to language output, the study of its morphology in developmental dyslexics appears to be especially relevant. Paradoxically, such studies are rather sparse. While Galaburda *et al.* (1985) have reported at the cytoarchitectonic level the presence of numerous ectopias and dysplasias bilaterally in the inferior frontal gyrus of developmental adult dyslexics, others using neuroimaging have shown macroscopic symmetry of the anterior speech region in dyslexic children (Hynd *et al.*, 1990). However, Jernigan *et al.* (1991) found no significant difference between language-disordered individuals and normal controls in the inferior frontal regions. Recently, Clark and Plante (1998) showed a relation between the sulcal morphology of the inferior posterior frontal gyrus and a family history of developmental language disorders, suggesting an increased risk factor for these disabilities when extra sulci are present in this frontal region (without, however, any lateralized effect).

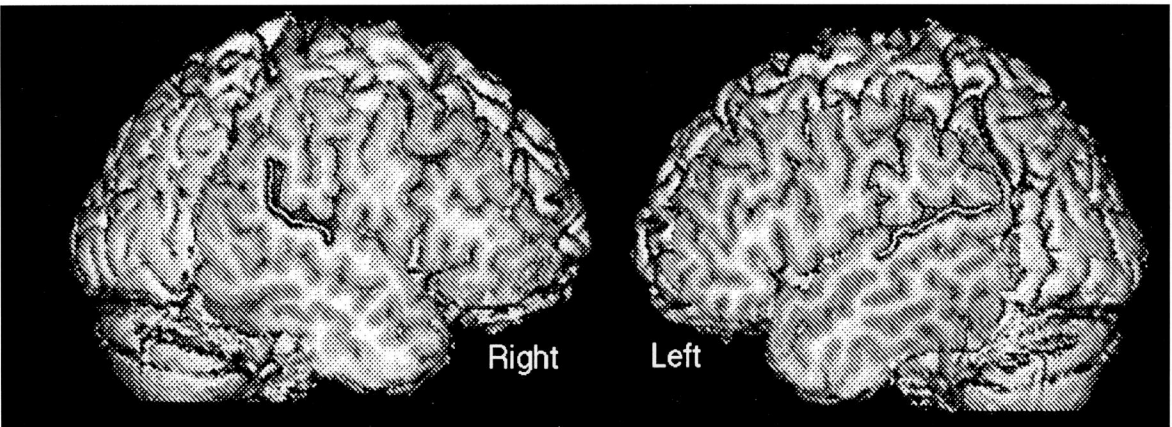

Figure 1 Morphological sagittal images obtained from 3D MRI reconstruction in a typical right-handed normal reader. The planum temporale (thick white line) is only slightly asymmetrical, but due to different orientation of the sylvian fissure, the parietal operculum, on the opposite bank of the fissure (fine black line), is clearly larger on the left side. In dyslexics, this asymmetry tends, on average, to disappear (Habib and Robichon, 1996).

Table I

Study	Method	Subjects	Chronological age (years)	Anatomical structure measured	Asymmetries and/or regional abnormalities — Dyslexics	Controls	Others
Rumsey et al., 1986	MRI 0.5 T	10 male dyslexics	22.6 (3.34)	Lateral ventricles	R < L: 20% / R > L: 40% / R = L: 90%		
Hynd et al., 1990	MRI 0.6 T	10 dyslexics (8 males + 2 females) / 10 ADD/H (8 males + 2 females) / 10 controls (8 males + 2 females)	9.9 (2.04) / 10.0 (3.36) / 11.8 (2.0)	Temporal lobes[a]			
				Length of the planum temporale[b]	R < L: 10%	R < L: 70%	R < L: 70%[k]
				Length of the insula	R < L	R > L	R < L
				Width of the frontal lobe	R = L	R > L	R = L
Larsen et al., 1990	MRI 1.5 T	19 dyslexics (ratio: 4 males/1 female) / 17 controls (ratio: 4 males/1 female)	15.1 (0.3) / 15.4 (0.4)	Surface of the planum temporale	R < L: 31.5% / R = L: 68.5%	R < L: 70.5% / R = L: 29.5%	
Duara et al., 1991	MRI 1.0 T	21 dyslexics (12 males + 9 females) / 29 controls (15 males + 14 females)	39.1 (11.0) / 35.3 (10.0)	Postcentral surface[a,c]	(R > L), NS[s]	(R = L), NS	
				Posterior surface	R > L, p = 0.007	(R < L), NS	
Jernigan et al., 1991	MRI 1.5 T	20 L/LI (13 males + 7 females) / 12 controls (8 males + 4 females)	8.9 (0.7) / 9.0 (0.7)	"Inferior–anterior" volume[d]	R > L,[l] $p < 0.01$	R > L, $p < 0.01$	
				"Superior–posterior" volume[e]	R < L, $p < 0.01$	R < L, $p < 0.01$	
				"Inferior–posterior" volume[a]	R < L: 45% / R > L: 50%; R = L: 5%		
Plante et al., 1991	MRI 0.5 T	8 SLI / 8 normal MRI scans from male subjects selected from a database	5.2 / —	Perisylvian region	R < L: 25%[m] / R > L: 37.5% / R = L: 37.5%	Not reported	
Kushch et al., 1993	MRI 1.5 T	17 dyslexics (9 males + 8 females) / 21 controls (8 males + 13 females)	26.2 (15.0) / 33.4 (15.0)	SSTL[f]			
				Anterior	(R < L), NS	R < L, $p < 0.001$	
				Posterior	(R > L), NS	R < L, $p < 0.001$	
				Total	(R = L), NS	R < L, $p < 0.001$	
Leonard et al., 1993	MRI 1.0 T	9 dyslexics (7 males + 2 females) / 10 unaffected siblings (4 males + 6 females) / 12 controls (5 males + 7 females)	36 (17.1) / 25.7 (20.3) / 37.1 (13.5)	Length of the planum temporale[g]	(R < L), NS	(R < L), NS	(R < L), NS[n]
				Temporal length (T)	R < L, $p < 0.001$	R < L, $p < 0.5$	R < L, $p < 0.01$
				Parietal length (P)	(R > L), NS	R > L, $p < 0.01$	R > L, $p < 0.5$
				Left intrahemispheric asymmetry[h]	T < P: 22% / T > P: 78%	T > P: 100%	T > P: 100%
				Right intrahemispheric asymmetry[h]	T < P: 55.5% / T < P: 44.5%	T > P: 100%	T < P: 40% / T > P: 50% / T = P: 10%

Parietal operculum[i]	LCH[o]	RCH[p]	Bil[q]	LCH	RCH	Bil	LCH	RCH	Bil
Type 3	67%			8%			40%		
Type 4	22%	11%	11%	8%			50%		
Multiple Heschl gyri	33%			11%			10%		10%

Study	Method	Subjects	Age	Region	Findings
Schultz et al., 1994	MRI 1.5 T	17 dyslexics (10 males + 7 females), 14 controls (7 males + 7 females)	8.68 (0.64), 8.94 (0.67)	Surface of planum temporale[i]	R < L: 76% R < L: 71%
Rumsey et al., 1997	MRI	16 right-handed dyslexic men, 14 matched controls	18–40	Planum temporale and ascending posterior ramus of SF ("planum parietale")	Both groups have 70–80% leftward planum temporale asymmetry and 50–60% rightward planum parietale asymmetry
Gauger et al., 1997	MRI	11 L/LI and 19 matched controls	5.6–13	Planum temporale; Broca's areas (pars triangularis)	Left pars triangularis significantly smaller in SLI children; More incidence of rightward asymmetry of language structures; Anomalous morphology in these language areas correlated with depressed language ability
Clark and Plante, 1998	MRI 1.5 T	41 normal adults, including 20 parents of L/LI children; among these, 15 probable (test-identified) dyslexics, 4 probable dyslexics in the nonparent population	30–51	Broca's area; seven types according to gyrification and sulcal patterns	Morphological types, including an extra sulcus in the inferior frontal gyrus of test-identified dyslexics (both hemispheres combined)
Dalby et al., 1998	MRI	17 dyslexics, 6 retarded readers, 12 normal controls		Three measures of temporal lobes	L < R or L = R : 82% for one of the three measures (temporal region or coronal slices, cortico-subcortical, including anterior insula) L > R: 72%[r]

[a] Including the planum temporale, among other regions. Volumetric measurements (study A) and surface measurements (study D).
[b] On axial slices.
[c] Measurements were made on an axial slice divided into six areas: anterior polar (prefrontal regions), anterior (premotor regions and Broca's area), anterior central (anterior part of the superior temporal gyrus), posterior central (posterior part of the superior temporal gyrus, including the planum temporale), posterior (including the angular gyrus), and posterior polar (lateral occipital cortex and the cuneus of the occipital lobe). Positive correlation between the severity of dyslexia and the surface of the right posterior polar brain segment.
[d] Prefrontal region below the frontal operculum, including orbitofrontal lobe bilaterally.
[e] Region including superior parietal lobe above the parietal operculum.
[f] SSTL: superior surface of the temporal lobe, defined as extending from the end of the sylvian fissure to the anterior border of the temporal lobe and divided into two equal anterior and posterior surfaces.
[g] Total length: temporal border (T) + parietal border (P).
[h] Intrahemispheric comparison of length of temporal border (T) and length of parietal border (P).
[i] See Steinmetz et al., 1990: four types of configuration of the parietal operculum.
[j] The planum temporale is defined as extending from the end of the sylvian fissure to the posterior border of the Heschl gyrus. Other measurements concern the superior surface of the temporal lobe and the volume of the temporal lobe; in order to account for individual variations in overall brain size, the images of brains were enlarged to reach a standard size but no significant difference between the two groups could be pointed out, even for the measurement of the surface of the planum temporale.
[k] ADD/H children.
[l] L/LI children.
[m] SLI children.
[n] Unaffected siblings.
[o] Left hemisphere (percentage of cases).
[p] Right hemisphere (percentage of cases).
[q] Bilateral abnormalities (percentage of cases).
[r] Normal controls and retarded readers combined.
[s] NS: not significant.

463

In our engineer population, we recently measured Broca's area asymmetry (unpublished results) and found a more frequent symmetrical pattern in areas 44 and 45 in dyslexics and a correlation between this pattern and the non-word reading performances of subjects. This result is consistent with functional imaging studies (see below) suggesting a role of the left inferior frontal gyrus in phonological aspects of reading (Fiez and Petersen, 1998; Price, 1998).

B. The Interhemispheric Deficit Theory of Dyslexia

Besides theories pointing to defective brain lateralization, another frequently proposed potential mechanism is abnormal collaboration and/or communication between the hemispheres. This hypothesis relies on well-documented evidence of impaired interhemispheric transfer of sensory or motor information in dyslexics (Gross-Glenn and Rothenberg, 1984; Best, 1985; Gladstone *et al.*, 1989; Moore *et al.*, 1995; Markee *et al.*, 1996). A few studies have looked for a structural concomitant of impaired callosal function, by measurement of the midsagittal surface of the corpus callosum on MRI scans. Duara *et al.* (1991) found, in 21 adult dyslexics compared to controls, a larger total callosal area in female but not male dyslexics and a larger posterior (splenial) area in male and female dyslexics. Conversely, Larsen *et al.* (1992) failed to demonstrate any difference in callosal measurements, for either total or splenial areas, between 17 dyslexic adolescents and 19 controls. Hynd *et al.* (1995) compared 16 dyslexic children to 16 age-matched controls and only found significant differences in the anteriormost region (genu), which was smaller in dyslexics. Finally, Rumsey *et al.* (1996) found, in 21 dyslexic men compared to 19 controls, a larger posterior third of the callosum that included the isthmus and splenium.

In our own study of 16 dyslexic men (Robichon and Habib, 1998), we found that (1) our dyslexics' corpus callosum displayed a more rounded and evenly thicker callosal shape and (2) only right-handed dyslexics had a larger midcallosal surface, especially in the isthmus. These findings are globally consistent with the fact that more symmetrical brains may possess more overall (right plus left) brain tissue in temporo-parietal regions connected through the posterior part of the callosum. Moreover, they raise the important issue of whether more callosal connections reflects lesser cortical asymmetry or has a special significance per se, for instance in terms of interhemispheric inhibition or collaboration. Finally, a changed callosal size in dyslexia may also result from intensive remedial therapy, since it has been shown that intensive training may affect callosal morphology (Schlaug *et al.*, 1995).

C. The Visual Theory of Dyslexia

It may seem obvious to seek the mechanism of reading disorders within the domain of visual perception, using tools primarily designed for investigations of visual function. Several lines of evidence argue in favor of this strategy. First, clinical studies have long reported that most dyslexics make errors that follow visual rather than phonetic laws, e.g., confusions between symmetrical (b/d) or visually close (m/n) letters, and that at least some of them may have purely perceptual impairments. Accordingly, the characterization of a "dyseidetic" subgroup of dyslexics (Boder, 1971) assumed a visual deficit at the origin of the disorder, with preferential use of a phonetic strategy when reading. A reverse dissociation was proposed for "dysphonetic" dyslexics. This dichotomy has regained favor in the light of cognitive models that consider dyseidetic dyslexia as the developmental equivalent of surface dyslexia. This notion is based on a larger number of errors with irregular words than nonwords (Hanley *et al.*, 1992; Valdois *et al.*, 1995). Up to 75% of dyslexic children may be affected by ophthalmologic problems that disturb binocular vision, ocular tracking, or motion perception, to the point that ophthalmologic remediation methods have even been proposed as treatment in the past. However, each of these problems can be interpreted as a consequence rather than a cause of reading impairment, if one considers that these various abilities develop in normal readers partly under the influence of reading itself.

Such an explanation does not hold, however, for more elementary, perceptual abnormalities repeatedly reported in dyslexic children. Globally, visual perceptual studies have shown that dyslexic children process visual information more slowly than normal readers. For instance, there are studies that show longer visual persistence at low spatial frequencies (Lovegrove *et al.*, 1980a,b) or a slower flicker fusion rate (Martin and Lovegrove, 1984). However, the best demonstration of a low-level visual deficit in dyslexics is that of altered contrast sensitivity. Dyslexics may need 10-fold lower spatial frequencies to perceive the same contrast as nondyslexic children. This contrast sensitivity deficit may affect 75% of dyslexics, especially those with evidence of an associated phonological deficit (Lovegrove *et al.*, 1980a,b, 1982, 1990; Eden *et al.*, 1996b).

Several researchers have suggested that deficits observed in psychophysical experiments may be accounted for by reference to the distinction between sustained and transient visual channels (see Stein, 1997, for a review). Since these channels can be distinguished by their preferred spatial frequencies, their temporal properties, and their contrast sensitivity, it has been suggested that the impairment observed in dyslexics both in contrast sensitivity and visual persistence may result from disturbance in the transient system, which mediates perception of global form, movement, and temporal resolution. As a confirmation of this hypothesis, Livingstone *et al.* (1991) have provided electrophysiological and neuroanatomical evidence of an alteration of the magnocellular component of the visual pathway (M-system) and shown the absence of specific electric responses to high spatial frequency and low-contrast visual targets in dyslexic children who responded normally to targets with greater contrast and lower spatial frequency. In the same article, they report that the dyslexic brains previously shown by Galaburda *et al.* to display cortical changes also display subtle abnormalities in the neuronal organization of the lateral geniculate nucleus (the thalamic relay of the retinocortical pathway). Consistent with the M-system theory, only neurons of the magnocellular part of the nucleus were abnormally atrophied, whereas the parvocellular part of the nucleus was intact. These neuropathological findings, however, have never been replicated since, a deficiency that represents an obvious limitation to any line of argument based on such neuropathological evidence.

D. The Phonological Theory of Dyslexia

One of the most robust discoveries in the domain of cognitive mechanisms leading to dyslexia is the repeated demonstration that the core deficit responsible for impaired learning to read is phonological in nature and has to do with oral language rather than visual perception. The deficit is in the ability to manipulate in abstract form the sound constituents of oral language, so-called phonological awareness (or metaphonology). Whereas most children are able to perform tasks requiring segmenting words into smaller units (syllables, and partly phonemes) well before reading age, dyslexic children are still unable to do so even after several months of reading and writing (Liberman, 1982). Lundberg *et al.* (1988) showed improved reading abilities in children previously trained in such exercises and these observations are the basis of the widespread use of oral language exercises for the rehabilitation of reading and spelling disorders.

An important concept of the phonological processing theory is that there is a deficit at the level of phoneme representation itself. For instance, several researchers have found that dyslexics are poorer than age-matched controls (and also than controls matched for reading age) at tasks that require processing of subtle differences between phonemes that are close to each other. This is best exemplified in tasks of categorical perception when children have to categorize as 'ba' or

'da' an artificial acoustic continuum between the two syllables. A number of studies (Godfrey *et al.*, 1981; Werker and Tees, 1987; Reed, 1989) have shown a deficit in this task in a proportion of dyslexics that is variable across different studies. The deficit is generally found for items situated close to the intercategorical boundary, especially articulatory oppositions (/ba/-/da/; /da/-/ga/), or less often voice-onset oppositions such as /ba/-/pa/ (Manis *et al.*, 1997). The latter authors showed such deficits are found specifically in a subgroup of dyslexic children with a phonological awareness deficit (as assessed in a task in which subjects had to pick out a phoneme within a nonword said aloud by the examiner). Manis *et al.* (1997) concluded that inadequate representations of phonemic units resulting from such perceptual deficits could prevent dyslexic children from using and normally manipulating phonological information, thus impairing their ability to acquire phonological prerequisites to learning to read.

Valuable contributions to the domain of phonological processing in dyslexia have also been made by electrophysiological techniques.

II. Event-Related Potentials and Developmental Language Disorders

Recent years have seen a growing use of electrophysiological techniques in research on the neurobiological mechanisms underlying language-learning disorders. This section reviews some of the results from studies focusing on different components of the event-related potentials (ERPs): MMN, P300, and N400. Since this is not a major component of this chapter, methodological considerations will not be dealt with in detail here. The ERPs of childhood have been described in detail elsewhere (see, for instance, Taylor (1995) for a review).

A. The Mismatch Negativity

The mismatch negativity (or MMN) is an ERP characterized by a negative deflection with a frontocentral distribution that peaks between 100 and 250 ms after stimulus onset. The MMN is thought to be generated in the supratemporal auditory cortex and is elicited in situations where any physically deviant auditory stimulus occurs randomly and infrequently in a series of homogeneous, or standard, stimuli (Näätänen *et al.*, 1978). It can be elicited independent of attention and by very small acoustic changes. The MMN therefore reflects an automatic "change-detection response" (Kraus *et al.*, 1995) and may be used to investigate, at a preattentive level, whether the auditory system has distinguished between two stimuli.

Using the MMN evoked response, Kraus and colleagues (1996) obtained evidence suggesting auditory deficits in certain children with learning problems. Using speech stimuli from two continua (/da/ to /ga/ and /ba/ to /wa/), these researchers found that learning-impaired children were poorer in speech discrimination than were normal controls and that impaired discrimination was correlated with diminished MMNs. In other words, children who were poor at discriminating certain speech contrasts also showed reduced or absent MMNs. The MMN is a correlate of auditory processing at preattentive levels, so these findings suggest discrimination deficits in some learning-impaired children originate in the auditory pathways before conscious perception. Several studies using a similar paradigm have consistently shown reduced MMN in learning-disordered children (for a review, see Leppänen and Lyytinen, 1997). For instance, Schulte-Körne et al. (1998) recently presented 12-year-old dyslexics and controls with either language stimuli (85% standard /da/, 15% deviant /ba/) or pure tones (standard 1000 Hz, deviant 1050 Hz).

The MMN response differed between the two groups only for the language stimuli. These results point to a specific deficit of preattentive mechanisms for language processes as a possible source of these children's difficulties in learning to read.

Finally, one potentially useful application of this method has been proposed by Leppänen and Lyytinen (1997), who compared the MMN in 6-month-old infants with and without familial risk of learning disorder. Children genetically at risk displayed reduced MMN amplitudes in the left electrodes only, suggesting a predictive value of this pattern for the later occurrence of dyslexia.

B. The P300

Another group of electrophysiological studies of developmental language impairments has compared the amplitude and/or latency of the P300 component. The P300 is elicited in "odd-ball" tasks where subjects must attend to a train of frequently occurring stimuli and respond at the presentation of a different, infrequent deviant stimulus. This event is related to conscious processing and evaluation of stimuli as well as memory updating. Several studies have reported a smaller or later P300 in developmental dyslexics (Taylor and Keenan, 1990) and in children with attention-deficit disorder (ADD) (Holcomb et al., 1985), suggesting inefficient processing of task-relevant stimuli. Duncan et al. (1994) found P300 abnormalities among adults with childhood dyslexia only in those also suffering from ADD. Since attention deficits are present in many dyslexics, it is difficult to determine the respective contribution of the two disorders to the ERP abnormalities (Taylor, 1995).

C. The N400

An anomalous N400 has been observed in many studies of developmental language disorders (Stelmack et al., 1988; Neville et al., 1993). The results, however, are inconsistent, making an interpretation difficult. Stelmack et al. (1988), for example, found a reduced N400 in dyslexics, which they interpreted as a "failure to engage long-term semantic memory." Other investigators (Neville et al., 1993), on the other hand, found an enhanced N400 in language-impaired children. More recent observations (M. Besson, personal communication) suggest that the N400, classically obtained when brain activity is recorded while normal subjects read incongruous sentences ("the mother holds the child in her nostrils") and compared to recordings during reading of sentences with congruous endings (". . . in her arms"), is elicited in dyslexics on congruous endings as well. This could mean that semantic integration is deficient or requires more effort in dyslexics or alternatively, that in reading, dyslexics use semantic strategies not used by normal readers.

III. Functional Brain Imaging in Dyslexia

Table II summarizes 13 studies published to date in which images of the functioning brain have been compared between a group of dyslexics and a group of matched nondyslexic controls. It must be noted, first, that most of these studies have involved adults with a past diagnosis of learning disorder mainly affecting reading abilities, so it is highly difficult to retrospectively ascertain the type and intensity of the disorder. Likewise, these studies did not take into account the diversity of clinical forms of dyslexia and thus are exposed to the possible pitfall of putting together cases with different pathophysiological mechanisms. The variety of imaging methods, from magnetoencephalography to functional MRI, as well as multiple techniques and experimental designs even across studies using one method render fragile any attempt to draw firm conclusions from this overview. However, some important information has been already obtained.

In the following review, after presenting some early results from older techniques, we will divide the contemporary contributions according to the kind of experimental paradigm rather than the imaging method used.

A. The Pioneering Studies with Xenon and FDG-PET

In the late 1980's, the ability to obtain images of the brain in function had just appeared and a few studies were devoted to dyslexia, with the objective of trying to

Table II

Study	Method	Subjects	Chronological age (years)	Imaging technique	Functional activation	Results
Rumsey et al., 1987	Xenon-133 inhalation	14 men with severe dyslexia 14 controls	22 (3.5)	Cerebral blood flow measurements 16 collimated probes on each side of the scalp	Semantic classification task Line orientation task	Increased L > R asymmetry Reduced anterior to posterior difference
Flowers et al., 1991	Xenon-133 inhalation	69 controls (39 males + 30 females) 83 adults with known past learning abilities 33 dyslexic adults 27 borderline 23 nondisabled	29.8 (7.5) 33.8 (5.7) 33.8 (4.3) 32.2 (5.3)	Cerebral blood flow 8 detectors on each hemisphere	Verbal memory task Auditory perception task Spelling analysis task (indicate for each heard word if it has four letters or not)	Spelling task: hypoactivation in the left superior temporal region Hyperactivation in a more temporoparietal region
Gross-Glenn et al., 1991	[^{18}F]FDG positron emission tomography	11 adult dyslexics (all males) 14 controls (all males)	30.3 (8.3) 27.6 (6.4)	Regional cerebral metabolic values 32 regions in each hemisphere	Serial word reading during 30 min, one every 5 s	Bilateral hyperactivation lingual gyrus (controls more leftward asymmetry) Relative right frontal hypoactivation, controls more rightward asymmetry
Hagman et al., 1992	[^{18}F]FDG positron emission tomography	10 dyslexic adults 10 matched controls	39.1 (11.0) 35.3 (10.0)	Regional cerebral metabolic values 32 regions in each hemisphere	Identify the target syllable 'da' within a series of six different stop consonant–vowel syllables	Bilateral increase in metabolism in medial temporal regions
Rumsey et al., 1992	^{15}O-labeled H_2O positron emission tomography	14 dyslexics 14 controls	27 (5) 26 (5)	Regions of interest (ROI) method 49 regions analyzed	Rhyme detection task (experimental task) Control task: discrimination of intensity of a simple tone	Failure to activate left posterior temporal and inferior parietal regions; hyperactivation in temporal cortex
Rumsey et al., 1994a,b	^{15}O-labeled H_2O positron emission tomography	15 dyslexics 20 controls	27 (5) Matched	Regions of interest (ROI) method 49 regions analyzed	Listening to pairs of sentences differing in grammatical construction, but having the same meaning or not Listening to paired sequences (3–4 tones); had to indicate which were identical	Reduced blood flow at rest in left temporoparietal cortex Activation in anterior language regions not different from controls Reduction of normal right frontotemporal hyperactivity

continues

Table II *continued*

Study	Method	Subjects	Chronological age (years)	Imaging technique	Functional activation	Results
Paulesu et al., 1996	^{15}O-labeled H$_2$O positron emission tomography	5 dyslexics 5 controls	25.2 (1.5) 27.2 (2.2)	Whole-head PET scanning Analysis with SPM software	Two experimental tasks involving subvocal rehearsal of series of letters (a) Rhyming task (subarticulation, no memory) Reference task: shape similarity with corean letters (b) Memory task (subarticulation, keeping sounds in immediate memory) Reference task: visual short-term memory with corean letters	(a) Normal: left perisylvian zone, including inferior precentral, areas 44 and 45, and superior temporal gyrus Dyslexics: anterior part only (b) Left perisylvian, wider zone in temporopatietal region, including parietal operculum Duslexics: posterior part only (Heschl's gyrus and parietal operculum)
Eden et al., 1996a	fMRI	6 dyslexics 8 controls	26.8 (6.2) 25.5 (6.2)	Blood oxygenation level dependent (BOLD) 1.5-T MRI apparatus 30 5-mm contiguous, coronal slices	Reference task: fixation of a central cross Motion task (Magno stimulus): low-contrast array of black dots on a gray background, all moving in the same direction Pattern task (Parvo stimulus): stationary, high-contrast patterned stimulus	Controls: bilateral motion sensivity in the region VS/MT Dyslexics: 5, no activation in this region; 1, unilateral activation No group difference for stationary pattern
Salmelin et al., 1996	Magnetoencephalography	6 dyslexics (3 males + 3 females) 8 controls (4 males + 4 females)	19–35 18–37	Whole-head MEG 122 SQUID sensors	Finish words and nonwords presented for 300 ms every 3 s	Dyslexics fail to activate a left inferior temporooccipital region normally responding 180 ms after the presentation of words Left inferior frontal lobe activates within 400 ms in 4 out of 6 dyslexics and none controls

468

Study	Method	Subjects	Age	Technique	Task	Results
Rumsey et al., 1997 H₂O	¹⁵O-labeled dyslexic men positron emmission tomography	17 right-handed 27 (8) 14 matched controls	18–40, mean scanning	Whole-head PET	One visual fixation control task and four experimental tasks divided into "phonological" and "orthographic" Two pronunciation tasks: read pseudowords (phonological) and read irregular words (orthographic) Two decision-making tasks: which one of two written pseudowords (e.g., 'bape–'balik') sounds like a real word (phonological) and which one of two written forms of a real word ('hoal' or 'hole') is correct (orthographic)	Globally 30% more voxels activated and 50% more voxels deactivated in dyslexics as in controls Pronunciation tasks: reduced activations and unusual deactivations in bilateral mid to posterior temporal cortex + left inferior parietal cortex Decision-making tasks: no significant differences between dyslexics and controls
Demb et al., 1997; 1998	fMRI	5 dyslexics (3 males + 2 females) 5 controls (3 males + 2 females)	22.2 (2.9) 26.8 (6.1)	Blood oxygenation level dependent (BOLD) on 1.5-T scanner 8 adjacent 4-mm slices centered on the occipital region Retinotopy measurements on computationally flattened representations of each brain + psychophysical measures of M pathway functioning (speed discrimination thresholds) + five reading measures	Visual stimulation in conditions "known to preferentially stimulate M pathways" (especially low luminance level and moving gratings) as opposed to control stimuli "designed to elicit strong responses from multiple pathways" + a separate study similar to moving-dots experiment in Eden et al. (1996)	fMRI responses in both V1 and MT⁺ are lower in dyslexics across the full range of contrasts explored, with larger differences at higher contrasts, especially in V1 Strong negative correlation between MT⁺ activity and discrimination thresholds; weaker but significant correlation for V1 (both dyslexics and controls) Strong correlation between MT⁺ activity in M condition and reading (both dyslexics and controls)

469

continues

Table II *continued*

Study	Method	Subjects	Chronological age (years)	Imaging technique	Functional activation	Results
Shaywitz et al., 1998	fMRI	29 dyslexics (14 males + 15 females) 32 normal readers (16 males + 16 females)	16–54 18–63	1.5-T echo-planar imaging 9-mm slices 17 brain regions of interest "that previous research had implicated in reading and language"	Five tasks ordered hierarchically: line orientation judgment, letter-case judgment, single-letter rhyme, nonword rhyme, nonword rhyme, semantic category judgment	Group–task interaction in four ROI [posterior STG (Wernicke's area, angular gyrus (BA 39), striate cortex (BA 17), IFG (Broca's area)] and marginally in two more regions [inferior lateral extrastriate and anterior inferior frontal] Relative increase of activation in phonological vs orthograpic tasks is greater in posterior regions (areas 21, 40, 39, and 37) in normal readers and in anterior regions (B4 44–47 and 11) in dyslexics
Horwitz et al., 1998	¹⁵O-labeled H₂O positron emission	17 right-handed dyslexics (all males) 14 normal readers (all males)	27 (8) 25 (5)	Spatial normalization into stereotactic space (Talairach and Tournoux) Calculation of interregional correlations within each condition between one voxel representative of the angular gyrus and and all other brain voxels	Pseudoword reading task (read aloud) Exception word reading task (read aloud) Self-paced on a computer screen	(1) Pseudoword reading: large correlations between left angular gyrus temporo-occipital extrastriate areas, including motion area (V5/MT), lingual and fusiform gyri, inferior frontal (BA-45), and superior temporal (BA 22) (2) Exception word reading: as above for extrastriate, just miss significance for Wernicke's and Broca's areas. Dyslexics: all correlations absent or much less marked

demonstrate abnormal activity in dysfunctional brain circuits. Using the xenon inhalation technique, Rumsey *et al.* (1987) gave dyslexics and controls two tasks designed to activate respectively the left and right hemisphere: a semantic classification task and a line orientation task. This first study was interesting from a methodological point of view as the researchers paid particular attention to matching several variables across the two tasks. The tasks were similar in that subjects first looked at a probe screen, displaying the target word or line, and then had to choose among four visual items either a semantically related word or the corresponding line orientation. Results were relatively modest, since they showed an increase in the expected asymmetry (leftward for the verbal task and rightward for the spatial task), and were interpreted in terms of difficulties in bihemispheric integration. The same xenon method was used by Flowers *et al.* (1991), who designed an interesting spelling task where subjects had to decide whether or not words delivered through ear phones had four letters. Dyslexics showed reduced activation compared to controls in the left superior temporal region and hyperactivation of a more posterior temporo-parietal region. The authors concluded that dyslexics may be constrained by a deficit of superior temporal regions to a different strategy than normals.

With better spatial resolution than provided by the xenon method, but still with poor temporal resolution, the fluorodeoxyglucose (FDG) method (the precursor of modern PET studies) was used for two studies of dyslexia by Gross-Glenn *et al.* (1991). Their subjects performed a serial word reading task for 30 min. The isotope was injected intravenously 2 min after the reading task had begun, and the scan was performed at the end of reading, which lasted 20 min. Results showed bilateral activation in the lingual gyrus and a rightward frontal deactivation relative to controls. Finally, using the same FDG method with PET, Hagman *et al.* (1992) reported increased glucose use in the medial temporal regions in 10 severely dyslexic adults during a task that required subjects to identify the target syllable 'da' within a series containing six stop-consonant/vowel syllables. The only relevant result was hyperactivity in the medial temporal regions, suggesting involvement of fibers connecting internal and lateral (auditory) aspects of the temporal lobes.

Although retrospectively these early results acquire relevance in the light of data from more recent studies, important technical limitations and ambiguities resulting from poorly designed cognitive tasks considerably weaken their impact. Contemporaneously with these studies, the first studies of cognitive neuroanatomy with oxygen-15 were published. This more accurate method soon generated a large body of studies, mainly on nor-

mal volunteers, but also on dyslexic subjects. Within a few years, the first functional MRI studies of normal language and reading appeared that are now supplanting the more invasive PET method, with its reliance on radioactive tracers. The main advantage of fMRI is that repeated examinations can be done on the same subject. However, fMRI has its own limitations, different from those of PET, so both methods remain pertinent in the study of dyslexia. Finally, another imaging method, with much finer temporal resolution, magnetoencephalography, has been used with success in the investigation of dyslexics.

B. Functional Imaging Investigations of the Phonological Impairment in Dyslexia

As stated above, most modern researchers consider that phonological difficulties are central to the reading and learning disorders of dyslexics. Therefore, investigation of the brain substrate of these cognitive functions makes particular sense. Several groups have contributed to this topic in PET and fMRI studies of normal volunteers. These results will be summarized first, before a detailed description of the fewer studies in dyslexic patients.

▲ Brain Functional Anatomy of Phonological Processes (Démonet *et al.*, 1992, 1994a,b; Sergent *et al.*, 1992; Zatorre *et al.*, 1992; Paulesu *et al.*, 1993; Démonet, 1995; Fiez *et al.*, 1995)

As summarized in Fig. 2, these studies have demonstrated several foci of activation during various phonological tasks in which subjects had to manipulate mentally the sound content of heard words. For instance, Zatorre *et al.* (1992) asked subjects to signal if auditory word pairs ended with the same consonant sound (`bag'/`big'). Démonet *et al.* (1992, 1994a) asked subjects to determine whether or not a nonword spoken aloud contains the phoneme /b/ after the phoneme /d/ in a previous syllable. Three distinct cortical regions showed foci of activation: superior temporal cortex, at the level of primary auditory area (BA 41 or 42), Broca's area (BA 44 or 45), and left inferior parietal cortex either in the upper part of the supramarginal gyrus (Zatorre *et al.*, 1992) or in the more rostral parietal operculum (Paulesu *et al.*, 1993; Démonet *et al.*, 1994b).

The work by Paulesu *et al.* (1993) has shed light on the possible significance of these foci and the respective role of each region in phoneme processing. When subjects have to remember six letters successively flashed on a screen ("memory" condition), they try to pronounce them subvocally, in order to put them in the auditory phonological store postulated in classical mod-

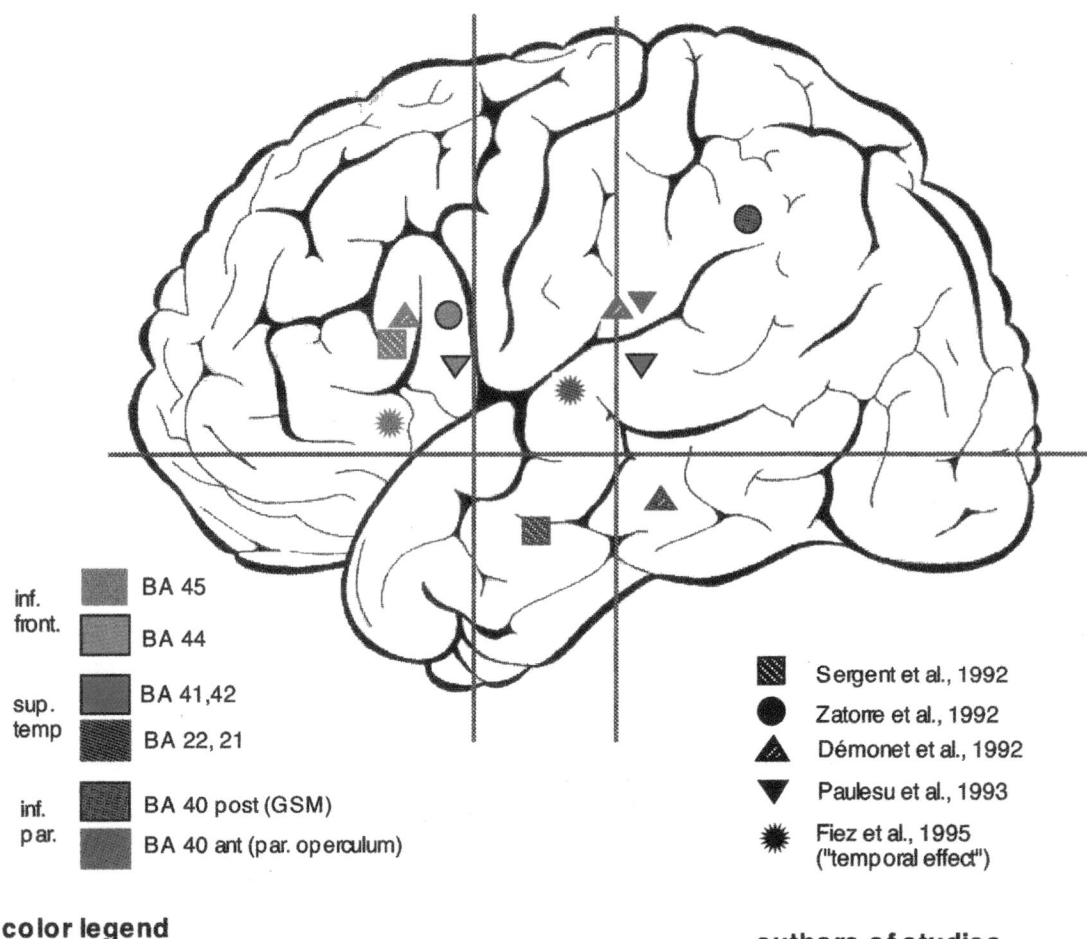

inf.
front.
 BA 45
 BA 44

sup.
temp
 BA 41,42
 BA 22, 21

inf.
par.
 BA 40 post (GSM)
 BA 40 ant (par. operculum)

color legend

 Sergent et al., 1992
 Zatorre et al., 1992
 Démonet et al., 1992
 Paulesu et al., 1993
 Fiez et al., 1995
 ("temporal effect")

authors of studies

Figure 2 Diagram summarizing, on the lateral surface of the left hemisphere, the main results of different studies of functional imaging (mainly ^{15}O-labeled-water PET) with phonological monitoring tasks in normal adults. Frontal, temporal, and parietal foci are represented with different color codes. Locations correspond to the foci of maximal activation in Talairach coordinates as reported by each author.

els of working memory (Baddeley, 1986). In another ("rhyming") condition, previously used by Sergent *et al.* (1992), subjects pronounce letters mentally but uniquely this task has no memory requirement and consists of a decision whether the name of a letter rhymes with a target letter 'b' ('b' rhymes with 'c', not with 'h'). Broca's area is activated in both conditions, whereas parietal operculum activation is specific to the memory condition. A similar activation of Broca's area has been found in a rhyming task by Sergent *et al.* (1992) and in tasks where subjects have to detect the presence of a phoneme within a series of heard syllables or words (Fiez *et al.*, 1995). Thus, Broca's area seems to be crucial to mental rehearsal of the articulatory form of phonemes, whether heard or transcoded from their visual form, whereas the parietal operculum is involved in the storage of phonological information. It is noteworthy that Broca's area may also activate without vol-

untary articulation, either when subjects listen passively to words or when they implicitly process them (Price *et al.*, 1996a,b). In these conditions, activation tends to be more anterior (BA 45) than when the same word is repeated aloud (BA 44).

Paulesu *et al.* (1993) and Démonet *et al.* (1994b) found activation in an almost identical location in the anterior part of the parietal operculum. The focus reported by Zatorre *et al.* is posterior to this location, probably because of the differential engagement of cognitive mechanisms. Zatorre and colleagues used short-word stimuli that did not engage working memory processes to the same extent as tasks used by the former authors. In a study where subjects had to maintain continuously subspan series of items in working memory, Fiez *et al.* (1996) did not find activation in the inferior parietal region. They hypothesized that this region was involved in encoding and retrieval aspects of the phonological loop rather than maintenance of information. Altogether, these results suggest that the "phonological store" postulated in neuropsychological models may depend on function of a restricted part of the parietal operculum (area SII and BA 43) rather than the supramarginal gyrus as a whole (BA 40). This detailed anatomical issue remains difficult to address in group studies because of the important variability of gyrification in the superior posterior perisylvian region (Talairach and Tournoux, 1988).

▲ Brain Activation during Phonological Tasks in Dyslexics

The first study to use PET and oxygen-15-labeled water in dyslexics was that of Rumsey *et al.* (1992; see Table II). However, this study (as well as two more from the same group; see Table II) used a region of interest method for analysis, which limits the scope of the results since it is always possible to overlook a peak of activation situated outside chosen regions with this method. Moreover, since the choice of regions of interest is dictated by preconceived ideas, the outcome of such studies is inevitably weakened by a risk of tautology. However, the results were important and have served as a basis for subsequent studies.

Fourteen adult dyslexics and 14 controls performed a rhyming task in which they had to press a button each time two words, within a pair presented binaurally through headphones, were judged to rhyme. Rhyming tasks are classically thought to explore some aspects of phonological awareness. To achieve such a task, subjects must concentrate on the sound form of the word endings, keep them in short-term memory, and compare them according to phonological similarity. Such a task is especially difficult to perform for adult dyslexics, who usually compensate by trying to resort to visual — orthographic mechanisms. For instance, non-rhyming pairs such as 'shoe'/'toe,' which are orthographically similar, and others such as 'head'/'said,' which rhyme but are orthographically dissimilar, are particularly puzzling for many dyslexics, showing high error rate in this task. In Rumsey *et al.*'s study, the task is made intentionally easy by avoiding such conflicting pairs. The main result is that dyslexics fail to activate a "left temporo-parietal region" activated in controls performing the task. Moreover, an interaction between group and condition for the left inferior and right anterior frontal regions was suggested in that dyslexics show a trend to relative deactivation in these regions where controls show nonsignificant increase in activity. Finally, dyslexics showed an activation compared to controls in a right middle temporal region. From these results, Rumsey *et al.* conclude that the main dysfunctional region in dyslexia is situated in the temporal cortex of the left hemisphere

and that by reference to previous PET studies showing that rhyming tasks rely upon activity of the posterior temporal region of the left hemisphere, this defect is related to their core phonological deficit.

The first ^{15}O PET study of dyslexics using whole-brain scanning and voxel-based analysis is that of Paulesu *et al.* (1996). These researchers used the same experimental paradigm as in their previous study (see above) comparing a rhyming and a memory condition. The rhyming condition is relevant to the issue of phonological awareness discussed by Rumsey *et al.* (1992) since, although the input is in the visual modality and stimuli are letter names rather than words, phonological segmentation and auditory comparisons are present in the tasks of both studies. However, in Paulesu's study, impact of the memory component is weakened by the presence of the target letter on the screen. On the other hand, in the memory task, the segmentation process is de-emphasized since a sequence of six letter names can be assimilated as a six-syllable word to be remembered, and thus the only segmentation required is at the syllabic, rather than phonemic, level.

Recent developments in psycholinguistic research concerning phonological awareness make a strong distinction between pre-reading phonological abilities (including syllable segmentation and rhyming abilities) and phonemic segmentation processes, explored by tasks such as phoneme deletion or discrimination (Goswami, 1993). Phonemic-level deficits are specifically impaired in most dyslexics, even though larger unit segmentation may provide an indicator of later reading abilities in younger children. In this connection, Paulesu's memory task may be less relevant to the phonologic deficit of dyslexia than is rhyming task. In any case, just as in Rumsey's study, Paulesu *et al.* assert that their dyslexics were as good as their controls at doing the tasks, which suggests that the differences in activation are independent of task difficulty.

Accordingly, Paulesu *et al.* found a specific pattern of activation in their five dyslexics, compared to five controls. In the memory task, dyslexics showed blood flow increases only in the posterior part (inferior parietal cortex) of the large perisylvian area activated in controls, whereas in the rhyming task, these patients only activated its anterior part (Broca's area). The common finding with both tasks was the absence of activation of the insular cortex. This led Paulesu *et al.* to an interpretation of dyslexic deficits in terms of disconnection between anterior and posterior zones of the language area. Since such impaired activation of the insular cortex has not been replicated by other functional imaging studies of dyslexics, this hypothesis awaits further evidence. Likewise, most studies published to date have shown increased rather than reduced activity close to

Broca's area during phonological tasks in dyslexics. However, the finding of reduced left superior temporal activation is consistent with other more recent studies (see below). Finally, the disconnection hypothesis has gained additional credence following a recent study (Horwitz *et al.,* 1998) showing that some areas, especially the angular gyrus, fail to coactivate in dyslexics.

C. Functional Imaging Studies of Reading in Dyslexia

▲ Brain Functional Anatomy of Normal Reading

Several studies, summarized in Fig. 3 and recently commented on by Fiez and Petersen (1998), have been performed with normal volunteers to understand the brain mechanisms of normal reading.

Following the pioneering studies of Petersen and collaborators (1989, 1990; see also Posner and Raichle, 1994), imaging the reading brain has become a challenge to neuroscientists. This interest started with some rather surprising findings reported in the early studies, which seemed to profoundly question classical concepts about the neuroanatomy of reading. Indeed, two of the main postulates of classical teaching in brain-reading relationships were disputed. In the classical view, as exposed in the late 1970's by Norman Geschwind (Geschwind, 1979), reading involved a crucial stage located in the angular gyrus, where the mental images of letters (Dejerine, 1892) are processed before being transferred to Wernicke's area for conversion into sounds (since Wernicke's area is the part of the auditory cortex thought to subserve auditory decoding of verbal messages). In the first PET studies, neither the angular gyrus nor Wernicke's area was activated while subjects read series of written words. Instead, activation of an unexpected area in the left visual cortex on the mesial side of the hemisphere (lingual gyrus) was found in a region since then referred to as left' extra-striate cortex. With time, these early results have been moderated since, under certain conditions, both Wernicke's area (Posner and Raichle, 1994) and the angular gyrus (Price *et al.,* 1994, 1996a) are activated by written visual stimuli. For instance, when subjects are presented with visual words and strings of consonants or pseudo-letters with an instruction to press a button when the letters of a word or pseudo-word contain ascending parts (such as d, b, l, and t), the comparison between words and nonsense strings (either made up of letters or pseudo-letters) reveals a large area of activation in the left hemisphere, encompassing all classical language areas (Price *et al.,* 1996a).

Figure 3 shows the different brain regions activated in various activation studies with reading tasks published to date. Each focus represents the mean location (the size of ovals representing standard deviations) obtained across nine different studies. Although all locations are not activated in all studies, there is high enough consistency to suggest that these foci are organized in a network involving several regions of the left and right hemispheres. One important consideration is that some of these foci are bilateral—BA 4 (motor cortex in the face area and opercular/insular region), BA 22/42 (primary and anterior association auditory cortex), and BA 22/21 (lateral temporal cortex)—whereas

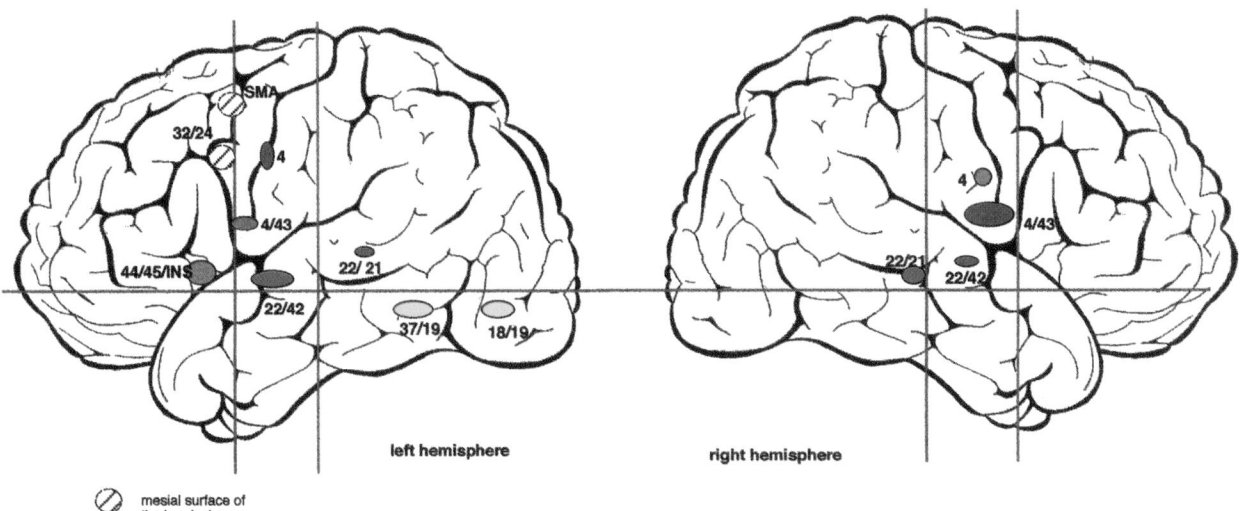

left hemisphere   right hemisphere

⊘ mesial surface of
  the hemisphere

Figure 3 Diagram summarizing the main data obtained in normal adults with PET activation studies during reading. Each oval is centered on the mean focus of maximal activation across nine studies, according to Talairach coordinates, and its size is indicative of standard deviations (modified, after Fiez and Petersen, 1998).

others are strictly unilaterally located to the left hemisphere—BA 44/45 (Broca's area), BA 37/19 (inferior temporo-occipital cortex), and BA 18/19 (extra-striate cortex). Several studies have resulted in a specification of the role of each region by contrasting different reading tasks. For example, as summarized by Fiez and Petersen (1998), both exception words (which break spelling-to-sound rules, such as 'pint' as compared to 'hint, mint, lint') and nonsense words seem to yield greater activation in Broca's area than consistent words. This apparently surprising observation can be explained if one refers to the special role of this region in prearticulatory phonological processing by assuming that pronounceable nonwords rely heavily on an orthography-to-phonology transformation procedure and that exception words require more intense or longer transformation between orthography and phonology. On the other hand, Pugh et al. (1996), using fMRI with ROI analysis, showed that left extrastriate areas are more activated in tasks requiring orthographic judgment, left inferior frontal regions are more activated in rhyming tasks, whereas postero-superior temporal regions are equally involved in phonological, orthographic, and semantic tasks. Activation of BA 22/42 is thought to reflect auditory stimulation by the reader's own voice on reading aloud. BA 22/21 activation, on the other hand, is probably associated with semantic processing.

One unexpected result of these studies is the consistency with which a small region of the basal aspect of the left temporal lobe, projecting within the limits of BA 37, has been activated in various studies involving the processing of written words. In a stimulating essay on the functional anatomy of language, Price (1998) proposed, on the basis of evidence of activation of this region in naming tasks contrasted with viewing objects (Price and Friston, 1997), that it is involved in phonological retrieval, i.e., the access to phonological forms of written words. Indeed, lesions in this area are classically associated with anomia and other naming deficits. However, another probably intuitively more satisfactory hypothesis would postulate a role of this basal temporal area in visual feature recognition specific to written material (Fiez and Petersen, 1998).

▲ Brain Functional Anatomy of Reading in Dyslexia

Three important functional imaging studies of reading in dyslexics have been reported in the past few years, one using the PET method with ^{15}O radiotracer, the second using fMRI, and the third (historically the first) using magnetoencephalography.

Comparing 17 right-handed dyslexics and 14 controls, Rumsey et al. (1997) performed a PET study that, unlike their previous reports, used a voxel-based whole-brain imaging technique. The main objective of their study was to try to contrast orthographic and phonologic processes in reading. This was done using two different paradigms. One, named "pronunciation," required participants to read aloud pseudowords (presumably relying on phonological processing) and irregular words (calling for orthographic processing). The other group of tasks, called "decision-making tasks," consisted of having participants read two words or nonwords with either a phonological instruction (decide which one of two nonwords sounds like a real word) or an orthographic instruction (decide which of two homophone forms of a same word is correctly spelled). Besides differences in the global volume of activation between the two groups (more brain tissue activated in dyslexics), the only significant changes observed in dyslexics were reduced activations and unusual deactivations in left posterior temporal/inferior parietal areas, in the pronunciation paradigm. The authors comment on the absence of any difference between the locations of brain activation with the word and nonword versions of the tasks, which could result, they suggest, from a common mechanism for both kinds of reading errors in dyslexia.

With fMRI using imaging parameters that were not optimal (9-mm thick slices, 17 regions of interest "that previous reseach had implicated in reading and language"), Shaywitz et al. (1998) proposed to 29 dyslexics and 32 controls a complete series of tasks tapping various levels of processing in reading: a line orientation judgment task, presumably exploring low-level perceptive mechanisms; a letter-case judgment, presumed to reflect orthographic mechanisms; a single-letter rhyme and a nonword rhyme task, exploring the phonological level of processing; and a semantic category judgment task. The main result is that in normals there is relative overactivation with phonological tasks (in comparison to orthographic ones) in posterior regions (posterior temporal, BA 21; supramarginal and angular gyri, BA 39 and 40; and inferior lateral temporal region, BA 37) and a reversed pattern (greater anterior than posterior activation when contrasting phonological and orthographic processing) in dyslexics. The authors propose a general explanation for posterior hypoactivation in dyslexics "due to actual disruption in a system in charge of phonological processes" whereas Broca's area hyperactivation reflects "increased effort required of dyslexics in carrying out phonological analysis."

Finally, probably the most informative study on brain functioning during reading tasks in dyslexics is that of Salmelin et al. (1996). These researchers used the method of magnetoencephalography (MEG) to compare six dyslexics and eight controls on various reading tasks involving Finnish words and nonwords presented for 300 ms every 3 s. Since this method provides only an approximate location for the presumed sources of recorded signal, one must remain cautious about information on the actual brain regions involved. However, the remarkable temporal resolution of the method makes it a valuable complement to other imaging methods, which can only provide information about processes occurring within a period of time of several seconds. Whereas normal controls activated a left inferior temporo-occipital region 180 ms after the presentation of words, dyslexics totally failed to activate this same region. Moreover, a left inferior frontal area activates within 400 ms in four out of six dyslexics but in none of the controls.

This result is of great importance because it clarifies data obtained by other methods whose interpretation had remained obscure. For instance, the inferior temporo-occipital activation is reminiscent of activation of BA 37 found in several studies. Since it appears as early as 180 ms after the presentation of a word, it is likely to represent early visual processing or immediate phonological processing occurring before any conscious recognition has occurred. That dyslexics do not activate this process suggests either an inability to achieve these early operations of global word-form perception or inefficient immediate phonological extraction. On the other hand, abnormal activity in Broca's area could result from compensatory silent articulation of a misperceived visual word form. These results thus provide information about the time course of reading in two areas already suspected to be dysfunctional in dyslexia, allowing a more complete discussion of their role.

A recent study by Brunswick et al. (1999) using ^{15}O PET to investigate explicit and implicit reading in dyslexic adults and normal readers came to similar conclusions. In both reading conditions, dyslexics activated two regions to a lesser degree than controls: left basal temporal lobe (area 37) and left frontal operculum. In the explicit reading condition, they overactivated a left pre-motor region situated 20 mm lateral to the area of reduced activation. This finding confirms that Broca's area is one of the most important brain regions at the origin of the learning impairment in dyslexics but that this important role is probably not unique, since different cortical zones within this anatomical region display different patterns of functional activation. Using the same inclusion criteria as in Brunswick's study, we recently conducted a parallel study in French-speaking dyslexics and matched controls (Chanoine et al., 1998). The only region to show greater activation in controls than in dyslexics on reading tasks was precisely the left inferior temporal region, exactly at the junction between lateral and mesial aspects of the temporal lobe (Fig. 4). It is noteworthy that this region is ideally lo-

cated to serve as an interface between areas associated with processing visual features of written words (especially the more mesial extra-striate cortex), other temporo-occipital regions involved in complex visual processing (including the motion area, which is in close vicinity; see below), and more dorsal language areas in the middle and superior temporal gyri, possibly subserving grapheme-to-phoneme transformation. One plausible role that could be attributed to this crucial region could be that of mediating the visual entry into the linguistic system, combining orthographic, lexical, and phonological information about words. Unpublished observations show that this region also activates with stimuli given in the auditory modality when subjects have to perform an orthographic computation on the heard words. Such data are also compatible with this formulation. Finally, the multimodal complexity of this region has been recently confirmed by Büchel et al. (1998), who show it activates equally in seeing persons reading words and blind persons reading Braille.

D. Functional Imaging Investigations of Visual Impairment in Dyslexia

▲ fMRI Study of Motion Perception in Dyslexia (Eden et al., 1996b)

By reference to theories based on psychophysical and electrophysiological evidence of a deficit in the magno-cellular component of the visual pathways, Eden et al. (1996b) designed an fMRI experiment contrasting two visual conditions that differentially activate the magno system and the parvo system. The experimental condition (M-stimulus) consisted of a moving-dot task, where participants had to look passively at an array of dots moving on a computer screen. In the reference task (P-stimulus), they had to look at a stationary pattern. The first condition is assumed to activate the magno system preferentially, especially the human area V5 (MT), which has been shown in earlier experiments to be in the posterior part of the inferior temporal sulcus. As expected, normal controls activated this area bilaterally with the M-stimulus, not with the P-stimulus, whereas dyslexics failed to activate this region even in the moving-dot condition. Eden et al.'s conclusion is that their data provide a direct demonstration of a magno-cellular deficit in dyslexia, which could be one manifestation of a basic disorder in the processing of temporal properties of stimuli.

▲ Relationship of Motion Perception Deficit to the Reading Disorder in Dyslexia

Demb et al. (1997, 1998) have replicated these findings in five adult dyslexics with a similar fMRI method,

sagittal

coronal

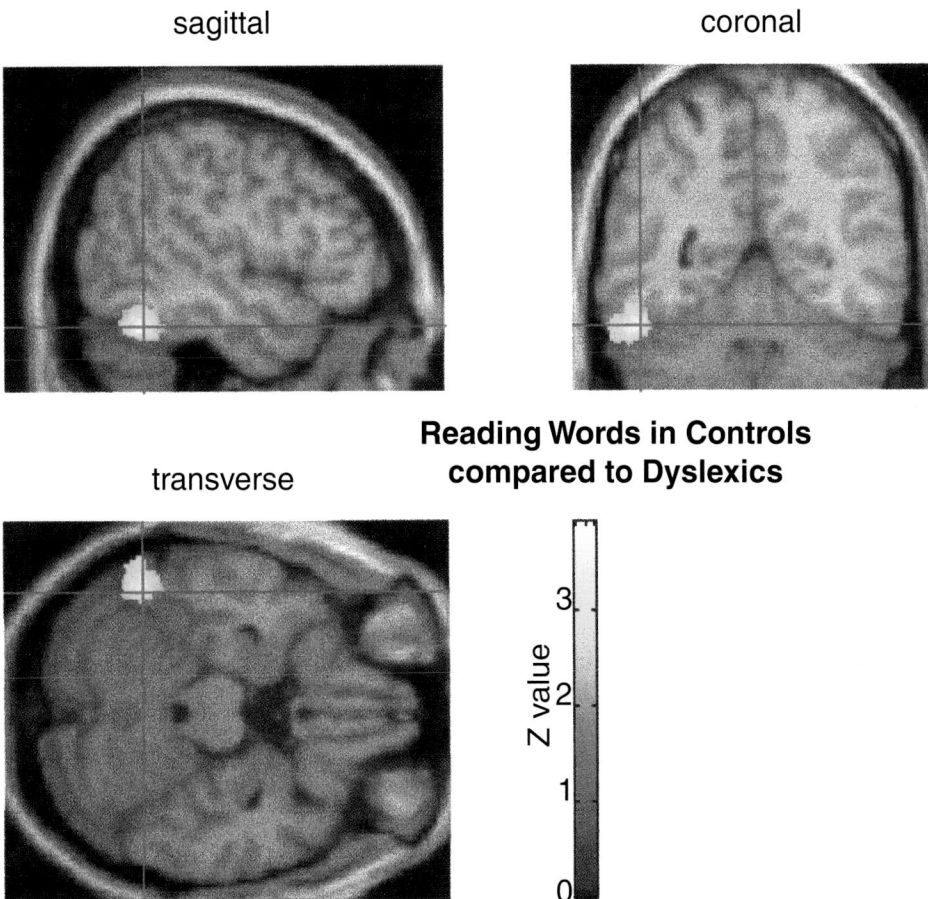

transverse

**Reading Words in Controls
compared to Dyslexics**

Figure 4 Three-dimensional representation of the main focus of over-activation obtained in six dyslexics compared to six controls during a word-reading task. The activated region corresponds to the basal temporal part of area 37 (junction between inferior temporal and fusiform gyri), a region thought to be crucial for multimodal interactions occurring in word recognition and/or orthographic processes (Chanoine *et al.*, 1998).

but with localization of the different areas by a more accurate technique that uses computational flattening of each brain. Speed discrimination thresholds were measured psychophysically for each participant to determine as accurately as possible the efficiency of the magno-cellular pathway. Finally, five reading tasks were administered to evaluate the impact of the magno defect on reading performance. Stimuli were low luminance level and moving gratings, known to preferentially stimulate M-pathways, as opposed to control stimuli "designed to elicit strong responses from multiple pathways." The results showed that fMRI responses in both V1 and MT^+ were lesser in dyslexics across the full range of contrasts explored, with larger differences at higher contrasts, especially in V1. Moreover, there was a strong negative correlation between MT^+ activity and discrimination thresholds, as well as a weaker but significant correlation for V1 (in both dyslexics and controls). Finally, a strong correlation was found between

MT^+ activity in the M-condition and reading speed (in both dyslexics and controls).

Altogether, these results obtained in the visual modality in dyslexics have been suggested as a potential brain marker for dyslexia. However, it must be noted that neither Eden *et al.* (1996b) nor Demb *et al.* (1998) have characterized their dyslexic adults by the neuropsychological form of dyslexia. For instance, it would have been interesting to know whether pure phonological dyslexics, with a prevalent grapheme-to-phoneme conversion deficit, are more or less impaired in their ability to activate their visual motion area. On the other hand, it would be important to know whether magnocellular impairment impinges specifically on the possibility of using global, whole-word strategies in reading, which could suggest a causal link between the perceptual deficit and learning disorder. On the contrary, if even purely phonological dyslexics fail to activate their motion area, this could mean that both visual deficit and phonological impairment stem from a common mechanism, as suggested by the temporal processing theory, for example.

▲ The Connection with the Temporal Processing Theory (Farmer and Klein, 1995; Fiez *et al.*, 1995; Tallal *et al.*, 1997) and Perspectives for Remediation

The notion that developmental dyslexia may result from a general, non-specific, deficit in perceiving rapidly changing auditory signals has been hotly debated in the past few years. This notion was first put forward by Tallal and collaborators (Tallal and Piercy, 1973; Tallal, 1980; Tallal *et al.*, 1997) following the observation that children with language-learning impairment failed to perform temporal order judgment tasks when inter-stimulus intervals were short (20–40 ms) but did not differ from normal controls for longer intervals (80–120 ms). Despite significant evidence accumulated since then in favor of this theory (for a review, see Farmer and Klein, 1995), others have objected that the problem is not acoustic but rather is linguistic in nature (Mody *et al.*, 1997). Finally, Tallal *et al.* (1996) proposed, based on the temporal processing theory, a training method for language-learning-impaired children using daily auditory exercises where the temporal features of stimuli were voluntarily and specifically altered. Recently, our group has obtained preliminary results suggesting that such auditory training, specifically aimed at facilitating perception of rapid acoustic changes in natural speech, may specifically improve phonological awareness in a subgroup of phonological dyslexic children, thus affecting the very crucial stage involved in learning disorder (Habib *et al.*, 1999).

In this regard, the temporal processing theory is probably one of the most powerful explanations available to date that accommodates the clinical and neuropsychological complexity and diversity of dyslexia and the only one able to reconcile data obtained from the neuropsychological approach, pointing to a phonological deficit, and those demonstrating a visual impairment. Difficulties for processing time in its different dimensions is an amazingly universal characteristic of the dyslexic child. Thus it is plausible that a general difficulty for the brain (most probably in the left hemisphere) is to integrate rapidly changing stimuli. Such a difficulty is responsible for (1) a deficit in generation or judgment of temporal order, (2) impaired perception of rapid unstable auditory stimuli, such as consonants, and (3) the deficit in all stages of reading that implies rapid processing, e.g., visuospatial letter arrangement, global word-form perception, and integration of successive relative word positions during oculo-motor scanning. Brain imaging methods may be useful in testing this hypothesis in normal readers and dyslexics. In normal

readers, Fiez *et al.* (1995) have shown that activation of Broca's area is similarly obtained when subjects have to listen to consonants and rapid nonverbal stimuli, but not when they listen to steady-state vowels without rapid acoustic changes. Recently, Belin *et al.* (1998) have used auditory activation with PET to show that nonverbal sounds containing rapid (40 ms) or extended (200 ms) frequency transitions yield different patterns of activation in the auditory cortex, bilateral symmetric for slow-transition stimuli and left unilateral for rapid transitions. Similar experiments in dyslexics would certainly provide valuable information about the temporal processing hypothesis in dyslexia and are currently in progress. Recent data obtained in our group (Cancès *et al.*, unpublished communication) using auditory ERPs show that a positive event peaking around 120 ms (part of the N1–P2 complex recorded following any acoustic stimulus) during automatic perception of the syllable /fi/ in a phonetic continuum /pi/-/fi/ is delayed for more than 50 ms in dyslexic subjects (Fig. 5). This systematic delay could represent the basic functional impairment in the temporal perception deficit of dyslexics. We are currently testing the electrophysiological responses of dyslexics with acoustically modified stimuli as a further evaluation of the temporal processing hypothesis.

Finally, such functional imaging studies in dyslexics, which has only been performed for the moment in adults, could be of great interest in children, especially in relation to the recently proposed training method using acoustically modified speech to improve temporal processing deficits in dyslexics (Tallal *et al.*, 1996; Habib *et al.*, 1999). In this context, functional imaging may serve to evaluate objectively the efficacy of therapy by measuring the nature and extent of changes in brain activity before and after training, either with natural or with acoustically modified speech. A related topic could be the possibility of imaging the effects of medications, for instance, the impact of methylphenidate on the associated hyperactivity disorder, which has been recently demonstrated with fMRI (Vaidya *et al.*, 1998).

IV. Conclusion

It is obviously unrealistic to expect a final conclusion in a chapter devoted to such a new and dramatically expanding topic as brain imaging in dyslexia. It seems only possible to roughly delineate, in light of the first results available to date, what could be interesting avenues to pursue in future studies.

Concerning morphological brain imaging, it is clear that, although sometimes contradictory, the results obtained strongly suggest that the dyslexic brain actually possesses morphological particularities whose better

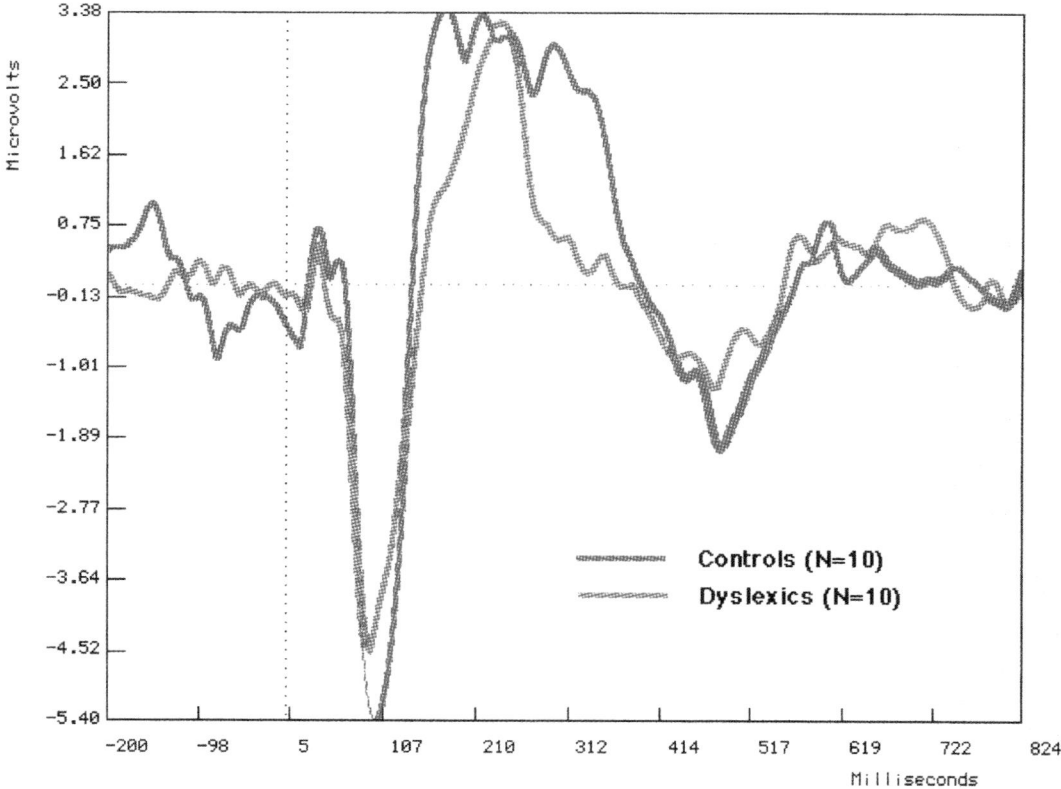

Figure 5 Average evoked potential (ERP) recorded on Cz scalp location for passive listening to syllable /fi/ among a continuum /fi/-/pi/ in dyslexics and controls (habituation-dehabituation paradigm). A delay in the early component P2 is observed in dyslexics compared to controls, a possible electrophysiological concomitant of temporal processing deficit in the auditory modality in dyslexia.

description should provide important indications toward possible pathophysiological mechanisms at their origin. For instance, specific changes in callosal morphology have already allowed discussion of the various factors that potentially influence the development of the dyslexic brain. It is probable that, along with rapid improvement of imaging techniques and morphometric tools, such information will become more accurate and finer in detail, allowing investigation of including regions not yet accessible to measurement.

Concerning functional imaging, the two objectives of studies of dyslexia and related disorders will be best tackled in parallel: (1) an attempt to understand the brain correlates of each dysfunctional system in order to contribute to an accurate characterization of the functioning dyslexic brain and (2) use of the dyslexic brain as a model to generate dissociations apt to enlighten the cognitive mechanisms subserving normal function, including reading and its oral language prerequisites. Such an approach has already proved fruitful, demonstrating, for instance, the multiple roles of Broca's area in reading and the importance of inferior temporal regions in orthographic processes.

One of the problems common to morphological and functional imaging is that of the heterogeneity of dyslexic populations studied as well as the multiple and diverse association between related entities. Presently, it is probable that imaging methods will not only benefit from improvement of clinical description and classification of scanned individuals but also contribute to them. For instance, it is conceivable that future clinical approaches of these conditions will include neuroimaging assessment for clinical purposes.

Finally, we wish to emphasize the fact that both morphological and functional imaging studies still have their role to play in future research on this fascinating topic and that, far from being mutually exclusive, their respective contribution should be combined in a complementary manner, oriented toward neurodevelopment with morphological imaging and to the neurocognitive side for functional methods.

References

Baddeley, A. (1986). "Working Memory." Oxford Univ. Press, Oxford.
Beaton, A. A. (1997). The relation of planum temporale asymmetry

and morphology of the corpus callosum to handedness, gender, and dyslexia: A review of the evidence. *Brain Lang.* **60**(2), 255–322.

Belin, P., Zilbovicius, M., Crozier, S., Thivard, L., Fontaine, A., Masure, M. C., and Samson, Y. (1998). Lateralization of speech and auditory temporal processing. *J. Cog. Neurosci.* **10**(4), 536–540.

Best, C. T. (1985). "Hemispheric Function and Collaboration in the Child." Academic Press, New York.

Binder JR Frost JA Hammeke TA Rao SM Cox RW (1996). Function of the left planum temporale in auditory and linguistic processing. *Brain* **119**(4), 1239–1247.

Boder, E. (1971). Developmental dyslexia: A diagnostic screening procedure based on three characteristic patterns of reading and spelling. *In* "Learning Disorders" (B. Bateman, Ed.), Vol. 4, pp. 297–343. Special Children Publications, Seattle.

Brunswick, N., McCrory, E., Price, C., Frith, C. D., and Frith, U. (1999). Explicit and implicit processing of words and pseudowords by adult developmental dyslexics: A search for Wernicke's Wortschatz? *Brain* **122**, 1901–1917.

Büchel, C., Price, C., and Friston, K. (1998). A multimodal language region in the ventral visual pathway. *Nature* **394**(6690), 274–277.

Celsis, P., Boulanouar, K., Doyon, B., Ranjeva, J. P., Berry, I., Nespoulous, J. L., and Chollet, F. (1998). Differential fMRI responses in the left posterior superior temporal gyrus and left supramarginal gyrus to habituation and change detection in syllables and tones. *NeuroImage* **9**, 135–144.

Clark, M. M., and Plante, E. (1998). Morphology of the inferior frontal gyrus in developmentally language-disordered adults. *Brain Lang.* **61**, 288–303.

Dalby, M. A., Elbro, C., and Stodkilde-Jorgensen, H. (1998). Temporal lobe asymmetry and dyslexia: An *in vivo* study using MRI. *Brain Lang.* **62**(1), 51–69.

Dejerine, J. (1892). Contribution à l'étude anatomo-clinique et clinique des différentes variétés de cécité verbale. *Mém. Soc. Biol.* **4**, 61–90.

Demb, J. B., Boynton, G. M., and Heeger, D. J. (1997). Brain activity in visual cortex predicts individual differences in reading performance. *Proc. Natl. Acad. Sci. U.S.A.* **94**, 13363–13366.

Demb, J. B., Boynton, G. M., and Heeger, D. J. (1998). Functional magnetic resonance imaging of early visual pathways in dyslexia. *J. Neurosci.* **18**(17), 6939–6951.

Démonet, J. F. (1995). Studies of language processes using positron emission tomography. *In* "Handbook of Neuropsychology" (R. Johnson, J.-C. Baron, J. Grafman, and J. Hendler, Eds.). Elsevier, Amsterdam.

Démonet, J. F., Chollet, F., Ramsay, S., Cardebat, D., Nespoulous, J.-L., Wise, R., Rascol, A., and Frackowiak, R. S. J. (1992). The anatomy of phonological and semantic processing in normal subjects. *Brain* **115**, 1753–1768.

Démonet, J. F., Price, C., Wise, R., and Frackowiak, R. S. J. (1994a). A PET study of cognitive strategies in normal subjects during language tasks. Influence of phonetic ambiguity and sequence processing on phoneme monitoring. *Brain* **117**, 671–682.

Démonet, J. F., Price, C., Wise, R., and Frackowiak, R. S. J . (1994b). Differential activation of right and left posterior sylvian regions by semantic and phonological tasks: A positron emission tomography study in normal human subjects. *Neurosci. Lett.* **182**, 25–28.

Duara, R., Kushch, A., Gross-Glenn, K., Barker, W. W., Jallad, B., Pascal, S., *et al.* (1991). Neuroanatomic differences between dyslexic and normal readers on magnetic resonance imaging scans. *Arch. Neurol.* **48**, 410–416.

Duncan, C. C., Rumsey, J. M., Wilkniss, S. M., Denckla, M. B., Hamburger, S. D., and Odou-Potkin, M. (1994). Developmental dyslexia and attention dysfunction in adults: Brain potential indices of information processing. *Psychophysiology* **31**, 386–401.

Eden, G. F., VanMeter, J. W., Rumsey, J. M., Maisog, J. M., Woods, R. P., and Zeffiro, T. A. (1996a). Abnormal processing of visual motion in dyslexia revealed by functional brain imaging. *Nature* **382**, 66–69.

Eden, G. F., VanMeter, J. W., Rumsey, J. M., Maisog, J. M., Woods, R. P., and Zeffiro, T. A. (1996b). The visual deficit theory of developmental dyslexia. *NeuroImage* **4**, S108–S117.

Farmer, M. E., and Klein, R. M. (1995). The evidence for a temporal processing deficit linked to dyslexia: A review. *Psychonom. Bull. Rev.* **2**(4), 460–493.

Fiez, J. A., and Petersen, S. E. (1998). Neuroimaging studies of word reading. *Proc. Natl. Acad. Sci. U.S.A.* **95**, 914–921.

Fiez, J. A., Tallal, P., Raichle, M. E., Miezin, F. M., Katz, W. F., and Petersen, S. E. (1995). PET studies of auditory and phonological processing: Effects of stimulus characteristics and task demands. *J. Cog. Neurosci.* **7**, 357–375.

Fiez, J. A., Raife, E. A., Balota, D. A., Schwarz, J. P., Raichle, M. E., and Petersen, S. E. (1996). A positron emission tomography study of the short-term maintenance of verbal information. *J. Neurosci.* **15**, 808–822.

Fisher, S. E., Marlow, A. J., Lamb, J., Maestrini, E., Williams, D. F., Richardson, A. J., Weeks, D. E., Stein, J. F., and Monaco, A. P. (1999). A quantitative trait locus on chromosome 6p influences different aspects of developmental dyslexia. *Am. J. Hum. Genet.* **64**, 146–156.

Flowers, D. L., Wood, F. B., and Naylor, C. E. (1991). Regional cerebral blood flow correlates of language processes in reading disability. *Arch. Neurol.* **48**, 637–643.

Galaburda, A. M., and Kemper, T. L. (1979). Cytoarchitectonic abnormalities in developmental dyslexia: A case study. *Ann. Neurol.* **6**, 94–100.

Galaburda, A. M., Sherman, G. F., Rosen, G. D., Aboitiz, F., and Geschwind, N. (1985). Developmental dyslexia: Four consecutive patients with cortical anomalies. *Ann. Neurol.* **18**, 222–233.

Gauger, L. M., Lombardino, L. J., and Leonard, C. M. (1997). Brain morphology in children with specific language impairment. *J. Speech Lang. Hear. Res.* **40**(6), 1272–1284.

Geschwind, N. (1979). Specializations of the human brain. *Sci. Am.* **241**(3), 158–168.

Geschwind, N., and Galaburda, A. M. (1985). Cerebral lateralization. Biological mechanisms, associations and pathology: I. A hypothesis and a program for research. *Arch. Neurol.* **42**, 428–459.

Geschwind, N., and Levitsky, W. (1968). Human brain: Left-right asymmetries in temporal speech region. *Science* **161**, 186–187.

Gladstone, M., Best, C. T., and Davidson, R. J. (1989). Anomalous bimanual coordination among dyslexic boys. *Dev. Psychol.* **25**, 236–246.

Godfrey, J. J., Syrdal-Lasky, A. K., Millay, K. K., and Knox, C. M. (1981). Performance of dyslexic children on speech perception tests. *J. Exp. Child Psychol.* **32**, 401–424.

Goswami, U. (1993). Phonological skills and learning to read. *Ann. N.Y. Acad. Sci.* **682**, 296–311.

Gross-Glenn, K., and Rothenberg, S. (1984). Evidence for deficit in interhemispheric transfer of information in dyslexic boys. *Int. J. Neurosci.* **24**, 23–25.

Gross-Glenn, K., Duara, R., Barker, W. W., Loewenstein, D., Chang, J. Y., Yoshii, F., *et al.* (1991). Positron emission tomography studies during serial word-reading by normal and dyslexic adults. *J. Clin. Exp. Neuropsychol.* **13**, 531–544.

Gross-Tsur, V., Shalev, R. S., Manor, O., and Amir, N. (1995). Developmental right-hemisphere syndrome: Clinical spectrum of the nonverbal learning disability. *J. Learn. Disabil.* **28**, 80–86.

Gross-Tsur, V., Manor, O., and Shalev, R. S. (1996). Developmental dyscalculia: Prevalence and demographic features. *Dev. Med. Child Neurol.* **38**, 25–33.

Habib, M., and Robichon, F. (1996). Parietal lobe morphology predicts phonological skills in developmental dyslexia. *Brain Cog.* **32**, 139–142.

Habib, M., Espesser, R., Rey, V., Giraud, K., Bruas, P., and Gres, C. (1999). Training dyslexics with acoustically modified speech: Evidence of improved phonological performance (abstract). *Brain Cog.* **40**, 143–146.

Hagman, J. O., Wood, F., Buchsbaum, M. S., Tallal, P., Flowers, L., and Katz, W. (1992). Cerebral brain metabolism in adult dyslexic subjects assessed with positron emission tomography during performance of an auditory task. *Arch. Neurol.* **49**, 734–739.

Hanley, R., Hastie, K., and Kay, J. (1992). Developmental surface dyslexia and dysgraphia: An orthographic processing impairment. *Q. J. Exp. Psychol.* **44A**(2), 285–319.

Hinshelwood, J. (1917). "Congenital Word Blindness." Lewis, London.

Holcomb, P. J., Dykman, R. A., and Ackerman, P. T. (1985). Cognitive event-related brain potentials in children with attention and reading deficits. *Psychophysiology* **22**, 656–667.

Horwitz, B., Rumsey, J. M., and Donohue, B. C. (1998). Functional connectivity of the angular gyrus in normal reading and dyslexia. *Proc. Natl. Acad. Sci. U.S.A.* **95**(15), 8939–8944.

Humphreys, P., Kaufmann, W. E., and Galaburda, A. M. (1990). Developmental dyslexia in women: Neuropathological findings in three patients. *Ann. Neurol.* **28**, 727–738.

Hynd, G. W., Semrud-Clikeman, M., Lorys, A. R., Novey, E. S., and Eliopoulos, D. (1990). Brain morphology in developmental dyslexia and attention deficit disorder/hyperactivity. *Arch. Neurol.* **47**, 919–926.

Hynd, G. W., Hall, J., Novey, E. S., Eliopulos, D., Black, K., Gonzales, J. J., *et al.* (1995). Dyslexia and corpus callosum morphology. *Arch. Neurol.* **52**, 32–38.

Jernigan, T. L., Hesselink, J. R., Sowell, E., and Tallal, P. A. (1991). Cerebral structure on magnetic resonance imaging in language- and learning-impaired children. *Arch. Neurol.* **48**, 529–545.

Karbe, H., Wurker, M., Herholz, K., Ghaemi, M., Pietrzyk, U., Kessler, J., and Heiss, W. (1995). Planum temporale and Brodmann's area 22. Magnetic resonance imaging and high-resolution positron emission tomography demonstrate functional left-right asymmetry. *Arch. Neurol.* **52**(9), 869–874.

Kraus, N., McGee, T., Carrell, T. D., and Sharma, A. (1995). Neurophysiologic bases of speech discrimination. *Ear Hear.* **16**, 19–37.

Kraus, N., McGee, T., Carrell, T. D., Zecker, S. G., Nicol, T. G., and Koch, D. B. (1996). Auditory neurophysiologic responses and discrimination deficits in children with learning problems. *Science* **273**, 971–973.

Kushch, A., Gross-Glenn, K., Jallad, B., Lubs, H., Rabin, M., Feldman, E., and Duara, R. (1993). Temporal lobe surface area measurements on MRI in normal and dyslexic readers. *Neuropsychologia* **31**, 811–821.

Larsen, J. P., Høien, T., Lundberg, I., and Ødegaard, H. (1990). MRI evaluation of the planum temporale in adolescents with developmental dyslexia. *Brain Lang.* **39**, 289–301.

Larsen, J. P., Høien, T., and Ødegaard, H. (1992). Magnetic resonance imaging of the corpus callosum in developmental dyslexia. *Cog. Neuropsychol.* **9**, 123–134.

Leonard, C. M., Voeller, K. K. S., Lombardino, L. J., Morris, M. K., Hynd, G. W., Alexander, A. W., Andersen, H. G., *et al.* (1993). Anomalous cerebral structure in dyslexia revealed with magnetic resonance imaging. *Arch. Neurol.* **50**, 461–469.

Leppänen, P. H. T., and Lyytinen, H. (1997). Auditory event-related potentials in the study of developmental language-related disorders. *Audiol. Neurootol.* **2**, 308–340.

Liberman, I. (1982). A language oriented view of reading and its disabilities. *In* "Progress in Learning Disabilities" (H. Myklebust, Ed.), Vol. 5. Grune & Stratton, New York.

Livingstone, M. S., Rosen, G. D., Drislane, F. W., and Galaburda, A. M. (1991). Physiological and anatomical evidence for a magnocellular defect in developmental dyslexia. *Proc. Natl. Acad. Sci. U.S.A.* **88**, 7643–7647.

Lovegrove, W. J., Bowling, A., Badcock, D., and Blackwood, M. (1980a). Specific reading disability: Differences in contrast sensitivity as a function of spatial frequency. *Science* **210**, 479–440.

Lovegrove, W. J., Heddle, M., and Slaghuis, W. (1980b). Reading disability: Spatial frequency deficits in visual information store. *Neuropsychologia* **18**, 111–115.

Lovegrove, W. J., Martin, F., Blackwood, M., Badcock, D., and Paxton, S. (1982). Contrast sensitivity functions and specific reading disability. *Neuropsychologia* **20**, 309–315.

Lovegrove, W. J., Garzia, R. P., and Nicholson, S. B. (1990). Experimental evidence for a transient system deficit in specific reading disability. *J. Optom. Assoc.* **61**, 137–146.

Lundberg, I., Frost, J., and Peterse, O. (1988). Effects of an extensive program for stimulating phonological awareness in preschool children. *J. Exp. Child Psychol.* **18**, 201–212.

Manis, F. R., McBride-Chang, C., Seidenberg, M. S., Keating, P., Doi, L. M., Munson, B., and Petersen, A. (1997). Are speech perception deficits associated with developmental dyslexia? *J. Exp. Child Psychol.* **66**, 211–235.

Markee, T. E., Brown, W. S., and Moore, L. H. (1996). Callosal function in dyslexia: Evoked potential interhemispheric transfer time and bilateral field advantage. *Dev. Neuropsychol.* **12**, 409–428.

Martin, F., and Lovegrove, W. J. (1984). The effect of size and luminance on contrast sensitivity differences between specifically reading disabled and normal children. *Neuropsychologia* **22**, 73–77.

Mody, M., Studdert-Kennedy, M., and Brady, S. (1997). Speech perception deficits in poor readers: Auditory processing or phonological coding? *J. Exp. Child Psychol.* **64**, 199–231.

Moore, L. H., Brown, W. S., Markee, T. E., Theberge, D. C., and Zvi, J. C. (1995). Bimanual coordination in dyslexic adults. *Neuropsychologia* **33**, 781–793.

Morgan, A. E., and Hynd, G. W. (1998). Dyslexia, neurolinguistic ability, and anatomical variation of the planum temporale. *Neuropsychol. Rev.* **8**(2), 79–93.

Morgan, W. P. (1896). A case of congenital word-blindness. *Br. Med. J.* **78**, 2–13.

Näätänen, R., Gaillard, A., and Mäntysalo, S. (1978). Early selective-attention effect on evoked potential reinterpreted. *Acta Psychol.* **42**, 313–329.

Neville, H. J., Coffey, S. A., Holcomb, P. J., and Tallal, P. (1993). The neurobiology of sensory and language processing in language-impaired children. *J. Cog. Neurosci.* **5**, 235–253.

Obrzut, J. E. (1988). Deficient lateralization in learning-disabled children: Developmental lag or abnormal cerebral organization? *In* "Brain Lateralization in Children: Developmental Implications" (D. L. Mofese and S. J. Segalowitz, Eds.), pp. 567–589. Guilford, New York.

Orton, S. T. (1937). "Reading, Writing and Speech Problems in Children." Norton, New York.

Paulesu, E., Frith, C. D., and Frackowiak, R. S. J. (1993). The neural correlates of the verbal component of working memory. *Nature* **362**, 342–344.

Paulesu, E., Frith, U., Snowling, M., Gallagher, A., Morton, J., Frackowiak, R. S., and Frith, C. D. (1996). Is developmental dyslexia a disconnection syndrome? Evidence from PET scanning. *Brain* **119**, 143–157.

Petersen, S. E., Fox, P. T., Posner, M. I., Mintun, M., and Raichle, M. E. (1989). Positron emission tomographic studies of the processing of single words. *J. Cog. Neurosci.* **1**, 153–170.

Petersen, S. E., Fox, P. T., Snyder, A. Z., and Raichle, M. E. (1990). Activation of extrastriate and frontal cortical areas by visual words and word-like stimuli. *Science* **249**, 1041–1044.

Plante, E., Swisher, L., Vance, R., and Rapcsak, S. (1991). MRI findings in boys with specific language impairment. *Brain Lang.* **41**, 52–66.

Price, C. J. (1998). The functional anatomy of language processing. *Trends Cog. Sci.* **2**, 281–288.

Price, C. J., and Friston, K. J. (1997). Cognitive conjuction: A new approach to brain activation experiments. *NeuroImage* **5**, 261–270.

Price, C. J., Wise, R. J. S., Watson, J. D. G., Patterson, K., Howard, D., and Frackowiak, R. S. J. (1994). Brain activity during reading. The effects of exposure duration and task. *Brain* **117,** 1255–1269.

Price, C. J., Wise, R. J., and Frackowiak, R. S. (1996a). Demonstrating the implicit processing of visually presented words and pseudowords. *Cereb. Cortex* **6**(1), 62–70.

Price, C. J., Wise, R. J. S., Warburton, E. A., Moore, C. J., Howard, D., Patterson, K., Frackowiak, R. S. J., and Friston, K. J. (1996b). Hearing and saying. The functional neuroanatomy of auditory word processing. *Brain* **119,** 919–931.

Price, C. J., Moore, C. J., Humphreys, G. W., and Wise, R. J. S. (1997a). Segregating semantic from phonological processes during reading. *J. Cog. Neurosci.* **9,** 727–733.

Price, C. J., Moore, C. J., and Friston, K. J. (1997b). Subtractions, conjunctions and interactions in experimental design of activation studies. *Hum. Brain Mapp.* **5,** 264–272.

Pugh, K. R., Shaywitz, B. A., Shaywitz, S. E., Constable, R. T., Skudlarski, P., Fulbright, R. K., Bronen, R. A., Shankweiler, D. P., Katz, L., Fletcher, J. M., and Gore, J. C. (1996). Cerebral organization of component processes in reading. *Brain* **119,** 1221–1238.

Rapin, I., and Allen, D. A. (1988). Syndromes in developmental dysphasia and adult aphasia. *In* "Language, Communication, and the Brain" (F. Plum, Ed.), pp. 57–75. Raven Press, New York.

Reed, M. A. (1989). Speech perception and the discrimination of brief auditory cues in reading disabled children. *J. Exp. Child Psychol.* **48,** 270–292.

Robichon, F., and Habib, M. (1998). Abnormal callosal morphology in male adult dyslexics: Relationships to handedness and phonological abilities. *Brain Lang.* **62,** 127–146.

Rumsey, J. M. (1996). Developmental dyslexia: Anatomic and functional neuroimaging. *Ment. Retard. Dev. Disabil. Res. Rev.* **2,** 28–38.

Rumsey, J. M., Dorwart, R., Vermess, M., Denckla, M. B., Kruesi, M. J. P., and Rapoport, J. L. (1986). Magnetic resonance imaging of brain anatomy in severe developmental dyslexia. *Arch. Neurol.* **43,** 1045–1046.

Rumsey, J. M., Berman, K. F., Denckla, M. B., Hamburger, S. D., Kruesi, M. J., and Weinberger, D. R. (1987). Regional cerebral blood flow in severe developmental dyslexia. *Arch. Neurol.* **44,** 1144–1150.

Rumsey, J. M., Andreason, P., Zametkin, A. J., Aquino, T., King, C., Hamburger, S. D., *et al.* (1992). Failure to activate the left temporoparietal cortex in dyslexia. An oxygen-15 positron emission tomography study. *Arch. Neurol.* **49,** 527–534.

Rumsey, J. M., Zametkin, A. J., Andreason, P., Hanahan, A. P., Hamburger, S. D., Aquino, T., *et al.* (1994a). Normal activation of frontotemporal language cortex in dyslexia as measured with oxygen-15 positron emission tomography. *Arch. Neurol.* **51,** 27–38.

Rumsey, J. M., Andreason, P., and Zametkin, A. J. (1994b). Right frontotemporal activation by tonal memory in dyslexia: An O-15 PET study. *Biol. Psychiatry* **36,** 171–180.

Rumsey, J. M., Casanova, M., Mannheim, G. B., Patronas, N., De Vaughn, N., Hamburger, S. D., and Aquino, T. (1996). Corpus callosum morphology, as measured with MRI, in dyslexic men. *Biol. Psychiatry* **39**(9), 769–775.

Rumsey, J. M., Donohue, B. C., Brady, D. R., Nace, K., Giedd, J. N., and Andreason, P. (1997). A magnetic resonance imaging study of planum temporale asymmetry in men with developmental dyslexia. *Arch. Neurol.* **54**(12), 1481–1489.

Salmelin, R., Service, E., Kiesilä, P., Uutela, K., and Salonen, O. (1996). Impaired visual word processing in dyslexia revealed with magnetoencephalography. *Ann. Neurol.* **40,** 157–162.

Schlaug, G., Jäncke, L., Huang, Y., Steiger, J. F., and Steinmetz, H. (1995). Increased corpus callosum size in musicians. *Neuropsychologia* **33,** 1047–1056.

Schulte-Körne, G., Deimel, W., Bartling, J., and Remschmidt, H. (1998). Auditory processing and dyslexia: Evidence for specific speech processing deficit. *NeuroReport* **9,** 337–340.

Schultz, R. T., Cho, N. K., Staib, L. H., Kier, L. E., Fletcher, J. M., Shaywitz, S. E., *et al.* (1994). Brain morphology in normal and dyslexic children: The influence of sex and age. *Ann. Neurol.* **35,** 732–742.

Sergent, J., Zuck, E., Levesque, M., and MacDonald, B. (1992). Positron emission tomography study of letter and object processing: Empirical findings and methodological considerations. *Cereb. Cortex* **2,** 68–80.

Shapleske, J., Rossell, S. L., Woodruff, P. W., and David, A. S. (1999). The planum temporale: A systematic, quantitative review of its structural, functional and clinical significance. *Brain Res. Brain Res. Rev.* **29**(1), 26–49.

Shaywitz, S. E., Shaywitz, B. A., Pugh, K. R., Fulbright, R. K., Constable, R. T., Mencl, W. E., Shankweiler, D. P., Liberman, A. M., Skudlarski, P., Fletcher, J. M., Katz, L., Marchione, K. E., Lacadie, C., Gatenby, C., Gore, J. C., (1998). Functional disruption in the organization of the brain for reading in dyslexia. *Proc. Natl. Acad. Sci. U.S.A.* **95**(5), 2636–2641.

Stein, J. (1997). To see but not to read: The magnocellular theory of dyslexia. *Trends Neurosci.* **20**(4), 147–152.

Steinmetz, H. (1996). Structure, functional and cerebral asymmetry: *In vivo* morphometry of the planum temporale. *Neurosci. Biobehav. Rev.* **20**(4), 587–591.

Stelmack, R. M., Saxe, B. J., Noldy-Cullum, N., Campbell, K. B., and Armitage, R. (1988). Recognition memory for words and event-related potentials: A comparison of normal and disabled readers. *J. Clin. Exp. Neuropsychol.* **10,** 185–200.

Talairach, J., and Tournoux, P. (1988). Co-planar stereotaxic atlas of the human brain. Thieme Medical Publications, New York.

Tallal, P. (1980). Auditory temporal perception, phonics, and reading disabilities in children. *Brain Lang.* **9,** 182–198.

Tallal, P., and Piercy, M. (1973). Defects of non-verbal auditory perception in children with developmental aphasia. *Nature* **241,** 468–469.

Tallal, P., Miller, S. L., Bedi, G., Byma, G., Wang, X., Nagarajan, S. S., Schreiner, C., Jenkins, W. M., and Merzenich, M. M. (1996). Language comprehension in language-learning impaired children improved with acoustically modified speech. *Science* **271,** 81–83.

Tallal, P., Miller, S., Jenkins, B., and Merzenich, M. (1997). The role of temporal processing in developmental language-based learning disorders: Research and clinical implications. *In* "Foundations of Reading Acquisition," pp. 343–356. Erlbaum, New York.

Taylor, M. J. (1995). The role of event-related potentials in the study of normal and abnormal cognitive development. *In* "Handbook of Neuropsychology" (F. Boller and J. Grafman, Eds.), Vol. 10, pp. 187–211. Elsevier, Amsterdam.

Taylor, M. J., and Keenan, N. K. (1990). Event-related potentials to visual and language stimuli in normal and dyslexic children. *Psychophysiology* **27,** 318–327.

Vaidya, C. J., Austin, G., Kirkorian, G., Ridlehuber, H. W., Desmond, J. E., Glover, G. H., and Gabrieli, J. D. (1998). Selective effects of methylphenidate in attention deficit hyperactivity disorder: A functional magnetic resonance study. *Proc. Natl. Acad. Sci. U.S.A.* **95,** 14494–14499.

Valdois, S., Gérard, C., Vanauld, P., and Dugas, M. (1995). Developmental dyslexia: A visual attentional account? *Cog. Neuropsychol.* **12,** 31–67.

Weintraub, S., and Mesulam, M.-M. (1983). Developmental learning disabilities of the right hemisphere. Emotional, interpersonal and cognitive components. *Arch. Neurol.* **40,** 463–468.

Werker, J. F., and Tees, R. C. (1987). Speech perception in severely disabled and average reading children. *Can. J. Psychol.* **41**(1), 48–61.

World Health Organization. (1993). "The International Classification of Diseases: Classification of Mental and Behavioural Disorders," Vol 10. WHO, Geneva, Switzerland.

Zatorre, R. J., Evans, A. C., Meyer, E., and Gjedde, A. (1992). Lateralization of phonetic and pitch discrimination in speech processing. *Science* **256,** 846–849.

IV

Psychiatric Disorders

19

Depression

Helen S. Mayberg

Rotman Research Institute, Departments of Psychiatry and Medicine (Neurology),
University of Toronto, Toronto, Ontario, Canada M6A 2E1

I. Introduction

II. Depression in Neurological Disease

III. Idiopathic Depression

IV. Parallel Studies of Normal Sadness

V. Working Model of Depression

References

I. Introduction

A. Diagnosis, Demographics, and Epidemiology

Disturbances of mood and affect are among the most prevalent of all behavioral disorders. Depressive symptoms are especially common, occurring as a clinical feature of various neurological, psychiatric, and medical illnesses and also as a normal response to external events or personal loss. The more specific diagnosis of a major depressive episode, however, whether idiopathic or occurring as a part of a defined neurological disorder, is based not only on the presence of a persistent negative mood state but on associated disturbances in attention, motivation, motor and mental speed, sleep, appetite, libido, as well as anhedonia, excessive or inappropriate guilt, recurrent thoughts of death with suicidal ideations and, in some cases, suicide attempts (American Psychiatric Association, 1994).

Depression has an average lifetime prevalence of about 15%, with a two-fold greater prevalence in women than men (Kaplan *et al.*, 1994). A biological etiology is suggested by family, adoption, and twin studies, where a high degree of heritability is reported (Golden and Gershon, 1988). Major depressive disorder generally begins after the age of 20 and before age 50. A later age of onset is associated with a higher incidence of structural brain lesions, including strokes and subcortical and periventricular white matter changes (Zubenko *et al.*, 1990; Coffey *et al.*, 1993; Greenwald *et al.*, 1996). Biological mechanisms for the increased vulnerability of women or the relative constancy in the age of onset are unknown. While single episodes are not rare, depression is generally considered a chronic, relapsing, and remitted illness. Periods of clinical normality are seen throughout the natural course of the disorder. However, recurrences occur with higher frequency and with greater intensity if episodes are not treated (Keller *et al.*, 1983; Maj *et al.*, 1992).

B. Foundation for Current Theoretical Models

A meaningful biological theory of depression must accommodate definitions of the general systems necessary for the processing and experience of emotions. To this end, most investigators agree that emotion can be defined as the specific autonomic, endocrine, and cognitive states that couple the perception of stimuli to an adaptive behavioral response (Darwin, 1872; James, 1884; Cannon, 1929; Schacter and Singer, 1962; Zajonc, 1980; Rolls, 1990; Lazarus, 1991; Izard, 1993; Lang *et al.,* 1993; Heilman, 1997). Fear, for example, would be defined as increased sympathetic activity, elevated glucocorticoids, and motor freezing seen in response to a threatening stimulus; sadness, the emotion most closely associated with depression, as the state of disrupted sleep, appetite, and libido, and withdrawal seen in response to loss.

Given these conceptualizations, disruption of pathways mediating normal emotional responses can be seen as the likely pathological substrates for psychiatric syndromes such as anxiety, panic, post-traumatic stress, and depression, where sustained changes in mood and affective state are coupled with disturbances in autonomic and motor behaviors as well as impairments in attention and cognitive performance (Livingston and Escobar, 1973; Mayberg, 1997). Using this model, one might further hypothesize that healthy emotional states arise from stereotypical stimuli and produce expected responses, whereas disease states involve the exaggeration or uncoupling of these stimulus-response events, manifesting as initiation or persistence of an emotional state in the absence of a provoking stimulus. As such, the delineation of normal pathways mediating the various components of the emotional experience becomes critical to the understanding of emotional disorders. Alternatively, the view that disease states and normal states are fundamentally non-overlapping can also be tested by the use of parallel studies in normal and abnormal subject groups. Functional and structural neuroimaging studies have taken on a unique role in testing these hypotheses.

A critical but non-exclusive role for limbic structures in the regulation of mood and emotional states is now considered a virtual axiom (Mesulam, 1985; Damasio, 1994; LeDoux, 1996a). To this end, these regions are considered central to the integration of exteroceptive and interoceptive inputs required for widespread motor, cognitive, and autonomic processes. Comparative cytoarchitectural, connectivity, and neurochemical studies have delineated reciprocal pathways linking various "limbic" structures with widely distributed brainstem, striatal, paralimbic, and neocortical sites (Nauta, 1986; Vogt and Pandya, 1987; Alexander *et al.,* 1990; Mesulam and Mufson, 1992; Morecraft *et al.,* 1993; Carmichael and Price, 1995, 1996; among others). Associations of specific regions and pathways to various aspects of motivational, affective, and emotional behaviors in animals have also been described. (MacLean, 1990; LeDoux, 1996b; Rolls, 1996; Tremblay and Schultz, 1999).

While definitive mechanisms for depression have yet to be identified, theories implicating focal lesions (Robinson, 1979, 1998; Starkstein and Robinson, 1993), specific neurochemical and neuropeptide systems (Schildkraut, 1965; Fibiger, 1984; Swerdlow and Koob, 1987), and selective dysfunction of known neural pathways (Drevets *et al.,* 1992; Mayberg, 1994b, 1997) have all been proposed, supported by a growing number of clinical and basic studies (Caldecott-Hazard *et al.,* 1988, 1991; Benca *et al.,* 1992; Nemeroff *et al.,* 1992; Overstreet, 1993; Petty *et al.,* 1997). These findings are complemented by parallel experiments of specific cognitive, motor, circadian, and affective behaviors mapped in healthy volunteers, which together suggest that depression is a systems-level disorder, affecting discrete but functionally linked pathways involving specific cortical, subcortical, and limbic sites and their associated neurotransmitter and peptide mediators. It is further postulated that depression is not simply dysfunction of individual regions or pathways but is failure of the coordinated interactions between them (Mayberg, 1997). More localized dysfunction of individual components might additionally explain comparable changes in motor, cognitive, and circadian behaviors seen in disorders where disturbances in mood are absent (Rogers *et al.,* 1987; Marin, 1990).

II. Depression in Neurological Disease

A. Lesion-Deficit Studies

Modern theories regarding the neuro-localization of depressive illness have evolved from several complementary directions. Kleist's early observations of mood and emotional sensations following direct stimulation of the ventral frontal lobes (Brodmann areas 47 and 11) focused attention on paralimbic brain regions (Kleist, 1937). Studies by Broca (1878), and later, Papez (1937), Yakovlev (1948), and MacLean (1990), elaborated many of the anatomical details of these cytoarchitecturally primitive regions of cortex as well as adjacent limbic structures, including the cingulate, amygdala, hippocampus, and hypothalamus. These studies were among the first to suggest the role of these regions in reward, motivation, and affective behaviors. Additional clinical observations in depressed patients have similarly identified a prominent role for the frontal and temporal lobes and the striatum in the expression and modulation of mood and affect (Bear, 1983; Damasio and Van Hoesen, 1983; Robinson *et al.,* 1984; Mesulam, 1985; Stuss and Benson, 1986).

Lesion-deficit studies of neurologically depressed patients have generally focused on three categories of disorders: (a) discrete lesions, as seen with trauma, ablative surgery, stroke, tumors, and focal seizures (reviewed in Starkstein and Robinson, 1993); (b) conditions where neurochemical or neurodegenerative changes are known, such as with Parkinson's disease, Huntington's disease, progressive supranuclear palsy, Fahr's disease, Wilson's disease, and carbon monoxide poisoning (Mayeux, 1983; Cummings, 1992; Mayberg and Solomon, 1995); and (c) diseases with generalized or randomly distributed pathologies, such as Alzheimer's disease, multiple sclerosis, and system illness with central nervous system involvement (Goodstein and Ferrel, 1977; Honer *et al.,* 1987; Zubenko and Moossy, 1988; Nemeroff, 1989; Cummings and Victoroff, 1990; Hirono *et al.,* 1998). These studies have consistently implicated frontal and temporal cortex and the striatum. The role of limbic regions is less clear, despite clear evidence that limbic structures are fundamentally involved in critical aspects of motivational, affective, and emotional behaviors (Kleist, 1937; Papez, 1937; Mesulam, 1985; Rolls, 1985; MacLean, 1990; Damasio, 1994; LeDoux, 1996a).

CT and MRI studies in stroke patients have demonstrated a high association of mood changes with infarctions of the frontal lobe and basal ganglia, particular those involving the head of the caudate (Robinson *et al.,* 1984; Starkstein *et al.,* 1987; Mendez *et al.,* 1989; Starkstein and Robinson, 1993; Robinson, 1998). Studies of trauma and tumor patients additionally suggest that dorsolateral rather than ventral frontal lesions are more commonly associated with depression and depressive-like symptoms such as apathy and psychomotor slowing. More precise localization has been hampered by the heterogeneity of these lesions (Blumer and Benson, 1975; Damasio and Van Hoesen, 1983; Stuss and Benson, 1986). Depression is also associated with lateralized temporal lobe seizures. However, anatomical studies have yet to define the critical sites within the temporal lobe most closely associated with these mood changes (Bear and Fedio, 1977; Altshuler *et al.,* 1990). Such is also the case in multiple sclerosis, where studies of plaque loci suggest an association of depression with lesions in the temporal lobes (Honer *et al.,* 1987).

There is also no clear consensus as to whether the left or right hemisphere is dominant in the expression of depressive symptoms. Reports of patients with traumatic frontal lobe injury indicate a high correlation between affective disturbances and right hemisphere pathology (Grafman *et al.,* 1986). Studies in stroke, on the other hand, suggest that left-sided lesions of both frontal cortex and the basal ganglia are more likely to result in depressive symptoms than right lesions, where displays of euphoria or indifference predominate, although there is still debate (Gainotti, 1972; Ross and Rush, 1981; Sinyor *et al.,* 1986; Starkstein *et*

al., 1987; Mendez *et al.,* 1989). Further evidence supporting the lateralization of emotional behaviors is provided in studies of pathological laughing and crying. Crying is more common with left hemisphere lesions, whereas laughter is seen in patients with right lesions (Sackeim *et al.,* 1982), consistent with reports of post-stroke mood changes. Lateralization of mood symptoms has also been examined in patients with temporal lobe epilepsy, although again, there is no consensus, as affective disorders have been described with left, right, and non-lateralized foci (Flor-Henry, 1969; Mendez *et al.,* 1986; Robertson *et al.,* 1987; Altshuler *et al.,* 1990).

B. Surgical Ablation Therapy

Supportive evidence for a neuro-localization of mood is additionally provided by observations of patients undergoing ablative surgery to alleviate refractory melancholia and unremitting emotional ruminations (Fulton, 1951; Livingston and Escobar, 1973). The mechanisms by which these destructive "lesions" improve mood are unknown as is the precise lesion site necessary for amelioration of depressive symptoms. Improved mood is, however, seen in many of the most severely ill patients following an anterior leukotomy or either a subcallosal or superior cingulotomy procedure (Cosgrove and Rauch, 1995; Malizia, 1997). These seemingly paradoxical effects suggest a more complicated interaction between cortical, subcortical, and limbic pathways in both normal and abnormal emotional processing than the lesion-deficit literature would intimate. Nonetheless, the overall regional convergence of published observations provides the necessary foundation for hypothesis-driven imaging studies of depressive disorders using functional techniques such as positron emission tomography (PET), single photon emission computed tomography (SPECT), and functional magnetic resonance imaging (fMRI).

C. Functional Imaging

Despite the many similarities among different neurological conditions, there is still significant variability in the location of identified lesions associated with acquired depression. This variability is due, in part, to the methodological and theoretical limitations of anatomical imaging techniques, which restrict lesion identification to those brain areas that are structurally damaged. Functional imaging (PET, SPECT, and fMRI in some cases) can complement structural imaging in that the consequences of anatomic or chemical lesions on global and regional brain function (metabolism, blood flow, transmitter) can also be assessed

(Cherry and Phelps, 1996). This approach has been an important tool for identifying previously unrecognized brain abnormalities and potential disease mechanisms. These methods additionally provide alternative strategies to test how similar mood symptoms occur with anatomically or neurochemically distinct disease states, or conversely, why comparable lesions do not always result in comparable behavioral phenomena. Parallel studies of primary affective disorder and patients with neurological depressions provide complementary perspectives.

One advantage of this approach is the accessibility of disease-specific controls—such as non-depressed patients with matched demographic and neurological characteristics. Furthermore, this design allows comparisons both within and between different disease groups. Methodological and interpretative considerations, however, have generally limited these studies to diseases where the primary pathology spares the frontal cortex, the region repeatedly implicated in the lesion-deficit literature.

Basal ganglia disorders such as Parkinson's disease, Huntington's disease, and caudate strokes have a particularly high incidence of affective disorder and, additionally, accommodate the stated methodological constraints. These diseases, therefore, have been the major focus of PET studies of this type. The overlap of clinical signs and symptoms between basal ganglia disease and primary affective disorder patients provides another important neurobiological clue. Motor and cognitive slowing, observed in many patients with unipolar depression, is common in certain basal ganglia disorders, even when depressive symptoms are absent (Starkstein et al., 1990a). In many cases, disease-specific motor and cognitive features may obscure recognition of concurrent mood changes when they are present. On the other hand, it is these clinical similarities among neurological and non-neurological depressions that suggest common mechanisms.

Mayberg, Starkstein, and colleagues, in a series of studies (1990, 1991c, 1992), focused on neurological diseases where functional abnormalities would not be confounded by gross cortical lesions. This approach allowed functional confirmation of lesion-deficit observations as well as characterization of functional changes remote from the site of primary injury or degeneration (Baron, 1989). As such, studies were restricted to those disorders with known or identifiable neurochemical, neurodegenerative, or focal changes and where the primary pathology spared frontal cortex (the region repeatedly implicated in the lesion-deficit literature). Parkinson's disease, Huntington's disease, and lacunar strokes of the basal ganglia best fit these criteria. Not only did clinical signs and symptoms in these depressed patients mirror

those seen in idiopathic depression, but several plausible biochemical mechanisms for mood symptoms had already been postulated (Robinson, 1979; Mayeux et al., 1988; Peyser and Folstein, 1990). The additional observation that motor and cognitive features present in these patients often obscured recognition of mood symptoms further suggested testable anatomical hypotheses. These clinical findings in combination with published animal and human studies of regional connectivity (Goldman-Rakic and Selemon, 1984; Alexander et al., 1990) provided additional foundation to postulate that regional dysfunction of specific frontal-subcortical pathways would discriminate depressed from nondepressed patients, independent of the underlying neurological disorder.

1. Parkinson's Disease

Several mechanisms have been proposed for the depression seen with Parkinson's disease (PD) (reviewed in Cummings, 1992; Mayberg and Solomon, 1995). A serotonergic etiology is strongly supported by reduced spinal fluid serotonin and serotonin metabolites in depressed, but not non-depressed, PD patients (Mayeux et al., 1988). A dopaminergic etiology, with differential involvement of the mesolimbic and mesocortical dopamine system, has also been proposed (Fibiger, 1984; Cantello et al., 1989). This hypothesis is supported by Torack and Morris' observation (1988) of selective cell loss in the ventral tegmental area of PD patients with prominent mood and cognitive features. These findings suggest that PD patients with preferential degeneration of VTA neurons may be more likely to develop depression than patients without VTA involvement, although clinical studies repeatedly demonstrate little effect of dopamine replacement or agonist therapy on mood symptoms (Marsh and Markham, 1973).

Decreases in whole-brain cortical glucose metabolism are present in patients with PD compared with healthy volunteers (Mayberg et al., 1990). Comparison of depressed and non-depressed PD patients further demonstrates selective hypometabolism involving the caudate, orbital frontal, and inferior prefrontal cortex in the depressed group (Fig. 1). In addition, frontal metabolism is inversely correlated with the severity of depressive symptoms, a finding repeatedly observed in patients with primary depression (Ketter et al., 1996). While frontal changes distinguish depressed from nondepressed patients, mood and cognitive deficits are not easily dissociated, suggesting a more complicated relationship between regional hypometabolism, depression, and behavior in PD.

To further explore this issue, metabolic changes specific to improved mood, motor, and cognitive perfor-

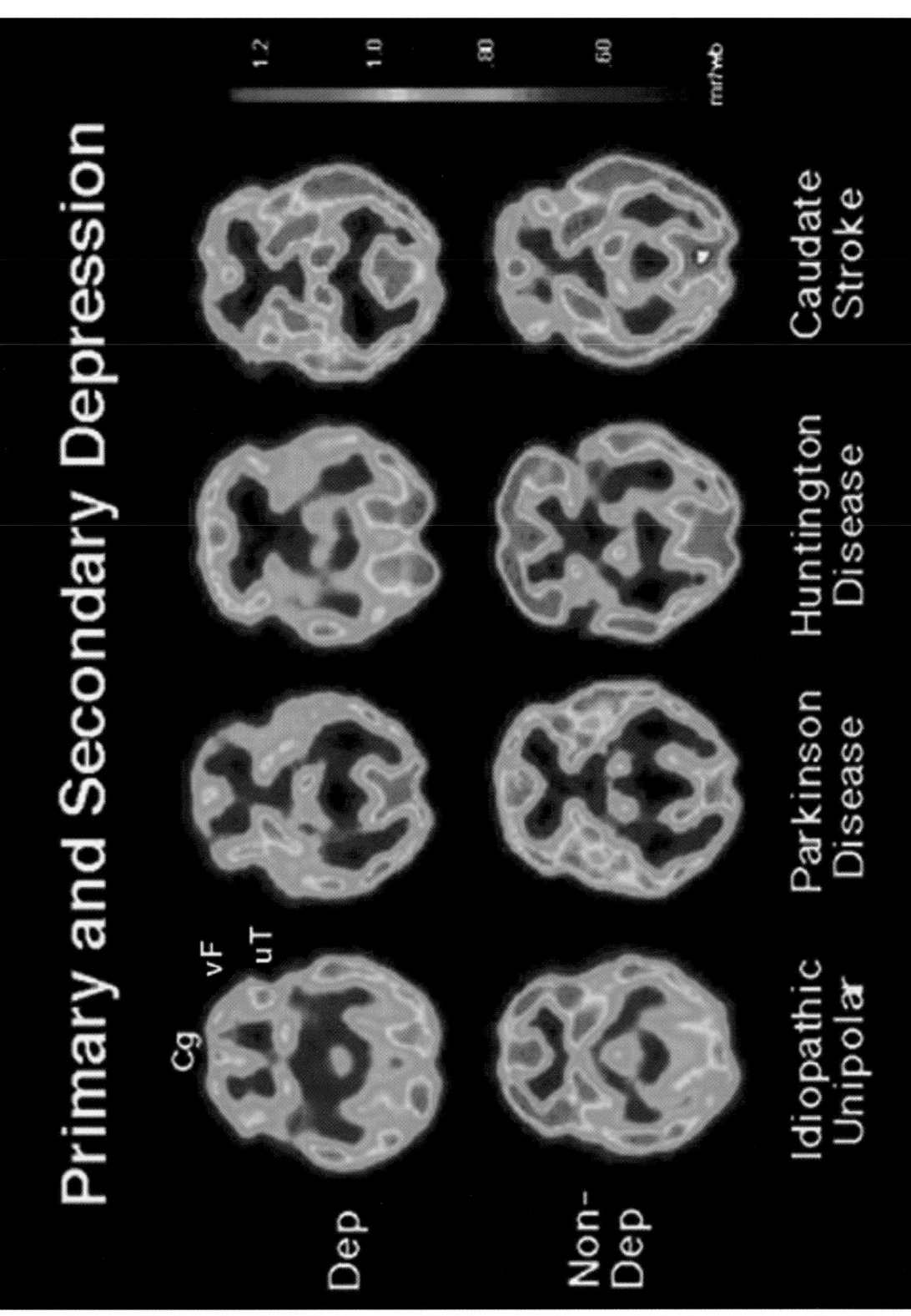

Figure 1 FDG PET in primary and secondary depression. Para-limbic hypometabolism common to patients with unipolar depression and depression associated with three basal ganglia disorders. FDG PET studies of primary and secondary depressed patients identify bilateral ventral frontal (vF), anterior temporal (uT), and anterior cingulate (Cg) hypometabolism that characterizes the depressive syndrome, independent of underlying disease etiology. Images are individual patients. Adapted from Mayberg, 1994.

mance in depressed PD patients treated with fluoxetine were subsequently examined (Mayberg *et al.*, 1995c). Drug treatment improved mood and performance on frontal-mediated cognitive tests (e.g., Stroop, Trails B, and verbal fluency tests), with no significant change in memory, visual–spatial perception, or motor deficits. Bilateral increases in metabolism were seen with treatment in ventral frontal and dorsal prefrontal cortex. Dorsal and ventral prefrontal increases correlated significantly with mood improvement; frontal-mediated cognitive performance tracked only with dorsal frontal changes. Neither area correlated with memory, visual–spatial, or motor performance. These findings suggest segregation of mood and cognitive behaviors, with some, but not all behaviors affected by changes in synaptic serotonin, a hypothesis supported by treatment trials using serotonin precursors (Coppen *et al.*, 1972; Sano and Taniguchi, 1972; Mayeux *et al.*, 1988; McCance-Katz *et al.*, 1992).

2. Huntington's Disease

Like PD, depression is the most prevalent mood disorder seen in Huntington's disease (HD), affecting about half of all patients and often preceding the motor abnormalities, even in people who may not recognize that they are genetically at-risk. (Folstein *et al.*, 1983). Unlike PD, mania also occurs, and impulsivity and suicide are common. Although the gene for HD is now known, it is still unclear how this defect translates into progressive loss of cells in the caudate nucleus and putamen with the eventual development of chorea, depression, and dementia that characterizes the illness.

Functional imaging studies (SPECT or PET) readily identify basal ganglia dysfunction–hypometabolism and hypoperfusion of the caudate and putamen–in both symptomatic patients and genetically at-risk subjects prior to the emergence of symptoms (Grafton *et al.*, 1990). Analogous to the findings in PD, when depressed and non-depressed HD patients are compared, the non-depressed patients have relatively normal cortical metabolism (Mayberg *et al.*, 1992). Depressed HD subjects, however, show decreased metabolism in the paralimbic orbital frontal and inferior prefrontal cortex, similar to that observed in the depressed PD patients (Fig. 1). The relationship between this paralimbic frontal hypometabolism and loss of caudate cells in HD is unclear, although disruption of frontolimbic–basal ganglia–thalamic pathways has been proposed. Neurochemical mechanisms are more obscure (Peyser and Folstein, 1990).

3. Basal Ganglia Strokes

Much of what is known about regional localization of mood has emerged from lesion-deficit studies of pa-

tients with stroke. While left frontal lesions are the most common, post-stroke depressions occur in other lesion locations, including sites in the right hemisphere (reviewed in Robinson, 1998). These clinical observations suggest a role for both hemispheres in mood regulation.

Clinical signs and symptoms seen with strokes generally correlate with the site of direct brain injury. However, anatomically uninjured brain regions functionally connected to, but anatomically removed from, the stroke lesion may also be affected (Baron, 1989). This phenomena, "remote diaschisis," likely explains the occurrence of frontal lobe deficits in patients with subcortical strokes, for example. Using a combination of structural and functional imaging methods, one can ask what pattern of cortical or subcortical dysfunction is common in patients with similar clinical findings and different brain lesions or, alternatively, what is different about patients with seemingly similar lesions but with discordant clinical symptoms.

This strategy has been used to identify the pattern of cortical hypometabolism specific to patients with secondary mood changes following unilateral lacunar strokes involving the head of the caudate (Starkstein *et al.*, 1990b; Mayberg *et al.*, 1991c; Mayberg, 1994b). While precise localization of the anatomical lesion in these studies was limited by the resolution of the available CT images, the pattern of cortical metabolic changes nonetheless differentiated depressed from euthymic patients. Temporal rather than frontal lobe changes discriminated the two groups, with bilateral hypometabolism characterizing the depressed patients. In contrast to the findings in PD and HD, frontal metabolism did not identify the patients with mood changes, as both depressed and nondepressed stroke patients showed bilateral frontal decreases (Fig. 1). These remote effects in orbital inferior frontal cortex may be lesion specific, disrupting orbital frontal–striatal–thalamic circuits in all patient subgroups, including a group of similarly lesioned patients with secondary mania (Starkstein *et al.*, 1990b). Temporal lobe changes, however, appear to be mood-state specific, implicating selective disruption of basotemporal limbic pathways in the patients with mood changes (Fulton, 1951; Nauta, 1971, 1986).

4. Common Findings across Patient Groups

The repeated observation from independent studies of depression associated with both degenerative and focal basal ganglia disease is the common involvement of paralimbic regions (ventral frontal and temporal cortex), independent of primary disease diagnosis (Mayberg, 1994b). These findings have been replicated in Parkinson's disease (Jagust *et al.*, 1992; Ring *et al.*, 1994) as well as in patients with temporal lobe epilepsy (Bromfield *et al.*, 1992), subcortical strokes (Laplane *et al.*, 1989; Grasso *et al.*, 1994), and Alzheimer's disease (Hirono *et al.*, 1998). The regional localization of changes is similar to (but not identical with) that seen in primary depression (shown in the next section) and is consistent with two known pathways: the orbital frontal-striatal-thalamic circuit (Goldman-Rakic and Selemon, 1984; Albin *et al.*, 1989, 1990; Alexander *et al.*, 1990) and the basotemporal limbic circuit which links orbital frontal cortex and anterior temporal cortex via the uncinate fasciculus (Papez, 1937; MacLean, 1949; Fulton, 1951; Nauta, 1971, 1986; Porrino *et al.*, 1981) (Fig. 2). Disease-specific disruption of converging pathways to these regions (Simon *et al.*, 1979; Dray, 1981; Azmitia and Gannon, 1986) best explains the presence of similar depressive symptoms in patients with distinctly different disease pathologies.

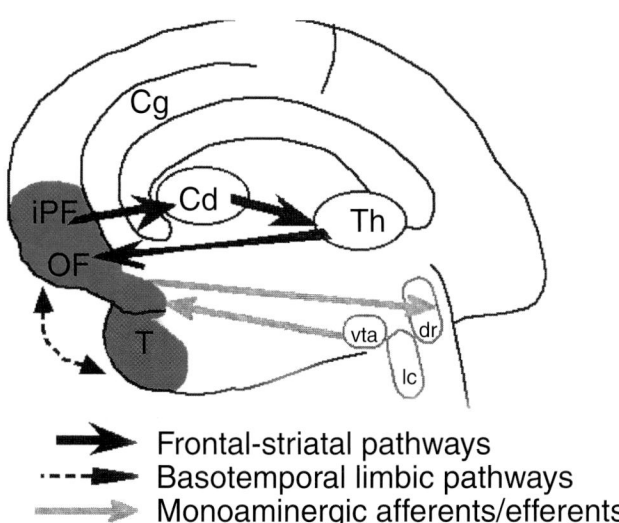

Figure 2 Depression in neurological disease. Postulated mechanisms for common paralimbic cortex hypometabolism (shown in blue) in secondary depression include anterograde or retrograde disruption of cortico-basal ganglia circuits from striatal degeneration or injury (solid black arrows); remote changes in basotemporal limbic regions (solid gray arrows); and degeneration of mesencephalic monoamine neurons (vta, dr, lc) and their cortical projections, with secondary involvement of serotonergic neurons, via disruption of orbital frontal outflow to the dorsal raphe (dashed arrows). Abbreviations: Cg, anterior cingulate; Cd, caudate; iPF, inferior prefrontal cortex; OF, orbital frontal cortex; T, temporal cortex; Th, thalamus; vta, ventral tegmental area; dr, dorsal raphe; lc, locus ceruleus. Adapted from Mayberg, 1994b.

III. Idiopathic Depression

A. Anatomical Studies

Anatomical studies of patients with primary affective disorders have been less consistent than those of depressed patients with neurological disorders (reviewed in Soars and Mann, 1997). Brain anatomy is grossly normal, and focal neocortical abnormalities have not been identified. Focal volume loss in subgenual medial orbital frontal cortex in both unipolar and bipolar depressed patients (Drevets *et al.*, 1997) has been identified, but not consistently (Botteron *et al.*, 1999). Reduced hippocampal and amygdala volumes have also been reported in patients with recurrent major depression (Sheline *et al.*, 1996, 1998), with a postulated mechanism of glucocorticoid neurotoxicity, consistent with both animal models (Sapolsky, 1994, 1996) and studies of patients with post-traumatic stress disorder (Bremner *et al.*, 1995). Nonspecific changes in ventricular size, and T_2-weighted MRI changes in subcortical gray and periventricular white matter have also been reported in some patient subgroups, most notably, elderly depressed patients (Zubenko *et al.*, 1990; Coffey *et al.*, 1993; Dupont *et al.*, 1995; Greenwald *et al.*, 1996; Hickie *et al.*, 1997; Steffens *et al.*, 1998). The parallels, if any, of these observations to the regional abnormalities described in lesion and neurological patients with depression are unclear. Studies of new-onset patients or preclinical at-risk subjects are needed to clarify if these changes reflect disease pathophysiology or are the consequence of chronic illness or treatment.

B. Resting-State Functional Imaging Studies

PET and SPECT studies in primary depression have, on the other hand, repeatedly reported frontal, cingulate, and less commonly temporal and parietal abnormalities, consistent with the general pattern seen in neurological depressions. The best replicated clinical studies of regional abnormalities in primary affective disorder have been those examining abnormal patterns of blood flow or metabolism under resting conditions, analogous to anatomical lesion-behavior correlations. To date, published studies have examined both young and old patients, drug-naive and medication-refractory disease, and a variety of patient subgroups. From a practical point of view, there is no evidence of physiological dissociation of flow from metabolism in depressed patients, although no explicit studies addressing this issue have been published. Similarly, differences in the sensitivity of PET versus SPECT in studies of comparable image quality have also not been directly examined.

Across studies, the most robust and consistent finding is decreased frontal lobe function (Buchsbaum *et al.*, 1986; Baxter *et al.*, 1989; Goodwin *et al.*, 1993; George *et al.*, 1994c; Lesser *et al.*, 1994; Mayberg *et al.*, 1994, 1997). The anatomical localization of frontal changes involves dorsolateral prefrontal cortex (Brodman areas 9, 10, and 46) as well as ventral prefrontal and orbital frontal cortex (Brodman areas 10, 11, and 47). Unlike the lesion-deficit literature, most of the studies report bilateral rather than left-lateralized abnormalities, although asymmetries have been reported. Of note, both right- and left-lateralized defects are seen in individual subjects, but to date, there are no behavioral correlates of this observation. In addition to frontal lobe changes, limbic (amygdala) (Drevets *et al.*, 1992; Hornig *et al.*, 1997), paralimbic (anterior temporal, cingulate) (Post *et al.*, 1987; Bench *et al.*, 1992; Wu *et al.*, 1992; Mayberg *et al.*, 1994; Bonne *et al.*, 1996; Ebert and Ebmeier, 1996; Drevets *et al.*, 1997), and subcortical (basal ganglia, thalamus) (Buchsbaum *et al.*, 1986; Drevets *et al.*, 1992) abnormalities have also been identified, but less consistently. Use of different analytic strategies (voxel-wise versus limited region-of-interest) likely accounts for some of these apparent inconsistencies (Mayberg *et al.*, 1994, 1997), with voxel-wise approaches generally identifying the non-frontal changes.

1. Sources of Variability

Despite the general consensus as to the regional localization of functional changes, there are some unresolved discrepancies, including contradictory reports as to whether depression is characterized by frontal and cingulate hypofunctioning (by example Baxter *et al.*, 1989; Mayberg *et al.*, 1994, 1997) or hyperfunctioning (Drevets *et al.*, 1992). Variability among experiments may be due in part to differences in scanner resolution and data analysis techniques (Bonne *et al.*, 1996). A more fundamental issue is how this variability reflects specific symptoms such as apathy, anxiety, psychomotor slowing, and executive cognitive dysfunction, present in varying combinations with dysphoric mood in individual depressed patients (Dolan *et al.*, 1992; Bench *et al.*, 1993a; Mayberg *et al.*, 1994; Osuch *et al.*, 1999). These relationships are less well studied.

Many PET and SPECT studies have demonstrated an inverse relationship between prefrontal activity and depression severity (reviewed by Ketter *et al.*, 1996), providing preliminary support for this argument. Significant correlations have also been shown for psychomotor speed (negative correlations with prefrontal and angular gyrus, Bench *et al.*, 1993a; negative correlation

with ventral frontal, Mayberg *et al.*, 1994), anxiety (positive correlation with inferior parietal lobule, Bench *et al.*, 1993a), and cognitive performance (positive correlation with medial frontal/cingulate, Dolan *et al.*, 1992; Bench *et al.*, 1993a). Complementary studies targeting these behaviors in normal subjects (Reivich *et al.*, 1983; Gottschalk *et al.*, 1991; Pardo *et al.*, 1991; Bench *et al.*, 1993a; George *et al.*, 1994b, 1997) as well as isolation of symptom-specific changes following treatment (Mayberg *et al.*, 1995a,b, 1999a) are necessary to clarify the many regional similarities among discordant behaviors seen with this correlational approach.

Other explanations for regional variability include medication status (drug naive vs drug washouts of varying duration), patient subgroups (familial vs depression spectrum), and transient fluctuations in mood at the time of the imaging study. Concerning medication status, several published studies (Bench *et al.*, 1992; Mayberg *et al.*, 1994) suggest that the clinical state of the patient at the time of the study (i.e., persistent clinical signs and symptoms of depression) drives the pattern of brain dysfunction, as no consistent difference in regional abnormalities could be discerned in acutely depressed patients on and off medication. The issue of patient subgroups is of additional clinical and diagnostic importance if depressed patients meeting different classification criteria actually have differing metabolic or flow imaging patterns, as suggested in some, but not all, studies. Most reports demonstrate comparable frontal hypometabolism in patients with depressions of varying types (unipolar depression, bipolar depression, and obsessive-compulsive disorder with depression) (Baxter *et al.*, 1989; Buchsbaum *et al.*, 1997). Hypermetabolism, on the other hand, has been demonstrated in patients with pure familiar unipolar depression—a finding not seen in depression spectrum disorder, where the more classical frontal hypometabolism is present (Drevets *et al.*, 1992). These findings require replication.

2. Cognitive Deficits in Depression

Cognitive deficits are a common feature of a major depressive episode. Attention, short-term memory, and psychomotor speed are the domains most affected (Hasher and Zacks, 1979; Weingartner *et al.*, 1981; Blaney, 1986; Calev *et al.*, 1986; Flint *et al.*, 1993; Brown *et al.*, 1994). Language, perception, and spatial abilities are generally preserved, although changes in these behaviors may be observed secondary to poor attention, motivation, or organizational abilities. Clinically significant anxiety, common in depression, may also impact cognitive efficiency (Rathus and Reber, 1994). Pseudodementia, also referred to as "depressive dementia," is encountered in a subset of depressed patients, primarily in the elderly (Stoudemire *et al.,* 1989; Emery and Oxman, 1992; Jones *et al.,* 1992).

The impaired performance of depressed patients on tasks testing these cognitive domains can be exploited using several functional strategies. As discussed in the previous section, co-variance analysis can be used to relate resting-state measures of flow or metabolism to performance on specific tests, measured separately (Dolan *et al.,* 1992; Bench *et al.,* 1993a; Mayberg, 1994). Mapping the task directly is an alternative approach, allowing direct comparisons of patients and healthy controls (Dolan *et al.,* 1993; George *et al.,* 1994a, 1997). With this type of design, one can quantify the neural correlates of the performance decrement as well as identify potential disease-specific sites of task reorganization. The advantage of this class of studies is that they can be performed with any of the available functional methods, including PET, fMRI, and ERP (Liotti *et al.,* 1997a, 1999).

Using this second strategy, George *et al.* (1994b, 1997) demonstrated blunting of an expected left anterior cingulate increase during performance of a Stroop task. A shift to the left dorsolateral prefrontal cortex, a region not normally recruited for this task in healthy subjects, was also observed. Elliott *et al.* (1997b), using the Tower of London test, described similar attenuation of an expected increase in dorsolateral prefrontal cortex and failure to activate anterior cingulate and caudate—regions recruited in controls. This group, in an additional set of experiments (Elliott *et al.,* 1997a,c), further demonstrated that, unlike healthy subjects, depressed patients also failed to activate the caudate in response to positive or negative feedback given while they performed this same task (e.g., "you were right" vs "you were wrong"). The fact that feedback valence influenced cognitive performance in normals and more dramatically in depressed individuals illustrates the highly interactive nature of mood and cognitive systems. These studies also underscore the critical importance of frontal–subcortical pathways in these behaviors.

C. Treatment Effects

1. Prognostic Markers

An untreated major depressive episode generally lasts 6–13 months, although treatment can significantly reduce this period (Meltzer, 1987; Bauer and Frazer, 1993). Antidepressants and interpersonal psychotherapy and cognitive behavioral therapy are generally effective in ameliorating depressive symptoms in patients with mild to moderate depression (Frank and Thase, 1999). Patients with a poor response to one antidepressant often respond well to another. Others will respond

to treatment augmentation strategies such as combinations of drugs with different pharmacological actions (Bauer and Frazer, 1993). Also highly effective is electroconvulsive therapy, although this approach is rarely used as a first-line therapy. The incidence of treatment resistance is reported to range anywhere from 20- to 40% (Keller *et al.,* 1983; Thase and Rush, 1995). Subtyping of patients for the purpose of treatment selection has been attempted, but at present there are no reliable stratification algorithms for guiding treatment. Furthermore, no clinical, neurochemical, or imaging markers can identify which patients will have a protracted disease course (Keller *et al.,* 1983; Coryell *et al.,* 1990; Maj *et al.* 1992), although this is a growing area of focused research.

Mayberg and colleagues (1997) reported that rostral anterior cingulate metabolism measured using PET uniquely predicts response to antidepressant medication. Hypermetabolism identified eventual treatment responders; hypometabolism characterized nonresponders (Fig. 3). Metabolism in no other region discriminated the two groups, nor did associated demographic, clinical, or behavioral measures. Wu and colleagues (Wu and Bunney, 1990; Wu *et al.,* 1992) described a similar hypermetabolic change in a nearby region of the dorsal anterior cingulate that predicts an antidepressant effect to one night of sleep deprivation. While replication of both findings is needed, metabolic signatures in individual patients may prove to be clinically useful in optimizing available treatment strategies or in identifying those patients at risk for a difficult disease course. The localization of these changes to rostral anterior cingulate is of particular significance as this region has unique reciprocal connections not only with dorsal anterior cingulate but also with dorsal neocortical (lateral prefrontal) and ventral paralimbic (insula, basal frontal) regions, previously discussed.

2. Symptom Remission

Recovery from depression is associated with normalization of certain of these regional abnormalities. Changes in cortical (prefrontal, ventral prefrontal, parietal), limbic—paralimbic (cingulate, amygdala, insula), and subcortical (caudate) regions have been described with different modes of antidepressant treatment, including drugs, psychotherapy, sleep deprivation, electroconvulsive therapy, and ablative surgery. Normalization of frontal and dorsal cingulate hypometabolism is the best replicated finding (Baxter *et al.,* 1989; Martinot *et al.,* 1990; Goodwin *et al.,* 1993; Bench *et al.,* 1995; Passero *et al.,* 1995; Ebert and Ebmeier, 1996; Buchsbaum *et al.,* 1997; Mayberg *et al.,* 1999a). Changes in associated limbic—paralimbic and subcortical regions are more variable (Drevets *et al.,* 1992; Nobler *et al.,* 1994; Bonne and Krausz, 1997; Malizia, 1997;

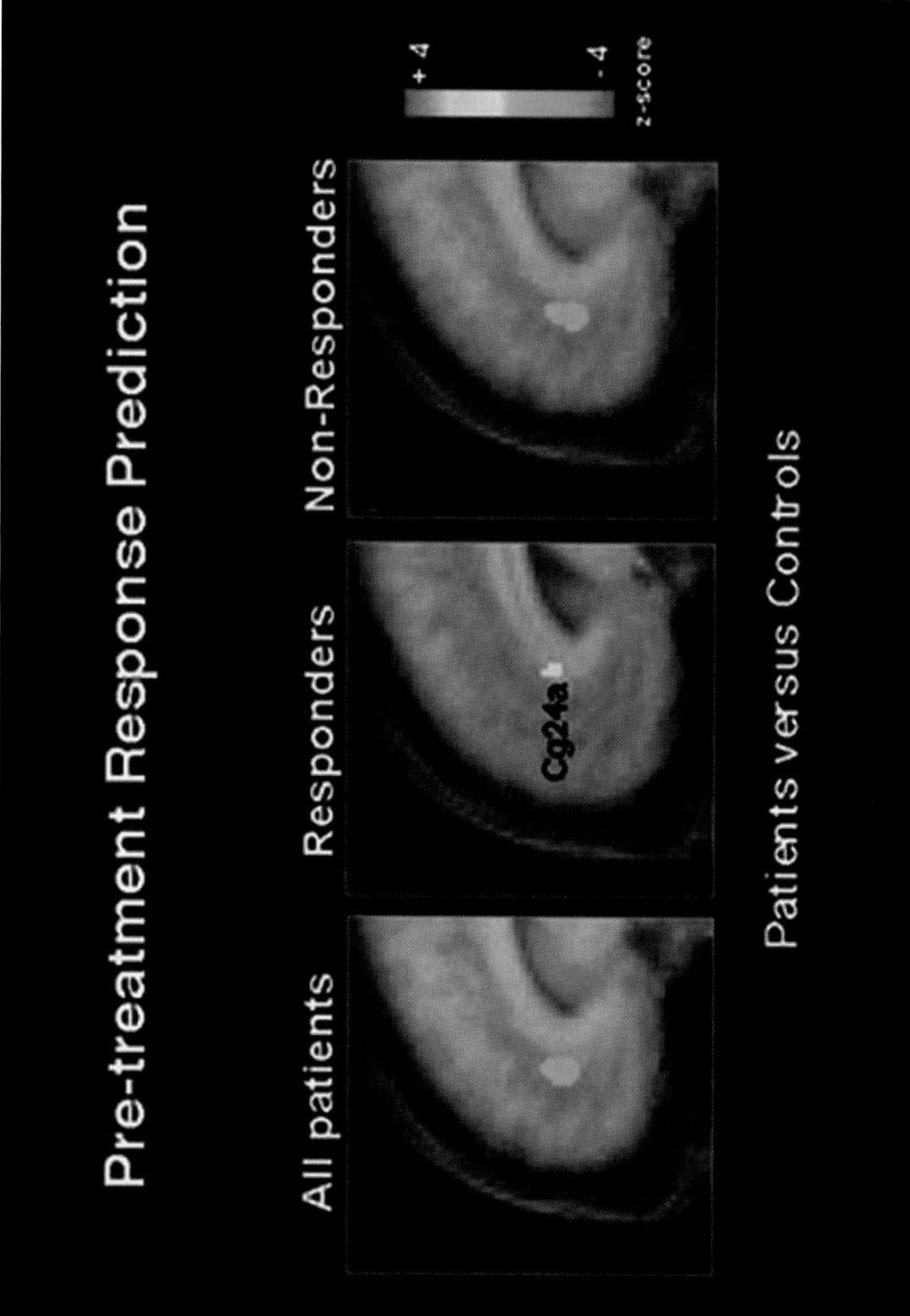

Figure 3 Value of rostral cingulate metabolism to predict eventual antidepressant response in unmedicated unipolar depressed patients. z-score maps demonstrating differences in direction, magnitude, and extent of changes in three depression groups compared to healthy controls. Rostral cingulate hypometabolism, relative to healthy controls (negative z-values, shown in green), characterizes nonresponders in contrast to hypermetabolism (positive z-values, shown in yellow) seen in treatment responders. Adapted from H. S. Mayberg, S. K. Brannan, R. K. Mahurin, P. A. Jerabek, J. S. Brickman, J. L. Tekell, J. A. Silva, S. McGinnis, T. G. Glass, C. C. Martin, and P. T. Fox (1997). Cingulate function in depression: A potential predictor of treatment response. *NeuroReport* **8**, 1057–1061. © Lippincott Williams & Wilkins.

Mayberg *et al.,* 1999a; Smith, G. S., *et al.,* 1999). A critical issue in sorting out contradictory results is to consider that drug-induced changes may be different in patients who respond compared to those that do not, as suggested by the pretreatment resting-state studies described above.

In support of this hypothesis, Mayberg and colleagues (Mayberg *et al.,* 1999b) demonstrated changes in dorsal cortical and ventral limbic-paralimbic regions in depressed patients treated for 6 weeks with fluoxetine (Fig. 4). Distinct patterns of change were seen at 1 week and 6 weeks of treatment, with the time course of metabolic changes reflecting the temporal delay in clinical response. Cortical increases (prefrontal and parietal) were seen in treatment responders and these increases were a normalization of the pre-treatment hypometabolic pattern. Ventral paralimbic areas, including subgenual cingulate, and hippocampus showed decreases, with resulting metabolism being less than that seen in healthy controls. Mood improvement was associated with both frontal increases and subgenual cingulate and hippocampal decreases; cognitive improvement only with cortical increases. The inverse pattern (cortical and posterior cingulate decreases; hippocampal increases, no change in cingulate) was seen in non-responders receiving identical treatment, and this nonresponse pattern was identical to metabolic changes seen in both groups with 1 week of treatment (Fig. 4).

These findings suggest not only an interesting relationship between limbic—paralimbic and neocortical systems and specific syndromal features but differences among patients in adaptation of target regions to chronic serotonergic modulation, a hypothesis supported by a growing literature targeting multiple neuroreceptor subtypes, second-messenger effects, and regionally specific regulatory mechanisms in both disease etiology and mechanisms of antidepressant action (Ballenger, 1988; Stancer and Cooke, 1988; Caldecott-Hazard *et al.,* 1991; Hyman and Nestler, 1996; Skolnick *et al.,* 1996; Vaidya *et al.,* 1997). However, specific neurochemical mechanisms that might account for these limbic, paralimbic, and neocortical metabolic changes remain speculative. Nonetheless, this growing body of data suggests that regional metabolic or blood flow pattern changes may serve as an important biomarker of clinical response in treatment trials of new medications with alternative mechanisms of action or purported faster onset of clinical effects.

D. Neurochemical Markers

Serotonergic and noradrenergic mechanisms have dominated the neurochemical literature on depression because most typical antidepressant drugs affect synaptic concentrations of these two transmitters (Bunney and Davis, 1965; Schildkraut, 1965; Vetulani and Sulser, 1975; Ballenger, 1988; Roy *et al.,* 1988a; Stancer and Cooke, 1988; Warsh *et al.,* 1988; Caldecott-Hazard *et al.,* 1991). Dietary restriction of tryptophan, resulting in an acute decrease in brain serotonin (the tryptophan depletion challenge), and catecholamines (the AMPT challenge) is associated with an abrupt relapse in remitted depressed patients (Delgado *et al.,* 1990) and changes in regional glucose metabolism and blood flow, further supporting a critical role for serotonin and norepinephrine in the regulation of depressive symptoms (Bremner *et al.,* 1997; K. A. Smith *et al.,* 1999). Changes in both serotonergic and noradrenergic metabolites have also been reported in subsets of depressed patients, but the relationship of these peripheral measures to changes in brain-stem nuclei or their cortical projections is unknown. Consistent with these findings, decreased serotonin transporter binding has been demonstrated in unipolar depressed patients using the SPECT ligand $[^{123}I]\beta$-CIT, an important observation directly implicating brainstem serotonergic dysfunction (Malison *et al.,* 1998). Postmortem studies of depressed suicide brains have additionally reported changes in serotonergic and noradrenergic receptors (Arango *et al.,* 1990). S_2-Serotonin receptor changes measured with PET have also been described in the temporal cortex of depressed stroke patients, with depressive symptoms negatively correlated with the magnitude of cortical receptor binding (Mayberg *et al.,* 1988; Mayberg *et al.,* 1991a). Studies of S_2 receptors in non-neurologically impaired depressed patients have also been reported but the findings are somewhat inconsistent. Massou and colleagues (1997) demonstrated upregulation of S_{2A}-receptor sites with antidepressant treatment, in line with the data in poststroke depression. In contrast, Meltzer *et al.* (1999) and Meyer *et al.* (1999) found no pretreatment abnormalities in S_2 receptors in late-life or mid-life depressed patients, respectively. Studies of other serotonin receptor subtypes are ongoing (Pike *et al.,* 1996; Drevets *et al.,* 1999).

While a primary dopaminergic mechanism for depression is generally considered unlikely, a role for dopamine in some aspects of the depressive syndrome is supported by several experimental observations (Fibiger, 1984; Rogers *et al.,* 1987; Cantello *et al.,* 1989; Zacharko and Anisman, 1991; Flint *et al.,* 1993). The mood-enhancing properties and clinical utility of methylphenidate in treating some depressed patients is well documented (Martin *et al.,* 1971), although dopaminergic stimulation alone does not generally alleviate all depressive symptoms. Dopaminergic pro-

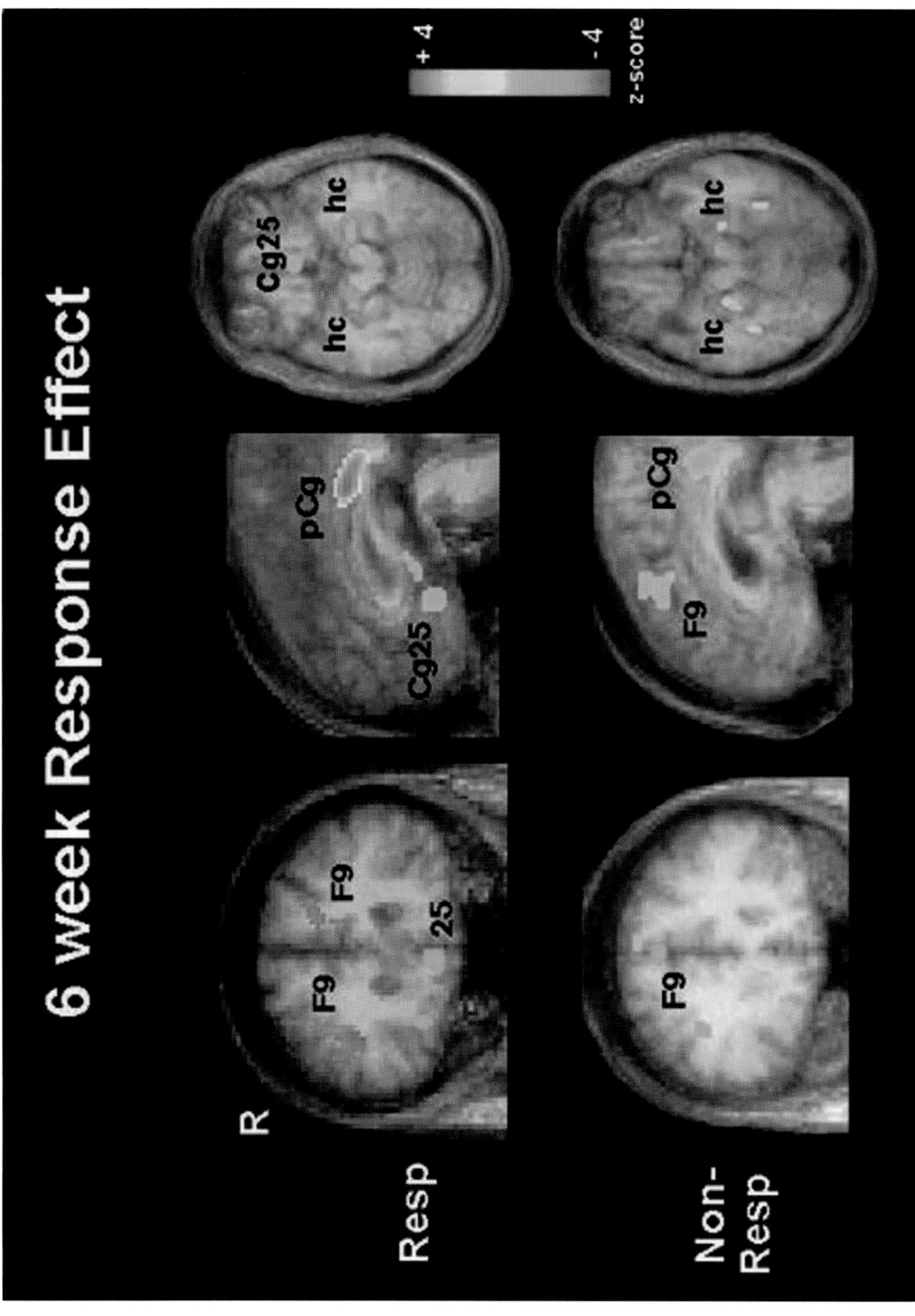

Figure 4 Common reciprocal changes in cortical and paralimbic function with shifts in mood state as measured with two different PET imaging techniques. Recovery from major depression is associated with increases in dorsal cortical (positive *z*-scores) and decreases in ventral paralimbic regional metabolism (negative *z*-scores). The reverse is seen with provocation of intense sadness in healthy volunteers, where dorsal decreases and ventral increases in blood flow accompany changes in mood state. F, frontal; ins, anterior insula; Cg25, subgenual cingulate; pCg31, posterior cingulate. Color scale: red, increases in flow or metabolism; green, decreases in flow or metabolism. Adapted from data in Mayberg *et al.*, 1999a.

jections from the ventral tegmental area (VTA) show regional specificity for the orbital/ventral prefrontal cortex, striatum, and anterior cingulate—areas repeatedly identified in functional imaging studies of primary and secondary depression (Simon *et al.*, 1979; Wise, 1980; Grabiel, 1990). Interestingly, a recent SPECT study has demonstrated D_2-receptor changes in the striatum and anterior cingulate with SSRI treatment (Larisch *et al.*, 1997). Despite the absence of pretreatment D_2-receptor abnormalities, these treatment effects suggest a potential role for serotonin-dopamine interactions in mechanisms of antidepressant action.

Studies of other transmitter and peptide systems, particularly those with known monoaminergic interactions, are the focus of increasing attention. Unfortunately, functional imaging ligands for many of the systems of interest either have not been tested (Janowski *et al.*, 1988; Petty *et al.*, 1992) or are not yet developed (Nemeroff *et al.*, 1984; Trullas and Skolnick, 1990; Duncan *et al.*, 1996; Hyman and Nestler, 1996; Nibuya *et al.*, 1996; Vaidya *et al.*, 1997). Increases in paralimbic mu-opiate receptors have been demonstrated with PET in refractory unipolar depressed patients (Mayberg *et al.*, 1991b), consistent with autoradiography studies in depressed suicide victims (Gross-Isseroff *et al.*, 1990) and regionally concordant with areas of hypoperfusion and hypometabolism seen in related studies (Mayberg, 1994). These increases are not seen in drug-naive patients, where mu binding is actually decreased compared to controls. Binding does, however, increase with treatment (Bencherif *et al.*, 1997). The full relationship of these findings to specific syndromal features awaits further investigation.

IV. Parallel Studies of Normal Sadness

Behaviorally, sadness is generally defined as the sustained state of withdrawal seen in response to loss. Observation of both animals and humans following the death of or separation from offspring confirms the universality of this state, with changes in body posture, disinterest in previously rewarding stimuli, and alterations in basic drive and circadian behaviors (i.e., feeding, reproduction, sleep, endocrine) (Harlow and Suomi, 1974; MacLean, 1990; Cowles, 1996; Levine *et al.*, 1997; Shively *et al.*, 1997) all readily apparent and commonly referred to as grief or bereavement (Zisook and DeVaul, 1985; Roy *et al.*, 1988b; Reynolds *et al.*, 1992; Zisook *et al.*, 1994). Despite the consistency of chronic behavioral changes, there are no good transient or short–term stimulus–response models to reliably study these phenom-

ena in animals (Willner, 1991; Thiebot *et al.*, 1992). People, on the other hand, can experience transient sadness in response to both personal internal cues and extrapersonal events. It is this capacity that has been exploited in functional imaging experiments of this emotion, to date.

A. Provocation Studies in Healthy Subjects

Methods to provoke sad mood are comparable to those used for other emotions, with recollection of past personal memories the most commonly employed tactic (Brewer *et al.*, 1980; Goodwin and Williams, 1982; Martin, 1990). Pardo *et al.* (1993) first described blood flow increases, using PET, in superior and inferior prefrontal cortex during spontaneous recollection of previous sad events. Gender differences were emphasized, as frontal effects were left lateralized in men and bilateral in women. This study was followed by George *et al.* (1995, 1996), who facilitated sad memory recollection in women with the simultaneous viewing of referential sad faces. Like Pardo, this paradigm identified bilateral superior frontal blood flow increases and, in addition, increases in anterior cingulate, basal ganglia, and thalamus. Findings in men neither replicated those of Pardo nor overlapped changes seen in the women.

Reiman *et al.* (1997) and Lane *et al.* (1997), in their combined studies of happiness, sadness, and disgust, reported similar increases in prefrontal cortex, basal ganglia, and thalamus using both self-generated autobiographical memories and viewing of sad film scenes. Changes not previously described by Pardo or George were seen in the hypothalamus and cerebellum. In contrast, Gemar *et al.* (1996), using a comparable autobiographical memory strategy, identified left frontal decreases, reminiscent of resting-state findings in clinically depressed patients (Baxter *et al.*, 1989). No significant increases were reported. The importance of the control task to these results was stressed; however, use of a neutral memory state was not unique to this study, having been employed by both George and Lane. The influence of gender is additionally unknown, as the experiment involved only men.

In each of these sad provocation studies (see also Partiot *et al.*, 1995; Schneider *et al.*, 1995; Baker *et al.*, 1997), mood-specific effects may have been confounded by changes due to the cognitive and memory strategies used to elicit the mood state. To address this, Mayberg *et al.* (1995b, 1999a) employed an autobiographical memory paradigm in which scans were acquired only after the sad state was reached and sustained with subjects no longer ruminating on the specifics of the personal situation used to provoke the mood. Ventral limbic and paralimbic increases (subgenual cingulate, anterior insula, and cerebellum) and neocortical decreases (right prefrontal, inferior parietal, and posterior cingulate) were seen with this strategy, replicating some but contradicting other findings previously reported. The anterior cingulate increases closely matched findings of George et al. (1995, 1996), Partiot et al. (1995), and Baker et al. (1997). In contrast, the right prefrontal decreases were the reverse of previous findings, perhaps due to scanning after rather than during active recollection of the sad memory (Pardo et al., 1993; George et al., 1995). In support of this assertion, these decreases also overlap areas shown to activate with sustained attentional tasks in the absence of mood manipulations (Posner and Petersen, 1990; Pardo et al., 1991; Corbetta et al., 1993; Paus et al., 1993)—behaviors commonly disturbed in depressed patients (Weingartner et al., 1981; Cohen et al., 1982; Roy-Bynre, 1986).

The critical importance of limbic and paralimbic regions in the mediation of negative emotions is further supported by a recent case of transient depression during deep-brain stimulation for treatment of intractable Parkinson's disease (Bejjani et al., 1999). Selective high-frequency stimulation to the left substantia nigra (2 mm below the subthalamic site that alleviated parkinsonian symptoms) provoked a reproducible and reversible depressive syndrome in a woman with no previous psychiatry history. Stimulation-induced mood changes were associated with focal blood flow increases in the left orbital frontal cortex, amygdala, globus pallidus, anterior thalamus, and right parietal cortex. This pattern of regional activation is similar to changes seen with memory-induced sadness in healthy volunteers, although there are clear differences requiring additional further studies. Nonetheless, this remarkable case provides important additional clues regarding regional circuits mediating normal and abnormal mood states and suggests a novel strategy for future research.

B. Similarities between Sadness and Depression

The localization of changes with transient normal sadness both matches resting-state abnormalities seen in depressed patients (Mayberg et al., 1997) and mirrors changes associated with remission of depressive symptoms (Mayberg et al., 1999a). More specifically, shifts in negative mood state in both patients and healthy volunteers involve a nearly identical set of ventral limbic-paralimbic (subgenual cingulate, anterior insula, cerebellum) and dorsal neocortical (prefrontal, parietal,

post-cingulate) regions (Fig. 5). Recovery from depression is associated with decreases in ventral paralimbic areas and increases in the dorsal neocortical regions. Transient sadness shows this identical pattern but in reverse—increases in ventral paralimbic regions and decreases in dorsal neocortex. The presence and maintenance of functional reciprocity between these regions with shifts in mood in either direction further suggest that these regional interactions are obligatory and likely mediate the well-recognized behavioral relationships between mood and cognition seen with both normal and pathological conditions (Liotti and Tucker, 1992; Ross et al., 1994; Barbas, 1995; Tucker et al., 1995; Heilman, 1997; Liotti et al., 1997a, 1999). The bidirectional nature of this limbic-cortical reciprocity provides additional evidence of potential mechanisms mediating cognitive ("top down"), pharmacological (mixed), and surgical ("bottom up") treatments of affective disorders.

V. Working Model of Depression

To facilitate the continued integration of clinical neuroimaging findings with complementary basic anatomical, chemical, and electrophysiological studies in the investigation of the pathogenesis of affective disorders, a working model of depression was recently proposed (Mayberg, 1997). This model (Fig. 6), implicating failure of the coordinated interactions of a distributed network of cortical-subcortical (cortical-limbic) pathways, is based on the convergence of findings from primary and neurological depressed patients. Brain regions with known anatomical interconnections that also show synchronized changes using PET in three behavioral states—normal transient sadness (controls), baseline depressed (patients), and posttreatment (patients)—are grouped into three compartments: dorsal, ventral, and rostral. The dorsal compartment (labeled attention-cognition) includes both neocortical and superior limbic elements and is postulated to mediate cognitive aspects of negative emotion such as apathy, psychomotor slowing, and impaired attention and executive function, based on complementary structural and functional lesion deficit correlational studies (Stuss and Benson, 1986; Bench et al., 1992; Dolan et al., 1993; Mayberg et al., 1994; Devinsky et al., 1995), symptom-specific treatment effects in depressed patients (Mayberg et al., 1999a,b), activation studies designed to explicitly map these behaviors in healthy volunteers (Pardo et al., 1991, 1993; George et al., 1995), and connectivity patterns in primates (Petrides and Pandya, 1984; Morecraft et al., 1993; Barbas, 1995).

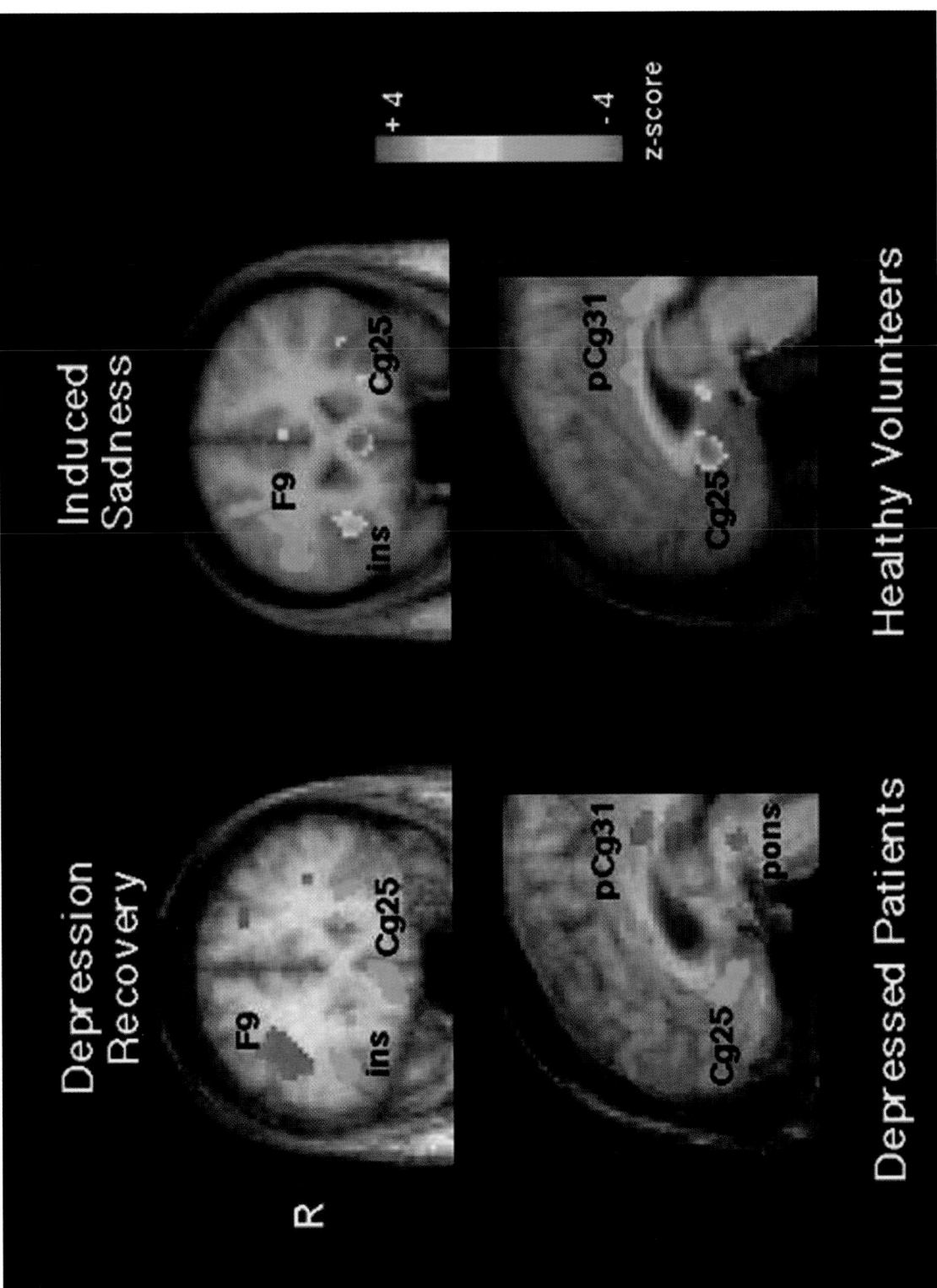

Figure 5 Response specific-effects of fluoxetine treatment in depression. Treatment response is associated with frontal and parietal cortex increases and subgenual cingulate, insula, posterior cingulate, and hippocampal decreases. Non-responders show metabolic changes in comparable brain regions, but in the opposite direction (frontal decreases, posterior cingulate and hippocampal increases; no change in subgenual cingulate). Comparisons are of before and after 6 weeks of treatment. Increases in metabolism with treatment are shown as positive z-scores, in red; decreases in metabolism as negative z-scores, in green. F9, prefrontal; Cg25, subgenual cingulate; pCg, posterior cingulate; hc, hippocampus. Adapted by permission of Elsevier Science from Early and late fluoxetine effects on regional glucose metabolism in depression. by H. S. Mayberg *et al. Biological Psychiatry,* **45**(S8), 111S, #357. Copyright 1999 by the Society of Biological Psychiatry.

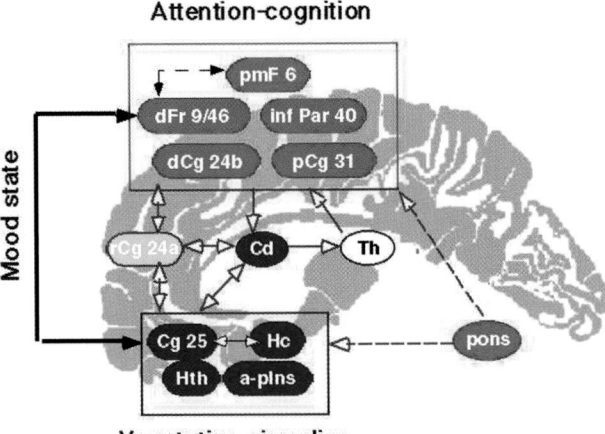

Figure 6 Depression model. Regions with known anatomical interconnections that also show synchronized changes using PET in three behavioral states—normal transient sadness (controls), baseline depressed (patients), and posttreatment (patients)—are grouped into three main compartments, dorsal (red box), ventral (blue box), and rostral (in yellow). The dorsal–ventral segregation additionally identifies those brain regions where an inverse relationship is seen across the different PET paradigms. Sadness and depressive illness are both associated with decreases in dorsal limbic and neocortical regions (red areas) and relative increases in ventral paralimbic areas (blue areas). The model, in turn, proposes that illness remission occurs when there is inhibition of the overactive ventral regions and activation of the previously hypofunctioning dorsal areas (solid black arrows), an effective facilitated by fluoxetine action in dorsal raphe and its projection sites (dotted lines). Integrity of the rostral cingulate (yellow) with its direct anatomical connections to both the dorsal and ventral compartments is postulated to be additionally required for the occurrence of these adaptive changes, as pretreatment metabolism in this region uniquely predicts antidepressant treatment response. Short white arrows identify segregated ventral and dorsal compartment afferents and efferents to and from the striatum and thalamus, although individual cortico-striatal-thalamic pathways are not delineated. Abbreviations: Red box, dFr, dorsolateral prefrontal; inf Par, inferior parietal; dCg, dorsal anterior cingulate; pCg, posterior cingulate; pmF, premotor. Blue box, Cg25, subgenual cingulate; a–pIns, anterior and posterior insula; Hc, hippocampus; Hth, hypothalamus. Yellow, rCg24a, rostral anterior cingulate; Cd, caudate; Th, thalamus. Numbers are Brodmann designations. Adapted from Mayberg (1997) and Mayberg *et al.* (1999a).

The ventral compartment (labeled vegetative–circadian) is composed of limbic, paralimbic, and subcortical regions known to mediate circadian and vegetative aspects of depression including sleep, appetite, libidinal, and endocrine disturbances, based on clinical and related animal studies (MacLean, 1949; Neafsey, 1990; Mesulam and Mufson, 1992; Augustine, 1996). The dorsal–ventral segregation additionally identifies those brain regions where an inverse relationship has been demonstrated in converging PET experiments as described in the previous section (see also Pandya and Kuypers, 1969; Chavis and Pandya, 1976; Petrides and Pandya, 1984; Pandya and Yeterian, 1996).

The rostral cingulate (rCg24a) is isolated from both the ventral and dorsal compartments based on its cytoarchitectural characteristics and reciprocal connections to both dorsal and ventral anterior cingulate (Nauta, 1971, 1986; Baleydier and Mauguiere, 1980; Vogt and Pandya, 1987; Morecraft *et al.*, 1993; Van Hoesen *et al.*, 1993; Carmichael and Price, 1994, 1995, 1996; Kunishio and Haber, 1994; Vogt *et al.*, 1995). Contributing to this position in the model are the additional observations that metabolism in this region uniquely predicts antidepressant response in acutely depressed patients (Mayberg *et al.*, 1997) and more recent evidence that this region is a principal site of abberent response during mood induction in remitted depressed patients (Liotti *et al.*, 1997b; Mayberg *et al.*, 1998). These anatomical and clinical distinctions suggest that the rostral anterior cingulate may serve an important regulatory role in the overall network by facilitating the interactions between the dorsal and ventral compartments (Livingston and Escobar, 1973; Crino, 1993). As such, dysfunction in this area could have significant impact on remote brain regions regulating a variety of behaviors, including the interaction between mood, cognitive, somatic, and circadian responses that characterize an emotional state. Additionally, this model proposes that this distributed cortical-limbic network is fundamental to the mediation of mood-cognitive interactions necessary for both acute and sustained maintenance of emotional homeostasis in health and disease. Testing of these hypotheses and refinement of the model are ongoing.

Acknowledgments

I thank my collaborators, Stephen Brannan, Mario Liotti, Roderick Mahurin, and Scott McGinnis, at the Research Imaging Center, Sergio Starkstein at the Raul Carrera Institute in Buenos Aires, and Robert Robinson at the University of Iowa, for their significant contributions to the research discussed in this chapter. This work was supported by National Institute of Mental Health MH49553, the National Alliance for Research on Schizophrenia and Depression (NARSAD), the Charles A. Dana Foundation, and Eli Lilly and Company.

References

Albin, R. L., Young, A. B., and Penney, J. B. (1989). The functional anatomy of basal ganglia disorders. *Trends Neurosci.* **12**, 366–375.

Albin, R. L., Young, A. B., Penney, J. B., Handelin, B., Balfour, R., Anderson, K. D., Markel, D. S., Tourtellotte, W. W., and Reiner, A. (1990). Abnormalities of striatal projection neurons and N-methyl-D-aspartate receptors in presymptomatic Huntington's disease. *N. Engl. J. Med.* **322**, 1293–1298.

Alexander, G. E., Crutcher, M. D., and De Long, M. R. (1990). Basal ganglia-thalamocortical circuits: Parallel substrates for motor, oculomotor, "prefrontal" and "limbic" functions. *Prog. Brain Res.* **85**, 119–146.

Altshuler, L. L., Devinsky, O., Post, R. M., and Theodore, W. (1990). Depression, anxiety, and temporal lobe epilepsy: Laterality of focus and symptoms. *Arch. Neurol.* **47**, 284–288.

American Psychiatric Association. (1994). "Diagnostic and Statistical Manual of Mental Disorders," 4th ed. American Psychiatric Association, Washington, DC.

Arango, V., Ernsberger, P., Marzuk, P. M., Chen, J. S., Tierney, H., Stanley, M., Reis, D. J., and Mann, J. J. (1990). Autoradiographic demonstration of increased serotonin 5-HT$_2$ and β-adrenergic receptor binding sites in the brain of suicide victims. *Arch. Gen. Psychiatry* **47,** 1038–1047.

Augustine, J. R. (1996). Circuitry and functional aspects of the insular lobe in primates including humans. *Brain Res. Brain Res. Rev.* **22,** 229–244.

Azmitia, E. C., and Gannon, P. J. (1986). Primate serotonergic system: A review of human and animal studies and a report on *Macaca fascicularis*. *In* "Advances in Neurology" (S. Fahn, Ed.), Vol. 43, pp. 407–468. Raven Press, New York.

Baker, S. C., Frith, C. D., and Dolan, R. J. (1997). The interaction between mood and cognitive function studied with PET. *Psychol. Med.* **27,** 565–578.

Baleydier, C., and Mauguiere, F. (1980). The duality of the cingulate gyrus in the rhesus monkey: Neuroanatomical study and functional hypotheses. *Brain* **103,** 525–554.

Ballenger, J. C. (1988). Biological aspects of depression: Implications for clinical practice. *In* "Review of Psychiatry" (A. J. Frances and R. E. Hales, Eds.), Vol. 7, pp. 169–187. American Psychiatric Press, Washington, DC.

Barbas, H. (1995). Anatomical basis of cognitive-emotional interactions in the primate prefrontal cortex. *Neurosci. Biobehav. Rev.* **19,** 499–510.

Baron, J. C. (1989). Depression of energy metabolism in distant brain structures: Studies with positron emission tomography in stroke patients. *Semin. Neurol.* **9,** 281–285.

Bauer, M., and Frazer, A. (1993). "Mood Disorders and Their Treatment in Biological Bases of Brain Function and Disease," 2nd ed. Raven Press, New York.

Baxter, L. R., Jr., Schwartz, J. M., Phelps, M. E., Mazziotta, J. C., Guze, B. H., Selin, C. E., Gerner, R. H., and Sumida, R. M. (1989). Reduction of prefrontal cortex glucose metabolism common to three types of depression. *Arch. Gen. Psychiatry* **46,** 243–250.

Bear, D. M. (1983). Hemispheric specialization and the neurology of emotion. *Arch. Neurol.* **40,** 195–202.

Bear, D., and Fedio, P. (1977). Quantitative analysis of interictal behavior in temporal lobe epilepsy. *Arch. Neurol.* **34,** 454–467.

Bejjani, B. P., Damier, P., Arnulf, I., Thivard, L., Bonnet, A.-M., Dormont, D., Cornu, P., Pidoux, B., Samson, Y., and Agid, Y. (1999). Transient acute depression induced by high-frequency deep-brain stimulation. *N. Engl. J. Med.* **340,** 1476–1480.

Benca, R. M., Obermeyer, W. H., Thisted, R. A., and Gillin, J. C. (1992). Sleep and psychiatric disorders: A meta-analysis. *Arch. Gen. Psychiatry* **49,** 651–668.

Bench, C. J., Friston, K. J., Brown, R. G., Scott, L. C., Frackowiak, R. S., and Dolan, R. J. (1992). The anatomy of melancholia: Focal abnormalities of cerebral blood flow in major depression. *Psychol. Med.* **22,** 607–615.

Bench, C. J., Friston, K. J., Brown, R. G., Frackowiak, R. S., and Dolan, R. J. (1993a). Regional cerebral blood flow in depression measured by positron emission tomography: The relationship with clinical dimensions. *Psychol. Med.* **23,** 579–590.

Bench, C. J., Frith, C. D., Grasby, P. M., Friston, K. J., Paulesu, E., Frackowiak, R. S., and Dolan, R. J. (1993b). Investigations of the functional anatomy of attention using the Stroop test. *Neuropsychology* **31,** 907–922.

Bench, C. J., Frackowiak, R. S. J., and Dolan, R. J. (1995). Changes in regional cerebral blood flow on recovery from depression. *Psychol. Med.* **25,** 247–251.

Bencherif, B., Treisman, G. J., Zubieta, J. K., *et al.* (1997). Mu opioid receptor binding correlates with symptoms and treatment response in unipolar depression. *Soc. Neurosci. Abstr.* **23**(2), 1207.

Blaney, P. H. (1986). Affect and memory: A review. *Psychol. Bull.* **99,** 229–246.

Blumer, D., and Benson, D. F. (1975). Personality changes with frontal and temporal lobe lesions. *In* "Psychiatric Aspects of Neurological Disease" (D. F. Benson and D. Blumer, Eds.), pp. 151–170. Grune & Stratton, New York.

Bonne, O., and Krausz, Y. (1997). Pathophysiological significance of cerebral perfusion abnormalities in major depression: Trait or state marker? *Eur. Neuropsychopharmacol.* **7,** 225–233.

Bonne, O., Krausz, Y., Gorfine, M., Karger, H., Gelfin, Y., Shapira, B., Chisin, R., and Lerer, B. (1996). Cerebral hypoperfusion in medication resistant, depressed patients assessed by Tc99m-HMPAO SPECT *J. Affect. Disord.* **41,** 163–171.

Botteron, K. M., Raichle, M. E., Heath, A. C., Price, J. L., Sternhell, K. E., Singer, T. M., and Todd, R. D. (1999). An epidemiological twin study of prefrontal neuromorphometry in early onset depression. *Biol. Psychiatry* **45**(8S), 59S, 188.

Bremner, J. D., Randall, P., Scott, T. M., Bronen, R. A., Delaney, R. C., Seibyl, J. P., Southwick, S. M., McCarthy, G., Charney, D. S., and Innis, R. B. (1995). MRI-based measurement of hippocampal volume in post-traumatic stress disorder. *Am. J. Psychiatry* **152,** 973–981.

Bremner, J. D., Innis, R. B., Salomon, R. M., Staib, L. H., Ng, C. K., Miller, H. L., Bronen, R. A., Krystal, J. H., Duncan, J., Rich, D., Price, L. H., Malison, R., Dey, H., Soufer, R., and Charney, D. S. (1997). Positron emission tomography measurement of cerebral metabolic correlates of tryptophan depletion-induced depressive relapse. *Arch. Gen. Psychiatry* **54**(4), 364–374.

Brewer, D., Doughtie, E. B., and Lubin, B. (1980). Induction of mood and mood shift. *J. Clin. Psychol.* **36,** 215–226.

Broca, P. (1878). Anatomie comparée des circonvolutions cérébrales. Le grant lobe limbique et la scissure limbique dans la série des mammiféres. *Rev. Anthropol.* **1,** 385–498.

Bromfield, E. B., Altschuler, L., Leiderman, D. B., Balish, M., Ketter, T. A., Devinsky, O., Post, R. M., and Theodore, W. H. (1992). Cerebral metabolism and depression in patients with complex partial seizures. *Arch. Neurol.* **49,** 617–623.

Brown, R. G., Scott, L. C., Bench, C. J., and Dolan, R. J. (1994). Cognitive function in depression: Its relationship to the presence and severity of intellectual decline. *Psychol. Med.* **24,** 829–847.

Buchsbaum, M. S., Wu, J., DeLisi, L. E., Holcomb, H., Kessler, R., Johnson, J., King, A. C., Hazlett, E., Langston, K., and Post, R. M. (1986). Frontal cortex and basal ganglia metabolic rates assessed by positron emission tomography with ^{18}F-2-deoxyglucose in affective illness. *J. Affect. Disord.* **10,** 137–152.

Buchsbaum, M. S., Someya, T., Wu, J. C., Tang, C. Y., and Bunney, W. E. (1997a). Neuroimaging bipolar illness with positron emission tomography and magnetic resonance imaging. *Psychiatr. Ann.* **27,** 489–495.

Buchsbaum, M. S., Wu, J., Siegel, B. V., Hackett, E., Trenary, M., Abel, L., and Reynolds, C. (1997b). Effect of sertraline on regional metabolic rate in patients with affective disorder. *Biol. Psychiatry* **41,** 15–22.

Bunney, W. E. J., and Davis, J. M. (1965). Norepinephrine in depressive reactions. *Arch. Gen. Psychiatry* **13,** 438–493.

Caldecott-Hazard, S., Mazziotta, J., and Phelps, M. (1988). Cerebral correlates of depressed behavior in rats, visualized using ^{14}C-2-deoxyglucose autoradiography. *J. Neurosci.* **8,** 1951–1961.

Caldecott-Hazard, S., Morgan, D. G., Deleon-Jones, F., Overstreet, D. H., and Janowsky, D. (1991). Clinical and biochemical aspects of depressive disorders. II. Transmitter/receptor theories. *Synapse* **9,** 251–301.

Calev, A., Korin, Y., Shapira, B., Kugelmass, S., and Lerer, B. (1986). Verbal and non-verbal recall by depressed and euthymic affective patients. *Psychol. Med.* **16,** 789–794.

Cannon, W. B. (1929). "Bodily Changes in Pain, Hunger, Fear, and Rage," Vol. 2. Appleton, New York.

Cantello, R., Aguaggia, M., Gilli, M., Delsedime, M., Chiardo Cutin, I., Riccio, A., and Mutani, R. (1989). Major depression in Parkinson's disease and the mood response to intravenous methylphenidate: Possible role of the "hedonic" dopamine synapse. *J. Neurol. Neurosurg. Psychiatry* **52,** 724–731.

Carmichael, S. T., and Price, J. L. (1994). Architectonic subdivision of the orbital and medial prefrontal cortex in the macaque monkey. *J. Comp. Neurol.* **346,** 366–402.

Carmichael, S. T., and Price, J. L. (1995). Limbic connections of the orbital and medial prefrontal cortex in macaque monkeys. *J. Comp. Neurol.* **363,** 615–641.

Carmichael, S. T., and Price, J. L. (1996). Connectional networks within the orbital and medial prefrontal cortex of macaque monkeys. *J. Comp. Neurol.* **371,** 179–207.

Chavis, D. A., and Pandya, D. N. (1976). Further observations on corticofrontal connections in the rhesus monkey. *Brain Res.* **117,** 369–386.

Cherry, S. R., and Phelps, M. E. (1996). Imaging brain function with positron emission tomography. *In* "Brain Mapping: The Methods" (A. W. Toga and J. C. Mazziotta, Eds.), pp. 191–222. Academic Press, San Diego.

Coffey, C. E., Wilkinson, W. E., Weiner, R. D., Parashos, L. A., Djang, W. T., Webb, M. C., Figiel, G. S., and Spritzer, C. E. (1993). Quantitative cerebral anatomy in depression: A controlled magnetic resonance imaging study. *Arch. Gen. Psychiatry* **50,** 7–16.

Cohen, R. M., Weingartner, H., Smallberg, S. A., *et al.* (1982). Effort and cognition in depression. *Arch. Gen. Psychiatry* **39,** 593–597.

Coppen, A., Metalve, M., Carroll, J. D., and Morris, J. G. L. (1972). Levodopa and L–tryptophan therapy in parkinsonism. *Lancet* **1,** 654–657.

Corbetta, M., Miezin, F. M., Shulman, G. L., *et al.* (1993). A PET study of visuospatial attention. *J. Neurosci.* **13,** 1020–1026.

Coryell, W., Endicott, J., and Keller, M. B. (1990). Outcome of patients with chronic affective disorders: A five year follow-up. *Am. J. Psychiatry* **147,** 1627–1633.

Cosgrove, G. R., and Rauch, S. L. (1995). Psychosurgery. *Neurosurg. Clin. N. Am.* **6,** 167–176.

Cowles, K. V. (1996). Cultural perspectives of grief: An expanded concept analysis. *J. Adv. Nurs.* **23,** 287–294.

Crino, P. B., Morrison, J. H., and Hof, P. R. (1993). Monoamine inervation of cingulate cortex. *In* "The Neurobiology of Cingulate Cortex and Limbic Thalamus: A Comprehensive Handbook" (B. A. Vogt and M. Gabriel, Eds.), pp. 285–310. Birkhauser, Boston.

Cummings, J. L. (1992). Depression and Parkinson's disease: A review. *Am. J. Psychiatry* **149,** 443–454.

Cummings, J. L., and Victoroff, J. I. (1990). Noncognitive neuropsychiatric syndromes in Alzheimer's disease. *Neuropsychiatry, Neuropsychol. Behav. Neurol.* **2,** 140–158.

Damasio, A. R. (1994). Descartes' Error. Putnam, New York.

Damasio, A. R., and Van Hoesen, G. W. (1983). Emotional disturbances associated with focal lesions of the limbic frontal lobe. *In* "Neuropsychology of Human Emotion" (K. M. Heilman and P. Satz, Eds.), pp. 85–110. Guilford, New York.

Darwin, C. (1872). The expression of the emotions in man and animals. Murray, London.

Delgado, P. L., Charney, D. S., Price, L. H., Aghajanian, G. K., Landis, H., and Heninger, G. R. (1990). Serotonin function and the mechanism of antidepressant action: Reversal of antidepressant-induced remission by rapid depletion of plasma tryptophan. *Arch. Gen. Psychiatry* **47,** 411–418.

Devinsky, O., Morrell, M. J., and Vogt, B. A. (1995). Contributions of anterior cingulate cortex to behavior. *Brain* **118,** 279–306.

Dolan, R. J., Bench, C. J., Brown, R. G., Scott, L. C., Friston, K. J., and Frackowiak, R. S. (1992). Regional cerebral blood flow abnormalities in depressed patients with cognitive impairment. *J. Neurol. Neurosurg. Psychiatry* **55,** 768–773.

Dolan, R. J., Bench, C. J., Liddle, P. F., Friston, K. J., Frith, C. D., Grasby, P. M., and Frackowiak, R. S. (1993). Dorsolateral prefrontal cortex dysfunction in the major psychoses: Symptom or disease specificity? *J. Neurol. Neurosurg. Psychiatry* **56**(12), 1290–1294.

Dray, A. (1981). Serotonin in the basal ganglia: Functions and interactions with other neuronal pathways. *J. Physiol. (Paris)* **77,** 393–403.

Drevets, W. C., Videen, T. O., Price, J. L., Preskorn, S. H., Carmichael, S. T., and Raichle, M. E. (1992). A functional anatomical study of unipolar depression. *J. Neurosci.* **12,** 3628–3641.

Drevets, W. C., Price, J. L., Simpson, J. R., Jr., Todd, R. D., Reich, T., Vannier, M., and Raichle, M. E. (1997). Subgenual prefrontal cortex abnormalities in mood disorders. *Nature* **386,** 824–827.

Drevets, W. C., Price, J. D., Kupfer, D. J., Frank, E., Holt, D., Huant, H., Proper, S. M., Gautier, C., and Mathis, C. (1999). PET imaging of serotonin 1A receptor binding in depression. *Biol. Psychiatry* **45**(8S), 118S, #382.

Duncan, G. E., Knapp, D. J., Johnson, K. B., and Breese, G. R. (1996). Functional classification of antidepressants based on antagonism of swim stress-induced fos-like immunoreactivity. *J. Pharmacol. Exp. Ther.* **277,** 1076–1089.

Dupont, R. M., Jernigan, T. L., Heindel, W., Butters, N., Shafer, K., Wilson, T., Hesselink, J., and Gillin, J. C. (1995). Magnetic resnonance imaging and mood disorders. Localization of white matter and other subcortical abnormalities. *Arch. Gen. Psychiatry* **52,** 747–755.

Ebert, D., and Ebmeier, K. (1996). Role of the cingulate gyrus in depression: from functional anatomy to depression. *Biol. Psychiatry* **39,** 1044–1050.

Elliott, R., Frith, C. D., and Dolan, R. J. (1997a). Differential neural response to positive and negative feedback in planning and guessing tasks. *Neuropsychologia* **35,** 1395–1404.

Elliott, R., Baker, S. C., Rogers, R. D., O'Leary, D. A., Paykel, E. S., Frith, C. D., Dolan, R. J., and Sahakian, B. J. (1997b). Prefrontal dysfunction in depressed patients performing a complex planning task: A study using positron emission tomography. *Psychol. Med.* **27**(4), 931–942.

Elliott, R., Sahakian, B. J., Herrod, J. J., Robbins, T. W., and Paykel, E. S. (1997c). Abnormal response to negative feedback in unipolar depression: Evidence for a diagnosis specific impairment. *J. Neurol. Neurosurg. Psychiatry* **63**(1), 74–82.

Emery, V. O., and Oxman, T. E. (1992). Update on the dementia spectrum of depression. *Am. J. Psychiatry* **149,** 305–317.

Fibiger, H. C. (1984). The neurobiological substrates of depression in Parkinson's disease: A hypothesis. *Can. J. Neurol. Sci.* **11**(1), 105–107.

Flint, A. J., Black, S. E., Campbell-Taylor, I., Gailey, G. F., and Levinton, C. (1993). Abnormal speech articulation, psychomotor retardation, and subcortical dysfunction in major depression. *J. Psychiatr. Res.* **27,** 309–319.

Flor-Henry, P. (1969). Psychosis and temporal lobe epilepsy. *Epilepsia* **10,** 363–395.

Folstein, S. E., Abbott, M. H., Chase, G. A., Jensen, B. A., and Folstein, M. F. (1983). The association of affective disorder with Huntington's disease in a case series and in families. *Psychol. Med.* **13:**537–542.

Frank, E., and Thase, M. E. (1999). Natural history and preventative treatment of recurrent mood disorders. *Annu. Rev. Med.* **50,** 453–468.

Fulton, J. F. (1951). "Frontal Lobotomy and Affective Behavior. A Neurophysiological Analysis." Chapman & Hall, London.

Gainotti, G. (1972). Emotional behavior and hemispheric side of the lesion. *Cortex* **8**, 41–55.

Gemar, M. C., Kapur, S., Segal, Z. V., *et al.* (1996). Effects of self-generated sad mood on regional cerebral activity: A PET study in normal subjects. *Depression* **4**, 81–88.

George, M. S., Ketter, T. A., Parekh, P. I., *et al.* (1994a). Spatial ability in affective illness: Differences in regional brain activation during a spatial matching task ($H_2^{15}O$ PET). *Neuro-psychiatry Neuropsychol. Behav. Neurol.* **7**, 143–153.

George, M. S., Ketter, T. A., Parekh, P. I., Rosinsky, N., Ring, H., Casey, B. J., Trimble, M. R., Horwitz, B., Herscovitch, P., and Post, R. M. (1994b). Regional brain activity when selecting a response despite interference: An $H_2^{15}O$ PET study of the Stroop and an emotional Stroop. *Hum. Brain Mapp.* **1**, 194–209.

George, M. S., Ketter, T. A., and Post, R. M. (1994c). Prefrontal cortex dysfunction in clinical depression. *Depression* **2**, 59–72.

George, M. S., Ketter, T. A., Parekh, P. I., Horwitz, B., Herscovitch, P., and Post, R. M. (1995). Brain activity during transient sadness and happiness in healthy women. *Am. J. Psychiatry* **152**(3), 341–351.

George, M. S., Ketter, T. A., Parekh, P. I., *et al.* (1996). Gender differences in regional cerebral blood flow during transient self-induced sadness or happiness. *Biol. Psychiatry* **40**(9), 859–871.

George, M. S., Ketter, T. A., Parekh, P. I., Rosinsky, N., Ring, H. A., Pazzaglia, P. J., Marangell, L. B., Callahan, A. M., and Post, R. M. (1997). Blunted left cingulate activation in mood disorder subjects during a response interference task (the Stroop). *J. Neuropsychiatry Clin. Neurosci.* **9**, 55–63.

Golden, L. R., and Gershon, E. S. (1988). The genetic epidemiology of major depressive illness. *In* "Review of Psychiatry" (A. J. Frances and R. E. Hales, Eds.), Vol. 7, pp. 148–168. American Psychiatry Press, Washington, DC.

Goldman-Rakic, P. S., and Selemon, L. D. (1984). Topography of corticostriatal projections in nonhuman primates and implications for functional parcellation of the neostriatum. *In* "Cerebral Cortex" (E. G. Jones and A. Peters, Eds.), pp. 447–466. Plenum, New York.

Goodstein, R. K., and Ferrel, R. B. (1977). Multiple sclerosis presenting as depressive illness. *Dis. Nerv. Syst.* **38**, 127–131.

Goodwin, A. M., and Williams, J. M. G. (1982). Mood-induction research: Its implications for clinical depression. *Behav. Res. Ther.* **20**, 373–382.

Goodwin, G. M., Austin, M. P., Dougall, N., Ross, M., Murray, C., O'-Carroll, R. E., Moffoot, A., Prentice, N., and Ebmeier, K. P. (1993). State changes in brain activity shown by the uptake of ^{99m}Tc-exametazime with single photon emission tomography in major depression before and after treatment. *J. Affect. Disord.* **29**, 243–253.

Gottschalk, L. A., Buchsbaum, M. S., Gillin, J. C., Wu, J., Reynolds, C. A., and Herrera, D. B. (1991). Positron emission tomographic studies of the relationship of cerebral glucose metabolism and the magnitude of anxiety and hostility experienced during dreaming and waking. *J. Neuropsychiatry Clin. Neurosci.* **3**, 131–142.

Grabiel, A. M. (1990). Neurotransmitters and neuromodulators in the basal ganglia. *Trends Neurosci.* **13**, 244–254.

Grafman, J., Vance, S. C., Weingartner, H., Salazar, A. M., and Amin, D. (1986). The effects of lateralized frontal lesions on mood regulation. *Brain* **109**, 1127–1148.

Grafton, S. T., Mazziotta, J. C., Pahl, J. J., St. George-Hyslop, P., Haines, J. L., Gusella, J., Hoffman, J. M., Baxter, L. R., and Phelps, M. E. (1990). A comparison of neurological, metabolic, structural and genetic evaluations in persons at risk for Huntington's. *Ann. Neurol.* **28**, 614–621.

Grasso, M. G., Pantano, P., Ricci, M., Intiso, D. F., Pace, A., Padovani, A., Orzi, F., Pozzilli, C., and Lenzi, G. L. (1994). Mesial temporal cortex hypoperfusion is associated with depression in subcortical stroke. *Stroke* **25**, 980–985.

Greenwald, B. S., Kramer-Ginsberg, E., Krishnan, R. R., Ashtari, M., Aupperle, P. M., and Patel, M. (1996). MRI signal hyperintensities in geriatric depression. *Am. J. Psychiatry* **153**, 1212–1215.

Gross-Isseroff, R., Dillon, K. A., Israeli, M., and Biegon, A. (1990). Regionally selective increases in mu opioid receptor density in the brains of suicide victims. *Brain Res.* **530**, 312-316.

Harlow, H. F., and Suomi, S. J. (1974). Induced depression in monkeys. *Behav. Biol.* 12, 273–296.

Hasher, L., and Zacks, R. T. (1979). Automatic and effortful processes in memory. *J. Exp. Psychol. Gen.* **108**, 356–388.

Heilman, K. M. (1997). The neurobiology of emotional experience. *J. Neuropsychiatry Clin. Neurosci.* **9**, 439–448.

Hickie, I., Scott, E., Wilhelm, K., and Brodaty, H. (1997). Subcortical hyperintensities on magnetic resonance imaging in patients with severe depression: A longitudinal evaluation. *Biol. Psychiatry* **42**(5), 367–374 .

Hirono, N., Mori, E., Ishii, K., Ikejire, Y., Imamura, T., Shimomura, T., Hashimoto, M., Yamashita, H., and Sasaki, M. (1998). Frontal lobe hypometabolism and depression in Alzheimer's disease. *Neurology* **50**, 380–383.

Honer, W. G., Hurwitz, T., Li, D. K., Palmer, M., and Paty, D. W. (1987). Temporal lobe involvement in multiple sclerosis patients with psychiatric disorders. *Arch. Neurol.* **44**, 187–190.

Hornig, M., Mozley, P. D., and Amsterdam, J. D. (1997). HMPAO SPECT brain imaging in treatment-resistant depression. *Prog. Neuropsychopharmacol. Biol. Psychiatry* **21**, 1097–1114.

Hyman, S. E., and Nestler, E. J. (1996). Initiation and adaptation: A paradigm for understanding psychotropic drug action. *Am. J. Psychiatry* **153**, 151–162.

Izard, C. E. (1993). Four systems for emotion activation: Cognitive and noncognitive processes. *Psychol. Rev.* **100**, 68–90.

Jagust, W. J., Reed, B. R., Martin, E. M., Eberling, J. L., and Nelson-Abbott, R. A. (1992). Cognitive function and regional cerebral blood flow in Parkinson's disease. *Brain* **115**, 521–537.

James, W. (1884). What is an emotion? *Mind* **9**, 188–205.

Janowski, D. S., Risch, S. C., and Gillin, J. C. (1988). Cholinergic involvement in affective illness. *In* "Receptors and Ligands in Psychiatry" (A. K. Sen and T. Lee, Eds.), pp. 228–244. Cambridge Univ. Press, New York.

Jones, R. D., Tranel, D., Benton, A., and Paulsen, J. (1992). Differentiating dementia from "pseudodementia" early in the clinical course: Utility of neuropsychological tests. *Neuropsychology* **6**, 13–21.

Kaplan, H. I., Sadock, B. J., and Grebb, J. A. (1994). Mood disorders. *In* "Synopsis of Psychiatry: Behavioral Sciences/Clinical Psychiatry," pp. 516–572. Williams & Wilkins, Baltimore.

Keller, M. B., Lavori, P. W., and Klerman, G. L. (1983). Predictors of relapse in major depressive disorder. *JAMA* **250**, 3299–3304.

Ketter, T. A., George, M. S., Kimbrell, T. A., Benson, B. E., and Post, R. M. (1996). Functional brain imaging, limbic function, and affective disorders. *Neuroscientist* **2**, 55–65.

Kleist, K. (1937). Bericht über die Gehirnpathologie in ihrer Bedeutung für Neurologie und Psychiatrie. *Z. Gesamte Neurol. Psychiatr.* **158**, 159–193.

Kunishio, K., and Haber, S. N. (1994). Primate cingulostriatal projection: Limbic striatal versus sensorimotor striatal input. *J. Comp. Neurol.* **350**, 337–356.

Lane, R. D., Reiman, E. M., Ahern, G. L., *et al.* (1997). Neuroanatomical correlates of happiness, sadness and disgust. *Am. J. Psychiatry* **154**, 926–933.

Lang, P. J., Greenwald, M. K., Bradley, M. M., and Hamm, A. O. (1993). Looking at pictures: Affective, facial, visceral, and behavioral reactions. *Psychophysiology* **30**, 261–273.

Laplane, D., Levasseur, M., Pillon, B., Dubois, B., Baulac, M., Mazoyer, B., Tran Dinh, S., Sette, G., Danze, F., and Baron, J. C. (1989). Obsessive-compulsive and other behavioural changes with bilateral basal ganglia lesions. A neuropsychological, magnetic resonance imaging and positron tomography study. *Brain* **12**, 699–725.

Larisch, R., Klimke, A., Vosberg, H., Loffler, S., Gaebel, W., and Muller-Gartner, H. W. (1997). *In vivo* evidence for the involvement of dopamine-D_2 receptors in striatum and anterior cingulate gyrus in major depression. *NeuroImage* **5**, 251–260.

Lazarus, R. S. (1991). Cognition and motivation in emotion. *Am. Psychologist* **46**, 352–367.

LeDoux, J. (1996a). "The Emotional Brain." Simon & Schuster, New York.

LeDoux, J. E. (1996b). In search of an emotional system in the brain: Leaping from fear to emotion and consciousness. *In* "The Cognitive Neurosciences" (M. S. Gazzaniga, Ed.), pp. 1049–1061. MIT Press, Cambridge, MA.

Lesser, I., Mena, I., Boone, K. B., Miller, B. L., Mehringer, C. M., and Wohl, M. (1994). Reduction of cerebral blood flow in older depressed patients. *Arch. Gen. Psychiatry* **51**, 677–686.

Levine, S., Lyons, D. M., and Schatzberg, A. F. (1997). Psychobiological consequences of social relationships. *Ann. N.Y. Acad. Sci.* **807**, 210–218.

Liotti, M., and Tucker, D. M. (1992). Right hemisphere sensitivity to arousal and depression. *Brain Cog.* **18**, 138–151.

Liotti, M., Mayberg, H. S., Ryder, K., *et al.* (1997a). An ERP study of mood provocation in remitted depression. *Soc. Neurosci. Abstr.* **23**(2), 1657.

Liotti, M., Mayberg, H. S., Brannan, S. K., McGinnis, S., Jerabek, P. A., Martin, C. C., and Fox, P. T. (1997b). Mood challenge in remitted depression: An ^{15}O-water PET Study. *NeuroImage* **5**(4), S60.

Liotti, M., Woldorff, M. G., Perez, R., and Mayberg, H. S. (1999). ERP correlates of attentional impairment during intense sadness. *Biol. Psychiatry* **45**(8S), 109S, #352.

Livingston, K. E., and Escobar, A. (1973). Tentative limbic system models for certain patterns of psychiatric disorders. *In* "Surgical Approaches in Psychiatry" (V. Laitinen and K. E. Livingstone, Eds.), pp. 245–252. Medical and Technical Publishing Co., Lancaster.

MacLean, P. D. (1949). Psychosomatic disease and the visceral brain. Recent developments bearing on the Papex theory of emotion. *Psychosom. Med.* **11**, 338–353.

MacLean, P. D. (1990). "The Triune Brain in Evolution: Role in Paleocerebral Function." Plenum, New York.

Maj, M., Veltro, F., Pirozzi, R., Lobrace, S., and Magliano, L. (1992). Patterns of recurrence of illness after recovery from an episode of major depression: A prospective study. *Am. J. Psychiatry* **149**, 795–800.

Malison, R. T., Price, L. H., Berman, R. M., van Dyck, C. H., Pelton, G. H., Carpenter, L., Sanacora, G., Owens, M. J., Nemeroff, C. B., Rajeevan, N., Baldwin, R. M., Seibyl, J. P., Innis, R. B., and Charney, D. S. (1998). Reduced midbrain serotonin transporter binding in depressed vs healthy subjects as measured by ^{123}I-β-CIT SPECT. *Biol. Psychiatry* **44**, 1090–1098.

Malizia, A. (1997). Frontal lobes and neurosurgery for psychiatric disorders. *J. Psychopharm.* **11**(2), 179–187.

Marin, R. S. (1990). Differential diagnosis and classification of apathy. *Am. J. Psychiatry* **147**, 22–30.

Marsh, G. G., and Markham, C. H. (1973). Does levodopa alter depression and psychopathology in parkinsonism patients? *J. Neurol. Neurosurg. Psychiatry* **36**, 925–935.

Martin, M. (1990). On the induction of mood. *Clin. Psychol. Rev.* **10**, 669–697.

Martin, W. R., Sloan, J. W., Sapira, J. D., and Jasinski, D. R. (1971). Physiologic subjective, and behavioural effects of amphetamine, metamphetamine, ephedrine, phenmetrazine, and methylphenidate in man. *Clin. Pharmacol. Ther.* **12**, 245–258.

Martinot, J. L., Hardy, P., Feline, A., Huret, J. D., Mazoyer, B., Attar-Levy, D., Pappata, S., and Syrota, A. (1990). Left prefrontal glucose hypometabolism in the depressed state: A confirmation. *Am. J. Psychiatry* **147**, 1313–1317.

Massou, J. M., Trichard, C., Attar-Levy, D., Feline, A., Corruble, E., Beaufils, B., and Martinot, J. L. (1997). Frontal 5-HT$_{2A}$ receptors studied in depressive patients during chronic treatment by selective serotonin reuptake inhibitors. *Psychopharmacology* **133**(1), 99–101.

Mayberg, H. S. (1994a). Clinical correlates of PET and SPECT defects in dementia. *J. Clin. Psychiatry* **55**(11), 12–21.

Mayberg, H. S. (1994b). Frontal lobe dysfunction in secondary depression. *J. Neuropsychiatry Clin. Neurosci.* **6**, 428–442.

Mayberg, H. S. (1997). Limbic-cortical dysregulation: A proposed model of depression. *J. Neuropsychiatry Clin. Neurosci.* **9**, 471–481.

Mayberg, H. S., and Solomon, D. H. (1995). Depression in PD: A biochemical and organic viewpoint. *In* "Behavioral Neurology of Movement Disorders, Advances in Neurology" (W. J. Weiner and A. E. Lang, Eds.), Vol. 65, pp. 49–60. Raven Press, New York.

Mayberg, H. S., Robinson, R. G., Wong, D. F., Parikh, R., Bolduc, P., Starkstein, S. E., Price, T., Dannals, R. F., Links, J. M., Wilson, A. A., Ravert, H. T., and Wagner, N. N. (1988). PET imaging of cortical S$_2$-serotonin receptors after stroke: Lateralized changes and relationship to depression. *Am. J. Psychiatry* **145**(8), 937–943.

Mayberg, H. S., Starkstein, S. E., Sadzot, B., Preziosi, T., Andrezejewski, P. L., Dannals, R. F., Wagner, H. N., Jr., and Robinson, R. G. (1990). Selective hypometabolism in the inferior frontal lobe in depressed patients with Parkinson's disease. *Ann. Neurol.* **28**, 57–64.

Mayberg, H. S., Parikh, R. M., Morris, P. L., and Robinson, R. G. (1991a). Spontaneous remission of post-stroke depression and temporal changes in cortical S$_2$-serotonin receptors. *J. Neuropsychiatry Clin. Neurosci.* **3**, 80–83.

Mayberg, H. S., Ross, C. A., Dannals, R. F., Wilson, A. A., Ravert, H. T., and Frost, J. J. (1991b). Elevated mu opiate receptors measured by PET in patients with depression. *J. Cereb. Blood Flow Metab.* **11**(S), 821.

Mayberg, H. S., Starkstein, S. E., Morris, P. L., Federoff, J. P., Price, T. R., Dannals, R. F., Wagner, H. N., and Robinson, R. G. (1991c). Remote cortical hypometabolism following focal basal ganglia injury: Relationship to secondary changes in mood. *Neurology* **41** (Suppl.), 266.

Mayberg, H. S., Starkstein, S. E., Peyser, C. E., Brandt, J., Dannals, R. F., and Folstein, S. E. (1992). Paralimbic frontal lobe hypometabolism in depression associated with Huntington's disease. *Neurology* **42**, 1791–1797.

Mayberg, H. S., Lewis, P. J., Regenold, W., and Wagner, H. N., Jr. (1994). Paralimbic hypoperfusion in unipolar depression. *J. Nucl. Med.* **35**, 929–934.

Mayberg, H. S., Brannan, S. K., Mahurin, R. K., Silva, J. A., Tekell, J. L., Jerabek, P. A., Glass, T. G., Martin, C. C., and Fox, P. T. (1995a). Functional correlates of mood and cognitive recovery in depression: An FDG PET study. *Hum. Brain Mapp.* **S1**, 428.

Mayberg, H. S., Liotti, M., Jerabek, P. A., Martin, C. C., and Fox, P. T. (1995b). Induced sadness: A PET model of depression. *Hum. Brain Mapp.* **S1**, 396.

Mayberg, H. S., Mahurin, R. K., Brannan, S. K., Glass, T. G., Solomon, D., New, P., Jerabek, P. A., Martin, C. C., and Fox, P. T. (1995c). Parkinson's depression: Discrimination of mood-sensitive and mood-insensitive cognitive deficits using fluoxetine and FDG PET. *Neurology* **45**(Suppl.), A166.

Mayberg, H. S., Brannan, S. K., Mahurin, R. K., Jerabek, P. A., Brickman, J. S., Tekell, J. L., Silva, J. A., McGinnis, S., Glass, T. G., Martin, C. C., and Fox, P. T. (1997). Cingulate function in depression: A potential predictor of treatment response. *NeuroReport* **8**, 1057–1061.

Mayberg, H. S., Liotti, M., Brannan, S. K., McGinnis, S., Jerabek, P. A., Martin, C. C., and Fox, P. T. (1998). Disease and state-specific effects of mood challenge on rCBF. *NeuroImage* **7**, S901.

Mayberg, H. S., Liotti, M., Brannan, S. K., McGinnis, S., Mahurin, R. K., Jerabek, P. A., Silva, J. A., Tekell, J. L., Martin, C. C., Lancaster, J. L., and Fox, P. T. (1999a). Reciprocal limbic-cortical function and negative mood: Converging PET findings in depression and normal sadness. *Am. J. Psychiatry* **156**(5), 675–682.

Mayberg, H. S., Brannan, S. K., Mahurin, R. K., McGinnis, S., Tekell, J. L., Silva, J. A., Jerabek, P. A., Martin, C. C., and Fox, P. T. (1999b). Early and late fluoxetine effects on regional glucose metabolism in depression. *Biol. Psychiatry* **45**(S8), 111S, #357.

Mayeux, R. (1983). Emotional changes associated with basal ganglia disorders. *In* "Neuropsychology of Human Emotion"(K. M. Heilman and P. Satz, Eds.), pp. 141–164. Guilford, New York.

Mayeux, R., Stern, Y., Sano, M., Williams, J. B., and Cote, L. J. (1988). The relationship of serotonin to depression in Parkinson's disease. *Mov. Disord.* **3**, 237–244.

McCance-Katz, E. F., Marek, K. L., and Price, L. H. (1992). Serotonergic dysfunction in depression associated with Parkinson's disease. *Neurology* **42**, 1813–1814.

Meltzer, C. C., Price, J., Mathis, C. A., Greer, P. J., Houck, P. R., Lopresti, B., Ben-Eliezer, D., Cantwell, M. C., Dekosky, S. T., and Reyolds, C. F. (1999). [F-18]-Altanserin binding to serotonin (5HT-2A) receptors in late-life depression and Alzheimer's disease. *Biol. Psychiatry* **45**(8S), 61S, #193.

Meltzer, H., Ed. (1987). "Psychopharmacology: The Third Generation of Progress," pp. 493–686. Raven Press, New York.

Mendez, M. F., Cummings, U. L., and Benson, D. F. (1986). Depression in epilepsy. *Arch. Neurol.* **43**, 766–770.

Mendez, M. F., Adams, N. L., and Lewandowski, K. S. (1989). Neurobehavioral changes associated with caudate lesions. *Neurology* **39**, 349–354.

Mesulam, M. M. (1985). Patterns in behavioral neuroanatomy: Association areas, the limbic system, and hemispheric specialization. *In* "Principles of Behavioral Neurology"(M. M. Mesulam, Ed.), pp. 1–70. Davis, Philadelphia.

Mesulam, M.-M., and Mufson, E. J. (1992). Insula of the Old World Monkey I, II, III. *J. Comp. Neurol.* **212**, 1–52.

Meyer, J. H., Kapur, S., House, S., DaSilva, J., Eisfeld, B., Brown, G. M., Wilson, A. A., and Kennedy, S. H. (1999). Prefrontal cortex 5-HT$_2$ receptors in depression. *Biol. Psychiatry* **45**(8S), 127S, 410.

Morecraft, R. J., Geula, C., and Mesulam, M. M. (1993). Architecture of connectivity within a cingulo-fronto-parietal neurocognitive network for directed attention. *Arch. Neurol.* **50**, 279–284.

Nauta, W. J. H. (1971). The problem of the frontal lobe: A reinterpretation. *J. Psychol. Res.* **8**, 167–187.

Nauta, W. J. H. (1986). Circuitous connections linking cerebral cortex, limbic system, and corpus striatum. *In* "The Limbic System: Functional Organization and Clinical Disorders" (B. K. Doane and K. E. Livingston, Eds.), pp. 43–54. Raven Press, New York.

Neafsey, E. J. (1990). Prefrontal cortical control of the autonomic nervous system: Anatomical and physiological observations. *Prog. Brain Res.* **85**, 147–66.

Nemeroff, C. B. (1989). Clinical significance of psychoneuroendocrinology in psychiatry: Focus on the thyroid and adrenal. *J. Clin. Psychiatry* **50**(Suppl.), 13–22.

Nemeroff, C. B., Ranga, K., and Krishnan, R. (1992). "Neuroendocrine alterations in psychiatric disorders in neuroendocrinology." CRC Press, Boca Raton, FL.

Nemeroff, C. B., Widerlov, E., Bissette, G., Walleus, H., Karlsson, I., Eklund, K., Kilts, C. D., Loosen, P. T., and Vale, W. (1984). Elevated concentrations of CSF corticotropin-releasing factor-like immunoreactivity in depressed patients. *Science* **226**, 1342–1344.

Nibuya, M., Nestler, E. J., and Duman, R. S. (1996). Chronic antidepressant administration increases the expression of cAMP response element binding protein (CREB) in rat hippocampus. *J. Neurosci.* **16**, 2365–2372.

Nobler, M. S., Sackeim, H. A., Prohovnik, I., Moeller, J. R., Mukherjee, S., Schnur, D. B., Prudic, J., and Devanand, D. P. (1994). Regional cerebral blood flow in mood disorders. III. Treatment and clinical response. *Arch. Gen. Psychiatry* **51**, 884–897.

Osuch, E. A., Ketter, T. A., Kimbrell, T. A., George, M. S., Hbenson, B. E., Willis, M. W., McCann, U., and Post, R. M. (1999). Regional cerebral metabolism unique to anxiety symptoms in affective disorder patients. *Biol. Psychiatry* **45**(8S), 129S, #417.

Overstreet, D. H. (1993). The Flinders sensitive line rats: A genetic animal model of depression. *Neurosci. Biobehav. Rev.* **17**, 51–68.

Pandya, D. N., and Kuypers, H. G. J. M. (1969). Cortico–cortical connections in the rhesus monkey. *Brain Res.* **13**, 13–36.

Pandya, D. N., and Yeterian, E. H. (1996). Comparison of prefrontal architecture and connections. *Philos. Trans. R. Soc. Lond.* **351**, 1423–1432.

Papez, J. W. (1937). A proposed mechanism of emotion. *Arch. Neurol. Psychiatry* **38**, 725–743.

Pardo, J. V., Raichle, M. E., and Fox, P. T. (1991). Localization of a human system for sustained attention by positron emission tomography. *Nature* **349**, 61–63.

Pardo, J. V., Pardo, P. J., and Raichle, M. E. (1993). Neural correlates of self-induced dysphoria. *Am. J. Psychiatry* **150**, 713–719.

Partiot, A., Grafman, J., Sadato, N., *et al.* (1995). Brain activation during the generation of non-emotional and emotional plans. *NeuroReport* **6**, 1397–1400.

Passero, S., Nardini, M., and Battistini, N. (1995). Regional cerebral blood flow changes following chronic administration of antidepressant drugs. *Prog. Neuropsychopharmacol. Biol. Psychiatry* **19**, 627–636.

Paus, T., Petrides, M., Evans, A. C., and Meyer, E. (1993). Role of the human anterior cingulate cortex in the control of oculomotor, manual and speech responses: A PET study. *J. Neurophysiol.* **70**, 453–469.

Petrides, M., and Pandya, D. N. (1984). Projections to the frontal cortex from the posterior parietal region in the rhesus monkey. *J. Comp. Neurol.* **228**, 105–116.

Petty, F., Kramer, G. L., Gullion, C. M., and Rush, A. J. (1992). Low plasma gamma-aminobutyric acid levels in male patients with depression. *Biol. Psychiatry* **32**, 354–63.

Petty, F., Kramer, G. L., Wu, J., and Davis, L. L. (1997). Posttraumatic stress and depression: A neurochemical anatomy of the learned helplessness animal model. *Ann. N.Y. Acad. Sci.* **821**, 529–532.

Peyser, C. E., and Folstein, S. E. (1990). Huntington's disease as a model for mood disorders: Clues from neuropathology and neurochemistry. *Mol. Chem. Neuropathol.* **12**, 99–119.

Pike, V. W., McCarron, J. A., Lammertsma, A. A., Osman, S., Hume, S. P., Sargen, P. A., Bench, C. J., Cliffe, I. A., Fletcher, A., and Grasby, P. M. (1996). Exquisite delineation of 5-HT$_{1A}$ receptors in human brain with PET and [*carbonyl*-^{11}C]WAY-100635. *Eur. J. Pharmacol.* **301**(1–3), R5–R7.

Porrino, L. J., Crane, A. M., and Goldman-Rakic, P. S. (1981). Direct and indirect pathways from the amygdala to the frontal lobe in rhesus monkeys. *J. Comp. Neurol.* **198**, 121–136.

Posner, M. I., and Petersen, S. E. (1990). The attention system of the human brain. *Annu. Rev. Neurosci.* **13**, 25–42.

Post, R. M., DeLisi, L. E., Holcomb, H. H., Uhde, T. W., Cohen, R., and Buchsbaum, M. S. (1987). Glucose utilization in the temporal cortex of affectively ill patients: Positron emission tomography. *Biol. Psychiatry* **22**, 545–553.

Rathus, J. H., and Reber, A. S. (1994). Implicit and explicit learning: Differential effects of affective states. *Percept. Mot. Skills* **79**, 163–184.

Reiman, E. M., Lane, R. D., Ahern, G. L., *et al.* (1997). Neuroanatomical correlates of externally and internally generated human emotion. *Am. J. Psychiatry* **154**, 918–925.

Reivich, M., Gur, R., and Alavi, A. (1983). Positron emission tomographic studies of sensory stimuli, cognitive processes and anxiety. *Hum. Neurobiol.* **2**, 25–33.

Reynolds, C. F., 3d, Hoch, C. C., Buysse, D. J., *et al.* (1992). Electroencephalographic sleep in spousal bereavement and bereavement-related depression of late life. *Biol. Psychiatry* **31**(1), 69–82.

Ring, H. A., Bench, C. J., Trimble, M. R., Brooks, D. J., Frackowiak, R. S., and Dolan, R. J. (1994). Depression in Parkinson's disease. A positron emission study. *Br. J. Psychiatry* **165**, 333–339.

Robertson, M. M., Trimble, M. R., and Townsend, H. R. A. (1987). Phenomenology of depression in epilepsy. *Epilepsia* **28**, 364–368.

Robinson, R. G. (1979). Differential behavioral effects of right versus left hemispheric cerebral infarction: Evidence for cerebral lateralization in the rat. *Science* **205**, 707–710.

Robinson, R. G. (1998). "The Clinical Neuropsychiatry of Stroke." Cambridge Univ. Press, Cambridge, UK.

Robinson, R. G., Kubos, K. L., Starr, L. B., Rao, K., and Price, T. R. (1984). Mood disorders in stroke patients: Importance of location of lesion. *Brain* **107**, 81–93.

Rogers, D., Lees, A. J., Smith, E., Trimble, M., and Stern, G. M. (1987). Bradyphrenia in Parkinson's disease and psychomotor retardation in depressive illness: An experimental study. *Brain* **110**, 761–776.

Rolls, E. T. (1985). Connections, functions and dysfunctions of limbic structures, the prefrontal cortex and hypothalamus. *In* "The Scientific Basis of Clinical Neurology" (M. Swash and C. Kennard, Eds.), pp. 201–213. Churchill Livingstone, London.

Rolls, E. T. (1990). A theory of emotion, and its application to understanding the neural basis of emotion. *Cog. Emot.* **4**, 161–190.

Rolls, E. T. (1996). The orbital frontal cortex. *Phil. Trans. R. Soc. London B.* **351**, 1433–1444.

Ross, E. D., and Rush, A. J. (1981). Diagnosis and neuroanatomical correlates of depression in brain-damaged patients. *Arch. Gen. Psychiatry* **39**, 1344–1354.

Ross, E. D., Homan, R. W., and Buck, R. (1994). Differential hemispheric lateralization of primary and social emotions. *Neuropsychiatry Neuropsychol. Behav. Neurol.* **7**, 1–19.

Roy, A., Pickar, D., De Jong, J., *et al.* (1988a). Norepinephrine and its metabolites in cerebrospinal fluid, plasma, and urine. *Arch. Gen. Psychiatry* **45**, 849–857.

Roy, A., Gallucci, W., Avgerinos, P., *et al.* (1988b). The CRH stimulation test in bereaved subjects with and without accompanying depression. *Psychiatry Res.* **25**(2), 145–156.

Roy-Byrne, P. P., Weingartner, H., Bierer, L. M., *et al.* (1986). Effortful and automatic cognitive processes in depression. *Arch. Gen. Psychiatry* **43**, 265–267.

Sackeim, H., Greenberg, M. S., Weiman, A. L., Gur, R. C., Hungerbuhler, J. P., and Geschwind, N. (1982). Hemispheric asymmetry in the expression of positive and negative emotions. *Arch. Neurol.* **39**, 210–218.

Sano, I., and Taniguchi, K. (1972). L-5-Hydroxytryptophan treatment of Parkinson's disease. *Munch. Med. Wochenschr.* **114**, 1717–1719.

Sapolsky, R. M. (1994). The physiological relevance of glucocorticoid endangerment of the hippocampus. *Ann. N.Y. Acad. Sci.* **746**, 294–304.

Sapolsky, R. M. (1996). Why stress is bad for your brain. *Science* **273**, 749–750.

Schacter, S., and Singer, J. E. (1962). Cognitive, social, and physiological determinants of emotional state. *Psychol. Rev.* **69**, 379–399.

Schildkraut, J. J. (1965). The catecholamine hypothesis of affective disorders: A review of supporting evidence. *Am. J. Psychiatry* **122**, 509–522.

Schneider, F., Gur, R. E., Mozley, L. H., *et al.* (1995). Mood effects on limbic blood flow correlate with emotional self-rating: A PET study with O-15 labeled water. *Psychiatry Res.* **61**, 265–283.

Sheline, Y. I., Wang, P. W., Gado, M. H., Csernansky, J. G., and Vannier, M. W. (1996). Hippocampal atrophy in recurrent major depression. *Proc. Natl. Acad. Sci. U.S.A.* **93**, 3908–3913.

Sheline, Y. I., Gado, M. H., and Price, J. L. (1998). Amygdala core nuclei volumes are decreased in recurrent major depression. *NeuroReport* **22**, 2023–2028.

Shively, C. A., Laber-Laird, K., and Anton, R. F. (1997). Behavior and physiology of social stress and depression in female cynomolgus monkeys. *Biol. Psychiatry* **41**, 871–882.

Simon, H., LeMoal, M., and Calas, A. (1979). Efferents and afferents of the ventral tegmental-A$_{10}$ region studied after local injection of [^{3}H]-leucine and horseradish peroxidase. *Brain Res.* **178**, 17–40.

Sinyor, D., Jacques, P., Kaloupek, D. G., Becker, R., Goldenberg, M., and Coopersmith, H. (1986). Post stroke depression and lesion location: An attempted replication. *Brain* **109**, 537–546.

Skolnick, P., Layer, R. T., Popik, P., Nowak, G., Paul, I. A., and Trullas, R. (1996). Adaptation of N-methyl-D-aspartate (NMDA) receptors following antidepressant treatment: Implications for the pharmacotherapy of depression. *Pharmacopsychiatry* **29**(1), 23–26.

Smith, G. S., Reynolds, C. F., Pollock, B., Derbyshire, S., Nofzinger, E., Dew, M. A., Houck, P. R., Milko, D., Meltzer, C. C., and Kupfer, D. J. (1999). Cerebral glucose metabolic response to combined total sleep deprivation and antidepressant treatment in geriatric depression. *Am. J. Psychiatry* **156**, 683–689.

Smith, K. A., Morris, J. S., Friston, K. J., Cowen, P. J., and Dolan, R. J. (1999). Brain mechanisms associated with depressive relapse and associated cognitive impairment following acute tryptophan depletion. *Br. J. Psychiatry* **174**, 525–529.

Soars, J. C., and Mann, J. J. (1997). The anatomy of mood disorders: Review of structural neuroimaging studies. *Biol. Psychiatry* **41**, 86–106.

Stancer, H. C., and Cooke, R. G. (1988). Receptors in affective illness. *In* "Receptors and Ligands in Psychiatry" (A. K. Sen and T. Lee, Eds.), pp. 303–326. Cambridge Univ. Press, New York.

Starkstein, S. E., and Robinson, R. G., Eds. (1993). "Depression in Neurologic Diseases." Johns Hopkins Press, Baltimore.

Starkstein, S. E., Robinson, R. G., and Price, T. R. (1987). Comparison of cortical and subcortical lesions in the production of post-stroke mood disorders. *Brain* **110**, 1045–1059.

Starkstein, S. E., Bolduc, P. L., Mayberg, H. S., *et al.* (1990a). Cognitive impairments and depression in Parkinson's disease: A follow-up study. *J. Neurol. Neurosurg. Psychiatry* **53**, 597–602.

Starkstein, S. E., Mayberg, H. S., Berthier, M. L., Fedoroff, P., Price, T. R., Dannals, R. F., Wagner, H. N., Leiguarda, R., and Robinson, R. G. (1990b). Mania after brain injury: Neuroradiological and metabolic findings. *Ann. Neurol.* **27**, 652–659.

Steffens, D. C., Tupler, L. A., Ranga, K., and Krishnan, R. (1998). Magnetic resonance imaging signal hypointensity and iron content of putamen nuclei in elderly depressed patients. *Psychiatry Res.* **83**, 95–103.

Stoudemire, A., Hill, C. D., and Gulley, L. R. (1989). Neuropsychological and biomedical assessment of depression—dementia syndromes. *J. Neuropsychiatry Clin. Neurosci.* **1**, 347–361.

Stuss, D. T., and Benson, D. F. (1986). "The Frontal Lobes." Raven Press, New York.

Swerdlow, N. R., and Koob, G. F. (1987). Dopamine, schizophrenia, mania and depression: Towards a unified hypothesis of cortico-striato-pallido-thalamic function. *Behav. Brain Sci.* **10**, 197–245.

Thase, M. E., and Rush, A. J. (1995). Treatment-resistant depression. *In* "Psychopharmacology: The Fourth Generation of Progress" (F. E. Blood and D. J. Kupfer, Eds.), pp. 1081–1097. Raven Press, New York.

Thiebot, M. H., Martin, P., and Puech, A. J. (1992). Animal behavioral studies in the evaluation of antidepressant drugs. *Br. J. Psychiatry* **15,** 44–50.

Torack, R. M., and Morris, J. C. (1988). The association of ventral tegmental area histopathology with adult dementia. *Arch. Neurol.* **45,** 211–218.

Tremblay, L., and Schultz, W. (1999). Relative reward preference in primate orbitofrontal cortex. *Nature* **398,** 704–708.

Trullas, R., and Skolnick, P. (1990). Functional antagonists at the NMDA receptor complex exhibit antidepressant actions. *Eur. J. Pharmacol.* **185**(1), 1–10.

Tucker, D. M., Luu, P., and Pribram, K. H. (1995). Social and emotional self-regulation. *Ann. N.Y. Acad. Sci.* **769,** 213-239.

Vaidya, V. A., Marek, G. J., Aghajanian, G. K., and Duman, R. S. (1997). 5-HT$_{2A}$ receptor-mediated regulation of brain-derived neurotrophic factor mRNA in the hippocampus and the neocortex. *J. Neurosci.* **17,** 2785–2795.

Van Hoesen, G. W., Morecraft, R. J., and Vogt, B. A. (1993). Connections of the monkey cingulate cortex. *In* "The Neurobiology of Cingulate Cortex and Limbic Thalamus: A Comprehensive Handbook" (B. A. Vogt and M. Gabriel, Eds.), pp. 249–284. Birkhauser, Boston.

Vetulani, J., and Sulser, F. (1975). Actions of various antidepressant treatments reduce reactivity of noradrenergic cyclic AMP-generating system in limbic forebrain. *Nature* **257,** 455–456.

Vogt, B. A., and Pandya, D. N. (1987). Cingulate cortex of the rhesus monkey: II. Cortical afferents. *J. Comp. Neurol.* **262,** 271–289.

Vogt, B. A., Nimchinsky, E. A., Vogt, L. J., and Hof, P. R. (1995). Human cingulate cortex: Surface features, flat maps, and cytoarchitecture. *J. Comp. Neurol.* **359,** 490–506.

Warsh, J. J., Chiu, A. S., and Li, P. P. (1988). Noradrenergic mechanisms in affective disorders: Contributions of receptor research. *In* "Receptors and Ligands in Psychiatry" (A. K. Sen and T. Lee, Eds.), pp. 271–302. Cambridge Univ. Press, New York.

Weingartner, H., Cohen, R. M., Murphy, D. L., Martello, J., and Gerdt, C. (1981). Cognitive processes in depression. *Arch. Gen. Psychiatry* **38,** 42–47.

Willner, P. (1991). Animal models as simulations of depression. *Trends Pharmacol. Sci.* **12,** 131–136.

Wise, R. A. (1980). The dopamine synapse and the notion of "pleasure centers" in the brain. *Trends Neurosci.* **4,** 91–95.

Wu, J. C., and Bunney, W. E. (1990). The biological basis of an antidepressant response to sleep deprivation and relapse: Review and hypothesis. *Am. J. Psychiatry* **147,** 14–21.

Wu, J. C., Gilin, J. C., Buchsbaum, M. S., Hershey, T., Johnson, J. C., and Bunney, W. E., Jr. (1992). Effect of sleep deprivation on brain metabolism of depressed patients. *Am. J. Psychiatry* **149,** 538–543.

Yakovlev, P. I. (1948). Motility, behavior, and the brain: Stereodynamic organization and neural coordinates of behavior. *J. Nerv. Ment. Dis.* **107,** 313–335.

Zacharko, R. M., and Anisman, H. (1991). Stressor-induced anhedonia in the mesocorticolimbic system. *Neurosci. Biobehav. Rev.* **15,** 391–405.

Zajonc, R. (1980). Feeling and thinking: Preferences need no inferences. *Am. Psychol.* **35,** 151–175.

Zisook, S., and DeVaul, R. (1985). Unresolved grief. *Am. J. Psychoanal.* **45**(4), 370–379.

Zisook, S., Shuchter, S. R., Sledge, P. A., *et al.* (1994). The spectrum of depressive phenomena after spousal bereavement. *J. Clin. Psychiatry* **55**(S), 29–36.

Zubenko, G. S., and Moossy, J. (1988). Major depression in primary dementia. *Arch. Neurol.* **45,** 1182–1186.

Zubenko, G. S., Sullivan, P., Nelson, J. P., Belle, S. H., Huff, F. J., and Wolf, G. L. (1990). Brain imaging abnormalities in mental disorders of late life. *Arch. Neurol.* **47,** 1107–1111.

20

The Neurobiology of Anxiety and Anxiety-Related Disorders: A Functional Neuroimaging Perspective

P. Chua[*] and R. J. Dolan[†,1]

[*]*Department of Neuropsychiatry and Department of Psychiatry, The Royal Melbourne Hospital, Parkville, Victoria 3050, Australia*
[†]*Wellcome Department of Cognitive Neurology, London WC1N 3BG, United Kingdom*

I. Introduction
II. Conceptual Issues in Neurobiological Accounts of Anxiety
III. Fear and Emotional Processing in the Brain
IV. Neuroimaging Studies of Fear Processing
V. Classical Conditioning as a Model of Fear Learning
VI. Processing Learned Fear Responses with and without Awareness
VII. Amygdala Interactions Related to Processing Aware and Nonaware Fear-Relevant Stimuli
VIII. Induction of Anxiety in Volunteer Subjects
IX. Psychopathological Studies of Anxiety Disorders
X. A Neuroanatomical Model of Anxiety
XI. Outstanding Issues in Understanding Mechanisms of Anxiety
XII. A Model of Emotional Processing
XIII. Conclusions
References

I. Introduction

Anxiety, can be conceptualized as a negative cognitive-affective-somatic state, the prominent feature being "a sense of uncontrollability focused on possible future threat, danger, or other upcoming, potentially negative events" (Barlow *et al.*, 1996). The central element of this negative affective state is a sense of fear (Izard and Youngstrom, 1996) and helplessness (Barlow *et al.*, 1996), accompanied by somatic components of central nervous

---

[1] To whom correspondence should be addressed.

system arousal. Its functional role, as manifest in normal anxiety, is to prepare for future threats, which may or may not be behaviorally relevant (Izard and Youngstrom, 1996).

Although anxiety is a normal psychological reaction to impending threat, or uncertainty, it also represents both the commonest psychiatric symptom and, in the form of general anxiety disorder (GAD), one of the most prevalent syndromes (Weissman *et al.*, 1978). Thus, in psychopathological contexts anxiety can occur either in isolation or as a component symptom of more widespread psychological morbidity. When anxiety is the predominant symptom, it is referred to as generalized anxiety disorder, a condition that reflects situations where patients experience pervasive anxiety in the absence of panic, agoraphobia, or other phobic symptoms. Consequently, core symptoms of fear, apprehension, and autonomic arousal occur in the absence of specific environmental precipitants. These objectless states can be contrasted with what are termed phobic disorders, which include agoraphobia and specific phobias, where the anxiety is bound to specific objects or environmental contexts. Examples of the former include snake or spider phobias whereas an example of the latter is agoraphobia, where anxiety is experienced in open or enclosed spaces. Anxiety is also a major component symptom in obsessive-compulsive disorder, where symptoms arise in response to recurrent and irrational fears, for example, a fear of contamination.

While many forms of anxiety arise without clear environmental precipitants, there are well-defined situations where anxiety arises directly out of individual experience. The paradigmatic example is what is termed post-traumatic stress disorder (PTSD), where anxiety is a consequence of exposure to traumatic events that are out of the range of everyday experience. Consequently, anxiety is re-experienced in the setting of evoked trauma-related memories or actual exposure to contexts that are similar to or identical with that of the precipitating traumatic event. Note again that a unifying feature in these disparate conditions is the presence of a psychological state of fear and apprehension associated with symptoms of autonomic arousal, including tachycardia, tachypnoea, pupil dilatation, enhanced reflexes, tremor, and increased startle response.

In this chapter, we provide a framework for understanding the neurobiology of clinical anxiety. The principal data source that we consider will be functional neuroimaging. A key assumption that we make is that anxiety is intimately related to the emotion fear. Fear represents what is termed a basic emotion, meaning an evolved pan-cultural state, that can be functionally defined as an adaptive response to danger or threat. This adaptive response is characterized by stereotyped behavioral patterns that involve co-ordinated motor, autonomic, and perceptual responses. It can be assumed that fear, and its associated behavioral repertoires, are necessary for survival (Griffiths, 1997). Its adaptive value accounts for the fact that fear, and its mediating neurobiological mechanisms, are conserved across phylogeny (LeDoux, 1996). This evolutionary perspective suggests that fear and anxiety are mechanisms whose prime function is to ensure danger avoidance. This reasoning implicitly suggests that fear-related responses are particularly likely to be engendered by situations historically related to scenarios that represented threat or danger in our environment of evolutionary adaptiveness. Empirical data that support this idea include observations that fear responses are acquired with greater facility to specific stimuli such as snakes or particular facial expressions of emotion (Orr and Lanzetta, 1980, 1984). This facilitation of fear-related learning is often referred to as learning preparedness (Seligman, 1971).

The relevance of these considerations for clinical phenomena is most obviously apparent in the case of phobias. Phobic responses to stimuli such as spiders and snakes not only are common but are seen in environments where snakes and hazardous spiders are relatively rare or even absent. This contrasts with responses to objects that represent greater danger, for example, cars, which are rarely the focus of phobic responses per se. It may be conjectured that an inherited facilitation of a learning disposition toward certain classes of stimuli may underpin some manifestations of anxiety.

II. Conceptual Issues in Neurobiological Accounts of Anxiety

A traditional approach to studying the neurobiology of psychiatric phenomena involves identifying target patient populations and comparing them with a control group in order to isolate differences with respect to some dependent variable. There is an extensive neuroimaging literature, including a literature on anxiety and related disorders, where this approach has highlighted regional differences in neuronal function between patients and controls, frequently under conditions of task-related activation. The major limitation of this approach is its atheoretical nature that invariably leaves unexplained the functional significance of any highlighted patient-control differences. We propose a different procedure that is critically dependent on formulating an a priori neurobiological hypothesis. This approach is strongly influenced by the ideas of David Marr, who argued that an adequate psychology, and one assumes an adequate psychopathology, must make use of three levels of description. In this scheme the highest level is the psychological; the intermediate level is the computational and describes what the system computes to give rise to the highest level; and the lowest level describes how these computational processes are implemented in brain function (Marr, 1982). An example of this approach is a recent study that addressed mechanisms of face processing (George *et al.*, 1999). The psychological issue of prime interest in this study was face recognition (level 1). We showed that face recognition depends upon a computational process that involves extraction of 3-D information from light and shading cues (level 2). The neurobiological instantiation of this process involved the midfusiform region of inferior temporal cortex (level 3).

Translating this approach to the context of psychopathology means that we have to first define the particular psychological phenomenon under investigation. A tradition in psychiatry is to take the syndrome of disease entity as the frame of reference, for example, schizophrenia or depression. This we believe is a mistaken assumption, insofar as specification of these entities is highly arbitrary. Instead we emphasize symptoms that can be defined unambiguously (Dolan *et al.*, 1993). Taking symptoms as the frame of reference allows reformulation of the nature of the symptom in terms of an underlying psychological process. This provides a link to Marr's second level that involves defining the computational or functional role of the specified psychological process. The third level of description involves providing an account of the neural implementation of these two other levels.

In this framework an analysis of anxiety and anxiety disorders, from the psychological perspective, may be seen as the inappropriate engagement of a fear system. The functional or computational role of a fear system is to detect stimuli that represent danger or threat to the organism. The goal of neurobiological research is to understand its implementation in terms of brain function. Pathological anxiety, as seen in anxiety disorders, by definition implies that anxiety is situationally inappropriate. It follows that the delineation between normal and pathological anxiety, at the phenomenological and neurobiological levels, remains arbitrary. This overlap between fear and anxiety, as well as their coexistence in a variety of psychopathological disorders, suggests closely related neurobiological processes. We will argue that understanding anxiety requires an account of how the human brain processes fear or danger. As already indicated, the functional role of fear is to detect danger and implement, in an adaptive manner, appropriate behavioral responses. In psychopatholgy such as anxiety disorders, the proposal is that this danger-detecting system is, for whatever reason, maladaptively engaged.

In terms of an analysis of the phenomenology of fear and anxiety, it is worth considering their commonalities. A common cognitive component is a predictive state of imminent danger. Both fear and anxiety involve a state of heightened autonomic arousal and motor expectancy. Finally, as a consequence of this altered motor and autonomic state, there is a change in the internal mileu or bodily state. In anxiety disorders the range of triggering stimuli include phobic-related stimuli and exposure to traumatic contexts. However, it is clear that triggering stimuli must also include a wide range of cognitions. The diversity of inputs that can elicit anxiety strongly suggests major epigenetic components derived from learning. By contrast to the diversity of triggering stimuli, the behavioral sequelae to engagement of anxiety and fear are highly stereotyped, involving, for example, autonomic arousal and enhanced startle, suggestive of hard-wired mechanisms.

In this chapter, we begin with an account of how the human brain processes stimuli that evoke or represent fear. We then consider the degree to which neurobiological systems identified in fear processing are also implicated in functional neuroimaging studies of anxiety and anxiety-related disorders. Next we ask whether there are characteristics of human fear processing that can account for core features of anxiety disorders. These core features include, first, the prominent disjunction between the cognitive appraisal of a situation, for example, the knowledge that spiders are harmless, and intense fear responses experienced in these same situations. Finally, we provide a neurobiological model of fear and ask how the function, or dysfunction, of this system might account for phenomena of anxiety.

III. Fear and Emotional Processing in the Brain

Emotional processing can be characterized in terms of input and output components. An emotional input refers to a sensory stimulus that elicits an emotional state. An emotional output refers to the behavioral sequelae to engagement of an emotional system. It should be noted that primary emotions, including fear, are relatively invariant in their expression (Griffiths, 1997). By contrast, there is a high degree of variance in the input stimuli that elicit primary emotions. The implication is that most emotion-eliciting stimuli are learned. There is, nevertheless, substantial evidence that emotional learning mechanisms are innately biased (Seligman, 1971; Ohman *et al.*, 1976).

A considerable literature indicates that the neurobiological implementation of this danger-detection system involves the amygdala and anatomically interconnected structures. One of the earliest observations was in monkeys where bilateral lesions of the anterior temporal cortex that included the amygdala, hippocampus, and surrounding cortical areas led to the Kluver-Bucy syndrome, a characteristic feature of which is substantial attenuation or obliteration in expression of fear to threatening stimuli (Kluver *et al.*, 1939). Other evidence is derived from elicitation of fear by electrical or chemical stimulation as well as observations on the effects of lesions and local infusion of drugs that impair amygdala function (Davis, 1997). For example, benzodiazepines injected into the amygdala have specific anti-anxiety effects (Thomas *et al.*, 1985). The anxiolytic effects of peripherally administered benzodiazepines are blocked by the injection of GABA-A and benzodiazepine antagonists into the amygdala (Sanders and Shekhar, 1995). Electrical stimulation of the amygdala in humans elicits feelings of fear and anxiety and autonomic arousal (Chapman *et al.*, 1950; Gloor *et al.*, 1982). Lesions of the amygdala in humans impair the perception of fear or danger, particularly as expressed in conspecifics (Adolphs *et al.*, 1995, 1998; Calder *et al.*, 1996). Consequently, a general conclusion can be drawn that animal and human lesion data provide a powerful case that the amygdala is essential for the acquisition of fear conditioning and the expression of innate and learned fear responses (Armony, 1997).

IV. Neuroimaging Studies of Fear Processing

Neuroimaging studies addressing the neurobiology of fear processing in human subjects have used two general approaches involving either the use of stimuli with innate fear value, such as facial expressions, or fear-conditioning models. Morris and colleagues provided evidence of differential neuronal responses in the amygdala when volunteer subjects processed facial expressions of fear as opposed to happy facial expressions (Morris *et al.*, 1996). This amygdala response increased as a linear function of increasing degree of actual facial expression of fear. The fact that the task requirement in this study was for subjects to make a gender classification in response to presentation of the faces, with no requirement for explicit emotional recognition, suggests a high degree of automaticity in processing fear-related stimuli. Breiter and colleagues presented subjects with fearful, happy, and neutral faces and demonstrated activation of the amygdala in response to fearful versus happy faces and happy versus neutral faces, suggesting a generalized response to stimuli with emotional valence (Breiter *et al.*, 1996a).

V. Classical Conditioning as a Model of Fear Learning

A prerequisite for successful adaptation is an ability to associate the meaning and behavioral relevance of sensory stimuli. This associative learning is driven primarily by changes in expectations about whether future events involve rewards or punishment. One of the best characterized is what is termed classical conditioning. In standard conditioning paradigms a sensory cue (CS+) comes to predict reward or punishment. For this type of learning to occur the cue must consistently precede the reward (the unconditioned stimulus, or US, which usually has innate value) in order for a predictive association to develop. In the standard paradigm, known as delay conditioning, the US onset precedes CS offset. With repeated pairings, the CS comes to elicit a response previously expressed to the US alone. Unlike explicit forms of learning, this type of learning can occur without any awareness of the relationship between the CS and US. In contrast, extinction refers to the process whereby responses to conditioned stimuli are attenuated, and abolished, through repeated presentation of the conditioned stimulus without reinforcement from an US. The ventromedial prefrontal and medial orbitofrontal cortices (Morgan *et al.*, 1997) are implicated in this process, which is also indicative of an active learning process (Gewitz *et al.*, 1997).

The optimal conditions for the study of classical conditioning require the use of mixed-trial paradigms where sensory stimuli that predict the occurrence (CS+) or absence (CS−) of reward or punishment can be presented in an interleaved and random manner. Thus, addressing how classical conditioned fear responses are acquired has benefited from the advent of event-related functional magnetic resonance imaging (fMRI) (Buckner *et al.*, 1996; Josephs *et al.*, 1997). Studies using PET and fMRI have previously been limited by a necessity to use blocked designs in which target stimuli are repeatedly presented. In a typical experiment subjects are conditioned during one block by presenting the CS with the US and the effect tested in a second block where the CS is presented alone (Morris *et al.*, 1997). However, this procedure is confounded by the fact that the second block, when the CS is presented alone, may reflect extinction as much as learning.

Event-related fMRI (Buckner *et al.*, 1996; Dale and Buckner, 1997; Josephs *et al.*, 1997) is homologous to event-related potentials in electrophysiology, where responses to different stimuli, presented repeatedly over time, are individually sampled and subsequently averaged. To study the acquisition of conditioning using event-related fMRI, we chose as our target stimuli four neutral faces, two male and two female, taken from the Ekman series (Ekman, 1982). Subjects were scanned during two distinct phases. During conditioning two (one male and one female) of the four faces were paired with an unpleasant tone, lasting 500 ms and adjusted to 10% above each subjects pain threshold, to become CS+. To assess the evoked hemodynamic response to the CS+ in the absence of the US (i.e., tone), we employed a 50% partial reinforcement strategy. In effect, only half the presentations of the two CS+ were paired with the tone. In total we presented 104 stimuli over a period of approximately 20 min. Fifty-two of these were CS−, 26 were CS+ paired with noise, and 26 were CS− not paired with noise.

We defined four event types; three were time-locked to the onset of the presentation of the face and the fourth to the onset of the tone. The three visual events were subdivided into (i) CS−, (ii) CS+ paired, and (iii) CS+ unpaired face stimuli. The comparisons of interest were between conditioning-evoked neural responses to the unpaired CS+ (i.e., no noise presentations) and the CS−. Differential activation of the anterior cingulate gyrus, bilateral anterior insula, and medial parietal cortex were seen specific to presentation of the CS+ conditions (see Fig. 1). Further differential responses were detected in the supplementary motor area (SMA), the left premotor cortex, and bilateral red nuclei at a lower level of significance. These data, acquired during learn-

ing acquisition, implicate an extended neural system that includes sensory processing regions, such as the insula, and motor output regions, such as the cingulate and SMA, in fear learning.

A predicted amygdala activation was not found in our predefined analysis involving categorical comparisons of event types that assumes time-invariant learning responses. However, there are neurobiological data that suggest fear-related responses in the amygdala rapidly habituate (Breiter *et al.*, 1996a). Consequently, we carried out an analysis that explicitly modeled habituation effects to characterize a response profile involving an initial increase and subsequent decrease in neuronal response to presentation of the CS+. With such an analysis, bilateral activations of amygdala were evident, consistent with rapid habituation. A similar habituation in response of the amygdala has also been reported by others (LaBar *et al.*, 1998). Thus, a major conclusion from neuroimaging studies of fear conditioning is that the amygdala has a time-limited role in fear learning. This might suggest that engagement of the amygdala initiates learning-related plastic changes in afferent systems that convey behaviorally relevant sensory inputs to the amygdala as well as in output systems that mediate adaptive behavioral responses. For example, it has been shown that thalamo-cingulate plasticity associated with conditioning mediates learned behavioral responses and that this plasticity is dependent

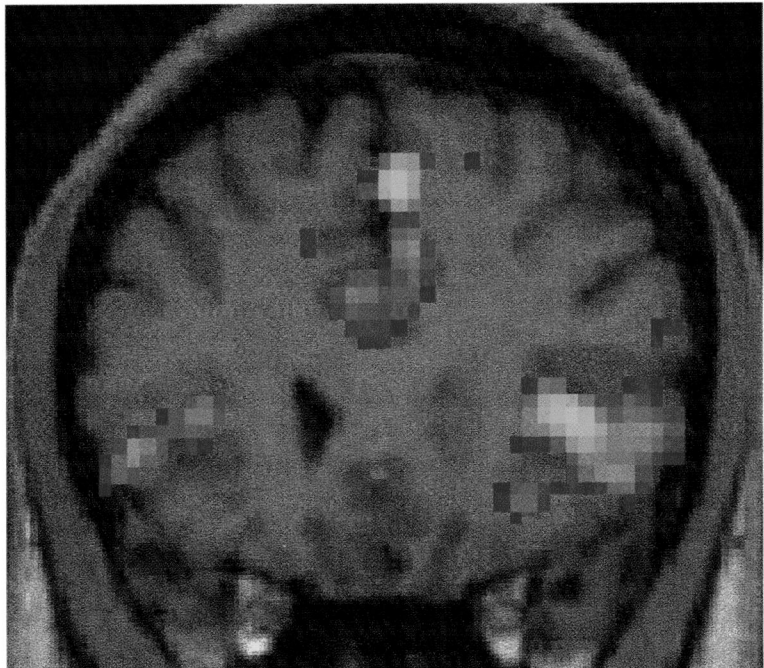

Figure 1 Illustration of activation in the anterior cingulate cortex and bilateral insulae in a single individual subject, for the categorical contrast of CS+ and CS− stimuli (see text).

upon an intact amygdala (Poremba *et al.*, 1997). However, it is important to acknowledge that the amygdala is not the only region implicated in fear conditioning. Consideration of the functional significance of activation in other key regions such as the insula and cingulate will be a subject matter in following sections.

VI. Processing Learned Fear Responses with and without Awareness

Fear-relevant stimuli can activate a danger-processing system with little or no conscious awareness of the elicitor. Important empirical research consistent with this conjecture includes evidence of discriminatory skin conductance responses (SCR's) to aversively conditioned stimuli that have been backwardly masked to prevent conscious awareness of their occurrence (Esteves *et al.*, 1994; Öhman and Soares, 1994; Parra *et al.*, 1997). Studies of patients with selective brain lesions also provide insights into the mechanisms of unconscious processing of behaviorally relevant stimuli. Patients with amygdala damage fail to acquire conditioned SCRs to stimuli paired with an aversive unconditioned stimulus (UCS) despite intact declarative knowledge concerning stimulus associations (Bechara *et al.*, 1995; LaBar *et al.*, 1995). Conversely, patients with intact amygdalae and damaged hippocampi acquire reliable discriminatory SCRs but do not report which stimuli have been paired with the UCS (Fried *et al.*, 1997).

Is the amygdala implicated in processing fearful but unseen behaviorally relevant stimuli in humans? To address this question, we used a backward masking procedure involving repeatedly presenting targets on a screen for 30 ms that were immediately followed by a masking stimulus for 45 ms. This masking procedure effectively prevents reportable awareness of the target fear-elicit-

ing stimulus. We used faces as stimuli, consisting of two angry and two neutral expressions, where in a conditioning study phase one of the angry faces (the CS+) was paired with a 1-s 100-dB white noise burst. None of the neutral faces was ever paired with the noise. During scanning, these two faces, a target, and mask were repeatedly shown at 5-s intervals for each experimental condition. In effect, for half the experimental conditions, the subject's awareness of target faces, conditioned and unconditioned, was prevented by masking with neutral faces.

The comparison of primary interest in this study was a contrast of neural responses elicited by masked CS+ and CS− scan conditions. This contrast revealed a highly differential effect in the region of the right amygdala driven by an enhanced response in the masked CS+ compared to the masked CS− condition (see Fig. 2a). When we subsequently contrasted unmasked CS+ and CS− faces, a significant neural response was evident in the region of the left amygdala. The interaction between conditioning and masking indicated that masking significantly enhanced the right amygdala response to CS+ faces. Conversely, unmasking enhanced the left amygdala response to the CS+. Note again that the only difference between the CS+ and CS− conditions was the subject's prior experience of the temporal association between the CS+ faces and the aversive noise.

Two general conclusions can be drawn from these data. First, they indicate that the amygdala discriminates between perceptually similar stimuli solely on the basis of their prior behavioral history. This discrimination (at least for the right amygdala) can occur independent of conscious awareness. The differential amygdala activation provides a neurobiological account of behavioral data, which indicate unconscious processing of aversive stimuli (Esteves *et al.*, 1994; Öhman and Soares, 1994; Parra *et al.*, 1997). They also extend other

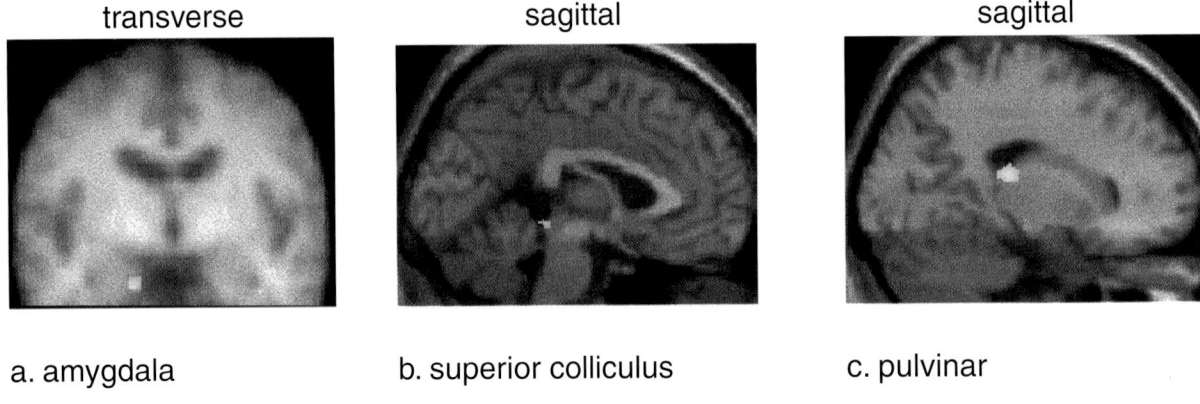

transverse	sagittal	sagittal
a. amygdala	b. superior colliculus	c. pulvinar

Figure 2 (a) Activation of the right amygdala (note that right is left in this image as per neuroradiological convention) during a comparison of neural responses associated with processing a masked CS+ and masked CS− (see text). Images (b) and (c) show the superior colliculus and pulvinar nucleus of the thalamus, which show maximal covariation in response with right amygdala during processing of masked unseen conditioned targets.

functional neuroimaging data that demonstrate that the amygdala can discriminate between innately salient, but perceptually distinct, stimuli (Whalen and Kapp, 1991). The possible clinical relevance of the finding is the implication that fear- and anxiety-eliciting stimuli do not have to be represented in conscious awareness to mediate their effects.

VII. Amygdala Interactions Related to Processing Aware and Nonaware Fear-Relevant Stimuli

The amygdala receives a major visual input from the anterior temporal lobe and this is the most likely pathway through which visual CS's access the amygdala (Amaral et al., 1992). This raises the question of what pathway mediates processing of masked or unseen behaviorally relevant stimuli. The disruption of normal visual processing by visual backward masking provides a temporary functional equivalent of the permanent loss of visual awareness seen with striate cortex lesions. Studies of patients with lesions to striate cortex, so-called "blindsight," provide evidence for parallel visual systems in the brain associated with different levels of conscious visual awareness (Weiskrantz et al., 1974). This has led to the suggestion that preserved abilities in these patients are mediated by a parallel secondary visual pathway comprising the superior colliculus and pulvinar nucleus of the posterior thalamus (Zihl and von Crammon, 1979; Gross, 1991).

We conjectured that the anatomical pathways proposed as providing a second visual pathway in blindsight (i.e., superior colliculus and pulvinar) might also be implicated in processing masked, fear-conditioned ("unseen") stimuli. It should be noted in this regard that animal models of auditory fear conditioning have proposed that a subcortical anatomical pathway involving the medial geniculate nucleus (MGN), and its projections to the lateral nucleus of the thalamus and amygdala, provide a rapid processing system for auditory CS's (LeDoux, 1996). Thus, in animal experiments involving auditory fear-conditioning paradigms, posttraining lesions of the auditory cortex result in loss of ability for discriminatory conditioned responding, though crude non-discriminatory responding is preserved (Jarrell et al., 1987). This suggests that an intact subcortical input to the amygdala can provide a sufficient input to mediate basic components of this form of associative learning.

To address the nature of the amygdala's functional interactions with cortical and subcortical regions, we extended the analysis of the neuroimaging data described in the previous section (Morris et al., 1998b, 1999). The

key question in this analysis related to whether the right amygdala showed a distinct pattern of interaction, with cortical or subcortical regions, under the context of processing masked conditioned stimuli compared to all other experimental conditions. This analysis is predicated on an assumption that brain regions constituting a functionally co-operative network should manifest context-specific patterns of interactions (Friston et al., 1997). In agreement with our a priori prediction, superior colliculus (Fig. 2b) and bilateral pulvinar (Fig. 2c) demonstrated positive covariation in reponse with activity in the right amygdala during masked (unaware) presentations of CS+ faces. By contrast, bilateral regions of fusiform, subsuming regions implicated in explicit processing of faces, as well as orbitofrontal cortices showed negative covariation of activity with right amygdala specific to the masked CS+ condition (Morris et al., 1999).

These results provide support for a hypothesis that behaviorally relevant features of the visual environment may be detected and processed, without conscious awareness, by a subcortical colliculo-pulvinar-amygdala pathway that also controls reflexive and autonomic responses. Explicit detection of face stimuli involves specialized cortical processing areas, for example, fusiform gyrus and temporal pole, regions associated with detailed analysis of the visual scene leading to object categorization (Dolan et al., 1997). The backward masking employed in this study disrupts processing in these cortical pathways while leaving the subcortical pathway unaffected. Unlike the retino-geniculo-striate pathway, which comprises relatively slowly conducting wavelength-sensitive neurons, the retino-collicular-pulvinar pathway is characterized by a rapidly conducting non-wavelength-sensitive transient response profile (Schiller and Malpeli, 1977). This response profile is in keeping with an idea that this pathway is suited to processing briefly presented visual stimuli and is less vulnerable to the influence of a backward mask. However, it needs to be borne in mind that the context-dependent interactions that we noted may be a downstream effect from the amygdala itself, reflecting the fact that a behaviorally relevant visual input through a cortical route is degraded due to masking.

These findings again have potential clinical relevance in that they raise the possibility that fear and anxiety-related environmental cues, most obviously phobic stimuli, can access the amygdala via a non-cortical route. Although speculative, these data may explain why phobic fears can be so irrational. Modulation of normal sensory processing by higher cognition may be relatively ineffective for stimuli subject to processing through this subsidiary route, a possibility that is clearly open to empirical testing using appropriate clinical populations.

VIII. Induction of Anxiety in Volunteer Subjects

Having considered the functional anatomy of fear processing, we next consider the degree to which functional systems described in studies of anxiety disorder map to neural systems mediating fear acquisition and expression. Recall that the most robust findings in the latter studies involve anterior insula, anterior cingulate cortex, and amygdala, regions that are likely to mediate separate components of the experience of fear.

One of the earliest reports of interest involved a study of anticipatory anxiety induced by threat of electric shocks (Reiman *et al.*, 1989). Here, induced anxiety produced significant increases in neuronal activity in bilateral temporal poles, although there was subsequent controversy as to whether the findings reflected muscle artifact from anxiety-associated teeth clenching (Drevets *et al.*, 1992).

In a more recent study, we used a similar paradigm involving the threat of mild electric shocks to the hand to induce states of anticipatory anxiety. Simultaneously, we manipulated the subject's attention by asking the subject to perform either an easy or difficult distracting motor learning task. Thus, the experiment involved a dual-task paradigm of electric shocks and motor performance designed to tease apart the neural substrates involved in cognitive aspects of anxiety from those responsible for affective and physiological components. We found that anticipatory anxiety, induced by the expectancy of electric shocks, was associated with increased neural activity in regions such as the left insula

and left anterior cingulate cortex (see Fig. 3). No interaction was detected between the shock and motor distraction conditions, possibly reflecting the fact that the latter was insufficiently attention demanding. When subjectively rated anxiety scores were used as a covariate of interest, a strong positive correlation emerged between high anxiety scores and activity in the left orbitofrontal cortex, left insula, and left anterior cingulate cortex. The latter finding is suggestive of a degree of specificity in these regions for the experience of anxiety.

A novel pharmacological symptom provocation approach to induction of anxiety was reported in a study of Benkelfat *et al.* (1995). These investigators used a double-blind design of intravenous injection of cholecystokinin tetrapeptide (CCK4) or saline placebo in healthy volunteers. CCK4 is known to induce panic or anxiety in healthy subjects. Relative activation was noted in the claustrum-insular-amygdala complex, left anterior cingulate, and cerebellar vermis specific to the CCK4-induced anxiety condition. In a similar approach, Ketter *et al.* used procaine to induce fear in healthy subjects and found that this manipulation increased activity in anterior paralimbic structures (Ketter *et al.*, 1996). Strikingly, a resulting psychological state of fear was positively correlated with activity in the left amygdala (Ketter *et al.*, 1996).

The consensus from these studies of induced anxiety is that the most robust activations are seen in anterior insula and to a lesser degree in anterior medial temporal lobe structures. Activation of the amygdala has mainly been observed during pharmacological induction of anxiety.

transverse

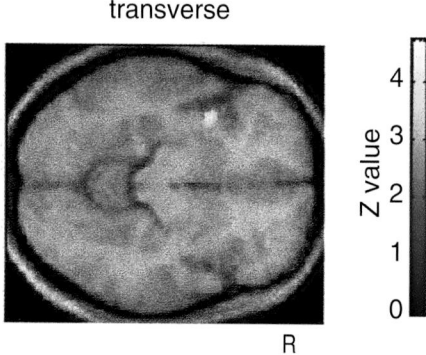

a. anterior insula

Z = 4.85, p = 0.000 (0.026 corrected)

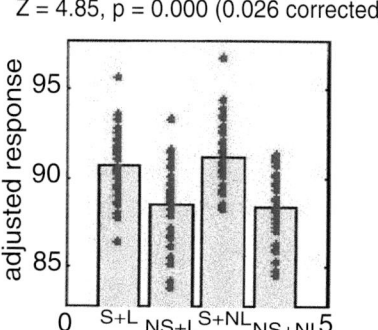

b. rCBF

Figure 3 The area of maximal activation in the left anterior insula during a state of anticipatory anxiety is illustrated on the left with condition-specific adjusted rCBF responses illustrated in the right-hand panel (see text). S,- shock condition; NS,- non-shock condition; L,- motor learning;- NL,- non-learning condition. Enhanced rCBF during the shock conditions can be seen in the histograms, with no modulation by the dual-task conditions.

IX. Psychopathological Studies of Anxiety Disorders

In this section, we consider selective studies of patient populations, focusing on general anxiety disorders (GAD), phobic disorders, obsessive-compulsive disorder (OCD), and post-traumatic stress disorder (PTSD).

A. Anxiety (GAD)

The studies described in the previous section used symptom provocation strategies in normal volunteer subjects to explore the functional anatomy of anxiety. A more conventional approach is to study patient populations. However, very few studies have addressed patient-control differences in GAD. One interesting approach has been reported by Rauch et al. (1997). These authors pooled data from three studies of OCD, simple phobia, and PTSD, respectively, all conditions in which anxiety is a prominent symptom. Using a diagnosis-independent measure of anxiety, they explored correlations with regional cerebral blood flow (rCBF). Significant correlations were found in right posterior medial orbitofrontal cortex, right inferior frontal cortex, bilateral insular cortices, and bilateral brainstem.

B. Obsessive–Compulsive Disorder (OCD)

Obsessive–compulsive disorder is characterized by recurrent thoughts and impulses and ritualistic compulsions. Patients with this condition have immense difficulty in suppressing thoughts and impulses, without engendering distressing levels of anxiety, despite retaining insight into the senseless nature of their ritualistic actions. Activation paradigms have capitalized on symptom provocation techniques to induce OCD symptoms, including anxiety, in this patient group. For example, McGuire et al. used a graded exposure paradigm by subjecting medication-free patients to an individualized hierarchy of contaminants that elicited the compulsion to handwash and associated anxiety (McGuire et al., 1994). Enhanced neural responses in the inferior frontal gyrus, caudate nucleus, putamen, globus pallidus, and thalamus on the right side and hippocampus, posterior cingulate gyrus, and cuneus on the left side correlated significantly with induced symptom intensity. Rauch et al., in a similar type of study that also involved a provoked symptomatic state, demonstrated a significant increase in relative activity during the symptomatic state in the right caudate nucleus, left anterior cingulate cortex, and bilateral orbitofrontal cortex (Rauch et al., 1994). Finally, Breiter et al., again using a symptom provocation technique, reported activation in medial orbitofrontal, lateral frontal, anterior temporal, and insular cortices in two-thirds of their patients (Breiter et al., 1996b).

C. Phobic Disorders

Phobias are characterized by clinically significant anxiety in response to objects or situations, often associated with active avoidance of the feared stimulus. One of the earliest studies of this disorder by Mountz et al. (1989) failed to find any group differences in global or regional activity, despite increased anxiety, in subjects with small-animal phobias when confronted by the phobic stimulus. Wik et al. (1993) measured activity during exposure to videotapes of snakes, aversive scenes, or neutral scenes in subjects with snake phobia. During exposure to the phobogenic visual stimuli, increased activity was noted in the secondary visual cortex and decreased activity in the hippocampus, orbitofrontal cortex, prefrontal cortex, temporopolar cortex, and posterior cingulate relative to a situation of viewing neutral videos. Similarily, Rauch et al. measured activity during symptom provocation in simple phobics (Rauch et al., 1994). Significant activation was observed during symptomatic states in the anterior cingulate, left insular cortex, right anterior temporal cortex, left somatosensory cortex, left posterior medial orbitofrontal cortex, and the left thalamus.

D. Post-traumatic Stress Disorder (PTSD)

PTSD is characterized by re-experiencing an extremely traumatic event accompanied by symptoms of increased arousal, persistent avoidance of stimuli associated with the event, and numbing of general responsiveness. Rauch et al. studied PTSD subjects and compared a traumatic imagery condition with a neutral control condition and reported increased activity in several right-sided structures, including amygdala, orbitofrontal cortex, insular cortex, anterior temporal pole, anterior cingulate cortex, and secondary visual cortex (Rauch et al., 1996). In a follow-up study, adult survivors of childhood sexual abuse with and without PTSD symptoms were compared using script-driven imagery conditions. Increased neural activity was observed in orbitofrontal cortex and anterior temporal poles in the PTSD group specific to a traumatic imagery condition (Shin et al., 1999). In a study of war veterans with PTSD, exposure to traumatic material resulted in deactivation in medial prefrontal cortex (Bremner et al., 1999).

A summary of the above studies, using symptom provocation techniques in volunteers and patients with anxiety-based conditions, does not provide a sense of consistency across the various studies. One of the difficulties with these studies is the lack of a clear model of brain systems implicated in anxiety and their functional role. We now consider a putative model.

X. A Neuroanatomical Model of Anxiety

As suggested, a difficulty in interpreting findings from patients is distinguishing anxiety-related activity from nonspecific activations related to experimental procedures, for example, pharmacological provocation or cooccurrence of symptoms such as traumatic imagery as, for example, in the case of PTSD-induced anxiety. Nevertheless, despite these interpretative limitations, the brain regions most commonly highlighted in studies of GAD, and anxiety-related disorders, are orbitofrontal cortex, anterior cingulate, and anterior insulae. Strikingly, activation of the amygdala is not a common finding in studies of GAD and anxiety-related conditions. In this section, we consider the functional significance of activation in these key regions during states of anxiety.

A. Amygdala

On the basis of studies of fear conditioning in animals and humans, a powerful argument can be made that the amygdala is a pivotal structure in the genesis of fear and anxiety (LeDoux, 1986, 1996; Davis, 1992; LaBar *et al.*, 1995; Morris *et al.*, 1996). The stimuli that trigger a response in the amygdala include incoming sensory signals, such as sight of phobic objects, interoceptive cues involving alteration in the internal bodily milieu, and cognition, for example, retrieved traumatic memories. The functional role of the amygdala in response to these inputs is derived from determining their behavioral relevance. More specifically, its functional role is to link sensory or cognitive stimuli to central representations of fear. When this linkage is to specific objects or contexts, then we propose the ensuing psychological state will have strong phobic elements. When the linkage is to an interoceptive or higher cognitive state, the ensuing psychological state will have strong, non-stimulus-bound, anxiety elements.

In terms of anatomical inputs to the amygdala, these are mainly to the lateral and basolateral nuclei (LeDoux *et al.*, 1990). These pathways convey highly processed sensory inputs from widespread cortical regions, including auditory and visual sensory systems, as well as the insular cortex. These lateral nuclei project in turn to the central nucleus of the amygdala, which then sends efferent projections to the hypothalamus and brainstem. There is overwhelming evidence from animal studies that these output pathways are critical in generating autonomic responses. Other major efferents

project to primary and association sensory cortices, prefrontal cortices, including orbitofrontal cortex, hippocampus, and olfactory cortex, ventral striatum, bed nucleus of the stria terminalis, and the thalamus (Amaral and Price, 1984; Amaral, 1987). These efferent pathways have been hypothesized to mediate other components of fear and anxiety, including perceptual, cognitive, and motoric components. For example, the basolateral amygdala has extensive connections to the frontal cortex (Amaral and Price, 1984) which may be important in the actual conscious perception of fear itself. The amygdalostriatal projections may be involved in affective and motoric response components of reward and conditioning (Russchen *et al.*, 1985).

One outstanding question is why amygdala activation is not generally highlighted in studies of anxiety and related disorders. In our fear-conditioning studies, and in other data, it is clear that the role of the amygdala in signaling danger is time-limited (Fried *et al.*, 1997; Buchel *et al.*, 1998). Recall that its profile of response in fear conditioning involves initial activation and subsequent response attenuation. In blocked experimental designs, an approach that characterizes all patient studies reported to date, this type of response pattern would not be evident due to the fact that critical anxiety-eliciting stimuli or contexts are present across the entire block. Thus, to adequately address the role of the amygdala requires experimental paradigms where it will be possible to assess Time × by Condition interactions, a fact that will necessitate the use of event-related designs (see above). Clearly, this approach is feasible for disorders such as specific phobias and PTSD but is likely to be more problematic in GAD, where phasic anxiety responses to provoking stimuli are less evident.

B. Insular Cortex

The anterior insular cortex is consistently activated in studies of anxiety as well as in fear-conditioning models. It receives dense sensory input, including inputs from the viscera. The richness of its sensory connections lends support to a notion that a key functional role is as both a visceral sensory and visceral motor area. Notably, the insula is a cortical region with anatomical projections to anterior cingulate, perirhinal, entorhinal, and periamygdaloid cortex and various amygdaloid nuclei, which has led to its conceptualization as an area functionally associated with emotional processing (Augustine, 1996). Several functional neuroimaging studies suggest a role for the anterior insula in processing emotional contexts, for example, pain (Casey *et al.*, 1994) or recollection of affect-laden autobiographical information (Fink *et al.*, 1996). Its strong connections to sensory regions, including the viscera, may also mean that its activation reflects interoceptive changes that anticipate or recapitulate those experienced during adversity, for example, an unpleasant noise (Mesulam and Mufson, 1982). The outstanding question is, what is its involvement in fear conditioning and anxiety? Recall that a critical component of anxiety and fear involves an autonomic and motor-generated change in visceral and somatic states. These changes are likely to be fed back to the cortex to provide critical information regarding the internal milieu. We propose that activation of the insula reflects a remapping of these interoceptive changes arising mainly out of the central generation of an autonomic response in both fear processing and anxiety disorders.

C. Anterior Cingulate Cortex

The cingulate is a functionally heterogeneous region with a known role in regulating context-dependent behaviors (Devinsky and Luciano, 1993). The anterior cingulate has been subdivided into areas specific for emotion (Brodmann areas 24, 25, and 33) and cognition (Brodmann areas 24″ and 32″) (Vogt and Pandya, 1987a,b; Vogt and Barbas, 1988; Devinsky et al., 1995; Vogt et al., 1995). The more ventral component of the anterior cingulate is primarily implicated in emotion and has extensive connections with amygdala, periaqueductal gray, and autonomic brainstem motor nuclei (Pandya et al., 1981; Devinsky et al., 1995). In humans, bilateral lesions of the anterior cingulate 24 induce complex emotional responses that include anxiety (Devinsky and Luciano, 1993). Activation in a similar region has been reported in association with processing the affective component of painful stimuli (Rainville et al., 1997). We have demonstrated activation of the anterior cingulate cortex in a functional neuroimaging study of fear conditioning (Buchel et al., 1998). Its activation in fear conditioning, and in the study of anxiety disorders, may reflect a predictive function with respect to the probability of occurrence of some future aversive event. A parallel function would involve automatic selection of context-appropriate adaptive behavioral outputs that include central engagement of autonomic mechanisms. We have shown that activation of anterior cingulate cortex is strongly related to the experience of cognitive effort and that activity in this region correlates with indices of peripheral autonomic arousal (Critchley et al., 2000). In this regard, we suggest that the cingulate is a key region in the co-ordination of cognition and autonomic function.

D. Orbitofrontal Cortex

Involvement of orbitofrontal cortex in anxiety and fear is not a consistent finding in neuroimaging studies. This is somewhat surprising in that orbitofrontal cortex is widely implicated in emotional processing in primates and humans. For example, orbitofrontal lesions disrupt feeding and social behavior in monkeys (Baylis and Gaffan, 1991). In humans, orbitofrontal lesions also impair social behavior and are associated with lack of affect (Damasio, 1994). Particularly relevant to the present discussion is the fact that orbitofrontal lesions cause deficits on reward and reward-related learning such as discrimination reversal learning and extinction (Butter, 1969; Jones and Mishkin, 1972). The strong implication is that learning processes involved in extinction of fear responses require the integrity of orbitofrontal cortex. Only one functional neuroimaging conditioning study in the literature has considered neural processes during extinction and it did not highlight involvement of amygdala (LaBar et al., 1998).

XI. Outstanding Issues in Understanding Mechanisms of Anxiety

A central tenet in fear conditioning is the observation that repeated exposure to a conditioned stimulus without reinforcement leads to extinction of fear responses. This raises the question as to why extinction is not seen in anxiety disorders. This might suggest that regions of the brain that are implicated in reversal learning are dysfunctional, for example, the orbitofrontal cortex. An alternate possibility is that iterative processes are engendered in those conditions where an unpleasant effect, or bodily state, associated with anxiety can act as a UCS that further reinforces the anxiety, be it focal in response to a discrete stimulus or more diffuse as in GAD. In other words, cognitive or somatic states themselves acquire, through associative mechanisms, an ability to act as reinforcers. This account would provide a generic mechanism for the maintenance of anxiety states and suggests that the most effective interventions would be those that are targeted at cognitive or altered somatic states. Indeed this suggestion gains explicit support from the therapeutic literature, where the principal target for effective cognitive therapy is dysfunctional cognition. However, it is clear that anxiety and fear responses may be engendered by stimuli that are not consciously perceived, which raises the possibility that cognitive therapies that require conscious access to antecedents of anxiety may be inadequate in a proportion of patients with anxiety disorders.

XII. A Model of Emotional Processing

A number of models of emotion and emotional learning have been articulated on the basis of animal

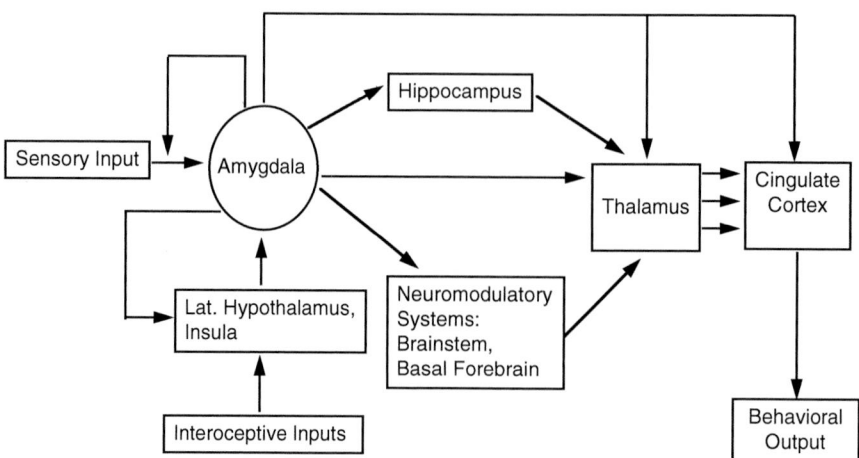

Figure 4 A model of emotional processing in the brain based upon a fear-conditioning model. The core value system involves the amygdala whose functional role is to detect situations that represent danger or threat to the organism. Inputs to the amygdala derive primarily from sensory channels, the internal mileu via the hypothalamus, or viscera via the insula. These inputs provide a continuously updated input. The primary outputs of the amygdala are to hippocampus, concerned with providing context-related information in the service of declarative knowledge, neuromodulatory systems such as the nucleus basalis of Meynert (NMB), involved in initiated learning-dependent plasticity, and brainstem, thalamus, and anterior cingulate cortex, responsible for adaptive behavioral responses.

data (LeDoux, 1996). A framework applied to human functional neuroimaging data, pertinent to understanding emotional disorders, is a modification of the model of value-dependent learning proposed by Friston *et al.* (1994). The model assumes the presence of evolutionary selected value systems that signal contexts that have survival value for the organism (see Fig. 4). Outputs from value systems reflect hard-wired responses to salient stimuli or contexts, such as food, sex, or danger, but also provide constraints, through selective consolidation of synaptic change, for the acquisition of novel adaptive behaviors. Thus, a key proposal of the model is that stimuli acquire value or behavioral relevance through consolidation of inputs to, and outputs from, the innately specified value systems. Note that this consolidation occurs under the influence of outputs from value systems and is expressed primarily when there is a change in the behavioral relevance of a stimulus. In terms of empirical neurobiology data that address this model, it is known that outputs from the amygdala to nucleus basalis of Meynert are capable of engendering this type of plasticity (Weinberger, 1995). Sensory afferent plasticity has been reported in neuroimaging studies of fear conditioning (Morris *et al.,* 1998). In anxiety and related disorders recursive mechanisms, as described in the previous section, may provide a mechanistic basis to the perpetuation of these disorders.

XIII. Conclusions

We have proposed a model of anxiety that is predicated on an assumption that a core psychological process is engagement of a fear system. Using fear conditioning as a model, we have provided an account of the relevant neurobiological systems in which the amygdala has a central role. Anxiety represents a complex state in which the experience of fear, a psychological state of expectancy, and enhanced autonomic arousal are core features. Our neurobiological model proposes that individual components of this state, involving fear, an altered autonomic state, and motor expectancy, are represented in activations within distinct cortical regions, namely, the amygdala, insula, and anterior cingulate cortex, respectively.

References

Adolphs, R., Tranel, D., Damasio, H., and Damasio, A. R. (1995). Fear and the human amygdala. *J. Neurosci.* **15,** 5879–5891.

Adolphs, R., Tranel, D., and Damasio, A. R. (1998). The human amygdala in social judgment. *Nature* **393,** 470–474.

Amaral, D. G. (1987). Memory: Anatomical organisation of candidate brain regions. *In* "Handbook of Physiology—The Nervous System V" (V. B. Mountcastle *et al.,* Eds.), pp. 211–294. American Physiological Society, Bethesda.

Amaral, D. G., and Price, J. L. (1984). Amygdalo-cortical projections in the monkey (*Macaca fascicularis*). *J. Comp. Neurol.* **230,** 465–496.

Amaral, D., Price, J. L., Pitkanen, A., and Carmichael, S. T. (1992). Anatomical organization of the primate amygdaloid complex. *In*

"The Amygdala: Neurobiological Aspects of Emotion, Memory and Mental Dysfunction" (J. P. Aggleton, Ed.), pp. 1–66. Wiley-Liss, New York.

Armony, J. L., and LeDoux, J. E. (1997). How the brain processes emotional information. *Ann. N.Y. Acad. Sci.* **821**, 259–270.

Augustine, A. R. (1996). Circuitry and functional aspects of the insular lobe in primates including humans. *Brain Res. Rev.* **22**, 229–244.

Barlow, D. H., Chorpita, B. F., and Turovsky, J. (1996). Fear, panic, anxiety, and disorders of emotion. *Nebr. Symp. Motiv.* **43**, 251–328.

Baylis, L. L., and Gaffan, D. (1991). Amygdalectomy and ventromedial prefrontal ablation produce similar deficits in food choice and in simple object discrimination learning for an unseen reward. *Exp. Brain Res.* **86**, 617–622.

Bechara, A., Tranel, D., Damasio, H., Adolphs, R., Rockland, C., and Damasio, A. R. (1995). Double dissociation of conditioning and declarative knowledge relative to the amygdala and hippocampus in humans. *Science* **269**, 115–118.

Benkelfat, C., Bradwejn, J., Ellenbogen, M., Milot, S., Gjedde, A., and Evans, A. (1995). Functional neuroanatomy of CCK4-induced anxiety in normal healthy volunteers. *Am. J. Psychiatry* **152**, 1180–1184.

Breiter, H. C., Ectoff, N. L., Whalen, P. J., *et al.* (1996a). Response and habituation of the human amygdala during visual processing of facial expression. *Neur* **2**, 875–887.

Breiter, H. C., RaBreiuch, S. L., Kwong, K. K., *et al.* (1996b). Functional magnetic resonance imaging of symptom provocation in obsessive-compulsive disorder. *Arch. Gen. Psychiatry* **53**, 595–606.

Bremner, D. J., Staib, L. H., Kaloupek, D., Southwick, S. M., Soufer, R., and Charney, D. S. (1999). Neural correlates of exposure to traumatic pictures and sound in Vietnam combat veterans with and without postraumatic stress disorder: A positron emission tomography study. *Biol. Psychiatry* **45**, 806–816.

Buchel, C., Morris, J., Dolan, R. J., and Friston, K. J. (1998). Brain systems mediating aversive conditioning: An event related fMRI study. *Neur* **20**, 947–957.

Buckner, R. L., Bandettini, P. A., Ocraven, K. M., *et al.* (1996). Detection of cortical activation during averaged single trials of a cognitive task using functional magnetic resonance imaging. *Proc. Natl. Acad. Sci. U.S.A.* **93**, 14878–14883.

Butter, C. M. (1969). Perseveration in extinction and in discrimination reversal tasks following selective prefrontal ablations in *Macaca mulatta*. *Physiol. Behav.* **4**, 163–171.

Calder, A. J., Young, A. W., Rowland, D., Perrett, D. I., Hodges, J. R., and Etcoff, N. L. (1996). Facial emotion recognition after bilateral amygdala damage: Differentially severe impairment of fear. *Cog. Neuropsychol.* **13**, 699–745.

Casey, K. L., Minoshima, S., Berger, K. L., Koeppe, R. A., Morrow, T. J., and Frey, K. A. (1994). Positron emission tomographic analysis of cerebral structures activated specifically by repetitive noxious heat stimuli. *J. Neurophysiol.* **71**, 802–807.

Chapman, W. P., Livingston, K. E., and Poppen, J. L. (1950). Effect upon blood pressure of electrical stimulation of tips of temporal lobes in man. *J. Neurophysiol.* **13**, 65–71.

Critchley, H. D., Corfield, D. R., Chandler, M. P., Mathias, C. J., and Dolan, R. J. (2000). Central correlate of peripheral autonomic cardiovascular arousal: A functional neuroimaging investigation. *J. Neurophysiol.* (in press).

Dale, A. M., and Buckner, R. L. (1997). Selective averaging of rapidly presented individual trials using fMRI. *Hum. Brain Mapp.* **5**, 329–340.

Damasio, A. (1994). "Descartes Error." Grosset/Putnam, New York.

Davis, M. (1992). The role of amygdala in conditioned fear. *In* "The Amygdala: Neurobiological Aspects of Emotion, Memory and Mental Dysfunction" (J. P. Aggleton, Ed.), pp. 255–306. Wiley-Liss, New York.

Davis, M. (1997). Neurobiology of fear responses: The role of the amygdala. *J. Neuropsychiatry Clin. Neurosci.* **9**, 382–402.

Devinsky, O., and Luciano, D. (1993). The contributions of the cingulate cortex to human behaviour. *In* "Neurobiology of the Cingulate Cortex and Limbic Thalamus" (B. A. Vogt *et al.*, Eds.), pp. 527–556. Birkhauser, Boston.

Devinsky, O., Morrell, M. J., and Vogt, B. A. (1995). Contributions of anterior cingulate cortex to behaviour. *Brain* **118**, 279–306.

Dolan, R. J., Bench, C. J., Liddle, P. F., *et al.* (1993). Dorsolateral prefrontal cortex dysfunction in the major psychoses: Symptom or disease specificity. *J. Neurol. Neurosurg. Psychiatry* **56**(12), 1292–1298.

Dolan, R. J., Fink, G. R., Rolls, E., *et al.* (1997). How the brain learns to see objects and faces in an impoverished context. *Nature* **389**, 596–599.

Drevets, W. C., Videen, T. Q., MacLeod, A. K., Haller, J. W., and Raichle, M. E. (1992). PET images of blood flow changes during anxiety: Correction. *Science* **256**, 1696.

Ekman, P. (1982). "Emotion in the Human Face." Cambridge Univ. Press, Cambridge, UK.

Esteves, F., Dimberg, U., and Öhman, A. (1994). Automatically elicited fear: Conditioned skin conductance responses to masked facial expressions. *Cog. Emot.* **9**, 99–108.

Fink, G. R., Markowitsch, H. J., Reinkemeier, M., Bruckbauer, T., Kessler, J., and Heiss W.-D. (1996). Cerebral representation of one's own past: Neural networks involved in autobiographical memory. *J. Neurosci.* **16**, 4275–4282.

Fried, I., Macdonald, K. A., and Wilson, C. L. (1997). Single neuron activity in hippocampus and amygdala during recognition of faces and objects. *Neur* **18**, 875–887.

Friston, K. J., Tononi, G., Reeke, G. N., Sporns, O., and Edelman, G. M. (1994). Value-dependent selection in the brain: Simulation in a synthetic neural model. *Neuroscience* **30**, 77–86.

Friston, K. J., Buechal, C., Fink, G., Morris, J. S., Rolls, E. T., and Dolan, R. J. (1997). Psychophysiological and modulatory interactions in neuroimaging. *NeuroImage* **6**, 218–229.

George, N., Dolan, R. J., Fink, G. R., Baylis, G., and Russell, C. (1999). Contrast polarity and face recognition in the human fusiform gyrus. *Nat. Neurosci.* **2**, 574–580.

Gewitz, J. C., Falls, W. A., and Davis, M. (1997). Normal conditioned inhibition and extinction of freezing and fear potentiated startle following electrolytic lesions of medial prefrontal cortex. *Behav. Neurosci.* **111**, 712–726.

Gloor, P., Olivier, A., Quesney, L. F., Andermann, F., and Horowitz, S. (1982). The role of the limbic system in experiential phenomena of temporal lobe epilepsy. *Ann. Neurol.* **12**, 129–144.

Griffiths, P. E. (1997). "What Emotions Really Are." Univ. of Chicago Press, Chicago.

Gross, C. G. (1991). Contribution of striate cortex and the superior colliculus to visual function in area MT, the superior temporal polysensory area and the inferior temporal cortex. *Neuropsychologia* **29**, 497–515.

Izard, C. E., and Youngstrom, E. A. (1996). The activation and regulation of fear and anxiety. *Nebr. Symp. Motiv.* **43**, 1–59.

Jarrell, T. W., Gentile, C. G., Romanski, L. M., McCabe, P. M., and Schneiderman, N. (1987). Involvement of cortical and thalamic auditory regions in retention of differential conditioning to acoustic conditioned stimuli in rabbits. *Brain Res.* **412**, 285–294.

Jones, B., and Mishkin, M. (1972). Limbic lesions and the problem of stimulus-reinforcement associations. *Exp. Neurol.* **36**, 362–377.

Josephs, O., Turner, R., and Friston, K. (1997). Event-related fMRI. *Hum. Brain Mapp.* **5**, 243–248.

Ketter, T. A., Andreason, P. J., George, M. S., *et al.* (1996). Anterior paralimbic mediation of procaine-induced emotional and psychosensory experience. *Arch. Gen. Psychiatry* **53**, 59–69.

Kluver, H., and Bucy, P. C. (1939). Preliminary analysis of the functions of the temporal lobes in monkeys. *Arch. Neurol. Psychiatry* **42,** 979–1000.

LaBar, K. S., LeDoux, J. E., Spencer, D. D., and Phelps, E. A. (1995). Impaired fear conditioning following unilateral temporal lobectomy. *J. Neurosci.* **15,** 6846–6855.

LaBar, K. S., Gatenby, J. C., Gore, J. C., LeDoux, J. E., and Phelps, E. A. (1998). Human amygdala activation during conditioned fear acquisition and extinction: A mixed-trial fMRI study. *Neur* **20,** 937–945.

LeDoux, J. E. (1986). Emotion. *In* "Handbook of Physiology. The Nervous System V" (V. B. Mountcastle *et al.,* Eds.), pp. 419–459. American Physiological Society, Bethesda.

LeDoux, J. (1996). "The Emotional Brain." Simon & Schuster, New York.

LeDoux, J. E., Cicchetti, P., Xagoraris, A., and Romanski, L. M. (1990). The lateral amygdaloid nucleus: Sensory interface of the amygdala in fear conditioning. *J. Neurosci.* **10,** 1062–1069.

Marr, D. (1982). "Vision." Freeman, San Francisco.

McGuire, P. K., Bench, C. J., Frith, C. D., Marks, I. M., Frackowiak, R. S. J., and Dolan, R. J. (1994). Graded activation of symptoms in obsessive compulsive disorder. *Br. J. Psychiatry* **164,** 459–468.

Mesulam, M.-M., and Mufson, E. J. (1982). Insula of the old world monkey. III. Efferent cortical output and comments on its function. *J. Comp. Neurol.* **242,** 38–52.

Morgan, M. A., Romanski, L. M., and LeDoux, J. E. (1997). Extinction of emotional learning: Contribution of medial prefrontal cortex. *Neurosci. Lett.* **163,** 109–113.

Morris, J., Frith, C. D., Perrett, D., *et al.* (1996). A differential neural response in the human amygdala to fearful and happy facial expressions. *Nature* **383,** 812–815.

Morris, J., Friston, K. J., and Dolan, R. J. (1997). Neural responses to salient visual stimuli. *Proc. R. Soc. London B* **264,** 769–775.

Morris, J. S., Friston, K. J., and Dolan, R. J. (1998a). Experience-dependent modulation of tonotopic neural responses in human auditory cortex. *Proc. R. Soc. London B* **265,** 649–657.

Morris, J. S., Ohman, A., and Dolan, R. J. (1998b). Conscious and unconscious emotional learning in the human amygdala. *Nature* **393,** 467–470.

Morris, J., Ohman, A., and Dolan, R. J. (1999). A subcortical pathway to the right amygdala mediating "unseen" fear. *Proc. Natl. Acad. Sci. U.S.A.* **96,** 1680–1685.

Mountz, J. M., Modell, J. G., Wilson, M. W., *et al.* (1989). Positron emission tomographic evaluation of cerebral blood flow during state anxiety in simple phobia. *Arch. Gen. Psychiatry* **46,** 501–504.

Öhman, A., and Soares, J. F. (1994). "Unconscious anxiety": Phobic responses to masked stimuli. *J. Abnorm. Psychol.* **103,** 231–240.

Ohman, A., Fredrikson, M., Hugdahl, K., and Rimmo, P. A. (1976). The premise of equipotentiality in human classical conditioning: Conditioned electrodermal responses to potentially phobic stimuli. *J. Exp. Psychol. [Gen.]* **105,** 313–337.

Orr, S. P., and Lanzetta, J. T. (1980). Facial expressions of emotion as conditioned human autonomic responses. *J. Pers. Soc. Psychol.* **38,** 278–282.

Orr, S. P., and Lanzetta, J. T. (1984). Extinction of an emotional response in the presence of facial expressions of emotion. *Motiv. Emot.* **8,** 55–66.

Pandya, D. N., Van Hoesen, G. W., and Mesulam, M.-M. (1981). Efferent connections of the cingulate gyrus in the rhesus monkey. *Exp. Brain Res.* **42,** 319–330.

Parra, C., Esteves, F., Flykt, A., and Öhman, A. (1997). Pavlovian conditioning to social stimuli: Backward masking and dissociation of implicit and explicit cognitve processes. *Eur. Psychol.* **2,** 106–117.

Poremba, A., and Gabriel, M. (1997). Amygdalar lesions block discriminative avoidance and cingulothalamic training-induced neuronal plasticity in rabbits. *J. Neurosci.* **17,** 5237–5244.

Rainville, P., Duncan, G. H., Price, D. D., Carrier, B., and Bushnell,

M. C. (1997). Pain affect encoded in human anterior cingulate but not somatosensory cortex. *Science* **277,** 968–971.

Rauch, S. L., Jenike, M. A., Alpert, N. M., *et al.* (1994). Regional cerebral blood flow measured during symptom provocation in obsessive-compulsive disorder using oxygen—15-labeled carbon dioxide and positron emission tomography. *Arch. Gen. Psychiatry* **51,** 62–70.

Rauch, S. L., van der Kolk, B. A., Fisler, R. E., *et al.* (1996). A symptom provocation study of posttraumatic stress disorder using positron emission tomography and script-driven imagery. *Arch. Gen. Psychiatry* **53,** 380–387.

Rauch, S. L., Savage, C. R., Alpert, N. M., Fischman, A. J., and Jenike, M. A. (1997). The functional neuroanatomy of anxiety: A study of three disorders using positron emission tomography and symptom provocation. *Biol. Psychiatry* **42,** 446–452.

Reiman, E. M., Fusselman, M. J., Fox, P. T., and Raichle, M. E. (1989). Neuroanatomical correlates of anticipatory anxiety. *Science* **243,** 1071–1074.

Russchen, F. T., Bakst, I., Amaral, D. G., and Price, J. L. (1985). The amygdalostriatal projections in the monkey. An anterograde tracing study. *Brain Res.* **329,** 241–257.

Sanders, S. K., and Shekhar, A. (1995). Anxiolytic effects of chlordiazepoxide blocked by injection of GABA-A and benzodiazepine receptor antagonists in the region of the anterior basolateral amygdala of rats. *Biol. Psychiatry* **37,** 473–476.

Schiller, P. H., and Malpeli, J. G. (1977). Properties and tectal projections of monkey retinal ganglion cells. *J. Neurophysiol.* **40,** 428–445.

Seligman, M. E. P. (1971). Phobias and preparedness. *Behav. Ther.* **2,** 307–320.

Shin, L. M., McNally, R. J., Kosslyn, S. M., *et al.* (1999). Regional cerebral blood flow during script-driven imagery in childhood sexual abuse-related PTSD: A PET investigation. *Am. J. Psychiatry* **156,** 575–584.

Thomas, S. R., Lewis, M. E., and Iversen, S. D. (1985). Correlation of [^{3}H]-diazepam density with anxiolytic locus in the amygdaloid complex of the rat. *Brain Res.* **342,** 85–90.

Vogt, B. A., and Pandya, D. N. (1987). Cingulate cortex of the rhesus monkey: II. Cortical afferents. *J. Comp. Neurol.* **262,** 271–289.

Vogt, B. A., Pandya, D. N., and Rosene, D. L. (1987). Cingulate cortex of the rhesus monkey: I. Cytoarchitecture and thalamic afferents. *J. Comp. Neurol.* **262,** 256–270.

Vogt, B. A., and Barbas, H. (1988). Structure and connections of the cingulate vocalization region in the rhesus monkey. *In* "The Physiological Control of Mammalian Vocalization" (J. D. Newman, Ed.), pp. 203–225. Plenum, New York.

Vogt, B. A., Nimchinsky, E. A., Vogt, L. J., and Hof, P. R. (1995). Human cingulate cortex: Surface feature, flat maps and cytoarchitecture. *J. Comp. Neurol.* **359,** 490–506.

Weinberger, N. M. (1995). Retuning the brain by fear conditioning. *In* "The Cognitive Neurosciences" (M. Gazzaniga, Ed.), pp. 1071–1090. MIT Press, Cambridge, MA.

Weiskrantz, L., Warrington, E. K., Sanders, M. D., and Marshall, J. (1974). Visual capacity in the hemianopic field following a restricted occipital ablation. *Brain* **97,** 709–728.

Weissman, M. M., Myers, J. K., and Harding, P. S. (1978). Psychiatric disorders in a US urban community. *Am. J. Psychiatry* **135,** 459–462.

Whalen, P. J., and Kapp, B. S. (1991). Contributions of the amygdaloid central nucleus to the modulation of the nictitating membrane reflex in the rabbit. *Behav. Neurosci.* **104,** 141–153.

Wik, G., Fredrickson, K., Stone-Elander, S., and Greitz, T. (1993). A functional cerebral response to frightening visual stimulation. *Psychiatr. Res.* **50,** 15–24.

Zihl, J., and von Crammon, D. (1979). The contribution of the "'second'" visual system to directed visual attention in man. *Brain* **102,** 835–856.

21

Functional Neuroimaging Studies of Schizophrenia

Sarah-J. Blakemore and Chris D. Frith

Wellcome Department of Cognitive Neurology, University College London, London WC1N 3BG, United Kingdom

I. Introduction

II. Neuropathology

III. Structural Studies

IV. Neuroreceptor Imaging of Antipsychotics Using PET

V. Cognitive Activation Studies

VI. Imaging Symptoms

VII. Obstacles to Functional Neuroimaging and Schizophrenia Studies

VIII. Conclusion

References

I. Introduction

Schizophrenia is a complex mental illness characterized by acute phases of delusions, hallucinations, and thought disorder and, chronically, by apathy, flat affect, and social withdrawal. Schizophrenia affects 1% of the world's population, independent of country or culture and constitutes a severe public health issue (WHO, 1975). Schizophrenia was originally named "dementia praecox" by Emil Kraepelin in 1896 (Kraepelin, 1919), emphasizing its chronic and deteriorating course and separating it from psychoses with a good prognosis (which he called "maniacal-depressive insanity") and dementia in the elderly. Later, in 1911, Bleuler noted the fragmentation of personality and cognition that is fundamental to the illness and gave it its current label (schizo = split; phrenia = mind; Bleuler, 1987). This label encouraged the common misperception that schizophrenia is characterized by a split personality. In fact, schizophrenia is a complex illness comprising a combination of different symptoms, for example, auditory hallucinations, delusions, and incoherent speech. Schizophrenic people normally start to display symptoms in their late teens to early twenties, but the time of onset and the course of the illness are very variable (American Psychiatry Association, 1987). Approximately a third of patients experience one chronic episode after which they make a more or less full recovery; another third are affected by the illness throughout their lives but their symptoms are to some extent alleviated by anti-psychotic drugs; and the remaining third are so chronically ill that they show little or no improvement even with medication (Johnstone, 1991).

A. General Intellectual Decline

There is a striking deterioration in intellectual and cognitive function in schizophrenia, a feature emphasized by Kraepelin in his original description of the disorder as dementia praecox. Several recent studies have confirmed Kraepelin's observations showing that the IQ of schizophrenic patients is significantly lower than that of controls (e.g., Johnstone, 1991). In most studies, premorbid IQ of the patients and the controls was matched, suggesting that the difference in IQ was due to a rapid postmorbid intellectual decline in the patients.

B. Specific Psychological Impairments

In addition to general intellectual decline, schizophrenic patients experience specific cognitive impairments. Findings demonstrating that intellectual impairment is more marked in some domains than others support the notion that circumscribed brain abnormalities may be associated with schizophrenia. Schizophrenic patients have difficulty in initiating and completing everyday tasks, being easily distracted and tending to give up when confronted by any obstacles. These deficits are similar to the problems in initiation, planning, and modification of behavior associated with frontal-lobe lesions (Shallice and Burgess, 1991). This has led many researchers to suggest that the core deficit of schizophrenia is a failure to activate frontal cortex appropriately during cognitive tasks involving planning and decision making ("task-related hypofrontality"), a notion supported in part by many functional imaging studies.

In addition to planning deficits, schizophrenic patients also show deficits in attention and memory tasks that engage prefrontal, hippocampal, and medial temporal systems (Green, 1996), even in drug-free populations (Saykin et al., 1994). However, these observations do not directly address the causes of the disorder.

C. Etiology

It is currently widely accepted that schizophrenia has a biological etiology. However, the shift toward this agreement is recent, and the etiology of schizophrenia has been the subject of lengthy and intense debate. The debate has been split between those who propose a biological etiology and those who postulate a psychodynamic origin of schizophrenia. In the latter camp, nonbiological factors such as family interaction and stressful life events (Kuipers and Bebbington, 1988), for example, have been proposed to play a causal role in the acquisition of schizophrenia. However, these theories have received little empirical support. In addition,

in the past 50 years, two main lines of evidence supporting a biological role in its etiology have become apparent. First, there was the discovery of antipsychotic drugs in the 1950s, and second, a significant hereditary contribution to the disorder has been demonstrated. These observations strongly suggest that schizophrenia has a biological basis.

D. Dopamine Hypothesis

In France in 1952, Delay, Deniker, and Harl found that dopamine-blocking drugs have a positive therapeutic effect on psychotic symptoms. This led in the 1960s to the dopamine (DA) hypothesis of schizophrenia, which posits that schizophrenia is caused by an overactivity of dopamine receptors (Van Rossum, 1966). This theory has received strong support since its conception: neuroleptic drugs produce some degree of antipsychotic or sedative action, or both, in most schizophrenic patients, and many studies have demonstrated a striking correlation between dosage of neuroleptic drug and its in vivo concentration for blocking DA receptors (Seeman, 1986). There are, however, some complications to the theory that have become apparent. First, there is a therapeutic lag. Whereas neuroleptics can alter the levels of DA activity on the first day of treatment, any distinct clinical improvement may take 10–50 days to appear (Straube and Oades, 1992). Second, the DA story is apparently more complicated, with some neuroleptics showing a similar but less marked blocking effect on the noradrenergic system (e.g., Iversen, 1986). In addition, many results on which the DA hypothesis is based have come from methodologies that are now believed to be problematic or confounded. In the 1960s and 1970s, techniques allowing the examination of cerebrospinal fluid (CSF) became available. However, CSF studies have produced weak or negative findings. Receptor binding studies have been performed on post-mortem brains. However, it has become clear that such studies will always be confounded by the effects of death, hypoxia, and neuroleptic treatment in life (Seeman, 1987).

The most objective way to investigate the dopamine hypothesis is the in vivo visualization of radioactive ligand binding to quantify dopamine receptor densities in drug-naive patients using positron emission tomography (PET). These studies will be reviewed later in the chapter.

E. Genetic Studies

That there is a significant genetic factor in schizophrenia is shown by studies demonstrating that monozygotic (MZ) twins (who share identical genes) have a 50% concordance rate for the illness (Gottesman and Shields, 1982). This finding supports the numerous studies demonstrating that a schizophrenic person is more likely than a non-schizophrenic person to have a blood relation who is also schizophrenic (e.g., McGue *et al.*, 1986). However, more recent studies have shown that the MZ twin concordance rate is lower than previously believed (e.g., 31%; Kendler and Robinette, 1983), and the lack of complete concordance between MZ twins indicates that non-genetic factors also play a part. In addition, the exact mode of inheritance is unclear. It is unknown whether what is inherited is a genetic predisposition to develop schizophrenia that is expressed in a direct biological way, or a vulnerability to express schizophrenic symptoms that depends on environmental or psychological factors.

Thus, current methodologies present a challenge for research into the genetic and dopamine theories of schizophrenia. As a consequence, research has become increasingly focused on attempts to elucidate some structural or functional brain abnormality since it is widely held that schizophrenia is a disease of the brain (e.g., Ron and Harvey, 1990). The theory that some gross brain lesion characterizes schizophrenia seems unlikely. Instead, it is generally accepted that schizophrenia is characterized not by structural damage, but by functional abnormalities. This is supported by the relapsing and remitting course of the illness, fluctuations in symptoms, and response to pharmacological interventions. Therefore the advent of functional neuroimaging has been important in the study of schizophrenia because it enables brain function and its abnormalities to be investigated. As stated by Weinberger *et al.* (1996): "Functional neuroimaging in psychiatry has had its broadest application and greatest impact in the study of schizophrenia."

A major problem is that images of brain function reflect the current mental state of the patient (i.e., symptoms) and these are very variable. Symptoms include disorders of inferential thinking (delusions), perception (hallucinations), goal-directed behavior (avolition), and emotional expression. Current thinking generally distinguishes symptoms that comprise the presence of something that should be absent (positive symptoms) and the absence of something that should be present (negative symptoms) (Crow, 1980). Factor-analytic studies of symptoms suggest that positive symptoms should be subdivided further into a psychotic group (comprising delusions and hallucinations) and a disorganized group (comprising disorganized speech, formal thought disorder, disorganized behavior, and inappropriate affect) (Liddle, 1987).

The original two-syndrome model proposed by Crow suggested that positive and negative symptoms might be due to a different pathophysiology. Positive symptoms, Crow suggested, were due to an overactivity of the dopamine system, whereas negative symptoms were due to diffuse neuronal loss. Even though Crow's delineation of symptoms is rarely used by clinicians today, pathophysiology and abnormal psychology specific to the two symptom types have been supported by numerous studies, which will be addressed later in the chapter. The diversity of symptoms in schizophrenia makes it unlikely that the pathophysiology can be accounted for by a single localized brain dysfunction. Instead, the strategy of attempting to localize specific symptoms to specific brain regions or connections between regions is encouraging, and this chapter will evaluate the results of such studies.

II. Neuropathology

Studies investigating structure in postmortem brains have shown evidence for a decrease in overall brain size (e.g., Brown *et al.*, 1986) and have linked schizophrenia to structural abnormalities in the prefrontal cortex and temporal lobe, especially the hippocampus and amygdala (Brown *et al.*, 1986; Benes *et al.*, 1991a). Several studies have found that schizophrenic brains tend to have enlarged lateral ventricles compared to nonschizophrenic brains (e.g., Brown *et al.*, 1986). The findings suggest that ventricular enlargement is associated with tissue loss in or abnormal development of the temporal lobes.

Histological studies have shown evidence for abnormal synaptic function in the cingulate (Benes *et al.*, 1991b; Honer *et al.*, 1997) and hippocampal pyramidal cells (Benes *et al.*, 1991a) in schizophrenia. The most reproducible positive anatomical finding in postmortem hippocampal formation has been the reduced size of neuronal cell bodies in schizophrenia. However, many studies have not revealed a clear change in hippocampal formation size, and recent studies of neuronal number and of cytoarchitecture have been largely negative. These inconsistent results might be due to the complexities surrounding this type of research, such as the effects of death, hypoxia, and neuroleptic treatment in life.

III. Structural Studies

A. Computerized Tomography (CT) Studies

The main finding from CT scan studies is that the lateral ventricles are enlarged in schizophrenic patients compared to normal controls. In the first study using this technique, Johnstone *et al.* (1976) found that 17 chronically hospitalized schizophrenic patients had significantly enlarged ventricles compared to normal controls. In a review of the CT literature, Andreasen *et al.* (1990) noted that 36 out of 49 subsequent studies have replicated this finding to some extent. There have also been two meta-analyses of the CT scan data (Raz and Raz, 1990; Van Horn and McManus, 1992), which supported the finding of enlarged ventricles in schizophrenic patients.

However, the extent of ventricular enlargement in schizophrenia is often small and within the control range (e.g., Weinberger *et al.*, 1979). Many further studies have found no evidence of ventricular enlargement. One problem contributing to the inconsistent results is the selection of controls. Smith and Iacano (1986) compared 21 studies with equivocal results. They plotted the mean ventricle : brain ratio (VBR) for the schizophrenic patients and the normal controls in each study. Normal control brains in the studies with positive findings had, on average, smaller VBRs than control brains in the studies with negative findings. These results suggest that positive findings might be due to the control group having smaller ventricles rather than the schizophrenic patients having larger ventricles. In support of this, Van Horn and McManus (1992) found in their meta-analysis that choice of control group was a contributing factor to the variability of control VBR.

Correlation between CT findings and symptoms has been investigated. Lewis *et al.* (1990) reviewed 41 CT scan studies that addressed the issue of heterogeneity of schizophrenic symptoms. Only 1 of 18 studies found any association between increased ventricular area and chronicity of illness. Five out of 18 found evidence of a relationship with negative symptoms. Poor treatment response was found to be associated with increased ventricular area in about half the studies. The presence of tardive dyskinesia (in 4 out of 6 studies) and neuropsychological impairment (11 out of 14 studies) were also found to be associated with large ventricles.

Chua and McKenna (1995) reviewed the CT literature and concluded that although the finding of increased ventricular area in schizophrenic patients is widespread and replicated, the difference between schizophrenic patients and normal controls is clearly small and depends significantly on the control group chosen. Overlap with the normal population is appreciable and might be complete if suitable controls were used (Smith and Iacano, 1986).

B. MRI Studies

MRI improves on CT and other scanning techniques due to its better spatial resolution and ability to differentiate white and gray matter. This allows the size of cortical and subcortical regions to be measured. There have been several replicated findings using structural MRI scans to investigate the structure of schizophrenics' brains. Studies vary significantly in the MRI methodology used, earlier studies tending to use a single or small number of slices, whereas more recent studies have been able to scan multiple slices, or the whole brain. In most studies, regions of interest have had to be identified by drawing round them manually (Coffman and Nasrallah, 1986; Kelsoe *et al.*, 1988). A number of studies have shown evidence for reductions in overall brain size, but most of these have used poor control groups and/or small numbers of schizophrenic patients (e.g., Andreasen *et al.*, 1986; Harvey *et al.*, 1993). In addition, this finding has not been replicated by other studies (DeMyer *et al.*, 1988; Andreasen *et al.*, 1990).

Many MRI studies provide evidence that schizophrenic patients have larger lateral ventricles than normal controls (e.g., Coffman and Nasrallah, 1986; Kelsoe *et al.*, 1988; Andreasen *et al.*, 1990; Suddath *et al.*, 1990; Woodruff *et al.*, 1997a). However, the results are conflicting, and there are several negative findings (e.g., Smith *et al.*, 1987; Johnstone *et al.*, 1989; Rossi *et al.*, 1994). Older studies were limited by poor control groups and small number of schizophrenic patients. Recent studies have used larger subject groups and multiple control groups. Sharma *et al.* (1998) compared MRI scans from 31 people with schizophrenia, 57 relatives, and 39 unrelated control subjects. Subjects with schizophrenia had larger lateral ventricles than both control groups. Relatives who were "presumed obligate carriers" had larger left lateral ventricles than other relatives and the control subjects.

Temporal lobe reductions in schizophrenia have been reported and are especially prominent in the hippocampus (Fukuzako *et al.*, 1997), parahippocampal gyrus, and the amygdala (Yurgelun-Todd *et al.*, 1996a). Suddath *et al.* (1990) using the identical twins of the schizophrenic patients as controls found evidence for reductions of the left temporal lobe, including the hippocampus. Kwon *et al.* (1999) recently replicated the finding of a reduction in volume of the left temporal lobe. Cannon *et al.* (1998) suggested that frontal and temporal structural changes might reflect genetic (or shared environmental) effects. In a study using a large group of patients and well-matched controls (their nonpsychotic siblings and a group of normal controls), they found that volume reductions of the frontal and temporal lobes were present in patients with schizophrenia and in some of their siblings without schizophrenia. However, temporal lobe reductions are equivocal, with some negative findings in the literature (e.g., Young *et al.*, 1991). There have been many conflicting and negative results in studies attempting to locate clinical correlates with temporal abnormalities. Many studies have found no association between chronicity and temporal lobe size (e.g., Kelsoe *et al.*, 1988; Young *et al.*, 1991), although one study found an inverse relationship between these two factors (DeLisi *et al.*, 1991). Size of the superior temporal gyrus has been associated with hallucinations (Barta *et al.*, 1990) and formal thought disorder (Shenton *et al.*, 1992). Bilateral reduction in volume of the hippocampal formation has been associated with the severity of disorganization syndrome (Fukuzako *et al.*, 1997).

Some studies have found frontal lobe reductions in schizophrenic patients (Harvey *et al.*, 1993; Cannon *et al.*, 1998). However, the results are inconsistent, especially with relation to the precise localization of the frontal abnormalities (Raine *et al.*, 1992). Buchanan *et al.* (1998) improved on this by subdividing the prefrontal cortex (PFC) into superior, middle, inferior, and orbital regions and found that patients with schizophrenia exhibited selective gray matter volume reductions in the inferior PFC bilaterally. There was no difference between the schizophrenic and control groups in any other region of the frontal lobes.

There are some MRI data that provide support for a hypothesis of disconnection between brain areas in schizophrenia. Breier *et al.* (1992) found that schizophrenic patients, compared with matched healthy controls, had reductions in the right and left amygdala, the left hippocampus, and prefrontal white matter. Moreover, the right prefrontal white matter volume in schizophrenic patients was significantly related to right amygdala/hippocampal volume, suggesting there might be abnormal connections between these areas. Buchsbaum *et al.* (1997) found evidence for a decreased left hemispheric volume in frontal and temporal regions in schizophrenic patients. This result was supported by Woodruff *et al.* (1997a), who found that inter-regional correlations were significantly reduced in schizophrenics between prefrontal and superior temporal gyrus volumes. The authors propose that these results support the existence of a relative "fronto-temporal dissociation" in schizophrenia.

Reversal or reduction of normal structural cerebral asymmetries may be related to the pathogenesis of schizophrenia (Crow, 1995). Lack of normal asymmetry has been especially associated with early onset of schizophrenia. For example, DeLisi *et al.* (1997) investigated MRI structural scans of 87 patients with a first episode of schizophrenia, 52 normal controls, and an independent group of 14 pairs of siblings with schizophrenia to evaluate evidence of heritability to cerebral asymmetries. Width asymmetries were reduced in patients compared with controls in the posterior and occipital regions. Brain horizontal length, on the other hand, was significantly more asymmetrical in patients (left >

right). In the 14 pairs of psychotic siblings, within-pair correlations for the horizontal sylvian fissure asymmetry were significantly greater than between-pair correlations. These findings are consistent with the early presence (possibly genetic) of anomalous cerebral asymmetry. Fukuzako *et al.* (1997) found a significant asymmetry in the hippocampal formation volume in control subjects, but not in patients, and a significant positive correlation between the asymmetry index and the patient's age at the onset of schizophrenia. Similarly, Maher *et al.* (1998) found that low levels of hemispheric asymmetry in the frontal and temporal areas were associated with early onset of schizophrenia, the association with frontal volume being more marked than with temporal volume. These findings are consistent with the hypothesis that failure to develop asymmetry is an important component of the pathology underlying some forms of schizophrenia.

1. Voxel-Based Morphometry

Voxel-based morphometry is a new approach of looking at structural brain abnormalities using MRI. It is a data-led technique in which the brain images are normalized and then differences between groups anywhere in brain are identified (Wright *et al.*, 1995). Andreasen *et al.* (1994) created normal and schizophrenic average brains, compared the latter with the former, and found decreased thalamus size in schizophrenic patients. Wolkin *et al.* (1998) used linear intersubject averaging of structural MR images to create a single averaged brain for the schizophrenic group ($n = 25$) and for the control group ($n = 25$). The signal intensity differences between these average images were consistent with cortical thinning/sulcal widening and ventricular enlargement.

C. Functional Imaging

1. PET Resting Studies

There are many PET metabolism studies in the literature, and this review is not exhaustive. Many of the earlier studies investigated small sample sizes, and we will concentrate on studies that have used large sample sizes. The main finding from PET metabolism studies is that there is lower metabolism in the frontal lobes of schizophrenic patients compared to normal controls. This became known as *hypofrontality*. The first study using isotope imaging was by Ingvar and Franzen (1974), who compared 15 normal controls with 11 patients with dementia, and two groups of schizophrenics (one group comprised nine chronic, elderly patients; the other comprised 11 younger patients). Whereas the demented patients showed a lower level of overall metabolism, both schizophrenic groups showed some evidence for reduced blood flow in anterior, relative to posterior, regions (Fig. 1).

However, disagreement about the definition of hypofrontality has caused problems and inconsistencies in the results. Studies vary on the frontal areas in which activity was measured. Some included all anterior regions, whereas others looked at frontal or prefrontal subdivisions only. The earlier studies tended to use the frontal:occipital ratio as a measure of hypofrontality, whereas recently most studies have used absolute frontal flow values with or without correction for mean total brain blood flow rates. The majority of studies, using any of these definitions, have found little evidence for statistically significant hypofrontality, and in several studies, hypofrontality was due to an elevated flow in posterior regions (Mathew *et al.*, 1988; Siegel *et al.*, 1993). In other studies, the differences between control and schizophrenic frontal cortex metabolism are small. It has been claimed that this is not due to the drug status of the patients at scanning (Waddington, 1990). However, recently anti-psychotic medication has been found to affect brain metabolism (Miller *et al.*, 1997). This latter study showed that subjects treated with antipsychotic medication had significantly higher rCBF in the left basal ganglia and left fusiform gyrus and lower rCBF in the anterior cingulate, left dorsolateral and inferior frontal cortices, and both cerebellar hemispheres, compared to when they were off medication.

Several studies have looked at clinical correlates associated with hypofrontality, but the results are inconsistent. Among those showing a positive relationship with hypofrontality are chronicity (Mathew *et al.*, 1988), negative symptoms (e.g., Ebmeier *et al.*, 1993), and neuropsychological task impairment (e.g., Paulman *et al.*,

1990). However, the interpretation of these findings is ambiguous, since it is impossible to attribute cause or effect.

Liddle *et al.* (1992) correlated rCBF levels with symptom severity scores for his three clinical subdivisions in 30 schizophrenic patients. They demonstrated that the psychomotor poverty syndrome, which has been shown to involve a diminished ability to generate words, was associated with decreased perfusion of the dorsolateral prefrontal cortex (DLPFC) at a locus that is activated in normal subjects during the internal generation of words (see Fig. 1). The disorganization syndrome, which has been shown to involve impaired suppression of inappropriate responses (e.g., in the Stroop test), was associated with increased perfusion of the right anterior cingulate gyrus at a location activated in normal subjects performing the Stroop test. The reality distortion syndrome, which might arise from disordered internal monitoring of activity, was associated with increased perfusion in the medial temporal lobe at a locus activated in normal subjects during the internal monitoring of eye movements. Therefore the abnormalities of brain metabolism underlying each of the three syndromes might be widely distributed over the brain.

Using data from the same patients, Friston *et al.* (1992) examined correlations between rCBF and a measure of psychopathology receiving equal contributions from each of Liddle's three subsystems. The degree of psychopathology correlated highly with rCBF in the left medial temporal region, and mesencephalic, thalamic, and left striatal structures. The highest correlation was in the left parahippocampal region, and the authors proposed that this might be a central deficit in schizophrenia. A canonical analysis of the same data highlighted the left parahippocampal region and left striatum (globus pallidus), in which rCBF increased with increasing severity of psychopathology. Friston and colleagues suggested that disinhibition of left medial temporal lobe activity mediated by fronto-limbic connections might explain these findings.

IV. Neuroreceptor Imaging of Antipsychotics Using PET

A. Dopamine Receptors

Neuroreceptor imaging of antipsychotics using PET has been carried out for over a decade. Many studies have shown evidence that DA (in particular D_2) receptors are increased in schizophrenic brains and that several different antipsychotics bind to these receptors (Wong *et al.*, 1986; Breier *et al.*, 1992). However, other studies using PET have produced contradictory results

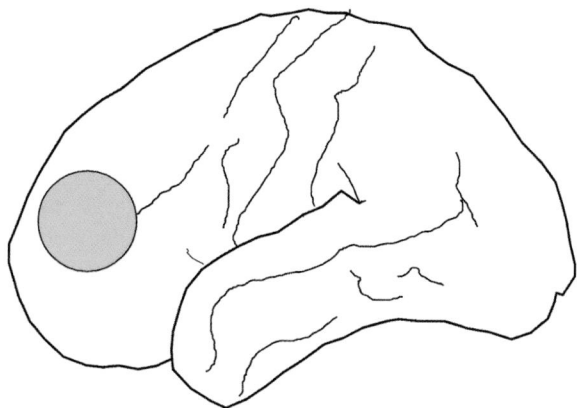

Figure 1 Approximate average localization of hypofrontality found by Liddle *et al.* (1992), Weinberger *et al.* (1986), Daniel *et al.* (1991), Volz *et al.* (1997), Ragland *et al.* (1998), Spence *et al.* (1998), Yurgelun-Todd *et al.* (1996), and Fletcher *et al.* (1998).

or only very weak evidence of an abnormality in DA receptor number (e.g., Crawley *et al.*, 1986; Farde *et al.*, 1987; Pilowsky *et al.*, 1994). These discrepant results might, in part, be due to the diversity of PET methodology used by these groups.

Meta-analytic methods were applied to post-mortem and PET and single photon emission computerized tomography (SPECT) neuroimaging data by Zakzanis and Hansen (1998). They found increases in D_2 receptor density to be approximately 70% in schizophrenia. That is, about 30% of patients with schizophrenia could not be discriminated from normal controls. Based on the findings, it was argued that D_2 receptor density increases in patients with schizophrenia, although a reliable finding in many patients, are not a specific or consistent marker of schizophrenia. Laruelle's (1998) meta-analysis of studies measuring D_2 receptor density parameters revealed that, compared to healthy controls, patients with schizophrenia showed a significant but mild increase and a larger variability of D_2 receptor density.

D_2 receptors are expressed most highly in the striatum, but many PET studies have failed to show any change in D_2 receptor density in the striatum of schizophrenics, raising the possibility that other receptors may also be involved. Okubo *et al.* (1997) used PET to examine the distribution of D_1 and D_2 receptors in the brains of drug-naive or drug-free schizophrenic patients. Although no differences were observed in the striatum relative to control subjects, binding of radioligand to D_1 receptors was reduced in the PFC of schizophrenics.

V. Cognitive Activation Studies

Functional neuroimaging is most frequently used to evaluate the regional cerebral responses (rCBF in PET or BOLD in fMRI) to a particular cognitive or sensorimotor process. Typically, subjects are scanned while performing an activation task, which engages the cognitive/sensorimotor process of interest, and a baseline task, which engages all components of the activation task except the cognitive/sensorimotor process of interest. Regions that show significantly more activity in the experimental state than in the baseline state are considered to be involved in the cognitive/sensorimotor process of interest. These task-specific activity patterns can then be compared between patients and control subjects to determine how the condition affects brain function.

Most of the cognitive processes on which schizophrenic patients' performance is impaired involve the frontal lobes, since patients with frontal lesions perform poorly on these tasks and these tasks activate various regions of the frontal lobes in normal controls. There has been considerable evidence for hypofrontality in schizophrenia from cognitive activation studies. In addition, there is increasing evidence that schizophrenics show abnormal functioning of the temporal lobes, and, less consistently, the parietal lobes and hippocampus, during cognitive tasks. Early studies used less sensitive imaging techniques, for example, ^{133}Xe inhalation or SPECT. In this review, we shall discuss some SPECT studies, but we concentrate on studies using more sensitive methods, specifically PET and functional magnetic resonance imaging (fMRI). There have been many negative findings, due to a number of possible factors, which will be discussed. Finally, in studies finding abnormal activity in the frontal and temporal lobes in schizophrenia, there have been inconsistencies in terms of precise localization. However, the evidence is now starting to point toward anomalous frontotemporal integration during cognitive tasks. In this review, we shall discuss several types of task that have been used to evaluate brain function in schizophrenic patients.

A. Task-Based Studies of Executive Function

In many behaviorial studies the most striking impairments are seen when schizophrenic patients perform the various complex executive tasks associated with the frontal lobes. Brain imaging makes it possible to identify the abnormal pattern of brain activity associated with this abnormal performance.

1. Wisconsin Card Sorting Task (WCST)

Schizophrenia is largely characterized by impairments in planning and execution and therefore tasks that involve this kind of planning and modification of behavior have been exploited in the scanner. Several researchers have investigated brain activity in schizophrenic patients while they perform a version of the Wisconsin Card Sorting Task (WCST). This task is known to activate the DLPFC in normal controls and is particularly sensitive to damage to DLPFC (Berman *et al.*, 1995; Nagahama *et al.*, 1996). In the typical computerized version of the WCST, subjects view a computer screen that displays a number of stimuli. These stimuli differ along three dimensions: color, shape, and number. On each of a series of trials, the subjects have to match a target stimulus with one of the four standard stimuli. However, the match is not exact, but has to made in terms of either color, shape, or form. Subjects are not informed of how to make the match, but are informed after each choice whether they are right or wrong. They have to determine from trial and error which dimension is correct. After subjects have made a series of correct responses, the rule is changed and subjects must determine a new rule for matching.

Weinberger *et al.* (1986) measured rCBF using ^{133}Xe inhalation SPECT in 20 medication-free patients with chronic schizophrenia and 25 normal controls during the WCST and a number-matching control task. During the WCST, but not during the control task, normal subjects showed increased DLPFC rCBF, whereas patients did not (patient performance was worse than that of the controls). Furthermore, in patients, DLPFC rCBF correlated positively with WCST performance. The authors suggest that this result shows that the better DLPFC was able to function, the better patients could perform the task. However, this conclusion is based on an ill-founded assumption. It is impossible to determine whether task-related underactivation *causes* or *reflects* poor task performance. This is a crucial issue in cognitive activation studies and will be discussed in more detail throughout the chapter.

Daniel *et al.* (1991) used SPECT to study the effect of amphetamine (a DA agonist) and a placebo on rCBF in 10 chronic schizophrenic patients while they performed a version of the WCST. They compared this to rCBF during a sensorimotor control task matched to the WCST in terms of visual stimulation and motor responses, but without the abstract reasoning and working memory components of the WCST. On placebo no significant activation was seen during the WCST compared with the control task. In contrast, significant activation of the left DLPFC occurred on the amphetamine trials. Daniel and colleagues point out that patients' performance improved with amphetamine relative to placebo and that with amphetamine, but not with placebo, a significant correlation was found between activation of DLPFC and performance on the WCST task. Again, this is a finding that is impossible to interpret: did amphetamine facilitate task performance, which then caused an increased rCBF in the PFC? Or did amphetamine cause PFC activity to increase, which facilitated task performance?

Volz *et al.* (1997) used fMRI to investigate activity during the WCST in 13 chronic schizophrenics on stable neuroleptic medication. They also showed evidence for lack of activation in the right PFC and a trend toward increased left temporal activity during the WCST compared to normal controls. However, again the task performance was different between the two groups and therefore the results remain ambiguous. In addition, this study was limited because a one-slice imaging technique was used so no information about the activation pattern in adjacent brain regions was obtained.

Ragland *et al.* (1998) found that better WCST performance correlated with rCBF increase in prefrontal

regions for controls and in the parahippocampal gyrus for patients. The results suggest that schizophrenia may involve a breakdown in the integration of a frontotemporal network that is responsive to executive and declarative memory demands in healthy individuals.

2. Tower of London Task

The Tower of London task involves high-level strategic planning among a number of other processes (Shallice, 1982). In this task, subjects have to rearrange a set of three balls presented on a computer screen so that their positions match a goal arrangement also presented on the screen (see Fig. 2). The complexity of the task, in terms of the number of moves necessary to complete the task, can be varied.

Andreasen *et al.* (1992) used SPECT during the Tower of London task in three different groups: 13 neuroleptic-naive schizophrenic patients; 23 non-naive schizophrenic patients who had been chronically ill but were medication free for at least 3 weeks; and 15 healthy normal volunteers. The Tower of London task activated the left mesial frontal cortex (probably including parts of the cingulate gyrus) in normal controls, but not in either patient group. Both patient groups also lacked activation of the right parietal cortex, representing the circuitry specifically activated by the Tower of London task in normal controls. Importantly, decreased activation occurred only in the patients with high scores

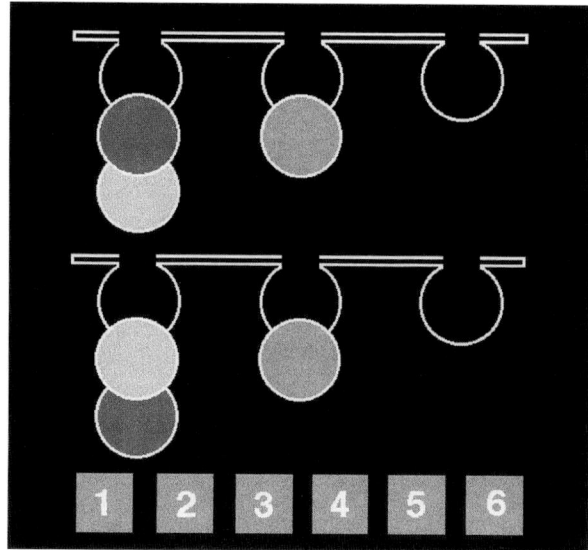

Figure 2 The Tower of London task involves high-level strategic planning among a number of other processes (Shallice, 1982). In this task, subjects have to rearrange a set of three balls presented on a computer screen so that their positions match a goal arrangement also presented on the screen. The complexity of the task, in terms of the number of moves necessary to complete the task, can be varied.

for negative symptoms. The authors therefore suggested that hypofrontality is related to negative symptoms and is not a long-term effect of neuroleptic treatment or of chronicity of illness. This is an important study because of the large number of neuroleptic-naive patients investigated, but it is limited in several ways, in particular the low-resolution imaging technique that was used. There was also the methodological problem that schizophrenic patients performed poorly on the tasks involved, so whether less activation of the PFC was due to poorer performance or vice versa cannot be resolved.

B. The Component Processes Underlying Executive Function

One problem with studies that used complex executive tasks is that these tasks involve many processes. For example, the WCST involves choosing a strategy, remembering the previous responses in order to learn by trial and error, attending to one dimension rather than another, and so on. In the absence of a series of carefully constructed comparison tasks, it is not possible to relate the various brain regions activated with each of the component processes. In the following section, we review studies in which simpler tasks were used with far fewer component processes.

1. Motor Tasks

Even tasks that require no more than the production of a simple sequence of movements can be associated with abnormal patterns of brain activity in schizophrenia.

Mattay *et al.* (1997) studied seven patients with schizophrenia and seven normal subjects while they performed a finger movement task of increasing complexity. Patients showed greater ipsilateral activation in the primary sensorimotor and lateral premotor regions and had a significantly lower laterality quotient than normal subjects. These functional abnormalities increased with the complexity of the task. The authors proposed that these results demonstrate a functional disturbance in the cortical motor circuitry of schizophrenic patients.

Schröder *et al.* (1999) asked 12 patients and 12 healthy controls to produce sequences of pronation and supination movements at three different speeds during fMRI scanning. Both groups showed increasing activity with increasing speed in sensorimotor cortex and supplementary motor area (SMA). However, the patients showed less overall activation than the controls. While the patients did not differ in the amplitude and frequency of their movements, they did show significantly greater variability in their movements than the controls. The differences were most marked in a subgroup of pa-

tients who were drug free at the time of testing. Both these studies raise the possibility of a fundamental, but subtle problem of motor control associated with schizophrenia.

2. Willed Action

Willed action involves a somewhat "higher" stage in the control of action. There is a fundamental distinction between actions elicited by external stimuli and actions elicited by internal goals (acts of will). Routine actions are specified by a stimulus. In contrast, in willed (or self-generated) acts, the response is open-ended and involves making a deliberate choice. Willed actions are a fundamental component of executive tasks. In normal subjects, willed acts in two response modalities (speaking a word or lifting a finger), relative to routine actions, were associated with increased blood flow in the DLPFC (Brodmann area 46; Frith *et al.,* 1991). Schizophrenic patients typically show abnormalities of willed behavior. In chronic patients, intentions of will are no longer properly formed and so actions are rarely elicited via this route. This gives rise to behavioral negative signs (e.g., poverty of speech and action).

Spence *et al.* (1998) performed a PET study in which subjects had to make voluntary joystick movements in the experimental condition, and stereotyped (routine) movements in the baseline condition, and do nothing in a control rest condition. They analyzed data from 13 schizophrenic patients, comparing two occasions when symptoms were severe and when they had subsided, and included data from a normal control group to clarify the role of the left DLPFC in volition. The DLPFC was activated by normal controls for the free-choice task only. However, it was not activated by schizophrenics with symptoms but became activated when their symptoms decreased. The authors noted that the DLPFC was also activated during the stereotyped joystick movement task in schizophrenic patients in remission, in contrast to a control group. Spence and colleagues concluded that, since hypofrontality was evident in schizophrenics who can perform the experimental task, the DLPFC is not necessary for that task. In addition, hypofrontality seems to depend on current symptoms. They suggested the reason for previous equivocal hypofrontality results is that schizophrenics with a varying amount and combination of symptoms were being compared with normal controls.

This study raises the puzzle of the precise role of the DLPFC in freely chosen movements. As the authors point out, DLPFC was activated by both the freely chosen and the stereotyped movement tasks in schizophrenics without symptoms. If patients do behave in a more random manner than controls, it follows that the patients might find making stereotyped movements harder than controls. In other words, it is possible that stereotyped movements are not truly stereotyped for schizophrenic patients. On the other hand, the production of random movements may be easier. Thus, although this study represents an advance on those in which patients show impaired task performance, as the authors acknowledge, caution is necessary when interpreting the results. What is the functional significance of DLPFC activation in normal controls and patients without symptoms if schizophrenic patients with symptoms can perform the task without recruiting the DLPFC?

3. Verbal Fluency

Verbal fluency tasks involve subjects having to generate words to a given cue. For example, subjects might have to produce a word beginning with a certain letter, a different letter being presented every 5 s. This can be seen as a task that involves willed action since subjects have to choose for themselves precisely which word to say. Verbal fluency tasks engage a distributed brain system similar to that engaged by motor response selection tasks associated with willed action (Frith *et al.*, 1991).

Yurgelun-Todd *et al.* (1996b) investigated verbal fluency using fMRI in 12 schizophrenic patients and 11 normal control subjects. They showed evidence for reduced left PFC activation and increased left temporal activation relative to control subjects during the verbal fluency task.

However, the lack of frontal activation by cognitive tasks in schizophrenic patients has not consistently been located in the PFC. Dolan *et al.* (1995) and Fletcher *et al.* (1996) used a factorial design to test the effect of apomorphine, a nonselective DA agonist, which when given in very low doses as in this experiment acts primarily on autoreceptors, thus decreasing the release of endogenous dopamine. Brain systems engaged by a paced verbal fluency task in unmedicated schizophrenic patients and normal controls were studied using PET. Activation of the DLPFC was normal, but they found a failure of task-related activation in anterior cingulate cortex and deactivation of the left superior temporal gyrus in the schizophrenic subjects (see Fig. 3). Fletcher and colleagues therefore suggested that schizophrenia is associated with both segregated (anterior cingulate) and integrative (fronto-temporal) functional abnormalities. Cingulate activation was restored by low-dose apomorphine in schizophrenics. Additionally, the abnormal fronto-temporal pattern of activation in schizophrenic subjects was normalized by this neuropharmacological intervention. Overall, in schizophrenic subjects

the effect of apomorphine was to modify the pattern of brain activity, making it more similar to that seen in control subjects. The interpretation of the apomorphine-induced reversal of the deactivation in the left temporal lobe in schizophrenic subjects is unclear. It might reflect a direct influence of apomorphine on the temporal lobe; alternatively, the reversal could be due to a "downstream" effect of the change in anterior cingulate function. The authors interpret the absence of the normal reciprocal interaction between the frontal and the superior temporal cortex in schizophrenia (the failure of task-related deactivation of the superior temporal gyrus in the schizophrenic group) to suggest the presence of impaired functional integration. This is an important concept, especially given the relatively large amount of evidence showing a lack of temporal deactivation in the presence of a lack of frontal activation in schizophrenia.

The finding of normal prefrontal activation found in this study is at variance with some previous functional neuroimaging studies investigating schizophrenia, as outlined above. It is in agreement, however, with a study in which task performance was optimized by pacing the task (Frith *et al.*, 1995). Using PET, these researchers investigated rCBF of 18 chronic schizophrenic patients and 6 normal controls matched for age, sex, and premorbid IQ while they performed (a) paced verbal fluency, (b) paced word categorization, and (c) paced word repetition. The schizophrenic patients were split into three groups according to their verbal fluency task performance level. All patient groups showed the *same* pattern of left PFC activation as control subjects, independent of their level of performance. However, in the left superior temporal cortex, all patient groups failed to show a normal deactivation when verbal fluency was compared with word repetition. Again, this result was interpreted to reflect abnormal functional connectivity between frontal and temporal cortex.

Friston and Frith (1995) performed an additional analysis of their data from the same three groups of schizophrenic patients according to the level of task performance: poverty (no words), odd (wrong words), and unimpaired. They used analytic techniques specifically to assess cortico-cortical interactions. Normal controls showed negative frontotemporal interactions whereas all the schizophrenic patients showed positive interactions, mostly between the left PFC and inferotemporal cortex. Friston and Frith suggested that this might represent a failure of the PFC to inhibit the temporal lobes. They postulated that the temporal lobes may be required to recognize the consequences of actions initiated by the frontal lobes in order to integrate action and perception.

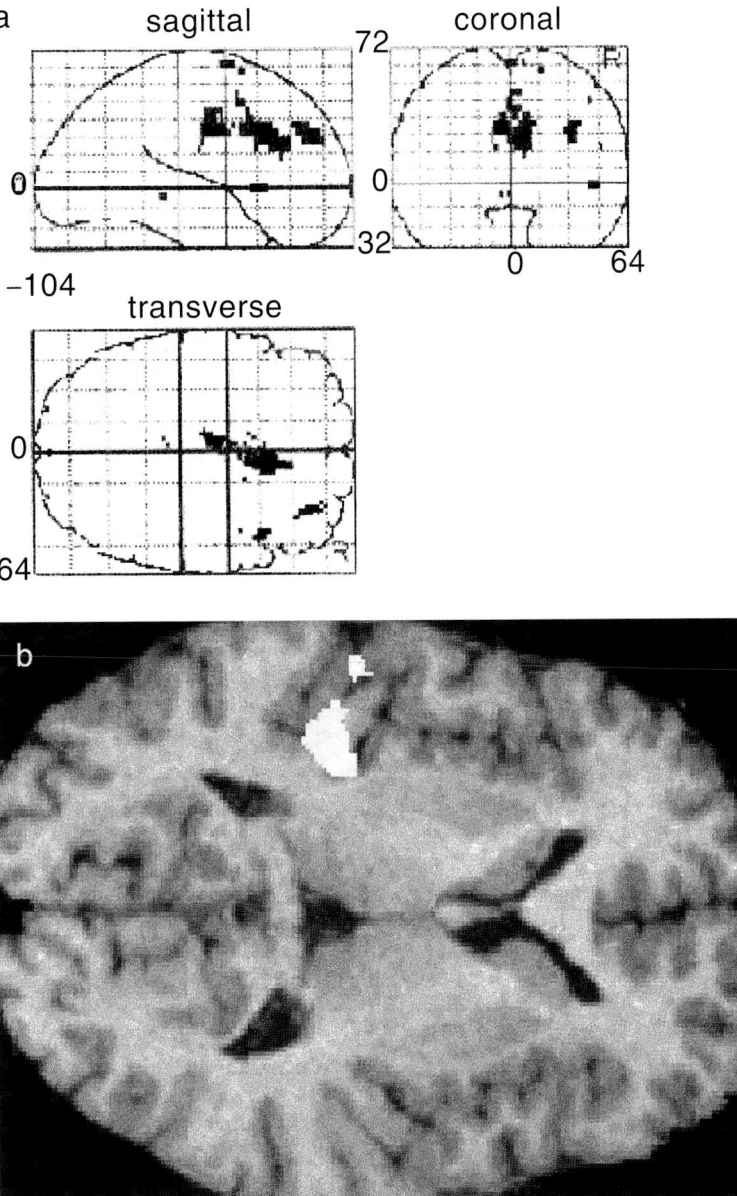

Figure 3 Reproduced with permission from Fletcher *et al.* (1996). (a) Statistical parametric maps (SPMs) showing brain regions where there was a significant ($p <$ 0.005) difference in drug (apomorphine)–task (verbal fluency) interaction between the schizophrenic group and the control group. The area in which there was an augmenting effect of the drug on the task-related activity occurring in schizophrenics compared to controls was the anterior cingulate gyrus. In other words, the impaired cingulate activation seen during the verbal fluency task in schizophrenic patients was significantly reversed by apomorphine. (b) Transverse section showing the verbal fluency task-related deactivation of the superior temporal gyrus that was absent in schizophrenic relative to control subjects. The authors interpret the failure of task-related deactivation of the superior temporal gyrus in the schizophrenic group to suggest the presence of impaired functional temporo-frontal integration.

C. Memory Tasks

Memory impairments are especially enduring symptoms in schizophrenia (Green, 1996), with memory storage particularly affected (Feinstein *et al.*, 1998). Working memory (WM) depends upon executive processes instantiated in frontal cortex and is involved in complex tasks such as the WCST and the Tower of London task. The neuroanatomy of memory in normal subjects has been mapped to a network of cortical and subcortical structures in the human brain (Ungerleider, 1995). This provides a model of memory with which to compare the functioning of memory systems in schizophrenia in order to investigate disturbances of specific memory components. In particular, the DLPFC and the hippocampal formation have been the subject of investigation in schizophrenia, as these are involved in various aspects of memory (Arnold, 1997; Goldman-Rakic and Selemon, 1997). The DLPFC is activated by semantic processing during encoding and retrieval, whereas hippocampal activation is associated with the detection of novelty and the creation of associations during encoding (Schacter *et al.*, 1996; Dolan and Fletcher, 1997). As for other executive and cognitive functions, impairment of WM in schizophrenia has been associated with decreased blood flow in the DLPFC (Goldman-Rakic and Selemon, 1997). Recent studies have used memory tasks on which the patients' memory is at normal levels and have shown evidence for hypofrontality.

Several functional neuroimaging studies have failed to find evidence for abnormal activation of temporal or frontal cortex in schizophrenia during memory tasks (Busatto *et al.*, 1994, using a verbal memory task with SPECT; Ragland *et al.*, 1998, using a paired associate recognition test with PET). Other studies have shown rCBF changes that overlapped in the schizophrenic and control groups, with a trend toward patients showing smaller activations than controls in frontal and superior temporal cortical regions (e.g., Ganguli *et al.*, 1997, using a verbal free-recall supraspan memory task). These differences may be due to the different type of memory tasks used by each group. Other groups have found evidence for hypofrontality during memory tasks, but these have often been confounded by performance levels. For example, Carter *et al.* (1998) used PET to evaluate rCBF associated with the "N-back" WM task, which activates the PFC as a function of WM load in normal subjects. Under low-working-memory-load conditions, the accuracy of both groups in the N-back task was equal, but when the memory load increased, the patients' performance deteriorated more than did that of control subjects. The rCBF response to increased WM load was significantly reduced in the patients' right DLPFC. Callicott *et al.* (1998) investigated BOLD sig-

nal changes in 10 patients with schizophrenia and 10 controls performing a novel N-back WM task, using fMRI. After confounds were removed and subjects were matched for signal variance (voxel stability), decreased DLPFC activity and a tendency for overactivation of parietal cortex were seen. However, these findings are difficult to interpret in the context of abnormal task performance in patients. There may be nothing inherently abnormal about the physiology of the frontal cortex in schizophrenia, but patients may be failing to select frontally mediated cognitive strategies because of abnormal connectivity between otherwise normal regions.

Wiser *et al.* (1998) used PET to measure rCBF during a long-term recognition memory task for words in schizophrenic patients and in healthy subjects. The task was designed so that performance scores were similar in the patient and control subjects. This memory-retrieval task did not activate PFC, precuneus, and cerebellum in patients as much as it did in the control group. This finding suggests that there is a dysfunctional cortico-cerebellar circuit in schizophrenia.

Hypofrontality has not always been found in studies using modified tasks to optimize the performance of schizophrenic subjects. Hypofrontality was not found in a study by Heckers and colleagues (1998) in which they used PET to evaluate 13 schizophrenic patients and a group of normal control subjects during memory-retrieval tasks. Prior to each memory-retrieval scan, subjects learned a list of written words. In half the trials, a "shallow" learning task was used, in which subjects were asked to count the number of right angles (T-junctions) in the letters of each word. This is known to result in poor subsequent memory for those words ("low recall"). In the other half of the trials, subjects performed a deep-encoding task in which words were learned by counting the number of meanings for each word, which is known to produce good subsequent memory for those words ("high recall"). Subjects were then scanned during high and low levels of memory recall. During the scans subjects were asked to retrieve studied words on the basis of three-letter (stem) cues, allowing a comparison between the effort of retrieval and the process of successful retrieval.

In this study, activation of the PFC correlated with effort of retrieval and hippocampal and parahippocampal activation occurred during successful retrieval in normal control subjects, a finding consistent with previous studies of memory in normals (Schacter *et al.*, 1996; Dolan and Fletcher, 1997). The schizophrenic patients recruited the PFC during the effort of retrieval but did not recruit the hippocampus during conscious recollection. This pattern of activation was associated with higher accuracy during low recall and lower accuracy during high recall in schizophrenics than in control subjects. In addition to the task-specific hippocampal underactivation, the authors observed a generally higher overall level of nonspecific hippocampal activity, supporting previous metabolic studies (e.g., Liddle et al., 1992). The authors suggested that high baseline hippocampal activity together with an absence of task-specific activation demonstrates abnormal cortico-hippocampal functional integration in schizophrenia.

The schizophrenic patients showed a more widespread activation of prefrontal areas and parietal cortex during recollection than controls, and the authors propose that this overactivation represents an "effort to compensate for the failed recruitment of the hippocampus." This interpretation again moves away from the simple notion of dysfunction in isolated brain regions explaining the cognitive deficits in schizophrenia and toward the idea that neural abnormality in schizophrenia reflects a disruption of integration between brain areas.

Other groups have found evidence for abnormal activation of different frontal regions. For example, Stevens et al. (1998) used fMRI to study the neural basis of the dissociation of auditory verbal and nonverbal WM in schizophrenia using the Word Serial Position Task and Tone Serial Position Task. Activation in the left inferior frontal gyrus (Brodmann areas 6, 44, and 45) was much reduced during the Word Serial Position Task in the patients and failed to show the same task-specific activation as in controls. Reduced activation in patients also extended to a medial area during the Tone Serial Position Task and to premotor and anterior temporal lobes during both tasks.

Fletcher et al. (1998) used PET to compare rCBF in memory-impaired and nonimpaired schizophrenic patients with normal controls during a parametrically graded memory task. They found that DLPFC activity correlates with memory task difficulty and performance in the control group. In contrast, for both schizophrenic groups, DLPFC activity levels plateaued as task difficulty increased despite a significant difference in performance between the two schizophrenic groups. The authors therefore suggested that hypofrontality in schizophrenics correlates with *task difficulty,* rather than task performance, since the memory-impaired schizophrenic group performed worse than the nonimpaired group even though both groups showed no increase in PFC activity as task difficulty increased.

Unlike the control group, there was no inferior temporal/parietal deactivation in either schizophrenic group. The authors suggest that the lack of deactivation of these areas might represent a temporofrontal disconnection in schizophrenics. Indeed they suggested that because temporal/parietal activations were not corre-

lated with performance, they therefore might represent a core pathology of schizophrenia. This study improves on previous cognitive activation studies in which the confound of non-matched task performance occurs. However, the function of the PFC in memory is difficult to interpret since the unimpaired group could still perform the task when the level of PFC rCBF had plateaued as task difficulty increased.

Further evidence for abnormal integration between brain areas in schizophrenia comes from a study that specifically investigated the functional integration between brain areas (Fletcher et al., 1999). Functional integration considers complex cognitive processes as emergent properties of interconnected brain area, building on the idea that simple cognitive processes can be localized in discrete anatomical modules (referred to as "functional segregation"). Brain areas A and B may be functionally connected if it can be shown that an increase (or decrease) of activity in area A is associated with an increase (or decrease) in area B, which can be shown empirically by analysis of covariance. In this case, activity in A might cause activity in B, or activations in A and B might be caused by changes in another area (C) that projects to A and B. Alternatively, areas A and B may be effectively connected if their relationship can be shown to be causal. This requires a more complex approach in which the anatomical components of a cognitive system are defined. Connections between these regions are designated on the basis of empirical neuroanatomy and the connections are allocated weights or path strengths by an iterative least-squares approach in such a way that the resultant functional model of inter-regional influences best accounts for the observed variance-covariance structure generated by the functional neuroimaging observations (Friston et al., 1993).

A simplified version of effective connectivity (Friston et al., 1997) was employed by Fletcher et al. (1999) to evaluate effective connectivity between regions in the data from their PET study of a graded memory task in schizophrenia. They demonstrated that in control subjects, but not in the schizophrenic patients, the product of PFC and ACG activity predicted a bilateral temporal and medial PFC deactivation. They interpreted these results as showing that in schizophrenia there is an abnormality in the way in which left PFC influences left superior temporal cortex, and this abnormality is due to a failure of the ACG to modulate the prefronto-temporal relationship (see Fig. 4).

D. Conclusion

There is a body of evidence suggesting that schizophrenic patients show abnormal interactions and influ-

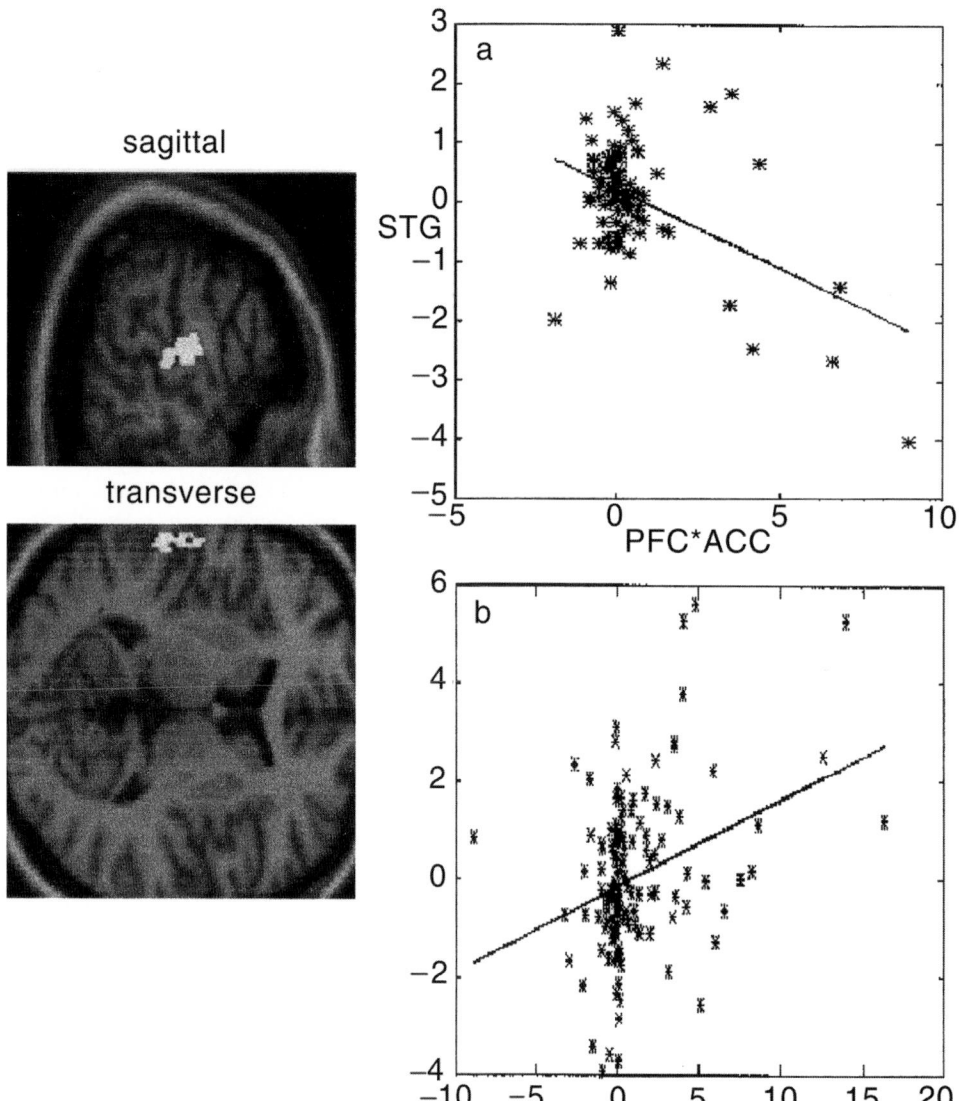

Figure 4 Reproduced with permission from Fletcher *et al.* (1999). Left panel: Region of left superior temporal cortex showing a significant difference between control and schizophrenic subjects ($p < 0.05$). A SPM rendered onto sagittal and transverse section of a stereotactically normalized structural MRI is shown. Right panel: Graphic representation of the relationship between activity in left superior temporal cortex (STG; *y* axis) and the combination of cingulo-prefrontal activity (PFC × ACG; *x* axis) in (a) controls and (b) schizophrenic subjects. The linear best fit is shown for both groups. The regression coefficient of each of the two groups is considered to provide a measure of Contribution of PFC × ACC to superior temporal cortical activity. Thus, in the controls an increase in PFC × ACC produces an inhibition of superior temporal activity. In the schizophrenic subjects, the line slopes upward, indicating the opposite effect.

ences among brain regions (or functional integration) during cognitive tasks. There is currently little *direct* evidence in favor of this hypothesis, and several regions have been found to function abnormally, with no unequivocal evidence for any particular region being involved. However, the majority of positive findings suggest that a disruption of fronto-temporal integration is a core feature of schizophrenia. However, findings have

been confounded by several factors, especially use of poor control tasks such as rest and nonmatched task performance in the schizophrenic and control groups. Future cognitive activation studies using improved methodologies should resolve issues such as whether abnormal frontal function causes or reflects poor task performance in schizophrenia. A question of clinical importance is whether different patterns of cortical interaction correlate with or predict schizophrenic symptoms or outcome.

VI. Imaging Symptoms

Functional neuroimaging is also useful for evaluating neural activity in patients experiencing psychotic symptoms.

A. Hallucinations

Hallucinations, perceptions in the absence of external stimuli, are prominent among the core positive symptoms of schizophrenia. Although auditory hallucinations are more common than any other type, hallucinations in other sensory modalities occur in a proportion of patients. Hallucinations in all modalities tend to be associated with activity in the neural substrate associated with that particular sensory modality. However, there is evidence that activation of the sensory cortex particular to the false perception is not in itself sufficient for the perception. Instead, research suggests that the interaction of a distributed cortico-subcortical neural network might provide a biological basis for schizophrenic hallucinations.

1. Auditory Hallucinations

The most common type of hallucination in schizophrenia is in the auditory domain and normally consists of spoken speech or voices (Hoffman, 1986). Functional neuroimaging studies of auditory hallucinations suggest that they involve neural systems dedicated to auditory speech perception as well as a distributed network of other cortical and subcortical areas. There are two distinct approaches to the study of the physiological basis of auditory hallucinations. The first, which we will call the state approach, asks what changes in brain activity can be observed at the time hallucinations are occurring. The second, which we call the trait approach, asks whether there is permanent abnormality of brain function present in patients who are prone to experience auditory hallucinations when they are ill. This abnormality will be observable even in the absence of current symptoms.

a. State Studies Silbersweig *et al.* (1995) used PET to study brain activity associated with the occurrence of hallucinations in six schizophrenic patients. Five patients with classic auditory verbal hallucinations demonstrated activation in subcortical (thalamic and striatal) nuclei, limbic structures (especially hippocampus), and paralimbic regions (parahippocampal and cingulate gyri and orbitofrontal cortex). Temporoparietal auditory-linguistic association cortex activation was present in each subject. One drug-naive patient had visual as well as auditory verbal hallucinations and showed activations in visual and auditory-linguistic association cortices. The authors propose that activity in deep-brain structures seen in all subjects may generate or modulate hallucinations, and the particular sensory cortical regions activated in individual patients may affect their specific perceptual content. Importantly, this study pointed to the possibility that hallucinations coincide with activation of the sensory and association cortex specific to the modality of the experience, a notion that has received support from several further studies.

David *et al.* (1996) used fMRI to scan a schizophrenic patient while he was experiencing auditory hallucinations and again when hallucination-free. The subject was scanned during presentation of exogenous auditory and visual stimuli and while he was on and off antipsychotic drugs. The BOLD signal in the temporal cortex to exogenous auditory stimulation (speech) was significantly reduced when the patient was experiencing hallucinating voices, regardless of medication. Visual cortical activation to flashing lights remained the same over all four scans, whether the subject was experiencing auditory hallucinations or not.

A similar result was obtained by Woodruff *et al.* (1997b), who used fMRI to study seven schizophrenic patients while they were experiencing severe auditory verbal hallucinations and again after their hallucinations had subsided. On the former occasion, these patients had reduced responses in temporal cortex, especially the right middle temporal gyrus, to external speech, compared to when their hallucinations were mild. The authors thus proposed that auditory hallucinations are associated with reduced responsivity in temporal cortical regions that overlap with those that normally process external speech, possibly due to competition for common neurophysiological resources.

Recently, Dierks *et al.* (1999) used event-related fMRI to investigate three paranoid schizophrenics who were able to indicate the on-set and off-set of their hallucinations as in the study by Silbersweig *et al.*. Using this design, they found that primary auditory cortex, including Heschl's gyrus, was associated with the presence of auditory hallucinations. Secondary auditory cortex, temporal lobe, and frontal operculum (Broca's area) were also activated during auditory hallucinations, supporting the notion that auditory hallucinations are related to inner speech. Finally, hallucinations were also associated with increased activity in the hippocampus and amygdala. The authors suggested that these activations could be due to retrieval from memory of the hallucinated material and emotional reaction to the voices, respectively.

Studies are divided on the issue of laterality of activity in the temporal lobe associated with schizophrenic hallucinations. Most studies report activity in the left temporal lobe (e.g., Cleghorn *et al.*, 1990; Suzuki *et al.*, 1993; McGuire *et al.*, 1996), consistent with the lateralization of language areas. Very few studies report activity in the right temporal lobe, and most that do have used a very small number of subjects, in some cases a single subject (e.g., Woodruff *et al.*, 1997b).

b. Trait Studies   The finding that auditory hallucinations are associated with activation of auditory and language association areas is consistent with the proposal that auditory verbal hallucinations arise from a disorder in the experience of inner speech (Frith, 1992). This was investigated by McGuire *et al.* (1996). They used PET to evaluate the neural correlates of tasks that engaged inner speech and auditory verbal imagery in schizophrenic patients with a strong predisposition to auditory verbal hallucinations (hallucinators), schizophrenic patients with no history of hallucinations (nonhallucinators), and

normal controls. There were no differences between hallucinators and controls in regional cerebral blood flow during thinking in sentences. However, when imagining sentences spoken in another person's voice, which entails both the generation and monitoring of inner speech, hallucinators showed reduced activation of the left middle temporal gyrus and the rostral supplementary motor area, regions activated by both normal subjects and nonhallucinators. Conversely, when nonhallucinators imagined speech, they differed from both hallucinators and controls in showing reduced activation in the right parietal operculum (see Fig. 5). McGuire and his colleagues suggest that the presence of verbal hallucinations is associated with a failure to activate areas concerned with the monitoring of inner speech.

2. Visual Hallucinations

The neural correlates of visual hallucinations also seem to be located in the neural substrate of visual perception. At least part of the activity in the brain associated with the experience of visual hallucinations is located in the visual cortex. For example, using SPECT, Hoksbergen *et al.* (1996) found that visual hallucinations were associated with hypoperfusion in the right occipito-temporal region, which showed partial normal-

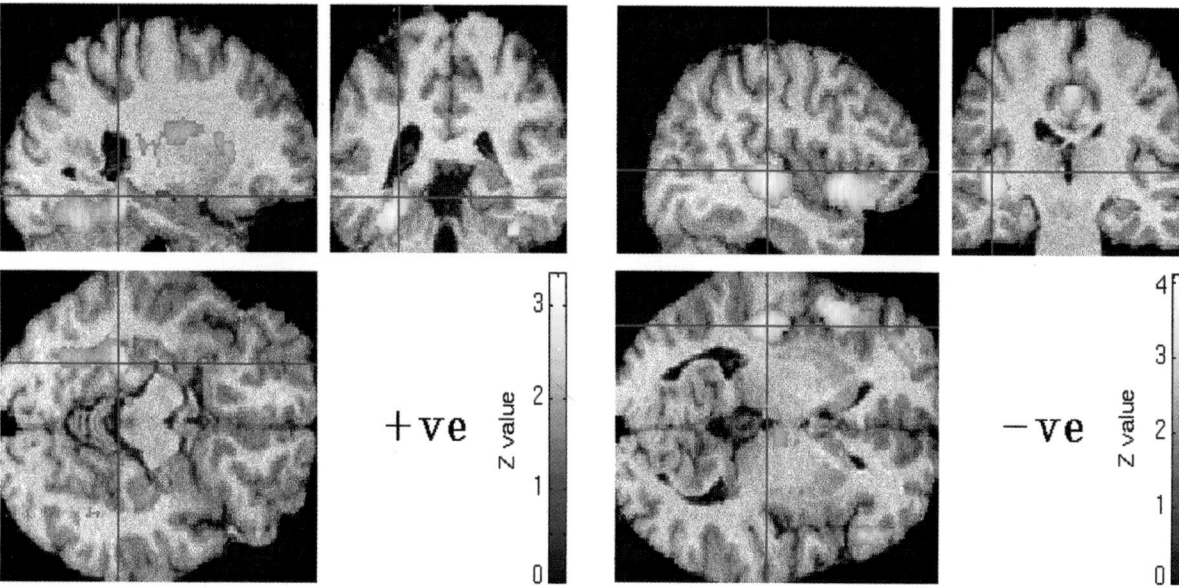

Figure 5 PET data in a group of patients with schizophrenia. Subjects were shown ambiguous pictures that provoked incoherent (thought-disordered) verbal responses when they were asked to describe what the pictures showed. Each subject was shown 12 different stimuli, which elicited a range of responses that varied with respect to how thought disordered they were. The severity of thought disorder was correlated with rCBF across the 12 scans, controlling for differences in the total number of words articulated. Positive correlations (areas that were more active the more disordered the speech) were evident in the left parahippocampal/anterior fusiform and right anterior fusiform gyri, whereas negative correlations were seen in the inferior frontal gyri, the cingulate gyrus, and the left superior temporal gyrus. Modified from McGuire *et al.* (1998).

ization after the visual hallucinations had subsided. Howard *et al.* (1997) used fMRI to investigate the visual cortical response to photic stimulation during and in the absence of continuous visual hallucinations. When visual hallucinations were absent, photic stimulation produced a normal bilateral activation in striate cortex. During hallucinations, very limited activation in striate cortex could be induced by exogenous visual stimulation. Similarly, Ffytche *et al.* (1998) found activity in ventral extrastriate visual cortex in patients with the Charles Bonnet syndrome when they experienced visual hallucinations. Moreover, the content of the hallucinations reflected the functional specialization of the region; for example, hallucinations of color activated V4 (the human color center) (Zeki *et al.*, 1991).

In conclusion, functional neuroimaging studies suggest that hallucinations involve an interaction between the neural systems dedicated to the particular sensory modality in which the false perception occurs and a widely distributed cortico-subcortical system, including limbic, paralimbic, and frontal areas. Intersubject variability in the specific location of the sensory activation associated with the hallucination could arise from differences between the patients in the sensory content and experience of their hallucinations.

B. Thought Disorder

McGuire *et al.* (1998) scanned six schizophrenic subjects with PET while they described a series of ambiguous pictures, which provoked different degrees of thought-disordered speech in each patient. The severity of thought disorder was correlated with rCBF across the 12 scans, controlling for differences in the total number of words articulated. Verbal disorganization (positive thought disorder) was inversely correlated with activity in the inferior frontal, cingulate, and left superior temporal cortex, areas implicated in the regulation and monitoring of speech production (see Fig. 5). The authors propose that this reduced activity might contribute to the articulation of the linguistic anomalies that characterize positive thought disorder. Verbal disorganization was positively correlated with activity in the parahippocampal/anterior fusiform region bilaterally, which may reflect this region's role in the processing of linguistic anomalies.

C. Passivity

Spence *et al.* (1997) performed a PET study in which subjects had to make voluntary joystick movements in the experimental condition, and stereotyped (routine) movements in the baseline condition, and do nothing in

the rest condition. They investigated a group of schizophrenic patients with passivity (delusions of control) and without (the same group in remission) and a group of normal controls. Schizophrenic patients with passivity showed hyperactivation of the inferior parietal lobe (Brodmann area 40), the cerebellum, and the cingulate cortex relative to schizophrenic patients without passivity (see Fig. 6). Similar results were found when schizophrenic patients with passivity were compared to normal controls. A comparison of all schizophrenics with normal controls revealed hypofrontality in the patients. When patients were in remission and no longer experienced passivity symptoms, a reversal of the hyperactivation of the parietal lobe and cingulate was seen. Hyperactivity in the parietal cortex may reflect the "unexpected" nature of the experienced movement in patients, as though it were being caused by an external force.

VII. Obstacles to Functional Neuroimaging and Schizophrenia Studies

A. Subject Matching

It is important that patients are matched to the control group on as many factors as possible. However, the best way to match IQ and education level is uncertain. It is not clear whether the control group's education level should be matched to the patient's parental or premorbid education level. However, both of these are su-

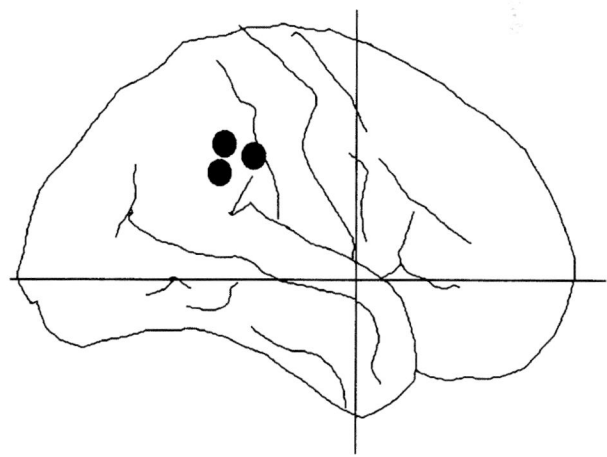

Figure 6 Diagram of overactivity in right parietal lobe (Brodmann area 40) in patients with passivity while making voluntary joystick movements (in the study by Spence *et al.*, 1997). When patients were in remission and no longer experienced passivity symptoms, a reversal of this hyperactivation of parietal lobe was seen. Hyperactivity in parietal cortex might reflect the patient's experiencing the movement as "unexpected," as if it were being caused by some external force.

perior to matching controls to patients' current IQ level, which may be considerably impaired by illness.

Another question is whether the control group should be normal or psychiatric. Normal controls are important in order to establish a baseline model of the neural circuitry involved in an experimental task. However, there are clearly several problems with using nonpsychiatric controls. These include the fact that normal controls are not taking medication, hospitalized, or affect-flattened, factors that might have a direct effect on rCBF or that might cause the patients to be more or less motivated or to attend or think more or less. Since psychiatric patients will be more matched on these factors, they may constitute a preferable control group. However, there are also problems with using psychiatric patients as controls. There is the question of which psychiatric population should be used. Should they be taking the same medication, or is hospitalization the most important factor? They might not be able to perform the experimental task for some reason that is different from that causing impairment in schizophrenic patients. Therefore, comparing schizophrenic to depressed patients, for example, may reveal activity specific to depression or to schizophrenia, rather than the task in question.

Some studies have used non-affected siblings, or even MZ twins, as controls for the schizophrenic patients. However, these are not matched for medication or hospitalization, even though they are usually matched for background. Furthermore, many studies investigate the non-affected MZ twins of schizophrenic patients as a high-risk group, since there is evidence and theory that this group may be affected in some way by predisposing factors (Crow, 1995).

It is questionable whether male and female schizophrenic patients should be evaluated as a single group since gender differences have been documented in schizophrenia (e.g., Andreasen *et al.,* 1994).

B. Experimental Tasks

As has been discussed at length throughout this chapter, there is a conceptual problem with applying the approach of cognitive activation studies to patient groups. It is impossible to interpret patterns of brain activity that differ between control and patient groups when the performance of the two groups on tasks differs in terms of degrees of efficiency and success. Many functional neuroimaging studies have used unmodified frontal tasks, in which the performance of schizophrenics falls below that of controls. It is unclear how to interpret increases or decreases of regionally specific activation in these studies. Any difference in brain activity between the two tasks could represent a critical

abnormality in schizophrenia and might cause poor task performance, or alternatively it might reflect poor performance. It is difficult to distinguish these two alternatives. Therefore, recent studies have employed tasks on which performance of the patient and control group is matched.

However, there is also a problem with interpreting the results of activation studies in which the task performance of the patient and control group is matched. What do differences in brain activity mean in the context of normal task performance? If an area is activated more in the controls than in the patient group during such a task, the functional significance of that activation is difficult to understand—it is clearly not necessary for performing the task. The most obvious interpretation is that patients and controls are using different strategies to achieve similar task performance. Therefore, interpretational difficulties remain: What is the nature of the relationship between differences in brain activity and behavior? How do these two variables relate to the schizophrenic state? These problems have not been resolved and remain when interpreting data from studies in which task performance of patient and control groups is matched.

Comparing an experimental condition with a very low-level baseline, or rest, is problematic because schizophrenics might show different brain activation during baseline than normal controls. In other words, schizophrenics' brains might not activate in the "normal" way during rest. Indeed there is much evidence from PET metabolism studies to indicate that blood flow in the brains of schizophrenic patients at rest is different from that in normal controls. Different rCBF at rest could be interpreted as activation (or deactivation) during the experimental task if rest is used as the control task.

C. Symptom-Specific Groups

Schizophrenia is a heterogeneous illness, comprising a variety of different symptoms. Using groups of patients defined by diagnosis (schizophrenia) may explain the inconsistent and equivocal results of functional imaging studies, since each symptom can be associated with a different brain pattern or functional abnormality. Attempts have been made to correlate cognitive brain activity with specific symptoms or clinical signs. However, as reviewed in this chapter, the results are inconsistent. Correlation studies should be hypothesis-driven, since correlations will always occur by chance if enough variables are evaluated. It would be an improvement if a hypothesis were made based on past data, for example that temporal lobe abnormality might contribute to auditory hallucinations because temporal lobe epileptics experience such symptoms. If, on the other hand, a trawling exercise is performed to identify symptom–brain activation pattern correlations, proper statistical thresholds with corrections for multiple comparisons need to be made to obtain any informative results.

A clear advantage of using symptom-specific schizophrenic groups is that the control group can comprise people with a diagnosis of schizophrenia, who are thus matched in terms of medication and hospitalization, but who do not have a particular symptom. Better still, the same group of schizophrenic patients can be used as their own control group if and when the symptom evaluated remits (see the study by Spence *et al.,* 1998). However, a clear shortcoming of using symptom-specific groups is that little can be discovered about schizophrenia as a syndrome.

VIII. Conclusion

Although there has been no one specific, unequivocal finding peculiar either to the syndrome of schizophrenia or to a particular symptom, there have been repeated findings of frontal and temporal abnormalities in schizophrenia, in both resting metabolism and cognitive activation studies. It is becoming increasingly evident that schizophrenic patients show abnormal interactions between brain regions during cognitive tasks. There is currently little *direct* evidence in favor of the hypothesis of functional disconnection, and several regions have been found to function abnormally, but with no unequivocal evidence for the consistent involvement of any particular region. However, the majority of positive findings suggest that a disruption of cortico-cortical (or cortico-subcortical) integration is a core feature of the schizophrenic syndrome. Future studies would do well to investigate this possibility, using methods of analyzing functional or effective connectivity to evaluate the influence of one brain region over another.

References

American Psychiatry Association. (1987). "Diagnostic and Statistical Manual of Mental Disorders," DSM-III-R, 3rd ed. American Psychiatry Association, Washington, DC.

Andreasen, N., Nasrallah, H. A., Dunn, V., Olson, S. C., Grove, W. M., Ehrhardt, J. C., Coffman, J. A., and Crossett, J. H. (1986). Structural abnormalities in the frontal system in schizophrenia. A magnetic resonance imaging study. *Arch. Gen. Psychiatry* **43**(2), 136–144.

Andreasen, N. C., Swayze, V. W., 2nd, Flaum, M., Yates, W. R., Arndt, S., and McChesney, C. (1990). Ventricular enlargement in schizophrenia evaluated with computed tomographic scanning. Effects of gender, age, and stage of illness. *Arch. Gen. Psychiatry* **47**(11), 1008–1015.

Andreasen, N. C., Rezai, K., Alliger, R., Swayze, V. W., Flaum, M., Kirchner, P., Cohen, G., and O'Leary, D. S. (1992). Hypofrontality

in neuroleptic-naive patients and in patients with chronic schizophrenia: Assessment with xenon-133 single-photon emission computed tomography and the Tower of London. *Arch. Gen. Psychiatry* **49**, 943–958.

Andreasen, N. C., Arndt, S., Swayze, V., 2nd, Cizadlo, T., Flaum, M., O'Leary, D. S., Ehrhardt, J. C., and Yuh, W. T. (1994). Thalamic abnormalities in schizophrenia visualised through magnetic resonance image averaging. *Science* **266**, 294–298.

Arnold, S. E. (1997). The medial temporal lobe in schizophrenia. *J. Neuropsychiatry Clin. Neurosci.* **9**, 460–470.

Barta, P. E., Pearlson, G. D., Powers, R. E., Richards, S. S., and Tune, L. E. (1990). Auditory hallucinations and smaller superior temporal gyral volume in schizophrenia. *Am. J. Psychiatry* **147**(11), 1457–1462.

Benes, F. M. (1991). Evidence for neurodevelopment disturbances in anterior cingulate cortex of post-mortem schizophrenic brain. *Schizophr. Res.* **5**(3), 187–188.

Benes, F. M., Sorensen, I., and Bird, E. D. (1991). Reduced neuronal size in posterior hippocampus of schizophrenic patients. *Schizophr. Bull.* **17**(4), 597–608.

Berman, K. F., Ostrem, J. L., Randolph, C., Gold, J., Goldberg, T. E., Coppola, R., Carson, R. E., Herscovitch, P., and Weinberger, D. R. (1995). Physiological activation of a cortical network during performance of the Wisconsin Card Sorting Test: A positron emission tomography study. *Neuropsychologia* **33**(8), 1027–1046.

Bleuler, E. (1987). Dementia praecox or the group of schizophrenias. *In* "The Clinical Routes of the Schizophrenia Concept" (J. Cutting and M. Shepherd, Eds.). Cambridge Univ. Press, Cambridge, UK.

Breier, A., Buchanan, R. W., Elkashef, A., Munson, R. C., Kirkpatrick, B., and Gellad, F. (1992). Brain morphology and schizophrenia. A magnetic resonance imaging study of limbic, prefrontal cortex, and caudate structures. *Arch. Gen. Psychiatry* **49**(12), 921–926.

Brown, R., Colter, N., Corsellis, J. A., Crow, T. J., Frith, C. D., Jagoe, R., Johnstone, E. C., and Marsh, L. (1986). Postmortem evidence of structural brain changes in schizophrenia. Differences in brain weight, temporal horn area, and parahippocampal gyrus compared with affective disorder. *Arch. Gen. Psychiatry* **43**(1), 36–42.

Buchanan, R. W., Vladar, K., Barta, P. E., and Pearlson, G. D. (1998). Structural evaluation of the prefrontal cortex in schizophrenia. *Am. J. Psychiatry* **155**(8), 1049–1055.

Buchsbaum, M. S., Yang, S., Hazlett, E., Siegel, B. V., Jr., Germans, M., Haznedar, M., O'Flaithbheartaigh, S., Wei, T., Silverman, J., and Siever, L. J. (1997). Ventricular volume and asymmetry in schizotypal personality disorder and schizophrenia assessed with magnetic resonance imaging. *Schizophr. Res.* **27**(1), 45–53.

Busatto, G. F., Costa, D. C., Ell, P. J., Pilowsky, L. S., David, A. S., and Kerwin, R. W. (1994). Regional cerebral blood flow (rCBF) in schizophrenia during verbal memory activation: A ^{99m}Tc-HMPAO single photon emission tomography (SPET) study. *Psychol. Med.* **24**(2), 463–472.

Callicott, J. H., Ramsey, N. F., Tallent, K., Bertolino, A., Knable, M. B., Coppola, R., Goldberg, T., van Gelderen, P., Mattay, V. S., Frank, J. A., Moonen, C. T., and Weinberger, D. R. (1998). Functional magnetic resonance imaging brain mapping in psychiatry: Methodological issues illustrated in a study of working memory in schizophrenia. *Neuropsychopharmacology* **18**(3), 186–196.

Cannon, T. D., van Erp, T. G., Huttunen, M., Lonnqvist, J., Salonen, O., Valanne, L., Poutanen, V. P., Standertskjold-Nordenstam, C. G., Gur, R. E., and Yan, M. (1998). Regional gray matter, white matter, and cerebrospinal fluid distributions in schizophrenic patients, their siblings, and controls. *Arch. Gen. Psychiatry* **55**(12), 1084–1091.

Carter, C. S., Perlstein, W., Ganguli, R., Brar, J., Mintun, M., and Cohen, J. D. (1998). Functional hypofrontality and working memory dysfunction in schizophrenia. *Am. J. Psychiatry* **155**(9), 1285–1287.

Chua, S. E., and McKenna, P. J. (1995). Schizophrenia, a brain disease? A critical review of structural and functional cerebral abnormality in the disorder. *Br. J. Psychiatry* **166**, 563–582.

Cleghorn, J. M., Garnett, E. S., Nahmias, C., Brown, G. M., Kaplan, R. D., Szechtman, H., Szechtman, B., Franco, S., Dermer, S. W., and Cook, P. (1990). Regional brain metabolism during auditory hallucinations in chronic schizophrenia. *Br. J. Psychiatry* **157**, 562–570.

Coffman, J. A., and Nasrallah, H. A. (1986). Magnetic brain imaging in schizophrenia. *In* "The Neurology of Schizophrenia" (H. A. Nasrallah and D. R. Weinberger, Eds.), pp. 251–266. Elsevier, Amsterdam.

Crawley, J. C., Owens, D. G., Crow, T. J., Poulter, M., Johnstone, E. C., Smith, T., Oldland, S. R., Veall, N., Owen, F., and Zanelli, G. D. (1986). Dopamine D_2 receptors in schizophrenia studied *in vivo*. *Lancet* **2**, 224–225.

Crow, T. J. (1980). Positive and negative schizophrenic symptoms and the role of dopamine. *Br. J. Psychiatry* **137**, 383–386.

Crow, T. J. (1995). Aetiology of schizophrenia: An evolutionary theory. *Int. Clin. Psychopharmacol.* **10**(Suppl. 3), 49–56.

Daniel, D. G., Weinberger, D. R., Jones, D. W., Zigun, J. R., Coppola, R., Handel, S., Bigelow, L. B., Goldberg, T. E., Berman, K. F., and Kleinman, J. E. (1991). The effect of amphetamine on regional cerebral blood flow during cognitive activation in schizophrenia. *J. Neurosci.* **11**, 1907–1917.

David, A. S., Woodruff, P. W., Howard, R., Mellers, J. D., Brammer, M., Bullmore, E., Wright, I., Andrew, C., and Williams, S. C. (1996). Auditory hallucinations inhibit exogenous activation of auditory association cortex. *NeuroReport* **7**(4), 932–936.

Delay, J., Deniker, P., and Harl, J. M. (1952). Traitement des etats d'exicitation et d'agitation par une methode medicamenteuse derives de l'hibernotherapie. *Ann. Med. Psychol.* **110**, 267–273.

DeLisi, L. E., Hoff, A. L., Schwartz, J. E., Shields, G. W., Halthore, S. N., Gupta, S. M., Henn, F. A., and Anand, A. K. (1991). Brain morphology in first-episode schizophrenic-like psychotic patients: A quantitative magnetic resonance imaging study. *Biol. Psychiatry* **29**(2), 159–175.

DeLisi, L. E., Sakuma, M., Kushner, M., Finer, D. L., Hoff, A. L., and Crow, T. J. (1997). Anomalous cerebral asymmetry and language processing in schizophrenia. *Schizophr. Bull.* **23**(2), 255–271.

DeMyer, M. K., Gilmor, R. L., Hendrie, H. C., DeMyer, W. E., Augustyn, G. T., and Jackson, R. K. (1988). Magnetic resonance brain images in schizophrenic and normal subjects: Influence of diagnosis and education. *Schizophr. Bull.* **14**(1), 21–37.

Dierks, T., Linden, D. E. J., Jandl, M., Formisano, E., Goebel, R., Lanfermann, H., and Singer, W. (1999). Activation of Heschl's gyrus during auditory hallucinations. *Neuron* **22**(3), 615–621.

Dolan, R. J., and Fletcher, P. C. (1997). Dissociating prefrontal and hippocampal function in episodic memory encoding. *Nature* **388**, 582–585.

Dolan, R. J., Fletcher, P., Frith, C. D., Friston, K. J., Frackowiak, R. S. J., and Grasby, P. J. (1995). Dopaminergic modulation of an impaired cognitive activation in the anterior cingulate cortex in schizophrenia. *Nature* **378**, 180–183.

Ebmeier, K. P., Blackwood, D. H., Murray, C., Souza, V., Walker, M., Dougall, N., Moffoot, A. P., O'Carroll, R. E., and Goodwin, G. M. (1993). Single-photon emission computed tomography with ^{99m}Tc-exametazime in unmedicated schizophrenic patients. *Biol. Psychiatry* **33**(7), 487–495.

Farde, L., Wiesel, F. A., Hall, H., Halldin, C., Stone-Elander, S., and Sedvall, G. (1987). No D_2 receptor increase in PET study of schizophrenia. *Arch. Gen. Psychiatry* **44**, 671–672.

Feinstein, A., Goldberg, T. E., Nowlin, B., and Weinberger, D. R. (1998). Types and characteristics of remote memory impairment in schizophrenia. *Schizophr. Res.* **30**(2), 155–163.

ffytche, D. H., Howard, R. J., Brammer, M. J., David, A., Woodruff, P., and Williams, S. (1998). The anatomy of conscious vision: An fMRI study of visual hallucinations. *Nat. Neurosci.* **1**(8), 738–742.

Fletcher, P. C., Frith, C. D., Grasby, P. M., Friston, K. J., and Dolan, R. J. (1996). Local and distributed effects of apomorphine on fronto-temporal function in acute unmedicated schizophrenia. *J. Neurosci.* **16**(21), 7055–7062.

Fletcher, P. C., McKenna, P. J., Frith, C. D., Grasby, P. M., Friston, K. J., and Dolan, R. J. (1998). Brain activations in schizophrenia during a graded memory task studied with functional neuroimaging. *Arch. Gen. Psychiatry* **55**(11), 1001–1008.

Fletcher, P. C., McKenna, P. J., Friston, K. J., Frith, C. D., and Dolan, R. J. (1999). Abnormal cingulate modulation of fronto-temporal connectivity in schizophrenia. *NeuroImage* **9**, 337–342.

Friston, K. J., and Frith, C. D. (1995). Schizophrenia: A disconnection syndrome? *Clin. Neurosci.* **3**, 89–97.

Friston, K. J., Liddle, P. F., Frith, C. D., Hirsch, S. R., and Frackowiak, R. S. (1992). The left medial temporal region and schizophrenia. A PET study. *Brain* **115** (Part 2), 367–382.

Friston, K. J., Frith, C. D., Liddle, P. F., and Frackowiak, R. S. (1993). Functional connectivity: The principal-component analysis of large (PET) data sets. *J. Cereb. Blood Flow Metab.* **13**(1), 5–14.

Friston, K. J., Buechel, C., Fink, G. R., Morris, J., Rolls, E., and Dolan, R. J. (1997). Psychophysiological and modulatory interactions in neuroimaging. *NeuroImage* **6**(3), 218–229.

Frith, C. D. (1992). "The Cognitive Neuropsychology of Schizophrenia." Lawrence Erlbaum, Hove.

Frith, C. D., Friston, K., Liddle, P. F., and Frackowiak, R. S. (1991). Willed action and the prefrontal cortex in man: A study with PET. *Proc. R. Soc. Lond. B. Biol. Sci.* **244**(1311), 241–246.

Frith, C. D., Friston, K. J., Herold, S., Silbersweig, D., Fletcher, P., Cahill, C., Dolan, R. J., Frackowiak, R. S. J., and Liddle, P. F. (1995). Regional brain activity in chronic schizophrenic patients during the performance of a verbal fluency task. *Br. J. Psychiatry* **167**, 343–349.

Fukuzako, H., Yamada, K., Kodama, S., Yonezawa, T., Fukuzako, T., Takenouchi, K., Kajiya, Y., Nakajo, M., and Takigawa, M. (1997). Hippocampal volume asymmetry and age at illness onset in males with schizophrenia. *Eur. Arch. Psychiatry Clin. Neurosci.* **247**(5), 248–251.

Ganguli, R., Carter, C., Mintun, M., Brar, J., Becker, J., Sarma, R., Nichols, T., and Bennington, E. (1997). PET brain mapping study of auditory verbal supraspan memory versus visual fixation in schizophrenia. *Biol. Psychiatry* **41**(1), 33–42.

Goldman-Rakic, P. S., and Selemon, L. D. (1997). Functional and anatomical aspects of prefrontal pathology in schizophrenia. *Schizophr. Bull.* **23**, 437–458.

Gottesman, I. I., and Shields, J. A. (1982). "Schizophrenia: The Epigenetic Puzzle." Cambridge Univ. Press, Cambridge, UK.

Green, M. F. (1996). What are the functional consequences of neurocognitive deficits in schizophrenia? *Am. J. Psychiatry* **153**, 321–330.

Harvey, I., Ron, M. A., Du Boulay, G., Wicks, D., Lewis, S. W., and Murray, R. M. (1993). Reduction of cortical volume in schizophrenia on magnetic resonance imaging. *Psychol. Med.* **23**(3), 591–604.

Heckers, S., Rauch, S. L., Goff, D., Savage, C. R., Schacter, D. L., Fischman, A. J., and Alpert, N. M. (1998). Impaired recruitment of the hippocampus during conscious recollection in schizophrenia. *Nat. Neurosci.* **1**(4), 318–323.

Hoffman, R. E. (1986). Verbal hallucinations and language production processes in schizophrenia. *Behav. Brain Sci.* **9**(3), 503–517.

Hoksbergen, I., Pickut, B. A., Marien, P., Slabbynck, H., Kunnen, J., and De Deyn, P. P. (1996). SPECT findings in an unusual case of visual hallucinosis. *J. Neurol.* **243** (8), 594–598.

Honer, W. G., Falkai, P., Young, C., Wang, T., Xie, J., Bonner, J., Hu, L., Boulianne, G. L., Luo, Z., and Trimble, W. S. (1997). Cingulate cortex synaptic terminal proteins and neural cell adhesion molecule in schizophrenia. *Neuroscience* **78**(1), 99–110.

Howard, R., David, A., Woodruff, P., Mellers, I., Wright, J., Brammer, M., Bullmore, E., and Williams, S. (1997). Seeing visual hallucinations with functional magnetic resonance imaging. *Dement. Geriatr. Cogn. Disord.* **8**(2), 73–77.

Ingvar, D. H., and Franzen, G. (1974). Distribution of cerebral activity in chronic schizophrenia. *Lancet* **2**, 1484–1486.

Iversen, S. D. (1986). Animal models of schizophrenia. *In* "The Psychopharmacology and Treatment of Schizophrenia" (P. B. Bradley and S. R. Hirsch, Eds.). Oxford Univ. Press, Oxford, UK.

Johnstone, E. C. (1991). Defining characteristics of schizophrenia. *Br. J. Psychiatry* **13** (Suppl.), 5–6.

Johnstone, E. C., Crow, T. J., Frith, C. D., Husband, J., and Kreel, L. (1976). Cerebral ventricular size and cognitive impairment in chronic schizophrenia. *Lancet* **2**, 924–926.

Johnstone, E. C., Owens, D. G., Crow, T. J., Frith, C. D., Alexandropolis, K., Bydder, G., and Colter, N. (1989). Temporal lobe structure as determined by nuclear magnetic resonance in schizophrenia and bipolar affective disorder. *J. Neurol. Neurosurg. Psychiatry* **52**(6), 736–741.

Kelsoe, J. R., Jr., Cadet, J. L., Pickar, D., and Weinberger, D. R. (1988). Quantitative neuroanatomy in schizophrenia. A controlled magnetic resonance imaging study. *Arch. Gen. Psychiatry* **45**(6), 533–541.

Kendler, K. S., and Robinette, C. D. (1983). Schziophrenia in National Academy of Sciences–National Research Council Twin Registry: A 16-year update. *Am. J. Psychiatry* **140**, 1551–1563.

Kraepelin, E. (1919). "Dementia Praecox and Paraphrenia." Livingstone, Edinburgh.

Kuipers, L., and Bebbington, P. (1988). Expressed emotion research in schizophrenia: Theoretical and clinical implications. *Psychol. Med.* **18**(4), 893–909.

Kwon, J. S., McCarley, R. W., Hirayasu, Y., Anderson, J. E., Fischer, I. A., Kikinis, R., Jolesz, F. A., and Shenton, M. E. (1999). Left planum temporale volume reduction in schizophrenia. *Arch. Gen. Psychiatry* **56**(2), 142–148.

Laruelle, M. (1998). Imaging dopamine transmission in schizophrenia. A review and meta-analysis. *Q. J. Nucl. Med.* **42**(3), 211–221.

Lewis, S. W. (1990). Computerised tomography in schizophrenia 15 years on. *Br. J. Psychiatry* **9**(Suppl.), 16–24.

Liddle, P. F. (1987). The symptoms of chronic schizophrenia. A re-examination of the positive-negative dichotomy. *Br. J. Psychiatry* **151**, 145–151.

Liddle, P. F., Friston, K. J., Frith, C. D., and Frackowiak, R. S. (1992). Cerebral blood flow and mental processes in schizophrenia. *J. R. Soc. Med.* **85**(4), 224–227.

Maher, B. A., Manschreck, T. C., Yurgelun-Todd, D. A., and Tsuang, M. T. (1998). Hemispheric asymmetry of frontal and temporal gray matter and age of onset in schizophrenia. *Biol. Psychiatry* **44**(6), 413–417.

Mathew, R. J., Wilson, W. H., Tant, S. R., Robinson, L., and Prakash, R. (1988). Abnormal resting regional cerebral blood flow patterns and their correlates in schizophrenia. *Arch. Gen. Psychiatry* **45**(6), 542–549.

Mattay, V. S., Callicott, J. H., Bertolino, A., Santha, A. K., Tallent, K. A., Goldberg, T. E., Frank, J. A., and Weinberger, D. R. (1997). Abnormal functional lateralization of the sensorimotor cortex in patients with schizophrenia. *NeuroReport* **8**(13), 2977–2984.

McGue, M., Gottesman, I. I., and Rao, D. C. (1986). The analysis of schizophrenia family data. *Behav. Genet.* **16**, 75–87.

McGuire, P. K., Silbersweig, D. A., Wright, I., Murray, R. M., Frackowiak, R. S., and Frith, C. D. (1996). The neural correlates of inner speech and auditory verbal imagery in schizophrenia: relationship to auditory verbal hallucinations. *Br. J. Psychiatry* **169**(2), 148–159.

McGuire, P. K., Quested, D. J., Spence, S. A., Murray, R. M., Frith,

C. D., and Liddle, P. F. (1998). Pathophysiology of "positive" thought disorder in schizophrenia. *Br. J. Psychiatry* **173**, 231–235.

Miller, D. D., Rezai, K., Alliger, R., and Andreasen, N. C. (1997). The effect of antipsychotic medication on relative cerebral blood perfusion in schizophrenia: Assessment with technetium-99m hexamethyl-propyleneamine oxime single photon emission computed tomography. *Biol. Psychiatry* **41**, 550–559.

Nagahama, Y., Fukuyama, H., Yamauchi, H., Matsuzaki, S., Konishi, J., Shibasaki, H., and Kimura, J. (1996). Cerebral activation during performance of a card sorting test. *Brain* **119**(5), 1667–1675.

Okubo, Y., Suhara, T., Suzuki, K., Kobayashi, K., Inoue, O., Terasaki, O., Someya, Y., Sassa, T., Sudo, Y., Matsushima, E., Iyo, M., Tateno, Y., and Toru, M. (1997). Decreased prefrontal dopamine D_1 receptors in schizophrenia revealed by PET. *Nature* **385**(6617), 634–636.

Paulman, R. G., Devous, M. D., Sr., Gregory, R. R., Herman, J. H., Jennings, L., Bonte, F. J., Nasrallah, H. A., and Raese, J. D. (1990). Hypofrontality and cognitive impairment in schizophrenia: Dynamic single-photon tomography and neuropsychological assessment of schizophrenic brain function. *Biol. Psychiatry* **27**(4), 377–399.

Pilowsky, L. S., Costa, D. C., Ell, P. J., Verhoeff, N. P., Murray, R. M., and Kerwin, R. W. (1994). D_2 dopamine receptor binding in the basal ganglia of antipsychotic-free schizophrenic patients. An ^{123}I-IBZM single photon emission computerised tomography study. *Br. J. Psychiatry* **164**(1), 16–26.

Ragland, J. D., Gur, R. C., Glahn, D. C., Censits, D. M., Smith, R. J., Lazarev, M. G., Alavi, A., and Gur, R. E. (1998). Frontotemporal cerebral blood flow change during executive and declarative memory tasks in schizophrenia: A positron emission tomography study. *Neuropsychology* **12**(3), 399–413.

Raine, A., Lencz, T., Reynolds, G. P., Harrison, G., Sheard, C., Medley, I., Reynolds, L. M., and Cooper, J. E. (1992). An evaluation of structural and functional prefrontal deficits in schizophrenia: MRI and neuropsychological measures. *Psychiatry Res.* **45**(2), 123–137.

Raz, S., and Raz, N. (1990). Structural brain abnormalities in the major psychoses: A quantitative review of the evidence from computerized imaging. *Psychol. Bull.* **108**(1), 93–108.

Ron, M. A., and Harvey, I. (1990). The brain in schizophrenia. *J. Neurol. Neurosurg. Psychiatry* **53**(9), 725–726.

Rossi, A., Stratta, P., Mancini, F., Gallucci, M., Mattei, P., Core, L., Di Michele, V., and Casacchia, M. (1994). Magnetic resonance imaging findings of amygdala–anterior hippocampus shrinkage in male patients with schizophrenia. *Psychiatry Res.* **52**(1), 43–53.

Saykin, A. J., Shtasel, D. L., Gur, R. E., Kester, D. B., Mozley, L. H., Stafiniak, P., and Gur, R. C. (1994). Neuropsychological deficits in neuroleptic naive patients with first-episode schizophrenia. *Arch. Gen. Psychiatry* **51**(2), 124–131.

Schacter, D. L., Alpert, N. M., Savage, C. R., Rauch, S. L., and Albert, M. S. (1996). Conscious recollection and the human hippocampal formation: Evidence from positron emission tomography. *Proc. Natl. Acad. Sci. U.S.A.* **93**, 321–325.

Schröder, J., Essig, M., Baudendistel, K., Jahn, T., Gerdsen, I., Stockert, A., Schad, L. R., and Knopp, M. V. (1999). Motor dysfunction and sensorimotor cortex activation changes in schizophrenia: A study with functional magnetic resonance imaging. *NeuroImage* **9**(1), 81–87.

Seeman, P. (1986). Dopamine/neuroleptic receptors in schizophrenia. *In* "Handbook on Studies of Schizophrenia" (G. D. Burrows, T. R. Norman, and G. Rubenstein, Eds.), Part 2. Elsevier, Amsterdam.

Seeman, P. (1987). Dopamine receptors and the dopamine hypothesis of schizophrenia. *Synapse* **1**, 133–152.

Shallice, T. (1982). Specific impairments of planning. *Philos. Trans. R. Soc. Lond. B. Biol. Sci.* **25**(298), 199–209.

Shallice, T., and Burgess, P. W. (1991). Deficits in strategy application following frontal lobe damage in man. *Brain* **114** (Part 2), 727–741.

Sharma, T., Lancaster, E., Lee, D., Lewis, S., Sigmundsson, T., Takei, N.,

Gurling, H., Barta, P., Pearlson, G., and Murray, R. (1998). Brain changes in schizophrenia. Volumetric MRI study of families multiply affected with schizophrenia: The Maudsley Family Study 5. *Br. J. Psychiatry* **173,** 132–138.

Shenton, M. E., Kikinis, R., Jolesz, F. A., Pollak, S. D., LeMay, M., Wible, C. G., Hokama, H., Martin, J., Metcalf, D., Coleman, M., *et al.* (1992). Abnormalities of the left temporal lobe and thought disorder in schizophrenia. A quantitative magnetic resonance imaging study. *N. Engl. J. Med.* **327**(9), 604–612.

Siegel, B. V., Jr., Buchsbaum, M. S., Bunney, W. E., Jr., Gottschalk, L. A., Haier, R. J., Lohr, J. B., Lottenberg, S., Najafi, A., Nuechterlein, K. H., Potkin, S. G., *et al.* (1993). Cortical–striatal–thalamic circuits and brain glucose metabolic activity in 70 unmedicated male schizophrenic patients. *Am. J. Psychiatry* **150**(9), 1325–1336.

Silbersweig, D. A., Stern, E., Frith, C., Cahill, C., Holmes, A., Grootoonk, S., Seaward, J., McKenna, P., Chua, S. E., Schnorr, L., *et al.* (1995). A functional neuroanatomy of hallucinations in schizophrenia. *Nature* **378,** 176–179.

Smith, G. N., and Iacono, W. G. (1986). Lateral ventricular size in schizophrenia and choice of control group. *Lancet* **1,** 1450.

Smith, R. C., Baumgartner, R., and Calderon, M. (1987). Magnetic resonance imaging studies of the brains of schizophrenic patients. *Psychiatry Res.* **20**(1), 33–46.

Spence, S. A., Brooks, D. J., Hirsch, S. R., Liddle, P. F., Meehan, J., and Grasby, P. M. (1997). A PET study of voluntary movement in schizophrenic patients experiencing passivity phenomena (delusions of alien control). *Brain* **120,** 1997–2011.

Spence, S. A., Hirsch, S. R., Brooks, D. J., and Grasby, P. M. (1998). Prefrontal cortex activity in people with schizophrenia and control subjects. Evidence from positron emission tomography for remission of "'hypofrontality'" with recovery from acute schizophrenia. *Br. J. Psychiatry* **172,** 316–323.

Stevens, A. A., Goldman-Rakic, P. S., Gore, J. C., Fulbright, R. K., and Wexler, B. E. (1998). Cortical dysfunction in schizophrenia during auditory word and tone working memory demonstrated by functional magnetic resonance imaging. *Arch. Gen. Psychiatry* **55**(12), 1097–1103.

Straube, E. R., and Oades, R. D. (1992). *In* "Schizophrenia: Empirical Research and Findings," Chapter 13. Academic Press, San Diego.

Suddath, R. L., Christison, G. W., Torrey, E. F., Casanova, M. F., and Weinberger, D. R. (1990). Anatomical abnormalities in the brains of monozygotic twins discordant for schizophrenia. *N. Engl. J. Med.* **322**(12), 789–794.

Suzuki, M., Yuasa, S., Minabe, Y., Murata, M., and Kurachi, M. (1993). Left superior temporal blood flow increases in schizophrenic and schizophreniform patients with auditory hallucination: A longitudinal case study using ^{123}I-IMP SPECT. *Eur. Arch. Psychiatry Clin. Neurosci.* **242**(5), 257–261.

Ungerleider, L. G. (1995). Functional brain imaging studies of cortical mechanisms for memory. *Science* **270,** 769–775.

Van Horn, J. D., and McManus, I. C. (1992). Ventricular enlargement in schizophrenia. A meta-analysis of studies of the ventricle : brain ratio. *Br. J. Psychiatry* **160,** 687–697.

Van Rossum, J. M. (1966). The significance of dopamine receptor blockade for the mechanism of action of neuroleptic drugs. *Arch. Int. Pharmacodyn. Ther.* **160,** 492–494.

Volz, H. P., Gaser, C., Hager, F., Rzanny, R., Mentzel, H. J., Kreitschmann-Andermahr, I., Kaiser, W. A., and Sauer, H. (1997). Brain activation during cognitive stimulation with the Wisconsin Card Sorting Test: A functional MRI study on healthy volunteers and schizophrenics. *Psychiatry Res.* **75**(3), 145–157.

Waddington, J. L. (1990). Sight and insight: Regional cerebral metabolic activity in schizophrenia visualised by positron emission tomography, and competing neurodevelopmental perspectives. *Br. J. Psychiatry* **156,** 615–619.

Weinberger, D. R., Torrey, E. F., Neophytides, A. N., and Wyatt, R. J. (1979). Lateral cerebral ventricular enlargement in chronic schizophrenia. *Arch. Gen. Psychiatry* **36**(7), 735–739.

Weinberger, D. R., Berman, K. F., and Zec, R. F. (1986). Physiologic dysfunction of dorsolateral prefrontal cortex in schizophrenia. I. Regional cerebral blood flow evidence. *Arch. Gen. Psychiatry* **43,** 114–124.

Weinberger, D. R., Mattay, V., Callicott, J., Kotrla, K., Santha, A., van Gelderen, P., Duyn, J., Moonen, C., and Frank, J. (1996). fMRI applications in schizophrenia research. *NeuroImage* **4**(3), S118–S126.

Wiser, A. K., Andreasen, N. C., O'Leary, D. S., Watkins, G. L., Boles Ponto, L. L., and Hichwa, R. D. (1998). Dysfunctional cortico-cerebellar circuits cause "cognitive dysmetria" in schizophrenia. *NeuroReport* **9**(8), 1895–1899.

Wolkin, A., Rusinek, H., Vaid, G., Arena, L., Lafargue, T., Sanfilipo, M., Loneragan, C., Lautin, A., and Rotrosen, J. (1998). Structural magnetic resonance image averaging in schizophrenia. *Am. J. Psychiatry* **155**(8), 1064–1073.

Wong, D. F., Wagner, H. N., Jr., Tune, L. E., Dannals, R. F., Pearlson, G. D., Links, J. M., Tamminga, C. A., Broussolle, E. P., Ravert, H. T., Wilson, A. A., Toung, J. K. T., Malat, J., Williams, J. A., O'Tuama, L. A., Snyder, S. H., Kuhar, M. J., and Gjedde, A. (1986). Positron emission tomography reveals elevated D_2 dopamine receptors in drug-naive schizophrenics. *Science* **234,** 1558–1563.

Woodruff, P. W., Wright, I. C., Shuriquie, N., Russouw, H., Rushe, T., Howard, R. J., Graves, M., Bullmore, E. T., and Murray, R. M. (1997a). Structural brain abnormalities in male schizophrenics reflect fronto-temporal dissociation. *Psychol. Med.* **27**(6), 1257–1266.

Woodruff, P. W., Wright, I. C., Bullmore, E. T., Brammer, M., Howard, R. J., Williams, S. C., Shapleske, J., Rossell, S., David, A. S., McGuire, P. K., and Murray, R. M. (1997b). Auditory hallucinations and the temporal cortical response to speech in schizophrenia: A functional magnetic resonance imaging study. *Am. J. Psychiatry* **154**(12), 1676–1682.

World Health Organization. (1975). "Schizophrenia: A Multinational Study." WHO, Geneva.

Wright, I. C., McGuire, P. K., Poline, J. B., Travere, J. M., Murray, R. M., Frith, C. D., Frackowiak, R. S., and Friston, K. J. (1995). A voxel-based method for the statistical analysis of gray and white matter density applied to schizophrenia. *NeuroImage* **2**(4), 244–252.

Young, A. H., Blackwood, D. H., Roxborough, H., McQueen, J. K., Martin, M. J., and Kean, D. (1991). A magnetic resonance imaging study of schizophrenia: Brain structure and clinical symptoms. *Br. J. Psychiatry* **158,** 158–164.

Yurgelun-Todd, D. A., Renshaw, P. F., Gruber, S. A., Ed, M., Waternaux, C., and Cohen, B. M. (1996a). Proton magnetic resonance spectroscopy of the temporal lobes in schizophrenics and normal controls. *Schizophr. Res.* **19**(1), 55–59.

Yurgelun-Todd, D. A., Waternaux, C. M., Cohen, B. M., Gruber, S. A., English, C. D., and Renshaw, P. F. (1996b). Functional magnetic resonance imaging of schizophrenic patients and comparison subjects during word production. *Am. J. Psychiatry* **153**(2), 200–205.

Zakzanis, K. K., and Hansen, K. T. (1998). Dopamine D_2 densities and the schizophrenic brain. *Schizophr. Res.* **32**(3), 201–206.

Zeki, S., Watson, J. D., Lueck, C. J., Friston, K. J., Kennard, C., and Frackowiak, R. S. (1991). A direct demonstration of functional specialization in human visual cortex. *J. Neurosci.* **11**(3), 641–649.

22

Addictive States

Frank W. Telang[*,1] **and Nora D. Volkow**[*,†]

Medical Department, Brookhaven National Laboratory, Upton, New York 11973
†*Department of Psychiatry, State University of New York at Stony Brook, Stony Brook, New York 11794*

I. Introduction

II. Pharmacological Properties of Drugs of Abuse in the Human Brain

III. Imaging and Addictive Processes: Evaluation of the Addicted Subject

IV. Studies with Abused Drugs

V. Conclusions

   References

I. Introduction

Illicit drug abuse is a social and medical affliction that touches the core of human existence. Estimates of the yearly cost to just the United States is in the range of $250–300 billion. In 1994, the RAND drug policy research study, initiated by the White House, found that treatment programs are seven times more cost-effective than domestic drug enforcement as a method to cut drug use and fifteen times better in reducing social costs of crime and lost productivity (Rydell, 1994). Yet, to improve treatment, we need to answer the question, What specific substance-related event causes human beings

to persist in this obsessive, self-destructive behavior? While clinicians must now focus interest mostly on behavior modification for symptom control, neuroscientists must look to the brain to identify etiologic mechanisms; that explain addiction, since improvements in drug abuse treatment require better understanding.

A. Substance Abuse: Reinforcement, Craving, Tolerance, and Withdrawal. Brief Outline of the Neuroanatomical Theory as Defined by Animal Studies

Brain reward mechanisms define reinforcement, as shown by brain stimulation reward experiments. An animal pressing a lever in an operant chamber will learn specific behaviors to receive a drug, and this associated electrophysiologic measurement can identify involved neuroanatomic structures. In 1972, using electrical stimulation experiments, Crow suggested that direct activation of the cell bodies or axons of the mesolimbic and mesostriatal dopamine (DA) systems was rewarding and that DA was the crucial neurotransmitter (Crow, 1973). It has since been noted that endogenous opioid release is seen in the ventral pallidum as a consequence of rewarding brain stimulation (Olive and Maidment, 1998; Panagis et al., 1998). GABAergic medium spinal neurons in the nucleus accumbens output pathway are also involved (Bardo, 1998). These studies support a

---

[1] To whom correspondence should be addressed.

three-stage pathway to account for the "hit," "rush," or "high" sought by the human substance abuser. As described by Gardner, the first component comprises descending, myelinated, moderately fast conducting neurons that run rostrally in the median forebrain bundle. These neurons synapse into the ventral mesencephalic nuclei, containing the cell bodies of the ascending mesotelencephalic DA system, whose axons run rostrally to limbic and cortical projection areas, to provide the second component. Abusing substances preferentially activates those neurons, especially whose fibers terminate in the shell of the nucleus accumbens. From the nucleus accumbens, the third-stage reward-relevant neurons, some of them enkephalinergic and/or GABAergic, carry the reward signal further to the ventral pallidum (Gardner and Lowinson, 1991).

Craving is defined as the "compelling urge" that intrudes upon the drug user's thoughts, affects the user's mood, and compels alteration in his or her behavior. Behavioral psychologists describe it under the Incentive-Sensitization Theory, in which addictive drugs share the ability to enhance mesotelencephalic DA neurotransmission. One function of this neural system is to attribute "incentive salience" to the perception and mental representation of events associated with activation of the system. Halikas and Kuhn suggested a neuroelectrical basis for cocaine craving as a form of "kindling" response (Halikas and Kuhn, 1990). Rather than Incentive-Sensitization Theory, newer concepts define craving as a incentive-motivational behavior, hypothesized to be determined by a continuous interaction between the hedonic rewarding properties of drugs (incentive) and the motivational state of the organism (organismic state). In drug-dependent individuals, the incentive-motivational value of drugs (i.e., drug craving) is greater compared to non-drug-dependent individuals due to the motivational state (i.e., withdrawal) developed with repeated drug administration (Markou et al., 1993).

Tolerance, or the need to increase dose to achieve the desired reinforcement, may, in at least the case of the opioid system, result from changes that occur at receptors, possibly from an uncoupling of receptors from G proteins. Symptoms of opiate withdrawal may be the result of interactions between opioid and other neurotransmitter systems, such as noradrenergic neurons in the locus ceruleus, as suggested by Nestler, who demonstrated that morphine activates an overshoot of cAMP signal transduction pathways in this region (Nestler, 1996). Using the model of spontaneous nicotine withdrawal in rats, Epping-Jordan and colleagues found that a decrease in brain reward as measured by increased brain reward thresholds persists for 4 days. Intervention with a nicotine-receptor antagonist also increases brain reward thresholds in a dose-dependent manner in rats

chronically treated with nicotine. Thus, poorly functioning brain reward systems, at least with nicotine, may constitute an important motivational factor that contributes to craving, relapse, and continued tobacco consumption in humans (Epping-Jordan et al., 1998).

II. Pharmacological Properties of Drugs of Abuse in the Human Brain

Investigation of the pharmacology of drugs of abuse entails studies of both pharmacokinetics and pharmacodynamics. Because neuroimaging studies can be conducted in alert human subjects, one can investigate the temporal relation between the effects of drugs and their uptake and clearance from the brain.

A. Pharmacokinetics

PET is well suited to studies of drugs of abuse because the availability of the positron emitter carbon-11 enables direct labeling of drugs without changing their pharmacological properties. Thus PET can be used to measure absolute uptake, regional distribution, and kinetics in the human brain. Whole-body PET can also be used to determine target organs for the drug and its labeled metabolites and thus provide information on potential organ toxicity as well as peripheral organ kinetics. Table I shows addictive drugs that have been labeled with positron emitters and whose distribution has been evaluated with PET methodology.

This strategy is illustrated by the investigation of the pharmacokinetics of cocaine in the brain and the human body (Fowler et al., 1989; Volkow et al., 1992a). The binding of cocaine in the human brain was highest in the basal ganglia, where it bound to the dopamine transporters, confirming previous animal studies. These studies also documented high brain uptake of cocaine (8–10% of injected dose) and very fast uptake (peaking 4–6 min after injection) and clearance (half-life 20 min) from the brain. Furthermore, for the first time, it was possible to evaluate the relation between the kinetics of an abused drug and the temporal relation of the behavioral effects. A parallelism was found between the kinetics of uptake and clearance of cocaine in the brain and cocaine-induced high.

Drugs can be directly compared. For example, [^{11}C]cocaine and [^{11}C]methylphenidate have almost identical brain uptake (7–10% of the injected dose) and regional distribution, yet their pharmacokinetics differed (Volkow et al., 1995a). They were both taken up rapidly (peak uptake 4–8 min) but methylphenidate cleared from the striatum much more slowly (half-life >90 min from peak uptake) than cocaine (half-life 20 min from peak uptake) (Fig. 3). For both drugs, their

Table I PET and SPECT Studies of the Pharmacokinetics and Pharmacodynamics of Substance of Abuse

Drug	Parameter measured	Labeled drug or tracer	Reference
Psychostimulants			
Cocaine	Pharmacokinetics	[^{11}C]Cocaine	Fowler *et al.*, 1989, 1992 Volkow *et al.*, 1992a, 1995a,b
	Metabolism	^{18}FDG	Volkow *et al.*, 1991a, 1992c, 1993 London *et al.*, 1990, 1996
	Blood flow	H$_2$^{15}O [^{99m}Tc]HMPAO [^{123}I]Amphetamine	Volkow *et al.*, 1988a Levin *et al.*, 1994, 1995 Weber *et al.*, 1993
	Dopamine receptors	[^{18}F]NMS [^{11}C]Raclopride	Volkow *et al.*, 1990 Volkow *et al.*, 1994
	Dopamine transporters	[11C]Cocaine [^{11}C]*d-threo*-MP; [^{123}I]β-CIT	Volkow *et al.*, 1996a,b Malison *et al.*, 1995
	Dopamine metabolism	[^{18}F]DOPA	Baxter *et al.*, 1988
Methylphenidate	Pharmacokinetics	[^{11}C]MP	Volkow *et al.*, 1995a
	Blood flow	H$_2$^{15}O	Wang *et al.*, 1994
	Dopamine responsivity	[^{11}C]Raclopride	Volkow *et al.*, 1994
Amphetamine	Metabolism	^{18}FDG	Wolkin *et al.*, 1987
	Dopamine responsivity	[^{123}I]IBZM	Laruelle *et al.*, 1995
Sedative hypnotics			
Alcohol	Metabolism	^{18}FDG	Volkow *et al.*, 1990, 1992a,c de Wit *et al.*, 1991; Adams *et al.*, 1995
		[^{11}C]Glucose	Wik *et al.*, 1988
	Blood flow	H$_2$^{15}O [^{99m}Tc]HMPAO	Volkow *et al.*, 1988b Nicolas *et al.*, 1993
	Benzodiazepine receptors	[^{11}C]Flumazenil	Litton *et al.*, 1993
	Dopamine receptors	[^{11}C]Raclopride	Hietala *et al.*, 1994 Volkow *et al.*, 1999a
Benzodiazepines	Benzodiazepine receptors	[^{123}I]Iomazenil	Sybirska *et al.*, 1993
	Metabolism	^{18}FDG	Volkow *et al.*, 1993, 1995c Buchsbaum *et al.*, 1987 de Wit *et al.*, 1991
Barbiturates	Metabolism	^{18}FDG	Theodore, 1986
Opiates			
Morphine and heroin	Metabolism	^{18}FDG	London *et al.*, 1989, 1990
Opiates	Pharmacokinetics	[^{11}C]Morphine [^{11}C]Heroin [^{11}C]Pethidine [^{11}C]Codeine	Hartvig *et al.*, 1984
	Binding site distribution	[^{11}C]Carfentanil	Frost *et al.*, 1990 (review)
Cannabinoids			
THC	Metabolism	^{18}FDG	Volkow *et al.*, 1991b, 1999a
Tobacco smoke			
Nicotine	Pharmacokinetics	[^{11}C]Nicotine	Maziere *et al.*, 1976 Bergstrom *et al.*, 1995
	Metabolism	^{18}FDG	Stapleton *et al.*, 1993
Other substances	MAO A and B	[^{11}C]Clorgyline and [^{11}C]L-Deprenyl-D$_2$	Fowler *et al.*, 1996a,b
Polysubstance abuse			
	Metabolism	^{18}FDG	Stapleton *et al.*, 1995

fast uptake in the striatum paralleled the experience of the high. However, whereas for cocaine the rate of clearance paralleled the decline in the high, for methylphenidate the high declined as fast as that for cocaine while there was still significant binding of the drug in the brain. Because it was the rate of uptake that was associated with the high for both drugs rather than the presence of the drug in the brain, it was postulated that the rate of clearance may affect the propensity of a drug to promote frequent repeated administration. Methylphenidate's relatively slow clearance from the brain may not facilitate the frequent repeated administration that occurs with cocaine.

The natural stereoisomerism of drugs of abuse in the interaction with the brain can also be examined with PET. It has been used to compare the regional distribution in the brain of the two enantiomers of methylphenidate. The study showed that [^{11}C]d-threo-methylphenidate, the more active, d-threo, enantiomer of methylphenidate, bound specifically to striatum but not [^{11}C]l-threo-methylphenidate (Ding et al., 1997). Another PET study showed that the active enantiomer of cocaine ((−)-cocaine) readily entered the brain whereas the inactive enantiomer ((+)-cocaine) was immediately and completely metabolized to inactive intermediates in the blood-stream (Gatley et al., 1990).

B. Pharmacodynamics

An extensive panel of PET radiotracers is now available to probe the biological processes involved in drug effects. With appropriate radiotracers, drug effects on metabolism and CBF, on neurotransmitter activity, and on enzyme activity can be assessed. Addicted subjects can be compared with normal controls to evaluate chronic effects of drug self-administration. For both chronic and acute studies, changes in rCMRglu and CBF have been most frequently assessed. These strategies allow identification of the brain regions most sensitive to drug effects. Because the studies are done in alert human subjects, they allow regional changes in metabolism or flow to be correlated with behavioral effects. Most of the common drugs of abuse have been studied. As an example, heroin-induced changes in scores related to euphoria were associated with heroin-induced changes in metabolic activity in lateral occipital and primary visual cortex (London, 1993). While most drugs of abuse decrease regional brain glucose metabolism, some drugs increase whereas others decrease CBF. Discrepancies between metabolism and CBF are probably indications of the vasoactive properties of many of these pharmacological agents. These properties are relevant to understanding their toxicity as it relates to cerebrovascular pathology. An important caveat for

these studies is that changes in metabolism reflect FDG uptake over a period of usually 35–45 min, whereas the temporal resolution of blood flow measurements is much higher (PET, usually 30–45 s; fMRI, usually 3–5 s). These differences in temporal resolution become particularly relevant when the relationship between metabolic or flow changes and the behavioral effects of drugs is investigated. The behavioral effects can be of short duration; hence the circuits involved in their evolution may not be discernible when activity is averaged over relatively long time periods.

The effects of drugs of abuse on neurotransmitter activity can also be investigated with PET and SPECT. For example, changes in synaptic DA concentration in response to psychostimulant drugs have been measured using DA receptor ligands, which are sensitive to the endogenous concentration of DA (Volkow et al., 1994; Laruelle et al., 1995). For this purpose, subjects are scanned twice, at baseline and after administration of the psychostimulant drug known to increase DA concentration. Changes in binding of the ligand between the two scans indicate relative changes in the synaptic concentration of DA. Responses between subjects are quite variable and decrease with age (Volkow et al., 1994). The intensity of the behavioral effects closely correlated with the magnitude of the estimated changes in DA concentration. Also, recent work in non-human primates has shown with PET that modulation of GABA transmission using a suicide inhibitor of GABA-transaminase (GVG) is effective in lowering nicotine- and cocaine-induced increases in extracellular DA in the nucleus accumbens. Parallel studies in rats showed that these effects were associated with a reduction of drug-induced conditioned place preference behavior (Dewey et al., 1998, 1999).

In vivo potency of psychostimulant drugs and their indirect actions can be investigated at the molecular level using PET. For example, PET can be used to measure dose/occupancy ratios of drugs (cocaine or methylphenidate) at the DA transporter (DAT) and correlated to assess abuse liability (Volkow et al., 1999c). DAT occupancies were measured with [^{11}C]cocaine, as a DAT ligand, in 8 normal controls for the methylphenidate study and in 17 active cocaine abusers for the cocaine study. The ratio of the distribution volume of [^{11}C]cocaine in striatum to that in the cerebellum, which corresponds to $B_{max}/K_d + 1$, was used as a measure of DAT availability. Methylphenidate and cocaine produced comparable dose-dependent blockade of DAT, with an estimated ED_{50} (dose required to block 50% of the DAT) for methylphenidate of 0.07 mg/kg and for cocaine of 0.13 mg/kg (Volkow, 1999a). PET can also be used to assess dose/occupancy ratios of drugs at their postsynaptic targets. In the case of psychostimulants, which increase DA indirectly, this allows for a measurement of the change in D_2 receptor occupancy by DA (Volkow et al., 1999b).

Radionuclide imaging technologies can also be used to investigate the relation between behavioral effects and degree of receptor or transporter occupancy. These studies use a radioligand for the receptor that is the target of the drug action. A paired study is performed at baseline and after the drug of interest is given. For example, SPECT was used to evaluate receptor occupancy by the benzodiazepine agonist lorazepam. This showed that only a very small fraction of the receptors are occupied at pharmacological doses, findings that corroborate in humans the existence of receptor reserve for benzodiazepine receptors (Sybirska, 1993).

The effects of drugs of abuse on enzyme activity have also been assessed with PET. The best-studied example involves monoamine oxidase B (MAO B) in brain and uses the radiotracer $[^{11}C]$L-deprenyl-D_2, which binds specifically to this enzyme. For example, one study showed that cigarette smokers have a reduction in brain MAO B of about 40% relative to nonsmokers and former smokers (Fowler et al., 1996). MAO B inhibition is associated with enhanced activity of DA as well as decreased production of hydrogen peroxide, a source of reactive oxygen species (Reiter, 1985). Inhibition of MAO by cigarette smoke could be one of the mechanisms accounting for the lower incidence of Parkinson's disease in cigarette smokers.

C. Drug Toxicity

PET/SPECT can be used to assess the toxicity from drugs toward brain and other organs. Because imaging technologies provide functional information, they are very sensitive to brain pathology secondary to drug use. Such sensitivity is illustrated by PET studies that documented for the first time abnormalities in CBF in cocaine abusers (Volkow et al., 1988a). Similar findings were subsequently reported by SPECT documenting perfusion defects in cocaine abusers (Holman et al., 1991). These findings corroborated the clinical reports of the occurrence of cerebral strokes and hemorrhages associated with cocaine consumption. The defects in CBF appear to be secondary to the vasoactive properties of cocaine. This early study points out the high incidence of vascular perfusion deficits in the poly-drug abuse addict, which can be seen regardless of structural MRI or CT imaging deficits.

Organ toxicity from drug use can be evaluated from several perspectives. In the heart, for example, functional imaging strategies can be used to quantify changes in myocardial blood flow and metabolism in substance abusers. Also, the ^{11}C-labeled drug can be used to measure its distribution in the various organs. This can indicate the likelihood of direct toxic effects on a given organ. For example, PET studies with $[^{11}C]$cocaine documented a significant accumulation in human heart (Volkow et al., 1992a). Also using $[^{18}F]$fluoro norepinephrine as a ligand to monitor the function of the peripheral norepinephrine transporter, Fowler et al. showed cocaine to inhibit the transporter in heart (Fowler et al., 1994). This is of particular interest because of the documented cardiotoxic properties of cocaine. Cocaine is a local anesthetic and its accumulation in the heart could result in direct myocardial toxicity. At the same time, inhibition of the norepinephrine transporter by cocaine interferes with a protective mechanism of the heart to remove circulating catecholamines.

III. Imaging and Addictive Processes: Evaluation of the Addicted Subject

PET and SPECT have been used to assess neurochemical and functional changes in the brain of addicted subjects and their changes as a function of withdrawal and/or drug treatment. For example, imaging studies recently documented that the decrements in CBF in cocaine abusers improved with the administration of buprenorphine (Levin et al., 1995). Functional imaging strategies can also be used to assess the pattern of brain activation during drug-related states triggered by behavioral interventions such as those that induce drug craving. For example, in cocaine abusers in who craving was elicited by showing a cocaine paraphernalia video, activation was induced and recorded in limbic brain regions (Childress et al., 1995).

MRI studies have potential advantages over PET/SPECT in some regards. They benefit from the ability to perform multiple studies in the same subject without the limitation of radiation exposure. BOLD activation studies during injection of abused substances/placebo are now being published and have better temporal and spatial resolution than PET. Methods for tagging RBC's to measure flow, FLAIR sequencing, and perfusion-weighted imaging to gain blood flow (blood volume) information are now being developed. Proton magnetic resonance spectroscopy (MRS) studies to measure the regional distribution of neurotransmitters such as GABA or alcohol directly have been completed or are in progress, in addition to creatine, choline, and N-acetylaspartate in the substance abuser. ^{31}P MRS studies for the assessment of energy metabolism in the polysubstance abuser have also been completed. The following sections review some studies in addiction research in more detail.

IV. Studies with Abused Drugs

A. Alcohol

The molecular sites of action of alcohol in the CNS are complex and not completely understood. Alcohol affects ligand gated ion channels, perhaps through changes in the membrane bilayer (Hunt, 1985). Yet the intoxicant nature of alcohol thought to be based on this ability to permeate and then alter membranes is not substantiated at doses that cause moderate to mild intoxication. Other studies have implicated alcohol in altering $GABA_A$, opioid, NMDA, and serotonin receptors. In fact, alcohol's effects on opiate receptors may account for naltrexone's therapeutic effectiveness in alcoholics (Herz, 1998). Genetic differences in these neurotransmitters in turn may affect an individual response to alcohol as evidenced by studies in transgenics, knockouts, and animals genetically bred for their preference to alcohol. Studies of genetic linkage in rats bred to prefer alcohol have identified a region that includes the neuropeptide Y (NPY) gene (Carr, 1998). NPY knockouts of mice show increased preference for alcohol; and are less sensitive to its sedative-hypnotic effects when compared to control animals. Overexpression of the NPY gene gave opposite effects, suggesting direct evidence that alcohol consumption and resistance are inversely related to NPY levels in the brain (Thiele *et al.*, 1998). A recent study showed that knockout dopamine D_2 receptor mice do not show conditioned place preference to alcohol, indicating an involvement of D_2 receptors in the reinforcing effects of alcohol (El-Ghundi *et al.*, 1998). In contrast, mutant mice lacking the dopamine D_4 receptor had an enhanced locomotive response to ethanol when compared to control wild-type strains (Rubinstein *et al.*, 1997). Also in mice, there is an association between the familial subunit variants of $GABA_A$ and the severity of acute physiological dependence to alcohol (Buck and Hood, 1998). Effects of alcohol on opioid receptors have been linked to craving (Froehlich *et al.*, 1998), which may explain naltrexone's effectiveness in diminishing craving in alcoholics (O'Malley, 1996). Other studies have shown an abnormal rate of serotonin turnover in the CNS in young males prone to early alcohol addiction that is coupled with early aggressiveness (conduct disorder) (Virkkunen and Linnoila, 1997). The serotonergic hypothesis is supported by studies of $5\text{-}HT_{1B}$ knockout

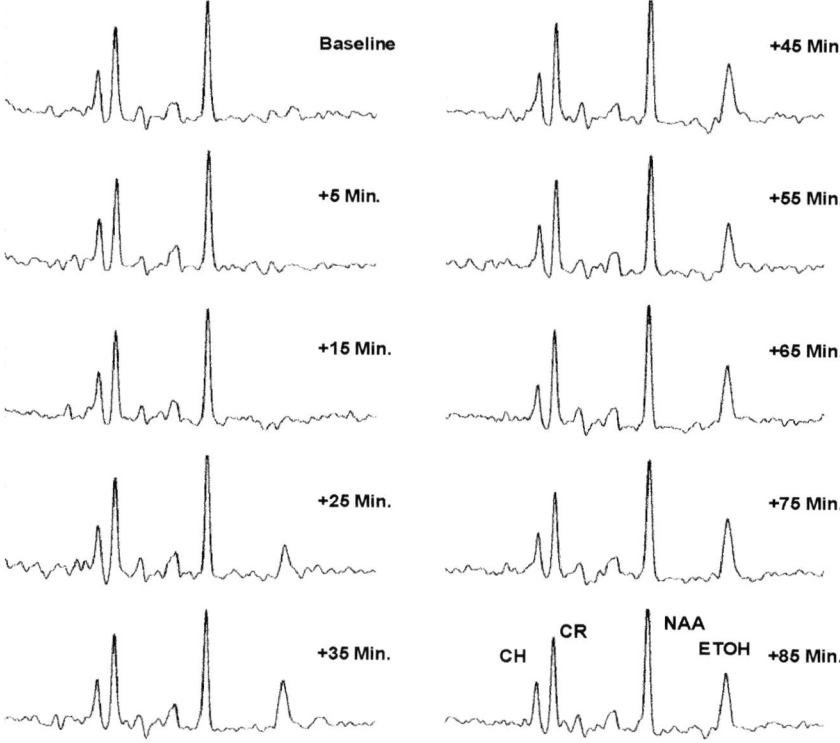

Figure 1 Spectrum acquired from the baseline (before drinking) spectroscopic image and spectra from the same spatial location as a function of time after drinking. The major observed resonances are choline (CH), creatine (CR), *N*-acetylaspartate (NAA), and ethanol (EtOH). The duration to the midpoint of the acquisition after drinking is listed to the right of each spectrum.

mice. They expressed enhanced aggression and drank twice as much ethanol as the wild-type strain, voluntarily ingesting solutions containing up to 20% ethanol in water (Crabbe *et al.*, 1996). These mice also have less of an ethanol-induced conditioned place preference response, which suggests reduced reinforcement in a dose-dependent manner (Risinger *et al.*, 1996).

1. PET/SPECT Studies

PET studies in alcoholics have been done to measure CBF and brain glucose metabolism, benzodiazepine receptors, DA D_2 receptors, and DA transporters. These studies have been done to determine if the brains of alcoholics differ from those of non-alcoholic normal controls and to assess if there are changes in the brains of subjects who have a genetic predisposition for alcoholism. PET has also been used to investigate brain metabolic changes in chronic alcoholics with and without neurological impairment (Volkow *et al.*, 1992b). Alcoholics with Korsakoff's encephalopathy showed decreased metabolism in prefrontal, parietal, and temporal cortices and alcoholics with neurological symptoms other than Korsakoff's showed decreased metabolism in frontal and parietal cortices. Studies in alcoholics with no evidence of neurological impairment have also shown PET evidence of frontal abnormalities. Frontal dysfunction in alcoholics has also been consistently found in non-PET studies of regional cerebral blood flow. The degree of brain metabolic recovery with detoxification has also been evaluated with PET in alcoholics monitored at different times after detoxification (Volkow *et al.*, 1993). This study showed that brain metabolism increased significantly during detoxification, predominantly during the first 30 days of detoxification. Regional increases in metabolism were larger in frontal regions. In alcoholics, metabolism in frontal, parietal, and left temporal cortices was negatively correlated with years of alcohol use and with age whereas in the comparison group they were not. Decrements in metabolism were most accentuated in the older alcoholics with the longer histories of alcohol consumption.

Acute alcohol administration decreased rCMRglu and CBF (Volkow *et al.*, 1988b, 1990). Alcoholics showed a larger metabolic response to ethanol than controls despite smaller subjective responses to the intoxicating properties of ethanol. In normals, intoxication was significantly correlated with metabolic response. The paradoxical metabolic and behavioral responsivity in alcoholics could reflect tolerance of the brain to ethanol-induced metabolic changes.

The acute (de Wit *et al.*, 1991; Volkow *et al.*, 1995c) and chronic (Buchsbaum *et al.*, 1987) effects of benzodiazepine drugs on rCMRglu have also been evaluated with PET. Similar to ethanol, benzodiazepines decrease brain glucose metabolism and the effects are more pronounced in the occipital cortex, the area of the brain with the highest density of benzodiazepine receptors. Other studies have shown that whereas normal and alcoholic subjects showed a similar response to lorazepam in occipital and cerebellar metabolism, the alcoholics showed a significantly lower response in the thalamus, basal ganglia, and orbitofrontal cortex. Thus, alcoholics have a blunted response to lorazepam, which in follow-up studies persisted after detoxification (Volkow *et al.*, 1997c). This is regionally specific and could reflect chronic alcohol administration and/or genetic differences.

To assess the contributions of genetic factors in the sensitivity to benzodiazepines, the regional brain metabolic response to lorazepam was evaluated in subjects with a positive family history for alcoholism (FHP) and compared with that of subjects without a family history for alcoholism (FHN). At baseline, FHP subjects showed lower cerebellar metabolism than FHN subjects and when challenged with lorazepam, they also showed a blunted response in the cerebellum and in the cingulate gyrus. These results corroborate electrophysiological and behavioral studies showing that subjects at risk have blunted responses to alcohol. Lorazepam-induced changes in cerebellar metabolism were significantly correlated with motor impairment. These changes could account for the decreased sensitivity to the motor effects of alcohol and benzodiazepine in FHP subjects and indicate regional brain metabolic differences in children of alcoholics (Volkow *et al.*, 1995c).

Imaging studies have also measured receptor and transporter densities in alcoholics. Benzodiazepine receptors were measured using [^{11}C]flumazenil. While an initial study did not show differences in receptor numbers between normals and alcoholics, transporters had a significantly larger variability for B_{max} measures than controls (Litton *et al.*, 1993). Others have shown significant reduction in benzodiazepine receptors in alcoholics compared with controls. Using SPECT and [^{123}I]iomazenil, region of interest analysis revealed that alcoholic patients had a significantly lower benzodiazepine distribution volume than comparison subjects in the frontal, anterior cingulate, and cerebellar cortices. SPM analysis revealed two areas in which the distribution volume in alcoholic patients was significantly lower than in comparison subjects: the anterior cingulate, extending into the right middle frontal gyrus, and the left occipital cortex (Abi-Dargham *et al.*, 1998; Lingford-Hughes *et al.*, 1998). The DA system has also been evaluated with PET. Studies measuring DA D_2 receptors have shown that alcoholics had a 20% decrease in B_{max}/K_d and a 12% decrease in D_2 receptor numbers when compared with controls (Hietala *et al.*, 1994;

Volkow *et al.*, 1997c). No significant correlations were found between days of detoxification (168 weeks) and D_2 receptor numbers. Because persistent reductions in D_2 receptor availability have also been reported in cocaine abusers, D_2 decrements may represent a variable related to vulnerability for addictive disorders. Studies measuring DA transporters have been inconsistent. One SPECT study reported that non-violent alcoholics had decreases in DA transporters (25%) when compared with non alcoholic controls (Tiihonen *et al.*, 1995). However, others have failed to document changes in DA transporters in alcoholics (Volkow *et al.*, 1997c). Since DA D_2 receptors in the striatum are mainly localized in GABA cells, decreases in D_2 receptors in alcoholics corroborate with involvement of GABA cells in alcoholics.

The influence of gender on the pharmacokinetic and pharmacodynamic distributions of alcohol in the human brain has yet to be determined. Wang reported no differences between the $[^{18}F]$FDG scans of female alcoholics and normal female controls, as a group. In his attempts to relate female to male scans, it was difficult to find women subjects that imbibed alcohol on an equal dose/weight basis to that of the male subjects for comparison (Wang *et al.*, 1998). Nonetheless, that study failed to show metabolic abnormalities in female alcoholics of moderate severity.

Abnormalities in the serotonin system have also been documented by studies showing decreases (a mean of about 30%) in the availability of brainstem serotonin transporters in alcoholics. The reductions correlated with lifetime alcohol consumption and with ratings of depression and anxiety during withdrawal (Heinz *et al.*, 1998).

2. Magnetic Resonance Imaging

Multiple structural studies have been done with CT and, more recently, with MRI to determine if there are volume changes in the brains of alcoholics that may evidence neurotoxicity. These studies have consistently documented ventricular enlargement and cortical atrophy in alcoholics (Eckardt *et al.*, 1988). Alcoholics have been shown to have a greater rate of age-related gray matter volume loss than non-alcoholics (Pfefferbaum *et al.*, 1997). The accentuation of the age effects has led some investigators to postulate that the neurotoxic effects of alcohol are due to an acceleration of age neurodegenerative changes. Interestingly, studies have documented that some of these changes may be reversible. Recently, a study documented that the rate of ventricular enlargement in alcoholic patients who maintained virtual sobriety was comparable to that in the control subjects (Pfefferbaum *et al.*, 1998). Numerous studies have also been done to assess if there are brain regions,

specifically the mamillary bodies, hippocampus, and cerebellum, that are more sensitive to the effects of chronic alcohol exposure (Sullivan *et al.*, 1995; Shear *et al.*, 1996). MRI may also change the clinical organization of brain diseases related to alcohol. One important study pointed out that thinning of the genu and body of the corpus callosum is seen with regularity in chronic alcoholics, suggesting a continuum of graded brain dysmorphology rather than classical alcoholic-related subsyndromes, such as the Marchiafava–Bignami syndrome, which may occur in this disease (Pfefferbaum *et al.*, 1996). MRI studies documenting specific areas in the brain, namely the cerebellum, basal ganglia, and corpus collosum, in children that are vulnerable to alcohol's teratogenic effects, i.e., fetal alcohol syndrome, have also been performed (Roebuck *et al.*, 1998).

Magnetic resonance spectroscopy (MRS) has been used to quantitate alcohol's pharmacologic effects, which include long- and short-term tolerance. An important technical issue that illustrates the difficulty inherent in this modality is the issue of NMR visibility of ethanol in the brain (Chiu *et al.*, 1994). Low visibility results from an anomalously short T_2, arising from the interaction of ethanol with membrane lipids. However, after nearly a decade of investigation, measures of the visibility of brain alcohol *in vivo* vary widely, ranging from 21 to 100%, depending on the pulse sequence and biological model used. For example, Moxon *et al.* (1991) reported an ethanol visibility of 23% in the dog brain. Similar visibilities in the human brain (21–26%) (Mendelson, 1990; Chiu, 1994) and in rhesus monkeys (23%) (Kaufman *et al.*, 1994) have been reported using STEAM methods. Other studies using spin–echo methods and refocusing pulses, which reduce J modulation, have yielded higher values. For example, Meyerhoff *et al.* (1996) reported an ethanol visibility of 66% in an *in vivo* rat brain model, which is similar to that reported in human gray matter by Spielman, 72% (Spielman, 1993). Finally, using a spectral editing sequence at 4.7 T in a rabbit model, Petroff *et al.* reported an ethanol visibility of nearly 100% (Petroff *et al.*, 1990). A recent study at 4.0 T in normal adults also indicates that brain alcohol visibility is approximately 100% (Hetherington, 1999). Advances in the basic understanding of MRS signals will facilitate this type of investigation.

Proton MRS has also been used to investigate the *N*-acetyl-aspartate (NAA) to choline ratios (Stein *et al.*, 1998) and total creatine in chronic alcoholic patients. As expected, the NAA/Cho ratio, thought to represent neuronal reserve, was reduced in the frontal lobe area, thalamus, and cerebellum. Creatine was reduced as well, suggesting a change in white matter constituents (Jagannathan *et al.*, 1996). NAA/Cho ratios were also seen to increase in the superior cerebellar vermis in alcoholics after 3–4 weeks of abstinence. Patients with alcoholic dementia had very low ratios (Martin *et al.*, 1995).

B. Phenethylamines

Amphetamine and methamphetamine, which act by releasing dopamine from nerve terminals and storage vesicles and, in some drugs or in selected patients, by damaging 5-HT neurons, have been investigated with imaging technologies.

1. PET/SPECT

Initial PET/FDG studies with oral amphetamine, 0.5 mg/kg, showed global decreases in CMR (Wolkin *et al.*, 1987; de Wit *et al.*, 1991). A later study, which used intravenous amphetamine and a continuous attention task, did not show significant increases in global CMRglu over baseline (Ernst *et al.*, 1994), although there were regional differences noted between normal controls and subjects with attention deficit hyperactivity disorder (ADHD) in frontal lobe areas. Amphetamine-induced dopamine release can be assessed by using [^{11}C]raclopride and PET. In a recent study, Carson *et al.* documented the decrease in binding in monkeys related to bolus vs bolus plus infusion intravenous dosing, and using a new plasma metabolite model to measure the total distribution volume, they showed a linear relationship between raclopride displacement and increases in extracellular dopamine, demonstrating a 22–42% reduction in the post-amphetamine study (Carson *et al.*, 1997).

Amphetamine abusers (21 cases, use ranged from 1 month to several years) studied early on with SPECT/[^{99m}Tc]HMPAO were found to have "perfusion defects," but these were not related to severity of symptoms, duration of abuse, or level of drug consumption (Kao *et al.*, 1994; Laruelle *et al.*, 1995). A recent [^{18}F]FDG study to record rCMRglu in normal volunteers given high euphorigenic doses of *d*-amphetamine (0.9–1.0 mg/kg p.o.) shows that the induced mania-like state correlates with increased metabolism in the anterior cingulate cortex, caudate nucleus, putamen, and thalamus (Vollenweider *et al.*, 1998).

3,4-Methylenedioxymethamphetamine (MDMA), also known as Ecstasy, used illicitly by millions of individuals, has also recently been studied with PET imaging. In rodents, this compound is known to damage serotonergic (5-HT) neurons, with toxicity limited to axon terminals. Initial studies in baboons used both enantiomers of [^{11}C]McN 5652, the (+) moiety being a potent 5-HT transporter ligand, as well as [^{11}C]βCIT, a cocaine analog that labels both the DAT and SERT. Baboons were given the active isomer of McN5652 as baseline scans with all three ligands and then received MDMA (5 mg/kg s.c.

twice daily for 4 consecutive days). PET studies at 13, 19, and 40 days post-MDMA revealed decreases in mean radioactivity levels in all brain regions when [^{11}C](+)-McN5652 was used, but not with [^{11}C](−)-McN5652 or [^{11}C]RTI-55. Reductions in specific [^{11}C](+)-McN5652 binding (calculated as the difference in radioactivity concentrations between (+)- and (−)-[^{11}C]McN5652) ranged from 44% in the pons to 89% in the occipital cortex. PET studies at 9 and 13 months showed regional differences in the apparent recovery of 5-HT transporters, with increases in some brain regions (e.g., hypothalamus) and persistent decreases in others (e.g., neocortex) (Scheffel *et al.*, 1998). This work was then extended to human trials, where they analyzed whether there were differences in 5-HT transporter binding, again using [^{11}C]McN5652, between abstinent MDMA users and participants in the control group. MDMA users showed decreased global and regional brain 5-HT transporter binding compared with controls. Decreases in 5-HT transporter binding positively correlated with the extent of previous MDMA use (McCann *et al.*, 1998).

2. Magnetic Resonance Imaging

A preliminary fMRI study involving fast echo-planar imaging was conducted in which three volunteers and two subjects with narcolepsy underwent activation by presentation of periodic auditory and visual stimuli, both before and after administration of amphetamine. The effect of amphetamine in controls was a small reduction in the extent of sensory-induced activation (Howard *et al.*, 1996). fMRI in the setting of a pharmacologic intervention has recently been performed using *d*-amphetamine and the cocaine analog 2β-carbomethoxy-3β-(4-fluorophenyl)tropane (CFT). Both compounds led to increased signal intensity in gradient-echo images in regions of the brain with high dopamine receptor density (frontal cortex, striatum, cingulate cortex > parietal cortex). Lesioning the animals with unilaterally administered 6-hydroxydopamine (6-OHDA) led to ablation of the MRI response on the ipsilateral side control measurements of rCBV and rCBF using bolus injections of Gd-DTPA showed that the baseline rCBV and rCBF values were intact on the lesioned side. The time course of the BOLD signal changes paralleled the changes observed by microdialysis measurements of dopamine release in the striatum for both amphetamine and CFT, peaking at 20–40 min after injection and returning to baseline at about 70–90 min. Signal changes did not correlate with heart rate, blood pressure, or pCO$_2$ (Chen *et al.*, 1997).

C. Caffeine

Caffeine is the most widely used psychoactive substance in the world. In Western society, at least 80% of the adult population consume caffeine in amounts large enough to affect the brain. It is an antagonist of G protein-linked adenosine receptors and has been reported to modify behavioral responses to a variety of centrally acting agents, including cocaine, amphetamines, benzodiazepines, alcohol, and morphine (Daly *et al.*, 1998).

1. Radioisotope Techniques

Early studies centered on use of the ^{133}Xe-inhalation technique. Two groups of normal volunteers underwent rCBF studies before and 30 min after 250 or 500 mg of caffeine (cup of coffee: 100–150 mg) given orally. rCBF was measured in a third group of subjects, twice, at a similar interval under identical laboratory conditions. Subjects who received caffeine showed significant decreases in rCBF whereas the others showed no rCBF change from the first to the second measurement. However, the two caffeine groups did not differ in degrees of rCBF reduction. There were no regional variations in the postcaffeine decrease in cerebral blood flow. The three groups did not show significant changes in end-tidal carbon dioxide, pulse rate, blood pressure, forehead skin temperature, and respiratory rate (Mathew *et al.*, 1983).

A PET study using [^{15}O]water quantified the effect of caffeine on whole-brain and regional cerebral blood flow (CBF) in humans. A mean dose of 250 mg of caffeine produced approximately a 30% decrease in whole-brain CBF; regional differences in caffeine effect were not observed. Precaffeine CBF strongly influenced the magnitude of the caffeine-induced decrease. Caffeine decreased $paCO_2$ and increased systolic blood pressure significantly; the change in $paCO_2$ did not account for the change in CBF. Smaller increases in diastolic blood pressure, heart rate, plasma epinephrine and norepinephrine, and subjectively reported anxiety were also observed (Cameron *et al.*, 1990). FDG/PET studies of caffeine do not appear to have been performed in humans. Rats were studied with the quantitative [^{14}C 2-deoxyglucose autoradiographic method to study the effects of acute intravenous injections (15 min prior to study) of caffeine on brain energy metabolism. With doses of 0.1 mg/kg, the effects of caffeine on cerebral glucose utilization were limited to the habenula, spinal trigeminal, and paraventricular nuclei. At the 1.0 mg/kg dose, significant increases were additionally seen in the caudate, ventral tegmental area, and medial septum. After the injection of 10 mg/kg of caffeine, average glucose utilization of the brain as a whole was increased by 15% (Nehlig *et al.*, 1984). A preliminary PET study using the direct ligand [^{11}C]caffeine in the baboon compared the kinetics of oral vs intravenous dosing and showed, as expected, a longer latency due to absorption differences.

A high degree of nonspecific binding was seen (Ding *et al.*, 1998).

2. Magnetic Resonance Imaging

Innovative proton MRS, using a rapid echo-planar technique, was applied to dynamically measure regional brain metabolic response to oral caffeine. Heavy caffeine users and caffeine-intolerant individuals were compared for changes in brain lactate, thought to be a combined result of stimulation of glycolysis and reduction in CBF. NAA/lactate ratios were used to compare groups. One potential problem is that the lactate peak acquired with this sequence technique may be obscured by residual lipid artifact and also affected by head movement. One would expect that these features were especially troublesome in the caffeine-intolerant individuals, yet the investigators were able to show significant increases in global and regionally specific brain lactate only among the caffeine-intolerant subjects. In addition, re-exposure of caffeine to heavy users after a drug holiday (4 weeks) resulted in elevations in brain lactate similar to those seen in the caffeine-intolerant group, suggesting that the caffeine-induced changes in lactate were not responsible for the differences in tolerance to caffeine (Dager *et al.*, 1999).

D. Cocaine/Methylphenidate

Abuse of intravenous and "crack" cocaine is a major societal problem. Cocaine's highly reinforcing and addictive properties are manifested to the extreme in laboratory animals, which, when given free access to intravenously administered cocaine, will self-administer until death (Koob and Bloom, 1988). This sets cocaine apart from other drugs of abuse, which are not self-administered at the expense of life-prolonging and life-preserving behaviors. The mesoaccumbens dopamine pathway, extending from the ventral tegmentum to the nucleus accumbens, appears to be central to the reinforcing effects of cocaine. Other areas receiving DA input, such as the basal forebrain and amygdaloid complex, have been tied to the craving response. Much evidence implicates inhibition of the DA transporter (DAT) and a resulting increase in synaptic DA as cocaine's primary site of action. However, cocaine also inhibits reuptake at the serotonin and norepinephrine transporters and directly inhibits the voltage-gated Na^+ channels. Recent studies in DAT or SERT knockout mice have demonstrated retention of the capacity to exhibit cocaine conditioned place preference. This may have substantial implications for understanding cocaine actions and for strategies to produce medications to combat cocaine addiction (Sora *et al.*, 1998).

1. PET/SPECT

In initial PET studies using [^{15}O]water, chronic cocaine users showed decreased relative CBF in the prefrontal cortex when compared with normal subjects. The repeated scans of some cocaine users, after 10 days of cocaine withdrawal, continued to show a regional pattern of decreased relative CBF, suggesting a sympathomimetic effect on vasculature, and an association has been made between this pattern and cognitive impairment (Volkow *et al.*, 1991c). SPECT has also been used in the identification of vascular perfusion deficits related to polydrug use. An initial early study looked at cocaine-dependent polydrug users. Ninety-five percent had abnormal perfusion characterized primarily as small focal defects involving inferoparietal, temporal, and anterofrontal cortex and basal ganglia. In this study, all cocaine-dependent subjects had abnormal psychometric testing, but no relationship was found between the severity of SPECT abnormalities and mode of administration or frequency or length of cocaine use (Holman *et al.*, 1991). This early study points out the high incidence of vascular perfusion deficits in the polydrug abuse addict, which can be seen regardless of structural MRI or CT imaging deficits. Vascular defects have been associated with cognitive impairment in drug abusers (Volkow *et al.*, 1988a).

In PET studies with polydrug users, intravenous administration of 40 mg of cocaine decreased rCMRglu in both cortical and subcortical structures, and these changes correlated with the subjective sense of intoxication (London *et al.*, 1996). An important question in studies of drug abusers is whether a particular parameter changes with time after drug withdrawal. In recently detoxified cocaine abusers (<1 week), brain glucose metabolism was significantly higher in orbitofrontal cortex and in striatum than in healthy non-abusing controls (Volkow *et al.*, 1991a). However, metabolic activity was highest in subjects tested during the initial 72 h after withdrawal and cocaine abusers who had the highest metabolic values had the highest subjective ratings for craving. In contrast, cocaine abusers tested between 1 and 4 months of detoxification showed significant reductions in metabolic activity in the prefrontal cortex, orbitofrontal cortex, temporal cortex, and cingulate gyrus.

Studies done in cocaine abusers during exposure to cocaine-related cues or to recall of their cocaine experiences have been done to identify the brain regions that get activated during craving. One such study that measured regional metabolic changes during a cocaine video that elicited craving showed activation in several cortical and limbic regions implicated in the craving response, the dorsolateral prefrontal cortex, medial temporal lobe (amygdala), and cerebellum (Grant

et al., 1996). Another study, which measured CBF also during a cocaine video that elicited craving, showed activation in limbic brain regions (amygdala and anterior cingulate) and CBF decreases in basal ganglia (Childress *et al.*, 1999). On the other hand, craving induced by recalling previous experiences with cocaine self-administration was associated with increases in metabolism in orbitofrontal cortex and temporal insula (Wang *et al.*, 1999).

PET studies using [^{11}C]cocaine have been hampered due to the fact that the specific to nonspecific binding ratio of cocaine is poor in areas that have a relatively low number of DAT. The detectability of weak signals from PET has been enhanced and the spatial and temporal resolution improved by averaging the scans of a number of normal subjects and coregistering the data to the Talairach–Tournoux Atlas. With this method, the highest concentration of [^{11}C]cocaine in the initial 5 min after injection occurred in the putamen, caudate, thalamus, and selective cortex, in particular, the posterior cingulate and the precuneus (Fig. 2). At 10 and 20 min, the highest concentration was also in the putamen, with intermediate concentrations in the caudate, nucleus accumbens, and dorsomedial thalamus. At 30 min, the only brain region where significant activity remained was in the striatum. The amygdala, hippocampus, and post-hippocampal gyrus were found to have relatively low distribution volume values, with concentrations roughly 10% of those in the putamen and 20% of those in the accumbens (Telang *et al.*, 1999). These values are lower than those found in *ex vivo* studies on primate brain with [^{3}H]cocaine and autoradiography (Madras and Kaufman, 1994). This is likely to reflect the poor specific to nonspecific binding of [^{11}C]cocaine binding studies performed *in vivo* when compared with those performed *in vitro*.

Recent work involving the coinjection of cocaine and [^{11}C]cocaine demonstrated that intravenous doses commonly abused (0.3–0.6 mg/kg) occupied 60–77% of the striatal binding sites; the degree of occupancy and its time course were directly related to the high. At least a 50% occupancy was required for the perception of the high (Volkow *et al.*, 1997a). The *O*-ethyl homolog of cocaine, cocaethylene, which is found *in vivo* in subjects abusing both cocaine and ethanol, has been studied with PET using the ^{11}C-labeled compound. Specific binding in the striatum was the same as that for cocaine but was significantly greater in the thalamus, cerebellum, and whole brain. The slower clearance may lead to accentuated physiological effects relative to cocaine (Fowler *et al.*, 1992).

Several imaging studies have been done to assess the acute effects of methylphenidate on cerebral blood flow and on metabolism. Methylphenidate (0.5 mg/kg i.v.)

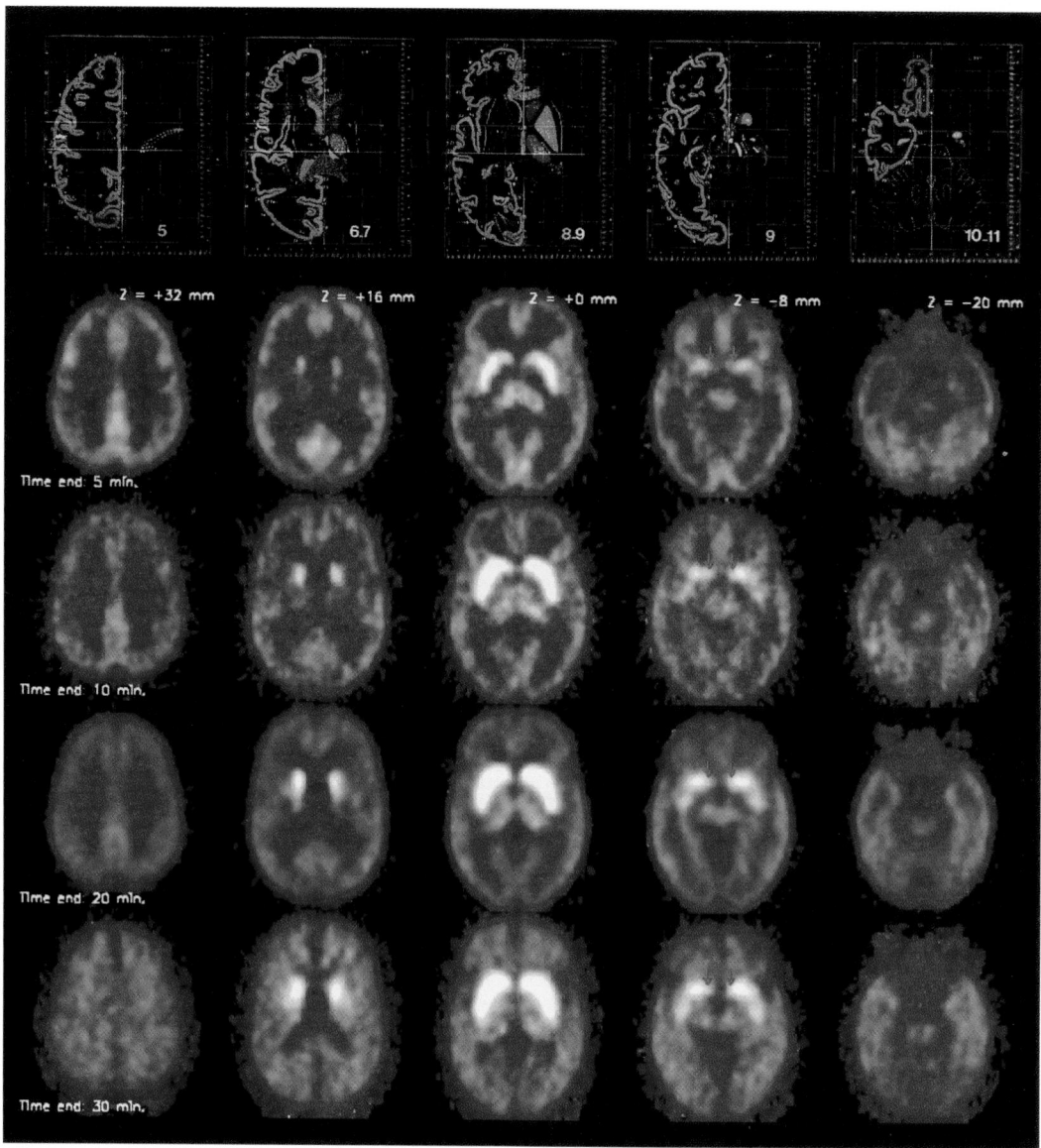

Figure 2 Brain images obtained with [^{11}C]cocaine at five different levels obtained at 5, 10, 20, and 30 min after radiotracer administration. Images have been normalized to highest activity in a given time frame. The top row illustrates representative Talairach–Tournoux images. The blue arrows indicate location of the nucleus accumbens.

consistently decreased global CBF in the human brain by 20–25% (Wang *et al.*, 1994). In contrast, the same dose of methylphenidate induced variable changes in brain glucose metabolism: brain metabolism was increased in six subjects, decreased in two, and not changed in seven. However, cerebellar metabolism was consistently increased. This variability in the metabolic responses was due, in part, to the variability in the levels of DA D_2 receptors, since for some of the regions, the metabolic changes were significantly correlated with D_2 levels. Frontal metabolism and temporal metabolism were increased in subjects with high D_2 receptors and decreased in subjects with low D_2 receptors (Volkow *et*

al., 1997b). A similar, but not identical, relationship was found in cocaine abusers in whom methylphenidate increased activity in the cingulate gyrus and thalamus (Volkow *et al.*, 1999b).

The relationship between the brain kinetics of [^{11}C]methylphenidate and the behavioral and cardiovascular effects of intravenous methylphenidate has been investigated in normal controls. The study showed that its kinetics in brain parallel the temporal course for self-reports of restlessness and changes in systolic blood pressure and heart rate, but not self-reports of high or anxiety or changes in diastolic blood pressure (Volkow *et al.*, 1995a).

PET has also been used to assess if further blocking of the DAT could be used to prevent the reinforcing effects of a DAT blocking drug such as cocaine. This investigation emphasizes one of the main strategies in drug development for cocaine addiction, finding drugs that would interfere with the binding of cocaine to the DAT. Using PET to measure DAT occupancy at different times after intravenous methylphenidate administration, Volkow et al. found the reinforcing effects of intravenous methylphenidate were equivalent whether given to subjects with no blockade of the DAT or to subjects in whom already 70–80% of the DATs were blocked (Volkow et al., 1995b). The study suggests that for DAT inhibitors to be therapeutically effective, occupancies of >80% may be required (Volkow et al., 1996b).

2. Magnetic Resonance Imaging

fMRI has been used to map the effects of acute cocaine administration (0.6 mg/kg i.v.) in the human brain using the BOLD technique. With better spatial resolution, cocaine induced focal signal increases in the nucleus accumbens/subcallosal cortex (Nac/Scc), insula, parahippocampal gyrus and hippocampus, and the cingulate and induced signal decreases in the amygdala, temporal pole, and medial frontal cortex. Saline produced only a few positive or negative activations, which were localized to the lateral prefrontal cortex and temporo-occipital cortex. The pattern was similar to that seen in a recent PET study of cue-induced cocaine craving in addicts (Grant et al., 1996) but fMRI has better spatial resolution. Brain regions that exhibited early and short-duration signal maxima showed a higher correlation with rush ratings. These included the ventral tegmentum, pons, basal forebrain, caudate, cingulate, and most regions of the lateral prefrontal cortex. In contrast, regions that demonstrated early but sustained signal maxima exhibited higher correlations with craving; these regions included the Nac/Scc, the right parahippocampal gyrus, and some regions of the lateral prefrontal cortex. Sustained negative signal changes were noted in the amygdala, which correlated with craving ratings, but there are signal artifacts in this region as a result of bone structures and the resultant imaging approach (Breiter et al., 1997).

To investigate the direct effects of cocaine on cerebral vasculature, dynamic susceptibility contrast MRI was used to determine whether a dose-effect relationship could be detected between cocaine administration and cerebral blood volume reduction in the human brain. Echo-planar images were collected with a spin-echo sequence (7-mm thickness, TR = 1000 ms, TE = 100 ms, FOV = 40 × 20 cm, matrix 128 × 64) to maximize blood volume measurement toward small vessels,

less than 100 μm. Subjects underwent measurements of relative cerebral blood volume (rCBV) at baseline and 10 min after administration of intravenous cocaine (0.2 or 0.4 mg/kg). Placebo administration resulted in reproducible results of ±4%. Cocaine doses induced CBV decreases to an average of 77% of baseline values at 10 min after injection (Kaufman et al., 1998). This is consistent with cocaine-induced reductions in flow noted in a similar SPECT study (Wallace et al., 1996).

Magnetic resonance spectroscopy has also been used to determine if there are persistent neurochemical abnormalities in asymptomatic chronic cocaine users (Fig. 3). Fifty-two African-American men (26 HIV-negative asymptomatic heavy cocaine users and 26 normal subjects) were studied and compared using proton MRS. Ventricle-to-brain ratios (VBR) and white matter lesions (WML) were quantified using MRI. N-Acetyl-containing compounds (NA), total creatine, choline-containing compounds, myoinositol, and glutamate + glutamine were then measured. VBR and WML were not significantly different in the cocaine users compared to the normal controls. Elevated creatine (+7%; $p = 0.05$) and myoinositol (+18%; $p = 0.01$) in the white matter were associated with cocaine use. N-Acetylaspartate, a measure of neuronal content, was normal (Chang et al., 1997).

Phosphorus magnetic resonance spectroscopy (^{31}P MRS) at 1.5 T was performed on nine polysubstance-abusing men scanned 2–7 days after admission to a drug treatment unit where they received methadone and were compared with 11 controls. In the brain, the phosphorus metabolite signal expressed as a percentage of total phosphorus signal was 15% higher for phosphomonoesters, 10% lower for nucleotide triphosphates (β-NTP), and 7% lower for total nucleotide phosphates in polydrug abusers compared with controls. Phosphodiesters, inorganic phosphate, phosphocreatine, total phosphorus, pH, and free magnesium concentration were unchanged. None of these parameters correlated with the methadone dose or number of days of abstinence, suggesting that cerebral high-energy phosphate and phospholipid metabolite changes result from long-term drug abuse and/or withdrawal and that these changes can be detected and studied by ^{31}P MRS (Christensen et al., 1996).

E. Marijuana

Cannibis sativa is a plant that currently elicits considerable controversy. With its widespread use documented among adolescents, there has been little perceived risk in its consumption. There has also been much interest in its potential use in management of certain diseases. Although THC, the main active ingredient

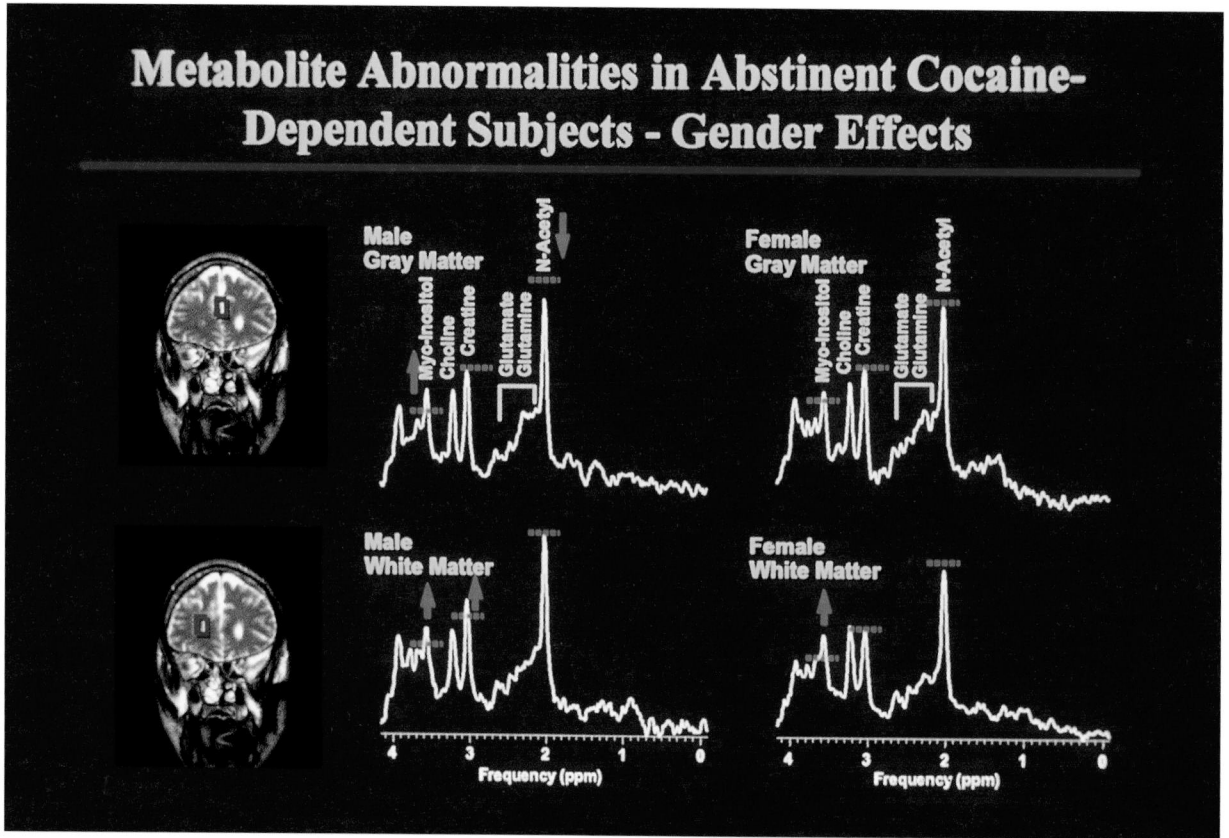

Figure 3 These ^{1}H MR spectra are from representative male and female crack cocaine users. The dashed lines indicate the mean peak height of normal comparison subjects for each corresponding peak. Note decreased *N*-acetyl compounds and increased myoinositol in the frontal gray matter and increased creatine and myoinositol in the white matter of the male user (left top and bottom spectra). In contrast, the female user showed mildly elevated myoinositol in the frontal gray matter and more significantly elevated myoinositol in the frontal white matter (top and bottom right). From *Biol. Psychiatry* **42,** 1105–1114, 1997. Copyright 1997, the American Psychiatric Association. Reprinted by permission.

of marijuana, can be legally prescribed for oral use, to improve appetite in AIDS/cancer wasting syndromes, and to relieve nausea, the smoked drug is preferred by some patients. There is ample evidence to support the view that cannabis is physically addictive, following the standard DSM-IV criteria for dependence. A recent study (Crowley *et al.*, 1998) evaluating a cohort of 340 patients showed that most patients claimed serious problems from cannabis, and 79% met standard adult criteria for cannabis dependence. Two-thirds of cannabis-dependent patients reported withdrawal. Progression from first to regular cannabis use was as rapid as tobacco progression and more rapid than that of alcohol, suggesting that cannabis is a reinforcer. The data indicated that for adolescents with conduct problems cannabis use is not benign and that the drug potently reinforces cannabis taking, producing both dependence and withdrawal.

THC acts at G protein-coupled receptors (CB1 receptors, primarily in the brain, and CB2 receptors, systemically). Anandamide, an endogenous antagonist of the cannabinoid system, has also been identified. The CB1 receptors are abundant in specific brain regions, including the cerebellum, hippocampus, and outflow nuclei of the basal ganglia. CNR1, a cannabinoid receptor gene, has also been identified, and a study looking at variants (additional trinucleotide repeats) in a cohort of 92 polydrug users determined that the gene might be associated with a susceptibility to dependence with alcohol or other drugs. In this study, the number of intravenous drugs used by the subject was significantly greater for those carrying the > or =/> or = 5 genotype than for other genotypes ($p = 0.005$). The association with specific types of drug dependence was greatest for cocaine, amphetamine, and cannabis dependence, consistent with a role of cannabinoid receptors in the modulation of dopamine and cannabinoid reward pathways (Comings *et al.*, 1997).

1. PET/SPECT

Studies evaluating effects of acute THC administration have shown that the most consistent observation in both normal controls and habitual marijuana users was increased rCMR in the cerebellum. The increases in cerebellar metabolism were positively correlated with plasma THC levels and subjective sense of intoxication (Volkow *et al.*, 1991b). In chronic users, but not controls, THC also increased metabolism in the orbitofrontal cortex and cingulate gyrus. In chronic THC use, the average increase in cortical and subcortical rCMR after administration was reduced in users compared with controls, and they had lower cerebellar metabolism at baseline (Volkow *et al.*, 1996a). This involvement of the cerebellum in the psychoactive effects of marijuana and in changes in rCMR is consistent with the view that THC interacts with CB1 receptors. Functions known to be associated with the cerebellum, such as motor coordination, proprioception, and learning, are adversely affected both during acute marijuana intoxication and in habitual users. In a recent [^{15}O]water study, chronic marijuana users received THC (intravenously) in a doubleblind fashion with saline. They showed dose-related rCBF increases in the frontal regions, insula and cingulate gyrus, and subcortical structures, with somewhat greater effects in the right hemisphere (Mathew *et al.*, 1997).

SPECT studies using AM281, a CB1 receptor antagonist, in non-human primates have been completed to determine its potential value as an *in vivo* imaging ligand for cannabinoid receptors (Fig. 4) (Gatley *et al.*, 1998). They demonstrated a rapid passage of [^{123}I]AM281 into the brain after intravenous injection, appropriate regional brain specificity of binding, and reduction of binding after treatment with SR141716A, a partial antagonist of CB1 (Lin *et al.* 1998). Human studies with this ligand are pending Investigational New Drug (IND) approval.

F. Opiates

Morphine and related drugs activate G protein-coupled receptors whose physiological ligands are peptides. Three classes of opioid receptors, each comprising two or more subtypes, are generally recognized, termed mu (μ), delta (δ), and kappa (κ) receptors. Data supporting the μ-receptor as the primary site for morphine reward come from structure-activity studies showing that μ-receptor affinities reasonably parallel the ability of opiate drugs to mediate reinforcement (Zubieta *et al.*, 1996). μ-Receptor knockout mice lose almost all the rewarding actions of morphine, even though levels of κ and δ receptors are relatively intact (Uhl *et al.*, 1999).

1. PET/SPECT

In an early PET study of 12 polydrug abusers, acute morphine (30 mg i.m.) reduced global CMR and rCMR by 10% in whole brain and by about 5–15% in telencephalic areas and the cerebellar cortex (London *et al.*, 1990). Effects on glucose utilization were not significantly related to measures of euphoria. To our knowl-

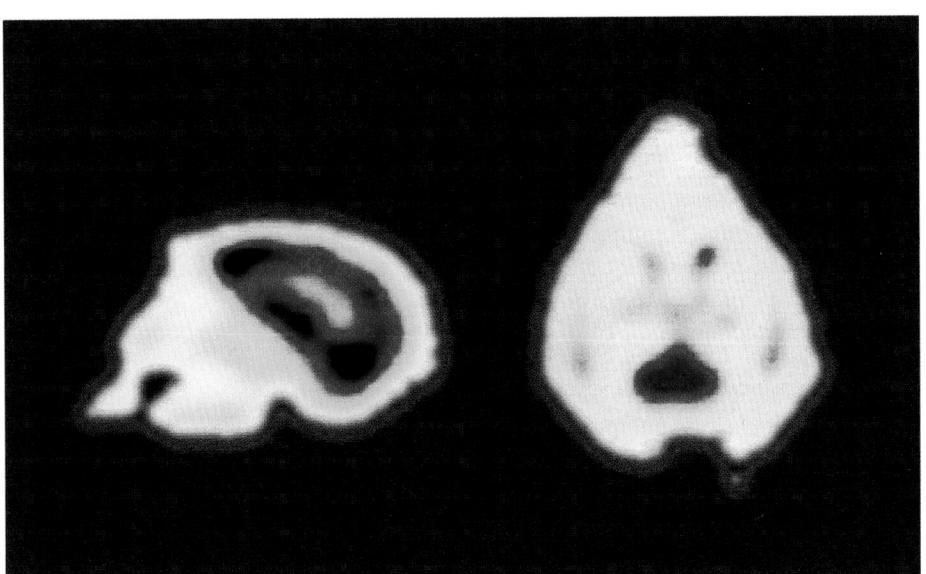

Figure 4 SPECT baboon brain images after intravenous injection of 8–10 mCi of [^{123}I]AM281 formulated in 1 ml of homologous blood plasma. AM281 is a CB1 cannabinoid ligand that shows promise for human studies since it crosses the blood–brain barrier with good affinity yet has poor systemic uptake. Activity appears increased in the cerebellum, cortex, and striatum, in agreement with the known distribution of the CB1 receptors in rats and mice. Courtesy of S. J. Gatley.

edge, there have been no published PET/SPECT studies of opioid abusers using labeled opioid drugs.

PET studies in primates using [^{11}C]morphine, heroin, codeine, and pethidine show uniform volumes of distribution, due to the lipophilicity and high nonspecific binding, a prolonged half-life of greater than 2 h in the case of morphine, and rapid formation of metabolites. Pethidine was found to have the quickest uptake kinetics, normalized to 6.3 that of morphine (Hartvig *et al.*, 1984). However, [^{11}C]carfentanil was used in a recent comparative nonimaging study to assess the duration of blockade of μ-opioid receptors by the antagonists naloxone and nalmefene. Using a simple (dual-detector) positron radiation detector system in normal volunteers, Kim *et al.* found that nalmefene had a prolonged kinetic profile in comparison to naloxone (Kim *et al.*, 1997). Future work is needed to address PET studies of opioid receptor occupancies by abused drugs and by medications designed to help recovering addicts. Buprenorphine (BPN) is a mixed-opiate agonist-antagonist, used as an analgesic and in the treatment of opiate addiction. The regional distribution of [^{11}C]BPN was measured with PET in the baboon brain and was striatum > thalamus cingulate gyrus > frontal cortex > parietal cortex > occipital cortex > cerebellum. This corresponds to opiate receptor density measured *in vitro* (Belcheva *et al.*, 1998). Naloxone administered 30–40 min after tracer injection at a dose of 1.0 mg/kg i.v. reduced [^{11}C]BPN binding in the thalamus, striatum, cingulate gyrus, and frontal cortex to 25–60% of control values. [^{11}C]Methadone has been developed as a novel ligand in this field of research; plans for a clinical study are ongoing (Ding *et al.*, 1996).

Two separate PET studies, using [^{15}O]water, have identified specific regions activated by nociceptive stimuli and the effects of fentanyl, a synthetic opiate, on these activation patterns. One study found that pain increased rCBF in the anterior cingulate, ipsilateral thalamus, prefrontal cortex, and contralateral supplementary motor area whereas fentanyl increased rCBF in the anterior cingulate and contralateral motor cortices and decreased rCBF in the thalamus and posterior cingulate. During combined pain stimulation and fentanyl administration, fentanyl significantly augmented pain-related rCBF increases in the supplementary motor area and prefrontal cortex. This activation pattern was associated with decreased pain perception, as measured on a visual analog scale (Adler *et al.*, 1997). In the other study, PET data were coregistered to MRI, and fentanyl administration correlated to significant increases in rCBF, consistent with regional neuronal activation in the cingulate, orbitofrontal and medial prefrontal cortices, and caudate nuclei. These areas are known to be

involved in avoidance learning, reward, and addiction (Firestone *et al.*, 1996).

Using PET and [^{11}C]raclopride, Wang *et al.* found that opioid-dependent subjects had lower D$_2$ receptor availability than controls (baseline measures for B_{max}/K_d). Naloxone induced withdrawal in dependent subjects but did not change [^{11}C]raclopride binding, indicating the withdrawal does not alter synaptic dopamine in the striatum as measured by this method (Wang *et al.*, 1997).

2. Magnetic Resonance Imaging

A recent MRI study using high-resolution volumetric analysis has suggested that individuals with polysubstance abuse disorder (mostly heroin and cocaine) exhibit structural deficits in the prefrontal cortex. They had been drug free (except for nicotine) for 2 weeks prior to scanning. Specific deficits in gray, but not white, matter was seen when the reduced volume was segmented and compared to normal age/gender-matched controls, suggesting that structural deficits in the prefrontal cortex contribute to the neuropathology of the functional impairments seen in this condition.

G. Nicotine

In the middle of the sixteenth century, tobacco (*Nicotiana tabacum*) was brought from Portugal to France by the French ambassador Jean Nicot, who gave it to the Queen Catherine de Medicis to relieve headache. Since that time, the widespread use of tobacco, whatever its administration (chewing, snuffing, smoking), has given rise to the appearance of neurophysiological and neuropsychological changes that make its use very difficult to give up once started. There are 45 million cigarette smokers in the United States alone, with 400,000 deaths/year associated with smoking, and medical costs amount to $50 billion a year. Surprisingly, little is known about the neurochemical actions of tobacco smoker exposure on the human brain and very few imaging studies have examined the smoker's brain. Nicotine, (*S*)-3-(1-methyl-2-pyrrolidinyl)pyridine, the main alkaloid extracted from this plant, has been found to be responsible for most of the effects sought after by tobacco users.

In 1914, Dale compared the actions of various synthetic and endogenous choline derivatives, including acetylcholine, to those of the plant alkaloids muscarine and nicotine. He postulated that acetylcholine was an autonomic nervous system neurotransmitter and that it had dual actions, i.e., muscarinic and nicotinic. Nicotine and acetylcholine can exist in similar molecular forms. After injection, [^{14}C] nicotine is distributed diffusely throughout the body; it concentrates in the nervous system. Schwartz and Kellar (1985) described the autoradiographic distribution of nAch' Rs in rat brain using [³H]nicotine. Both [³H]nicotine and [³H]Ach binding is highest in the interpeduncular nucleus, most thalamic nuclei, superior colliculus, medial habenula, presubiculum, layers I, III, and IV of the cerebral cortex, substantia nigra pars compacta, and the ventral tegmental area.

In 1976, the first *in vivo* study of $(-)$-[¹¹C]nicotine was performed in baboons with a conventional γ camera (Maziere *et al.*, 1976). Since 1979, all studies in baboons and humans have been done with PET. There is an increased uptake of [¹¹C]nicotine in young tobacco smokers compared to control nonsmokers (Nyback, 1994), in concordance with an increase in the number of high-affinity nicotinic receptors found postmortem in smoker's brain tissue (Benwell *et al.*, 1988). Yet, until recently, attempts to model the ligand-receptor interaction of this compound based on a three-compartment model were unsuccessful; therefore the model used included a blood compartment and only a single tissue compartment (which integrates the free and bound ligand compartments). Therefore, the exact contribution relative to flow was uncertain. In 1998, Muzic *et al.* did a quantification study using PET and both enantiomers of [¹¹C]nicotine in a challenge fashion with the (*S*) isomer. They determined that there was very little change in distribution volume (DV), depending on the isomer chosen in the three-compartment model, and concluded that very little specific binding is attributable to [¹¹C]nicotine; that is, [¹¹C]nicotine is a relatively poor PET tracer (Muzic *et al.*, 1998).

Glucose metabolic activity has been compared in a relatively small number of smokers and nonsmokers, with one study reporting slight elevations in metabolism and a second reporting no significant differences (Fowler *et al.*, 1996a). The acute administration of intravenous nicotine has also been reported to reduce brain metabolism (Stapleton, 1993; Villemagne *et al.*, 1997). More recently, monoamine oxidase (MAO) has been examined in the human brain (Fowler *et al.*, 1998a,b). MAO breaks down neurotransmitter amines like dopamine, serotonin, and norepinephrine as well as amines from exogenous sources. It occurs in two subtypes, MAO A and MAO B, which can be imaged *in vivo* using [¹¹C]clorgyline and [¹¹C]L-deprenyl. Using these ligands, Fowler and colleagues showed that cigarette smokers have a reduction in brain MAO B of about 40% relative to nonsmokers or former smokers and that smokers have a 28% reduction in brain MAO A relative to nonsmokers. It is known that nicotine does not inhibit MAO B at physiologically relevant concen-

trations. MAO A and MAO B inhibition is associated with enhanced activity of dopamine, a neurotransmitter involved in reinforcing and motivating behaviors and in movement as well as decreased production of hydrogen peroxide, a source of reactive oxygen species. Inhibition of MAO by cigarette smoke could be one of the mechanisms accounting for the lower incidence of Parkinson disease in cigarette smokers. MAO A and MAO B inhibition by smoke may also account for some of the epidemiological features of smoking, which include a higher rate of smoking in individuals with depression and addiction to other substances. In this regard, it is noted that MAO A inhibitors are effective in the treatment of depression.

Studies have also evaluated the complicated relationship of nicotine on rCBF, independent of changes in PaCO₂, which is known to increase with cigarette smoking. Using [¹⁵O]water, Nagata was able to show increases in the cerebellum and frontal lobes due to smoking the first cigarette of the day and a subdued response that was thought to represent acute tolerance with the second smoking session. Yet nicotine's effect on the brain includes sympathetic stimulation to the nerves innervating blood vessels, which in itself is a biphasic response related to dose; also carbon monoxide and carboxy hemoglobin are produced, impairing oxygen delivery and consequent cerebral vasodilation. A recent SPECT study confirms that chronic exposure to cigarette smoke alters the brain response to the blood flow tracer [¹²³I]iodoamphetamine (IMP), with a global decrease in activation to stimulus in comparison to control subjects (Rourke *et al.*, 1997). Attempts to label nicotinic acetylcholine receptor radioligands for human studies are also ongoing. Studies with epibatidine were slowed by its inherent toxicity; [¹¹C]A-84543, or 3-[(1-[¹¹C]methyl-2(*S*)-pyrrolidinyl)methoxy]pyridine, has been produced and a study in mice shows a biodistribution consistent with that of nAch' Rs (Kassiou *et al.*, 1998).

1. Magnetic Resonance Imaging

fMRI has been used to determine the acute CNS effects of intravenous nicotine in 16 active cigarette smokers. Saline and three doses of nicotine (0.75, 1.50, and 2.25 mg/70 kg of body weight) were sequentially administered intravenously over 1-min periods in an ascending, cumulative-dosing paradigm while whole-brain gradient-echo, echo-planar images were acquired every 6 s during consecutive 20-min trials. Nicotine induced a dose-dependent increase in neuronal activity in a distributed system of brain regions, including the nucleus accumbens, amygdala, cingulate, and frontal lobes (Stein *et al.*, 1998).

V. Conclusions

Imaging technologies have started to document mechanisms of reinforcement of addictive substances and to delineate neurochemical changes in the brain of the addicted subject. Though findings are mostly of a preliminary nature, they provide an indication of the potential of imaging in the area of substance abuse. The capabilities include the following:

1. Assessment of the neurobiological behavior of drugs of abuse in the human brain. This is important since drug pharmacokinetics and pharmacodynamics may vary across animal species. It also enables the assessment of drug behavior directly in the drug addict.

2. Determination of a relationship, in awake human subjects, between behavioral and regional brain effects both in neurotransmitters and in function as assessed with measures of glucose metabolism or cerebral blood flow. Studies can also be done to assess the relation between pharmacokinetics of a given drug and the time course of its pharmacological effects.

3. The ability to view neurochemical and functional changes from many perspectives directly in the addicted individual.

4. The ability to carry out studies of kinetics of the effects of drugs on regional brain activity using the fast imaging capabilities of fMRI.

5. Further applications in the development of new therapeutic interventions.

Acknowledgments

This research was supported in part by the U.S. Department of Energy (Office of Health and Environmental Research) under Contract DE-AC02-98CH10886, NIDA under Grant DA06891, and the NIAAA, NIDA under Grant AA 09481.

References

Abi-Dargham, A., Krystal, J. H., Anjilvel, S., Scanley, B. E., Zoghbi, S., Baldwin, R. M., Rajeevan, N., Ellis, S., Petrakis, I. L., Seibyl, J. P., Charney, D. S., Laruelle, M., and Innis, R. B. (1998). Alterations of benzodiazepine receptors in type II alcoholic subjects measured with SPECT and [^{123}I]iomazenil. *Am. J. Psychiatry* **155**, 1550–1555.

Adams, K. M., Gilman, S., Johnson-Greene, D., Koeppe, R. A., Junck, L., Kluin, K. J., Martorello, S., Johnson, M. J., Heumann, M., and Hill, E. (1998). The significance of family history status in relation to neuropsychological test performance and cerebral glucose metabolism studied with positron emission tomography in older alcoholic patients. *Alcohol Clin. Exp. Res.* **22**, 105–110.

Adler, L. J., Gyulai, F. E., Diehl, D. J., Mintun, M. A., Winter, P. M., and Firestone, L. L. (1997). Regional brain activity changes associated with fentanyl analgesia elucidated by positron emission tomography [published erratum appears in *Anesth. Analg.* **84**(5), 949 (1997)]. *Anesth. Analg.* **84**, 120–126.

Bardo, M. T. (1998). Neuropharmacological mechanisms of drug reward: Beyond dopamine in the nucleus accumbens. *Crit. Rev. Neurobiol.* **12**, 37–67.

Baxter, L. R., Jr., Schwartz, J. M., Phelps, M. E., Maziotta, J. C., Barrio, J., Rawson, R. A., Engel, J., Guze, B. H., Selin, C., and Sumida, R. (1988). Localization of neurochemical effects of cocaine and other stimulants in the human brain. *J. Clin. Psychiatry* **49**, 23–26.

Belcheva, M. M., Bohn, L. M., Ho, M. T., Johnson, F. E., Yanai, J., Barron, S., and Coscia, C. J. (1998). Brain opioid receptor adaptation and expression after prenatal exposure to buprenorphine. *Brain Res. Dev.* **111**, 35–42.

Benwell, M. E., Balfour, D. J., and Anderson, J. M. (1988). Evidence that tobacco smoking increases the density of ($-$)-[^{3}H]nicotine binding sites in human brain. *J. Neurochem.* **50**, 1243–1247.

Bergstrom, M., Nordberg, A., Lunell, E., Antoni, G., and Langstrom, B. (1995). Regional deposition of inhaled ^{11}C-nicotine vapor in the human airway as visualized by positron emission tomography. *Clin. Pharmacol. Ther.* **57**, 309–317.

Breiter, H. C., Gollub, R. L., Weisskoff, R. M., Kennedy, D. N., Makris, N., Berke, J. D., Goodman, J. M., Kantor, H. L., Gastfriend, D. R., Riorden, J. P., Mathew, R. T., Rosen, B. R., and Hyman, S. E. (1997). Acute effects of cocaine on human brain activity and emotion. *Neuron* **19**, 591–611.

Buchsbaum, M. S., Wu, J., Haier, R., Hazlett, E., Ball, R., Katz, M., Sokolski, K., Lagunas-Solar, M., and Langer, D. (1987). Positron emission tomography assessment of effects of benzodiazepines on regional glucose metabolic rate in patients with anxiety disorder. *Life Sci.* **40**, 2393–2400.

Buck, K. J., and Hood, H. M. (1998). Genetic association of a GABA(A) receptor gamma2 subunit variant with severity of acute physiological dependence on alcohol. *Mamm. Genome* **9**, 975–978.

Cameron, O. G., Modell, J. G., and Hariharan, M. (1990). Caffeine and human cerebral blood flow: A positron emission tomography study. *Life Sci.* **47**, 1141–1146.

Carr, L. (1998). A quantitative trait locus for alcohol consumption in selectively bred rat lines. *Alcohol. Clin. Exp. Res.* **22**, 884–887.

Carson, R. E., Breier, A., de Bartolomeis, A., Saunders, R. C., Su, T. P., Schmall, B., Der, M. G., Pickar, D., and Eckelman, W. C. (1997). Quantification of amphetamine-induced changes in [^{11}C]raclopride binding with continuous infusion. *J. Cereb. Blood Flow Metab.* **17**, 437–447.

Chang, L., Mehringer, C. M., Ernst, T., Melchor, R., Myers, H., Forney, D., and Satz, P. (1997). Neurochemical alterations in asymptomatic abstinent cocaine users: A proton magnetic resonance spectroscopy study. *Biol. Psychiatry* **42**, 1105–1114.

Chen, Y. C., Galpern, W. R., Brownell, A. L., Matthews, R. T., Bogdanov, M., Isacson, O., Keltner, J. R., Beal, M. F., Rosen, B. R., and Jenkins, B. G. (1997). Detection of dopaminergic neurotransmitter activity using pharmacologic MRI: Correlation with PET, microdialysis, and behavioral data. *Magn. Reson. Med.* **38**, 389–398.

Childress, A., Mozley, D., Fitzgerald, J., Reivich, M., Jaggi, J., and O'Brien, C. P. (1995). Limbic activation during cue-induced cocaine craving. *Abstr. Soc. Neurosci.* **767**, 1.

Childress, A. R., Mozley, P. D., McElgin, W., Fitzgerald, J., Reivich, M., and O'Brien, C. P. (1999). Limbic activation during cue-induced cocaine craving. *Am. J. Psychiatry* **156**, 11–18.

Chiu, T. M., Mendelson, J. H., Woods, B. T., Teoh, S. K., Levisohn, L., and Mello, N. K. (1994). *In vivo* proton magnetic resonance spectroscopy detection of human alcohol tolerance. *Magn. Reson. Med.* **32**, 511–516.

Christensen, J. D., Kaufman, M. J., Levin, J. M., Mendelson, J. H., Holman, B. L., Cohen, B. M., and Renshaw, P. F. (1996). Abnormal cerebral metabolism in polydrug abusers during early withdrawal: A ^{31}P MR spectroscopy study. *Magn. Reson. Med.* **35**, 658–663.

Comings, D. E., Muhleman, D., Gade, R., Johnson, P., Verde, R., Saucier, G., and MacMurray, J. (1997). Cannabinoid receptor gene (CNR1): Association with i.v. drug use. *Mol. Psychiatry* **2**, 161–168.

Crabbe, J. C., Phillips, T. J., Feller, D. J., Hen, R., Wenger, C. D., Lessov, C. N., and Schafer, G. L. (1996). Elevated alcohol consumption in null mutant mice lacking 5-HT$_{1B}$ serotonin receptors. *Nat. Genet.* **14**, 98–101.

Crow, T. J. (1973). Catecholamine-containing neurones and electrical self-stimulation. 2. A theoretical interpretation and some psychiatric implications. *Psychol. Med.* **3**, 66–73.

Crowley, T. J., Macdonald, M. J., Whitmore, E. A., and Mikulich, S. K. (1998). Cannabis dependence, withdrawal, and reinforcing effects among adolescents with conduct symptoms and substance use disorders. *Drug Alcohol Depend.* **50**, 27–37.

Dager, S. R., Layton, M. E., Strauss, W., Richards, T. L., Heide, A., Friedman, S. D., Artru, A. A., Hayes, C. E., and Posse, S. (1999). Human brain metabolic response to caffeine and the effects of tolerance. *Am. J. Psychiatry* **156**, 229–237.

Daly, J. W., Holmen, J., and Fredholm, B. B. (1998). [Is caffeine addictive? The most widely used psychoactive substance in the world affects same parts of the brain as cocaine]. *Lakartidningen* **95**, 5878–5883.

de Wit, H., Metz, J., and Cooper, M. (1991). Effects of ethanol, diazepam and amphetamines on cerebral metabolic rate: PET studies using FDG. *NIDA Res. Monogr.* **105**, 61–67.

Dewey, S. L., Morgan, A. E., Ashby, C. R., Jr., Horan, B., Kushner, S. A., Logan, J., Volkow, N. D., Fowler, J. S., Gardner, E. L., and Brodie, J. D. (1998). A novel strategy for the treatment of cocaine addiction. *Synapse* **30**, 119–129.

Dewey, S. L., Brodie, J. D., Gerasimov, M., Horan, B., Gardner, E. L., and Ashby, C. R., Jr. (1999). A pharmacologic strategy for the treatment of nicotine addiction. *Synapse* **31**, 76–86.

Ding, Y. S., Fowler, J., Volkow, N. D., Studebaker, R., and Pipolo, F. (1996). Synthesis and PET baboon study of [^{11}C]methadone and (−)-[^{11}C]methadone. *Proc. 43rd Annu. Meet. Soc. Nucl. Med.* **37**, 40P.

Ding, Y. S., Fowler, J. S., Volkow, N. D., Dewey, S. L., Wang, G. J., Logan, J., Gatley, S. J., and Pappas, N. (1997). Chiral drugs: Comparison of the pharmacokinetics of [^{11}C]*d-threo-* and *l-threo*-methylphenidate in the human and baboon brain. *Psychopharmacology (Berl.)* **131**, 71–78.

Ding, Y., Volkow, N. D., Yang, G. J., and Dewey, S. (1998). Distribution of [^{11}C]caffeine in the baboon brain. Brookhaven National Laboratory. Personal communication.

Eckardt, M. J., Rohrbaugh, J. W., Rio, D., Rawlings, R. R., and Coppola, R. (1988). Brain imaging in alcoholic patients. *Adv. Alcohol Subst. Abuse* **7**, 59–71.

El-Ghundi, M., George, S. R., Drago, J., Fletcher, P. J., Fan, T., Nguyen, T., Liu, C., Sibley, D. R., Westphal, H., and O'Dowd, B. F. (1998). Disruption of dopamine D$_1$ receptor gene expression attenuates alcohol-seeking behavior. *Eur. J. Pharmacol.* **353**, 149–158.

Epping-Jordan, M. P., Watkins, S. S., Koob, G. F., and Markou, A. (1998). Dramatic decreases in brain reward function during nicotine withdrawal. *Nature* **393**, 76–79.

Ernst, M., Zametkin, A. J., Matochik, J. A., Liebenauer, L., Fitzgerald, G. A., and Cohen, R. M. (1994). Effects of intravenous dextroamphetamine on brain metabolism in adults with attention-deficit hyperactivity disorder (ADHD). Preliminary findings. *Psychopharmacol. Bull.* **30**, 219–225.

Firestone, L. L., Gyulai, F., Mintun, M., Adler, L. J., Urso, K., and Winter, P. M. (1996). Human brain activity response to fentanyl imaged by positron emission tomography. *Anesth. Analg.* **82**, 1247–1251.

Fowler, J. S., Volkow, N. D., Wolf, A. P., Dewey, S. L., Schlyer, D. J., Macgregor, R. R., Hitzemann, R., Logan, J., Bendriem, B., Gatley, S. J., *et al.* (1989). Mapping cocaine binding sites in human and baboon brain *in vivo*. *Synapse* **4**, 371–377.

Fowler, J. S., Volkow, N. D., MacGregor, R. R., Logan, J., Dewey, S. L., Gatley, S. J., and Wolf, A. P. (1992). Comparative PET studies of the kinetics and distribution of cocaine and cocaethylene in baboon brain. *Synapse* **12**, 220–227.

Fowler, J. S., Ding, Y. S., Volkow, N. D., Martin, T., MacGregor, R. R., Dewey, S., King, P., Pappas, N., Alexoff, D., Shea, C., *et al.* (1994).

PET studies of cocaine inhibition of myocardial norepinephrine uptake. *Synapse* **16**, 312–317.

Fowler, J. S., Volkow, N. D., Wang, G. J., Pappas, N., Logan, J., Shea, C., Alexoff, D., MacGregor, R. R., Schlyer, D. J., Zezulkova, I., and Wolf, A. P. (1996a). Brain monoamine oxidase A inhibition in cigarette smokers. *Proc. Natl. Acad. Sci. U. S. A.* **93**, 14065–14069.

Fowler, J. S., Volkow, N. D., Wang, G.-J., Pappas, N., Logan, J., MacGregor, R., Alexoff, D., Shea, C., Wolf, A. P., Warner, D., Zezulkova, I., and Cilento, R. (1996b). Inhibition of MAO B in the brains of smokers. *Nature* **379**, 733–736.

Fowler, J. S., Volkow, N. D., Logan, J., Pappas, N., King, P., MacGregor, R., Shea, C., Garza, V., and Gatley, S. J. (1998a). An acute dose of nicotine does not inhibit MAO B in baboon brain *in vivo*. *Life Sci.* **63**, L19-L23.

Fowler, J. S., Volkow, N. D., Wang, G. J., Pappas, N., Logan, J., MacGregor, R., Alexoff, D., Wolf, A. P., Warner, D., Cilento, R., and Zezulkova, I. (1998b). Neuropharmacological actions of cigarette smoke: Brain monoamine oxidase B (MAO B) inhibition. *J. Addict. Dis.* **17**, 23–34.

Froehlich, J. C., Badia-Elder, N. E., Zink, R. W., McCullough, D. E., and Portoghese, P. S. (1998). Contribution of the opioid system to alcohol aversion and alcohol drinking behavior. *J. Pharmacol. Exp. Ther.* **287**, 284–292.

Frost, J. J., Mayberg, H. S., Sadzot, B., Dannals, R. F., Lever, J. R., Ravert, H. T., Wilson, A. A., Wagner, H. N., Jr., and Links, J. M. (1990). Comparison of [^{11}C]diprenorphine and [^{11}C]carfentanil binding to opiate receptors in humans by positron emission tomography. *J. Cereb. Blood Flow Metab.* **10**, 484–492.

Gardner, E. L., and Lowinson, J. H. (1991). Marijuana's interaction with brain reward systems: Update 1991. *Pharmacol. Biochem. Behav.* **40**, 571–580.

Gatley, S. J., MacGregor, R. R., Fowler, J. S., Wolf, A. P., Dewey, S. L., and Schlyer, D. J. (1990). Rapid stereoselective hydrolysis of (+)-cocaine in baboon plasma prevents its uptake in the brain: Implications for behavioral studies. *J. Neurochem.* **54**, 720–723.

Gatley, S. J., Lan, R., Volkow, N. D., Pappas, N., King, P., Wong, C. T., Gifford, A. N., Pyatt, B., Dewey, S. L., and Makriyannis, A. (1998). Imaging the brain marijuana receptor: Development of a radioligand that binds to cannabinoid CB1 receptors *in vivo*. *J. Neurochem.* **70**, 417–423.

Grant, S., London, E. D., Newlin, D. B., Villemagne, V. L., Liu, X., Contoreggi, C., Phillips, R. L., Kimes, A. S., and Margolin, A. (1996). Activation of memory circuits during cue-elicited cocaine craving. *Proc. Natl. Acad. Sci. U.S.A.* **93**, 12040–12045.

Halikas, J. A., and Kuhn, K. (1990). Possible neurophysiologic basis of cocaine craving. *Ann. Clin. Psychiatry* **2**, 79–83.

Hartvig, P., Bergstrom, K., Lindberg, B., Lundberg, P. O., Lundqvist, H., Langstrom, B., Svard, H., and Rane, A. (1984). Kinetics of ^{11}C-labeled opiates in the brain of rhesus monkeys. *J. Pharmacol. Exp. Ther.* **230**, 250–255.

Heinz, A., Ragan, P., Jones, D. W., Hommer, D., Williams, W., Knable, M. B., Gorey, J. G., Doty, L., Geyer, C., Lee, K. S., Coppola, R., Weinberger, D. R., and Linnoila, M. (1998). Reduced central serotonin transporters in alcoholism. *Am. J. Psychiatry* **155**, 1544–1549.

Herz, A. (1998). Opiate reward mechanisms: A key role in drug abuse? *Can. J. Physiol. Pharmacol.* **76**(3), 252–258.

Hetherington, H. P., Telang, F., Pan, J. P., Sammi, M., Schuhlein, D., Molina, P., and Volkow, N. D. (1999). Spectroscopic imaging of the uptake kinetics and visibility of human brain ethanol *in vivo* at 4 T. *Magn. Reson. Med.* **42**, 1019–1026.

Hietala, J., West, C., Syvalahti, E., Nagren, K., Lehikoinen, P., Sonninen, P., and Ruotsalainen, U. (1994). Striatal D$_2$ dopamine receptor binding characteristics *in vivo* in patients with alcohol dependence. *Psychopharmacology (Berl.)* **116**, 285–290.

Holman, B. L., Carvalho, P. A., Mendelson, J., Teoh, S. K., Nardin, R., Hallgring, E., Hebben, N., and Johnson, K. A. (1991). Brain perfusion is abnormal in cocaine-dependent polydrug users: A study using technetium-99m-HMPAO and SPECT. *J. Nucl. Med.* **32**, 1206–1210.

Howard, R. J., Ellis, C., Bullmore, E. T., Brammer, M., Mellers, J. D., Woodruff, P. W., David, A. S., Simmons, A., Williams, S. C., and Parkes, J. D. (1996). Functional echoplanar brain imaging correlates of amphetamine administration to normal subjects and subjects with the narcoleptic syndrome. *Magn. Reson. Imaging* **14**, 1013–1016.

Hunt, W. A. (1985). "Alcohol and Biological Membranes." Guilford Press, New York.

Jagannathan, N. R., Desai, N. G., and Raghunathan, P. (1996). Brain metabolite changes in alcoholism: An *in vivo* proton magnetic resonance spectroscopy (MRS) study. *Magn. Reson. Imaging* **14**, 553–557.

Kao, C. H., Wang, S. J., and Yeh, S. H. (1994). Presentation of regional cerebral blood flow in amphetamine abusers by ^{99m}Tc-HMPAO brain SPECT. *Nucl. Med. Commun.* **15**, 94–98.

Kassiou, M., Scheffel, U. A., Ravert, H. T., Mathews, W. B., Musachio, J. L., London, E. D., and Dannals, R. F. (1998). Pharmacological evaluation of [^{11}C]A-84543: An enantioselective ligand for *in vivo* studies of neuronal nicotinic acetylcholine receptors. *Life Sci.* **63**, L13–L18.

Kaufman, M. J., Chiu, T. M., Mendelson, J. H., Woods, B. T., Mello, N. K., Lukas, S. E., Fivel, P. A., and Wighton, L. G. (1994). *In vivo* proton magnetic resonance spectroscopy of alcohol in rhesus monkey brain. *Magn. Reson. Imaging* **12**, 1245–1253.

Kaufman, M. J., Levin, J. M., Ross, M. H., Lange, N., Rose, S. L., Kukes, T. J., Mendelson, J. H., Lukas, S. E., Cohen, B. M., and Renshaw, P. F. (1998). Cocaine-induced cerebral vasoconstriction detected in humans with magnetic resonance angiography. *JAMA* **279**, 376–380.

Kim, S., Wagner, H. N., Jr., Villemagne, V. L., Kao, P. F., Dannals, R. F., Ravert, H. T., Joh, T., Dixon, R. B., and Civelek, A. C. (1997). Longer occupancy of opioid receptors by nalmefene compared to naloxone as measured *in vivo* by a dual-detector system. *J. Nucl. Med.* **38**, 1726–1731.

Koob, G. F., and Bloom, F. E. (1988). Cellular and molecular mechanisms of drug dependence. *Science* **242**, 715–723.

Laruelle, M., Abi-Dargham, A., van Dyck, C. H., Rosenblatt, W., Zea-Ponce, Y., Zoghbi, S. S., Baldwin, R. M., Charney, D. S., Hoffer, P. B., Kung, H. F., *et al.* (1995). SPECT imaging of striatal dopamine release after amphetamine challenge. *J. Nucl. Med.* **36**, 1182–1190.

Levin, J. M., Holman, B. L., Mendelson, J. H., Teoh, S. K., Garada, B., Johnson, K. A., and Springer, S. (1994). Gender differences in cerebral perfusion in cocaine abuse: Technetium-^{99}m-HMPAO SPECT study of drug-abusing women. *J. Nucl. Med.* **35**, 1902–1909.

Levin, J., Mendleson, J. H., Holman, L. B., Teoh, S. K., Garada, B., Schwartz, R. B., and Mello, N. K. (1995). Improved regional cerebral blood flow in chronic cocaine polydrug users treated with buprenorphine. *J. Nucl. Med.* **36**, 1211–1215.

Lin, S., Khanolkar, A. D., Fan, P., Goutopoulos, A., Qin, C., Papahadjis, D., and Makriyannis, A. (1998). Novel analogues of arachidonylethanolamide (anandamide): Affinities for the CB1 and CB2 cannabinoid receptors and metabolic stability. *J. Med. Chem.* **41**, 5353–5361.

Lingford-Hughes, A. R., Acton, P. D., Gacinovic, S., Suckling, J., Busatto, G. F., Boddington, S. J., Bullmore, E., Woodruff, P. W., Costa, D. C., Pilowsky, L. S., Ell, P. J., Marshall, E. J., and Kerwin, R. W. (1998). Reduced levels of GABA-benzodiazepine receptor in alcohol dependency in the absence of grey matter atrophy. *Br. J. Psychiatry* **173**, 116–122.

Litton, J. E., Neiman, J., Pauli, S., Farde, L., Hindmarsh, T., Halldin, C., and Sedvall, G. (1993). PET analysis of [^{11}C]flumazenil binding to benzodiazepine receptors in chronic alcohol-dependent men and healthy controls. *Psychiatry Res.* **50**, 1–13.

London, E. D. (1989). The effects of drug abuse on glucose metabolism. *J. Neuropsychiatry Clin. Neurosci.* **1**, S30–S36.

London, E. D. (1993). Studies of sigma receptors and metabolic responses to sigma ligands in the brain. *NIDA Res. Monogr.* **133**, 55–68.

London, E. D., Broussolle, E. P., Links, J. M., Wong, D. F., Cascella, N. G., Dannals, R. F., Sano, M., Herning, R., Snyder, F. R., Rippetoe, L. R., *et al.* (1990). Morphine-induced metabolic changes in human brain. Studies with positron emission tomography and [fluorine 18]fluorodeoxyglucose. *Arch. Gen. Psychiatry* **47**, 73–81.

London, E. D., Stapleton, J. M., Phillips, R. L., Grant, S. J., Villemagne, V. L., Liu, X., and Soria, R. (1996). PET studies of cerebral glucose metabolism: Acute effects of cocaine and long-term deficits in brains of drug abusers. *NIDA Res. Monogr.* **163**, 146–158.

Madras, B. K., and Kaufman, M. J. (1994). Cocaine accumulates in dopamine-rich regions of primate brain after i.v. administration: Comparison with mazindol distribution. *Synapse* **18**, 261–275.

Malison, R. T., Best, S. E., Wallace, E. A., McCance, E., Laruelle, M., Zoghbi, S. S., Baldwin, R. M., Seibyl, J. S., Hoffer, P. B., Price, L. H., *et al.* (1995). Euphorigenic doses of cocaine reduce [123I]beta-CIT SPECT measures of dopamine transporter availability in human cocaine addicts. *Psychopharmacology (Berl.)* **122**, 358–362.

Markou, A., Weiss, F., Gold, L. H., Caine, S. B., Schulteis, G., and Koob, G. F. (1993). Animal models of drug craving. *Psychopharmacology (Berl.)* **112**, 163–182.

Martin, P. R., Gibbs, S. J., Nimmerrichter, A. A., Riddle, W. R., Welch, L. W., and Willcott, M. R. (1995). Brain proton magnetic resonance spectroscopy studies in recently abstinent alcoholics. *Alcohol. Clin. Exp. Res.* **19**, 1078–1082.

Mathew, R. J., Barr, D. L., and Weinman, M. L. (1983). Caffeine and cerebral blood flow. *Br. J. Psychiatry* **143**, 604–608.

Mathew, R. J., Wilson, W. H., Coleman, R. E., Turkington, T. G., and DeGrado, T. R. (1997). Marijuana intoxication and brain activation in marijuana smokers. *Life Sci.* **60**, 2075–2089.

Maziere, M., Comar, D., Marazano, C., and Berger, G. (1976). Nicotine-^{11}C: Synthesis and distribution kinetics in animals. *Eur. J. Nucl. Med.* **30**, 255–258.

McCann, U. D., Szabo, Z., Scheffel, U., Dannals, R. F., and Ricaurte, G. A. (1998). Positron emission tomographic evidence of toxic effect of MDMA ("Ecstasy") on brain serotonin neurons in human beings. *Lancet* **352**, 1433–1437.

Mendelson, J. H., Woods, B. T., Chiu, T. M., Mello, N. K., Lukas, S. E., Teoh, S. K., Sintavanarong, P., Cochin, J., Hopkins, M. A., and Dobrosielski, M. (1990). *In vivo* proton magnetic resonance spectroscopy of alcohol in human brain. *Alcohol* **7**, 443–447.

Meyerhoff, D. J., Rooney, W. D., Tokumitsu, T., and Weiner, M. W. (1996). Evidence of multiple ethanol pools in the brain: An *in vivo* proton magnetization transfer study. *Alcohol. Clin. Exp. Res.* **20**, 1283–1288.

Moxon, L. N., Rose, S. E., Haseler, L. J., Galloway, G. J., Brereton, I. M., Bore, P., and Doddrell, D. M. (1991). The visibility of the ^{1}H NMR signal of ethanol in the dog brain [published erratum appears in *Magn. Reson. Med.* 21(2), 329 (1991)]. *Magn. Reson. Med.* **19**, 340–348.

Muzic, R. F., Jr., Berridge, M. S., Friedland, R. P., Zhu, N., and Nelson, A. D. (1998). PET quantification of specific binding of carbon-11-nicotine in human brain. *J. Nucl. Med.* **39**, 2048–2054.

Nehlig, A., Lucignani, G., Kadekaro, M., Porrino, L. J., and Sokoloff, L. (1984). Effects of acute administration of caffeine on local cerebral glucose utilization in the rat. *Eur. J. Pharmacol.* **101**, 91–100.

Nestler, E. J. (1996). Under siege: The brain on opiates. *Neuron* **16**, 897–900.

Nicolas, J. M., Catafau, A. M., Estruch, R., Lomena, F. J., Salamero, M., Herranz, R., Monforte, R., Cardenal, C., and Urbano-Marquez, A. (1993). Regional cerebral blood flow-SPECT in chronic alcoholism: Relation to neuropsychological testing. *J. Nucl. Med.* **34**, 1452–1459.

Nyback, H., Halldin, C., Ahlin, A., Curvall, M., and Eriksson, L. (1994). PET studies of the uptake of (S)- and (R)-[^{11}C]nicotine in the human brain: Difficulties in visualizing specific receptor binding *in vivo*. *Psychopharmacology (Berl.)* **115**, 31–36.

Olive, M. F., and Maidment, N. T. (1998). Repeated heroin administration increases extracellular opioid peptide-like immunoreactivity in the globus pallidus/ventral pallidum of freely moving rats. *Psychopharmacology (Berl.)* **139**, 251–254.

O'Malley, S. S. (1996). Opioid antagonists in the treatment of alcohol dependence: Clinical efficacy and prevention of relapse. *Alcohol Alcohol. Suppl.* **1**, 77–81.

Panagis, G., Kastellakis, A., and Spyraki, C. (1998). Involvement of the ventral tegmental area opiate receptors in self-stimulation elicited from the ventral pallidum. *Psychopharmacology (Berl.)* **139**, 222–229.

Petroff, O. A. C., Novotny, E. J., Ogino, T., Avison, M., and Prichard, J. W. (1990). *In vivo* measurements of ethanol concentration in rabbit brain by ^{1}H magnetic resonance spectroscopy. *J. Neurochem.* **54**, 1188–1195.

Pfefferbaum, A., Lim, K. O., Desmond, J. E., and Sullivan, E. V. (1996). Thinning of the corpus callosum in older alcoholic men: A magnetic resonance imaging study. *Alcohol. Clin. Exp. Res.* **20**, 752–757.

Pfefferbaum, A., Sullivan, E. V., Mathalon, D. H., and Lim, K. O. (1997). Frontal lobe volume loss observed with magnetic resonance imaging in older chronic alcoholics. *Alcohol. Clin. Exp. Res.* **21**, 521–529.

Pfefferbaum, A., Sullivan, E. V., Rosenbloom, M. J., Mathalon, D. H., and Lim, K. O. (1998). A controlled study of cortical gray matter and ventricular changes in alcoholic men over a 5-year interval. *Arch. Gen. Psychiatry* **55**, 905–912.

Reiter, R. (1985). Oxidative processes and antioxidative defense mechanisms in the aging brain. *FASEB J.* **9**, 526–533.

Risinger, F. O., Bormann, N. M., and Oakes, R. A. (1996). Reduced sensitivity to ethanol reward, but not ethanol aversion, in mice lacking 5-HT$_{1B}$ receptors. *Alcohol. Clin. Exp. Res.* **20**, 1401–1405.

Roebuck, T. M., Mattson, S. N., and Riley, E. P. (1998). A review of the neuroanatomical findings in children with fetal alcohol syndrome or prenatal exposure to alcohol. *Alcohol. Clin. Exp. Res.* **22**, 339–344.

Rourke, S. B., Dupont, R. M., Grant, I., Lehr, P. P., Lamoureux, G., Halpern, S., and Yeung, D. W. (1997). Reduction in cortical IMP-SPECT tracer uptake with recent cigarette consumption in a young group of healthy males. San Diego HIV Neurobehavioral Research Center. *Eur. J. Nucl. Med.* **24**, 422–427.

Rubinstein, M., Phillips, T. J., Bunzow, J. R., Falzone, T. L., Dziewczapolski, G., Zhang, G., Fang, Y., Larson, J. L., McDougall, J. A., Chester, J. A., Saez, C., Pugsley, T. A., Gershanik, O., Low, M. J., and Grandy, D. K. (1997). Mice lacking dopamine D$_4$ receptors are supersensitive to ethanol, cocaine, and methamphetamine. *Cell* **90**, 991–1001.

Rydell, C. P., Everingham, S. S. (1994). The National Report on Substance Abuse, pp. 2–6. RAND, MR-331-ONDCP/A/DPRC, Santa Monica.

Scheffel, U., Szabo, Z., Mathews, W. B., Finley, P. A., Dannals, R. F., Ravert, H. T., Szabo, K., Yuan, J., and Ricaurte, G. A. (1998). *In vivo* detection of short- and long-term MDMA neurotoxicity: A positron emission tomography study in the living baboon brain. *Synapse* **29**, 183–192.

Schwartz, R. D., and Kellar, K. J. (1985). *In vivo* regulation of [^{3}H]acetylcholine recognition sites in brain by nicotinic cholinergic drugs. *J. Neurochem.* **45**, 427–433.

Shear, P. K., Sullivan, E. V., Lane, B., and Pfefferbaum, A. (1996). Mammillary body and cerebellar shrinkage in chronic alcoholics with and without amnesia. *Alcohol. Clin. Exp. Res.* **20**, 1489–1495.

Sora, I., Wichems, C., Takahashi, N., Li, X. F., Zeng, Z., Revay, R., Lesch, K. P., Murphy, D. L., and Uhl, G. R. (1998). Cocaine reward models: Conditioned place preference can be established in dopamine- and in serotonin-transporter knockout mice. *Proc. Natl. Acad. Sci. U.S.A.* **95**, 7699–7704.

Spielman, D. M., Glover, G. H., Macovski, A., and Pfefferbaum, A. (1993). Magnetic resonance spectroscopic imaging of ethanol in the human brain: A feasibility study. *Alcohol. Clin. Exp. Res.* **17**, 1072–1077.

Stapleton, J. M., Henningfield, J. E., Wong, D. F., Phillips, R. L., Grayson, R. F., Dannals, R. F., and London, E. D., Eds. (1993). "Nicotine Reduces Cerebral Glucose Utilization in Humans." DHHS, Washington, DC.

Stapleton, J. M., Morgan, M. J., Phillips, R. L., Wong, D. F., Yung, B. C., Shaya, E. K., Dannals, R. F., Liu, X., Grayson, R. L., and London, E. D. (1995). Cerebral glucose utilization in polysubstance abuse. *Neuropsychopharmacology* **13**, 21–31.

Stein, E. A., Pankiewicz, J., Harsch, H. H., Cho, J. K., Fuller, S. A., Hoffmann, R. G., Hawkins, M., Rao, S. M., Bandettini, P. A., and Bloom, A. S. (1998). Nicotine-induced limbic cortical activation in the human brain: A functional MRI study. *Am. J. Psychiatry* **155**, 1009–1015.

Sullivan, E. V., Marsh, L., Mathalon, D. H., Lim, K. O., and Pfefferbaum, A. (1995). Anterior hippocampal volume deficits in nonamnesic, aging chronic alcoholics. *Alcohol. Clin. Exp. Res.* **19**, 110–122.

Sybirska, E., Seibyl, J. P., Bremner, J. D., Balswin, R. M., Al-Tikriti, M. S., Bradberry, C., Malison, R. T., Zea Ponce, Y., Zoghbi, S., During, M., Goddard, A. W., Woods, S. W., Hoffer, P. B., and Charney, D. S. (1993). [^{123}I]Iomazenil SPECT imaging demonstrates significant benzodiazepine receptor reserve in human and non human primate brain. *Neuropharmacology* **32**, 671–680.

Telang, F. W., Volkow, N. D., Levy, A., Logan, J., Fowler, J. S., Felder, C., Wong, C., and Wang, G. J. (1999). Distribution of tracer levels of cocaine in the human brain as assessed with averaged [^{11}C]cocaine images. *Synapse* **31**, 290–296.

Theodore, W. H., DiChiro, G., Margolin, R., Fishbein, D., Porter, R. J., and Brooks, R. A. (1986). Barbiturates reduce human cerebral glucose metabolism. *Neurology* **36**, 60–64.

Thiele, T., Marsh, D. J., Ste. Marie, L., Bernstein, I. L., and Palmiter, R. D. (1998). Ethanol consumption and resistance are inversely related to neuropeptide Y levels. *Nature* **396**, 366–369.

Tiihonen, J., Kuikka, J., Bergstrom, K., Hakola, P., Karhu, J., Ryynanen, O. P., and Fohr, J. (1995). Altered striatal dopamine re-uptake site densities in habitually violent and non-violent alcoholics [see comments]. *Nat. Med.* **1**, 654–657.

Uhl, G. R., Sora, I., and Wang, Z. (1999). The mu opiate receptor as a candidate gene for pain: Polymorphisms, variations in expression, nociception, and opiate responses. *Proc. Natl. Acad. Sci. U.S.A.* **96**(14), 7752–7755.

Villemagne, V. L., Horti, A., Scheffel, U., Ravert, H. T., Finley, P., Clough, D. J., London, E. D., Wagner, H. N., Jr., and Dannals, R. F. (1997). Imaging nicotinic acetylcholine receptors with fluorine-18-FPH, an epibatidine analog. *J. Nucl. Med.* **38**, 1737–1741.

Virkkunen, M., and Linnoila, M. (1997). Serotonin in early-onset alcoholism. *Recent Dev. Alcohol.* **13**, 173–189.

Volkow, N. D., Mullani, N., Gould, K. L., Adler, S., and Krajewski, K. (1988a). Cerebral blood flow in chronic cocaine users: A study with positron emission tomography. *Br. J. Psychiatry* **152**, 641–648.

Volkow, N. D., Mullani, N., Gould, L., Adler, S. S., Guynn, R. W., Overall, J. E., and Dewey, S. (1988b). Effects of acute alcohol intoxication on cerebral blood flow measured with PET. *Psychiatry Res.* **24**, 201–209.

Volkow, N. D., Hitzemann, R., Wolf, A. P., Logan, J., Fowler, J. S., Christman, D., Dewey, S. L., Schlyer, D., Burr, G., Vitkun, S., *et al.* (1990). Acute effects of ethanol on regional brain glucose metabolism and transport. *Psychiatry Res.* **35**, 39–48.

Volkow, N. D., Fowler, J. S., Wolf, A. P., Hitzemann, R., Dewey, S., Bendriem, B., Alpert, R., and Hoff, A. (1991a). Changes in brain glucose metabolism in cocaine dependence and withdrawal [see comments]. *Am. J. Psychiatry* **148,** 621–626.

Volkow, N. D., Gillespie, H., Mullani, N., Tancredi, L., Grant, C., Ivanovic, M., and Hollister, L. (1991b). Cerebellar metabolic activation by delta-9-tetrahydro-cannabinol in human brain: A study with positron emission tomography and ^{18}F-2-fluoro-2-deoxyglucose. *Psychiatry Res.* **40,** 69–78.

Volkow, N. D., Mullani, N., Gould, L. K., Adler, S. S., and Gatley, S. J. (1991c). Sensitivity of measurements of regional brain activation with oxygen-15- water and PET to time of stimulation and period of image reconstruction. *J. Nucl. Med.* **32,** 58–61.

Volkow, N. D., Fowler, J. S., Wolf, A. P., Wang, G. J., Logan, J., MacGregor, R., Dewey, S. L., Schlyer, D., and Hitzemann, R. (1992a). Distribution and kinetics of carbon-11-cocaine in the human body measured with PET. *J. Nucl. Med.* **33,** 521–525.

Volkow, N. D., Hitzemann, R., Wang, G. J., Fowler, J. S., Burr, G., Pascani, K., Dewey, S. L., and Wolf, A. P. (1992b). Decreased brain metabolism in neurologically intact healthy alcoholics. *Am. J. Psychiatry* **149,** 1016–1022.

Volkow, N. D., Hitzemann, R., Wang, G. J., Fowler, J. S., Wolf, A. P., Dewey, S. L., and Handlesman, L. (1992c). Long-term ftrontal brain metabolic changes in cocaine abusers. *Synapse* **11,** 184–190. [published erratum appears in *Synapse* (1992) **12**(1), 86.]

Volkow, N. D., Wang, G. J., Hitzemann, R., Fowler, J. S., Wolf, A. P., Pappas, N., Biegon, A., and Dewey, S. L. (1993). Decreased cerebral response to inhibitory neurotransmission in alcoholics. *Am. J. Psychiatry* **150,** 417–422.

Volkow, N. D., Wang, G. J., Fowler, J. S., Logan, J., Schlyer, D., Hitzemann, R., Lieberman, J., Angrist, B., Pappas, N., MacGregor, R., *et al.* (1994). Imaging endogenous dopamine competition with [^{11}C]raclopride in the human brain. *Synapse* **16,** 255–262.

Volkow, N. D., Ding, Y. S., Fowler, J. S., Wang, G. J., Logan, J., Gatley, J. S., Dewey, S., Ashby, C., Liebermann, J., Hitzemann, R., *et al.* (1995a). Is methylphenidate like cocaine? Studies on their pharmacokinetics and distribution in the human brain. *Arch. Gen. Psychiatry* **52,** 456–463.

Volkow, N. D., Ding, Y. S., Fowler, J. S., Wang, G. J., Logan, J., Gatley, S. J., Schlyer, D. J., and Pappas, N. (1995b). A new PET ligand for the dopamine transporter: Studies in the human brain. *J. Nucl. Med.* **36,** 2162–2168.

Volkow, N. D., Wang, G. J., Begleiter, H., Hitzemann, R., Pappas, N., Burr, G., Pascani, K., Wong, C., Fowler, J. S., and Wolf, A. P. (1995c). Regional brain metabolic response to lorazepam in subjects at risk for alcoholism. *Alcohol. Clin. Exp. Res.* **19,** 510–516.

Volkow, N. D., Gillespie, H., Mullani, N., Tancredi, L., Grant, C., Valentine, A., and Hollister, L. (1996a). Brain glucose metabolism in chronic marijuana users at baseline and during marijuana intoxication. *Psychiatry Res.* **67,** 29–38.

Volkow, N. D., Wang, G. J., Fowler, J. S., Gatley, S. J., Ding, Y. S., Logan, J., Dewey, S. L., Hitzemann, R., and Lieberman, J. (1996b). Relationship between psychostimulant-induced "high" and dopamine transporter occupancy. *Proc. Natl. Acad. Sci. U.S.A.* **93,** 10388–10392.

Volkow, N. D., Wang, G. J., Fischman, M. W., Foltin, R. W., Fowler, J. S., Abumrad, N. N., Vitkun, S., Logan, J., Gatley, S. J., Pappas, N., Hitzemann, R., and Shea, C. E. (1997a). Relationship between subjective effects of cocaine and dopamine transporter occupancy. *Nature* **386,** 827–830.

Volkow, N. D., Wang, G. J., Fowler, J. S., Logan, J., Angrist, B., Hitzemann, R., Lieberman, J., and Pappas, N. (1997b). Effects of methylphenidate on regional brain glucose metabolism in humans: Relationship to dopamine D$_2$ receptors. *Am. J. Psychiatry* **154,** 50–55.

Volkow, N. D., Wang, G. J., Overall, J. E., Hitzemann, R., Fowler, J. S., Pappas, N., Frecska, E., and Piscani, K. (1997c). Regional brain metabolic response to lorazepam in alcoholics during early and late alcohol detoxification. *Alcohol. Clin. Exp. Res.* **21,** 1278–1284.

Volkow, N. D., Fowler, J. S., Gatley, S. J., Dewey, S. L., Wang, G. J., Logan, J., Ding, Y. S., Franceschi, D., Gifford, A., Morgan, A., Pappas, N., and King, P. (1999a). Comparable changes in synaptic dopamine induced by methylphenidate and by cocaine in the baboon brain. *Synapse* **31,** 59–66.

Volkow, N. D., Wang, G. J., Fowler, J. S., Gatley, S. J., Logan, J., Ding, Y. S., Dewey, S. L., Hitzemann, R., Gifford, A. N., and Pappas, N. R. (1999b). Blockade of striatal dopamine transporters by intravenous methylphenidate is not sufficient to induce self-reports of "high." *J. Pharmacol. Exp. Ther.* **288,** 14–20.

Volkow, N. D., Wang, G., Fowler, J., Fischman, M., Foltin, R., Abumrad, N. N., Gatley, S. J., Logan, J., Wong, C., Gifford, A., Ding, Y. S., Hitzemann, R., and Pappas, N. (1999c). Methylphenidate and cocaine have a similar *in vivo* potency to block dopamine transporters in the human brain. *Life Sci.* **65,** 7–12.

Vollenweider, F. X., Maguire, R. P., Leenders, K. L., Mathys, K., and Angst, J. (1998). Effects of high amphetamine dose on mood and cerebral glucose metabolism in normal volunteers using positron emission tomography (PET). *Psychiatry Res.* **83,** 149–162.

Wallace, E. A., Wisniewski, G., Zubal, G., van Dyck, C. H., Pfau, S. E., Smith, E. O., Rosen, M. I., Sullivan, M. C., Woods, S. W., and Kosten, T. R. (1996). Acute cocaine effects on absolute cerebral blood flow. *Psychopharmacology* (*Berl.*) **128,** 17–20.

Wang, G. J., Volkow, N. D., Fowler, J. S., Ferrieri, R., Schlyer, D. J., Alexoff, D., Pappas, N., Lieberman, J., King, P., Warner, D., *et al.* (1994). Methylphenidate decreases regional cerebral blood flow in normal human subjects. *Life Sci.* **54,** 143–146.

Wang, G. J., Volkow, N. D., Fowler, J. S., Logan, J., Abumrad, N. N., Hitzemann, R. J., Pappas, N. S., and Pascani, K. (1997). Dopamine D$_2$ receptor availability in opiate-dependent subjects before and after naloxone-precipitated withdrawal. *Neuropsychopharmacology* **16,** 174–182.

Wang, G. J., Volkow, N. D., Fowler, J. S., Pappas, N. R., Wong, C. T., Pascani, K., Felder, C. A., and Hitzemann, R. J. (1998). Regional cerebral metabolism in female alcoholics of moderate severity does not differ from that of controls. *Alcohol. Clin. Exp. Res.* **22,** 1850–1854.

Wang, G. J., Volkow, N. D., Fowler, J. S., Cervany, P., Hitzemann, R. J., Pappas, N. R., Wong, C. T., and Felder, C. (1999). Regional brain metabolic activation during craving elicited by recall of previous drug experiences. *Life Sci.* **64,** 775–784.

Weber, D. A., Franceschi, D., Ivanovic, M., Atkins, H. L., Cabahug, C., Wong, C. T., and Susskind, H. (1993). SPECT and planar brain imaging in crack abuse: Iodine-123-iodoamphetamine uptake and localization. *J. Nucl. Med.* **34,** 899–907.

Wik, G., Borg, S., Sjogren, I., Wiesel, F. A., Blomqvist, G., Borg, J., Greitz, T., Nyback, H., Sedvall, G., Stone-Elander, S., *et al.* (1988). PET determination of regional cerebral glucose metabolism in alcohol-dependent men and healthy controls using ^{11}C-glucose. *Acta Psychiatr. Scand.* **78,** 234–241.

Wolkin, A., Angrist, B., Wolf, A., Brodie, J., Wolkin, B., Jaeger, J., Cancro, R., and Rotrosen, J. (1987). Effects of amphetamine on local cerebral metabolism in normal and schizophrenic subjects as determined by positron emission tomography. *Psychopharmacology* (*Berl.*) **92,** 241–246.

Zubieta, J. K., Gorelick, D. A., Stauffer, R., Ravert, H. T., Dannals, R. F., and Frost, J. J. (1996). Increased mu opioid receptor binding detected by PET in cocaine-dependent men is associated with cocaine craving. *Nat. Med.* **2,** 1225–1229.

V

Therapeutics and Recovery of Function

23

Plasticity

Mark Hallett

*Human Motor Control Section, National Institute of Neurological Disorders and Stroke,
National Institutes of Health, Bethesda, Maryland 20892*

I. Introduction

II. Methods

III. Peripheral Lesions

IV. Spinal Cord Lesions

V. Brain Lesions

VI. Activity/Learning

VII. Blind, Cross-Modal Plasticity

VIII. Severe Sensory Neuropathy

IX. Maladaptive Plasticity

References

I. Introduction

The brain has millions of neurons connected in a highly specific fashion by billions of synapses. The components are put into place during the complex process of development. It is difficult to imagine that the brain could be reorganized, and for many years that was the generally held belief. Now it is clear that the brain is not only capable of reorganizing itself, but is in fact constantly doing so. An obvious example is learning; new knowledge or a new motor skill must mean that the brain has changed. Plastic changes are also induced in a number of pathological conditions, including injury to

the peripheral nervous system and brain injury. Reorganization is now recognized to be the principal process responsible for spontaneous recovery of function after stroke (see Chapter 24).

There are a number of mechanisms for plasticity in humans. Their physiology has been studied in model systems, and we have only incomplete knowledge as to which mechanism applies to which phenomenon.

First, a change in the balance of excitation and inhibition can happen very quickly. This process depends on the fact that neurons or neural pathways have a much larger region of anatomical connectivity than their usual territory of functional influence. Some zones may be kept in check by tonic inhibition. If the inhibition is removed, the region of influence can be quickly increased or unmasked; and this process is often called unmasking. For example, following the application of the GABA antagonist bicuculline to the forelimb area of the rat motor cortex, stimulation of the adjacent vibrissa area led to forelimb movements; this strongly suggests that GABAergic neurons are crucial to the maintenance of cortical motor representations (Jacobs and Donoghue, 1991).

A second process that can also be relatively fast is strengthening or weakening of existing synapses, in processes such as long-term potentiation (LTP) or long-term depression (LTD). LTP or LTD occurs following specific patterns of synaptic activity and may last for long periods of time. Such processes can occur in the motor cortex (Hess and Donoghue, 1996; Hess *et al.*, 1996).

A third process is a change in neuronal membrane excitability. One possible mechanism for this is an alteration of sodium channels in neuronal membranes, which has been demonstrated to occur with operantly conditioned changes in the H-reflex (Halter *et al.*, 1995).

A fourth process is anatomical changes, which must need a longer period of time. Specific anatomical changes include sprouting of new axon terminals and formation of new synapses. For example, there can be increases in synaptic density that would strengthen a pre-existing, but previously weak connection. After long-term thalamic stimulation in adult cats, synaptic proliferation can be identified in the motor cortex (Keller *et al.*, 1992). In extreme circumstances, there can be growth of new connections into new regions. This was demonstrated by Pons *et al.* (1991) after long-term deafferentation of a limb in adult monkeys.

II. Methods

Information about plasticity in humans has been growing rapidly with the use of a number of noninvasive methods. The basic approach is to localize and characterize the brain representation area for specific functions and then to see if this changes in different circumstances. This has been done most extensively for the motor system, which has a well-understood brain topography. Moreover, physiological methods for studying the motor system allow measurements of its excitability.

Functional neuroimaging with PET and fMRI depends on changes in rCBF to demarcate a functional area. They are indirect measures of synaptic activity, depending on well-behaved blood flow responses to metabolic needs. Generally, they have good spatial resolution and poor temporal resolution. One problem with PET is that the magnitude of the blood flow change in a region is confounded by the size of the responsive region. For example, an apparent enlargement of a region may just be an increase in metabolic activity. This is less of a problem with fMRI.

EEG and MEG are also measures of synaptic activity, but they are direct measures in real time. The problem is that the sources of the EEG are ambiguous. This has been enough of a problem that these methods are not often used. In the future, the merger of EEG/MEG with PET/fMRI will be stronger than either type of method alone.

Transcranial magnetic stimulation (TMS) has been a valuable tool for studying plasticity. This method can map muscle or movement representations of the primary motor cortex (M1) with a high degree of precision. Similar to PET, however, assessing representation size is confounded with excitability. A change in the representation area is best demonstrated by a shift of the location of the map rather than a change in concentric size. In attempting to define the"central"position of a map, one possible position is the point of producing the motor evoked potential (MEP) of the largest amplitude. Typically, this is one point, although it could be two positions (or even more). Another choice, often better, is the center of gravity. This gives a position that takes all positions into account and gives an average weighted by the amplitude at each position. There is excellent concordance of TMS mapping of muscle representations and PET and fMRI (Wassermann *et al.*, 1996; Krings *et al.*, 1997).

Various TMS measures can evaluate different aspects of cortical excitability. *Threshold* for producing a MEP reflects the excitability of a central core of neurons that arises from the excitability of individual neurons and their local density. The *recruitment curve* is the growth of MEP size as a function of stimulus intensity and background contraction force. This measurement is less well understood but must involve neurons in addition to the core region activated at threshold. These neurons have a higher threshold for activation, either because they are intrinsically less excitable or because they are spatially further from the center of activation by the magnetic stimulus. These neurons would be part of the"subliminal fringe."*Intracortical inhibition* (*ICI*) and *intracortical facilitation* (*ICF*) are obtained with paired pulse studies and reflect interneuron influences in the cortex (Kujirai *et al.*, 1993; Ziemann *et al.*, 1996). ICI, for example, is likely largely a GABAergic effect.

TMS can also temporarily deactivate a region of the brain, and this effect can be used for mapping or studies of function. While single pulses can be used for this purpose, many effects of this type require repetitive TMS (rTMS).

These methods have been utilized to gather considerable information about the organization of M1 in humans. Different parts of the body such as the arm, leg, and face have clearly separate representations, but muscles within the same body part are intermingled (Wassermann *et al.*, 1992). This differs from the sensory cortex, which is organized in a simple topographic fashion. Thus, for example, maps in the motor cortex for the muscles moving the fingers are essentially the same, while the topography in the sensory cortex for the different fingers is laid out like the keys of a piano.

Also with these methods, it is possible to show in normal individuals a role of ipsilateral pathways for motor control. Although in normal subjects it is rare to record responses to TMS in resting ipsilateral muscles, ipsilateral silent periods (pauses in EMG activity) may be readily observed in voluntarily activated muscles (Wassermann *et al.*, 1991). Ipsilateral MEPs on a background of voluntary activation are also seen in some subjects, and these are usually more prominent in proximal muscles. When MEPs are present ipsilaterally, their map is not identical to that of the homologous muscle contralaterally (Wassermann *et al.*, 1994). For example, when mapping the ipsilateral first dorsal interosseous (FDI) the largest ipsilateral MEPs occurred with TMS at more sites more lateral than the contralateral FDI, near the representation of the contralateral risorius. With functional imaging, ipsilateral motor activation is seen, but generally only when the movement is complex. With EEG, ipsilateral activation of the motor cortex region is only slightly less than contralateral activation even with simple, self-paced movements (Toro *et al.*, 1993). The role of normal ipsilateral activation is not always clear, and some of the activity could be inhibitory in nature, to "suppress" mirror movements.

III. Peripheral Lesions

The pioneering work of Merzenich and colleagues in primates demonstrated that the organization of the sensory cortex changed after deafferentation of a limb (Merzenich *et al.*, 1984; Merzenich and Jenkins, 1993). Similar observations were made by Donoghue *et al.* for the motor cortex (Donoghue *et al.*, 1990; Donoghue, 1995). These studies demonstrated that the deafferented (or deefferented) cortex did not stay idle, but was taken over by body representations adjacent to the deafferented body part. Our group has investigated this situation in the motor system of humans (Hallett *et al.*, 1993). We studied patients with traumatic, surgical, or congenital amputations of the arm at about the level of the elbow (Cohen *et al.*, 1991a). MEPs were recorded from the biceps and deltoid muscles immediately proximal to the stump and from the same muscles contralaterally. Motor representation areas targeting muscles ipsilateral and immediately proximal to the stump were larger than those for muscles contralateral to the stump (Fig. 1). The threshold for activation of muscles proximal and ipsilateral to the stump was decreased compared with muscles on the intact side. A decrease in threshold, increase in MEP size, and increase in map size are all related, and may be different manifestations of the same phenomenon. These results are consistent with the idea that the motor cortex for the muscles proximal to the amputation had expanded into the territory of the amputated part. Kew *et al.* (1994b) reported studies with PET confirming enlargement of the motor territory of muscles proximal to amputation (Fig. 2) and sensory region adjacent to the amputation (Kew *et al.*, 1997).

Findings from intraoperative rolandic cortex mapping during awake craniotomy for a tumor in a patient with a contralateral upper extremity amputation verified that this reorganization takes place (Ojemann and Silbergeld, 1995). This patient sustained a traumatic amputation at the mid-humerus 24 years previously. The motor representation of the face and jaw extended more superiorly than the sensory representation. Shoulder movements were evoked more laterally than usual

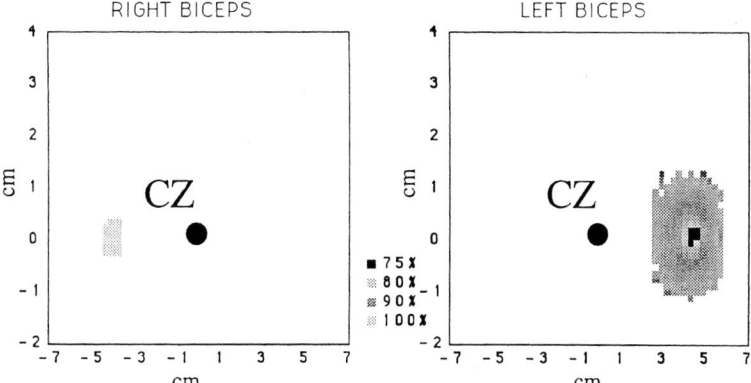

Figure 1 TMS mapping of left and right biceps from a 78-year-old subject with a left upper limb amputation. Testing was performed 11 months after amputation. Thresholds for activation at different scalp locations are indicated. Note that the muscle representation area for the left biceps is larger than that for the right biceps and that the threshold for activation of the left biceps was lower than that for the right biceps. Reproduced from Hallett *et al.* 1993, with permission from Springer-Verlag.

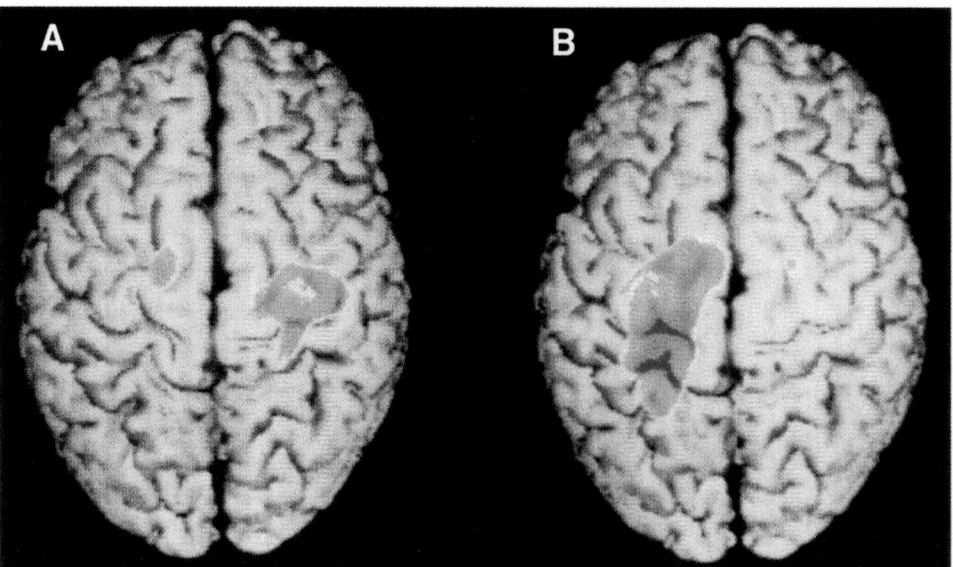

Figure 2 rCBF changes measured with PET in a traumatic above-elbow amputee. Areas of cortex in which significant increases in rCBF occurred during movement of the intact (A) and the amputated (B) arms are displayed on the cortical surface of the subject's T_1-weighted MRI volume. Significant increases in rCBF are in yellow and red. In the contralateral primary sensorimotor cortex increases were more significant and widespread during movement of the amputated arm than the intact arm. Reproduced from Kew *et al.*, 1994b, with permission of the American Physiological Society.

at the superior aspect of the craniotomy. A small region of the precentral gyrus, between the jaw and shoulder representations, elicited no detectable effect when stimulated. While somatosensory mapping showed a similar topographical distribution of face and mouth cortex, right arm and hand (phantom) sensations were evoked in areas posterior and inferior to the shoulder motor cortex. At least in this case, while reorganization was present in the motor cortex, it was not in the sensory cortex.

Similar findings are seen in Bell's palsy (Rijntjes *et al.*, 1997). With TMS, the size of the area producing MEPs of the abductor pollicis brevis (APB) muscle, the sum of MEP amplitudes within this area, and the volume over the mapping area were compared between both hemispheres in eight patients. With PET, increases in rCBF were compared between six patients and six healthy volunteers during sequential finger opposition. With both methods, a lateral enlargement of the hand area contralateral to the facial palsy was found, extending into the site of the presumed face area.

Two studies were performed to see how fast and at which level along the motor pathways modulation of motor outputs could occur following reversible deafferentation of a limb segment in humans. Reversible deafferentation was accomplished by using a blood pressure cuff. In the first study, a pneumatic double tourniquet was placed just below the elbow in normal

subjects; anesthesia was produced both by the ischemia and by a regional lidocaine infusion (Brasil-Neto *et al.*, 1992). Anesthesia was determined by the absence of voluntary finger and wrist movements, disappearance of MEPs from APB ipsilateral to the cuff, and absence of position sense and tactile sensation below the cuff. The amplitudes of MEPs to magnetic stimulation from muscles immediately proximal to the temporarily anesthetized forearm increased in minutes after the onset of anesthesia and returned to control values after the anesthesia subsided (Fig. 3).

In the second study, in order to determine the level at which the early modulation of motor outputs took place, we recorded maximal H reflexes, peripheral M responses, MEPs to TMS, and MEPs to transcranial electrical stimulation (TES) and spinal electrical stimulation (SES) from a muscle immediately proximal to a limb segment made ischemic by a pneumatic tourniquet (Brasil-Neto *et al.*, 1993). A pneumatic tourniquet was placed just below the subject's right knee. The amplitudes of MEPs to TMS, but not to TES and SES, were larger during ischemia. Reflecting the same process, the map, obtained with TMS, of cortical representation areas for the muscle proximal to the ischemic limb segment was also enlarged. The maximal H/M ratios were unaffected by ischemia, indicating that alpha-motoneuron excitability to segmental Ia inputs remained unchanged. TES appears to have much of its ef-

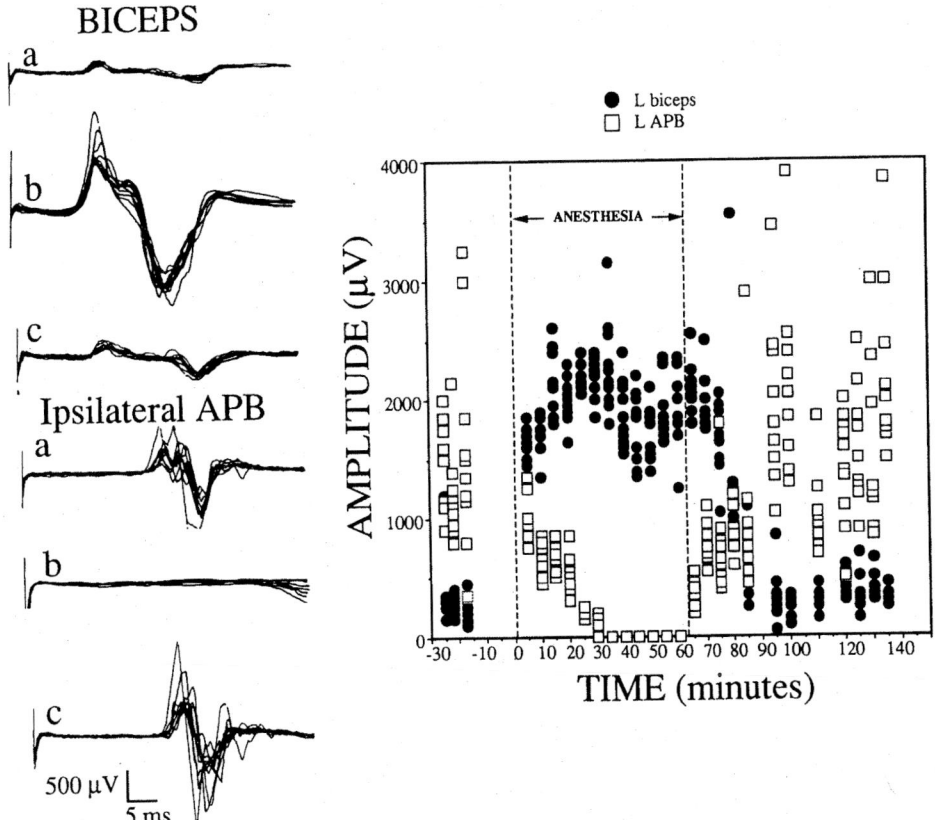

Figure 3 Effect of anesthesia below the elbow on biceps MEPs. (Left) Ten superimposed MEPs from biceps and the ipsilateral abductor pollicis brevis before anesthesia (a), during anesthetic block (b), and after anesthesia (c) in one subject. (Right) Amplitudes of MEPs from biceps and APB as a function of the time course of the experiment in the same subject. Reproduced from J. P. Brasil-Neto, L. G. Cohen, A. Pascual-Leone, F. K. Jabir, R. T. Wall, and M. Hallet (1992). Rapid reversible modulation of human motor outputs after transient deafferentation of the forearm: A study with transcranial magnetic stimulation. *Neurology* **42,** 1302–1306.

fect by direct stimulation of axons of corticospinal tract neurons, whereas TMS appears to have much of its effect due to intracortical influences upon the excitability of the neuronal cell body (or initial segment of the axon). The fact that MEPs changed after TMS, but not TES, strongly suggests that cortical neuronal nets targeting corticospinal tract neurons are the sites where most of the modulation occurs.

Sadato *et al.* confirmed cortical changes in this ischemic model using PET (Sadato *et al.,* 1995). Transient anesthesia of the right forearm caused an increase of rCBF in the primary sensorimotor area bilaterally at rest, but there was no change of rCBF with movement, indicating that the movement-related change in CBF was reduced (Fig. 4). These findings are consistent with increased excitability of neurons as a result of deafferentation.

Mano and colleagues showed that projections from the biceps region of the motor cortex can be directed to the spinal cord neurons of intercostal nerves in patients with brachial plexus avulsion after the intercostal nerve is anastomosed to the musculocutaneous nerve (Mano *et al.,* 1995). Moreover, they have shown that eventually the biceps can be controlled separately from respiration, demonstrating that control of the spinal neurons has been completely altered as a result of the brain plasticity. This process took a year or more to occur.

It is likely that the physiology of the acute and chronic states differs. We sought to understand these differences by using similar TMS methods of assessment in the situation of amputation and ischemic limb deafferentation. Ziemann *et al.* studied seven normal subjects with ischemic nerve block (Ziemann *et al.,* 1998). Baseline excitability measures were obtained in the biceps brachii muscle. Then a tourniquet was inflated above systolic blood pressure across the elbow distal to the biceps brachii. The pressure level was kept constant until complete block to the APB was achieved

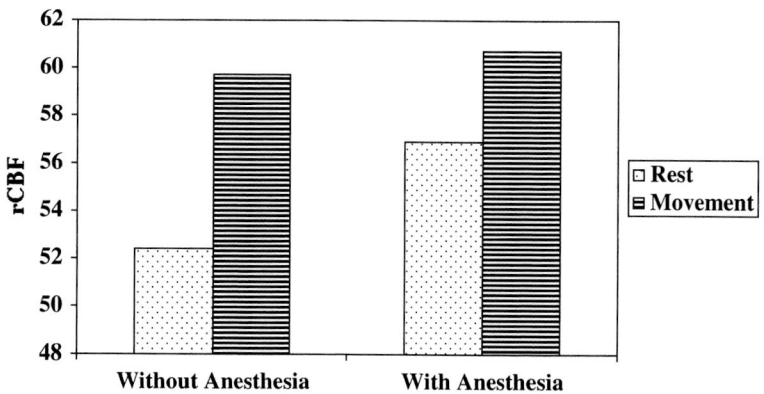

Figure 4 Effect of anesthesia at the level of the elbow on rCBF assessed with PET in the contralateral primary sensorimotor area. Measurements are made at rest and with repetitive movements of the elbow. Note a large increase in rCBF at rest caused by the anesthesia, but little change with movement. From N. Sadato, T. A. Zeffiro, G. Campbell, J. Konishi, H. Shibasaki, and M. Hallet (1995). Regional cerebral blood flow changes in motor cortical areas after transient anesthesia of the forearm. *Annals of Neurology* **37**, 74-81.

$(31.7 \pm 3.8 \text{ min})$. After an additional 5 min, the excitability parameters were re-measured. Immediately thereafter, the tourniquet was gradually released over a period of 2–3 min. The excitability parameters were remeasured 20, 40, and 60 min after tourniquet deflation was started. Chen *et al.* investigated 14 subjects aged 25 to 76 years with unilateral amputation and two subjects aged 48 and 60 years with bilateral amputation of the lower limbs (Chen *et al.,* 1998). The amputations were above the knee in nine subjects and below the knee in seven subjects. Amputation had occurred 7 months to 53 years earlier, and a prosthesis was used by 13 of the 16 subjects. None of the subjects had any neurological disorders. The quadriceps femoris was investigated.

With ischemia, there was no change in MEP threshold, ICI, and ICF. MEP amplitude was increased. There was no significant effect of stimulation intensity and the MEPs from the three intensities were pooled. There was a 39% increase in amplitude which lasted for the 60-min period of observation following cuff deflation.

With amputation, there was a lowering of threshold. For the patients with unilateral amputation, the threshold averaged 76% of maximal stimulator output on the amputated side and 89% on the intact side. The maximum MEP, measured in terms of the percentage of maximum compound muscle action potential, was increased. Given the way the measurement was made, the increase in MEP amplitude may well be due to the decreased threshold (since the stimulus for the amputated side would be a higher percentage above the motor threshold). On the other hand, it has been demonstrated previously that there is a more rapid rise in

MEP amplitude with increasing TMS intensities (Ridding and Rothwell, 1997). The results of the paired TMS studies for the normal subjects and the intact side of the amputees were similar, both showing inhibition at interstimulus intervals of 2 and 4 ms, followed by facilitation at interstimulus intervals of 10 to 30 ms (Fig. 1). The amputated side showed more facilitation at all intervals from 2 to 30 ms compared with the intact side and with normal subjects. For example, for the inhibitory interstimulus intervals of 2 and 4 ms, the test MEP amplitude on the amputated side was 240% of control compared with 60% both on the intact side and for normal subjects. At the facilitatory interstimulus intervals of 10 and 15 ms, the facilitation of the test MEP was 384% on the amputated side compared with 285% on the intact side and with 296% for normal subjects (but these latter differences were not formally significant).

Thus, with acute deafferentation, we detected changes only for MEP size, which reflects a change in the recruitment curve. Since there was no change in threshold, only the higher threshold cortical neurons are affected. Given the rapidity of the change, this must be due to unmasking of these neurons. With chronic deafferentation, there are additional changes in threshold and intracortical inhibition and facilitation, indicating changes in the low-threshold neurons, including level of excitability and synaptic changes. These changes, which have additional time to occur, must be due to processes like LTP and new synapse formation. Large-scale sprouting of cortical connections has been seen in a primate study (Florence *et al.,* 1998). It is important to realize that while many of the measurements are made on the cortex, subcortical changes are also possible and would be reflected in cortical measurements (Jones and Pons, 1998).

IV. Spinal Cord Lesions

Using TMS, investigators have looked for plastic changes in muscles proximal to the level of spinal cord injury. For both cervical (Levy *et al.,* 1990) and thoracic spinal injury (Topka *et al.,* 1991), there is increased excitability of these muscles identified by larger MEP amplitude and larger map size (Fig. 5).

V. Brain Lesions

A. Hemispherectomy

After hemispherectomy, motor function in the limb contralateral to the excised hemisphere experiences a substantial degree of recovery, particularly when surgery is performed at early age. To understand the mechanisms underlying this recovery of function, we studied patients with hemispherectomy (Cohen *et al.,* 1991b). TMS of the remaining hemisphere induced bilateral activation of deltoid and biceps. Similar findings were obtained by Benecke *et al.* (1991). Evaluation of MEPs indicated that muscles ipsilateral to the preserved hemisphere were activated by stimulation of

scalp positions anterior and lateral to those activating muscles on the normal side. Similarly, ipsilateral elbow movements were associated with CBF increases in an area centered slightly anterior and lateral to that activated by the same movements on the normal side (Fig. 6). These results indicate that ipsilateral and contralateral representations in the remaining hemisphere are topographically differentiated, with ipsilateral representations having a more anterior and lateral scalp distribution. The anatomy suggests that the normal ipsilateral representation has become more influential in these patients and has likely contributed to the recovery.

B. Tumors

PET was used to map the rCBF changes related to voluntary finger movements in patients with tumors in the hand area of motor cortex (Seitz *et al.,* 1995). Compared with the unaffected side, the activations were shifted by 9–43 mm either along the mediolateral body representation of motor cortex or into premotor or parietal somatosensory cortex (Fig. 7). These results provided evidence that slowly developing lesions can induce large-scale reorganization that is not confined to changes within the somatotopic body representation in motor cortex. An additional six patients with gliomas of the precentral gyrus were studied by the same group with both PET and TMS mapping (Wunderlich *et al.,*

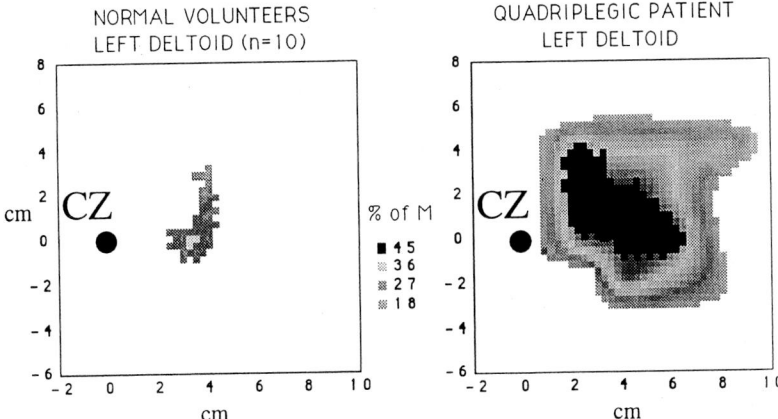

Figure 5 TMS maps showing views of the top of the head for a group of 10 normal volunteers and a patient with complete spinal cord injury at the C5 level that occurred 4 months before testing. The sizes of MEPs from the left deltoid (expressed as percentage of the alpha-motoneuron pool) evoked by TMS at different scalp positions are indicated. The group map displays the maximal size of the MEP evoked in the 10 control subjects after stimulation of each scalp location. The patient map displays the average size of three MEPs evoked after stimulation of each scalp position. Note that the MEP and motor representation area was larger for the left deltoid in the patient than the largest values of any individual in the control group. Reproduced from Hallett *et al.,* 1993, with permission from Springer-Verlag.

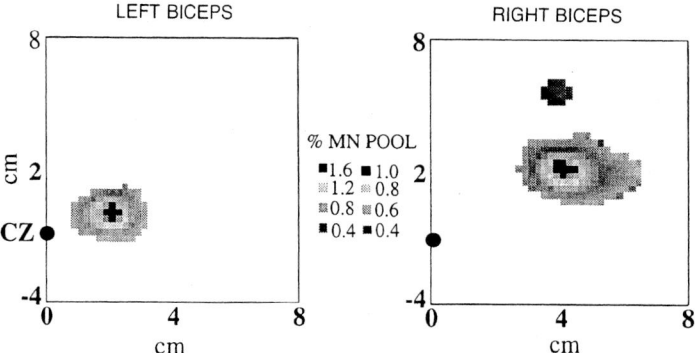

Figure 6 TMS maps of the motor representation areas for the left and right biceps on the right hemisphere of a patient who had a left hemispherectomy. The amplitudes of MEPs evoked by magnetic stimulation (expressed as percentage of the alpha-motoneuron pool) are indicated. Note that the cortical representation for the right biceps (on the ipsilateral hemisphere) was clearly differentiated and anterolateral to that of the left biceps. Reproduced from Hallett *et al.*, 1993, with permission from Springer-Verlag.

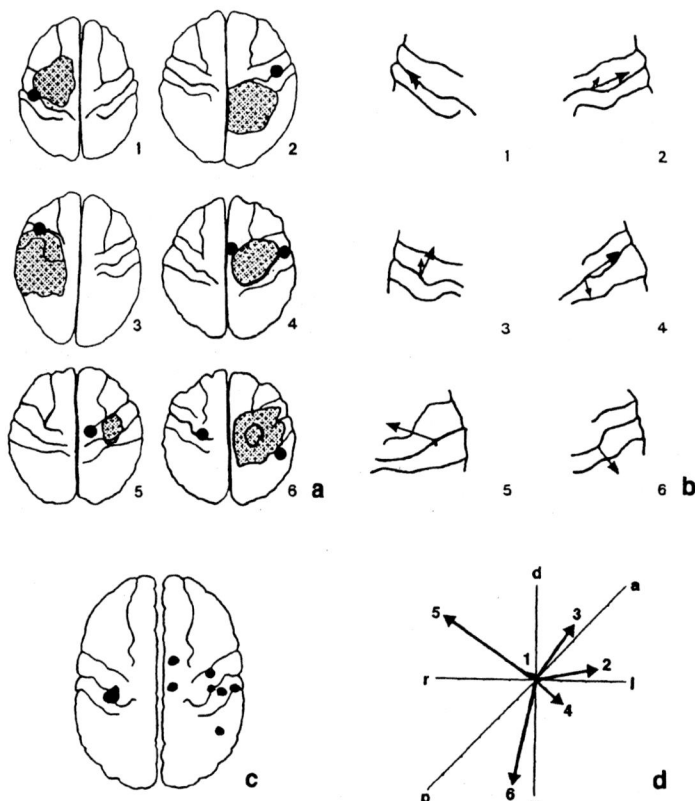

Figure 7 PET cortical motor activation sites produced by finger movements in patients with brain tumors. (a) Six individual subjects. The tumor is indicated by shading, and the point of maximal rCBF increase by a black dot. (b) More detail from the central sulcus region in the same six patients. A small arrow indicates the vectorial displacement of the central sulcus by the tumor, and a large arrow indicates the direction of shift of the activation site. (c) Superimposition of the sites from all six patients (right side of figure) compared with the normal position of activation (left side of figure). (d) Vectorial displacements of the rCBF increases in the affected hemisphere. The origin corresponds to the normal motor hand area ($x = 35$, $y = 20$, $z = 50$) (a, = anterior; p, = posterior; d, = dorsal; v, = ventral; r, = right; l, = left; unit = 10 mm). Reproduced from R. J. Seitz, Y. Huang, U. Knorr, L. Tellmann, H. Herzog, and H. J. Freund (1995). Large-scale plasticity of the human motor cortex. *NeuroReport* **6**, 742–744.

1998). Again movement-related increases of rCBF, compared with the contralateral side, were shifted by 20 ± 13 mm within the dorsoventral dimension of the precentral gyrus. Dorsal tumor growth resulted in ventral displacement of motor hand representation, leaving the motor cortical output system unaffected, whereas ventral tumor growth leading to dorsal displacement of motor hand representation compromised the motor cortical output, as evident from TMS.

In patients affected by intracerebral gliomas, ipsilateral pathways may play some role activated by the impairment of contralateral control. TMS was used to examine this idea in seven patients (Caramia *et al.*, 1998). While ipsilateral MEPs were generally absent in normal subjects, they were found in the patients, although with higher thresholds and longer latencies.

C. Amyotrophic Lateral Sclerosis (ALS)

PET was used to measure rCBF in ALS patients while they performed stereotyped and freely selected movements of a joystick with their right hand (Kew *et al.*, 1993). Compared with normal subjects, rCBF at rest was significantly reduced in the primary sensorimotor cortex, the lateral premotor cortex, the supplementary motor area, the anterior cingulate cortex, the paracentral lobule, and the superior and inferior parietal cortex. Comparison of the increase in rCBF caused by freely selected joystick movements over the resting state between the two groups of subjects showed significantly greater activation in ALS patients in the ventral third (face area) of the contralateral primary sensorimotor cortex and in the adjacent contralateral ventral premotor and parietal association cortices; significantly greater activation of the contralateral anterior insula and the ipsilateral anterior cingulate cortex (dorsocaudal area 24) was also present in ALS patients. The pattern of reduced rCBF at rest in ALS patients probably reflects a combination of neuronal loss in all areas of the cortex projecting through the pyramidal tract together with loss of projections from the sensorimotor cortex to the motor association areas. The expansion of the upper limb output zone of the sensorimotor cortex in ALS patients during contralateral upper limb movement may represent cortical reorganization in response to Betz cell loss or corticospinal tract disruption. Abnormal recruitment of nonprimary motor areas may also represent functional adaptation to a corticospinal tract lesion. In a subsequent study, patients with progressive lower motor neuron degeneration (LMND) were also studied (Kew *et al.*, 1994a). Patients with LMND had normal rCBF at rest in the primary sensorimotor cortex. During joystick movement, LMND patients showed significantly greater rCBF increases than controls only in the anterior insular cortex bilaterally.

The finding of reduced rCBF at rest, together with abnormal bilateral activation and altered somatotopy during movement, in the sensorimotor cortex of ALS but not LMND patients suggests that these abnormalities reflect loss of pyramidal neurons and resultant reorganization. Abnormal activation of perisylvian areas (insular and SII cortices) during limb movement in both LMND and ALS patients suggests that these may be accessory sensorimotor areas that are recruited nonspecifically in response to limb weakness.

D. Parkinson's Disease

Patients with Parkinson's disease have difficulty in performing long sequences of movement, presumably as one result of their nigrostriatal loss of dopamine and its resultant effects on cortical function. We used PET to assess the rCBF associated with the performance of well-learned sequences of finger movements of varying length (Catalan *et al.*, 1999). Sequential finger movements in the Parkinson's disease patients were associated with an activation pattern similar to that found in normal subjects, but they showed relative overactivity in the precuneus, premotor, and parietal cortices. Increasing the complexity of movements resulted in increased rCBF in the premotor and parietal cortices of normal subjects; the Parkinson's disease patients showed greater increases in these same regions and had additional significant increases in the anterior supplementary motor area (SMA)/cingulate. This last point has particular interest since performance of self-selected movements induces significant activation of the anterior SMA/cingulate in normal subjects but not in Parkinson's disease patients. Thus, in Parkinson's disease patients more cortical areas are recruited to perform sequential finger movements; this may be the result of increasing corticocortical activity to compensate for their striatal dysfunction.

VI. Activity/Learning

Cortical changes also result from changes in the patterns of behavior. Indeed, it appears that there is a constant battle for the control of each neuron among its various inputs. The function of each neuron or neuron pool will be determined by the dominant inputs resulting from several dynamic processes (Donoghue, 1995). To investigate this issue, we performed detailed mapping of the motor cortical areas targeting the FDI and the adductor digiti minimi (ADM) bilaterally in Braille readers and blind controls with focal TMS (Pascual-Leone *et al.*, 1993). Subjects and controls were matched for age and age of blindness. Subjects had learned Braille at age 8 to 14 years and used it daily for 5–10 h.

Controls had not learned Braille until age 17 to 21 years and used it daily for less than 1 h. Motor threshold of subjects and controls was similar. In the controls, motor representations of right and left FDI and ADM were not significantly different. However, in the proficient Braille readers, the representation of the FDI in the reading hand was significantly larger than that in the nonreading hand or in either hand of the controls (Fig. 8). Conversely, representation of the ADM in the reading hand was significantly smaller than that in the nonreading hand or in either hand of the controls. These results suggest that the cortical representation for the reading finger in proficient Braille readers is enlarged at the expense of the representation of other fingers.

We also studied six Braille proofreaders, comparing the maps obtained on a day in which they worked reading Braille for approximately 6 h with the maps obtained on a day after having been off from work for 2 days (Pascual-Leone *et al.*, 1995c). The maps for the FDI of the reading hand were significantly larger in the former case than in the latter case. These results illustrate the rapid modulation in motor cortical outputs in relation to preceding activity.

On the other hand, Leipert and colleagues studied cortical plasticity in 22 patients who had a unilateral immobilization of the ankle joint without any peripheral nerve lesion (Liepert *et al.*, 1995). The motor cortex area of the inactivated tibial anterior muscle diminished compared to the unaffected leg, without changes in spinal excitability or motor threshold (Fig. 9). The area reduction was correlated to the duration of immobilization and could be quickly reversed by voluntary muscle contraction.

As in the setting of amputation, we can look at how fast such changes can occur, and for this purpose we have studied motor learning. Pascual-Leone *et al.*, looked at the motor cortical representation of the hand over a 5-day period in normal subjects as they learned a skilled task with their hand (Pascual-Leone *et al.*, 1995a). As subjects became more skilled at a five-finger exercise on a piano, the size of the motor representation of the hand increased (Fig. 10). This result is supported by a number of studies, including those of Grafton *et al.*, who showed that rCBF increased in primary motor area as subjects learned a pursuit rotor task (Grafton *et al.*, 1992, 1994). Enlargement of the representation of the hand has also been seen with fMRI for subjects who learned a finger-tapping task (Karni *et al.*, 1995).

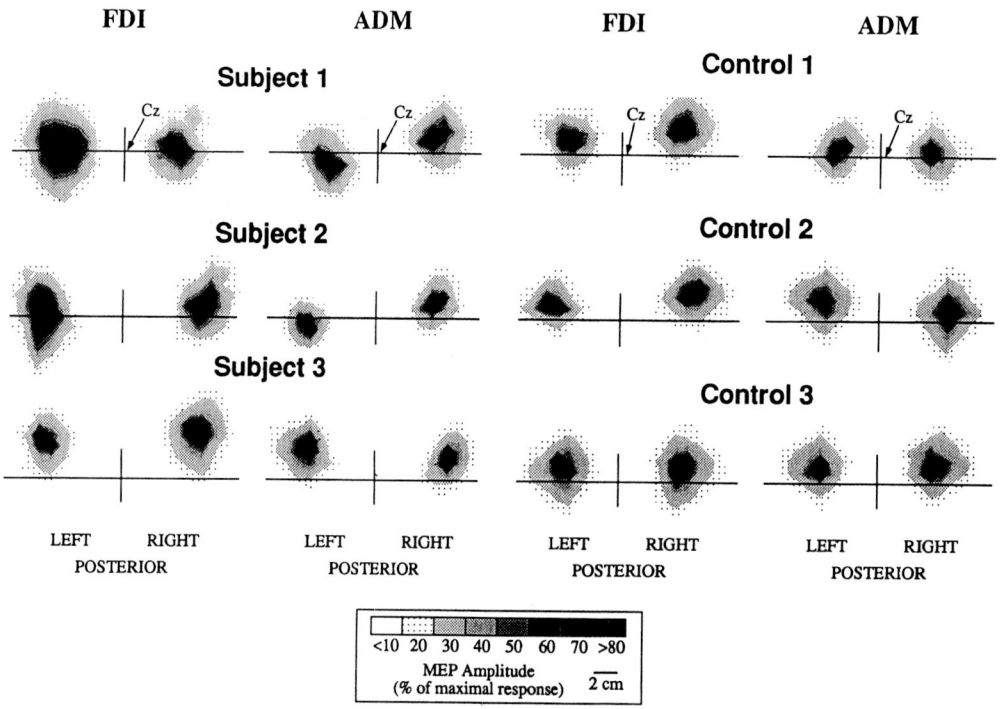

Figure 8 TMS maps of right and left first dorsal interosseous and abductor digiti minimi muscles in three proficient Braille readers and three blind controls. Braille readers 1 and 2 and control subjects 1 and 2 are right-handed, whereas the third Braille reader and control subject are left-handed. Contour maps represent the amplitude of the MEPs induced by focal TMS as a function of scalp position. Reproduced from A. Pascual-Leone, A. Cammarota, E. M. Wassermann, J. P. Brasil-Neto, L. G. Cohen, and M. Hallet (1993). Modulation of motor cortical outputs to the reading hand of Braille readers. *Annals of Neurology* **34,** 33–37.

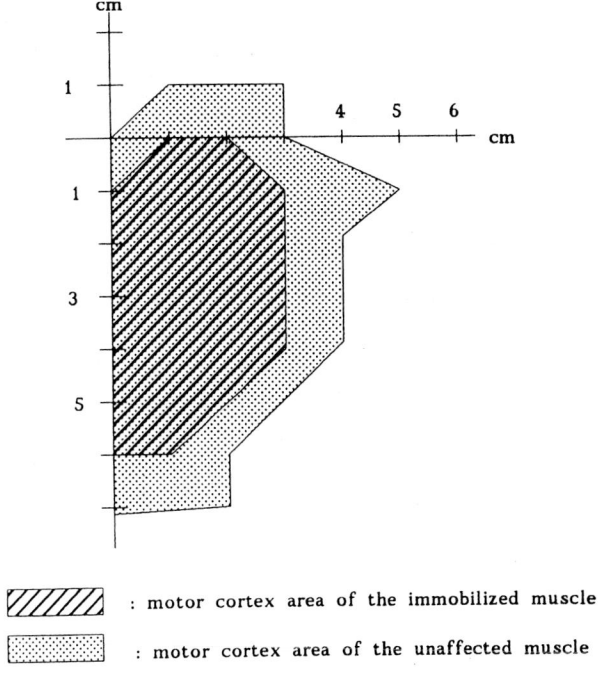

Legend:

///// : motor cortex area of the immobilized muscle

::::: : motor cortex area of the unaffected muscle

Figure 9 TMS maps of the tibialis anterior muscles bilaterally in a patient with unilateral immobilization of the ankle joint for 8 weeks. Reprinted from *Electroencephalogr. Clin. Neurophysiol.* (now *Clinical Neurophysiology*) **97**; J. Liepert, M. Tegenthoff, and J. P. Malin; Changes of cortical motor area size during immobilization, pp. 382–386. Copyright 1995, with permission from Elsevier Science.

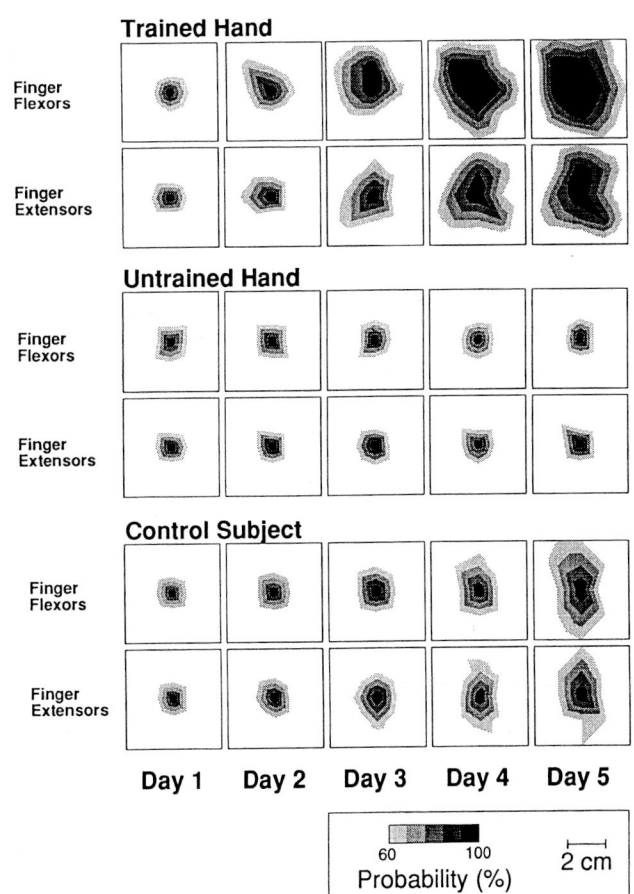

Figure 10 TMS maps of the long finger flexor and extensor muscles centered over the left sensorimotor cortex of single subjects over a series of 5 successive days. The trained hand practiced a five-finger exercise on a piano for 2 h each day. The untrained hand was the contralateral hand of the same subject. Maps for the control subject were for the hand that played individual notes on the piano for 2 h each day, but not a specific sequence. Each map is based on 25 measured points. Reproduced from Pascual-Leone *et al.*, 1995a, with permission of the American Physiological Society.

A. Adaptation Learning

Motor learning is a complex phenomenon with a number of different components (Hallett *et al.*, 1996; Hallett and Grafman, 1997). One component can be defined as a change in motor performance with practice, and it is composed of several different phenomena, including adaptation and skill learning. Adaptation is simply a change in the nature of the motor output whereas skill learning is the development of a new capability. Adaptation learning involves the cerebellum as demonstrated by several experiments that show deficits in adaptation in patients with cerebellar lesions. Such experimental paradigms include adaptation to lateral displacement of vision as produced by prism glasses and tasks with a change in the visual–motor gain.

Eye-blink conditioning is a form of motor learning and might fit the definition proposed here for adaptation learning. In nonhuman animal studies, eye-blink conditioning seems to require the cerebellum, at least for the expression and timing of the response. A number of groups have studied eye-blink conditioning in patients with cerebellar lesions and found them to be markedly deficient (Daum *et al.*, 1993; Topka *et al.*, 1993). Eye-blink conditioning has also been studied with neuroimaging. Several groups have shown activation of the cerebellum associated with learning, as well

as a number of other structures (Molchan *et al.*, 1994; Logan and Grafton, 1995; Blaxton *et al.*, 1996). In one study, the well-trained state was compared with an unconditioned state and was characterized by increased activity in the inferior cerebellar cortex/deep nuclei bilaterally, anterior cerebellar vermis, contralateral cerebellar cortex and pontine tegmentum, ipsilateral inferior thalamus/red nucleus, ipsilateral hippocampal formation, ipsilateral lateral temporal cortex, and bilateral ventral striatum (Logan and Grafton, 1995). Thus, among other structures, there are changes in the cerebellum with adaptation learning.

B. Skill Learning

As pointed out earlier, one area of brain associated with motor learning is the primary sensorimotor region.

This has been identified in a number of different paradigms, including finger-tapping sequences (Grafton *et al.*, 1992, 1994; Seitz and Roland, 1992; Schlaug *et al.*, 1994; Karni *et al.*, 1995). Other regions are also activated, depending in part on the task. These regions included, in a hand trajectory task, the premotor cortical regions bilaterally (Seitz *et al.*, 1994), in a pursuit rotor task, the supplementary motor area (SMA), thalamus, contralateral cingulate area, precuneate cortex, and the ipsilateral anterior cerebellum (Grafton *et al.*, 1992, 1994), and in a sequence of keypresses, the prefrontal, premotor, and parietal cortices and the cerebellum, all bilaterally (Jenkins *et al.*, 1994).

The serial reaction time test (SRTT) is a nice paradigm to study motor learning of sequences (Nissen and Bullemer, 1987; Pascual-Leone *et al.*, 1995b). The task is a choice reaction time, with typically four possible responses. The responses can be carried out by keypresses with four different fingers. A visual stimulus indicates which is the appropriate response. The completion of one response triggers the next stimulus. Each movement is simple and separate from the others so that the movement aspect of this task is different (and easier) from that of the other tasks considered previously such as finger tapping or piano playing. The trick in this task is that unbeknownst to the naive subject, the stimuli are a repeating sequence. With practice at this task, the responses get faster even though the subject has no conscious recognition that the sequence is repetitive. This is called implicit learning. With continuing practice and improvement, there is recognition that there is a sequence, but it may not be possible to specify what it is. Now knowledge is becoming explicit. With even more practice, the sequence can be specified and it has become declarative as well as procedural. Performance gets even better at this stage, but the subject's strategy can change since the stimuli can be anticipated.

We have examined the dynamic involvement of different brain regions in the SRTT using PET (Honda *et al.*, 1998). Test sessions consisted of 10 cycles of the same 10-item sequence. During the implicit learning phase, improvement of the reaction time was associated with increased activity only in the contralateral primary sensorimotor cortex (Fig. 11). Explicit learning, shown as a positive correlation of rCBF with the correct recall of the sequence, was associated with increased activity in the posterior parietal cortex, precuneus and premotor cortex bilaterally, as well as the supplementary motor area predominantly in the left anterior part, left thalamus, and right dorsolateral prefrontal cortex (Fig. 12). During the explicit learning phase, the reaction time was significantly correlated with activity in a part of the frontoparietal network. Similar results with a slightly different paradigm were seen by Grafton *et al.* (Grafton *et al.*, 1995; Hazeltine *et al.*, 1997). If after learning an SRTT, the body part performing the movements changed, activity in the inferior parietal cortex remained unchanged, suggesting that this area had encoded the sequence at an abstract level (Grafton *et al.*, 1998).

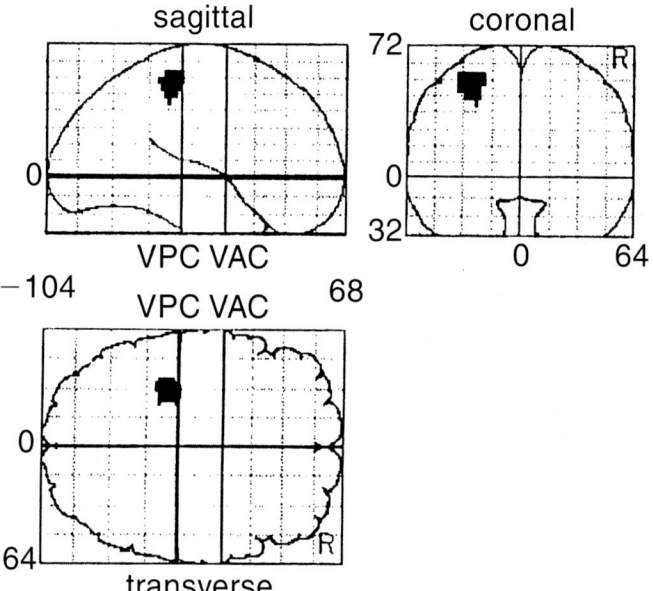

Figure 11 Statistical parametric map of PET study showing areas activated in the implicit learning phase of a serial reaction time task. The map shows regions of negative correlation of rCBF with normalized reaction time prior to any explicit knowledge of the sequence. Reproduced from Honda *et al.*, 1998, by permission of Oxford University Press.

In relation to the question of the involvement of the primary motor cortex in implicit learning, we mapped the motor cortex with TMS contralateral to the hands of normal subjects performing the SRTT (Pascual-Leone *et al.*, 1994). Mapping was done at intervals while the subjects were at rest between blocks of the SRTT. The map gradually enlarged during the implicit and explicit learning phase, but as soon as full explicit learning was achieved, the map size returned to baseline.

Thus in the SRTT, it appears that multiple structures in the brain are involved, but the involvement comes at different stages. The primary motor cortex appears to play a role in implicit learning. Premotor and parietal cortical areas play a role in explicit learning, and the parietal cortex specifically may store the sequence.

VII. Blind, Cross-Modal Plasticity

Sadato *et al.* studied rCBF in a group of subjects blind since early infancy and a group of sighted volunteers performing the same tactile discrimination task (Sadato *et al.*, 1996). We were seeking confirmation with PET of the findings mentioned earlier that blind subjects who read Braille have a larger motor representation of the reading fingers. We found what we were looking for (Sadato *et al.*, 1998) but also to our surprise that this task activated visual primary and association areas in the blind but not in normal volunteers. Primary visual cortex activation was verified using fMRI (Sadato

and Hallett, 1999) (Fig. 13). These results were interpreted as indicating that cortical areas normally reserved for vision may be activated by other sensory modalities. The functional neuroimaging study by itself could not prove that the activation was functionally useful in the discrimination process. To address this question we used repetitive transcranial magnetic stimulation (rTMS) applied over different scalp positions while blind and sighted subjects performed tactile identification of Braille and Roman letters (Cohen *et al.*, 1997). The hypothesis was that disruption of the reading task by rTMS to specific cortical regions would imply that that region was involved in performing the specific task. We found that in the blind subjects stimulation of occipital regions induced more errors in the reading task than stimulation of any other region or in control subjects. Therefore, not only are occipital areas in the blind activated in association with Braille reading tasks but they are an essential component of the network involved in Braille reading in the blind. These findings indicate the functional relevance of cross-modal plasticity in the blind.

Subsequently, we determined that this cross-modal plasticity depended on age of blindness (Cohen *et al.*, 1999). Individuals blinded after age 14 ("late blind") did not activate visual cortex when reading Braille, nor did rTMS influence Braille reading. The late blind subjects were not able to use the visual cortex to process somatosensory information. Thus, this form of dramatic plasticity is apparently only available to the young.

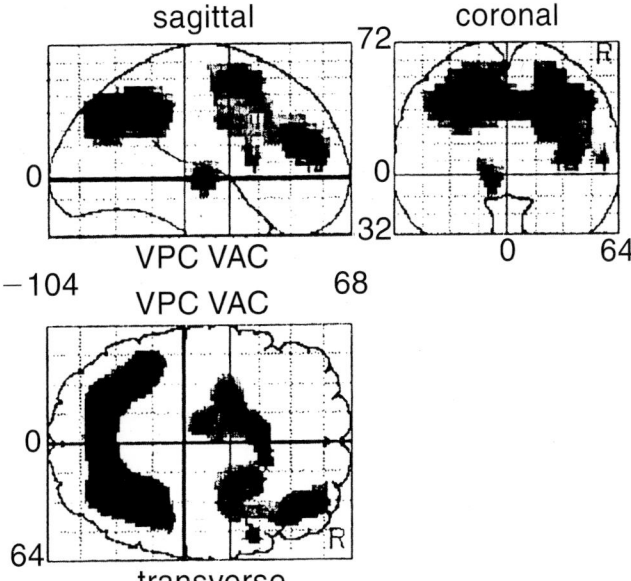

Figure 12 Statistical parametric map of PET study showing areas activated in the explicit learning phase of a serial reaction time task. The map shows regions of positive correlation of rCBF with report accuracy of the sequence. Reproduced from Honda *et al.*, 1998, by permission of Oxford University Press.

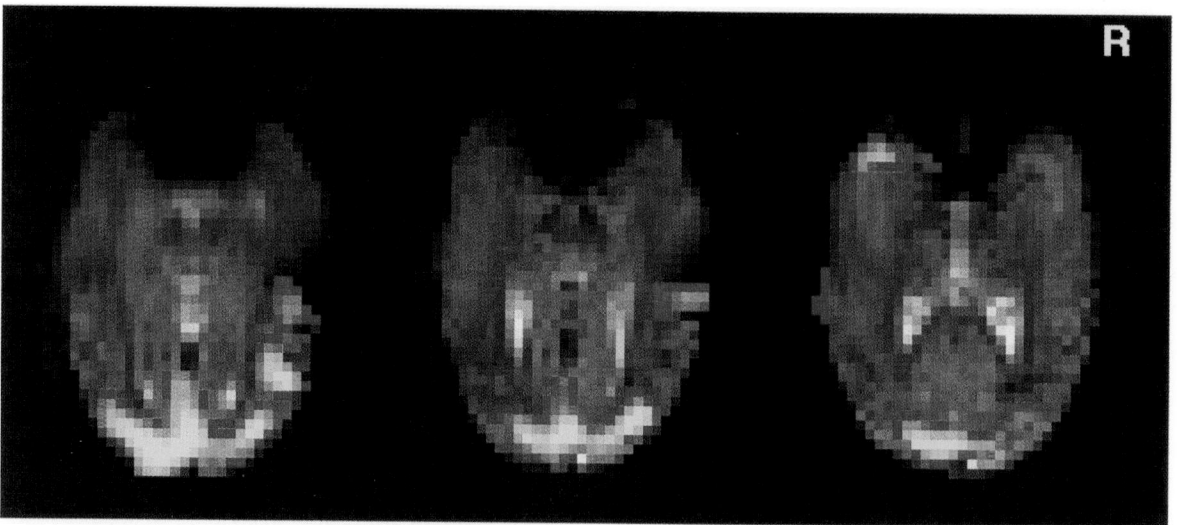

Figure 13 fMRI showing activation of the visual cortex when a congenitally blind person performs shape discrimination tasks with the right hand. The functional scans are superimposed on transaxial echo-planar MRI sections. Reproduced from N. Sadato and M. Hallett (1999). fRMI occipital activation by tactile stimulation in a blind man. *Neurology* **52,** 423.

VIII. Severe Sensory Neuropathy

We have examined patients with severe sensory neuropathies and normal muscle power, looking for evidence of transmodal plasticity in this circumstance (Weeks *et al.,* 1997). In addition, since the disabling symptoms of these patients can be substantially improved by visually monitoring their movements, we hypothesized that during visually guided movements, these patients would show overactivity of regions specialized for visuomotor control. We used PET to evaluate five conditions with the right hand: a sequential finger movement task under visual guidance; the same motor task without observation of the hand; monitoring a video of the same sequential finger movement; a passive visual task observing a reversing checkerboard; and an unconstrained rest condition. When the patient group was compared to the control group, activation was not deficient in any brain areas of the patient cohort in any of the contrasts tested. In particular, in the nonvisually guided movement task, in which meaningful visual and proprioceptive input was absent, the patient group activated primary motor, premotor, and cerebellar regions. This suggests that these areas are involved in motor processing independent of sensory input. In all conditions involving visual observation of hand movements, there was highly significant overactivity of the left parietal operculum (SII) and right parieto-occipital cortex (PO) in the patient group. Recent nonhuman primate studies have suggested that the PO region contains a visual representation of hand movements. Overactivity of this area and the activation of SII

by visual input appear to indicate that both compensatory overactivity of visual areas and cross-modal plasticity of somatosensory areas occur in deafferented patients.

IX. Maladaptive Plasticity

The genesis of occupational cramps such as writer's cramp or pianist's cramp has been extensively studied. There is common agreement that these disorders are focal dystonias (Cohen and Hallett, 1988; Berardelli *et al.,* 1998; Hallett, 1998a,b). There are likely several causes or contributing factors, and repetitive activity may be one. A possible animal model of dystonia was created in nonhuman primates with synchronous, widespread sensory stimulation to the hand during a repetitive motor task (Byl *et al.,* 1996, 1997). Over a period of months, the animals' motor performance deteriorated. After development of the movement disorder, the primary somatosensory cortex was mapped, and each cell was analyzed for the region of the body that activated the cell, its "receptive field." Receptive fields in area 3b were increased 10- to 20-fold, often extending across the surface of two or more digits. The investigators suggested that synchronous sensory input over a large area of the hand can lead to remapping of the receptive fields and subsequently to a movement disorder. However, these tasks also involve repetitive movements, which can lead to remapping of the motor system directly.

If this situation holds in human dystonia, then the somatosensory cortex might also be abnormal. Specifically, if the receptive fields have enlarged, they should overlap more than in the normal situation. Then the centers of the sensory receptive field maps from the different fingers in patients with dystonia should be closer together. We mapped the human cortical hand somatosensory area of six patients with focal dystonia of the hand, using noninvasive, high-resolution electrophysiologic and magnetic resonance imaging techniques (Bara-Jimenez *et al.,* 1998). The center of the receptive field from each finger was approximated by

calculating the electrical dipole source for the somatosensory evoked potential produced by cutaneous stimulation. We found degradation of the normal homuncular organization of the finger representations in the primary somatosensory cortex. The cortical finger representations in these patients were closer to each other than in normal subjects and often had an abnormal somatotopic arrangement (Fig. 14). These findings support the concept that abnormal plasticity is involved in the development of dystonia. Our findings have been

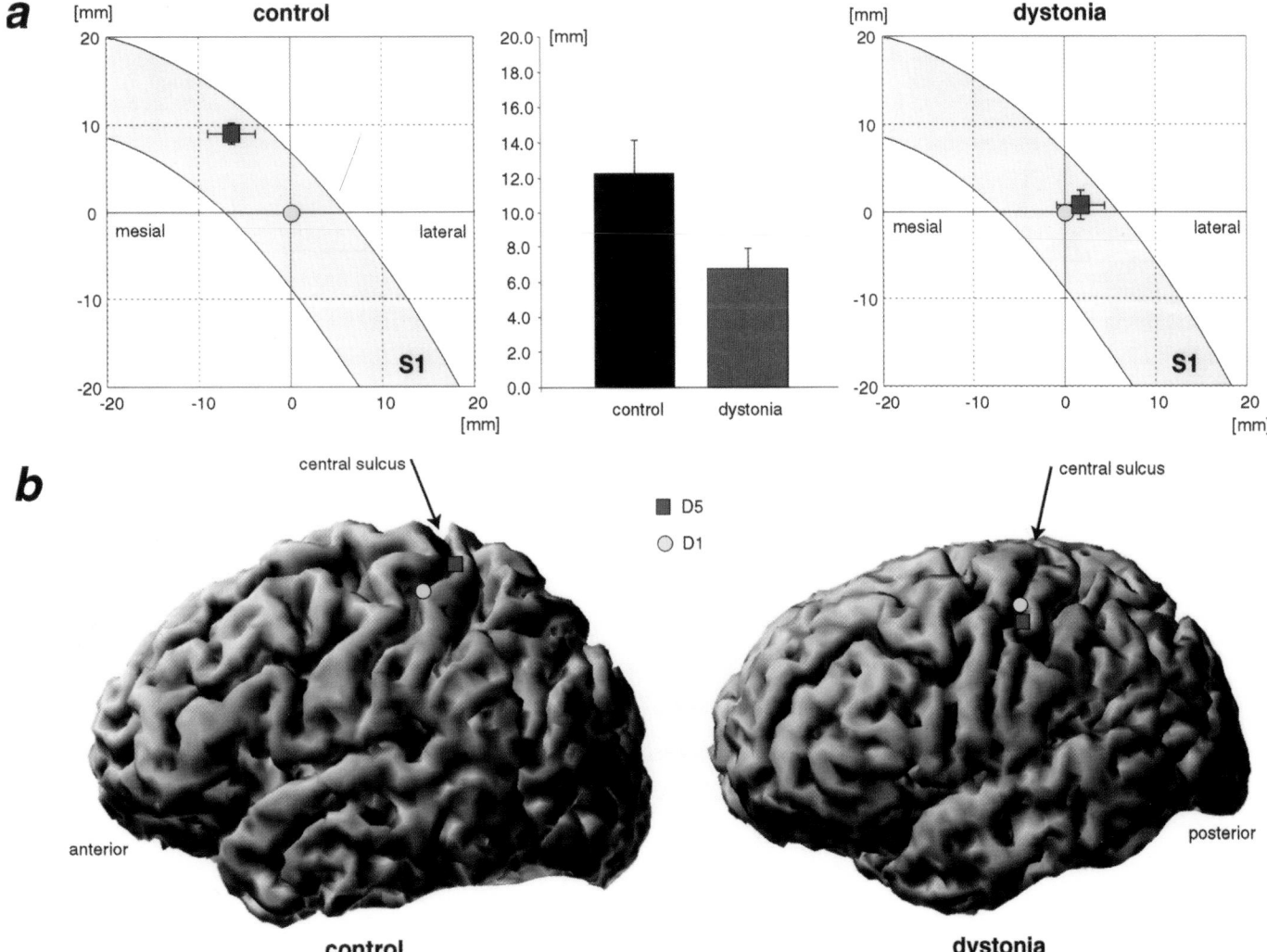

Figure 14 (a) Schematic drawings of the left postcentral gyrus (primary somatosensory cortex, S1) of six normal subjects (left) and six patients with hand dystonia (right) showing the group mean of the N20 source localization of the little finger (D5) relative to the thumb (D1) in the coronal plane. For spatial normalization across subjects, D1 served as the origin (0/0) of the coordinate system in each subject. Error bars indicate one standard error of the mean. Finger representations in the patients are located closer to each other than in the normal subjects ($p = 0.0163$). The bar graph in the middle shows group means and standard errors of the absolute distance between D1 and D5 in the coronal plane. Unlike the schematic drawings, the graph represents the absolute distance irrespective of the fingers' relative mediolateral position to each other. (b) Three-dimensional reconstruction of the brain of a normal subject (left) and a patient with hand dystonia (right). N20 source localizations for D1 and D5 are displayed. The normal cortical homuncular arrangement of the fingers is degraded in the patient. Reproduced from W. Bara-Jimenez, M.-J. Catalan, M. Hallett, and C. Gerloff (1998). Abnormal somatosensory homunculus in dystonia of the hand. *Annals of Neurology* **44,** 828–831.

recently reproduced using MEG (Elbert *et al.*, 1998). When doing thalamotomies in patients with dystonia, Lenz *et al.* (1999) have found enlarged receptive fields in thalamic neurons. These results are also consistent.

A similar situation is a possible human model of blepharospasm proposed in a case report of blepharospasm-like symptoms developing contralateral to an eyelid weakened by facial nerve palsy (Chuke *et al.*, 1996). With a weak lid, the central nervous system must increase the neural signal to the muscle to achieve eyelid closure. This is accomplished by increasing the output of the brainstem signal for a given command for eyelid closure. Such an increase in gain also affects the normal eyelid since changes in gain are always bilateral. Thus, hyperexcitability of the normal eyelid might be a maladaptive consequence of the weakness of the affected lid. This theory is supported by the observation that the eyelid spasms were eliminated by the implantation of a gold weight to assist closure of the paretic eyelid. Excitability of the brainstem controller can be assessed using the blink reflex recovery cycle. In this test, the amplitude of a second blink reflex is compared with a first one at short intervals. Ordinarily, there is inhibition of the second blink reflex for a period lasting several seconds. We have added additional support to the concept of brainstem hyperexcitability in facial palsy by finding reduced inhibition in the blink reflex recovery cycle, similar to that seen in patients with blepharospasm (Syed *et al.*, 1999).

It is also possible that phantom pain comes from maladaptive plasticity. Several studies have indicated that the amount of plasticity correlates with the subjective amount of pain (Knecht *et al.*, 1995, 1996; Flor *et al.*, 1998).

References

Bara-Jimenez, W., Catalan, M. J., Hallett, M., and Gerloff, C. (1998). Abnormal somatosensory homunculus in dystonia of the hand. *Ann. Neurol.* **44**, 828–831.

Benecke, R., Meyer, B. U., and Freund, H.-J. (1991). Reorganization of descending motor pathways in patients after hemispherectomy and severe hemispheric lesions demonstrated by magnetic brain stimulation. *Exp. Brain Res.* **83**, 419–426.

Berardelli, A., Rothwell, J. C., Hallett, M., Thompson, P. D., Manfredi, M., and Marsden, C. D. (1998). The pathophysiology of primary dystonia. *Brain* **121**, 1195–1212.

Blaxton, T. A., Zeffiro, T. A., Gabrieli, J. D., Bookheimer, S. Y., Carrillo, M. C., Theodore, W. H., *et al.* (1996). Functional mapping of human learning: A positron emission tomography activation study of eyeblink conditioning. *J. Neurosci.* **16**, 4032–4040.

Brasil-Neto, J. P., Cohen, L. G., Pascual-Leone, A., Jabir, F. K., Wall, R. T., and Hallett, M. (1992). Rapid reversible modulation of human motor outputs after transient deafferentation of the forearm: A study with transcranial magnetic stimulation. *Neurology* **42**, 1302–1306.

Brasil-Neto, J. P., Valls-Solé, J., Pascual-Leone, A., Cammarota, A., Amassian, V. E., Cracco, R., *et al.* (1993). Rapid modulation of human cortical motor outputs following ischemic nerve block. *Brain* **116**, 511–525.

Byl, N., Merzenich, M. M., and Jenkins, W. M. (1996). A primate genesis model of focal dystonia and repetitive strain injury: I. Learning-induced dedifferentiation of the representation of the hand in the primary somatosensory cortex in adult monkeys. *Neurology* **47**, 508–520.

Byl, N. N., Merzenich, M. M., Cheung, S., Bedenbaugh, P., Nagarajan, S. S., and Jenkins, W. M. (1997). A primate model for studying focal dystonia and repetitive strain injury: Effects on the primary somatosensory cortex. *Phys. Ther.* **77**, 269–284.

Caramia, M. D., Telera, S., Palmieri, M. G., Wilson-Jones, M., Scalise, A., Iani, C., *et al.* (1998). Ipsilateral motor activation in patients with cerebral gliomas. *Neurology* **51**, 196–202.

Catalan, M. J., Ishii, K., Honda, M., Samii, A., and Hallett, M. (1999). A PET study of sequential finger movements of varying length in patients with Parkinson's disease. *Brain* **122**, 483–495.

Chen, R., Corwell, B., Yaseen, Z., Hallett, M., and Cohen, L. G. (1998). Mechanisms of cortical reorganization in lower-limb amputees. *J. Neurosci.* **18**, 3443–3450.

Chuke, J. C., Baker, R. S., and Porter, J. D. (1996). Bell's Palsy-associated blepharospasm relieved by aiding eyelid closure. *Ann. Neurol.* **39**, 263–268.

Cohen, L. G., and Hallett, M. (1988). Hand cramps: Clinical features and electromyographic patterns in a focal dystonia. *Neurology* **38**, 1005–1012.

Cohen, L. G., Bandinelli, S., Findley, T. W., and Hallett, M. (1991a). Motor reorganization after upper limb amputation in man. *Brain* **114**, 615–627.

Cohen, L. G., Roth, B. J., Wassermann, E. M., Topka, H., Fuhr, P., Schultz, J., *et al.* (1991b). Magnetic stimulation of the human cerebral cortex, an indicator of reorganization in motor pathways in certain pathological conditions. *J. Clin. Neurophysiol.* **8**, 56–65.

Cohen, L. G., Celnik, P., Pascual-Leone, A., Corwell, B., Falz, L., Dambrosia, J., *et al.* (1997). Functional relevance of cross-modal plasticity in blind humans. *Nature* **389**, 180–183.

Cohen, L. G., Weeks, R. A., Sadato, N., Celnik, P., Ishii, K., and Hallett, M. (1999). Period of susceptibility for cross-modal plasticity in the blind. *Ann. Neurol.* **45**, 451–460.

Daum, I., Schugens, M. M., Ackermann, H., Lutzenberger, W., Dichgans, J., and Birbaumer, N. (1993). Classical conditioning after cerebellar lesions in humans. *Behav. Neurosci.* **107**, 748–756.

Donoghue, J. P. (1995). Plasticity of adult sensorimotor representations. *Curr. Opin. Neurobiol.* **5**, 749–754.

Donoghue, J. P., Suner, S., and Sanes, J. N. (1990). Dynamic organization of primary motor cortex output to target muscles in adult rats. II. Rapid reorganization following motor nerve lesions. *Exp. Brain Res.* **79**, 492–503.

Elbert, T., Candia, V., Altenmuller, E., Rau, H., Sterr, A., Rockstroh, B., *et al.* (1998). Alteration of digital representations in somatosensory cortex in focal hand dystonia. *NeuroReport* **9**, 3571–3575.

Flor, H., Elbert, T., Muhlnickel, W., Pantev, C., Wienbruch, C., and Taub, E. (1998). Cortical reorganization and phantom phenomena in congenital and traumatic upper-extremity amputees. *Exp. Brain Res.* **119**, 205–212.

Florence, S. L., Taub, H. B., and Kaas, J. H. (1998). Large-scale sprouting of cortical connections after peripheral injury in adult macaque monkeys. *Science* **282**, 1117–1121.

Grafton, S. T., Mazziotta, J. C., Presty, S., Friston, K. J., Frackowiak, R. S. J., and Phelps, M. E. (1992). Functional anatomy of human procedural learning determined with regional cerebral blood flow and PET. *J. Neurosci.* **12**, 2542–2548.

Grafton, S. T., Woods, R. P., and Tyszka, M. (1994). Functional imaging of procedural motor learning: Relating cerebral blood flow with individual subject performance. *Hum. Brain Mapp.* **1,** 221–234.

Grafton, S. T., Hazeltine, E., and Ivry, R. (1995). Functional mapping of sequence learning in normal humans. *J. Cog. Neurosci.* **7,** 497–510.

Grafton, S. T., Hazeltine, E., and Ivry, R. B. (1998). Abstract and effector-specific representations of motor sequences identified with PET. *J. Neurosci.* **18,** 9420–9428.

Hallett, M. (1998a). The neurophysiology of dystonia. *Arch. Neurol.* **55,** 601–603.

Hallett, M. (1998b). Physiology of dystonia. *In* "Dystonia 3. Advances in Neurology" (S. Fahn, C. D. Marsden, and M. DeLong, Eds.), Vol. 78, pp. 11–18. Lippincott-Raven, Philadelphia.

Hallett, M., and Grafman, J. (1997). Executive function and motor skill learning. *In* "The Cerebellum and Cognition"(J. D. Schmahmann, Ed.), Vol. 41, pp. 297–323. Academic Press, San Diego.

Hallett, M., Cohen, L. G., Pascual-Leone, A., Brasil-Neto, J., Wassermann, E. M., and Cammarota, A. N. (1993). Plasticity of the human motor cortex. *In* "Spasticity: Mechanisms and Management" (A. F. Thilmann, D. J. Burke, and W. Z. Rymer, Eds.), pp. 67–81. Springer-Verlag, Berlin.

Hallett, M., Pascual-Leone, A., and Topka, H. (1996). Adaptation and skill learning. Evidence for different neural substrates. *In* "Acquisition of Motor Behavior in Vertebrates" (J. R. Bloedel, T. J. Ebner, and S. P. Wise, Eds.), pp. 289–301. MIT Press, Cambridge.

Halter, J. A., Carp, J. S., and Wolpaw, J. R. (1995). Operantly conditioned motoneuron plasticity: Possible role of sodium channels. *J. Neurophysiol.* **73,** 867–871.

Hazeltine, E., Grafton, S. T., and Ivry, R. (1997). Attention and stimulus characteristics determine the locus of motor-sequence encoding. A PET study. *Brain* **120,** 123–140.

Hess, G., and Donoghue, J. P. (1996). Long-term depression of horizontal connections in rat motor cortex. *Eur. J. Neurosci.* **8,** 658–665.

Hess, G., Aizenman, C. D., and Donoghue, J. P. (1996). Conditions for the induction of long-term potentiation in layer II/III horizontal connections of the rat motor cortex. *J. Neurophysiol.* **75,** 1765–1778.

Honda, M., Deiber, M. P., Ibanez, V., Pascual-Leone, A., Zhuang, P., and Hallett, M. (1998). Dynamic cortical involvement in implicit and explicit motor sequence learning. A PET study. *Brain* **121,** 2159–2173.

Jacobs, K. M., and Donoghue, J. P. (1991). Reshaping the cortical motor map by unmasking latent intracortical connections. *Science* **251,** 944–947.

Jenkins, I. H., Brooks, D. J., Nixon, P. D., Frackowiak, R. S. J., and Passingham, R. E. (1994). Motor sequence learning: A study with positron emission tomography. *J. Neurosci.* **14,** 3775–3790.

Jones, E. G., and Pons, T. P. (1998). Thalamic and brainstem contributions to large-scale plasticity of primate somatosensory cortex. *Science* **282,** 1121–1125.

Karni, A., Meyer, G., Jezzard, P., Adams, M., Turner, R., and Ungerleider, L. G. (1995). Functional MRI evidence for adult motor cortex plasticity during motor skill learning. *Nature* **377,** 155–158.

Keller, A., Arissian, K., and Asanuma, H. (1992). Synaptic proliferation in the motor cortex of adult cats after long-term thalamic stimulation. *J. Neurophysiol.* **68,** 295–308.

Kew, J. J., Leigh, P. N., Playford, E. D., Passingham, R. E., Goldstein, L. H., Frackowiak, R. S., *et al.* (1993). Cortical function in amyotrophic lateral sclerosis. A positron emission tomography study. *Brain* **116,** 655–680.

Kew, J. J., Brooks, D. J., Passingham, R. E., Rothwell, J. C., Frackowiak, R. S., and Leigh, P. N. (1994a). Cortical function in progressive lower motor neuron disorders and amyotrophic lateral sclerosis: A comparative PET study. *Neurology* **44,** 1101–1110.

Kew, J. J., Ridding, M. C., Rothwell, J. C., Passingham, R. E., Leigh, P. N., Sooriakumaran, S., *et al.* (1994b). Reorganization of cortical blood flow and transcranial magnetic stimulation maps in human subjects after upper limb amputation. *J. Neurophysiol.* **72,** 2517–2524.

Kew, J. J., Halligan, P. W., Marshall, J. C., Passingham, R. E., Rothwell, J. C., Ridding, M. C., *et al.* (1997). Abnormal access of axial vibrotactile input to deafferented somatosensory cortex in human upper limb amputees. *J. Neurophysiol.* **77,** 2753–2764.

Knecht, S., Henningsen, H., Elbert, T., Flor, H., Hohling, C., Pantev, C., *et al.* (1995). Cortical reorganization in human amputees and mislocalization of painful stimuli to the phantom limb. *Neurosci. Lett.* **201,** 262–264.

Knecht, S., Henningsen, H., Elbert, T., Flor, H., Hohling, C., Pantev, C., *et al.* (1996). Reorganizational and perceptional changes after amputation. *Brain* **119,** 1213–1219.

Krings, T., Buchbinder, B. R., Butler, W. E., Chiappa, K. H., Jiang, H. J., Cosgrove, G. R., *et al.* (1997). Functional magnetic resonance imaging and transcranial magnetic stimulation: Complementary approaches in the evaluation of cortical motor function. *Neurology* **48,** 1406–1416.

Kujirai, T., Caramia, M. D., Rothwell, J. C., Day, B. L., Thompson, P. D., Ferbert, A., *et al.* (1993). Corticocortical inhibition in human motor cortex. *J. Physiol. (Lond.)* **471,** 501–519.

Lenz, F. A., Jaeger, C. J., Seike, M. S., Lin, Y. C., Reich, S. G., DeLong, M. R., and Vitek, J. L. (1999). Thalamic single neuron activity in patients with dystonia: Dystonia-related activity and somatic sensory reorganization. *J. Neurophysiol.* **82,** 2372–2392.

Levy, W. J., Jr., Amassian, V. E., Traad, M., and Cadwell, J. (1990). Focal magnetic coil stimulation reveals motor cortical system reorganized in humans after traumatic quadriplegia. *Brain Res.* **510,** 130–134.

Liepert, J., Tegenthoff, M., and Malin, J. P. (1995). Changes of cortical motor area size during immobilization. *Electroencephalogr. Clin. Neurophysiol.* **97,** 382–386.

Logan, C. G., and Grafton, S. T. (1995). Functional anatomy of human eyeblink conditioning determined with regional cerebral glucose metabolism and positron-emission tomography. *Proc. Natl. Acad. Sci. U.S.A.* **92,** 7500–7504.

Mano, Y., Nakamuro, T., Tamura, R., Takayanagi, T., Kawanishi, K., Tamai, S., *et al.* (1995). Central motor reorganization after anastomosis of the musculocutaneous and intercostal nerves in patients with traumatic cervical root avulsion. *Ann. Neurol.* **38,** 15–20.

Merzenich, M. M., and Jenkins, W. M. (1993). Reorganization of cortical representations of the hand following alterations of skin inputs induced by nerve injury, skin island transfers, and experience. *J. Hand Ther.* **6,** 89–104.

Merzenich, M. M., Nelson, R. J., Stryker, M. P., Cynader, M. S., Schoppmann, A., and Zook, J. M. (1984). Somatosensory cortical map changes following digit amputation in adult monkeys. *J. Comp. Neurol.* **224,** 591–605.

Molchan, S. E., Sunderland, T., McIntosh, A. R., Herscovitch, P., and Schreurs, B. G. (1994). A functional anatomical study of associative learning in humans. *Proc. Natl. Acad. Sci. U.S.A.* **91,** 8122–8126.

Nissen, M. J., and Bullemer, P. (1987). Attentional requirements of learning: Evidence from performance measures. *Cog. Psychol.* **19,** 1–32.

Ojemann, J. G., and Silbergeld, D. L. (1995). Cortical stimulation mapping of phantom limb rolandic cortex. Case report. *J. Neurosurg.* **82,** 641–644.

Pascual-Leone, A., Cammarota, A., Wassermann, E. M., Brasil-Neto, J. P., Cohen, L. G., and Hallett, M. (1993). Modulation of motor cortical outputs to the reading hand of Braille readers. *Ann. Neurol.* **34,** 33–37.

Pascual-Leone, A., Grafman, J., and Hallett, M. (1994). Modulation of cortical motor output maps during development of implicit and explicit knowledge. *Science* **263,** 1287–1289.

Pascual-Leone, A., Dang, N., Cohen, L. G., Brasil-Neto, J. P., Cammarota, A., and Hallett, M. (1995a). Modulation of muscle responses evoked by transcranial magnetic stimulation during the acquisition of new fine motor skills. *J. Neurophysiol.* **74,** 1037–1045.

Pascual-Leone, A., Grafman, J., and Hallett, M. (1995b). Procedual learning and prefrontal cortex. *In* "Structure and Function of the Human Prefrontal Cortex" (J. Grafman, F. Boller, and K. J. Holyoak, Eds.), pp. 61–70. The New York Academy of Sciences, New York.

Pascual-Leone, A., Wassermann, E. M., Sadato, N., and Hallett, M. (1995c). The role of reading activity on the modulation of motor cortical outputs to the reading hand in Braille readers. *Ann. Neurol.* **38,** 910–915.

Pons, T. P., Garraghty, P. E., Ommaya, A. K., Kaas, J. H., Taub, E., and Mishkin, M. (1991). Massive cortical reorganization after sensory deafferentation in adult macaques. *Science* **252,** 1857–1860.

Ridding, M. C., and Rothwell, J. C. (1997). Stimulus/response curves as a method of measuring motor cortical excitability in man. *Electroencephalogr. Clin. Neurophysiol.* **105,** 340–344.

Rijntjes, M., Tegenthoff, M., Liepert, J., Leonhardt, G., Kotterba, S., Muller, S., *et al.* (1997). Cortical reorganization in patients with facial palsy. *Ann. Neurol.* **41,** 621–630.

Sadato, N., and Hallett, M. (1999). fMRI occipital activation by tactile stimulation in a blind man. *Neurology* **52,** 423.

Sadato, N., Zeffiro, T. A., Campbell, G., Konishi, J., Shibasaki, H., and Hallett, M. (1995). Regional cerebral blood flow changes in motor cortical areas after transient anesthesia of the forearm. *Ann. Neurol.* **37,** 74–81.

Sadato, N., Pascual-Leone, A., Grafman, J., Ibanez, V., Deiber, M. P., Dold, G., *et al.* (1996). Activation of the primary visual cortex by Braille reading in blind subjects. *Nature* **380,** 526–528.

Sadato, N., Pascual-Leone, A., Grafman, J., Deiber, M. P., Ibanez, V., and Hallett, M. (1998). Neural networks for Braille reading by the blind. *Brain* **121,** 1213–1229.

Schlaug, G., Knorr, U., and Seitz, R. J. (1994). Inter-subject variability of cerebral activations in acquiring a motor skill: A study with positron emission tomography. *Exp. Brain Res.* **98,** 523–534.

Seitz, R. J., and Roland, P. E. (1992). Learning of sequential finger movements in man: A combined kinematic and positron emission tomography (PET) study. *Eur. J. Neurosci.* **4,** 154–165.

Seitz, R. J., Canavan, A. G., Yaguez, L., Herzog, H., Tellmann, L., Knorr, U., *et al.* (1994). Successive roles of the cerebellum and premotor cortices in trajectorial learning. *NeuroReport* **5,** 2541–2544.

Seitz, R. J., Huang, Y., Knorr, U., Tellmann, L., Herzog, H., and Freund, H. J. (1995). Large-scale plasticity of the human motor cortex. *NeuroReport* **6,** 742–744.

Syed, N. A., Delgado, A., Sandbrink, F., Schulman, A. E., Hallett, M., and Floeter, M. K. (1999). Blink reflex recovery in facial weakness: An electrophysiologic study of adaptive changes. *Neurology* **52,** 834-838.

Topka, H., Cohen, L. G., Cole, R. A., and Hallett, M. (1991). Reorganization of corticospinal pathways following spinal cord injury. *Neurology* **41,** 1276–1283.

Topka, H., Valls-Solé, J., Massaquoi, S., and Hallett, M. (1993). Deficit in classical conditioning in patients with cerebellar degeneration. *Brain* **116,** 961–969.

Toro, C., Matsumoto, J., Deuschl, G., Roth, B. J., and Hallett, M. (1993). Source analysis of scalp-recorded movement-related electrical potentials. *Electroencephalogr. Clin. Neurophysiol.* **86,** 167–175.

Wassermann, E. M., Fuhr, P., Cohen, L. G., and Hallett, M. (1991). Effects of transcranial magnetic stimulation on ipsilateral muscles. *Neurology* **41,** 1795–1799.

Wassermann, E. M., McShane, L. M., Hallett, M., and Cohen, L. G. (1992). Noninvasive mapping of muscle representations in human motor cortex. *Electroencephalogr. Clin. Neurophysiol.* **85,** 1–8.

Wassermann, E. M., Pascual-Leone, A., and Hallett, M. (1994). Cortical motor representation of the ipsilateral hand and arm. *Exp. Brain Res.* **100,** 121–132.

Wassermann, E. M., Wang, B., Zeffiro, T. A., Sadato, N., Pascual-Leone, A., Toro, C., *et al.* (1996). Locating the motor cortex on the MRI with transcranial magnetic stimulation and PET. *NeuroImage* **3,** 1–9.

Weeks, R. A., Gerloff, C., and Hallett, M. (1997). PET study of visually and non-visually guided finger movements in patients with severe pan-sensory neuropathies and healthy controls. *Soc. Neurosci. Abstr.* **23** 1949.

Wunderlich, G., Knorr, U., Herzog, H., Kiwit, J. C., Freund, H. J., and Seitz, R. J. (1998). Precentral glioma location determines the displacement of cortical hand representation. *Neurosurgery* **42,** 18–26; discussion 26–17.

Ziemann, U., Rothwell, J. C., and Ridding, M. C. (1996). Interaction between intracortical inhibition and facilitation in human motor cortex. *J. Physiol. (Lond.)* **496,** 873–881.

Ziemann, U., Corwell, B., and Cohen, L. G. (1998). Modulation of plasticity in human motor cortex after forearm ischemic nerve block. *J. Neurosci.* **18,** 1115–1123.

24

Recovery of
Neurological Function

François Chollet*,[1] and Cornelius Weiller†

*INSERM U455 and Department of Neurology, Hôpital Purpan, Toulouse 31059, France
†Department of Neurology, Universitätskrankenhaus Hamburg-Eppendorf, D-20246 Hamburg, Germany*

I. Evidence for Spontaneous Cerebral
   Reorganization during Recovery in Stroke
   Patients
II. Learning Processes and Recovery of
   Neurological Function
III. Pharmacological Approach to Recovery of
   Neurological Function: Neuromodulation of
   Brain Networks
   References

Imaging recovery of neurological function cannot be considered a trivial challenge for many reasons. The main one may be the fact that recovery corresponds to improvement of clinical function in patients and not necessarily to any changes in cerebral activation patterns as they can be observed with modern neuroimaging techniques. Although recent studies confirm the relationship between training-induced improvement of impaired function and changes in the brain, we have to admit that there are few studies in the literature that

correlate behavioral and neuroimaging data in the field of recovery. On the other hand, there are many lines of evidence that indicate a capacity of the adult human brain to reorganize following disease and it is likely that a relationship exists between cerebral reorganization, time course of disease, and clinical features in patients. Stroke is an interesting model in this respect because it can manifest as an acute single lesion of the brain. The natural history of acute stroke is for the neurological deficit to recover spontaneously, though usually partially, in the first few months after its onset. Stroke provides an opportunity to investigate spontaneous cerebral reorganization and to correlate it with clinical improvement. However, though such correlations can be expected in patients with stroke, this is not the case with chronic degenerative neurological diseases. In this model, cerebral reorganization occurs in the context of disease progression and a deterioration of clinical performance.

Recovery of function after a lesion of the central nervous system and particularly after stroke is also a public health issue but remains very difficult to predict in individual cases. Until now, no validated treatment has improved clinical outcome in patients. Controversy exists about the effectiveness of rehabilitation and it seems from many series that rehabilitation may be

---

[1] To whom correspondence should be addressed.

Brain Mapping: The Disorders

effective but there is ignorance about the best and most valid procedures and their neural substrates (Nakayama *et al.*, 1994; Jorgensen *et al.*, 1995a,b). On the other hand, pharmacological evidence from experimental and clinical studies indicates drugs that may be effective in improving the magnitude of clinical recovery. Neuroimaging techniques may allow us to select patients for such treatment and to identify and monitor its cerebral effects.

Imaging recovery of function is difficult because of the technical limitations to performing technically valid longitudinal studies with tomographic techniques. It is not easy to repeat PET studies because of regulations governing exposure to radiation and behavioral and technical limitations are unresolved when using fMRI. The only solution to these problems at present is to include a separate control group. This makes studies more time-consuming and of limited validity for single subjects. These technical issues need to be addressed and solved urgently.

Neuroimaging studies have given the opportunity to investigate cerebral reorganization of the human brain under various circumstances and have illustrated the reality of plastic change. Modern brain mapping techniques are perfectly suited to re-test *in vivo* classical hypotheses about brain function determined by the clinico-anatomic approach. It is an obligation of neurologists to use these new findings to improve the care of patients. The relative dearth of effective interventions in neurorehabilitation may be attributable to relatively weak contributions from basic science and behavioral psychology in their evaluation. Now that techniques are available to study the human brain *in-vivo*, there is an awareness of this deficit and an increased effort to bridge the gap between clinical practice and an understanding of recovery mechanisms. We expect that this effort shall yield new physical rehabilitative therapies in the future.

There are two issues important to an understanding of the mechanisms underlying reorganization during recovery of function after cerebral lesions:

1. Brain functions are localized, either in distinct brain regions (e.g., perception of visual motion in V5 (MT) (Watson *et al.*, 1993)) or in extended, connected, overlapping, highly parallel and reciprocal processing networks, the modular parts of which may substitute for each other. Such networks have been proposed for attention (Mesulam, 1990), motor function (Rizzolatti *et al.*, 1998), object localization and identification (Ungerleider *et al.*, 1998), and language (Howard *et al.*, 1992).

2. Functional localization is not fixed as even the adult human brain retains "plastic" potential.

I. Evidence for Spontaneous Cerebral Reorganization during Recovery in Stroke Patients

We have identified, from neuroimaging studies of patients with stroke over the past 10 years, four main features of cerebral reorganization:

1. *Spontaneous reorganization* Spontaneous reorganization can be seen in the lesioned brain under basal conditions by a comparison of cerebral metabolic activity with normal unlesioned controls. In a series of eight patients with striatocapsular infarcts compared with PET to a population of controls, Weiller and colleagues showed that the resting activity in patients was lower in a wide area spreading from the lesion sites to involve insular cortex and brainstem (Weiller *et al.*, 1992). They also identified, in the contralateral, nonlesioned hemisphere areas where CBF was higher in patients than in controls. Thus lesions could lead to both relative activations and deactivations. Thus measurements of brain metabolism in basal conditions (at rest) show evidence of modification of resting metabolic patterns after brain injury. Modifications of metabolic patterns appear early after stroke and correspond to cerebral rearrangements consequent on a lesion. These metabolic alterations should be compared with the diaschisis phenomenon initially described by Baron between cerebral and cerebellar hemispheres (Baron *et al.*, 1980; Feeney and Baron, 1986). Crossed cerebellar diaschisis was later described in many supratentorial lesions, including those of frontal cortex (Martin and Raichle, 1983), parietal cortex (Lenzi *et al.*, 1982), thalamic and basal ganglia areas and internal capsule (Rougement *et al.*, 1983; Meneghetti *et al.*, 1984). It has also been described in transient lesions (Perani *et al.*, 1987) and in nonvascular injuries (Duncan *et al.*, 1987). Interruption of cerebro-cerebellar loops is thought to be the most likely mechanism of this remote trans-neuronal metabolic depression. Diaschisis follows neuroanatomical pathways spreading from the site of injury. Despite extensive studies in animal models and numerous descriptions in humans, the correlation between diaschisis and clinical symptoms remains unclear. Likewise, the precise nature of the correlation between changes of metabolic pattern and recovery of function is uncertain, but it seems unlikely that the metabolic rearrangements have nothing to do with the recovery processes.

2. *Recruitment of remote areas* The most direct evidence for the participation of remote areas and particularly of the ipsilateral motor system in recovery

processes comes from PET studies. Chollet *et al.* studied a series of six patients recovering from their first hemiplegic stroke with PET and an activation paradigm (Chollet *et al.*, 1991). They noted that when patients used a recovered hand to perform a finger-to-thumb sequential opposition task at a fixed rate, there was bilateral activation of the motor cortices. On the other hand, movements of the normal hand were associated only with activation of contralateral motor cortex. Bilateral activation was seen not only in primary motor cortex but also in cerebellum, premotor cortex, area 40, and probably SMA. Essentially, bilateral activation of the whole motor network was observed. Other techniques have contributed to the characterization of the capacity of the lesioned human brain to reorganize, to recruit remote cortical areas, and to participate in the recovery of function. Motor system studies in patients with stroke again provide an illustrative example.

The usefulness of the recruitment of remote areas in the recovery process remains somewhat controversial because longitudinal studies designed to find a correlation between both phenomena are still lacking (Chollet and Weiller, 1994). Several arguments can be articulated that suggest recruitment of remote areas is a nonspecific phenomenon, but none demonstrate that recruitment of remote areas is independent of recovery. First, clinical studies indicate that many motor representations are bilateral and indirectly suggest the participation of ipsilateral motor pathways in recovery after stroke. They also suggest the existence of ipsilateral corticofugal pathways (Jones *et al.*, 1989; Adams and Gandevia, 1990; Desrosiers *et al.*, 1996; Marque *et al.*, 1997; Prigatano and Wong, 1997). Second, recruitment of remote areas in patients recovering from stroke is not specific to the motor system. It has been shown by Weiller that the right hemisphere is activated by linguistic tasks in patients with good recovery from Wernicke's aphasia (Weiller *et al.*, 1995). Moreover, recruitment of remote areas is not specific to stroke patients and is not even specific to disease. It has been shown by Shibasaki *et al.*, with PET scanning, that recruitment of ipsilateral primary sensorimotor cortex and of contralateral cerebellum occurs in normal subjects performing a complex motor task even though no recruitment is observed when a simple motor task is performed (Shibasaki *et al.*, 1993). These studies raise the question of the functional significance of ipsilateral motor pathways and indirectly ask what are the cellular mechanisms that mediate such cerebral reorganization.

Electrophysiological studies and those using transcranial magnetic stimulation have contributed to an evaluation of the significance of ipsilateral motor

pathways in the recovery process (Benecke *et al.*, 1991; Cohen *et al.*, 1991a, 1993; Fries *et al.*, 1991; Palmer *et al.*, 1992; Carr *et al.*, 1993; Caramia *et al.*, 1996; Bastings *et al.*, 1997). Finally, it is clear that very different individual patterns of activation have been found in patients recovering from a vascular hemiplegia (Weiller *et al.*, 1993). In some patients, premotor cortices are activated bilaterally; in others, the ipsilateral premotor cortex does not contribute. This may be because of an underlying variability in the neuronal hard-wiring, the functional anatomy, the lesion site, or the extent and the time course of the process of recovery. Individual neuronal hard-wiring variability is supported by findings from activation studies of linguistic function in normals. These show that for the same task under the same experimental conditions, right hemisphere linguistic areas may or may not be activated in different individuals (Frackowiak, 1997).

3. *Extension of specialized areas*   Weiller was the first to demonstrate that plastic changes in primary motor cortex could be found in patients recovering from a motor deficit after stroke (Weiller *et al.*, 1993). In a selected series of eight patients with capsulothalamic stroke, he showed that during movements of fingers of a recovered hand, contralateral activation of primary motor cortex differed in different subjects. In four patients with lesions affecting the posterior limb of the internal capsule, the peak activation in primary motor cortex was shifted ventrally in Brodmann area 4. In these patients, the cortical motor field associated with finger movements had enlarged significantly into an area normally associated with the motricity of the face. Evidence is accumulating in the literature for plastic changes to cortical maps in adult animals following various types of peripheral injury. For example, Pons *et al.* demonstrated a 10- to 14-mm extension of the face area into the area of the upper limb in the primary sensory cortex of adult macaques after upper limb de-afferentation (Pons *et al.*, 1991). Magnetic brain stimulation studies in man have shown that following upper limb amputation, surviving limb muscles may be excited from a larger number of scalp positions than in normal subjects (Cohen *et al.*, 1991b). Focal (stroke like) cortical lesions in monkeys and further retraining are known to induce dramatic changes in cortical motor representations. In particular, area 3b, area 3a, and area 4 have shown marked plastic change (Merzenich *et al.*, 1996; Nudo *et al.*, 1996). Weiller also clearly showed that in a recovering human brain, activation of contralateral primary motor cortex is more important than in control subjects (Weiller *et al.*, 1992). Very few studies have tried to correlate synaptic activity (indexed by blood flow changes) and magni-

tude of recovery. Dettmers and colleagues described the methodological difficulties with such studies (Dettmers *et al.*, 1997). They correlated activity in motor cortex with force exerted by the hand in stroke patients and controls. They found a binomial relationship between force and activity changes that was very different from that observed in normals. This result suggests that residual activity is correlated with the magnitude of the remaining deficit (Dettmers *et al.*, 1997). In other words, a brain lesion can affect the dynamics of the responses of the executive motor system.

4. *Increased activity in nonspared areas* Disruption of the primary motor system leads to increased activity in secondary or higher order brain areas (Weiller *et al.*, 1992). Increased activity in lateral premotor cortex and SMA during motor tasks in stroke patients as compared to normal subjects suggests that these regions of the frontal lobe are involved in the recovery process. In a conventional view, the motor system is hierarchically organized, with primary motor cortex in control of limb movements, executing its influence via direct cortico-spinal fibers. Additional motor areas outside area 4, in the frontal lobe, are in this formulation responsible for "planning," preparation, initiation, and maintenance of movement. It could be argued that patients use such strategies and hence these areas to execute the required motor task. However, the additional motor areas, lateral premotor cortex, SMA, and cingulate cortex also contribute substantial numbers of fibers directly to the corticospinal tract. One of the most consistent findings in all stroke studies is increased and bilateral activation of inferior parietal cortices (BA 40) and the anterior parts of the insula/frontal operculum. These two areas seem to constitute an additional system tightly interconnected with the unimodal association cortices (SII and lateral premotor cortex). We suggest that the increased activity in spared areas and the recruitment of additional areas may reflect a mechanism by which the brain adapts to a lesion.

II. Learning Processes and Recovery of Neurological Function

It is likely that close links exist between recovery mechanisms and general mechanisms of learning. In some neurobiological models (Hebb, 1949; Melzac, 1990), representations are specified in terms of patterns of neuronal activity. These are determined genetically or by early learning and modulated or differentiated by use and activity. Individual neurons participate in various neuronal assemblies with different representations and functions. The function is defined by the timely interac-

tion between various parts of the active network (Singer, 1993, 1999). During the developmental phase, groups of neurons that are simultaneously active strengthen their connections and develop into assemblies with strong internal connections (Braitenberg and Schuz, 1994).

Learning effects resulting in changes in the anatomical somatotopy of the primary cortices can be differentiated from functional effects in higher order cortices. There are numerous examples of learning-induced or use-dependent changes in primary somatosensory or primary motor cortex within the representation of *one* extremity. Representations extend spatially with exaggerated use in Braille readers or players of string instruments (Pascual-Leone and Torres, 1993; Elbert *et al.*, 1995) or shrink in patients with forced immobilization of a leg (Liepert *et al.*, 1995). These changes can occur rapidly, within minutes (Sanes *et al.*, 1992; Classen *et al.*, 1998). A recent transcranial magnetic stimulation study showed learning-induced effects on the motor cortex *across limbs* (Liepert *et al.*, 1999). Learning-induced transfer of motor function from one extremity to another may be accomplished via secondary cortices. In a recent study, it was shown that limb-independent functional movement parameters are stored in unimodal, sensorimotor association cortices in parietal and frontal areas at the level of the extremity that habitually performs a given movement (e.g., the left lateral cortices for signing with the right hand) (Rijntjes *et al.*, 1997b). With some stereotyped movements, it is possible to perform them with limbs other than the habitual one. For example, most people can sign their name with the dominant and non-dominant hand and foot, even though these limbs have never learned to perform this movement. They are able to access and use the movement parameters for such a typical hand movement by activating the same parietal and frontal cortical structures as when the dominant hand is used to sign (Rijntjes *et al.*, 1997; Dettmers *et al.*, 1998). When the non-dominant hand learns to sign (e.g., after a week of training), homologous right hemisphere areas are now activated. It seems as if the left hand has stored the necessary motor parameters in its own unimodal association cortices by modifying functional connections within existing networks. Thus, inter-hemispheric transfer of motor programs associated with one extremity or learning may both take place via secondary cortices. An example of cross-modal plasticity induced by early learning is the activation of extrastriate visual and parietal association cortices by somatosensory stimulation by Braille reading in congenitally blind subjects (Büchel *et al.*, 1998; Sadato *et al.*, 1998).

Learning can be seen as a strengthening of connections between neural assemblies within an existing network (Gustafsson *et al.*, 1987; Ahissar *et al.*, 1992). What is the neurophysiological correlate of this? In the normal brain, associative learning is reflected in suppression of activation with repetition and an increase in effective connectivity between brain regions (Büchel *et al.*, 1999). After brain lesions, changes in representations induced by active learning or mediated by use may have different electrophysiological correlates. For example, instead of repetition *suppression* a temporary learning-induced *augmentation* of activity may be found, which gradually declines when effective connectivity is enhanced.

"Plastic" changes occur under very different conditions in the intact as well as in the lesioned brain (Merzenich *et al.*, 1982; Kaas, 1991). They may occur as a result of learning or adaptation, with or without concomitant change in behavior. After lesions, they may be found as a consequence of a structural lesion (e.g., diaschisis) or due to active therapy (e.g., rehabilitation). With recovery, peri-lesional extensions of motor representations shift from primary to secondary cortical systems and recruitment of contralateral homologous cortical areas in the unaffected hemispheric has been described in animals (Fries *et al.*, 1993; Nudo *et al.*, 1996; Rouiller *et al.*, 1998; Darian-Smith *et al.*, 1999; Liu and Rouiller, 1999) and in human stroke victims (Chollet *et al.*, 1991; DiPiero *et al.*, 1992; Weiller *et al.*, 1992, 1993, 1995; Chollet and Weiller, 1994; Seitz *et al.*, 1994, 1998; Weder and Seitz, 1994; Binkofski *et al.*, 1996; Dettmers *et al.*, 1997; Weiller, 1998). This reorganization is seen as a determinant of recovery of function in the presence of persistent structural tissue damage. However, the exact relation between reorganization and restitution of function is unknown. Plastic changes may sometimes occur without obvious teleological reason, i.e., without any improvement of function (Kaas, 1997). For example, patients with peripheral facial nerve palsy (Bell's Palsy) may show a large lateral extension of the hand representation into the face representation. This has been determined by assessment of activation extent with PET during finger tapping and by demonstration of increases in the area over the skull from which responses in abductor pollicis brevis can be elicited by transcranial magnetic stimulation (Rijntjes *et al.*, 1997). The representation of the hand in M1 extends into neighboring face areas, which are deafferented from their target facial muscles. This may indicate competition between adjoining brain areas.

We recently performed two studies in which we related the training-induced improvement of lost function to reorganizational changes in the brain. In the first study, the effect of constraint-induced (CI) movement therapy was examined in the brain of chronic stroke patients (Liepert *et al.*, 1998). This therapy was proposed on the basis of basic research with monkeys subjected to somatosensory de-afferentation and is based on a be-

havioral theory of "learned non-use" of the affected limb (Taub *et al.*, 1978, 1980). CI therapy reverses learned non-use by constraining the movement of the healthy upper extremity and intensely training the paretic arm. It has been shown to be effective in a subset of stroke patients (Taub *et al.*, 1993, 1994). We charted the effects of this training on the motor cortex with transcranial magnetic stimulation. After 2 weeks of CI therapy, motor performance had improved substantially in all patients. This improvement was accompanied by an increase in the area and a shift of the center of gravity of the motor output zone and of the amplitudes of motor-evoked potentials. This result indicates preserved neuronal excitability of the motor cortex in the damaged hemisphere despite chronic lack of movement of the plegic hand, raising the possibility of reactivation by rehabilitation. Although this study examined the primary motor cortex, it is possible that other, secondary areas may have mediated the changes. The results are in line with an early PET study (Weiller *et al.*, 1993) in which a large lateral extension of the "hand representation" was found with finger movements in stroke patients with posterior internal capsular infarcts and spared function of the facial musculature. It seems that fractionated finger movements, which are highly dependent on the integrity of the M1-cortex and the corticospinal tract, are so critical that the hand representation achieves privileged access to a larger area of M1 at the expense of neighboring representations, thereby providing a basis for recovery.

In a recent fMRI study, we compared a control group to a series of normal volunteers who underwent 4 weeks of proprioceptive training. For 20 min a day over 28 days, their wrists were repeatedly passively extended to provide "chronic" proprioceptive stimulation. Passive movements were calibrated and paced. fMRI was done before and after training using the same passive repetitive extension of the wrist contrasted with rest. In the trained group, we found increased activity in sensorimotor cortex and the supplementary motor (Carel *et al.*, 2000). Thus chronic proprioceptive stimulation, which is part of most rehabilitation procedures, induces a reorganization of sensorimotor activation patterns. Moreover, proprioceptive training results in a redistribution of brain motor activation patterns toward cortical structures involved in motor execution. Application of this paradigm to stroke patients is imminent.

In another study aimed to assess the relationship between reorganization and recovery of function, we investigated whether language training induces cortical reorganization in aphasic stroke (Musso *et al.*, 1999). We studied four patients with Wernicke's aphasia and a left perisylvian lesion. Language comprehension, a function that is often abolished in aphasia but tends to recover quickly, was assessed by an expert from a short version of the Token Test during 12 consecutive PET scans. Between scans, patients underwent intense comprehension training sessions of 8 min each. The training-induced improvement of language comprehension during scanning correlated with activity changes in language-related areas in the right hemisphere only (middle and superior temporal gyrus and supramarginal gyrus). This study strongly supports the notion that the right hemisphere plays a role in recovery from aphasia and adds further evidence that reorganization is actually beneficial.

III. Pharmacological Approach to Recovery of Neurological Function: Neuromodulation of Brain Networks

A considerable scientific literature exists about the effects of drugs on the cerebral distribution of neuroreceptors but most reports are of static studies using SPECT or PET and radioactive ligands with little or no reference to neurological function (Montastruc *et al.*, 1987).

The influence of drugs on cerebral functional networks and on specific neuroreceptors can now be addressed. Basically, a dynamic approach assesses the effect a pharmacological agent has on a specific neural network during recording of task performance and allows identification of the neuromodulatory effect of the drug on the relevant functional network.

Neuromodulation was initially demonstrated in human brains by Friston *et al.* and Grasby *et al.* Using PET, they performed measurements of brain activity during the performance of memory tasks before and after the administration of apomorphine (a dopamine agonist), buspirone (5-HT1A partial antagonist), and placebo. Increases of brain activity in response to memory challenge were attenuated by apomorphine in dorsolateral prefrontal cortex. Conversely, buspirone attenuated activity increases in the retrosplenial region. These interactions between drug and a cognitive challenge can be interpreted as neuromodulatory effects (Friston *et al.*, 1992; Grasby *et al.*, 1992). Similarly, serotonin effects on prefrontal cortex have also been shown to exert a neuromodulatory effect (Mann *et al.*, 1996).

A. Parkinson's Disease: Nigrostriatal Pathway Disruption

Parkinson's disease is a well-defined chronic neurological disorder characterized by a dysfunction of the extrapyramidal motor system. Parkinson's disease is associated with a disruption of the dopaminergic system,

particularly the nigrostriatal pathway. It offers a model of dysfunction of an anatomically well-characterized cortico-subcortical loop mediated by loss of a specific neuromodulator, dopamine. Akinesia is the major symptom found in patients with Parkinson's disease. A major output of the basal ganglia is to the premotor areas. There is evidence that directly implicates the supplementary motor area in the impairment of patients with Parkinson's disease. When patients are required to make self-paced repetitive movements without external cue, a decreased amplitude of the initial phase of the readiness potential is recorded electrophysiologically over frontal areas (Dick *et al.*, 1989). Patients with a lesion of SMA have no readiness potential and also an akinesia very similar to Parkinson's disease patients (Deecke *et al.*, 1987). PET and SPECT studies have shown that the subcortical dopamine deficit induces a de-afferentation of the SMA (Playford *et al.*, 1992; Rascol *et al.*, 1992). In a series of six selected patients with Parkinson's disease, Playford and colleagues used an internally generated motor task to show impaired activation of supplementary motor area compared with a series of normal controls.

1. Functional Re-afferentation of Motor Cortex

Rascol *et al.*, using SPECT, showed that primary motor cortex is also hypoactive in Parkinson patients in the akinetic "off" condition. He demonstrated that administration of a subcutaneous dose of apomorphine reversed the abnormality and resulted in normal activation of SMA and primary motor cortex. Jenkins and colleagues found very similar results with PET and intravenously administered apomorphine (Jenkins *et al.*, 1992). Apomorphine also changed the clinical status of the patients. SMA was considered functionally de-afferented in patients with Parkinson's disease and reafferented following drug administration. This finding represents an illustrative example of a direct interaction between the motor network and an externally administered drug. Apomorphine induced a clear-cut reorganization of the motor system. Levodopa is also active in normalizing SMA activation in patients undergoing long-term treatment (Rascol *et al.*, 1994).

B. Working Memory and Pharmacological Modulation

Another illustrative example of the influence of an externally administered drug is given by modulation of cholinergic transmission. Modulation of the cholinergic system results in changes in memory performance in animals and in patients with Alzheimer's disease. For example, tacrine, a long-acting acetyl-cholinesterase inhibitor, is used clinically to treat the cognitive symptoms of Alzheimer's disease. Furey and colleagues (Furey *et al.*, 1997), in order to identify associated changes in functional brain responses, studied performance measures and regional cerebral activity with PET in healthy subjects during the performance of a working memory task. Eight controls received an infusion of saline throughout the study and 13 other volunteers were given continuous infusion of physostigmine. Physostigmine is an acetylcholinesterase inhibitor that increases the duration of action of acetylcholine at synapses. It improves performance of a working memory task in animals and in patients with Alzheimer's disease. Physostigmine improved working memory efficiency as indicated by faster reaction times and reduced working memory task-related activity in anterior and posterior regions of the middle frontal gyrus, a region previously shown to be associated with working memory. This study shows that enhancement of cholinergic transmission results in an improvement of working memory efficiency that is correlated with an alteration of brain activity in a cortical region known to play a central role in working memory. It demonstrates a cholinergic modulation of functional brain responses. The mechanism of interaction between the drug and working memory network is unknown and the reduced activation of middle frontal regions was unexpected. It may reflect the shorter processing time during task performance.

C. Neuromodulation of the Motor System and Stroke Recovery

The literature gives many examples of the influence of drugs on recovery from central nervous system lesions. Briefly, it is possible from animal experiments to distinguish between drugs that exert a positive effect on recovery from experimental lesions and drugs that have a negative effect and prevent recovery. For example, benzodiazepines and neuroleptic agents (e.g., haloperidol) have a negative effect on recovery whereas amphetamine and some antidepressant drugs are assumed to have a beneficial effect (Goldstein, 1993, 1995). Clinical trials have provided evidence for the effectiveness of some of these drugs on recovery of motor function in patients after stroke. Fluoxetine and dextroamphetamine facilitate motor recovery in post-stroke patients undergoing rehabilitation (Walker-Batson *et al.*, 1995; Dam *et al.*, 1996). Fluoxetine is known to inhibit the reuptake of serotonin and dextroamphetamine and to enhance presynaptic release of monoamines (noradrenaline, dopamine, and serotonin). Many lines of evidence suggest that amphetamine-promoted recovery of function may be mainly mediated through the noradrenergic system (Goldstein, 1991, 1993; Boyeson, 1994). As significant cerebral reorganization occurs after stroke in

man (Chollet *et al.*, 1991; Weiller *et al.*, 1992, 1993; Chollet and Weiller, 1994), it is reasonable to ask whether drugs can interact or modulate them. Despite the demonstration in animals of a beneficial effect of noradrenaline (Goldstein, 1993, 1995; Small, 1994), and to a lesser extent dopamine (Feeney and Law, 1982), and the suggestive clinical studies, the influence of serotonergic modulation on motor recovery is even more uncertain (Boyeson *et al.*, 1994; Boyeson, 1996; Dam *et al.*, 1996). The beneficial effect of noradrenaline and dopamine on motor behavior is generally recognized (McMillen, 1983; Kuczenski *et al.*, 1995) but the role of serotonin on motor function is still controversial (Elliott, 1992; Jacobs, 1993; Mangan *et al.*, 1994; Veasey *et al.*, 1995; Geyer, 1996; Hasbroucq *et al.*, 1997; Porrino *et al.*, 1997). Although unclear, a role for serotonin in motor control cannot be discarded and may also be primarily modulatory (Jacobs, 1993, 1997). In addition, since post-stroke depression is a common sequel of stroke, serotonin involvement in motor and non-motor cognitive functions and their recovery seems warranted.

In recent work, we used fMRI to test the effect of a single dose of fluoxetine, a serotonin reuptake inhibitor, and of fenozolone, an amphetamine-like drug on motor activation, in a series of 12 normal subjects (Loubinoux *et al.*, 1999). Neither fluoxetine nor fenozolone altered the mean arterial blood pressure in healthy volunteers. On functional images no effect of fluoxetine or fenozolone was observed in areas that were not activated or deactivated by the sensorimotor tasks. The main effect of application of a single dose of either fluoxetine or fenozolone was in the cerebellum and contralateral sensorimotor cortices. Both drugs elicited a highly significant and widespread attenuation of motor-induced activation in bilateral cerebellum and vermis and an augmentation in primary motor cortex, premotor cortex, cingulate cortex, and SMA. It seems to us that cerebral motor activity can be modulated by a single low dose of fluoxetine or fenozolone in healthy subjects and that the sensitivity of fMRI is sufficient to demonstrate such changes. Our results show that both drugs induce similar profound changes in the entire motor pathway, including primary and secondary cortices, thalamus, and cerebellum. Moreover, the pattern of changes is in accordance with the known neuroanatomy of monoaminergic transmitter systems (Geyer and Zilles, 1997). The densities of adrenergic and serotonergic sites are high in sensorimotor cortex. The results are also in accordance with the similar distributions in sensorimotor cortex of [3H]prazosin and [3H]5-HT ligands, which bind to adrenergic and serotonergic sites, respectively (Geyer and Zilles, 1997). Drug effects interacting with cerebral motor activity suggest a specific action of these pharmacological compounds and a direct or indirect involvement of monoamines, including serotonin, in the regulation of cerebral motor activity. It is worth noting that fluoxetine, though not a pure selective serotonin re-uptake inhibitor (Wong *et al.,* 1990), induces enhanced activation focused on motor executive areas: motor and premotor cortices and posterior supplementary motor area. This result provides clear additional evidence for the postulated role of serotonin in motor control. The modulation of brain activation induced by fenozolone and fluoxetine constitutes a redistribution of brain acti-vation toward the executive motor system. An augmen-tation of activation is seen in the presence of active drug that is centered on the knob of the central sulcus, which is known to correspond to the motor representation of the hand (Yousry *et al.,* 1997). It is surrounded by an at-tenuation of activation in contralateral sensorimotor cortex. Similarly, SMA activity was redistributed toward its posterior part whereas the anterior part of the SMA showed attenuated activation. This result suggests that enhanced monoaminergic transmission favors execu-tion rather than programming of movements (Deiber *et al.,* 1996) in-so-far as such functions can be attributed to these structures. There is growing evidence that devel-opment of precision and automaticity of a motor task results in similar changes in the SMA (Jenkins *et al.,* 1994; Picard and Strick, 1996).

All together, these data are in line with the hypothe-sis that the brain monoamine systems, including the 5-HT system, facilitate motor output. Accumulating ev-idence indicates that the serotonin link to motor activ-ity is related both to co-activation of the noradrenergic nervous system and to modulation of afferent inputs of this system (Jacobs, 1997). It seems that the primary function of the 5-HT system in the brain is to facilitate motor output and concurrently inhibit sensory informa-tion processing (Jacobs, 1993).

These three examples (Parkinson's disease, working memory, and stroke patients) demonstrate some of the interactions between drugs and the central nervous system in the field of recovery from neurological def-icit. They show different mechanisms of action that all converge toward the improvement of clinical perfor-mance in patients. Parkinson's disease offers the oppor-tunity to characterize functional de-afferentation and re-afferentation of motor cortices and particularly of the supplementary motor area as a consequence of the ad-ministration of dopamine or of a dopamine agonist. The role as neuromodulators of cholinergic agents in working memory and of monoamines in stroke patients is well il-lustrated by these models. Modulation may improve or interfere with recovery and so studies are needed to demonstrate the relationship of modulation of activity in functional brain systems with behavioral outcome.

Acknowledgment

The work presented herein was partly supported by the BMBF, the DFG, and the ARC-program of the DAAD (to C.W.), Germany.

References

Adams, R., and Gandevia, S. C. (1990). The distribution of muscular weakness in upper motor neuron lesions affecting the lower limb. *Brain* **113,** 1459–1476.

Ahissar, E., Vaadia, E., Ahissar, M., Bergman, H., Arieli, A., and Abeles, M. (1992). Dependence of cortical plasticity on correlated activity of single neurons and on behavioural context. *Science* **257,** 1412–1415.

Baron, J., Bousser, M. G., Comar, D., and Castaigne, P. (1980). "Crossed cerebellar diaschisis" in human supratentorial brain in-farction. *Trans. Am. Neurol. Assoc.* **105,** 459–461.

Bastings, E., Rapisarda, G., Pennisi, G., Maerten de Noordhout, A., Lenaerts, M., Good, D. C., and Delwaide, P. J. (1997). Mechanisms of hand motor recovery after stroke: An electrophysiologic study of central motor pathways. *J. Neurol. Rehabil.* **11,** 97–108.

Benecke, R., Meyer, B. U., and Freund, H. J. (1991). Reorganization of descending motor pathways in patients with hemispherectomy and severe hemispheric lesions demonstrated by magnetic brain stimu-lation. *Exp. Brain Res.* **83,** 419–426.

Binkofski, F., Seitz, R., Arnold, S., Classen, J., Benecke, R., and Fre-und, H. J. (1996). Thalamic metabolism and corticospinal tract in-tegrity determine motor recovery in stroke. *Ann. Neurol.* **39,** 460–470.

Boyeson, M. (1996). Effects of fluoxetine and maprotiline on func-tional recovery in poststroke hemiplegic patients undergoing reha-bilitation therapy. *Stroke* **27,** 2145–2146.

Boyeson, M., Hamron, R. L., and Jones, J. (1994). Comparative effects of fluoxetine, amitryptiline and serotonin on functional motor re-covery after sensorimotor cortex injury. *Am. J. Phys. Med. Rehabil.* **73,** 76–83.

Braitenberg, V., and Schuz, A. (1994). Basic features of cortical con-nectivity and some considerations on language. *In* "Language Ori-gin: A Multidisciplinary Approach" (J. Wind, B. Bichahjiyn, A. Nocentini, and A. Jonker, Eds.), pp. 89–102. Kluwer Academic, Dordrecht/Norwell, MA.

Büchel, C., Price, C., Frackowiak, R. S., and Friston, K. (1998). Differ-ent activation patterns in the visual cortex of late and congenitally blind subjects. *Brain* **121,** 409–419.

Büchel, C., Coull, J. T., and Friston, K. J. (1999). The predictive value of changes in effective connectivity for human learning. *Science* **283** (5407), 1538–1541.

Caramia, M., Iani, C., and Bernardi, G. (1996). Cerebral plasticity af-ter stroke as revealed by ipsilateral responses to magnetic stimula-tion. *NeuroReport* **7,** 1756–1760.

Carel, C., Loubinoux, I., Boulanouar, K., Manelfe, C., Rascol, O., Cel-sis, P., and Chollet, F. (2000). *J. Cereb. Blood Flow Metabol.* (in press).

Carr, L. J., Harrison, L. M., Evans, A. L., and Stephens, J. A. (1993). Patterns of central motor reorganization in hemiplegic cerebral palsy. *Brain* **116,** 1223–1247.

Chollet, F., DiPiero, V., Wise, R. J. S., Brooks, D. J., Dolan, R., and Frackowiak, R. S. J. (1991). The functional anatomy of motor re-covery after stroke in humans: A study with positron emission to-mography. *Ann. Neurol.* **29**(1), 63–71.

Chollet, F., and Weiller, C. (1994). Imaging recovery of function fol-lowing brain injury. *Curr. Opin. Neurobiol.* **4**(2), 226–230.

Classen, J., Liepert, J., Wise, S. P., Hallett, M., and Cohen, L. G. (1998). Rapid plasticity of human cortical movement representation induced by practice. *J. Neurophysiol.* **79,** 1117–1123.

Cohen, L., Roth, B. J., Wassermann, E. M., Topka, H., Fuhr, P., Schultz, J., and Hallett, M. (1991a). Magnetic stimulation of the human cerebral cortex, an indicator of reorganization in motor pathways in certain pathological conditions. *J. Clin. Neurophysiol.* **8,** 56–65.

Cohen, L., Bandinelli, S., Findley, T. W., and Hallett, M. (1991b). Motor reorganization after limb amputation in man. *Brain* **114,** 615–627.

Cohen, L., Brasil-Neto, J. P., Pascual-Leone, A., and Hallett, M. (1993). Plasticity of cortical motor output organization following deafferentation, cerebral lesions and skill acquisition. *In* "Electrical and Magnetic Stimulations of the Brain and Spinal Cord" (O. Devinsky, A. Beric, and M. Dogali, Eds.), pp. 187–200. Raven Press, New York.

Dam, M., Tonin, P., DeBoni, A., Pizzolato, G., Casson, S., Ermani, M., Freo, U., Piron, L., and Battistin, L. (1996). Effects of fluoxetine and maprotiline on functional recovery in post-stroke hemiplegic patients undergoing rehabilitation therapy. *Stroke* **27,** 1211–1214.

Darian-Smith, I., Burman, K., and Darian-Smith, C. (1999). Parallel pathways mediating manual dexterity in the macaque. *Exp. Brain Res.* **128**(1/2), 101–108.

Deecke, L., Lang, W., Heller, H. J., *et al.* (1987). Bereitschaftspotential in patients with unilateral lesions of the supplementary motor area. *J. Neurol. Neurosurg. Psychiatry* **50,** 1430–1434.

Deiber, M. P., Sadato, N., and Hallett, M. (1996). Cerebral structures participating in motor preparation in humans: A positron emission tomography study. *J. Neurophysiol.* **75,** 233–247.

Desrosiers, J., Bourbonnais, D., Bravo, G., Roy, P. M., and Guay, M. (1996). Performance of the unaffected upper extremity of elderly stroke patients. *Stroke* **27,** 1564–1570.

Dettmers, C., Stephan, K. M., Lemon, R. N., and Frackowiak, R. S. J. (1997). Reorganization of the executive motor system after stroke. *Cerebrovasc. Dis.* **7,** 187–200.

Dettmers, C., Rijntjes, M., Rzanny, R., Gaser, C., Kiebel, S., Kaiser, W., and Weiller, C. (1998). Evidence for transhemispheric access to motor programs. *NeuroImage* **7,** S959.

Dick, J., Rothwell, J. C., Day, B. L., *et al.* (1989). The Bereitschaftspotential is abnormal in Parkinson's disease. *Brain* **112,** 233–244.

DiPiero, V., Chollet, F., MacCarthy, P., Lenzi, G. L., and Frackowiak, R. S. J. (1992). Motor recovery after ischemic stroke: A metabolic study. *J. Neurol. Neurosurg. Psychiatry* **55,** 990–996.

Duncan, R., Patterson, J., Boe, I., and Wyper, D. J. (1987). Reversible cerebellar diaschisis in focal epilepsy. *Lancet* **2,** 625–626.

Elbert, T., Pantev, C., Wienbruch, C., Rockstroh, B., and Taub, E. (1995). Increased cortical representation of the fingers of the left hand in string players. *Science* **270,** 305–307.

Elliott, P. (1992). Serotonin and L-norepinephrine as mediators of altered excitability in neonatal rat motoneurons studied *in vitro*. *Neuroscience* **47,** 533–544.

Feeney, D., and Baron, J. C. (1986). Diaschisis. *Stroke* **17,** 817–830.

Feeney, D., and Law, W. (1982). Amphetamine, haloperidol and experience interact to affect rate of recovery after motor cortex injury. *Science* **217,** 855–857.

Frackowiak, R. (1997). The cerebral basis of functional recovery. *In* "Human Brain Function": (R. Frackowiak, K. J. Friston, C. D. Frith, R. J. Dolan, and J. C. Mazziotta, Eds.), pp. 275–300. Academic Press, London.

Fries, W., Danek, A., and Thomas, N. W. (1991). Motor responses after transcranial electrical stimulation of cerebral hemispheres with a degenerated pyramidal tract. *Ann. Neurol.* **29,** 646–650.

Fries, W., Danek, A ., Scheidtmann, K., and Hamburger, C. (1993). Motor recovery following capsular stroke. *Brain* **116,** 369–382.

Friston, K., Grasby, P. M., Bench, C. J., Frith, C. D., Cohen, P. J., Liddle, P. F., Frackowiak, R. S. J., and Dolan, R. (1992). Measuring the neuromodulatory effects of drugs in man with positron emission tomography. *Neurosci. Lett.* **141,** 106–110.

Furey, M., Pietrini, P., Haxby, J. V., Alexander, G. E., Lee, H. C., VanMeter, J., Grady, C. L., Shetty, U., Rapoport, S. I., Schapiro, M. B., and Freo, U. (1997). Cholinergic stimulation alters performance and task-specific cerebral blood flow during working memory. *Proc. Natl. Acad. Sci. U.S.A.* **94,** 6512–6516.

Geyer, M. (1996). Serotoninergic functions in arousal and motor activity. *Behav. Brain Res.* **73,** 31–35.

Geyer, S., and Zilles, K. (1997). The somatosensory cortex of human: Cytoarchitecture and regional distributions of receptor-binding sites. *NeuroImage* **6,** 27–45.

Goldstein, L. (1991). Pharmacologic modulation of recovery after stroke: Clinical data. *J. Neurol. Rehabil.* **5,** 129–140.

Goldstein, L. (1993). Basic and clinical studies of pharmacologic effects on recovery from brain injury. *J. Neural Transpl. Plast.* **4,** 175–192.

Goldstein, L. (1995). Common drugs may influence motor recovery after stroke. *Neurology* **45,** 865–871.

Grasby, P., Friston, K. J., Bench, C., Cohen, P. J., Frith, C. D., Liddle, P. F., Frackowiak, R. S. J., and Dolan, R. J. (1992). Effects of the 5-HT1A partial antagonist buspirone on regional cerebral blood flow in man. *Psychopharmacology* **108,** 380–386.

Gustafsson, B., Abraham, W., and Huang, Y. (1987). Long term potentiation in the hippocampus using depolarizing current pulses as the conditioning stimulus to single volley synaptic potential. *J. Neurosci.* **7,** 774–780.

Hasbroucq, T., Blin, O., and Possamai, C. A. (1997). Serotonin and human information processing: Fluvoxamine can improve reaction time performance. *Neurosci. Lett.* **229,** 204–208.

Hebb, D. (1949). "The Organisation of Behaviour." Wiley, New York.

Howard, D., Patterson, K., Wise, R. J. S., Brown, W. D., Friston, K. J., and Weiller, C. (1992). The cortical localisation of the lexicons: PET evidence. *Brain* **115,** 1779–1782.

Jacobs, B. L. (1993). 5-HT and motor control: A hypothesis. *Trends Neurosci.* **16,** 346–352.

Jacobs, B. L. (1997). Serotonin and motor activity. *Curr. Opin. Neurobiol.* **7,** 820–825.

Jenkins, I., Fernandez, W., Playford, E. D., Lees, A. J., Frackowiak, R. S. J., Passingham, R. E., and Brooks, D. J. (1992). Impaired activation of the supplementary motor area in Parkinson's disease is reversed when akinesia is treated with apomorphine. *Ann. Neurol.* **32,** 749–757.

Jenkins, I., Brooks, D. J., Nixon, P. D., Frackowiak, R. S. J., and Passingham, R. E. (1994). Motor sequence learning: A study with positron emission tomography. *J. Neurosci.* **14,** 3775–3790.

Jones, R., Donaldson, I. M., and Parkin, P. J. (1989). Impairment and recovery of ipsilateral sensori-motor function following cerebral unilateral infarction. *Brain* **112,** 113–132.

Jorgensen, H. S., Nakayama, H., Raaschou, H. O., Vive-Larsen, J., Stoier, M., and Skyhoj Olsen, T. S. (1995a). Outcome and time course of recovery in stroke. Part II: Time course of recovery. The Copenhagen stroke study. *Arch. Phys. Med. Rehabil.* **76,** 406–412.

Jorgensen, H. S., Nakayama, H., Raaschou, H. O., and Skyhoj Olsen, T. (1995b). Recovery of walking function in stroke patients: The Copenhagen stroke study. *Arch. Phys. Med. Rehabil.* **76,** 27–32.

Kaas, J. (1991). Plasticity of sensory and motor maps in adult mammals. *Annu. Rev. Neurosci.* **14,** 137–167.

Kaas, J., and Florence Jain, N. (1997). Reorganization of sensory systems of primates after injury. *Neuroscientist* **3,** 123–130.

Kuczenski, R., Segal, D. S., Cho, A. K., and Melega, W. (1995). Hippocampus norepinephrine, caudate dopamine and serotonin, and behavioral responses to the stereoisomers of amphetamine and methamphetamine. *J. Neurosci.* **15,** 1308–1317.

Lenzi, G., Frackowiak, R. S. J., and Jones, T. (1982). Cerebral oxygen metabolism and blood flow in human cerebral ischemic infarction. *J. Cereb. Blood Flow Metab.* **2,** 321–335.

Liepert, J., Terborg, C., and Weiller, C. (1999). Motor plasticity induced by synchronized thumb and foot movements. *Exp. Brain Res.* **125**(4), 435–439.

Liepert, J., Tegenthoff, M., and Malin, J. P. (1995). Changes of cortical motor area size during immobilization. *Electroencephalogr. Clin. Neurophysiol.* **97,** 382–386.

Liepert, J., Miltner, M. W., Bauder, H., Sommer, M., Dettmers, C., and Taub, E. (1998). Motor cortex plasticity during constraint-induced movement therapy in stroke. *Neurosci. Lett.* **250,** 5–8.

Liu, Y., and Rouiller, E. (1999). Mechanisms of recovery of dexterity following unilateral lesion of the sensorimotor cortex in adult monkeys. *Exp. Brain Res.* **128**(1/2), 149–159.

Loubinoux, I., Boulanouar, K., Ranjeva, J. P., Berry, I., Rascol, O., Celsis, P., and Chollet, F. (1998). Effect of a single dose of fluoxetine and fenozolone on fMRI measurement of brain activity during hand sensorimotor tasks. *NeuroImage.*

Loubinoux, I., Boulanouar, K., Ranjeva, J. P., Carel, C., Berry, I., Rascol, O., Celsis, P., and Chollet, F. (1999). Cerebral functional magnetic resonance imaging activity modulated by a single dose of the monoamine neurotransmission enhancer fluoxetine and fenozolone during hand sensorimotor task. *J. Cerebr. Blood. Flow. Metabol.* **19,** 1365–1375.

Mangan, P. S., Cometa, A. K., and Friesen, W. O. (1994). Modulation of swimming behavior in the medicinal leech. IV. Serotonin-induced alteration of synaptic interactions between neurons of the swim circuit. *J. Comp. Physiol.* **175,** 723–736.

Mann, J. J., Malone, K. M., Diehl, D. J., Perel, J., Nichols, T. E., and Mintun, M. A. (1996). Positron emission tomographic imaging of serotonin activation effects on prefrontal cortex in healthy volunteers. *J. Cereb. Blood Flow Metab.* **16,** 418–426.

Marque, P., Felez, A., Puel, M., Demonet, J. F., Guiraud-Chaumeil, B., Roques, C. F., and Chollet, F. (1997). Impairment and recovery of left motor function in patients with right hemiplegia. *J. Neurol. Neurosurg. Psychiatry* **62,** 77–81.

Martin, W., and Raichle, M. E. (1983). Cerebellar blood flow and metabolism in cerebral hemisphere infarction. *Ann. Neurol.* **14,** 168–176.

McMillen, B. (1983). CNS stimulants: Two distinct mechanisms of action for amphetamine-like drugs. *Trends Pharmacol. Sci.,* 429–432.

Melzac, R. (1990). Phantom limbs and the concept of a neuromatrix. *Trends Neurosci.* **13,** 88–92.

Meneghetti, G., Vorstrup, S., Mickey, B., Lindewald, H., and Lassen, N. A. (1984). Crossed cerebellar diaschisis in ischemic stroke: A study of regional cerebral blood flow by ^{133}Xe inhalation and single photon computerized tomography. *J. Cereb. Blood Flow Metab.* **4,** 235–240.

Merzenich, M. M., Nelson, R., Stryker, M. P., Cynader, M. S., Shoppmann, A., and Zook, J. M. (1982). Somatosensory cortical map changes following digital amputation in adult monkey. *J. Comp. Neurol.* **224,** 591–605.

Merzenich, M., Wright, W., Xerri, C., Byl, N., Miller, S., and Tallal, P. (1996). Cortical plasticity underlying perceptual motor and cognitive skill development: Implications for neurorehabilitation. *Cold Spring Harbor Symp. Quant. Biol.* **61,** 1–8.

Mesulam, M. (1990). Large-scale neurocognitive networks and distribution processing for attention, language and memory. *Ann. Neurol.* **28,** 597–603.

Montastruc, J., Celsis, P., Agniel, A., *et al.* (1987). Levodopa-induced regional cerebral blood flow changes in normal volunteers and in patients with Parkinson's disease: Lack of correlation with clinical or neuropsychological improvement. *Mov. Disord.* **2,** 279–289.

Musso, M., Weiller, C., Kiebel, S., Müller, S. P., Bülau, P., and Rijntjes, M. (1999). Training-induced brain plasticity in aphasia. *Brain* **122** (Part 9), 1781–1790.

Nakayama, H., Jorgensen, H. S., Raaschou, H. O., and Skyhoj Olsen, T. (1994). Recovery of upper extremity function in stroke patients: The Copenhagen stroke study. *Arch. Phys. Med. Rehabil.* **75,** 394–398.

Nudo, R., Wise, B. M., SiFuentes, F., and Miliken, G. W. (1996). Neural substrates for the effects of rehabilitative training on motor recovery after ischemic infarct. *Science* **272,** 1791–1794.

Palmer, E., Ashby, P., and Hajek, V. E. (1992). Ipsilateral fast corticospinal pathways do not account for recovery in stroke. *Ann. Neurol.* **2,** 519–525.

Pascual-Leone, A., and Torres, F. B. (1993). Plasticity of the sensorimotor cortex representation of the reading finger in Braille readers. *Brain* **116,** 39–52.

Perani, D., Gerundini, P., and Lenzi, G. L. (1987). Cerebral hemispheric and contralateral cerebellar hypoperfusion during a transient ischemic attack. *J. Cereb. Blood Flow Metab.* **7,** 507–509.

Picard, N., and Strick, P. L. (1996). Motor areas of the medial wall: A review of their location and functional activation. *Cereb. Cortex* **6,** 342–353.

Playford, E. D., Jenkins, I. H., Passingham, R. E., Nutt, J., Frackowiak, R. S. J., and Brooks, D. J. (1992). Impaired mesial frontal and putamen activation in Parkinson's disease: A positron emission tomography study. *Ann. Neurol.* **32,** 151–161.

Pons, T., Garraghty, P. E., Ommaya, A. K., Kaas, J. H., Taub, E., and Miskin, M. (1991). Massive cortical reorganization after sensory deafferentation in adult macaques. *Science* **252,** 1857–1860.

Porrino, L. J., Miller, M., Hedgecock, A. A., Thornley, C., Matasi, J. J., and Davies, H. M. (1997). Local cerebral metabolic effects of the novel cocaine analog, WF-31: Comparisons to fluoxetine. *Synapse* **27,** 26–35.

Prigatano, G., and Wong, J. L. (1997). Speed of finger tapping and goal attainment after unilateral cerebral vascular accident. *Arch. Phys. Med. Rehabil.* **78,** 847–852.

Rascol, O., Sabatini, U., Chollet, F., Celsis, P., Montastruc, J. L., Marc-Vernes, J. P., and Rascol, A. (1992). Supplementary and primary sensory motor area activity in Parkinson's disease. Regional cerebral blood flow changes during finger movements and effects of apomorphine. *Arch. Neurol.* **49,** 144–148.

Rascol, O., Sabatini, U., Chollet, F., Celsis, P., Montastruc, J. L., Marc-Vergnes, J. P., and Rascol, A. (1994). Normal activation of supplementary motor area in patients with Parkinson's disease undergoing long-term treatment with levodopa. *J. Neurol. Neurosurg. Psychiatry* **57,** 567–571.

Rijntjes, M., Tegenthoff, M., Liepert, J., Leonhardt, G., Kotterba, S., Muller, S., Kiebel, S., Malin, J. P., Diener, H. C., and Weiller, C. (1997a). Cortical reorganization in patients with facial palsy. *Ann. Neurol.* **41,** 621–630.

Rijntjes, M., Dettmers, C., Rzanny, R., Kiebel, S., and Weiller, C. (1997b). A blueprint for movement in the human motor cortex. *NeuroImage* **5,** S229.

Rizzolatti, G., Luppino, G., and Matelli, M. (1998). The organization of the cortical motor system: New concepts. *Electroencephalogr. Clin. Neurophysiol.* **106,** 283–296.

Rougement, D., Baron, J. C., Lebrun-Grandie, P., Bousser, M. G., Cabanis, E., and Laplane, D. (1983). Debit sanguin cérébral et extraction d'oxygène dans les hémiplégies lacunaires. *Rev. Neurol.* **139,** 277–282.

Rouiller, E. M., Yu, X., Moret, V., Tempini, A., Wiesendanger, M., and Liang, F. (1998). Dexterity in adult monkeys following early lesion of the motor cortical hand area: The role of cortex adjacent to the lesion. *Eur. J. Neurosci.* **10,** 729–740.

Sadato, N., Pascual-Leone, A., Grafman, J., Deiber, M. P., Ibanez, V., and Hallett, M. (1998). Neural networks for Braille reading by the blind. *Brain* **121,** 1213–1229.

Sanes, J. N., Wang, J., and Donoghue, J. P. (1992). Immediate and delayed changes of rat motor cortical output representation with new forelimb configurations. *Cereb. Cortex* **2,** 141–152.

Seitz, R., Schlaug, G., Kleinschmidt, A., Knorr, U., Nebeling, B., and Wirrwar, A. (1994). Remote depressions of cerebral metabolism in hemiparetic stroke: Topography and relation to motor and somatosensory functions. *Hum. Brain Mapp.* **1,** 81–100.

Seitz, R. J., Höflich, P., Binkofski, F., Tellmann, L., Herzog, H., and Freund, H. J. (1998). Role of the premotor cortex in recovery from middle cerebral artery infarction. *Arch. Neurol.* **55,** 1081–1088.

Shibasaki, H., Sadato, N., and Lyshkow, H. (1993). Both primary motor cortex and supplementary motor area play a role in complex finger movements. *Brain* **116,** 1387–1398.

Singer, W. (1993). Synchronization of cortical activity and its putative role in information processing and learning. *Annu. Rev. Physiol.* **55,** 349–374.

Singer, W. (1999). Neurobiology. Striving for coherence [news; comment]. *Nature* **397,** 391–393.

Small, S. (1994). Pharmacotherapy of aphasia: A critical review. *Stroke* **25,** 1282–1289.

Taub, E., Williams, M., Barro, G., and Steiner, S. S. (1978). Comparison of the performance of deafferented and intact monkeys on continuous and fixed ration schedules of reinforcement. *Exp. Neurol.* **58,** 1–13.

Taub, E., Harger, M., Grier, H. C., and Hodos, W. (1980). Some anatomical observations following chronic dorsal rhizotomy in monkeys. *Neuroscience* **5,** 389–401.

Taub, E., Miller, N., Novack, T. A., Cook, E. W., Fleming, W. C., and Nepomuceno, C. S. (1993). Technique to improve chronic motor deficit after stroke. *Arch. Phys. Med. Rehabil.* **74,** 347–354.

Taub, E., Crago, J., Burgio, L. D., Groomes, T. E., Cook, E. W., and DeLuca, S. C. (1994). An operant approach to rehabilitation medicine: Overcoming learned nonuse by shaping. *J. Exp. Anal. Behav.* **61,** 281–293.

Ungerleider, L. G., Coutney, S., and Haxby, J. V. (1998). A neural system for human visual working memory. *Proc. Natl. Acad. Sci. U.S.A.* **95,** 883–890.

Veasey, S. C., Fornal, C. A., Metzler, C. W., and Jacobs, B. L. (1995). Response of serotonergic caudal raphe neurons in relation to specific motor activities in freely moving cats. *J. Neurosci.* **15,** 5346–5359.

Walker-Batson, D., Smith, P., Curtis, S., Unwin, H., and Greenlee, R. (1995). Amphetamine paired with physical therapy accelerates motor recovery after stroke. *Stroke* **26,** 2254–2259.

Watson, J. D. G., Myers, R., Frackowiak, R. S. J., Hajnal, J. V., Woods, R. P., and Mazziotta, J. C. (1993). Area V5 of the human brain: Evidence from a combined study using positron emission tomography and magnetic resonance imaging. *Cereb. Cortex* **3,** 79–94.

Weder, B., and Seitz, R. (1994). Deficient cerebral activation pattern in stroke recovery. *NeuroReport* **5,** 457–460.

Weiller, C. (1998). Imaging recovery from stroke. *Exp. Brain Res.* **123,** 13–17.

Weiller, C., Chollet, F., Friston, K. J., Wise, R. J. S., and Frackowiak, R. S. J. (1992). Functional reorganisation of the brain in recovery from striatocapsular infarction in man. *Ann. Neurol.* **31,** 463–472.

Weiller, C., Ramsay, S., Wise, R. J. S., Friston, K. J., and Frackowiak, R. S. J. (1993). Individual patterns of functional reorganization in the human cerebral cortex after capsular infarction. *Ann. Neurol.* **33,** 181–189.

Weiller, C., Insensee, C., Rijntjes, M., *et al.* (1995). Recovery from Wernicke's aphasia: A positron emission tomography study. *Ann. Neurol.* **37,** 723–732.

Wong, D. T., Fuller, R. W., and Robertson, D. W. (1990). Fluoxetine and its two enantiomers as selective serotonin uptake inhibitors. *Acta Pharm. Nord.* **2,** 171–180.

Yousry, T. A., Schmid, U. D., Alkadhi, H., Schmidt, D., Peraud, A., Buettner, A., and Winkler, P. (1997). Localization of the motor hand area to a knob on the precentral gyrus. A new landmark. *Brain* **120,** 141–157.

25

Therapeutics: Pharmacologic

William H. Theodore

National Institutes of Health, Bethesda, Maryland 20892

I. Introduction

II. Imaging Studies of the Effects of Antiepileptic Drugs

III. Drugs in Basal Ganglia and Movement Disorders

IV. Cognitive Dysfunction and Dementia

V. Imaging Studies of "Antipsychotic" Drugs

VI. Conclusion: The Possibility of Individualizing Therapy

References

I. Introduction

A. Clinical Pharmacology and Neurology

Until fairly recently, neurology was a relatively contemplative specialty, showing marked parallels with disciplines such as archaeology. Like the successive layers of destruction uncovered during excavation of an ancient city, neurologic dysfunction occurs in levels, and one could do about as much for most patients as archaeologists could for the political prospects of the Trojans.

Luckily, the situation has improved, and a wide range of therapeutic alternatives has become available. Half a dozen new antiepileptic drugs have been approved for use in the United States since 1995, for example, as opposed to only one from 1978 to 1994; additional agents are available in other countries. Several entirely new treatments for Parkinson's disease are now available.

B. Value of Imaging for Pharmacologic Studies

Of course, the plethora of new drugs presents pitfalls as well as opportunities. Clinical experience remains limited for many. Much less information on pharmacokinetics and pharmacodynamics is available on new drugs like lamotrigine (LTG) than older agents such as phenytoin (PHT). For example, studies showing the relation of plasma levels to clinical effect still need to be done for all the new antiepileptic drugs. Unfortunately, it is becoming more and more difficult to perform controlled trials to obtain these data after a drug has been approved for clinical use.

Pharmacodynamic measures are even more difficult to obtain than pharmacokinetic data (Macdonald and Meldrum, 1995). Moreover, it is more difficult to interpret the effect of drugs on the nervous system than on other organs, like the liver, whose function can be followed closely by chemical parameters or biopsied with relative ease and safety. Cerebrospinal fluid has been a disappointing marker for brain function.

Neuroimaging studies have the potential to provide physiologic data predicting drug efficacy and toxicity as well as hints on mechanisms of action and central pharmacokinetics. Direct observations of therapeutic or adverse drug effects may have more obvious clinical relevance. Clinical trials, however, are complex and costly, may be difficult to interpret, and only rarely can provide information on mechanisms of drug action. In addition, it is very difficult to use trial results to guide therapy in a particular case by trying to predict individual patient responses to drugs. Imaging studies may be a "shortcut" to these goals.

C. General Considerations (Table I)

Drug effects may depend on a number of pharmacokinetic factors, such as dose, route of administration, volume of distribution, half-life, clearance, and metabolic pathways. Subject-related factors, such as age, illness, other drugs being taken, and possibly genetic differences in pharmacodynamic parameters such as CNS receptors might play a role as well. It is also important to make sure that doses used in imaging studies are clinically relevant. In addition to direct CNS effects, drugs may have system actions that can influence brain function. Alterations in pulse, blood pressure, and cardiac output may affect cerebral blood flow (CBF), for example. This is more likely to be a factor during acute, intravenous administration than in more chronic dosing studies. While some studies image discrete receptor systems, many measure effects on cerebral glucose metabolism (CMRGlc) or CBF. To interpret the clinical implications of the study, it is important to have a sense of the relation of those global markers to disease phenomena or drug toxicity.

Several imaging modalities are available for drug studies (Table II). At present, PET and MRS can provide the best data for formal studies. The potential clinical role of the various techniques will be discussed below.

Table I   Imaging Studies of Pharmacologic Agents: Important Parameters

Route of administration
Chronic versus acute administration
Peripheral pharmacokinetics
Central pharmacokinetics
Individual subject variables
Systemic effects
Clinical relevance of doses used

II. Imaging Studies of the Effects of Antiepileptic Drugs

A. PET Studies of CMRGlc and CBF

Differences in the effect of antiepileptic drugs (AEDs) on cerebral metabolic rate for glucose (CMRGlc) may reflect different mechanisms of drug action and be related to adverse effects on neuropsychological performance. The predominant effect of phenobarbital (PB) is enhancement of chloride channel responses to γ-aminobutyric acid (GABA), although decreased glutamate responses and blockade of voltage-activated calcium channels also occur (Crane et al., 1978; MacDonald and Schulz, 1981; Heyer and MacDonald, 1982). In patients with seizures who had a mean blood level of 27 mg/L, CMRGlc was 37% lower on PB than their baseline AEDs (Theodore et al., 1986a; Theodore, 1988). In a study of barbiturate coma, gray matter CMRGlc was reduced to about 33% of resting conscious levels (Blacklock et al., 1987) (Fig. 1). Benzodiazepines, which also interact with the GABA receptor complex, have also been shown to reduce global CMRGlc by up to 20% (Foster et al., 1987). This effect may occur when anxiolytic but not adverse motor effects are evident (de Wit et al., 1991).

The effect on brain metabolism may be associated with the significant impairment in cognitive function that is associated with PB treatment (Vining et al., 1987). Transient exposure of immature rats to barbiturates leads to reduced CMRGlc when measured during adult life as well as adverse neurodevelopmental and behavioral effects (Pereira de Vasconcelos et al., 1990). These observations parallel results from a long-term study of febrile seizure prophylaxis. After 2 years on PB, mean IQ was 7.03 points lower in the PB than the placebo group, and the difference persisted even 6 months after drug had been stopped (Farwell et al., 1990).

In contrast to PB and benzodiazepines, PHT and carbamazepine (CBZ) exert their antiepileptic action by use-dependent inhibition of sodium channels (Woodbury and Kemp, 1971; McLean and MacDonald, 1986b). They both decrease CMRGlc by 10–15%, considerably less than PB (Theodore et al., 1986b, 1989a). At therapeutic levels, CBZ and PHT have only mild effects on rapid motor movements, and adverse cognitive effects are difficult to demonstrate (Thompson and Trimble, 1982; Trimble, 1987; Meador et al., 1990).

PHT, however, is particularly associated with cerebellar function impairment, and some evidence suggests that cerebellar atrophy may occur rarely after prolonged exposure at high levels (Theodore et al., 1987). Several studies have reported reduced CMRGlc in patients on PHT (Theodore et al., 1987; Seitz et al., 1996)

Table II   Advantages and Disadvantages of Imaging Techniques for Drug Studies

	PET	SPECT	MRI
Parameter	CMRGlc, CBF, receptors	CBF (receptors)	Metabolite levels, activation/CBF
Resolution	4–6 mm	8–10 mm	~1 mm to 1 cm
Advantages	Reliable, quantitative, versatile	Uses longer half-life isotopes, standard cameras	Standard hardware, no radiation, easy to repeat
Disadvantages	Need on-site cyclotron, radiation exposure	Radiation exposure, poor quantitation, low resolution	Complex software, labor intense, techniques in evolution

(Fig. 2). In a subgroup of patients with a history of drug intoxication in infancy, cerebellar CMRGlc was particularly decreased, suggesting a developmental vulnerability for AED toxicity (Seitz *et al.,* 1996). A study of the effects of phenobarbital on infantile rats suggested that the cerebellum was more sensitive than other structures to drug-induced developmental impairment (Pereira de Vasconcelos and Nehlig, 1987). These studies, in conjunction with the data for prolonged reduction in IQ scores, show the vulnerability of the developing nervous system to adverse effects of therapeutic drugs.

Valproate (VPA) may act either by raising wholebrain GABA levels (Loscher, 1981), an effect that may not be important at clinically relevant human doses, or by blocking voltage-dependent sodium channels (McLean and Macdonald, 1986a). In normal volunteers, at a mean blood level of 82 mg/L, CMRGlc decreased by only 9%, consistent with its negligible effect on cognitive performance (Gaillard *et al.,* 1996) (Fig. 3). Global CBF was reduced by 15%, most prominently in the thalamus (Fig. 4). This may be related to the drug's efficacy against primary generalized seizures.

When VPA was added to CBZ monotherapy, however, the effect was slightly greater than for either drug alone (Fig. 5). It is conceivable that this could be due, in part, to enhanced formation of the active CBZ metabolite, CBZ 10,11 -epoxide, which is equipotent with CBZ, but less protein bound, and so achieves higher brain levels (Theodore *et al.,* 1989b).

Vigabatrin (VGB) is an investigational drug in the United States but marketed in a number of countries around the world. It is particularly interesting as it has a clearly defined mechanism of action, blocking GABA-transaminase, the main GABA degradative enzyme (Holland *et al.,* 1992; Spanaki *et al.,* 1999). Added to baseline CBZ, it reduced global CMRGlc by 8.1 ± 6.5% and global CBF by 13.1 ± 10.4%. There was a significant relation between the rise in CSF GABA and fall in CMRGlc (Fig. 6) (Spanaki *et al.,* 1999). Patients on VGB had lower Zung and Bunney–Hamburg anxiety scores, which also decreased as CSF GABA levels rose (Theodore *et al.,* 1998).

It is interesting that VGB reduces CMRGlc so much less than PB or benzodiazepines, despite its action at the

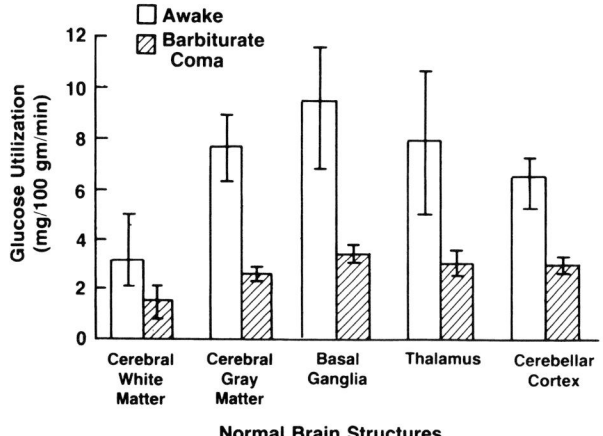

Figure 1 During barbiturate coma, cortical CMRGlc is reduced to about 33% of normal. Adapted from Blacklock *et al.* (1987).

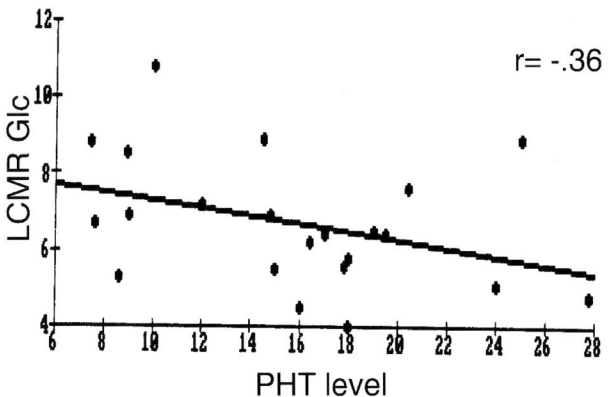

Figure 2 There is a significant relationship between PHT serum levels and cerebellar glucose metabolism. Adapted from Theodore *et al.* (1986b). Effect of phenytoin on human cerebral glucose metabolism. *J. Cereb. Blood Flow Metab.* **6,** 315–320.

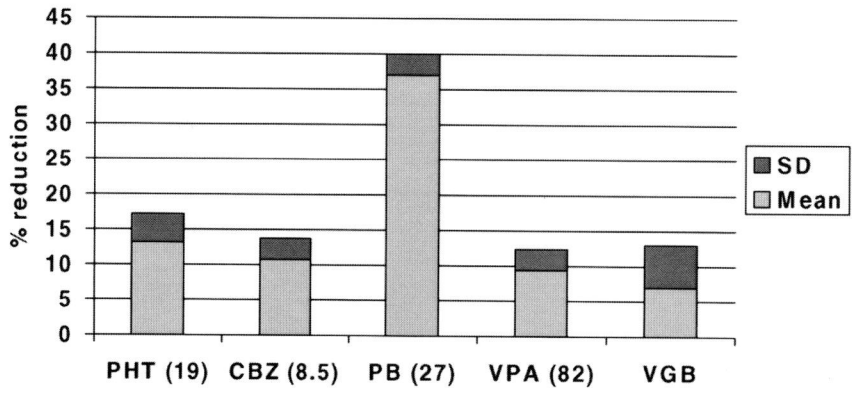

Figure 3 Barbiturates reduce CMRGlc to a greater extent than other AEDs.

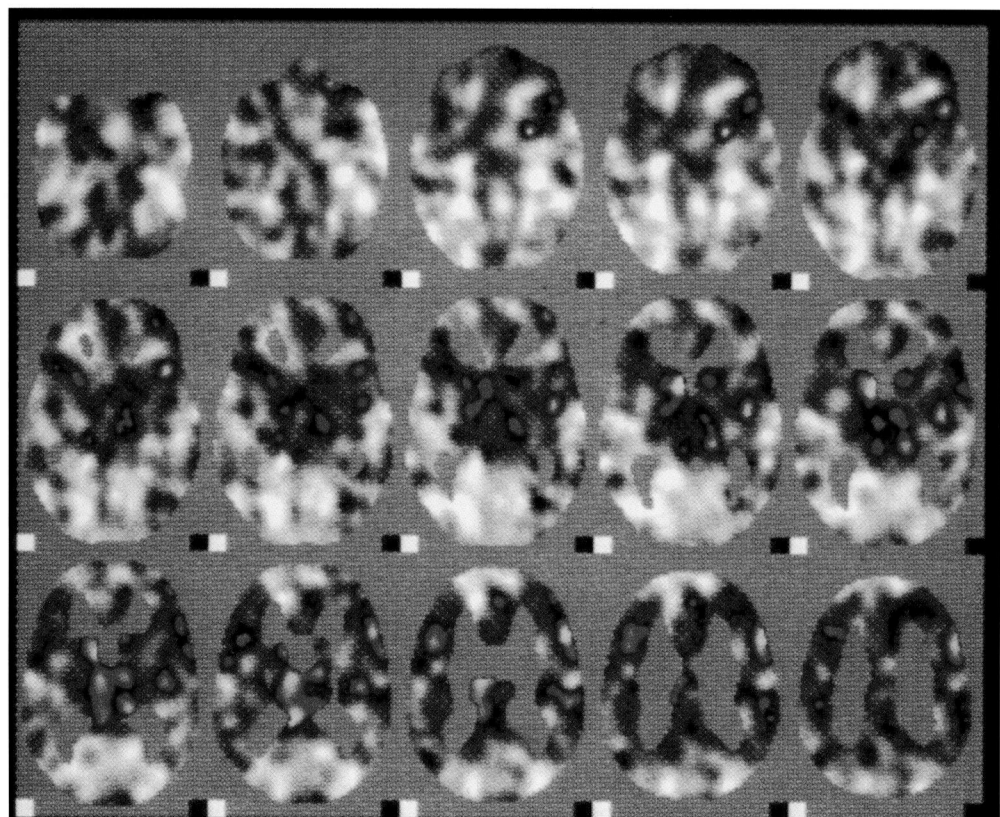

Figure 4 Valproic acid had its greatest effect on CBF in the thalamus, an effect that might be related to its clinical role in the treatment of absence seizures. Reproduced from Gaillard *et al.* (1996). Effect of valproate on cerebral metabolism and blood flow: An [18]F 2-deoxyglucose and [15]O water positron emission tomography study. *Epilepsia* **37**(6), 515–521.

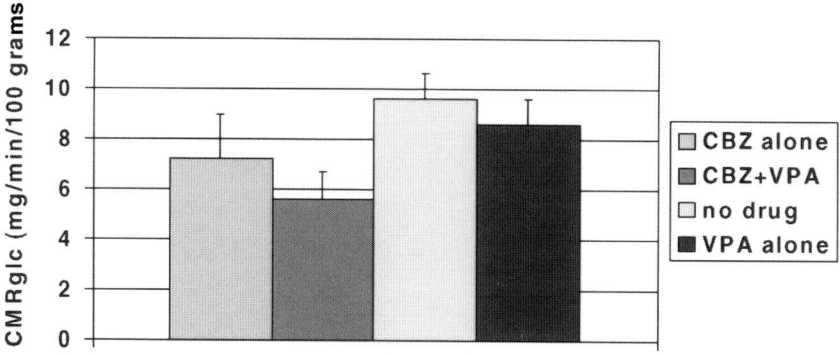

Figure 5 Valproic acid and carbamazepine together reduced CMRGlc more than either agent alone. This could be related to either pharmacokinetic or pharmacodynamic interactions. Adapted from Leiderman *et al.* (1989) and Gaillard *et al.* (1996).

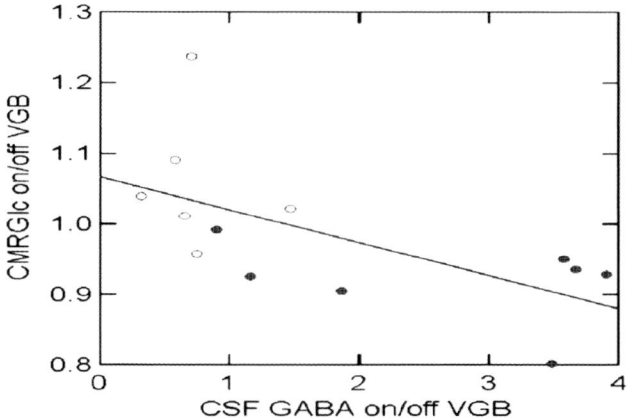

Figure 6 There was a significant relation between the level of CMRGlc and the relative rise in CSF GABA when patients were given the AED vigabatrin.

same receptor site. At GABAergic synapses themselves, feedback inhibition might limit neuronal firing. VGB's effect on GABA uptake and release may be reversible through a mechanism of action dependent on tissue concentration of the compound (Loscher, 1980). Recent work has suggested that VGB may cause a "use" or frequency-dependent enhancement of GABA-mediated transmission (Jackson *et al.*, 1994). These mechanisms might help account for the mild effect of VGB on CMRGlc and CBF.

Spanaki *et al.* (1999) found that inferior medial and lateral temporal cortex showed a more consistent statistically significant reduction than other regions (Fig. 7). It is interesting that one animal study found VGB-related increases in GABA were greater in the hippocampus than other brain regions (Pericic, 1980). PB, PHT, and CBZ, drugs effective against focal epilepsies, did not show regionally specific effects on CMRGlc or CBF (Theodore *et al.*, 1986a,b, 1989a). This is consistent

with a study by Baron *et al.* (1983), using [^{11}C]PHT, which found a uniform distribution of the drug without excess accumulation in epileptic foci. Interestingly, Peyron *et al.* (1994) reported that the direct GABA-A agonist THIP (4,5,6,7-tetrahydroisoxazolo[5,4-*c*]pyridin-3-ol) increased CMRGlc in temporal lobe epileptic foci to a greater degree than in the brain as a whole. Other THIP studies have usually found decreased metabolism, however (Theodore, 1988).

B. Receptor Studies

In contrast to basal ganglia disorders, relatively little work has been done on the interaction of AEDs with specific receptor systems. The role for any receptor system in epilepsy is not nearly as well established as is that of the dopaminergic system in Parkinson's disease. However, several studies have found reduced benzodiazepine receptor binding in epileptic foci. Investigations of the effect of vigabatrin have been performed as well. Weber *et al.* (1999) found no change in benzodiazepine receptor density in patients on VGB even though brain GABA levels measured by MRS rose. Interestingly, the potency of diazepam in displacing [^{11}C]flumazenil binding in baboons was enhanced in vigabatrin–pretreated animals, contrasting with the reduced anticonvulsant effects of diazepam in those animals in the pentylenetetrazole model, suggesting that increasing GABA levels reduced diazepam intrinsic efficacy (Schmid *et al.*, 1996). Staedt *et al.* (1995) reported that successful control of seizures with CBZ eliminated left temporal focal EEG slowing and returned BZP receptor density, measured on [^{123}I]iomazenil SPECT, to normal.

The availability of drugs with clear mechanisms of action allows investigation of receptor system interactions. Dewey *et al.* (1992) found that striatal [^{11}C]raclopride, a D_2-dopamine receptor ligand, binding increased

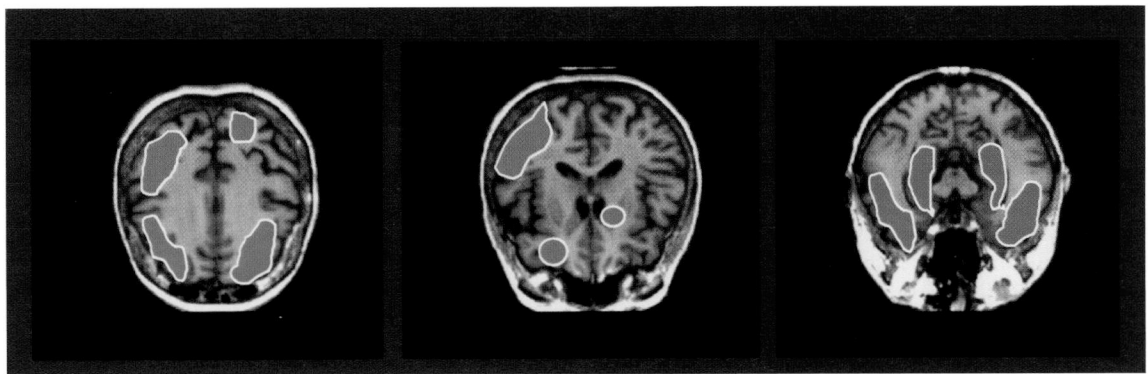

Figure 7 Vigabatrin led to a global reduction in CMRGlc, which was most prominent in bilateral temporal regions.

following both acute VGB and lorazepam administration to baboons, suggesting that drug-induced potentiation of GABAergic transmission affects D_2 receptor occupancy or binding potential. However, Ring *et al.* (1992) reported that a month of treatment with VGB in man was associated with a decrease in specific binding of SPECT [^{123}I]IBZM to basal ganglia D_2 receptors. The difference between these two studies could be due to acute versus chronic administration.

BZPs are effective AEDs but tolerance occurs fairly rapidly. The BZP alprazolam induced sedation at 16% receptor occupancy measured on [^{123}I]iomazenil SPECT. During chronic dosing, unoccupied receptors decreased 10% over 1 week but returned to baseline values in 2 weeks (Fujita *et al.*, 1999).

Surgical therapy might also affect receptor distribution. Savic *et al.* (1998) reported that temporal lobectomy reversed the preoperative reduction in BZP receptor density in primary projection areas of the epileptogenic zone.

Several studies have reported that opiate receptor number or binding potential is upregulated in epileptic foci (Frost *et al.*, 1988). Endogenous opiates are involved in regulation of human CBF and possibly in hypoperfusion in epileptic foci. In patients with CPS, naloxone had no effect on glucose metabolism but reduced blood flow 7–12% 45–60 min after infusion of 1 mg/kg (Theodore *et al.*, 1993). Interestingly, the degree of lateral temporal CBF asymmetry in patients with >10% baseline hypoperfusion was reduced, consistent with the suggestion that opiates might play a role in the hypoperfusion seen in epileptic foci.

C. Magnetic Resonance Spectroscopy

Magnetic resonance spectroscopy (MRS) has been used by a number of investigators to study the effects of AEDs. Petroff *et al.* (1995) reported that occipital GABA in 10 controls was 1 μmol/g, compared with 2.6 μmol/g in patients taking 3–6 g of vigabatrin. Glutamate was increased by 1.9 and glutamine decreased by 0.8 μmol/g in patients on VGB compared to those on other drugs. Spectra were acquired using a 2-T magnet, 8-cm coil, and 14-mm^3 volume in occipital cortex. Increasing vigabatrin doses up to 60 mg/kg/day correlated with increasing GABA levels and decreasing seizure frequency (Petroff *et al.*, 1996a). Patients on 3300–3600 mg/day had higher GABA levels than those on 1200–2400 mg/day (Petroff *et al.*, 1996b). Doubling the dose to 6 g/day did not further increase levels. Serial measurements, moreover, showed a gradual decline in GABA, despite constant doses, over 1–2 years (Petroff *et al.*, 1996b).

Patients with partial seizures had lower occipital GABA levels than controls, and there was a significant correlation between higher levels and lower seizure frequency. Petroff *et al.* (1996c) found that patients with complex partial seizures had GABA levels of 1.03 vs 1.18 mmol/kg for subjects without epilepsy ($p < 0.02$). In the patient group, low GABA levels were correlated with recent seizures. Patients who had had seizures within a day of measurement had mean levels of 0.92 mmol/kg, whereas those with none for 5 years had 1.28 mmol/kg.

Vigabatrin increases CSF GABA, and there was a significant relation between CSF and MRS GABA measurements in patients taking a mean vigabatrin dose of 3750 mg/day (Theodore *et al.*, 1997).

Part of the effect on seizures may be due to increased homocarnosine as well as GABA. Homocarnosine, a dipeptide of GABA and histidine, is thought to be an inhibitory neuromodulator synthesized in subclasses of GABAergic neurons. Daily low-dose (2 g) vigabatrin increased both homocarnosine and GABA. Larger doses of vigabatrin (4 g) further increased homocarnosine but changed GABA levels minimally. Patients whose seizure control improved with the addition of vigabatrin had higher mean homocarnosine concentrations than, but the same mean GABA concentrations as, those whose seizure control did not improve (Petroff *et al.*, 1998).

The effect of several other AEDs has been studied as well. Topiramate (TOP) is another new drug that probably has multiple mechanisms of action. Brain GABA concentrations were elevated by 64% at 6 h after a dose (Kuzniecky *et al.*, 1998). Patients taking chronic TOP therapy had elevated mean brain GABA, homocarnosine, and pyrrolidinone concentrations. Two patients already on vigabatrin had further GABA increases when TOP was added (Petroff *et al.*, 1999).

Gabapentin is a relatively new AED that was designed to mimic the effect of GABA but originally thought to have little GABAergic action, based on animal studies. In 14 patients, however, GABA was elevated compared with 14 others matched for antiepileptic drug treatment, except gabapentin, and appeared to be higher in patients taking 3300–3600 mg/day than in those taking standard doses of 1200–2400 mg/day (Petroff *et al.*, 1996d).

ACTH, which is used to treat infantile spasms, a severe epileptic encephalopathy in children, decreased brain *N*-acetylaspartate content. This effect may be related to the brain atrophy noted on serial MRI scans in patients taking the drug (Gee *et al.*, 1988; Kokate *et al.*, 1994; Maeda *et al.*, 1997).

D. EEG and Evoked Potentials

The effect of AEDs on the EEG varies (Bazil and Pedley, 1995). Drugs interacting with the GABA–benzodiazepine receptor complex may be more likely to affect background frequencies. BZPs increase activity in the 25- to 35-Hz range and reduce alpha voltage. Computerized EEG analysis showed a relation between BZP plasma levels and frontal beta. At higher doses, slow activity becomes prominent. Similar effects have also been shown for the experimental drug tiagabine, which inhibits GABA reuptake (Coenen *et al.,* 1995). Most studies of barbiturates have used IV drug: increased 18- to 25-Hz activity (most prominent frontally) is followed by "barbiturate spindles" and rhythmic delta as levels increase (Stanski *et al.,* 1984). High PB levels increase visual-evoked response latencies in children (Brinciotti, 1994). Phenytoin, carbamazepine, and valproic acid have little overt effect at therapeutic doses, although slow activity may appear at toxic levels. Some computerized studies showed decreased "power" in low and increased power in high-frequency bands at therapeutic PHT levels (Fink *et al.,* 1979). Others suggested that PHT and CBZ may increase theta and delta activity and decrease alpha activity (Salinsky *et al.,* 1994).

EEG measures might be used as surrogates for clinical effects. In rats, EC_{50} for EEG effects (increased amplitude in 11.5- to 30-Hz band) correlates with GABA receptor affinity measured by $[^3H]$-flumazenil displacement as well as activity in the PTZ seizure model (Mandema *et al.,* 1991). Peak velocity of saccadic and smooth-pursuit eye movements are reduced by BZPs and can be used to derive concentration–effect relations (van Stevenick *et al.,* 1992). Alterations in delta range activity may predict hypnotic, and in alpha, anxiolytic, and anticonvulsant properties of BZPs (Mandema and Danhof, 1992). EEG power spectra variables correlated linearly with the decrease in CMRGlc during propofol and isoflurane anesthesia (Alkire, 1998).

Some AEDs, particularly BZPs and barbiturates, reduce interictal spike frequency. PHT and CBZ show no consistent effect; VPA reduces generalized but not focal discharges (Sato *et al.,* 1982). Preliminary data suggest that computerized analysis of spike morphology can be used to predict clinical drug effects. Frost *et al.* (1986) found that decreases in spike amplitude, duration, and "sharpness" were associated with achievement of seizure control in children given CBZ or PB.

Drug toxicity may be associated with electrophysiological abnormalities. The competitive NMDA antagonist (carboxy piperazin-4yl)-propyl-l-phosphonic acid (CPP-ene) led to poor concentration, sedation, ataxia, depression, and memory impairment as well as background slowing (Sveinbjornsdottir *et al.,* 1993). In rats, both competitive and noncompetitive NMDA antagonists induce three dose-dependent stages of EEG patterns: (1) increase in cortical desynchronization periods; (2) increase in amplitude of cortical high-frequency (20–30 Hz), low-voltage (30–50 μV) background activity; (3) appearance of cortical slow (2–3 Hz) wave-sharp

wave complexes. These EEG changes are accompanied by stimulatory–depressive behavioral effects such as stereotypy (circling, head weaving) and ataxia (Popoli *et al.,* 1994). Frost *et al.* (1995) performed a prospective study of the effect of CBZ on EEG background frequency in 16 previously untreated patients aged 5–14. Patients whose performance at 1 year showed deterioration on the Wisconsin Intelligence Scale for Children (WISC) arithmetic and picture completion tests had had significantly greater decreases in alpha frequency when tested soon after initiation of therapy. Quantitative measures of beta range activity using fast Fourier transform distinguished the effect of a BZP agonist from a non-BZP agonist anxiolytic; the former increased beta activity and impaired digit symbol substitution test (DSST) performance and word list recall, whereas the latter impaired recall but not EEG or DSST (Greenblatt *et al.,* 1994).

Drugs that interact directly with the GABA–benzodiazepine receptor complex may have greater effects on cognitive function than channel-blocking agents such as PHT or CBZ, although differences can be hard to detect and test-specific. In one study, for example, only the digit symbol test revealed differences between PB, CBZ, and PHT (Meador *et al.,* 1990). Effects may be task-specific as well as agent-specific. PHT and CBZ increased theta-band activity and impaired performance on story recall but did not affect complex figure recall or selective reminding performance (Meador *et al.,* 1990, 1993).

BZPs reduce saccadic eye movement velocity and increase duration (Glue, 1991). Oral barbiturates and alcohol decrease peak velocity. Other AEDs such as CBZ and PHT, but not lamotrigine or vigabatrin, may impair smooth pursuit, adaptive tracking, and saccades (Glue, 1991; Peck, 1991; Zaccara *et al.,* 1992).

E. Transcranial Magnetic Stimulation

A new noninvasive technique, transcranial magnetic stimulation (TMS), can be used to evaluate central neurophysiologic parameters in man (Jacobs *et al.,* 1991; Kujirai *et al.,* 1993; Ziemann *et al.,* 1996, 1998). Thresholds for induction of a motor-evoked potential using TMS depend, at least in part, on membrane excitability. Since TMS activates predominantly corticocortical connections targeting pyramidal tract neurons, motor thresholds reflect the membrane excitability of parallel fibers and their terminations in the motor cortex and, in addition, the excitability of the postsynaptic targets (corticospinal neurons). Intracortical facilitation/inhibition measured using paired-pulse paradigms is thought to be related to GABAergic mechanisms, whereas intracortical excitability is thought to reflect the synaptic excitability of the interneuronal circuitry within the motor cortex regulated by ion channel permeability (Ziemann *et al.,* 1996). A single dose of lamotrigine or carbamazepine, both blockers of voltage-gated Na channels, increases the thresholds for induction of motor-evoked potentials without affecting intracortical inhibition. Paired-pulse intracortical inhibition is increased (and facilitation decreased) by drugs such as the benzodiazepine antianxiety agent lorazepam, ethanol, and gabapentin that enhance the activity of GABA (Ziemann *et al.,* 1996, 1998). Drugs such as dextromethorphan (Ziemann *et al.,* 1998) and riluzole (Liepert *et al.,* 1997) that antagonize the action of glutamate produce similar effects. This method offers a relatively simple approach to investigate drug mechanisms of action and potency. AEDs may have multiple effects and their relative weight can be assessed by TMS.

III. Drugs in Basal Ganglia and Movement Disorders

Imaging studies can be used for detailed study of the interaction of pharmacologic, anatomic, and physiologic mechanisms in Parkinson's disease. CBF activation of the putamen, supplementary motor area, and cingulate cortex, measured with inhaled oxygen-15 CO_2 PET, is impaired in patients with Parkinson's disease compared to normal controls when they are off treatment and performing motor tasks in the "off" state (Jenkins *et al.,* 1992). At rest, the dopaminergic agonist apomorphine had no effect on CBF. Activation of the supplementary motor area significantly improved when the akinesia was reversed with apomorphine, showing the role of the dopaminergic system in facilitating freely generated motor tasks. The level of dopaminergic "tone" may affect CBF activation of supplementary motor and dorsolateral prefrontal cortex during voluntary motor activity in patients with Parkinson's disease, correlating with difficulty in performing patterned motor activity (Brooks, 1998). Activation is reduced after 12 h off medication and improved by acute apomorphine administration.

It may be possible to use imaging studies to predict response to therapy. In patients with early, but not advanced, PD, apomorphine infusion decreased the L-[^{11}C]dopa influx rate, particularly in the dorsal putamen, suggesting that upregulated presynaptic inhibitory feedback in the putamen is lost with chronic dopamine agonist treatment as disease progresses (Ekesbo *et al.,* 1999). In the superior and inferior putamen, superior and middle caudate, ventral striatum, and inferior thalamus, relative reduction in metabolism on the side contralateral to predominant Parkinsonian signs was associated with L-dopa unresponsiveness (Dethy *et al.,*

1998). Mean BG/FC ratio uptake of [^{123}I]iodobenzamide ([^{123}I]IBZM), a dopamine receptor ligand, on SPECT in basal ganglia contralateral to predominant Parkinsonian symptoms was higher in medication responders than nonresponders (Hertel *et al.,* 1997).

IV. Cognitive Dysfunction and Dementia

Patients with SDAT (senile dementia of the Alzheimer's type) have reduced CBF and CMRGlc, particularly in the frontal and temporal neocortex, paralleling neuropathological findings and patterns of cholinergic innervation (Bryant and Jackson, 1998). Physostigmine, an anticholinesterase, increased CBF measured with [^{99m}Tc]HMPAO SPECT in the right frontal cortex of patients with SDAT. In another study, posterior parietal CBF increases were seen in patients but not controls (Bryant and Jackson, 1998). Increased CMRGlc has been reported as well. Physostigmine reduces the CBF activation during a "frontal lobe" working memory task in normal volunteers, suggesting increased processing efficiency. Reports of the effect of tacrine, the only drug approved for SDAT treatment, on CBF have been contradictory, and the link between blood flow modulation and cognitive improvement is weak. In one study, tacrine was given to three patients with mild SDAT in doses between 80 and 160 mg daily for 13–31 months. Improvement in [^{11}C]nicotine binding, CBF, EEG, trail making, and block design tests occurred earlier after initiation of tacrine treatment compared with the glucose metabolism, which did not increase until after several months of tacrine treatment (Nordberg *et al.,* 1998). After treatment with donepezil, improvement was noted on neuropsychological testing and occipital perfusion on SPECT (Warren *et al.,* 1998). Although some investigators have tried to use EEG to monitor the effect of therapy for SDAT, the results have been inconsistent (Hegerl and Moller, 1997).

Several studies have examined the relation between drugs acting on the dopaminergic system and cognitive activation paradigms. Patients with Korsakoff's psychosis scanned during a verbal fluency task showed increased frontal lobe blood flow measured by [^{99m}Tc] HMPAO SPECT after clonidine infusion compared with saline control (Moffoot *et al.,* 1994). This increased CBF was associated with improved task performance. Increased CBF in the posterior cingulate was also observed. On the basis of these results, the authors suggested that modulation of performance by the drug might be related to effects on intentional systems. Coull *et al.* (1997) found that in normal volunteers, clonidine reduced thalamic CBF during attention and working memory tasks as well as at rest; the effect was greater at rest, suggesting that arousal may interact with pharmacologic effects on CBF. Increased CBF was also found in the anterior cingulate and superior temporal cortex in all the conditions. These findings underline the complexity of drug effects, showing task, region, and state interactions.

V. Imaging Studies of "Antipsychotic" Drugs

PET can be used to measure central receptor occupancy and estimate the effect of pharmacologic treatments. D_2 dopamine receptor occupancy measurements may predict antipsychotic response and extrapyramidal side effects (Table III). There is a high occupancy of central D_2-dopamine receptors in patients that are treated with clinically effective doses of classical antipsychotic drugs (Farde, 1997). For "typical" neuroleptics, therapeutic response is associated with occupancy levels as low as 60–70%, which may be achieved at lower doses than routinely used; extrapyramidal side effects, but not therapeutic response, increases with higher occupancy levels (Heinz et al., 1996). A dose of 6 mg/day of risperidone led to mean D_2 receptor occupancy of 82% and development of extrapyramidal side effects, whereas a dose of 3 mg/day, equally clinically effective, led to an occupancy of 72% and reduction in extrapyramidal side effects (Nyberg et al., 1999).

Using [^{11}C]raclopride and [^{18}F]setoperone PET to measure D_2 and 5-HT$_2$ occupancies, Kapur et al. (1999) compared three antipsychotics, clozapine, risperidone, and olanzapine, that have fewer extrapyramidal side effects than older drugs. Clozapine showed a much lower

Table III Antipsychotic Doses and D_2 Receptor Occupancy[a]

Drug	Dose	Occupancy[b] (%)
Chlorpromazine	100	80
Thioridazine	100	75
Trifluperazine	5	80
Perphenazine	4	79
Perphenazine	30	88
Flupenthixol	5	74
Haloperidol	2	81
Haloperidol	3	85
Haloperidol	6	86
Clozapine	150	40
Clozapine	300	65

[a] Adapted from Farde et al., 1992.
[b] Occupancy levels of ~80% are associated with extrapyramidal side effects.

D_2 occupancy (16–68%) than risperidone (63–89%) and olanzapine (43–89%). Risperidone and olanzapine at their usual clinical doses gave the same level of D_2 occupancy as low-dose typical antipsychotics, possibly accounting for their lower toxicity. Interestingly, clozapine receptor occupancy may not correlate well with drug serum levels (Nordstrom et al., 1995). One study using [^{123}I]IBZM SPECT reported a range of receptor occupancy of 18–80% in patients with sustained, favorable responses to clozapine, although patients with more symptoms showed lower [^{123}I]IBZM specific binding, consistent with competition of endogenous dopamine for D_2 binding sites (Pickar et al., 1996).

Some currently available antipsychotic drugs induce significant occupancy of D_1 dopamine receptors (Sedvall et al., 1995). However, the experimental selective D_1 antagonist SCH-39166, in doses inducing a more than 70% occupancy of D_1 dopamine receptors, failed to induce an antipsychotic action, suggesting that, in contrast to D_2 blockade, selective D_1 antagonism does not mediate antipsychotic action in schizophrenia. Some, but not all, antipsychotic drugs also induced high occupancy of neocortical 5-HT$_{2A}$ receptors. Because selective 5-HT$_{2A}$ antagonism does not appear to be an efficient treatment for schizophrenia, it seems most likely that 5-HT$_{2A}$ receptors and, perhaps, D_1 receptors act in concert to modify aspects of the mandatory D_2 blockade to induce antipsychotic actions (Farde, 1997). Other data suggest that haloperidol exerts its primary antidopaminergic action in the basal ganglia and that thalamic and cortical sites are secondary to this primary site of drug action. Holcomb et al. (1996) reported a decrease in glucose metabolism in the caudate and putamen 30 days after withdrawal, indicating that haloperidol treatment enhanced glucose utilization in these areas. The thalamus, bilaterally but only in anterior areas, showed the same response to haloperidol. Only in the frontal cortex and in the anterior cingulate had metabolism increased 30 days after withdrawal, indicating that in those two cortical areas haloperidol depressed glucose metabolism.

A number of studies of drugs that can exacerbate symptoms have been performed as well. An acute psychotic state, particularly conceptual disorganization, induced by subanesthetic ketamine doses in normal volunteers was associated with focal increases in metabolic activity in the prefrontal cortex, suggesting that the prefrontal cortex may be involved in mediating NMDA receptor-induced psychosis (Breier et al., 1997a). The glutamatergic N-methyl-D-aspartate (NMDA) receptor antagonist ketamine produced decreases in specific [^{11}C]raclopride binding similar to those caused by amphetamine and significantly correlated with induction of schizophrenia-like symptoms (Breier et al., 1998). Patients with schizophrenia had significantly greater amphetamine-related reductions in [¹¹C]raclopride specific binding than healthy volunteers, suggesting an overactive dopaminergic response (Breier *et al.*, 1997b) (Fig. 8). Perhaps paradoxically, one study reported that dextroamphetamine improved performance on the Wiscon-

sin Card Sorting Task in patients with schizophrenia; the drug led to a greater increase of CBF measured by ¹³³Xe SPECT in the left lateral prefrontal cortex, right anterior cingulate, and posterior cingulate than when the task was performed after placebo (Daniel *et al.*, 1991). Interestingly, the atypical agents clozapine and risperidone do

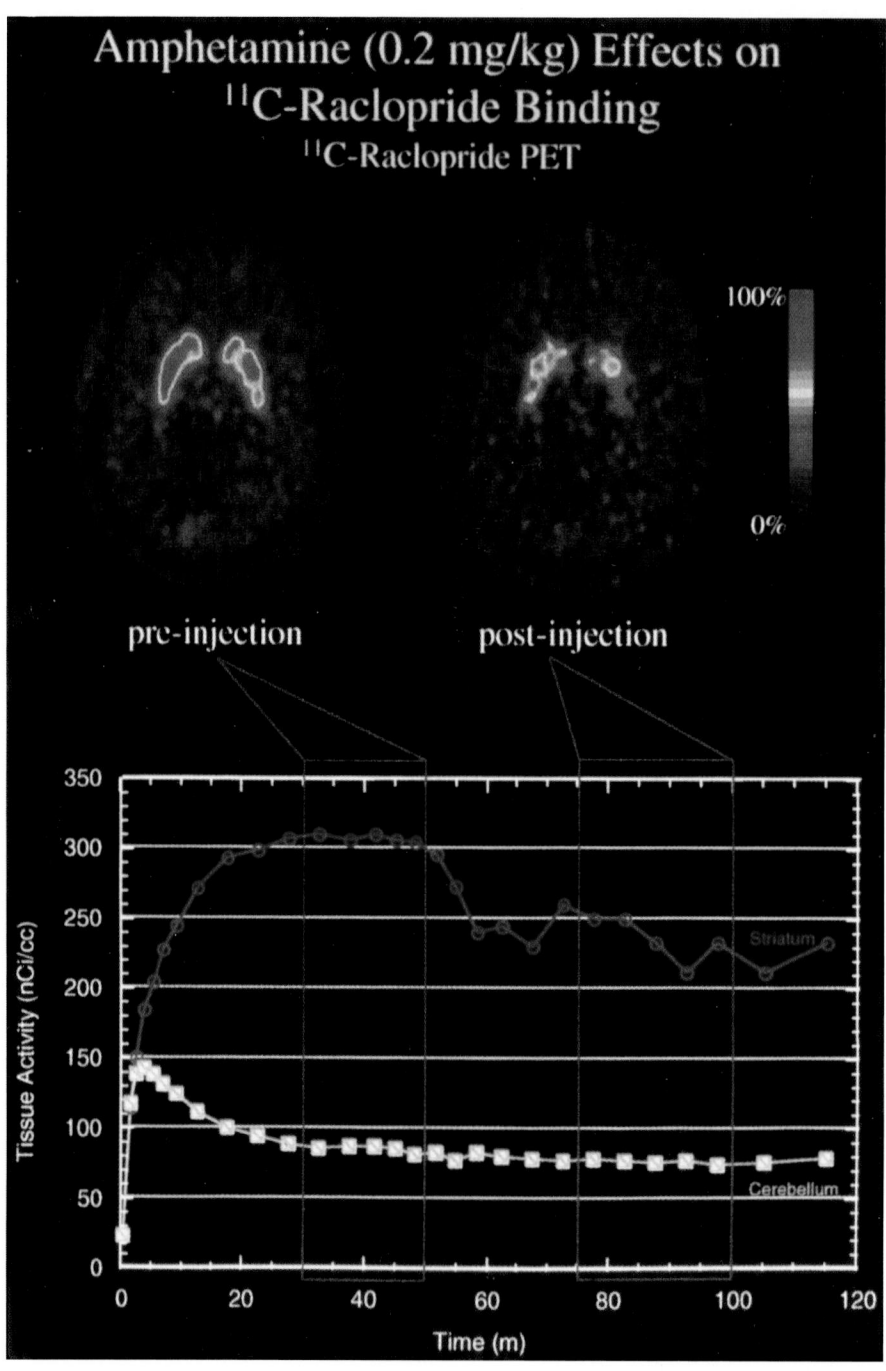

Figure 8 Patients with schizophrenia had significantly greater amphetamine-related reductions in [¹¹C]raclopride specific binding than healthy volunteers, probably due to greater endogenous dopamine release, suggesting an overactive dopaminergic response. Reproduced from Breier *et al.* (1997b). Copyright 1997 National Academy of Sciences, U.S.A.

not effect amphetamine-related striatal dopamine release (Breier *et al.,* 1999).

VI. Conclusion: The Possibility of Individualizing Therapy

This chapter has reviewed the role of imaging in pharmacologic studies for several neurologic and psychiatric disorders. The clinical impact has been greatest for the major psychoses, and data from "pharmacoimaging" studies appear to have immediate relevance for treatment. In principle, physicians can adjust antipsychotic drug therapy individually based on imaging studies. Unfortunately, PET, while possessing the greatest versatility and potential for providing quantitative data (essential for formal studies of drug-receptor interactions), is also the most expensive and difficult modality to use. It might be possible to use SPECT, however, to gain clinically useful estimates of receptor occupancy for individual patients and tailor drug doses appropriately.

The appearance of AEDs with specific mechanisms of action suggests the possibility of individualizing therapy based on receptor studies. Basic mechanisms are less well understood for epilepsy than Parkinson's disease or even schizophrenia. Epilepsy is probably a relatively heterogeneous condition. However, testable hypotheses could be formed based on existing data. For example, patients with reduced GABA on MRS, or decreased BZP-receptor binding, might benefit more from a GABAergic drug like gabapentin than a sodium channel agent like phenytoin.

References

Alkire, M. T. (1998). Quantitative EEG correlations with brain glucose metabolic rate during anesthesia in volunteers. *Anesthesiology* **89,** 323–333.

Baron, J. C., Roeda, D., Munari, C., *et al.* (1983). Brain regional pharmacokinetics of ^{11}C-labeled diphenylhydantoin: Positron emission tomography in humans. *Neurology* **33,** 580–585.

Bazil, C., and Pedley, T. (1995). Neurophysiological effects of AEDs. *In* "Antiepileptic Drugs" (R. H. Levy, R. H. Mattson, and B. S. Meldrum, Eds.), 4th ed., pp. 79–89. Raven Press, New York.

Blacklock, J. B., Oldfield, E. H., Di Chiro, G., *et al.* (1987). Effect of barbiturate coma on glucose utilization in normal brain versus gliomas. Positron emission tomography studies. *J. Neurosurg.* **67,** 71–75.

Breier, A., Malhotra, A. K., Pinals, D. A., Weisenfeld, N. I., and Pickar, D. (1997a). Association of ketamine-induced psychosis with focal activation of the prefrontal cortex in healthy volunteers. *Am. J. Psychiatry* **154**(6), 805–811.

Breier, A., Su, T. P., Saunders, R., Carson, R. E., Kolachana, B. S., de Bartolomeis, A., Weinberger, D. R., Weisenfeld, N., Malhotra, A. K., Eckelman, W. C., and Pickar, D. (1997b). Schizophrenia is associated with elevated amphetamine-induced synaptic dopamine concentrations: Evidence from a novel positron emission tomography method. *Proc. Natl. Acad. Sci. U.S.A.* **94**(6), 2569–2574.

Breier, A., Adler, C. M., Weisenfeld, N., Su, T. P., Elman, I., Picken, L., Malhotra, A. K., and Pickar, D. (1998). Effects of NMDA antagonism on striatal dopamine release in healthy subjects: Application of a novel PET approach. *Synapse* **29**(2), 142–147.

Breier, A., Su, T. P., Malhotra, A. K., Elman, I., Adler, C. M., Weisenfeld, N. I., and Pickar, D. (1999). Effects of atypical antipsychotic drug treatment on amphetamine-induced striatal dopamine release in patients with psychotic disorders. *Neuropsychopharmacology* **20**(4), 340–345.

Brinciotti, M. (1994). Effects of chronic high serum levels of phenobarbital on evoked potentials in epileptic children. *Electroencephalogr. Clin. Neurophysiol.* **92,** 11–16.

Brooks, D. J. (1998). Positron emission tomography studies in movement disorders. *Neurosurg. Clin. N. Am.* **9,** 263–281.

Bryant, C. A., and Jackson, S. H. D. (1998). Functional imaging of the brain in the evaluation of drug response and its application to the study of aging. *Drugs Aging* **13,** 211–222.

Coenen, A. M., Blezer, E. H., and van Luijtelaar, E. L. (1995). Effects of the GABA-uptake inhibitor tiagabine on electroencephalogram, spike-wave discharges and behaviour of rats. *Epilepsy Res.* **21,** 89–94.

Coull, J. T., Frith, C. D., Dolan, R. J., Frackowiack, R. S. J., and Grasby, P. M. (1997). The neural correlates of the noradrenergic modulation of human attention, arousal, and learning. *Eur. J. Neurosci.* **9,** 589–598.

Crane, P. D., Braun, L. D., Cornford, E. M., Cremer, J. E., Glass, J. M., and Oldendorf, W. H. (1978). Dose dependent reduction of glucose utilization by phenobarbital in rat brain. *Stroke* **9,** 12–18.

Daniel, D. G., Weinberger, D. R., Jones, D. W., Zigun, J. R., *et al.* (1991). The effect of amphetamine on regional cerebral blood flow during cognitive activation in schizophrenia. *J. Neurosci.* **11,** 1907–1917.

Dethy, S., Van Blercom, N., Damhaut, P., Wikler, D., Hildebrand, J., and Goldman, S. (1998). Asymmetry of basal ganglia glucose metabolism and dopa responsiveness in parkinsonism. *Mov. Disord.* **13**(2), 275–280.

Dewey, S. L., Smith, G. S., Logan, J., Brodie, J. D., Yu, D. W., Ferrieri, R. A., King, P. T., MacGregor, R. R., Martin, T. P., Wolf, A. P., *et al.* (1992). GABAergic inhibition of endogenous dopamine release measured *in vivo* with ^{11}C-raclopride and positron emission tomography. *J. Neurosci.* **12**(10), 3773–3780.

de Wit, H., Metz, J., Wagner, N., and Cooper, M. (1991). Effects of diazepam on cerebral metabolism and mood in normal volunteers. *Neuropsychopharmacology* **5,** 33–41.

Ekesbo, A., Rydin, E., Torstenson, R., Sydow, O., Laengstrom, B., and Tedroff, J. (1999). Dopamine autoreceptor function is lost in advanced Parkinson's disease. *Neurology* **52**(1), 120–125.

Farde, L. (1997). Brain imaging of schizophrenia: The dopamine hypothesis. *Schizophr. Res.* **28** (2–3), 157–162.

Farde, L., Nordstrom, A. L., Wiesel, F. A., Pauli, S., Halldin, C., and Sedvall, G. (1992). Positron emission tomographic analysis of central D_1 and D_2 dopamine receptor occupancy in patients treated with classical neuroleptics and clozapine. Relation to extrapyramidal side effects. *Arch. Gen. Psychiatry* **49**(7), 538–544.

Farwell, J. R., Lee, Y. J., Hirtz, D. G., Sulzbacher, S. I., Ellenberg, J. H., and Nelson, K. B. (1990). Phenobarbital for febrile seizures: Effects on intelligence and on seizure recurrence. *N. Engl. J. Med.* **322**(6), 364–369.

Fink, M., Irwin, P., Sannita, W., Papakostas, Y., and Green, M. A. (1979). Phenytoin: EEG effects and plasma levels in volunteers. *Ther. Drug Monit.* **1,** 93–103.

Foster, N. L., VanDerSpek, A. F., Aldrich, M. S., *et al.* (1987). The effect of diazepam sedation on cerebral glucose metabolism in Alzheimer's disease as measured using positron emission tomography. *J. Cereb. Blood Flow Metab.* **7,** 415–420.

Frost, J. D., Kellaway, P., Hrachovy, R. A., Glaze, D. G., and Mizrahi, E. M. (1986). Changes in epileptic spike configuration associated with attainment of seizure control. *Ann. Neurol.* **20,** 723–726.

Frost, J. D., Hrachovy, R. A., Glaze, D. G., and Rettig, G. M. (1995). Alpha rhythm slowing during initiation of carbamazepine therapy: Implications for future cognitive performance. *J. Clin. Neurophysiol.* **12,** 57–63.

Frost, J. D., Mayberg, H. S., Fisher, R. S., *et al.* (1988). Mu-opiate receptors measured by positron emission tomography are increased in temporal lobe epilepsy. *Ann. Neurol.* **23,** 231–237.

Fujita, M., Woods, S. W., Verhoeff, N. P., *et al.* (1999). Changes of benzodiazepine receptors during chronic benzodiazepine administration in humans. *Eur. J. Pharmacol.* **368,** 161–172.

Gaillard, W. D., Zeffiro, T., Fazilat, S., DeCarli, C., and Theodore, W. H. (1996). Effect of valproate on cerebral metabolism and blood flow: An ^{18}F 2-deoxyglucose and ^{15}O water positron emission tomography study. *Epilepsia* **37**(6), 515–521.

Gee, K. W., Bolger, M. B., Brinton, R. E., Coirini, H., and McEwen, B. S. (1988). Steroid modulation of the chloride ionophore in rat brain: Structure-activity requirements, regional dependence and mechanism of action. *J. Pharmacol. Exp. Ther.* **246,** 803–812.

Glue, P. (1991). The pharmacology of saccadic eye movements. *J. Psychopharmacol.* **5,** 377–387.

Greenblatt, D. J., Harmatz, J. S., Gouthro, T. A., Locke, J., and Shader, R. I. (1994). Distinguishing a benzodiazepine agonist (triazolam) from a nonagonist anxiolytic (buspirone) by electroencephalography: Kinetic–dynamic studies. *Clin. Pharmacol. Ther.* **56,** 100–111.

Hegerl, U., and Moller, H. J. (1997). Electroencephalography as a diagnostic instrument in Alzheimer's disease: Reviews and perspectives. *Int. Psychogeriatr.* **9** (Suppl. 1), 237–246.

Heinz, A., Knable, M. B., and Weinberger, D. R. (1996). Dopamine D$_2$ receptor imaging and neuroleptic drug response. *J. Clin. Psychiatry* **57** (Suppl. 11), 84–88.

Hertel, A., Weppner, M., Baas, H., Schreiner, M., Maul, F. D., Baum, R. P., Fischer, P. A., and Hor, G. (1997). Quantification of IBZM dopamine receptor SPET in *de novo* Parkinson patients before and during therapy. *Nucl. Med. Commun.* **18**(9), 811–822.

Heyer, E. S., and MacDonald, R. L. (1982). Barbiturate reduction of calcium dependent action potentials: Correlation with anesthetic action. *Brain Res.* **236,** 157–171.

Holcomb, H. H., Cascella, N. G., Thaker, G. K., Medoff, D. R., Dannals, R. F., and Tamminga, C. A. (1996). Functional sites of neuroleptic drug action in the human brain: PET/FDG studies with and without haloperidol. *Am. J. Psychiatry* **153**(1), 41–49.

Holland, K. D., McKeon, A. C., Canney, D. J., Covey, D. F., and Ferrendelli, J. A. (1992). Relative anticonvulsant effects of GABAmimetic and GABA modulatory agents. *Epilepsia* **33,** 981–986.

Jackson, M. F., Dennis, T., Esplin, B., and Capek, R. (1994). Acute effects of γ-vinyl GABA (vigabatrin) on hippocampal GABAergic inhibition *in vitro. Brain Res.* **651,** 85–91.

Jacobs, K. M., and Donoghue, J. P. (1991). Reshaping the cortical motor map by unmasking latent intracortical connections. *Science* **251,** 944–947.

Jenkins, I. H., Fernandez, W., Playford, E. D., Lees, A. J., Frackowiak, R. S., Passingham, R. E., and Brooks, D. J. (1992). Impaired activation of the supplementary motor area in Parkinson's disease is reversed when akinesia is treated with apomorphine. *Ann. Neurol.* **32,** 749–797.

Kapur, S., Zipursky, R. B., and Remington, G. (1999). Clinical and theoretical implications of 5-HT$_2$ and D$_2$ receptor occupancy of clozapine, risperidone, and olanzapine in schizophrenia. *Am. J. Psychiatry* **156**(2), 286–293.

Kokate, T. G., Svensson, B. E., and Rogawski, M. A. (1994). Anticonvulsant activity of neurosteroids: Correlation with γ-aminobutyric

acid evoked chloride current potentiation. *J. Pharmacol. Exp. Ther.* **27,** 1223–1229.

Kujirai, T., Caramia, M. D., Rothwell, J. C., Day, B. L., Thompson, P. D., Ferbert, A., Wroe, S., Asselman, P., and Marsden, C. D. (1993). Corticocortical inhibition in human motor cortex. *J. Physiol. (Lond.)* **471,** 501–519.

Kuzniecky, R., Hetherington, H., Ho, S., Pan, J., Martin, R., Gilliam, F., Hugg, J., and Faught, E. (1998). Topiramate increases cerebral GABA in healthy humans. *Neurology* **51**(2), 627–629.

Liepert, J., Schwenkreis, P., Tegenthoff, M., and Malin, J. P. (1997). The glutamate antagonist riluzole suppresses intracortical facilitation. *J. Neural Transm.* **104,** 1207–1214.

Loscher, W. (1980). Effects of inhibitors of GABA transaminase on synthesis binding uptake and metabolism of GABA. *J. Neurochem.* **34,** 1603–1608.

Loscher, W. (1981). Valproate induced changes in GABA metabolism at the subcellular level. *Biochem. Pharmacol.* **30,** 1363–1366.

MacDonald, R., and Meldrum, B. (1995). Priniciples of antiepileptic drug action. *In* "Antiepileptic Drugs" (R. H. Levy, R. H. Mattson, and B. S. Meldrum, Eds.), 4th ed., pp. 61–77. Raven Press, New York.

MacDonald, R. L., and Schulz, D. W. (1981). Barbiturate enhancement of GABA mediated inhibition and activation of chloride ion conductance: Correlation with anticonvulsant and anesthetic actions. *Brain Res.* **209,** 177–188.

Maeda, H., Furune, S., Nomura, K., Kitou, O., Ando, Y., Negoro, T., and Watanabe, K. (1997). Decrease of *N*-acetylaspartate after ACTH therapy in patients with infantile spasms. *Neuropediatrics* **28**(5), 262–267.

Mandema, J., and Danhof, M. (1992). Electroencephalogram effect measures and relationships between pharmacokinetics and pharmacodynamics of centrally acting drugs. *Clin. Pharmacokinet.* **23,** 191–215.

Mandema, J. W., Sansom, L. N., Dios-Vieitez, C., Hollander-Jansen, M., and Danhof, M. (1991). Pharmacokinetic–pharmacodynamic modelling of the electroencephalographic effects of benzodiazepines. Correlation with receptor binding and anticonvulsant activity. *J. Pharmacol. Exp. Ther.* **257,** 472–478.

McLean, M. J., and MacDonald, R. L. (1986a). Sodium valproate, but not ethosuximide, produces use and voltage-dependent limitation of high frequency repetitive firing of action potentials of mouse central neurons in cell culture. *J. Pharmacol. Exp. Ther.* **237,** 1001–1011.

McLean, M. J., and MacDonald, R. L. (1986b). Carbamazepine and 10,11-epoxycarbamazepine produce use and voltage dependent limitation of rapidly firing action potentials of mouse central neurons in cell culture. *J. Pharmacol. Exp. Ther.* **228,** 727–738.

Meador, K., Loring, D. W., Huh, K., Gallagher, B. B., and King, D. W. (1990). Comparative cognitive effects of anticonvulsants. *Neurology* **40,** 391–394.

Meador, K. J., Loring, D. W., Abney, O. L., Allen, M. E., Moore, E. E., Zamrini, E. Y., and King, D. W. (1993). Effects of carbamazepine and phenytoin on EEG and memory in healthy adults. *Epilepsia* **34,** 153–157.

Moffoot, A., O'Carroll, R. E., Dougall, N., Ebmeier, K., and Goodwin, G. M. (1994). Clonidine infusion increases uptake of ^{99m}Tc-exametazine in anterior cingulate cortex in Korsakoff's psychosis. *Psychol. Med.* **24,** 53–61.

Nordberg, A., Amberla, K., Shigeta, M., Lundqvist, H., Viitanen, M., Hellstrom-Lindahl, E., Johansson, M., Andersson, J., Hartvig, P., Lilja, A., Langstrom, B., and Winblad, B. (1998). Long-term tacrine treatment in three mild Alzheimer patients: Effects on nicotinic receptors, cerebral blood flow, glucose metabolism, EEG, and cognitive abilities. *Alzheimer Dis. Assoc. Disord.* **12**(3), 228–237.

Nordstrom, A. L., Farde, L., Nyberg, S., Karlsson, P., Halldin, C., and Sedvall, G. (1995). D_1, D_2, and 5-HT_2 receptor occupancy in relation to clozapine serum concentration: A PET study of schizophrenic patients. *Am. J. Psychiatry* **152**(10), 1444–1449.

Nyberg, S., Eriksson, B., Oxenstierna, G., Halldin, C., and Farde, L. (1999). Suggested minimal effective dose of risperidone based on PET-measured D_2 and 5-HT_{2A} receptor occupancy in schizophrenic patients. *Am. J. Psychiatry* **156**(6), 869–875.

Peck, A. W. (1991). Clinical pharmacology of lamotrigine. *Epilepsia* **32** (Suppl. 2), S9–S12.

Pereira de Vasconcelos, A., and Nehlig, A. (1987). Effects of early chronic phenobarbital treatment on the maturation of energy metabolism in the developing rat brain. I. Incorporation of glucose carbon into amino acids. *Brain Res.* **433**, 219–229.

Pereira de Vasconcelos, A., Boyet, S., and Nehlig, A. (1990). Consequences of chronic phenobarbital treatment on local cerebral glucose utilization in the developing rat. *Brain Res. Dev. Brain Res.* **53**, 168–178.

Pericic, D. (1980). Effects of γ-vinyl GABA on the enzymes of GABA system in specific brain regions. *Period. Biol.* **82**, 19–23.

Petroff, O. A. C., Rothman, D. L., Behar, K. L., and Mattson, R. H. (1995). Initial observations on effect of vigabatrin on *in vivo* ^{1}H spectroscopic measurements of γ-aminobutyric acid, glutamate, and glutamine in human brain. *Epilepsia* **36**, 457–464.

Petroff, O. A., Behar, K. L., Mattson, R. H., and Rothman, D. L. (1996a). Human brain gamma-aminobutyric acid levels and seizure control following initiation of vigabatrin therapy. *J. Neurochem.* **67**, 2399–2404.

Petroff, O. A., Rothman, D. L., Behar, K. L., and Mattson, R. H. (1996b). Human brain GABA levels rise after initiation of vigabatrin therapy but fail to rise further with increasing dose. *Neurology* **46**, 1459–1463.

Petroff, O. A., Rothman, D. L., Behar, K. L., and Mattson, R. H. (1996c). Low brain GABA level is associated with poor seizure control. *Ann. Neurol.* **40**, 908–911.

Petroff, O. A., Rothman, D. L., Behar, K. L., Lamoureux, D., and Mattson, R. H. (1996d). The effect of gabapentin on brain gamma-aminobutyric acid in patients with epilepsy. *Ann. Neurol.* **39**(1), 95–99.

Petroff, O. A., Mattson, R. H., Behar, K. L., Hyder, F., and Rothman, D. L. (1998). Vigabatrin increases human brain homocarnosine and improves seizure control. *Ann. Neurol.* **44**(6), 948–952.

Petroff, O. A., Hyder, F., Mattson, R. H., and Rothman, D. L. (1999). Topiramate increases brain GABA, homocarnosine, and pyrrolidinone in patients with epilepsy. *Neurology* **52**(3), 473–478.

Peyron, R., Cinotti, L., Le Bars, D., *et al.* (1994). Effects of $GABA_A$ receptors activation on brain glucose metabolism in normal subjects and temporal lobe epilepsy (TLE) patients. A positron emission tomography (PET) study. Part II: The focal hypometabolism is reactive to $GABA_A$ agonist administration in TLE. *Epilepsy Res.* **19**, 55–62.

Pickar, D., Su, T. P., Weinberger, D. R., Coppola, R., Malhotra, A. K., Knable, M. B., Lee, K. S., Gorey, J., Bartko, J. J., Breier, A., and Hsiao, J. (1996). Individual variation in D_2 dopamine receptor occupancy in clozapine-treated patients. *Am. J. Psychiatry* **153**(12), 1571–1578.

Popoli, P., Pezzola, A., and Sagratella, S. (1994). Diphenylhydantoin potentiates the EEG and behavioural effects induced by N-methyl-D-aspartate antagonists in rats. *Psychopharmacol. (Berl.)* **113**, 471–475.

Ring, H. A., Trimble, M. R., Costa, D. C., George, M. S., Verhoeff, P., and Ell, P. J. (1992). Effect of vigabatrin on striatal dopamine receptors: Evidence in humans for interactions of GABA and dopamine systems. *J. Neurol. Neurosurg. Psychiatry* **55**(9), 758–761.

Salinsky, M. C., Okun, B. S., and Morehead, L. (1994). Intraindividual analysis of antiepileptic drug effects on EEG background rhythms. *Electroencephalogr. Clin. Neurophysiol.* **90**, 186–193.

Sato, S., White, B. G., Perry, J. K., Dreifuss, F. E., Sackellares, J. C., and Kupferberg, H. J. (1982). Valproic acid versus ethosuximide in the treatment of absence seizures. *Neurology* **32**, 157–163.

Savic, I., Persson, A., Roland, P., Pauli, S., Sedvall, G., and Widen, L. (1988). *In-vivo* demonstration of reduced benzodiazepine receptor binding in human epileptic foci. *Lancet* **2**, 863–866.

Savic, I., Blomqvist, G., Halldin, C., Litton, J. E., and Gulyas, B. (1998). Regional increases in [^{11}C]flumazenil binding after epilepsy surgery. *Acta Neurol. Scand.* **97**, 279–286.

Schmid, L., Bottlaender, M., Brouillet, E., Fuseau, C., and Maziere, M. (1996). Vigabatrin modulates benzodiazepine receptor activity *in vivo*: A positron emission tomography study in baboon. *J. Pharmacol. Exp. Ther.* **276**(3), 977–983.

Sedvall, G., Farde, L., Hall, H., Halldin, C., Karlsson, P., Nordstrom, A. L., Nyberg, S., and Pauli, S. (1995). Utilization of radioligands in schizophrenia research. *Clin. Neurosci.* **3**(2), 112–121.

Seitz, R. J., Piel, S., Arnold, S., *et al.* (1996). Cerebellar hypometabolism in focal epilepsy is related to age of onset and drug intoxication. *Epilepsia* **37**, 1194–1199.

Spanaki, M. V., Siegel, H., Kopylev, L., Fazilat, S., Dean, A., Liow, K., Ben-Menachem, E., Gaillard, W. D., and Theodore, W. H. (1999). The effect of vigabatrin (γ-vinyl GABA) on cerebral blood flow and metabolism. *Neurology* **53**(7), 1518–1522.

Staedt, J., Stoppe, G., Kogler, A., and Steinhoff, B. J. (1995). Changes of central benzodiazepine receptor density in the course of anticonvulsant treatment in temporal lobe epilepsy. *Seizure* **4**, 49–52.

Stanski, D. R., Hudson, R. J., Homer, T. D., Saidman, L. J., and Meathe, E. (1984). Pharmacodynamic modeling of thiopental anesthesia. *J. Pharmacokinet. Biopharmaceut.* **12**, 223–240.

Sveinbjornsdottir, S., Sander, J. W. A. S., Upton, D., *et al.* (1993). The excitatory amino acid antagonist d-CPP-ene (SDZ EAA-494) in patients with epilepsy. *Epilepsy Res.* **16**, 165–174.

Theodore, W. H., DiChiro, G., Margolin, R., Fishbein, D., Porter, R. J., and Brooks, R. A. (1986a). Barbiturates reduce human cerebral glucose metabolism. *Neurology* **36**, 60–64.

Theodore, W. H., Bairamian, D., Newark, M. E., *et al.* (1986b). Effect of phenytoin on human cerebral glucose metabolism. *J. Cereb. Blood Flow Metab.* **6**, 315–320.

Theodore, W. H., Fishbein, D., Deitz, M., and Baldwin, P. (1987). Complex partial seizures: Cerebellar metabolism. *Epilepsia* **28**, 319–323.

Theodore, W. H. (1988). Antiepileptic drugs and cerebral glucose metabolism. *Epilepsia* **29** (Suppl. 2), S48–S55.

Theodore, W. H., Bromfield, E., and Onorati, L. (1989a). The effect of carbamazepine on cerebral glucose metabolism. *Ann. Neurol.* **25**, 516–520.

Theodore, W. H., Narang, P. K., Holmes, M. D., Reeves, P., and Nice, F. J. (1989b). Carbamazepine and its epoxide: Relation of plasma levels to toxicity and clinical control. *Ann. Neurol.* **25**, 194–196.

Theodore, W. H., Leiderman, D., Gaillard, W., Khan, I., Reeves, P., and Lloyd-Hontz, K. (1993). The effect of naloxone on cerebral blood flow and glucose metabolism in patients with complex partial seizures. *Epilepsy Res.* **16**(1), 51–54.

Theodore, W. H., Levy, L., Ben-Menachem, E., Nystrom, L., Oletsky, H., Hunter, K., and Dean, A. (1997). Vigabatrin increases both CSF GABA and brain GABA measured by MRS. *Ann. Neurol.* **42**, 393.

Theodore, W. H., Ketter, T. A., Kimbrell, T., Dean, A., Siegel, H., Benson, B., and Ben-Menachem, E. (1998). The effect of vigabatrin on psychiatric rating scales. *Epilepsia* **39** (Suppl. 6), 240.

Thompson, P. J., and Trimble, M. R. (1982). Anticonvulsant drugs and cognitive functions. *Epilepsia* **23**, 531–544.

Trimble, M. R. (1987). Anticonvulsant drugs and cognitive function: A review of the literature. *Epilepsia* **28** (Suppl. 3), S37–S45.

Twyman, R. E., and MacDonald, R. L. (1992). Neurosteroid regulation of GABA$_A$ receptor single-channel kinetic properties of mouse spinal cord neurons in culture. *J. Physiol. (Lond.)* **456**, 215–245.

van Stevenick, A. L., Verver, S., Schoemaker, H. C., Pieters, M. S. M., Kroon, R., Breimer, D. D., and Cohen, A. F. (1992). Effects of temazepam on saccadic eye movements: Concentration-effect relationships in individual volunteers. *Clin. Pharmacol. Ther.* **52**, 402–408.

Vining, E. P. G., Mellits, E. D., Dorsen, M. M., *et al.* (1987). Psychologic and behavioral effects of antiepileptic drugs in children: A double-blind comparison between phenobarbital and valproic acid. *Pediatrics* **80**, 165–174.

Warren, S., Hier, D. B., and Pavel, D. (1998). Visual form of Alzheimer's disease and its response to anticholinesterase therapy. *J. Neuroimaging* **8**(4), 249–252.

Weber, O. M., Verhagen, A., Duc, C. O., Meier, D., Leenders, K. L., and Boesiger, P. (1999). Effects of vigabatrin intake on brain GABA activity as monitored by spectrally edited magnetic resonance spectroscopy and positron emission tomography. *Magn. Reson. Imaging* **17**(3), 417–425.

Woodbury, D. M., and Kemp, J. M. (1971). Pharmacology and mechanisms of action of diphenylhydantoin. *Psychiatr. Neurol. Neurochir.* **74**, 91–117.

Zaccara, G., Gangemi, P. F., Messori, A., Parigi, A., Massi, S., Valenza, T., and Monza, G. C. (1992). Effects of oxcarbazepine and carbamazepine on the central nervous system: Computerized analysis of saccadic and smooth-pursuit eye movements. *Acta Neurol. Scand.* **85**, 425–429.

Ziemann, U., Lonnecker, S., Steinhoff, B. J., and Paulus, W. (1996). Effects of various antiepileptic drugs on motor cortex excitability in man. A transcranial magnetic stimulation study. *Ann. Neurol.* **40**, 367–378.

Ziemann, U., Steinhoff, B. J., Tergau, F., and Paulus, W. (1998). Transcranial magnetic stimulation: Its current role in epilepsy research. *Epilepsy Res.* **30**, 11–30.

26

Therapeutics: Surgical

Robert S. Turner, Thomas Henry, and Scott T. Grafton[1]

Departments of Neurology and Radiology, Emory University School of Medicine, Atlanta, Georgia 30322

I. Introduction

II. Surgical Therapies for Parkinson's Disease

III. Surgical Therapies for Tremor

IV. Surgical Therapies for Epilepsy

V. Summary

References

This chapter focuses on the application of positron emission tomography and magnetic resonance imaging for investigating the effects of surgery on functional brain circuits. Particular attention is given to imaging studies of Parkinson's disease and epilepsy surgery. Imaging can be used to identify changing activation patterns before and after surgery that suggest possible pathophysiologic mechanisms for the cardinal symptoms of a disease. It can also be used to examine possible functions of basal ganglia by comparing activation patterns before and after subcortical surgery or with deep-brain stimulation. Patterns of cortical reorganization after removal of cortical or subcortical tissue can be identified. Probable mechanisms of action for peripheral devices such as the vagus nerve stimulator can be determined. Ultimately, these imaging studies may suggest new targets for functional surgery and improve control of symptoms in intractable neurological diseases.

I. Introduction

In the past decade there have been remarkable advances in neurosurgical techniques for the control of nervous system disorders that were once considered intractable. Examples include refined stereotaxic ablation of tumors previously considered to be unresectable, improved accuracy in the removal of epileptogenic tissue, the reemergence of revascularization for intracranial ischemic vascular disease, and functional surgery for controlling neurologic symptoms associated with progressive neurodegenerative disorders such as Parkinson's disease. Other chapters in this book examine the utility of functional brain imaging for identifying appropriate candidates for neurosurgical procedures, predicting outcome of surgery, and identifying eloquent brain areas as part of presurgical planning. In this chapter we address issues that emerge when effects of these evolving neurosurgical methods are studied with functional imaging. For this purpose it is useful to categorize broadly neurosurgical methods into five general approaches.

A. Resective Surgery

These methods involve the total or subtotal removal of tissue by open craniotomy or stereotaxic surgery.

---

[1] To whom correspondence should be addressed.

Common indications are the treatment of brain tumor and the resection of epileptogenic tissue. An obvious question that arises whenever resective surgery is performed is how the brain adapts or accommodates to the removal of potentially functional tissue. This is analogous to asking what happens during functional recovery after stroke, when the tissue damage is due to a vascular event rather than surgery. After resection, does the brain return to a more normal state or are new systems recruited to preserve function? By studying patients with functional brain imaging, before and at multiple time points after surgery, this question can be examined in detail. When combined with behavioral assessments, resective surgery can also be used to determine the degree to which particular brain areas, activated during functional imaging, are essential for a particular neurobehavioral task. For example, if a task such as confrontational naming activates a cortical site that is located within a planned temporal lobe resection, will that task be disrupted after surgery? If not, it suggests that the activation site is an auxiliary member of a functional circuit. These questions will be discussed in the section of this chapter that examines functional imaging of epilepsy patients.

B. Transective Surgery

These surgeries involve the manipulation or ablation of specific functional circuits within the brain. Historic examples include frontal leukotomy for the manipulation of mood and voluntary behavior, dorsal rhizotomy of the spinal cord for control of pain, and more recently, stereotaxic ablation of basal ganglia nuclei for modification of motor behavior. All methods of transective surgery rely on a fundamental assumption that a particular tract or nucleus to be ablated carries a particular type of functional information. For the case of thalomotomy, it is assumed that excessive rhythmic activity from the cerebellum to the thalamus is culminating in the complex facilitation of rhythmic activity between the thalamus and motor cortical areas. After a selective lesion of the appropriate thalamic nucleus, this inappropriate rhythmic activity can be disrupted without significant loss of other thalamic functions. Functional imaging can be used to substantiate these assumptions by examining the activity of these circuits across a variety of behaviors. Once transective surgery is performed, functional effects in downstream circuits can be identified and reorganization within these circuits can be examined by serial studies. Functional brain imaging can also be used to examine conceptual paradoxes that arise when brain circuits are eliminated. For example, if basal ganglia output from the globus pallidus to the thalamus is essential for modifying motor behavior, why does ab-

lation of the globus pallidus in Parkinson's disease lead to a restitution of motor function without the development of new motor deficits? These issues will be examined in detail in this chapter.

C. Transplantation

These surgeries involve stereotaxic placement of fetal, adrenal, carotid body, xenograft, or cell culture derived tissue into the brain for the generation of specific neurochemical substrates. Well-known examples include the use of fetal substantia nigra cells for the treatment of Parkinson's disease. Functional imaging after transplantation surgery has been used primarily to examine the viability and growth of grafted tissue after surgery. It has been assumed that regional dopamine-producing capacity of the brain following fetal dopamine transplants will be predicted by the degree of F-18 dopa influx into striatum. It should be noted that the contribution of other processes that might lead to increased F-18 dopa influx, such as host tissue reaction secondary to surgery, have yet to be characterized. There is another potential process that has yet to be investigated with functional imaging. It would be useful to know if functional circuits downstream from a transplanted area demonstrate restitution of normal activation patterns after reintroduction of a neural modulator such as dopamine. For example, in Parkinson's disease, functional imaging could be used to determine if transplanted patients demonstrate normal patterns of activation during movement after functioning dopamine cells are implanted in targets such as the putamen.

D. Implantation

These surgeries include the placement of stimulating electrodes within the peripheral or central nervous system. Driven by programmable pacemakers, these electrodes can be used to modulate specific sensory or motor circuits. The rapid development of implantable devices for the control of seizures and movement disorders raises a number of important questions that can be addressed with functional imaging. Most importantly, what is the mechanism of action for a stimulating device. For peripherally placed devices such as the vagal nerve stimulator, used for intractable epilepsy, an immediate question is what central circuits are modified by stimulation. For centrally placed stimulators, additional questions arise. Stimulators are currently being placed in the thalamus, subthalamic nucleus, and globus pallidus for control of parkinsonian symptoms. In the future there is also the possibility that stimulators could be placed in epileptogenic regions for the interruption of electrical activity heralding a seizure. For central stimulators, brain imaging can define which circuits are involved and the functional specificity of these circuits. Are the patterns of activation during stimulation specific to particular behaviors or behavior independent? In addition, it may be possible to map functional systems associated with cardinal symptoms of a disease by stimulating different targets within a circuit that differentially modulate these symptoms. These questions will be examined in detail.

E. Revascularization

In addition to traditional approaches such as carotid endarterectomy, revascularization also refers to more complex vascular techniques such as angioplasty, extracranial-intracranial bypass surgery for intracranial stenosis, and embolization procedures. Although it has yet to be performed, functional imaging could be used to examine reorganization of brain activity after revascularization.

In the remainder of this chapter, we focus on two nervous system disorders where neurosurgical treatment is being used increasingly and most of the above issues are being explored with functional imaging. First, we focus on Parkinson's disease, giving particular attention to current models of the functional circuitry of motor behavior. Then we examine brain mapping studies of surgically-treated epileptic patients. The ultimate lessons learned from these functional imaging studies may improve our current understanding of the systems-level pathophysiologic substrates of disease and lead to improved surgical targets, better rational for surgical therapy, and more accurate selection of candidate patients for those procedures.

II. Surgical Therapies for Parkinson's Disease

A. Overview of Basal Ganglia Organization and Parkinson's Disease

The classic clinical features of Parkinson's disease (PD) include akinesia, bradykinesia, rigidity, rest tremor, postural instability, and gait disturbance (Parkinson, 1817). Akinesia is characterized by "poverty and slowness of initiation and execution of willed and associated movements and difficulty in changing one motor pattern to another in the absence of paralysis" (Lakke, 1981). Thus, the term akinesia refers to various aspects of disordered movement. In contrast, the term bradykinesia refers exclusively to slowness in the execution of movements, measurable as prolonged movement durations or reduced movement velocities as distinct from prolonged reaction times. This distinction between akinesia and

bradykinesia is substantiated by the studies of Evarts *et al.*, who demonstrated that PD patients can have impairments of movement initiation (prolonged reaction times) without prolonged movement times (bradykinesia) and vice versa (Evarts *et al.*, 1981).

One of the major goals of functional imaging in PD is to determine if the clinically distinct deficits of PD are related to common or distinct abnormalities in the activation or inhibition of motor circuits. As will become evident, nearly all functional imaging studies to date have focused on understanding the pathophysiological basis for parkinsonian akinesia. With the employment of a greater variety of motor tasks in recent years, we are beginning to discover that the abnormalities of task-related brain activity associated with PD are characterized more accurately as distributed patterns of under- and over-activation rather than just focal deficits in brain activity. Which brain areas are involved is influenced strongly by the type of motor task employed. Moreover, the effects of surgical therapy for PD appear to be a task-specific normalization of the abnormal patterns. These observations are consistent with an emerging view that dissociable neural systems underlie the different symptoms of PD.

Interpretation of function imaging studies of PD requires a clear understanding of basal ganglia (BG) circuitry. Current definitions of the BG include a set of anatomically-interconnected subcortical and brainstem nuclei (DeLong *et al.*, 1984). These include the striatum (caudate and putamen), the globus pallidus (internal and external segments, GPi and GPe, respectively), the substantia nigra (compacta and reticulata, SNc and SNr, respectively), and the subthalamic nucleus (STN). The major connections within the BG are summarized in Fig. 1. As shown, the striatum is the primary receptive nucleus of the BG and it receives inputs from nearly all areas of the neocortex. These cortical projections are topographically organized, and there is clear evidence for a parallel, anatomically segregated organization of motor, associative, and limbic cortical inputs to the striatum (Alexander *et al.*, 1990).

Numerous studies have substantiated the presence of both direct and indirect (via GPe) pathways through the BG, ultimately converging in GPi and SNr (Parent and Hazrati, 1995; Wichmann and DeLong, 1996). The primary outputs from the BG are inhibitory projections from GPi and SNr to thalamic nuclei (Yoshida *et al.*, 1972; DeVito and Andersen, 1982) and to midbrain nuclei, including the pedunculopontine nucleus (PPN) and superior colliculus (SC) (Rye *et al.*, 1996). (Recent observations in nonhuman primates indicate the possibility of BG brainstem projections to precerebellar nuclei, thereby providing an anatomical substrate for direct BG influences on cerebellar activity (Rye, personal communication).)

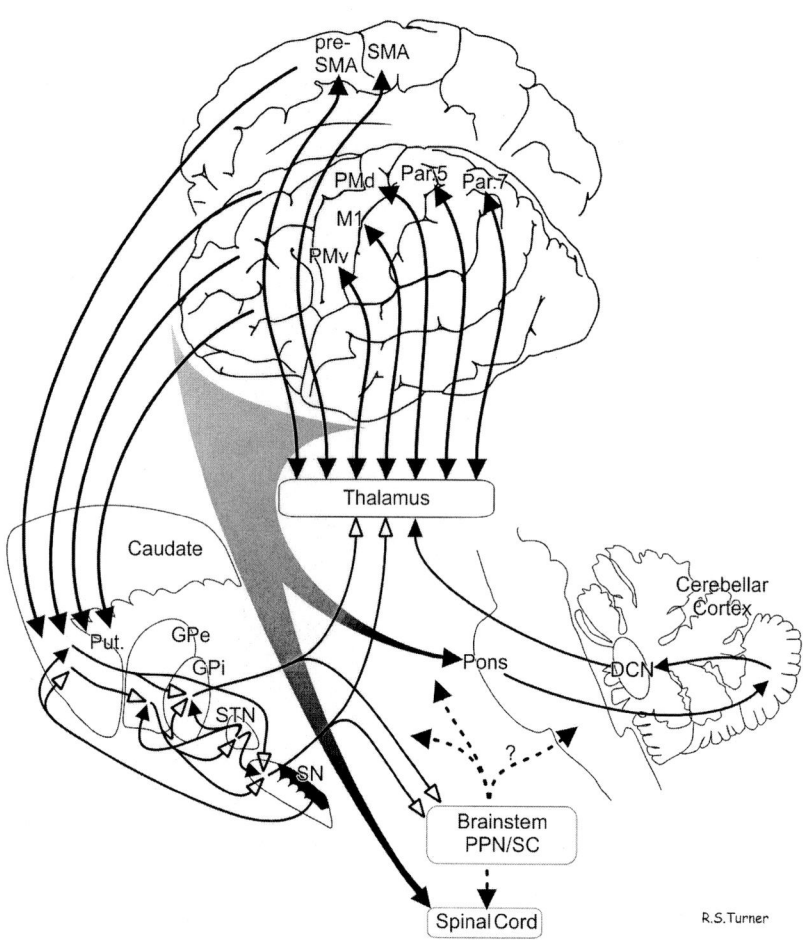

Figure 1 Basal ganglia- and cerebello-thalamocortical circuitry for motor control. Excitatory connections are indicated by closed arrowheads and inhibitory connections by open arrowheads. Multiple motor and parietal cortical areas send parallel segregated excitatory afferents to the putamen (Put., the primary input nucleus of basal ganglia) and to the pontine nuclei (Pons, a primary input nucleus for the cerebellum). Efferents from the basal ganglia originate in the internal segment of the globus pallidus and substantia nigra (GPi and SN, respectively) and project to the thalamus and brainstem nuclei: pedunculopontine and superior colliculus nuclei (PPN and SC, respectively). Efferents from the cerebellum directed to the thalamus originate in the deep cerebellar nuclei (DCN). Basal ganglia and cerebellar efferents to the thalamus terminate in separate thalamic territories, but basal ganglia and cerebellar pathways converge in the cortical motor areas: primary motor cortex (M1), dorsal premotor (PMd), ventral premotor (PMv), supplementary motor area (SMA), and pre-SMA. Portions of the mesial aspect of the left hemisphere are shown as if reflected in a mirror. For a description of the intrinsic basal ganglia circuitry, see Fig. 2. GPe, external segment of the globus pallidus.

Basal ganglia and cerebellar efferents terminate in anatomically-segregated regions of the thalamus but these pathways converge at the cortical level. Recent anatomical evidence reveals that the BG output to cortex (via the thalamus) is spatially organized into discrete channels (Hoover and Strick, 1993; Middleton and Strick, 1997b). Even within the motor circuit of the BG, there appear to be anatomically-segregated subcircuits that serve different frontal motor areas. For instance, GPi neurons that project via the thalamus to the sup-

plementary motor area (SMA) are located dorsomedially, whereas those projecting to the ventral premotor area (PMv) are in the ventral GPi and those influencing the motor cortex are in an intermediate location (Hoover and Strick, 1993). A similar segregated organization also appears in the cerebellar output nuclei and a further somatotopic segregation appears within each of the motor subcircuits (Middleton and Strick, 1997a; Hoover and Strick, 1999). These and other recent anatomic studies have clearly shown that BG and cerebellar efferent pathways are segregated at the thalamic level but that they converge in most, if not all, of the frontal motor areas, albeit with different weightings for different cortical areas (Rouiller *et al.,* 1994; Inase and Tanji, 1995; Inase *et al.,* 1996b; Hoover and Strick, 1999; Sakai *et al.,* 1999). The parallel anatomic channels from the BG and cerebellum could subserve different aspects of motor control, and differential involvement of these channels in movement disorders such as PD could explain the variability of clinical symptomatology.

Models for the roles of the BG in normal motor control remain speculative (Graybiel *et al.,* 1994; Houk, 1995; Mink, 1996). It is clear that activity in the output nucleus of the BG motor circuit often reflects both parameters of movement (Georgopoulos *et al.,* 1983; Turner and Anderson, 1997; Turner *et al.,* 1998) and the behavioral context in which movement is performed (Brotchie *et al.,* 1991b; Mushiake and Strick, 1995). It has been found consistently that pallidal movement-related discharge has a late onset, too late to contribute to movement initiation, but at a time appropriate to influence an ongoing movement or subsequent movements (DeLong and Georgopoulos, 1979; Anderson and Horak, 1985; Brotchie *et al.,* 1991a; Mink and Thach, 1991; Turner and Anderson, 1997). Reductions in the tonic discharge of BG output neurons probably disinhibit recipient neurons and thereby facilitate activity in select thalamocortical and brainstem circuits (Deniau and Chevalier, 1985). Increases in neuronal discharge,

which occur more commonly than decreases within the skeletomotor circuit, likely suppress the activation of thalamocortical circuits. It has been hypothesized that the inhibitory outflow from the BG motor circuit acts on frontal motor areas to modulate the scale of movement (Georgopoulos *et al.,* 1983; Turner and Anderson, 1997; Turner *et al.,* 1998), to serve as an internal trigger for subsequent motor acts (Brotchie *et al.,* 1991a), or to suppress competing or "inappropriate" reflexes (Mink and Thach, 1991). It may be possible to subsume many, if not all, of these functions within the general concept that the BG "select" the frontal thalamocortical circuits whose activation is predicted to be optimal for the present behavioral context (Cools *et al.,* 1984; Marsden and Obeso, 1994; Redgrave *et al.,* 1999). Although numerous experimental results are consistent with an involvement of the BG in selection, the details of the theory remain sketchy and much remains to be tested concerning any of the proposed normal motor functions of the BG.

Pathophysiological models of BG disorders, in contrast, have solid experimental support. They have proved a rational substrate for the development of innovative surgical therapies, and the efficacy of those therapies may in fact allow us to reject certain hypotheses concerning the normal motor functions of the BG (DeLong, 1990; Wichmann and DeLong, 1996). The loss of nigrostriatal dopamine in PD is known to alter the balance of activity in direct and indirect efferent pathways from the striatum (Fig. 2). The net effect is a rela-

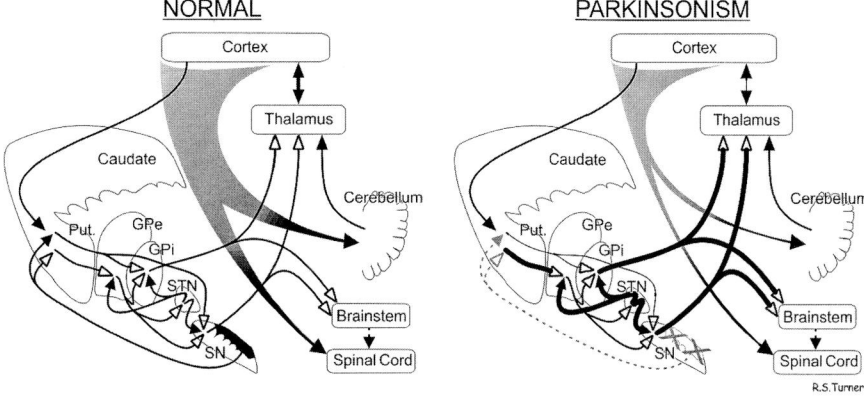

Figure 2 Basal ganglia circuitry and parkinsonian pathophysiology. Parkinsonism arises from a loss of dopaminergic neurons in the pars compacta segments of the substantia nigra (black area of SN for Normal and gray Xs for Parkinsonism) and subsequent differential changes in striato-pallidal and striato-nigral activity. (The relative increases and decreases in tonic discharge rates associated with parkinsonism are denoted by wider and thinner lines, respectively.) Reduced inhibition via the direct pathway (putamen to internal globus pallidus and substantia nigra, e.g., Put. → GPi and SN) combined with increased excitation from the subthalamic nucleus (STN) leads to increased discharge rates and altered discharge patterns in the basal ganglia projection neurons in the GPi and SN. This abnormal activity may cause motor impairment by suppressing or altering thalamocortical facilitation and by altering the normal function of brainstem circuits. The cause of the increased discharge of STN neurons remains controversial (Wichmann and De-Long, 1996; Levy *et al.,* 1997; Obeso *et al.,* 1997).

tive overactivity of the inhibitory BG output neurons (GPi and SNr) (Filion *et al.*, 1988; Miller and DeLong, 1988), which, in turn, causes an excessive inhibition (or deactivation) of BG-recipient thalamic and brainstem regions. Under normal conditions, reciprocal cortico-thalamocortical excitatory projections are thought to facilitate activity in the associated cortical area (Jones, 1985). It is reasonable to predict, then, that the excessive inhibition of motor-related thalamic areas in PD culminates in an activation deficit in associated frontal motor areas such as the SMA. In the recent past, it was thought that outflow from the BG motor circuit terminates nearly exclusively in SMA-related thalamic areas (Schell and Strick, 1984). A BG influence on SMA activity has been of particular interest as this cortical region is associated with movement selection, movement initiation, movement sequencing, and, in particular, generation of internally guided movements (Goldberg, 1985; Picard and Strick, 1996). Abnormalities of these functions in PD, arising from excessive pallidal inhibition of SMA-related thalamic areas, could culminate in akinesia. Basal ganglia influences on the other motor cortical areas and on brainstem circuits, although well established anatomically (Hoover and Strick, 1993; Rye *et al.*, 1996), have not been integrated fully into models of parkinsonian pathophysiology. For instance, the behavioral correlates of an excessive pallidal inhibition (via the thalamus) of primary motor cortex versus lateral or ventral premotor areas remain to be explored.

It is important to point out that not all aspects of the pathophysiological model outlined above are on equally firm footing (Wichmann and DeLong, 1996; Levy *et al.*, 1997; Obeso *et al.*, 1997). Data are sparse concerning the physiological effects on cortical activity of excessive thalamic inhibition, and particularly so for effects on *task-related* changes in cortical activity. Recent studies of cortical glucose metabolism in parkinsonian patients are in relative agreement on a pattern of reduced frontal metabolism in the resting state (Eidelberg *et al.*, 1994; Piert *et al.*, 1996; see also Chapter 10). Studies using metabolic markers in animal models of parkinsonism, however, have failed to demonstrate the changes in cortical activity predicted by the model (Crossman *et al.*, 1985; Mitchell *et al.*, 1989; Blandini and Greenamyre, 1995; Delfs *et al.*, 1995). Moreover, the few single-unit recording studies of frontal cortical activity in the MPTP monkey model of PD found no significant change in tonic discharge rates and relatively subtle changes in task-related discharge (Doudet *et al.*, 1990; Watts and Mandir, 1992; Turner and DeLong, in preparation). Because of the complex adaptive responses to changes in tonic synaptic input observed at multiple levels in the CNS (Koch, 1997), it would in fact be surprising if the thalamus conducted to cortex an exact replica

of the tonic inhibitory signal it receives from the BG (Anderson and Turner, 1991; Inase *et al.*, 1996a). Increasing attention is also being drawn to the abnormal oscillatory activity and synchronization seen in the GPi and SNr of dopamine-depleted animals (Filion and Tremblay, 1991; Nini *et al.*, 1995; Rohlfs *et al.*, 1997; Wichmann *et al.*, 1999). Some of these observations may be taken to indicate that abnormal patterns of tonic activity and abnormal responsiveness to sensory stimuli in fact dominate the changes in neural discharge associated with PD (Filion and Tremblay, 1991; Nini *et al.*, 1995; Wichmann *et al.*, 1999). Thus, it is reasonable to conjecture that abnormal discharge patterns play a critical role in disturbing the normal operations of the thalamocortical system—perhaps by injecting erroneous information into the system or simply by adding "nonspecific noise" (Wichmann and DeLong, 1996). In short, BG outflow in the parkinsonian state may cause abnormal *patterns* of thalamocortical activity, patterns that could not result from excessive tonic inhibition alone. As will become evident below, recent neuroimaging results are also incompatible with a model of parkinsonian pathophysiology based solely on the excessive tonic inhibition of select thalamocortical circuits.

B. PET Studies of PD: The Akinesia Model

Positron emission tomography (PET) and functional magnetic resonance (fMRI) studies are beginning to investigate activity within the BG circuitry in normal subjects and PD patients while they perform a variety of motor tasks. For example, PET and fMRI studies of simple movement can detect activation in almost all of the nuclei of the cortico-subcortical motor circuit (Turner *et al.*, 1998). Imaging studies are beginning to link specific parkinsonian signs (bradykinesia, akinesia, etc.) with changes of motor circuit activity. The majority of published PET studies of PD have focused on the influence of basal ganglia dysfunction on SMA activity as a model for akinesia. In principle, this is a reasonable approach as the SMA is one of the main cortical receiving areas of the BG motor circuit (Schell and Strick, 1984) and the SMA has been linked to a variety of motor behaviors that are impaired in PD, including, most notably, the selection and generation of internally-guided movements. Numerous psychophysical studies have demonstrated that PD patients are particularly impaired on tasks requiring internal selection or guidance of movement (Goldberg, 1985; Dick *et al.*, 1989; Jahanshahi *et al.*, 1995) and that the motor performance of PD patients improves when sensory feedback is enhanced (Dietz *et al.*, 1990; Glickstein and Stein, 1991; Georgiou *et al.*, 1994).

Thus, tasks that require repeated internal selection and initiation of discrete movements should provide a good substrate for testing the association between parkinsonian akinesia and SMA activity. As predicted, PD patients show a smaller than normal increase in CBF in the SMA during movement tasks that require selection and execution of unidirectional ballistic joystick movements (Fig. 3) (Playford *et al.,* 1992). The widely employed "internal generation" task used in this experiment, however, requires subjects to select movement directions *at random* from four allowed directions (forward, backward, left, or right). The task is fairly complex, therefore, in that it requires conceptualization

of randomization and working memory of previous movements in addition to internal selection and generation of simple movements. The added cognitive demands of the task confound the utility of this task for studying akinesia. In a critical follow-up experiment, a more carefully designed task was used to compare internally- and externally-generated movements in normals and PD patients (Jahanshahi *et al.,* 1995). Subjects were trained to make simple index finger extensions every 3 s by self-initiation or external triggering, yoked to the same rate. The tasks required minimal working memory or other cognitive demands. As before, PD patients had a smaller than normal activation of SMA for

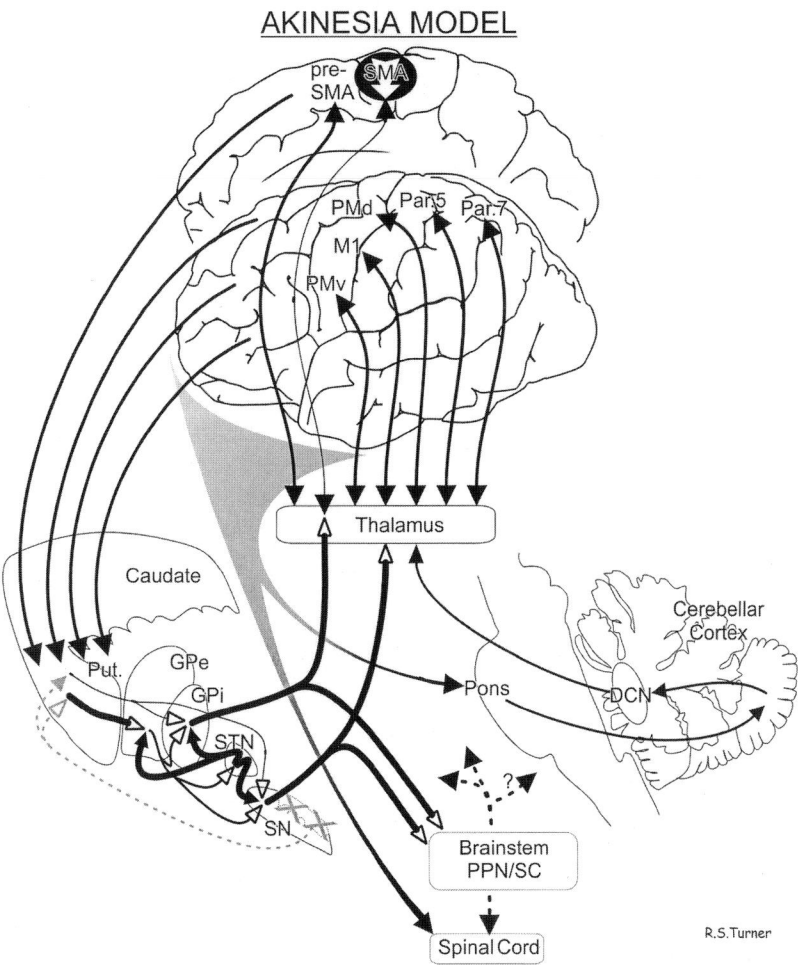

Figure 3 The traditional model for parkinsonian akinesia. This model hinges on the assumption that outflow from the basal ganglia motor circuit (GPi → Thalamus) terminates preferentially in the thalamic areas that interconnect with mesial premotor areas such as the SMA. In parkinsonism, excessive inhibitory outflow from the GPi inhibits or interrupts the facilitatory thalamocortical relationship, thereby leading to a disfacilitation of activity in the SMA. Sub-normal activation of the SMA has been associated with impaired movement selection and initiation under internally guided and sequential-task conditions. Therapeutic interventions for Parkinson's disease (dopamine replacement, pallidotomy, or DBS) both restore normal levels of SMA activity and improve motor performance when assessed under internally guided and sequential-task conditions.

self-initiated movements. It is noteworthy that no differences in brain activity between normal subjects and PD patients were found in this study when they performed similar movements under an externally triggered condition (Jahanshahi *et al.*, 1995).

When PD patients performing the internal generation task are treated with dopamine agonists (apomorphine), there is a "normalization" of the movement-related activation of SMA accompanied by a reduction in reaction times (Jenkins *et al.*, 1992). A similar effect of dopamine replacement therapy was observed by Rascol *et al.* in PD patients performing a sequential movement task that requires frequent initiation of self-generated discrete finger-to-thumb movements (Rascol *et al.*, 1992). They showed, using single photon emission computed tomography (SPECT), that the SMA is under activated in PD patients during this task (i.e., that the SMA had a smaller than normal task-related increase in CBF) and that the SMA defect normalized with apomorphine therapy (Rascol *et al.*, 1992). These results provide additional evidence that SMA activation is modulated by the BG motor circuit and that dopamine replacement therapy can reduce the inadequate thalamocortical facilitation of the SMA.

The imaging experiments outlined so far are consistent with a simple model that accounts for the "akinesia" associated with hypokinetic movement disorders in which BG dysfunction culminates in an inadequate recruitment of SMA neurons, resulting in impaired movement initiation. Dopamine replacement therapy, by releasing thalamocortical facilitation, restores normal SMA activation patterns and movement initiation improves. We next review evidence that this model may be an oversimplification and that it certainly does not account for symptoms other than akinesia that are common in PD.

C. Functional Reorganization and Task-Specific Compensation in PD

Recent functional imaging studies have detected patterns of CBF in PD patients that may reflect adaptive changes, some of which may be closely linked to the particular motor task being performed. Using SPECT, Rascol *et al.* (1997) found that untreated PD patients demonstrated an abnormally high activation of the cerebellum ipsilateral to the moving arm when they performed sequential finger-to-thumb movements. (That is, when scans from PD patients were compared with those from a matched group of neurologically normal subjects, task-related increases in CBF in the cerebellum were significantly greater for the PD subjects.) Coincident with the cerebellar overactivation was a smaller-than-normal activation of the SMA, as predicted by the akinesia model (Section IIB). It is unlikely that the exaggerated cerebellar activity in PD subjects could be attributed to differences in motor performance because the extent and frequency of movements during scanning were controlled carefully across subject groups (Rascol *et al.*, 1997). Furthermore, the increased activity in the cerebellum was not seen in a separate group of PD subjects who were studied when on their normal dopamine replacement therapy. The authors conclude that cerebellar overactivation in untreated PD patients may be part of a compensatory recruitment of alternate motor circuits in the parkinsonian brain (including the visually-driven cortico-ponto-cerebellar loop (Glickstein and Stein, 1991)) in an attempt to overcome impaired function of the mesial frontal cortical circuits normally used to perform this task (Rascol *et al.*, 1997).

Other recent studies also provide evidence of abnormal increased cerebral activity (CBF) in PD patients and indicate, additionally, that the specific patterns of under- and over-activation hinge on what behavioral task is used. Using PET, Samuel *et al.* found a bilateral task-related increase in CBF in dorsolateral premotor and inferior parietal cortices in untreated PD subjects performing a sequential finger tapping task (Fig. 4) (Samuel *et al.*, 1997a). These areas were not activated in normal subjects performing the same task. Again, as predicted from previous studies (Section IIB), Samuel *et al.* also found a task-related underactivation of mesial frontal and prefrontal areas in the PD subjects. These observations have been confirmed and extended recently by Catalan *et al.* (1999) in a PET study of PD and normal subjects performing either sequential finger movements of increasing complexity or an internal generation task (similar to the internal generation task used by Playford *et al.* (1992) described above). During sequential finger movements, they found a relative overactivation (i.e., a greater task-related increase in CBF than observed in normals) of bilateral parietal cortices, lateral premotor areas, and precuneus. Interestingly, Catalan *et al.* observed that mesial frontal areas (anterior SMA/cingulate cortex) were activated during motor sequence performance in both PD and normal subjects but that CBF increased progressively with more complex sequences only in the PD subjects. In contrast, when the same PD subjects performed the internal generation task, no parietal or premotor overactivations were observed and the mesial frontal areas, including the SMA, were underactive, as previous studies predicted (Playford *et al.*, 1992).

Although some of the results described thus far can be interpreted within the model for parkinsonian akinesia, other results call for a revised or expanded model. The contrasting results for sequential movement

TASK-SPECIFIC OVERACTIVITY

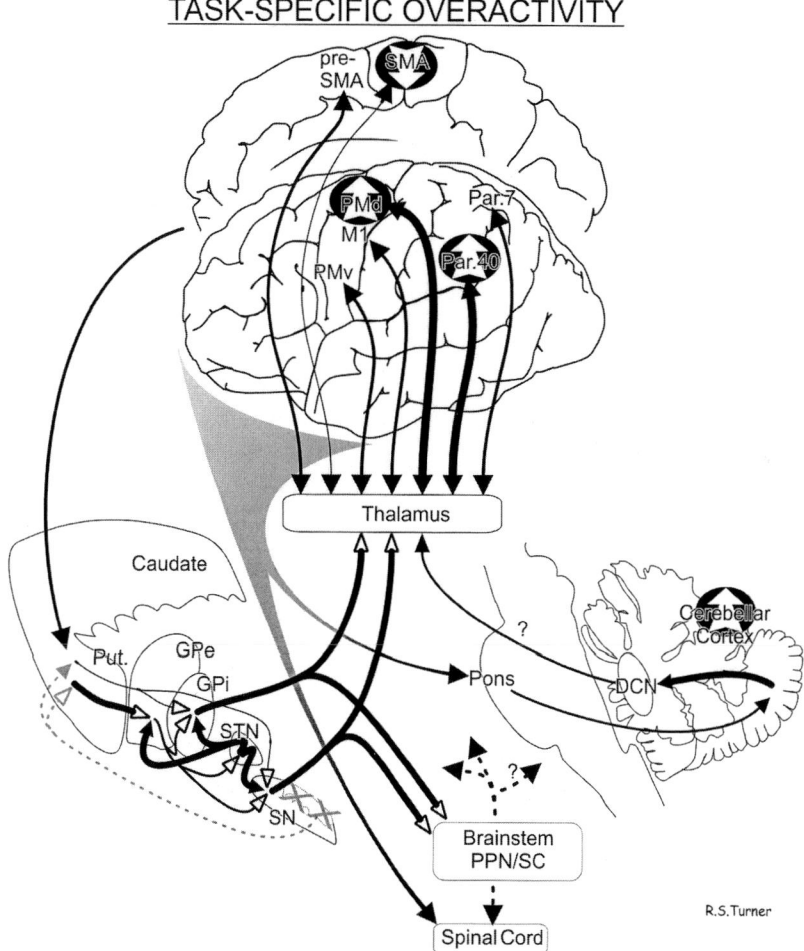

R.S.Turner

Figure 4 Areas of relative increased cerebral activity observed in parkinsonian subjects performing sequential movement tasks. Compared with normal subjects, parkinsonian patients had greater task-related increases in CBF in the dorsal premotor and parietal cortex (PMd and Par. 40) and in the cerebellar cortex. One of the sites of overactivation (PMd) is now known to receive basal ganglia outflow directly via the thalamus. Other sites of overactivation lie far from a known direct influence of basal ganglia output (Par. 40 and Cerebellar Cortex). These areas of relative overactivity may reflect a primary facet of parkinsonian pathophysiology (e.g., the inability to select or focus activation to the cerebral areas appropriate for the task) or may instead reflect the adoption of compensatory strategies. Note that increased activation of the cerebellar cortex under these conditions is most likely accompanied by altered activity in the efferent pathways from the deep cerebellar nuclei such as the cerebello-thalamocortical projections to motor and parietal areas.

and internal generation tasks in the Catalan *et al.* study, for instance, indicate that the specific differences in brain activity between PD and normal subjects depend critically on the nature of the behavioral task being performed. The use of tasks that accentuate different facets of parkinsonian motor impairment may expand our understanding of the functional substrates of parkinsonian symptoms other than akinesia. Recent functional imaging results from our group support this view.

We performed PET scans on PD patients (off their normal antiparkinsonian medication) and age-matched normal adults while they performed a smooth pursuit tracking task designed to probe the neural substrates of bradykinesia. (Six of the PD patients were studied again after they received a left hemisphere pallidotomy. See Section IIE.)] The task required subjects to follow the smooth movement of a visual target across a video screen using an on-screen cursor whose position was controlled by a joystick held in the subject's right hand. On different scans, the target movement followed a sinusoidal position–time function at one of three rates (0.1, 0.4, and 0.7 Hz (Fig. 5); see Turner *et al.* (1998) for a detailed description of the tasks.) In this way, the task manipulated across scans the mean velocity of movement and frequency of movement reversals while holding movement extent constant. Parkinsonian patients showed clear signs of bradykinesia when performing these tasks, being unable to increase their hand velocity to the level required to follow the full extent of target movement. This slowing of movement and hypometria were especially evident for the faster target rates (Figs. 5A and 5B, 0.7 Hz). The timing of tracking movements, however, was only slightly altered for PD subjects, therefore suggesting that signs of akinesia were not evident in this task.

For the PD patients, movement-related increases in CBF were found over a wider extent of the cortical surface but at fewer subcortical sites when compared with normals. Activations associated with arm tracking in normal subjects (that is, increases in CBF for all movement rates relative to that during an eye-tracking "rest" condition) were found at a collection of loci implicated by many previous functional imaging studies in the control of arm movements (Stephan *et al.*, 1995; Winstein *et al.*, 1996). Briefly, these included sensorimotor, dorsal premotor, and parietal (area 5) cortical areas (blue areas in Fig. 6A), the left BG and thalamus (Fig. 6B), and the right cerebellum (Fig. 6C). Additional smaller activations were observed in the right precentral gyrus,

right superior parietal lobule, and right BG. Movement-related increases in CBF in the PD group were found at several cortical loci that were not activated in normals, including the ventral premotor and parietal cortices (area 7) of the left hemisphere and the dorsal premotor cortex of the right hemisphere (compare blue areas in Figs. 6A and 6D). At the primary sites of normal movement-related activity, however, the PD group showed attenuated activations or no movement-related activation at all. Smaller than normal increases in CBF were observed in both cortical (sensorimotor, dorsal premotor, and parietal area 5) and subcortical (VL thalamus and right cerebellum) regions.

The reduction or loss of movement-related CBF in the BG, associated thalamic areas, and cerebellum of parkinsonian subjects may arise from the abnormal neural discharge patterns known to be exiting the BG in PD (Miller and DeLong, 1987; Filion and Tremblay, 1991; Hutchison *et al.*, 1994; Sterio *et al.*, 1994; Hayase *et al.*, 1998; Hurtado *et al.*, 1999). An attenuation of *relative* task-related increases in CBF at the locations of GPi terminal fields in the thalamus, for example, may result from the abnormally high tonic discharge of GPi neurons in PD. (Recall that regional CBF is linked closely to the level of local presynaptic activity (Jueptner and Weiller, 1995).) One factor contributing to this relative attenuation could be the simple fact that the same movement-related change in discharge will amount to a smaller *relative* change when it is riding on top of a higher tonic discharge rate. Higher tonic discharge rates may also attenuate the task-related signal

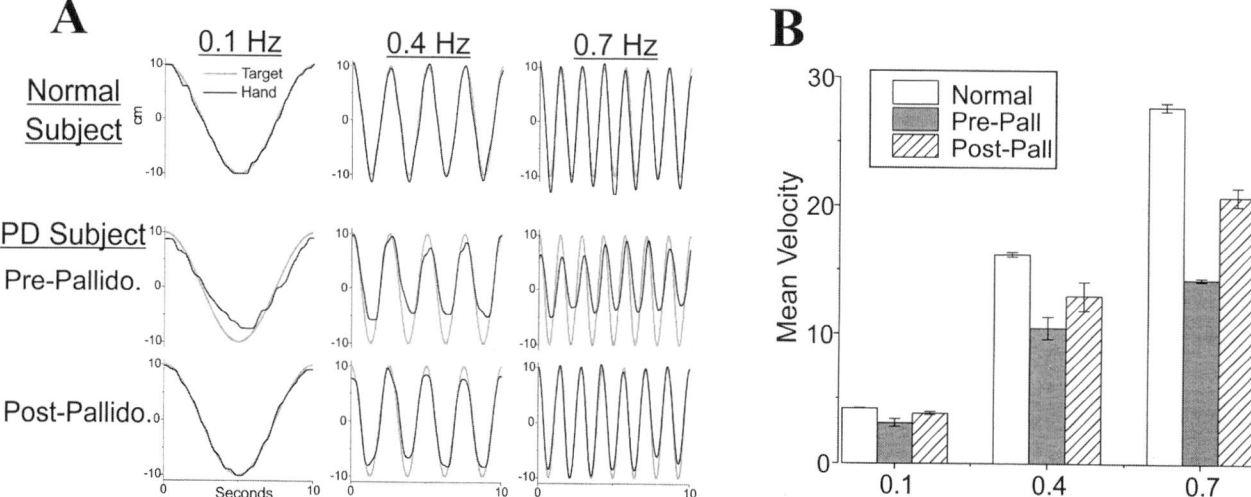

Figure 5 Performance of normal and PD subjects during tracking at three target rates. (A) Examples for individual subjects. For the PD subject, data are shown for equivalent studies before and after pallidotomy. Only 10 s of each 100-s long trial is shown. (B) Mean tangential velocity across all subjects studied. Tracking performance was degraded in the parkinsonian subjects pre-pallidotomy, showing a reduction in movement amplitude and mean movement velocity that was exacerbated at faster target rates. Following a clinically effective pallidotomy, the performance of patients improved.

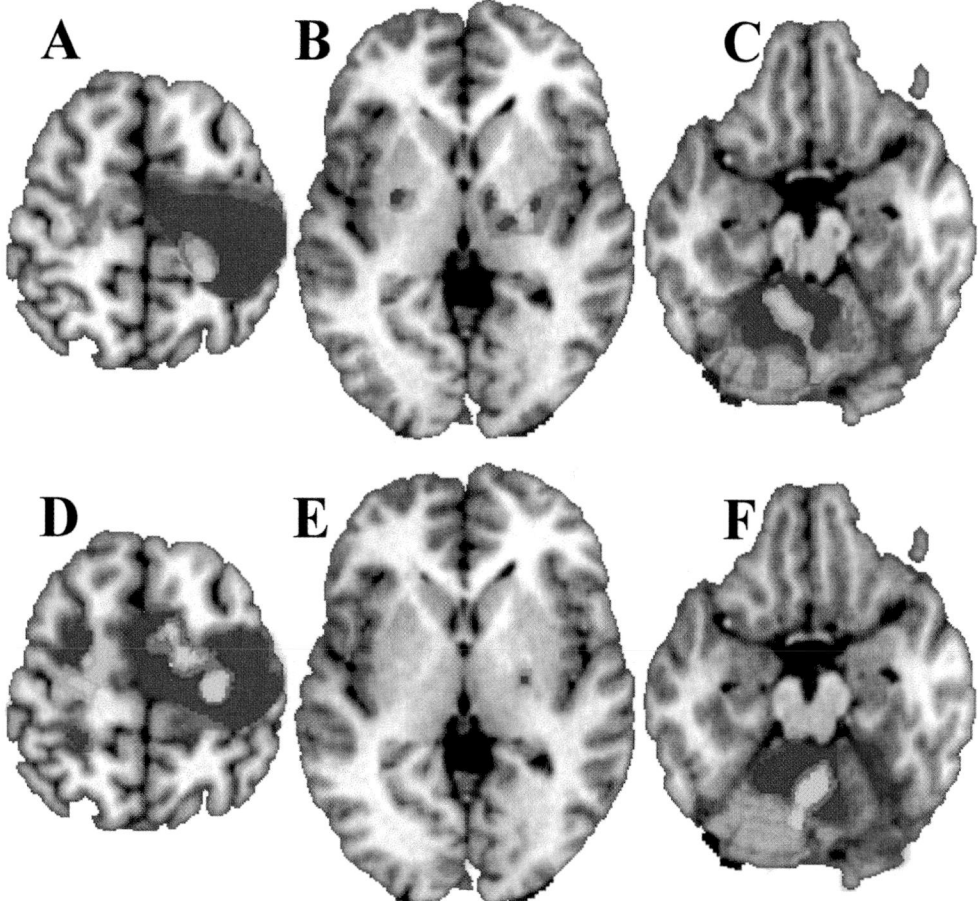

Figure 6 Significant movement- and rate-related CBF (blue and yellow regions) for normal (A–C) and parkinsonian (D–F) subjects. Axial views of *t*-maps for PET comparisons (*p* < 0.005) are shown superimposed in Talairach coordinates on an MRI from one subject. (Right hemisphere is to the left.) A and D: cortex. Task-related activation of cortex was greater for parkinsonian subjects, including a greater activation of mesial frontal cortex and bilateral lateral premotor and parietal areas. Rate-related activations in parkinsonian subjects expanded beyond primary sensorimotor (yellow in A) to include bilateral mesial and lateral premotor cortices (white in D). B and E: basal ganglia/thalamus were underactivated in PD subjects and showed no velocity relations. C and F: Movement- and velocity-related activations of the cerebellum both were attenuated in the PD subjects. Panels A–C adapted with permission from Turner *et al.*, 1998.

in GPi terminals through ceiling effects or depolarization block if GPi discharge is driven beyond the maximum rates possible. The reduced movement-related CBF in the cerebellum of PD subjects may be a product of similar effects working in the newly described pathway from basal ganglia to precerebellar brainstem nuclei (Rye, personal communication) or via polysynaptic corticopontine pathways. The regions of reduced cortical CBF observed in our parkinsonian subjects may be accounted for using a standard explanation; namely, that excessive inhibitory outflow from the parkinsonian BG causes a loss of normally excitatory thalamocortical loops. Areas of abnormally increased cortical CBF, however, cannot be easily accounted for with well-established models of parkinsonian pathophysiology.

We will discuss possible mechanisms contributing to areas of abnormally increased CBF after briefly reviewing the effects of movement rate on CBF.

Rate-related activations for the normal subjects (identified as areas where CBF increased monotonically with the rate of target movement) were detected at only three sites: the left sensorimotor cortex, the motor territory of the left GP, and the right anterior cerebellum (yellow regions in Figs. 6A, 6B, and 6C, respectively). The rate-related activations of motor cortex and cerebellum are consistent with previous reports of cerebral activity associated with increasing movement frequency (VanMeter *et al.*, 1995; Blinkenberg *et al.*, 1996; Sadato *et al.*, 1996; Jenkins *et al.*, 1997), but frequency-related activations in the basal ganglia have not been described previously. For the PD subjects, rate-related activity was found at a far greater number of cortical sites, including left mesial (SMA), lateral, and ventral premotor areas and lateral premotor cortex ipsilateral to the moving arm (compare yellow areas in Figs. 6A and 6D). At the three loci modulated with rate in the normal subjects, however, the effects of movement rate in the PD group were either reduced (left sensorimotor and cerebellum, Fig. 7) or absent (left GP, Fig. 6E).

Differences in motor performance between the normal and PD subjects could not account for these different patterns of movement- and rate-related activity. For instance, for loci that were normally rate-related, the

slope of the relation between CBF and actual movement velocity was reduced for the PD subjects (cerebellum and sensorimotor cortex in Fig. 7), whereas, for cortical sites that were rate-related only in PD patients, the CBF/velocity slopes were increased (SMA and ventral premotor, Fig. 7). A post hoc multiple linear regression analysis confirmed the statistical significance of these observations.

An appealing explanation for the movement rate results is that the velocity, force, or overall scale of movement is normally regulated by a small subset of areas, including the BG motor circuit, and that disorders of this subcircuit lead to bradykinesia. The rate-related activation of the posterior GP in normals is consistent with previous observations that neural activity in the BG motor loop is often correlated with parameters of movement scale (Georgopoulos *et al.*, 1983; Turner and Anderson, 1997). The three structures that were strongly modulated with velocity in normal subjects, including the motor portion of the GP, showed *reduced* velocity relations in the parkinsonian subjects. This result complements many previous observations that the disruption of neural activity in the BG motor circuit results in bradykinesia or hypometria (Hallett and Khoshbin, 1980; Hore and Villis, 1980; Horak and Anderson, 1984a; Stelmach *et al.*, 1989; Alamy *et al.*, 1996; Berardelli *et al.*, 1996).

In summary, the results discussed in this section are consistent with an emerging view that the specific pattern of abnormally attenuated or enhanced brain activity associated with PD is strongly influenced by the behavioral task being used. When tasks require the internal selection, sequencing, or initiation of movement, mesial premotor areas including the SMA are hypoactivated (Playford *et al.*, 1992; Rascol *et al.*, 1992; Jahanshahi *et al.*, 1995; Samuel *et al.*, 1997a; Catalan *et al.*, 1999) and lateral premotor, parietal, and cerebellar areas may have greater than normal task-related increases in CBF (Rascol *et al.*, 1997; Samuel *et al.*, 1997a; Catalan *et al.*, 1999). When, instead, a task emphasizes online visuomotor control and the scaling of movement velocity, there is an underactivation of contralateral primary motor cortex, dorsal premotor cortex, parietal area 5, BG, thalamus, and anterior cerebellum, accompanied by greater than normal changes in CBF (movement- and/or rate-related) at a variety of premotor sites, including the SMA. It is far too early to say conclusively how the interaction between the demands of a task and parkinsonian pathophysiology determines which areas will be under- and over-activated. It is reasonable to hypothesize, however, that the areas underactivated during a specific task may indicate which brain areas are associated with the parkinsonian symptoms that are manifested during that task. The variety of

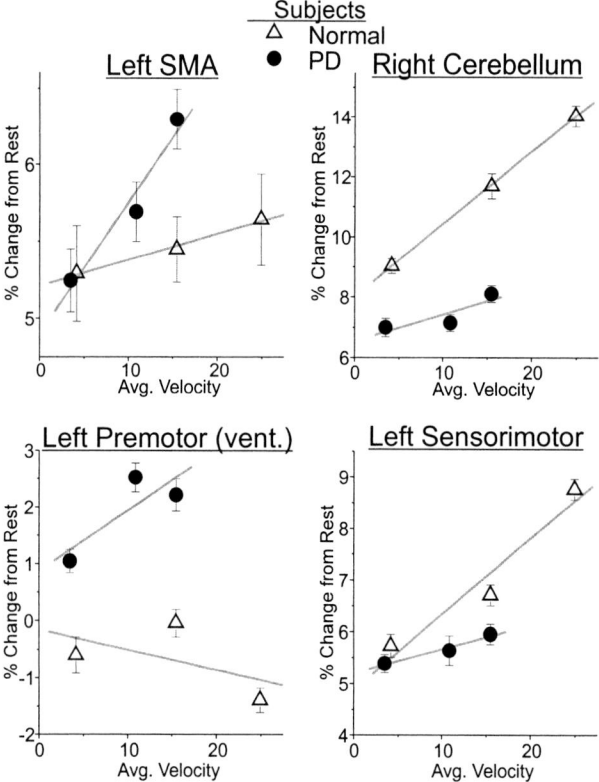

Figure 7 The differences in task-related CBF between normal and parkinsonian subjects could not be attributed to differences in task performance. For four representative areas, the mean percentage change in CBF (relative to the eye-only rest condition, ±SEM) is plotted versus actual mean hand velocities for normal and parkinsonian subjects. Multiple cortical premotor areas including left supplementary motor (L. SMA) and ventral premotor (L. Premotor vent.) areas had increased movement and rate-related activation in the parkinsonian subjects. These effects were evidenced in a regression analysis by significant increases in *y*-intercepts and slopes, respectively, for the parkinsonian group. Reduced *y*-intercepts and slopes at cerebellar and primary sensorimotor cortex sites supported the view that parkinsonian subjects were unable to adequately modulate activity in primary motor execution areas. Shaded lines, linear regressions for normal and parkinsonian groups.

data emerging from these recent functional imaging studies, however, are rather uninterpretable if the pathophysiologic model of PD is limited to that of an attenuated contribution of SMA to movement initiation.

Areas of abnormally enhanced CBF observed in PD have been interpreted by most investigators as a recruitment of visually competent motor circuits in an effort (probably automatic or subconscious) to compensate for dysfunction of the BG–mesial frontal circuits (Rascol *et al.,* 1997; Samuel *et al.,* 1997a; Catalan *et al.,* 1999). It is unlikely, however, that such an explanation can account for the progressively increasing CBF observed in mesial premotor areas both by Catalan *et al.* with increasing sequence difficulty (Catalan *et al.,* 1999) and by ourselves with increasing movement rate. Furthermore, if the areas of increased CBF do reflect some type of compensatory response to parkinsonian motor impairment, the effectiveness of the compensation is questionable given the simple fact that parkinsonian motor performance remained impaired in spite of the increased activity. The areas of abnormally-increased CBF may, alternatively, reflect a facet of the primary pathophysiology of PD such as an inability to suppress the activation of contextually inappropriate motor circuits. An impaired ability to control or focus regional activity could arise from dysfunction of the basal ganglia, per se (Mink, 1996), or, equally likely, from dysfunction of the thalamus and/or frontal lobes secondary to abnormal BG outflow. The observation of a similar recruitment, or faulty suppression, of areas not normally activated with a task after functional recovery from subcortical stroke (Weiller *et al.,* 1992) suggests that task-dependent changes like these are a relatively common correlate of CNS insult. Similar task-specific alterations in brain activity were also observed as a correlate of cognitive decline associated with aging (Esposito *et al.,* 1999). Although much remains to be discovered concerning areas of abnormally increased CBF in PD, it is clear that the patterns are task-specific and, thus, may provide information about the functional substrates of various parkinsonian symptoms. For instance, it will be shown below (Section IIE) that a therapeutic intervention that reduces bradykinesia (pallidotomy) also results in a relative normalization of the abnormally expanded pattern of rate-related brain activity.

D. Rationale for Surgical Intervention in PD: Pallidotomy

The current model of PD pathophysiology provides a clear rationale for surgical treatment of PD by stereotaxic ablation of the posteroventral GPi (pallidotomy) (Fig. 8). (The motor territory of the pallidum is located posteroventrally.) Both in PD patients and in primate models of PD, pallidotomy can reduce significantly the cardinal symptoms of PD while producing no overt side-effects (Laitinen *et al.,* 1992; Dogali *et al.,* 1995; Baron *et al.,* 1996). The presumed mechanism of action for pallidotomy is an elimination of excessive pallidothalamic inhibition and a subsequent recovery of

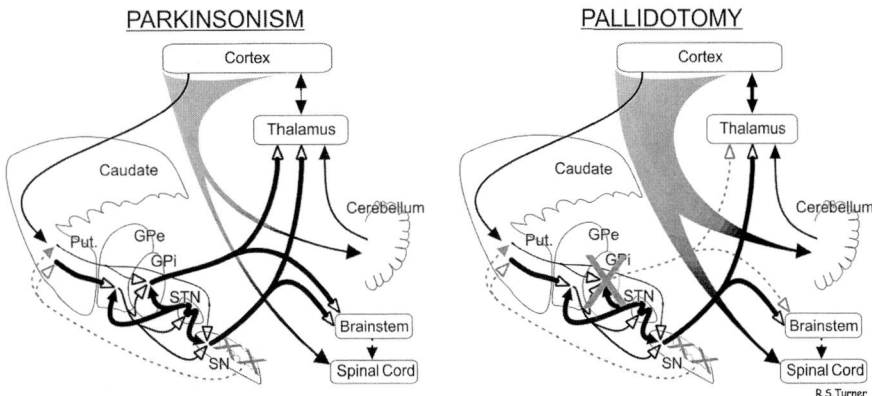

Figure 8 The hypothesized mechanism by which pallidotomy results in a normalization in thalamocortical activity. Parkinsonism is associated with abnormally high tonic discharge and altered discharge patterns in the projection neurons of the basal ganglia (neurons projecting from GPi to SN). This abnormal inhibitory input to the thalamus (and brainstem) interferes with normal thalamocortical function and culminates in disordered movement. Pallidotomy (thermocoagulation of the motor-related portion of the GPi) eliminates one source of abnormal activity, thereby allowing a relative normalization of thalamocortical function. Note that pallidotomy leaves intact the substantia nigra (SN), a potentially important source of abnormal basal ganglia outflow and a possible target for therapeutic intervention.

function in the previously over-inhibited frontal cortical areas. The efficacy of pallidotomy as a treatment for PD points clearly to the conclusion that most of the symptoms of PD arise from the impaired function of cortical motor areas secondary to excessive inhibitory outflow from the pallidum and not, as might be assumed, from impaired BG function per se (Wichmann and DeLong, 1996). In other words, many of the symptoms associated with PD, and very likely those associated with other striatal disorders, may tell us more about the functions of the frontal areas affected by BG outflow than about the BG itself.

If the BG plays an important role in motor control, then why doesn't ablation of the primary output nucleus of this structure lead to an overt impairment of motor performance? As far as has been tested, naturally-occurring pallidal lesions in humans lead to subtle, short-lived, or no detectable deficits of motor performance (DeLong and Coyle, 1979; DeLong and Georgopoulos, 1981, review; Marsden and Obeso, 1994; Haaxma et al., 1995). The disruption of normal BG outflow in non-human primates is reported to cause, at most, modest changes in the speed and/or the metrics of trained arm movements, and sometimes changes in postural reflexes, while not impairing the initiation or the directional accuracy of movement (Hore and Villis, 1980; Horak and Anderson, 1984a,b; Mink and Thach, 1991; Kato and Kimura, 1992; Alamy et al., 1996; Inase et al., 1996a). A parsimonious explanation of these observations is that the BG plays a minimal role in the actual control of movement execution.

Some have proposed that the BG motor circuit, instead, plays a specialized role in higher level motor functions such as the automatic execution of complex overlearned motor actions or the learning of new motor skills (Berridge and Whishaw, 1992; Marsden and Obeso, 1994; Cromwell and Berridge, 1996; Aldridge and Berridge, 1998). If true, then it is reasonable to suppose that these functions will be impaired in pallidal-lesioned normal animals and in parkinsonian patients who have received a pallidotomy. Surprisingly, no studies to date have examined in normal animals the effects of pallidal inactivation on the performance of complex motor skills such as sequential movement or on motor learning. Pallidotomy for PD has been, in fact, reported to improve the performance of some complex movements and of overlearned motor sequences (Baron et al., 1996; Kimber et al., 1999; Limousin et al., 1999). A recent preliminary report suggests, however, that pallidotomy impairs patients' ability to learn new motor sequences under implicit conditions (Brown et al., 1998). Although many studies have examined the behavioral effects of striatal inactivations on complex motor skills and motor learning (Packard et al., 1989; Berridge and

Whishaw, 1992; Cromwell and Berridge, 1996; Miyachi et al., 1997), interpretation of these studies is complicated by uncertainty about the net downstream effects of striatal inactivation. One might recall at this point an important lesson gleaned from the therapeutic success of pallidotomy: that striatal disorders may give rise to motor impairments that are only indirectly related to the actual functions of the BG.

More complex explanations postulate that compensatory mechanisms may obscure the true negative aspects of pallidotomy. For instance, the GPi–thalamic projection is not the only output pathway for the BG and other pathways (SNr to the thalamus, GPe to the reticular nucleus of the thalamus, and STN projections to the brainstem (PPN)) may also contribute to motor control in a significant manner (Parent and Hazrati, 1993; Hazrati and Parent, 1994; Joel and Weiner, 1994). These other projections could compensate for GPi lesions in otherwise normal subjects (Alamy et al., 1994). Finally, the possible compensatory utilization of other brain systems for controlling movement (e.g., the cortico-ponto-cerebellar pathway) should not be neglected (Glickstein and Stein, 1991). It has long been argued that cerebellar pathways in particular may be used by PD patients and subjects with BG lesions to control movements relying on visually driven strategies (Hore et al., 1977; Majsak et al., 1998). These pathways and others may also be of increased importance for motor control following pallidotomy.

In summary, although pallidotomy has proved to be an effective therapy for PD, the behavioral correlates of pallidotomy are only now being explored. It is quite possible that the full impact of pallidotomy on motor control systems is not evident at the behavioral level because of compensatory neural mechanisms. Recent neuroimaging results, some reviewed below, indicate that the effects of pallidotomy can be characterized, in large part, as a relative normalization of the abnormal parkinsonian pattern of brain activity but that the specific effects of pallidotomy depend upon the motor task being performed.

E. Functional Imaging Studies of Pallidotomy for PD

Many of the functional imaging studies of pallidotomy have provided results consistent with the akinesia model of PD pathophysiology (Ceballos-Baumann et al., 1994; Grafton et al., 1994, 1995; Samuel et al., 1997b). Ceballos-Baumann and colleagues were the first to show, using the internal generation task and PET in one parkinsonian patient, that pallidotomy can cause a marked improvement in the movement-related activation of the SMA and dorsolateral prefrontal areas

(Ceballos-Baumann *et al.*, 1994). (That is, the task-related increase in CBF was significantly greater following pallidotomy.) In a detailed report that followed, Samuel *et al.* (1997b) studied six parkinsonian subjects "off" medication both before and 3–4 months following a right posteroventral pallidotomy. (All of the subjects showed a marked amelioration of their parkinsonian symptoms following pallidotomy.) The subjects used their contralateral (left) arm to perform an internal generation task. For the patients, movement-related increases in CBF in the SMA and dorsolateral prefrontal area were of greater magnitude following surgery, just as the akinesia model of PD would predict. Greater increases in CBF were also observed in visual association areas and the anterior insula ipsilateral to the moving arm. In this study, the only brain area with a significant pallidotomy-related reduction in CBF was at the site of the lesion in the right pallidum (Samuel *et al.*, 1997b).

Grafton *et al.* also used PET to study movement-related changes in CBF in PD subjects before and after pallidotomy (Grafton *et al.*, 1994, 1995). In those studies, six PD patients reached and grasped illuminated Plexiglas dowels. The task, therefore, required stimulus-driven movement initiation and on-line visually-guided reaching. The frequency of movement was held constant for pre- and post-pallidotomy scans. Following a pallidotomy in the hemisphere contralateral to the moving arm, subjects showed a greater movement-related increase in CBF (compared to that prior to pallidotomy) in the SMA, lateral premotor, and anterior insular cortex (Fig. 9). Once again, smaller movement-related changes in CBF after pallidotomy were observed only around the site of the lesion in the basal ganglia (Grafton *et al.*, 1995).

A feature common to the motor tasks used in these pallidotomy imaging studies was a strong emphasis on movement selection and initiation, paced by visual or auditory stimuli. When such tasks are used, functional imaging of post-pallidotomy patients provides further support for the well-established model of parkinsonian akinesia. The effects of pallidotomy on areas of significant overactivity, however, were not addressed in the first studies. We will next review some of our own data showing that the functional correlates of pallidotomy depend strongly on what task is used.

For our subjects, pallidotomy had an overall effect of reversing the aberrant patterns of movement- and rate-related cerebral activations seen preoperatively. Six of the parkinsonian subjects that participated in the study of visuomotor tracking (Section IIC) were scanned again at 3–4 months after receiving a left hemisphere pallidotomy. Five of the six subjects showed a significant improvement in tracking performance during post-pallidotomy scans (Fig. 5B). For those five patients, movement-related increases in CBF were of greater magnitude after pallidotomy in the dorsal parietal and cingulate motor cortices ipsilateral to the lesion (Figs. 10A and 10B, respectively) and in the primary sensorimotor, dorsolateral premotor, and dorsal parietal cortices contralateral to the lesioned hemisphere (Fig. 10A). Greater task-related activations were also seen in the lateral anterior cerebellar cortex and midbrain tegmentum (Fig. 10D). Movement-related increases in CBF were *reduced* in magnitude following pallidotomy in the SMA bilaterally (Fig. 10A), in the ventral premotor cortex (PMv, Fig. 10B), and in the anterior putamen (Fig. 10C). The attenuated activations of the SMA and ventral premotor (relative to the task-related CBF seen in these areas in PD patients prior to pallidotomy) and the increased task-related activation of the cerebellum could be interpreted as a pallidotomy-related normalization of movement-related activity. (Recall that in PD subjects before pallidotomy, the SMA and ventral premotor areas had greater-than-normal movement-related increases in CBF whereas the cerebellum was underactivated; Section IIC.) The emergence after

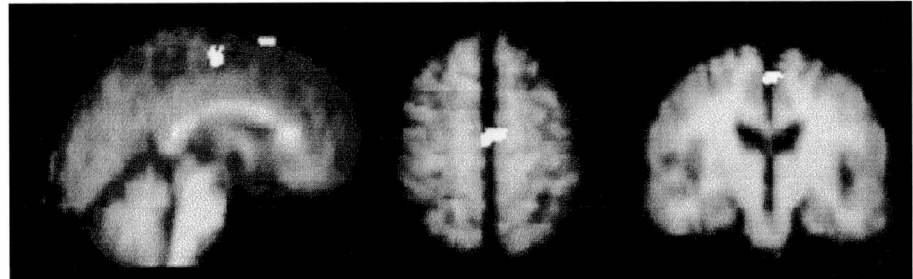

Figure 9 Interaction of pallidotomy and movement-related responses in supplementary motor cortex. A key site with a significant increase of movement-related activity after pallidotomy is shown in white, superimposed on the same subjects' average MRI scans. The three sections show the location of the SMA site in sagittal (left), transaxial (middle), and coronal (right) view. Adapted from S. T. Grafton, C. Waters, J. Sutton, M. F. Lew, and W. Couldwell (1995). Pallidotomy increases activity of motor association cortex in Parkinson's disease: A PET study. *Ann. Neurol.* **37,** 776–783.

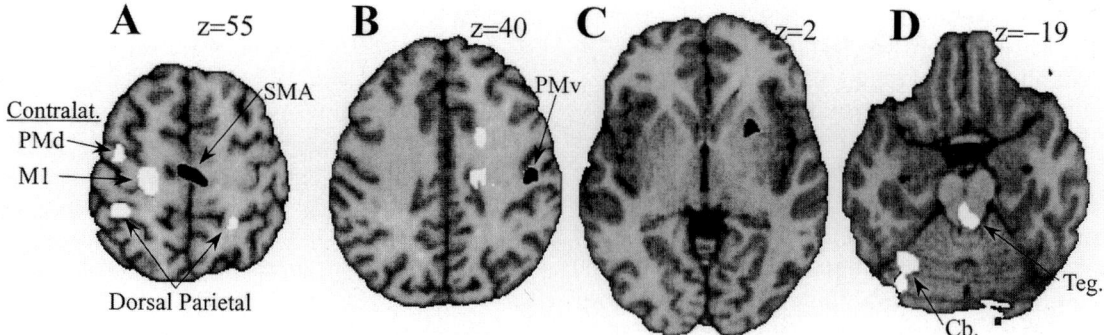

Figure 10 The effects of pallidotomy on movement-related activity in five PD patients during smooth-pursuit tracking. White: regions where movement-related activity was greater postpallidotomy. Black: regions where movement-related activity was reduced postpallidotomy. Clinically effective pallidotomy resulted in increased movement-related activation of cingulate motor areas (B), contralateral primary sensorimotor cortex and dorsal premotor cortex (M1 and PMd), and bilateral dorsal parietal areas (A). Reduced movement-related activations were observed in the supplementary motor area bilaterally (SMA in A), ventral premotor cortex (PMv in B), and the basal ganglia (C). Contralat.: right hemisphere contralateral to the pallidotomy and ipsilateral to the moving arm. (All areas significant at $p < 0.05$ corrected for multiple comparisons.)

pallidotomy of task-related increases in CBF in cingulate motor areas and in motor areas ipsilateral to the moving arm, however, must be taken as evidence that the postpallidotomy motor system employs a new motor control strategy because these areas were not activated by normal subjects performing the same task.

CBF correlated with the rate of movement was increased after pallidotomy in the motor thalamus and in the cerebellum while it was reduced in multiple cortical motor areas. Increased relations between CBF and movement rate were seen in left thalamic motor areas (VA) and visual association areas (white areas in Fig. 11C) and in the cerebellum (Fig. 11D; see also Fig. 12). Decreased CBF-rate relations were seen in cortical motor structures both contralateral and ipsilateral to the

moving arm (black areas in Figs. 11A and 11B). It is noteworthy that most of the cortical motor areas that had increased CBF-rate relations in PD patients compared to normals (Fig. 6) showed reduced CBF-rate relations in the pre- versus post-pallidotomy comparison. Although the left pallidum was a prominent site for rate-related activity in normal subjects, it is not surprising that rate-related CBF did not appear there following pallidotomy because of the lesion. The enhanced relation of CBF to movement rate in the thalamus after pallidotomy may be due to a combination of factors: an increase in the velocity-related inflow from cerebellothalamic afferents, a loss after pallidotomy of excessive non-velocity-related inflow from pallidothalamic afferents, and, perhaps, increased corticothalamic in-

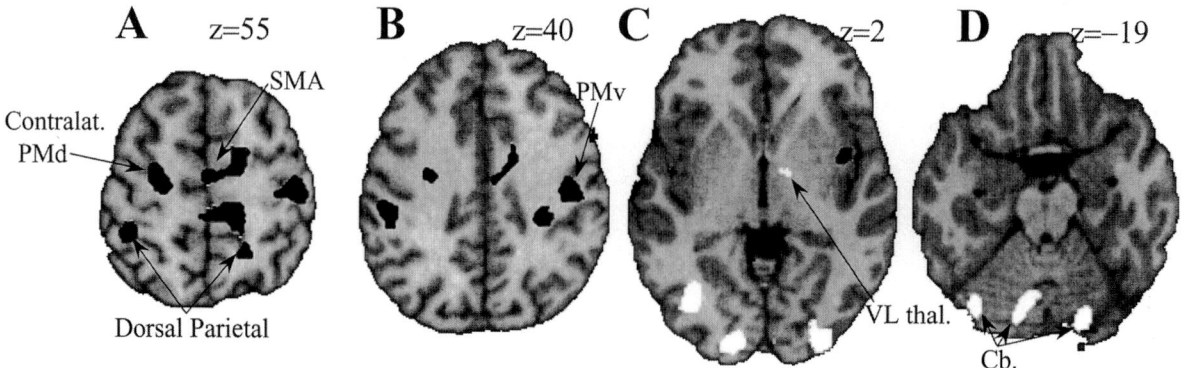

Figure 11 The effects of pallidotomy on rate-related activity. White: regions where a monotonic relation between CBF and movement rate increased postpallidotomy. Black: regions where the rate-related activity was decreased postpallidotomy. Note that the cortical areas that had abnormally increased rate relations in parkinsonian subjects pre-pallidotomy (Fig. 6) showed reduced CBF/rate slopes postpallidotomy (A and B). Conversely, areas that were underactive pre-pallidotomy, such as the cerebellum (Cb. in D), had increased rate relations after pallidotomy. Contralat.: right hemisphere contralateral to the pallidotomy and ipsilateral to the moving arm. (All areas significant at $p < 0.05$ corrected for multiple comparisons.)

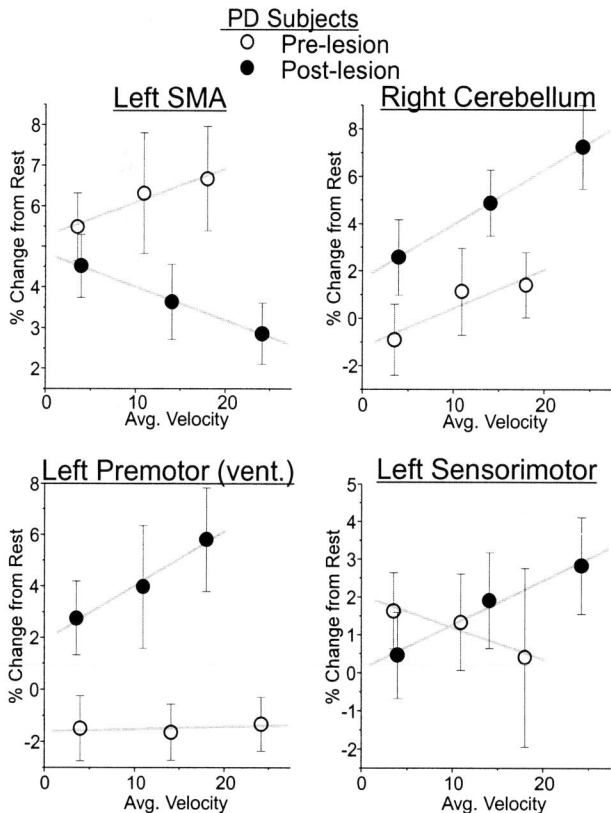

PD Subjects
○ Pre-lesion
● Post-lesion

Figure 12 Pallidotomy effects on CBF were independent of treatment-related changes in performance. Mean percentage change in CBF (relative to the eye-only rest condition, ± SEM) versus actual movement velocities for five PD subjects pre- and postpallidotomy. Cortical premotor areas showed reduced movement- and rate-related activity (L. SMA and L. Premotor vent.) whereas the cerebellum and thalamus had increased activity. Shaded lines, group linear regressions.

flow. The effects of pallidotomy on CBF could not be accounted for by pallidotomy-related changes in task performance (Fig. 12). This was true for both the reduced movement- and rate-related activations of cortical areas (SMA and ventral premotor, Fig. 12) and the increased movement- and rate-related activations of the thalamus and cerebellum.

To recap, neuroimaging studies have shown that the effects of pallidotomy on brain activity can be described, in many respects, as a restoration of normal patterns of task-related activity. Clinically-effective pallidotomy is associated with an enhanced task-related activation of brain areas that are normally engaged during performance of a task and, at the same time, reduced activation of brain areas that are not normally activated with the task. The effects of pallidotomy are task-specific in a manner that mirrors the effects of PD on brain activity. For tasks that emphasize movement selection and initiation, an impaired activation of the

SMA (and dorsolateral prefrontal, in some studies) is observed in PD and that activity is restored following pallidotomy (Grafton *et al.*, 1995; Samuel *et al.*, 1997b). For our task, which emphasized visuomanual control and the scaling of velocity, many of the motor execution areas that are underactivated in PD (sensorimotor, parietal area 5, motor thalamus, and right cerebellum) have movement- and/or rate-related increases in CBF of greater magnitude following pallidotomy. We found, additionally, that the areas of abnormally high movement- and rate-related CBF observed in PD were suppressed following pallidotomy.

Furthermore, by studying rate-related CBF before and after pallidotomy, we have identified a potential functional correlate of parkinsonian bradykinesia. Bradykinesia appears to be associated with an underactivation of the motor execution areas and an expansion of activity into multiple premotor cortical sites (areas not normally sensitive to movement rate). Amelioration of bradykinesia by pallidotomy is accompanied by increased task-related CBF in some of the motor areas that normally scale their activity with the rate of movement and by a loss of the abnormal rate-related activation of premotor cortices. In a few important respects, however, the post-pallidotomy pattern of movement- and rate-related activation was different from that seen in normals, thereby suggesting that pallidotomy itself may induce a task-specific pattern of functional compensation.

F. Deep-Brain Stimulation as a Treatment for PD

A relative drawback of surgical pallidotomy is the potential morbidity (acute and chronic) resulting from a permanent brain lesion. The recent introduction of high-frequency deep-brain stimulation (DBS) is an exciting alternative to ablation because the electrode can be introduced without producing significant brain damage and, by adjusting stimulation sites and parameters, the optimal response can be obtained. Initial reports of clinical response to DBS have been very encouraging (Siegfried and Lippitz, 1994; Limousin *et al.*, 1995; Krack *et al.*, 1998). As shown in Fig. 13, the stimulating electrode can be positioned at several nodes of the subcortical motor circuit, including the GPi, subthalamic nucleus (STN), and the motor thalamus. Prospective trials are being implemented both to determine the optimal target for DBS individualized to a patient's clinical deficits and to establish the long-term safety and clinical efficacy of the procedure.

The mechanism by which DBS achieves therapeutic results remains speculative. In the structures that have been studied, the clinical effect of stimulation is similar to the effect of inactivation of neural activity in the tar-

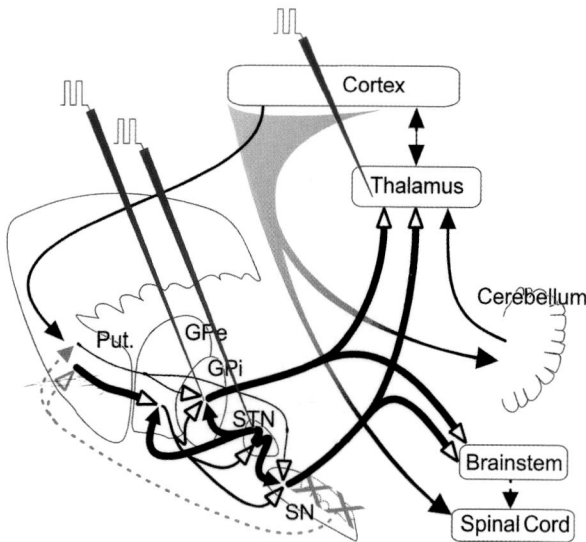

Figure 13 Potential sites for therapeutic deep-brain stimulation (DBS) in Parkinson's disease. The effects of excessive and abnormal outflow from the parkinsonian basal ganglia can be reduced or eliminated by high-frequency electrical stimulation (depicted as red arrows) of the subthalamic nucleus (STN), internal globus pallidus (GPi), or motor-related thalamic nuclei. Note how the connectivity of the STN may allow STN-DBS to reduce abnormal neural activity in both of the primary basal ganglia output nuclei (GPi and SN). DBS of the thalamus is an effective treatment for tremor but appears to be less effective than GPi- or STN-DBS for other parkinsonian symptoms.

get nucleus (Bergman *et al.*, 1990a,b; Wichmann *et al.*, 1994a; Dogali *et al.*, 1995; Lozano *et al.*, 1995; Baron *et al.*, 1996), thereby suggesting that DBS inactivates or at least reduces abnormally high neural discharge in the target nucleus. DBS could produce this effect by recruitment of inhibitory afferents at the site of stimulation, direct depolarization block, entrainment, or some other process. Thus, it is thought that continuous high-frequency electrical stimulation in the GPi has an effect similar to pallidotomy in interrupting the abnormal GPi outflow, thereby freeing the relatively intact frontal thalamocortical structures to function more normally. With DBS of the STN, GPi outflow may be normalized by a reduction of the excitatory drive from the STN (Wichmann *et al.*, 1994b).

G. PET Studies of Deep-Brain Stimulation in PD

When combined with functional imaging, DBS can be a powerful technique for assessing the systemwide effects of focal disruption at different nodes of the BG motor circuit. Task-specific changes in activity and the expected responses based on the current models of BG functional organization can then be compared.

There are two published studies using PET to examine the effects of therapeutic DBS on CBF. In the first, Limousin *et al.* explored the effects on cerebral blood flow of DBS in the GPi and STN using the internal generation task described in Section IIB (Limousin *et al.*, 1997). Clinically effective levels of stimulation in the STN led to a greater task-related increase in CBF in the SMA and dorsolateral prefrontal cortex compared to ineffective stimulation. The authors interpreted this result in the context of the simple akinesia model. Of concern, however, clinically-effective stimulation of the GPi produced no significant changes in CBF. As noted in our ongoing study of DBS, large task-specific changes in CBF can occur with DBS during a control condition; thus a study design such as Limousin's that only tests for differences between movement and control scans with DBS could miss important treatment-by-task interactions.

In the second study, Davis *et al.* examined the effect of GPi DBS on brain activity during a "rest" condition (Davis *et al.*, 1997). Clinically beneficial stimulation in the GPi was associated with a CBF increase, relative to CBF during scans performed with no stimulation, in mesial frontal cortex anterior to the SMA. This result suggests that DBS altered the inhibitory GPi output in a manner analogous to ablation and thereby disinhibited the frontal thalamocortical circuit. The authors proposed that the increased CBF in the mesial cortical areas, although observed under a "resting" condition, could be responsible for a reduction of akinesia. The "rest" condition in this study required subjects to silently count to themselves and to make no movement. It is noteworthy that the area rostral to the SMA, the "pre-SMA," showed the greatest increase in CBF with DBS during this silent counting task. Studies in normal subjects report that the pre-SMA area has task-related increases in CBF during imagined movement and speech (Paus *et al.*, 1993). (See Tanji (1994) for a thorough review of the important anatomical and functional differences between the SMA-proper and pre-SMA.) Therefore, the frontal areas showing increased CBF with GPi DBS during silent counting are not the same as those (e.g., SMA-proper) activated during performance of arm movement initiation tasks in normal subjects and in PD patients treated with dopamine agonists, pallidotomy, or DBS in the STN.

A recent case report illustrates the analytic power of combining DBS and functional imaging. Bejjani *et al.* studied the behavioral and functional imaging correlates of DBS in the substantia nigra in a patient with an acute depressive syndrome who responded to such stimulation (1999). This patient received bilateral STN stimulators for treatment of medically intractable Parkinson's disease. Stimulation through electrode contacts located within the STN, as determined by MRI, yielded relief from parkinsonian symptoms with no accompanying change in mood. Stimulation at an electrode contact below the clinically-effective site, however, at a location corresponding with the central portion of the left substantia nigra reticulata (SNr), evoked marked and reproducible dysphoria that lasted as long as the stimulation was applied. This stimulation-induced mood disorder met nearly all of the criteria for major depression and could not be explained by parallel changes in the patient's motor symptoms, levodopa therapy, or pain perception. A PET study of this patient, comparing CBF during SNr stimulation with that during no stimulation, found that DBS in the SNr evoked an increase in CBF in the left orbitofrontal cortex, left amygdala, left globus pallidus and thalamus, and right parietal lobe. The left orbitofrontal cortex and amygdala are areas that previous studies have associated with negative affect (Adolphs *et al.*, 1995; Zald and Pardo, 1997). DBS in the SNr most likely affected cerebral activity in these areas via the well-established SNr projection to the orbitofrontal cortex via the MD thalamus (Ilinsky *et al.*, 1985; Goldman-Rakic, 1995). This case report illustrates how the use of DBS in conjunction with neuroimaging can be used to identify both the behavioral and CBF correlates of altered neural activity in just one of the parallel segregated pathways through the basal ganglia (Alexander *et al.*, 1990).

In an ongoing study of DBS in the STN, we have also found that stimulation at clinically efficacious levels increases CBF in a task-specific manner. To illustrate our results, we present data from one representative subject. We performed PET scans of a parkinsonian patient who had stimulators implanted in the STN bilaterally. The subject performed visuomanual and eye-only tracking tasks (described in Section IIC) at an intermediate target rate (0.4 Hz). The two tasks were performed while DBS was delivered at voltages that were either just above or just below the threshold for clinical effectiveness. Stimulation at therapeutic levels led to a dramatic improvement in performance of the visuomotor tracking task.

The dominant effect of clinically effective stimulation was to increase CBF in motor and premotor cortical areas. Which areas were affected depended on the task being performed. When the subject was tracking the target with eye movements alone, therapeutic stimulation caused increases in CBF (>10% relative to that during DBS at a subtherapeutic level) in dorsal mesial cortex bilaterally (Fig. 14A) and in the right precentral gyrus (Fig. 14B). These areas of increased CBF correspond fairly well with previously identified oculomotor control areas: the supplementary eye areas on the dorsomesial cortical surface and the frontal eye field in the precentral gyrus. In contrast, when the subject was performing tracking movements also with the arm, thera-

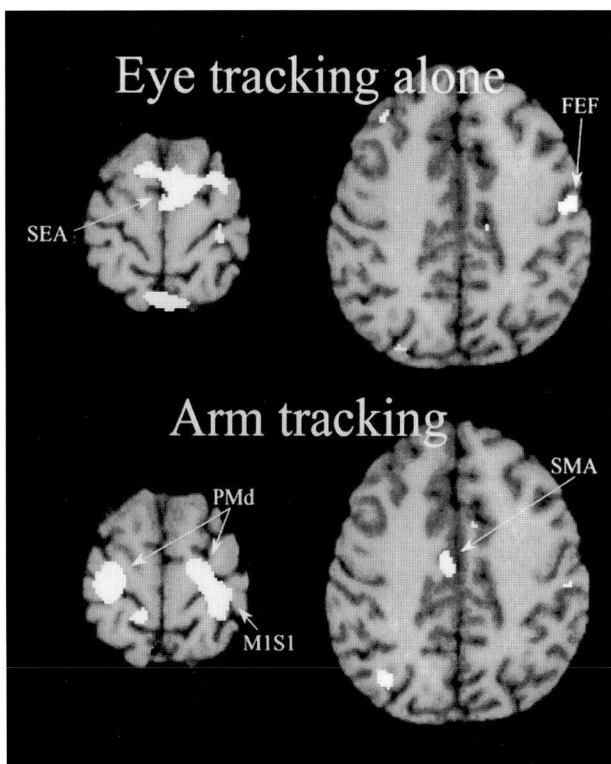

Figure 14 Effects of clinically effective DBS in STN in one parkinsonian patient. White: regions where CBF increased >10% relative to that during stimulation at a voltage just below the threshold for clinical effect. When the subject tracked target movements with only the eyes, effective stimulation caused increased activation of oculomotor control areas such as bilateral supplementary eye areas (SEA in A) and left frontal eye field (FEF in B). In contrast, during visuomanual tracking, clinically effective stimulation caused increased CBF in primary sensorimotor cortex (M1S1 in C), dorsal premotor cortex bilaterally (PMd in C), and supplementary motor area (SMA in D).

peutic stimulation increased CBF bilaterally (>10%) in primary sensorimotor and dorsal premotor cortices (Fig. 14C) and in the SMA (Fig. 14D), all of which are areas associated with the control of arm movement.

That the effects of DBS depend on the behavioral task being performed points out the difficulty of selecting an appropriate rest condition when studying task-related activations in the context of neurologic disease. The traditional approach for measuring both movement-related activations and the effects of therapeutic manipulations on movement-related activations requires that scans obtained under some rest condition be subtracted from movement condition scans (i.e., by computing the significance of the movement-by-stimulation interaction). There is little evidence, however, to support the primary assumption of that approach—namely, that different behaviors (e.g., eye movement and arm movement) will have additive effects on brain activity—and analyses based solely on that approach can produce misleading results. If the traditional approach had been used with our STN DBS data, for instance, the movement-by-stimulation interaction would have indicated that effective stimulation caused a *decrease* in the movement-related CBF for the supplementary eye areas, whereas those areas actually had selectively increased CBF under the eye-only condition. It is quite likely that both of the published PET studies of DBS also included effects that were selective for a "rest" condition (Davis *et al.*, 1997; Limousin *et al.*, 1997). It may be possible to avoid the potential confound of task-selective effects of therapy by studying instead the effects of DBS on parametric relations between CBF and an experimentally manipulated task variable (like the rate of movement as used in our study).

The combined results of Limousin *et al.* (1997), Davis *et al.* (1997), and our studies can be taken as further evidence that therapeutic interventions for PD lead to increased cerebral activity at locations that are contingent on the particular task being performed. The influence of DBS on areas of abnormally increased activity remains to be determined.

H. Conclusion

The studies reviewed here lead beyond a simple model of parkinsonian pathophysiology in which a few select frontal areas (e.g., the SMA) have impaired task-related activation because those areas are differentially affected by excessive inhibitory outflow from the parkinsonian basal ganglia. Recent neuroimaging studies of PD and of surgical interventions for PD are consistent with alternate pathophysiological models such as a general impairment in the ability to select the appropriate pattern of brain activation for the task at hand. Such a model would account for the results of recent PET studies showing that some of the brain areas normally activated with a task have smaller-than-normal increases in CBF in PD patients whereas other cortical areas, including some not normally activated with the task, have greater-than-normal increases in CBF. Which areas show smaller-than-normal and greater-than-normal activation depends on the nature of the task being performed. This model of impaired selection of the task-appropriate activation pattern could also explain many of the task-specific alterations in brain activity observed following surgical interventions for PD. The anatomical substrate for this disorder of selection remains unknown. Although a role in the context-dependent facilitation and suppression of frontal thalamocortical circuits has been proposed for the BG (Cools *et al.*, 1984; Marsden and Obeso, 1994; Mink, 1996; Redgrave *et al.*, 1999), it remains unclear if this focusing-like function can account

for the expansion and pallidotomy-related contraction of rate-related activity in premotor areas in our studies, for example. It may be just as likely, if not more so, that the abnormal patterns of CBF observed in PD arise from thalamic and/or cortical dysfunctions that are caused secondarily by the abnormal outflow from the parkinsonian BG. The fact that these abnormal activity patterns normalize following ablation or DBS of the primary output nucleus of the basal ganglia furthers the view that most parkinsonian symptoms and their correlates in abnormal patterns of CBF are linked only indirectly to the normal functions of the BG.

III. Surgical Therapies for Tremor

Thermal ablation of the thalamic ventral intermedial nucleus (VIM) has long been recognized as an effective treatment for tremor associated with a variety of neurologic disorders (Burchiel, 1995; Shahzadi *et al.*, 1995; Brophy *et al.*, 1997). The functional correlates of thalamotomy, however, have not been investigated extensively. Boecker *et al.* (1997) examined the effects of VIM thalamotomy on changes in CBF associated with performance of the internal generation task (described in Section IIB). Two parkinsonian patients were studied prior to and following thalamotomy for their tremor-dominant symptoms. After thalamotomy there was a reduction in both task-related and resting CBF in multiple motor cortical areas, including primary motor cortex and lateral premotor and parietal cortices in the hemisphere ipsilateral to the lesion. These results can be seen as consistent with established understanding in which reciprocal thalamocortical interactions facilitate cortical activity, perhaps to a pathologic degree in cases of tremor (Lenz *et al.*, 1993). Interpretation of these results is complicated, however, by the confounding effects of thalamotomy on tremor itself. The pattern of reduced CBF observed in this study might be attributed equally well to an interruption of the central tremor-generating circuitry or to an elimination of the sensory consequences of tremor. A deeper understanding of the functional correlates of therapeutic thalamotomy will require future studies in which there are control conditions for the presence of overt tremor.

IV. Surgical Therapies for Epilepsy

A. Surgical Approaches to Epilepsy Management

At least one-fourth of individuals with epilepsy cannot gain full seizure control with available antiepileptic drugs (AEDs). Furthermore, many medication-treated individuals sustain complete or improved seizure control only when using AED regimens that impose significant toxicities. Systemic AED administration requires that nonepileptogenic brain tissues be exposed to undesired drug effects, as the unavoidable consequence of delivering the agent to sites that participate in seizure initiation and propagation.

Disordered synchrony at large numbers of cerebral synapses underlies the deficient or deviant information processing that occurs during seizures (Engel *et al.*, 1998). During "generalized-onset" seizures, paroxysmal thalamocortical dysfunction drives the entire cortex simultaneously into a hypersynchronous ictal state, with manifestations ranging from impaired information processing that is clinically evident only in association cortex (during absence seizures) to unconsciousness with widespread, convulsive discharges of corticospinal and other pathways (during generalized tonic-clonic seizures). "Partial-onset" seizures begin focally in one cerebral region and then spread to variable extents, with manifestations ranging from electrographic seizures with no detectable signs (subclinical partial seizures), through nonconvulsive seizures without or with impaired awareness (simple or complex partial seizures), to secondarily generalized seizures (generalized tonic-clonic seizures of focal onset). A variety of cerebral insults can induce epileptogenic synaptic reorganization and perhaps other types of epileptogenic neuronal dysfunction, which support the pathological states of interneuronal synchrony that characterize seizures. Numerous genetic conditions also can predispose to pathological interneuronal hypersynchrony.

Epilepsy surgery can offer complete seizure control for some individuals whose seizures have proved refractory to all reasonable medication trials. Epilepsy surgery is most often efficacious in particular epilepsy syndromes. Approximately 80% of individuals gain complete seizure control following anterior temporal lobectomy for refractory limbic temporal lobe epilepsy (TLE), which is the single most common form of epilepsy, and following extratemporal resections in other partial epilepsies when an associated lesion is present. Seizure control is also excellent following multilobar resection in the less common syndrome of infantile spasms with unilateral cortical dysplasia. Corpus callosotomies can arrest epileptic drop attacks (either tonic or atonic seizures that cause injurious falls), but affected individuals usually have generalized tonic-clonic seizures or other types of seizures that do not respond to callosotomy. Extratemporal resections are considerably less efficacious if no lesion is present than if a lesion is found on MRI. Epilepsy can be caused by lesions that expose adjacent cortex to mass effect, is-

chemia, or hemosiderin. In some cases, lesion resection may partially or fully reverse the tendency of insulted cortex to generate seizures, but lesionectomy combined with resection of epileptogenically-reorganized cortex achieves a higher rate of complete seizure control (Fried and Cascino, 1998). In addition to mapping lesions with structural and functional imaging techniques, functional imaging can map alterations of structurally normal cortex that is epileptogenically-reorganized. Localization of epileptogenic lesions and of the ictal onset zone constitutes a unique role for imaging applications in resective epilepsy surgery.

Surgical resections, and the less-often-performed transections of white matter bundles, impose significant risk of irreversible injury to neuronal assemblies essential for particular motor, sensory, cognitive, and other processes. Functional imaging can offer localization of primary motor, sensory, and language cortex, thereby reducing surgical risks. Brain mapping also is useful in evaluating the effects of epilepsy surgery. Brain MRI is required to determine the site and extent of resection or transection, which are major determinants of surgical success in eliminating seizures. Mechanisms of post-surgical improvements in cerebral function and of iatrogenic declines in cerebral function can be studied with functional imaging techniques.

Despite advancements in presurgical evaluation, many individuals cannot undergo epilepsy surgery due to excessive risks of iatrogenic disability or simply because no site for efficacious resection or transection can be found. For such individuals, current research focuses on chronically implanted devices that deliver electrical stimulation or drugs to highly localized neural sites and on neural tissue transplantation. To date, only the neural electrostimulation approach has been widely tested in humans. Mapping of acute and chronic cerebral blood flow has been used to study mechanisms of action of vagus nerve stimulation and should be useful in studying intracranial electrostimulation.

B. Resective Epilepsy Surgery

1. Electrophysiological Mapping of Ictal Onset Zones

Refractory partial-onset seizures most often originate in the limbic temporal lobe but also may begin in the temporal neocortex and essentially any extratemporal cortical site. Sites of earliest ictal electrophysiological changes, and anatomical patterns of ictal propagation, have often been studied with intracranially placed macroelectrodes (Spencer, 1981; Wieser, 1983; So *et al.*, 1989). Surgically placed intracerebral and subdural electrodes permit recording of electrophysiological seizure activity with high spatial resolution and exceedingly high temporal resolution. By contrast, extracranial electrodes provide limited sampling of electrophysiological activity generated primarily at crests of cortical gyri, with much degradation of signal by volume conduction effects (Pacia and Ebersole, 1997). In some cases, unusual pathways of ictal propagation can cause a seizure that begins in one hippocampus to spread more rapidly to neocortex of the contralateral temporal lobe than to ipsilateral temporal neocortex; this will cause the scalp EEG to show earliest ictal changes over temporal electrodes of the side opposite the actual ictal onset zone (Sammaritano *et al.*, 1987). In years past, essentially all resective epilepsy surgery could be planned only after recording of seizures with intracranially placed electrodes. Despite limitations of extracranial EEG in detailed anatomical localization of ictal onset zones, these noninvasive recordings provide an essential starting point in presurgical evaluation. Ictal recordings with extracranial electrodes can distinguish partial-onset from generalized-onset seizures, distinguish psychogenic nonepileptic seizures ("pseudoseizures," which often generate ictal behaviors that resemble those of complex partial seizures) from epileptic seizures, and provide localizing information that helps to limit the high-probability sites of seizure origin (Risinger *et al.*, 1989; Henry and Drury, 1997).

2. Concordant Imaging and Electrophysiological Abnormalities

Engel developed the concept that concordance in localization of the ictal onset zone, by extracranial ictal EEG recordings and by specific neuroimaging abnormalities, can accurately predict that intracranial ictal EEG recordings would show all electrographic seizure onsets over one mesial temporal region and that in such cases anterior temporal lobectomy can be performed with a high degree of confidence even though intracranial EEG recording has not been performed (Engel *et al.*, 1982). Engel described specific criteria for performing temporal lobectomy without prior invasive electrophysiological studies, based on technically satisfactory extracranial EEG recording that showed all earliest ictal changes over one temporal region, interictal regional hypometabolism on fluorodeoxyglucose (FDG) PET over the same temporal lobe, and absence of contradictory localizing information, in patients who have the clinical syndrome of refractory limbic temporal lobe epilepsy (Engel *et al.*, 1982). This premise has gathered considerable support and now is recognized by nearly all epilepsy surgery programs.

The variety of useful imaging abnormalities has expanded beyond the temporal lobe hypometabolism on interictal FDG PET originally used by Engel to include brain MRI abnormalities and ictal SPECT abnormali-

ties. Current efforts aim to find functional imaging modalities that are more sensitive and that detect more focal amygdalar-hippocampal abnormalities for use in this role. Anterior temporal lobectomies and amygdalo-hippocampectomies are intended to remove the entire amygdala and hippocampal formation on the side of limbic temporal ictal onset, so definition of a margin of resection is not required. Extratemporal resections require definition of a margin of resection. Extratemporal epilepsies therefore will always require intracranial electrode placement, at least to permit ictal recordings that define the margin of resection and sometimes to better define the lobe of ictal onset. Structural and functional imaging abnormalities are highly useful in selecting intracranial electrode sites in extratemporal epilepsies and in complex cases of TLE.

The other aspect of presurgical localization, that of localizing essential cortical processing zones that should not be resected, has for many years been accomplished with mapping by direct cortical electrical stimulation (Penfield and Bradley, 1937; Ojemann, 1979) and the intracarotid amobarbital procedures (Jones-Gotman *et al.*, 1998). Regional blood flow activation imaging provides a less invasive alternative to the older forms of presurgical functional mapping.

3. Structural Imaging in Presurgical Localization of Ictal Onset Zones

Structural brain imaging with MRI is essential in localization of alien-tissue lesions, sites of brain injury, and sites of disturbed cerebral development. Such lesions are detected with MRI, and not with X-ray CT, in the majority of individuals with refractory partial-onset seizures (Bronen *et al.*, 1996; Duncan *et al.*, 1997). Such lesions are often, but not always, located at or near the sites of ictal onsets (Holmes *et al.*, 1999).

The most common lesion in TLE is hippocampal sclerosis, and this is usually detectable with MRI. The major MRI feature of hippocampal sclerosis is hippocampal atrophy, which is usually associated with hippocampal T_2 signal increase, and sometimes with hippocampal T_1 signal decrease, or with loss of definition of internal hippocampal structure (Bronen *et al.*, 1997). Unilateral hippocampal sclerosis is highly associated with a single ictal onset zone that may be efficaciously resected. A single region of hippocampal sclerosis does not exclude the possibility of two independent foci of ictal onset, nor even the possibility that all seizures originate at a site distant from the MR-detected unilateral hippocampal sclerosis (King *et al.*, 1997). Quantification of atrophy by hippocampal volumetry is a useful research tool, which may or may not increase detection of hippocampal sclerosis beyond expert scan interpretation (Jack *et al.*, 1992). Quantification of T_2 signal increase by T_2 relaxometry (Jackson *et al.*, 1993) has yet to be studied fully.

Brain MRI provides considerable information concerning water content, hemosiderin-related signal voids, and other sources of internal inhomogeneities of solid or cystic lesion structure, such that specificity of lesion diagnosis is much greater with MRI than with other imaging modalities. Potential epileptogenic lesions that are readily detected and often specifically characterized by MRI include astrocytoma, oligodendroglioma, ganglioglioma, gangliocytoma, dysembryoplastic neuroepithelial tumor, meningioma, cerebral metastasis, arteriovenous malformation, cavernous angioma, glial hamartoma (isolated or multiple in tuberous sclerosis), encephalomalacia (due to trauma, infarction, or infection), chronic intracranial hemorrhage, porencephaly, ulegyria, schizencephaly, cysticercosis, Rasmussen's encephalitis, focal cortical dysplasia, regional and band heterotopias, polymicrogyria, hemimegalencephaly, lissencephaly, and the lesions of neurofibromatosis and Sturge-Weber syndrome (Kuzniecky and Jackson, 1995; Bronen *et al.*, 1997).

Brain MRI or CT guides stereotaxic localization for planning intracranial electrode implantation (Ross *et al.*, 1996). Post-implantation MRI is useful in determining the actual sites of the electrodes (Zhang *et al.*, 1993; Meiners *et al.*, 1996). Frameless MR systems also can display information on lesion and ictal onset zone locations in the operating room at the time of resection, using moveable markers in the resection field that are automatically localized within the patient's preoperative MR images (Chabrerie *et al.*, 1997).

4. Functional Imaging in Presurgical Localization of Ictal Onset Zones

Partial and generalized epilepsies often are not associated with cerebral, macrostructural abnormalities detectable with MRI or are associated with "nonspecific" MRI abnormalities such as subcortical white matter puncta of T_2 signal increase, which have no established correlation with localized ictal onset zones. While some believe that normal MRI bodes poorly for efficacious epilepsy surgery (Scott *et al.*, 1999), it is clear that some individual patients with normal MRI benefit from ablative surgery of electrophysiologically defined ictal onset zones. Detection of localized abnormalities of certain brain functions can be highly useful when MRI is normal or non-specifically abnormal.

The functional imaging techniques currently utilized for localization of ictal onset zones are ictal SPECT with [^{99m}Tc]hexamethylpropyleneamineoxime ([^{99m}Tc] HMPAO) or [^{99m}Tc]ECD and interictal PET with FDG (Berkovic, 1993; Henry *et al.*, 1993a-c). Both techniques provide information that is nonredundant with MRI. Both techniques permit

1. Lateralization and regionalization of functional abnormality in TLE, for resection without prior intracranial EEG monitoring, based on concordance with other noninvasive data

2. Regionalization of functional abnormality in TLE and other partial epilepsies, for direction of intracranial electrode placement

Additionally, interictal FDG PET has proved useful in

3. Lateralization and regionalization of functional abnormality in infantile spasms, for resection without prior intracranial EEG monitoring, based on concordance with other noninvasive data

4. Prognostication of surgical outcome with regard to seizure control

Ictal SPECT may be shown to have similar applications in the future.

Ictal SPECT detects focal hyperperfusion in patients with normal or abnormal MRI (Lee *et al.*, 1988; Rowe *et al.*, 1989; Marks *et al.*, 1992; Newton *et al.*, 1992; Berkovic, 1993; Ho *et al.*, 1995; Arroyo *et al.*, 1997; O'Brien *et al.*, 1999; Spanaki *et al.*, 1999). Ictal SPECT is most sensitive and specific when the patient's interictal SPECT images are coregistered with and subtracted from the ictal images, with superimposition of the perfusion difference images on the patient's MRI (O'Brien *et al.*, 1999). Patterns of ictal hyperperfusion on ictal SPECT studies are much larger than the small foci of ictal onset that usually are detected with intracerebral EEG recordings, but the focus of earliest intracerebrally recorded ictal discharges is contained within the region of ictal hyperperfusion (Spanaki *et al.*, 1999). Thus, it appears likely that ictal SPECT is usually demonstrating the site of ictal onset and sites of early or most intense ictal propagation. Ictal SPECT is most accurate in regionalizing the ictal onset zone when radiopharmaceutical injection takes place during the electrographic seizure itself (Newton *et al.*, 1992). Short complex partial seizures, which often occur in frontal lobe epilepsies, are more difficult to image with ictal SPECT (Marks *et al.*, 1992). Nonetheless, when properly acquired and analyzed, ictal SPECT is highly reliable in temporal and extratemporal epilepsies.

Interictal FDG PET detects focal hypometabolism in patients with normal or abnormal MRI (Theodore *et al.*, 1983, 1992; Swartz *et al.*, 1989; Engel *et al.*, 1990; Henry *et al.*, 1993a-c; Ho *et al.*, 1995) (Fig. 15). Interictal FDG PET probably is most sensitive and specific when the patient's FDG images are compared statistically with normal FDG activity and the voxels with significantly reduced FDG activity are displayed superimposed on the patient's MRI (Faber *et al.*, 1996). Regions of inter-

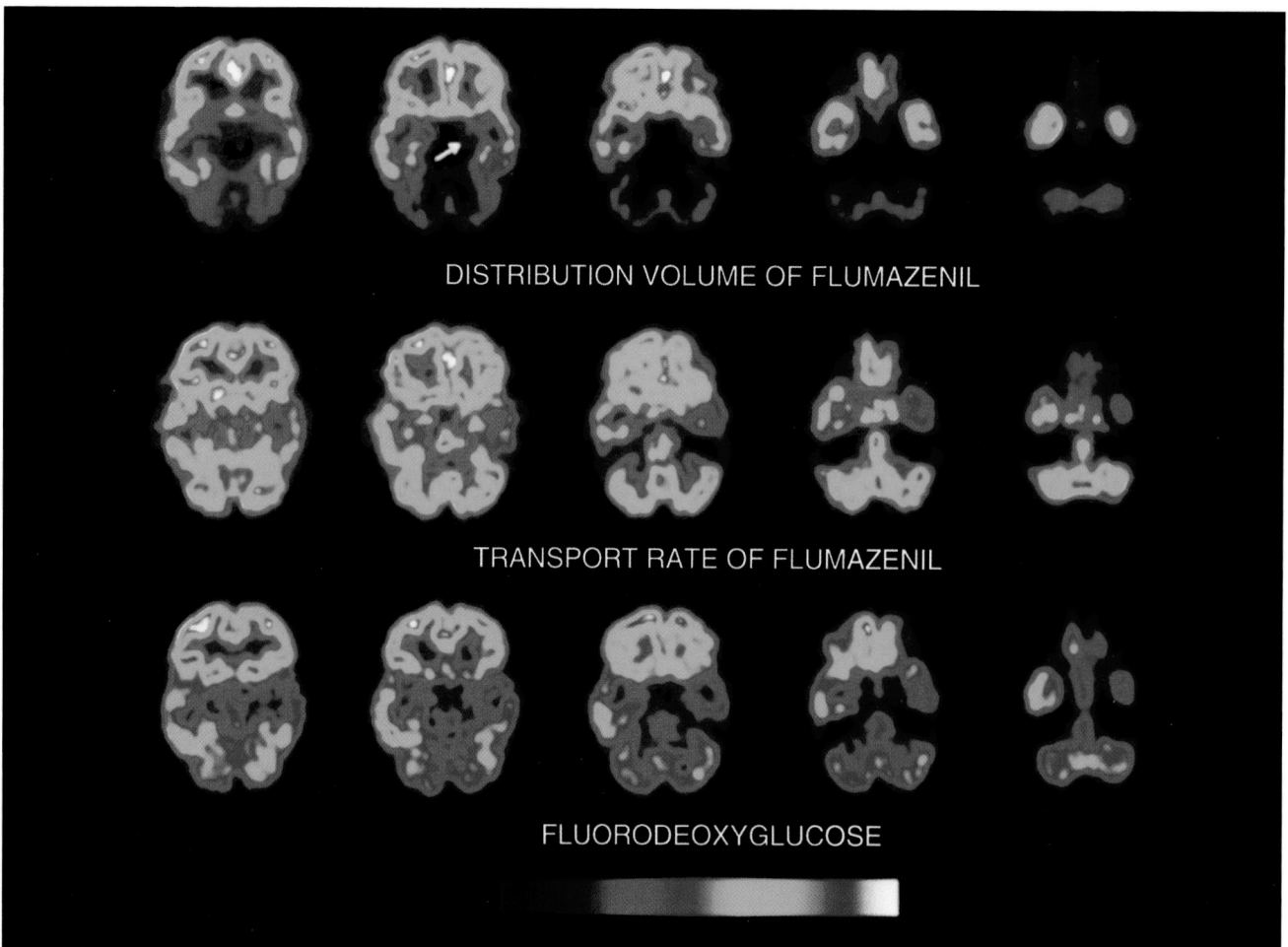

Figure 15 Selected coregistered flumazenil distribution volume and transport rate images and fluorodeoxyglucose images of a patient with epilepsy. Five adjacent image planes are displayed for each modality. The more superior image plane is to the left of each row. The subject's left is on the right of each image. An arrow indicates the site of greatest decrease in FMZ binding as determined by visual inspection. Adapted with permission from Henry *et al.* (1993b). *In vivo* cerebral metabolism and central benzodiazepine-receptor binding in temporal lobe epilepsy. *Neurology* **43,** 1998–2006.

ictal hypometabolism on FDG PET studies typically are much larger than the small foci of ictal onset that usually are detected with intracerebral EEG recordings, but the focus of earliest intracerebrally recorded ictal discharges is contained within the region of ictal hyperperfusion (Engel *et al.*, 1990). Ablative structural lesions clearly contribute to localized decreases in metabolism. Neuronal loss and diaschisis could have been the cause of the widespread interictal hypometabolism in TLE patients with hippocampal sclerosis, but this hypothesis was refuted by correlation of quantified preoperative FDG PET and quantitative neuronal volumetric densitometry of resected temporal tissue (Henry *et al.*, 1994). Currently, the pathophysiology of interictal glucose hy-

pometabolism is fully explained by macrostructural and microstructural alterations in temporal lobe epilepsy (Henry, 1996; Reutens *et al.*, 1998). Interictal FDG PET is less often abnormal in nonlesional extratemporal epilepsies than in partial epilepsies with lesions and in TLE. False lateralization is rarely observed with either interictal FDG PET or ictal SPECT, but as with MRI, these functional imaging techniques are rather insensitive to detection of bilateral independent epileptic foci. The relative sensitivities of MRI, peri-ictal SPECT, and interictal FDG PET in presurgical evaluation of partial epilepsies remain unclear (Spencer, 1994).

Large, unilateral cortical areas of decreased or increased FDG activity often occur in patients with refractory infantile spasms (Chugani *et al.*, 1990). The metabolically dysfunctional cerebral regions are highly associated with the volume of cortex that can be resected to control the seizures and prevent subsequent development of other forms of epilepsy. Specimens from these multilobar resections or functional hemispherectomies usually include sites of gross or microscopic malformations of cortical development or other structural abnormalities. The normal absence of myelination of subcortical white matter in infants may decrease the contrast of ectopic gray matter and adjacent white matter for MRI, explaining the low sensitivity of MRI to malformations of cortical development in infants, despite its high sensitivity in older children and adults. Resection of large volumes of cortex can result in cessation of seizures, and normal or near-normal cognitive development, when performed at early ages (Chugani *et al.*, 1993).

Functional MRI and [^{15}O]water PET have been performed during status epilepticus (Hajek *et al.*, 1991; Jackson *et al.*, 1994; Detre *et al.*, 1995). Both techniques demonstrate sites of maximal ictal involvement. The superior temporal resolution of fMRI permits further distinction of earlier regional hyperperfusion from sites of intense ictal propagation. Functional MRI and [^{15}O]water PET are not practical for ictal imaging in most surgical cases, however, due to the unpredictability and relative infrequency of complex partial and secondarily generalized seizures.

Proton MR spectroscopy detects reduced *N*-acetylaspartate (NAA), a neuron-specific metabolite, and NAA-to-creatine/choline ratios in the ictal onset zone in many cases of TLE (Cendes *et al.*, 1994; Connelly *et al.*, 1994). It is not clear that the association of NAA decreases with mesial temporal epileptogenesis provides sufficient sensitivity and specificity for clinical application. Phosphorus-31 MR spectroscopy detects reduced phosphocreatine/inorganic-phosphorus and adenosinetriphosphate/inorganic-phosphorus ratios in the ictal onset zone in many cases of TLE (Kuzniecky *et al.*, 1992; Laxer *et al.*, 1992). Conflicting reports of decreased pH and normal pH in the mesial temporal lobe on phosphorus-31 MR spectroscopy remain unresolved (Kuzniecky *et al.*, 1992; Laxer *et al.*, 1992; Chu *et al.*, 1998).

Reduced density of central benzodiazepine receptors (at the GABA$_A$ receptor-chloride ionophore complex) is highly associated with the ictal onset zone in TLE and other partial epilepsies (Henry *et al.*, 1993a-c; Savic *et al.*, 1995; Koepp *et al.*, 1996) (Fig. 15). Central benzodiazepine imaging with [^{11}C]flumazenil PET has not yet been fully validated as a presurgical evaluation tool. Other PET radioligands, such as α-[^{11}C]methyltryptophan (Chugani *et al.*, 1998), also may demonstrate localized functional ab-normalities that add information concerning regional epileptogenesis to that provided by MRI.

5. Presurgical Localization of Essential Cortical Processing Zones

Resection of primary motor cortex causes irreversible loss of volitional motor function in patients who do not have severe preoperative deficits in motor performance. Iatrogenic motor deficits are most readily demonstrated in movements that involve independent finger control. Mapping of primary motor cortex can be reliably obtained with direct cortical electrical stimulation (DCES). The primary motor cortex is characterized by a lower threshold to obtain movements of contralateral face or appendages than the lowest current required to induce movements on stimulation of secondary motor areas (Penfield and Bradley, 1937). Mapping with DCES can be performed during general anesthesia, with exposure of dorsolateral frontal cortex. Both fMRI and CBF PET obtained with motor tasks generate maps of primary motor cortex that are highly concordant with maps generated by DCES (Vinas *et al.*, 1997; Yetkin *et al.*, 1997). Functional imaging can be used as a noninvasive tool to avoid iatrogenic injury to primary motor cortex during epilepsy surgery (Duncan, 1997; Duncan *et al.*, 1997).

Resection of primary sensory cortex can cause loss of conscious sensory perception for specific sensory systems. For example, resection of calcarine cortex can result in visual processing deficits, resection of the postcentral gyrus in somatosensory deficits, and resection of Heschl's (transverse temporal) gyrus in auditory deficits. It is more difficult to demonstrate primary sensory cortices in humans for olfaction, gustation, and viscerosensation, in part because candidate areas are relatively difficult to resect due to overlying branches of the middle cerebral artery (insula) or are less often resected due to infrequent surgical indications (olfactory bulb). Mapping with DCES reveals low-threshold regions of primary sensory cortices, whose stimulation results in reported "elementary" (unformed) sensations (Penfield and Bradley, 1937). Sensory mapping with DCES must be performed in conscious patients, either with chronically implanted subdural grid electrodes or with acute electrocorticography and local anesthesia at the craniotomy site. Primary sensory cortex is concordantly localized with activation PET or DCES (Vinas *et al.*, 1997; Bittar *et al.*, 1999). Calcarine cortex and Heschl's gyrus rarely are resected for epilepsy, and comparative studies of DCES and functional imaging to localize primary visual and auditory cortices are lacking. As for primary motor cortex, functional imaging can be used to avoid iatrogenic injury to primary sensory cortex during epilepsy surgery (Duncan, 1997; Duncan *et al.*, 1997).

Resection of Broca's area (inferior portion of the left inferior frontal gyrus, in most individuals) or of Wernicke's area (the left angular gyrus or posterior portions of the left superior temporal gyrus, in most individuals) can cause irreversible loss of expressive or receptive language functions. Mapping with DCES has been used to localize left dorsolateral frontal and temporoparietal sites, which must be spared resection to avoid naming and other language deficit (Ojemann *et al.*, 1989). Such studies reveal great interindividual variability in size and location of left dorsolateral frontal and dorsolateral temporoparietal sites of language functions (Ojemann, 1979; Lesser *et al.*, 1986; Ojemann *et al.*, 1989); male–female differences partially explain this variability (Ojemann *et al.*, 1989). These techniques also defined an additional left basal temporal language area (Luders *et al.*, 1986).

The "Wada" test, or intracarotid amobarbital infusion with behavioral testing, usually anesthetizes the frontotemporoparietal cortex unilaterally. The Wada test is the standard method for determination of left, right, and mixed hemispheric language dominance (Rausch *et al.*, 1993). Early injuries, particularly those occurring before 5 years of age, can be associated with development of right hemisphere dominance or mixed hemispheric dominance for language (Rasmussen and Milner, 1977). In such individuals the resection of eloquent cortex of the right hemisphere might result in permanent language deficits. Language processing is widely distributed over the left hemisphere and also occurs within the right hemisphere, even in individuals with left hemisphere language dominance (Binder *et al.*, 1996; Damasio *et al.*, 1996; Martin *et al.*, 1996; Henry *et al.*, 1998a,b). Word generation, visual confrontation naming, and other language protocols for fMRI and CBF PET can be used to determine hemispheric dominance for language (Binder *et al.*, 1997; Worthington *et al.*, 1997; Yetkin *et al.*, 1997; Hunter *et al.*, 1999).

Dysnomia is highly prevalent in left TLE, and naming and word-finding impairments often occur after left temporal lobectomy (Stafiniak *et al.*, 1990; Langfitt and Rausch, 1997). Many patients experience significant improvements in early postoperative naming declines over the next 6–12 months, but some are left with persisting iatrogenic hyponomia (Stafiniak *et al.*, 1990; Langfitt and Rausch, 1997). The DCES-defined basal temporal language area may be injured during left temporal lobectomy, producing new or increased dysnomia. The basal temporal language area lies mainly on the anterior fusiform gyrus, although adjacent regions of the parahippocampal gyrus and inferior temporal gyrus also may function in this area. Naming activates blood flow increases in the fusiform gyri, left inferior frontal

gyrus, and other regions on PET studies averaged in normal groups (Fig. 16) (Damasio *et al.*, 1996; Martin *et al.*, 1996). Many individuals with left TLE have lost the ability to achieve normal activation of the left anterior fusiform gyrus during visual confrontation naming, and these individuals also have impaired naming performance (Henry *et al.*, 1998a,b). Naming activation CBF PET and DCES provide highly concordant maps of the basal temporal language area within individual TLE patients (Bookheimer *et al.*, 1997).

Working memory, which supports "immediate" recall of verbally and nonverbally encoded material, is particularly dependent on normal function of dorsolateral prefrontal cortex (Grasby *et al.*, 1993; Petrides *et al.*, 1993a,b). Presumably, resection of critical frontal sites might result in permanent deficits of working memory, although frontal lobe epilepsies without demonstrable structural lesions are less often surgically treated due to lesser efficacy in seizure control. Additionally, frontal resections for epilepsy are unilateral, and both frontal lobes appear to support working memory. Bilateral dorsolateral prefrontal areas (and to a lesser degree, several other areas) were shown to increase FDG activity significantly during continuous-performance memory tasks that could be sustained for long periods in normal volunteers (Swartz *et al.*, 1995). The same FDG activation PET protocol did not result in any significant frontal or other regional CMRGlu increases in patients with unilateral frontal lobe epilepsy (Swartz *et al.*, 1996). These observations support the general principle that interictal dysfunctions of localization-related epilepsies can alter task-specific activation patterns at disseminated sites of parallel cortical processing, even though a single region is maximally involved in seizure generation. These observations do not support a role for functional imaging in avoidance of iatrogenic deficits of attention and working memory due to unilateral frontal resection.

Resection of the hippocampal formations bilaterally causes persisting global amnesia for declarative memories stored over about 5 min, although some ability to learn motor procedures is spared. While bilateral hippocampectomy is never considered following several famous iatrogenic catastrophes, an ongoing concern is the risk of global amnesia following resection of an epileptogenic hippocampus located contralateral to an injured hippocampus that cannot support processing for delayed recall. A more commonly occurring concern in epilepsy surgery is of significant new or exacerbated old memory impairments, without complete amnesia, following temporal lobectomy (Novelly *et al.*, 1984; Ojemann and Dodrill, 1985; Saykin *et al.*, 1989). Hippocampal DCES has shown evidence for verbal memory processing in the left hippocampus (Lee *et al.*, 1990), but

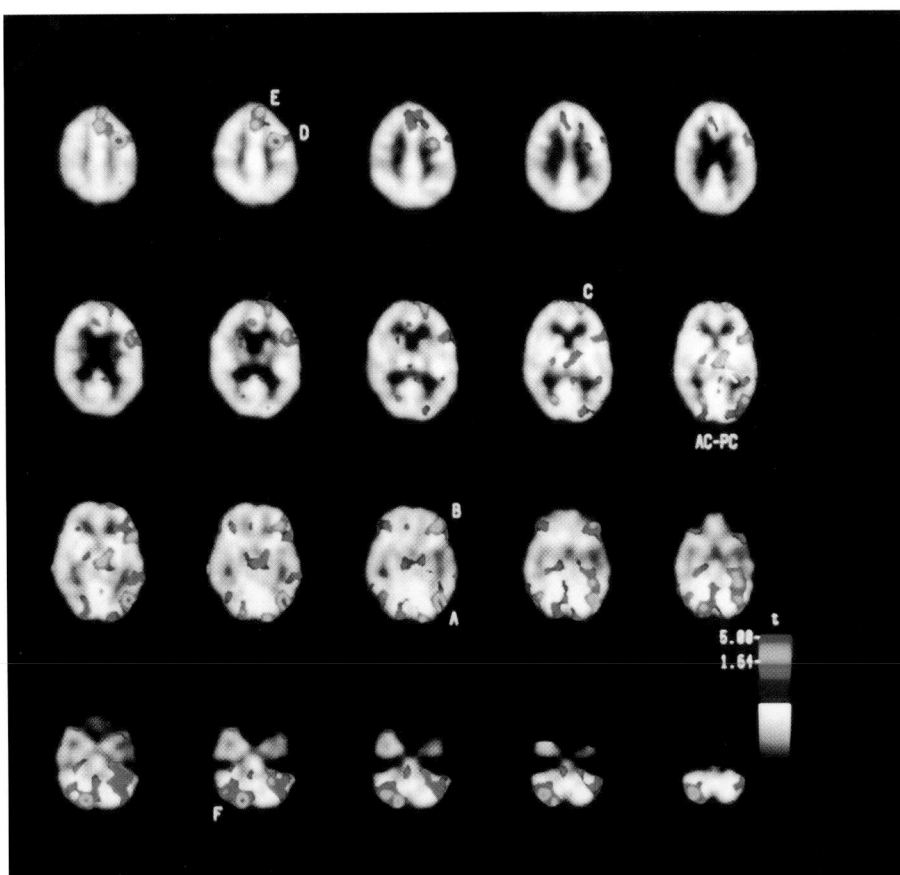

Figure 16 Naming activation images of a patient with left temporal lobe epilepsy (TLE). Activation $H_2^{15}O$ PET during visual confrontation naming in a patient with left TLE (T2). Seven areas of significant cerebral blood flow (CBF) increase were detected: A, posterior left fusiform gyrus; B, left inferior frontal gyrus; C, left superior frontal gyrus, inferiorly; D, left superior frontal gyrus, superiorly; E, left cingulate gyrus; F, right cerebellar hemisphere; arrow, left thalamus. Areas of CBF increases are shown in color, superimposed on the patient's summated CBF PET images in grayscale; significant increases are represented by orange or red pixels. Images are in the axial plane, in descending order from the upper left-hand image. Image spacing is 4.5 mm. The image falling on the line through the anterior and posterior commissures is labeled "AC–PC." The patient's right side is displayed on the left side of each image. With permission from Henry *et al.* (1998b). Regional cerebral blood flow increases during naming in left temporal lobe epilepsy. *Neurology* **50,** 787–790.

it is difficult to stimulate long enough to interfere with the consolidation process. Currently, the Wada test is used to predict severe amnesic states that might occur after unilateral temporal lobectomy, when memory failure occurs after amobarbital injection on the side of intended resection (Rausch *et al.,* 1993). Hippocampal blood flow can be activated with complex viewing-retrieval tasks (Nyberg *et al.,* 1996; Tulving *et al.,* 1996). It remains to be seen whether differential right and left hippocampal activation protocols will permit specific mapping of long-term, declarative memory processing.

6. Postoperative Evaluation of Surgical Effects

Postoperative brain imaging offers a unique opportunity to fully evaluate the structural and functional effects of surgery. Prior to the availability of high-resolution MRI, the extent of hippocampal resection at temporal lobectomy could not be adequately determined. It is surprisingly difficult for neurosurgeons to accurately estimate the length of hippocampal resection by visualization at resection or by measurement of the tissue specimen (even for the minority of surgeons who remove the anterior temporal tissues en bloc, without fragmentation or frank suctioning of the tissue). In some patients who experience seizures after anterior temporal lobectomy, MRI reveals that only a small volume of hippocampus was resected, and a subsequent, more extensive hippocampal resection results in full seizure control. Volume of temporal lobe resection on MRI correlates significantly with postoperative seizure control following anterior temporal lobectomy; extent of mesiobasal resection is a stronger predictor of seizure outcome than is lateral temporal resection volume (Awad *et al.*, 1989; Nayel *et al.*, 1991). As might be expected, reoperation to resect additional hippocampal tissue usually results in an excellent outcome among patients with seizures after temporal lobectomy who are shown on MRI to have had insufficient initial hippocampal resection (Wyler *et al.*, 1989). Care must be taken not to image the brain too early after surgery, if the purpose of imaging is evaluation of resection volume. In particular, transient postoperative extra-axial fluid collections will artifactually alter determinations of volume of resection on comparisons of pre- and postoperative images. Within the first 1–2 days after temporal lobectomy, essentially all patients have extra-axial fluid, air, or blood on MRI (Henegar *et al.*, 1996) and the extra-axial collections resolve gradually, reportedly with complete absence by 1–2 months (Sato *et al.*, 1997). Transient cerebral and persisting dural enhancement have been reported on MRI after temporal lobectomy (Hajek *et al.*, 1994; Sato *et al.*, 1997).

C. Transective Epilepsy Surgery

Corpus callosotomy can eliminate or markedly improve control of generalized tonic seizures and generalized atonic seizures, which constitute the highly injurious forms of epileptic drop attacks (Gates *et al.*, 1993). Most individuals with these symptomatic generalized epilepsies in addition have generalized tonic-clonic seizures or other types of seizures that do not respond to callosotomy. Early postoperative MRI effectively localizes sites of hemorrhage and edema after callosotomy (Harris *et al.*, 1989). Corpus callosotomy can result in mutism and disconnection syndromes, and the extent of callosal transection to some extent correlates with the types of interhemispheric information transfer that is impaired (Risse *et al.*, 1989). Assessment of extent of callosal resection is best accomplished with T_1-weighted, parasagittal MR images (Bogen *et al.*, 1988; Risse *et al.*, 1989). A number of ingenious tasks have been devised to explore the behavioral consequences of callosal transection in light of impaired interhemispheric information transfer (Gazzaniga, 1985). Activation fMRI and CBF PET studies might further clarify anatomical correlates of these behavioral alterations.

D. Implanted Devices for Epilepsy Therapy

1. Nonablative Surgical Interventions

Resective or transective epilepsy surgery often cannot be used to treat medically refractory partial and generalized epilepsies with adequate efficacy and safety. In some cases, the predicted benefit in seizure control is too low to consider ablative surgery. In other cases, the potential for iatrogenic disability due to cognitive or other permanent deficits is too great. Proposed therapies for such patients include (1) chronic electrical stimulation of neural structures by implanted devices (Fisher *et al.*, 1998), (2) chronic brain instillation of antiepileptic chemicals by implanted cannula (Mirski *et al.*, 1997), and (3) brain transplantation of neural tissue that produces useful neurotransmitters, such as GABA, or other beneficial effects (Schachter and Saper, 1998). Nonablative surgical therapies are conceptually attractive but far less developed in clinical application than are resective approaches to intractable seizures. Among these approaches, only neural electrostimulation has been widely tested in humans to date. Uncontrolled trials of cerebellar stimulation, intended to activate widespread inhibitory cortical projections, were extensive but inconclusive (Fisher *et al.*, 1998). Smaller trials of thalamic stimulation at the centromedian nucleus produced somewhat divergent results with regard to efficacy (Fisher *et al.*, 1992; Velasco *et al.*, 1995). The anterior nucleus of the thalamus, the locus coeruleus, and several other brain structures have been proposed as targets for further investigation of chronic electrostimulation (Feinstein *et al.*, 1989; Fisher *et al.*, 1998). Left cervical vagus nerve stimulation (VNS) is the only form of chronic, intermittent neuroelectrical stimulation that has shown sufficient efficacy and safety in prospective, randomized, controlled trials to achieve international regulatory acceptance as established epilepsy therapy at this time. VNS can induce EEG desynchronization in cats (Bailey and Bremer, 1938). This observation, and an extensive literature demonstrating hypersynchronous states of cortical and thalamocortical neuronal interactions during seizures (Lothman *et al.*, 1991; Steriade and Contreras, 1995), led to the proposal that VNS-induced electrocerebral desynchronization might prevent seizures (Zabara, 1985). Seizures were prevented or attenuated in several experimental models of epilepsy (Lockard *et al.*, 1990; Woodbury and Woodbury, 1990; Zabara, 1992; Takaya *et al.*, 1996). These studies demonstrated that the antiseizure effects of VNS outlast the period during which the nerve is stimulated. Experiments also showed that unilateral VNS activates multiple brain structures in higher order, central vagal pathways. There was involvement of forebrain structures, including the posterior cortical amygdaloid nucleus, as well as cingulate and retrosplenial cortex and ventromedial and arcuate hypothalamic nuclei (Naritoku *et al.*, 1995). Further studies showed antiseizure effects of VNS could be blocked by bilateral locus coeruleus lesions (Krahl *et al.*, 1998). Animal studies have not definitively established mechanisms of action of VNS, however.

Randomized, double-blind, controlled trials in refractory partial epilepsies did establish that VNS induces significant seizure reduction, with acceptable safety and tolerability (Salinsky and Group, 1995; Handforth *et al.*, 1998). Intermittent electrical stimulation of the left cervical segment of the vagus nerve also appears to be effective for generalized-onset seizures, in reports of primary and symptomatic generalized epilepsies (Tecoma *et al.*, 1995; Labar *et al.*, 1998). A pacemaker-like device is implanted below the clavicle, and this externally programmable device delivers electrical pulses to a lead wire that is coiled around the vagus nerve. During periods of VNS, stimulation occurs automatically at set intervals (usually every 1–5 min), during waking and sleep. The stimuli are delivered in pulse trains (usually lasting 7–30 s) as high-frequency (about 30 Hz), brief pulses of alternating polarity, with pauses between trains. The complexities of therapeutic VNS have been reviewed in detail (Schachter and Saper, 1998).

2. Imaging Studies of Vagus Nerve Electrical Stimulation

Brain mapping of VNS effects in humans offers an alternative to experimental epilepsy models for investigating therapeutic mechanisms of VNS. Electrical stimulation of the peripheral vagus nerve almost certainly requires synaptic transmission to mediate its antiseizure activity. Regional alterations in overall synaptic activity can be mapped as regional cerebral blood flow (rCBF) changes with activation PET and with fMRI techniques. Each vagus nerve has particularly dense, bilateral efferent projections onto the nuclei of the tractus solitarius (Rhoton *et al.*, 1966; Beckstead, 1979), so left cervical VNS is likely to cause increased transsynaptic neurotransmission in the nucleus of the tractus solitarius (NTS) bilaterally. At the cervical level, the left vagus nerve in primates is composed of approximately 80% afferent fibers. In addition to NTS projections, the vagal afferents synapse in the area postrema, the nucleus ambiguus, the dorsal motor nucleus of the vagus, the nucleus of the spinal trigeminal tract, and the medullary reticular regions (Rhoton *et al.*, 1966; Beckstead, 1979). The NTS and other nuclei that receive vagal afferents are located predominantly in dorsal, central, rostral medullary regions, as "the dorsal medullary vagal complex" (see Fig. 18). These nuclei project with numerous intramedullary synapses and descending efferents but also have major ascending projections that can influence pontomesencephalic, diencephalic, and cortical activities. The NTS projects densely onto the parabrachial nucleus, and the parabrachial nucleus projects heavily to multiple nuclei of the thalamus and the anterior insula; higher-order projections of the anterior insula are

particularly dense in inferior and inferolateral frontal cortex (Saper and Loewy, 1980; Parent, 1996). The NTS also projects to the inferior cerebellar hemispheres and vermis, the paraventricular and other hypothalamic nuclei, the parvicellular portion of the ventral posteromedial nucleus and other thalamic nuclei, the amygdala and hippocampus (and thereby to entorhinal cortex and anterior temporal cortex), and the cingulate gyrus (Rhoton *et al.*, 1966; Beckstead, 1979; Beckstead *et al.*, 1979; Somana and Walberg, 1979; Saper and Loewy, 1980; Nieuwenhuys *et al.*, 1988; Rutecki, 1990; Parent, 1996). The nucleus of the spinal trigeminal tract projects to somatosensory thalamic neurons, which project to the inferior postcentral gyrus and inferior parietal lobule (Parent, 1996). These polysynaptic higher pathways of cervical vagus afferents support major functional roles for vagal input in the central autonomic network, the visceral and somatic sensory systems, the reticular activating system, and the limbic system (Benarroch, 1997).

Vagal afferents have access to two special neuromodulatory systems of the brain and spinal cord, via bulbar noradrenergic and serotonergic projections (see Fig. 17). The locus coeruleus (LC) is a collection of dorsal pontine neurons that provide extremely widespread noradrenergic innervation of the entire cortex, diencephalon, and many other brain structures. Most afferents to the LC arise from two medullary nuclei, the nucleus paragigantocellularis (PGi) and the nucleus prepositus hypoglossi (PrH). The NTS projects to the LC through two major disynaptic pathways, one via the PGi and the other via the PrH (Aston-Jones *et al.*, 1991). The PGi afferent synapses use excitatory amino acids, which have their effects via a non-NMDA receptor as well as other neurotransmitters. The PrH afferent synapses are inhibitory, using GABA as the transmitter. Thus, VNS effects on the LC might variably be excitatory (PGi predominating), inhibitory (PrH predominating), or "neutral." Unlike the relatively compact LC, the raphe nuclei are distributed among midline reticular neurons from inferior medulla through the mesencephalon. The raphe nuclei provide extremely widespread serotonergic innervation of the entire cortex, diencephalon, and other brain structures. The NTS projects to multiple raphe nuclei, as do other nuclei of the dorsal medullary vagal complex, but the complexity of NTS-raphe pathways and transmitters is greater than for NTS-LC interactions (Nieuwenhuys *et al.*, 1988; Parent, 1996). Based on chemical anatomy alone, it is unclear whether VNS significantly alters LC and raphe nuclear functions. Vagal-LC and vagal-raphe interactions are potentially relevant to VNS mechanisms, because the LC and raphe are the major sources of norepinephrine and serotonin, respectively, in most of the brain, and norepinephrine and serotonin exert antiseizure effects (Stanton *et al.*, 1992; Dailey *et al.*, 1996) in addition to modulating many normal thalamic and cortical activities. Comparative mapping of cerebral serotonin synthesis before and after VNS might be accomplished

with α-[^{11}C]methyl-L-tryptophan and PET (Chugani *et al.*, 1998), and noradrenergic or serotonergic ligand displacement studies of VNS effect also would be of considerable interest. Comparative mapping of cerebral GABA concentrations might be performed with MRS before and after VNS, similarly to MRS studies of AEDs (Petroff *et al.*, 1996; Kuzniecky *et al.*, 1998). To date, such studies have not been reported.

Mapping of VNS effects on rCBF has been investigated with activation PET techniques (Garnett *et al.*, 1992; Ko *et al.*, 1996; Henry *et al.*, 1998a,b, 1999). In such studies, it is possible to acquire [^{15}O]water PET data over a period of time that is identical with the duration of trains of VNS pulses in standard therapeutic applications, for periods of 30–60 s. The net effect of VNS on regional CBF over such periods of stimulation can be determined by acquiring three or more stimulator-on scans and three or more stimulator-off scans in the same subject, with image subtraction and statistical analysis. These studies can be performed acutely (when VNS has been used for less than 24 h) or chronically (after a month or longer of ongoing VNS). The therapeutic effects of VNS often increase over months and may not plateau for up to 12 months or longer in some individuals (Salinsky *et al.*, 1996). It, therefore, would not be surprising should VNS-induced alterations in rCBF differ between acute and chronic states of VNS.

In the only reported functional imaging study of acute VNS effects in humans, acute VNS induced significant brainstem and bilateral cerebellar, diencephalic, and hemispheric blood flow changes (Henry *et al.*, 1998a) (Fig. 18). This investigation used averaged data of two groups of patients with refractory complex partial and secondarily generalized seizures. The two groups, each with five patients, differed only in the power of stimulation applied to the vagus nerve during the stimulator-on PET acquisitions. The low-stimulation group received VNS individually adjusted to the lowest threshold of perception (as left cervical sensations). The high-stimulation group received VNS individually titrated up to the highest comfortable level of stimula-

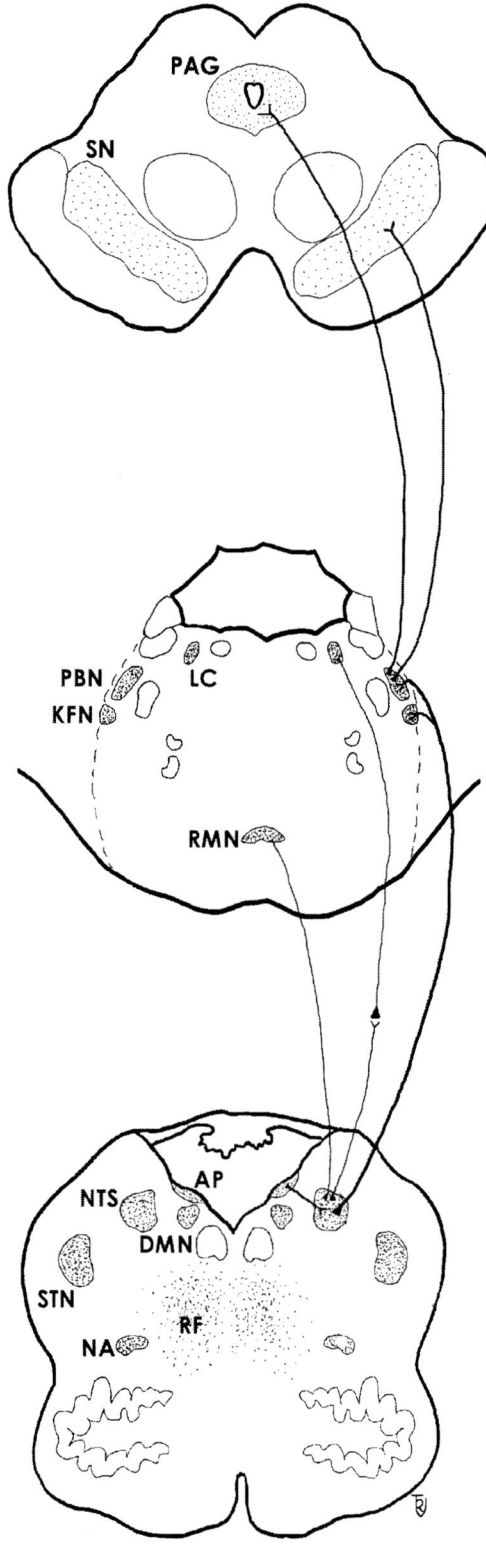

Figure 17 Schema of major vagus nerve polysynaptic brainstem connections. The locations of relevant nuclei are shown in axial cross-sections of the rostral mesencephalon (upper), mid pons (middle), and rostral medulla oblongata (lower). Each vagus nerve has afferent synapses on six ipsilateral medullary nuclei and on the contralateral nucleus of the tractus solitarius. The indicated projections from the nucleus of the tractus solitarius to higher levels are monosynaptic, except for a disynaptic projection to the locus coeruleus. Abbreviations: AP, area postrema; DMN, dorsal motor nucleus of the vagus; KFN, Kolliker–Fuse nucleus; LC, locus coeruleus; NA, nucleus accumbens; NTS, nucleus of the tractus solitarius; PAG, periaqueductal gray; PBN, parabrachial nucleus; RF, reticular formation; RMN, raphe magnus nucleus; SN, substantia nigra; STN, spinal trigeminal nucleus.

tion. The mean VNS power was 68 times greater in the high- than in the low-stimulation group. In general, the high-stimulation group demonstrated larger volumes of significant activations and deactivations than did the low-stimulation group. Acute VNS activated the dorsal, rostral medullary region, which is the location of the dorsal medullary vagal complex. Acute VNS activated sites in the somatosensory pathways subserving left cervical sensation, with right inferior postcentral gyrus activation (not observed in contralateral homologous areas), consistent with the left cervical sensations that patients consistently report during left cervical VNS. Most of the other cortical and subcortical activation and deactivation sites were bilaterally homologous, consistent with left vagus nerve projections to the NTS bilaterally and NTS projections to higher levels. Acute VNS induced right and left inferior cerebellar hemispheric and vermian CBF increases. Acute VNS also induced bilateral rCBF increases in the thalami, hypothalami, and insular and inferior frontal regions, but induced bilateral rCBF decreases in the amygdalae, posterior hippocampi, and cingulate gyri. It was concluded that left cervical VNS acutely alters synaptic activities in a widespread and bilateral distribution over brain structures that receive polysynaptic projections from the left vagus nerve.

Chronic VNS activation PET studies showed much smaller volumes of significant rCBF alterations than were found on PET studies of acute VNS (Garnett et al., 1992; Ko et al., 1996; Henry et al., 1998a,b). The three such investigations showed quite variable results, but patient selection and several technical aspects of PET studies differed among the chronic VNS-activation PET studies. Garnett and colleagues reported that left VNS activated blood flow in the left thalamus and left anterior cingulate gyrus on averaged PET data of five patients (Garnett et al., 1992). In this study, two of five patients had seizures during image acquisition. The results of this VNS-activation PET study are rendered more difficult to interpret by the inclusion of ictal scans, because partial-onset seizures are known to cause complex, multiregional alterations in CBF (Newton et al., 1992). Ko and colleagues reported that left VNS activated blood flow in the right thalamus, right posterior temporal cortex, left putamen, and left inferior cerebellum on averaged PET data of three patients (Ko et al., 1996). These results may have been influenced by prior epilepsy surgery, with right anterior temporal lobectomy in one case and left frontal resection in another of the three subjects. The Emory group restudied the 10 patients of the acute VNS activation PET studies (Henry et al., 1998a,b) using the same PET techniques after 3 months of ongoing VNS. Mean seizure frequency decrease was 38% between the acute and chronic PET studies. These chronic VNS activation PET studies

showed that some acute VNS-induced CBF changes declined, whereas other VNS-induced CBF changes persisted. VNS-induced CBF increases had the same distributions over the right postcentral gyrus and bilateral thalami, hypothalami, inferior cerebellar hemispheres, and inferior parietal lobules during chronic compared with acute VNS activation studies. The acute VNS-induced alterations of bilateral insular, hippocampal, amygdala, and cingulate CBF did not occur during the chronic studies. The chronic VNS activation studies did not show sites of activation or deactivation that had not occurred during the acute VNS studies. These studies suggest that subcortical effects of VNS may be more important in antiseizure mechanisms than are cortical effects of VNS. Alternatively, the loss of significant VNS-induced synaptic activity alterations in the hippocampi, amygdalae, and cingulate and insular cortices may be due to cortical processes that act to attenuate VNS effects on cerebral synaptic activities.

Regional VNS-induced activations and deactivations in synaptic activities may or may not represent sites where the activity alterations act to attenuate seizure onset or propagation. For this reason, a further analysis was performed to determine which, if any, acute VNS-induced rCBF changes correlated with subsequent changes in seizure frequency. The PET data were acquired at the beginning of a 3-month period of VNS, performed in the double-blind, randomized, active-controlled study of VNS. The active-control group received low stimulation (as a true control could not be designed, given the sensory stimuli that occur during VNS), and the treatment group received high stimulation (as described above). Antiepileptic drugs and other conditions were not changed during these 3 months, so changes in seizure frequency should be attributable to VNS effects. The "active-control" group showed a decline in seizures during the initial 3 months of VNS. The treatment group showed a significantly greater mean seizure reduction than did the active-control group, at $p = 0.02$, t-test, among the 194 patients of the multicenter study (Handforth et al., 1998). All 11 participants at the Emory site also had acute VNS-activation CBF PET at initiation of VNS, and all completed the 3-month treatment period. The Emory patients experienced seizure frequency changes ranging from 71% decrease to 2% increase in the high-stimulation group and from 59% decrease to 12% increase in the low-stimulation group. Percent changes in seizure frequency were rank ordered across the 11 patients. The rank order across patients was also established for each region's CBF changes during acute VNS-activation PET, using the 25 regions that showed significant effects in the group-averaged studies (Henry et al., 1998a,b). Regional CBF changes for these volumes of interest were measured in

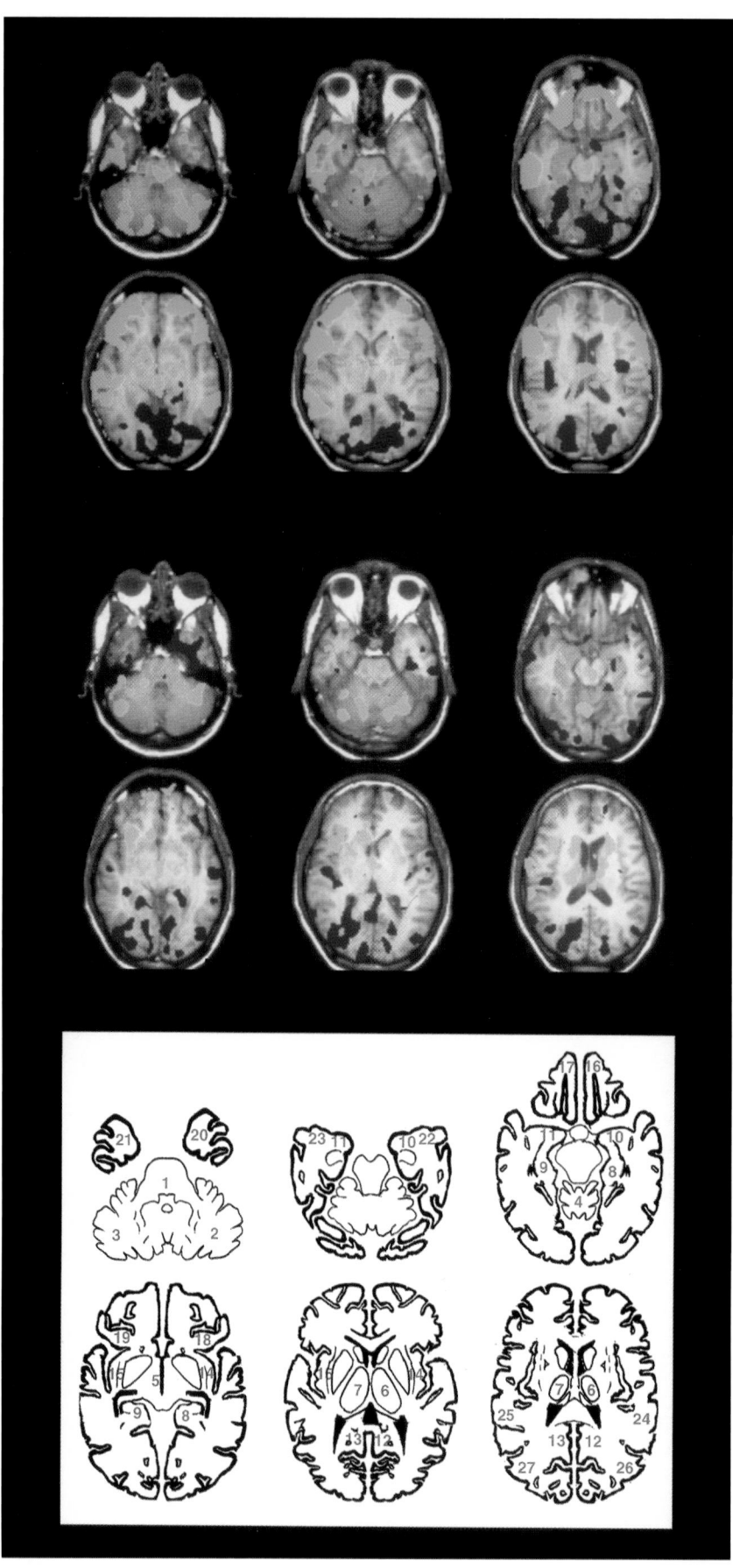

two ways: (1) the number of voxels that met particular *T*-value thresholds and (2) the maximal *T*-value within the volume of interest (Henry *et al.*, 1999). Spearman correlations determined associations between the 3-month change in seizure frequency and each region's volume of activation during acute VNS and between the 3-month change in seizure frequency and each region's maximal single-voxel *T*-values during acute VNS. Various regions showed obvious trends toward association of activation with subsequent improvement in seizure control, but after correction for multiple comparisons within the data set, only the right and left thalami showed significant associations. Patients who had greater improvement in seizure frequency had greater volumes of VNS-induced rCBF increase in the right and left thalami, with Spearman correlations of 0.88 and 0.81, respectively, both significant at $p < 0.05$ after Bonferroni correction. Patients who had greater improvement in seizure frequency also had higher maximal *T*-values in the right and left thalami, with Spearman correlations of 0.88 and 0.85, respectively, both significant at $p < 0.05$ after Bonferroni correction. Similar analyses were performed to compare the power of VNS applied during the PET acquisitions with the VNS-induced rCBF change; power of VNS during PET showed no significant correlation with altered thalamic CBF, but greater increases in right postcentral gyrus CBF were significantly associated with higher VNS power. It was concluded that increased thalamic synaptic activities probably mediate some anti-seizure effects of VNS.

Thalamocortical interactions modulate cortical epileptic excitability, as discussed above, so correlations of altered thalamic activities during acute VNS with therapeutic efficacy are not surprising. Future studies will be required to determine particular neurotransmitter-receptor alterations that VNS induces in reticular and specific thalamic nuclei in order to attenuate seizures. Evolution in VNS-related synaptic activities across time, determined with chronic activation imaging, and their correlation with therapeutic efficacy also could be useful in determining VNS anti-seizure mechanisms. Future studies of larger groups may show that some extrathalamic sites also have blood flow alterations during acute VNS, which are significantly associated with improvement in seizure control. For example, bilateral inferior cerebellar hemispheric activation during acute VNS tended to be greater in VNS responders,

although this association did not quite reach significance in the Emory study (Henry *et al.*, 1999). One interesting possibility is that patients who have greater initial thalamic activation by VNS represent a subpopulation of VNS responders. If so, activation PET or fMRI might identify VNS responders before device implantation, based on extracorporeal vagal stimulation, and implantation of VNS non-responders could be avoided. Alternatively, it might be shown that alterations in VNS settings that cause greater thalamic activation could lead to greater seizure reduction. Based on this principle, activation PET or fMRI might rapidly identify VNS settings that are more likely to cause greater seizure reduction, thereby obviating long periods of trial and error in adjustment of VNS settings.

Future rCBF activation studies of VNS will use fMRI. Safe and technically satisfactory fMRI is more difficult to obtain than is activation PET during VNS. An early safety concern was the possibility that radiofrequency pulses during MRI might induce electrical currents in the coiled vagus nerve leadwire, causing excessive tissue heating (Henry, 1996). Cervical tissue heating has not been observed in studies that use appropriately configured head coils (personal communication, R. Terry). Another question was whether the electrical fields generated by VNS would have sufficient spatial distribution so as to interfere with MR signal acquisition. This has not been observed to be a problem in early studies of VNS during fMRI (personal communication, R. Terry). Early studies did reveal a tendency for the VNS generator to be shut off on exposure to the high-strength static field of the MR instrument, however. It later became apparent that the "reed" switch of the generator is not closed by the unidirectional (static) magnetic field in all orientations of the generator. In other words, the generator can be positioned such that it is not shut off when the patient enters the MR scanner. The superior temporal resolution of fMRI may make it possible to trace the anatomical sequence of polysynaptic VNS effects, among other types of information that are obscured by the lesser temporal resolution of activation PET.

3. Imaging Studies of Brain Electrical Stimulation

The considerable information provided by functional imaging for investigations of VNS suggests similar application to intracranial stimulation. In a recent study,

---

Figure 18 *t* statistical mapping of regional blood flow increases and decreases induced by vagus nerve stimulation (VNS) in high- and low-stimulation groups, with schemata for locations of predetermined structures of interest. Volumes of BF increases (in yellow) and decreases (in blue) which have *t*-values of >5 are superimposed on gray-scale magnetic resonance images. The subjects' left is displayed on image right. The first two rows consist of axial images with ~1-cm spacing, arranged from inferior (top left) to superior (bottom right), of the high-stimulation group. The middle two rows are similarly arranged data for the low-stimulation group. The bottom two rows are axial brain schemata at the same levels as the other rows. With permission from Henry *et al.* (1998a). Brain blood flow alterations induced by therapeutic vagus nerve stimulation in partial epilepsy: I. Acute effects at high and low levels of stimulation. *Epilepsia* **39,** 983–990.

activation PET detected local CBF increases during after-discharges induced by electrical stimulation of macroelectrodes implanted for presurgical localization (Kahane *et al.,* 1995). Such studies may permit mapping of preferential pathways of ictal propagation, among numerous other possible investigations of cerebral electrochemical conductivity and of therapeutic mechanisms. Activation PET is readily compatible with electrical stimulation of the brain, but fMRI requires no exposure to ionizing radiation and has several other potential advantages. Concurrent intracranial electrical stimulation poses technical difficulties for fMRI, because generation of electrical fields to stimulate brain tissue can interfere excessively with magnetic signal acquisition. Perhaps properly timed brain MR signal acquisition will avoid electromagnetic signals due to electrical stimulation pulses, using uncoiled extracranial wires leading to intracranial electrodes, with pulse generation by devices that are compatible with the MR environment.

V. Summary

In this chapter the effects of surgery for movement disorders and epilepsy were examined in detail. Functional imaging provides insights into the mechanism of action for surgical therapies and can be used to identify major anatomic regions that are modulated by resection or implantable stimulator devices. Most striking is the modulation of "downstream" cortical areas after manipulation of peripheral or subcortical circuits. In combination with clinical assessments, the imaging results are also beginning to yield insight into functional anatomic correlates of cardinal symptoms, such as akinesia and bradykinesia. Ultimately, these observations may guide the selection of new targets for functional surgery.

References

Adolphs, R., Tranel, D., Damasio, H., and Damasio, A. R. (1995). Fear and the human amygdala. *J. Neurosci.* **15,** 5879–5891.

Alamy, M., Trouche, E., Nieoullon, A., and Legallet, E. (1994). Globus pallidus and motor initiation: The bilateral effects of unilateral quisqualic acid-induced lesion on reaction times in monkeys. *Exp. Brain Res.* **99,** 247–258.

Alamy, M., Pons, J., Gambarelli, D., and Trouche, E. (1996). A defective control of small-amplitude movements in monkeys with globus pallidus lesions: An experimental study on one component of pallidal bradykinesia. *Behav. Brain Res.* **72,** 57–62.

Aldridge, J. W., and Berridge, K. C. (1998). Coding of serial order by neostriatal neurons: A "natural action" approach to movement sequence. *J. Neurosci.* **18,** 2777–2787.

Alexander, G. E., Crutcher, M. D., and DeLong, M. R. (1990). Basal ganglia thalamo-cortical circuits: Parallel substrates for motor, oculomotor, "prefrontal" and "limbic" functions. *Prog. Brain Res.* **85,**

119–146.

Anderson, M. E., and Horak, F. B. (1985). Influence of the globus pallidus on arm movements in monkeys. III. Timing of movement-related information. *J. Neurophysiol.* **54,** 433–448.

Anderson, M. E., and Turner, R. S. (1991). Activity of neurons in cerebellar-receiving and pallidal-receiving areas of the thalamus of the behaving monkey. *J. Neurophysiol.* **66,** 879–893.

Arroyo, S., Santamaria, J., Sanmarti, F., Lomena, F., Catafau, A., Casamitjana, R., Setoain, J., and Tolosa, E. (1997). Ictal laughter associated with paroxysmal hypothalamopituitary dysfunction [published erratum appears in *Epilepsia* **38,** 1363 (1997)]. *Epilepsia* **38,** 114–117.

Aston-Jones, G., Shipley, M. T., Chouvet, G., Ennis, M., van Bockstaele, E., Pieribone, V., Shiekhattar, R., Akaoka, H., Drolet, G., Astier, B., *et al.* (1991). Afferent regulation of locus coeruleus neurons: Anatomy, physiology and pharmacology. *Prog. Brain Res.* **88,** 47–75.

Awad, I. A., Katz, A., Hahn, J. F., Kong, A. K., Ahl, J., and Lüders, H. (1989). Extent of resection in temporal lobectomy for epilepsy. I. Interobserver analysis and correlation with seizure outcome. *Epilepsia* **30,** 756–762.

Bailey, P., and Bremer, F. (1938). A sensory cortical representation of the vagus nerve with a note on the effects of low pressure on the cortical electrogram. *J. Neurophysiol.* **1,** 405–412.

Baron, M. S., Vitek, J. L., Bakay, R. A. E., Green, J., Kaneoke, Y., Hashimoto, T., Turner, R. S., Woodard, J. L., Cole, S. A., McDonald, W. M., and DeLong, M. R. (1996). Treatment of advanced Parkinson's disease by posterior GPi pallidotomy: 1-year results of a pilot study. *Ann. Neurol.* **40,** 355–366.

Beckstead, R. M. (1979). An autoradiographic examination of corticocortical and subcortical projections of the mediodorsal-projection (prefrontal) cortex in the rat. *J Comp Neurol* **184,** 43–62.

Beckstead, R. M., Domesick, V. B., and Nauta, W. J. H. (1979). Efferent connections of the substantia nigra and ventral tegmental area in the rat. *Brain Res.* **175,** 191–217.

Bejjani, B. P., Damier, P., Arnulf, I., Thivard, L., Bonnet, A. M., Dormont, D., Cornu, P., Pidoux, B., Samson, Y., and Agid, Y. (1999). Transient acute depression induced by high-frequency deep-brain stimulation [see comments]. *N. Engl. J. Med.* **340,** 1476–1480.

Benarroch, E. E. (1997). "Central Autonomic Network: Functional Organization and Clinical Correlations." Futura, Armonk, NY.

Berardelli, A., Hallett, M., Rothwell, J. C., Agostino, R., Manfredi, M., Thompson, P. D., and Marsden, C. D. (1996). Single-joint rapid arm movements in normal subjects and in patients with motor disorders. *Brain* **119,** 661–674.

Bergman, H., Wichmann, T., and DeLong, M. R. (1990a). Amelioration of Parkinsonian symptoms by inactivation of the subthalamic nucleus (STN) in MPTP treated green monkeys. First International Congress of Movement Disorders, Abstracts.

Bergman, H., Wichmann, T., and DeLong, M. R. (1990b). Reversal of experimental parkinsonism by lesions of the subthalamic nucleus. *Science* **249,** 1436–1438.

Berkovic, S. F. (1993). Single photon emission tomography. *In* "Surgical Treatment of the Epilepsies" (J. J. Engel, Ed.), pp. 223–244. Raven Press, New York.

Berridge, K. C., and Whishaw, I. Q. (1992). Cortex, striatum, and cerebellum: Control of serial order in a grooming sequence. *Exp. Brain Res.* **90,** 275–290.

Binder, J. R., Swanson, S. J., Hammeke, T. A., Morris, G. L., Mueller, W. M., Fischer, M., Benbadis, S., Frost, J. A., Rao, S. M., and Haughton, V. M. (1996). Determination of language dominance using functional MRI: A comparison with the Wada test. *Neurology* **46,** 978–984.

Binder, J. R., Frost, J. A., Hammeke, T. A., Cox, R. W., Rao, S. M., and Prieto, T. (1997). Human brain language areas identified by functional magnetic resonance imaging. *J. Neurosci.* **17,** 353–362.

Bittar, R. G., Olivier, A., Sadikot, A. F., Andermann, F., Comeau,

R. M., Cyr, M., Peters, T. M., and Reutens, D. C. (1999). Localization of somatosensory function by using positron emission tomography scanning: A comparison with intraoperative cortical stimulation. *J. Neurosurg.* **90,** 478–483.

Blandini, F., and Greenamyre, J. T. (1995). Effect of subthalamic nucleus lesion on mitochondrial enzyme activity in rat basal ganglia. *Brain Res.* **669,** 59–66.

Blinkenberg, M., Bonde, C., Holm, S., Svarer, C., Andersen, J., Paulson, O. B., and Law, I. (1996). Rate dependence of regional cerebral activation during the performance of a repetitive motor task: A PET study. *J. Cereb. Blood Flow Metab.* **16,** 794–803.

Boecker, H., Wills, A. J., Ceballos-Baumann, A., Samuel, M., Thomas, D. G., Marsden, C. D., and Brooks, D. J. (1997). Stereotactic thalamotomy in tremor-dominant Parkinson's disease: An H₂(15)O PET motor activation study. *Ann. Neurol.* **41,** 108–111.

Bogen, J. E., Schultz, D. H., and Vogel, P. J. (1988). Completeness of callosotomy shown by magnetic resonance imaging in the long term. *Arch. Neurol.* **45,** 1203–1205.

Bookheimer, S. Y., Zeffiro, T. A., Blaxton, T., Malow, B. A., Gaillard, W. D., Sato, S., Kufta, C., Fedio, P., and Theodore, W. H. (1997). A direct comparison of PET activation and electrocortical stimulation mapping for language localization. *Neurology* **48,** 1056–1065.

Bronen, R. A., Fulbright, R. K., Spencer, D. D., Spencer, S. S., Kim, J. H., Lange, R. C., and Sutilla, C. (1996). Refractory epilepsy: Comparison of MR imaging, CT, and histopathologic findings in 117 patients. *Radiology* **201,** 97–105.

Bronen, R. A., Fulbright, R. K., Kim, J. H., Spencer, S. S., and Spencer, D. D. (1997). A systematic approach for interpreting MR images of the seizure patient. *AJR, Am. J. Roentgenol.* **169,** 241–247.

Brophy, B. P., Kimber, T. J., and Thompson, P. D. (1997). Thalamotomy for parkinsonian tremor. *Stereotact. Funct. Neurosurg.* **69,** 1–4.

Brotchie, P., Iansek, R., and Horne, M. K. (1991a). Motor function of the monkey globus pallidus. 1. Neuronal discharge and parameters of movement. *Brain* **114,** 1667–1683.

Brotchie, P., Iansek, R., and Horne, M. K. (1991b). Motor function of the monkey globus pallidus. 2. Cognitive aspects of movement and phasic neuronal activity. *Brain* **114,** 1685–1702.

Brown, R., Jahanshahi, M., Limousin, P., Quinn, N., and Rothwell, J. (1998). Incidental sequence learning in patients with Parkinson's disease (PD): The impact of stereotaxic posteroventral pallidotomy (PVP). *Mov. Disord.* **13** (Suppl.), 199.

Burchiel, K. J. (1995). Thalamotomy for movement disorders. *Neurosurg. Clin. N. Am.* **6,** 55–71.

Catalan, M. J., Ishii, K., Honda, M., Samii, A., and Hallett, M. (1999). A PET study of sequential finger movements of varying length in patients with Parkinson's disease. *Brain* **122,** 483–495.

Ceballos-Baumann, A. O., Obeso, J. A., Vitek, J. L., DeLong, M. R., Bakay, R., Linaasoro, G., and Brooks, D. J. (1994). Restoration of thalamocortical activity after posteroventrolateral pallidotomy in Parkinson's disease. *Lancet* **344,** 814.

Cendes, F., Andermann, F., Preul, M. C., and Arnold, D. L. (1994). Lateralization of temporal lobe epilepsy based on regional metabolic abnormalities in proton magnetic resonance spectroscopic images. *Ann. Neurol.* **35,** 211–216.

Chabrerie, A., Ozlen, F., Nakajima, S., Leventon, M. E., Atsumi, H., Grimson, E., Keeve, E., Helmers, S., Riviello, J., Jr., Holmes, G., Duffy, F., Jolesz, F., Kikinis, R., and Black, P. M. (1997). Three-dimensional reconstruction and surgical navigation in pediatric epilepsy surgery. *Pediatr. Neurosurg.* **27,** 304–310.

Chu, W. J., Hetherington, H. P., Kuzniecky, R. I., Simor, T., Mason, G. F., and Elgavish, G. A. (1998). Lateralization of human temporal lobe epilepsy by ³¹P NMR spectroscopic imaging at 4.1 T. *Neurology* **51,** 472–479.

Chugani, H. T., Shields, W. D., Shewmon, D. A., Olson, D. M., Phelps, M. E., and Peacock, W. J. (1990). Infantile spasms: I. PET identifies focal cortical dysgenesis in cryptogenic cases for surgical treatment. *Ann. Neurol.* **27,** 406–413.

Chugani, H. T., Shewmon, D. A., Shields, W. D., Sankar, R., Comair, Y., Vinters, H. V., and Peacock, W. J. (1993). Surgery for intractable infantile spasms: Neuroimaging perspectives. *Epilepsia* **34,** 764–771.

Chugani, D. C., Chugani, H. T., Muzik, O., Shah, J. R., Shah, A. K., Canady, A., Mangner, T. J., and Chakraborty, P. K. (1998). Imaging epileptogenic tubers in children with tuberous sclerosis complex using alpha-[¹¹C]methyl-L-tryptophan positron emission tomography. *Ann. Neurol.* **44,** 858–866.

Connelly, A., Jackson, G. D., Duncan, J. S., King, M. D., and Gadian, D. G. (1994). Magnetic resonance spectroscopy in temporal lobe epilepsy. *Neurology* **44,** 1411–1417.

Cools, A. R., Jaspers, R., Schwarz, M., Sontag, K. H., Vrijmoed-de-Vries, M., and van den Bercken, J. (1984). Basal ganglia and switching motor programs. *In* "The Basal Ganglia. Structure and Function" (J. S. McKenzie *et al.*, Eds.), pp. 513–544. Plenum, New York.

Cromwell, H. C., and Berridge, K. C. (1996). Implementation of action sequences by a neostriatal site: A lesion mapping study of grooming syntax. *J. Neurosci.* **16,** 3444–3458.

Crossman, A. R., Mitchell, I. J., Clarke, C. E., and Boyce, S. (1985). Regional brain uptake of 2-deoxyglucose in *N*-methyl-4-phenyl-1,2,3,6-tetrahydropyridine (MPTP)-induced parkinsonism in the macaque monkey. *Neuropharmacology* **24,** 587–591.

Dailey, J. W., Yan, Q. S., Adams-Curtis, L. E., Ryu, J. R., Ko, K. H., Mishra, P. K., and Jobe, P. C. (1996). Neurochemical correlates of antiepileptic drugs in the genetically epilepsy-prone rat (GEPR). *Life Sci.* **58,** 259–266.

Damasio, H., Grabowski, T. J., Tranel, D., Hichwa, R. D., and Damasio, A. R. (1996). A neural basis for lexical retrieval [published erratum appears in *Nature* **381,** 810 (1996)]. *Nature* **380,** 499–505.

Davis, K. D., Taub, E., Houle, S., Lang, A. E., Dostrovsky, J. O., Tasker, R. R., and Lozano, A. M. (1997). Globus pallidus stimulation activates the cortical motor system during alleviation of parkinsonian symptoms. *Nat. Med.* **3,** 671–674.

Delfs, J. M., Ciaramitaro, V. M., Parry, T. J., and Chesselet, M. F. (1995). Subthalamic nucleus lesions: Widespread effects on changes in gene expression induced by nigrostriatal dopamine depletion in rats. *J. Neurosci.* **15,** 6562–6575.

DeLong, M. R. (1990). Primate models of movement disorders of basal ganglia origin. *Trends Neurosci.* **13,** 281–285.

DeLong, M. R., and Coyle, J. T. (1979). Globus pallidus lesions in the monkey produced by kainic acid: Histologic and behavioral effects. *Appl. Neurophysiol.* **42,** 95–97.

DeLong, M. R., and Georgopoulos, A. P. (1979). Motor functions of the basal ganglia as revealed by studies of single cell activity in the behaving primate. *In* "Advances in Neurology" (L. J. Poirier *et al.*, Eds.), Vol. 24, pp. 131–140. Raven Press, New York.

DeLong, M. R., and Georgopoulos, A. P. (1981). Motor functions of the basal ganglia. *In* "Handbook of Physiology. The Nervous System. Motor Control" (J. M. Brookhart *et al.*, Eds.), Sect. 1, Vol. II, Part 2, pp. 1017–1061. American Physiological Society, Bethesda, MD.

DeLong, M. R., Alexander, G. E., Georgopoulos, A. P., Crutcher, M. D., Mitchell, S. J., and Richardson, R. T. (1984). Role of basal ganglia in limb movements. *Hum. Neurobiol.* **2,** 235–244.

Deniau, J. M., and Chevalier, G. (1985). Disinhibition as a basic process in the expression of striatal functions. II. The striato-nigral influence on thalamocortical cells of the ventromedial thalamic nucleus. *Brain Res.* **334,** 227–233.

Detre, J. A., Sirven, J. I., Alsop, D. C., O'Connor, M. J., and French, J. A. (1995). Localization of subclinical ictal activity by functional magnetic resonance imaging: Correlation with invasive monitoring. *Ann. Neurol.* **38,** 618–624.

DeVito, J. L., and Andersen, M. E. (1982). An autoradiographic study of the efferent connections of the globus pallidus in *Macaca mullata*. *Brain Res.* **46,** 107–117.

Dick, J. P. R., Rothwell, J. C., Day, B. L., Cantello, R., Buruma, O., Gioux, M., Benecke, R., Berardelli, A., Thompson, P. D., and Marsden, C. D. (1989). The bereitschaftspotential is abnormal in Parkinson's disease. *Brain* **112,** 233–244.

Dietz, M. A., Goetz, C. G., and Stebbins, G. T. (1990). Evaluation of a modified inverted walking stick as a treatment for parkinsonian freezing episodes. *Mov. Disord.* **5,** 243–247.

Dogali, M., Fazzini, E., Kolodny, E., Eidelbert, D., Sterio, D., Devinsky, O., and Beric, A. (1995). Stereotactic ventral pallidotomy for Parkinson's disease. *Neurology* **45,** 753–761.

Doudet, D. J., Gross, C., Arluison, M., and Bioulac, B. (1990). Modifications of precentral cortex discharge and EMG activity in monkeys with MPTP-induced lesions of DA nigral neurons. *Exp. Brain Res.* **80,** 177–188.

Duncan, J. D., Moss, S. D., Bandy, D. J., Manwaring, K., Kaplan, A. M., Reiman, E. M., Chen, K., Lawson, M. A., and Wodrich, D. L. (1997). Use of positron emission tomography for presurgical localization of eloquent brain areas in children with seizures. *Pediatr. Neurosurg.* **26,** 144–156.

Duncan, J. S. (1997). Imaging and epilepsy. *Brain* **120,** 339–377.

Eidelberg, D., Moeller, J. R., Dhawan, V., Spetsieris, P., Takikawa, S., Ishikawa, T., Chaly, T., Robeson, W., Margouleff, D., Przedborski, S., *et al.* (1994). The metabolic topography of parkinsonism. *J. Cereb. Blood Flow Metab.* **14,** 783–801.

Engel, J., Kuhl, D. E., Phelps, M. E., and Crandall, P. H. (1982). Comparative localization of epileptic foci in partial epilepsy by PCT and EEG. *Ann. Neurol.* **12,** 529–537.

Engel, J., Jr., Henry, T. R., Risinger, M. W., Mazziotta, J. C., Sutherling, W. W., Levesque, M. F., and Phelps, M. E. (1990). Presurgical evaluation for partial epilepsy: Relative contributions of chronic depth-electrode recordings versus FDG-PET and scalp-sphenoidal ictal EEG. *Neurology* **40,** 1670–1677.

Engel, J. J., Dichter, M. A., and Schwartzkroin, P. A. (1998). Basic mechanism of human epilepsy. *In* "Epilepsy: A Comprehensive Textbook" (J. J. Engel and T. A. Pedley, Eds.), pp. 449–512. Lippincott-Raven, Philadelphia.

Esposito, G., Kirkby, B. S., Van Horn, J. D., Ellmore, T. M., and Berman, K. F. (1999). Context-dependent, neural system-specific neurophysiological concomitants of ageing: Mapping PET correlates during cognitive activation. *Brain* **122,** 963–979.

Evarts, E. V., Teravainen, H., and Calne, D. (1981). Reaction time Parkinson's disease. *Brain* **104,** 167–186.

Faber, T. L., Hoffman, J. M., Henry, T. R., Votaw, J. R., Brummer, M. E., and Garcia, E. V. (1996). Identifying hypometabolism in PET images of the brain: Application to epilepsy. *In* "Visualization in Biomedical Computing" (K. H. Hohne and R. Kikinis, Eds.), pp. 457–465. Springer-Verlag, Berlin.

Feinstein, B., Gleason, C. A., and Libet, B. (1989). Stimulation of locus coeruleus in man. Preliminary trials for spasticity and epilepsy. *Stereotact. Funct. Neurosurg.* **52,** 26–41.

Filion, M., and Tremblay, L. (1991). Abnormal spontaneous activity of globus pallidus neurons in monkeys with MPTP-induced parkinsonism. *Brain Res.* **547,** 142–151.

Filion, M., Tremblay, L., and Bedard, P. J. (1988). Abnormal influences of passive limb movement on the activity of globus pallidus neurons in parkinsonian monkeys. *Brain Res.* **444,** 165–176.

Fisher, R. S., Uematsu, S., Krauss, G. L., Cysyk, B. J., McPherson, R., Lesser, R. P., Gordon, B., Schwerdt, P., and Rise, M. (1992). Placebo-controlled pilot study of centromedian thalamic stimulation in treatment of intractable seizures. *Epilepsia* **33,** 841–851.

Fisher, R. S., Mirski, M., and Krauss, G. L. (1998). Brain stimulation. *In*

"Epilepsy: A Comprehensive Textbook" (J. Engel and T. A. Pedley, Eds.), pp. 1867–1875. Lippincott-Raven, Philadelphia.

Fried, I., and Cascino, G. D. (1998). Lesionectomy. *In* "Epilepsy: A Comprehensive Textbook" (J. J. Engel and T. A. Pedley, Eds.), pp. 1841–1850. Lippincott-Raven, Philadelphia.

Garnett, E. S., Nahmias, C., Scheffel, A., Firnau, G., and Upton, A. R. (1992). Regional cerebral blood flow in man manipulated by direct vagal stimulation. *Pacing Clin. Electrophysiol.* **15,** 1579–1580.

Gates, J. R., Wada, J. A., and Reeves, A. G. (1993). Reevaluation of corpus callostomy. *In* "Surgical Treatment of Epilepsy" (J. J. Engel, Ed.), pp. 637–648. Raven Press, New York.

Gazzaniga, M. S. (1985). Some contributions of split-brain studies to the study of human cognition. *In* "Epilepsy and the Corpus Callosum" (A. G. Reeves, Ed.), pp. 196–230. Plenum Press, New York.

Georgiou, N., Bradshaw, J. L., Iansek, R., Phillips, J. G., Mattingley, J. B., and Bradshaw, J. A. (1994). Reduction in external cues and movement sequencing in Parkinson's disease. *J. Neurol. Neurosurg. Psychiatry* **57,** 368–370.

Georgopoulos, A. P., DeLong, M. R., and Crutcher, M. D. (1983). Relations between parameters of step-tracking movements and single cell discharge in the globus pallidus and subthalamic nucleus of the behaving monkey. *J. Neurosci.* **3,** 1586–1598.

Glickstein, M., and Stein, M. (1991). Paradoxical movement in Parkinson's disease. *Trends Neurosci.* **14,** 480–483.

Goldberg, G. (1985). Supplementary motor area structure and function: Review and hypotheses. *Behav. Brain Sci.* **8,** 567–616.

Goldman-Rakic, P. S. (1995). Toward a circuit model of working memory and the guidance of voluntary motor action. *In* "Models of Information Processing in the Basal Ganglia" (J. C. Houk *et al.*, Eds.), pp. 131–148. MIT Press, Cambridge, MA.

Grafton, S. T., Sutton, J., Couldwell, W., Lew, M., and Waters, C. (1994). Network analysis of motor system connectivity in Parkinson's disease: Modulation of thalamocortical interactions after pallidotomy. *Hum. Brain Mapp.* **2,** 45–55.

Grafton, S. T., Waters, C., Sutton, J., Lew, M. F., and Couldwell, W. (1995). Pallidotomy increases activity of motor association cortex in Parkinson's disease: A PET study. *Ann. Neurol.* **37,** 776–783.

Grasby, P. M., Frith, C. D., Friston, K. J., Bench, C., Frackowiak, R. S., and Dolan, R. J. (1993). Functional mapping of brain areas implicated in auditory–verbal memory function. *Brain* **116,** 1–20.

Graybiel, A. M., Aosaki, T., Flaherty, A. W., and Kimura, M. (1994). The basal ganglia and adaptive motor control. *Science* **265,** 1826–1831.

Haaxma, R., van Boxtel, A., Brouwer, W. H., Goeken, L. N., Denier van der Gon, J. J., Colebatch, J. G., Martin, A., Brooks, D. J., Noth, J., and Marsden, C. D. (1995). Motor function in a patient with bilateral lesions of the globus pallidus. *Mov. Disord.* **10,** 761–777.

Hajek, M., Antonini, A., Leenders, K. L., and Wieser, H. G. (1991). Epilepsia partialis continua studied by PET. *Epilepsy Res.* **9,** 44–48.

Hajek, M., Wieser, H. G., Khan, N., Antonini, A., Schrott, P. R., Maguire, P., Beer, H. F., and Leenders, K. L. (1994). Preoperative and postoperative glucose consumption in mesiobasal and lateral temporal lobe epilepsy. *Neurology* **44,** 2125–2132.

Hallett, M., and Khoshbin, S. (1980). A physiological mechanism of bradykinesia. *Brain* **103,** 301–314.

Handforth, A., DeGiorgio, C. M., Schachter, S. C., Uthman, B. M., Naritoku, D. K., Tecoma, E. S., Henry, T. R., Collins, S. D., Vaughn, B. V., Gilmartin, R. C., Labar, D. R., Morris, G. L., 3rd, Salinsky, M. C., Osorio, I., Ristanovic, R. K., Labiner, D. M., Jones, J. C., Murphy, J. V., Ney, G. C., and Wheless, J. W. (1998). Vagus nerve stimulation therapy for partial-onset seizures: A randomized active-control trial. *Neurology* **51,** 48–55.

Harris, R. D., Roberts, D. W., and Cromwell, L. D. (1989). MR imaging of corpus callosotomy. *AJNR, Am. J. Neuroradiol.* **10**, 677–680.

Hayase, N., Miyashita, N., Endo, K., and Narabayashi, H. (1998). Neuronal activity in GP and Vim of parkinsonian patients and clinical changes of tremor through surgical interventions. *Stereotact. Funct. Neurosurg.* **71**, 20–28.

Hazrati, L.-N., and Parent, A. (1994). Projection from the external pallidum to the reticular thalamic nucleus in the squirrel monkey. *Brain* **550**, 142–146.

Henegar, M. M., Moran, C. J., and Silbergeld, D. L. (1996). Early postoperative magnetic resonance imaging following nonneoplastic cortical resection. *J. Neurosurg.* **84**, 174–179.

Henry, T. R. (1996). Progress in epilepsy research: Functional neuroimaging with positron emission tomography. *Epilepsia* 1141–1154.

Henry, T. R., and Drury, I. (1997). Non-epileptic seizures in temporal lobectomy candidates with medically refractory seizures. *Neurology* **48**, 1374–1382.

Henry, T. R., Chugani, H. T., Abou Khalil, B. W., Theodore, W. H., and Swartz, B. E. (1993a). Positron emission tomography in presurgical evaluation of epilepsy. *In* "Surgical Treatment of the Epilepsies" (J. J. Engel, Ed.), pp. 221–232. Raven Press, New York.

Henry, T. R., Frey, K. A., Sackellares, J. C., Gilman, S., Koeppe, R. A., Brunberg, J. A., Ross, D. A., Berent, S., Young, A. B., and Kuhl, D. E. (1993b). *In vivo* cerebral metabolism and central benzodiazepine-receptor binding in temporal lobe epilepsy. *Neurology* **43**, 1998–2006.

Henry, T. R., Mazziotta, J. C., and Engel, J. (1993c). Interictal metabolic anatomy of mesial temporal lobe epilepsy. *Arch. Neurol.* **50**, 582–589.

Henry, T. R., Babb, T. L., Engel, J., Jr., Mazziotta, J. C., Phelps, M. E., and Crandall, P. H. (1994). Hippocampal neuronal loss and regional hypometabolism in temporal lobe epilepsy. *Ann. Neurol.* **36**, 925–927.

Henry, T. R., Bakay, R. A., Votaw, J. R., Pennell, P. B., Epstein, C. M., Faber, T. L., Grafton, S. T., and Hoffman, J. M. (1998a). Brain blood flow alterations induced by therapeutic vagus nerve stimulation in partial epilepsy: I. Acute effects at high and low levels of stimulation. *Epilepsia* **39**, 983–990.

Henry, T. R., Buchtel, H. A., Koeppe, R. A., Pennell, P. B., Kluin, K. J., and Minoshima, S. (1998b). Regional cerebral blood flow increases during naming in left temporal lobe epilepsy. *Neurology* **50**, 787–790.

Henry, T. R., Votaw, J. R., Pennell, P. B., Epstein, C. M., Bakay, R. A., Faber, T. L., Grafton, S. T., and Hoffman, J. M. (1999). Acute blood flow changes and efficacy of vagus nerve stimulation in partial epilepsy. *Neurology* **52**, 1166–1173.

Ho, S. S., Berkovic, S. F., and Berlangieri, S. U. (1995). Comparison of ictal SPECT and interictal PET in the presurgical evaluation of temporal lobe epilepsy. *Ann. Neurol.* **37**, 738–745.

Holmes, M. D., Wilensky, A. J., Ojemann, G. A., and Ojemann, L. M. (1999). Hippocampal or neocortical lesions on magnetic resonance imaging do not necessarily indicate site of ictal onsets in partial epilepsy. *Ann. Neurol.* **45**, 461–465.

Hoover, J. E., and Strick, P. L. (1993). Multiple output channels in the basal ganglia. *Science* **259**, 819–821.

Hoover, J. E., and Strick, P. L. (1999). The organization of cerebellar and basal ganglia outputs to primary motor cortex as revealed by retrograde transneuronal transport of herpes simplex virus type 1. *J. Neurosci.* **19**, 1446–1463.

Horak, F. B., and Anderson, M. E. (1984a). Influence of globus pallidus on arm movements in monkeys. I. Effects of kainic-induced lesions. *J. Neurophysiol.* **52**, 290–304.

Horak, F. B., and Anderson, M. E. (1984b). Influence of globus pallidus on arm movements in monkeys. II. Effects of stimulation. *J. Neurophysiol.* **52**, 305–322.

Hore, J., and Villis, T. (1980). Arm movement performance during re-versible basal ganglia lesions in the monkey. *Exp. Brain Res.* **39**, 217–228.

Hore, J., Meyer-Lohmann, J., and Brooks, V. B. (1977). Basal ganglia cooling disables learned arm movements of monkeys in the absence of visual guidance. *Science* **195**, 584–586.

Houk, J. C. (1995). Information processing in modular circuits linking basal ganglia and cerebral cortex. *In* "Models of Information Processing in the Basal Ganglia" (J. C. Houk *et al.*, Eds.), pp. 3–10. MIT Press, Cambridge, MA.

Hunter, K. E., Blaxton, T. A., Bookheimer, S. Y., Figlozzi, C., Gaillard, W. D., Grandin, C., Anyanwu, A., and Theodore, W. H. (1999). (15)O water positron emission tomography in language localization: A study comparing positron emission tomography visual and computerized region of interest analysis with the Wada test. *Ann. Neurol.* **45**, 662–665.

Hurtado, J. M., Gray, C. M., Tamas, L. B., and Sigvardt, K. A. (1999). Dynamics of tremor-related oscillations in the human globus pallidus: A single case study. *Proc. Natl. Acad. Sci. U.S.A.* **96**, 1674–1679.

Hutchison, W. D., Luhn, M. A., and Schmidt, R. F. (1994). Responses of lateral thalamic neurons to algesic chemical stimulation of the cat knee joint. *Exp. Brain Res.* **101**, 452–464.

Ilinsky, I. A., Jouandet, M. L., and Goldman-Rakic, P. S. (1985). Organization of the nigrothalamocortical system in the rhesus monkey. *J. Comp. Neurol.* **236**, 315–330.

Inase, M., and Tanji, J. (1995). Thalamic distribution of projection neurons to the primary motor cortex relative to afferent terminal fields from the globus pallidus in the macaque monkey. *J. Comp. Neurol.* **353**, 415–426.

Inase, M., Buford, J. A., and Anderson, M. E. (1996a). Changes in the control of arm position, movement, and thalamic discharge during local inactivation in the globus pallidus of the monkey. *J. Neurophysiol.* **75**, 1087–1104.

Inase, M., Tokuno, H., Nambu, A., Akazawa, T., and Takada, M. (1996b). Origin of thalamocortical projections to the presupplementary motor area (pre-SMA) in the macaque monkey [published erratum appears in *Neurosci. Res.* **26**(4), 401–402 (1996)]. *Neurosci Res* **25**, 217–227.

Jack, C. R., Sharbrough, F. W., and Cascino, G. D. (1992). MRI-based hippocampal volumetry correlation with outcome after temporal lobectomy. *Ann. Neurol.* **31**, 138–146.

Jackson, G. D., Connelly, A., Duncan, J. S., Grunewald, R. A., and Gadian, D. G. (1993). Detection of hippocampal pathology in intractable partial epilepsy: Increased sensitivity with quantitative magnetic resonance T_2 relaxometry. *Neurology* **43**, 1793–1799.

Jackson, G. D., Connelly, A., Cross, J. H., Gordon, I., and Gadian, D. G. (1994). Functional magnetic resonance imaging of focal seizures. *Neurology* **44**, 850–856.

Jahanshahi, M., Jenkins, I. H., Brown, R. G., Marsden, C. D., Passingham, R. E., and Brooks, D. J. (1995). Self-initiated versus externally triggered movements: I. An investigation using measurement of regional cerebral blood flow with PET and movement-related potentials in normal and Parkinson's disease subjects. *Brain* **188**, 913–933.

Jenkins, I. H., Fernandez, W., Playford, E. D., Lees, A. J., Frackowiak, R. S. J., Passingham, R. E., and Brooks, D. J. (1992). Impaired activation of the supplementary motor area in Parkinson's disease is reversed when akinesia is treated with apomorphine. *Ann. Neurol.* **32**, 749–757.

Jenkins, I. H., Passingham, R. E., and Brooks, D. J. (1997). The effect of movement frequency on cerebral activation: A positron emission tomography study. *J. Neurol. Sci.* **151**, 195–205.

Joel, D., and Weiner, I. (1994). The organization of the basal ganglia-thalamocortical circuits: Open interconnected rather than closed segregated. *Neuroscience* **63**, 363–379.

Jones, E. G. (1985). "The Thalamus." Plenum, New York.

Jones-Gotman, M., Smith, M. L., and Wieser, H. G. (1998). Intra-arterial amobarbital procedures. *In* "Epilepsy: A Comprehensive Textbook." Lippincott-Raven, Philadelphia.

Jueptner, M., and Weiller, C. (1995). Does measurement of regional cerebral blood flow reflect synaptic activity? Implications for PET and fMRI. *NeuroImage* **2**, 148–156.

Kahane, P., Merlet, I., Gregoire, M. C., Munari, C., Perret, J., and Mauguiere, F. (1999). An H_2^{15} O-PET study of cerebral blood flow changes during focal epileptic discharges induced by intracerebral electrical stimulation. *Brain* **122**, 1851–1865.

Kato, M., and Kimura, M. (1992). Effects of reversible blockade of basal ganglia on a voluntary arm movement. *J. Neurophysiol.* **68**, 1516–1534.

Kimber, T. E., Tsai, C. S., Semmler, J., Brophy, B. P., and Thompson, P. D. (1999). Voluntary movement after pallidotomy in severe Parkinson's disease. *Brain* **122**, 895–906.

King, D., Spencer, S. S., McCarthy, G., and Spencer, D. D. (1997). Surface and depth EEG findings in patients with hippocampal atrophy. *Neurology* **48**, 1363–1367.

Ko, D., Heck, C., Grafton, S., Apuzzo, M. L., Couldwell, W. T., Chen, T., Day, J. D., Zelman, V., Smith, T., and DeGiorgio, C. M. (1996). Vagus nerve stimulation activates central nervous system structures in epileptic patients during PET $H_2(15)O$ blood flow imaging. *Neurosurgery* **39**, 426–430; discussion 430–431.

Koch, C. (1997). Computation and the single neuron [News]. *Nature* **385**, 207–210.

Koepp, M. J., Richardson, M. P., Brooks, D. J., Poline, J. B., Van Paesschen, W., Friston, K. J., and Duncan, J. S. (1996). Cerebral benzodiazepine receptors in hippocampal sclerosis. An objective *in vivo* analysis. *Brain* **119**, 1677–1687.

Krack, P., Pollak, P., Limousin, P., Hoffmann, D., Xie, J., Benazzouz, A., and Benabid, A. L. (1998). Subthalamic nucleus or internal pallidal stimulation in young onset Parkinson's disease. *Brain* **121**, 451–457.

Krahl, S. E., Clark, K. B., Smith, D. C., and Browning, R. A. (1998). Locus coeruleus lesions suppress the seizure-attenuating effects of vagus nerve stimulation. *Epilepsia* **39**, 709–714.

Kuzniecky, R. I., and Jackson, G. D. (1995). "Magnetic Resonance in Epilepsy." Raven Press, New York.

Kuzniecky, R., Elgavish, G. A., Hetherington, H. P., Evanochko, W. T., and Pohost, G. M. (1992). *In vivo* ^{31}P nuclear magnetic resonance spectroscopy of human temporal lobe epilepsy. *Neurology* **42**, 1586–1590.

Kuzniecky, R., Hetherington, H., Ho, S., Pan, J., Martin, R., Gilliam, F., Hugg, J., and Faught, E. (1998). Topiramate increases cerebral GABA in healthy humans. *Neurology* **51**, 627–629.

Labar, D., Nikolov, B., Tarver, B., and Fraser, R. (1998). Vagus nerve stimulation for symptomatic generalized epilepsy: A pilot study. *Epilepsia* **39**, 201–205.

Laitinen, L. V., Bergenheim, A. T., and Hariz, M. I. (1992). Leksell's posteroventral pallidotomy in the treatment of Parkinson's disease. *J. Neurosurg.* **76**, 53–61.

Lakke, P. W. F. (1981). Classification of extrapyramidal disorders. *J. Neurol. Sci.* **51**, 311–327.

Langfitt, J. T., and Rausch, R. (1997). Word-finding deficits persist after left anterotemporal lobectomy. *Arch. Neurol.* **53**, 72–76.

Laxer, K. D., Hubesch, B., Sappey-Marinier, D., and Weiner, M. W. (1992). Increased pH and inorganic phosphate in temporal seizure foci demonstrated by [^{31}P]MRS. *Epilepsia* **33**, 618–623.

Lee, G. P., Loring, D. W., Smith, J. R., and Flanigin, H. F. (1990). Material specific learning during electrical stimulation of the human hippocampus. *Cortex* **26**, 433–442.

Lee, H. J., Rye, D. B., Hallanger, A. E., Levey, A. I., and Wainer, B. H.

(1988). Cholinergic vs. noncholinergic efferents from the mesopontine tegmentum to the extrapyramidal motor system nuclei. *J. Comp. Neurol.* **275**, 469–492.

Lenz, F. A., Vitek, J. L., and DeLong, M. R. (1993). Role of the thalamus in parkinsonian tremor: Evidence from studies in patients and primate models. *Stereotact. Funct. Neurosurg.* **60**, 94–103.

Lesser, R. P., Luders, H., Morris, H. H., Dinner, D. S., Klem, G., Hahn, J., and Harrison, M. (1986). Electrical stimulation of Wernicke's area interferes with comprehension. *Neurology* **36**, 658–663.

Levy, R., Hazrati, L. N., Herrero, M. T., Vila, M., Hassani, O. K., Mouroux, M., Ruberg, M., Asensi, H., Agid, Y., Feger, J., Obeso, J. A., Parent, A., and Hirsch, E. C. (1997). Re-evaluation of the functional anatomy of the basal ganglia in normal and Parkinsonian states. *Neuroscience* **76**, 335–343.

Limousin, P., Pollak, P., Benazzouz, A., Hoffmann, D., Broussolle, E., Perret, J. E., and Benabid, A. L. (1995). Bilateral subthalamic nucleus stimulation for severe Parkinson's disease. *Mov. Disord.* **10**, 672–674.

Limousin, P., Greene, J., Pollak, P., Rothwell, J., Benabid, A.-L., and Frackowiak, R. S. J. (1997). Changes in cerebral activity pattern due to subthalamic nucleus or internal pallidum stimulation in Parkinson's disease. *Ann. Neurol.* **42**, 283–291.

Limousin, P., Brown, R. G., Jahanshahi, M., Asselman, P., Quinn, N. P., Thomas, D., Obeso, J. A., and Rothwell, J. C. (1999). The effects of posteroventral pallidotomy on the preparation and execution of voluntary hand and arm movements in Parkinson's disease. *Brain* **122**, 315–327.

Lockard, J. S., Congdon, W. C., and DuCharme, L. L. (1990). Feasibility and safety of vagal stimulation in monkey model. *Epilepsia* **31**, S20–S26.

Lothman, E. W., Bertram, E. H. D., and Stringer, J. L. (1991). Functional anatomy of hippocampal seizures. *Prog. Neurobiol.* **37**, 1–82.

Lozano, A. M., Lang, A. E., Galvez-Jimenez, N., Miyasaki, J., Duff, J., Hutchinson, W. D., and Dostrovsky, J. O. (1995). Effect of GPi pallidotomy on motor function in Parkinson's disease. *Lancet* **346**, 1383–1387.

Luders, H., Lesser, R. P., Hahn, J., Dinner, D. S., Morris, H., Resor, S., and Harrison, M. (1986). Basal temporal language area demonstrated by electrical stimulation. *Neurology* **36**, 505–510.

Majsak, M. J., Kaminski, T., Gentile, A. M., and Flanagan, J. R. (1998). The reaching movements of patients with Parkinson's disease under self-determined maximal speed and visually cued conditions. *Brain* **121**, 755–766.

Marks, D. A., Katz, A., Hoffer, P., and Spencer, S. S. (1992). Localization of extratemporal epileptic foci during ictal single photon emission computed tomography. *Ann. Neurol.* **31**, 250–255.

Marsden, C. D., and Obeso, J. A. (1994). The functions of the basal ganglia and the paradox of stereotaxic surgery in Parkinson's disease. *Brain* **117**, 877–897.

Martin, A., Wiggs, C. L., Ungerleider, L. G., and Haxby, J. V. (1996). Neural correlates of category-specific knowledge. *Nature* **379**, 649–652.

Meiners, L. C., Bakker, C. J., van Rijen, P. C., van Veelen, C. W., van Huffelen, A. C., van Dieren, A., Jansen, G. H., and Mali, W. P. (1996). Fast spin echo MR of contact points on implanted intracerebral stainless steel multicontact electrodes. *AJNR, Am. J. Neuroradiol.* **17**, 1815–1819.

Middleton, F. A., and Strick, P. L. (1997a). Cerebellar output channels. *Int. Rev. Neurobiol.* **41**, 61–82.

Middleton, F. A., and Strick, P. L. (1997b). New concepts about the organization of basal ganglia output. *In* "The Basal Ganglia and New Surgical Approaches for Parkinson's Disease" (J. A. Obeso *et al.*, Eds.), pp. 57–68. Lippincott-Raven, Philadelphia.

Miller, W. C., and DeLong, M. R. (1987). Altered tonic activity of neurons in the globus pallidus and subthalamic nucleus in the primate MPTP model of parkinsonism. *In* "The Basal Ganglia II" (M. B. Carpenter and A. Jayaraman, Eds.), pp. 415–427. Plenum, New York.

Miller, W. C., and DeLong, M. R. (1988). Parkinsonian symptomatology: An anatomical and physiological analysis. *Ann. N.Y. Acad. Sci.* **515,** 287–302.

Mink, J. W. (1996). The basal ganglia: Focused selection and inhibition of competing motor programs. *Prog. Neurobiol.* **50,** 381–425.

Mink, J. W., and Thach, W. T. (1991). Basal ganglia motor control. III. Pallidal ablation: Normal reaction time, muscle cocontraction, and slow movement. *J. Neurophysiol.* **65,** 330–351.

Mirski, M. A., Rossell, L. A., Terry, J. B., and Fisher, R. S. (1997). Anticonvulsant effect of anterior thalamic high frequency electrical stimulation in the rat. *Epilepsy Res.* **28,** 89–100.

Mitchell, I. J., Clarke, C. E., Boyce, S., Robertson, R. G., Peggs, D., Sambrook, M. A., and Crossman, A. R. (1989). Neural mechanisms underlying parkinsonian symptoms based upon regional uptake of 2-deoxyglucose in monkeys exposed to 1-methyl-4-phenyl-1,2,3,6-tetrahydropyridine. *Neuroscience* **32,** 213–226.

Miyachi, S., Hikosaka, O., Miyashita, K., Karadi, Z., and Rand, M. K. (1997). Differential roles of monkey striatum in learning of sequential hand movement. *Exp. Brain Res.* **115,** 1–5.

Mushiake, H., and Strick, P. L. (1995). Pallidal neuron activity during sequential arm movements. *J. Neurophysiol.* **74,** 2754–2758.

Naritoku, D. K., Terry, W. J., and Helfert, R. H. (1995). Regional induction of fos immunoreactivity in the brain by anticonvulsant stimulation of the vagus nerve. *Epilepsy Res.* **22,** 53–62.

Nayel, M. H., Awad, I. A., and Luders, H. (1991). Extent of mesiobasal resection determines outcome after temporal lobectomy for intractable complex partial seizures. *Neurosurgery* **29,** 55–60; discussion 60–61.

Newton, M. R., Berkovic, S. F., Austin, M. C., Rowe, C. C., McKay, W. J., and Bladin, P. F. (1992). Postictal switch in blood flow distribution and temporal lobe seizures. *J. Neurol. Neurosurg. Psychiatry* **55,** 891–894.

Nieuwenhuys, R., Voogd, J., and van Huijzen, C. (1988). "The Human Central Nervous System: A Synopsis and Atlas." Springer-Verlag, New York.

Nini, A., Feingold, A., Slovin, H., and Bergman, H. (1995). Neurons in the globus pallidus do not show correlated activity in the normal monkey, but phase-locked oscillations appear in the MPTP model of Parkinsonism. *J. Neurophysiol.* **74**(4), 1800–1805.

Novelly, R. A., Augustine, E. A., Mattson, R. H., Glaser, G. H., Williamson, P. D., Spencer, D. D., and Spencer, S. S. (1984). Selective memory improvement and impairment in temporal lobectomy for epilepsy. *Ann. Neurol.* **15,** 64–67.

Nyberg, L., McIntosh, A. R., Houle, S., Nilsson, L. G., and Tulving, E. (1996). Activation of medial temporal structures during episodic memory retrieval. *Nature* **380,** 715–717.

Obeso, J. A., Rodriguez, M. C., and DeLong, M. R. (1997). Basal ganglia pathophysiology. A critical review. *Adv. Neurol.* **74,** 3–18.

O'Brien, T. J., So, E. L., Mullan, B. P., Hauser, M. F., Brinkmann, B. H., Jack, C. R., Jr., Cascino, G. D., Meyer, F. B., and Sharbrough, F. W. (1999). Subtraction SPECT co-registered to MRI improves postictal SPECT localization of seizure foci. *Neurology* **52,** 137–146.

Ojemann, G. A. (1979). Individual variability in cortical localization of language. *J. Neurosurg.* **50,** 164–169.

Ojemann, G. A., and Dodrill, C. B. (1985). Verbal memory deficits after left temporal lobectomy for epilepsy. Mechanism and intraoperative prediction. *J. Neurosurg.* **62,** 101–107.

Ojemann, G., Ojemann, J., Lettich, E., and Berger, M. (1989). Cortical language localization in left, dominant hemisphere. An electrical stimulation mapping investigation in 117 patients. *J. Neurosurg.* **71,**

316–326.

Pacia, S. V., and Ebersole, J. S. (1997). Intracranial EEG substrates of scalp ictal patterns from temporal lobe foci. *Epilepsia* **38,** 642–654.

Packard, M. G., Hirsch, R., and White, N. M. (1989). Differential effects of fornix and caudate nucleus lesions on two radial maze tasks: Evidence for multiple memory systems. *J. Neurosci.* **9,** 1465–1472.

Parent, A. (1996). "Carpenter's Human Neuroanatomy." Williams & Wilkins, Baltimore.

Parent, A., and Hazrati, L.-N. (1993). Anatomical aspects of information processing in primate basal ganglia. *Trends Neurosci.* **16**(3), 111–116.

Parent, A., and Hazrati, L.-N. (1995). Functional anatomy of the basal ganglia. I. The cortico-basal ganglia-thalamo-cortical loop. *Brain Res.* **20,** 91–127.

Parkinson, J. (1817). "An Essay on the Shaking Palsy." Wittingham & Rowland, London.

Paus, T., Petrides, M., Evans, A. C., and Meyer, E. (1993). Role of the human anterior cingulate cortex in the control of oculomotor, manual, and speech responses: A positron emission tomography study. *J. Neurophysiol.* **70,** 453–469.

Penfield, W. J., and Bradley, E. (1937). Somatic motor and sensory representation in the cerebral cortex of man as studied by electrical stimulation. *Brain* **60,** 389–443.

Petrides, M., Alivisatos, B., Evans, A. C., and Meyer, E. (1993a). Dissociation of human mid-dorsolateral from posterior dorsolateral frontal cortex in memory processing. *Proc. Natl. Acad. Sci. U.S.A.* **90,** 873–877.

Petrides, M., Alivisatos, B., Meyer, E., and Evans, A. C. (1993b). Functional activation of the human frontal cortex during the performance of verbal working memory tasks. *Proc Natl Acad Sci* **90,** 878–882.

Petroff, O. A., Rothman, D. L., Behar, K. L., and Mattson, R. H. (1996). Human brain GABA levels rise after initiation of vigabatrin therapy but fail to rise further with increasing dose. *Neurology* **46,** 1459–1463.

Picard, N., and Strick, P. L. (1996). Motor areas of the medial wall: A review of their location and functional activation. *Cereb. Cortex* **6,** 342–353.

Piert, M., Koeppe, R. A., Giordani, B., Minoshima, S., and Kuhl, D. E. (1996). Determination of regional rate constants from dynamic FDG-PET studies in Parkinson's disease. *J. Nucl. Med.* **37,** 1115–1122.

Playford, E. D., Jenkins, I. H., Passingham, R. E., Nutt, J., Frackowiak, R. S. J., and Brooks, D. J. (1992). Impaired mesial frontal and putamen activation in Parkinson's disease: A positron emission tomography study. *Ann. Neurol.* **32,** 151–161.

Rascol, O., Sabatini, U., Chollet, F., Celsis, P., Montastruc, J.-L., Marc-Vergnes, J.-P., and Rascol, A. (1992). Supplementary and primary sensory motor area activity in Parkinon's disease. Regional cerebral blood flow changes during finger movements and effects of apomorphine. *Arch. Neurol.* **49,** 144–148.

Rascol, O., Sabatini, U., Fabre, N., Brefel, C., Loubinoux, I., Celsis, P., Senard, J. M., Montastruc, J. L., and Chollet, F. (1997). The ipsilateral cerebellar hemisphere is overactive during hand movements in akinetic parkinsonian patients. *Brain* **120,** 103–110.

Rasmussen, T., and Milner, B. (1977). The role of early left-brain injury in determining lateralization of cerebral speech functions. *Ann. N.Y. Acad. Sci.* **299,** 355–369.

Rausch, R., Silfvenius, H., Wieser, H. G., Dodrill, C. B., Meador, K. J., and Jones Gotman, M. (1993). Internal amobarbital procedures. *In* "Surgical Treatment of the Epilepsies" (J. J. Engel, Ed.), pp. 341–357. Raven Press, New York.

Redgrave, P., Prescott, T. J., and Gurney, K. (1999). Is the short-latency dopamine response too short to signal reward error? *Trends Neurosci.* **22,** 146–151.

Reutens, D. C., Gjedde, A. H., and Meyer, E. (1998). Regional lumped constant differences and asymmetry in fluorine-18-FDG uptake in temporal lobe epilepsy. *J. Nucl. Med.* **39,** 176–180.

Rhoton, A. L., Jr., O'Leary, J. L., and Ferguson, J. P. (1966). The trigeminal, facial, vagal, and glossopharyngeal nerves in the monkey. Afferent connections. *Arch. Neurol.* **14,** 530–540.

Risinger, M. W., Engel, J., Jr., Van Ness, P. C., Henry, T. R., and Crandall, P. H. (1989). Ictal localization of temporal lobe seizures with scalp/sphenoidal recordings. *Neurology* **39,** 1288–1293.

Risse, G. L., Gates, J., Lund, G., Maxwell, R., and Rubens, A. (1989). Interhemispheric transfer in patients with incomplete section of the corpus callosum. Anatomic verification with magnetic resonance imaging. *Arch. Neurol.* **46,** 437–443.

Rohlfs, A., Nikkah, G., Rosenthal, C., Rundfeldt, C., Brandis, A., Samii, M., and Loscher, W. (1997). Hemispheric asymmetries in spontaneous firing characteristics of substantia nigra pars reticulata neurons following a unilateral 6-hydroxydopamine lesion of the rat nigrostriatal pathway. *Brain Res* **761,** 352–356.

Ross, D. A., Brunberg, J. A., Drury, I., and Henry, T. R. (1996). Intracerebral depth electrode monitoring in partial epilepsy: The morbidity and efficacy of placement using magnetic resonance image-guided stereotactic surgery. *Neurosurgery* **39,** 327–333; discussion 333–334.

Rouiller, E. M., Liang, F., Babalian, A., Moret, V., and Wiesendanger, M. (1994). Cerebellothalamocortical and pallidothalamocortical projections to the primary and supplementary motor cortical areas: A multiple tracing study in macaque monkeys. *J. Comp. Neurol.* **345,** 185–213.

Rowe, C. C., Berkovic, S. F., Sia, S. T., Austin, M., McKay, W. J., Kalnins, R. M., and Bladin, P. F. (1989). Localization of epileptic foci with postictal single photon emission computed tomography. *Ann. Neurol.* **26,** 660–668.

Rutecki, P. (1990). Anatomical, physiological, and theoretical basis for the antiepileptic effect of vagus nerve stimulation. *Epilepsia* **31,** S1–S6.

Rye, D. B., Turner, R. S., Vitek, J. L., Bakay, R. A. E., Crutcher, M. D., and DeLong, M. R. (1996). Anatomical investigations of the pallidotegmental pathway in monkey and man. *In* "The Basal Ganglia V" (H. Ohye, J. S. McKenzie, and M. Kimura, Eds.), pp. 59–75. Plenum Press, New York.

Sadato, N., Zeffiro, T. A., Campbell, G., Konishi, J., Shibasaki, H., and Hallett, M. (1996). Frequency-dependent changes of regional cerebral blood flow during finger movements. *J. Cereb. Blood Flow Metab.* **16,** 23–33.

Sakai, S. T., Inase, M., and Tanji, J. (1999). Pallidal and cerebellar inputs to thalamocortical neurons projecting to the supplementary motor area in *Macaca fuscata*: A triple-labeling light microscopic study. *Anat. Embryol. (Berl.)* **199,** 9–19.

Salinsky, M. C., and Group, V. N. S. S. (1995). A randomized controlled trial of chronic vagus nerve stimulation for treatment of medically intractable seizures. *Neurology* **45,** 224–230.

Salinsky, M. C., Uthman, B. M., Ristanovic, R. K., Wernicke, J. F., and Tarver, W. B. (1996). Vagus nerve stimulation for the treatment of medically intractable seizures. Results of a 1-year open-extension trial. Vagus Nerve Stimulation Study Group. *Arch. Neurol.* **53,** 1176–1180.

Sammaritano, M., de Lotbiniere, A., Andermann, F., Olivier, A., Gloor, P., and Quesney, L. F. (1987). False lateralization by surface EEG of seizure onset in patients with temporal lobe epilepsy and gross focal cerebral lesions. *Ann. Neurol.* **21,** 361–369.

Samuel, M., Ceballos-Baumann, A. O., Blin, J., Uema, T., Boecker, H., Passingham, R. E., and Brooks, D. J. (1997a). Evidence for lateral premotor and parietal overactivity in Parkinson's disease during sequential and bimanual movements. A PET study. *Brain* **120,** 963–976.

Samuel, M., Ceballos-Baumann, A. O., Turjanski, N., Boecker, H.,

Gorospe, A., Linazasoro, G., Holmes, A. P., DeLong, M. R., Vitek, J. L., Thomas, D. G. T., Quinn, N. P., Obeso, J. A., and Brooks, D. J. (1997b). Pallidotomy in Parkinson's disease increases supplementary motor area and prefrontal activation during performance of volitional movements: An H$_2$^{15}O PET study. *Brain* **120,** 1301–1313.

Saper, C. B., and Loewy, A. D. (1980). Efferent connections of the parabrachial nucleus in the rat. *Brain Res.* **197,** 291–317.

Sato, N., Bronen, R. A., Sze, G., Kawamura, Y., Coughlin, W., Putman, C. M., and Spencer, D. D. (1997). Postoperative changes in the brain: MR imaging findings in patients without neoplasms. *Radiology* **204,** 839–846.

Savic, I., Thorell, J. O., and Roland, P. (1995). [^{11}C]Flumazenil positron emission tomography visualizes frontal epileptogenic regions. *Epilepsia* **36,** 1225–1232.

Saykin, A. J., Gur, R. C., Sussman, N. M., O'Connor, M. J., and Gur, R. E. (1989). Memory deficits before and after temporal lobectomy: Effect of laterality and age of onset. Brain Cogn. **9,** 191–200.

Schachter, S. C., and Saper, C. B. (1998). Vagus nerve stimulation. *Epilepsia* **39,** 677–686.

Schell, G. R., and Strick, P. L. (1984). The origin of thalamic inputs to the arcuate premotor and supplementary motor areas. *J. Neurosci.* **4,** 539–560.

Scott, C. A., Fish, D. R., Smith, S. J., Free, S. L., Stevens, J. M., Thompson, P. J., Duncan, J. S., Shorvon, S. D., and Harkness, W. F. (1999). Presurgical evaluation of patients with epilepsy and normal MRI: Role of scalp video-EEG telemetry. *J. Neurol. Neurosurg. Psychiatry* **66,** 69–71.

Shahzadi, S., Tasker, R. R., and Lozano, A. (1995). Thalamotomy for essential and cerebellar tremor. *Stereotact. Funct. Neurosurg.* **65,** 11–17.

Siegfried, J., and Lippitz, B. (1994). Bilateral chronic electrostimulation of ventroposterolateral pallidum: A new therapeutic approach for alleviating all parkinsonian symptoms. *Neurosurgery* **35,** 1126–1129; discussion 1129–1130.

So, N., Gloor, P., Quesney, L. F., Jones-Gotman, M., Olivier, A., and Andermann, F. (1989). Depth electrode investigations in patients with bitemporal epileptiform abnormalities. *Ann. Neurol.* **25,** 423–431.

Somana, R., and Walberg, F. (1979). Cerebellar afferents from the nucleus of the solitary tract. *Neurosci. Lett.* **11,** 41–47.

Spanaki, M. V., Zubal, I. G., MacMullan, J., and Spencer, S. S. (1999). Periictal SPECT localization verified by simultaneous intracranial EEG. *Epilepsia* **40,** 267–274.

Spencer, S. S. (1981). Depth electroencephalography in selection of refractory epilepsy for surgery. *Ann. Neurol.* **9,** 207–214.

Spencer, S. S. (1994). The relative contributions of MRI, SPECT, and PET imaging in epilepsy. *Epilepsia* **35,** S72–S89.

Stafiniak, P., Saykin, A. J., Sperling, M. R., Kester, D. B., Robinson, L. J., O'Connor, M. J., and Gur, R. C. (1990). Acute naming deficits following dominant temporal lobectomy: Prediction by age at 1st risk for seizures. *Neurology* **40,** 1509–1512.

Stanton, P. K., Mody, I., Zigmond, D., Sejnowski, T., and Heinemann, U. (1992). Noradrenergic modulation of excitability in acute and chronic model epilepsies. *Epilepsy Res. Suppl.* **8,** 321–334.

Stelmach, G. E., Teasdale, N., Phillips, J., and Worringhan, C. J. (1989). Force production characteristics in Parkinson's disease. *Exp. Brain Res.* **76,** 165–172.

Stephan, K. M., Fink, G. R., Passingham, R. E., Silbersweig, D., Ceballos-Baumann, A. O., Frith, C. D., and Frackowiak, R. S. J. (1995). Imaging the execution of movements. *J. Neurophysiol.* **73,** 373–386.

Steriade, M., and Contreras, D. (1995). Relations between cortical and thalamic cellular events during transition from sleep patterns to paroxysmal activity. *J. Neurosci.* **15,** 623–642.

Sterio, D., Beric, A., Dogali, M., Fazzini, E., Alfaro, G., and Devinsky,

O. (1994). Neurophysiological properties of pallidal neurons in Parkinson's disease. *Ann. Neurol.* **35,** 586–591.

Swartz, B. E., Halgren, E., Delgado-Escueta, A. V., Mandelkern, M., Gee, M., Quinones, N., Blahd, W. H., and Repchan, J. (1989). Neuroimaging in patients with seizures of probable frontal lobe origin. *Epilepsia* **30,** 547–558.

Swartz, B. E., Halgren, E., Fuster, J. M., Simpkins, E., Gee, M., and Mandelkern, M. (1995). Cortical metabolic activation in humans during a visual memory task. *Cereb. Cortex* **5,** 205–214.

Swartz, B. E., Halgren, E., Simpkins, F., Fuster, J., Mandelkern, M., Krisdakumtorn, T., Gee, M., Brown, C., Ropchan, J. R., and Blahd, W. H. (1996). Primary or working memory in frontal lobe epilepsy: An ¹⁸FDG-PET study of dysfunctional zones. *Neurology* **46,** 737–747.

Takaya, M., Terry, W. J., and Naritoku, D. K. (1996). Vagus nerve stimulation induces a sustained anticonvulsant effect. *Epilepsia* **37,** 1111–1116.

Tanji, J. (1994). The supplementary motor area in the cerebral cortex. *Neurosci. Res.* **19,** 251–268.

Tecoma, E. S., Iragui, V. J., and Alksne, J. I. (1995). Refractory primary generalized epilepsy: Treatment with intermittent vagus nerve stimulation. *Epilepsia* **36** (Suppl. 3), 228.

Theodore, W. H., Newmark, M. E., Sato, S., Brooks, R., Patronas, N., De La Paz, R., DiChiro, G., Kessler, R. M., Margolin, R., Manning, R. G., *et al.* (1983). [¹⁸F]Fluorodeoxyglucose positron emission tomography in refractory complex partial seizures. *Ann. Neurol.* **14,** 429–437.

Theodore, W. H., Sato, S., Kufta, C., Balish, M. B., Bromfield, E. B., and Leiderman, D. B. (1992). Temporal lobectomy for uncontrolled seizures: The role of positron emission tomography. *Ann. Neurol.* **32,** 789–794.

Tulving, E., Markowitsch, H. J., Craik, F. E., Habib, R., and Houle, S. (1996). Novelty and familiarity activations in PET studies of memory encoding and retrieval. *Cereb. Cortex* **6,** 71–79.

Turner, R. S., and Anderson, M. E. (1997). Pallidal discharge related to the kinematics of reaching movements in two dimensions. *J. Neurophysiol.* **77,** 1051–1074.

Turner, R. S., Grafton, S. T., Hoffman, J. M., Votaw, J. R., and DeLong, M. R. (1998). Motor subcircuits mediating the control of movement velocity: A PET study. *J. Neurophysiol.* **80,** 2162–2176.

VanMeter, J. W., Maisog, J. M., Zeffiro, T. A., Hallett, M., Herscovitch, P., and Rapoport, S. I. (1995). Parametric analysis of functional neuroimages: Application to a variable-rate motor task. *NeuroImage* **2,** 273–283.

Velasco, F., Velasco, M., Velasco, A. L., Jimenez, F., Marquez, I., and Rise, M. (1995). Electrical stimulation of the centromedian thalamic nucleus in control of seizures: Long-term studies. *Epilepsia* **36,** 63–71.

Vinas, F. C., Zamorano, L., Mueller, R. A., Jiang, Z., Chugani, H., Fuerst, D., Muzik, O., Mangner, T. J., and Diaz, F. G. (1997). [¹⁵O]-Water PET and intraoperative brain mapping: A comparison in the localization of eloquent cortex. *Neurol. Res.* **19,** 601–608.

Watts, R. L., and Mandir, A. S. (1992). The role of motor cortex in the pathophysiology of voluntary movement deficits associated with parkinsonism. *Neurol. Clin. N. Am.* **10,** 451–469.

Weiller, C., Chollet, F., Friston, K. J., Wise, R. J. S., and Frackowiak, R. S. J. (1992). Functional reorganization of the brain in recovery from striatocapsular infarction in man. *Ann. Neurol.* **31,** 1463–

1472.

Wichmann, T., and DeLong, M. R. (1996). Functional and pathophysiological models of the basal ganglia. *Curr. Opin. Neurobiol.* **6,** 751–758.

Wichmann, T., Baron, M. S., and DeLong, M. R. (1994a). Local inactivation of the sensorimotor territories of the internal segment of the globus pallidus and the subthalamic nucleus alleviates parkinsonian motor signs in MPTP treated monkeys. *In* "The Basal Ganglia IV: New Ideas and Data on Structure and Function" (G. Percheron *et al.,* Eds.), pp. 357–363. Plenum, New York.

Wichmann, T., Bergman, H., and DeLong, M. R. (1994b). The primate subthalamic nucleus. III. Changes in motor behavior and neuronal activity in the internal pallidum induced by subthalamic inactivation in the MPTP model of parkinsonism. *J. Neurophysiol.* **72,** 521–530.

Wichmann, T., Bergman, H., Starr, P. A., Subramanian, T., Watts, R. L., and DeLong, M. R. (1999). Comparison of MPTP-induced changes in spontaneous neuronal discharge in the internal pallidal segment and in the substantia nigra pars reticulata in primates. *Exp. Brain Res.* **125,** 397–409.

Wieser, H. G. (1983). Electroclinical features of the psychomotor seizure. A stereoencephalographic study of ictal symptoms and chronotopographical seizure patterns including clinical effects of intracerebral stimulation.

Winstein, C. J., Grafton, S. T., and Pohl, P. S. (1996). Motor task difficulty and brain activity: An investigation of goal-directed reciprocal aiming using positron emission tomography (PET). *J. Neurophysiol.* **77,** 1581–1594.

Woodbury, D. M., and Woodbury, J. W. (1990). Effects of vagal stimulation on experimentally induced seizures in rats. *Epilepsia* **31,** S7–S19.

Worthington, C., Vincent, D. J., Bryant, A. E., Roberts, D. R., Vera, C. L., Ross, D. A., and George, M. S. (1997). Comparison of functional magnetic resonance imaging for language localization and intracarotid speech amytal testing in presurgical evaluation for intractable epilepsy. Preliminary results. *Stereotact. Funct. Neurosurg.* **69,** 197–201.

Wyler, A. R., Hermann, B. P., and Richey, E. T. (1989). Results of reoperation for failed epilepsy surgery. *J. Neurosurg.* **71,** 815–819.

Yetkin, F. Z., Mueller, W. M., Morris, G. L., McAuliffe, T. L., Ulmer, J. L., Cox, R. W., Daniels, D. L., and Haughton, V. M. (1997). Functional MR activation correlated with intraoperative cortical mapping. *AJNR, Am. J. Neuroradiol.* **18,** 1311–1315.

Yoshida, M., Rabin, A., and Anderson, M. E. (1972). Monosynaptic inhibition of pallidal neurons by axon collaterals of caudatonigral fibers. *Exp. Brain Res.* **15,** 333–347.

Zabara, J. (1985). *Electroencephalogr. Clin. Neurophysiol.* **61,** 162.

Zabara, J. (1992). Inhibition of experimental seizures in canines by repetitive vagal stimulation. *Epilepsia* **33,** 1005–1012.

Zald, D. H., and Pardo, J. V. (1997). Emotion, olfaction, and the human amygdala: Amygdala activation during aversive olfactory stimulation. *Proc. Natl. Acad. Sci. U.S.A.* **94,** 4119–4124.

Zhang, J., Wilson, C. L., Levesque, M. F., Behnke, E. J., and Lufkin, R. B. (1993). Temperature changes in nickel–chromium intracranial depth electrodes during MR scanning. *AJNR, Am. J. Neuroradiol.* **14,** 497–500.

Index

A

Ablations
  lesions, 67–70
  surgical, therapy, 487
  thermal, 121–123
N-Acetylaspartate
  -aspartate ratio, 552–553
  cerebral neoplasia, 408–409
  -cho ratio, 552–553
  -creatine ratio
    MS, clinical correlations, 374
    MS pathology, 379–380
    PD, resective surgery, 637
    SCA1, 423
Acetyl-cholinesterase inhibitor, 592–593
Acquired immunodeficiency syndrome lymphomas, 392
AD, see Alzheimer's disease
Addiction, see Substance abuse
Adenosine triphosphate, 340–341
Adrenocorticotrophic hormone, 604
Akinesia, 592, 606, 618–620
Alcohol studies
  MRI, 552–553
  PET/SPECT, 551–552
  sites of action, 550–551
ALS, see Amyotrophic lateral sclerosis
Alzheimer's disease, see also Dementia
  atlases, 23
  average image templates, 165–166
  biological markers, 217
  functional imaging
    activation studies, 232
    early, 226–227
    methodological issues, 227–228
    overview, 208–209
    resting studies, 230–332
  pharmacological modulation, 592–593
  structural imaging
    cortical morphometry, 224–226
    segmentation, 218–219
    volumetrics, 219–221, 223–224
Amino acid tracers, 405–406
γAminobutyric acid
  drug abuse studies, 549–550
  epilepsy studies, 339–340
  function, 323–324
  pharmacological response, 604–606
  plasticity, 569
Amphetamine, 553
Amygdala
  AD structures, 220–221
  anxiety, 518
  cocaine effects, 555
  emotions, 511–512
  fear conditioning, 513
Amygdalohippocampectomy, 320
Amyotrophic lateral sclerosis, 577
Anatomical data, 37
Anatomical models, 143–144
Anterior cingulate cortex, 519
Antiepileptic drugs
  EEG, 604–605
  evoked potentials, 604–605
  MRS, 604
  PET, 600–603
  receptor studies, 603–604
  surgery vs., 632–633
  TMS, 605–606
Antipsychotic drugs, 607–609

655

Anxiety, *see also* Fear
  characterization, 509–510
  disorders, 517–518
  emotional processing, 519–520
  induction, 516
  mechanisms, 519
  neuroanatomical model, 518–519
  neurobiology, 510–511
Aphasia, *see also* Language
  cognitive studies, 182–184
  lesion-deficit model, 181–182
  neuroimaging
    brain systems, 188–190
    dyslexics, 192–195
    expectations, 184–185
    implications, 196, 198
    language recovery, 192–196
    limitations, 185–186
    task-dependent lesions, 192
    types, 185
    undamaged region effects, 186–188
  subcortical, 310
Apomorphine, 606
Arteriovenous visualization, 115
Asynchrony, 48, 50–51
Atlases
  algorithms, 139–140
  anatomical models, 143–144
  anatomic templates, 132, 136
  anatomic variations, 138
  average image templates, 164–168
  brain asymmetry, 147
  brain data analysis, 138–139
  challenges, 131–132
  continuum-mechanical, 140
  coordinate systems, 134, 136
  corpus callosum, 147
  cortical
    asymmetries, 159–160, 164
    averaging, 157
    modeling, 155–156
    tensor maps, 157, 159
    variability, 157
  cortical patterns, 132
  demographics, 134
  digital manipulations, 136
  digital templates, 138
  disease-specific, 132, 166
  displacement maps, 145–146
  dynamic, 168
  encoding anatomic variability, 146
  generation, 20, 22–24
  image distortion, 166–167
  individualizing, 141

  multimodality, 136–137
  parallel implementations, 141, 143
  parametric mesh modeling, 146–147
  pathology, 134
  pathology-based
    anisotropic Gaussian fields, 151–152
    bayesian pattern, 153–154
    deformable probabilistic, 150
    morphometry, 152–153
    pattern-theoretic, 153
    subject comparing, 150–151
    variation encoding, 150
  postmortem, 137
  registration accuracy, 166
  schizophrenia, 147, 149
  spatial normalization, 137–138
  statistical templates, 140–141
  surface parameterization, 144–145
ATP, *see* Adenosine triphosphate
Atrophy, *see* Multiple-system atrophy;
    Olivopontocerebellar atrophy
Attention deficit hyperactivity disorder, 553
Atypical language lateralization theory, 460, 463–464
Autoregulation, 301

B

Basal ganglia
  disorders, 606
  motor circuit, 626
  organization, 615–618
Bayesian patterns, 153–154
Benzodiazepine receptors
  central, 329
  epilespy-associated, 323–327
  peripheral, 328–329
Benzodiazepines, 600
Blood flow, *see* Cerebral blood flow
Blood oxygen level dependent techniques, 12, 86–87
Blood perfusion
  analysis methods, 299–300
  early spontaneous, 305
  imaging studies, 306–308
  luxury, 301
  misery, 301–302
Braille, 577–581
Brain
  arterial obstruction, long-standing, 301–303
  asymmetry, 147
  cancer
    clinical features, 393–394
    diagnosis, 394–395
    epidemiology, 388
    etiology, 393

genetics, 393
histological features, 388–391
incidence, 391–393
intracranial
    diffusion MRI, 410–411
    MRS, 408–410
    PET FDG, 403–405
    SPECT, 403–405
management
    chemotherapy, 400–402
    investigational therapies, 402–403
    neurosurgery, image-guided, 397–399
    radiation therapy, image-guided, 399–400
    survival factors, 395–397
survival, 391–393
data analysis, 138–139
deep, stimulation, 253, 629–632
differentiation, 19–20
dopaminergic therapy, 250–251
emotional processing, 511–512
fear processing, 511–512
hemodynamics, 299–300
ictal onset zones, 633–634
integration, 34
metabolism, 300–301
nonlinear coupling, 54
perfusion, 299–301
plasticity
    ALS, 577
    conditions, 591
    cross-modal, 581
    learning
        adaptation, 579
        mechanisms, 577–578
        skills, 579–581
    lesions, 571–575
    maladaptive, 582–584
    noninvasive studies, 570–571
    Parkinson's disease, 577
    processes, 569–570
    severe sensory neuropathy, 582
    tumors, 577
segregation, 34–35
spatial normalization, 95–96, 137–138
specialization, 34–35
systems, regions vs., 210–212
target sites, 67–70
variation, 150
white matter, 218–219, 311
Broca's area, 474–475
Buprenorphine, 560
Burden of disease analysis
    clinical correlations, 373–374
    MRI burden, 377–379

C

Caffeine, 553–554
Cannibis sativa, see Marijuana
Carbamazepine, 600–603
Carotid amytal testing, 333
Categorical designs, 42
Cavernomas, 320
CBD, see Corticobasal degeneration
CBZ, see Carbamazepine
Cerebral blood flow
    acute ischemic stroke, 303–305
    antiepileptic drugs, 602–603
    apomorphine effect, 606
    caffeine effects, 554
    cerebral neoplasia, 398
    cognitive dysfunction, 606–607
    drug effects, 548–549
    epilepsy, 322–323, 330–332
    imaging techniques, 10–13
    lesions, 62–66
    measurement, 8, 10
    Parkinson's disease
        deep-brain stimulation, 630–632
        pallidotomy, 626–629
        PET, 245–246
        reorganization, 620, 622–625
    peripheral lesions, 573
    PSP studies, 270–271
    schizophrenia, 530
    tremor, 632
Cerebral spinal fluid
    age factors, 218
    antiepileptic drugs, 604
    imaging issues, 228
Cerebrospinal fluid studies, 358–3359
Children, 79, 81
Choreas, 292, 294, see also specific diseases
Cocaine
    MRI, 557
    PET, 546–547
    PET/SPECT, 555–556
    sites of action, 554–555
Cognition
    aphasic studies, 182–184
    depression, 492–493
    derived ESM, 83–84
    dysfunction, 606–607
    epilepsy studies, 335–336
    mild impairment
        Alzheimer-association, 217
        3D mapping, 221
        ROI scans, 218
Cognitive conjunctions, 42
Cognitive subtraction, 42

Conditioning, 512–515
Conditioning, blink, 579
Confounds, spatially coherent, 47
Conjunction analyses, 51–53
Connectivity, 53–54
Constraint-induced movement therapy, 591
Coordinate reference systems, 95–96
Cortex
    anterior cingulate, 519
    asymmetry
        abnormal, 164
        dyslexia, 460, 463–464
        emerging, 159–160
    averaging, 157
    disease patterns, 132
    dorsolateral prefrontal, 249–250, 528–530
    focal, hypometabolism, 321
    insular, 518–519
    lesions, 70–72
    magnetic stimulation, 419–420
    modeling, 155–156
    orbitofrontal, 519
    prefrontal
        language, 532
        memory, 534–537
        responses, 205–206
        schizophrenia, 527
        task-related activation, 209–210
    rolandic, 78–79
    sensory, 571–572
    stimulation, 637–638
    subcortical, 310
    tensor maps, 157, 159
    variability, 157
Corticobasal degeneration, 275–278
Creatine/N-acetylaspartate
    MS, clinical correlations, 374
    pathology in vivo, 379–380
    SCA1, 423
CSF, Cerebral spinal fluid
Cystokinin tetrapeptide, 516

D

DCES, see Direct cortical electrical stimulation
Deep-brain stimulation, 629–632
Dementia, see also Alzheimer's disease
    pharmacology, 606–607
    semantic, 191
    subcortical, 311
Dentatorubropallidoluysian, 430–432
Depression
    basal ganglia disorders, 488, 490–491
    diagnosis, 485

epidemiology, 485
    functional imaging, 487–488, 490–491
    idiopathic, 491–493
    lesion-deficit studies, 486–487
    long-term (see Long-term depression)
    neurochemical markers, 495, 497
    recovery, 493
    sadness vs., 497–498
    surgical ablation therapy, 487
    theoretical models, 485–486
    treatment, 493, 495
    working model, 498, 500
Designs, imaging, 42–45, 50–51
Diaschisis
    crossed cerebellar, 309
    definition, 226
    thalamo-cortical, 310
Diffusion imaging, 380–381
Direct cortical electrical stimulation, 637–638
Disconnection syndromes
Displacement maps, 145–146
Dopa decarboxylase, 242
Dopamine
    cell implantation, 254–256
    depression, 495, 497
    drug reinforcement, 545–546
    dysfunction, 266–267
    induced-dyskinesias, 251
    ligands, 244–245
    receptors
        alcohol abuse, 550–553
        schizophrenia, 528–529
        tardive dyskinesia, 294–295
    schizophrenia hypothesis, 524
    system imaging, 242–245
    therapy, 250–251
    transporter
        cocaine at, 548–549, 554–557
        function, 242–243
        methylphenidate, 554–557
    uptake, 273–274, 277–278
Dorsolateral prefrontal cortex, 528–530
DRPLA, see Dentatorubropallidoluysian
Dyskinesias, 251
Dyslexia
    atypical language lateralization theory, 460, 463–464
    cortical asymmetry, 460, 463–464
    definition, 460
    developmental, 192–193
    event-related potentials, 465–466
    FDG-PET, 466, 471
    interhemispheric deficit theory, 464
    phonological impairment, 471–474

phonological theory, 465
reading, 474–476
right hemisphere reading, 193–194
temporal processing theory, 478
visual impairment, 476–478
visual theory, 464–465
xenon imaging, 466, 471
Dysnomia, 638
Dystonias, *see also specific diseases*
acquired hemidystonia, 289
description, 287
dopa-responsive, 278–279, 289
idiopathic, 287–289, 288
maladaptive, 582–584
paroxysmal, 289–290

E

EAE, *see* Experimental allergic encephalomyelitis
EEG, *see* Electroencephalography
Eigenimage analysis, 53–54
Electrical stimulation mapping, *see also specific
  techniques*
cognition derived, 83–84
description, 78
fMRI *vs.*, 83
functional imaging *vs.*, 83
language identification, 79–83
rolandic cortex, 78–79
Electrocorticogram, 77–78, 83–84
Electrocorticography, 90
Electroencephalography
advantages, 16–17
antiepileptic drugs, 603–605
application, 3, 5
brain metabolism, 300
DRPLA, 432
epilepsy
  generalized, 322
  resective surgery, 634
  source localization, 342
plasticity studies, 570–571
Emotional processing, 519–520
Epilepsies
diffusion MRI, 336
electrophysiological imaging, 342–343
fMRI studies, 333–334
generalized, 322–323
idiopathic generalized, 329
imaging history, 317
intractable, 115–116
juvenile myoclonic, 323
localization-related
  activation studies, 322

benzodiazepine receptors, 323–327
cortical malformations, 321, 338–339
drug therapy, 321–322
extra temporal lobe, 338
frontal lobe, 320–321
histamine receptors, 328
imaging techniques, 318
monoamine oxidase type B, 327–328
NMDA receptors, 328
opioid receptors, 327
peripheral benzodiazepine receptors, 328–329
serotoninergic neurons, 328
syndromes, 321
temporal lobe, 318–320, 337–338
vagal nerve stimulation, 322
MRS
  carbon spectroscopy, 342
  proton spectroscopy, 336–339
  u31UPspectroscopy, 340–341
pharmacology
  EEG, 604–605
  evoked potentials, 604–605
  MRS, 604
  PET, 600–603
  receptor studies, 603–604
  TMS, 605–606
SPECT, 330–334
surgery
  approaches, 632–633
  imaging, 641–643, 645
  nonablative, 640–641
  postoperative evaluation, 639–640
  transective, 640
ERP, *see* Event-related potentials
Event-related potentials
advantages, 16–17
applications, 3, 5
dyslexia, 465–466
Evoked potentials
antiepileptic drugs, 604–605
motor, 570–573
multiple sclerosis, 363
SCA1, 423–424
somatosensory (*see* Somatosensory evoked
  potentials)
Experimental allergic encephalomyelitis, 375
Eye-blink conditioning, 579

F

Factorial designs, 44–45
Fast-spin-echo studies, 359–360
F-DOPA tracers, 405

Fear, *see also* Anxiety
  learning, 512–515
  processing
    amygdala interactions, 515
    brain mechanism, 511–512
    imaging studies, 512
Fenozolone, 593
Fields, theory of, 40–41
Finite impulse response models, 39–40
Flumazenil, 311–312
Fluorodexyglucose positron emission tomography
  caffeine effects, 554
  cerebral neoplasia, 403–405
  dementia studies, 226–227, 230–232
  epilepsies
    activation studies, 322
    cortical malformations, 321
    drug therapy, 321–322
    frontal lobe, 320–321
    general focus, 288
    generalized, 322–323
    resective surgery, 634–637
    syndromes, 321
    temporal lobe, 318–320
  essential tremors, 286
  Friedreich's ataxia, 419
  idiopathic dystonia, 288
  MSA, 264–265, 443–447
  SCA1, 422
  SCA2, 424
  SCA3, 426
  SCA6, 426
  sporadic olivopontocerebellar atrophy,
    438–439
Fluoxetine, 593
Friedreich's ataxia, 418–420
Functional magnetic resonance imaging
  Alzheimer's disease, 232
  anatomical normalization, 36–37
  application, 5
  cerebral neoplasia, 397–399
  cocaine effects, 557
  connectivity, 53–54
  dementia, 213
  depression, 487–488, 490–491
  design/analysis issues, 33–34
  dyslexia
    phonological impairment, 471–474
    reading, 475
    visual impairment, 476–478
  electrical stimulation, 646
  electrophysiological correlates, 84
  epilepsy studies, 334–336
  epoch-related studies, 48–50

  ESM *vs.*, 83
  event-related studies, 48–50
  experiments, designing, 42–48
  fear conditioning, 512–514
  integration, 53–54, 54–56
  lesions, 69
  lesions flow, 65
  memory impairment
    brain response patterns, 205–206
    brain systems, 210–212
    as diagnostic tool, 207–208
    group comparisons, 208–209
    major difficulties, 204–207
    single events, 212–213
    structural, 202–204
    task-related activation, 209–210
  methylphenidate effects, 557
  neurological recovery, 591
  nicotine effects, 561
  optical imaging *vs.*, 86–87
  phenethylamines, 553
  population inference, 51–53
  restless legs syndrome, 295
  schizophrenia
    executive function, 531–535
    hallucinations, 537–539
    tasks studies, 529
  spatial realignment, 36–37
  specialization, 54–56
  stroke recovery, 593
  structural maps, 95
  subject inference, 51–53
  tics, 292
  tremors, essential, 286–287

G

GABA, *see* γAminobutyric acid
Gaussian fields, 151–152
General linear model, 38–40
Glial proliferation, 312
Glioblastoma multiforme
  chemotherapy, 400–401
  incidence, 392–393
Glioma
  low-grade, 114–115
  primary, 396–397
  resection, 125–126
Global normalization, 47
Globus pallidus
  functional alterations, 245–246
  outflow, 251
  stimulation, 253
Glutamate, 340

Glutamine, 340
GMB, Glioblastoma mulitforme

H

Hallucinations, 537–539
Hemianopia, 310
Hemiparkinson-hemiatrophy syndrome, 242
Hemodynamic failure, 301
Hemodynamic reserve, 301–302
Hemodynamic response function, 45–46
Histamine receptors, 328
Homocarnosine, 604
HPHA, *see* Hemiparkinson-hemiatrophy syndrome
Huntington's disease
    akinetic-rigid, 279
    depression, 488, 490
    imaging studies, 292
Hypermetabolism, 493
Hyperperfusion, 305
Hyperthermia, 126

I

ICA, *see* Independent component analysis
ILT, *see* Interstitial laser treatment
Image-guided therapy, 397–399, 399–400
Implantation, 614–615
Independent component anaylsis, 54
Insular cortex, 518–519
Integration
    brain, 34
    functional, 53–54
    lesion-deficit model, 54–56
Interhemispheric deficit theory, 464
Interstitial laser treatment, 123
Intraoperative brain mapping
    development, 77–78
    electrical stimulation
        cognition derived, 83–84
        description, 78
        functional imaging *vs.*, 83
        language identification, 79–83
        rolandic cortex, 78–79
    optical imaging, 84–88, 88, 90
    structural maps
        advantages, 90, 92–93
        coordinate reference systems, 95–96
        creating, 93–94
        morphology, 98
        preoperative, 94–95
        registration, 96–98
        spatial normalization, 95
        stereotaxy, 95–96, 98–100
        visualization, 98–100

Intraoperative mapping, 20
Inversion-recovery scans, 358

K

Kernohan system, 389
Korsakoff's psychosis, 606

L

Lactate, 339
Lamotrigine, 600
Landau-Kleffner syndrome, 321
Language
    aphasia models
        brain systems, 188–189
        dyslexics, 192–195
        processing, 196, 198
        recovery, 195–196
        task-dependent lesions, 192
        undamaged regions, 186–188
    atypical lateralization theory, 460, 463–464
    developmental disorders, 465–466
    identification, 79–83
    lesion-deficit model, 181–182
    optical imaging, 90
Learning
    activation studies, 249–250
    neurological recovery, 590–592
    plasticity
        adaptation, 579
        mechanisms, 577–578
        skills, 579–581
Lennox-Gastaut syndrome, 321
Lesch Nyhan syndrome, 292, 294
Lesioning, 19
Lesions
    blood flow, 62–66
    cortical mapping, 70–72
    -deficit model
        depression, 486–487
        description, 54–56
        language, 181–182
    neuropharmacology *in vivo*, 66–67
    neurotransmitters, 66–67
    peripheral, plasticity, 571–575
    recovery
        evidence, 588–590
        imaging, 587–588
        learning process, 590–592
    structural data, 60–61
    targeting, 67–70
    task-dependent, 192
    vascular anatomy, 61–62

Ligands
  idiopathic dystonia, 288–289
  PET, 14–16
  tracers, 333–334
Lincoln, Abraham, 427
Linear time-invariant systems, 39–40
Long-term depression, 569
Long-term potentiation, 569
LTG, *see* Lamotrigine
Luxury perfusion, 301

M

Machado-Joseph disease, 424–426
Magnetic brain stimulation, 589–590
Magnetic resonance angiograpy
  application, 5
  description, 6–7
  lesions, 61–62
  surgical visualization, 110
Magnetic resonance imaging
  alcohol studies, 552–553
  Alzheimer's disease, 229
  application, 5
  caffeine effects, 554
  CBD studies, 275–276
  cerebral neoplasia, 392, 394–395
  cocaine effects, 557
  cortical asymmetry, 460, 463–464
  description, 6
  diffusion-weighted, 336, 410–411
  disease-based atlases, 138
  DRPLA, 432
  drug abuse studies, 549
  epilepsy
    generalized, 322–323
    history, 317
    postoperative evaluation, 639–640
    resective surgery, 635, 637
  Friedreich's ataxia, 419
  functional (*see* Functional magnetic resonance
    imaging)
  hippocampal sclerosis, 634–635
  intraoperative techniques
    ablations, 121–123
    advantages, 116–117
    cases, 125–126
    procedures, 120–121
    real-time, 123–125
    technology, 118–119
    visualization, 100
  lesions
    blood flow, 65
    structural data, 60–61
    targeting, 68–69

methylphenidate effects, 557
MSA, 441, 443
multiple sclerosis
  atrophy, 381
  clinical trials
    acute-phase analysis, 376–377
    advantages, 375–376
    burden of disease, 377–379
    conventional approaches, 357–363
    natural history studies
      clinical correlations, 371, 373–374
      combined activity, 367–368
      gadolinium enhancement, 365–366
      morphological changes, 364–365
      placebo group behavior, 369–370
    new technologies, 379–381
    pathological correlation, 375
opiate effects, 560
perfusion studies, 11–13
PSP, 269–270
SCA1, 422
SCA2, 424
SCA3, 426
SCA6, 428
SCA7, 429
schizophrenia, 526–527
sporadic olivopontocerebellar atrophy, 435
structural maps, 94–95
surgical visualization, 109–110
Magnetic resonance spectroscopy
  alcohol studies, 552
  antiepileptic drugs, 604
  brain metabolism, 300
  cerebral neoplasia, 408–410
  DRPLA, 432
  drug abuse studies, 549
  epilepsy studies, 336–339, 342
  metabolism, 13–14
  multiple sclerosis *in vivo*, 379–380
  proton, 266–267
  SCA1, 423
  sporadic olivopontocerebellar atrophy, 439
Magnetic stimulation, 419–420
Magnetization transfer imaging, 374, 380
Magnetoencephalography
  advantages, 18
  application, 3, 5
  brain metabolism, 300
  dyslexia, 475–476
  epilepsy studies, 342–343
  plasticity studies
Mapping, *see specific techniques*
Marijuana, 557–559, 558–559
MEG, *see* Magnetoencephalography

Memory
    functional imaging
        brain response patterns, 205–206
        brain systems, 210–212
        as diagnostic tool, 207–208
        group comparisons, 208–209
        major difficulties, 204–207
        single events, 212–213
        structural studies, 202–204
        task-related activation, 209–210
    impairment, 220–221, 321
    schizophrenia, 534
    short-term, 84
    working, 592–593
Metabolism
    CBD studies, 276–277
    FDOPA/PET studies, 245–246
    glucose, 602–603
    hypo, forcal cortical, 321
    imaging techniques, 13–14
    lesions, 65–66
    measurement, 300
    network analysis, 246–249
    PSP studies, 270–272
Metal ion tracers, 406–407
Methamphetamine, 553
N-Methyl-D-aspartate receptors, 328, 605, 607
3,4-Methylenedioxymethamphetamine, 553
Methylphenidate
    MRI, 557
    PET, 546–547
    PET/SPECT, 555–556
    sites of action, 554–555
Misery perfusion, 310–303
Mismatch negativity, 465–466
Monoamine oxidase, 561
Monoamine oxidase type B, 327–328, 549
Motor areas
    evoked potentials, 570–573
    neuromodulation, 593–594
    response, 78–79
    supplementary (see Supplementary motor area)
Movement, 36, 591
MS, see Multiple sclerosis
MSA, see Multiple-system atrophy
MTI, see Magnetization transfer imaging
Multimodal volume registration, 110–111
Multiple modality imaging, 18–19
Multiple sclerosis
    burden of disease, 371–374
    MRI
        atrophy, 381
        clinical trials
            acute-phase analysis, 376–377

        advantages, 375–376
        burden of disease, 377–379
    conventional imaging, 357–363
    natural history studies
        clinical correlations, 371, 373–374
        combined activity, 367–368
        gadolinium enhancement, 365–366
        morphological changes, 364–365
        placebo group behavior, 369–370
    new technologies, 379–381
    pathological correlation, 375
    neurophysiological techniques, 363–364
Multiple-system atrophy
    description, 263–264
    dopaminergic dysfunction, 266–267
    functional imaging, 438
    metabolic studies, 264–265
    MRI studies, 264
    neuropathology, 441
    opioid dysfunction, 267–268
    proton MRS, 266–267
    structural imaging, 434, 441, 443
multiple-system atrophy, 440
Multiple time graphical analysis, 242
MVR, see Multimodal volume registration

N

Neoplasia, cerebral
    clinical features, 393–394
    diagnosis, 394–395
    epidemiology, 388
    etiology, 393
    genetics, 393
    histological features, 388–391
    incidence, 391–393
    intracranial
        diffusion MRI, 410–411
        MRS, 408–410
        PET FDG, 403–405
        SPECT, 403–405
    management
        chemotherapy, 400–402
        investigational therapies, 402–403
        neurosurgery, image-guided, 397–399
        radiation therapy, image-guided, 399–400
        survival factors, 395–397
    survival, 391–393
Nervous system disorders, see specific disorders
Neurodegenerative disorders, see specific disorders
Neurofibrillary tangle patterns, 218
Neurological function, recovery
    evidence, 588–590
    imaging, 587–588

Neurological function, recovery (*continued*)
   learning process, 590–592
   pharmacological approach, 592
Neuro-oncology, 288
Neuropathy, 582
Neuropharmacology, 66–67
Nicotine, 560–561
Noise spectrum, 46–47
Nucleic acid transport tracers, 406

O

Obsessive-compulsive disorder, 517
OIS, *see* Optical intrinsic signal imaging
Olivopontocerebellar atrophy, 263–264, 268
OPCA, *see* Olivopontocerebellar atrophy
Opiates, 559–560
Opioids
   dysfunction, 267–268
   PSP binding
   receptors, 327, 329
Optical imaging
   description, 84–88
   intraoperative, 88, 90
Optical intrinsic signal imaging
   advantages, 18
   language, 90
   measures, 85–86
Orbitofrontal cortex, 519

P

P300, 466
PAF, *see* Pure autonomic failure
Pallidotomy
   description, 251–253, 625–626
   functional imaging, 626–629
Palsy, *see* Progressive supranuclear palsy
Parameterization, surface, 144–145
Parametric designs, 42–44
Parametric mesh modeling, 146–147
Parkinson's disease
   atlases, 23
   basal ganglia organization, 615–618
   brain activation studies, 249–250
   deep-brain stimulation, 629–630, 630–632
   depression, 488, 490
   dopamine system imaging, 242–245
   functional imaging
      blood flow studies, 245–246
      metabolic network, 246–249
      neuroprotective therapies, 256–257
      neurorestorative therapies, 254–256
      stereotaxic surgical therapies, 251–254
      symptomatic pharmacology, 250–251

   metabolic studies, 265
   MSA
      clinical presentation, 440
      functional imaging, 443–447
      structural imaging, 441, 443
   nigrostriatal pathway disruption, 592
   pharmacology, 606
   plasticity, 577
   rating scale, 256
   resting tremor, 285
   surgery, 626–629
   task-specific compensation, 620–625
Passivity, 539
PCA, *see* Principal component analysis
PCr, *see* Phosphocreatine
PD, *see* Parkinson's disease
PDE, *see* Phosphodiesters
Penumbral tissue, 303–306
Pharmacology, *see also specific drugs*
   antiepileptic drugs
      EEG, 604–605
      evoked potentials, 604–605
      MRS, 604
      PET, 600–603
      receptor studies, 603–604
      surgery *vs.,* 632–633
      TMS, 605–606
   antipsychotic drugs, 607–609
   basal ganglia disorders, 606
   cognitive dysfunction, 606–607
   dementia, 606–607
   imaging studies, 599–600
   movement disorders, 606
Phenethylamines, 553
Phenobarbital, 600
Phenytoin
   EEG, 605
   evoked potentials, 605
   PET, 600, 603
Phobic disorders, 517
Phonological theory of dyslexia, 465
Phosphocreatine, 340–341
Phosphodiesters, 340–341
Phosphomonoesters, 340–341
PHT, *see* Phenytoin
Physostigmine, 606
Plasticity
   ALS, 577
   conditions, 591
   cross-modal, 581
   learning
      adaptation, 579
      mechanisms, 577–578
      skills, 579–581

lesions, 575
   maladaptive, 582–584
   mapping, 24, 26
   noninvasive studies, 570–571
   Parkinson's disease, 577
   peripheral lesions, 571–575
   processes, 569–570
   severe sensory neuropathy, 582
   tumors, 577
PME, *see* Phosphomonoesters
PMR, *see* Proton magnetic resonance
Population inferences, 51–53
Positron emission tomography
   acute ischemic stroke, 303–306
   alcohol studies, 551–552
   Alzheimer's disease
      activation studies, 232
      methodological issues, 227–229
      pharmacological modulation, 593
      resting studies, 230–332
   amino acid tracers, 405–406
   anatomical normalization, 36–37
   antiepileptic drugs, 600–601
   antipsychotic drugs, 607–609
   application, 5
   blood flow, 8, 10
   cerebral neoplasia, 398
   connectivity, 53–54
   dementia, 213
   depression
      brain damage, 487–488
      Huntington's associated, 490–491
      working model, 498, 500
   design/analysis issues, 33–34
   disease-based atlases, 138
   dopa-responsive dystonia, 279
   drug abuse studies, 546–549
   dyslexia, 471–476
   electrical stimulation, 646
   electrophysiological correlates, 84
   epilepsy studies
      benzodiazepine receptors, 323–327
      histamine receptors, 328
      idiopathic generalized, 329
      monoamine oxidase type B, 327–328
      NMDA receptors, 328
      opioid receptors, 327, 329
      serotoninergic neurons, 328
   epoch-related studes, 48–50
   ESM *vs.*, 83
   event-related studies, 48–50
   experimental designs, 42–45
   FDG, *see* Fluorodexyglucose positron emission
      tomography

-FDOPA
   blood flow studies, 245–246
   cerebral neoplasia, 405
   DAT imaging, 242–243
   metabolic network analysis, 246–249
   neuroprotective therapies, 256–257
   neurorestorative therapies, 254–256
Huntington's disease, 292
integration, 53–54, 54–56
intraoperative, 100
lesions
   cortical mapping, 70–72
   neurotransmitters, 66–67
   peripheral, 571–572
ligand imaging, 14–16
marijuana effects, 558–559
memory impairment
   brain response patterns, 205–206
   brain systems, 210–212
   as diagnostic tool, 207–208
   group comparisons, 208–209
   major difficulties, 204–207
   single events, 212–213
   structural, 202–204
   task-related activation, 209–210
metabolism, 13
metal ion tracers, 406–407
MSA, 443–447
multiple sclerosis, 363
neurological recovery, 591, 592
neuroreceptor imaging, 14–16
novel tracers, 405–407
nucleic acid transport tracers, 406
opiate effects, 559–560
pallidotomy, 627
Parkinson's disease
   akinesia model, 618–620
   pharmacology, 606
   SMA, 592
   symptomatic pharmacology, 250–251
peripheral lesions, 571–573
phenethylamines, 553
population inference, 51–53
PSP studies, 271
resting-state, 491–493
resting studies, 527–529
restless legs syndrome, 295
SCA3, 426
schizophrenia
   executive function, 531–535
   hallucinations, 537–539
   resting studies, 527–529
severe sensory neuropathy, 582
skills learning, 580

Positron emission tomography (*continued*)
  spatial realignment, 36–37
  specialization, 54–56
  sporadic olivopontocerebellar atrophy, 436–438
  stroke recovery, 588–589
  structural maps, 95
  subject inference, 51–53
  tremors, 286–287
  tumors, 575, 577
Posttraumatic stress disorder, 510, 517–518
Prefrontal cortex
  language, 532
  memory, 534–537
  responses, 205–206
  schizophrenia, 527
  task-related activation, 209–210
Preoperative mapping
  basic types, 59–60
  lesions
    blood flow, 62–66
    ligands, 66–67
    metabolism, 65–66
    neuropharmacology *in vivo,* 66–67
    neurotransmitters, 66–67
    structural data, 60–61
    vascular anatomy, 61–62
  strategies, 19–20
  structural maps, 94–95
  surgical planning, 72–73
  targeting, 67–70
Principal component analysis, 53–54, 246–249
Progressive supranuclear palsy
  cholinergic binding, 274
  description, 242, 268–269
  dopaminergic system, 273–274
  metabolic studies, 270–272
  MRI studies, 269–270
  opioid binding, 274
Proton magnetic resonance, 272–273, 277
PSP, *see* Progressive supranuclear palsy
Pure autonomic failure, 263–264, 268

R

Radiation therapy, 399–400
Random-effect analyses, 51, 53
Realignment, 36
Recovery
  neurological function
    imaging, 587–588
    learning process, 590–592
    pharmacological approach, 592
  stroke, 588–590, 593–594
Relaxation measurements *vivo,* 381

Resective surgery, 613–614, 633–634
Restless legs syndrome, 295
Revascularization, 615
Ringherz system, 389
RMP-7, 406–407

S

Sadness, 497–498
SCA, *see* Spinocerebellar ataxia
Scaled subprofile model, 246–249
Schizophrenia
  atlases, 147, 149
  characterization, 523–524
  cognitive activation studies, 529
  cortical asymmetries, 159–160
  CT studies, 525–526
  dopamine hypothesis, 524
  etiology, 524
  executive function
    component processes, 531–535
    task-based studies, 529–530
  functional studies, 527–528
  genetics, 524–525
  imaging symptoms, 537–539
  MRI, 526–527
  neuropathology, 525
  neuroreceptor imaging, 528–529
  pharmacology, 607–608
  prefrontal interaction, 534–537
  research obstacles, 539–541
Segmentation, 111–112
Segregation, 34–35
Sensory cortex, 571–572
Serial reaction time test, 580
Serotonin, 328
SES, *see* Spinal electrical stimulation
Shy-Drager syndrome, *see* Multiple-system atrophy
Single photon emission computed tomography
  alcohol studies, 551–552
  Alzheimer's disease, 230–332
  application, 5
  cerebral neoplasia, 395, 403–405, 407–408
  depression, 487–488, 490–491
  DRPLA, 432
  drug abuse studies, 549
  epilepsy
    cerebral blood flow tracers, 330–332
    ligand tracers, 333–334
    resective surgery, 634, 636
  lesions, 65–67
  ligands, 16
  marijuana effects, 558–559
  MSA, 264–267, 446–447

multiple sclerosis, 364
neurological recovery, 592
opiate effects, 559–560
perfusion, 10–11, 306–308
pharmacology, 606–607
phenethylamines, 553
resting-state, 491–493
restless legs syndrome, 295
schizophrenia, 529–531
SMA, 592
sporadic olivopontocerebellar atrophy, 439
Tourette syndrome, 290
Sleep disorders, 440
SND, *see* Striatonigral degeneration
SOA, *see* Stimulus onset asynchrony
Somatosensory evoked potentials
intraoperative
language, 90
optical activity, 86–87
rolandic cortex, 79
multiple sclerosis, 363
Spatially coherent confounds, 47
Spatial normalization, 36–37
Spatial smoothing, 37
Specialization, 34–35, 54–56
Spen-echo techniques, 359
Spinal electrical stimulation, 572
Spinocerebellar ataxia
genotype classification, 417
type 1, 421–424
type 2, 421
type 3, 421–426
type 4, 426
type 5, 427–428
type 6, 428
type 7, 428–429
type 8, 429–430
type 9, 430
type 10, 430
SPM, *see* Statistaical parametric mapping
Sporadic olivopontocerebellar atrophy
clinical presentation, 433–434
description, 432–433
functional imaging, 436–439
neuropathology, 434
structural imaging, 434–435
SSEPs, *see* Somatosensory evoked potentials
SSM, *see* Scaled subprofile model
St. Anne-Mayo system, 389
Stationary stochastic designs, 50–51
Statistaical parametric mapping, 37–40
Statistical inference, 40–41
Statistical parametric mapping, 34
Stereotaxy, 95–96, 251–254

Stimulation
deep-brain, 629–632
electrical, 645–646
magnetic, 419–420
vagus nerve, 640–643, 645
Stimulus onset asynchrony, 48
STN, *see* Subthalamic nucleus
Striatonigral degeneration
description, 263–264
distinguishing, 242
MRI studies, 264
MSA overlap, 268
PET studies, 265–266
Stroke
acute ischemic
metabolic effects, 309–311
PET studies, 303–306
therapy, 308–309
basal ganglia, 490
depression link, 487
recovery, 588–590, 593–594
Structural maps
advantages, 90, 92–93
coordinate reference systems, 95–96
creating, 93–94
morphology, 98
preoperative, 94–95
registration, 96–98
spatial normalization, 95–96
stereotaxy, 95–96, 98–100
visualization, 98–100
Substance abuse
alcohol studies
MRI, 552–553
PET/SPECT, 551–552
sites of action, 550–551
caffeine, 553–554
cocaine
MRI, 557
PET, 546–547
PET/SPECT, 555–556
sites of action, 554–555
costs, 545
marijuana, 557–559
methylphenidate, 554–557
neuronanatomical theory, 545–546
nicotine, 560–561
opiates, 559–560
pharmacodynamics, 548–549
pharmacokinetics, 546–547
phenethylamines, 553
tolerance, 546
withdrawal, 545–546
Subthalamic nucleus, 245, 254

Supplementary motor area
    fear conditioning, 513
    idiopathic dystonia, 287–288
    learning, 249–250
    lesions, 592
    localization, 616
    pallidotomy, 627, 629
    PET, 618–620
    schizophrenia, 531
    skill learning, 580
    stereotaxic surgical therapies, 251–253
    stimulation, 630–631
    stroke recovery, 594
Surgery
    ablation therapy, 487
    DCES, 637–638
    epilepsy
        approaches, 632–633
        postoperative evaluation, 639–640
        resective, 633–634
        transplantation, 640–643, 645
    image-guided, 397–399
    implantation, 614–615
    intraoperative (see Intraoperative brain mapping)
    pallidotomy, 625–629
    Parkinson's disease
        deep-brain stimulation, 629–632
        pallidotomy, 625–626
        rationale, 625–626
    planning, 72–73
    preoperative (see Preoperative mapping)
    resective, 613–614, 633–634
    revascularization, 615
    stimulation, deep-brain, 629–632
    transective, 614, 640
    transplantation
        description, 614
        epilepsy, 640–643, 645
    tremor, 632
    visualization
        cases, 114–116
        data acquisition, 110
        3D preoperative modeling, 109–110
        filtering, 110
        image-guided therapy, 108
        limits, 108
        MRI
            ablations, 121–123
            advantages, 116–117
            cases, 125–126
            procedures, 120–121
            real-time, 123–125
            technology, 118–119
        MVR, 110–111

    navigation, 113–114
    planning, 108–109
    registration model, 112–113
    segmentation, 111–112
    TMS, 111

T

Talairach spaces, 224
Tardive dyskinesia, 294–295
Targeting, 19
Tasks
    based-studies, 529–530
    -compensation, PD, 620–625
    -dependent lesions associated, 192
    related activation impairment, 209
    Wisconsin card sorting, 529–530
Temporal basis functions, 39–40
Temporal filtering, 47
Temporal processing theory, 478
TES, see Transcranial electrical stimulation
Thalamotomy, 251–253
Thalamus stimulation, 82–83
THC, see Marijuana
Theory of Gaussian fields, 40–41
Thought disorders, 539
Tics, 290, 292
Time series, 45
TMS, see Transcranial magnetic stimulation
TOA, see Trial onset asynchrony
TOP, see Topiramate
Topiramate, 604
Tourette syndrome, 290
Tower of London task, 530–531
Transcranial electrical stimulation, 572–573
Transcranial magnetic stimulation
    advantages, 18
    antiepileptic drugs, 605–606
    neurological recovery, 589
    noninvasive studies, 570–571
    plasticity, 581
    sporadic olivopontocerebellar atrophy,
        439–440
    surgical visualization, 111
    tumors, 575, 577
Transective surgery, 614, 640
Transplantation
    description, 614
    epilepsy, 640–643, 645
Tremors
    description, 285–287
    Parkinson's disease, 286
    surgery, 632
Trial onset asynchrony, 50–51

Tumors
    margins, 92–93
    plasticity, 575, 577

V

Vagus nerve stimulation
    description, 322
    imaging, 641–643, 645
    induction, 640–641
Valproate, 601, 605
Verbal fluency tasks, 532
Vesicular monoamine transporter, 243–244
Vigabatrin, 601–604
Visual impairment
    dyslexia, 464–465, 476–478
    learning, 577–581
Visualization
    surgical
        cases, 114–116
        data acquisition, 110
        3D preoperative modeling, 109–110
        filtering, 110
        image-guided therapy, 108
        limits, 108
        MRI
            ablations, 121–123
            advantages, 116–117
            cases, 125–126
            procedures, 120–121
            real-time, 123–125
            technology, 118–119
        MVR, 110–111

    navigation, 113–114
    planning, 108–109
    registration model, 112–113
    segmentation, 111–112
    structural maps, 98–100
    TMS, 111
VMAT, see Vesicular monoamine transporter
Voxel-based morphometry, 527
VPA, see Valproate

W

West's syndrome, 321
Willed action, 531–532
Wilson's disease, 289
Wisconsin card sorting task, 529–530

X

Xenon-inhalation technique, 466, 471, 554
X-ray computed tomography
    application, 5
    cerebral neoplasia, 391–392, 395
    description, 6
    DRPLA, 432
    lesions, 60–62, 65
    MSA, 441
    multiple sclerosis, 358
    schizophrenia, 525–526
    spinal, 7–8
    sporadic olivopontocerebellar atrophy, 434
    surgical visualization, 109–110
    xenon-enhanced, 11